PROGRESS IN BRAIN RESEARCH

VOLUME 162

NEUROBIOLOGY OF HYPERTHERMIA

Other volumes in PROGRESS IN BRAIN RESEARCH

Volume 125: Transmission Revisited, by L.F. Agnati, K. Fuxe, C. Nicholson and E. Syková (Eds.) – 2000, ISBN 0-444-50314-5.
Volume 126: Cognition, Emotion and Autonomic Responses: the Integrative Role of the Prefrontal Cortex and Limbic Structures, by H.B.M. Uylings, C.G. Van Eden, J.P.C. De Bruin, M.G.P. Feenstra and C.M.A. Pennartz (Eds.) – 2000, ISBN 0-444-50332-3.
Volume 127: Neural Transplantation II. Novel Cell Therapies for CNS Disorders, by S.B. Dunnett and A. Björklund (Eds.) – 2000, ISBN 0-444-50109-6.
Volume 128: Neural Plasticity and Regeneration, by F.J. Seil (Ed.) – 2000, ISBN 0-444-50209-2.
Volume 129: Nervous System Plasticity and Chronic Pain, by J. Sandkühler, B. Bromm and G.F. Gebhart (Eds.) – 2000, ISBN 0-444-50509-1.
Volume 130: Advances in Neural Population Coding, by M.A.L. Nicolelis (Ed.) – 2001, ISBN 0-444-50110-X.
Volume 131: Concepts and Challenges in Retinal Biology, by H. Kolb, H. Ripps and S. Wu (Eds.) – 2001, ISBN 0-444-50677-2.
Volume 132: Glial Cell Function, by B. Castellano López and M. Nieto-Sampedro (Eds.) – 2001, ISBN 0-444-50508-3.
Volume 133: The Maternal Brain. Neurobiological and Neuroendocrine Adaptation and Disorders in Pregnancy and Post Partum, by J.A. Russell, A.J. Douglas, R.J. Windle and C.D. Ingram (Eds.) – 2001, ISBN 0-444-50548-2.
Volume 134: Vision: From Neurons to Cognition, by C. Casanova and M. Ptito (Eds.) – 2001, ISBN 0-444-50586-5.
Volume 135: Do Seizures Damage the Brain, by A. Pitkänen and T. Sutula (Eds.) – 2002, ISBN 0-444-50814-7.
Volume 136: Changing Views of Cajal's Neuron, by E.C. Azmitia, J. DeFelipe, E.G. Jones, P. Rakic and C.E. Ribak (Eds.) – 2002, ISBN 0-444-50815-5.
Volume 137: Spinal Cord Trauma: Regeneration, Neural Repair and Functional Recovery, by L. McKerracher, G. Doucet and S. Rossignol (Eds.) – 2002, ISBN 0-444-50817-1.
Volume 138: Plasticity in the Adult Brain: From Genes to Neurotherapy, by M.A. Hofman, G.J. Boer, A.J.G.D. Holtmaat, E.J.W. Van Someren, J. Verhaagen and D.F. Swaab (Eds.) – 2002, ISBN 0-444-50981-X.
Volume 139: Vasopressin and Oxytocin: From Genes to Clinical Applications, by D. Poulain, S. Oliet and D. Theodosis (Eds.) – 2002, ISBN 0-444-50982-8.
Volume 140: The Brain's Eye, by J. Hyönä, D.P. Munoz, W. Heide and R. Radach (Eds.) – 2002, ISBN 0-444-51097-4.
Volume 141: Gonadotropin-Releasing Hormone: Molecules and Receptors, by I.S. Parhar (Ed.) – 2002, ISBN 0-444-50979-8.
Volume 142: Neural Control of Space Coding, and Action Production, by C. Prablanc, D. Pélisson and Y. Rossetti (Eds.) – 2003, ISBN 0-444-509771.
Volume 143: Brain Mechanisms for the Integration of Posture and Movement, by S. Mori, D.G. Stuart and M. Wiesendanger (Eds.) – 2004, ISBN 0-444-513892.
Volume 144: The Roots of Visual Awareness, by C.A. Heywood, A.D. Milner and C. Blakemore (Eds.) – 2004, ISBN 0-444-50978-X.
Volume 145: Acetylcholine in the Cerebral Cortex, by L. Descarries, K. Krnjević and M. Steriade (Eds.) – 2004, ISBN 0-444-51125-3.
Volume 146: NGF and Related Molecules in Health and Disease, by L. Aloe and L. Calzà (Eds.) – 2004, ISBN 0-444-51472-4.
Volume 147: Development, Dynamics and Pathology of Neuronal Networks: From Molecules to Functional Circuits, by J. Van Pelt, M. Kamermans, C.N. Levelt, A. Van Ooyen, G.J.A. Ramakers and P.R. Roelfsema (Eds.) – 2005, ISBN 0-444-51663-8.
Volume 148: Creating Coordination in the Cerebellum, by C.I. De Zeeuw and F. Cicirata (Eds.) – 2005, ISBN 0-444-51754-5.
Volume 149: Cortical Function: A View from the Thalamus, by V.A. Casagrande, R.W. Guillery and S.M. Sherman (Eds.) – 2005, ISBN 0-444-51679-4.
Volume 150: The Boundaries of Consciousness: Neurobiology and Neuropathology, by Steven Laureys (Ed.) – 2005, ISBN 0-444-51851-7.
Volume 151: Neuroanatomy of the Oculomotor System, by J.A. Büttner-Ennever (Ed.) – 2006, ISBN 0-444-51696-4.
Volume 152: Autonomic Dysfunction after Spinal Cord Injury, by L.C. Weaver and C. Polosa (Eds.) – 2006, ISBN 0-444-51925-4.
Volume 153: Hypothalamic Integration of Energy Metabolism, by A. Kalsbeek, E. Fliers, M.A. Hofman, D.F. Swaab, E.J.W. Van Someren and R. M. Buijs (Eds.) – 2006, ISBN 978-0-444-52261-0.
Volume 154: Visual Perception, Part 1, Fundamentals of Vision: Low and Mid-Level Processes in Perception, by S. Martinez-Conde, S.L. Macknik, L.M. Martinez, J.M. Alonso and P.U. Tse (Eds.) – 2006, ISBN 978-0-444-52966-4.
Volume 155: Visual Perception, Part 2, Fundamentals of Awareness, Multi-Sensory Integration and High-Order Perception, by S. Martinez-Conde, S.L. Macknik, L.M. Martinez, J.M. Alonso and P.U. Tse (Eds.) – 2006, ISBN 978-0-444-51927-6.
Volume 156: Understanding Emotions, by S. Anders, G. Ende, M. Junghofer, J. Kissler and D. Wildgruber (Eds.) – 2006, ISBN 978-0-444-52182-8.
Volume 157: Reprogramming of the Brain, by A.R. Møller (Ed.) – 2006, ISBN 978-0-444-51602-2.
Volume 158: Functional Genomics and Proteomics in the Clinical Neurosciences, by S.E. Hemby and S. Bahn (Eds.) – 2006, ISBN 978-0-444-51853-8.
Volume 159: Event-Related Dynamics of Brain Oscillations, by C. Neuper and W. Klimesch (Eds.) – 2006, ISBN 978-0-444-52183-5.
Volume 160: GABA and the Basal Ganglia: From Molecules to Systems, by J.M. Tepper, E.D. Abercrombie and J.P. Bolam (Eds.) – 2007, ISBN 978-0-444-52184-2.
Volume 161: Neurotrauma: New Insights into Pathology and Treatment, by J.T. Weber and A.I.R. Maas (Eds.) – 2007, ISBN 978-0-444-53017-2.

PROGRESS IN BRAIN RESEARCH

VOLUME 162

NEUROBIOLOGY OF HYPERTHERMIA

EDITED BY

HARI SHANKER SHARMA

*Laboratory of Cerebrovascular Research, Department of Surgical Sciences, Division of Anaesthesiology
and Intensive Care Medicine,
University Hospital, Uppsala University, SE-75185 Uppsala, Sweden*

ELSEVIER

AMSTERDAM – BOSTON – HEIDELBERG – LONDON – NEW YORK – OXFORD
PARIS – SAN DIEGO – SAN FRANCISCO – SINGAPORE – SYDNEY – TOKYO
2007

Elsevier
Radarweg 29, PO Box 211, 1000 AE Amsterdam, The Netherlands
Linacre House, Jordan Hill, Oxford OX2 8DP, UK

First edition 2007

Copyright © 2007 Elsevier B.V. All rights reserved

No part of this publication may be reproduced, stored in a retrieval system
or transmitted in any form or by any means electronic, mechanical, photocopying,
recording or otherwise without the prior written permission of the publisher

Permissions may be sought directly from Elsevier's Science & Technology Rights
Department in Oxford, UK: phone (+44) (0) 1865 843830; fax (+44) (0) 1865 853333;
e-mail: permissions@elsevier.com. Alternatively you can submit your request online by
visiting the Elsevier web site at http://www.elsevier.com/locate/permissions, and selecting
Obtaining permission to use Elsevier material

Notice
No responsibility is assumed by the publisher for any injury and/or damage to persons
or property as a matter of products liability, negligence or otherwise, or from any use
or operation of any methods, products, instructions or ideas contained in the material
herein. Because of rapid advances in the medical sciences, in particular, independent
verification of diagnoses and drug dosages should be made

Library of Congress Cataloging-in-Publication Data
A catalog record for this book is available from the Library of Congress

British Library Cataloguing in Publication Data
A catalogue record for this book is available from the British Library

ISBN: 978-0-444-51926-9 (this volume)
ISSN: 0079-6123 (Series)

For information on all Elsevier publications
visit our website at books.elsevier.com

Printed and bound in The Netherlands

07 08 09 10 11   10 9 8 7 6 5 4 3 2 1

Working together to grow
libraries in developing countries

www.elsevier.com | www.bookaid.org | www.sabre.org

ELSEVIER   BOOK AID International   Sabre Foundation

List of Contributors

E.R. Adair, US Air Force, 50 Deepwood Drive, Hamden, CT 06517, USA

O. Ahlers, Department of Anesthesiology and Intensive Care Medicine, Charité Medical School, University Medicine Berlin, Campus Virchow Clinic, D-13353 Berlin, Germany

I.S. Aneja, Department of Physiology, Faculty of Medicine, Kuwait University, P.O. Box 24923, Safat 13110, Kuwait

D.M. Aronoff, Division of Infectious Diseases, Department of Internal Medicine, The University of Michigan Health Systems, 6323 MSRB III/Box 0642, 1150 W. Medical Center Drive, Ann Arbor, MI 48109-0642, USA

E.F. Barrett, Department of Physiology and Biophysics and the Neuroscience Program, Miller School of Medicine, University of Miami, Miami, FL 33136, USA

J.N. Barrett, Department of Physiology and Biophysics and the Neuroscience Program, University of Miami School of Medicine, Room 5025 Rosenstiel Medical Science Building (RMSB) 1600 N.W. 10th Avenue, Miami, FL 33101, USA

C.M. Blatteis, Department of Physiology, College of Medicine, The University of Tennessee Health Science Center, 894 Union Avenue, Memphis, TN 38163, USA

H.M. Bramlett, Department of Neurological Surgery, Miami Project to Cure Paralysis, University of Miami, Miller School of Medicine, 1095 NW 14th Terrace (R-48), Miami, FL 33136, USA

M.A. Buccellato, Department of Veterinary Biosciences, The Ohio State University, 1925 Coffey Road, Columbus, OH 43210, USA

T. Carsillo, Department of Molecular Virology, Immunology and Medical Genetics, The Ohio State University, 333 West 10th Avenue, Columbus, OH 43210, USA

C.-K. Chang, Department of Surgery, Mackay Memorial Hospital, and Graduate Institute of Injury Prevention and Control, Taipei Medical University and Municipal Wan-Fan Hospital, Taipei, Taiwan

C.-P. Chang, Department of Medical Research, Chi-Mei Medical Center, Tainan, Taiwan

S.S. Cheung, Environmental Ergonomics Laboratory, School of Health and Human Performance, Dalhousie University, 6230 South Street, Halifax, NS B3H 3J5, Canada

O.L. Cremer, Department of Intensive Care Medicine, University Medical Center, Q04.460, PO Box 85500, 3508 GA Utrecht, The Netherlands

J.A. D'Andrea, Directed Energy Bioeffects Laboratory, Naval Health Research Center Detachment, 8315 Navy Road, Brooks City-Base, TX 78235-5365, USA

A. Dieing, Department of Oncology and Hematology, Charité Campus Mitte, University Medicine Berlin, Schumannstrasse 20/21, D-10117 Berlin, Germany

W.D. Dietrich, Department of Neurological Surgery, Miami Project to Cure Paralysis, University of Miami, Miller School of Medicine, 1095 NW 14th Terrace (R-48), Miami, FL 33136, USA

M. Frosini, Dipartimento di Scienze Biomediche, Sezione di Farmacologia, Fisiologia e Tossicologia Polo Scientifico di S. Miniato, Università di Siena, Viale A. Moro 2, lotto C, 53100 Siena, Italy

C.J. Gordon, Neurotoxicology Division, National Health and Environmental Effects Research Laboratory, US Environmental Protection Agency, B105-04, 109 S.T.W. Alexander Drive, Research Triangle Park, NC 27711, US

M. Hallberg, Department of Pharmaceutical Biosciences, Division of Biological Research on Drug Dependence, Uppsala University, S-751 24 Uppsala, Sweden

B. Hildebrandt, Medical Clinic, Department of Hematology and Oncology, Charité Medical School, University Medicine Berlin, Campus Virchow Clinic, D-13353 Berlin, Germany

M. Horowitz, Laboratory of Environmental Physiology, Faculty of Dental Medicine, Hadassah Medical School, The Hebrew University, P.O. Box 12272, Jerusalem 91120, Israel

B. Hu, Department of Neurology and the Neuroscience Program, Miller School of Medicine, University of Miami, Miami, FL 33136, USA

T. Iwamura, Department of Organic Chemistry, Gifu Pharmaceutical University, 5-6-1 Mitahora-higashi, Gifu 502-8585, Japan

C.E. Johanson, Department of Neurosurgery, Brown Medical School, Rhode Island Hospital, 593 Eddy Street, Providence, RI 02903, USA

C.J. Kalkman, Division of Perioperative Care and Emergency Medicine, University Medical Center E03.511, P.O. 85500, 3508 GA Utrecht, The Netherlands

T. Kerner, Department of Anesthesiology and Intensive Care Medicine, Charité Medical School, University Medicine Berlin, Campus Virchow Clinic, D-13353 Berlin, Germany

E.A. Kiyatkin, Cellular Neurobiology Branch, National Institute on Drug Abuse – Intramural Research Program, National Institutes of Health, DHHS, 5500 Nathan Shock Drive, Baltimore, MD 21224, USA

L.R. Leon, US Army Research Institute of Environmental Medicine, Thermal and Mountain Medicine Division, Kansas Street, Building 42, Natick, MA 01760-5007, USA

C.M. Lilly, University of Massachusetts, School of Medicine, Worcester, MA, USA

M.-T. Lin, Department of Medical Research, Chi-Mei Medical Center, Yung Kang, Tainan, Taiwan 710

S.-Y. Liu, Department of Oral and Maxillofacial Surgery, National Defense Medical Center and Taipei Medical University, Taipei, Taiwan

L.E. Luca, Department of Physiology and Biophysics and the Neuroscience Program, Miller School of Medicine, University of Miami, Miami, FL 33136, USA

K. Nisijima, Department of Psychiatry, Jichi Medical University, Minamikawachi-Machi, Kawachi-Gun, Tochigi-Ken 329-0498, Japan

D. Nonner, Department of Physiology and Biophysics, Miller School of Medicine, University of Miami, Miami, FL 33136, USA

F. Nyberg, Department of Pharmaceutical Biosciences, Division of Biological Research on Drug Dependence, Uppsala University, S-751 24 Uppsala, Sweden

L. Nybo, Department of Human Physiology, Institute of Exercise and Sport Sciences, August Krogh Institute, Universitetsparken 13, DK-2100 Copenhagen, Denmark

M. Oglesbee, Department of Veterinary Biosciences, The Ohio State University, 1925 Coffey Road, Columbus, OH 43210, USA

A. Pavlik, Department of Physiology, Faculty of Medicine, Kuwait University, P.O. Box 24923, Safat 13110, Kuwait

K. Possinger, Department of Oncology and Hematology, Charité Campus Mitte, University Medicine Berlin, Schumannstrasse 20/21, D-10117 Berlin, Germany

S.D.M. Robinson, Laboratory of Environmental Physiology, Faculty of Dental Medicine, Hadassah Medical School, The Hebrew University, P.O. Box 12272, Jerusalem, 91120, Israel

A.A. Romanovsky, Systemic Inflammation Laboratory, Trauma Research, St. Joseph's Hospital and Medical Center, 350 W. Thomas Road, Phoenix, AZ 85013, USA

O. Saleh, Department of Physiology and Biophysics, Miller School of Medicine, University of Miami, Miami, FL 33136, USA

M.N. Sawka, Thermal and Mountain Medicine Division, US Army Research Institute of Environmental Medicine, Natick, MA 01760, USA

A. Sharma, Laboratory of Cerebrovascular Research, Department of Surgical Sciences, Anaesthesiology and Intensive Care Medicine, Uppsala University Hospital, Uppsala University, SE-75185 Uppsala, Sweden

H.S. Sharma, Laboratory of Cerebrovascular Research, Department of Surgical Sciences, Anaesthesiology and Intensive Care Medicine, Uppsala University Hospital, Uppsala University, SE-75185 Uppsala, Sweden

K. Shioda, Department of Psychiatry, Jichi Medical University, Minamikawachi-Machi, Kawachi-Gun, Tochigi-Ken 329-0498, Japan

L.A. Sonna, Division of Pulmonary and Critical Care Medicine, University of Maryland School of Medicine, Baltimore, MD 21201, USA

I. Tamm, Medical Clinic, Department of Hematology and Oncology, Charité Medical School, University Medicine Berlin, Campus Virchow Clinic, D-13353 Berlin, Germany

Z. Traylor, Department of Veterinary Biosciences, The Ohio State University, 1925 Coffey Road, Columbus, OH 43210, USA

M.G. White, Department of Physiology and Biophysics and the Neuroscience Program, Miller School of Medicine, University of Miami, Miami, FL 33136, USA

P. Wust, Department of Radiation Medicine, Charité Medical School, University Medicine Berlin, Campus Virchow Clinic, D-13353 Berlin, Germany

J.M. Ziriax, Directed Energy Bioeffects Laboratory, Naval Health Research Center Detachment, 8315 Navy Road, Brooks City-Base, TX 78235-5365, USA

Preface

Hyperthermia and heat-related illnesses pose a serious medical problem in our society that has still not been addressed well either in clinical situations or at the basic research level (Centers for Disease Control and Prevention (CDC), 2006; Hajat et al., 2006). With the increase in global warming, the frequency and intensities of heat-waves have increased in the last five years (Keatinge and Donaldson, 2004). Massive heat-related deaths in Europe and in the United States of America in recent years necessitated new research on the mechanisms of heat-induced morbidity and mortality to find a suitable therapeutic strategy (Keatinge and Donaldson, 2004; Centers for Disease Control and Prevention (CDC), 2006; Hajat et al., 2006). Although, sporadic reports of heat-induced death are reported in the literature since Biblical times (Judith:8), the importance of the central nervous system (CNS) in heat-related injuries is still not well understood (see Sharma and Westman, 1998). Recent research clearly suggests that the CNS is the most vulnerable organ in heat-related illnesses (Sharma, 1999, 2005; Sharma and Hoopes, 2003). However, the detailed cellular and molecular mechanisms of hyperthermia-induced brain dysfunction are not known. Thus, it is still unclear whether heat-related illnesses could adversely influence the human populations suffering from cardiovascular, endocrine, or other metabolic diseases. Moreover, whether environmental factors, e.g., presence of ultrafine particles in the air, pollution, or other infectious agents will worsen the pathological outcome of heat-stress induced symptoms (Sharma and Dey, 1987; Sharma et al., 1992). Therapeutic hyperthermia is a common procedure for cancer therapy that also precipitates neurological symptoms (Sharma and Westman, 1998; Sharma and Hoopes, 2003). Thus, an increased understanding of heat-related illnesses is necessary to develop suitable therapeutic strategies to attenuate hyperthermia-induced morbidity and mortality. Research on thermoregulation and body temperature control in healthy conditions is well known since the last several decades (Feldberg, 1965; Satinoff, 1978; Blatteis, 1988); however, our knowledge on the pathological aspects of temperature regulation in relation to brain function is still rudimentary. Recent trends in basic and clinical research indicate that temperature control in physiological and pathological conditions is regulated by the CNS (Dascombe and Milton, 1976; Gordon and Heath, 1986; Sharma et al., 1992; Sharma and Hoopes, 2003; Kiyatkin, 2005). However, information pertaining to this aspect has never been assembled in any single volume or treatise before. This volume is the first to address the problems of CNS regulation of hyperthermia in which neural regulation of fever, therapeutic hyperthermia, gene expression, brain and cerebrospinal fluid (CSF) dysfunction, neurochemical transmission, as well as brain hyperthermia in relation to CNS injuries and repair are covered.

This volume is a refereed collection of 25 chapters written by leading experts in the field. In each chapter, the authors highlight cutting edge facts of their research in the light of currently available literature on the subject and suggest new areas for future research.

The volume is divided into X sections. Section I deals with hyperthermia associated with fever and contains two chapters. It appears that the classical theory of fever production involving prostaglandins (PGs) and cytokine interaction in the hypothalamus requires new insights. Blatteis (Chapter 1) revisited this hypothesis and presented new ideas based on his innovative experiments on the lack of a febrile response of circulating cytokines following intravenously administered bacterial endotoxin, lipopolysaccharide (LPS). New evidence clearly shows that the peripheral febrigenic message is conveyed to the hypothalamus via a neural route, specifically by the vagus through PGs that express PGE_2 receptors (EP_3). In Chapter 2,

Aronoff and Romanovsky describe the importance of eicosanoids viz., prostaglandins, leukotrienes, and other lipoxygenase metabolites in regulating hyperthermia in febrile illness and in normal healthy conditions. The authors review the new signaling pathways for eicosanoids, i.e., endocannabinoids/endovanilloids system in fever, and find that although the PGE_2 is the principal mediator of fever, it is not involved in the maintenance of normal core body temperature.

Section II describes hyperthermia caused by physiological stimulation, i.e., performing exercise or psychological tasks, and is divided into two chapters. Nybo (Chapter 3) summarizes the current knowledge on cerebral metabolic, thermodynamic, and neurohumoral responses during prolonged exercise in the heat and their potential relations with hyperthermia-induced fatigue. New data presented by the author show that hyperthermia-induced fatigue is related to the CNS function, known as "central fatigue" that depends on heat accumulation in the brain. In this context, Cheung's contribution (Chapter 4) further suggests that several psychological safeguards protect individuals prior to the catastrophic effects of hyperthermia and attenuate central drive for exercise before reaching the critical limit of CNS hyperthermia. Thus, the face and head are highly thermal–sensitive regions that primarily detect the early physiological heat strain in order to take effective measures to improve ventilation, airflow, or active cooling in order to prevent mental impairment in heat stress.

In Section III, the effect of drug metabolism in relation to hyperthermia is described in two chapters. In general, hyperthermia exacerbates the toxicity of many types of drugs and environmental toxicants. In Chapter 5, Gordon describes the effects of hyperthermic drugs in animal models in relation to thermoregulatory ability and the body mass. The data suggest that the area-to-mass ratio and the ambient temperature play important roles in determining hyperthermia-induced toxicity. Nisijima et al. (Chapter 6) review hyperthermic complications arising out of neuroleptic malignant syndrome (NMS) and serotonin syndrome (SS). These syndromes show profound hyperthermia and other clinical symptoms, e.g., diaphoresis, tachycardia, and muscle rigidity. The authors find very similar profiles of monoamine metabolism in the CSF of the NMS and SS cases indicating that these syndromes could be very similar in nature. Thus, use of selective serotonin $_{2A}$ receptor antagonists may be more effective for treating these syndromes, as they are also successful in containing hyperthermia.

Therapeutic hyperthermia and its consequences are presented in Section IV that comprises three chapters. D'Andrea et al. (Chapter 7) presents the current safe exposure recommendations to radio frequency electromagnetic radiation (RF-EMF). The scientific bases on which these recommendations are made, e.g., dosimetry, thermoregulatory, and behavioral responses are also described in detail. In Chapter 8, Dieing et al. discuss current immunological aspects of therapeutical hyperthermia used for cancer therapy in patients. Since fever activates a variety of immunological reactions, therapeutic hyperthermia is also likely to alter the immune system. The neurological symptoms precipitated by therapeutic hyperthermia are described by Cremer and Kalkman (Chapter 9). The most frequent symptoms when the temperature exceeds 40°C include demyelating peripheral neuropathy, nausea, delirium, apathy, stupor, and coma. In addition, transient vasoparalysis, cerebral metabolic uncoupling and loss of pressure-flow autoregulation are also seen in cancer patients. It appears that these symptoms are related to the development of brain edema, intracerebral hemorrhage, and intracranial hypertension. The consequences and ways to avoid these problems based on the author's new data are discussed in this review.

Section V deals with problems of hyperthermia and brain pathology and is divided into four chapters. To understand the clinical problems of whole body hyperthermia with special regard to brain pathology, a good animal model is necessary. Sharma (Chapter 10) describes a new model of whole body hyperthermia that induces brain edema and brain pathology very similar to that seen in clinical situations. For comparison, alternative methods to produce brain hyperthermia and key measurable parameters of brain pathology, e.g., breakdown of the blood–brain barrier permeability and brain edema formation is also described in detail. In Chapter 11, Dietrich and Bramlett discuss the harmful effects of hyperthermia, e.g., fever following brain or spinal cord injuries. This necessitates maintaining normal body temperature during

CNS injury in neurosurgical care units. Small elevations in temperature during or following an insult to the brain or spinal cord worsen the histopathological and behavioral outcomes. Based on the author's own investigations in this area, it is clear that a mild hyperthermia ($>37°C$) aggravates multiple pathomechanisms, including excitotoxicity, free radical generation, inflammation, apoptosis, and genetic responses to injury in which gender differences also play important roles. In this context, Kiyatkin (Chapter 12) points out an important role of brain temperature in regulating various physiological and pathological mechanisms in animal models. Since hyperthermia exacerbates drug-induced toxicity and is destructive to neural cells and brain functions, use of these drugs in these that restrict heat loss conditions may pose a significant health risk, resulting in both acute life-threatening complications and chronic destructive CNS changes. The effects of hyperthermia are aggravated in regions showing high pollution in the environment. It is likely that ultrafine particles, nowadays known as nanoparticles, suspended in the air influence the pathomechanisms of hyperthermia. Although, our knowledge regarding influence of nanoparticles on brain function *in vivo* during normal or hyperthermic conditions is still lacking, few reports indicate that nanoparticles when entering into the CNS may induce neurotoxicity. However, the ability of nanoparticles to influence body temperature in normal or in hyperthermic conditions is not known. Sharma and Sharma presented their novel data in Chapter 13 showing that nanoparticles derived from metals (e.g., Cu, Ag, or Al, ≈ 50 to $60\,nm$) are capable of inducing brain dysfunction in normal animals and aggravating brain pathology and cognitive impairments caused by whole body hyperthermia.

The role of neurochemicals in hyperthermia is discussed in Section VI in two chapters. Nyberg and Hallberg (Chapter 14) review the role of various neuropeptides in relation to hyperthermia and heat-related disorders. Their data suggests that the peptidergic systems are affected during thermal challenges and neuropeptides are involved in the control of heat production. In Chapter 15, Sharma provides new information showing an interaction between amino acid neurotransmitters and opioid receptors in hyperthermia-induced brain damage. Thus, increased excitotoxicity (glutamate and aspartate) and a decrease in inhibitory amino acid neurotransmission (GABA and glycine) in heat stress are associated with brain pathology and sensory–motor disturbances. Blockade of multiple opioid receptors with naloxone restored the whole body hyperthermia-induced decline in inhibitory amino acids and thwarted the elevation of excitotoxic amino acids in the CNS and attenuated cognitive dysfunction and brain pathology.

Section VII is devoted to hyperthermia-induced gene expression and contains three chapters. Sonna et al. (Chapter 16) present new evidence showing expression of several genes in humans following exertional heat illnesses (EHI). New data show that EHI produces time-dependent changes in gene expression that are easily detectable in peripheral blood mononuclear cells (PBMCs). The factors that determine pathway-specific components of gene expression in heat-related injuries are also discussed. In Chapter 17, White et al. describe gene expression in the developing brain that is especially sensitive to hyperthermia. Employing *in vivo* and *in vitro* approaches, the authors demonstrate hyperthermia-induced expression of several proteins in the endoplasmic reticulum probably due to denaturing of nascent polypeptide chains as well as nuclear and cytoskeletal damage including mitochondrial distortion and neuronal depolarization. The global genomic response, detected using gene-chips and cluster analyses, Horowitz (Chapter 18) provides new evidence implying the upregulation of genes encoding ion channels, pumps, and transporters (markers for neuronal excitability) in the hypothalamus at the onset of heat acclimation and downregulation of metabotropic genes during long-term adaptation to hyperthermia.

Expression of heat shock proteins in hyperthermia is presented in Section VIII and includes three chapters. In Chapter 19, Buccellato et al. discuss an important role of HSPs in the maintenance of cellular homeostasis in hyperthermia that adversely affect normal cellular structure and function. New data show that expression of HSPs plays a role in the maintenance and defense of cellular viability as well as in the preservation of neuron-specific functions that support synaptic activity. Pavlik and Aneja (Chapter 20) review the distribution of Hsp70 in neural and non-neural cell types in whole body hyperthermia. Their data show that expression of Hsp70 in oligodendrocytes, microglia, and vascular cells in hyperthermia far exceeds in Hsp70 induction in astrocytes and in most neurons. The neuronal non-responsiveness of Hsp70 in heat stress is discussed. Horowitz and Robinson (Chapter 21) present the new role of HSP as molecular chaperones and their possible functions in cellular repair mechanisms in heat stroke-induced cerebral ischemia and hypotension. New results suggest that heat acclimation increases tissue reserves of HSP70 and thus accelerates the heat shock response indicating their role in central integrative systems in hyperthermia.

Influence of hyperthermia on the CSF is discussed in Section IX and is divided into two chapters. The work of Frosini (Chapter 22) suggests that the CSF composition is altered during heat stress and fever. A decrease in CSF taurine and GABA levels following hyperthermia without any change in aspartate and glutamate levels indicate the role of GABA and taurine as endogenous cryogens influencing central mechanisms of body heat production in hyperthermia and in fever. However, alterations in CSF composition following heat stress or fever denotes impairment of blood–CSF-barrier (BCSFB). Chapter 23 (Sharma and Johanson) describes the breakdown of the BCSFB in whole body hyperthermia in animal model as evidenced by degeneration of choroidal epithelium (anatomical site of BCSFB) and the CSF-bordering ependymal cells. New evidence indicates a crucial role of BSCFB disruption in hyperthermia in edema formation cell and tissue injuries in the CNS.

Problems of heat stroke following hyperthermia are presented in Section X that comprises two chapters. Leon (Chapter 24) reviews the role of cytokines in heat stroke that are known to be important modulators of acute stress response, infection, and inflammation. Thus, antagonism of cytokine may attenuate morbidity and mortality in heat stroke. In Chapter 25, Chang et al. provide new insight on the role of oxidative stress in heat stroke-induced brain injuries. Hyperthermia, hypotension, and cerebral ischemia during heat stroke is associated with increased production of free radicals (specifically hydroxyl radicals and superoxide anions), higher lipid peroxidation, lower enzymatic anti-oxidant defenses, and higher enzymatic pro-oxidants in the brain. New data show that the conventional hydroxyl radical scavengers (e.g., mannitol or α-tocopherol) and traditional medicine, e.g., Shengmai San or magnolol (Chinese herbal medicines) or hypervolemic hemodilution (produced by intravenous infusion of 10% human albumin) are equally effective for prevention and repair of ischemic and oxidative damage in the brain during heat stroke. These new findings may have important implications for management and therapy of heat stroke patients.

There are reasons to believe that the basic mechanisms of brain damage and repair mechanisms in hyperthermia or following different CNS insults, like hypoxia, ischemia, trauma, and neurodegenerative diseases etc., are quite similar in nature. Thus, the new therapeutic advances presented in this volume could also be used for the development of suitable therapeutic strategies against brain diseases in general and neurodegeneration in particular. It is hoped that this volume will provide tremendous new input to both basic and clinical neuroscientists, especially, neurobiologists, neuropharmacologists, neurologists, neuropsychiatrists, neuropathologists, neurotraumatologists, neurosurgeons, and researchers to use the current knowledge for the benefit of mankind and to improve the current health care system for the victims of heat-related disorders.

I am grateful to Hilary Rowe (Elsevier, San Diego, USA) and Maureen Twaig (Elsevier, Amsterdam, The Netherlands) for their tireless help, encouragement and support throughout the development of this book project. Without their continuous support at every moment, it would have been difficult to compile this volume in time. Editorial assistance of Erin Labonte-McKay, Cindy Minor (Elsevier, San Diego, USA), and Aruna Sharma (Uppsala, Sweden) is gratefully acknowledged with thanks.

I hope that the volume will stimulate new research in heat-related illnesses leading to further expansion of our knowledge in basic science and consequently development of suitable therapeutic strategies to improve health care and quality of life.

Hari Shanker Sharma
Uppsala

References

Centers for Disease Control and Prevention (CDC). (2006) Heat-related deaths — United States, 1999–2003. MMWR Morb. Mortal. Wkly. Rep., 55(29): 796–798.

Hajat, S., Armstrong, B., Baccini, M., Biggeri, A., Bisanti, L., Russo, A., Paldy, A., Menne, B. and Kosatsky, T. (2006) Impact of high temperatures on mortality: Is there an added heat wave effect? Epidemiology, 17(6): 632–638.

Sharma, H.S. and Westman, J. (1998) Brain Function in Hot Environment. Progress in Brain Research, Vol. 115, pp. 1–617; Elsevier Science Publishers, Amsterdam, The Netherlands.

Keatinge, W.R. and Donaldson, G.C. (2004) The impact of global warming on health and mortality. South Med. J., 97(11): 1093–1099. Review

Sharma, H.S. and Hoopes, P.J. (2003) Hyperthermia induced pathophysiology of the central nervous system. Int. J. Hyperthermia, 19(3): 325–354. Review

Sharma, H.S. (1999) Pathophysiology of blood-brain barrier, brain edema and cell injury following hyperthermia: new role of heat shock protein, nitric oxide and carbon monoxide. An experimental study in the rat using light and electron microscopy. Acta Universitatis Upsaliensis, 830: 1–94.

Sharma, H.S. (2005) Editorial. Heat-related deaths are largely due to brain damage. Indian J. Med. Res., 121(5): 621–623.

Sharma, H.S. and Dey, P.K. (1987) Influence of long-term acute heat exposure on regional blood-brain barrier permeability, cerebral blood flow and 5-HT level in conscious normotensive young rats. Brain Res., 424(1): 153–162.

Sharma, H.S., Zimmer, C., Westman, J. and Cervos-Navarro, J. (1992) Acute systemic heat stress increases glial fibrillary acidic protein immunoreactivity in brain: experimental observations in conscious normotensive young rats. Neuroscience, 48(4): 889–901.

Satinoff, E. (1978) Neural organization and evolution of thermal regulation in mammals. Science, 201(4350): 16–22. Review

Blatteis, C.M. (1988) Neural mechanisms in the pyrogenic and acute-phase responses to interleukin-1. Int. J. Neurosci., 38(1–2): 223–232.

Feldberg, W. (1965) A new concept of temperature control in the hypothalamus. Proc. R Soc. Med., 58: 395–404.

Gordon, C.J. and Heath, J.E. (1986) Integration and central processing in temperature regulation. Annu. Rev. Physiol., 48: 595–612. Review

Dascombe, M.J. and Milton, A.S. (1976) Cyclic adenosine $3',5'$-monophosphate in cerebrospinal fluid during thermoregulation and fever. J. Physiol., 263(3): 441–463.

Kiyatkin, E.A. (2005) Brain hyperthermia as physiological and pathological phenomena. Brain Res. Brain Res. Rev., 50(1): 27–56. Review

Contents

List of Contributors v

Preface ix

Section I. Fever and Hyperthermia

1. The onset of fever: new insights into its mechanism
   C.M. Blatteis (Memphis, TN, USA). 3

2. Eicosanoids in non-febrile thermoregulation
   D.M. Aronoff and A.A. Romanovsky (Ann Arbor, MI and Phoenix, AZ, USA). . . 15

Section II. Physiological Mechanisms in Hyperthermia

3. Exercise and heat stress: cerebral challenges and consequences
   L. Nybo (Copenhagen, Denmark) 29

4. Neuropsychological determinants of exercise tolerance in the heat
   S.S. Cheung (Halifax, NS, Canada) 45

Section III. Drugs and Hyperthermia

5. Thermophysiological responses to hyperthemic drugs: extrapolating from rodent to human
   C.J. Gordon (Research Triangle Park, NC, USA). 63

6. Neuroleptic malignant syndrome and serotonin syndrome
   K. Nisijima, K. Shioda and T. Iwamura (Tochigi-Ken and Gifu, Japan). 81

Section IV. Therapeutic Hyperthermia and Consequences

7. Radio frequency electromagnetic fields: mild hyperthermia and safety standards
   J.A. D'Andrea, J.M. Ziriax and E.R. Adair (Brooks City-Base, TX and Hamden, CT, USA). 107

8. The effect of induced hyperthermia on the immune system
   A. Dieing, O. Ahlers, B. Hildebrandt, T. Kerner, I. Tamm, K. Possinger and P. Wust (Berlin, Germany)......... 137

9. Cerebral pathophysiology and clinical neurology of hyperthermia in humans
   O.L. Cremer and C.J. Kalkman (Utrecht, The Netherlands)......... 153

Section V. Hyperthermia and Brain Pathology

10. Methods to produce hyperthermia-induced brain dysfunction
    H.S. Sharma (Uppsala, Sweden)......... 173

11. Hyperthermia and central nervous system injury
    W.D. Dietrich and H.M. Bramlett (Miami, FL, USA)......... 201

12. Physiological and pathological brain hyperthermia
    E.A. Kiyatkin (Baltimore, MD, USA)......... 219

13. Nanoparticles aggravate heat stress induced cognitive deficits, blood–brain barrier disruption, edema formation and brain pathology
    H.S. Sharma and A. Sharma (Uppsala, Sweden)......... 245

Section VI. Neurochemicals and Hyperthermia

14. Neuropeptides in hyperthermia
    F. Nyberg and M. Hallberg (Uppsala, Sweden)......... 277

15. Interaction between amino acid neurotransmitters and opioid receptors in hyperthermia-induced brain pathology
    H.S. Sharma (Uppsala, Sweden)......... 295

Section VII. Hyperthermia and Gene Expression

16. Exertional heat illness and human gene expression
    L.A. Sonna, M.N. Sawka and C.M. Lilly (Baltimore, MD, Natick and Worcester, MA, USA)......... 321

17. Cellular mechanisms of neuronal damage from hyperthermia
    M.G. White, L.E. Luca, D. Nonner, O. Saleh, B. Hu, E.F. Barrett and J.N. Barrett (Miami, FL, USA)......... 347

18. Heat acclimation and cross-tolerance against novel stressors: genomic–physiological linkage
    M. Horowitz (Jerusalem, Israel)......... 373

Section VIII. Heat Shock Proteins in Hyperthermia

19. Heat shock protein expression in brain: a protective role spanning intrinsic thermal resistance and defense against neurotropic viruses
    M.A. Buccellato, T. Carsillo, Z. Traylor and M. Oglesbee (Columbus, OH, USA) ... 395

20. Cerebral neurons and glial cell types inducing heat shock protein Hsp70 following heat stress in the rat
    A. Pavlik and I.S. Aneja (Safat, Kuwait) ... 417

21. Heat shock proteins and the heat shock response during hyperthermia and its modulation by altered physiological conditions
    M. Horowitz and S.D.M. Robinson (Jerusalem, Israel) ... 433

Section IX. Hyperthermia and Cerebrospinal Fluid

22. Changes in CSF composition during heat stress and fever in conscious rabbits
    M. Frosini (Siena, Italy) ... 449

23. Blood–cerebrospinal fluid barrier in hyperthermia
    H.S. Sharma and C.E. Johanson (Uppsala, Sweden and Providence, RI, USA) ... 459

Section X. Heat Stroke and Hyperthermia

24. Heat stroke and cytokines
    L.R. Leon (Natick, MA, USA) ... 481

25. Oxidative stress and ischemic injuries in heat stroke
    C.-K. Chang, C.-P. Chang, S.-Y. Liu and M.-T. Lin (Taipei, Taiwan) ... 525

Subject Index ... 547

SECTION I

Fever and Hyperthermia

CHAPTER 1

The onset of fever: new insights into its mechanism

Clark M. Blatteis*

Department of Physiology, College of Medicine, The University of Tennessee Health Science Center, Memphis, TN 38163, USA

Abstract: The classical view of fever production is that it is modulated in the ventromedial preoptic area (VMPO) in response to signaling by pyrogenic cytokines elaborated in the periphery by mononuclear phagocytes and the consequent induction of cyclooxygenase (COX)-2-dependent prostaglandin (PG)E_2 in the VMPO. This mechanism has, however, been questioned, in particular because the appearance of circulating cytokines lags the onset of the febrile response to intravenously (iv) injected bacterial endotoxic lipopolysaccharide (LPS), an exogenous pyrogen. Moreover, COX-2, in this case, is itself an inducible enzyme, the de novo synthesis of which similarly lags significantly the onset of fever. Issues also exist regarding the accessibility of the POA to blood-borne cytokines. New data adduced over the past 10 years indicate that the peripheral febrigenic message is conveyed to the VMPO via a neural rather than a humoral route, specifically by the vagus to the nucleus tractus solitarius (NST), and that the peripheral trigger is PGE_2, not cytokines; vagal afferents express PGE_2 receptors (EP_3). Thus, the initiation of the febrile responses to both iv and intraperitoneal (ip) LPS is temporally correlated with the appearance of LPS in the liver's Kupffer cells (Kc), its arrival immediately activating the complement (C) cascade and the consequent production of the anaphylatoxin C5a; the latter is the direct stimulus for PGE_2 production, catalyzed non-differentially by constitutive COX-1 and -2. From the NST, the signal proceeds to the VMPO via the ventral noradrenergic bundle, causing the intrapreoptic release of norepinephrine (NE) which then evokes two distinct core temperature (T_c) rises, viz., one α_1-adrenoceptor (AR)-mediated, rapid in onset, and PGE_2-independent, and the other α_2-AR-mediated, delayed, and COX-2/PGE_2-dependent, i.e., the prototypic febrile pattern induced by iv LPS. The release of NE is itself modulated by nitric oxide contemporaneously released in the VMPO.

Keywords: pyrogenic agents; preoptic-anterior hypothalamus; vagal afferents; Kupffer cells; prostaglandin E_2; complement 5a; norepinephrine; nitric oxide

Introduction

The conventional view of the mechanism by which infectious fevers are produced holds that infectious noxa (e.g., Gram-negative bacteria and/or their products [bacterial endotoxic lipopolysaccharides, LPS]) that invade the body activate mononuclear phagocytes that then produce and release pyrogenic cytokines. These, in turn, are transported by the bloodstream to the ventromedial preoptic area (VMPO) of the anterior hypothalamus, the "fever-producing center", where they act (for reviews, see Saper, 1998; Roth and De Souza, 2001; Dunn, 2002; Dinarello, 2004). It is generally agreed that evidence of their central action is the depression of

*Corresponding author. Tel.: +1-901-448-5845; Fax: +1-901-448-1673; E-mail: blatteis@physio1.utmem.edu

DOI: 10.1016/S0079-6123(06)62001-3

warm-sensitive neurons in this region, leading, in accordance with Hammel's classical model (1965), to the conservation of heat and, hence, to a rise in body core temperature (T_c) (Boulant, 2000). It is, however, uncertain how cytokines, as hydrophilic peptides, could penetrate the brain and, indeed, whether they or other, secondarily elaborated factors mediate the febrile response. To wit, it is generally believed that, rather than acting directly, the cytokines induce the local generation and release of prostaglandin (PG)E_2, a lipid mediator that is clearly thermogenic when injected centrally (for reviews, see Blatteis, 1997; Ivanov and Romanovsky, 2004). Its production in this instance has been demonstrated to be dependent on the activation of two enzymes, cyclooxygenase (COX)-2 and microsomal PGE synthase (mPGES)-1, which catalyze its conversion from arachidonic acid (AA) present in the membranes of cells (Ivanov et al., 2002).

Questions have arisen over the past 10 years, however, about the validity of this concept. They are based, in particular, on findings that low-to-moderate doses of LPS injected intravenously (iv) elicit in many species, e.g., guinea pigs (Sehic et al., 1996a, b), significant elevations of T_c and VMPO PGE_2 levels within 10 min after administration, whereas the first cytokine to appear in the blood, tumor necrosis factor (TNF)α, is not detectable until, at the earliest, 30 min after injection (Jansky et al., 1995). This would make it unlikely, therefore, that TNFα or other pyrogenic cytokines released subsequently could play a major role in the quick induction of these responses. This is not completely surprising, however, since cytokines are not generally expressed constitutively in mononuclear phagocytes, but are induced de novo in response to these cells' activation by LPS, a synthetic process that requires some time (Conti et al., 2004). By the same token, it has been demonstrated that both the COX-2 and mPGES-1 that catalyze the production of preoptic PGE_2 in response to peripheral LPS are also not constitutive, but rather are upregulated after a ca. 60 plus-min delay (Inoue et al., 2002). Hence, it would also seem improbable that PGE_2 elaborated by this mechanism could mediate the initiation of the febrile response. A different mechanism must therefore be operating to account for the prompt onset of fever after iv LPS.

Pyrogen activation of the brain: the role of the vagus

The first part of the old concept that can probably be rejected is the notion that the cytokines are transported to the brain by the circulation for their action. If they are not detectable in the bloodstream before or coincidently with the beginning of fever, their signaling of the brain is most likely not related to their physical arrival at this site. Indeed, if they had a role at all in initiating fever, the only pathway that would allow the rapid and direct transmission of their message from the periphery to the brain would be neural. Moreover, since the liver contains the body's largest population of mononuclear phagocytes (Kupffer cells, Kc), it is the primary clearinghouse of circulating LPS and therefore also the principal source of LPS-induced cytokines; the pyrogenic signal should therefore originate there. Indeed, the essential role of Kc in fever production was recently verified (Feleder et al., 2003; Li and Blatteis, 2004; Li et al., 2004). Furthermore, a number of studies have now established that the subdiaphragmatic section of both vagal trunks, and its hepatic branches in particular, significantly attenuates the febrile response to LPS (Watkins et al., 1995; Sehic and Blatteis, 1996; Romanovsky et al., 1997a, b; Fleshner et al., 1998; Gaykema et al., 1998; Simons et al., 1998; Wieczorek et al., 2005), albeit that it is still controversial whether this effect is conditional on the route and dose of LPS administered (Goldbach et al., 1997; Hansen et al., 2001). Biotinylated IL-1-receptor antagonist has also been shown to bind to glomus cells in hepatic vagal paraganglia (Goehler et al., 1997), suggesting that this could be the mechanism by which IL-1β activates the vagal afferents. In support, the intraportal vein administration of IL-1β increases the discharge rate of vagal afferents (Niijima, 1996) and the expression of c-*fos* in the nucleus of the solitary tract (NTS, the primary projection area of the vagal nerves) is enhanced after iv and intraperitoneal (ip) IL-1β and iv LPS (Wan et al., 1994). Subdiaphragmatic vagotomy abrogates this effect and electrolytic lesions of the NTS attenuate the febrile response to ip LPS (Wan et al., 1994). However, since Kc do not express cytokines constitutively, they cannot be the factors that rapidly stimulate these terminals after LPS administration. Consequently, we should look for another factor that is quickly liberated in response to circulating LPS and capable of binding to vagal afferents to mediate the febrile response.

PGE_2, not cytokines, is the peripheral fever trigger

Such a factor could be PGE_2. It is produced by Kc activated by LPS and its level rises in venous blood and, to a lesser extent, in arterial blood very quickly after the peripheral administration of both exogenous (e.g., LPS) and endogenous (e.g., IL-1β) pyrogens (Skarnes et al., 1981; Rotondo et al., 1988; Perlik et al., 2005; Li et al., 2006). This raises the possibility that, as was conceived earlier in regard to the cytokines, the PGE_2 that acts in the VMPO could enter it from the blood: being lipophilic, it could either cross the blood-brain barrier (BBB) or diffuse to this site through "leaky" ports in the barrier, e.g., the organum vasculosum laminae terminalis (Blatteis and Sehic, 1997). It is, however, controversial whether PGE_2 can actually pass from the blood into the brain and, especially, whether PGE_2 entering the brain in this manner can raise T_c (Morimoto et al., 1992; Sehic et al., 1996a, b; Abul et al., 1997; Romanovsky et al., 1999). Indeed, although the rise of T_c and preoptic PGE_2 levels induced by iv LPS are both abrogated by peripherally injected antipyretics, findings that they are also prevented by the intraVMPO microinjection of COX-2 inhibitors would indicate that it is more likely that the applicable PGE_2 is generated inside rather than outside the BBB (Steiner et al., 2001). Hence, again, it would seem more probable that its message should be rapidly conveyed from the liver to the brain neurally rather than humorally. In support, abundant PGE_2 receptors of the EP_3 subtype occur in nodose ganglion vagal sensory neurons receiving information from the abdominal compartment (Ek et al., 1998).

There is a problem, however, with the notion that PGE_2 produced by Kc in quick response to LPS is the candidate mediator of the febrile response, because LPS is actually a weak trigger of AA release, the activation of group IV cytosolic phospholipase A_2 ($cPLA_2$) by LPS being very slow (Ambs et al., 1995). Moreover, as indicated earlier, the increased biosynthesis of PGE_2 induced by LPS is selectively catalyzed by inducible COX-2 and mPGES-1, and, in conscious rats at least, their mRNAs are not detectable in liver until 30 min after iv LPS (Ivanov et al., 2002). The delay imposed by this synthetic process consequently implies that the prompt elevation of plasma PGE_2 levels observed in response to iv LPS cannot be accounted for by its COX-2/mPGES-1-mediated production in Kc. A different factor must therefore be involved.

This mediator is the anaphylatoxic complement (C) component C5a (for review, see Blatteis et al., 2004a, b). The C cascade is activated on contact by LPS via the alternative pathway, resulting in the very quick production of all its components, including C5a. Kc express its cognate receptor, $C5aR_1$ (Schieferdecker et al., 1997, 2001; Schlaf et al., 2003). The production of PGE_2 by Kc is initiated within minutes after the addition of C5a, both *in vitro* and *in vivo*; C depletion inhibits this release (Schieferdecker et al., 2001). PGE_2, under these conditions, could be generated by the hydrolysis of membrane-associated phosphoinositide (PI, which has a high arachydonoyl chain content) by PI-specific phospholipase C (PI-PLC); indeed, AA liberation by PI-PLC is 10-fold more rapid (within seconds) than that mediated by $cPLA_2$ (Rhur, 1994). Moreover, PI-PLC is activated by C, but not by LPS or IL-1β, and the subsequent conversion of this AA to PGE_2 is unselectively catalyzed by COX-1 and/or COX-2 (Schütze and Krönki, 1994), which are both constitutive in Kc. Hence, the initial peripheral fever trigger could indeed be PGE_2 released by Kc stimulated by LPS-activated C5a and binding to EP_3 receptors on vagal afferents.

This hypothesis was recently substantiated in a series of studies (for review, see Blatteis et al., 2004a, b; Perlik et al., 2005). Thus, the intraportal vein injection into anesthetized guinea pigs of LPS or cobra venom factor (CVF, another immediate activator of the C cascade that quickly elevates the levels of all the C components, but, unlike LPS, that eventually depletes the C substrate, thereby ultimately causing hypocomplementemia) induced similar increases of PGE_2 in the inferior vena cava (ivc, near its confluence with the hepatic veins) within the first 5 min after treatment. The rises in PGE_2 due to

LPS ⟶ Liver ⟶ C5a ⟶ Kupffer cells ⟶ PGE$_2$

Fig. 1. Depiction of the sequence of steps that occur in the periphery following the iv or ip bolus injection of a small-to-moderate dose of LPS into conscious guinea pigs, culminating in the rapid production and release by Kc of PGE$_2$, the endogenous trigger of the febrile response. Pyrogenic cytokines do not play a role in this phase of fever onset.

CVF returned to control levels within 15 min, whereas those due to LPS increased further 30–45-min later, then stabilized. LPS given to the same animals 3 h after CVF (when the pool of C5a was exhausted) also elevated ivc PGE$_2$, but after a 30–45-min delay. CVF *per se* did not affect the basal level of ivc PGE$_2$. It also did not alter basal ivc TNFα, IL-1β, and IL-6 levels nor their responses to LPS; in the latter case, their rise lagged significantly that of PGE$_2$. These results thus confirm that the PGE$_2$ that appears in the ivc immediately after an LPS challenge originates in the liver and is probably due to the postulated activation of C5a by LPS. The secondary elevation of PGE$_2$ 30–45 min after LPS is presumably due to the Toll-like receptor 4-mediated activation of Kc by LPS and the consequent, delayed upregulation of COX-2/mPGES-1. The essential role of Kc as the cellular source of this PGE$_2$ had been shown earlier (Blatteis et al., 2004a) and was reconfirmed in this study (Perlik et al., 2005). The association between the uptake by Kc of iv or ip injected LPS and the febrile response and that between the latter and the rise in plasma PGE$_2$ were also demonstrated in parallel studies in conscious animals (Blatteis et al., 2004a; Li et al., 2006). Especially important in this regard was the finding that PGE$_2$ antiserum administered iv 10 min before iv LPS prevented the febrile response (Li et al., 2006). In sum, these results strongly support the view that PGE$_2$ is generated by C5a-activated Kc in immediate response to LPS and that it, rather than pyrogenic cytokines, triggers the febrile response. In view of the rapid onset of fever following the iv administration of LPS and the likelihood reviewed earlier that the pyrogenic signal is conveyed to the VMPO neurally, it is probable that the released PGE$_2$ acts on hepatic vagal afferents for this transmission, although its binding to these terminals has not yet been specifically demonstrated. Figure 1 summarizes the sequence of events triggered in the periphery by the iv or ip injection of LPS that initiate the febrile response.

Contribution of preoptic PGE$_2$ to fever production

It is well established that the pyrogenic signal of peripheral LPS activates specific neural systems in the VMPO region. The enhanced expression of c-*fos* in the NTS and its blockade by subdiaphragmatic vagotomy were already mentioned. In addition, the central noradrenergic system is activated, causing the release of norepinephrine (NE), an effect that is also blocked by vagotomy (Fleshner et al., 1995; Ishizuka et al., 1997); the increase in the activity of the noradrenergic system produced by proinflammatory stimuli is well documented (for review, see Dunn, 2001). NE microinjected into the VMPO of conscious guinea pigs raises their T_c (Zeisberger, 1987; Quan and Blatteis, 1989), and electrical stimulation of their ascending noradrenergic system in the NTS produces the same effect (Szelenyi et al., 1976, 1977) whereas chemical sympathectomy prevents this response (Zeisberger, 1987). NE also stimulates the release of PGE$_2$ in brain tissue *in vitro* (Hori et al., 1987), and the microdialysis of NE into the VMPO of conscious guinea pigs augments the local production of PGE$_2$ (Sehic et al., 1996a, b; Feleder et al., unpublished observation); in turn, the released PGE$_2$ inhibits the further presynaptic release of NE (Bergstrom et al., 1973; for review, see Hedqvist, 1977). Hence, NE locally released in the VMPO consequent to the vagally transmitted hepatic PGE$_2$ signal could induce the production of preoptic PGE$_2$ and thereby account for its presence and traditional pyrogenic function there. Indeed, the release of NE in the VMPO following the systemic administration of LPS or IL-1β has been demonstrated in several species (Linthorst et al., 1995; Wieczorek et al., 2005; Feleder et al., unpublished observation; for review, see Dunn, 2001). It has further been reported that pretreatment with the non-selective COX inhibitor aspirin prevents the T_c rise produced by PGE$_2$ microinjected into a lateral ventricle (Navarro et al., 1988).

If NE were the direct stimulus for the increased production of PGE$_2$ in the VMPO in response to peripheral LPS-stimulated, vagally transmitted signals, the induction of PGE_2 should begin promptly and, according to current dogma, be mediated by COX-2/mPGES-1. Indeed, the LPS-induced rise in preoptic PGE_2 occurs promptly following the peripheral injection of LPS (Sehic et al., 1996a, b; Feleder et al., unpublished observation; for review, see Blatteis, 1997). However, the increase in preoptic NE that attends the course of the febrile response to LPS is associated with the early rather than the late phase of fever (Linthorst et al., 1995; Feleder et al., unpublished observation). Hence, the temporal discrepancy between the occurrence of NE and PGE_2 in the VMPO would seem to argue against their functional interaction. There is another challenge: as mentioned earlier, although, in brain, both COX-2 and mPGES-1 are expressed constitutively, but rather minimally, in endothelial cells and, to a lesser extent, in neuronal cell bodies, dentritic spines, and glia, their role in fever production is attributed to their upregulation in endothelial and/or glial cells, but not in neurons (Breder and Saper, 1996; Matsumura et al., 1997; Cao et al., 1998; Quan et al., 1998). Moreover, their mRNAs become detectable in the VMPO between 0.5 and 4 h after the iv or ip administration of LPS or a cytokine (Inoue et al., 2002; Ivanov et al., 2002). The prompt elevation of preoptic PGE_2 could therefore not be due to its COX-2/mPGES-1-mediated production. Consequently, its production should be mediated either via COX-1 or via a COX-independent pathway.

In a series of experiments designed to clarify the mechanism that could underlie the production of PGE_2 in the VMPO in response to local NE and, at the same time, to identify the adrenoceptor (AR) subtype(s) that could be involved in this effect, NE and specific α_1- and α_2-AR agonists and antagonists were microdialyzed into the VMPO of conscious guinea pigs pretreated intraVMPO with selective COX-1 and COX-2 inhibitors (Feleder et al., 2004). The results unexpectedly revealed that NE mediates not one, but two successive T_c rises, each associated with a different mechanism. Thus, the intraVMPO microdialysis of the selective α_1-AR agonist cirazoline induced a prompt T_c rise without, remarkably, affecting basal preoptic PGE_2 levels, whereas the selective α_2-AR agonist clonidine caused a significantly delayed T_c rise that followed an initial T_c fall; both these responses were accompanied by parallel changes in the levels of VMPO PGE_2. The elevation of PGE_2, however, was not associated under these conditions with a demonstrable upregulation of COX-2 (Feleder et al., unpublished observation), implying, therefore, that it was mediated by the activation of this enzyme, not its de novo synthesis. The thermal effects of both agonists were validated by their blockade by their respective, selective antagonists, prazosin and yohimbine. Furthermore, both the increases in T_c and preoptic PGE_2 levels caused by clonidine were prevented by the intraVMPO microdialysis of a selective COX-2 inhibitor, MK-0663; they were unaffected by that of the selective COX-1 inhibitor SC-560, although it unexpectedly suppressed the initial decreases of both these variables caused by this α_2-AR agonist. The intraVMPO microdialysis of NE reproduced the early and the late T_c and preoptic PGE_2 level changes induced by clonidine, but not the early T_c rise evoked by cirazoline; both the clonidine-mediated effects were inhibited by yohimbine and MK-0663, but not by prazosin. Cirazoline and clonidine microdialyzed together replicated the late T_c rises but not the early T_c falls elicited by clonidine and NE, and also not the early T_c rises caused by cirazoline; on the other hand, their co-microdialysis induced both the VMPO PGE_2 falls and rises associated with clonidine and NE. In further support of the non-involvement of PGE_2 in the cirazoline-induced T_c rise and, in contrast, the involvement of COX-2-dependent PGE_2 in that caused by clonidine, conscious, wild-type, and COX-1$^{-/-}$ mice respond similarly to the intracerebroventricular (icv) microinjection of these two α-AR agonists as the correspondingly treated guinea pigs; and, most relevant to this context, the late clonidine-induced T_c rise is absent in COX-2$^{-/-}$ mice (Blatteis et al., 2004a).

The finding that the specific α_1-AR agonist cirazoline evoked very quickly a rise in T_c without the intermediation of PGE_2 was indeed remarkable and unexpected. It suggested that the α_1-AR activated by NE are located on postsynaptic warm-sensitive or thermoinsensitive neurons in the VMPO and that NE directly reduces or augments, respectively, the activities of these neurons; according to the classical model of Hammel (1965); both responses promote heat conservation. Since these neurons, moreover, are thought to inhibit synaptically connected cold-sensitive neurons, these are concomitantly facilitated, stimulating heat production; i.e., in combination, these effector mechanisms raise T_c. Support for the direct, PGE_2-independent involvement of α_1-AR is also provided by preliminary data from single-unit extracellular recordings of the discharge rates of thermally characterized, individual neurons in horizontal slices of rat hypothalami. In the experiments to date, cirazoline inhibited three of four warm-sensitive neurons and excited two of three thermoinsensitive neurons (Boulant and Blatteis, unpublished observation). The specific α_1-AR subtype involved in this hyperthermic effect remains to be identified.

The findings that, when the microdialysis of clonidine was continued for another 3 h, the initial hypothermic response was followed by a protracted T_c rise and an associated increase in the animals' levels of preoptic PGE_2 were both also novel. Since both these responses were also inhibited by yohimbine pretreatment, they were therefore also α_2-AR-mediated; and since, moreover, they were prevented by the prior microdialysis of the selective COX-2 inhibitor MK-0663, this late hyperthermic response was specifically mediated by COX-2-dependent PGE_2. The brain cell type expressing COX-2 in response to NE in this context remains to be determined. But, since the increase of COX-2 following the peripheral administration of LPS has been observed, as already mentioned, in glial and cerebromicrovascular endothelial cells, but only irregularly in neurons, we conjecture that the PGE_2 collected in the microdialysate effluents from the VMPO interstitial space in these experiments was generated by astrocytic processes contacting noradrenergic synaptic regions rather than by postsynaptic (warm-sensitive) neurons. The subsequent effect of the thus released PGE_2 on the electrical activities of VMPO warm-sensitive and thermoinsensitive neurons is presumptively similar to that indicated earlier for α_1-AR-mediated responses, viz., a reduction and an increase in their discharge rates, respectively. The PGE_2-sensitive receptor involved in these neuronal effects is probably the EP_3 subtype; it has been linked to the development of fever and is present in the VMPO (Ushikubi et al., 1998; Ek et al., 2000; for review, see Oka, 2004). The receptor implicated in the inhibition of presynaptic NE release by PGE_2 has also been previously identified as the EP_3 subtype (Schlicker and Gothert, 1998). These results thus demonstrated that the induction of PGE_2 by NE in the VMPO is modulated by α_2-AR and indeed catalyzed by COX-2. The identity of the α_2-AR subtype involved in this effect remains to be elucidated.

Taken together, the preceding data would suggest, therefore, that the NE released in the VMPO in response to the vagally conveyed pyrogenic message of Kc-generated PGE_2 could mediate the febrile response of guinea pigs in the following two, successive ways: (1) it could induce the first of the characteristic two T_c rises evoked in conscious guinea pigs by iv LPS by rapidly activating α_1-AR without the intermediation of PGE_2, and (2) it could cause the second of the two T_c rises and also the delayed, single rise produced by ip LPS (due to the late arrival of LPS in the liver [Li and Blatteis, 2004]) by also stimulating at the same time α_2-AR, consequently activating (after the delay imposed by the de novo synthesis of the enzymes) the production and release of COX-2/mPGES-1-dependent PGE_2 in the POA. This hypothesis was tested (Feleder et al., unpublished observation) by measuring over short intervals the febrile response and the levels of NE and PGE_2 in the interstitial fluid of the VMPO of conscious guinea pigs pretreated with intraVMPO prazosin or yohimbine (or their vehicles) to the iv or Ip injection of *Salmonella enteritidis* LPS, following the same protocol as in the preceding study. In guinea pigs, pyrogenic doses of iv injected LPS rapidly and characteristically evoke two successive T_c rises associated with concurrent elevations in preoptic PGE_2 levels (Sehic et al., 1996a, b); however, the second rise is attenuated significantly more than the first by COX-2 blockade (Steiner et al., 2001). Ip injected LPS, on the other hand, induces coincident monophasic T_c and PGE_2 responses after some delay (Inoue et al., 2002). The iv injection of a low-to-moderate dose of LPS (2 µg/kg) promptly induced the appearance of NE in the VMPO interstitial fluid; its level culminated in 30 min, then gradually returned toward its control value over the following 2 h. T_c and preoptic PGE_2 levels also both increased promptly, first rapidly, then more slowly, and in correspondence with each other; the first T_c peak was reached in ~60 min and the second in ~150 min. Pretreatment with the α_1-AR antagonist prazosin significantly slowed the rate of rise of the early phase of fever and eliminated the first T_c peak; but it did not affect the onset latency of the febrile response nor the magnitude and time of the second peak of fever. It also did not affect the initial, LPS-induced elevation of preoptic PGE_2; but, importantly, this increase was not sustained: PGE_2 levels returned from their first highs at 30 min to their control values at 60 min. They then resumed their rise, culminating not differently than their untreated counterparts and coincidentally with the second peak of fever (Feleder et al., unpublished observations). These results thus confirmed that the onset of fever evoked in conscious guinea pigs by iv LPS is associated with the intraVMPO release of NE and accompanied by coincident increases in T_c and preoptic PGE_2 levels, in conformity with previous findings (Linthorst et al., 1995; Sehic et al., 1996a, b; Wieczorek et al., 2005). They further demonstrated that the initial T_c rise is indeed mediated by the NE-induced activation of α_1-AR and intimated that, its very onset apparently excepted, the first phase of fever was maintained by the direct noradrenergic activation of the relevant neurons in the VMPO, without the intermediation of PGE_2. Pretreatment with the α_2-AR antagonist yohimbine, by contrast, did not change the onset and rate of rise of the early febrile response, the first T_c peak reaching the same value and occurring at the same time as that of the untreated controls; but it completely abrogated the second peak, consequently significantly reducing the overall magnitude of the febrile response by comparison with that of untreated controls. The fever, moreover, abated more slowly in this group than in its controls. Remarkably, this treatment completely suppressed the LPS-induced rise in preoptic PGE_2 levels (Feleder et al., unpublished observations). These findings thus indicated that the LPS fever of these α_2-AR-blocked guinea pigs is initiated, maintained, and even extended in the total absence of corresponding increases in VMPO PGE_2 levels. Hence, by deduction, it is mediated entirely by PGE_2-independent, α_1-AR activation. Neither the microdialysis of prazosin nor of yohimbine *per se* affected the animals' T_c and preoptic PGE_2 levels (Feleder et al., unpublished observations). In sum, it would appear that the early phase of LPS fever is indeed mediated independently of PGE_2 by α_1-AR stimulation whereas the late phase is dependent on PGE_2 induced consequent to α_2-AR stimulation, as hypothesized. But then what accounts for the initial rise in preoptic PGE_2 of all the animals, including the yohimbine-pretreated ones?

The first possibility is that COX-1 could mediate the initial, LPS-induced PGE_2 rise. But this possibility was refuted by findings that the intraVMPO microdialysis of the selective COX-1 inhibitor SC-560 had absolutely no effect on the onset, height, and course of T_c and the associated changes in preoptic PGE_2 levels induced in conscious guinea pigs by iv LPS (Feleder et al., unpublished observation). Acetaminophen, a putatively selective inhibitor of COX-1 variant retaining intron 1 (for review, see Botting and Ayoub, 2005), also did not affect the T_c and intraVMPO PGE_2 responses to iv LPS (Feleder et al., unpublished observation). Administration of the selective COX-2 inhibitor MK-0663, by contrast, did not alter the early T_c and preoptic PGE_2 rises, but prevented both their late rises (Feleder et al., unpublished observation). These results, therefore, documented again that the late, but not the early phase of LPS fever is mediated by COX-2-dependent PGE_2, and that COX-1 evidently plays no role in the febrile response (except, presumptively in part, in the initial induction of PGE_2 by C5a-stimulated Kc, as mentioned earlier).

Since the intraVMPO microdialysis of selective COX-1, -2, and -3 inhibitors do not prevent the prompt, initial increase in PGE_2 levels observed in the POA/VMPO following the peripheral administration of LPS, it may be surmised that this rise is induced by a COX-independent pathway, i.e., presumptively by the non-enzymatic isoprostane pathway of free radical-catalyzed peroxidation of AA (for review, see Basu and Helmersson, 2005). Indirect support for the involvement of free radicals in the observed, initial PGE_2 elevation comes from the finding that the intraVMPO microdialysis of the antioxidant catechin throughout the febrile course had the same effect as pretreatment with the α_2-AR antagonist yohimbine, i.e., suppression of both the LPS-induced early and late rises of preoptic PGE_2 and of the second peak of fever; the latency of fever onset and its first peak were not affected, the fever continuing at its early phase high level until it abated normally (Feleder et al., unpublished observation). The free radicals in this case could be generated by the auto-oxidation of NE and/or nitric oxide (NO). The latter is also released locally in the VMPO after LPS administration (see below); however, no direct measurement of biomarkers of oxidative stress was made in these studies. It is not clear at this writing whether catechin also inhibited the expression of COX-2 and, hence, the production of COX-2-dependent PGE_2 (Wang et al., 2004). But it has been reported previously that fever is prevented when the production of free radicals is blocked. The effect was ascribed then not to the inhibition by antioxidants of isoprostane production, but rather to their reduction of the thiol groups attached to N-methyl-D-aspartate receptors, thereby depressing glutamate-mediated neuronal excitability and, hence, limiting fever (Riedel, 1997; Riedel and Maulik, 1999; Riedel et al., 2003); the activation by peripheral pyrogens of glutaminergic pathways both in the NTS and the OVLT has been reported (Mascarucci et al., 1998; Huang et al., 2001, 2004). It was therefore suggested that NO and oxygen radicals, but not PGE_2, modulate fever (Riedel et al., 2003). Indeed, since the initial T_c rise provoked by the peripheral administration of LPS is evidently mediated by PGE_2-independent α_1-AR activation in the VMPO and since, moreover, the febrile rise can evidently be sustained by this mechanism alone until its normal abatement, it raises the heretic possibility that, in contrast to PGE_2 generated in the liver, PGE_2 in the VMPO, however it may occur there, may not be material to the febrile response!

Finally, reports that the gaseous transmitter NO stimulates the biosynthesis of PGE_2 by increasing the activities of both isoforms of COX prompted us to investigate whether NO could also have a pyretic function in the central mediation of the febrile response. Indeed, the various isoforms of NO synthase (endothelial, neural, and inducible NOS), the enzyme that converts L-arginine into citrulline and NO, occur in the hypothalamus (for review, see Mollace et al., 2005), and circulating LPS and cytokines stimulate the release of NO in the VMPO (for review, see Schmid et al., 1998; Simon, 1998; Gerstberger, 1999). Furthermore, as reviewed above, NE induces the production of COX-2/mPGES-1-dependent PGE_2 in the VMPO via an α_2-AR-mediated mechanism, and others have shown that NE activates NOS (Canteros et al., 1996). Hence, the existence of a pyrogenic NE-NO-PGE_2 cascade in the POA in response to iv LPS seems plausible. But the testing of this hypothesis revealed that, quite to the contrary, NO donors microdialyzed into the POA of conscious

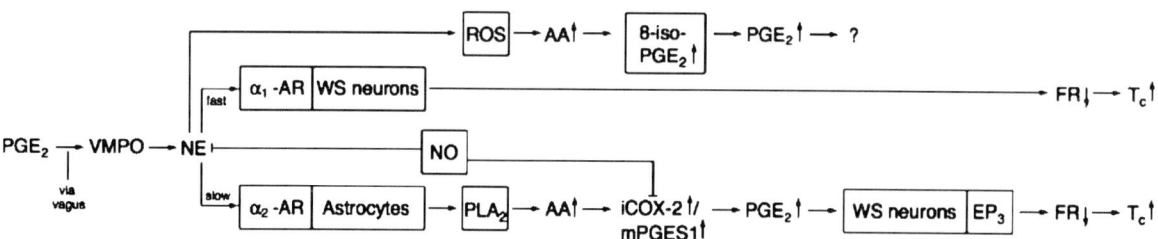

Fig. 2. Schematic illustration of the dual mechanisms, one fast, α_1-AR-mediated and PGE_2-indenpendent, the other slow, α_2-AR-mediated and COX-2/mPGES-1-derived PGE_2-dependent, activated by NE liberated in the VMPO. Both mechanisms sequentially lead to the depression of the firing rates (FR) of local warm-sensitive (WS) neurons, thereby causing T_c to rise. NO inhibits the release of NE and may also inhibit the upregulation of COX-2. Reactive oxygen species (ROS) due to the auto-oxidation of NE or derived from NO may also induce isoprostanes (8-iso-PGE_2) that are quickly converted to PGE_2; it, however, does not appear to contribute to the initial T_c rise. Circulating pyrogenic cytokines and cerebromicrovascular endothelial cells-generated PGE_2 probably contribute to sustaining the course of the late phase of fever.

guinea pigs inhibit rather than promote the febrile response to iv LPS and that they do so by inhibiting the LPS-induced release of NE in the VMPO and consequently preventing the α_2-AR-mediated activation of COX-2-dependent PGE_2 synthesis (Feleder et al., unpublished observation). NO scavengers microdialyzed into the POA have exactly the opposite effects (Feleder et al., unpublished observation). Indeed, previous data regarding a potential role of NO in fever have been conflicting, some indicating pyretic and others antipyretic effects (Riedel, 1997; Gerstberger, 1999; Kozak and Kozak, 2003). The present results would indicate, therefore, that NO, presumptively released in the VMPO coincidentally with or very shortly after NE, serves as a local negative-regulatory factor, i.e., it is a central, endogenous antipyretic mediator of LPS fever. Figure 2 illustrates the sequence of events occurring in the VMPO following the arrival of the peripheral PGE_2-induced, vagally transmitted, pyrogenic signal.

Conclusions

LPS-induced fever arises as the result of a complex, phased sequence of interactions among soluble factors and cells that is initiated in the periphery and then transmitted neurally to the VMPO, which modulates the febrile response. Thus, novel evidence accumulated over the past 10 years suggests that the febrigenic process is initiated by the arrival of LPS in the liver and its uptake by Kc, causing virtually immediately the activation of the C cascade and, hence, the generation of the anaphylatoxin C5a. It, in turn, stimulates the Kc very rapidly to release constitutive COX-1- and COX-2-dependent PGE_2. The released PGE_2 activates local sensory vagal terminals that project to the NTS. From the NTS, the input of PGE_2 is transmitted to the VMPO via the ventral noradrenergic bundle. NE consequently secreted in the VMPO activates both local α_1- and α_2-AR. The stimulation of the first rapidly evokes an initial rise in T_c which is associated with a decrease in the firing rates of preoptic warm-sensitive neurons, inhibiting heat loss and stimulating heat production, but is not accompanied by any change in the levels of preoptic PGE_2. Stimulation of the second causes, after a significant delay, a second, more prolonged T_c rise that is associated with a concurrent increase in preoptic COX-2/mPGES-1-dependent PGE_2 levels. Hence, two distinctly produced PGE_2s would appear to mediate the febrile response: one, generated in the liver, is the immediate distal trigger of the febrile response, and the other, produced in the VMPO, is its subsequent proximal modulator. The late phase is probably also supported by meanwhile produced circulating cytokines and PGE_2 as well as PGE_2 generated by cerebral endothelial and/or other cells, all acting presumably as originally proposed for their roles.

In summary, the key, new findings regarding the initiation of LPS fever are: (1) peripheral PGE_2 rather than pyrogenic cytokines initiates the febrile response, (2) NE propagates the pyrogenic signal forward within the VMPO, (3) NO modulates its release in the VMPO and, hence, the intensity of the febrile response, and (4) COX-2/mPGES-1-dependent PGE_2 mediates the late, but not the early phase of fever; the latter appears to be independent of COX-derived PGE_2.

Acknowledgments

This research was supported by National Institute of Health grants Nos. NS-34857 and NS-38594. I gratefully acknowledge the invaluable contributions of my co-workers in this research, Drs. Elmir Sehic, Shuxin Li, Carlos Feleder, Vit Perlik, Zhonghua Li, and Ying Tang, without whose hard work and unswerving dedication these studies could not have been performed.

References

Abul, H.T., Davidson, J., Milton, A.S. and Rotondo, D. (1997) Prostaglandin E_2 enters the brain following stimulation of the acute phase immune response. Ann. N.Y. Acad. Sci., 813: 287–295.

Ambs, P., Baccarini, M., Fitzke, E. and Dieter, P. (1995) Role of cytosolic phospholipase A2 in arachidonic acid release of rat-liver macrophages: regulation by Ca^{2+} and phosphorylation. Biochem. J., 311: 189–195.

Basu, S. and Helmersson, J. (2005) Factors regulating isoprostane formation in vivo. Antioxid. Redox Signal, 7: 221–235.

Bergstrom, S., Farnebo, L.O. and Fuxe, K. (1973) Effect of prostaglandin E_2 on central and peripheral catecholamine neurons. Eur. J. Pharmacol., 21: 362–368.

Blatteis, C.M. (1997) Prostaglandin E_2: a putative fever mediator. In: Mackowiak P.A. (Ed.), Fever: Basic Mechanisms and Management (2nd ed.). Lippincott-Raven, Philadelphia, PA, pp. 117–145.

Blatteis, C.M., Li, S., Li, Z., Perlik, V. and Feleder, C. (2004a) Complement is required for the induction of endotoxic fever in guinea pigs and mice. J. Thermal Biol., 9: 369–381.

Blatteis, C.M., Li, S., Li, Z., Perlik, V. and Feleder, C. (2004b) Signaling the brain in systemic inflammation: the role of complement. Front. Biosci., 9: 915–931.

Blatteis, C.M. and Sehic, E. (1997) Fever: how may circulating pyrogens signal the brain? News Physiol. Sci., 12: 1–7.

Botting, R. and Ayoub, S.A. (2005) COX-3 and the mechanism of action of paracetamol/acetaminophen. Prostaglandins Leukot. Essent. Fatty Acids, 72: 85–87.

Boulant, J.A. (2000) Role of the preoptic-anterior hypothalamus in thermoregulation and fever. Clin. Infect. Dis., 31(Suppl 5): S157–S161.

Breder, C.D. and Saper, C.B. (1996) Expression of inducible cyclooxygenase mRNA in the mouse brain after systemic administration of bacterial lipopolysaccharide. Brain Res., 713: 64–69.

Canteros, G., Rettori, V., Genaro, A., Suburo, A., Gimeno, M. and McCann, S.M. (1996) Nitric oxide synthase content of hypothalamic explants; increased by norepinephrine and inactivated by NO and cGMP. Proc. Natl. Acad. Sci. U.S.A., 93: 4246–4250.

Cao, C., Matsumura, K., Yamagata, K. and Watanabe, Y. (1998) Cyclooxygenase-2 is induced in brain blood vessels during fever evoked by peripheral or central administration of tumor necrosis factor. Brain Res. Mol. Brain Res., 56: 45–56.

Conti, B., Tabarean, I., Andrei, C. and Bartfai, T. (2004) Cytokines and fever. Front. Biosci., 9: 1433–1449.

Dinarello, C.A. (2004) Infection, fever, and exogenous and endogenous pyrogens: some concepts have changed. J. Endotoxin Res., 10: 201–222.

Dunn, A.J. (2001) Effects of cytokines and infections on brain neurochemistry. In: Ader R., Felten D.L. and Cohen N. (Eds.), Psychoneuroimmunology. Academic Press, New York, pp. 645–666.

Dunn, A.J. (2002) Mechanisms by which cytokines signal the brain. Int. Rev. Neurobiol., 52: 43–65.

Ek, M., Arias, C., Sawchenko, P. and Ericsson-Dahlstrand, A. (2000) Distribution of the EP3 prostaglandin E_2 receptor subtype in the rat brain: relationship to sites of interleukin-1-induced cellular responsiveness. J. Comp. Neurol., 428: 5–20.

Ek, M., Kurosowa, M., Lundeberg, T. and Ericsson, A. (1998) Activation of vagal afferents after intravenous injection of interleukin-1β: role of endogenous prostaglandins. J. Neurosci., 18: 9471–9479.

Feleder, C., Li, Z., Perlik, V., Evans, A. and Blatteis, C.M. (2003) The spleen modulates the febrile response of guinea pigs to LPS. Am. J. Physiol., 284: R1466–R1476.

Feleder, C., Perlik, V. and Blatteis, C.M. (2004) Preoptic α_1- and α_2-noradrenergic agonists induce, respectively, PGE_2-independent and PGE_2-dependent hyperthermic responses in guinea pigs. Am. J. Physiol., 286: R1156–R1166.

Fleshner, M., Goehler, L.E., Hermann, J., Relton, J.K., Maier, S.F. and Watkins, L.R. (1995) Interleukin-1β-induced corticosterone elevation and hypothalamic NE depletion is vagally mediated. Brain Res. Bull., 37: 605–610.

Fleshner, M., Goehler, L.E., Schwartz, B.A., McGorry, M., Martin, D., Maier, S.F. and Watkins, L.R. (1998) Thermogenic and corticosterone responses to intravenous cytokines (IL-1β and TNF-α) are attenuated by subdiapgragmatic vagotomy. J. Neuroimmunol., 86: 134–141.

Gaykema, R.P.A., Goehler, L.E., Tilders, F.J.H., Bol, J.G.J.M., McGorry, M., Fleshner, M., Maier, S.F. and Watkins, L.R. (1998) Bacterial endotoxin induces Fos immunoreactivity in primary afferent neurons of the vagus nerve. Neuroimmunomodulation, 5: 234–240.

Gerstberger, R. (1999) Nitric oxide and body temperature control. News Physiol. Sci., 14: 30–36.

Goehler, L.E., Relton, J.K., Dripps, D., Kiechle, R., Tartaglia, N., Maier, S.F. and Watkins, L.R. (1997) Vagal paraganglia bind biotinylated interleukin-1 receptor antagonists (IL-1ra) in the rat: a possible mechanism of immune-to-brain communication. Brain Res. Bull., 43: 357–364.

Goldbach, J.M., Roth, J. and Zeisberger, E. (1997) Fever suppression by subdiaphragmatic vagotomy in guinea pigs depends on the route of pyrogen administration. Am. J. Physiol., 272: R675–R681.

Hammel, H.T. (1965) Neurons and temperature regulation. In: Yamamoto W.S. and Brobeck J.R. (Eds.), Physiological Controls and Regulations. W.B. Saunders, Philadelphia, PA, pp. 71–97.

Hansen, M.K., O'Connor, K.A., Goehler, L.E., Watkins, L.R. and Maier, S.F. (2001) The contribution of the vagus nerve in interleukin-1beta-induced fever is dependent on dose. Am. J. Physiol., 280: R929–R934.

Hedqvist, P. (1977) Basic mechanisms of prostaglandin action on autonomic neurotransmission. Annu. Rev. Pharmacol. Toxicol., 17: 259–279.

Hori, Y., Blatteis, C.M. and Nasjletti, A. (1987) Production of PGE_2 by brain slices stimulated by various thermoactive agents. Fed. Proc., 46: 683.

Huang, W.T., Tsai, S.M. and Lin, M.T. (2001) Involvement of brain glutamate in pyrogenic fever. Neuropharmacology, 41: 811–818.

Huang, W.T., Wang, J.J. and Lin, M.T. (2004) Cyclooxygenase inhibitors attenuate augmented glutamate release in organum vasculosum laminae terminalis and fever induced by staphylococcal enterotoxin A. J. Pharmacol. Sci., 94: 192–196.

Inoue, W., Matsumura, K., Yamagata, K., Takemiya, T., Shiraki, T. and Kobayashi, S. (2002) Brain-specific endothelial induction of prostaglandin E_2 synthesis enzymes: its temporal relation to fever. Neurosci. Res., 44: 51–61.

Ishizuka, Y., Ishida, Y., Kunitake, T., Kato, K., Hanamori, T., Mitsuyama, Y. and Kannan, H. (1997) Effects of area postrema lesion and abdominal vagotomy on interleukin-1β-induced norepinephrine release in the hypothalamic paraventricular nucleus region in the rat. Neurosci. Lett., 223: 57–60.

Ivanov, A.I., Pero, R.S., Scheck, A.C. and Romanovsky, A.A. (2002) Prostaglandin E (2)-synthesizing enzymes in fever: differential transcriptional regulation. Am. J. Physiol., 283: R1104–R1117.

Ivanov, A.I. and Romanovsky, A.A. (2004) Prostaglandin E_2 as a mediator of fever: synthesis and catabolism. Front. Biosci., 9: 1977–1993.

Jansky, L., Vybiral, S., Pospisilova, D., Roth, J., Dornand, J. and Zeisberger, E. (1995) Production of systemic and hypothalamic cytokines during the early phase of endotoxin fever. Neuroendocrinology, 62: 55–61.

Kozak, W. and Kozak, A. (2003) Genetic models in applied physiology: selected contributions: differential role of nitric oxide synthase isoforms in fever of different etiologies: studies using Nos gene-deficient mice. J. Appl. Physiol., 94: 2534–2544.

Li, Z. and Blatteis, C.M. (2004) Fever onset is linked to the appearance of lipopolysaccharide in the liver. J. Endotoxin Res., 10: 1–15.

Li, Z., Feleder, C. and Blatteis, C.M. (2004) Lipopolysaccharide challenge causes exaggerated fever and increased hepatic lipopolysaccharide uptake in vinblastine-induced leukopenic guinea pigs. Crit. Care Med., 32: 2131–2134.

Li, Z., Perlik, V., Feleder, C., Tang, Y. and Blatteis, C.M. (2006) Kupffer cell-generated PGE2 triggers the febrile response of guinea pigs to intravenously injected LPS. Am. J. Physiol., 290: R1262–R1270.

Linthorst, A.C., Flachskamm, C., Holsboer, F. and Reul, J.M. (1995) Intraperitoneal administration of bacterial endotoxin enhances noradrenergic neurotransmission in the rat preoptic area: relationship with body temperature and hypothalamic–pituitary–adrenocortical axis activity. Eur. J. Neurosci., 7: 2418–2430.

Mascarucci, P., Perego, C., Terrazzino, S. and De Simoni, M.G. (1998) Glutamate release in the nucleus tractus solitarius induced by peripheral lipopolysaccharide and interleukin-1β. Neuroscience, 86: 1285–1290.

Matsumura, K., Cao, C. and Watanabe, Y. (1997) Possible role of cyclooxygenase-2 in the brain vasculature in febrile response. Ann. N.Y. Acad. Sci., 813: 302–306.

Mollace, V., Muscoli, C., Masini, E., Cuzzocrea, S. and Salvemini, D. (2005) Modulation of prostaglandin biosynthesis by nitric oxide and nitric oxide donors. Pharmacol. Rev., 57: 217–252.

Morimoto, A., Morimoto, K., Watanabe, T., Sakata, Y. and Murakami, N. (1992) Does an increase in prostaglandin E_2 in the blood circulation contribute to a febrile response in rabbits? Brain Res. Bull., 29: 189–192.

Navarro, E., Romero, S.D. and Yaksh, T.L. (1988) Release of prostaglandin E_2 from brain of cat: II. In vivo studies on the effects of adrenergic, cholinergic and dopaminergic agonists and antagonists. Neuropharmacology, 27: 1067–1072.

Niijima, A. (1996) The afferent discharges from sensors for interleukin-1β in the hepatoportal system in the anesthetized rat. J. Autonom. Nerv. Syst., 61: 287–291.

Oka, T. (2004) Prostaglandin E_2 as a mediator of fever: the role of prostaglandin E (EP) receptors. Front. Biosci., 9: 3046–3057.

Perlik, V., Li, Z., Goorha, S., Ballou, L.R. and Blatteis, C.M. (2005) LPS-activated complement, not LPS per se, triggers the early release of PGE2 by Kupffer cells. Am. J. Physiol., 289: R332–R339.

Quan, N. and Blatteis, C.M. (1989) Intrapreoptically microdialyzed and microinjected norepinephrine evokes different thermal responses. Am. J. Physiol., 257: R816–R821.

Quan, N., Whiteside, M. and Herkenham, M. (1998) Cyclooxygenase-2 mRNA expression in rat brain after peripheral injection of lipopolysaccharide. Brain Res., 802: 189–197.

Rhur, S.G. (1994) Regulation of phosphoinositide-specific phospholipase C by G protein. In: Liscovitch M. (Ed.), Signal-Activated Phospholipases. RG Landes, Austin, TX, pp. 1–12.

Riedel, W. (1997) Antipyretic role of nitric oxide during endotoxin-induced fever in rabbits. Int. J. Tissue React., 19: 171–178.

Riedel, W., Lang, U., Oetjen, U., Schlapp, U. and Shibata, M. (2003) Inhibition of oxygen radical formation by methylene blue, aspirin, or α-lipoic acid, prevents bacterial-lipopolysaccharide-induced fever. Mol. Cell Biochem., 247: 83–94.

Riedel, W. and Maulik, G. (1999) Fever: an integrated response of the central nervous system to oxidative stress. Mol. Cell Biochem., 196: 125–132.

Romanovsky, A.A., Ivanov, A.I. and Karman, E.K. (1999) Blood-borne, albumin-bound prostaglandin E_2 may be involved in fever. Am. J. Physiol., 276: R1840–R1844.

Romanovsky, A.A., Kulchitsky, V.A., Simons, C.T., Sugimoto, N. and Szekely, M. (1997b) Febrile responsiveness of vagotomized rats is suppressed even in the absence of malnutrition. Am. J. Physiol., 273: R777–R783.

Romanovsky, A.A., Simons, C.T., Szekely, M. and Kulchitsky, V.A. (1997a) The vagus nerve in the thermoregulatory response to systemic inflammation. Am. J. Physiol., 273: R407–R413.

Roth, J. and De Souza, G.E. (2001) Fever induction pathways: evidence from responses to systemic or local cytokine formation. Braz. J. Med. Biol. Res., 34: 301–314.

Rotondo, D., Abul, H.T., Milton, A.S. and Davidson, J. (1988) Pyrogenic immunomodulators increase the level of prostaglandin E_2 in the blood simultaneously with the onset of fever. Eur. J. Pharmacol., 154: 145–152.

Saper, C.B. (1998) Neurobiological basis of fever. Ann. N.Y. Acad. Sci., 856: 90–94.

Schieferdecker, H.L., Rothermel, E., Timmermann, A., Gotze, O. and Jungermann, K. (1997) Anaphylatoxin C5a receptor mRNA is strongly expressed in Kupffer and stellate cells and weakly in sinusoidal endothelial cells but not in hepatocytes of normal rat liver. FEBS Lett., 406: 305–308.

Schieferdecker, H.L., Schlaf, G., Jungermann, K. and Gotze, O. (2001) Functions of anaphylatoxin C5a in rat liver: direct and indirect actions on nonparenchymal and parenchymal cells. Int. Immunopharmacol., 1: 469–481.

Schlaf, G., Schmitz, M., Rothermel, E., Jungermann, K., Schieferdecker, H.L. and Gotze, O. (2003) Expression and induction of anaphylatoxin C5a receptors in the rat liver. Histol. Histopathol., 18: 299–308.

Schlicker, E. and Gothert, M. (1998) Interactions between the presynaptic alpha2-autoreceptor and presynaptic inhibitory heteroreceptors on noradrenergic neurones. Brain Res. Bull., 47: 129–132.

Schmid, H.A., Riedel, W. and Simon, E. (1998) Role of nitric acid in temperature regulation. Prog. Brain Res., 115: 25–49.

Schütze, S. and Krönki, M. (1994) Activation of phosphatidylcholine-specific phospholipase C by cytokines. In: Liscovitch M. (Ed.), Signal-Activated Phospholipases. RG Landes, Austin, TX, pp. 101–123.

Sehic, E. and Blatteis, C.M. (1996) Blockade of lipopolysaccharide-induced fever by subdiaphragmatic vagotomy in guinea pigs. Brain Res., 726: 160–166.

Sehic, E., Szekely, M., Ungar, A.L., Oladehin, A. and Blatteis, C.M. (1996a) Hypothalamic PGE_2 during lipopolysaccharide-induced fever in guinea pigs. Brain Res. Bull., 39: 391–399.

Sehic, E., Ungar, A.L. and Blatteis, C.M. (1996b) Interaction between norepinephrine and prostaglandin E_2 in the preoptic area of guinea pigs. Am. J. Physiol., 271: R528–R536.

Simon, E. (1998) Nitric oxide as a peripheral and central mediator in temperature regulation. Amino Acids, 14: 87–93.

Simons, C.T., Kulchitsky, V.A., Sugimoto, N., Homer, L.D., Szekely, M. and Romanovsky, A.A. (1998) Signaling the brain in systemic inflammation: which vagal branch is involved in fever genesis? Am. J. Physiol., 275: R63–R68.

Skarnes, R.C., Brown, S.K., Hull, S.S. and McCracken, J.A. (1981) Role of prostaglandin E in the biphasic fever response to endotoxin. J. Exp. Med., 154: 1212–1224.

Steiner, A.A., Li, S., Llanos-Q, J. and Blatteis, C.M. (2001) Differential inhibition by nimesulide of the early and late phases of intravenous- and intracerebroventricular-LPS-induced fever in guinea pigs. Neuroimmunomodulation, 9: 263–275.

Szelenyi, Z., Zeisberger, E. and Bruck, K. (1976) Effects of electrical stimulation in the lower brainstem on temperature regulation in the unanaesthetized guinea-pig. Pflugers Arch., 364: 123–127.

Szelenyi, Z., Zeisberger, E. and Bruck, K. (1977) A hypothalamic alpha-adrenergic mechanism mediating the thermogenic response to electrical stimulation of the lower brainstem in the guinea pig. Pflugers Arch., 370: 19–23.

Ushikubi, F., Segi, E., Sugimoto, Y., Murata, T., Matsuoka, T., Kobayashi, T., Hizaki, H., Tuboi, K., Katsuyama, M., Ishikawa, A., Tanaka, T., Yoshida, N. and Narumiya, S. (1998) Impaired febrile response in mice lacking the prostaglandin E receptor subtype EP3. Nature, 395: 281–284.

Wan, W., Wetmore, L., Sorensen, C.M., Greenberg, A.H. and Nance, D.M. (1994) Neural and biochemical mediators of endotoxin and stress-induced c-*fos* expression in the rat brain. Brain Res. Bull., 34: 7–14.

Wang, T., Qin, L., Liu, B., Liu, Y., Wilson, B., Eling, T.E., Langenbach, R., Taniura, S. and Hong, J.-S. (2004) Role of reactive oxygen species in LPS-induced production of prostaglandin E_2 in microglia. J. Neurochem., 88: 939–947.

Watkins, L.R., Goehler, L.E., Reston, J.K., Tartaglia, N., Gilbert, L., Martin, D. and Maier, S.F. (1995) Blockade of interleukin-1-induced hyperthermia by subdiaphragmatic vagotomy: evidence for vagal mediation of immune-brain communication. Neurosci. Lett., 183: 27–31.

Wieczorek, M., Swiergiel, A.H., Pournajafi-Nazarloo, H. and Dunn, A.J. (2005) Physiological and behavioral responses to interleukin-1β and LPS in vagotomized mice. Physiol. Behav., 85: 500–511.

Zeisberger, E. (1987) The roles of monoaminergic neurotransmitters in thermoregulation. Can. J. Physiol. Pharmacol., 65: 1395–1401.

CHAPTER 2

Eicosanoids in non-febrile thermoregulation

David M. Aronoff[1,*] and Andrej A. Romanovsky[2]

[1]Division of Infectious Diseases, Department of Internal Medicine, The University of Michigan Health Systems, 6323 MSRB III/Box 0642, 1150 W. Medical Center Dr., Ann Arbor, MI 48109-0642, USA
[2]Systemic Inflammation Laboratory, Trauma Research, St. Joseph's Hospital and Medical Center, 350 W. Thomas Road, Phoenix, AZ 85013, USA

Abstract: Eicosanoids are a large group of oxygenated fatty acids [viz., ω-3 (*n*-3) and ω-6 (*n*-6) C_{20} polyunsaturated fatty acids], the most important source being the ω-6 cell membrane-derived arachidonic acid (AA). Eicosanoids are produced by many different cell types; through their ligation and activation of specific membrane-bound and intracellular receptors, they regulate myriad physiological and pathological functions, including body temperature (T_b). However, the thermoregulatory role of eicosanoids has mainly been associated with fever, i.e., with T_b changes induced during illness; their importance in maintaining T_b during health remains unclear. In this review, we address the question of whether AA-derived mediators (viz., prostaglandins, leukotrienes and other lipoxygenase metabolites, and the endocannabinoids/endovanilloids) are involved in normal (non-febrile) thermoregulation. We conclude that although prostaglandin E_2 is a principal mediator of fever, it is unlikely to be involved in the maintenance of normal T_b. Other eicosanoids reviewed also seem to have no major role in non-febrile thermoregulation. Newly discovered signaling pathways for eicosanoids, such as the endovanilloid system, may participate in thermoregulation, but further studies are required before definitive conclusions can be made.

Keywords: thermoregulation; prostaglandins; fever; endocannabinoids; endovanilloids; cyclooxygenase

Introduction

Eicosanoids are a large group of oxygenated fatty acids that are derived from ω-3 (*n*-3) and ω-6 (*n*-6) C_{20} polyunsaturated fatty acids and include: prostaglandins (PGs); leukotrienes (LTs), lipoxins, and monohydroxy fatty acids produced via lipoxygenase pathways; epoxy and dihydroxy fatty acids formed by cytochrome P-450s; conjugated arachidonic acid (AA) compounds including endocannabinoids and endovanilloids; and several non-enzymatically generated products of fatty acid oxidation including isoprostanes and isoleukotrienes.

AA is an ω-6 fatty acid that serves as the major substrate for the production of eicosanoids (Smith et al., 2000). Eicosanoids are produced by many different cell types. Through their ligation and activation of specific membrane-bound and intracellular receptors, eicosanoids regulate myriad physiological and pathological functions, including body temperature (T_b) regulation. However, the thermoregulatory role of eicosanoids has mainly been associated with fever, i.e., with T_b changes induced during illness; their importance in

*Corresponding author. Tel.: +1 734 936 5205; Fax: +1 734 764 4556; E-mail: daronoff@umich.edu

maintaining T_b during health remains unclear. In this review, we address the question of whether AA-derived mediators are involved in normal (non-febrile) thermoregulation.

For a compound to be implicated as an *endogenous* thermoregulatory mediator, it should be possible to precipitate aberrant thermoregulatory responses, such as hypothermia, hyperthermia, or impaired diurnal T_b variation, by inhibiting its production or signaling *in vivo*. With this somewhat simple criterion in mind, the role of specific eicosanoids in thermoregulation is discussed in the following sections. The synthetic pathways of the various AA metabolites addressed below are depicted in Fig. 1.

Prostaglandins

Overview and involvement in febrile pathogenesis

Among the eicosanoids, no group of compounds has been so intensively studied for its role in thermoregulation as the PGs, especially PGE_2. These molecules are derived via the oxygenation and cyclization of AA by cyclooxygenase (COX) enzymes, which exist in two isoforms termed COX-1 and COX-2 (Smith et al., 2000). COX-1 is constitutively expressed in a broad variety of tissue and cell types, whereas COX-2 (which is constitutively expressed in a limited number of tissues, e.g., brain, lungs, and kidney) is readily inducible in

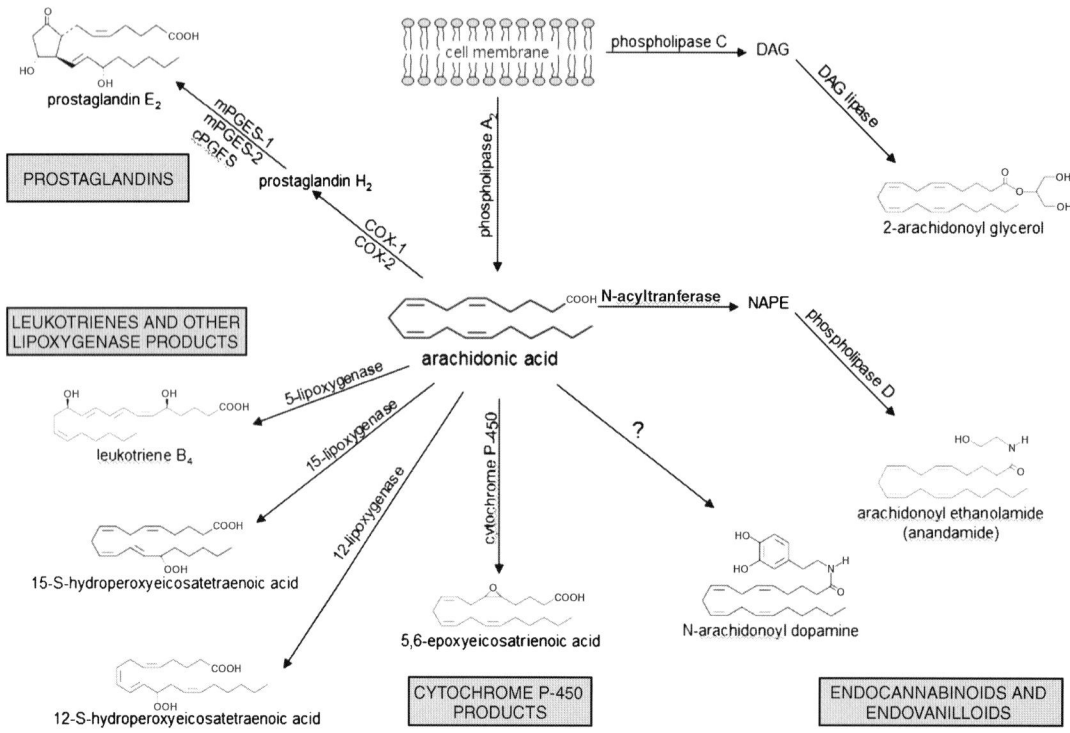

Fig. 1. The biosynthetic pathways of eicosanoids implicated in thermoregulation. Arachidonic acid is liberated from cell membrane phospholipids by various phospholipase A_2 enzymes. With the exception of the endocannabinoid molecule 2-arachidonoyl glycerol, the eicosanoids depicted are synthesized from the liberated arachidonic acid substrate. The enzymatic pathways involved in the production of *N*-arachidonoyl dopamine are unknown. COX, cyclooxygenase; cPGES, cytosolic prostaglandin E_2 synthase; DAG, diacyl glycerol; NAPE, *N*-acyl-phosphatidylethanolamine; mPGES, microsomal prostaglandin E_2 synthase. Structure diagrams courtesy of Cayman Chemical (Ann Arbor, MI).

many cells by a variety of stimuli including inflammatory. The product of COX metabolism of AA is the unstable endoperoxide PGH_2, which is isomerized to the physiologically relevant prostanoids by terminal synthases. PGE_2, for example, can be generated by any of three PGE synthases (PGES), including the inducible microsomal (m)PGES-1, constitutive mPGES-2, and constitutive cytosolic (c)PGES (Helliwell et al., 2004).

A critical role for PGE_2 has been established in the pathogenesis of fever (reviewed in Ivanov and Romanovsky, 2004). Endothelial and perivascular cells within the hypothalamus serve as the main sources of the febrigenic PGE_2 (Schiltz and Sawchenko, 2003; Matsumura and Kobayashi, 2004), which reflects the up-regulation of several PGE_2-synthesizing enzymes, COX-2 and mPGES-1 (Ivanov and Romanovsky, 2004). The involvement of PGE_2 in fever pathogenesis is highlighted by the fact that most antipyretic agents alleviate fever by inhibiting COX (Aronoff and Neilson, 2001).

Given its key role in fever pathophysiology, it was hypothesized that PGE_2 might be important in (i) the maintenance of normal T_b (throughout the diurnal cycle); (ii) the circadian increase in core T_b (that occurs in the late afternoon in humans); and (iii) the defense of T_b against the cold (low ambient temperature, T_a). Furthermore, since naturally occurring COX inhibitors (e.g., salicylates) and later synthetic agents have been used for centuries to reduce fever (Mackowiak, 2000), there has long been an interest in the ability of these compounds to affect normal T_b. A detailed review of the numerous publications exploring the T_b-altering effects of COX inhibitors in healthy (non-febrile) humans and animals extends beyond the scope of the current review, and the reader is referred to two comprehensive reviews by Clark (Clark and Clark, 1981; Clark, 1987). However, we will present a succinct discussion of what is known and unknown regarding PG-mediated thermoregulation, with particular regard to the three hypotheses stated above.

Hypothesis 1. *PGE_2 is involved in the maintenance of normal T_b*: As alluded to above, PGE_2 plays a fundamental role in the febrile response to bacterial infection. In this light, it is notable that treatment of *uninfected* animals with oral antibiotics, which eliminated the flora of bacteria colonizing the gastrointestinal tract, reduced their basal T_b (Conn et al., 1991; Fuller and Mitchell, 1999). One inference from these data is that host-derived mediators (such as PGE_2), perhaps provoked by resident gut microbes, may be contributing to normal T_b. In other words, normal, "non-febrile" T_b may have a febrile component, due to constant stimulation of the immune system with bacterial products of normal intestinal flora.

However, the administration of antipyretic drugs to humans, at doses sufficient to reduce febrile temperatures, has not provided convincing evidence to support a role of PGs in the maintenance of normal T_b (Clark and Clark, 1981; Clark, 1987). Sée (1877) reported that salicylate did not significantly alter T_b in healthy animals or man. Nearly a century later, in a study of 12 healthy volunteers, Rosendorff and Cranston (1968) administered either oral or intravenous salicylates and recorded T_b at three sites (aural, oral, and rectal) before and during thermal (heat) stress. Neither oral aspirin nor parenteral salicylate altered basal T_b or the thermoregulatory response to a heat load.

These results do not surprise the clinician, who will note that the vast majority of patients treated either chronically or acutely with antipyretic medications do not develop noticeable hypothermia. Rarely, however, COX inhibitors (such as acetaminophen and indomethacin) have been noted to produce a fall in T_b in previously afebrile humans or animals, particularly when administered at high doses (Ayoub et al., 2004). Interestingly, this is a property not generally shared by the salicylates (Milton, 1976). Although some antipyretics uncommonly reduce T_b in the *febrile* host below the usual basal T_b (Milton, 1976), the low frequency at which these hypothermic effects occur and the inconsistent ability of structurally distinct COX inhibitors to produce such effects argue against a physiological role.

Animal models have yielded somewhat contradictory results (Milton, 1973; Ayoub et al., 2004). Milton (1973) found that indomethacin and acetaminophen both caused T_b to fall when administered intraperitoneally to the conscious cat, although aspirin had no effect on T_b. Milton (1976) concluded that the effects of the two drugs to activate heat loss mechanisms (vasodilatation and panting) were non-specific, i.e., not mediated through the inhibition of PG synthesis.

A more recent study examined the effect of acetaminophen on T_b in normal C57/BL6 mice and mice genetically deficient in COX-1 or COX-2 (Ayoub et al., 2004). Although antipyretic doses of acetaminophen ($\sim$6–25 mg/kg) were previously not found to alter murine T_b (Cashin and Heading, 1968), a high (100–300 mg/kg) dose of acetaminophen administered intraperitoneally caused hypothermia in these experiments (Ayoub et al., 2004). In the wild-type mice, this acetaminophen-induced hypothermia correlated with the suppression of basal PGE_2 levels in the CNS. In COX-1 null (but not COX-2 null) mice, the hypothermic effect of acetaminophen was blunted, suggesting that COX-1 derived PGE_2 is important for T_b maintenance in the mouse. However, the lack of COX-1 did not completely prevent acetaminophen from lowering basal T_b, suggesting to the authors that, "another mechanism may also operate in acetaminophen hypothermia, perhaps unrelated to COX inhibition" (Ayoub et al., 2004). Further supporting such a conclusion, basal T_b of untreated COX-1 and COX-2 knockout mice appeared to be similar to the wild-type mice (Li et al., 1999, 2001; Ayoub et al., 2004).

A few studies have explored thermoregulation in mice lacking any of the known PGES enzymes. It is unclear how the different terminal synthases contribute to basal PGE_2 synthesis within the CNS, although mPGES-1 deficient mice have reduced PGE_2 levels in the brain (Boulet et al., 2004). Both mPGES-1 and cPGES are expressed in the hypothalamus under normal conditions (Ivanov et al., 2002), whereas mPGES-2 (has not been studied in the hypothalamus) is expressed at many extrahypothalamic sites (Engblom et al., 2003). In a recent knockout study, absence of mPGES-1 was found to affect neither basal T_b nor diurnal T_b variation (recorded over a 48 h observation period) (Saha et al., 2005). However, hypothalamic PGE_2 levels were not measured to confirm a reduction in basal concentration in the mPGES-1 knockout mice. Similar studies in cPGES and mPGES-2 deficient mice have not been published to date.

Hypothesis 2. *PGE_2 is involved in the circadian increase in T_b*: Several studies have addressed the role of PGE_2 in circadian T_b variation (Scales et al., 1988; Murphy et al., 1996; Baker et al., 2002). Scales et al. (1988) treated healthy, ambulatory human volunteers with antipyretic doses of aspirin (650 mg every 4 h by mouth) and continuously monitored T_b over 2 days with indwelling colonic probes. Despite achieving therapeutic salicylate concentrations in the blood, no effects were seen on the circadian rhythm of T_b (Fig. 2). More recently, a study in 13 healthy women demonstrated no effect of moderate oral doses of acetaminophen (1300 mg by mouth every 6 h) on T_b over a 24 h period (Baker et al., 2002). The same experiments also showed that acetaminophen did not affect the normal increase in T_b experienced by women during the luteal phase of the menstrual cycle.

A curious result was reported by Murphy et al. (1996), who showed that human volunteers administered aspirin or ibuprofen at 11:00 P.M. experienced less of a decrease in T_b over the subsequent 4 h than did untreated controls. However, T_b of the treated subjects was only marginally (by $\sim$0.2°C) different than that of controls at the 3:00 A.M. timepoint (Murphy et al., 1996); furthermore, no effect of the COX inhibitors was observed during the daytime hours of 3:00 P.M.–5:00 P.M. This study was limited by the brief and discontinuous periods of observation. Nevertheless, the data raise the possibility that products of COX participate in the decrease in T_b observed during night time in human beings. Given the established ability of PGE_2 to increase T_b, these results seem paradoxical and have not been reproduced by others.

Several studies exploring the effects of COX inhibition on the circadian rhythm of T_b have been performed in small animals maintained in a thermoneutral or nearly thermoneutral environment (thereby reducing the confounding effects of the hot or cold). Scales and Kluger (1987) performed a study in rats at T_a of $\sim$26°C, which was likely to be near-neutral in their setup

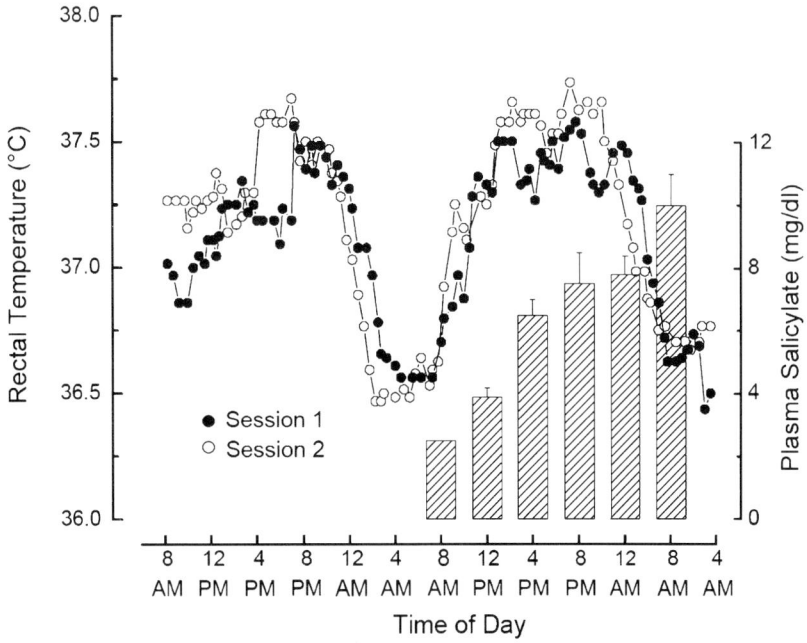

Fig. 2. Circadian rhythm of rectal temperature in healthy male volunteers treated with aspirin. Each point is a mean ($n = 4$). During Session 1, aspirin was given as an oral dose of 650 mg every 4 h for a total of 20 h, as indicated by the arrows. During Session 2, no aspirin was administered. Plasma salicylate concentration (mean ± SE) was measured over time during Session 1 (bars). Adapted with permission from Scales et al. (1988).

(Romanovsky et al., 2002). These authors initially reported that high doses of salicylate or aspirin (300 mg/kg intraperitoneally) given to rats at 5:00 P.M. (the time at which basal T_b begins its evening ascent) resulted in a dramatic fall in T_b. Administration of these medications at 9:00 A.M. did not affect T_b. Similar experiments with indomethacin yielded a dose-dependent depression of nocturnal T_b regardless of the time of drug administration. The authors suggested that the initiation and maintenance of the nocturnal T_b rise in rats required PGE_2, a hypothesis supported by the finding that hypothalamic PGE_2 levels are increased in both mice and rats during the evening hours (Fraifeld et al., 2001). In contrast to this, the same group did not find any alteration in circadian T_b when they repeated these experiments with a lower, yet antipyretic, dose of salicylate, 66 mg/kg (Scales et al., 1988). The fact that the effect was observed only at very high salicylate doses raises the possibility of COX-independent actions of the drug (Aronoff and Neilson, 2001; Tegeder et al., 2001).

Recent studies of PGE_2 receptor inhibition cast further doubt on the premise that PGE_2 plays a role in the circadian increase in T_b. PGE_2 signals via four distinct G-protein-coupled receptors, termed E-prostanoid (EP) 1–4 (Breyer et al., 2001). Rat studies reveal expression of each EP receptor except EP2 within the preoptic area of the anterior hypothalamus (Oka et al., 2000), and both EP1 and EP3 have been found to mediate the febrile response in murine models (Ushikubi et al., 1998; Oka et al., 2003). Genetic knockout of each of the four EP receptor subtypes has been accomplished in mice. A recent study of EP1 and EP3 null mice performed at 22° to 24°C (below thermoneutrality, Gordon, 1993) failed to demonstrate any difference in T_b or diurnal temperature changes in these mice compared with the wild-type strain (Oka et al., 2003). Similar studies of EP2 and EP4 knockouts have not been reported, although a pharmacological study in rats involving the peripheral and CNS administration of the EP1 and EP2 receptor antagonist AH-6809 did not elicit T_b abnormalities or alterations in diurnal variations of T_b (Matsumura et al., 1989).

Hypothesis 3. *PGE_2 is involved in the defense of core T_b against the cold*: Because the physiological and behavioral efforts employed in defending core T_b against cold are the same as those involved in the generation of fever (i.e., increasing heat production and reducing heat loss), it is reasonable to suspect an involvement of PGE_2 in such processes. Such an involvement was examined nearly 40 years ago in rats exposed to cold (T_a of 22 or 2°C) (Bizzi et al., 1965). Doses of salicylate as high as 300 mg/kg (intraperitoneally) did not affect T_b when T_a was 22°C; however, salicylate did result in a gradual, dose-dependent drop in T_b in rats held at 2°C (Bizzi et al., 1965). Satinoff (1972) later demonstrated quite similar results in rats treated with salicylate and maintained at 5°C; she also found that large doses of salicylate (180–300 mg/kg, intraperitoneally) reduced the T_b of rats kept at 23°C. Although the thermoneutral zone is different for the same animal in different experimental setups, the temperatures employed in the above-mentioned studies (2° to 23°C) are clearly subneutral for rats (Romanovsky et al., 2002). These experiments suggest that PGs might participate in the generation of cold-defense responses, at least in the rat.

This specific possibility was further investigated by Solomonovich and Kaplanski (1985). The authors performed intricate experiments using either salicylate or indomethacin administered intraperitoneally to rats kept at 4, 23, or 34°C. In addition to monitoring rectal T_b, the authors analyzed hypothalamic PGE_2 concentrations. In control rats, hypothalamic PGE_2 levels did not increase when rats were moved from 23 to 4°C, which argues against a defensive increase in PGE_2 to counter the reduced T_a. Indomethacin treatment (50 mg/kg) potently depressed PGE_2 production in rats at each T_a but did not affect T_b. Salicylate (300 mg/kg) lowered T_b at both 23 and 4°C but provoked hyperthermia when T_a was 34°C; hypothalamic PGE_2 levels were significantly reduced by the drug only at 23°C. The authors concluded that the changes in T_b in rats treated with salicylate resulted from a pharmacological effect unique to salicylate and independent of its action on hypothalamic PGE_2 production. As noted above, many COX-independent effects of high doses of salicylates have been identified (Aronoff and Neilson, 2001; Tegeder et al., 2001), including the suppression of plasma free fatty acid levels in rats exposed to the cold (thus limiting substrate available for metabolism) (Bizzi et al., 1965).

In addition, studies involving larger animals, such as monkeys (Barney and Elizondo, 1981) and cats (Cranston et al., 1975), further suggest that prostanoids do not play a key role in thermal defense against cold. This point is underscored by reports that cerebrospinal fluid PGE levels remained unchanged in both cats and rabbits during cold exposure (reviewed in Barney and Elizondo, 1981).

In summary, there are no definitive data to support a role for PGE_2 (or other prostanoids) in regulating T_b of experimental animals or humans under non-febrile conditions, including during cold exposure. Results of experiments with salicylate compounds administered at high doses show definite effects on T_b, but these might be due to actions independent of COX inhibition.

Leukotrienes and other lipoxygenase products of AA

LTs are potent mediators of inflammation derived from the 5-lipoxygenase-catalyzed oxygenation of AA (reviewed in Peters-Golden et al., 2005). 5-lipoxygenase, in concert with the 5-lipoxygenase activating protein, generates the unstable epoxide LTA_4, which can be hydrolyzed to LTB_4 or conjugated with glutathione to form the cysteinyl LTs (LTC_4, LTD_4, and LTE_4). Few studies have been done to explore the role of LTs in non-febrile T_b regulation.

A murine study (male CD/1 strain) showed that treatment with MK-886 (an inhibitor of 5-lipoxygenase activating protein) at doses up to 1 mg/kg provoked a transient increase (0.4° to 0.6°C) in T_b (Paul et al., 1999). The same investigators later reported that MK-886 (1 mg/kg) prevented a normal drop in T_b observed in CD/1 mice during the early afternoon (Fraifeld et al., 2001). However, these experiments were conducted at T_a of 22° to 24°C, which are below the neutral T_a of the mouse (Gordon, 1993). Thus it is possible that LTs are involved in the thermoregulatory responses to cold. Data regarding the role of LTs in T_b homeostasis in a thermoneutral environment are lacking, and the involvement of LTs in non-murine species remains unknown.

AA can be mono-oxygenated by lipoxygenase enzymes other than 5-lipoxygenase. 12-(S)- and 15-(S)-hydroperoxyeicosatetraenoic acids (12S- and 15S-HpETE) are the respective products of the 12- and 15-lipoxygenase enzymes. Although neither of these enzymes nor their products have been implicated in thermoregulation, both 12S- and 15S-HpETE are capable of activating vanilloid receptors (Van Der Stelt and Di Marzo, 2004) that may be involved in T_b regulation (described in the following section).

Conjugates of AA

The α-carbon carboxylic acid moiety of AA facilitates the generation of ether, ester, and amide conjugates of AA, some of which have been found to occur *in vivo* and appear to have physiological importance. The best studied of such compounds, particularly with respect to thermoregulation, are the endocannabinoids and endovanilloids (Van Der Stelt and Di Marzo, 2004; Rodriguez de Fonseca et al., 2005).

(-)-Δ^9-Tetrahydrocannabinol, the main psychoactive component of *Cannabis sativa* has long been known to exert hypothermic effects in mammals (Holtzman et al., 1969). Two types of cannabinoid receptor, CB_1 (Matsuda et al., 1990) and CB_2 (Munro et al., 1993), have been identified. Whereas the CB_1 receptor is present in the CNS and some peripheral tissues, the CB_2 receptor is mainly expressed by immune cells (Pertwee and Ross, 2002). Several AA conjugates have been shown to serve as endogenous cannabinoid receptor ligands, the best characterized being anandamide (arachidonoylethanolamide, AEA) and 2-arachidonoylglycerol (2-AG) (Devane et al., 1992; Mechoulam et al., 1995; Sugiura et al., 1995).

A role for endocannabinoids and the CB_1 receptor in normal thermoregulation is suggested by the findings that (1) AEA and 2-AG are produced in detectable quantities within the CNS (Sugiura et al., 1995) and their concentrations vary in a diurnal fashion (Valenti et al., 2004); (2) the peripheral injection of AEA provokes hypothermia in animal models (Fride and Mechoulam, 1993); and (3) highly selective, synthetic cannabinoid receptor agonists induce hypothermia in rats when delivered directly into the hypothalamus (Rawls et al., 2002). The availability of both pharmacological CB_1 inhibitors and CB_1 knockout mice has facilitated the investigation of the role of endogenously produced cannabinoids in T_b homeostasis.

A recent study in rats demonstrated that the specific CB_1/CB_2 agonist WIN 55212-2 could induce profound hypothermia whether administered intramuscularly or directly into the preoptic area of the anterior hypothalamus (Rawls et al., 2002). These effects were blocked with the selective CB_1 antagonist SR141716A but not the CB_2 antagonist SR144528 (Rawls et al., 2002), consistent with other studies demonstrating that cannabinoid-induced hypothermia is CB_1-mediated (Compton et al., 1996; Nava et al., 2000). However, treatment of the rats with either antagonist alone did not alter baseline T_b over the 5 h study duration. Similarly, mice treated with SR141716A did not display altered T_b compared with untreated animals (Compton et al., 1996). In addition, CB_1 knockout mice did not exhibit a different T_b than wild-type mice (Zimmer et al., 1999). These data suggest that the endocannabinoid system does not tonically modulate T_b. However, detailed investigations of the effects of CB_1 deficiency on either circadian T_b variation or defense against thermal stress have not been published.

AEA and several other AA derivatives have been shown to bind the capsaicin receptor, also known as the transient receptor potential of the vanilloid family (TRPV)1 receptor (O'Neil and Brown, 2003). This receptor, expressed in small-diameter primary afferent fibers and in CNS neurons, is a temperature and pH-sensitive calcium channel that can also be activated by capsaicin, the pungent ingredient of hot peppers (reviewed in O'Neil and Brown, 2003). The observation that the treatment of animals either peripherally or centrally with capsaicin resulted in profound hypothermia and impaired defense against an elevated T_a (Jancso-Gabor et al., 1970a, b), suggested that TRPV1 might play a role in thermoregulation.

Structural similarities between capsaicin and AEA led to studies of the effects of AEA on TRPV1 signaling (Di Marzo et al., 1998; Zygmunt et al., 1999). Indeed, AEA was shown to be capable of activating TRPV1, though much less potently than capsaicin (Zygmunt et al., 1999). Since then, other AA derivatives, including 12S- and 15S-HpETE and the novel compound N-arachidonoyldopamine (NADA), have been reported to activate TRPV1 (reviewed in Van Der Stelt and Di Marzo, 2004). NADA is the most potent endogenous lipid TRPV1 ligand discovered to date and is almost equipotent to capsaicin (Van Der Stelt and Di Marzo, 2004). The presence of NADA in brain tissue suggests that it may be an important endogenous ligand in the CNS (Huang et al., 2002). Very little is known about the relevance of 12S- and 15S-HpETE as TVPR1 ligands *in vivo*.

Although TRPV1 appears essential for defense against external heat (at least in rodents), a role for TRPV1 in normal thermoregulation has not been firmly established. An early study of TRPV1 deficient mice demonstrated no differences in T_b compared with wild-type mice over a 7-day observational period (Caterina et al., 2000). Similarly, two recent studies (Szelenyi et al., 2004; Iida et al., 2005) of daily T_b rhythm (at a neutral T_a) and heat tolerance in TRPV1 knockout mice showed that mean T_b and mean diurnal changes in T_b of these mice were similar to those of wild-type animals (although the TRPV1 null mice in one of these studies demonstrated a higher amplitude of the usual "ultradian" T_b oscillations that occur throughout the day (Szelenyi et al., 2004)). Iida et al. (2005) also observed a reduced febrile response of TRPV1 knockout mice to bacterial lipopolysaccharide, but an extensive pharmacological study by Dogan et al. (2004) conducted in rats did not reveal any involvement of TRPV1 in fever.

Another member of the TRPV calcium channel family, TRPV4, although not activated by capsaicin, has been shown to respond to 5,6-epoxyeicosatrienoic acid (5,6-EET), a cytochrome P-450 metabolite of AA (Watanabe et al., 2003). TRPV4 is thermosensitive and is expressed by hypothalamic neurons (Guler et al., 2002). TRPV4 knockout mice exhibited similar basal level and circadian fluctuations of T_b, as well as similar thermoregulatory responses to both cold and warm environments compared with wild-type mice (Liedtke and Friedman, 2003; Lee et al., 2005). Although the TRPV4-deficient animals preferred a somewhat higher T_a than the genetically normal controls (Lee et al., 2005), the physiological basis for this difference is unknown.

Thus, to date there are few data to suggest that cannabinoid or vanilloid receptor signaling pathways are important in non-febrile thermoregulation and even fewer data to implicate endocannabinoid or endovanilloid AA derivatives in normal T_b homeostasis.

Conclusions

Although eicosanoids such as PGE_2 have been found to be involved in thermoregulation during illness (i.e., fever), few data exist to support a role for AA derivatives in the maintenance of normal T_b. Newly discovered signaling pathways for eicosanoids, such as the endovanilloid system, may participate in thermoregulation, but further studies are required before definitive conclusions can be made.

Abbreviations

AA	arachidonic acid
CB	cannabinoid receptor
CNS	central nervous system
COX	cyclooxygenase
EP	E-prostanoid receptor
HpETE	hydroperoxyeicosatetraenoic acid
LT	leukotriene

NADA	*N*-arachidonoyldopamine
PG	prostaglandin
PGES	prostaglandin E_2 synthase
T_a	ambient temperature
T_b	body temperature
TRPV	transient receptor potential vanilloid

Acknowledgments

Funding for this work was provided by the National Institutes of Health Grant HL078727 (D.M.A.). We thank Dr. Nicolas Flamand for his critical review of an earlier draft of this manuscript.

References

Aronoff, D.M. and Neilson, E.G. (2001) Antipyretics: mechanisms of action and clinical use in fever suppression. Am. J. Med., 111: 304–315.

Ayoub, S.S., Botting, R.M., Goorha, S., Colville-Nash, P.R., Willoughby, D.A. and Ballou, L.R. (2004) Acetaminophen-induced hypothermia in mice is mediated by a prostaglandin endoperoxide synthase 1 gene-derived protein. Proc. Natl. Acad. Sci. U.S.A., 101: 11165–11169.

Baker, F.C., Driver, H.S., Paiker, J., Rogers, G.G. and Mitchell, D. (2002) Acetaminophen does not affect 24-h body temperature or sleep in the luteal phase of the menstrual cycle. J. Appl. Physiol., 92: 1684–1691.

Barney, C.C. and Elizondo, R.S. (1981) Prostaglandins and temperature regulation in the rhesus monkey. J. Appl. Physiol., 50: 1248–1254.

Bizzi, A., Garattini, S. and Veneroni, E. (1965) The action of salicylate in reducing plasma free fatty acids and its pharmacological consequences. Br. J. Pharmacol., 25: 187–196.

Boulet, L., Ouellet, M., Bateman, K.P., Ethier, D., Percival, M.D., Riendeau, D., Mancini, J.A. and Methot, N. (2004) Deletion of microsomal prostaglandin E2 (PGE2) synthase-1 reduces inducible and basal PGE2 production and alters the gastric prostanoid profile. J. Biol. Chem., 279: 23229–23237.

Breyer, R.M., Bagdassarian, C.K., Myers, S.A. and Breyer, M.D. (2001) Prostanoid receptors: subtypes and signaling. Annu. Rev. Pharmacol. Toxicol., 41: 661–690.

Cashin, C.H. and Heading, C.E. (1968) The assay of antipyretic drugs in mice, using intracerebral injection of pyretogenins. Br. J. Pharmacol., 34: 148–158.

Caterina, M.J., Leffler, A., Malmberg, A.B., Martin, W.J., Trafton, J., Petersen-Zeitz, K.R., Koltzenburg, M., Basbaum, A.I. and Julius, D. (2000) Impaired nociception and pain sensation in mice lacking the capsaicin receptor. Science, 288: 306–313.

Clark, W.G. (1987) Changes in body temperature after administration of antipyretics, LSD, delta 9-THC and related agents: II. Neurosci. Biobehav. Rev., 11: 35–96.

Clark, W.G. and Clark, Y.L. (1981) Changes in body temperature after administration of antipyretics, LSD, delta 9-THC, CNS depressants and stimulants, hormones, inorganic ions, gases, 2,4-DNP and miscellaneous agents. Neurosci. Biobehav. Rev., 5: 1–136.

Compton, D.R., Aceto, M.D., Lowe, J. and Martin, B.R. (1996) In vivo characterization of a specific cannabinoid receptor antagonist (SR141716A): inhibition of delta 9-tetrahydrocannabinol-induced responses and apparent agonist activity. J. Pharmacol. Exp. Ther., 277: 586–594.

Conn, C.A., Franklin, B., Freter, R. and Kluger, M.J. (1991) Role of gram-negative and gram-positive gastrointestinal flora in temperature regulation of mice. Am. J. Physiol., 261: R1358–R1363.

Cranston, W.I., Hellon, R.F. and Mitchell, D. (1975) Is brain prostaglandin synthesis involved in responses to cold? J. Physiol., 249: 425–434.

Devane, W.A., Hanus, L., Breuer, A., Pertwee, R.G., Stevenson, L.A., Griffin, G., Gibson, D., Mandelbaum, A., Etinger, A. and Mechoulam, R. (1992) Isolation and structure of a brain constituent that binds to the cannabinoid receptor. Science, 258: 1946–1949.

Di Marzo, V., Bisogno, T., Melck, D., Ross, R., Brockie, H., Stevenson, L., Pertwee, R. and De Petrocellis, L. (1998) Interactions between synthetic vanilloids and the endogenous cannabinoid system. FEBS Lett., 436: 449–454.

Dogan, M.D., Patel, S., Rudaya, A.Y., Steiner, A.A., Szekely, M. and Romanovsky, A.A. (2004) Lipopolysaccharide fever is initiated via a capsaicin-sensitive mechanism independent of the subtype-1 vanilloid receptor. Br. J. Pharmacol., 143: 1023–1032.

Engblom, D., Saha, S., Engstrom, L., Westman, M., Audoly, L.P., Jakobsson, P.J. and Blomqvist, A. (2003) Microsomal prostaglandin E synthase-1 is the central switch during immune-induced pyresis. Nat. Neurosci., 6: 1137–1138.

Fraifeld, V., Paul, L., Kaplanski, J., Kozak, W. and Kluger, M.J. (2001) Evidence for the involvement of eicosanoids in the regulation of normal body temperature. J. Thermal Biol., 26: 295–297.

Fride, E. and Mechoulam, R. (1993) Pharmacological activity of the cannabinoid receptor agonist, anandamide, a brain constituent. Eur. J. Pharmacol., 231: 313–314.

Fuller, A. and Mitchell, D. (1999) Oral antibiotics reduce body temperature of healthy rabbits in a thermoneutral environment. J. Basic Clin. Physiol. Pharmacol., 10: 1–13.

Gordon, C.J. (1993) Temperature Regulation in Laboratory Rodents. Cambridge University Press, Cambridge.

Guler, A.D., Lee, H., Iida, T., Shimizu, I., Tominaga, M. and Caterina, M. (2002) Heat-evoked activation of the ion channel, TRPV4. J. Neurosci., 22: 6408–6414.

Helliwell, R.J., Adams, L.F. and Mitchell, M.D. (2004) Prostaglandin synthases: recent developments and a novel hypothesis. Prostaglandins Leukot. Essent. Fatty Acids, 70: 101–113.

Holtzman, D., Lovell, R.A., Jaffe, J.H. and Freedman, D.X. (1969) 1-delta9-Tetrahydrocannabinol: neurochemical and behavioral effects in the mouse. Science, 163: 1464–1467.

Huang, S.M., Bisogno, T., Trevisani, M., Al-Hayani, A., De Petrocellis, L., Fezza, F., Tognetto, M., Petros, T.J., Krey, J.F., Chu, C.J., Miller, J.D., Davies, S.N., Geppetti, P., Walker, J.M. and Di Marzo, V. (2002) An endogenous capsaicin-like substance with high potency at recombinant and native vanilloid VR1 receptors. Proc. Natl. Acad. Sci. U.S.A., 99: 8400–8405.

Iida, T., Shimizu, I., Nealen, M.L., Campbell, A. and Caterina, M. (2005) Attenuated fever response in mice lacking TRPV1. Neurosci. Lett., 378: 28–33.

Ivanov, A.I., Pero, R.S., Scheck, A.C. and Romanovsky, A.A. (2002) Prostaglandin E(2)-synthesizing enzymes in fever: differential transcriptional regulation. Am. J. Physiol. Regul. Integr. Comp. Physiol., 283: R1104–R1117.

Ivanov, A.I. and Romanovsky, A.A. (2004) Prostaglandin E2 as a mediator of fever: synthesis and catabolism. Front. Biosci., 9: 1977–1993.

Jancso-Gabor, A., Szolcsanyi, J. and Jancso, N. (1970a) Irreversible impairment of thermoregulation induced by capsaicin and similar pungent substances in rats and guinea-pigs. J. Physiol., 206: 495–507.

Jancso-Gabor, A., Szolcsanyi, J. and Jancso, N. (1970b) Stimulation and desensitization of the hypothalamic heat-sensitive structures by capsaicin in rats. J. Physiol., 208: 449–459.

Lee, H., Iida, T., Mizuno, A., Suzuki, M. and Caterina, M.J. (2005) Altered thermal selection behavior in mice lacking transient receptor potential vanilloid 4. J. Neurosci., 25: 1304–1310.

Li, S., Ballou, L.R., Morham, S.G. and Blatteis, C.M. (2001) Cyclooxygenase-2 mediates the febrile response of mice to interleukin-1beta. Brain Res., 910: 163–173.

Li, S., Wang, Y., Matsumura, K., Ballou, L.R., Morham, S.G. and Blatteis, C.M. (1999) The febrile response to lipopolysaccharide is blocked in cyclooxygenase-2(−/−), but not in cyclooxygenase-1(−/−) mice. Brain Res., 825: 86–94.

Liedtke, W. and Friedman, J.M. (2003) Abnormal osmotic regulation in trpv4−/− mice. Proc. Natl. Acad. Sci. U.S.A., 100: 13698–13703.

Mackowiak, P.A. (2000) Brief history of antipyretic therapy. Clin. Infect. Dis., 31(Suppl 5): S154–S156.

Matsuda, L.A., Lolait, S.J., Brownstein, M.J., Young, A.C. and Bonner, T.I. (1990) Structure of a cannabinoid receptor and functional expression of the cloned cDNA. Nature, 346: 561–564.

Matsumura, H., Honda, K., Choi, W.S., Inoue, S., Sakai, T. and Hayaishi, O. (1989) Evidence that brain prostaglandin E2 is involved in physiological sleep-wake regulation in rats. Proc. Natl. Acad. Sci. U.S.A., 86: 5666–5669.

Matsumura, K. and Kobayashi, S. (2004) Signaling the brain in inflammation: the role of endothelial cells. Front. Biosci., 9: 2819–2826.

Mechoulam, R., Ben-Shabat, S., Hanus, L., Ligumsky, M., Kaminski, N.E., Schatz, A.R., Gopher, A., Almog, S., Martin, B.R., Compton, D.R., et al. (1995) Identification of an endogenous 2-monoglyceride, present in canine gut, that binds to cannabinoid receptors. Biochem. Pharmacol., 50: 83–90.

Milton, A.S. (1973) Prostaglandin E1 and endotoxin fever, and the effects of aspirin, indomethacin, and 4-acetamidophenol. Adv. Biosci., 9: 495–500.

Milton, A.S. (1976) Modern views on the pathogenesis of fever and the mode of action of antipyretic drugs. J. Pharm. Pharmacol., 28: 393–399.

Munro, S., Thomas, K.L. and Abu-Shaar, M. (1993) Molecular characterization of a peripheral receptor for cannabinoids. Nature, 365: 61–65.

Murphy, P.J., Myers, B.L. and Badia, P. (1996) Nonsteroidal anti-inflammatory drugs alter body temperature and suppress melatonin in humans. Physiol. Behav., 59: 133–139.

Nava, F., Carta, G. and Gessa, G.L. (2000) Permissive role of dopamine D(2) receptors in the hypothermia induced by delta(9)-tetrahydrocannabinol in rats. Pharmacol. Biochem. Behav., 66: 183–187.

Oka, T., Oka, K., Kobayashi, T., Sugimoto, Y., Ichikawa, A., Ushikubi, F., Narumiya, S. and Saper, C.B. (2003) Characteristics of thermoregulatory and febrile responses in mice deficient in prostaglandin EP1 and EP3 receptors. J. Physiol., 551: 945–954.

Oka, T., Oka, K., Scammell, T.E., Lee, C., Kelly, J.F., Nantel, F., Elmquist, J.K. and Saper, C.B. (2000) Relationship of EP(1–4) prostaglandin receptors with rat hypothalamic cell groups involved in lipopolysaccharide fever responses. J. Comp. Neurol., 428: 20–32.

O'Neil, R.G. and Brown, R.C. (2003) The vanilloid receptor family of calcium-permeable channels: molecular integrators of microenvironmental stimuli. News Physiol. Sci., 18: 226–231.

Paul, L., Fraifeld, V. and Kaplanski, J. (1999) Evidence supporting involvement of leukotrienes in LPS-induced hypothermia in mice. Am. J. Physiol., 276: R52–R58.

Pertwee, R.G. and Ross, R.A. (2002) Cannabinoid receptors and their ligands. Prostaglandins Leukot. Essent. Fatty Acids, 66: 101–121.

Peters-Golden, M., Canetti, C., Mancuso, P. and Coffey, M.J. (2005) Leukotrienes: underappreciated mediators of innate immune responses. J. Immunol., 174: 589–594.

Rawls, S.M., Cabassa, J., Geller, E.B. and Adler, M.W. (2002) CB1 receptors in the preoptic anterior hypothalamus regulate WIN 55212-2 [(4,5-dihydro-2-methyl-4(4-morpholinylmethyl)-1-(1-naphthalenyl-carbonyl)-6H-pyrrolo[3,2,1ij]quinolin-6-one]-induced hypothermia. J. Pharmacol. Exp. Ther., 301: 963–968.

Rodriguez de Fonseca, F., Del Arco, I., Bermudez-Silva, F.J., Bilbao, A., Cippitelli, A. and Navarro, M. (2005) The endocannabinoid system: physiology and pharmacology. Alcohol Alcohol, 40: 2–14.

Romanovsky, A.A., Ivanov, A.I. and Shimansky, Y.P. (2002) Selected contribution: ambient temperature for experiments in rats — a new method for determining the zone of thermal neutrality. J. Appl. Physiol., 92: 2667–2679.

Rosendorff, C. and Cranston, W.I. (1968) Effects of salicylate on human temperature regulation. Clin. Sci., 35: 81–91.

Saha, S., Engstrom, L., Mackerlova, L., Jakobsson, P.J. and Blomqvist, A. (2005) Impaired febrile responses to immune challenge in mice deficient in microsomal prostaglandin E synthase-1. Am. J. Physiol. Regul. Integr. Comp. Physiol., 288(5): R1100–R1107.

Satinoff, E. (1972) Salicylate: action on normal body temperature in rats. Science, 176: 532–533.

Scales, W.E. and Kluger, M.J. (1987) Effect of antipyretic drugs on circadian rhythm in body temperature of rats. Am. J. Physiol., 253: R306–R313.

Scales, W.E., Vander, A.J., Brown, M.B. and Kluger, M.J. (1988) Human circadian rhythms in temperature, trace metals, and blood variables. J. Appl. Physiol., 65: 1840–1846.

Schiltz, J.C. and Sawchenko, P.E. (2003) Signaling the brain in systemic inflammation: the role of perivascular cells. Front. Biosci., 8: s1321–s1329.

Sée, G. (1877) Etudes sur l'acide salicylique et les salicylates; traitement du rhumatisme aigu et chronique, de la goutte, et de diverse affections du systeme nerveux sensitif par les slicylates. Bull. Acad. Med. (Paris), 26: 689–706.

Smith, W.L., DeWitt, D.L. and Garavito, R.M. (2000) Cyclooxygenases: structural, cellular, and molecular biology. Annu. Rev. Biochem., 69: 145–182.

Solomonovich, A. and Kaplanski, J. (1985) Effects of salicylate and indomethacin in nonfebrile rats at different ambient temperatures. Prostaglandins Leukot. Med., 19: 161–165.

Sugiura, T., Kondo, S., Sukagawa, A., Nakane, S., Shinoda, A., Itoh, K., Yamashita, A. and Waku, K. (1995) 2-Arachidonoylglycerol: a possible endogenous cannabinoid receptor ligand in brain. Biochem. Biophys. Res. Commun., 215: 89–97.

Szelenyi, Z., Hummel, Z., Szolcsanyi, J. and Davis, J.B. (2004) Daily body temperature rhythm and heat tolerance in TRPV1 knockout and capsaicin pretreated mice. Eur. J. Neurosci., 19: 1421–1424.

Tegeder, I., Pfeilschifter, J. and Geisslinger, G. (2001) Cyclooxygenase-independent actions of cyclooxygenase inhibitors. FASEB J., 15: 2057–2072.

Ushikubi, F., Segi, E., Sugimoto, Y., Murata, T., Matsuoka, T., Kobayashi, T., Hizaki, H., Tuboi, K., Katsuyama, M., Ichikawa, A., Tanaka, T., Yoshida, N. and Narumiya, S. (1998) Impaired febrile response in mice lacking the prostaglandin E receptor subtype EP3. Nature, 395: 281–284.

Valenti, M., Vigano, D., Casico, M.G., Rubino, T., Steardo, L., Parolaro, D. and Di Marzo, V. (2004) Differential diurnal variations of anandamide and 2-arachidonoyl-glycerol levels in rat brain. Cell. Mol. Life Sci., 61: 945–950.

Van Der Stelt, M. and Di Marzo, V. (2004) Endovanilloids. Putative endogenous ligands of transient receptor potential vanilloid 1 channels. Eur. J. Biochem., 271: 1827–1834.

Watanabe, H., Vriens, J., Prenen, J., Droogmans, G., Voets, T. and Nilius, B. (2003) Anandamide and arachidonic acid use epoxyeicosatrienoic acids to activate TRPV4 channels. Nature, 424: 434–438.

Zimmer, A., Zimmer, A.M., Hohmann, A.G., Herkenham, M. and Bonner, T.I. (1999) Increased mortality, hypoactivity, and hypoalgesia in cannabinoid CB1 receptor knockout mice. Proc. Natl. Acad. Sci. U.S.A., 96: 5780–5785.

Zygmunt, P.M., Petersson, J., Andersson, D.A., Chuang, H., Sorgard, M., Di Marzo, V., Julius, D. and Hogestatt, E.D. (1999) Vanilloid receptors on sensory nerves mediate the vasodilator action of anandamide. Nature, 400: 452–457.

SECTION II

Physiological Mechanisms in Hyperthermia

CHAPTER 3

Exercise and heat stress: cerebral challenges and consequences

Lars Nybo*

Department of Human Physiology, Institute of Exercise and Sport Sciences, August Krogh Institute, Universitetsparken 13, DK-2100 Copenhagen Ø, Denmark

Abstract: This review deals with new aspects of exercise in the heat as a challenge that not only influences the locomotive and cardiovascular systems, but also affects the brain. Activation of the brain during such exercise is manifested in the lowering of the cerebral glucose to oxygen uptake ratio, the elevated ratings of perceived exertion and increased release of hypothalamic hormones. While the slowing of the electroencephalographic (EEG), the decreased endurance and hampered ability to activate the skeletal muscles maximally during sustained isometric and repeated isokinetic contractions appear to relate to central fatigue arising as the core/brain increases, the central fatigue during exercise with hyperthermia thus can be considered as the ultimate safety break against catastrophic hyperthermia. This would force the subject to stop exercising or decrease the internal heat production. It appears that the dopaminergic system is important, but several other factors may interact and feedback from the skeletal muscles and internal temperature sensors are probably also involved. The complexity of brain fatigue response is discussed based on our own investigations and in the light of recent literature.

Keywords: heat balance; central fatigue; exercise; cerebral blood flow and metabolism; voluntary muscle activation

Introduction: the hot and hard working brain

While a moderate rise of the internal body temperatures as observed during prolonged exercise in thermoneutral environments will be beneficial for various metabolic and enzymatic processes, it has become clear that endurance is markedly reduced when exercise results in excessive increases of the body core temperatures, i.e. hyperthermia (Gonzalez-Alonso et al., 1999; Nybo et al., 2001). The reduced work capacity may relate to more than one factor (see Chapter 4/Cheung for discussion),

*Corresponding author. Tel.: +45 35321620;
Fax: +45 35321600; E-mail: lnnielsen@aki.ku.dk

but exhaustion during prolonged exercise in the heat seems to coincide with the attainment of high internal temperatures (Nielsen et al., 1993; Fuller et al., 1998; Gonzalez-Alonso et al., 1999; Walters et al., 2000); and experiments with goats indicate that brain temperature may be a dominant factor affecting the ability to maintain motor activity (Caputa et al., 1986).

Furthermore, the idea that hyperthermia-induced fatigue relates to so-called "central fatigue" is supported by the observation that exercise-induced hyperthermia is associated with a reduced voluntary activation of the α-motor neurons during sustained maximal muscle contractions (Nybo and Nielsen, 2001a). A reduced level of

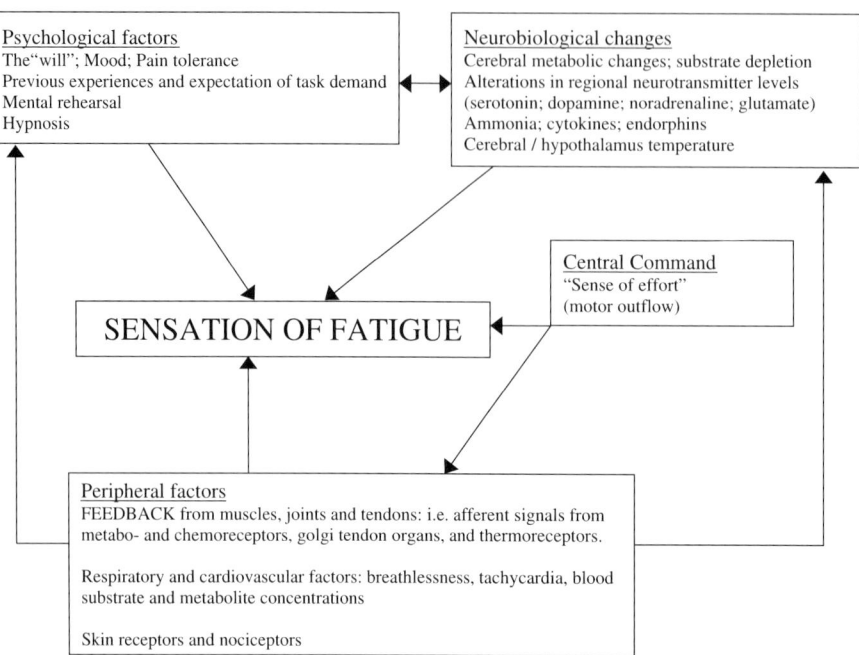

Fig. 1. Simplified model of physiological and psychological factors affecting "the sensation of fatigue". (Adapted from Nybo and Secher, 2004 with permission.)

central activation is also observed following passively induced hyperthermia (Morrison et al., 2004; Todd et al., 2005), and the lower voluntary activation level has been demonstrated via both electrical stimulation of the peripheral motor neuron (Nybo and Nielsen, 2001a) and transcranial magnetic stimulation of the motor cortex (Todd et al., 2005). Hyperthermia-induced activation deficit is most pronounced during prolonged sustained muscle contractions and may not always appear during brief maximal isometric contractions (Nybo and Nielsen, 2001a; Saboisky et al., 2003), which indicates that hyperthermia impairs the ability to sustain an adequate α-motor drive rather than the ability to activate the skeletal muscles maximally for a very brief period (see Central fatigue and critical internal temperatures for discussion).

Declining isometric strength is one of the characteristics of skeletal muscle fatigue and it is obvious that such deterioration of the contractile force indeed may involve factors located in the skeletal muscles, as demonstrated by stimulation of isolated muscles *in vitro* or electrically evoked activation of skeletal muscles *in vivo* (Merton, 1954; West et al., 1996; Westerblad et al., 1998). Furthermore, it should be acknowledged that the development of fatigue is complex and influenced by the interplay between peripheral and central factors including psychological aspects such as motivation and the will to succeed (see Fig. 1). However, this chapter is focused on cerebral metabolic, thermodynamic and neurohumoral responses during prolonged exercise in the heat and the potential relations between hyperthermia-induced fatigue and the cerebral perturbations provoked during such exercise.

Central fatigue and critical internal temperatures

Several studies support the notion that there is an internal temperature above which animals and humans will not continue to exercise voluntarily (MacDougal et al., 1974; Nielsen et al., 1993; Fuller et al., 1998; Gonzalez-Alonso et al., 1999; Walters et al., 2000); and even in experiments where dehydration is superimposed and markedly impairs cardiovascular function in hyperthermic athletes and reduces the perfusion of the exercising skeletal muscles (González-Alonso et al., 1997, 1998), exhaustion seems to relate to attainment of a high core temperature, rather than altered muscle metabolism (González-Alonso et al., 1999). It may be debated whether this internal temperature (most likely the brain/hypothalamic temperature) is "critical" and represents a definitive safety break against catastrophic heat injury, as supported by the observation that trained subjects during repeated trials with different starting temperatures or rates of heat storage stop exercising at similar body core temperatures (of ~40°C), but after dissimilar duration of the exercise. However, the consistency of the core temperatures at voluntary exhaustion in laboratory experiments both in trained (Gonzalez-Alonso et al., 1999) and untrained subjects (Cheung and McLellan, 1998) may also relate to the study designs, where low to moderate intensity exercise is usually combined with a large external (uncompensable) heat stress. Muscle metabolite accumulation, substrate depletion and other factors influencing peripheral fatigue may be of minor importance under such conditions, and the gradually increasing inhibition from a high brain/hypothalamic temperature may become the major factor dictating the point of exhaustion. However, fatigue is a complex phenomenon and the body core temperature at exhaustion may be influenced by factors such as training status, exercise intensity/mode and motivation. For example, differences in motivation between laboratory experiments and sports competitions combined with the influence of the subjects personality and training status could explain why untrained subjects during hot exercise conditions become exhausted at core temperatures between 38 and 39°C (Sawka and Wenger, 1988), whereas trained subjects may attain core temperatures as high as 41°C during sports competitions (Pugh et al., 2002), although they, as described above, become exhausted or unwilling to continue exercising, when their core temperature exceeds ~40°C in laboratory settings (Gonzalez-Alonso et al., 1999; Nybo and Nielsen, 2001a).

Experimental support for the involvement of central fatigue is provided by the data presented in Fig. 2, which demonstrates that exercise-induced hyperthermia reduces the level of voluntary activation during a sustained maximal knee extension. The maximal contractions were performed immediately after bicycle exercise, which in the hyperthermic trial increased the core temperature to ~40°C and exhausted the subjects after 50 min of exercise, whereas during the control trial the core temperature stabilized at ~38°C and exercise was maintained for 1 h without exhausting the subjects. Of note, although the hyperthermic exercise trial exhausted the subjects, it did not impair the knee extensors' ability to generate force (similar force when electrical stimulation was superimposed), and voluntary force production was also similar during the initial phase of the maximal voluntary contraction (MVC). However, during hyperthermia the subjects were unable to sustain the same activation, as during the control trial, and the voluntary force production as well as the rectified integrated surface electromyogram (EMG) from *m. vastus lateralis* became low. In addition, following a resembling bicycle protocol, force development during a sustained handgrip contraction followed a similar pattern of response as for the knee extensors, indicating that the attenuated ability to activate the skeletal muscles did not depend on whether the muscle group had been active or inactive during the preceding exercise bout (Nybo and Nielsen, 2001a). Conversely, hyperthermia did not affect maximal force development or central activation during brief maximal knee extensions (2 s duration) even if the MVCs were repeated 40 times and interspaced by only 3 s of rest (see Fig. 3). This may indicate that during exercise conditions where central fatigue is enhanced, the central nervous system (CNS) regains the ability to activate the skeletal muscles within a short period of recovery (Nybo and Nielsen, 2001a). Thus, if we compare the effect of hyperthermia with that of hypoglycemia on the development of fatigue during prolonged exercise and the activation pattern during a sustained MVC (cf. Nybo and Nielsen, 2001a; Nybo, 2003), it appears that both conditions are associated with central fatigue. However, in both studies the voluntary force production was unaffected during the initial phase of the sustained MVCs, and the level of voluntary activation was not reduced until the contraction

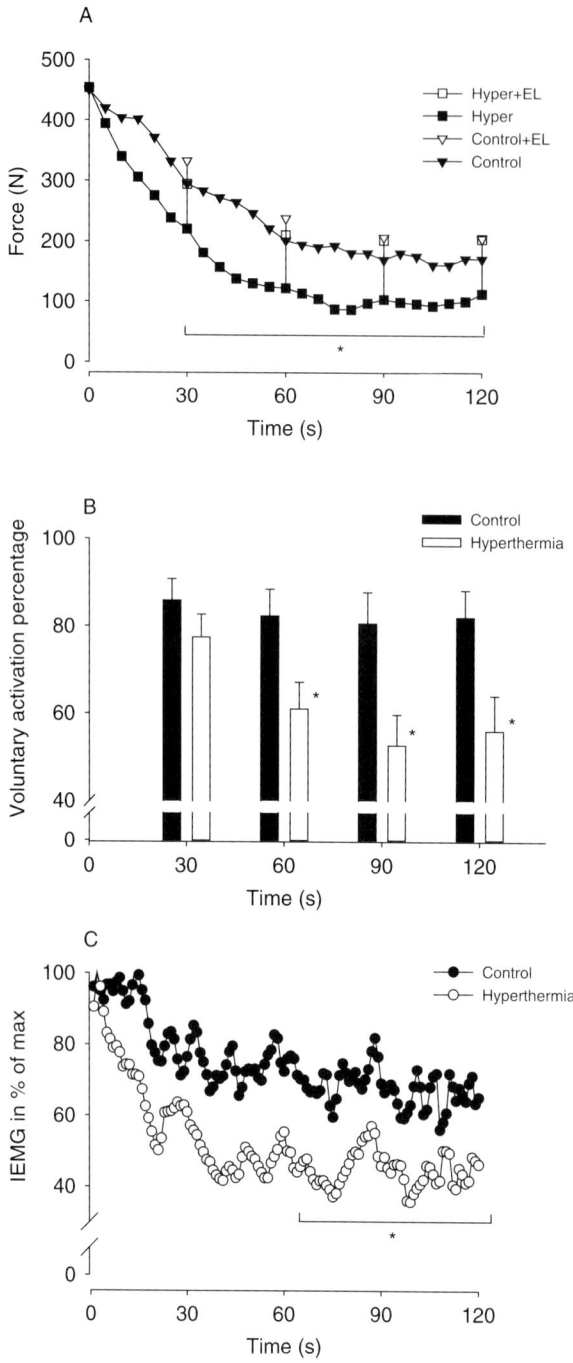

Fig. 2. (A) Force production, (B) voluntary activation level and (C) rectified integrated surface electromyography (IEMG) from m. vastus lateralis during 2 min of sustained maximal knee extension during hyperthermia (core temperature of ~40°C) and control (core temperature of 38°C). The subjects were instructed and verbally encouraged to make a maximal effort throughout the contraction and electrical stimulation (EL) was superimposed every 30 s to assess the level of voluntary activation, which was calculated as voluntary force divided by the force elicited when EL was superimposed. Data are means ± SE for eight subjects (error bars not included in A). * Indicates that all values in this period are significantly lower than control, $P<0.05$. (Adapted with permission from Nybo and Nielsen, 2001a.)

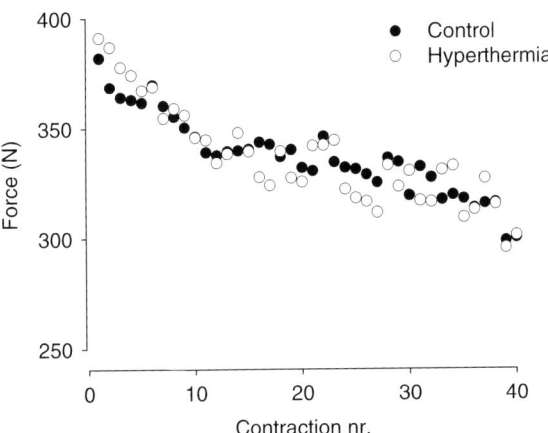

Fig. 3. Mean maximal force development in seven subjects during 40 consecutive MVCs with the knee extensors. The brief maximal contractions (duration of ~2 s) were repeated every 5 s and the two MVC protocols were performed immediately (~30 s) after cycling to exhaustion in a 40°C hot environment (exercise time of 50 ± 3 min and a core temperature of ~40°C) or following 1 h of non-exhaustive cycling at a similar exercise intensity in a thermoneutral environment (core temperature of ~38°C). (Adapted with permission from Nybo and Nielsen, 2001a.)

had been sustained for some time. Depletion of substrates within the CNS, inadequate oxygen delivery and/or alterations in the level of certain neurotransmitters are potential mechanisms underlying the decline in central activation during the sustained muscle contraction (see Cerebral blood flow and metabolism and Neurohumoral responses in relation to hyperthermia-induced fatigue), but sensory feedback from the contracting muscles could also be a major factor influencing the pattern of CNS activation. Inhibitory feedback from muscle chemo- and metaboreceptors may be of minor importance for the activation level during the initial phase of isometric contractions, whereas it may inhibit motor activation when the contraction is sustained and muscle metabolites accumulate (Kent-Braun, 1999). In accordance with the notion that fatigue is composed of many factors, it appears that central activation becomes markedly impaired when hyperthermia is combined with inhibitory signals from the skeletal muscles, whereas inhibition from a high brain/hypothalamus temperature (Caputa et al., 1986) may be overridden providing inhibitory feedback from chemo- and metaboreceptors, which is low. In addition, heating will decrease time to peak twitch force as well as the half-relaxation time of the skeletal muscles. Consequently hyperthermia may increase the firing frequency necessary to sustain maximal activation of the motor units and make it more difficult (or impossible) for the central nervous system to maintain/produce maximal force (Morrison et al., 2004).

Based on the discussion above, it seems logical to suggest that central fatigue is also the main factor underlying the reduced endurance during prolonged dynamic exercise in the heat. However, evaluation of the relative importance of peripheral and central factors' contribution to fatigue during ongoing dynamic exercise is much more difficult than during isometric contractions. Therefore, at present we only have indirect/circumstantial evidence for a central component of the fatigue that develops during prolonged work in hot environments (Fritzsche et al., 2000; Kay et al., 2001; Nybo and Nielsen, 2001c; Pitsiladis et al., 2002). On the other hand, it seems unlikely that hyperthermia-induced central fatigue should influence only isometric contractions and not dynamic exercise, and in support of this notion Martin et al. (2005) report that exercise-induced hyperthermia also lowers voluntary drive to the skeletal muscles in an exercise protocol with repeated maximal isokinetic contractions.

Difficulties in retaining power output during exercise with progressive hyperthermia are reflected in the subjects' rating of perceived exertion, and it appears that the increase in rating of perceived exertion (RPE) is closely related to an electroencephalographic (EEG) frequency shift that may reflect decreased arousal and impeded ability of the brain to sustain motor activity (Nielsen et al., 2001). Thus, two studies by Nielsen and co-workers (Nielsen et al., 2001; Nybo and Nielsen, 2001c) have demonstrated that perceived exertion is highly associated with increases in core temperature and frequency changes of the EEG obtained over the prefrontal cortex (specifically a rise in the A_α/A_β index — i.e. the ratio between EEG activity in the 8–13 Hz band and 13–30 Hz band). A similar slowing of the EEG is observed during sleep; however, it seems far too simple to conclude that exercise with hyperthermia reduces arousal, as systemic norepinephrine levels are higher during such exercise indicating that sympathetic activity is increased rather than reduced. It is possible that the altered EEG simply reflects the sensation of the increasing temperature or that it responds to other signals arising secondarily to the increase in core temperature. On the other hand, the EEG-RPE relation appears to be robust to ventilatory and circulatory changes (Rasmussen et al., 2004) and it is striking that the gradual increase in perceived exertion is so closely correlated to the EEG frequency shift, and while there is no correlations between perceived exertion and the electromyographic response of the exercising muscles, it indicates that altered activity within the central nervous system rather than fatigue-induced changes in motor-unit recruitment and/or discharge rates are involved in the development of fatigue during prolonged exercise in hot environments.

Indirect evidence for a centrally mediated component of fatigue is also provided by the observation that the systemic serum concentration of prolactin is markedly elevated after prolonged exercise in the heat compared to exercise in cool environments (Frewin et al., 1976; Radomski et al., 1998; Vigas et al., 2000; Pitsiladis et al., 2002). Hypothalamic activity cannot be directly assessed in exercising humans, but systemic hyperprolactinemia may be a marker for activation of the serotonergic system, as serotonin is a prominent excitatory neurotransmitter for the release of prolactin from the pituitary gland (Freeman et al., 2000; Weicker and Strüder, 2001). In addition, serotonergic neurones are involved with thermoregulation (Komiskey and Rudy, 1977; Hillegaart, 1991) and it seems likely that hyperthermia may enhance serotonergic activity, although during prolonged exercise an elevated core temperature does not alter the uptake of tryptophan, the amino acid precursor for serotonin synthesis (Nybo et al., 2003b). However, prolactin release from the pituitary gland is influenced by other factors than serotonin (e.g. dopaminergic D_2 activity, which inhibits prolactin secretion; Ben-Jonathan and Hnasko, 2001), and hyperthermia-induced hyperprolactinemia seems to have a significant non-serotonergic component (Bridge et al., 2003). Also, prolactin clearance from the circulation could be impaired during exercise with heat stress, as hyperthermia reduces the perfusion of the internal organs including the kidneys (Rowell et al., 1965), which extracts substantial amounts of low and medium molecular weight polypeptide hormones (Katz and Emmanouel, 1978; Emmanouel et al., 1981). In addition, it is not clear whether an enhancement of serotonergic activity during exercise with hyperthermia relates only to thermoregulatory factors (Hillegaart, 1991) or if it is also associated with central fatigue (cf. Neurohumoral responses in relation to hyperthermia-induced fatigue; Davis and Bailey, 1997; Nybo et al., 2003a).

Cerebral thermodynamic responses during exercise

At rest the average brain temperature remains stable at ~37°C, implying that heat removal from the brain is equal to the global cerebral metabolic heat production. However, temperature is heterogeneously distributed in the brain (Mariak et al., 1999), and the local temperature depends on the proportion between the regional cerebral blood flow and the metabolic rate (Yablonskiy et al., 2000). Thus, the net chemical reaction of oxygen and glucose liberates 470 kJ·mol^{-1} of oxygen (Zuntz and Schumburg, 1901), and in the brain almost all this energy ends up as heat, since no mechanical work is performed (Yablonskiy et al., 2000). The cerebral energy production is covered mainly by oxidation, and therefore the cerebral heat production is practically proportional to the cerebral metabolic rate (CMR_{oxygen}), while heat removal from the brain is the product of the specific heat capacity of blood, cerebral blood flow (CBF) and the arterio-venous blood temperature difference. At rest the human brain has an average metabolic rate of $\sim 1.5\,\mu mol\, O_2 \cdot g^{-1} \cdot min^{-1}$ (Lassen, 1985; Madsen et al., 1993), which corresponds to a cerebral heat production of $\sim 0.6\, j \cdot g^{-1} \cdot min^{-1}$ and heat balance is established with a jugular venous to arterial temperature difference (v-aD_{temp}) of $\sim 0.3\,°C$ and a CBF of $\sim 0.50\, ml \cdot g^{-1} \cdot min^{-1}$ (Yablonskiy et al., 2000; Nybo et al., 2002b).

However, during cerebral stimulation the regional oxygen consumption usually increases to a lesser degree than the perfusion and this uncoupling of the regional CBF and regional oxygen consumption may cause a lowering of the local brain temperature, because heat removal from the activated brain area will increase in proportion to the regional CBF and exceed the rate of heat production, which depends on the CMR_{oxygen}. Accordingly, data obtained with MRI in humans indicate that the tissue temperature in the visual cortex is lowered by $\sim 0.2\,°C$ in response to reversing checkerboard stimulation (Yablonskiy et al., 2000), and a corresponding decrease in local brain temperature is observed in monkeys after amygdala stimulation (Hayward and Baker, 1968). Of note, the opposite response is observed in rats, where sensory stimulation or other means of functional neuronal activation will cause an increase in the local brain temperature (Kiyatkin, 2005). This difference between species may relate to anatomical differences and to the large difference in head size or specifically the difference in the ratio between head surface and volume (see Zhu et al., 2004; Kiyatkin, 2005 for discussion).

In opposition to sensory stimulation, which in humans may have little or no effect on the average brain temperature and even cause a lowering of the temperature in the activated brain areas, exercise will increase the global cerebral temperature (Nybo et al., 2002b). Physical activity is associated with a major increase in the heat production by the exercising muscles, which elevates the core temperature and thus also the arterial blood. This implicates that the temperature of the blood supplied to the brain will increase and subsequently challenge the cerebral heat balance. During exercise in a thermoneutral environment, the core temperature increases rapidly at the onset of exercise and the v-aD_{temp} across the brain is narrowed for the first 10–15 min resulting in storage of heat in the brain; but a new balance between cerebral heat release and heat production is then developed. Therefore, during the first 15 min of moderate intense exercise the average brain temperature increases by $\sim 1\,°C$, but it then stabilizes at this new level (Fig. 5).

In contrast, during combined exercise and heat stress the body core and arterial blood temperatures keep increasing, and together with reductions of the CBF and increases in the CMR_{oxygen} it disrupts the cerebral heat balance. As illustrated in Fig. 4, heat removal via the jugular venous blood is reduced during exercise with hyperthermia compared to exercise in a thermoneutral environment, and the impaired heat removal is primarily a result of the reduced CBF (see Cerebral blood flow and metabolism), because the v-aD_{temp} is almost similar during hyperthermia compared to normothermia. Also, the elevated CMR_{oxygen} during exercise with hyperthermia increases the metabolic heat production in the brain and, combined with the impaired heat removal via the circulation, it results in continuous storage of heat in the brain. During hyperthermic exercise the cerebral temperature therefore rises in parallel with the core temperature, and this is unaltered by facial fanning (Fig. 5; see also Jessen and Kuhnen, 1992; Nielsen and Jessen, 1992; Corbett and Laptook, 1998).

Selective brain cooling defined as a lowering of the average brain temperature below the aortic/arterial temperature (The Commission for Thermal Physiology of the International Union of Physiological Sciences, 1987) has been a matter of large controversy (Jessen and Kuhnen, 1992; Nielsen and Jessen, 1992; Brengelmann, 1993; Cabanac, 1993). However, based on the data presented in Fig. 5, it appears that hyperthermic humans, in contrast to some animal species (McConaghy et al., 1995; Mitchell et al., 2002), both at rest and during exercise, have a brain

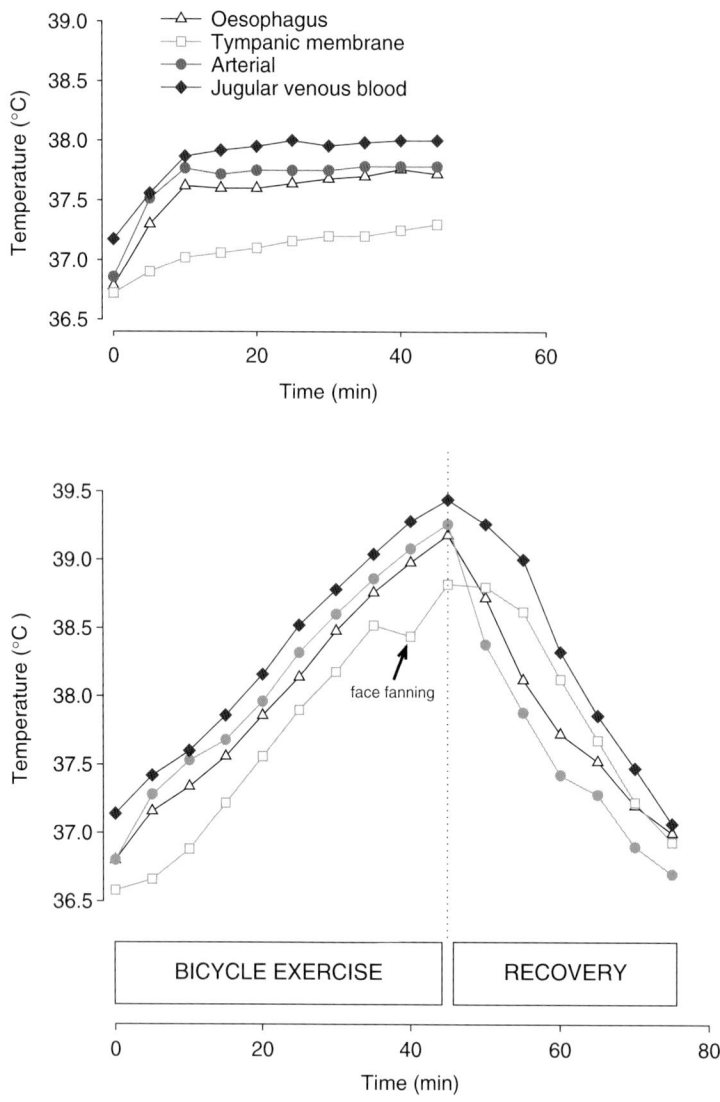

Fig. 4. Oesophageal, tympanic, arterial and jugular venous temperature responses during cycling with a normal core temperature response (top panel; control trial) and during a similar exercise bout with progressive hyperthermia (lower panel). Values are means of seven subjects. Standard deviations are omitted for simplicity, but the SD of all temperatures were in the range of 0.1° to 0.3°C. (Adapted with permission from Nybo et al., 2002b.)

temperature that is higher than that of the trunk. During prolonged exercise in the heat, exhaustion coincides with the attainment of a critical internal temperature (Nielsen et al., 1993; Fuller et al., 1998; Gonzalez-Alonso et al., 1999; Walters et al., 2000) and the thermodynamic response of the brain is, as mentioned, of special interest because the brain temperature appears to be a dominant factor affecting motor activity. Thus, during treadmill exercise goats reduce their speed, or refuse to move, when the brain temperature is independently increased to >42°C (Caputa et al., 1986). With brain temperature as a main factor influencing motor activation during exercise in hot environments, a thermal limit may be reached more quickly in species that have weak or no selective

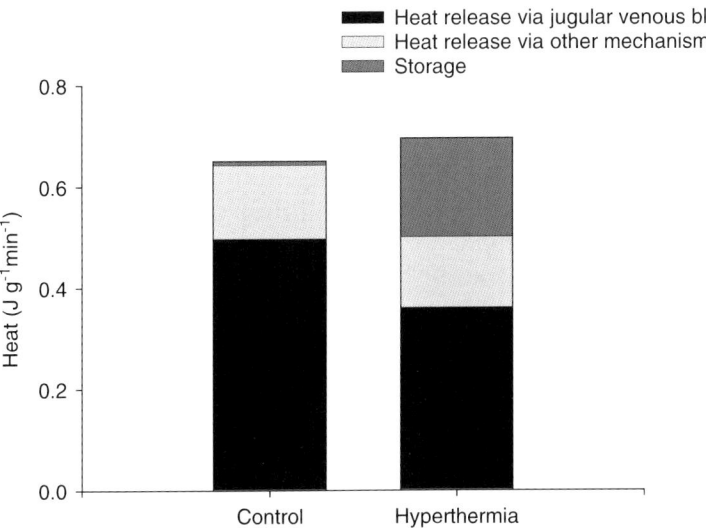

Fig. 5. Rate of cerebral heat production, heat removal via the jugular venous blood (lower, black segment) and heat storage (upper segment of the bar) during exercise with a normal temperature response (control) and exercise with progressive hyperthermia (see Fig. 5). The rate of cerebral heat production, which is represented by the total height of the bars, was calculated on the basis of the Kety-Schmidt-determined values for cerebral oxygen uptake and lactate release, while heat removal via the jugular venous blood was determined on the basis of the global CBF and the arterial to internal jugular venous temperature difference. Storage is the average rate of heat storage from 30 to 45 min of exercise, as determined from the change in cerebral venous blood temperature, and "heat removal via other mechanisms" (than convective removal via the jugular venous blood) is the difference between the rate of cerebral heat production and the computed rate of storage in the brain summed with the rate of heat removal by the jugular venous blood. The values represent means of three subjects. (Adapted with permission from Nybo et al., 2002b.)

brain cooling. But are humans capable of "selectively" cooling the brain? Selective brain cooling is observed in some species that lack a carotid rete (McConaghy et al., 1995), but it is best developed in animals with a carotid rete, such as the artiodactyls and felids (Jessen, 2001). Although humans do not have a carotid rete, the basis for cooling of the arterial blood on its passage from the heart to the brain exists (Rubenstein et al., 1960) and especially so during exercise when the pulmonary ventilation is markedly increased (Hanson, 1974). However, the transit time in the carotid artery appears to be too short for the blood temperature to equilibrate with the temperature of the surrounding tissue (Crezee and Lagendijk, 1992) and during the passage from the heart to brain the arterial blood temperature is lowered by less than 0.1°C (Nybo et al., 2002b). During mild hyperthermia, the subdural brain temperature may be slightly reduced in response to facial fanning (Mariak et al., 2003) or respiratory cooling of the upper airways (Mariak et al., 1999), but the lowering of the brain surface temperature in the studies by Mariak et al. (1999, 2003) seems to be influenced largely by the concomitant reduction of the core temperature in the resting postsurgical patients. Conversely, when the core temperature remains stable, face fanning fails to reduce the average brain temperature (Shiraki et al., 1988; Jessen and Kuhnen, 1992; Nielsen and Jessen, 1992), indicating that any cerebral cooling is restricted to the superficial layers of the brain. Furthermore, in the study by Nybo et al. (2002b) during exercise with hyperthermia the jugular venous blood remained ~0.2°C warmer than the arterial blood in all subjects (individual peak core temperatures were in the range from 39.0° to 40.1°C with corresponding jugular blood temperatures of 39.2° to 40.4°C) and neither the cerebral venous blood temperature nor the v-aD$_{temp}$ were affected by fan cooling of the head, which lowered the average head skin temperature by 5°C. It therefore appears that humans have a limited ability to cool the arterial/carotid blood that supplies the brain and that selective head cooling is insufficient to lower the global/average brain temperature beneath that of the aortic blood.

Cerebral blood flow and metabolism

Strenuous exercise appears to be associated with an overall activation of the brain as supported by the observation that the ratio between the global cerebral oxygen and glucose uptake becomes reduced during recovery from maximal exercise (Ide et al., 2000), following exercise with partial curarization (Dalsgaard et al., 2002), after exercise made difficult by obstructing blood flow to the exercising limbs (Dalsgaard et al., 2003) and also following exercise with hyperthermia (Nybo et al., 2003b). In addition, the global CMR_{oxygen} is increased by ~7% when exercise with hyperthermia (core temperature of 39.5°C and maximal RPE) is compared to exercise with a core temperature of 38°C and perceived exertion expressed as fairly light. Although the rise in the cerebral oxygen uptake probably relates to the Q_{10} effect (the Van't Hoff effect on tissue energy turnover; Busija et al., 1988), it could be influenced by the level of cerebral neuronal activity associated with exertion (Nybo et al., 2002a). Also, a functional link between the postexercise lowering of the cerebral metabolic uptake ratio, depletion of brain glycogen stores and central fatigue has been suggested (Ide et al., 2000; Dalsgaard et al., 2002). The notion that a reduced oxygen to carbohydrate uptake ratio relates to a lowering of the brain glycogen stores is supported by the observations in rats that sensory activation of the brain is associated with a reduced cerebral glycogen content (Madsen et al., 1995, 1999), and that the cerebral metabolic ratio remains low for several minutes following such activation (Swanson et al., 1992).

Although exercise with hyperthermia induces an overall activation of the brain and although the cerebral metabolic rate increases by ~7%, the global cerebral blood flow actually decreases by ~20% during such exercise (Nybo et al., 2002a). The reduced CBF is primarily caused by hyperthermia-induced hyperventilation, which lowers arterial CO_2 tension (P_aCO_2) and consequently also the global CBF. Thus, a lowering of the arterial CO_2 tension appears to be a stronger regulator of the global CBF than an increased CMR_{oxygen} — at least during conditions where oxygen delivery to the brain does not restrict the cerebral metabolism, as increased oxygen extraction can compensate for the lower perfusion.

It would not be obscure to suggest that the fatigue arising during hyperthermic exercise could relate inadequate oxygen delivery to the brain, since the blood supply to the brain is reduced at a point in time where the need for oxygen is elevated as the cerebral metabolic rate increases (Nybo and Nielsen, 2001b; Nybo et al., 2002a). Estimations of the average mitochondrial oxygen tension (according to the calculations described by Gjedde et al., 2005) reveals that during the worst-case scenarios of hyperthermic exercise, the mitochondrial oxygen tension may approach zero (Gjedde and Nybo, unpublished observations). On the other hand, *in vitro* studies have shown that mitochondrial respiration/function is not deteriorated even at very low oxygen tensions (Gnaiger et al., 1995; Gnaiger and Kuznetsov, 2002) and the observation that lactate spillover from the brain remains unchanged (Nybo et al., 2002a) also argues against the notion that oxygen delivery becomes insufficient during exercise with hyperthermia — at least in laboratory settings, where subjects seem to stop within a safe limit before the cerebral perfusion becomes critically low, whereas during sport competitions they may push themselves beyond that limit (Nielsen and Nybo, 2003). In addition, during exercise with hyperthermia the decline in middle cerebral artery mean blood velocity (MCA V_{mean}) (indicating reduced regional and global CBF) does not explain the fatigue that develops during such exercise (Nybo and Nielsen, 2001a). Thus, perceived exertion is unchanged, or slightly increased, when MCA V_{mean} via CO_2 inhalation is restored to the same level as during control exercise (Rasmussen et al., 2004). Inadequate oxygen delivery therefore does not explain the fatigue arising during exercise in the heat, but since the CBF appears to keep decreasing for as long as the core temperature increases (see Nybo and Nielsen, 2001b) it is likely that critically low perfusion of the brain in part explains the commonly observed presyncopal signs and occasional black outs observed during extreme conditions of hyperthermia.

Neurohumoral responses in relation to hyperthermia-induced fatigue

Several hypotheses have connected central fatigue with alterations in the cerebral level of different neurotransmitters with special attention to the synthesis and metabolism of serotonin (5-hydroxytryptamine), because of its role in arousal, sleepiness and mood (Newsholme et al., 1987; Davis et al., 1992, 2000; Blomstrand, 2001). As previously mentioned, there is some evidence for elevated serotonergic activity during exercise with hyperthermia (see Central fatigue and critical internal temperatures). However, it seems too simple to ascribe central fatigue to changes in the global cerebral level of a single neurotransmitter as fatigue is much more complex and most likely influenced by multiple factors including several neurotransmitter systems (cf. Fig. 1; Davis and Bailey, 1997; Meeusen et al., 1997; Meeusen and Piacentini, 2003). Thus, it seems clear that several neurotransmitter systems are activated during exercise (Meeusen and De Meirleir, 1995) and that some of these systems affect the preoptic area and anterior hypothalamus, which is of major importance for thermoregulation (see Chapter 1 by Blatteis). Consequently, the separate influence of the different neurotransmitter systems may be difficult to evaluate, because inhibition or promotion of a single specific neurotransmitter or its receptor(s) may exert various effects on different brain regions and concomitantly influence thermoregulation, mood, the drive to continue exercising and other factors of importance for exercise performance. For example, inhibition of the preoptic area and anterior hypothalamus deteriorates thermoregulatory function in exercising rats (Hasegawa et al., 2005), and inhibition of the serotonergic system via administration of pizotifen (a serotonin [5-HT$_{2C}$ receptor] antagonist) increase in the rectal temperature at rest and tends to induce a greater rate of core temperature increase during exercise in humans (Strachan et al., 2005). However, the 5-hydroxytryptamine (5-HT$_{2C}$) receptor blockade did not influence exercise performance, plasma prolactin or cortisol, and these parameters were also unchanged following administration of paroxetine, a selective 5-HT re-uptake inhibitor, which in addition failed to influence the core temperature response during exercise (Meeusen et al., 2001).

Two relatively new studies revealed that the exercise capacity in hot environments is not affected by branched-chain amino acid supplementation (Cheuvront et al., 2004; Watson et al., 2004), although such supplementation has been proposed to benefit performance during exercise by reducing serotonergic activity. In contrast, Mittleman et al. (1998) observed that branched-chain amino acid supplementation extended time to exhaustion in both men and women exercising in a warm environment (34°C and ~40% relative humidity). However, in that study the exercise intensity was quite low and the subjects were not hyperthermic by the end of the exercise trials (core temperatures below 38°C), and exhaustion was probably related to other factors than hyperthermia-induced central fatigue. Thus, while the rationale for the "serotonin-fatigue hypothesis" is clear and although it is supported by results from animal studies, the experimental evidence is not convincing in humans (see van Hall et al., 1995; Davis et al., 2000; Nybo and Secher, 2004 for details and discussion).

There has also been major interest in the relation between dopamine and central fatigue, and the study by Bridge et al. (2003) indicate that a high activity in the dopaminergic system is associated with a higher tolerance to exercise in the heat. Watson et al. (2005) recently investigated the effect of a dual dopamine/noradrenalin reuptake inhibitor (bupropion) on time trial (TT) performance in both temperate and warm conditions, and while they observed no influence of the drug in the control trial/temperate condition (TT performance ~31 min in both placebo and bupropion trials), a significant improvement from ~40 min to ~36 min was apparent following administration of bupropion in the heat. These results indicate that dopamine/noradrenalin reuptake inhibition enables subjects to dampen or override inhibitory signal arising from the central nervous system to cease exercise due to hyperthermia (Watson et al., 2005).

Administration of bupropion may not only postpone fatigue and enhance exercise performance in the heat, it may also interfere with thermoregulatory functions as adrenergic and dopaminergic neurons are richly represented in the preoptic area and anterior hypothalamus. Therefore, dopamine/noradrenalin reuptake inhibition may increase the risk of overheating as it may allow the subjects to push themselves beyond the safe limit of internal body temperatures. Especially administration of the drug to highly motivated athletes (e.g. during competitions) may jeopardize health, as these subjects already exercise with very high core temperatures.

Conclusions

Exercise in the heat is a challenge, not only for the locomotive and cardiovascular systems, but also for the brain. Activation of the brain during such exercise is manifested in the lowering of the cerebral glucose to oxygen uptake ratio, the elevated ratings of perceived exertion and increased release of hypothalamic hormones. While the slowing of the EEG, the decreased endurance and hampered ability to activate the skeletal muscles maximally during sustained isometric and repeated isokinetic contractions appear to relate to central fatigue arising as the core/brain increases, the central fatigue during exercise with hyperthermia may be considered as the ultimate safety break against catastrophic hyperthermia, forcing the subject to stop exercising or decrease the internal heat production, when heat loss mechanisms — in the given environment and with the prearranged exercise intensity — are insufficient to secure heat homeostasis. Elevated brain/hypothalamic temperature appears to be the main factor affecting motor activation, but the neurological factors underlying hyperthermia-induced fatigue are not apparent. The cerebral perfusion is reduced, but oxygen delivery to the brain does not appear to be critically low during laboratory experiments. The dopaminergic system seems to be important, but several other factors may interact and feedback from the skeletal muscles and internal temperature sensors are probably also involved. The complexity of fatigue as well as the complexity of the brain makes it difficult to obtain a definitive understanding of central fatigue, but exercise with hyperthermia provides an excellent model for studying the mechanisms involved with fatigue, as it is clear that central factors are of major importance for motor performance during such exercise.

Abbreviations

5-HT	5-hydroxytryptamine (serotonin)
CBF	cerebral blood flow
CMR	cerebral metabolic rate
CNS	central nervous system
EL	electrical stimulation
EEG	electroencephalogram
EMG	electromyogram
IEMG	integrated surface electromyography
MCA V_{mean}	middle cerebral artery mean blood velocity
MVC	maximal voluntary contraction
P_aCO_2	arterial CO_2 tension
RPE	rating of perceived exertion

References

Ben-Jonathan, N. and Hnasko, R. (2001) Dopamine as a prolactin (PRL) inhibitor. Endocr. Rev., 22: 724–763.

Blomstrand, E. (2001) Amino acids and central fatigue. Amino Acids, 20: 25–34.

Brengelmann, G.L. (1993) Specialized brain cooling in humans? FASEB J., 7: 1148–1153.

Bridge, M., Weller, A., Rayson, M. and Jones, D. (2003) Responses to exercise in the heat related to measures of hypothalamic serotonergic and dopaminergic function. Eur. J. Appl. Physiol., 89: 451–459.

Busija, D.W., Leffler, C.W. and Pourcyrous, M. (1988) Hyperthermia increases cerebral metabolic rate and blood flow in neonatal pigs. Am. J. Physiol., 255: H343–H346.

Cabanac, M. (1993) Selective brain cooling in humans: "fancy" or fact? FASEB J., 7: 1143–1146.

Caputa, M., Feistkorn, G. and Jessen, C. (1986) Effect of brain and trunk temperatures on exercise performance in goats. Pflugers Arch. Physiol., 406: 184–189.

Cheung, S.S. and McLellan, T.M. (1998) Heat acclimation, aerobic fitness, and hydration effects on tolerance during uncompensable heat stress. J. Appl. Physiol., 84: 1731–1739.

Cheuvront, S.N., Carter III, R., Kolka, M.A., Lieberman, H.R., Kellogg, M.D. and Sawka, M.N. (2004) Branched-chain amino acid supplementation and human performance when hypohydrated in the heat. J. Appl. Physiol., 97: 1275–1282.

Corbett, R.J. and Laptook, A.R. (1998) Failure of localized head cooling to reduce brain temperature in adult humans. Neuroreport, 9: 2721–2725.

Crezee, J. and Lagendijk, J.J. (1992) Temperature uniformity during hyperthermia: the impact of large vessels. Phys. Med. Biol., 37(6): 1321–1337.

Dalsgaard, M., Ide, K., Cai, Y., Quistorff, B. and Secher, N.H. (2002) The intent to exercise influences the cerebral O_2/carbohydrate uptake ratio in humans. J. Physiol., 540: 681–689.

Dalsgaard, M., Nybo, L., Cai, Y. and Secher, N.H. (2003) Cerebral metabolism is influenced by muscle ischaemia during exercise in humans. Exp. Physiol., 88: 297–302.

Davis, J.M., Alderson, N. and Welsh, R. (2000) Serotonin and central nervous system fatigue: nutritional considerations. Am. J. Clin. Invest., 72: 573S–578S.

Davis, J.M. and Bailey, S.P. (1997) Possible mechanisms of central nervous system fatigue during exercise. Med. Sci. Sports Exerc., 29: 45–57.

Davis, J.M., Bailey, S.P., Woods, J., Galiano, F., Hamilton, M. and Bartoli, W. (1992) Effects of carbohydrate feedings on plasma free-tryptophan and branched-chain amino acids during prolonged cycling. Eur. J. Appl. Physiol., 65: 513–519.

Emmanouel, D., Fang, V. and Katz, A. (1981) Prolactin metabolism in the rat: role of the kidney in degradation of the hormone. Am. J. Physiol., 240: F437–F445.

Freeman, M., Kanyicska, B., Lerant, A. and Nagy, G. (2000) Prolactin: structure, function, and regulation of secretion. Physiol. Rev., 80: 1523–1631.

Frewin, D., Frantz, A. and Downey, J. (1976) The effect of ambient temperature on the growth hormone and prolactin response to exercise. Aust. J. Exp. Biol. Med. Sci., 54: 97–101.

Fritzsche, R., Switzer, T., Hodgkinton, B., Lee, S., Martin, J.C. and Coyle, E.F. (2000) Water and carbohydrate ingestion during prolonged exercise increase maximal neuromuscular power. J. Appl. Physiol., 88: 730–737.

Fuller, A., Carter, R.N. and Mitchell, D. (1998) Brain and abdominal temperatures at fatigue in rats exercising in the heat. J. Appl. Physiol., 84: 877–883.

Gjedde, A., Johannsen, P., Cold, G. and Østergaard, L. (2005) Cerebral metabolic response to low blood flow: possible role of cytochrome oxidase inhibition. J. Cereb. Blood Flow Metab., 25: 1–14.

Gnaiger, E. and Kuznetsov, A.V. (2002) Mitochondrial respiration at low levels of oxygen and cytochrome c. Biochem. Soc. Transact., 30: 252–258.

Gnaiger, E., SteinlechnerMaran, R., Mendez, G., Eberl, T. and Margreiter, R. (1995) Control of mitochondrial and cellular respiration by oxygen. J. Bioenerg. Biomemb., 27: 583–596.

González-Alonso, J., Calbet, J.A. and Nielsen, B. (1998) Muscle blood flow is reduced with dehydration during prolonged exercise in humans. J. Physiol., 513: 895–905.

González-Alonso, J., Calbet, J.A. and Nielsen, B. (1999) Metabolic and thermodynamic responses to dehydration-induced reductions in muscle blood flow in exercising humans. J. Physiol., 520: 577–589.

González-Alonso, J., Mora-Rodríguez, R., Below, P.R. and Coyle, E.F. (1997) Dehydration markedly impairs cardiovascular function in hyperthermic endurance athletes during exercise. J. Appl. Physiol., 82: 1229–1236.

Gonzalez-Alonso, J., Teller, C., Andersen, S., Jensen, F., Hyldig, T. and Nielsen, B. (1999) Influence of body temperature on the development of fatigue during prolonged exercise in the heat. J. Appl. Physiol., 86: 1032–1039.

van Hall, G., Raaymakers, J.S., Saris, W.H. and Wagenmakers, A.J. (1995) Ingestion of branched-chain amino acids and tryptophan during sustained exercise in man: failure to affect performance. J. Physiol., 486: 789–794.

Hanson, R.D.G. (1974) Respiratory heat loss at increased core temperature. J. Appl. Physiol., 37: 103–107.

Hasegawa, H., Ishiwata, T., Saito, T., Yazawa, T., Aihara, Y. and Meeusen, R. (2005) Inhibition of the preoptic area and anterior hypothalamus by tetrodotoxin alters thermoregulatory functions in exercising rats. J. Appl. Physiol., 98: 1458–1462.

Hayward, J. and Baker, M. (1968) Role of cerebral arterial blood in the regulation of brain temperature in the monkey. Am. J. Physiol., 215: 389–403.

Hillegaart, V. (1991) Functional topography of brain serotonergic pathways in the rat. Acta Physiol. Scand. Suppl., 598: 1–54.

Ide, K., Schmalbruch, I.K., Quistorff, B., Horn, A. and Secher, N.H. (2000) Lactate, glucose and O_2 uptake in human brain during recovery from maximal exercise. J. Physiol., 522: 159–164.

Jessen, C. (2001) Selective brain cooling in mammals and birds. Jpn. J. Physiol., 51: 291–301.

Jessen, C. and Kuhnen, G. (1992) No evidence for brain stem cooling during face fanning in humans. J. Appl. Physiol., 72: 664–669.

Katz, A. and Emmanouel, D. (1978) Metabolism of polypeptide hormones by the normal kidney and in uremia. Nephron, 22: 69–80.

Kay, D., Marino, F., Cannon, J., St Clair Gibson, A., Lambert, M. and Noakes, T. (2001) Evidence for neuromuscular fatigue during high-intensity cycling in warm, humid conditions. Eur. J. Appl. Physiol., 84: 115–121.

Kent-Braun, J.A. (1999) Central and peripheral contributions to muscle fatigue in humans during sustained maximal effort. Eur. J. Appl. Physiol., 80: 57–63.

Kiyatkin, E.A. (2005) Brain hyperthermia as physiological and pathological phenomena. Brain Res. Brain Res. Rev., 50: 27–56.

Komiskey, H. and Rudy, T. (1977) Serotonergic influences on brain stem thermoregulation mechanisms in the cat. Brain Res., 134: 297–315.

Lassen, N.A. (1985) Normal average value of CBF in young adults is 50 ml/100 g/min (editorial). J. Cereb. Blood Flow Metab., 5: 347–349.

MacDougal, J.D., Reddan, W.G., Layton, C.R. and Dempsey, J.A. (1974) Effects of metabolic hyperthermia on performance during heavy prolonged exercise. J. Appl. Physiol., 36: 538–544.

Madsen, P.L., Cruz, N.F. and Dienel, G.A. (1995) Metabolic flux to and from brain and levels of metabolites in brain of rats during rest, sensory stimulation, and recovery. J. Cereb. Blood Flow Metab., 15: S77.

Madsen, P.L., Cruz, N.F., Sokoloff, L. and Dienel, G.A. (1999) Cerebral oxygen/glucose ratio is low during sensory stimulation and rises above normal during recovery: excess glucose consumption during stimulation is not accounted for by lactate efflux from or accumulation in brain tissue. J. Cereb. Blood Flow Metab., 19: 393–400.

Madsen, P.L., Sperling, B.K., Warming, T., Schmidt, J.F., Secher, N.H., Wildschiødtz, G., Holm, S. and Lassen, N.A. (1993) Middle cerebral artery blood velocity and cerebral blood flow and O_2 uptake during dynamic exercise. J. Appl. Physiol., 74: 245–250.

Mariak, Z., White, M.D., Lewko, J., Lyson, T. and Piekarski, P. (1999) Direct cooling of the human brain by heat loss from the upper respiratory tract. J. Appl. Physiol., 87: 1609–1613.

Mariak, Z., White, M., Lyson, T. and Lewko, J. (2003) Tympanic temperature reflects intracranial temperature changes in humans. Pflugers Arch., 446: 279–284.

Martin, P.G., Marino, F.E., Rattey, J., Kay, D. and Cannon, J. (2005) Reduced voluntary activation of human skeletal muscle during shortening and lengthening contractions in whole body hyperthermia. Exp. Physiol., 90: 225–236.

McConaghy, F., Hales, J., Rose, R. and Hodgson, D. (1995) Selective brain cooling in the horse during exercise and environmental heat stress. J. Appl. Physiol., 79: 1849–1854.

Meeusen, R. and De Meirleir, K. (1995) Exercise and brain neurotransmission. Sports Med., 20: 160–188.

Meeusen, R. and Piacentini, M.F. (2003) Exercise, fatigue, neurotransmission and the influence of the neuroendocrine axis. Dev. Tryptophan Serotonin Metab., 527: 521–525.

Meeusen, R., Piacentini, M., Van Den Eynde, S., Magnus, L. and De Meirleir, K. (2001) Exercise performance is not influenced by a 5-HT reuptake inhibitor. Int. J. Sports Med., 22: 329–336.

Meeusen, R., Roeykens, J., Magnus, L., Keizer, H. and De Meirleir, K. (1997) Endurance performance in humans: the effect of a dopamine precursor or a specific serotonin (5-HT2A/2C) antagonist. Int. J. Sports Med., 18: 571–577.

Merton, P. (1954) Voluntary strength and fatigue. J. Physiol., 123: 553–564.

Mitchell, D., Maloney, S., Jessen, C., Laburn, H., Kamerman, P., Mitchell, G. and Fuller, A. (2002) Adaptive heterothermy and selective brain cooling in arid-zone mammals. Comp. Biochem. Physiol. B Biochem. Mol. Biol., 131: 571–585.

Mittleman, K., Ricci, M. and Bailey, S. (1998) Branched-chain amino acids prolong exercise during heat stress in men and women. Med. Sci. Sports Exerc., 30: 83–91.

Morrison, S., Sleivert, G.G. and Cheung, S.S. (2004) Passive hyperthermia reduces voluntary activation and isometric force production. Eur. J. Appl. Physiol., 91: 729–736.

Newsholme, E.A., Acworth, I. and Blomstrand, E. (1987) Amino-acids, brain neurotransmitters and a functional link between muscle and brain that is important in sustained exercise. In: Benzi G. (Ed.), Advances in Biochemistry. John Libbey Eurotext Ltd., London, pp. 127–133.

Nielsen, B., Hales, J.R.S., Strange, N.J., Christensen, N.J., Warberg, J. and Saltin, B. (1993) Human circulatory and thermoregulatory adaptations with heat acclimation and exercise in a hot, dry environment. J. Physiol., 460: 467–485.

Nielsen, B., Hyldig, T., Bidstrup, F., Gonzalez-Alonso, J. and Christoffersen, G. (2001) Brain activity and fatigue during prolonged exercise in the heat. Pflugers Arch. Physiol., 442: 41–48.

Nielsen, B. and Jessen, C. (1992) Evidence against brain stem cooling by face fanning in severely hyperthermic humans. Pflugers Arch. Physiol., 422: 168–172.

Nielsen, B. and Nybo, L. (2003) Cerebral changes during exercise in the heat. Sports Med., 33: 1–11.

Nybo, L. (2003) CNS fatiue and prolonged exercise – effect of glucose supplementation. Med. Sci. Sports Exerc., 35: 589–594.

Nybo, L., Jensen, T., Nielsen, B. and Gonzalez-Alonso, J. (2001) Effects of marked hyperthermia with and without dehydration on VO2 kinetics during intense exercise. J. Appl. Physiol., 90: 1057–1064.

Nybo, L., Møller, K., Volianitis, S., Nielsen, B. and Secher, N.H. (2002a) Effects of hyperthermia on cerebral blood flow and metabolism during prolonged exercise in humans. J. Appl. Physiol., 93: 58–64.

Nybo, L. and Nielsen, B. (2001a) Hyperthermia and central fatigue during prolonged exercise in humans. J. Appl. Physiol., 91: 1055–1060.

Nybo, L. and Nielsen, B. (2001b) Middle cerebral artery blood flow velocity is reduced with hyperthermia during prolonged exercise in humans. J. Physiol., 534: 279–286.

Nybo, L. and Nielsen, B. (2001c) Perceived exertion during prolonged exercise with progressive hyperthermia is associated with an altered electrical activity of the brain. J. Appl. Physiol., 91: 2017–2023.

Nybo, L., Nielsen, B., Blomstrand, E., Moller, K. and Secher, N. (2003) Neurohumoral responses during prolonged exercise in humans. J. Appl. Physiol., 95: 1125–1131.

Nybo, L., Secher, N.H. and Nielsen, B. (2002b) Inadequate heat release from the human brain during prolonged exercise with hyperthermia. J. Physiol., 545: 697–704.

Nybo, L. and Secher, N.H. (2004) Cerebral perturbations provoked by prolonged exercise. Prog. Neurobiol., 72: 223–261.

Pitsiladis, Y., Strachan, A., Davidson, I. and Maughan, R. (2002) Hyperprolactinaemia during prolonged exercise in the heat: evidence for a centrally mediated component of fatigue in trained cyclists. Exp. Physiol., 87: 215–226.

Pugh, L., Corbett, J. and Johnson, R. (2002) Rectal temperatures, weight losses, and sweat rates in marathon running. J. Appl. Physiol., 23: 347–352.

Radomski, M., Cross, M. and Buguet, A. (1998) Exercise-induced hyperthermia and hormonal responses to exercise. Can. J. Physiol. Pharmacol., 76: 547–552.

Rasmussen, P., Stie, H., Nybo, L. and Nielsen, B. (2004) Heat induced fatigue and changes of the EEG is not related to reduced perfusion of the brain during prolonged exercise in humans. J. Thermal Biol., 29: 731–737.

Rowell, L.B., Blackmon, J., Martin, R., Mazzarella, J. and Bruce, R.A. (1965) Hepatic clearance of indocyanine green in man under thermal and exercise stresses. J. Appl. Physiol., 20: 384–394.

Rubenstein, E., Meub, D. and Eldridge, F. (1960) Common carotid blood temperature. J. Appl. Physiol., 15: 603–604.

Saboisky, J., Marino, F.E., Kay, D. and Cannon, J. (2003) Exercise heat stress does not reduce central activation to non-exercised human skeletal muscle. Exp. Physiol., 88: 783–790.

Sawka, M.N. and Wenger, C.B. (1988) Physiological responses to acute exercise-heat stress. In: Pandolf K.B., Sawka M.N. and Gonzalez R.R. (Eds.), Human Performance Physiology and Environmental Medicine at Terrestrial Extremes. pp. 97–151.

Shiraki, K., Sagawa, S., Tajima, F., Yokota, A., Hashimoto, M. and Brengelmann, G.L. (1988) Independence of brain and tympanic temperatures in an unanesthetized human. J. Appl. Physiol., 65: 482–486.

Strachan, A.T., Leiper, J.B. and Maughan, R.J. (2005) Serotonin(2C) receptor blockade and thermoregulation during exercise in the heat. Med. Sci. Sports Exerc., 37: 389–394.

Swanson, R.A., Morton, M.M., Sagar, S.M. and Sharp, F.R. (1992) Sensory stimulation induces local cerebral glycogenolysis: demonstration by autoradiography. Neuroscience, 51: 451–461.

The Commission for Thermal Physiology of the International Union of Physiological Sciences. (1987) Glossary of thermal physiology. Pflugers Arch., 410: 567–587.

Todd, G., Butler, J.E., Taylor, J.L. and Gandevia, S.C. (2005) Hyperthermia: a failure of the motor cortex and the muscle. J. Physiol., 563: 621–631.

Vigas, M., Celko, J. and Koska, J. (2000) Role of body temperature in exercise-induced growth hormone and prolactin release in non-trained and physically fit subjects. Endocr. Regul., 34: 175–180.

Walters, T.J., Ryan, K.L., Tate, L.M. and Mason, P.A. (2000) Exercise in the heat is limited by a critical internal temperature. J. Appl. Physiol., 89: 799–806.

Watson, P., Hasegawa, H., Roelands, B., Piacentini, M.F., Looverie, R. and Meeusen, R. (2004) Acute dopamine/noradrenaline reuptake inhibition enhances exercise performance in warm, but not temperate conditions. Eur. J. Appl. Physiol., 93: 306–314.

Watson, P., Shirreffs, S.M. and Maughan, R.J. (2004) The effect of acute branched-chain amino acid supplementation on prolonged exercise capacity in a warm environment. Eur. J. Appl. Physiol.

Weicker, H. and Strüder, H. (2001) Influence of exercise on serotonergic neuromodulation in the brain. Amino Acids, 20: 35–47.

West, W., Hicks, A., McKelvie, R. and O'Brien, J. (1996) The relationship between plasma potassium, muscle membrane excitability and force following quadriceps fatigue. Pflugers Arch., 432: 43–49.

Westerblad, H., Allen, D., Bruton, J., Andrade, F. and Lannergren, J. (1998) Mechanisms underlying the reduction of isometric force in skeletal muscle fatigue. Acta Physiol. Scand., 162: 253–260.

Yablonskiy, D.A., Ackerman, J. and Raichle, M.E. (2000) Coupling between changes in human brain temperature and oxidative metabolism during prolonged visual stimulation. PNAS, 97: 7603–7608.

Zhu, M.M., Nehra, D., Ackerman, J.J.H. and Yablonskiy, D.A. (2004) On the role of anesthesia on the body/brain temperature differential in rats. J. Thermal Biol., 29: 599–603.

Zuntz, N. and Schumburg, W.A.E. (1901) Physiologie des Marsches. Pflugers Arch. Physiol., 83: 557.

CHAPTER 4

Neuropsychological determinants of exercise tolerance in the heat

Stephen S. Cheung*

Environmental Ergonomics Laboratory, School of Health & Human Performance, Dalhousie University, 6230 South Street, Halifax, NS, Canada

Abstract: Traditionally, exercise in the heat has been assumed to be primarily limited by cardiovascular constraints. However, an evolutionary perspective suggests that psychological safeguards should also protect individuals prior to catastrophic hyperthermia, and exposure to hot environments or elevated body temperature may directly attenuate central drive for exercise even well before the attainment of a critical limiting central temperature. Voluntary exercise tolerance or pacing may be influenced by a complex integration of peripheral and central thermal afferents, with regional differences in thermosensitivity across the skin surface and individual variability due to age and fitness. Despite the risk of accidents from impairments in mental function, heat exposure guidelines are commonly driven by physiological parameters, and the incorporation of a psychological component should be an important focus in occupational health and safety. In directly counteracting the effects of heat stress, the face and head is a region of high sudomotor and thermal sensitivity, and may thereby serve as an effective site for reducing perceptual and/or physiological heat strain via improvements in ventilation, airflow, or active cooling.

Keywords: head; cognition; hyperthermia; heat exposure; perception; head cooling; heat strain index; pacing

Introduction

Climatic warming, along with a growing participation in extreme or ultra-endurance sports and other outdoor activities, has contributed to the increased incidence of serious exertional heat illnesses worldwide (Armstrong et al., 1996; Epstein et al., 1999; Coris et al., 2004). Hyperthermia increases the physiological strain on the body, and can result in a marked decrease in exercise capacity and potentially heat injury and death. Exercise in hot and even temperate environments can severely impair exercise capacity, with significant decreases in tolerance time to exhaustion (Galloway and Maughan, 1997). Traditionally, exercise in the heat has been assumed to be primarily limited by cardiovascular constraints, with reduced blood volume along with reduced venous return from increased dilation of peripheral vascular beds for thermoregulation leading to cardiovascular collapse. However, as heat stress or hyperthermia may also impact cognitive responses (Hancock and Vasmatzidis, 2003; see Chapters 13 and 15 in this volume), an evolutionary perspective suggests that psychological safeguards should protect individuals prior to catastrophic hyperthermia.

Various animals will cease exercise when their core temperatures exceed safe limits, and some

*Corresponding author. Tel.: +1-905-688-5550 × 4957; Fax: +1-905-688-8364; E-mail: stephen.cheung@brocku.ca

research indicates there may be a similar behavioral response in humans to reduce metabolic heat production and ultimately protect physiological integrity. One hypothesis suggests that exhaustion occurs upon the attainment of a critical internal temperature, based on observations of consistent core and/or muscle temperatures at the point of voluntary termination of exercise despite different starting core temperatures and/or rates of heat storage (MacDougall et al., 1974; Gonzalez-Alonso et al., 1999; Walters et al., 2000). An alternate perspective proposes that exercising humans can anticipate the intensity of heat stress that they will be, or are, exposed to, and will regulate their workload accordingly to minimize heat storage (Marino et al., 2004; Tucker et al., 2004). This as-yet controversial proposal of teleo-anticipation, arguing that exercise starts and ends in the brain (Kayser, 2003) even before the initiation of exercise, seemingly argues against the traditional perceived importance of cardiovascular and oxygen supply limitations to exercise, and highlights the need for a better understanding of the effects of hyperthermia and heat stress on cognitive function and neuropsychological responses.

Thus, this review will survey the recent literature on the neuropsychological responses to exercise in the heat in the hopes of stimulating further research into the relationship between body temperature and voluntary exhaustion during exercise in the heat. For a detailed review of the neurophysiological responses to heat stress and hyperthermia, see Chapter 3 by Nybo, in this volume.

Neuropsychological responses to hyperthermia

Cognitive effects of hyperthermia

Studies of occupational settings generally report greater subjective fatigue and discomfort when working in hot environments (Chen et al., 2003) and also increased frequency of unsafe behaviors (Ramsey et al., 1983), with increasing decrements with duration of exposure. Much of the research into the effects of human exposure to heat stress has focused on the core temperature responses or the underlying physiological (e.g., cardiovascular, neuromuscular, immunological) mechanisms (Cheung et al., 2000). As a result, the development of health and occupational exposure limits are heavily weighted toward physiological determinants (Parsons, 1993, 1995). In contrast, the effects of thermal stress on cognitive function and complex task performance are less intensively explored. Yet, it is critical to understand how elevations in body temperature impact on cognitive function because: (1) cognitive performance may be much more sensitive to environmental stress than physiological markers (Hancock and Vasmatzidis, 1998; Pilcher et al., 2002), and (2) errors in judgment and decision making can have severe consequences for the health and safety of the individual. Given the complexity of modern work environments and the dangers from even minor errors, this leads to concerns that occupational safety limits should be supplemented or indeed driven by psychological rather than physiological parameters (Hancock and Vasmatzidis, 1998).

In altered thermal environments, performance impairments in complex tasks could reflect impairments in one or more stages of information processing, from signal detection through to central integration and motor response. However, one major problem in defining the nature of this relationship is in establishing cognitive measures that are both scientifically and extrinsically valid. These sometimes-competing validity requirements has led to a wide variety of cognitive measures and tests being employed, along with differing magnitudes and durations of heat stress and subsequent heat strain, making it difficult to compare results across different studies and to establish clear patterns. Indeed, different cognitive processes (e.g., perception, decision making, motor planning and execution) might have different thresholds for impairment under thermal stress, and the reader is referred to excellent recent reviews comprehensively surveying the available literature on the effects of both hot and cold stress on information processing (Pilcher et al., 2002; Hancock and Vasmatzidis, 2003), and physical fitness and exercise on cognitive functioning (Etnier et al., 1997). This review will focus on some of the major implications of heat stress and cognitive functioning relating to exercise performance.

Brain activity and hyperthermia

Brain activity, specifically the ratio of low frequency ($\alpha = 8$–$13\,Hz$) and high frequency ($\beta = 13$–$30\,Hz$) brain waves as an indicator of arousal, during hyperthermia and exercise has been explored in humans cycling at 60% aerobic power in both a hot ($\sim 40°C$) and a cool ($\sim 19°C$) environment (Nielsen et al., 2001). A progressive reduction in β waves in the hot exercise condition was evident, such that the ratio of α/β waves was increased. This is similar to what happens during sleep, so it may reflect a reduced state of arousal in hyperthermic subjects. Furthermore, the magnitude of increase in the α/β ratio was strongly correlated to elevated core temperature ($r^2 = 0.94$–0.98) (Nielsen et al., 2001). Similarly, with passive hyperthermia a reduction in electroencephalographic (EEG) frequency has been reported in primates, but not until a brain (epidural) temperature of $\sim 41.5°C$ (Eshel and Safar, 2002). The functional significance of altered EEG activity remains to be determined but it is worth noting that the altered brain activity was associated with changes in ratings of perceived exertion (Nielsen et al., 2001) during exercise in humans. Subjects continually rated their effort higher during hyperthermic trials, with the best predictor of the rate of perceived exertion being a reduction in EEG frequency in the frontal cortex of the brain (Nielsen et al., 2001) (Fig. 1).

Problems in the literature

One fundamental problem in defining the role of thermal stress on cognitive impairment and task performance is that few studies have systematically and completely tracked thermal stress with changes in physiological and/or perceptual thermal strain through to cognitive and ultimately task performance impairments. For example, Neave et al. (2004) reported impaired attention and vigilance along with slower reaction times, but no changes in either physiological or perceptual responses, in young cricket batsmen. While it is intuitive that these cognitive abilities are critical determinants of batting performance, the lack of batting data preclude

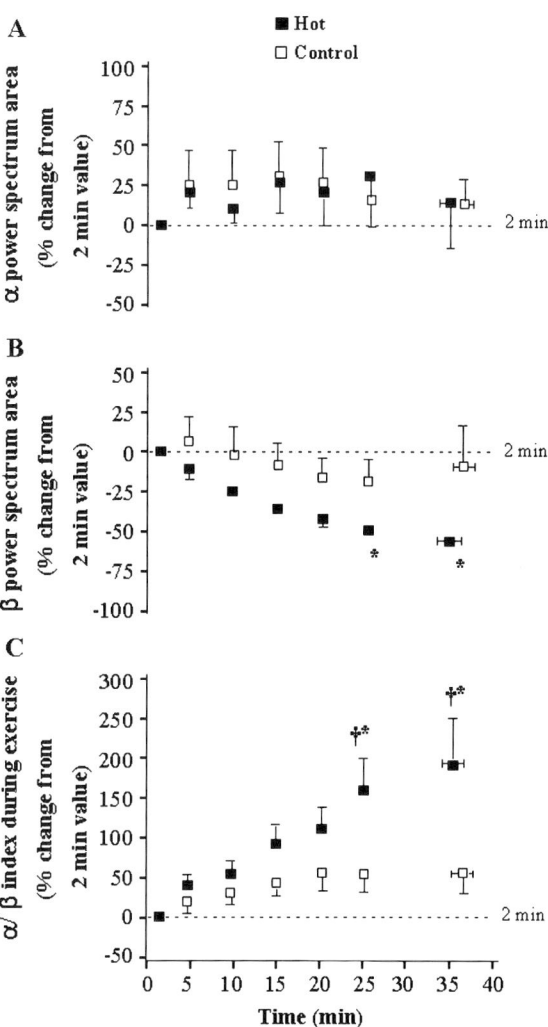

Fig. 1. α Power spectrum areas (A), β power spectrum areas (B), and exercise α/β index (C), all as a percentage of the first measurement made 2 min of exercise in the hot and control conditions. Mean $\pm$ SE for seven subjects. *Significantly different from 2-min value ($P<0.05$). †Significantly different from control ($P<0.05$). (Adapted with permission from Nielsen et al., 2001.)

conclusive determination of this connection. Other studies expose subjects to heat for a set and often short period of time and assume that subjects experience similar levels of heat strain without any physiological quantification (Chase et al., 2003). Cognitive impairment, while sensitive to thermal stress, may also be negated or minimized by other compensatory mechanisms, such that performance outcomes may be much less susceptible to thermal impairment than presumed from cognitive changes alone (Hancock and Vasmatzidis, 2003). For example, mild and stable levels of whole body heating or cooling, the latter including significant cooling of hand temperatures, produced subjective thermal discomfort but actually improved some aspects of both precision and accuracy amongst trained marksmen in a shooting simulator (Tikuisis et al., 2002a), but unfortunately the lack of cognitive-based test measures in this study precludes an exploration of the underlying mechanisms.

Hocking et al. (2001) used steady-state probe topography to demonstrate an increase in amplitude and a decrease in latency of visually evoked potentials with increased core temperature, suggesting that hyperthermia systematically altered electrical responses of the brain. Hocking's research (2001) highlights the ability of neural potential recording and analysis to quantify cognitive output. Similarly, event-related potential (ERP) recording allows comparison of potentials to a timed sequence of events, offering specific details on a given temporal or spatial aspect of neural functioning. Paired with appropriate neuropsychological tests, ERP analysis has proven to be both a valuable assessment tool of cognitive performance in psychological research (Picton, 1992) and also more recently as a clinical diagnostic tool (Polich, 1998; Connolly and D'Arcy, 2000; Marchand et al., 2002). While potentially promising, research applications for ERP analysis in thermal physiology are yet to be explored in detail. The use of ERP assessment may allow for accurate determination of cognitive performance regardless of physiological deficits such as loss of manual dexterity from shivering, as measurable ERP events can occur before, after, or in the absence of behavioral responses, and do not require behavioral observations to fully correlate the relative timing of processing to a specific event (Connolly and D'Arcy, 2000).

Perception of heat stress

The perception of the magnitude and the all-esthesial quality of thermal stress plays a major role in human behavioral thermoregulation and possibly exercise capacity. Preferred self-adjusted ambient temperature for a group of European males and females averaged $26.6 \pm 2.6°C$, though substantial inter-individual and diurnal variability were observed (Grivel and Candas, 1991). Exercise in the heat at a constant workload resulted in higher cardiovascular and thermal strain while eliciting a greater thermal discomfort and perception of effort (Maw et al., 1993). The perception of thermal stimuli and its subsequent influence on exercise performance and tolerance in the heat remains a complex issue. While the dominant center for overall thermal stimuli integration and response resides within the hypothalamus, a hierarchic system of less complex thermal integration sites also exist elsewhere within the central nervous system (Satinoff, 1983). This is informed in part by cold and warm receptors under human skin existing at an average depth of 0.15–0.17 mm and 0.3–0.6 mm, respectively (Hensel, 1981). The overall integration of these signals, and the role of individual factors, play a significant role in modulating the perception of thermal stress, and potentially the behavioral response to heat stress and also possibly the capacity for exercise.

Peripheral and regional sensitivity

The relative role of central versus peripheral afferents in perceptual responses to thermal stimuli is made difficult by methodological problems in manipulating one variable without altering the other. For example, Frank et al. (1999) used a combination of a thermally controlled mattress (at 14, 34, or 42°C) to passively alter and maintain mean skin temperature to cool, neutral, and hot, followed by 4°C intravenous saline infusion at $70 \, \text{mL min}^{-1}$ to achieve rapid core cooling. A clear separation of thermal perception and physiological response was observed, with multiple linear regression analyses demonstrating that core and skin temperature contributed about equally to perceptions of perceived temperature, but that core temperature dominated in driving vasomotor tone, metabolic heat production with core cooling, and epinephrine and norepinephrine response (Frank et al.,

1999). While such research is important in understanding the relative weighting of thermal afferents, care must be taken to avoid assuming a philosophical construct of central and peripheral afferents as two distinct entities with minimal interaction. Rather, to develop models of mean body temperature inputs into thermal stimuli integration, it would be interesting to extend such work to different levels and rates of core temperature alterations along with exploring regional skin temperature manipulations during exercise in the heat.

In addition to the interactions between central and peripheral thermoreceptors, thermal sensation may be determined in part by regional variability in thermosensitivity of different skin surface regions with local heating or cooling. Local thermal stimulation of different skin surface regions plays a major role in thermal perception of heat stress independent of core temperature, and thermal comfort may be strongly tied to the alliesthesial effects of sweating rate and skin wettedness (Mower, 1976). The existence of thermosensitivity variability across different skin regions, notably the face, has been both supported (Crawshaw et al., 1975) and rejected (Libert et al., 1984). However, one systematic limitation of these studies is the closed-loop approach to thermal manipulation, such that one region is thermally stimulated without thermal clamping of the other skin regions or core temperature, thus altering the overall thermal afferent input apart from the stimulated site. Recently, Cotter and Taylor (2005) employed a water-perfused suit enabling an open-loop approach, whereby skin regions could be stimulated while thermal clamping of non-stimulated regions could be maintained. Using this design, cooling of the face was demonstrated to be two to five times more effective in suppressing sweating response and thermal discomfort than an equivalent skin surface area elsewhere (Cotter and Taylor, 2005). The neurological origins of this increased sensitivity in humans is unclear but, based on animal models, it may be due to a higher facial thermoreceptor density (Dickenson et al., 1979) or minimal thermoafferent convergence (Dawson and Hellon, 1979). Overall, though the existence and importance of selective brain cooling in humans remain contentious (see Chapter by Nybo), both the behavioral and autonomic (e.g., sudomotor) sensitivity of the face suggests an effective potential site for targeted local cooling to lower perceived heat stress and possibly prolonged exercise capacity during hyperthermia (see Heat Stress Countermeasures) (Fig. 2).

Individual variability in thermal perception

Along with differing physiological thermoregulatory strategies and abilities, thermal perception appears to differ throughout the human lifespan. Prepubertal children have been proposed to lose heat faster due to a higher surface area to volume ratio, but both the onset threshold and sensitivity of their sweating response (Falk et al., 1991, 1992; Anderson and Mekjavic, 1996), along with thirst sensation (Meyer and Bar-Or, 1994; Bar-Or and Wilk, 1996), appear to be diminished compared to young adults. This is accompanied and possibly compensated by a heightened subjective sensitivity to increases in core temperature (Anderson and Mekjavic, 1996). With aging, there also appears to be a gradual decrement in heat dissipation and conservation capacities (Natsume et al., 1992), with greater core temperature changes required to initiate sweating and shivering (Ogawa et al., 1993; Anderson et al., 1996). This is accompanied by decreased perception to changes in thermal stimuli or thirst (Miescher and Fortney, 1989; Natsume et al., 1992; Ogawa et al., 1993), thus requiring a greater thermal stimulus to elicit behavioral responses (Taylor et al., 1995). In contrast, others argue that aging minimally influences thermoregulatory ability, and that functional capacity (e.g., aerobic fitness, body composition) and health status are the primary determinants of physiological thermoregulatory capacity with aging (Kenney and Havenith, 1993; Pandolf, 1997).

Concomitant with maturational issues, aerobic fitness and training history are other inter-individual variables that may also alter individual perception of thermal stress. During exercise in an uncompensable heat stress environment, Tikuisis et al. (2002b) compared the linkage between subjective ratings of perceived exertion and thermal discomfort as behavioral analogs for the respective

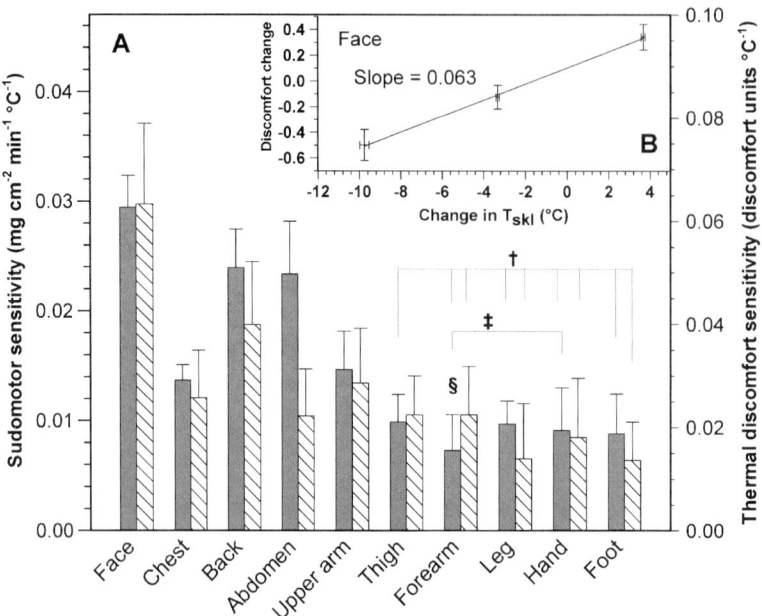

Fig. 2. Local cutaneous thermosensitivities for sudomotor control and whole-body thermal discomfort (alliesthesia). (A) Data are site-specific; mean sudomotor (shaded bars) and alliesthesial thermosensitivities ($\pm$S.E.M.) derived across three localized thermal treatments ($N = 12$): mild warming (4°C), mild cooling (−4°C), and moderate cooling (−11°C). Significant differences ($P<0.05$) between local sites were only apparent with respect to the face (†), abdomen (‡), and back (§). The inset (B) displays the relation between changes in whole-body thermal discomfort and local skin temperature (T_{skl}), and represents the alliesthesial thermosensitivity for these treatments at the face. (Adapted with permission from Cotter and Taylor, 2005.)

physiological parameters of heart rate and rectal temperature. While aerobically untrained subjects were generally very close in matching the modeled perceptual strain with its physiological counterpart, aerobically trained subjects consistently underestimated their modeled physiological strain during exercise in the heat while wearing semi-permeable chemical protective clothing (Tikuisis et al., 2002b). This difference between perceived and physiological strain during exercise in the heat could contribute to the significantly greater endpoint core temperatures consistently attained by aerobically fit individuals (Cheung and McLellan, 1998; Selkirk and McLellan, 2001), and also the ability of highly fit runners to sustain highly elevated core temperatures throughout prolonged competition (Noakes et al., 1991). Overall, this suggests the somewhat counterintuitive notion that highly fit individuals, due to their higher metabolic rate — and therefore heat production — along with their perceptual attenuation, may be at a greater risk for developing heat exhaustion (Noakes et al., 1991). Epidemiological evidence argues against this though, with low running ability and high body mass index (kg m^{-2}) being strong predictors of heat illness in Marine Corp recruits (Gardner et al., 1996). Regardless, occupational monitoring and the awareness of heat illness risks and symptoms appear critical in all worker and athletic populations exposed to heat stress.

Pacing strategies in the heat

Well before the attainment of physiological impairment of exercise via cerebral or neuromuscular mechanisms (see chapter by Nybo), the physiological impact of hyperthermia may be circumvented by voluntary control of effort at the commencement of and during heat stress. From studies modeling heat storage during running in the heat (Dennis and Noakes, 1999) and investigating self-selected running intensities under a variety of thermal environments (Marino et al., 2000, 2004), an alternative explanation to the direct effect of a high core temperature accelerating fatigue has recently been proposed (Marino, 2004). This hypothesis proposes that it is not so much a fatigue process at a critical physiological threshold that limits exercise performance in the heat, but rather an anticipatory regulation process influenced by rates of heat storage, presumably providing some level of feedback to the brain, that is activated the moment exercise begins. This would prevent excessive heat production in hot conditions and ensure that a critical limiting body temperature is not reached prematurely during voluntary exercise.

In the first study of a series leading toward this hypothesis, Dennis and Noakes (1999) calorimetrically modeled the effects of body size on maximal running speed in a warm and humid (30°C, 60% relative humidity) environment prior to attaining a situation of uncompensable heat stress, where the evaporative heat exchange required to maintain thermoneutrality matches the maximal evaporative capacity of the environment. Lower body mass and size provides a higher surface area to volume ratio and also results in lower absolute heat production at a given running speed, such that 45 kg runners were calculated as being capable of running at 19.1 km h^{-1} at the point of uncompensable heat stress as opposed to only 12.2 km h^{-1} for 75 kg runners (Dennis and Noakes, 1999). This modeling study was followed up by a validation study utilizing highly trained runners with similar aerobic capacity but a wide range of body sizes (Marino et al., 2000). Subjects ran in 15, 25, and 35°C for 30 min at a set pace, followed by an 8 km self-paced run. Running speeds progressively increased and were similar during the 8 km run at 15 and 25°C (17.8 and 17.5 km^{-1}, respectively), in marked contrast to a progressively slower pace averaging 15.8 km h^{-1} in 35°C conditions. Furthermore, body mass inversely and significantly correlated with running speed at 35°C, such that larger body masses were concluded to result in greater imbalances between heat production and dissipation, leading to a lower self-selected pace to maintain thermal balance during exercise in hot environments (Marino et al., 2000).

The common methodological paradigm of fixed intensity exercise to exhaustion, used in some studies on pre-cooling efficacy (Lee and Haymes, 1995) and in postulating a critically limiting temperature (Cheung and McLellan, 1998; Gonzalez-Alonso et al., 1999), has been challenged as artificially clamping exercise intensity and thus negating the potential contributions of voluntary control (Marino, 2004). Rather, the use of time trial efforts with self-paced intensity permits voluntary input and is also more realistic to real-life performance (Booth et al., 1997; Marino, 2004). In a comparison of African and Caucasian runners, a reduction in running speed was observed in the larger Caucasian athletes from the moment they were exposed to running in hot conditions, in which they could not maintain heat balance, as well as in cool conditions (Marino et al., 2004). Thus they slowed their running speed long before they became hyperthermic or reached some critical limiting core temperature. This study may be criticized for using a cross-sectional design with relatively non-elite subjects, such that observed differences may have been due to differences in fitness across subject groups. Additionally, levels of relative strain during the 30 min steady state run that preceded the 8 km time-trial were higher in Caucasian athletes, even in cool conditions. Although these differences were not significant, this is likely a result of poor statistical power ($n = 6$ each group), and the higher relative strain in the Caucasians may have contributed to the performance differences observed.

A survey of existing literature investigating the effects of thermal stress on exercise performance and using self-paced intensity exercise provides some support to the hypothesis of voluntary anticipation of heat stress. Core pre-cooling by 0.7°C resulted in ~300 m greater running distance over 30 min in trained runners, suggesting, though no interval timepoints were analyzed, that running speed could have been higher throughout exercise (Booth et al., 1997). In our laboratory, Olympic-caliber rowers performing a 1500 m ergometer test produced higher power outputs throughout each 500 m interval when torso pre-cooling was provided during passive exposure and 30 min warm-up (Johnson et al., 2005). Elite cyclists demonstrated a similar rectal temperature throughout a 30 min time trial in both hot (32°C) and temperate

(23°C) conditions, but the power output, albeit during only the final 10 min, was lower in hot conditions (Tatterson et al., 2000). More tellingly, a ~60 min pre-cooling period of water immersion, where skin temperature was decreased 5° to 6°C without any changes in rectal temperature prior to exercise (albeit ~0.3°C lower from 15–25 min of exercise), resulted in greater cycling distances over a 30 min self-paced time trial (Kay et al., 1999) (Fig. 3).

Understanding or defining the signal, or signals, that becomes integrated to produce self-paced effort is a difficult problem to elucidate, as it is most likely an amalgam of physiological and psychological sources and further mediated by individual factors (e.g., training status, hydration, age). Core temperature itself would appear to be an obvious signal, and the progressive decline in central neuromuscular activation during isometric leg flexion with passive heating from 37.4° to 39.4°C (Morrison et al., 2004) has been argued to provide a neuromuscular basis for this idea (Marino, 2004). Recent data from our lab also showed the same progressive impairment with passive heating independent of local soleus temperature during plantar flexion (Thomas et al., 2006). However, analysis of dynamic (isokinetic) exercise using the same passive heating model did not find any progressive impairment with core temperature increases (Cheung and Sleivert, 2004), and the nature and magnitude of impairment observed during isolated joint movements with hyperthermia appears dependent on the type (isometric versus dynamic) and duration of contractions. The role of absolute core temperature is also argued against by self-paced studies. For example, Tatterson et al. (2000) reported higher power output in cyclists in temperate (23°C) environments

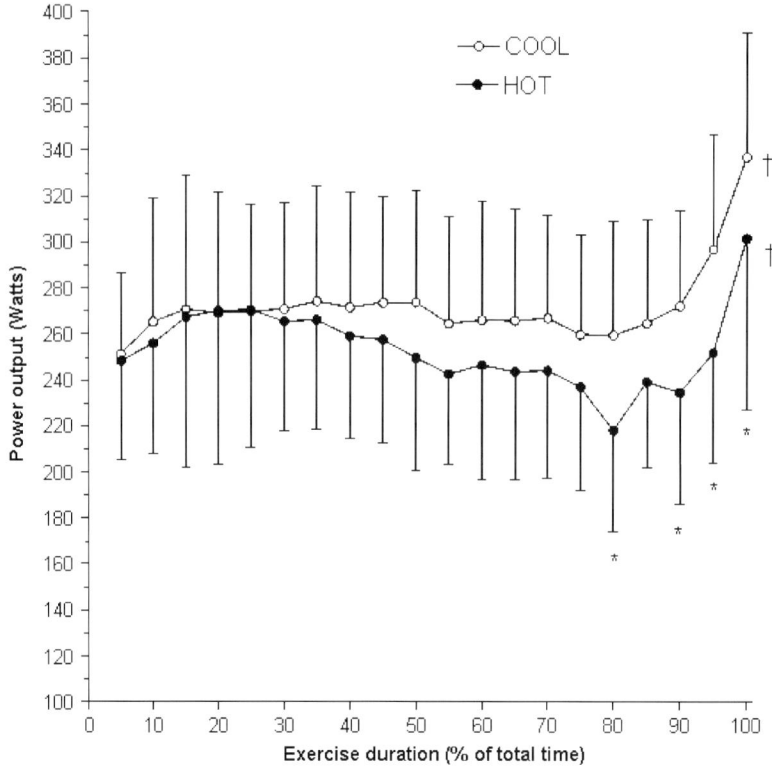

Fig. 3. Power output at intervals of 5% of the total duration of a self-paced, 20-km cycling time-trial at 35°C (HOT) or 15°C (COOL). Means±SD for 10 subjects. *$P<0.05$ vs. COOL; †$P<0.05$ vs. preceding time intervals within the same environmental condition. (Adapted with permission from Tucker et al., 2004.)

despite similar rectal temperature compared to hot (32°C) conditions. Similarly, both Kay et al. (1999) and Cotter et al. (2001) reported higher power outputs following pre-cooling even at points where absolute core temperatures were similar across conditions. The rate of core temperature increase is also unlikely to be a primary signal, as Cotter et al. (2001) reported similar rates with or without pre-cooling. Clearly, further work is required to directly determine whether there is an anticipatory process in the brain controlling work output through feedback loops involving thermal or other inputs, along with interactions from perceptual and psychological parameters.

Heat stress countermeasures

Psychological heat stress/strain indices

A great store of safety standards around acceptable exposure limits to heat stress have been developed and used for a variety of occupations from mining, firefighting, to military purposes (Parsons, 1995). These standards are typically based around the environmental heat stress imposed on the individual, using heat stress indices such as wet bulb globe temperature (WBGT) to quantify the four potential avenues (radiative, conductive, convective, and evaporative) for heat exchange between the individual and the environment. However, one limitation of such indices is the underlying assumption that all individuals respond similarly to an imposed level of external heat stress. Furthermore, to safely accommodate the widest range of individual responses, these standards are designed to be highly conservative in estimating thermal response and tolerance. Other indices (e.g., ISO 7933) incorporate more individual responses by being based on heat balance calculations and required sweat rates, but may be too cumbersome to calculate in real-time.

In response to these limitations in heat stress indices, a Physiological Strain Index (PSI) incorporating an equal weighting to cardiovascular and core temperature responses relative to resting and peak values has been modeled (Moran et al., 1998b; Moran, 2000). The theoretical advantage of such strain-based indices is the incorporation of individual responses to heat stress and work intensity, and thus the potential to individually tailor exposure durations. Relying on measures (heart rate and rectal temperature) that are relatively easy to monitor real-time in practical settings, the PSI has been validated against previous experimental data to be sensitive to variables such as hydration status (Moran et al., 1998a) and sex (Moran et al., 1999), and appears promising as an index during exercise in the heat. However, work is required to validate the PSI against criterion measures of heat strain and exercise performance/tolerance, and potentially to model and incorporate a perceptual component.

$$PSI = 5(T_{ret} - T_{re0}) \cdot (39.5 - T_{re0})^{-1} + 5(HR_t - HR_0) \cdot (180 - HR_0)^{-1}$$

where T_{re} = rectal temperature, t = current value, 0 = baseline value, HR = heart rate (Moran et al., 1998b).

$$PeSI = 5\frac{(TS_t - 7)}{6} + 5\frac{PE_t}{10}$$

where TS is a modified version of the Gagge et al. (1967) rating of thermal sensation from comfortable to intolerably hot (on a scale from 7 to 13, respectively), and PE is a modified version of the Borg (1970) rating of physical exertion (on a scale from 0 to 10) (Tikuisis et al., 2002b).

One critical factor that the indices above, and the vast majority of safety standards in general, do not incorporate are the cognitive or behavioral responses to heat stress and hyperthermia. In a majority of modern occupational settings with their heavy reliance on complex technological tools and systems, effective, efficient, and error-free task completion, rather than the length of time an individual can sustain a set physical workload, is the ultimate criterion for successful performance. In addition, surveys of the literature generally acknowledge that degradations in cognitive responses are much more sensitive to environmental and especially thermal perturbations than to physiological tolerance (Pilcher et al., 2002; Hancock and Vasmatzidis, 2003). Therefore, as argued by Hancock and Vasmatzidis (1998), "...*continuing work exposure after work performance efficiency begins to fail, but before current physiological limits are reached, is inappropriate for both the safety and productivity of the worker, their colleagues, and the systems within which they operate."* As an alternative, they propose that occupational exposure standards should incorporate both physiological and psychomotor performance factors. This paradigm shift gains extra currency given the importance of thermal perception on both behavioral and exercise response to heat stress. However, the development of such standards is presently still in its infancy, due to the numerous neuropsychological factors underlying task performance and also difficulty in developing valid and practical field measures of such factors.

Helmet use and design

In military, recreational and occupational settings, a helmet is often worn for protection from impact or collision. Helmet use for the purpose of reducing collision impact is generally accepted as standard practice in many sports such as motor racing, equestrian, American football, and cricket. There have been notable exceptions, such as the initial attempt by the Union Cyclistes Internationale to make protective helmets mandatory for professional bike racing in 1991 being met with united resistance by the athletes, primarily over the issue of perceived heat stress. Indeed, it was not until after the death of a pro cyclist during racing in 2003 that a mandatory helmet law was enforced, albeit until 2005 with a provision for the removal of helmets at the finish of mountain stages. While the wearing of protective headgear has clear importance in the prevention of trauma injuries and advances in material technology has greatly reduced their weight and increased their ventilation in some cases, their use generally remains at the expense of extra thermal insulation over an area of high potential heat dissipation (Nunneley, 1989; McCullough and Kenney, 2003). Coupled with increased understanding of the role of head temperature on perception of effort and mental performance, it is reasonable to question the impact of helmet use on heat storage, risk of mental errors, and overall exercise capacity.

Impact protection and other protective characteristics are typically the primary design characteristics of most helmets, and their performance in these issues has generally been very well investigated. In contrast, ventilation and thermal characteristics have, for the most part, been a secondary concern (Liu and Holmer, 1997), even though ventilation ports on top of occupational safety helmets resulted in less thermal strain and maintained work performance better compared to helmets with either no ventilation (Holland et al., 2002; Kim and Park, 2004) or with side ventilation (Holland et al., 2002). In some instance, such as with bicycle helmets marketed to the public, ventilation acquires major importance in acceptance (Bruhwiler et al., 2004), along with secondary characteristics such as general appearance. While commercially-available bicycle helmets vary significantly in ventilation characteristics, especially when tilt angle to air flow was considered (Bruhwiler et al., 2004), the use of a cycling helmet in either a hot-dry or hot-humid lab environment did not alter either physiological (T_{re}, HR, head temperature) or perceived magnitude of heat load when cycling compared to a bare head (Sheffield-Moore et al., 1997). Therefore, helmet use cannot be excluded on the basis of heat storage issues, although much work remains on optimizing ventilation.

It is possible that, despite a lack of subjective or physiological strain, cognitive impairment remains highly sensitive to helmet use. Neave et al. (2004) investigated the effects of helmet use on teenage cricket batsmen during a 30-min batting session and reported a clear disparity between physiological and subjective responses versus cognitive performance. Oral temperature increased and body weight decreased equally whether a helmet was worn or not, suggesting, despite limited reliability of these measures, no differences in thermal strain or sweating response. In addition, subjective sensations of fatigue, thermal stress, thirst, and alertness were similar with or without a helmet. However, reaction time, attention, and vigilance were significantly impaired following the 30 min batting session when helmets were worn, versus minimal changes when no helmets were used. Unfortunately, batting data were not reported, so it is unclear whether the cognitive impairment was merely incidental or systematically impaired actual performance.

Active head cooling

With the face as a site of high sudomotor and alliesthesial thermosensitivity during heat stress (Crawshaw et al., 1975; Cotter and Taylor, 2005), targeted active cooling of the head during exercise has long been recognized as efficacious in improving subjective tolerance and/or physiological response to heat stress (Shvartz, 1970; Nunneley et al., 1971; Greenleaf et al., 1980). Although it only comprises ~8–10% of the body surface area (Brown and Williams, 1982), it can account for the dissipation of 30% of the resting heat load and ~20% during moderate-intensity exercise (Nunneley et al., 1971). Furthermore, a lack of vasomotor response in the head region ensures a constantly high blood flow and thermal gradient for heat transfer (Hertzman and Roth, 1942). Therefore, unlike other areas in which vasoconstriction occurs during cooling to conserve heat, the cooling potential of the head is not limited by the cooling process itself. While whole body cooling with air or water underneath protective clothing has clear improvements in physiological response and exercise capacity during exercise in the heat (McLellan et al., 1999; Selkirk et al., 2004), the efficacy and most efficient pathway for head cooling remains unclear.

Fanning of the face with either ambient or cool air is a relatively low-energy alternative to targeted cooling around the head. 10°C air blown at 12 m s^{-1} into a hood decreased facial skin temperature in a 40°C environment during both rest and 90 W cycling, and also decreased sweat rate without decreasing either esophageal temperature or skin temperature in other body regions (Desruelle and Candas, 2000). Mannino and Washburn (1987) also reported decreased heart rates and forearm blood flow during 10 min of facial cooling with 4°C air at 5 m s^{-1} in resting subjects, while Armada-da-Silva et al. (2004) found that face cooling was able to slightly attenuate overall perception of effort during exercise in the heat, with lesser effect when body temperature was passively elevated prior to exercise.

If greater power generation is present, another method of active head cooling may be the use of ice packs or cooling hoods around the head and the neck regions. In clinical settings, cooling helmets have been employed for inducing brain hypothermia rapidly following cardiac arrest (Hachimi-Idrissi et al., 2001; Hachimi-Idrissi and Huyghens, 2004; Wang et al., 2004; Holzer et al., 2005) to minimize neural damage. In these situations, with shivering pharmacologically blocked, head cooling to 34°C was achieved within 60 min and whole-body (bladder) temperature within 180 min (Hachimi-Idrissi et al., 2001). Cooling helmets have also been used to reduce heat stress and alleviate symptoms in individuals with multiple sclerosis (Ku et al., 1996). Studies on subjects simulating helicopter or airplane flying conditions have reported significantly improved physiological (Nunneley et al., 1982; Nunneley and Maldonado, 1983) and cognitive response on reaction time and performance accuracy (Nunneley et al., 1982) when active head cooling using water-perfused hoods was provided. In particular, head + torso cooling was claimed to be two to three times more efficient than torso cooling alone, despite the small additional surface area of coverage, (Nunneley and Maldonado, 1983). When using a wearable neck-cooling device, rectal temperatures (0.2°C) and sweat rates (92 mL h^{-1}) were slightly but significantly decreased in trained runners during a constant-speed track run in temperate (21°C) conditions compared to no cooling, with no changes in heart rate and perceived exertion (Gordon et al., 1990). Furthermore, pre-cooling solely of the head or head-cooling during exercise with a liquid-conditioned hood, both, resulted in a higher self-selected pace in trained runners during running in hot conditions compared to no cooling (Palmer et al., 2001).

The efficacy of either face fanning or head cooling to influence either brain temperature or physiological response and performance in the heat is not universally evident in the literature, and the continuing debate of whether humans exhibit selective brain cooling during heat stress is covered in the Chapter 3 by Nybo. Modeling studies of the human head indicate that brain temperature closely follows arterial temperature and is controlled through systemic thermoregulation independent of head surface temperature (Nelson and Nunneley, 1998; Dennis et al., 2003). Specifically, the primary limitation behind head cooling may be due to the dense capillary network around the brain and large degree of warm blood perfusion from the uncooled carotid artery (Nelson and Nunneley, 1998; Dennis et al., 2003), as a modeled brain temperature of 33°C within 30 min could only be successfully achieved with a major reduction in carotid blood flow (Dennis et al., 2003). This is supported by proton magnetic resonance spectroscopy in human subjects, where neither any cooling nor any difference in superficial cortex or thalamus temperature over 50 min with a scalp temperature of either 15.8 or 34.7°C were observed (Corbett and Laptook, 1998). Using a similar system of a liquid-cooled hood and pilot clothing as Nunneley and Maldonado (1983), Frim (1989) reported improved subjective comfort but no physiological or performance improvements with using head + torso cooling compared to torso-cooling by itself. Similarly, neither a refrigerated headpiece nor an ice-packet vest reduced heat strain during wheelchair exercise in the heat compared to no cooling. (Armstrong et al., 1995).

Summary

Overall, hyperthermia impacts both physiological and psychological response to exercise. The head and face is an area of high thermosensitivity for both sudomotor and perceptual response to heat stress, but the physiological mechanism by which active head cooling appears to benefit exercise performance remains unclear. The connection between head ventilation and cooling on mental performance argues for improved cognitive abilities, but studies connecting these underlying cognitive changes to actual differences in realistic and relevant outcome measures remain rare. Despite modeling studies that demonstrate minimal effect on actual brain or whole body temperature on conscious and healthy humans, increasing ventilation in protective helmets and active head cooling during passive exposure or active exercise in the heat appears to promote an improved subjective perception of heat stress. In turn, ventilation and cooling generally reduces whole-body thermal and physiological strain during exercise performance, leading to higher self-selected pacing during prolonged exercise. This would tend to support the idea that afferent feedback from the skin and especially the head, irrespective of actual brain temperature, may play an important role in regulating exercise intensity and pacing, and would argue for maximizing ventilation and cooling whenever possible or practicable.

One significant caveat to the use of active cooling during heavy or prolonged exercise in extreme heat, however, is the risk of over-riding the body's physiological capacity for tolerating heat stress by artificially dampening the perceptual response to heat stress, resulting in greater exposure to heat strain with its consequent immunological and endotoxemic perturbations and risk of heat stroke. This is especially relevant in elite performance with highly aerobically trained individuals, who can already tolerate a higher terminal core temperature prior to voluntary exhaustion than non-trained individuals (Cheung and McLellan, 1998; Selkirk and McLellan, 2001) along with greater tolerance to passive heating (Morrison et al., 2006), and who consistently perceptually underestimate their estimated physiological strain (Tikuisis et al., 2002b). Therefore, targeted head cooling should only be employed with real-time physiological monitoring or thermal and cardiovascular strain, and strict cutoff parameters should be adhered to.

Acknowledgments

The author was supported by a Discovery Grant from the Natural Sciences and Engineering Research Council (NSERC) of Canada.

References

Anderson, G.S. and Mekjavic, I.B. (1996) Thermoregulatory responses of circum-pubertal children. Eur. J. Appl. Physiol., 74: 404–410.

Anderson, G.S., Meneilly, G.S. and Mekjavic, I.B. (1996) Passive temperature lability in the elderly. Eur. J. Appl. Physiol., 73: 278–286.

Armada-da-Silva, P.A., Woods, J. and Jones, D.A. (2004) The effect of passive heating and face cooling on perceived exertion during exercise in the heat. Eur. J. Appl. Physiol., 91: 563–571.

Armstrong, L.E., Epstein, Y., Greenleaf, J.E., Haymes, E.M., Hubbard, R.W., Roberts, W.O. and Thompson, P.D. (1996) American College of Sports Medicine position stand: heat and cold illnesses during distance running. Med. Sci. Sports Exerc., 28: i–x.

Armstrong, L.E., Maresh, C.M., Riebe, D., Kenefick, R.W., Castellani, J.W., Senk, J.M., Echegaray, M. and Foley, M.F. (1995) Local cooling in wheelchair athletes during exercise heat stress. Med. Sci. Sports Exerc., 27: 211–216.

Bar-Or, O. and Wilk, B. (1996) Water and electrolyte replenishment in the exercising child. Int. J. Sport Nutr., 6: 93–99.

Booth, J., Marino, F. and Ward, J.J. (1997) Improved running performance in hot humid conditions following whole body precooling. Med. Sci. Sports Exerc., 29: 943–949.

Borg, G. (1970) Perceived exertion as an indicator of somatic stress. Scand. J. Rehabil. Med., 2: 92–98.

Brown, G.A. and Williams, G.M. (1982) The effect of head cooling on deep body temperature and thermal comfort in man. Aviat. Space Environ. Med., 53: 583–586.

Bruhwiler, P.A., Ducas, C., Huber, R. and Bishop, P.A. (2004) Bicycle helmet ventilation and comfort angle dependence. Eur. J. Appl. Physiol., 92: 698–701.

Chase, B., Karwowski, W., Benedict, M.E., Quesada, P.M. and Irwin-Chase, H.M. (2003) A study of computer-based task performance under thermal stress. Int. J. Occup. Saf. Ergon., 9: 5–15.

Chen, M.L., Chen, C.J., Yeh, W.Y., Huang, J.W. and Mao, I.F. (2003) Heat stress evaluation and worker fatigue in a steel plant. AIHA J. (Fairfax, VA), 64: 352–359.

Cheung, S.S. and McLellan, T.M. (1998) Influence of heat acclimation, aerobic fitness, and hydration effects on tolerance during uncompensable heat stress. J. Appl. Physiol., 84: 1731–1739.

Cheung, S.S., McLellan, T.M. and Tenaglia, S. (2000) The thermophysiology of uncompensable heat stress: physiological manipulations and individual characteristics. Sports Med., 29: 329–359.

Cheung, S.S. and Sleivert, G.G. (2004) Lowering of skin temperature decreases isokinetic maximal force production independent of core temperature. Eur. J. Appl. Physiol., 91: 723–728.

Connolly, J.F. and D'Arcy, R.C. (2000) Innovations in neuropsychological assessment using event-related brain potentials. Int. J. Psychophysiol., 37: 31–47.

Corbett, R.J. and Laptook, A.R. (1998) Failure of localized head cooling to reduce brain temperature in adult humans. Neuroreport, 9: 2721–2725.

Coris, E.E., Ramirez, A.M. and Van Durme, D.J. (2004) Heat illness in athletes: the dangerous combination of heat, humidity and exercise. Sports Med., 34: 9–16.

Cotter, J.D., Sleivert, G.G., Roberts, W.S. and Febbraio, M.A. (2001) Effect of pre-cooling, with and without thigh cooling, on strain and endurance exercise performance in the heat. Comp. Biochem. Physiol. A Mol. Integr. Physiol., 128: 667–677.

Cotter, J.D. and Taylor, N.A.S. (2005) The distribution of cutaneous sudomotor and alliesthesial thermosensitivity in mildly heat-stressed humans: an open-loop approach. J. Physiol. (Lond.), 565: 335–345.

Crawshaw, L.I., Nadel, E.R., Stolwijk, J.A. and Stamford, B.A. (1975) Effect of local cooling on sweating rate and cold sensation. Pflugers Arch., 354: 19–27.

Dawson, N.J. and Hellon, R.F. (1979) Facilitation and suppression of antidromic invasion by orthodromic impulses in the cat [proceedings]. J. Physiol., 289: 57P–58P.

Dennis, B.H., Eberhart, R.C., Dulikravich, G.S. and Radons, S.W. (2003) Finite-element simulation of cooling of realistic 3-D human head and neck. J. Biomech. Eng., 125: 832–840.

Dennis, S.C. and Noakes, T.D. (1999) Advantages of a smaller bodymass in humans when distance-running in warm, humid conditions. Eur. J. Appl. Physiol., 79: 280–284.

Desruelle, A.V. and Candas, V. (2000) Thermoregulatory effects of three different types of head cooling in humans during a mild hyperthermia. Eur. J. Appl. Physiol., 81: 33–39.

Dickenson, A.H., Hellon, R.F. and Taylor, D.C. (1979) Facial thermal input to the trigeminal spinal nucleus of rabbits and rats. J. Comp. Neurol., 185: 203–209.

Epstein, Y., Moran, D.S., Shapiro, Y., Sohar, E. and Shemer, J. (1999) Exertional heat stroke: a case series. Med. Sci. Sports Exerc., 31: 224–228.

Eshel, G.M. and Safar, P. (2002) The role of the central nervous system in heatstroke: reversible profound depression of cerebral activity in a primate model. Aviat. Space Environ. Med., 73: 327–332 discussion 333–334.

Etnier, J.L., Salazar, W., Landers, D.M., Petruzzello, S.J., Han, M. and Nowell, P. (1997) The influence of physical fitness and exercise upon cognitive functioning: a meta-analysis. J. Sport Exerc. Psychol., 19: 249–277.

Falk, B., Bar-Or, O. and MacDougall, J.D. (1992) Sweat gland response to exercise in the heat among pre-, mid-, and late-pubertal boys. Med. Sci. Sports Exerc., 24: 313–319.

Falk, B., Bar-Or, O., MacDougall, J.D., McGillis, L., Calvert, R. and Meyer, F. (1991) Sweat lactate in exercising children and adolescents of varying physical maturity. J. Appl. Physiol., 71: 1735–1740.

Frank, S.M., Raja, S.N., Bulcao, C.F. and Goldstein, D.S. (1999) Relative contribution of core and cutaneous temperatures to thermal comfort and autonomic responses in humans. J. Appl. Physiol., 86: 1588–1593.

Frim, J. (1989) Head cooling is desirable but not essential for preventing heat strain in pilots. Aviat. Space Environ. Med., 60: 1056–1062.

Gagge, A.P., Stolwijk, J.A. and Hardy, J.D. (1967) Comfort and thermal sensations and associated physiological responses at various ambient temperatures. Environ. Res., 1: 1–20.

Galloway, S.D. and Maughan, R.J. (1997) Effects of ambient temperature on the capacity to perform prolonged cycle exercise in man. Med. Sci. Sports Exerc., 29: 1240–1249.

Gardner, J.W., Kark, J.A., Karnei, K., Sanborn, J.S., Gastaldo, E., Burr, P. and Wenger, C.B. (1996) Risk factors predicting exertional heat illness in male Marine Corps recruits. Med. Sci. Sports Exerc., 28: 939–944.

Gonzalez-Alonso, J., Teller, C., Andersen, S.L., Jensen, F.B., Hyldig, T. and Nielsen, B. (1999) Influence of body temperature on the development of fatigue during prolonged exercise in the heat. J. Appl. Physiol., 86: 1032–1039.

Gordon, N.F., Bogdanffy, G.M. and Wilkinson, J. (1990) Effect of a practical neck cooling device on core temperature during exercise. Med. Sci. Sports Exerc., 22: 245–249.

Greenleaf, J.E., Van Beaumont, W., Brock, P.J., Montgomery, L.D., Morse, J.T., Shvartz, E. and Kravik, S. (1980) Fluid-electrolyte shifts and thermoregulation: rest and work in heat with head cooling. Aviat. Space Environ. Med., 51: 747–753.

Grivel, F. and Candas, V. (1991) Ambient temperatures preferred by young European males and females at rest. Ergonomics, 34: 365–378.

Hachimi-Idrissi, S., Corne, L., Ebinger, G., Michotte, Y. and Huyghens, L. (2001) Mild hypothermia induced by a helmet device: a clinical feasibility study. Resuscitation, 51: 275–281.

Hachimi-Idrissi, S. and Huyghens, L. (2004) Resuscitative mild hypothermia as a protective tool in brain damage: is there evidence? Eur. J. Emerg. Med., 11: 335–342.

Hancock, P.A. and Vasmatzidis, I. (1998) Human occupational and performance limits under stress: the thermal environment as a prototypical example. Ergonomics, 41: 1169–1191.

Hancock, P.A. and Vasmatzidis, I. (2003) Effects of heat stress on cognitive performance: the current state of knowledge. Int. J. Hyperthermia, 19: 355–372.

Hensel, H. (1981) Thermoreception and temperature regulation. Monogr. Physiol. Soc., 38: 1–321.

Hertzman, A.B. and Roth, L.W. (1942) The absence of vasoconstrictor reflexes in the forehead circulation. Am. J. Physiol., 136: 692–697.

Hocking, C., Silberstein, R.B., Lau, W.M., Stough, C. and Roberts, W. (2001) Evaluation of cognitive performance in the heat by functional brain imaging and psychometric testing. Comp. Biochem. Physiol. A Mol. Integr. Physiol., 128: 719–734.

Holland, E.J., Laing, R.M., Lemmon, T.L. and Niven, B.E. (2002) Helmet design to facilitate thermoneutrality during forest harvesting. Ergonomics, 45: 699–716.

Holzer, M., Bernard, S.A., Hachimi-Idrissi, S., Roine, R.O., Sterz, F. and Mullner, M. (2005) Hypothermia for neuroprotection after cardiac arrest: systematic review and individual patient data meta-analysis. Crit. Care Med., 33: 414–418.

Johnson, E.A., Sleivert, G.G., Cheung, S.S. and Wenger, H. (2005) Pre-cooling decreases psychophysical strain during steady-state rowing and enhances self-paced performance in elite rowers. Med. Sci. Sports Exerc., 37: S170.

Kay, D., Taaffe, D.R. and Marino, F.E. (1999) Whole-body pre-cooling and heat storage during self-paced cycling performance in warm humid conditions. J. Sports Sci., 17: 937–944.

Kayser, B. (2003) Exercise starts and ends in the brain. Eur. J. Appl. Physiol., 90: 411–419.

Kenney, W.L. and Havenith, G. (1993) Heat stress and age: skin blood flow and body temperature. J. Thermal Biol., 18: 341–344.

Kim, H.E. and Park, S.J. (2004) The effect of safety hat on thermal responses and working efficiency under a high temperature environment. J. Physiol. Anthropol. Appl. Human Sci., 23: 149–153.

Ku, Y.T., Montgomery, L.D. and Webbon, B.W. (1996) Hemodynamic and thermal responses to head and neck cooling in men and women. Am. J. Phys. Med. Rehabil., 75: 443–450.

Lee, D.T. and Haymes, E.M. (1995) Exercise duration and thermoregulatory responses after whole body precooling. J. Appl. Physiol., 79: 1971–1976.

Libert, J.P., Candas, V., Sagot, J.C., Meyer, J.P., Vogt, J.J. and Ogawa, T. (1984) Contribution of skin thermal sensitivities of large body areas to sweating response. Jpn. J. Physiol., 34: 75–88.

Liu, X. and Holmer, I. (1997) Evaluation of evaporative heat transfer characteristics of helmets. Appl. Human Sci., 16: 107–113.

MacDougall, J.D., Reddan, w.G., Layton, C.R. and Dempsey, J.A. (1974) Effects of metabolic hyperthermia on performance during heavy prolonged exercise. J. Appl. Physiol., 36: 538–544.

Mannino, J.A. and Washburn, R.A. (1987) Cardiovascular responses to moderate facial cooling in men and women. Aviat. Space Environ. Med., 58: 29–33.

Marchand, Y., D'Arcy, R.C. and Connolly, J.F. (2002) Linking neurophysiological and neuropsychological measures for aphasia assessment. Clin. Neurophysiol., 113: 1715–1722.

Marino, F.E. (2004) Anticipatory regulation and avoidance of catastrophe during exercise-induced hyperthermia. Comp. Biochem. Physiol. B Biochem. Mol. Biol., 139: 561–569.

Marino, F.E., Lambert, M.I. and Noakes, T.D. (2004) Superior performance of African runners in warm humid but not in cool environmental conditions. J. Appl. Physiol., 96: 124–130.

Marino, F.E., Mbambo, Z., Kortekaas, E., Wilson, G., Lambert, M.I., Noakes, T.D. and Dennis, S.C. (2000) Advantages of smaller body mass during distance running in warm, humid environments. Pflugers Arch., 441: 359–367.

Maw, G.J., Boutcher, S.H. and Taylor, N.A. (1993) Ratings of perceived exertion and affect in hot and cool environments. Eur. J. Appl. Physiol., 67: 174–179.

McCullough, E.A. and Kenney, W.L. (2003) Thermal insulation and evaporative resistance of football uniforms. Med. Sci. Sports Exerc., 35: 832–837.

McLellan, T.M., Frim, J. and Bell, D.G. (1999) Efficacy of air and liquid cooling during light and heavy exercise while wearing NBC clothing. Aviat. Space Environ. Med., 70: 802–811.

Meyer, F. and Bar-Or, O. (1994) Fluid and electrolyte loss during exercise: the paediatric angle. Sports Med., 18: 4–9.

Miescher, E. and Fortney, S.M. (1989) Responses to dehydration and rehydration during heat exposure in young and older men. Am. J. Physiol., 257: R1050–R1056.

Moran, D.S. (2000) Stress evaluation by the physiological strain index (PSI). J. Basic Clin. Physiol. Pharmacol., 11: 403–423.

Moran, D.S., Montain, S.J. and Pandolf, K.B. (1998a) Evaluation of different levels of hydration using a new physiological strain index. Am. J. Physiol., 44: R854–R860.

Moran, D.S., Shapiro, Y., Laor, A., Izraeli, S. and Pandolf, K.B. (1999) Can gender differences during exercise-heat stress be assessed by the physiological strain index? Am. J. Physiol., 276: R1798–R1804.

Moran, D.S., Shitzer, A. and Pandolf, K.B. (1998b) A physiological strain index to evaluate heat stress. Am. J. Physiol. Regul. Integr. Comp. Physiol., 44: R129–R134.

Morrison, S., Sleivert, G.G. and Cheung, S.S. (2004) Passive hyperthermia reduces voluntary activation and isometric force production. Eur. J. Appl. Physiol., 91: 729–736.

Morrison, S.A., Sleivert, G.G. and Cheung, S.S. (2006) Aerobic influence on neuromuscular function and tolerance during passive hyperthermia. Med. Sci. Sports Exerc., 38: 1754–1761.

Mower, G.D. (1976) Perceived intensity of peripheral thermal stimuli is independent of internal body temperature. J. Comp. Physiol. Psychol., 90: 1152–1155.

Natsume, K., Ogawa, T., Sugenoya, J., Ohnishi, N. and Imai, K. (1992) Preferred ambient-temperature for old and young men in summer and winter. Int. J. Biometeorol., 36: 1–4.

Neave, N., Emmett, J., Moss, M., Ayton, R., Scholey, A. and Wesnes, K. (2004) The effects of protective helmet use on physiology and cognition in young cricketers. Appl. Cogn. Psychol., 18: 1181–1193.

Nelson, D.A. and Nunneley, S.A. (1998) Brain temperature and limits on transcranial cooling in humans: quantitative modeling results. Eur. J. Appl. Physiol. Occup. Physiol., 78: 353–359.

Nielsen, B., Hyldig, T., Bidstrup, F., Gonzalez-Alonso, J. and Christoffersen, G.R. (2001) Brain activity and fatigue during prolonged exercise in the heat. Pflugers Arch. Eur. J. Physiol., 442: 41–48.

Noakes, T.D., Myburgh, K.H., Du Plessis, J., Lang, L., Lambert, M., Van der Riet, C. and Schall, R. (1991) Metabolic rate, not percent dehydration, predicts rectal temperature in marathon runners. Med. Sci. Sports Exerc., 23: 443–449.

Nunneley, S.A. (1989) Heat stress in protective clothing: interactions among physical and physiological factors. Scand. J. Work Environ. Health, 15(Suppl 1): 52–57.

Nunneley, S.A. and Maldonado, R.J. (1983) Head and/or torso cooling during simulated cockpit heat stress. Aviat. Space Environ. Med., 54: 496–499.

Nunneley, S.A., Reader, D.C. and Maldonado, R.J. (1982) Head-temperature effects on physiology, comfort, and performance during hyperthermia. Aviat. Space Environ. Med., 53: 623–628.

Nunneley, S.A., Troutman Jr., S.J. and Webb, P. (1971) Head cooling in work and heat stress. Aerosp. Med., 42: 64–68.

Ogawa, T., Ohnishi, N., Imai, K. and Sugenoya, J. (1993) Thermoregulatory responses of old men to gradual changes in ambient temperature. J. Thermal Biol., 18: 345–348.

Palmer, C.D., Sleivert, G.G. and Cotter, J.D. (2001) The effects of head and neck cooling on thermoregulation, pace selection, and performance. Int. Thermal Physiol. Symposium, 32: 122P.

Pandolf, K.B. (1997) Aging and human heat tolerance. Exp. Aging Res., 23: 69–105.

Parsons, K.C. (1995) International heat stress standards: a review. Ergonomics, 38: 6–22.

Picton, T.W. (1992) The P300 wave of the human event-related potential. J. Clin. Neurophysiol., 9: 456–479.

Pilcher, J.J., Nadler, E. and Busch, C. (2002) Effects of hot and cold temperature exposure on performance: a meta-analytic review. Ergonomics, 45: 682–698.

Polich, J. (1998) P300 clinical utility and control of variability. J. Clin. Neurophysiol., 15: 14–33.

Ramsey, J.D., Burford, C.L., Beshir, M.Y. and Jensen, R.C. (1983) Effects of workplace thermal conditions on safe work behavior. J. Safety Res., 14: 105–114.

Satinoff, E. (1983) A reevaluation of the concept of the homeostatic organization of temperature regulation. In: Satinof E. and Teitelbaum P. (Eds.), Handbook of Behavioral Neurobiology, Vol. 6. Plenum Press, New York, pp. 443–472.

Selkirk, G.A. and McLellan, T.M. (2001) Influence of aerobic fitness and body fatness on tolerance to uncompensable heat stress. J. Appl. Physiol., 91: 2055–2063.

Selkirk, G.A., McLellan, T.M. and Wong, J. (2004) Active versus passive cooling during work in warm environments while wearing firefighting protective clothing. J. Occup. Environ. Hyg., 1: 521–531.

Sheffield-Moore, M., Short, K.R., Kerr, C.G., Parcell, A.C., Bolster, D.R. and Costill, D.L. (1997) Thermoregulatory responses to cycling with and without a helmet. Med. Sci. Sports Exerc., 29: 755–761.

Shvartz, E. (1970) Effect of a cooling hood on physiological responses to work in a hot environment. J. Appl. Physiol., 29: 36–39.

Tatterson, A.J., Hahn, A.G., Martin, D.T. and Febbraio, M.A. (2000) Effects of heat stress on physiological responses and exercise performance in elite cyclists. J. Sci. Med. Sport, 3: 186–193.

Taylor, N.A., Allsopp, N.K. and Parkes, D.G. (1995) Preferred room temperature of young vs. aged males: the influence of thermal sensation, thermal comfort, and affect. J. Gerontol. A Biol. Sci. Med. Sci., 50: M216–M221.

Thomas, M.M., Cheung, S.S., Elder, G.C. and Sleivert, G.G. (2006) Voluntary muscle activation is impaired by core temperature rather than local muscle temperature. J. Appl. Physiol., 100: 1361–1369.

Tikuisis, P., Keefe, A.A., Keillor, J., Grant, S. and Johnson, R.F. (2002a) Investigation of rifle marksmanship on simulated targets during thermal discomfort. Aviat. Space Environ. Med., 73: 1176–1183.

Tikuisis, P., McLellan, T.M. and Selkirk, G. (2002b) Perceptual versus physiological heat strain during exercise-heat stress. Med. Sci. Sports Exerc., 34: 1454–1461.

Tucker, R., Rauch, L., Harley, Y.X. and Noakes, T.D. (2004) Impaired exercise performance in the heat is associated with an anticipatory reduction in skeletal muscle recruitment. Pflugers Arch., 448: 422–430.

Walters, T.J., Ryan, K.L., Tate, L.M. and Mason, P.A. (2000) Exercise in the heat is limited by a critical internal temperature. J. Appl. Physiol., 89: 799–806.

Wang, H., Olivero, W., Lanzino, G., Elkins, W., Rose, J., Honings, D., Rodde, M., Burnham, J. and Wang, D. (2004) Rapid and selective cerebral hypothermia achieved using a cooling helmet. J. Neurosurg., 100: 272–277.

ics
Drugs and Hyperthermia

CHAPTER 5

Thermophysiological responses to hyperthermic drugs: extrapolating from rodent to human

Christopher J. Gordon*

Neurotoxicology Division, National Health and Environmental Effects Research Laboratory, B105-04, US Environmental Protection Agency, 109 S.T.W. Alexander Drive, Research Triangle Park, NC 27711, USA

Abstract: This chapter focuses on the effects of hyperthermia on drug and chemical toxicity. In general, hyperthermia exacerbates the toxicity of many types of drugs and environmental toxicants. Using rodents to model the potential responses of humans to hyperthermic drugs is hampered by the unique differences in thermoregulatory ability and body mass. Because of their relatively large surface area:mass ratio, ambient temperature has a more profound influence on the potential hyperthermic effect of a drug in rodents. The relative increase in heat production (i.e., as a percentage of their basal metabolic rate) required to raise core temperature by 1°C will increase with a decrease in body mass. The thermoregulatory response to methylenedioxymethamphetamine (MDMA) is used to illustrate the differences in thermoregulatory responses of rats and humans to a hyperthermic drug. Overall, the interaction between ambient temperature and drug-induced changes in body temperature is critical in the evaluation of hyperthermic-induced toxicity in rodent models.

Keywords: temperature regulation; hyperthermia; extrapolation; MDMA

Introduction

Hyperthermia generally exacerbates the physiological and pathological responses of drugs, environmental contaminants, and other insults. Thermal kinetics provides a fundamental explanation of hyperthermia and drug efficacy. That is, the chemical reactions of all life processes depend on temperature, as expressed by the principles of the Arrhenius equation (Burton and Edholm, 1955).

The Arrhenius equation is based on thermal kinetics, and states that the rate of chemical reactions increases exponentially with a rise in temperature (Fig. 1). Most molecular, cellular, and physiological processes have a positive temperature coefficient, meaning their activity increases in a manner similar to that predicted by the Arrhenius equation. Thermal biologists often use the Q_{10} to describe the effects of temperature. The Q_{10} of most biological processes ranges between 2 and 3, which equates to a doubling and tripling of the reaction rate with a 10°C increase in temperature.

Cellular and molecular mechanisms of toxicity generally have positive temperature coefficients (Gordon, 2005). Processes such as receptor binding, lipid peroxidation, metabolic deactivation of a toxicant, and oxidative phosphorylation generally

---

☆ This paper has been reviewed by the National Health and Environmental Effects Research Laboratory, US Environmental Protection Agency, and approved for publication. Mention of trade names or commercial products does not constitute endorsement or recommendation for use.

*Corresponding author. Tel.: +1-919-541-1509;
Fax: +1-919-541-4416; E-mail: gordon.christopher@epa.gov

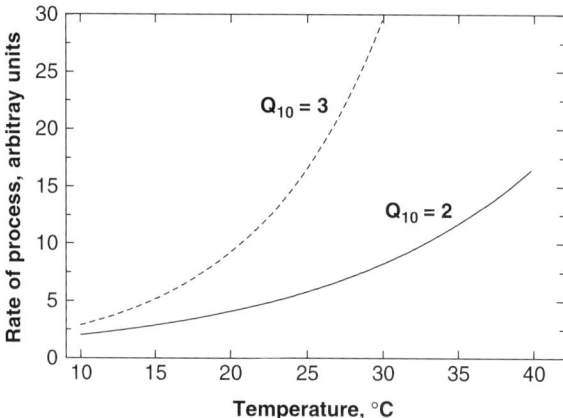

Fig. 1. A diagrammatic representation of the effect of temperature on the rate of a chemical reaction or physiological process. Theoretical functions show the effects of a doubling ($Q_{10} = 2$) or tripling ($Q_{10} = 3$) the rate of the process with a 10°C increase in temperature. Adapted from Schmidt-Nielsen (1975).

increase with temperature but there are some notable exceptions. Insecticides such as DDT and pyrethroids that induce depolarization of nerve tissue and induction of some protective proteins, exhibit a negative temperature coefficient. In most cases, the toxic efficacy of a chemical is proportional to tissue temperature. Thus, in an environment where the thermoregulatory system is challenged such that a hyperthermic state exists, there will be increased susceptibility to a drug or toxic chemical.

This chapter has three main goals: review the literature on the role of hyperthermia on the toxic efficacy of drugs and xenobiotic agents, explore the mechanism of action of selected amphetamine-derivatives on the thermoregulatory system, and emphasize the importance of extrapolating thermoregulatory data from rodent to human.

Thermoregulatory profile and hyperthermic efficacy of drugs

Rodents have been and will continue to be a primary test species in the study of the toxicity of drugs and other chemical agents. Many investigators may not be fully aware of the interaction between ambient temperature, thermoregulatory requirements of rodents, and the hyperthermic efficacy of drugs and other agents. A thermoregulatory profile is essentially the interrelationship between ambient temperature, body temperature, skin temperature, and activity of thermoeffectors (Fig. 2). Measuring metabolic rate, evaporative water loss, and skin temperature (or skin blood flow) over a range of ambient temperature reveals a pattern of thermoeffector activity that is typical for rodents and most other mammals (Fig. 2). There is a range of ambient temperatures termed the *thermoneutral zone*, where metabolic rate is at or near basal levels. In this zone, temperature regulation is achieved by control of sensible heat loss, meaning without regulatory changes in metabolic rate or evaporative water loss. The control of body temperature in the thermoneutral zone is achieved with modulations in skin blood flow, which controls the rate of heat loss with no additional metabolic requirements. As ambient temperature decreases below the thermoneutral zone, the blood flow to the skin is minimal as a result of peripheral vasoconstriction. As ambient temperature decreases, metabolism must increase above basal levels by shivering and non-shivering thermogenesis in order for heat production to match heat loss to the environment. The ambient temperature where metabolic rate increases is termed the *lower critical temperature*. As ambient temperature decreases below the lower critical temperature, skin temperature falls passively but may increase with extreme cold exposure as a result of cold-induced vasodilation (CIVD) of the peripheral blood vessels. This is a protective response to keep

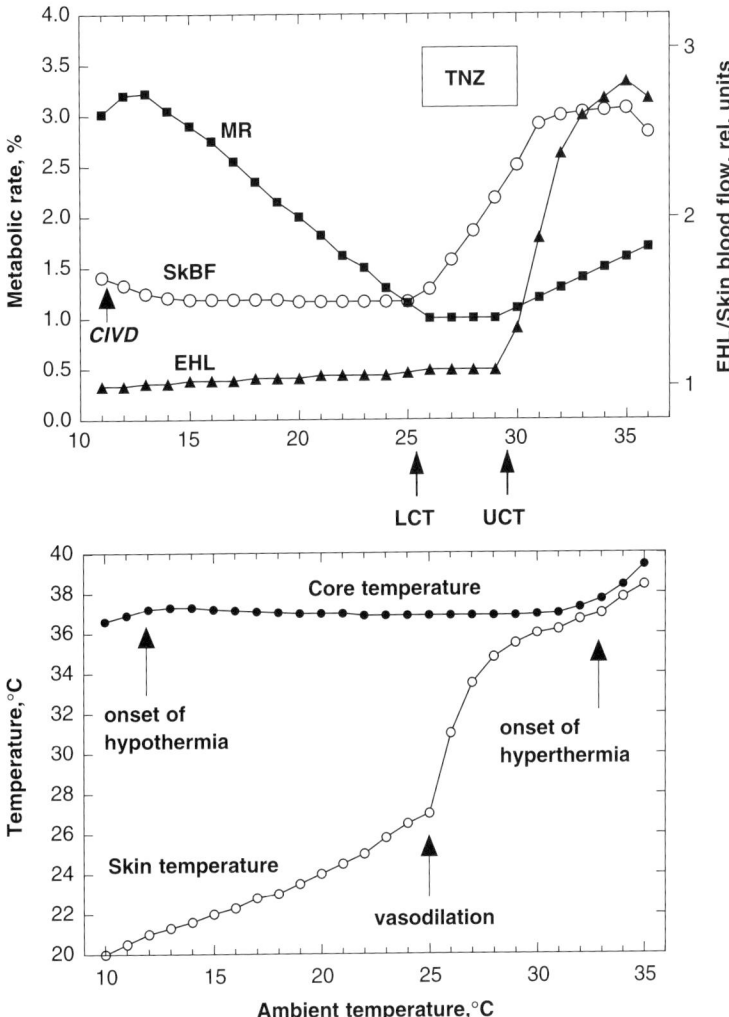

Fig. 2. General pattern of core and skin temperature and activity of autonomic thermoeffectors as a function of ambient temperature in a homeotherm. SkBF, skin blood flow; EHL, evaporative heat loss; MR, metabolic rate; LCT, lower critical temperature; UCT, upper critical temperature.

exposed tissues from freezing. Eventually, a temperature is reached where metabolic rate cannot maintain the pace of heat loss, and the animal becomes hypothermic.

As ambient temperature increases through the thermoneutral zone, skin blood flow increases and there is a disproportionate rise in skin temperature. However, if skin blood flow is maximal, skin temperature cannot increase above the internal core temperature. This means that as ambient temperature increases, the gradient between the skin and ambient temperature becomes smaller, thus limiting the rate of heat loss by convection, conduction, and radiation. Hence, at temperatures above the thermoneutral zone, evaporative heat loss mechanisms (i.e., panting, sweating, saliva grooming) must be activated to maintain thermal balance. This ambient temperature is termed the *upper critical temperature.* It is also identified with the point where core temperature and metabolism begin to rise (see IUPS Thermal Commission, 2001). At around the point of the upper critical temperature, skin temperature has increased to a level that is just below core temperature, reflecting maximal redistribution of warm blood from core to the periphery. With further increase in ambient temperature, skin and core temperature parallel each other until the point of thermoregulatory failure. At this point, evaporation is ineffective to maintain sufficient heat loss and core temperature spirals upward leading to hyperthermic death.

It is obvious that the hyperthermic efficacy of a drug will be exacerbated when rodents are housed at a warm temperature that exceeds thermoneutrality. But why should researchers studying the efficacy and responses of hyperthermic drugs be concerned with the characteristics of the thermoregulatory system below thermoneutrality? Due to their small size, rodents have a relatively large surface area:mass ratio, meaning that they lose body heat faster and must rely more on a high metabolic rate rather than adjustments in peripheral vasomotor tone to thermoregulate below the lower critical temperature. Mice and rats have a lower critical temperature of 28° to 31°C, which is notably much warmer than the standard temperature for housing in most laboratory settings. In other words, under standard conditions they are essentially cold stressed and thermoregulate by maintaining a metabolic rate above basal levels. Hence, in studies where the efficacy of hyperthermic drugs is studied at standard room temperature, the mouse and rat are in an environment that is relatively cool and this will impact on the thermoregulatory response to a drug. Most importantly, it will affect the manner in which one wants to extrapolate the effects of the drug to that of a human (see below).

One can argue that rodents maintained under standard housing conditions of 22°C are able to acclimate to the cool conditions and therefore are not "stressed". However, when one puts rats, mice, and other rodents in a temperature gradient, they select temperatures associated with metabolic thermoneutrality (approximately 28° to 30°C), a temperature that is warmer than that of their housing conditions (Gordon, 1993). Hence, one can infer from this behavior that rodents are often housed under conditions that are likely to be a mild cold stress. Of course, the housing conditions, including number of animals per cage, bedding type, air flow, and other factors will affect the rodent's thermoregulatory response under so-called standard conditions. For example, wood shaving bedding that allows for burrowing will minimize the effects of ambient temperature on heat loss while a wire-screen floor accelerates heat loss. Groups of mice housed in cages with wood shavings maintain a significantly warmer core temperature during the day than that of mice housed on wood chip bedding which does not allow for burrowing (Gordon, 2004).

Ambient temperature and potential hyperthermic response

Whether a toxicant will cause an increase, decrease, or have no effect on body temperature will depend largely on ambient temperature and the thermoeffector systems affected (Table 1). The potential changes in temperature, indicated by the number of arrows in each block of the table, is dependent on whether the animal is housed in a cool, warm, or thermoneutral environment. For example, a drug that blocks metabolic thermogenesis will manifest the most effective change in body temperature when exposure occurs at a relatively cool ambient temperature. A toxicant that induces peripheral vasoconstriction, thus restricting blood flow and heat loss from the skin will have relatively minor effects in a cold environment because the animal is in a state of peripheral vasoconstriction but would lead to hyperthermia in a thermoneutral and warm environment. On the other hand, an agent that causes peripheral vasodilation would be mostly ineffective in a warm environment because skin blood flow is already elevated and an additional vasodilatory action should have little effect on total heat loss. Blocking salivation in rodents or sweating in humans would have little effect in the cold but would lead to dramatic hyperthermia if the blocking agents are administered in a warm environment.

If the toxicant impairs one thermoeffector without affecting CNS thermoregulatory control, then one would expect the animal to utilize other thermoeffectors to maintain thermal homeostasis. For example, if skin blood flow was stimulated in a

Table 1. Relative hyperthermic (↑) and hypothermic (↓) effects of a toxicant or drug that stimulates or blocks thermoeffectors at cool, thermoneutral, and warm ambient temperature

	Thermogenesis increased	Skin blood flow increased	Evaporation increased	Thermogenesis blocked	Skin blood flow blocked	Evaporation blocked
Cool temperature	↑	→		➡		
Thermoneutral temperature		→	→	→	⬅	
Warm temperature	⬅	→	→	→	⬅	⬅

Size of arrow indicates relative magnitude of change in core temperature. Adapted from Gordon (2005).

cold environment, then metabolic thermogenesis could increase to counter for the increased heat loss. Stimulation of metabolism in a warm environment, such as, occurs by exposure to chemical agents that uncouple oxidative phosphorylation, is accompanied by a marked increase in evaporation (Wood et al., 1983). Overall, these are idealized situations and drugs and toxicants generally affect the function of more than one thermoeffector system.

Patterns of drug toxicity as function of temperature

Fuhrman and Fuhrman (1961) first developed a general scheme to describe the toxicity of drugs as a function of ambient temperature (Fig. 3). The term "toxic potency" in the ordinate can describe both lethal and non-lethal endpoints. In general, there are three basic functions that describe the toxicity of drugs in homeotherms exposed over a wide range of ambient temperatures: (A) a V- or U-shaped function with minimum toxicity (i.e., high LD_{50}) at a narrow range of temperatures usually below the metabolic thermoneutral zone and increased toxicity as temperature increases or decreases from the nadir of the curve; (B) a linear function with toxicity increasing with temperature; and (C) a function with a constant toxicity over a wide range of cool ambient temperatures and then sudden increase in toxicity at a threshold temperature around or above the upper portion of the thermoneutral zone. The Type C response is questionable in the sense that if temperature is reduced to sufficient levels, a toxicant would eventually respond in a Type A response. Like the efficacy of many drugs, the toxicity of xenobiotics can follow one of the three responses depicted in Fig. 3. In general, the *in vivo* toxicity of mice and rats as a function of temperature exhibits the Type A response (V-shaped). Examples of a Type A response include chlorpromazine and toluene. Toxicants such as ethanol and dinitrophenol display a Type B response, while DDT shows a Type C response (Fig. 3). Whether a toxicant fits the Type A, B, or C function is dependent in part on the mechanism of action (Weihe, 1973). For example, agents that are sympathomimetics and stimulate metabolic thermogenesis would be expected to take on a Type B response. That is, the effectiveness of a compound to cause a dangerous elevation in core temperature, such as oxidative phosphorylation uncouplers, would increase with ambient temperature.

The Type A response is likely the most relevant function for most drugs and toxicants. The V- or U-shaped function is thought to be a result of agents that interfere with the CNS control of body temperature. Although Weihe (1973) suggested that Type A drugs would not be involved with a re-setting of the thermoregulatory set-point, the evidence with many toxicants would suggest otherwise. This function is a result of the combined

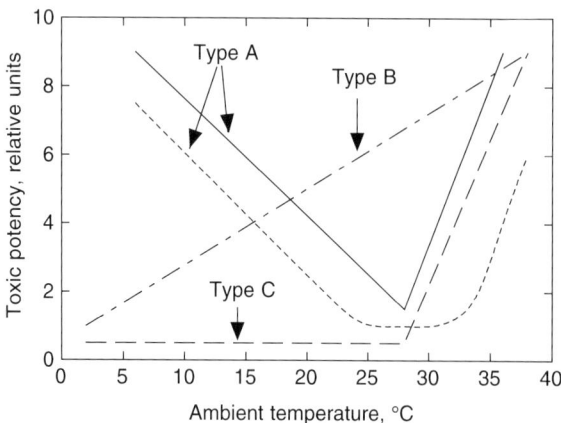

Fig. 3. General pattern of drug and chemical toxicity as a function of ambient temperature. Adapted from Fuhrman and Fuhrman (1961).

effects of the toxicant and ambient temperature on thermoregulatory control. That is, many toxic chemicals induce a regulated hypothermic response in rats and mice. There is an ideal range of ambient temperatures where the treated rodent can maintain a moderate hypothermic core temperature (Gordon, 2005). If ambient temperature changes above or below optimal levels, the thermoregulatory effectors are overwhelmed and are unable to maintain core temperature, possibly resulting in death from thermoregulatory failure.

Core temperature: its effect on magnitude versus duration of toxicity

The response or sensitivity to a toxicant is characterized by both the magnitude and length of time that the toxicant persists in the body (Doull, 1972). In most cases, hyperthermia is going to reduce the duration but increase the magnitude of toxicity. That is, when body temperature is raised, the mechanism(s) responsible for metabolizing and excreting the toxicants are going to be accelerated because the temperature coefficients for these processes are positively affected by temperature. Doull (1972) developed a general prediction of the response of biological systems to toxic levels of chemicals that can be applied to most drugs and toxic chemicals. "Temperature is directly correlated with the magnitude and inversely correlated with duration of drug response in biological systems". In other words, while the concentration of a toxicant will persist longer during hypothermia, the toxicity of the agent is reduced. This is a favorable situation for the efficacy of chemotherapeutic agents and the treatment of cancer because it allows for one to use lower doses of chemotherapeutics by raising the local temperature of the tissue to be treated (Falk and Issels, 2001).

The tenet of Doull's was stated over 30 years ago and remains a critical issue in toxicology and pharmacology. The magnitude and duration of a toxicity has a bearing on experimental and clinical toxicological studies. First, rodents are the primary species for toxicological studies and their thermoregulatory system is extremely sensitive to toxic insults. Their integrated thermoregulatory responses will have a marked impact on the pharmacokinetics and overall toxicity of the chemical or drug. Second, if hyperthermia exacerbates toxicity and hypothermia affords protection to toxicants, then this should have important implications in the way in which humans and domestic species are treated following episodes of poisoning (Gordon, 2005).

Extrapolating from rodent to human

An assumption of some degree of similarity between rodents and humans is the mainstay of biomedical research. Drug development and safety, toxicology, and other responses are studied in rodents with the notion that the effects can be extrapolated to that of humans. There can be tremendous uncertainty in the extrapolation process. Thermoregulatory sensitivity is a critical factor in the extrapolation from rodent to human that is often not considered. It is extremely important in the understanding of the hyperthermic drugs.

In addition to ambient temperature (e.g., Table 1), it is also important to consider the species' body mass when attempting to predict how a change in thermoeffector function will affect the control of body temperature. For example, species such as rodents with a relatively small body mass and large surface area:body mass ratio rely mostly on metabolic thermogenesis to thermoregulate, whereas peripheral vasomotor tone becomes more critical in species with large body mass. This also means that the amount of heat production needed for a given elevation in core temperature is going to increase with a reduction in body mass. This can be best illustrated by a plot of species body mass versus the amount of heat from microwave radiation needed to raise core temperature by $1°C$ (Fig. 4). In these studies, mice, rats, golden hamsters, and rabbits were exposed to RF radiation for 90 min while maintained at ambient temperatures of 20 and $30°C$. An ambient temperature of $20°C$ is relatively cool for the mouse, hamster, and rat but near thermoneutral for the rabbit. A temperature of $30°C$ is near thermoneutrality for the rodents but a mild to moderate heat stress for the rabbit.

At both ambient temperatures there was clear inverse relationship between body mass and the

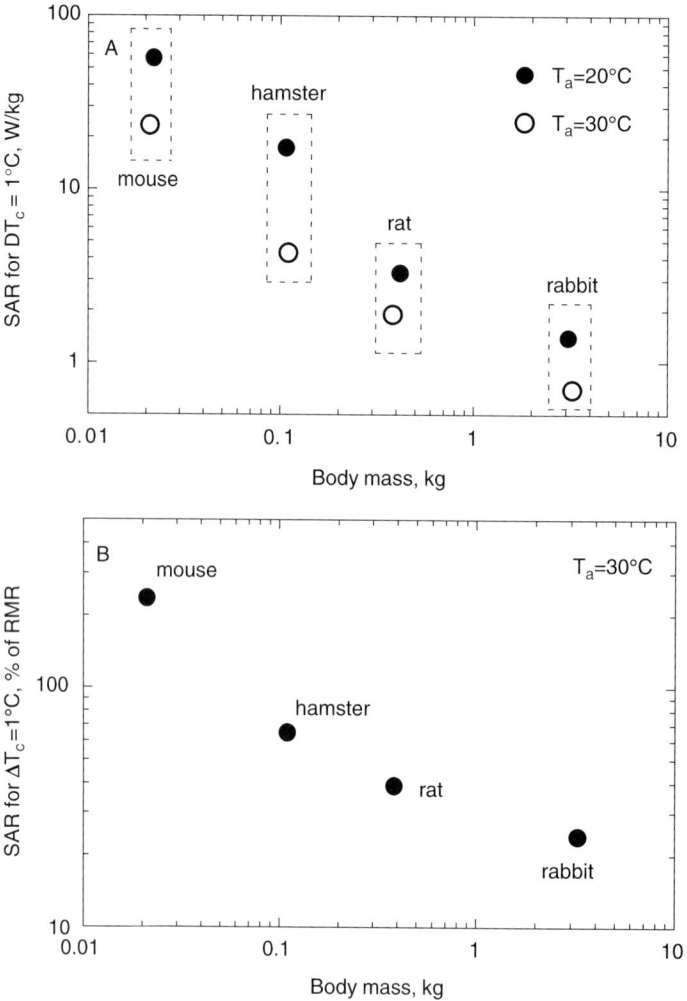

Fig. 4. (A) Relationship between body mass and the specific absorption rate (SAR) required to raise core temperature by 1°C in various species exposed to radiofrequency radiation for 90 min at ambient temperatures of 20 and 30°C. (B) Calculation of SAR as percentage of resting metabolic rate (RMR) required to raise core temperature by 1°C at an ambient temperature of 30°C. Data adapted from Gordon et al. (1986a, b).

amount of RF energy needed to raise core temperature by a given amount. It is important to note that the dosimetry units for exposure to RF radiation are the same as for metabolic rate normalized to body weight (i.e., W/kg). This provides an ideal means of comparing thermoregulatory responses across species. By calculating the SAR as a percentage of the species' metabolic rate at an ambient temperature of 30°C, one sees a marked decrease in the relative heat production that would be needed to raise body temperature with increase in body mass (Fig. 4).

Body mass also affects the manner in which heat is dissipated by physiological mechanisms. The contribution of peripheral vasomotor tone for heat dissipation is also linked to body mass. Assessing if the drug or toxicant causes vasodilation or vasoconstriction is a common approach in rodent studies with the intent that the vasomotor response can be extrapolated to that of a human response.

To make comparisons in the vasomotor responses of rodents and humans, one must also consider the impact of body size. Phillips and Heath (1995) utilized infrared thermography in a quantitative study of the impact of body mass on the role of peripheral vasomotor tone in thermoregulation. They noted that a small mammal such as a mouse or rat, by having a relatively large surface area:volume ratio, are "metabolic specialists". That is, homeotherms of small size rely on changes in their metabolic rate as a primary means to regulate body temperature. Large animals do not change metabolic rate as much with changes in ambient temperature and thermoregulate by controlling their surface temperature. The vasomotor index, a value reflecting the ability of an animal to regulate heat exchange through control of its skin temperature, was shown to be directly proportional to body mass. For example, the vasomotor index of a 80 kg human was predicted to be more than 6-fold greater than that of a 300 g rat. Could this mean that a vasomotor effect of a drug in a small mammal is magnified in a larger species that relies more on peripheral vasomotor tone to thermoregulate? Further work in comparative thermoregulatory responses to drugs and toxicants will help answer this question.

One should not infer from this discussion that rodents are better adapted to resist heat stress. In fact, the opposite is true when viewing heat resistance in terms of exposure to high ambient temperatures and/or radiant heat loads (Adolph, 1947). That is, due to their large surface area:mass ratio, mice and rats heat up quickly when exposed to high ambient temperatures relative to that of large mammals. With a smaller size, rodents must quickly dissipate heat by evaporation to prevent an explosive rise in core temperature. Rodents are capable of dissipating heat by evaporation through grooming of saliva and urine on their fur but they cannot maintain this response for long periods of time.

Recovery from acute elevations in core temperature is another important facet of extrapolation that is relevant to the study of hyperthermic drugs and toxic agents. The extrapolation of studies pertaining to lethal body temperature in mammals should be less problematic because the denaturation of proteins and cellular breakdown is universal, regardless of species. Overall, lethal body temperature is nearly equal among eutherian mammals although there are some peculiar species differences (Adolph, 1947). On the other hand, if one is interested in exploring the effects of transient exposure to high body temperatures, then it is important to understand that rodents will recover much faster than humans and other large species when exposed to near lethal heat stress. In the study of the thermoregulatory effects of exercise, it is recognized that small mammals (i.e., $< 1\,kg$) are capable of dissipating most of the heat burden by non-evaporative mechanisms because of their large surface area:mass relationship (Taylor, 1977). In large mammals ($> 10\,kg$), non-evaporative mechanisms are insufficient and greater reliance is placed on sweating to dissipate the heat load from exercise. This characteristic of body mass is important to consider in the study of hyperthermic drugs and the extrapolation of their effects from rodents to humans.

To summarize, in the extrapolation from rodent to human, one would expect the required increase in heat production to raise core temperature as a percentage of the basal metabolic rate would decrease. Small rodents are as not in much danger of overheating from a drug when exposed at a standard room temperature because they can easily dump heat by passive mechanisms due to their large surface area:volume. Humans are well adapted to dissipate heat by evaporation and vasomotor mechanisms but can overheat quickly if a drug or toxicant impairs heat loss mechanisms. These relationships are simplified and could certainly be affected by a variety of biological and environmental factors. One must be cautious in extrapolating a potential thermoregulatory effect from rodent-to-human or human-to-rodent.

Methylenedioxymethamphetamine (MDMA)

Rodent response

MDMA (street name: Ecstasy) is an ideal drug for discussion in this chapter for several reasons: (i) there is a substantial database on its thermoregulatory effects in both experimental animals and humans; (ii) it is a heavily used drug of abuse with thorough studies on its use and abuse; (iii) it is also considered a potential candidate for therapeutic treatment of psychoses; and (iv) overdose exposures to MDMA and resultant mortality has been attributed to its disabling of thermoregulatory mechanisms and death by hyperthermia.

MDMA stimulates the release of serotonin in the CNS resulting in a euphoric-like state. Serotonergic pathways in the CNS are also critical in the regulation of body temperature. It was recognized years ago that the abuse of MDMA in humans while engaged in dancing in clubs with poor ventilation and warm temperatures led in some cases to severe cases of hyperthermia and occasional deaths (Henry et al., 1992). The thermoregulatory effects of this drug in rats and mice have since been extensively studied.

Our laboratory was one of the first to study the thermoeffector responses of rats when dosed with MDMA at different ambient temperatures (Gordon et al., 1991). We found that the hyperthermic effects of MDMA were markedly dependent on ambient temperature (Fig. 5A). Rats and mice become hypothermic with MDMA given at relatively cool ambient temperatures, whereas the same dose give at warm temperatures results in hyperthermia. For example, a dose of 30 mg/kg MDMA in the rat led to hypothermia at 10°C, no effect at 20°C, and near lethal hyperthermia at 30°C (Fig. 5A). It is important to note that these same ambient temperatures have little effect on baseline core temperature of the rat. The hypothermic and hyperthermic effects are attributed to the combined effects of MDMA on heat production and heat loss. MDMA led to significant elevation in metabolic rate at ambient temperature of 20 and 30°C and evaporative water loss at ambient temperatures of 10, 20, and 30°C when housed in a metal chamber. MDMA also caused vasoconstriction regardless of ambient temperature (see below). MDMA stimulates motor activity due to its effect on serotonin release but its hyperthermic effects were only manifested at warmer temperatures. Vasoconstriction of skin blood flow in the cold will have little impact on a potential hyperthermic effect of a drug but the increase in evaporation was more likely a key factor explaining the hypothermia in the cold. At 20°C the increase in metabolic rate was essentially balanced by the increase in evaporative water loss resulting in no change in core temperature. At 30°C, the increase in metabolic rate combined with tail vasoconstriction was more than enough to overwhelm any heat loss from evaporation; the net result being near lethal hyperthermia.

The floor of the cage was critical in the thermoregulatory outcome of MDMA. When kept at room temperature in a cage with a metal floor that easily conducts heat away from the rat, the hyperthermic effects were less apparent. On a floor of low thermal conduction such as plastic or wood shavings, the hypothermic effects were most profound (Gordon and Fogelson, 1994). One of the most amazing effects on thermoregulation in the rat was the inability of the rat to vasodilate its tail in spite of encountering lethal hyperthermia (Fig. 5B). For example, 3 h after dosing with 20 mg/kg MDMA at ambient temperature of 22°C, the core temperature is at near lethal level of 41.5°C while tail skin temperature was 2.2°C below that of animals dosed with saline. By comparison, rats exposed to a hot ambient temperature of 37°C for 1 h had a core temperature of 40.5°C and a tail skin temperature of 31.7°C, reflecting a marked rise in skin blood flow. Note that an ambient temperature of 26° to 28°C is considered the threshold ambient temperature for tail vasodilation in the rat (Gordon, 1993). In rabbits, MDMA elicits profound vasoconstriction of blood flow to the ears, resulting in hyperthermia (Pedersen and Blessing, 2001). On the other hand, there was no indication of vasoconstriction of skin blood flow in humans administered MDMA (see below).

Jaehne et al. (2005) have recently measured behavioral thermoregulatory responses of rats in a temperature gradient when dosed with MDMA. They found that behavioral thermoregulatory mechanism to regulate body temperature were apparently unaffected by MDMA administration. For example, rats given MDMA and maintained in a warm environment quickly selected cool temperatures and brought their core temperature back to normothermic levels. Rats made hypothermic by administering MDMA and exposed in a cold environment quickly selected warmer temperatures and brought core temperature back to

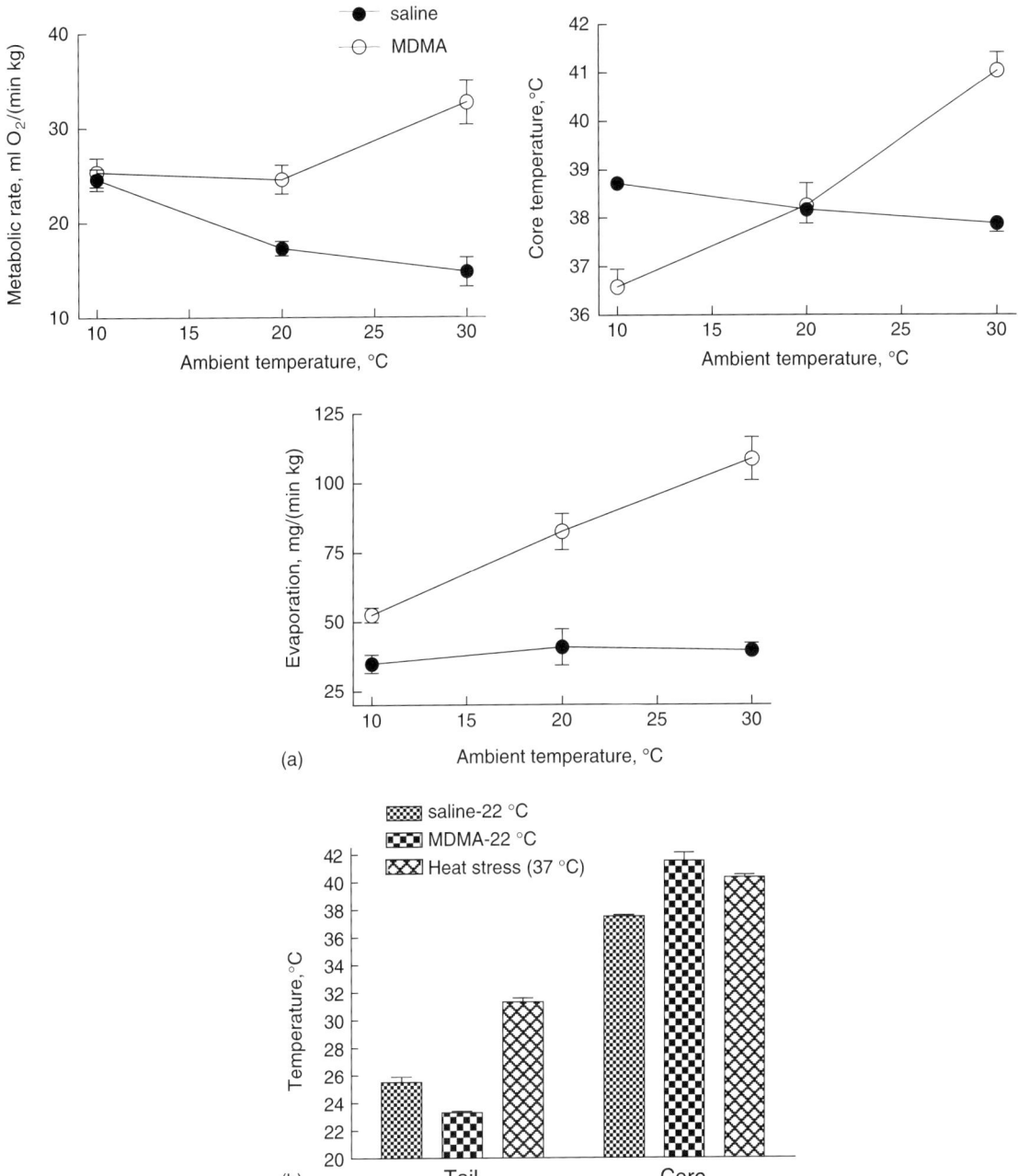

Fig. 5. (a) Effect of ambient temperature on the metabolic rate, evaporative water loss, and body temperature of rats dosed intraperitoneally with 30 mg/kg MDMA. Relationship between core and tail skin temperature in the rat dosed with 20 mg/kg MDMA when housed at an ambient temperature of 22°C. Also shown is the effects of heat stress (1 h at 37°C ambient temperature) on tail and core temperature. (b) Note relatively cool tail skin temperature in spite of hyperthermic body temperature in MDMA-treated rats as compared to warm tail (i.e., vasodilation) in heat-stressed rats. Data adapted from Gordon et al. (1991).

normal. Hence, autonomic thermoeffectors are clearly dysfunctional in MDMA-treated rats whereas behavioral thermoregulation appears to be unaffected. The response of the rat in thermal gradient provides an interesting contrast to the perceived thermal environment of humans given MDMA (see below).

Human response

In its use as a recreational drug MDMA is often ingested in crowded dance clubs ("raves") that can be quite warm. The primary toxic effect of MDMA in humans is hyperthermia with core temperatures as high as 43°C reported (Henry et al., 1992). It is clear that the thermoregulatory effects of MDMA combined with the impact of heat production from dancing in a warm environment will interact to compound the hyperthermic effects of MDMA. Until recently, there was little known on the thermoregulatory effects of MDMA in humans. Freedman et al. (2005) reported on the effects of MDMA ingestion in human volunteers (previous MDMA users) maintained at a relatively cold (18°C) and warm (30°C) ambient temperature. Core temperature measured by a radiotelemetric pill, metabolic rate, sweating, skin temperature, and subjective feelings of warm and cold were assessed. Three hours after ingesting 2 mg/kg MDMA, core temperature began to rise above control levels, increasing by 0.6 and 0.3°C in subjected exposed to warm and cold ambient temperatures, respectively (Fig. 6). The rise in temperature was associated with a mild rise in metabolic rate in the warm temperature and a sharper rise in metabolism in the cold. Interestingly, there was no indication of vasoconstriction as measured by skin temperature. While exposure to the warm temperature elicited a sweating response, it is interesting to note that sweating in the MDMA-exposed group was delayed by approximately 30 min in subjects given MDMA in the warm temperature. Subjects in the heat given MDMA had a greater perception of warmth compared to controls but their autonomic effectors (sweating, vasodilation) were apparently not activated to lower core temperature.

The response of humans and rodents to MDMA is similar in that autonomic effectors are activated to raise core temperature and there appears to be a normal behavioral sensation of warm temperatures. On the other hand, there are distinct differences between species. The effects of ambient temperature on the thermoregulatory response to MDMA were not as profound in humans. Rodents become hypothermic in the cold while the human subjects exposed to a cold environment that was sufficient to increase metabolic rate became hyperthermic with MDMA ingestion. Moreover, peripheral vasoconstricition is a key thermoeffector in the rat exposed to MDMA whereas this response seemed to be lacking in the human subjects. These comparisons have to be viewed with caution in view of the differences in doses and route of administration and the previous exposure to MDMA in the human subjects.

Interaction between hyperthermia and drug toxicity

Toxic mechanisms attenuated by hypothermia

Hypothermia affords protection to a variety of drugs and toxicants while hyperthermia augments toxicity. Systemic toxicological studies have shown that mild hypothermia improves the recovery of organ function and chances of survival following exposure to a variety of toxicants but there is relatively little known on the mechanisms of action (Gordon, 2005). Hyperthermia augments toxicity of most chemicals and drugs. Based on the Arrhenius relationship, it is reasonable to assume that hypothermia is most likely protecting tissues by slowing the rate at which the toxicant exerts its damaging effects at the cellular and intracellular level. That is, the kinetics of cell death as a function of temperature and toxicant exposure is often studied using an Arrhenius plot. When the relationship between temperature and cell killing is linear then it could be assumed that the toxicant follows a Arrhenius type of temperature dependence (see Gordon, 2005). On the other hand, a sudden departure from linearity would indicate a non-additive cytotoxic of the chemical with temperature. For example, Arrhenius plots for *in vitro* cell inactivation by the genotoxicants bleomycin and paraquat show a distinct, non-linear break while other anti-cancer agents show a linear

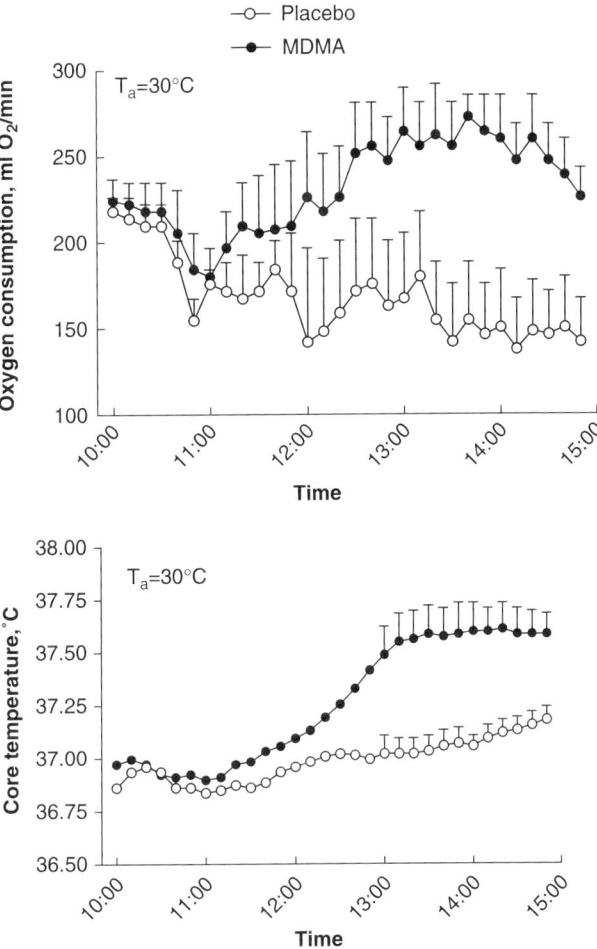

Fig. 6. Time-course of core temperature and metabolic rate in human volunteers administered MDMA. Data modified from Freedman et al. (2005).

temperature-dependent cytotoxicity (Takayoshi et al., 1987). A non-linear Arrhenius plot could result from simultaneous activation of the toxicant of two or more processes that have different activation energies.

There are other mechanisms to consider that go beyond Arrhenius type responses to explain the beneficial effects of mild hypothermia and damaging effects of hyperthermia. Recently, there have been numerous studies designed to understand the mechanisms of hypothermic protection to the CNS of rodents treated with substituted amphetamines such as MDMA (O'Callaghan and Miller, 2002). As discussed earlier, methamphetamines are drugs of abuse and are also well-known neurotoxicants that cause depletions in striatal levels of dopamine and serotonin. The database for the methamphetamines is considerable and a review of the mechanism of action of the amphetamines and their interaction with body temperature is relevant to understand the mechanisms of hypothermic protection exhibited by many of the other drugs and xenobiotics.

Blocking the hyperthermic effects or allowing the rodents to become hypothermic imparts remarkable protection on the neurotransmitter-depleting actions of the amphetamines. These observations have spurred numerous investigations on the mechanisms

Table 2. A summary of studies with plausible mechanisms to explain how hypothermia protects tissues in the CNS from pathological insults

Insult	Mechanism	Reference
Amphetamine toxicity	Attenuated 5-HT and DA release from striatum	Bowyer et al. (1992), Malberg et al. (1996)
Amphetamine[a] toxicity	Attenuated DA release and GFAP accumulation in striatum	Miller and O'Callaghan (1994)
Amphetamine toxicity	Prevented Ca^{+2} accumulation in nigrostriatum	Corbett et al. (1990)[a]
CNS ischemia	Reduced free radical formation	Globus et al. (1995)
CNS ischemia	Blocked translocation of PKC in striatum	Cardell et al. (1991)
CNS ischemia	Reduced ATP depletion in hippocampus	Zeevalk and Nicklas (1993)
CNS ischemia	Reduced glutamate and glycine release in hippocampus	Illievich et al. (1994)
Trimethyl tin	Reduced accumulation of GFAP in hippocampus	Gordon and O'Callaghan (1995)
CNS ischemia	Reduce oxidative stress metabolite (MDA) production	Katz et al. (2004)

Key: 5-HT, serotonin; DA, dopamine; PKC, phosphocreatine kinase C; GFAP, glial fibrillary acid protein; MDA, malondialdehyde.
[a]Various substituted amphetamines studied.

of amphetamine-induced striatal dopamine depletion and the protection afforded by hypothermia (e.g., Malberg and Seiden, 1998). For example, after treating rats with MDMA there is a marked depletion of serotonin in the striatum, frontal cortex, and other CNS sites when measured 14 days after treatment (Malberg et al., 1996). Preinjection with ketanserin (KET) or α-methyl-p-tyrosine (AMPT) with MDMA causes a profound hypothermia and no depletion in serotonin levels. Blocking the hypothermia in MDMA-treated animals given KET or AMPT led to marked serotonin depletion suggesting that the reduction in core temperature afforded neuroprotection by preventing loss of neurotransmitters. On the other hand, fluoxetine given with MDMA also blocks serotonin depletion but this combination of drugs has no effect on body temperature. Overall, these experiments serve as just one example to illustrate the difficulty in separating the protective actions of a drug or toxicant that is directly dependent on a thermoregulatory response or completely independent of body temperature. Researchers studying the methamphetamines have been frustrated by the difficulty to separate a true neuroprotective mechanism of a chemical treatment from the general protective effects of hypothermia. This frustration was best summed up in a review on the mechanisms of amphetamine neurotoxicity by O'Callaghan and Miller (2002) "Although painful to consider, what these data suggest is that the results of all 'mechanistic' studies of amphetamine neuropharmacology or neurotoxicity are compromised unless temperature can be ruled out as a contributing factor".

There has also been a resurgence to understand how hypothermia affords protection to the CNS following stroke as well as other types of traumatic brain injuries (Dietrich, 1992; Ginsberg and Busto, 1998). Like the amphetamine studies, there has been tremendous success in connecting a specific cellular or molecular endpoint that is manifested with hypothermia but it has been difficult to identify whether the mechanism is a direct result of hypothermia (Table 2). The potential impact of hyperthermia on the recovery from CNS ischemia is extremely relevant because fever is commonly observed in stroke patients and has been shown to be a key factor that impedes recovery (Ginsberg and Busto, 1998; Hanchaiphiboolkul, 2005). Many of the protective effects of hypothermia can likely be applied to toxicological studies because many toxicants exert the same sequelae as that seen from ischemia, including lipid peroxidation, free radical formation, Ca^{+2} imbalance, and many others (Stohs and Bagchi, 1995; Gultekin et al., 2000). The stroke and ischemia literature may well be helpful to the endeavors of toxicologists and pharmacologists to understand the protective role of hypothermia.

Reactive oxygen species (ROS)

Formation of ROS and the protective role of hypothermia has emerged as a key mechanism to understanding the mechanisms of toxicant-induced degeneration of the CNS and other tissues and organs (Halliwell, 1992). ROS cause damage to proteins, lipids, and nucleic acids leading to cell damage and death. Slikker et al. (2001) found that the percentage of surviving Chinese hamster ovary cells subjected to oxidative stress by exposure to hydrogen peroxide was inversely related to the temperature of incubation. This is not surprising in view of the studies that have shown how hyperthermia accentuates lipid peroxidation and free radical formation (Table 3). Furthermore, it was shown that reducing the incubation temperature led to greater expression of *bcl-2*, an anti-apoptotic protein that affords marked protection against oxidative stress. This experiment shows how hypothermia protects cells by more than just a simple Arrhenius effect and hyperthermia would most likely suppress the protective effects (Fig. 7). That is, lower temperatures induce and higher temperatures suppress the expression of a protein that is critical for cell survival. In this scheme, hyperthermia accentuates the activity of damaging processes that have a positive temperature coefficient (e.g., lipid peroxidation, ROS formation). Hypothermia activates expression of *bcl-2* that further protects cells from oxidative stress.

Table 3. A summary of studies with plausible mechanisms of how hyperthermia exacerbates the toxicity of drugs and other agents

Insult	Mechanism	Reference
MDMA	Accentuate production of lipid peroxidation products in liver	Carvalho et al. (2002)
Carbon monoxide	Accentuate production of lipid peroxidation products in brain	Kudo et al. (2001)
Turpentine fever	Accentuated lipid peroxidation (MDA) in kidney and liver	Brzezinska-Slebodzinska (2001)
Aging	Accentuates liver peroxidation and reduce catalase activity	Ando et al. (1997)
L-ephedrine	CNS neurodegeneration	Bowyer et al. (2001)

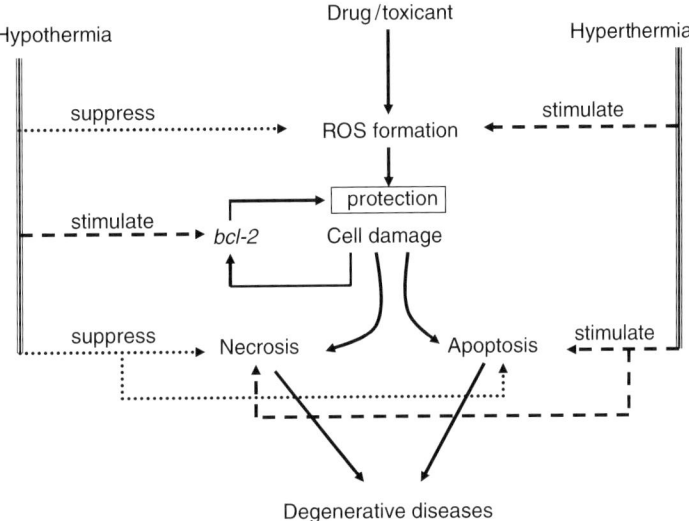

Fig. 7. A general mechanism of how hypothermia and hyperthermia can modulate the toxic-induced cell damage and the manifestation of neurodegenerative disease. Damaging processes with a positive temperature coefficient or $Q_{10}1.0$, including formation of reactive oxygen species (ROS), are exacerbated with an increase in temperature. Expression of proteins that protect cells from ROS formation such as *bcl-2* is a process that has a negative temperature coefficient and is activated with hypothermia. Hypothermia suppresses cell necrosis and apoptosis. Adapted from Gordon (2005) also see Slikker et al. (2001).

Conclusions

Hyperthermia is going to exacerbate the toxicity of many drugs and other chemicals. Ambient temperature and species mass are critical issues in the study of the toxicity of drugs and other chemicals in animals that are hyperthermic. The efficacy of a drug to raise body temperature is obviously dependent on ambient temperature. But the specie's body mass will also impart a significant impact on the effect of the drug on body temperature. One must be aware that, while rodents are excellent thermoregulators, they are more thermally labile when their thermoregulatory system is faced with a toxicant or other insult. Hyperthermic responses to a drug could be more transient and difficult to see in rodents compared to larger species. Radiotelemetry is thus the ideal method for monitoring the thermoregulatory responses to hyperthermic drugs in small rodents. Moreover, using laboratory rodents to model the responses of hyperthermic drugs in humans requires a firm understanding of comparative physiology and principles of scaling.

Acknowledgements

I am appreciative for the assistance provided by Peggy Becker in the preparation of this chapter. I thank Drs. Leon Lisa and Amir Rezvani for their review of the manuscript.

References

Adolph, E.F. (1947) Tolerance to heat and dehydration in several species of mammals. Am. J. Physicol., 151: 564–575.

Ando, M., Katagiri, K., Yamamoto, S., Wakamatsu, K., Kawahara, I., Asanuma, S., Usuda, M. and Sasaki, K. (1997) Age-related effects of heat stress on protective enzymes for peroxides and microsomal monooxygenase in rat liver. Environ. Health Perspect., 105: 726–733.

Bowyer, J.F., Hopkins, K.J., Jakab, R. and Ferguson, S.A. (2001) L-ephedrine-induced neurodegeneration in the parietal cortex and thalamus of the rat is dependent on hyperthermia and can be altered by the process of in vivo brain microdialysis. Toxicol. Lett., 125: 151–166.

Bowyer, J.F., Tank, A.W., Newport, G.D., Slikker Jr., W., Ali, S.F. and Holson, R.R. (1992) The influence of environmental temperature on the transient effects of methamphetamine on dopamine levels and dopamine release in rat striatum. J. Pharmacol. Exp. Ther., 260: 817–824.

Brzezinska-Slebodzinska, E. (2001) Fever induced oxidative stress: the effect on thyroid status and the 5′-monodeiodinase activity, protective role of selenium and vitamin E. J. Physiol. Pharmacol., 52: 275–284.

Burton, A.C. and Edholm, O.G. (1955) Man in a Cold Environment. Edward Arnold, London, UK, pp. 273.

Cardell, M., Boris-Moller, F. and Wieloch, T. (1991) Hypothermia prevents the ischemia-induced translocation and inhibition of protein kinase C in the rat striatum. J. Neurochem., 57: 1814–1817.

Carvalho, M., Carvalho, F., Remiao, F., de Lourdes Pereira, M., Pires-das-Neves, R. and de Lourdes Bastos, M. (2002) Effect of 3,4-methylenedioxymethamphetamine ("ecstasy") on body temperature and liver antioxidant status in mice: influence of ambient temperature. Arch. Toxicol., 76: 166–172.

Corbett, D., Evans, S., Thomas, C., Wang, D. and Jonas, R.A. (1990) MK-801 reduced cerebral ischemic injury by inducing hypothermia. Brain Res., 514: 300–304.

Dietrich, W.D. (1992) The importance of brain temperature in cerebral injury. J. Neurotrauma, 9: S475–S485.

Doull, J. (1972) The effect of physical environmental factors on drug response. Essays Toxicol., 3: 37–63.

Falk, M.H. and Issels, R.D. (2001) Hyperthermia in oncology. Int. J. Hyperthermia, 17: 1–18.

Freedman, R.R., Johanson, C.E. and Tancer, M.E. (2005) Thermoregulatory effects of 3,4-methylenedioxymethamphetamine (MDMA) in humans. Psychopharmacology (Berl.), 15: 1–9.

Fuhrman, G.J. and Fuhrman, F.A. (1961) Effects of temperature on the action of drugs. Ann. Rev. Pharmacol., 1: 65–78.

Ginsberg, M.D. and Busto, R. (1998) Combating hyperthermia in acute stroke: a significant clinical concern. Stroke, 29: 529–534.

Globus, M.Y.-T., Busto, R., Lin, B., Schnippering, H. and Ginsberg, M.D. (1995) Detection of free radical activity during transient global ischemia and recirculation: effects of intraischemic brain temperature modulation. J. Neurochem., 65: 1250–1256.

Gordon, C.J. (1993) Temperature Regulation in Laboratory Rodents. Cambridge University Press, New York, NY.

Gordon, C.J. (2004) Effect of cage bedding on temperature regulation and metabolism of group-housed female mice. Comp. Med., 54: 63–68.

Gordon, C.J. (2005) Temperature and Toxicology: An Integrative, Comparative, and Environmental Approach. CRC Press, Boca Raton, FL.

Gordon, C.J. and O'Callaghan, J.P. (1995) Trimethyltin-induced neuropathy in the rat: interaction with thermoregulation. Neurotoxicology, 16: 319–326.

Gordon, C.J. and Fogelson, L. (1994) Metabolic and thermoregulatory responses of the rat maintained in acrylic or wire-screen cages: implications for pharmacological studies. Physiol. Behav., 56: 73–79.

Gordon, C.J., Long, M.D. and Fehlner, K.S. (1986a) Temperature regulation in the unrestrained rabbit during exposure to 600 MHz radiofrequency radiation. Int. J. Radiat. Biol., 49: 987–997.

Gordon, C.J., Merrit, D., Long, M.D., Fehlner, K.S. and Stead, A.G. (1986b) Body temperature in the mouse, hamster, and rat exposed to radiofrequency radiation: an interspecies comparison. J. Therm. Biol., 11: 59–65.

Gordon, C.J., Watkinson, W.P., O'Callaghan, J.P. and Miller, D.B. (1991) Effects of 3,4-methylenedioxymethamphetamine on autonomic thermoregulatory responses of the rat. Pharm. Biochem. Behav., 38: 339–344.

Gultekin, F., Ozturk, M. and Akdogan, M. (2000) The effect o organophosphate insecticide chlorpyrifos-ethyl on lipid peroxidation and antioxidant enzymes (in vitro). Arch. Toxicol., 74: 533–538.

Halliwell, B. (1992) Reactive oxygen species and the central nervous system. J. Neurochem., 59: 1609–1632.

Hanchaiphiboolkul, S. (2005) Body temperature and mortality in acute cerebral infarction. J. Med. Assoc. Thai., 88: 26–31.

Henry, J.A., Jeffreys, K.J. and Dawling, S. (1992) Toxicity and deaths from 3,4-methylenedioxymethamphetamine ("ecstasy"). Lancet, 340: 384–387.

Illievich, U.M., Zornow, M.H., Choi, K.T., Strnat, M.A. and Scheller, M.S. (1994) Effects of hypothermia or anesthetics on hippocampal glutamate and glycine concentrations after repeated transient global cerebral ischemia. Anesthesiology, 80: 177–186.

IUPS Thermal Commission. (2001) Glossary of terms for thermal physiology. Third edition. Revised by The Commission for Thermal Physiology of the International Union of Physiological Sciences. Jpn. J. Physiol., 51, 245–280.

Jaehne, E.J., Salem, A. and Irvine, R.J. (2005) Effects of 3,4-methylenedioxymethamphetamine and related amphetamines on autonomic and behavioral thermoregulation. Pharmacol. Biochem. Behav., 81: 485–496.

Katz, L.M., Young, A.S., Frank, J.E., Wang, Y. and Park, K. (2004) Regulated hypothermia reduces brain oxidative stress after hypoxic-ischemia. Brain Res., 1017: 85–91.

Kudo, R., Adachi, J., Uemura, K., Maekawa, T., Ueno, Y. and Yoshida, K. (2001) Lipid peroxidation in the rat brain after CO inhalation is temperature dependent. Free Radic. Biol. Med., 31: 1417–1423.

Malberg, J.E., Sabol, K.E. and Seiden, L.S. (1996) Co-administration of NMDA with drugs that protect against MDMA neurotoxicity produces different effects on body temperature in the rat. J. Pharmacol. Exp. Ther., 278: 258–267.

Malberg, J.E. and Seiden, L.S. (1998) Small changes in ambient temperature cause large changes in 3,4-methylenedioxymethamphetamine (MDMA)-induced serotonin neurotoxicity and core body temperature in the rat. J. Neurosci., 18: 5086–5094.

Miller, D.B. and O'Callaghan, J.P. (1994) Environment-, drug- and stress-induced alterations in body temperature affect the neurotoxicity of substituted amphetamine in the C57Bl/6J mouse. J. Pharmacol. Exp. Therap., 270: 752–760.

O'Callaghan, J.P. and Miller, D.B. (2002) Neurotoxic effects of substituted amphetamines in rats and mice: challenges to the current dogma. In: Massaro E.J. (Ed.), Handbook of Neurotoxicology, Vol. 2. Humana Press, Inc., Totowa, NJ, pp. 269–301.

Pedersen, N.P. and Blessing, W.W. (2001) Cutaneous vasoconstriction contributes to hyperthermia induced by 3,4-methylenedioxymethamphetamine (ecstasy) in conscious rabbits. J. Neurosci., 21: 8648–8654.

Phillips, P.K. and Heath, J.E. (1995) Dependency of surface temperature regulation on body size in terrestrial mammals. J. Thermal Biol., 20: 281–289.

Schmidt-Nielsen, K. (1975) Animal Physiology: Adaptation and Environment. Cambridge University Press, London, UK.

Slikker III, W., Desao, V.G., Duhart, H., Feuers, R. and Iman, S.Z. (2001) Hypothermia enhances bcl-2 expression and protects against oxidative stress-induced cell death in Chinese hamster ovary cells. Free Radic. Biol. Med., 31: 405–411.

Stohs, S.J. and Bagchi, D. (1995) Oxidative mechanisms in the toxicity of metal ions. Free Radic. Biol. Med., 18: 321–336.

Takayoshi, S., Kohda, K. and Kawazoe, Y. (1987) Temperature-dependence of cytotoxicity of several genotoxicants in Chinese hamster V79 cells: bleomycin, paraquat, and some N-alkyl-N-nitrosureas Biochem. Biophys. Res. Commun., 146: 67–71.

Taylor, C.R. (1977) Exercise and environmental heat loads: different mechanisms for solving different problems? In: Robertshaw R. (Ed.), International Review of Physiology, Environmental Physiology II, Vol. 15. University Park Press, Baltimore, MD, pp. 119–146.

Weihe, W.H. (1973) The effect of temperature on the action of drugs. Ann. Rev. Pharmacol., 13: 409–425.

Wood, S., Rom, W.N., White, G.L. and Logan, D.C. (1983) Pentachlorophenol poisoning. J. Occup. Med., 25: 527–530.

Zeevalk, G.D. and Nicklas, W.J. (1993) Hypothermia, metabolic stress and NMDA-mediated excitotoxicity. J. Neurochem., 61: 1445–1453.

CHAPTER 6

Neuroleptic malignant syndrome and serotonin syndrome

Koichi Nisijima[1,*], Katsutoshi Shioda[1] and Tatsunori Iwamura[2]

[1]*Department of Psychiatry, Jichi Medical University, Minamikawachi-Machi, Kawachi-Gun, Tochigi-Ken 329-0498, Japan*
[2]*Department of Organic Chemistry, Gifu Pharmaceutical University, 5-6-1 Mitahora-higashi, Gifu 502-8585 Japan*

Abstract: This chapter is focused on drug-induced hyperthermia with special regard to use of antipsychotics and antidepressants for the treatment of schizophrenia and major depression, respectively. Neuroleptic malignant syndrome (NMS) develops during the use of neuroleptics, whereas serotonin syndrome is caused mainly by serotoninergic antidepressants. Although both syndromes show various symptoms, hyperthermia is the main clinical manifestation. In this review we describe the historical background, clinical manifestations, diagnosis, and differential diagnosis of these two syndromes based on our observations on the experimental and clinical data.

Keywords: neuroleptic malignant syndrome; serotonin syndrome; dantrolene; bromocriptine; cyproheptadine; 5-HT$_{2A}$ receptor antagonist; ECT; animal model

Introduction

Antipsychotics and antidepressants are indispensable drugs for the treatment of schizophrenia and major depression, respectively. Nevertheless, these agents induce numerous side effects. Neuroleptic malignant syndrome (NMS) develops during the use of neuroleptics, whereas serotonin syndrome is caused mainly by serotoninergic antidepressants. Although both syndromes show various symptoms, hyperthermia is a main clinical manifestation. The components of these two syndromes are similar, and some researchers consider them to exist on a spectrum of the same disorder (Demirkiran et al., 1986; Fink, 1996). Hyperthermia sometimes occurs in patients who are simultaneously being treated with both a neuroleptic and an antidepressant: in such cases it is often difficult to differentiate which syndrome it is. As described below, the pharmacotherapy for these two syndromes is slightly different and thus it is important to differentiate them. More than 40 years have elapsed since NMS was first reported, and 20 years have passed since serotonin syndrome was first reported, but not much is known about the pathophysiology or pathogenesis of either of them even now. In this chapter we describe the historical background, clinical manifestations, diagnosis, and differential diagnosis of these two syndromes. The clinical data and basic data we have collected this far are presented, and the pathophysiology and pathogenesis of the two syndromes are discussed.

*Corresponding author. Tel.: +81-285-58-7364; Fax: +81-285-44-6198; E-mail: psychiat@jichi.ac.jp

Neuroleptic malignant syndrome

Historical background

Although similar cases had been reported in the late 1950s (Preston, 1959), when neuroleptics began to be used clinically, NMS was first reported in 1960 by Delay et al. (1960) in France as "*syndrome malin des neuroleptiques*"; and the French term was translated into English as "*neuroleptic malignant syndrome*neuroleptic malignant syndrome" (Delay and Deniker, 1968) and became familiar in English-speaking countries around the world. However, the syndrome was reported only occasionally in the 1960s and 1970s, and did not arouse much interest. Caroff (1980) summarized the previous studies, and his paper renewed interest worldwide. Before 1980, mortality was reported to be 20% or more, but greater awareness of the syndrome and the development of therapeutic methods have reduced mortality to 10% or less.

Clinical manifestations and diagnosis

NMS should be suspected whenever a fever of unknown origin at 37° to 38°C occurs after administration of a neuroleptic. It should also be considered whenever extrapyramidal symptoms are exacerbated during the administration of a neuroleptic. Serum creatine phosphokinase (CPK) elevations and leukocytosis are found in as many as more than 90% of NMS cases (Caroff and Mann, 1993); however, there is no pathognomonic finding that leads to the diagnosis of NMS, and definitive diagnosis depends on the clinical manifestations. Four manifestations — high fever, altered consciousness, a variety of autonomic symptoms, and severe extrapyramidal symptoms — are observed in typical cases of NMS. However, many cases have not exhibited sufficient clinical manifestations because of better awareness and early diagnosis in recent years. Several diagnostic criteria for NMS have been proposed. Levenson's criteria (1985), which consist of three major manifestations (fever, muscle rigidity, and increased serum CPK level) and six minor manifestations (tachycardia, abnormal blood pressure, tachypnea, altered consciousness, diaphoresis, and leukocytosis), are regarded as the most useful for early detection of NMS. NMS can be diagnosed when the three major manifestations are detected, or when two major manifestations and four minor manifestations are detected. These criteria are highly sensitive but less specific, because cases of NMS without fever can be recognized. Caroff and Mann (1993) proposed that hyperthermia ($\geqq 38.0°C$), muscle rigidity, and five of the following 9 manifestations (change in mental status, tachycardia, tachypnea or hypoxia, diaphoresis or sialorrhea, tremor, incontinence, CPK elevation or myoglobinuria, leukocytosis, and metabolic acidosis) could be used to support the diagnosis when other disorders have been ruled out. Their diagnostic criterion is useful in making a definitive diagnosis.

Risk factors

NMS has been reported to develop in ~0.2% of patients treated with neuroleptics (Caroff and Mann, 1993). It tends to occur in patients with central nervous system defects such as mental retardation, organic brain disorders, etc. More than 80% of NMS patients have been suggested to experience dehydration, physical exhaustion, or psychomotor activity due to aggravation of a psychiatric disorder before they develop the syndrome (Yamawaki et al., 1990a). It is reported that, only in some cases, the syndrome showed familial occurrence (Otani et al., 1991), but hereditary cause is not generally found. Contributing factors to developing NMS include ambient heat and humidity, but these factors need not always be present for NMS to occur.

Among the antipsychotic drugs that cause NMS, it tends to develop when drugs with a potent dopamine D_2 receptor blocking effect, such as haloperidol, are administered. However, it has recently been reported that NMS is also induced by atypical antipsychotics with reduced D_2 receptor affinity, such as olanzapine, quetiapine, etc. (Ananth et al., 2004). Thus, whether NMS develops depends on the condition of the patient irrespective of the strength of the blocking effect of dopamine D_2 receptors. Furthermore, NMS does not always develop when the same neuroleptic is administered to a patient with a history of NMS. As stated above, not much is known about the etiology of NMS.

Pathogenesis

Hypotheses to explain the development of NMS

Because the clinical manifestations of NMS resemble those of the malignant hyperthermia that develops after administration of anesthetic gases, a hypothesis based on a skeletal muscle abnormality was proposed before 1980 to explain the etiology of NMS. In malignant hyperthermia, when inhalational anesthetics exert effects on an individual with an abnormality in skeletal muscle, calcium from the sarcoplasmic reticulum flows out, leading to abnormal contraction of skeletal muscle and causing fever (Denborough, 1998). Based on this mechanism, it was considered that antipsychotic drugs exert an effect on the skeletal muscle of patients with NMS and fever occurs. However, many NMS patients have been safely anesthetized with inhalational anesthetics. Also, while caffeine halothane contracture tests were found to be positive in the skeletal muscle of malignant hyperthermia patients, the results of the tests were inconclusive in NMS cases. These findings suggest that NMS and malignant hyperthermia have different pathogenetic mechanisms. Henderson and Wooten (1981) and Toru et al. (1981), on the other hand, hypothesized that NMS is caused by central dopamine hypofunction or dopamine receptor blockade because NMS-like symptoms sometimes occur after discontinuation of anti-parkinsonian drugs in patients with Parkison's disease or parkinsonism. Their hypothesis has been supported by the fact that causative drugs of NMS have dopamine D_2 receptor blocking effect, and that dopamine agonists, including bromocriptine, are effective in treating NMS.

Pathophysiology of NMS

The clinical manifestations of NMS are very diverse, and it seems impossible to explain all of them simply by a decrease in central dopamine function. To elucidate the pathophysiology of NMS, we performed a study on monoamine metabolism in the cerebrospinal fluid (CSF) of NMS patients. Age in the patient group and healthy control group was not matched in the initial study (Nisijima and Ishiguro, 1990), and a subsequent comparative study was performed on CSF monoamines and their metabolites in age-matched groups of patients and healthy control subjects (Nisijima and Ishiguro, 1995a). In the present study we studied CSF monoamine metabolism in seven other patients with NMS. Table 1 shows the general clinical symptoms in these patients, and all of them met the diagnostic criteria for NMS proposed by Caroff. We measured their CSF levels of homovanillic acid (HVA), a major metabolite of dopamine, 5-hydroxyindoleacetic acid (5-HIAA), a major metabolite of serotonin, and noradrenaline (NA), and the results are shown in Table 2. The results were basically consistent with those of our previous two studies, and the HVA levels in the active phase of NMS were significantly lower than in the control group. These findings seem to support the dopamine-hypofunction hypothesis of NMS. The HVA levels were lower after improvement of the clinical symptoms. Lumbar puncture was performed within two weeks after improvement. The low HVA levels after clinical improvement showed that dopamine function did not return to normal even when the clinical symptoms improved. NMS sometimes recurs when neuroleptics are given immediately after improvement of the clinical symptoms. The results of this study suggest that decreased dopamine function persists even after the clinical symptoms disappear, and that there is a risk of recurrence of NMS if administration of an antipsychotic is resumed under such conditions. The low HVA values we found in the NMS cases have been confirmed by other researchers (Ueda et al., 2001). No significant difference in CSF 5-HIAA values was found between the patients and controls. In the active phase of NMS the NA levels were significantly higher than in the control group, but they returned to the normal range after recovery. These results demonstrate the existence of central noradrenergic hyperactivity during NMS. On the other hand,

Table 1. Summary of seven cases of neuroleptic malignant syndrome

Case	Age	Sex	Underlying diseases	Causes of NMS	Body temperature (°C)	Muscle rigidity	Serum CPK (IU/l)	Pulse rate (beats/min)	Mental status change	Diaphoresis	Outcome
1	59	Female	Major depression, Parkinson's syndrome	Withdrawal of amantadine	38.0	+	9580	110	+	+	Recovered
2	44	Male	Mental retardation	HLP	41.0	+	1688	130	+	+	Recovered
3	25	Female	Mood disorder, mental retardation	Zotepine, lithium	40.0	+	44,880	140	+	+	Recovered
4	31	Female	Schizophrenia (undifferentiated)	Risperidone	39.8	+	22,423	130	+	+	Recovered
5	25	Female	Mood disorder, mental retardation	Thioridazine	38.2	+	1558	120	+	+	Recovered
6	35	Male	Schizophrenia (undifferentiated)	HLP, CP, perphenazine	40.6	+	2238	130	+	+	Recovered
7	43	Male	Schizophrenia (paranoid)	HLP, zotepine, lithium	39.5	+	10,110	122	+	+	Recovered

Notes: CPK, creatine phosphokinase; CP, chlorpromazine; HLP, haloperidol; the clinical course of Cases 1 and 2 was published in 1999.

Table 2. CSF levels of HVA, 5-HIAA, NA, and GABA in NMS patients before and after recovery and in controls

CSF controls (n = 8) (age: 39.1±8.9 y.o.)	HVA 52.9±9.3 (ng/ml)		5-HIAA 20.8±2.8 (ng/ml)		NA 0.08±0.2 (ng/ml)		GABA 380.8±96.8 (pmol/ml) (n = 6) (age: 30.5±9.4 y.o.)	
	During the active phase	After recovery	During the active phase	After recovery	During the active phase	After recovery	During the active phase	After recovery
NMS 1	16.1	/	13.7	/	0.18	/	24	/
NMS 2	34.5	15.7	24.7	18.7	0.175	0.07	125	312
NMS 3	42.4	39.2	27.2	25.8	0.339	0.228	167	459
NMS 4	48.5	/	23.1	/	0.358	/	178	/
NMS 5	31.9	37.6	17.7	13.3	0.217	0.09	409	595
NMS 6	5.7	9	20.1	7.3	0.59	0.098	306	277
NMS 7 (age: 37.4±12.2 y.o.)	26.7	30.2	19.1	18.4	0.297	0.224	/	/
Mean±S.D.	29.4±14.8*	26.3±13.4**	20.8±4.6	16.7±6.9	0.31±0.14***	0.14±0.08	201.5±136.4****	410.8±146.0

Note: NMS, neuroleptic malignant syndrome; HVA, homovanillic acid; 5-HIAA, 5-hydroxyindoleacetic acid; NA, noradrenaline; GABA, gamma-aminobutyric acid. The data were analyzed by the Mann-Whitney U-test.
*$p = 0.0078$ (vs. controls).
**$p = 0.0045$ (vs. controls).
***$p = 0.0012$ (vs. controls).
****$p = 0.0547$ (vs. controls).

Table 3. Catecholamines levels in the urine and plasma of NMS patients

Controls	Urine			Plasma	
	A (3–15 μg/day)	NA (26–121 μg/day)	VMA (1.3–5.1 mg/day)	A (100 pg/ml<)	NA (100–450 pg/ml)
NMS ($n = 7$) Mean±S.D.	63.2±72.2↑	331.9±216.6↑	6.3±2.7↑	117±73.4↑	1155±379.8↑

Note: NMS, neuroleptic malignant syndrome; A, adrenaline; NA, noradrenaline; VMA, vanillylmandelic acid.

increased peripheral levels of catecholamines and their metabolites have been reported in NMS (Gurrera and Romero, 1992; Gregorakos et al., 2000; Spivak et al., 2000). We also measured catecholamines levels in the urine and plasma of NMS patients we previously encountered. As shown in Table 3, the levels of catecholamines including adrenaline, NA, and vanillylmandelic acid were higher than in the healthy controls (unpublished data). These findings suggest the occurrence of peripheral sympathetic nervous system hyperactivity in NMS. Diverse autonomic nervous symptoms were observed in NMS, and the increased catecholamine levels in CSF, urine, and plasma may have maintained the symptoms.

We also measured gamma-aminobutyric acid (GABA) levels in the CSF and found that they tended to be low in the active phase of NMS. An abnormality in the GABAergic system is thought to occur during NMS, and there have been several reports that benzodiazepines, GABA mimetic agents, are effective against NMS. The results of our study seem to provide further evidence for the effectiveness of benzodiazepines.

Other methods have been used to elucidate the pathogenesis of NMS. Kish et al. (1990) analyzed monoamines at various sites in the autopsy brains of three patients thought to have NMS. The HVA concentration in the striatum was decreased in two of the cases, and a marked reduction in hypothalamic NA was found in all three cases. The reduced HVA levels in the Kish study may be related to the low CSF HVA levels in our study, and the NA depletion demonstrated by Kish may be secondary to hyperthermia and related to the increased CSF NA levels in our study.

DeReuck et al. (1991) performed positron emission tomography (PET) in three NMS patients and found increased metabolism in the striatum, cerebellum, and occipital cortex, suggesting that other parts of the central nervous systems in addition to the dopamine system are involved in the pathogenesis of NMS. We performed single-photon emission computed tomography (SPECT) with ^{123}I-N-isopropyl-p-iodoamphetamine (^{123}I-IMP) in three NMS patients and found asymmetrical ^{123}I-IMP accumulation in the basal ganglia during the active phase (Nisijima et al., 1995b). SPECT in two other cases revealed decreased ^{123}I-IMP accumulation in the parietal and occipital regions during the active phase of NMS, but these changes were no longer seen after recovery from NMS (unpublished data) (Fig. 1). Although the significance of these findings is uncertain, they suggest abnormalities in other brain regions besides the striatum. Jauss et al. (1996) performed SPECT with ^{123}I-iodo-benzamide (^{123}I-IZM) from the active phase to improvement in one case of NMS. ^{123}I-IZM is a benzamide-type ligand and allows imaging of dopamine D_2 receptors. The results showed that the dopamine D_2 receptors in the striatum were completely occupied by neuroleptics in the active phase of NMS, and that the occupation rate gradually decreased after improvement. They claimed that the persistent extrapyramidal symptoms after improvement of NMS could be explained by these results.

Other studies

Malignant hyperthermia is considered to be inherited as an autosomal dominant trait. Molecular biological studies around 1990 demonstrated mutations in the ryanodine receptor type 1 in the skeletal muscle of patients with malignant

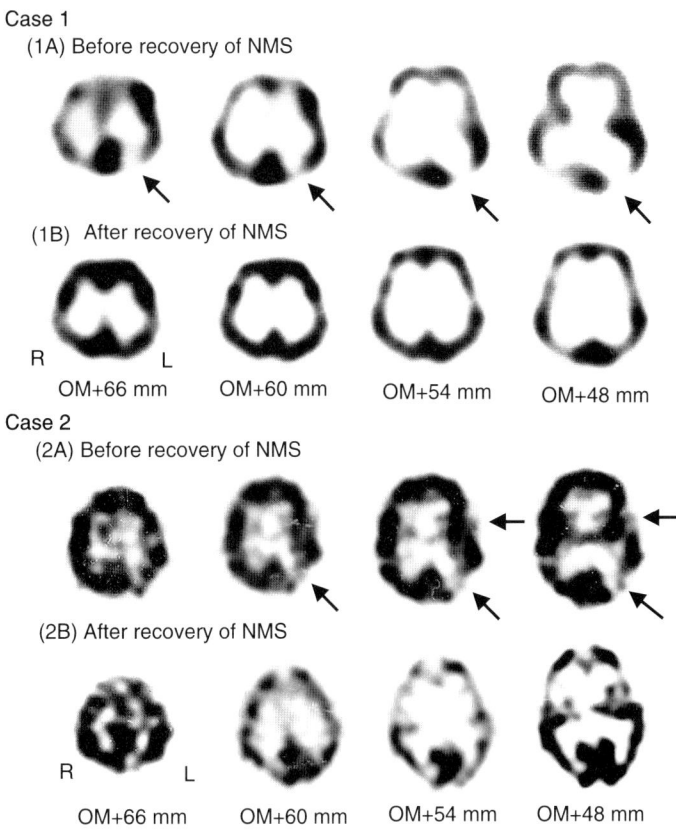

Fig. 1. (1A, 1B, 2A, 2B) Transverse SPECT scans of two NMS patients. Case 1. A 28-year-old woman with schizophrenia. Case 1 corresponds to Case 11 in Table 1 in our paper published in 1995. The early image demonstrated decreased accumulation of ^{123}I-IMP in the left parietal region (arrows) before improvement of NMS (1A), but no regions of abnormal accumulation were observed after improvement of NMS (1B). Case 2. A 35-year-old man with schizophrenia. Case 2 corresponds to Case 6 in Table 1. The early image demonstrated decreased accumulation of ^{123}I-IMP in the left temporo-parietal region (arrows) before improvement of NMS (2A), but no regions of abnormal accumulation were observed after improvement of NMS (2B). The numbers under each transverse image indicate the level of the slices from the orbitomeatal line.

hyperthermia (Denborough, 1998). Molecular studies on patients with a history of NMS have been performed to determine whether they had any variations, such as mutations in the cytochrome P450 genes related to the metabolism of psychotropic drugs. Other molecular studies were performed to analyze the dopamine D_2 receptor gene, serotonin receptor gene, and ryanodine receptor gene. Variations were reported in some of the studies (Iwahashi et al., 1999; Kawanishi, 2003; Dettling et al., 2004), but the number of cases was limited and no consistent results have been obtained.

Treatment of NMS

Basic treatment

When NMS is suspected, all neuroleptics must be discontinued immediately. Dehydration occurs in many cases, and fluid replacement may result in improvement in mild cases. Antipyretic drugs are generally ineffective against the fever, and complications, such as respiratory failure, rhabdomyolysis, renal failure, and disseminated intravascular coagulation (DIC), may adversely affect the outcome. A high fever of more than 40°C, in particular, is a risk factor for multiple organ failure and to lead to a poor outcome. Thus, rapid cooling of the whole body is important. Dysphagia and sialorrhea often occur in NMS patients, and since they may increase the risk of pneumonia, frequent aspiration is necessary.

Pharmacotherapy

Dantrolene is a peripheral muscle relaxant that attenuates calcium release from the sarcoplasmic reticulum and has often been described as effective in the treatment of NMS. It was first used clinically in the 1970s as a specific drug in the treatment of malignant hyperthermia (Krause et al., 2004). The efficacy of dantrolene against NMS was first reported by Delacour et al. in France in 1981, and a case of efficacy was subsequently reported by Coons et al. in the United States (1982). Since the etiology of NMS is thought to lie in the central nervous system, it is unknown why dantrolene's peripheral action is effective against NMS. Yamawaki et al. (1986) reported an NMS patient whose EEG abnormality improved after dantrolene administration and suggested that it may have central effects. We retrospectively studied changes in CSF HVA and 5-HIAA in a group treated with dantrolene (Group I) and a group treated with supportive therapy alone (Group II) (Nisijima and Ishiguro, 1993). The low HVA levels tended to become normal in the active phase of NMS in the dantrolene group, suggesting that dantrolene has effects on the dopamine system (Fig. 2). Oyamada et al. (1998) later reported that the dantrolene affects serotonin and dopamine metabolism based on the results of an animal study.

Dopamine agonists are also widely used to treat NMS, and bromocriptine is particularly widely used. Mueller and Vester (1983) was the first to report that bromocriptine is effective in the treatment of NMS. Similar reports followed, and bromocriptine is now recognized as being as effective as dantrolene. Bromocriptine is only available for oral use, and must be given via a nasogastric tube if the patient cannot ingest it orally. L-dopa is a dopamine precursor that is also used to treat NMS, although there have been fewer reports of the effectiveness of L-dopa than of bromocriptine. The advantage of L-dopa is that it can be given intravenously, because an injection preparation is available, the same for dantrolene. We treated three cases of NMS with intravenous L-dopa (50–100 mg/day) and obtained satisfactory results (Nisijima et al., 1997).

Amantadine, a dopamine agonist and an antagonist at the N-methyl-D asparatate (NMDA) type of the glutamate receptor, is also used to treat NMS. Weller and Kornhuber (1992) have claimed that NMS may be related to a relative glutamatergic transmission excess as a consequence of a dopaminergic blockade. Amantadine may indirectly reinforce the dopamine system via the glutamatergic system, but the precise mechanism of action of amantadine is unknown.

Benzodiazepines, which enhance GABAergic function, are also used in the treatment of NMS (Lew and Tollefson, 1983). Although they are recommended to control agitated patients being treated for NMS, the efficacy of benzodiazepines is moderate or transient (Caroff et al., 1998). They may be effective in treating a mild form of NMS (Kontaxakis et al., 1988).

Electroconvulsive therapy

Some patients exhibit psychiatric symptoms during the course of NMS, and cases that exhibit hyperthermia and psychomotor excitation have been dubbed "lethal catatonia" or "malignant catatonia." Discussion has focused on whether these states are different entities from NMS. Electroconvulsive therapy (ECT) is effective against both (Mann et al., 1990). Jessee and Anderson (1983) first reported that ECT is effective in the treatment of NMS, and many reports describing the efficacy of ECT in the treatment of NMS have been published subsequently (Davis et al., 1991; Trollor and Sachdev, 1999). We also used ECT in five cases of NMS associated with psychiatric symptoms and confirmed its effectiveness (Nisijima and Ishiguro, 1999). Pharmacotherapy takes ~10 days to improve NMS (Rosebush et al., 1991), whereas the mean time from the initial ECT to complete resolution was 6 days in our five cases.

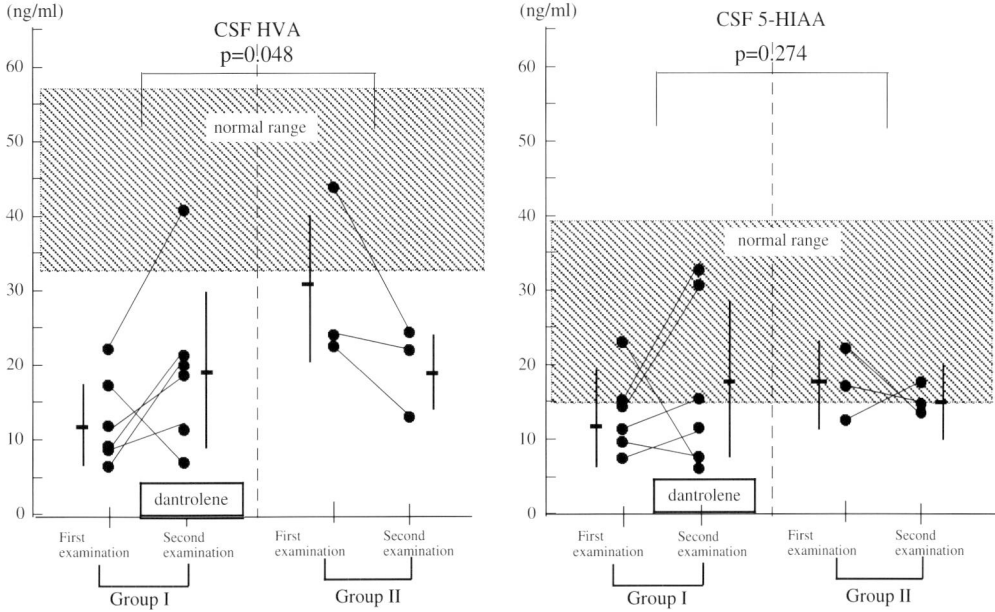

Fig. 2. Changes in CSF HVA and 5-HIAA levels in Groups I and II during the active phase of NMS. Dantrolene was administered to Group I, but not to Group II. HVA, homovanillic acid; 5-HIAA, 5-hydroxyindoleacetic acid. Normal range: mean ± S.D. The data were analyzed by Mann-Whitney U-test (Nisijima and Ishiguro, 1993)

Thus, ECT offers the advantage of treating not only the NMS but also the underlying psychosis, and may shorten the interval between the onset and resolution of NMS.

Animal models of NMS

As stated above, hypofunction of the dopamine system is thought to occur in the region of the striatum and hypothalamus in NMS. However, NMS develops only in 0.2% of cases treated with neuroleptics, and it also occurs in patients treated with atypical antipsychotics that have a weak dopamine receptor blocking effect (Hasan and Buckley, 1998). Moreover, it does not always recur when neuroleptics are given after improvement in patients with a history of NMS. Thus, the pathogenesis of NMS is not fully known. On the other hand, study on the pathogenesis of malignant hyperthermia is in marked contrast to those on NMS. These two diseases were first reported in 1960 (Denborough and Lovell, 1960), although the study of malignant hyperthermia is more advanced than that of NMS. This may be attributable to the fact that malignant hyperthermia is a hereditary disease and an adequate animal model has been found for malignant hyperthermia. The animal model was called "porcine stress syndrome (PSS)" and was developed in the 1960s (Hall et al., 1966). The animal model has been used to answer many questions about malignant hyperthermia, and for this reason it seems necessary to prepare an animal model for NMS. Several investigators have attempted to develop animal models of NMS. Yamawaki et al. (1990b) hypothesized dopamine-serotonin imbalance theory as the etiology of NMS, asserting that hyperfunction occurs in the serotonin system and hypofunction in the dopamine system, and they attempted to prepare an animal model of NMS based on this hypothesis. When haloperidol was administered to rats, and a small quantity of veratrine, which releases serotonin, was later injected into the rat preoptic anterior hypothalamus (PO/AH), which contains the thermoregulation center, the body temperature of the rats rose. They considered that animals with a fever induced by the above-mentioned manipulation was an animal model of NMS, but the veratrine injection into the PO/AH is not natural from the viewpoint of developing NMS. Keck et al. (1990) used PSS, i.e., an animal model of malignant hyperthermia, and investigated whether it could also be used as an animal model of NMS. They administered haloperidol to PSS, but did not detect any manifestations of NMS and reported that PSS is inadequate as an animal model of NMS. Parada et al. (1995) injected sulpiride into the perifornical lateral hypothalamus of rats, and their body temperature increased. Although they reported this as an animal model of NMS, they merely demonstrated that the dopamine D_2 receptor is involved in the regulation of body temperature. Since the increased body temperature in the rats they induced is not part of the course of development of NMS, it cannot be regarded as a valid animal model. Tanii et al. (1996) administered haloperidol and atropine to rabbits in an environment maintained at 35°C and reported that their body temperature increased by 2.3°C. We have already reported that a high temperature environment is a risk factor for the development of NMS, and the animal model prepared by Tanii is a better model for NMS. However, a problem is that the body temperature of animals treated with atropine alone in a high temperature environment also increased. There do not appear to have been any reports of new animal models of NMS since the study by Tanii. Further study is needed to prepare an adequate animal model of NMS.

Serotonin syndrome

Historical background

In the 1970s, the pharmacological effects of serotonin by administering various serotonin agonists to animals and observing their abnormal behavior were studied. The animals exhibited forepaw treading, straub tail, tremor, flat body posture, head weaving, and wet-dog shake. This combination of abnormal behaviors was called "serotonin behavioral syndrome" (Grahame-Smith, 1971; Jacobs, 1976). Thus, the term "serotonin syndrome" was initially used in the field of animal behavioral pharmacology. Insel et al. (1982) encountered two patients who exhibited restlessness, myoclonus, fever, hyperreflexia, etc., when a monoamine oxidase (MAO) inhibitor clorgyline and an antidepressant clomipramine were simultaneously administered, and they reported them as cases of serotonin syndrome in humans. Although this appears to be the first time the term "serotonin syndrome" was used in relation to humans, a case report, in which patients could be diagnosed as serotonin syndrome, was published in the 1950s (Mitchell, 1955). The patient exhibited restlessness, excitation, tremors, and hyperreflexia during simultaneous administration of iproniazid (an antituberculosis drug and an MAO inhibitor) and meperidine, and the manifestations were consistent with the current criteria for serotonin syndrome. Occasional cases of serotonin syndrome have been reported since the report by Insel et al. Sternbach (1991) conducted the first comprehensive clinical review of serotonin syndrome. At that time selective serotonin reuptake inhibitors (SSRIs), including fluoxetine, had begun to be used to treat many depressive patients, and his article appeared to be timely, because the increased use of SSRIs had increased the incidence of the syndrome. Interest in serotonin syndrome has rapidly grown since his review, and many reports on this subject have been published.

Clinical symptoms and diagnosis of serotonin syndrome

In his review, Sternbach evaluated 38 cases of serotonin syndrome in 12 articles and proposed a definition of the syndrome. The criteria for the diagnosis of serotonin syndrome include the following 10 symptoms: (1) mental status change (confusion, hypomania); (2) agitation; (3) myoclonus; (4) hyperreflexia; (5) diaphoresis; (6) shivering; (7) tremor; (8) diarrhea; (9) incoordination; and (10) fever. He suggested that the presence of at least three manifestations indicated a probability of serotonin syndrome if supported by the clinical history. This was the first diagnostic criterion for serotonin syndrome, and it is now widely used.

However, since according to this criterion, some patients may have all 10 symptoms and some may have only three, Sternbach's criterion does not provide any method of evaluating the severity of the syndrome in individual cases. Birmes et al (2003) have recently proposed stricter diagnostic criteria. Mills (1997), on the other hand, analyzed the symptoms in 127 published cases of serotonin syndrome and identified a total of 34 symptoms in cases, including muscle rigidity (49%), hypertension (33%), tachypnea (28%), hypotension (14%), and salivation (5%). These symptoms considerably overlap in NMS and in serotonin syndrome cases, although there are differences in their frequency. As described below, it is important to differentiate serotonin syndrome from NMS.

Causative drugs

Serotonin syndrome generally develops when several serotoninergic agents are used simultaneously, and it rarely develops as the result of administration of a single drug. Combinations of L-tryptophan and MAO inhibitors were frequent causes from the 1960s to the 1980s (Oates and Sjoerdsma, 1960; Baloh et al., 1982). Now, however, serotonin syndrome often develops during simultaneous use of SSRIs and MAO inhibitors. Some tricyclic antidepressants have a potent serotonin reuptake inhibiting effect. Clomipramine, for example, has a strong serotonin reuptake inhibiting effect comparable to that of the SSRIs. Imipramine and amitryptyline are antidepressants that have a serotonin reuptake inhibiting effect as well as an NA reuptake inhibitory effect (Richelson and Pfenning, 1984). Serotonin syndrome may develop as a result of the administration of high doses or simultaneous use of these drugs. Lithium carbonate does not cause serotonin syndrome directly, but serotonin neurotransmission may be promoted (Price et al., 1990) and the risk of developing serotonin syndrome may increase when used simultaneously with the above antidepressants. In fact, there have been reports of cases in which lithium carbonate has been involved in the development of serotonin syndrome (Mills, 1997).

In many of the cases of serotonin syndrome reported, MAO inhibitors have been simultaneously administered to a patient with a poor prognosis (Hilton et al., 1997). A reversible MAO inhibitor, moclobemide that has fewer adverse effects, began to be used in recent years and cases of serotonin syndrome related to moclobemide have also been reported (Neuvonen et al., 1993). Thus, special care should be taken with regard to simultaneous use of other antidepressants with MAO inhibitors.

Serotonin syndrome has also developed when analgesics, including pethidine, pentazocine, and tramadol, have been simultaneously used with MAO inhibitors or SSRIs, and these analgesics have been reported to promote the serotonin reuptake inhibitory effect or the release of serotonin (Larsen and Hyttel, 1985). 3,4-Methylenedioxymethamphetamine (MDMA), which has recently been used as a stimulant drug among young people, has a serotonin releasing effect (Green et al., 1995), and cases exhibiting clinical symptoms similar to those of serotonin syndrome have been reported (Mueller and Korey, 1998).

Pathogenesis of serotonin syndrome

Several subtypes of serotonin (5-HT) receptors have been identified in recent years, and $5\text{-}HT_{1A}$ receptors have been to be closely involved in the development of serotonin behavioral syndrome in animals. Based on the results of past animal experiments, Sternbach speculated that hyperstimulation of $5\text{-}HT_{1A}$ receptors in the brainstem and spinal cord may play an important role in the development of serotonin syndrome in humans. However, fever occurs in many human cases of serotonin syndrome. Stimulation of the $5\text{-}HT_{1A}$ receptor is thought to lead to a decrease in body temperature (O'Connell et al., 1992), whereas an increase in body temperature is reported to be induced by activation of the $5\text{-}HT_{2A}$ receptor (Mazzola Pomietto et al., 1995). Therefore, it is uncertain that human cases of serotonin syndrome develop as a result of excessive $5\text{-}HT_{1A}$ receptor stimulation alone. Extrapyramidal symptoms, such as muscle rigidity, are observed in nearly 50% of the cases (see "Clinical symptoms and diagnosis of serotonin syndrome"), suggesting hypofunction of the dopamine system; and autonomic instability, such as diaphoresis and tachycardia, are found in half of the cases of serotonin syndrome, indicating hyperactivity of the noradrenergic system. Thus, the abnormalities in serotonin syndrome may spread to other parts of the nervous system in addition to the serotonin system. In view of this, we evaluated monoamines and their metabolites in the CSF in the active phase and after the improvement in four cases of serotonin syndrome, the same as in the NMS cases (Nisijima et al., 2003a) (Table 4). The results showed that the CSF NA level during the active phase of serotonin syndrome was significantly higher than in the control group and that it returned to within the normal range after improvement. These findings suggest that the sympathetic nervous system is in an activated state in serotonin syndrome. The mean HVA value in the serotonin syndrome group was lower than in the control group, but the difference did not reach statistical significance. The CSF 5-HIAA level, however, was significantly lower during the active phase of serotonin syndrome than in the control group. Antidepressants with a potent serotonin reuptake inhibiting effect had been administered in the four serotonin syndrome cases (see Table 4). These antidepressants have inhibited serotonin reuptake to presynapses, and conversion of serotonin to 5-HIAA by MAO was inhibited. In this way, CSF 5-HIAA levels may decrease when antidepressants with a serotonin reuptake inhibiting effect are administered. In addition, the low CSF 5-HIAA level may be pathognomonic of serotonin syndrome. We also measured the CSF GABA levels in these four cases. The GABA levels in the serotonin syndrome group and the control group were 210 ± 231 pmol/ml and 434 ± 205 pmol/ml, respectively (unpublished data). Although the difference was not significant, the level tended to be lower in the serotonin syndrome group ($p = 0.089$). The serotonin syndrome cases were characterized by myoclonus, agitation, and mental status changes. Benzodiazepine derivatives are effective in treating such symptoms (Gillman, 1999). Since benzodiazepines have a potentiating effect on the GABA system, the tendency for CSF GABA to decrease appears to explain the effectiveness of benzodiazepines in serotonin syndrome. We found that the results of our study of CSF monoamine metabolism in the serotonin syndrome cases were very similar to our results in the NMS cases. These findings show that the two syndromes are similar to each other from the standpoint of CSF monoamine and GABA metabolism. However, few serotonin syndrome cases have been studied thus far, and more cases need to be studied.

Pathophysiology of serotonin syndrome

Because serotonin syndrome sometimes develops during administration of SSRIs and MAO inhibitors, which enhance central serotoninergic neurotransmission, the enhancement of serotonin activity in the brain appears to be closely related to the development of the syndrome. However, serotonin syndrome exhibits diverse clinical symptoms, and abnormalities in other parts of the central nervous systems besides the serotonin system may be involved. In fact, high CSF NA values were noted in the active phase in our CSF study of serotonin syndrome cases, and we used animal models of serotonin syndrome to evaluate this in greater detail. As already described above, under historical background, various types of serotonin-enhancing agents have been used to animal models of serotonin syndrome, and have included: 8-hydroxy-2-(di-n-propylamino)tetralin (8-OH-DPAT), a 5-HT$_{1A}$ receptor agonist; 5-HT precursors, such as L-tryptophan; potent serotonin reuptake inhibitors, such as SSRIs and clomipramine; and MAO inhibitors, such as tranylcypromine and clorgyline. Because high fever occurs in human cases of serotonin syndrome and sometimes leads to death, we regarded the development of high fever as an important condition when selecting drugs to prepare animal models of serotonin syndrome. Since the hypothalamus regulates body temperature, we prepared rat models of serotonin syndrome with a combination of 5-hydroxy-L-tryptophan (5-HTP, a precursor of serotonin) and clorgyline (an MAO-A inhibitor) and a combination of fluoxetine (an SSRI) and tranylcypromine (an MAO-B inhibitor), and evaluated

Table 4. Summary of four patients with serotonin syndrome and their CSF levels of NA, HVA, 5-HIAA, and GABA

Case	Age	Gender	Underlying diseases	Causative drugs of SS	Body temperature (°C)	Symptoms of SS	Suicidal ideation	HAM-D	CSF NA (ng/ml)	HVA (ng/ml)	5-HIAA (ng/ml)	GABA (pmol/ml)
1	43	M	Depression	Clomipramine	38.7	Mental status changes, hyperreflexia, diaphoresis, flushed skin, tachycardia	±	26	0.17 (0.10)	12.9 (40.6)	12.2 (19.2)	65 (142)
2	62	M	Depression	Imipramine	39.4	Mental status changes, myoclonus, diaphoresis, hyperreflexia, rigidity, tachycardia, incoordination	−	29	0.22 (−)	35.7 (−)	9.3 (−)	93 (−)
3	56	M	Depression	Fluvoxamine, clomipramine	37.8	Mental status changes, myoclonus, diaphoresis, incoordination, tachycardia	+	35	0.14 ( )	48.1 (−)	10.6 (−)	125 ( )
4	64	F	Depression	Fluvoxamine, milnacipran	37.2	Mental status changes, incoordination, tremor, diaphoresis, tachycardia	−	18	0.20 (0.11)	29.8 (35.4)	13.3 (11.2)	555 (728)
Mean±S.D.	56.3±9.5 y.o.								0.18±0.04* (0.105)	31.6±14.6 (38.0)	11.4±1.8** (15.2)	210±231*** (435)
Controls ($n=7$)	50.6±10.0 y.o.								0.07±0.02	49.7±13.5	21.2±5.1	434±205

Note: The levels of the four parameters after recovery from the 5-HT syndrome are in parenthesis. The GABA level data have not been published. SS, serotonin syndrome; HAM-D, the 17-item Hamilton Depression Rating Scale before development of SS.
* $p=0.0082$ (vs. controls) (Mann-Whitney U-test) (Nisijima et al., 2003a).
** $p=0.0233$ (vs. controls) (Mann-Whitney U-test) (Nisijima et al., 2003a).
*** $p<0.1$ (vs. controls) (Mann-Whitney U-test).

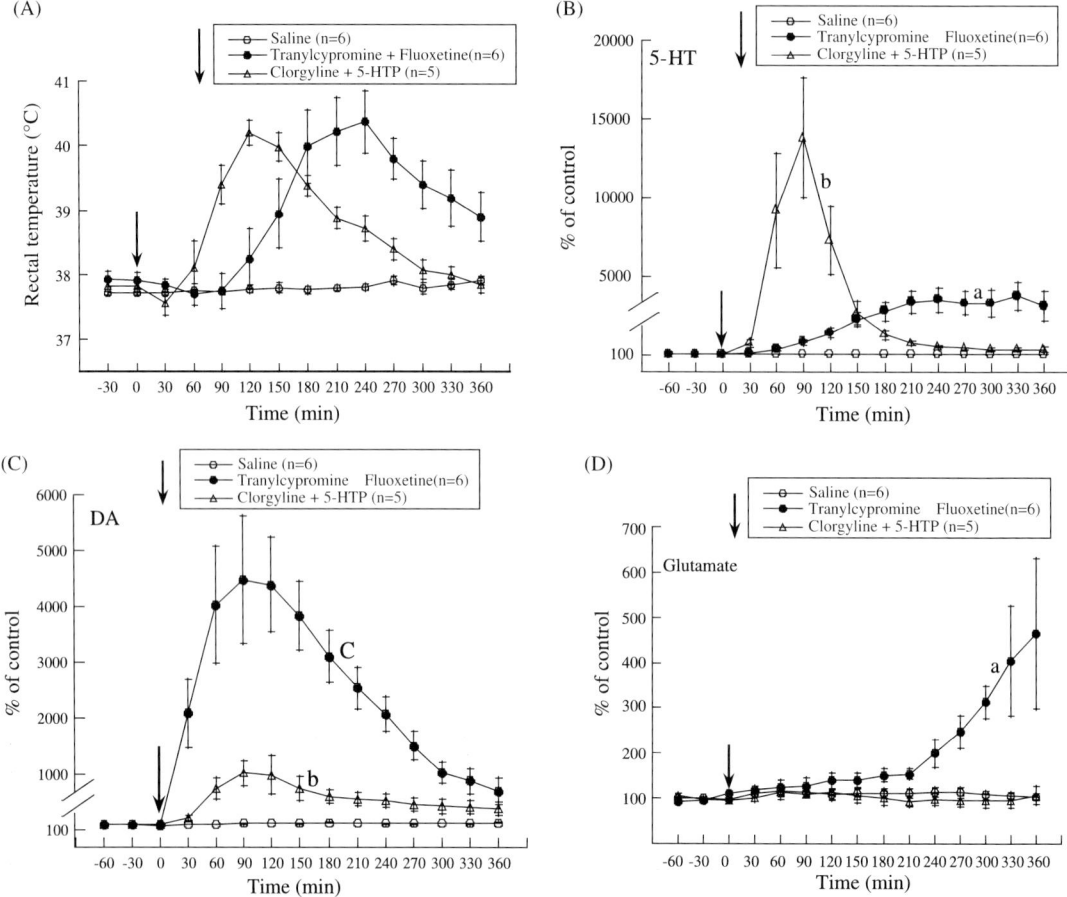

Fig. 3. (A–D) Time course of the rectal temperature of the rats (A) and the time course of serotonin (B), dopamine (C), and glutamate (D) levels in the anterior hypothalamus in the three groups. Tranylcypromine (3.5 mg/kg) + fluoxetine (10 mg/kg) or 5-HTP (80 mg/kg) + clorgyline (1.2 mg/kg) was administered intraperitoneally at time zero. Data are shown as the mean ± S.E. The areas under the curve (AUC) of the concentration vs. time plots for dialysates serotonin, dopamine, and glutamate were calculated, and statistical comparisons of the AUC data between the groups were made by the Mann-Whitney's U-test. a: $p = 0.0039$ (tranylcypromine + fluoxetine vs. saline), b: $p = 0.0062$ (clorgyline + 5-HTP vs. saline), c: $p = 0.0032$ (tranylcypromine + fluoxetine vs. saline) (Shioda et al., 2004).

changes in the concentration of extracellular neurotransmitters in the hypothalamus by a microdialysis method (Shioda et al., 2004). As shown in Fig. 3, the serotonin and dopamine concentration in the hypothalamus increased in both animal models of serotonin syndrome, although the extent of the increases differed. The concentration of NA also increased in two different animal models of serotonin syndrome (Fig. 4) (Nisijima et al., 2003b). These results suggest that high fever develops in response to combinations of various types of 5-HT receptor agonists and that dopamine and NA concentrations increase in the hypothalamus as well as the serotonin concentration. In the cases of serotonin syndrome induced by fluoxetine and tranylcypromine, a delayed increase in glutamate in the hypothalamus was also found (Fig. 3). These results suggest that various systems in the central nervous system, including the serotonin system, dopamine system, NA system, and glutamate system, are affected in the serotonin syndrome. The increase in dopamine concentration was found to be higher in the cases of serotonin syndrome cases induced by fluoxetine and

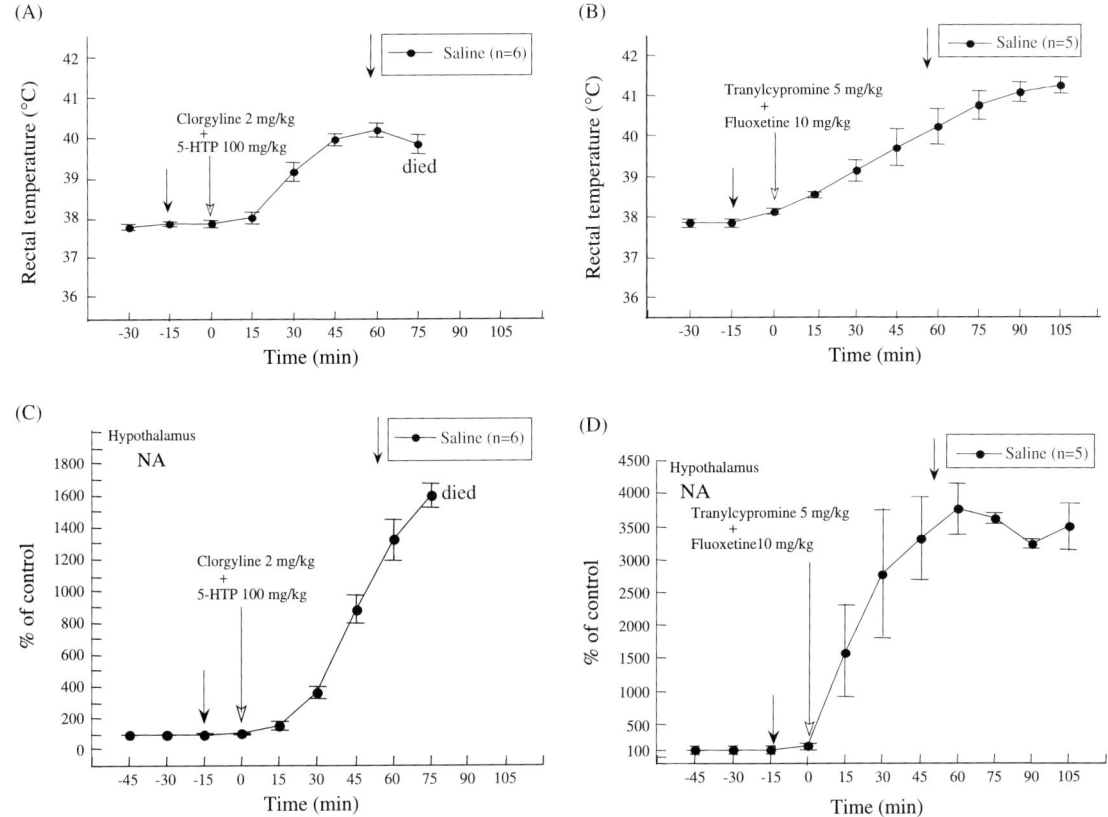

Fig. 4. (A–D) Time course of rectal temperature of the rats (A, B) and changes in NA levels in the anterior hypothalamus in the two different animal models of serotonin syndrome (C, D). 5-HTP (100 mg/kg)+clorgyline (2 mg/kg) or tranylcypromine (5 mg/kg)+fluoxetine (10 mg/kg) was administered intraperitoneally at time zero. Data are shown as the mean±S.E. (Nisijima et al., 2003b).

tranylcypromine than in the cases induced by 5-HTP and clorgyline, and monoamine and glutaminergic metabolism in the central nervous system were found to differ with the drug combination. Thus, the central neurochemical metabolism in serotonin syndrome may differ according to the nature of the causative drug.

Treatment of serotonin syndrome

Current treatment

The basic treatment of serotonin syndrome is discontinuation of the causative drugs and supportive therapy, including fluid replacement and cooling, the same as in NMS. Serotonin syndrome is usually self-limited, with an uneventful resolution, once the offending drug has been discontinued. However, there are cases in which hyperthermia, dyspnea, renal failure, and DIC develop and ultimately lead to death, and thus treatment of the complications is necessary in these severe cases. Several drugs have been used to treat serotonin syndrome. The most frequently used drugs are non-specific 5-HT receptor antagonists, such as cyproheptadine (McDaniel, 2001). Propranolol (a β-blocker) blocks the $5\text{-}HT_{1A}$ receptor and has been reported to be effective against serotonin syndrome in humans (Gillman, 1999), but few cases have been studied, and the efficacy of propranolol remains unclear. There have been reports that chlorpromazine, an antipsychotic drug, is effective in the treatment of serotonin syndrome

(Graham, 1997; Gillman, 1999), perhaps because it has a relatively potent 5-HT$_{2A}$ receptor blocking effect. Benzodiazepines, including diazepam, are sometimes used to treat myoclonus and anxiety in serotonin syndrome, although it is uncertain whether benzodiazepines are essential to the treatment of serotonin syndrome. Dantrolene has been reported to be effective for the treatment of serotonin syndrome induced by MDMA (Singarajah and Lavies, 1992), but only a few cases have been reported, and study of additional cases is needed. We recently reported a case of serotonin syndrome in which ECT were effective (Nisijima et al., 2002). Fink (1996) had previously described the efficacy of ECT for serotonin syndrome, but since only two cases have been reported, further study in a large number of cases will be needed to assess the effectiveness of ECT.

Evaluation of pharmacotherapy in an animal model of serotonin syndrome

Cyproheptadine, a nonspecific serotonin antagonist, has been the most consistently effective treatment in case reports of serotonin syndrome (Graudins et al., 1998). However, serotonin syndrome in humans often includes hyperthermia, and hyperthermia appears to be associated with activation of the 5-HT$_{2A}$ receptor. We hypothesized that 5-HT$_{2A}$ receptor antagonist2A receptor antagonist>s would be more effective than cyproheptadine. Also, the effectiveness of other drugs, including diazepam, chlorpromazine, propranolol, etc., has not yet been established for the treatment of serotonin syndrome. We therefore evaluated which drugs were therapeutic options for the treatment of serotonin syndrome in an animal model. An animal model of serotonin syndrome was prepared by intraperitoneally injecting rats with 5-HTP (100 mg/kg) and clorgyline (2 mg/kg) (Nisijima et al., 2000, 2001, 2003b, 2004). Rectal temperature began to rise 15 min after administration of 5-HTP and clorgyline, and reached 40°C by 60 min. All the rats died within 90 min of administration. By contrast, the increase in rectal temperature and subsequent death was completely averted in the rats pre-treated with potent 5-HT$_{2A}$ receptor antagonist2A receptor antagonist>s, i.e., including risperidone, ketanserin, ritanserin and pipamperone (Fig. 5). Cyproheptadine and chlorpromazine were effective only at high doses. Although propranolol suppressed the increase in rectal temperature, all the rats died within 120 min of administration of the two drugs. Rectal temperature in the group pretreated with WAY 100635 (a selective 5-HT$_{1A}$ receptor antagonist) increased in the same manner as in the saline group (Fig. 6). Diazepam suppressed the body temperature increase in the model of the serotonin syndrome, but did not prevent death as effectively as the 5-HT$_{2A}$ receptor blockers. Dantrolene did not exert as strong an effect in the model of serotonin syndrome as 5-HT$_{2A}$ receptor blockers, but it suppressed body temperature to some extent. The NMDA receptor antagonists memantine and MK 801 also suppressed the increase in body temperature in the model of serotonin syndrome (Fig. 7). In view of their inhibitory effects of the rectal temperature increase and mortality in each group (Table 5), the drugs with a potent 5-HT$_{2A}$ receptor blocking effect were the most effective, and were followed by the nonspecific serotonin receptor antagonists, such as cyproheptadine and chlorpromazine. Diazepam, memantine, and dantrolene also showed some effects. MK 801 appears to be effective for the treatment of serotonin syndrome, but it has adverse effects that include ataxia, amnesia, and psychotomimetic symptoms, and it is only used experimentally.

It is still risky to apply these results to the treatment of serotonin syndrome in human; however, the results of our experiment may be helpful in regard to the treatment of cases with life-threatening hyperthermia of over 40°C.

Effects of dantrolene on hyperthermia induced by MDMA

3,4-Methylenedioxymethamphetamine (MDMA) is now widely used by young people in Western countries, and serotonin syndrome sometimes develops when excessive doses are ingested. MDMA has a structural formula similar to that of

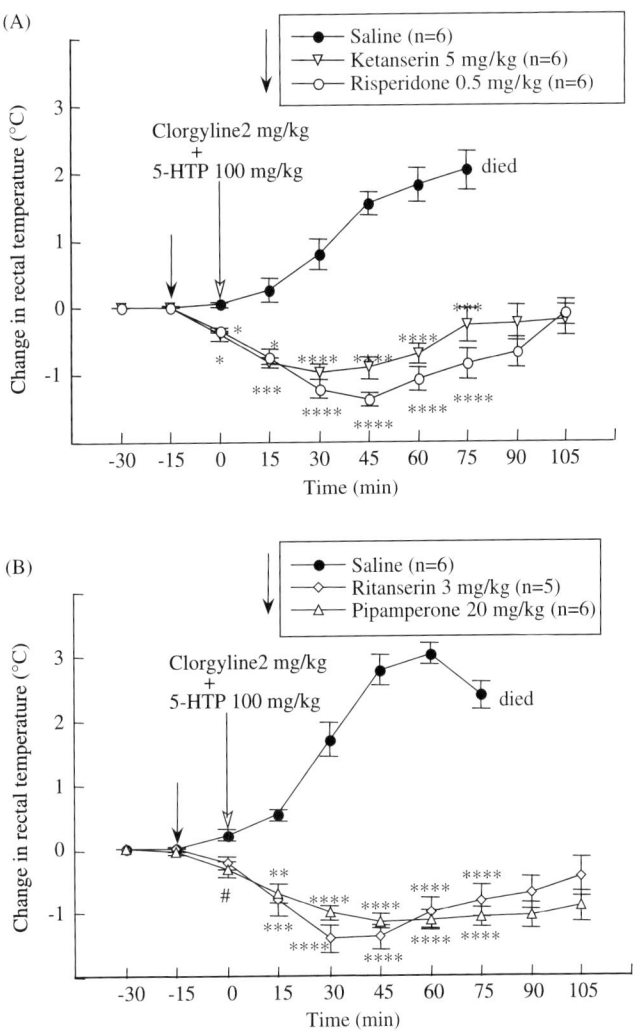

Fig. 5. (A, B) Time course of the effect of ketanserin, risperidone, ritanserin, and pipamperone on the rectal temperature of rats. 5-HTP (100 mg/kg) and clorgyline (2 mg/kg) were given at time zero. Saline, ketanserin, risperidone, ritanserin, or pipamperone was given intraperitoneally to rats 15 min before administration of 5-HTP and clorgyline. Data are shown as the mean ± S.E. The statistical differences between the groups were: #$P<0.05$ (vs. saline); ##$P<0.01$ (vs. saline); *$P<0.005$ (vs. saline); ***$P<0.0005$ (vs. saline); ****$P<0.0001$ (vs. saline) (Scheffe's test) (Nisijima et al., 2001, 2004).

amphetamine (Green et al., 1995), but has been reported to promote greater serotonin secretion than dopamine. This may explain why serotonin syndrome can be induced by MDMA. Dantrolene has been reported to be effective in treating serotonin syndrome induced by MDMA (Logan et al., 1993; Mallick and Bodenham, 1997). However, few cases have been reported, and it is unclear whether dantrolene is actually effective. We are currently evaluating the effectiveness of dantrolene using animals. Although the experiment is ongoing, we will present our preliminary results in this article. Male Wistar rats weighing 200–250 g were used in this study. Dantrolene was purchased from Sigma Chemical Co. (USA). MDMA was synthesized by our collaborating investigator T. Iwamura — a specialist on synthetic chemistry. In other studies, 10 mg/kg or 20 mg/kg of dantrolene were

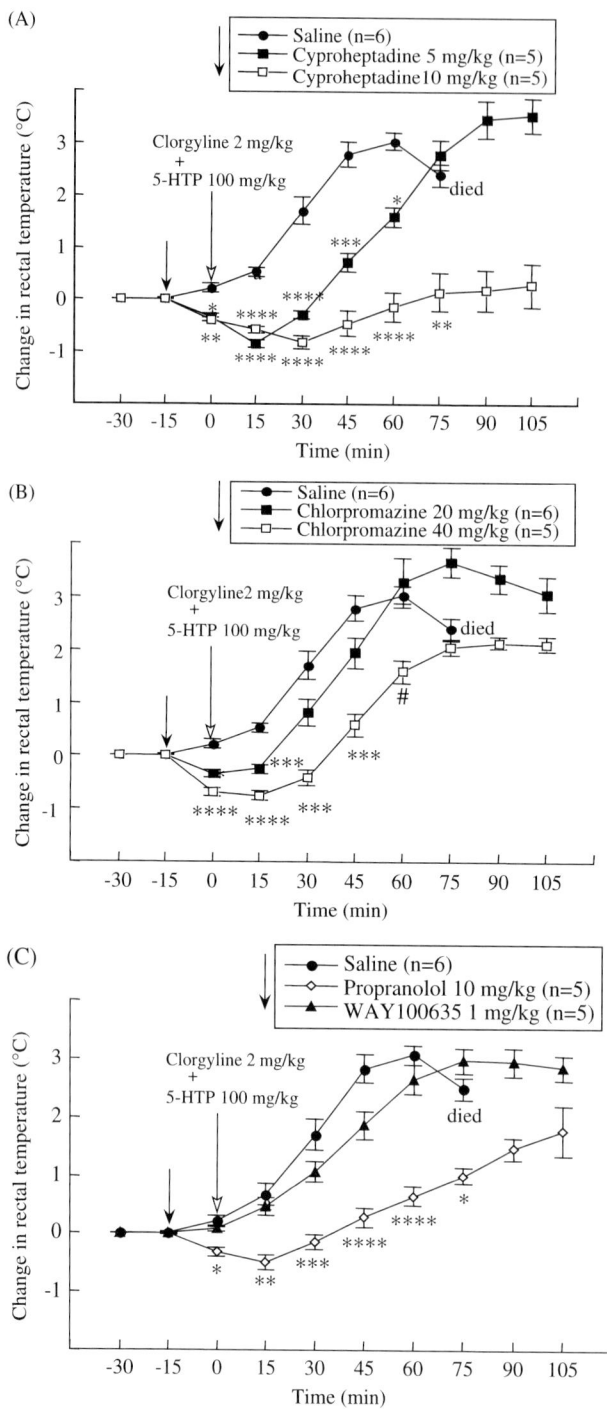

Fig. 6. (A–C) Time course of the effect of cyproheptadine, chlorpromazine, propranolol, and WAY 100635 on the rectal temperature of rats. Data are shown as the mean±S.E. The statistical differences between the groups were: #$P<0.05$ (vs. saline); *$P<0.005$ (vs. saline); **$P<0.0001$ (vs. saline); ***$P<0.0005$ (vs. saline); ****$P<0.0001$ (vs. saline) (Scheffe's test) (Nisijima et al., 2001).

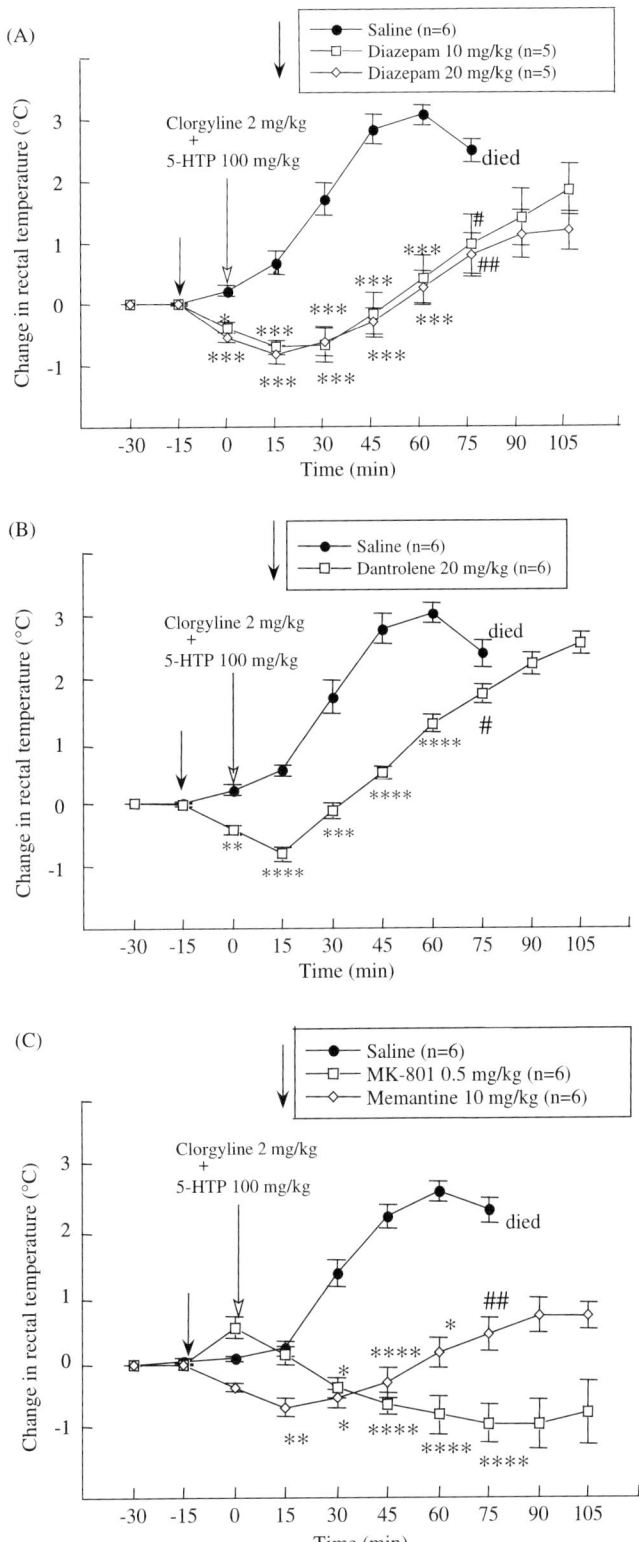

Fig. 7. (A–C) Time course of the effect of diazepam, dantrolene, MK 801, and memantine on the rectal temperature of rats. Data are shown as the mean ± S.E. The statistical differences between the groups were: #$P<0.05$ (vs. saline); ##$P<0.01$ (vs. saline); *$P<0.005$ (vs. saline); **$P<0.001$ (vs. saline); ***$P<0.0005$ (vs. saline); ****$P<0.0001$ (vs. saline) (Scheffe's test) (Nisijima et al., 2001, 2004).

Table 5. Mortality of rats in different groups 120 min after administration of 5-HTP and clorgyline

Drugs	Dose (mg/kg)	Deaths/total no.
Saline		6/6
Risperidone	0.5	0/6
Ketanserin	5	0/6
Ritanserin	3	0/5
Pipamperone	20	0/6
Cyproheptadine	5	5/5
	10	0/5
Chlorpromazine	20	6/6
	40	0/5
Propranolol	10	5/5
WAY 100135	1	5/5
Diazepam	10	2/5
	20	2/5
Dantrolene	20	6/6
Memantine	10	0/6
MK 801	0.5	0/6

Source: Nisijima et al. (2000, 2001, 2003b, 2004).

administered via an intraperitoneal route. Because dantrolene is only weakly water-soluble, we selected an oral route and a dose of 40 mg/kg. The rectal temperature was measured every 30 min using a thermocouple probe connected to a digital thermometer. MDMA (15 mg/kg) was subcutaneously injected to rats, 2 h later. As shown in Fig. 8, the rectal temperature reached 41°C at 60 min after MDMA administration in the group pretreated with physiological saline, and then gradually decreased. In the group that was pretreated with dantrolene, the highest temperature after MDMA administration was not lower than in the saline group, but the temperature rise from 150 min to 210 min after MDMA administration was significantly suppressed. These results suggest that dantrolene can, to some extent, suppress the hyperthermia caused by MDMA. Rusyniak et al. (2004) performed animal experiments and reported that dantrolene has no role in MDMA-mediated hyperthermia. However, their experimental design differed from that of ours. They intraperitoneally injected dantrolene (2.5 mg/kg) in rats and later administered 40 mg/kg of MDMA. Recently, dantrolene has been reported to be effective in a patient with MDMA-induced body temperatures of over 43°C (Kunitz et al., 2003). Since dantrolene may have some effect in humans, further animal studies should be performed. Also, since potent 5-HT$_{2A}$ receptor antagonist2A receptor antagonist > s were highly effective in animal models of serotonin syndrome induced with clorgyline and 5-HTP, we hypothesized that agents with a potent 5-HT$_{2A}$ receptor blocking effect may have a greater effect than dantrolene on MDMA-induced hyperthermia. In the future, we plan to study the suppression of MDMA-induced hyperthermia using 5-HT$_{2A}$ receptor antagonist2A receptor antagonist > s like risperidone or ritanserin.

Differentiation of serotonin syndrome from NMS

Diverse clinical symptoms, including hyperthermia, diaphoresis, and tachycardia, occur in both serotonin syndrome and NMS. As is evident from the comparison of the diagnostic criteria for these two symptoms, their clinical symptoms are very similar. As mentioned in regard to the treatment of serotonin syndrome, while some drugs can be used to both serotonin syndrome and NMS, other drugs should be used to treat only one of them. Thus, differentiation between the two syndromes is very important. First, the causative drugs of serotonin syndrome are primarily serotoninergic agents, whereas NMS is caused by dopamine receptor antagonists. The clinical symptoms in serotonin syndrome cases are characterized by psychotic symptoms, such as anxiety, irritation, excitation, and hypomania. Autonomic nervous symptoms, such as tachycardia, diaphoresis, labile blood pressure, etc., are often observed in both syndromes, whereas extrapyramidal symptoms, such as muscle rigidity, are more common in NMS. Myoclonus and hyperreflexia are specific to serotonin syndrome, and rarely occur in NMS. Blood tests show CPK elevation and leukocytosis in both syndromes, but these abnormalities are more frequent in NMS. Some cases of NMS have been reported in patients receiving SSRIs or serotonin noradrenaline reuptake inhibitors (SNRIs). Before 1990, cases that should have been diagnosed as serotonin syndrome may have been diagnosed as NMS (Brennan et al., 1988; Baca and Martinelli, 1990), because the concept of

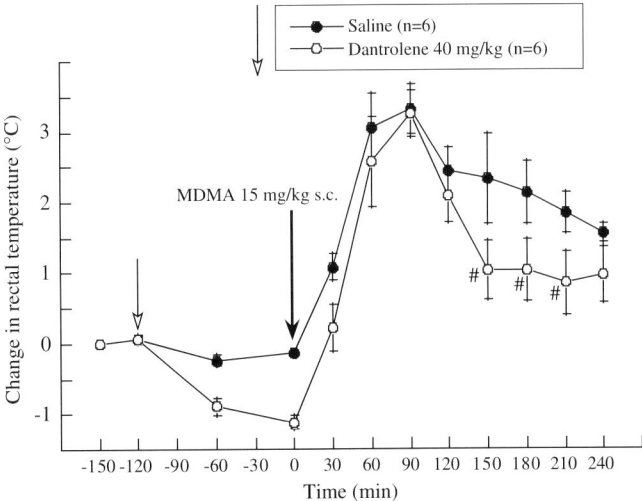

Fig. 8. Effect of dantrolene on MDMA-induced hyperthermia. MDMA (15 mg/kg) was subcutaneously injected at time zero. Saline or dantrolene had been orally preadministered 2 h before MDMA administration. The data were analyzed using an ANOVA with repeated measures, followed by a Fisher test. Data are shown as the mean ± S.E. #$P < 0.05$ (vs. saline).

serotonin syndrome was not widely recognized at the time. However, in recent years, cases of NMS caused by antidepressants have also been reported (Assion et al., 1998; Nimmagadda et al., 2000). The causative drugs are antidepressants in the above cases, but differentiation between NMS and serotonin syndrome are very difficult on the basis of the clinical symptoms. As already described in the evaluation of the CSF study in both syndromes, the abnormalities in CSF monoamines and GABA levels in these two syndromes are very similar. Based on these findings, we speculate that the pathology in serotonin syndrome may start with enhanced activity of serotonin system and then spread to other parts of the central nervous systems. When serotonin syndrome becomes severe, the two syndromes may exhibit very similar clinical symptoms.

Relationship between serotonin syndrome and NMS

The qualifier "toxic," i.e., "toxic serotonin syndrome," has sometimes been used when cases of serotonin syndrome have been reported (Fink, 1996; Graham, 1997). Thus, serotonin syndrome is considered to be caused by increased serotonin concentrations in the brain. On the other hand, NMS does not develop when high doses of neuroleptics are administered, although extrapyramidal symptoms often occur. Even a low dose of neuroleptics may induce NMS when patients are in such physiological states as dehydration, exhaustion, or psychomotor activity. NMS does not appear to result simply from overdose of neuroleptics. Thus, the two syndromes differ in their mode of development. However, in regard to findings in CSF monoamine and GABAergic metabolism in our study, there are similarities between the two syndromes. Also, there are many overlapping aspects of the clinical presentation between both syndromes. There is probably hypofunction of the dopamine system in regions including the striatum, hypothalamus, etc. in NMS cases, and the pathology in NMS may spread to other parts of the central nervous system. As a result, abnormalities of several divisions of the nervous systems may lead to diverse clinical manifestations. On the other hand, simultaneous use of several serotoninergic agents increases serotonin activity in the brain in serotonin syndrome, and the excessive serotoninergic activity may influence other parts of the central nervous system, such as the NA system and the dopamine system. These mechanisms may result in clinical symptoms similar to those of NMS. In other words, the dopamine system is affected first in NMS, whereas the serotonin system is affected first in serotonin syndrome. However, the clinical symptoms of the two syndromes may become very similar at the clinical end point.

Since there have been fewer reports on serotonin syndrome than on NMS, study of additional cases is needed to determine whether patient factors influence the development of serotonin syndrome. The syndrome is usually produced by combinations or high doses of serotoninergic medications. However, serotonin syndrome sometimes develops even after antidepressant doses within the therapeutic range are administered (Voirol et al., 2000; Pan and Shen, 2003). What factor contributes to the development of the serotonin syndrome in these cases should be elucidated.

References

Ananth, J., Parameswaran, S., Gunatilake, S., Burgoyne, K. and Sidhom, T. (2004) Neuroleptic malignant syndrome and atypical antipsychotic drugs. J. Clin. Psychiatry, 65: 464–470.

Assion, H.J., Heinemann, F. and Laux, G. (1998) Neuroleptic malignant syndrome under treatment with antidepressants? A critical review. Eur. Arch. Psychiatry Clin. Neurosci., 248: 231–239.

Baca, L. and Martinelli, L. (1990) Neuroleptic malignant syndrome: a unique association with a tricyclic antidepressant. Neurology, 40: 1797–1798.

Baloh, R.W., Dietz, J. and Spooner, J.W. (1982) Myoclonus and ocular oscillations induced by L-tryptophan. Ann. Neurol., 11: 95–97.

Birmes, P., Coppin, D., Schmitt, L. and Lauque, D. (2003) Serotonin syndrome: a brief review. C.N.A.J., 168: 1439–1442.

Brennan, D., MacManus, M., Howe, J. and McLoughlin, J. (1988) Neuroleptic malignant syndrome without neuroleptics. Br. J. Psychiatry, 152: 578–579.

Caroff, S.N. (1980) The neuroleptic malignant syndrome. J. Clin. Psychiatry, 41: 79–83.

Caroff, S.N. and Mann, S.C. (1993) Neuroleptic malignant syndrome. Med. Clin. North Am., 77: 185–202.

Caroff, S.N., Mann, S.C. and Keck, P.E. (1998) Specific treatment of the neuroleptic malignant syndrome. Biol. Psychiatry, 44: 378–381.

Coons, D.J., Hillman, F.J. and Marshall, R.W. (1982) Treatment of neuroleptic malignant syndrome with dantrolene sodium: a case report. Am. J. Psychiatry, 139: 944–945.

Davis, J.M., Janicak, P.G., Sakkas, P., Gilmore, C. and Wang, Z. (1991) Electroconvulsive therapy in the treatment of the neuroleptic malignant syndrome. Convuls. Ther., 7: 111–120.

Delacour, J.L., Daoudal, P., Chapoutot, J.L. and Rocq, B. (1981) Therapy of neuroleptic malignant syndrome with dantrolene. Nouv. Presse Med., 10: 3572–3573.

Delay, J., Pichot, P., Lemperiere, T., Elissalde, B. and Peigne, F. (1960) Un neuroleptique majeur non phenothiazinique et non reserpinique, l'haloperidol, dans le traitement des psychoses. Ann. Med. Psychol., 118: 145–152.

Delay, J. and Deniker, P. (1968) Drug-induced extrapyramidal syndromes. In: Vinken, P.J. and Bruyn G.W. (Eds.), Handbook of Clinical Neurology, Vol. 6. Diseases of the Basal Ganglia. North-Holland, Amsterdam, pp. 248–266.

Demirkiran, M., Jankovic, J. and Dean, J.M. (1986) Ecstasy intoxication: an overlap between serotonin syndrome and neuroleptic malignant syndrome. Clin. Neuropharmacol., 19: 157–164.

Denborough, M.A. and Lovell, R.R.H. (1960) Anaesthetic deaths in a family. Lancet, 2: 45.

Denborough, M.A. (1998) Malignant hyperthermia. Lancet, 352: 1131–1136.

DeReuck, J., Van Aken, J., Van Landegem, W. and Colardyn, F. (1991) Positron emission tomographic studies of changes in cerebral blood flow and oxygen metabolism in neuroleptic malignant syndrome. Eur. Neurol., 31: 1–6.

Dettling, M., Sander, T., Weber, M. and Steinlein, O.K. (2004) Mutation analysis of the ryanodine receptor gene isoform 3 (RYR3) in recurrent neuroleptic malignant syndrome. J. Clin. Psychopharmacol., 24: 471–473.

Fink, M. (1996) Toxic setotonin syndrome or neuroleptic malignant syndrome? Pharmacopsychiatry, 29: 159–161.

Gillman, P.K. (1999) The serotonin syndrome and its treatment. J. Psychopharmacol., 13: 100–109.

Grahame-Smith, D.G. (1971) Studies in vivo on the relationship between brain tryptophan, brain 5-HT synthesis and hyperactivity in rats treated with a monoamine oxidase inhibitor and L-tryptophan. J. Neurochem., 18: 1053–1066.

Graham, P.M. (1997) Successful treatment of the toxic serotonin syndrome with chlorpromazine. Med. J. Aust., 166: 166–167.

Graudins, A., Stearman, A. and Chan, B. (1998) Treatment of the serotonin syndrome with cyproheptadine. J. Emerg. Med., 16: 615–619.

Green, A.R., Cross, A.J. and Goodwin, G.M. (1995) Review of the pharmacology and clinical pharmacology of 3,4-methylenedioxymethamphetamine (MDMA or "Ecstasy"). Psychopharmacology, 119: 247–260.

Gregorakos, L., Thomaides, T., Stratouli, S. and Sakayanni, E. (2000) The use of clonidine in the management of autonomic overactivity in neuroleptic malignant syndrome. Clin. Auton. Res., 10: 193–196.

Gurrera, R.J. and Romero, J.A. (1992) Sympathoadrenomedullary activity in the neuroleptic malignant syndrome. Biol. Psychiatry, 32: 334–343.

Hall, L.W., Woolf, N., Bradley, J.W.P. and Jolly, D.W. (1966) Unusual reaction to suxamethonium chloride. Br. Med. J., 4: 1305–1306.

Hasan, S. and Buckley, P. (1998) Novel antipsychotics and the neuroleptic malignant syndrome: a review and critique. Am. J. Psychiatry, 155: 1113–1116.

Henderson, V.W. and Wooten, G.F. (1981) Neuroleptic malignant syndrome: a pathogenetic role for dopamine receptor blockade? Neurology, 31: 132–137.

Hilton, S.E., Maradit, H. and Moller, H.J. (1997) Serotonin syndrome and drug combinations: focus on MAOI and RIMA. Eur. Arch. Psychiatry Clin. Neurosci., 247: 113–119.

Insel, T.R., Roy, B.F., Cohen, R.M. and Murphy, D.L. (1982) Possible development of the serotonin syndrome in man. Am. J. Psychiatry, 139: 954–955.

Iwahashi, K., Yoshihara, E., Nakamura, K., Amano, K., Watanabe, M., Tsuneoka, Y., Ichikawa, Y. and Igarashi, K. (1999) CYP2D6 HhaI genotype and the neuroleptic malignant syndrome. Neuropsychobiology, 39: 33–37.

Jacobs, B.L. (1976) An animal behavior model for studying central serotonergic synapses. Life Sci., 19: 777–786.

Jauss, M., Krack, P., Franz, M., Klett, R., Bauer, R., Gallhofer, B. and Dorndorf, W. (1996) Imaging of dopamine receptors with ^{123}I-iodobenzamide single-photon emission-computed tomography in neuroleptic malignant syndrome. Mov. Disord., 11: 726–728.

Jessee, S.S. and Anderson, G.F. (1983) ECT in the neuroleptic malignant syndrome: case report. J. Clin. Psychiatry, 44: 186–188.

Kawanishi, C. (2003) Genetic predisposition to neuroleptic malignant syndrome: implications for antipsychotic therapy. Am. J. Pharmacogenomics, 3: 89–95.

Keck, P.E., Seeler, D.C., Pope, H.G. and Mcelroy, S.L. (1990) Porcine stress syndrome: an animal model for the neuroleptic malignant syndrome? Biol. Psychiatry, 28: 58–62.

Kish, S.J., Kleinert, R., Minauf, M., Gilbert, J., Walter, G.F., Slimovitch, C., Maurer, E., Rezvani, Y., Myers, R. and Hornykiewicz, O. (1990) Brain neurotransmitter changes in three patients who had a fatal hyperthermia syndrome. Am. J. Psychiatry, 147: 1358–1363.

Kontaxakis, V.P., Christodoulou, G.N., Markidis, M.P. and Havaki-Kontaxaki, B.J. (1988) Treatment of a mild form of neuroleptic malignant syndrome with oral diazepam. Acta Psychiatr. Scand., 78: 396–398.

Krause, T., Gerbershagen, M.U., Fiege, M., Weisshorm, R. and Wappler, F. (2004) Dantrolene — a review of its pharmacology, therapeutic use and new developments. Anaesthesia, 59: 364–373.

Kunitz, O., Ince, A., Kuhlen, R. and Rossaint, R. (2003) Hyperpyrexia and rhabdomyolysis after ecstasy (MDMA) intoxication. Anaesthesist, 52: 511–515.

Larsen, J.J. and Hyttel, J. (1985) 5-HT-uptake inhibition potentiates antinoception induced by morphine, pethidine, methadone and ketobemidone in rats. Acta Pharmacol. Toxicol., 57: 214–218.

Levenson, J.L. (1985) Neuroleptic malignant syndrome. Am. J. Psychiatry, 144: 1137–1145.

Lew, T.Y. and Tollefson, G. (1983) Chlorpromazine-induced neuroleptic malignant syndrome and its response to diazepam. Biol. Psychiatry, 18: 1441–1446.

Logan, A.S., Stickle, B., O'Keefe, N. and Hewitson, H. (1993) Survival following 'Ecsatsy' ingestion with a peak temperature of 42 degrees C. Anaesthesia, 48: 1017–1018.

Mallick, A. and Bodenham, A.R. (1997) MDMA induced hyperthermia: a survivor with an initial body temperature of 42.9 degrees C. J. Accid. Emerg. Med., 14: 336–338.

Mann, S.C., Caroff, S.N., Bleier, H.R., Antelo, R.E. and Un, H. (1990) Electroconvulsive therapy of the lethal catatonia syndrome. Convuls. Ther., 6: 239–247.

Mazzola Pomietto, P., Aulakh, C.S., Wozniak, K.M., Hill, J.L. and Murphy, D.L. (1995) Evidence that 1-(2,5-dimethoxy-4-iodophenyl)-2-amino-propane (DOI)-induced hyperthermia in rats is mediated by stimulation of 5-HT$_{2A}$ receptor. Psychopharmacology, 117: 193–199.

McDaniel, W.W. (2001) Serotonin syndrome: early management with cyproheptadine. Ann. Pharmacother., 35: 870–873.

Mills, K.C. (1997) Serotonin syndrome. A clinical update. Crit. Care Clin., 13: 763–783.

Mitchell, R.S. (1955) Fatal toxic encephalitis occurring during iproniazid therapy in pulmonary tuberculosis. Ann. Intern. Med., 42: 417–424.

Mueller, P.D. and Korey, W.S. (1998) Death by "Ecstasy": the serotonin syndrome? Ann. Emerg. Med., 32: 377–380.

Mueller, P.S. and Vester, J.W. (1983) Neuroleptic malignant syndrome: successful treatment with bromocriptine. JAMA, 249: 386–388.

Neuvonen, P.J., Pohjola-Sintonen, S., Tacke, U. and Vuori, E. (1993) Five fatal cases of serotonin syndrome after moclobemide-citalopram or moclobemide-clomipramine overdoses. Lancet, 342: 1419.

Nimmagadda, S.R., Ryan, D.H. and Atkin, S.L. (2000) Neuroleptic malignant syndrome after venlafaxine. Lancet, 355: 289–290.

Nisijima, K. and Ishiguro, T. (1990) Neuroleptic malignant syndrome: a study of CSF monoamine metabolism. Biol. Psychiatry, 27: 280–288.

Nisijima, K. and Ishiguro, T. (1993) Does dantrolene influence central dopamine and serotonin metabolism in the neuroleptic malignant syndrome? A retrospective study. Biol. Psychiatry, 33: 45–48.

Nisijima, K. and Ishiguro, T. (1995a) Cerebrospinal fluid levels of monoamine metabolites and gamma-aminobutyric acid in neuroleptic malignant syndrome. J. Psychiat. Res., 29: 233–244.

Nisijima, K., Matoba, M. and Ishiguro, T. (1995b) Single photon emission computed tomography with ^{123}I-IMP in three cases of the neuroleptic malignant syndrome. Neuroradiology, 36: 281–284.

Nisijima, K., Noguti, M. and Ishiguro, T. (1997) Intravenous injection of levodopa is more effective than dantrolene as

therapy for neuroleptic malignant syndrome. Biol. Psychiatry, 41: 913–914.
Nisijima, K. and Ishiguro, T. (1999) Electroconvulsive therapy for the treatment of neuroleptic malignant syndrome with psychotic symptoms: a report of five cases. J. ECT, 15: 158–163.
Nisijima, K., Yoshino, T. and Ishiguro, T. (2000) Risperidone counteracts lethality in an animal model of the serotonin syndrome. Psychopharmacology, 150: 9–14.
Nisijima, K., Yoshino, T., Yui, K. and Katoh, S. (2001) Potent serotonin (5-HT)$_{2A}$ receptor antagonists completely prevent the development of hyperthermia in an animal model of the 5-HT syndrome. Brain Res., 890: 23–31.
Nisijima, K., Nibuya, M. and Kato, S. (2002) Toxic serotonin syndrome successfully treated with electroconvulsive therapy. J. Clin. Psychopharmacol., 22: 338–339.
Nisijima, K., Nibuya, M. and Sugiyama, H. (2003a) Abnormal CSF monoamine metabolism in 5-HT syndrome. J. Clin. Psychopharmacol., 23: 528–531.
Nisijima, K., Shioda, K., Yoshino, T., Takano, K. and Kato, S. (2003b) Diazepam and chlormethiazole attenuate the development of hyperthermia in an animal model of the serotonin syndrome. Neurochem. Int., 43: 155–164.
Nisijima, K., Shioda, K., Yoshino, T., Takano, K. and Kato, S. (2004) Memantine, an NMDA antagonist, prevents the development of hyperthermia in an animal model of the serotonin syndrome. Pharmacopsychiatry, 37: 57–62.
Oates, J.A. and Sjoerdsma, A. (1960) Neurologic effects of tryptophan in patients receiving a monoamine oxidase inhibitor. Neurology, 10: 1076–1078.
O'Connell, M.T., Sarna, G.S. and Curzon, G. (1992) Evidence for postsynaptic medication of the hypothermic effect of 5-HT$_{1A}$ receptor activation. Br. J. Pharmacol., 106: 603–609.
Otani, K., Hariguti, M., Kondo, T. and Fukushima, Y. (1991) Is the predisposition to neuroleptic malignant syndrome genetically transmitted? Br. J. Psychiatry, 158: 850–853.
Oyamada, T., Hayashi, T., Kagaya, A., Yokota, N. and Yamawaki, S. (1998) Effect of dantrolene on K(+)- and caffeine-induced dopamine release in rat striatum assessed by in vivo microdialysis. Neurochem. Int., 32: 171–176.
Pan, J.J. and Shen, W.W. (2003) Serotonin syndrome induced by low-dose venlafaxine. Ann. Pharmacother., 37: 209–211.
Parada, M.A., de Parada, M.P., Rada, P. and Hernandez, L. (1995) Sulpiride increases and dopamine decreases intracranial temperature in rats when injected in the lateral hypothalamus: an animal model for the neuroleptic malignant syndrome? Brain Res., 674: 117–121.
Preston, J. (1959) Central nervous system reactions to small doses of tranquilizers. Am. Prac. Dig. Tr., 10: 627–630.
Price, L.H., Charney, D.S., Delgado, P.L. and Heninger, G.R. (1990) Lithium and serotonin function: implications for the serotonin hypothesis of depression. Psychopharmacology, 100: 3–12.
Richelson, E. and Pfenning, M. (1984) Blockade by antidepressants and related compounds of biogenic amine uptake into rat brain synaptosomes: most antidepressants selectively block norepinephrine uptake. Eur. J. Pharmacol., 104: 277–286.

Rosebush, P.I., Stewert, T. and Mazurek, M. (1991) The treatment of neuroleptic malignant syndrome: are dantrolene and bromocriptine useful adjuncts to supportive care? Br. J. Psychiatry, 159: 709–712.
Rusyniak, D.E., Banks, M.L., Mills, E.M. and Sprague, J.E. (2004) Dantrolene use in 3,4-methylenedioxymethamphetamine ("ecstasy")-mediated hyperthermia. Anesthesiology, 101: 263–264.
Shioda, K., Nisijima, K., Yoshino, T. and Kato, S. (2004) Extracellular serotonin, dopamine and glutamate levels are elevated in the hypothalamus in a serotonin syndrome animal model induced by tranylcypromine and fluoxetine. Prog. Neuro-Psychopharmacol. Biol. Psychiatry, 28: 633–640.
Singarajah, C. and Lavies, N.G. (1992) An overdose of ecstasy. A role for dantrolene. Anaesthesia, 47: 686–687.
Spivak, B., Maline, D.I., Vered, Y., Kozyrev, V.N., Mester, R., Neduva, S.A., Ravilov, R.S., Graff, E. and Weizman, A. (2000) Prospective evaluation of circulatory levels of catecholamines and serotonin in neuroleptic malignant syndrome. Acta Psychiatr. Scand., 102: 226–230.
Sternbach, H. (1991) The serotonin syndrome. Am. J. Psychiatry, 148: 705–913.
Tanii, H., Taniguchi, N., Niigawa, H., Hosono, T., Ikura, Y., Sakamoto, S., Kudo, T., Nishimura, T. and Takeda, M. (1996) Development of an animal model for neuroleptic malignant syndrome: heat-exposed rabbits with haloperidol and atropine administration exhibit increased muscle activity, hyperthermia, and high serum creatine phosphokinase level. Brain Res., 743: 263–270.
Toru, M., Matsuda, O., Makiguchi, K. and Sugano, K. (1981) Neuroleptic malignant syndrome-like state following a withdrawal of antiparkinsonian drugs. J. Nerv. Ment. Dis., 169: 324–327.
Trollor, J.N. and Sachdev, P.S. (1999) Electroconvulsive treatment of neuroleptic malignant syndrome: a review and report of cases. Aust. N.Z. J. Psychiatry, 33: 650–659.
Ueda, M., Hamamoto, M., Nagayama, H., Okubo, S., Amemiya, S. and Katayama, Y. (2001) Biochemical alterations during medication withdrawal in Parkinson's disease with and without neuroleptic malignant-like syndrome. J. Neurol. Neurosurg. Psychiatry, 71: 111–113.
Voirol, P., Hodel, P.F., Zullino, D. and Baumann, P. (2000) Serotonin syndrome after small doses of citalopram or sertraline. J. Clin. Psychopharmacol., 20: 713–714.
Weller, M. and Kornhuber, J. (1992) A rationale for NMDA receptor antagonist therapy of the neuroleptic malignant syndrome. Med. Hypotheses, 38: 329–333.
Yamawaki, S., Yanagawa, K., Morio, M. and Mori, K. (1986) Possible central effect of dantrolene sodium in neuroleptic malignant syndrome. J. Clin. Psychopharmacol., 6: 378–379.
Yamawaki, S., Yano, E. and Uchitomi, Y. (1990a) Analysis of 497 cases of neuroleptic malignant syndrome in Japan. Hiroshima J. Anesth., 26: 35–44.
Yamawaki, S., Kato, T. and Yano, E. (1990b) Studies on pathogenesis of neuroleptic malignant syndrome: effect of dantrolene on serotonin release in rat hypothalamus. Hiroshima J. Anesth., 26: 45–52.

SECTION IV

Therapeutic Hyperthermia and Consequences

CHAPTER 7

Radio frequency electromagnetic fields: mild hyperthermia and safety standards

John A. D'Andrea[1,*], John M. Ziriax[1] and Eleanor R. Adair[2]

[1]Directed Energy Bioeffects Laboratory, Naval Health Research Center Detachment, 8315 Navy Road, Brooks City-Base, TX 78235, USA
[2]US Air Force, 50 Deepwood Drive, Hamden, CT 06517, USA

Abstract: This chapter is a short review of literature that serves as the basis for current safe exposure recommendations by ICNIRP (International Commission on Non-Ionizing Radiation Protection, 1998) and the IEEE C95.1 (IEEE Standard for Safety Levels with Respect to Human Exposure to Radio Frequency Electromagnetic Fields, 3 kHz to 300 GHz, 2005) for exposure to radio frequency electromagnetic radiation (RF-EMF). Covered here are topics on dosimetry, thermoregulatory responses, behavioral responses, and how these have been used to derive safe exposure limits for humans to RF-EMF. Energy in this portion of the electromagnetic spectrum, 3 kHz–300 GHz, can be uniquely absorbed and is different from ionizing radiation both in dosimetry and effects. The deposition of thermalizing energy deep in the body by exposure to RF-EMF fields provides a unique exception to the energy flows normally encountered by humans. Behavioral effects of RF-EMF exposure range from detection to complete cessation of trained behaviors. RF-EMF is detectable and can in most cases, presumably by thermal mechanisms, support aversion and disruption or complete cessation (work stoppage) of behavior. Safety standards are based on behavioral responses by laboratory animals to RF-EMF, enhanced by careful studies of human thermoregulatory responses at four specific RF frequencies, thereby providing a conservative level of protection from RF-EMF for humans.

Keywords: behavioral; dosimetry; electromagnetic fields; mild hyperthermia; radio frequency; safety standards; thermoregulatory

Introduction

Exposure of living organisms to radio frequency electromagnetic radiations (RF-EMFs) produces heat within tissues primarily from molecular rotation. The heat within tissues is distributed throughout the body by convection (blood flow) and eventually lost to the environment. If the rate of heating is high it may lead to an overall body temperature increase. If the heating is severe it can overwhelm the organism's thermoregulatory responses and lead to hyperthermia. The heating and hyperthermia, of course, will depend on the strength of the microwave field, the duration of exposure, and other characteristics of the exposed organism and the environment, which might

*Corresponding author. Tel.: +1-210-536-6527; Fax: +1-210-536-6537; E-mail: john.dandrea@navy.brooks.af.mil
Reprinted with permission from IEEE C95.1 2005, IEEE Standard for Safety Levels with Respect to Human Exposure to Radio Frequency Electromagnetic Fields, 3 kHz to 300 GHz, Copyright 2006, by IEEE. The IEEE disclaims any responsibility or liability resulting from the placement and use in the described manner.

influence either microwave absorption or the loss of heat from the organism.

Personnel in occupational settings operate many different RF-EMF emitting devices. Such devices operate in the frequency range of 100 kHz–300 GHz and are used for heating, communications, medical, and many other purposes. Considering the number and kind of RF-EMF emitting devices in use today, there are many opportunities for accidental human exposure from many different exposure configurations and durations. Recent technological achievements have increased the power output of RF-EMF devices many times during the past 40 years, enhancing concerns over inadvertent human exposure.

From a safety viewpoint this topic has been of some concern for many years, thereby leading to the development of maximum permissible exposure levels (MPEs). The concern has been for both the possible hazardous effects of exposure from high power RF-EMF exposures as well as low-level chronic exposures. However, low-level chronic exposures have not produced a consistent database from which to derive safe exposure levels. The determination of safe exposure levels requires an extensive scientific database of biomedical information and knowledge of the physical factors that control the deposition of RF-EMF energy in the organism. Developing standards for safe exposure to RF-EMF has required a multidisciplinary approach with biologists, physicists, engineers, psychologists, and medical personnel working together to develop the MPEs. Recent reviews of the scientific literature evaluated the results of many studies conducted at all levels of exposure, including those studies that involve low-level exposures where increases in temperature could not be measured or were not expected (ICNIRP, 1998; Adair and Black, 2003; D'Andrea et al., 2003a, b; Black and Heynick, 2003; Elder, 2003a, b; Elder and Chou, 2003; Elwood, 2003; Heynick and Merritt, 2003; Heynick et al., 2003; Meltz, 2003; NRPB, 2004).

The scientific body of research upon which the safety standards are based is varied and ranges from basic research to applied medical applications. This research has shown, for example, that exposure to RF-EMF can produce mild whole-body hyperthermia and cause behavioral changes in man or laboratory animals ranging from detection of RF-EMFs to performance disruption. A variety of studies have documented the thresholds for detection of RF-EMF exposures in both animals and humans. In the study of behavior, laboratory animals will sometimes escape and subsequently avoid RF-EMF exposure but not in all circumstances. They have also learned to obtain a brief burst of RF-EMF exposure for a warming effect in a cold environment. Most importantly, food motivated behaviors can be either disrupted or stopped outright by RF-EMF exposure, presumably because of the ensuing mild hyperthermia. The stoppage of behavior has been used as the threshold level of absorption (4 W/kg) at which harmful effects might occur during exposures of <1 h. This chapter is devoted to a review and discussion of the functional, thermoregulatory, and behavioral changes produced by RF-EMF exposure and temperature change in tissues. Related topics of RF-EMF dosimetry, absorption, and thermoregulation are discussed relative to safety standards.

Dosimetry of radio frequency electromagnetic fields

Absorption of energy in the RF-EMF range of 100 kHz–300 GHz results primarily in tissue heating. The heating is produced by movement and oscillations of dipole molecules resulting in transfer of energy from the RF-EMFs to the biological tissues. A water molecule is small in size but has a large dipole moment such that it rapidly oscillates in RF-EMFs and produces heat generated by friction (Lee et al., 2000). In the RF-EMF frequency range, human and laboratory animals will scatter and absorb energy depending on factors such as wavelength, body size, body shape, and orientation in electric and magnetic field vectors (Gandhi, 1974; Gandhi et al., 1977; Durney et al., 1978). In comparison, for an organism exposed to much higher frequencies such as visible light and ionizing radiation, the absorption is directly related to the cross-sectional area of the organism (NCRP, 1986). In the RF-EMF region of the electromagnetic spectrum, however, these factors will strongly control absorption that occurs independently of cross-sectional area and may result in resonant absorption in both man and animal. Thus, measurement of the incident fields in the environment may not necessarily reflect the nature of the internal (*in situ*) fields within the tissues of the organism and the subsequent dose-rate. In contrast to much lower frequencies, at RF-EMF frequencies the epidermis of the skin is relatively transparent to fields and capacitive coupling allows energy to reach water found more deeply in the tissues (Lee et al., 2000). The rate of heating is not only dependent on the amplitude of the fields but also on the density of water molecules and dielectric characteristics of the various tissues. For example, the fatty layers between tissues heat much more slowly. Thus, understanding and measurement of the absorbed energy within the organism is important because this may determine not only how much energy is absorbed but where in the body it is absorbed.

Empirical dosimetry

To determine the toxicity of chemicals the knowledge of the dose of the toxic chemical is important to understanding the observed biological effects. In fact, this relationship between dose and the resultant biological effect is a cornerstone of toxicology as a science and has given birth to the concept of dose–response effects in biology. Formally defined, dose–response refers to the relationship between the dose of an agent and its effect on a biological system. The agent in a "dose" of RF-EMF exposure is the heat generated within tissues. While other mechanisms are postulated, tissue heating has been the predominant factor for any ensuing biological effects (Adair and Black, 2003). While the measurement of RF-EMF in the environment is readily accomplished with a densitometer, the measurement of RF-EMF fields within the tissues and subsequent heating of a living organism is not so easily accomplished. One method employed experimentally has been to measure the rate of temperature change during exposure in either the actual organism or in a physical model of the organism.

Some of the earliest estimates of the absorbed dose in a biological organism exposed to RF-EMF energy were accomplished in the 1970s by Don Justesen and colleagues, by using cylindrical models filled with saline (Justesen and King, 1970; Justesen et al., 1971, 1974). Justesen and colleagues were the first to define units of measurement for dose (J/kg) and dose-rate (W/kg) and quantify the amount of energy absorbed from the RF-EMFs (Justesen, 1975). These investigators were conducting experiments with rats exposed in a closed cavity system and required a measurement of absorbed energy to correlate with observed changes in rodent behavior. As later measurements would show, the saline models were within 10% of the dose-rate measured by more sophisticated and accurate methods developed by Hunt and Phillips (1972) and Phillips et al. (1975). They devised a twin-well calorimeter to make very precise measurements of the absorbed dose in rodents exposed to microwave fields. The efforts of these researchers led to the use of the specific absorption (SA) in J/kg and specific absorption rate (SAR) in W/kg as the method of reporting the dose rate of exposure (Justesen, 1975). These were defined in the IEEE C95.1 (2005) safety standard[1] as:

Specific absorption: The quotient of the incremental energy (dW) absorbed by (dissipated in) an incremental mass (dm) contained in a volume (dV) of a given density (ρ).

$$\text{SA} = \frac{dW}{dm} = \frac{dW}{\rho dV}$$

The specific absorption is expressed in units of joules per kilogram (J/kg).

Specific absorption rate: The time derivative of the incremental energy (dW) absorbed by (dissipated in) an incremental mass (dm) contained in a volume element (dV) of given density (ρ).

$$\text{SAR} = \frac{d}{dt}\left(\frac{dW}{dm}\right) = \frac{d}{dt}\left(\frac{dW}{\rho dV}\right)$$

where SAR is expressed in units of watts per kilogram (W/kg).

In the laboratory the measurement of SAR, while laborious, can be accomplished using

---
[1] From IEEE C95.1 2005, IEEE Standard for Safety Levels with Respect to Human Exposure to Radio Frequency electromagnetic Fields, 3 kHz to 300 GHz. Copyright 2006, by IEEE. All rights reseved.

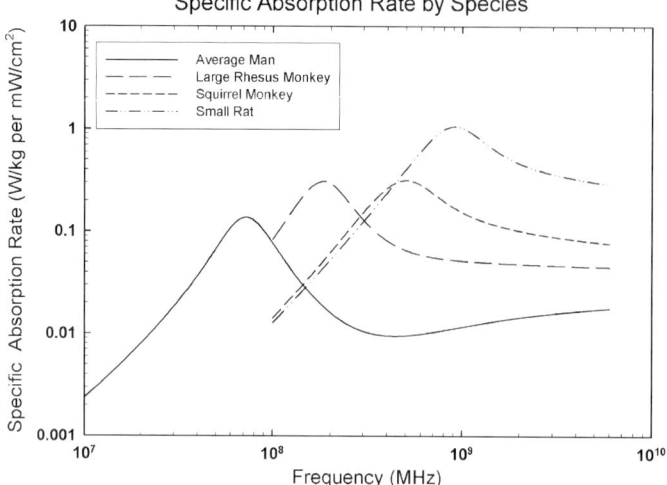

Fig. 1. Whole-body specific absorption rate profiles for man and several laboratory animals derived from the empirical equation by Hurt. (Adapted from Durney et al. (1978).)

physical models of the exposed organism or the organism itself. One compilation of empirical dosimetry data from animal experiments was given by Durney et al. (1986). From that compilation William Hurt formulated an empirical equation, which could predict the whole-body absorption in a prolate spheroid model representing animals of different sizes (software implementation of empirical equation by Hatcher and D'Andrea, 1992). Several examples are shown in Fig. 1 and are relatively accurate predictions for animals exposed to different frequencies of RF-EMF. For the average man (height 175 cm) maximal absorption occurs around a frequency of 70 MHz and can result in a sevenfold increase in SAR compared to an exposure at 2450 MHz (Gandhi and Chatterjee, 1982).

Computational dosimetry

In the laboratory the measurement of SAR, while laborious, can be accomplished using physical models of the exposed organism or the organism itself. In the field, however, such measurements are difficult if not impossible due to environmental influences on the measurements. To provide SAR estimates a variety of sophisticated computational methods have been used in recent years but the fidelity of the finite-difference time-domain (FDTD) method has made this the leader. Gandhi and Kang (2002) provide a good review of the various methods (see also Gandhi, 1994, 1995).

Anatomical models of both the human body and of laboratory animals have been developed by several different research groups in recent years (Stuchly et al., 1987; Gandhi, 1995; Dimbylow, 1997; Wang and Fujiwara, 2000; Mason et al., 2000a; Nagaoka et al., 2004). The FDTD method is widely used to model RF-EMF absorption by a variety of targets including the human body (Shlager and Schneider, 1995; and http://www.fdtd.org). Briefly, the FDTD method, developed by Yee (1966), allows the simultaneous computation of propagation of the electric and magnetic fields. The method has been applied to problems in three dimensions (Kunz and Luebbers, 1993; Taflove, 1995). In this case, the volume to be modeled is described by a three-dimensional grid. At each location in the grid or voxel, a material or tissue type is identified, and each tissue type has associated physical and electrical properties. The FDTD method simulates the propagation of the electrical and magnetic fields associated with an RF-EMF exposure through this volume by solving Maxwell's equations at finite increments of time at each voxel location.

Empirically, the results of the FDTD calculations are used to guide the placement of temperature probes inside a phantom or animal carcass. Measurement results, taken with these probes or at the surface with an infrared video camera, are compared to FDTD predictions to increase confidence in the validity of the model predictions. Thus, FDTD calculations are typically performed before empirical data is collected as a guide to the placement of temperature probes. Finally, FDTD may be used at the end of the project as a quantitative means of extrapolating laboratory results to the real world application.

The interaction RF-EMF with the body is complex and undoubtedly occurs on multiple levels simultaneously. The computational analysis of this interaction must be approached piecemeal. In short, the models will necessarily be simplifications of the real interaction. Thus, the FDTD method requires: (1) a software implementation of the FDTD method; (2) a frequency-dependent description of the physical electrical properties of the tissues to be modeled such as were reported by Gabriel and Gabriel (1996); and finally (3) an anatomical model (Gandhi et al., 1992).

High resolution anatomical models

The development of a number of anatomical models has been underway for some time by researchers at Brooks City-Base, TX. The Brooks anatomical models include both man (Fig. 2) and several laboratory animals (Mason et al., 2000a, b). The models are constructed from sets of serial images, such as a whole-body magnetic resonance image (MRI). Each image is manually color-coded by tissue type, and the resulting stack of coded images is converted to a single three-dimensional model. The models have voxel resolutions down to $0.7\,mm^3$ for the 70 kg man and have been used extensively with the FDTD method to solve non-ionizing electromagnetic radiation absorption at higher frequencies as well as contact current problems at low frequencies. The Brooks anatomical models are unique in that they are freely available on the Internet, making them the de facto standard for realistic anatomical models.

The Brooks Man anatomical model is presently the only human anatomical model, which has been made generally available (contact: john.ziriax@brooks.af.mil) and with a 0.7 mm resolution, it is one of the most detailed as well. The model was constructed from a single human male cadaver that had been thinly sliced and each slice photographed [National Library of Medicine, Visible Human Project]. Brooks researchers acquired the photographic dataset and coded each image by tissue type (Mason et al., 2000a). This database has come to be known as the "Brooks Man" model. In addition, recent software developments make it possible to pose the voxel anatomical models in a variety of poses to allow a wider variety of exposures to be simulated.

Detailed anatomical models have also been developed for several laboratory animals including a rat, rhesus monkey, and goat. Two ongoing projects in the second author's laboratory include an ultra-high resolution ($0.1\,mm^3$) baboon-eye model (Fig. 3) that is nearing completion, and a $2\text{-}mm^3$ model of a domestic swine. Both models are based on MRIs giving precise anatomical details.

The role of computational modeling in this effort is intimately entwined with the laboratory efforts. First, recognizing that empirical efforts require more time and effort, computational models will help suggest critical experiments and exposure parameters. Second, computational models can be used to extend empirical results, filling in the gaps between experiments and extrapolating from animal to human, and from the laboratory to real-world exposures. In both of these roles, the computational model does not stand alone; rather, it acts in partnership with the more expensive, time-consuming empirical efforts. The result of this partnership will be an understanding of the mechanisms and variables determining the biological effect of RF-EMF exposures.

Dosimetry is the technology used to attach a dose of radiation to a bioeffect. In demonstrating bioeffects resulting from RF-EMF exposures, the careful and accurate assessment of RF-EMF dose is not only critical in determining the exposure level at which a given effect occurs, but also in relating that exposure to other exposure conditions including potential human exposures. It is not surprising, therefore, that dosimetry is a part of every well-done experiment.

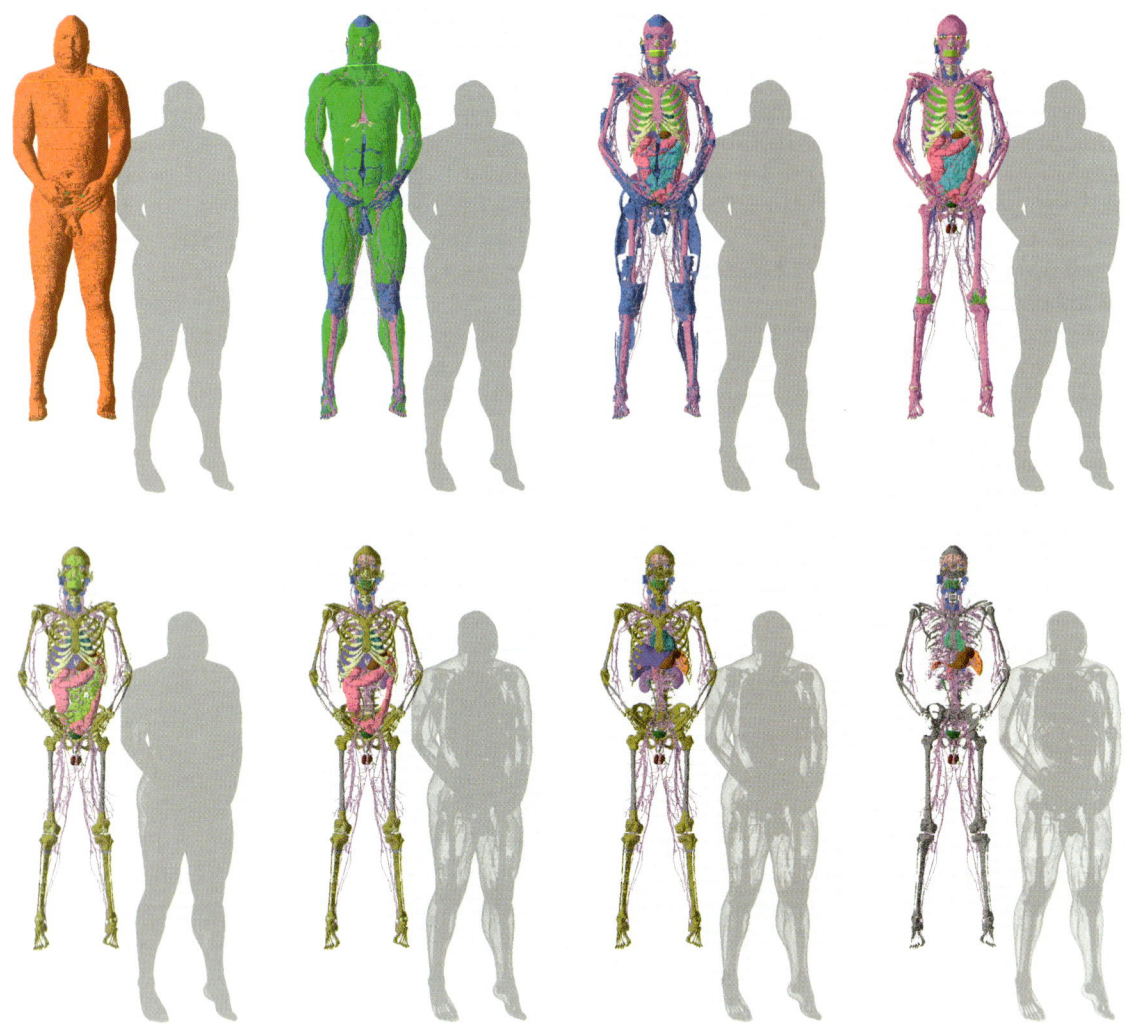

Fig. 2. Views of the Brooks Man anatomical model with progressively more tissues made transparent to show the internal structures in the model.

Thermoregulation[2]

Thermoregulation is the term that describes the maintenance of the body temperature within a prescribed range under conditions in which the thermal load on the body may vary. In humans, thermal loads come from alterations in ambient conditions (temperature, ambient vapor pressure, air velocity, clothing, and other environmental variables that may alter the temperature of the skin) and from changes in heat production within the body. The deposition of thermalizing energy deep in the body by exposure to RF-EMFs provides a unique exception to the energy flows normally encountered by humans, although metabolic

---

[2]From IEEE C95.1 2005, IEEE Standard for Safety Levels with Respect to Human Exposure to Radio Frequency electromagnetic Fields, 3 kHz to 300 GHz. Copyright 2006, by IEEE. All rights reseved.

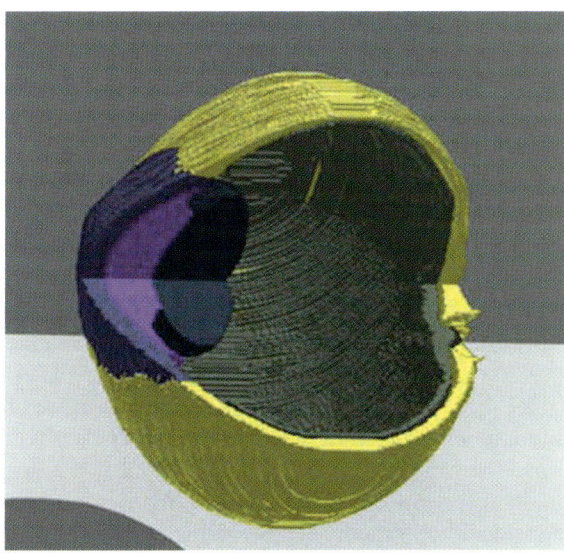

Fig. 3. Color-coded high-resolution baboon eye model. The final model will be placed in an enhanced-resolution head model from the Brooks Man.

activity in the muscles during exercise can also deposit large amounts of thermal energy directly into deep tissues.

The pattern of thermoregulation in humans and many animals is called endothermy, in which the body temperature depends on a high and regulated metabolic heat production and the dissipation of that heat to the environment is closely regulated. Most of the vital internal organs of endotherms function most efficiently when they are held at a relatively constant temperature that is characteristic of the species. For humans, this characteristic temperature is $\sim 37 \pm 0.5°C$ ($98.6 \pm 1.0°F$). While the temperature of individual body tissues may vary slightly, significant departures are associated with vigorous exercise, disease states, or possibly lethal conditions. The usual range of body temperatures extends from 35.5 to 40°C in humans and includes circadian variation, vigorous exercise, variations in ambient conditions (temperature, humidity, air velocity), sequelae of food intake, age factors, menstrual variation in women, emotional factors, and assorted effects of drugs and alcohol.

Body tissues are extremely vulnerable to excessive changes in temperature, particularly to overheating. Humans have an elaborate system of mechanisms for regulating internal body temperature. In endotherms, two distinct control systems are available for thermoregulation. (1) Behavioral thermoregulation involves conscious, voluntary acts that adjust the characteristics of the air–skin interface. (2) Autonomic (or physiological) thermoregulation involves the involuntary responses of the body that generate and dissipate body heat. In humans, autonomic thermoregulation, emphasized in this section, provides for the fine control of body temperature in the resting state and is the principal control during exercise, environmental heating, and RF-EMF exposure.

Thermoregulatory overload from deep heating[3]

Hyperthermia refers to the general condition where body temperatures are above normal. An elevated core temperature increases metabolism and certain other functions, such as heart rate, respiration rate, and nerve conduction velocity. Central nervous system function deteriorates at temperatures above 42° to 43°C and convulsions may occur. At this level, protein denaturation may begin and cells may be damaged. This is particularly dangerous for the brain, since lost neurons are not replaced. Thermoregulatory responses of sweating and vasodilation cease at approximately 43°C, after which body temperatures may rise very rapidly if external cooling is not imposed. Other events that occur at this level include confusion or unconsciousness, elevated enzyme levels, and damage to heart and kidneys. The conditions just described are those of heat stroke, a true hazard to human beings. The hallmark of this hazard is the increase in body temperatures above 42° to 43°C.

Three main factors have been identified that lead to the breakdown of heat loss mechanisms. These include (a) dehydration, which perturbs the cutaneous circulation and sweat secretion; (b) poor acclimatization to heat; and (c) poor physical fitness. Other factors that have been identified as potentially contributing to the problem include

---

[3]From IEEE C95.1 2005, IEEE Standard for Safety Levels with Respect to Human Exposure to Radio Frequency electromagnetic Fields, 3 kHz to 300 GHz. Copyright 2006, by IEEE. All rights reseved.

alcoholism, chronic illness, fatigue, lack of sleep, obesity, and restrictive clothing.

Heat stroke occurs primarily in workers or athletes exposed for long periods to very warm, humid environments. It is not clear that prolonged human exposure to RF-EMFs can produce comparable symptoms. Deeply penetrating RF-EMF energy at high field strengths is seldom encountered, except by accident. Surface RF-EMF energy deposition at high field strengths will be perceived immediately and escape maneuvers will result. These possibilities are further elaborated below.

Effects on essential physiological functions and survival: heat stroke

Essential physiological functions and survival depend critically on minimizing the overheating of the human body. Heat stroke is a set of symptoms that are produced by excessive body heating. The impact of such heat is determined not only by the absolute level of body temperature but also by the duration of heat exposure. While heat stroke is a medical emergency, it is not always recognized by physicians because it presents in different ways. Further, standard clinical thermometers do not record body temperatures above 41.5°C. Nevertheless, this condition requires rapid physiological treatment, the success of which depends upon an understanding of the physiology of heat stroke.

Disturbances of the CNS are always present in heat stroke, and the level of consciousness is often depressed, for example, coma, sleep, or delirium. Edema in the brain tissue and meninges with a flattening of the brain convolutions can be found during a pathological examination (see Chapters 10 and 13 in this volume). These facts infer that the temperature of the CNS tissue is critical to the occurrence of heat stroke and that the defense of the brain temperature seems to be of paramount importance. Cabanac (1983) estimated that the highest tolerated brain temperature is approximately 40.5°C, but it is also true that the actual brain temperature runs lower than that of the body core under normal conditions. Whether this is also the case during heat stroke is unknown, especially under prolonged steady-state conditions when thermoregulation fails. Those patients who have survived heat stroke with core temperatures of 45° to 47°C have had associated neurological complications or permanent deficits. Selective brain cooling has been demonstrated in several animal species (gazelle, goat, sheep, and dog) by counter current cooling of arterial blood as it passes through the carotid rete in the cavernous sinuses. Humans do not possess a carotid rete, however, and there is no comparable mechanism for significant brain cooling, despite contentions by Cabanac (1983) that such a mechanism exists.

RF-EMF energy deposited on the skin surface of humans (e.g., at millimeter wave frequencies) will raise the irradiated skin temperature but will minimally affect the core temperature. In environments considered thermally comfortable (e.g., 20° to 30°C), the average human skin temperature is ~33° to 34°C. The threshold of thermal pain is 44° to 45°C (Hardy et al., 1952). Data of Blick et al. (1997) and Riu et al. (1997) provided detection thresholds for warmth sensation (2.45–94 GHz) that involved increases in surface temperature of 0.05° to 0.08°C for 3 or 10-s exposures of small areas of skin. The corresponding range of thresholds was 63.1 mW/cm^2 at 2.45 GHz to 5.34 mW/cm^2 at 94 GHz. Larger areas of skin will presumably have lower perception thresholds (Michaelson, 1972). Human volunteers undergoing partial-body RF exposure can immediately detect spatial peak SARs of 6.6 W/kg, equivalent to peak power densities of 27 mW/cm^2 at 2450 MHz and 18 mW/cm^2 at 450 MHz (Adair et al., 1999a). Much higher peak power densities at 2450 MHz (50 and 70 mW/cm^2) produce either an asymptotic increase in temperature of the irradiated skin of ~2.0°C at the end of a 45-min RF-EMF exposure or a fall in skin temperature due to evaporative cooling (Adair et al., 2001a). RF-EMF exposure at these power densities is described by the subjects as highly directional and very warm and would serve as an appropriate stimulus to escape behavior. As is the case with exposure to strong sunlight, individuals exposed to strong RF-EMFs at frequencies at or above 1 GHz will detect it immediately, determine its directionality, and take measures promptly to get out of the field. For this reason, it is not considered necessary to recommend any guideline based on an increase in skin temperature in the new C95.1 standard. The surface pain threshold at 44° to 45°C is the upper limit, but an irradiated individual will never voluntarily allow this level to be reached if escape from the field is possible. Only if the rate of increase in skin temperature equals or exceeds 1.0°C/s would pain or thermal damage be possible.

Resonance and partial body thermal tissue response: animal studies[4]

The maximal absorption of RF-EMF energy during whole-body exposure occurs when the long axis of the body is parallel to the electric field vector (E-polarization) and the longest dimension of the body is approximately 0.4 of the free space wavelength (resonant frequency) (Durney et al., 1986). RF-EMF exposure of non-human primates at resonance produces somewhat less efficient thermoregulation than does exposure to sub-resonant or supra-resonant frequencies (Krupp, 1983; Lotz, 1985; Lotz and Saxton, 1988; Adair et al., 1992). Although the threshold for a reduction in metabolic heat production (M) may be lower at resonance, the magnitude of the response change may be less for a given SAR than at non-resonance and the colonic temperature (T_{co}) may rise. However, the hyperthermia is modest and well regulated even at SARs equivalent (in W/kg) to the level of resting M at thermoneutrality. The situation is similar to that occurring in humans during exercise (Adair, 1996). Some have expressed concern that human exposure at resonance may pose a greater hazard than exposure at other frequencies. The non-human primate studies are reassuring because, even though thermosensors in the skin (necessary for thermal perception and avoidance behavior) may be inefficiently stimulated, there is solid evidence that autonomic mechanisms are rapidly mobilized to dissipate heat generated deep in the body. Experiments recently completed, in which seated human adults undergo 45-min RF-EMF whole-body exposures at resonance (100 MHz), demonstrate this prediction exactly; no increase in core temperature occurs even at a power density that is eight times the IEEE C95.1 (1982) standard at 100 MHz (Adair et al., 2002).

If only part of the body is exposed to RF-EMF energy, the magnitude of the change in M reflects the total absorbed energy, as though it were integrated over the whole body (Adair, 1988). If an endotherm is exposed to RF-EMF energy at SARs greater than those that reduce M to the resting level, thermoregulation will be accomplished by mobilization of the next response in the thermoregulatory hierarchy, i.e., changes in blood flow (vasomotor state or conductance) (Adair, 1985; Candas et al., 1985; Lotz and Saxton, 1987). In humans, the threshold for initiation of a heat loss response, either vasodilation or sweating, will be a complex function of the core and skin temperatures (Adair and Black, 2003).

Organs with greater thermal sensitivity: animal studies

Unfortunately, for many organs, there is very little information about thresholds for thermal damage. There are physiological responses that occur in response to heat stress that may or may not be related to thermal damage, *per se*. For example, blood flow in the skin of both rodents and humans increases at temperatures near 40°C. This increase in blood flow may not reflect damage; it is a normal physiological response to increased temperature. A summary of data available on tissue thresholds for damage is provided in Table 3 in Dewhirst et al. (2003). It is important to note that these thresholds are affected by the type of endpoint used. The endpoints are indicated for each tissue listed. A preponderance of the data are from rodent studies, so they may not be directly applicable to humans. However, it is likely that the relative ranking of organ thermal sensitivities is similar between species. In other words, those organs most thermally sensitive in rodents are likely to be the most thermally sensitive in humans.

Many studies on humans and animals confirm that, even when no changes can be measured in the deep or peripheral temperatures of the body,

---

[4] From IEEE C95.1 2005, IEEE Standard for Safety Levels with Respect to Human Exposure to Radio Frequency electromagnetic Fields, 3 kHz to 300 GHz. Copyright 2006, by IEEE. All rights reseved.

sensitive thermoregulatory mechanisms are mobilized to dissipate heat generated in body tissues. This is especially true when the body absorbs RF-EMF energy. Well documented are the locus of the thermosensors in the mammalian body, their operating characteristics and neural network, and how they may be selectively stimulated by RF-EMF energy. In this analysis, RF-EMF frequency plays a prominent role; the lower the frequency, the deeper the energy penetrates. When seated human volunteers undergo whole-body RF-EMF exposure at resonance, for example, 100 MHz, no change is measured in core (esophageal) temperature despite maximal energy penetration (Adair et al., 2003). Minimal temperature change is recorded in the skin, locus of peripheral thermosensors. Nevertheless, vigorous heat loss responses of sweating and increased blood flow are rapidly mobilized, even though subjects have no sensation of warmth. FDTD modeling of a seated adult human exposed dorsally at 100 or 220 MHz shows high rates of energy absorption in deep neural tissues that harbor thermosensitive cells (Allen et al., 2005). Anatomical loci have been identified as several brainstem nuclei (preoptic and anterior hypothalamic areas) and the spinal cord, all of which are bathed by cerebrospinal fluid. A model of the neural elements involved in thermoregulation can explain the generation of autonomic heat loss responses (Adair, 2000). The evidence is strong that minute local temperature increases ($\Delta T = 0.03°C/s$) trigger the central thermosensitive cells to initiate heat loss.

Thermographic studies on tissue-equivalent models of humans and animals (Guy, 1971; Guy et al., 1974) indicate regions of high local SAR during exposure of the whole body to planewave RF-EMFs. Wrists, ankles, and neck (also the base of animal tails) are predicted to be foci of enhanced local SAR, where excessive elevations of tissue temperature may occur. Krupp (1983) studied anesthetized rhesus monkeys, equilibrated to $T_a = 23°C$ and exposed for 1–4 h to planewave 210 MHz RF-EMF energy (power density (PD) = 5–27 mW/cm^2). Substantial increases in T_{co} occurred but no evidence for localized regions of greatly elevated temperature (i.e., tissue "hot spots") in wrist, ankle, thigh, or biceps. Similar experiments at 2.06 GHz (Krupp, 1981) found different results. At PD = 15 mW/cm^2, 1 h of RF-EMF exposure produced no increase in T_{co} or local temperatures of neck and groin. Wrist and ankle temperatures rose slowly over the 1-h exposure but never reached the level of T_{co}. Krupp concluded that increased convective heat transfer by the blood, coupled with lower set point temperatures in the limbs, could protect individual tissues from overheating.

Several factors can influence the thermal sensitivity of tissues. Some of these could occur from occupational or accidental exposure to RF-EMFs. The critical factors include thermal tolerance, pH and pressure effects, and a phenomenon referred to as "step down" heating. The effects of these factors on thermal sensitivity are well characterized and can be described quantitatively based on Arrhenius analysis (Dewhirst et al., 2003).

A few assorted temperatures may be mentioned, for what they are worth. Core temperatures of 40° to 42°C have historically been considered dangerous because of their association with tissue damage and mortality. But this may be just the middle of a larger range. The intestines contain a large quantity of highly toxic lipopolysaccharide (LPS, endotoxin) sloughed from the walls of Gram-negative bacteria. Studies have shown that hyperthermia to 42° to 43°C leads to damage of the gut wall and entry of LPS into the circulation (see Fig. 4; Hales et al., 1996). At about this same level, there is a sharp increase in heart rate and decrease in mean arterial pressure in monkeys (Gathiram et al., 1988), but also evidence for leakage of LPS at temperatures as low as 40°C. On the other hand, it has been shown that heating only the abdominal viscera of anesthetized sheep to 45°C for 30–60 min can be fatal (Rawson and Quick, 1970). It is reasonable to expect that an organ as critical as the brain should be either less sensitive to temperature changes or protected from the high temperatures occurring in the rest of the body. The critical brain temperature range measured by Britt et al. (1984) was 42° to 43°C and included values at which changes in brainstem auditory evoked potentials changed abruptly. On the other hand, critical changes in brain cells *in vitro* occur at lower levels of 40° to

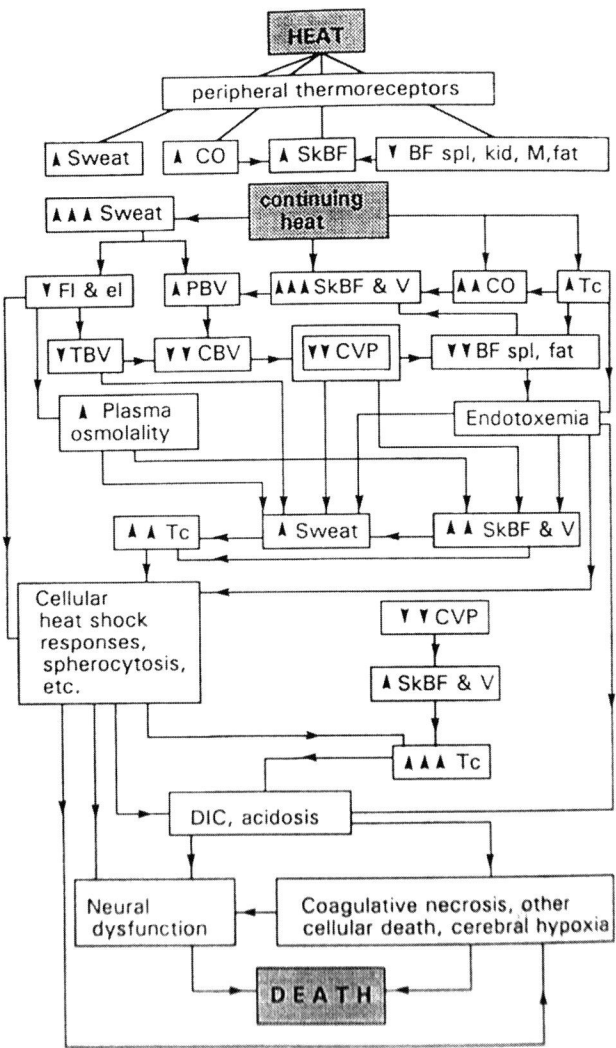

Fig. 4. A scheme of the interacting sequence of events occurring from the beginning of exposure to a hot environment to death from heat stroke, with known or proposed interrelationships. Arrows indicate increased (↑) or decreased (↓) parameter, but note, for example, in the eighth line ↑↑ SkSF & V shows that SkBF and volume are well above control levels but are reduced compared with ↑↑↑ in the fifth line. "Cellular" box includes, for example, 80–90% protein inhibition, synthesis of HSPs, and increased (150–200%) membrane ion flux rate. BF, blood flow; CO, cardiac output; CBV, central blood volume; CVP, central venous pressure; DIC, disseminated intravascular coagulation; el, electrolytes; Fl, fluids; kid, kidney; M, muscle; PBV, peripheral blood volume; Sk, skin; spl, splanchnic; TBV, total blood volume; T_c, core temperature; V, volume. (Adapted from Hales et al. (1996) with permission.)

41°C (Shen and Schwartzkroin, 1988). It does not make sense that this range should indicate that the brain is sensitive to a significantly lower level of temperature than the 42° to 45°C often quoted for other tissues. It is generally agreed that the human brain, and that of some other species, does have limited protection from general body hyperthermia but only to the extent that its temperature will often exceed 42°C in heat stroke victims (Hales et al., 1996).

Age factors: human exposures

Both ends of the age spectrum exhibit poorer heat tolerance than healthy young adults do (Makrides et al., 1987). Children have a higher ratio of surface area to body mass than adults. While this means they can more rapidly absorb heat from the environment, it should also mean that they can lose body heat more rapidly. However, in very young children this is not the case because maximal sweating capacity normally takes some years to develop. In addition, children are said to have a relatively low capacity for circulatory transport of heat from core to skin because cardiovascular reserves are limited in the growing body. Added disadvantages for young children are a higher metabolic rate per unit body mass and a slower rate of acclimation (Webster, 1987).

Old age, depending on whose paper you read, starts between 45 and 65 years of age. Clear reductions in thermoregulatory efficiency have been most widely documented for people above that age range. However, it appears that neither sensitivity (onset) nor capacity for sweating is diminished in old age. Therefore, the overall reduction in thermoregulatory efficiency must be related to a compromised cardiovascular performance, most probably conforming with the well-established decline in the capacity for oxygen uptake as one ages. However, Yousef's extensive studies (1987) showed no significant differences in heart rate or cardiac output responses of elderly and young people subjected to heat stress. On the other hand, forearm blood flow increased much less in the elderly, which may be attributed to factors such as reduced capacity to dilate skin blood vessels or to redistribute cardiac output. In younger people, the increase in cardiac output is directed pretty much to the skin, whereas it seems not to be serving that purpose in the elderly. Thus, as skin blood flow (SkBF) is already low, the fluids and electrolyte necessary to maintain sweating during severe heat stress may not be available to certain elderly persons.

Thermoregulatory responses of humans exposed to RF-EMF energy

The ultimate goal of research involving whole-body RF-EMF exposure of intact organisms is the prediction of such exposure on human beings. Most of the published research on physiological thermoregulation has been conducted on laboratory animals, with a heavy emphasis on laboratory rodents. Because they lack a physiological heat loss mechanism, small furred animals are very poor models for human beings. Basic information about the thermoregulatory capabilities of animal models relative to human capability is essential for the appropriate evaluation and extrapolation of animal data to humans. In general, reliance on data collected on humans and non-human primates, however fragmentary, yields a more accurate understanding of how RF-EMFs interact with humans. Such data are featured in a paper published in the BEMS Journal, Supplement 6 (Adair and Black, 2003) and include data from both clinic and laboratory. Featured topics include thermal sensation, human RF-EMF overexposures, exposures attending MRI, predictions based on simulation models, and laboratory studies of human volunteers.

Studies of human volunteers exposed to RF-EMF energy

More than 20 years ago, Tell and Harlen (1979) stated that no laboratory investigations of human beings exposed to RF-EMF fields had yet been conducted. Progress is finally being made on this front. Adair et al. (1998) reported the first study in a series in which human volunteers were exposed in the far field to RF-EMF at controlled power densities in highly controlled thermal environments. Thermoregulatory responses of heat production and heat loss were measured in seven adult volunteers (males and females) during 45-min dorsal exposures of the head and trunk to 450 MHz continuous-wave (CW) RF fields. Two peak power densities (PD = 18 and 24 mW/cm^2) were tested at each of three ambient temperatures ($T_a = 24$, 28, and 31°C) plus T_a controls (no RF-EMF). A standardized protocol (30-min baseline, 45-min RF-EMF or sham exposure, 10-min baseline) was always used. The normalized peak surface SAR, measured at the location of the subject's center back, was SAR = 5.8 and 7.7 W/kg. No change in metabolic heat production (M) occurred under any exposure conditions at either frequency. The magnitude of increase in those skin temperatures under direct irradiation was directly related to frequency, but local sweating rates on back and chest were related more to T_a and SAR. The efficient sweating contributed to the regulation of the deep body (esophageal) temperature (T_{esoph}) to within $\pm 0.1°C$ of the baseline level. At this frequency, normalized peak SARs, in excess of IEEE C95.1 (1991) guidelines, were easily counteracted by normal thermophysiological mechanisms. Even under ambient conditions that were often judged uncomfortable and very warm, T_{esoph} was regulated with precision because appropriate autonomic heat loss responses, principally sweating, were mobilized.

A second study (Adair et al., 1999b) compared the results described above with those collected on a second group of volunteers, exposed to 2450 MHz CW RF fields. The basic protocol was identical, as were the three T_a and response measures, with the addition of local SkBF at three sites on the body. The normalized peak surface SAR was the same for comparable PD at both frequencies. Again, no change in M occurred under any exposure condition. The magnitude of increase in the skin temperatures (T_{sk}) under direct irradiation was much greater at 2450 MHz than at 450 MHz, but local sweating rates were related more to T_a and SAR. Efficient sweating and increased local SkBF regulated T_{esoph} to within $\pm 0.1°C$ of the baseline level.

During whole-body exposure of squirrel monkeys at 2450 MHz CW and pulsed-wave (PW) fields, heat production and heat loss responses were nearly identical (Adair et al., 1993). To explore this question in humans, we exposed two different groups of human volunteers at 2450 MHz CW (two females, five males) and PW (65 μs pulse width, 10 kHz pulses/s (pps); three females, three males) RF-EMFs. We measured thermophysiological responses of heat production and heat loss (esophageal and six skin temperatures, M, local SkBF, and local sweat rate). The basic protocol was identical, as were the three T_a and response measures. At each T_a, the average PD studied were 0, 27, and 35 mW/cm². Mean data for each group showed minimal changes in core temperature and M for all test conditions and no reliable differences between CW and PW exposure. Local skin temperatures showed similar trends for CW and PW exposure that were PD dependent; only the skin temperature of the upper back (facing the antenna) showed a reliably greater increase ($p = 0.05$) during PW exposure than during CW exposure. Local sweat rate and SkBF were both T_a and PD dependent and showed greater variability than other measures between CW and PW exposure; this variability was attributable primarily to the composition of the two subject groups. With the one noted exception, no clear evidence for a differential response to CW and PW fields was found.

At 2450 MHz, two additional peak PD were tested (50 and 70 mW/cm²) (Adair et al., 2001b). The higher of these PD (normalized peak local SAR = 15.4 W/kg) is well outside the IEEE C95.1 (1999) guidelines for partial-body exposure, as is the estimated whole-body SAR ~1.0 W/kg. Seven volunteers (four males, three females) were tested at each PD in three ambient temperatures and under our standard protocol. The thermophysiological data collected were combined with comparable data at PD = 0, 27, and 35 mW/cm² from the 1999 study to generate response functions across PD. No change in T_{esoph} or M was recorded at any PD in any T_a. At PD = 70 mW/cm², skin temperature on the upper back (irradiated directly) increased 4.0°C in $T_a = 24°C$, 2.6°C in $T_a = 28°C$, and 1.8°C in $T_a = 31°C$. These differences were primarily due to the increase in local sweat rate, which was greatest in $T_a = 31°C$. Also at PD = 70 mW/cm², local SkBF on the back increased 65% over baseline levels in $T_a = 31°C$, but only 40% in $T_a = 24°C$. Although T_a becomes an important variable when RF-EMF exposure exceeds the C95.1 partial-body exposure limits, vigorous heat loss responses of blood flow and sweating maintain thermal homeostasis efficiently. It is also clear that strong sensations of heat and thermal discomfort will motivate a timely retreat from a strong RF-EMF field long before these physiological responses are exhausted.

At 100 MHz, RF-EMF energy penetrates the human body maximally, bypassing skin thermoreceptors. Twelve tests were conducted on each of seven adult volunteers seated inside an anechoic chamber in the far field of a dipole antenna (Adair et al., 2003). The standard protocol was used for three power densities (PD = 4, 6, and 8 mW/cm²) presented in each of the three ambient temperatures. T_{esoph} and six T_{sk}, M, local sweat rate at two sites, and local SkBF at three sites were measured. Theoretical dosimetry indicated high RF-EMF energy absorption in the legs, so ankle skin temperature (T_{ankle}) was also measured. Results showed no change in M and little or no change in local T_{sk}, including those sites on the back that were irradiated directly. T_{ankle} rose up to 4.0°C in some subjects at PD = 8 mW/cm². During RF-EMF exposure, T_{esoph} changed little (range = −0.15° to 0.13°C), a result of increased sweating and SkBF. These efficient physiological responses must have been stimulated by thermoreceptors deep in the body, not by those in the skin.

Since initiating this research program in 1994, we have learned much about how human beings respond in the presence of RF-EMF fields at levels that may heat the body tissues. We noted that there was a significant gap in the range of frequencies we had tested. Therefore, we measured human thermoregulatory responses to RF-EMF fields at a frequency in the critical transition range from deep body heating to more superficial energy deposition at 220 MHz (Adair et al., 2005). Thermoregulatory responses of heat loss and heat production were measured in six adult volunteers seated inside an anechoic chamber in the far field of a dipole antenna. The standard protocol was again used for three power densities (PD = 9, 12, and 15 mW/cm²), whole body average normalized SAR = 0.045 (W/kg)/(mW/cm²), and tested at three T_a plus T_a controls (no RF). Measured responses included T_{esoph}, seven T_{sk}, M, local sweat rate, and local SkBF. Derived measures included heart rate, respiration rate, and total evaporative water loss. FDTD modeling of a seated 70 kg human exposed to 220 MHz predicted six localized 'hot spots' at which local temperatures were also measured. No changes in M occurred under any test condition, while T_{esoph} showed small changes ($\leqslant 0.35$°C) but never exceeded 37.3°C. As with similar exposures at 100 MHz, local T_{sk} changed little and modest increases in SkBF were recorded. At 220 MHz, vigorous sweating occurred at PD = 12 and 15 mW/cm², with sweating levels higher than those observed for equivalent PD at 100 MHz. Predicted 'hot spots' were confirmed by local temperature measurements. The FDTD model showed the local SAR in deep neural tissues that harbor temperature-sensitive neurons (e.g., brainstem, spinal cord) to be greater at 220 than at 100 MHz. Human exposure at both 220 and 100 MHz results in far less skin heating than occurs during exposure at 450 MHz. However, the exposed subjects thermoregulate efficiently because of increased heat loss responses, particularly sweating. It is clear that these responses are controlled by neural signals from thermosensors deep in the brainstem and spinal cord, rather than those in the skin.

Supporting data from animal studies

Supporting data from animal studies include the thermoregulatory profile, response thresholds, physiological responses of heat production and heat loss, intense or prolonged exposure, RF-EMF effects on early development, circadian variation, and additive drug–RF-EMF interactions. Abundant details for these topics may be found in the comprehensive paper by Adair and Black (2003). The overall conclusion from the studies described above is inescapable: that human beings demonstrate far superior thermoregulatory ability over other organisms during RF-EMF exposure at, or even well above current human exposure guidelines.

Behavioral effects of RF-EMF exposure

Behavioral changes in man and laboratory animals that range from perceptions of warmth and sound to high body temperatures during RF-EMF exposure can produce disruption and complete cessation of ongoing behavior. Animals will escape and subsequently avoid RF-EMF exposure under the appropriate conditions, but they can also learn to work for and obtain a brief burst of RF-EMF exposure in a cold environment (Marr et al., 1988). Research over the past two decades has been devoted to gaining an understanding of these behavioral effects and the mechanisms that produce them. Research has also been directed toward investigating the relationship between whole-body SAR and the harmful effects of RF-EMF exposure. Evaluations of the exposure effects on performance of well-learned operant tasks by laboratory animals have been the primary avenue for determining this relationship. At present, safe exposure standards ICNIRP (1998) and IEEE C95.1 (2005) are based on *in vivo* exposures of rodents and non-human primates. Specifically, SAR thresholds to disrupt simple operant behaviors and produce work stoppage during acute whole-body exposure were determined. The threshold for behavioral change has been established by using the heating effects produced by absorption of RF-EMFs. The absorption of RF-EMF is complicated. Several studies have examined the RF-EMF characteristics that control SAR, such as frequency and polarization and the consequent different rates of heating. The factors of animal shape and size also interact to govern SAR (Schrot and Hawkins, 1976). Simple test protocols have been used to first establish a stable behavioral baseline of performance, and then to determine the effects of RF-EMF on the baseline performance. Generally, the result of microwave exposure has been a body temperature rise followed by either an increase or a reduction in behavioral response rates.

Detection

People have reported hearing pulsed RF-EMFs. The first reports of this phenomenon, which drew skepticism, came from workers standing next to a radar antenna who claimed they could hear the radar pulses (Airborne Instruments Laboratory, 1956). Frey (1961, 1962) conducted the first scientific studies and described the effect as "clicking," "knocking," or "buzzing" located behind the head and required pulsed energy for its occurrence. The initial conclusion was that this effect represented direct neural stimulation by RF-EMF energy. More recent evaluations have shown the effect to be a thermally mediated mechanism. It is thought to be a heat-induced tissue expansion created by the pulsed RF-EMFs (Foster and Finch, 1974). This thermoelastic expansion, initiated by a RF-EMF pulse, propagates as an acoustic wave through the brain. The acoustic wave is thought to be detected in the cochlea of the inner ear by producing minute movement of the natural hair cells, which initiates the auditory perception. Other early reports have verified the hearing sensation (Sharp et al., 1974) and measured cochlear microphonics produced by pulsed RF-EMFs (Chou et al., 1977). Lin (1978, 1990) reviewed the evidence for this effect and concluded that there is "little likelihood that RF-EMF auditory phenomena arise from direct interaction of RF-EMF pulses with the cochlear nerve or neurons at higher structures" (Lin, 1990, p. 313). Elder and Chou (2003) provide a more recent detailed review of the auditory hearing effect.

Some early studies were designed to evaluate the hearing effect in animals to determine whether the auditory sensation could be used as a cue to control behavior. For example, Johnson et al. (1976) trained rats to nosepoke for food reward when a 7.5-kHz acoustic click presented at 10 Hz was presented. The rats learned to respond during the acoustic stimulus and stop responding when the acoustic stimulus was terminated and food reward was no longer given. The investigators then substituted a pulsed 918-MHz (10 pps, 10 s duration) RF-EMF signal during the absence of the acoustic stimulus. During the RF-EMF signal the rats began responding immediately and continued to perform correctly with the RF-EMF stimulus. They showed that the RF-EMF pulses were nearly as effective as the auditory click cues in controlling rat behavior. A different study by Hjeresen et al. (1979) found that pulsed RF-EMFs were detected by rats and could be used to control their behavior. These studies supported the human perception studies of pulsed RF-EMFs. Animals have also been shown to detect very low levels of sinusoidally modulated CW RF-EMFs at 2450 MHz. In a classic study, King et al. (1971) determined the threshold for detection of sinusoidally modulated CW RF-EMFs to be at a whole-body SAR of 0.6 W/kg.

Adair and Black (2003) recently summarized much of the archival research on detection of RF-EMF by human subjects. From these studies they noted that when an area of forehead (37 cm^2) was irradiated for 4 s, the mean absolute threshold of warmth was 33.5 mW/cm^2 at 3 GHz, 12.6 mW/cm^2 at 10 GHz, and 4.2 mW/cm^2 at far IR frequencies. They noted further that small areas of irradiation on the skin and short irradiation durations required more energy to be detectable (temporal summation).

Table 1. Detection thresholds for human subjects (Blick et al., 1997)

Frequency (GHz)	Threshold (mW/cm^2) (mean $\pm$ S.E.M.)	Tissue penetration ($1/e^2$ depth, mm)	Wavelength (mm)
2.45	63.1 $\pm$ 6.7	32	122
7.5	19.5 $\pm$ 2.9	6.3	40
10	9.6 $\pm$ 2.9	3.9	30
35	8.8 $\pm$ 1.3	0.8	8.6
94	4.5 $\pm$ 0.6	0.4	3.2
(IR) $1-3 \times 10^5$	5.34 $\pm$ 1.1	<0.1	~0.002

A recent study by Blick et al. (1997) measured detection thresholds of several frequencies on the skin in the middle of the back of human subjects. They used long duration (10-s), large area (327-cm^2) stimuli to minimize temporal or spatial summation. They tested frequencies of 2.45, 7.5, 10, 35, and 94 GHz plus IR. They found that the thresholds dropped more than an order of magnitude from 2.45 to 94 GHz. As is shown in Table 1, the 2450 MHz threshold (63.1 mW/cm^2) and the IR detection threshold (5.3 mW/cm^2) were roughly 2.5 times the thresholds found by Justesen et al. (1982) of 27 and 1.7 mW/cm^2 at 2450 MHz.

Aversion

Early research investigating the possible aversive properties of RF-EMF had shown that laboratory rats would avoid pulsed RF-EMFs at low average power densities but not CW RF-EMFs at similar average power densities (Frey and Feld, 1975). Other studies found pulsed RF-EMF more disrupting of behavioral performance than CW RF-EMF (Thomas et al., 1982) while other studies did not find this difference (Lebovitz, 1983). At the present time, the relative aversive properties of pulsed versus CW RF-EMFs remains unclear. Stern (1980) suggested that the RF-EMF induced the auditory effect, which may be an aversive stimulus for the rat, with PW RF-EMFs that complicate simple interpretations of the aversion data.

The dosimetry studies performed on the laboratory rat suggested that electrical and thermal hot spots existed in different regions of the body (Lin et al., 1977; D'Andrea et al., 1985, 1987). The electric fields induced in the rat body during RF-EMF exposure are non-uniform and result in electrical field concentrations called electrical hotspots (Gandhi et al., 1979). Recent FDTD modeling studies predict hotspots in the human body with the ratio of peak SAR to whole-body average SAR at 100:1 (Bernardi et al., 2003). As described above, D'Andrea et al. (1985) found thermal hot spots in the tail and rectum of euthanatized rats during exposure to 360 and 2450 MHz fields. These were not seen during exposures at frequencies close to whole-body resonance, 700 MHz, for medium-sized rats. The most intense hot spots were found to exceed the whole body average SAR by 50 and 18 times for exposures at 360 and 2450 MHz, respectively. D'Andrea et al. (1988) theorized that these hot spots were aversive and would influence the behavior of live rats. Blood flow and convective heat transfer through areas of thermal hotspots in tissues should greatly reduce temperature elevations of localized thermal hot spots. D'Andrea et al. (1988) designed a novel behavioral paradigm to evaluate the aversive properties of the hotspots. The rats were tested in a long Plexiglas runway (see upper panel of Fig. 5), which generally kept them parallel to the electric field of the applied RF-EMFs. Without RF-EMF exposure each rat over several test sessions selected one (preferred) end of the runway in which to rest. On subsequent tests, an RF exposure was given when the rat selected the preferred end of the runway. If the rat moved to the non-preferred end of the runway the exposure was stopped. Three RF frequencies were tested, two (360 and 2450 MHz) that produced localized hot spots and one (700 MHz) that did not. The field power density for each frequency was adjusted to yield equivalent whole body SARs of 1, 2, 6, and 10 W/kg. The results showed (see lower panel of Fig. 5) that the rats vacated the preferred (irradiated) end of

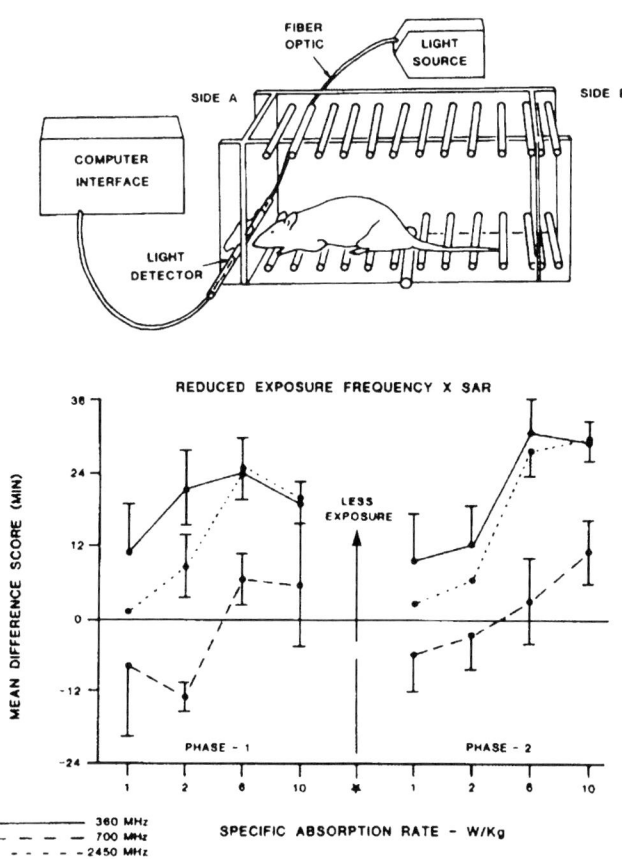

Fig. 5. (Upper panel) Schematic representation of the Plexiglas rat cage illustrating the fiber optic light system to determine the occupied side of the cage by detecting vertical tilt. (Lower panel) Results of rat exposures at 360, 700, and 2450 MHz. Increased difference score indicates aversion to RF-EMFs. During Phase 1, exposures occurred when the rat occupied the preferred end of the runway. During Phase 2, exposures occurred when the rat occupied the opposite end of the runway. (Adapted with permission from D'Andrea et al. (1988).)

the runway much sooner during 360 and 2450 MHz exposures than they did during 700 MHz exposures. The authors interpreted this to indicate an aversion to the frequencies that generated hot spots. Unfortunately, it was not possible to measure the local tissue temperature in the location of presumed hot spots during this experiment. A second experiment demonstrated for longer exposures that the whole body SAR threshold for aversion of 2450 MHz was 2.1–2.8 W/kg.

The detection of RF-EMF depends on the frequency. As described above (Blick et al., 1997) there is a several order of magnitude difference between the detection of 2.45 GHz fields and of 94 GHz fields for human subjects. This is presumably so because the 94 GHz energy is deposited in the first one-third millimeter of the skin while 2.45 GHz RF-EMF penetrates 2–3 cm. Most of the thermal sensors and pain receptors are found near the skin surface, a design that is optimal for detecting solar and terrestrial sources of infrared radiation. This too may explain the relative aversiveness of RF-EMF. Surface detected RF-EMF energy may be more likely to support aversion because it is readily detected. The inverse of this concept has been described by Justesen (1988) who has argued that RF-EMF fields at 918 MHz and 2450 MHz are inadequate to serve either as a negative reinforcement or a discriminative stimulus for operant behavior. He found that mice and rats failed to learn an escape response by moving to an area marked on the floor of the closed-space cavity in which they were irradiated by 918 MHz or 2450 MHz RF-EMF, respectively (Carroll et al., 1980; Levinson et al., 1982, 1985; Justesen, 1983; Justesen et al., 1985). These studies showed that naive animals fail to learn an escape response, even at lethal field strengths. All of these studies were conducted in multimode cavities that featured a "safe" area marked on the floor. Thermal receptors are known to exist in the spinal cord and hypothalamic area of the brainstem but from the experiments described, these did not support establishment of escape or avoidance behaviors with the "safe area" in the cavity. Presumably the RF-EMF fields penetrate deeply and the heating is not readily detected by skin receptors in the rat and mouse. In these experiments cessation of foot shock was always a potent reinforcer, while cessation of either a bright light or intense RF-EMFs alone was not. Justesen has pointed out that there may be specific situations in industry or the military where accidental exposure could be especially thermally hazardous to unsuspecting workers because of deeply penetrating RF-EMF frequencies that may not be readily detected by heating in the skin.

Performance disruption

As can be seen from the descriptions above, RF-EMF is detectable and can in most cases, presumably by thermal mechanisms, support aversion to exposure. In this section are descriptions of RF-EMF effects on behavior that are most clearly associated with thermal mechanisms and result in disruption or complete cessation (work stoppage) of ongoing behavior. Characteristics of several studies are shown in Table 2. The characteristics of the RF-EMF exposure (frequency, orientation, etc.) and the SAR are very important and were explained in the dosimetry section above. Performance disruption can be operationally defined as a significant change in behavior from a well-defined baseline (D'Andrea, 1999; D'Andrea et al., 2003a). Work stoppage is simply the point at which the animal ceases emitting the trained behavior for a predetermined time period.

For most studies of work stoppage a simple test protocol has been followed: establish a stable behavioral performance baseline and then evaluate the effects of RF-EMFs on the baseline performance. The typical outcome of RF-EMF exposure has been a mild body temperature rise followed by a reduction in behavioral responding. What happens when the animal slows or stops responding? Stern (1980) pointed out that the reduction of responding on a learned task during exposure may reflect the animals' attempts to engage in other thermoregulatory behaviors (i.e., escape), which are incompatible with behaviors such as lever pressing are required by the performance task (D'Andrea, 1991).

Behavioral changes have been used to evaluate different physical aspects of the exposure. For example, D'Andrea et al. (1977) evaluated the effect of RF-EMF frequency on a stable variable interval (VI) baseline of lever-pressing for food reinforcement. They studied four different frequencies (400, 500, 600, and 700 MHz all at 20 mW/cm^2). The whole-body resonance frequency was predicted then by Gandhi (1974) and Durney et al. (1978) to be 600 MHz for a prolate spheroidal model of this size rat. An example of resonance for this rat size is shown in Fig. 1. Rats were exposed to each frequency on different days, and the primary dependent variable was the exposure duration necessary to meet a criterion of 33% reduction below the baseline response rate. Rats quickly stopped lever pressing when exposed to 20 mW/cm^2 at 600 MHz. However, significantly longer exposure durations were needed to achieve work stoppage at lower or higher frequencies (see Fig. 6). Rats exposed to 600 MHz exhibited the greatest rate of temperature rise and the shortest exposure duration to work stoppage. The other frequencies required longer exposure times and showed lower temperature elevations. These results confirmed previous analytical and empirical predictions (Gandhi, 1974; Durney et al., 1978) that whole-body resonance was an important factor in determining the rate of heating, and subsequently duration of exposure necessary to disrupt behavior.

Table 2. Behavioral performance disruption

	Behavioral effect	CW effect	Pulsed effect	Exposure data	References
Albino rat	DRL response timing disrupted	No, 0.2–3.6 W/kg	Yes, 2.5 and 3.6 W/kg	2.8-GHz; 2 s pulse width at 500 pps; 1–15 mW/cm^2	Thomas et al. (1982)
Long-Evans rat	Multicomponent (fixed ratio timeout (TO)) task	Yes, 5.8 W/kg	Yes, 6.7 W/kg	1.3-GHz; 1 ms pulse width at 600 pps	Lebovitz (1983)
Rhesus monkeys	Disruption of lever pressing	Yes, see Tables 3 and 4	Yes, see Tables 3 and 4	225 MHz; 1.3, 2.45, 5.7 GHz	de Lorge (1983)
Squirrel monkeys	Disruption of lever pressing			2.45, 5.7 GHz	
Rats	Disruption of lever pressing			1.3, 2.45, 5.7 GHz	
Rats	Disruption of lever pressing	NA	Yes, 6.0 W/kg	1.28 GHz; 5.5–15 mW/cm^2 5.60 GHz; 7.5–48.5 mW/cm^2	de Lorge and Ezell (1980)
Rats	Disruption of lever pressing	Yes, 9.5 W/kg	NA	600 MHz	D'Andrea et al. (1977)
Rhesus monkeys	Disruption of lever pressing	NA	Yes, 4 and 6 W/kg	5.6 GHz; 100 pps; 2.8 ms; 50 ns PD	D'Andrea et al. (1994)

Another example of performance disruption was a study by de Lorge and Ezell (1980) who trained rats to lever press for food reinforcement each day during 40 min test sessions. The rats were exposed to RF-EMF at 5.6 GHz (7.5–48.5 mW/cm^2) and 1.28 GHz (5.5–15.0 mW/cm^2), both of which are thermalizing levels of exposure. At 5.6 GHz exposure, behavior was disrupted by power densities of 26–38.5 mW/cm^2 while at 1.28 GHz significant behavioral changes were measured at 15 mW/cm^2 in all rats, with some changes noted as low as 10 mW/cm^2. At 1.28 GHz, which is twice that of the resonant frequency of 600 MHz, much less power density was needed to achieve significant changes in behavior. That is because the coupling of RF-EMF energy to the rat body is better at 600 MHz than 1.28 GHz, thus less field strength is needed to produce heating.

De Lorge and his colleagues conducted a number of other studies designed to investigate performance disruption in rats, squirrel monkeys (*Saimiri sciureus*), and rhesus monkeys (*Macaca mulatta*) (de Lorge, 1976, 1979, 1983, 1984; de Lorge and Ezell, 1980). These were summarized by de Lorge (1983, 1984) in two articles. Table 3 (adapted from de Lorge, 1983, 1984) gives the basic characteristics of the experiments listed above. The three species tested are listed at the left side of the table, and the RF-EMF frequencies evaluated, 225 MHz, 1.3 GHz, and 5.8 GHz, are shown across the top. Both PW and CW RF-EMF fields were used. The operant tasks are listed in each cell and were either an observing response with two levers (B_1); a fixed interval schedule with one lever (B_2); or a repeated acquisition task with three levers (B_3). Colonic temperature was measured continuously during the test sessions (T) or estimated (t) from other measurements. The orientation of the subject's long body axis to the electric-field vector is shown as X, Y, and Z. Also indicated is whether the subject was facing the antenna (frontal), was below the antenna (dorsal), or had one side toward the antenna (lateral). Note that squirrel monkeys were not exposed to 1.3 GHz and that all exposures were not oriented to the same polarization.

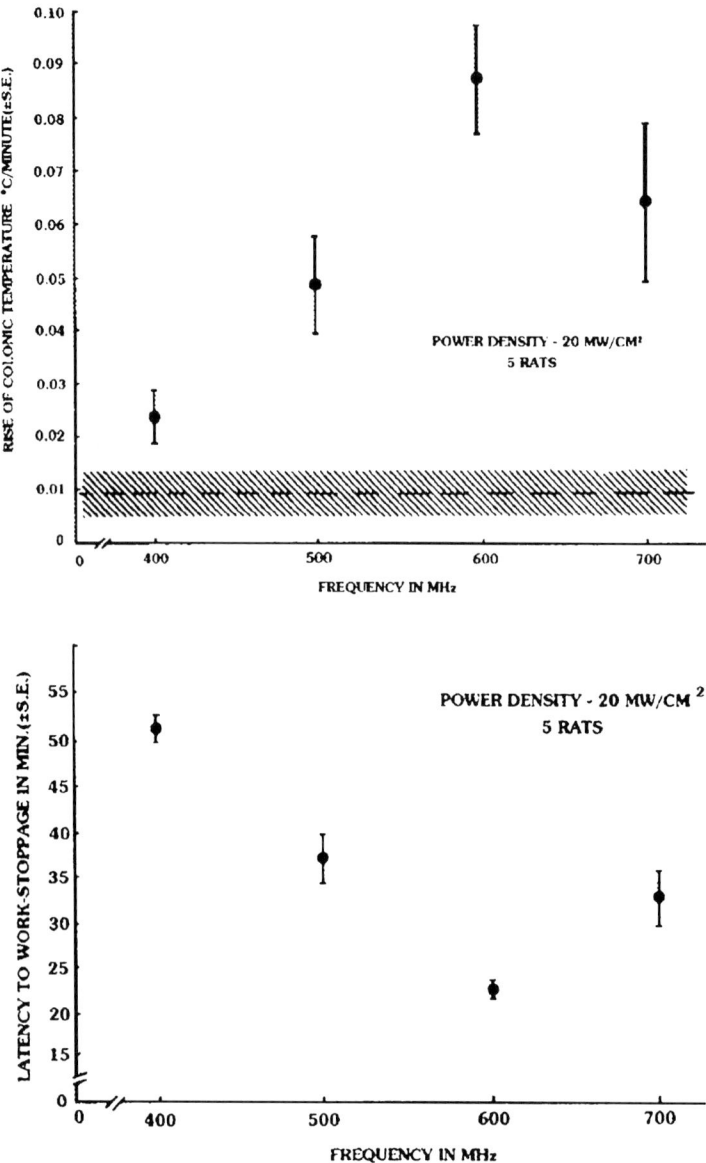

Fig. 6. Elevations of colonic temperature and latencies to work stoppage. (Adapted with permission from D'Andrea et al. (1977).)

Behavioral performance was characterized as ratios of response rates during RF-EMF exposure compared to sham exposure. Changes in colonic temperature were also recorded. Figure 7, modified from de Lorge (1984), shows typical changes in the response ratios at different power densities. These data are for five rhesus monkeys exposed to RF-EMF during 1-h sessions of observing response performance. Each point represents the mean of means from five monkeys exposed three times. The results show that a significant decline in behavioral responding occurred during RF-EMF exposure at each frequency. At 225 MHz there was a reliable effect at 8 mW/cm². However, maximal effect and actual work stoppage occurred at a power density of 10 mW/cm². For the 1.3 GHz

Table 3. Characteristics of contributing experiments: behavioral disruption

	225 MHz CW	1.3 GHz PF	2.45 GHz CW	5.8 GHz PF
Rat	–	B_1–T_{est}–$X+Y$–lateral	B_2–T_{est}–$X+Y$–dorsal B_1	$X+Y$–lateral
Saimiri	–	–	B_1–T–$Z+Y$–dorsal	B_3–T–Y–frontal
Macaca	B_1 T–Y–frontal	B_1–T–Y–frontal	B_1–T–Y–frontal	B_1–T–Y–frontal

Symbol identification: B_1, observing response; B_2, fixed interval schedule; B_3, repeated acquisition; X, Y, Z, long axis orientation to E field; T, colonic temperature measured; T_{est}, colonic temperature estimated; PF, pulsed field; CW, continuous wave field.

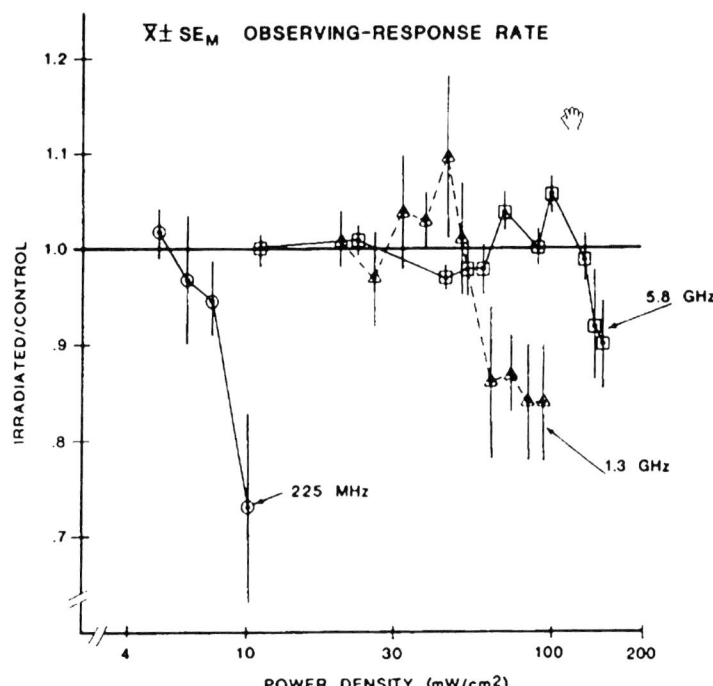

Fig. 7. Mean ratios of observing-responses emitted by rhesus monkeys during RF-EMF exposure sessions to responses made during sessions of sham exposure. The ratio for RF-EMF exposure sessions and sham sessions, which had no effect would be 1.0. Increased responses during exposure would result in a ratio >1.0 while decreased responding during exposure would produce a ratio <1.0. The horizontal line at 1.0 indicates no difference between the two conditions. Vertical bars indicate the standard error of the mean. (Adapted with permission from de Lorge (1984).)

exposures de Lorge described a more abrupt effect, which occurred at 56 mW/cm² and increased only slightly for 93 mW/cm² exposures. At 5.8 GHz behavioral changes were more gradual with an occasional increase in responding at the lower power densities. A reliable decrease in responding, however, did not occur until the monkeys were exposed to 140 mW/cm². As shown in Fig. 7, as frequency of the RF-EMF field was increased, the power density required also increased. Generally, behavior was not reliably affected until a colonic temperature increased by 1°C or more (Table 4).

Examination of temperature increases in rhesus monkeys across the three frequencies showed that 225 MHz exposures produced the highest temperature increases at relatively low power densities. This is because 225 MHz is very near the whole-body resonant frequency for the rhesus monkey and couples very well to this sized animal (see Fig. 1). In comparison, exposures to 1.3 and 5.8 GHz produced much less temperature increase even at much higher power densities than 225 MHz. This result from de Lorge (1984) is shown in Fig. 8 as temperature increase above the baseline colonic temperature in the monkeys. Whole-body SARs were either calculated based on the physical dimensions of each animal or measured in saline-filled models using procedures developed by Olsen et al. (1980). Only at 2.45 GHz were the SARs consistent across species where they averaged between 4 and 5 W/kg. The rat at 5.8 GHz and *Macaca* at 1.3 GHz also absorbed between 4 and 5 W/kg. However, *Saimiri* and *Macaca* at 5.8 GHz had higher whole-body SARs and the rat at 1.3 GHz had a lower whole-body SAR at the behavioral disruption threshold.

Basis of safety standards

The biological effects of electromagnetic fields have been studied for more than a century but studies on the effects of RF-EMF on organisms began shortly after World War II. The focus was on radar technology that involved increasing power output during and after the war (Osepchuk and Petersen, 2003). During the 1950s numerous meetings were held to evaluate known thermal effects and propose safe exposure limits. A limit of 100 W/m^2, first proposed by Hermann Schwan and recommended by the U.S. Navy, was adopted, and was designed to limit temperature rise in the body. This was a continuous exposure limit. Later standards, developed in the late 1960s and early 1970s, began using a 6 min time averaging concept such that more

Table 4. Power density (PD) thresholds (mW/cm^2) for behavioral disruption and temperature increase

	225 MHz CW	1.3 GHz PF	2.45 GHz CW	5.8 GHz PF
Rat	–	$B_1 = 10$ $T_{est} = 10$	$B_2 = 28$ $T_{est} = 32$	$B_1 = 20$
Saimiri	–	–	$B_1 = 45$ $T = 50$	$B_3 = 40$ $T = 40$
Macaca	$B_1 = 7.5$ $T = 7.5$	$B_1 = 53$ $T = 42$	$B_1 = 67$ $T = 67$	$B_1 = 145$ $T = 140$

Symbol identification: B_1, observing response; B_2, fixed interval schedule; B_3, repeated acquisition; T, colonic temperature rise of 1°C measured; T_{est}, colonic temperature rise of 1°C estimated; PF, pulsed field; CW, continuous wave field.

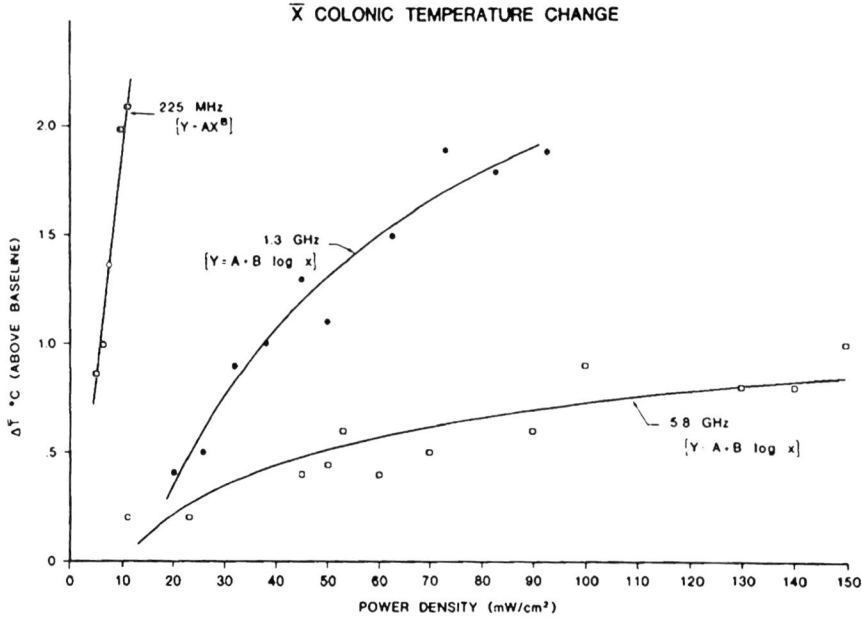

Fig. 8. Rhesus monkey colonic temperature by power density at three different frequencies. (Adapted with permission from de Lorge (1984).)

intense fields could be tolerated for periods shorter than 6 min. Six minutes was chosen because it was the thermal time constant for such tissues as the eyes and testes (Osepchuk and Petersen, 2003). The IEEE C95.1 (1982) standard was the first to recognize the importance of dosimetry and incorporated SAR as the basis for the standard. This standard also took into account the newly discovered resonant frequencies for both whole-body and for limb resonances in the range of 100–300 MHz (IEEE C95.1, 1982). Thus, MPEs within this frequency range are set low.

Both the IEEE C95.1 (2005) and the ICNIRP (International Commission on Non-Ionizing Radiation Protection) (1998) standards provide guidelines to limit RF-EMF exposure and give protection from known adverse health effects. Adverse health effects have been defined as "*detectable impairment of the health of the exposed individual or of his or her offspring*" (ICNIRP, 1998). Both standards distinguish biological effects from adverse effects and do not protect against biological effects. Both standards acknowledge tissue heating as the basis for adverse effects and set limits of exposure to prevent excessive heating. For human exposure in controlled environments to RF-EMFs from 3 kHz to 300 GHz, the MPE, in terms of rms electric (E) and magnetic (H) field strengths and the equivalent planewave free-space power densities (S), is given in Fig. 9 as a function of frequency. Controlled environment includes exposure by persons who are aware of the potential for exposure as a concomitant of employment. Uncontrolled exposure is where members of the public, who are unaware of exposure, may not exceed the MPE shown in Fig. 10. Limits of human exposure to induced currents associated with exposure to RF-EMF are also included in the IEEE C95.1 (2005) and ICNIRP (1998) standards. In addition each standard incorporates an appropriate time period over which an exposure is averaged to determine compliance with the MPE. For example the IEEE C95.1 (2005) standard incorporates a 6 and 30 min averaging time for controlled and uncontrolled exposures, respectively and promulgates an RF safety program. At higher frequencies (> 15 GHz controlled; > 3000 MHz uncontrolled) the averaging times are reduced

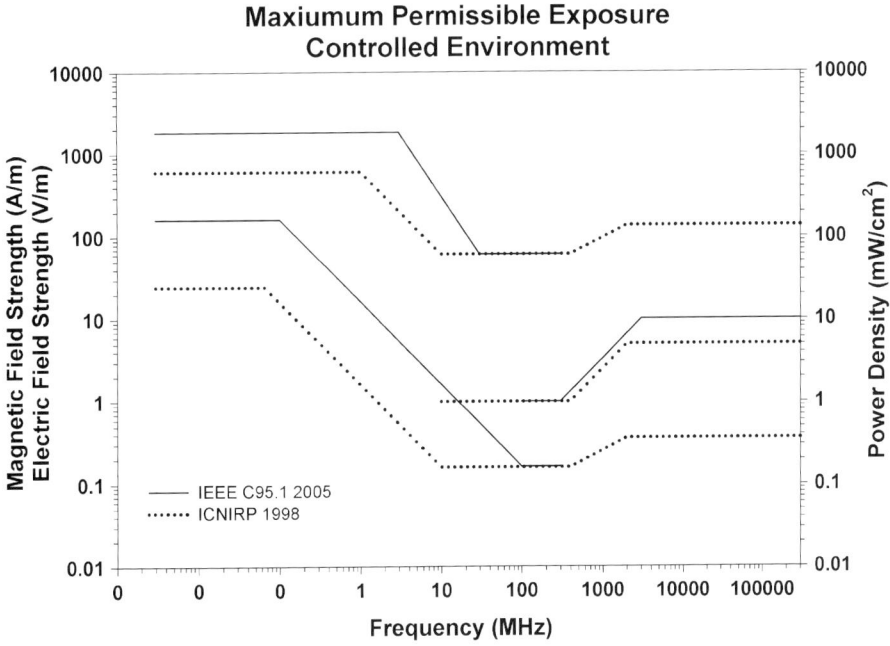

Fig. 9. Graphic representation of the IEEE C95.1 (2005) (solid line) and ICNIRP (International Commission on Non-Ionizing Radiation Protection) (1998) (dashed line) MPEs for exposures in controlled (occupational) environments from 3 kHz to 300 GHz.

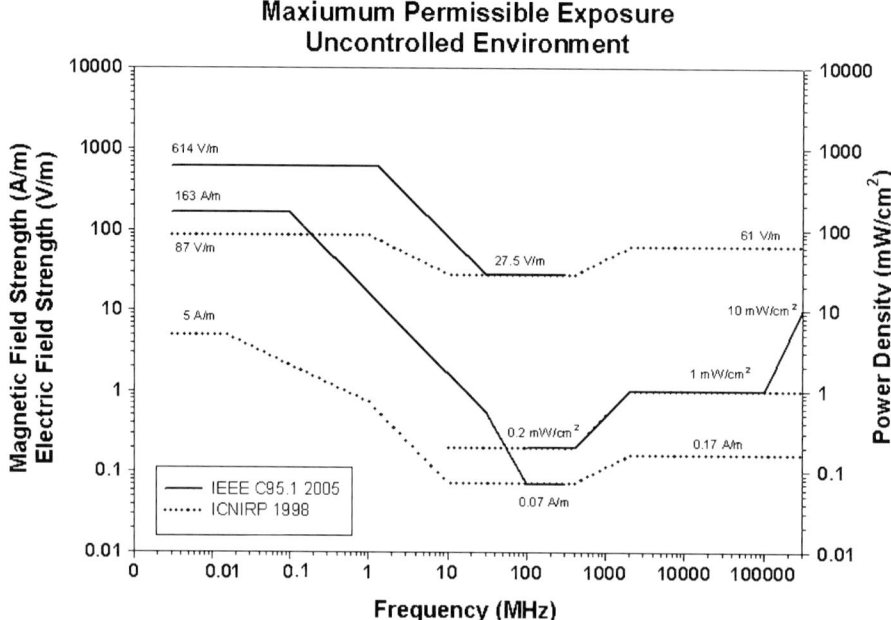

Fig. 10. Graphic representation of the IEEE C95.1 (2005) (solid line) and ICNIRP (International Commission on Non-Ionizing Radiation Protection) (1998) (dashed line) MPEs for exposures in uncontrolled (public) environments from 3 kHz to 300 GHz.

because of the more superficial absorption at the higher frequencies.

The current safety standards are based on reviews of the scientific literature and "weight of evidence" evaluations of the data. Currently, the behavioral effects described above have served as the basis for the safety standard whole-body SAR limits (IEEE C95.1, 1982, 1991, 1999, 2005; ICNIRP, 1998). Thermal effects appear to account for all of the reported behavioral disruption effects of RF-EMF energy exposures. As described above, the studies that show changes in behavioral performance involve some level of tissue heating, perhaps mild heat stress, and alternate behaviors that are thermoregulatory in nature. The threshold for the disruption of behavioral performance is near 4 W/kg (range of 3.2–8.4 W/kg). This information provides a scientific database from which safe exposure standards can be derived. Thus, one-tenth of this level has been set as the maximum permissible safe human exposure (0.4 and 0.08 W/kg) by two standard-setting organizations (ICNIRP, 1998; IEEE C95.1, 2005). The margin of safety for the IEEE C95.1 (1999) standard was a factor of 10 (occupational) and 50 (public) below the exposures that are disruptive of laboratory animal behavior, presumably due to mild heat stress.

Conclusions

The exposure standards offer protection from RF-EMF thermalizing effects by limiting the whole-body and localized SAR. The protection is based on limiting the whole-body SAR at a level (0.4 and 0.08 W/kg) that is well within the thermoregulatory capability of humans and that will prevent temperature rise in the body. While specifying SAR as the quantity that is regulated, the standards directly regulate the levels of electric, magnetic, and electromagnetic fields to which humans can be exposed because SAR is not a quantity that can be measured at the worksite. Thus, SAR is indirectly controlled by RF-EMF limits to prevent adverse thermal effects from occurring.

The whole-body SAR limit is based on thermal effects in laboratory species and is one-tenth of an adverse effect for laboratory species. The literature cited above leads to the conclusion that exposure to RF-EMF at a whole-body dose-rate of 4 W/kg will produce thermal changes in laboratory animals, leading to behavioral changes. These effects occur across species (rat, *Saimiri*, and *Macaca*) and occur within a narrow range of SAR. The behavioral effects have been thought of as predictive of an adverse thermal effect in human beings, but tests of human responses to exposure at 4 W/kg have never been conducted. The 0.4 W/kg limit may be quite conservative because the behavioral and physiological responses of laboratory animals to thermal insult are not necessarily predictive of human responses. Humans are far superior in managing thermal challenges than are the laboratory species tested (Adair and Black, 2003). Thus, as stated above, the exposure standards may be overly conservative in providing protection for humans from RF-EMF exposure. These standards are periodically revised as new information is made available identifying the adverse health effects of electric, magnetic, and electromagnetic fields.

Disclaimer

The views expressed in this article are those of the authors and do not reflect the official policy or position of the Department of the Navy, Department of Defense, nor the U.S. Government. Trade names of materials and/or products of commercial or non-government organizations are cited as needed for precision. These citations do not constitute official endorsement or approval of the use of such commercial materials and/or products. This review of the literature was sponsored by the Naval Health Research Center, Science Project Work Units (WU-60583 and WU-60584).

References

Adair, E.R. (1985) Thermal physiology of RFR interactions in animals and humans. In: Mitchell J.C. (Ed.), Proceedings of a Workshop on Radiofrequency Radiation Bioeffects. USAF Report SAM-TR-85-14. USAF School of Aerospace Medicine, Brooks AFB, TX, pp. 37–54.

Adair, E.R. (1988) Microwave challenges to the thermoregulatory system. In: O'Connor M.E. and Lovely R.F. (Eds.), Electromagnetic Waves and Neurobehavioral Function. Alan R. Liss, Inc., New York, NY, pp. 179–201.

Adair, E.R. (1996) Thermoregulation in the presence of microwave fields. In: Polk C.K. and Postow E. (Eds.), CRC Handbook of Biological Effects of Electromagnetic Fields (2nd ed.). CRC Press, Boca Raton, FL, pp. 403–434.

Adair, E.R. (2000) Thermoregulation: it's role in microwave exposure. In: Klauenberg B.J. and Miklavcic D. (Eds.), Radiofrequency Radiation Dosimetry. Kluwer Academic Publishers, Netherlands, pp. 345–356.

Adair, E.R., Adams, B.W. and Hartman, S.K. (1992) Physiological interaction processes and radiofrequency energy absorption. Bioelectromagnetics, 13: 497–512.

Adair, E.R., Adams, B.W. and Kelleher, S.A. (1993) Resonant CW and pulsed RF fields of identical average power density provoke similar thermoregulatory adjustments by squirrel monkeys. Bioelectromagnetics Society Fifteenth Annual Meeting Abstracts. Frederick, MD: BEMS, p. 63.

Adair, E.R. and Black, D.R. (2003) Thermoregulatory responses to RF energy absorption. Bioelectromagnetics, 24(Suppl 6): S17–S38.

Adair, E.R., Blick, D.W., Allen, S.J., Mylacraine, K.S., Ziriax, J.M. and Scholl, D.M. (2005) Thermophysiological responses of human volunteers to whole body RF exposure at 220 MHz. Bioelectromagnetics, 26(6): 448–461.

Adair, E.R., Cobb, B.L., Mylacraine, K.S. and Kelleher, S.A. (1999b) Human exposure at two radio frequencies (450 and 2450 MHz): similarities and differences in physiological response. Bioelectromagnetics, 20(Suppl 4): 12–20.

Adair, E.R., Kelleher, S.A., Berglund, L.G. and Mack, G.W. (1999a) Physiological and perceptual responses of human volunteers during whole-body RF exposure at 450 MHz. In: Bersani F. (Ed.), Electricity and Magnetism in Biology and Medicine. Kluwer Academic/Plenum, New York, NY, pp. 613–616.

Adair, E.R., Kelleher, S.A., Mack, G.W. and Morocco, T.S. (1998) Thermophysiological responses of human volunteers during controlled whole-body radiofrequency exposure at 450 MHz. Bioelectromagnetics, 19: 232–245.

Adair, E.R., Mylacraine, K.S. and Allen, S.J. (2002) Human thermophysiological responses to whole-body RF exposure (100 MHz CW) regulate the body temperature efficiently. BEMS, 24th Annual Meeting Abstracts, p. 34.

Adair, E.R., Mylacraine, K.S. and Allen, S.J. (2003) Thermophysiological consequences of whole-body resonant RF exposure (100 MHz) in human volunteers. Bioelectromagnetics, 24: 489–501.

Adair, E.R., Mylacraine, K.S. and Cobb, B.L. (2001a) Partial-body exposure of human volunteers to 2450 MHz pulsed or CW fields provokes similar thermoregulatory responses. Bioelectromagnetics, 22: 246–259.

Adair, E.R., Mylacraine, K.S. and Cobb, B.L. (2001b) Human exposure to RF energy at levels outside the C95.1 standard does not increase core temperature. Bioelectromagnetics, 22: 429–439.

Airborne Instruments Laboratory. (1956) An observation on the detection by the ear of microwave signals. Proc. IRE, 44: 2–5.

Allen, S.J., Adair, E.R., Mylacraine, K.S., Hurt, W. and Ziriax, J. (2005) Empirical and theoretical dosimetry in support of whole body radio frequency (RF) exposure in seated human volunteers at 220 MHz. Bioelectromagnetics, 26: 440–447.

D'Andrea, J.A. (1991) Microwave radiation absorption: behavioral effects. Health Phys., 61: 29–40.

D'Andrea, J.A. (1999) Behavioral evaluation of microwave irradiation. Bioelectromagnetics, 20: 64–74.

D'Andrea, J.A., Adair, E.R. and de Lorge, J.O. (2003a) Behavioral and cognitive effects of microwave exposure. Bioelectromagnetics, 24(Suppl 6): S39–S62.

D'Andrea, J.A., Chou, C.T., Johnson, S. and Adair, E.R. (2003b) Microwave effects on the nervous system. Bioelectromagnetics, 24(Suppl 6): S107–S147.

D'Andrea, J.A., DeWitt, J.R., Portuguez, L.M. and Gandhi, O.P. (1988) Reduced exposure to microwave radiation by rats: frequency specific effects. In: O'Connor M.E. and Lovely R.H. (Eds.), Electromagnetic Fields and Neurobehavioral Function. Alan R. Liss, Inc., New York, NY, pp. 289–308.

D'Andrea, J.A., Emmerson, R.Y., Bailey, C.M., Olsen, R.G. and Gandhi, O.P. (1985) Microwave radiation absorption in the rat: frequency dependent SAR distribution in body and tail. Bioelectromagnetics, 6: 199–206.

D'Andrea, J.A., Emmerson, R.Y., DeWitt, J.R. and Gandhi, O.P. (1987) Absorption of microwave radiation by the anesthetized rat: electromagnetic and thermal hotspots in body and tail. Bioelectromagnetics, 8: 385–396.

D'Andrea, J.A., Gandhi, O.P. and Lords, J.L. (1977) Behavioral and thermal effects of microwave radiation at resonant and non-resonant wavelengths. Radio Sci., 12: 251–256.

D'Andrea, J.A., Thomas, A. and Hatcher, D.J. (1994) Rhesus monkey behavior during exposure to high-peak-power 5.62-GHz microwave pulses. Bioelectromagnetics, 15: 163–176.

Bernardi, P., Cavagnaro, M., Pisa, S. and Piuzzi, E. (2003) Specific absorption rate and temperature elevation in a subject exposed in the far-field of radio-frequency sources operating in the 10–900-MHz range. IEEE Trans. Biomed. Eng., 50: 295–304.

Black, D. and Heynick, L.N. (2003) Radiofrequency (RF) effects on blood cells, cardiac, endocrine, and immunological functions. Bioelectromagnetics, 24: S187–S195.

Blick, D.W., Adair, E.R., Hurt, W.D., Sherry, C.J., Walters, T.J. and Merritt, J.H. (1997) Thresholds of microwave evoked warmth sensations in human skin. Bioelectromagnetics, 18: 403–409.

Britt, R.H., Lyons, B., Ryan, T., Saxer, E., Obana, W. and Rossi, G. (1984) Effect of whole-body hyperthermia on auditory brainstem and somatosensory and visual-evoked potentials. In: Hales J.R.S. (Ed.), Thermal Physiology. Raven, New York, NY, pp. 519–524.

Cabanac, M. (1983) Face fanning: a possible way to prevent or cure brain hyperthermia. In: Kogali M. and Hales J.R.S. (Eds.), Heat Stroke and Temperature Regulation. Academic Press, Sydney, Australia, pp. 213–222.

Candas, V., Adair, E.R. and Adams, B.W. (1985) Thermoregulatory adjustments in squirrel monkeys exposed to microwaves at high power densities. Bioelectromagnetics, 6: 221–234.

Carroll, D.R., Levinson, D.M. and Justesen, D.R. (1980) Failure of rats to escape from a potentially lethal microwave field. Bioelectromagnetics, 1: 101–115.

Chou, C.K., Guy, A.W. and Galambos, R. (1977) Characteristics of microwave-induced cochlear microphonics. Radio Sci., 12(6S): 221–227.

Dewhirst, M.W., Viglianti, B.L., Lora-Michiels, M., Hanson, M. and Hoopes, P.J. (2003) Basic principles of thermal dosimetry and thermal thresholds for tissue damage from hyperthermia. Int. J. Hyperthermia, 19(3): 267–294.

Dimbylow, P.J. (1997) FDTD calculations of the whole-body averaged SAR in an anatomically realistic voxel model of the human body from 1 MHz to 1 GHz. Phys. Med. Biol., 42: 479–490.

Durney, C.H., Johnson, C.C., Barber, P.W., Massoudi, H., Iskander, M.F., Lords, J.L., Ryser, D.K., Allen, S.J. and Mitchell, J.C. (1978) Radio-Frequency Radiation Dosimetry Handbook. 2nd ed., Report USAFSAM-TR-78-22, USAF School of Aerospace Medicine, Brooks Air Force Base, TX.

Durney, C.H., Massoudi, H. and Iskander, M.F. (1986) Radiofrequency Radiation Dosimetry Handbook [Fourth Edition], USAF School of Aerospace Medicine, Brooks AFB, TX, Report USAFSAM-TR-85-73.

Elder, J.A. (2003a) Ocular effects of radiofrequency energy. Bioelectromagnetics, 24(Suppl 6): S148–S161.

Elder, J.A. (2003b) Survival and cancer in laboratory mammals exposed to radiofrequency energy. Bioelectromagnetics, 24(Suppl 6): S101–S106.

Elder, J.A. and Chou, C.K. (2003) Auditory response to pulsed radiofrequency energy. Bioelectromagnetics, 24(Suppl 6): S162–S173.

Elwood, M.J. (2003) Epidemiological studies of radiofrequency exposures and human cancer. Bioelectromagnetics, 24(Suppl 6): S63–S73.

Foster, K.R. and Finch, E.D. (1974) Microwave hearing: evidence for thermoacoustic auditory stimulation by pulsed microwaves. Science, 185: 256–258.

Frey, A.H. (1961) Auditory system response to radio-frequency energy. Aerospace Med., 32: 1140–1142.

Frey, A.H. (1962) Human auditory system response to modulated electromagnetic energy. J. Appl. Physiol., 17: 689–692.

Frey, H.A. and Feld, S.R. (1975) Avoidance by rats of illumination with low power nonionizing electromagnetic energy. J. Comp. Physiol. Psychol., 89: 183–188.

Gabriel, C. and Gabriel, S. (1996) Compilation of the Dielectric Properties of Body Tissues at RF and Microwave Frequencies (internet document); URL: http://www.brooks.af.mil/AFRL/HED/hedr/reports/dielectric/home.html

Gandhi, O.P. (1974) Polarization and frequency effects on whole animal energy absorption of RF energy. Proc. IEEE, 62: 1171–1175.

Gandhi, O.P. (1994) Some numerical methods for dosimetry: ELF to microwave frequencies. Radio Sci., 29(November).

Gandhi, O.P. (1995) Some numerical methods for dosimetry: extremely low frequencies to microwave frequencies. Radio Sci., 30: 161–177.

Gandhi, O.P. and Chatterjee, I. (1982) Radio-frequency hazards in the VLF to MF band. Proc. IEEE, 70: 1462–1464.

Gandhi, O.P., Hagmann, M.J. and D'Andrea, J.A. (1979) Partbody and multibody effects on absorption of radio frequency electromagnetic energy by animals and by models of man. Radio Sci., 14(6S): 15–22.

Gandhi, O.P., Hunt, E.L. and D'Andrea, J.A. (1977) Deposition of electromagnetic energy in animals and in models of man with and without grounding and reflector effects. Radio Sci., 12(6S): 39–48.

Gandhi, O.P., Gu, Y.G., Chen, J.Y. and Bassen, H.I. (1992) Specific absorption rates and induced current distributions in an anatomically based human model for planewave exposures. Health Phys., 63(3): 281–290.

Gandhi, O.P. and Kang, G. (2002) Some present problems and a proposed experimental phantom for SAR compliance testing of cellular telephones at 835 and 1900 MHz. Phys. Med. Biol., 47: 1501–1518.

Gathiram, P.G., Brock-Utne, J.G., Wells, M.T. and Gaffm, S.L. (1988) Prophylactic corticosteroid prevents endotoxemia in heat stressed primates. Aviat. Space Environ. Med., 59: 142–145.

Guy, A.W. (1971) Analyses of electromagnetic fields in biological tissues by thermographic studies on equivalent phantom models. IEEE Trans. Microw. Theory Tech., 19: 205–214.

Guy, A.W., Lehmann, J.F. and Stonebridge, J.B. (1974) Therapeutic applications of electromagnetic power. Proc. IEEE, 62: 55–75.

Hales, J.R.S., Hubbard, R.W. and Gaffm, S.L. (1996) Limitation of heat tolerance. In: Fregly M.J. and Blatteis C.M. (Eds.), Handbook of Physiology, Section 4: Environmental Physiology, Vol. I. Oxford, New York, NY, pp. 285–355.

Hardy, J.D., Wolff, H.G. and Goodell, H. (1952) Pain Sensations and Reactions. Williams and Wilkins Co., Baltimore, MD.

Hatcher, D.J. and D'Andrea, J.A. (1992) A Computer Program to Calculate Planewave Average Specific Absorption Rate in a Prolate Spheroidal Model, NAMRL Technical Memorandum 92-3, Naval Aerospace Medical Research Laboratory, Pensacola, FL, August (AD A258 197).

Heynick, L.N., Johnston, S.A. and Mason, P.A. (2003) Radio frequency electromagnetic fields: cancer, mutagenesis, and genotoxicity. Bioelectromagnetics, 24(Suppl 6): S74–S100.

Heynick, L.N. and Merritt, J.H. (2003) Radiofrequency fields and teratogenesis. Bioelectromagnetics, 24(Suppl 6): S174–S186.

Hjeresen, D.L., Doctor, S.K. and Shelton, R.L. (1979) Shuttlebox side preference as mediated by pulsed microwave and conventional auditory cues. In: Stuchley S. (Ed.), Proceedings of Symposium on Electromagnetic Fields in Biological Systems. International Microwave Power Institute, Ottawa, Canada, pp. 194–214.

Hunt, E.L. and Phillips, R.D. (1972) Absolute physical dosimetry for whole animal experiments. In: Joint U.S. Army/ Georgia Institute of Technology Microwave Dosimetry Workshop, Digest of Papers. Walter Reed Army Institute of Research, Washington, DC, pp. 74–77.

ICNIRP (International Commission on Non-Ionizing Radiation Protection). (1998) Guidelines on limits of exposure to laser radiation of wavelengths between 180 nm and 1 mm. Health Phys., 71: 804–819.

IEEE C95.1. (1982) ANSI Standard for Safety Levels with Respect to Human Exposure to Radio Frequency Electromagnetic Fields, 3 kHz to 100 GHz, American National Standards Institute, New York, NY, p. 24. http://standards.ieee.org

IEEE C95.1. (1991) IEEE Standard for Safety Levels with Respect to Human Exposure to Radio Frequency Electromagnetic Fields, 3 kHz to 300 GHz. http://standards.ieee.org

IEEE C95.1. (1999) IEEE Standard for Safety Levels with Respect to Human Exposure to Radio Frequency Electromagnetic Fields, 3 kHz to 300 GHz. http://standards.ieee.org

IEEE C95.1. (2005) IEEE Standard for Safety Levels with Respect to Human Exposure to Radio Frequency Electromagnetic Fields, 3 kHz to 300 GHz. http://standards.ieee.org

Johnson, R.B., Myers, D.E., Guy, A.W. and Lovely, R.H. (1976) Discriminative control of appetitive behavior by pulsed microwave radiation in rats. In: Johnson C.C. and Shore M.L. (Eds.), Biological Effects of Electromagnetic Waves. Vol. II, Selected Papers of the USNC/URSI Annual Meeting, October 20–23, 1975, Boulder, CO. HEW publication (FDA) 77-8011, U.S. Government Printing Office, Washington, DC, pp. 238–247.

Justesen, D.R. (1975) Toward a prescriptive grammar for the radiobiology of non-ionising radiations: quantities, definitions, and units of absorbed electromagnetic energy — an essay. J. Microw. Power, 10(4): 343–354.

Justesen, D.R. (1983) Sensory dynamics of intense microwave irradiation: a comparative study of evasive behaviors by mice and rats. In: Adair E.R. (Ed.), Microwaves and Thermoregulation. Academic Press, New York, NY, pp. 203–230.

Justesen, D.R. (1988) Microwave and infrared radiations as sensory, motivational, and reinforcing stimuli. In: O'Connor M.E. and Lovely R.H. (Eds.), Electromagnetic Fields and Neurobehavioral Function. Alan R. Liss, Inc., New York, NY, pp. 235–264.

Justesen, D.R., Adair, E.R., Stevens, J.C. and Bruce-Wolfe, V. (1982) A comparative study of human sensory thresholds: 2450 MHz vs. far-infrared radiation. Bioelectromagnetics, 3: 117–125.

Justesen, D.R. and King, N.W. (1970) Behavioral effects of low level microwave irradiation in the closed space situation. In: Cleary S.F. (Ed.), Biological effects and health implications of microwave radiation — Symposium proceedings (U.S. Public Health Service Publication No. BRH/DBE 70-2). Washington, DC: U.S. Government Printing Office.

Justesen, D.R., Levinson, D.M., Clarke, R.L. and King, N.W. (1971) A microwave oven for behavioral and biological research: electrical and structural modifications, calorimetric dosimetry, and functional evaluation. J. Microw. Power, 6: 237–258.

Justesen, D.R., Levinson, D.M. and Justesen, L.R. (1974) Psychogenic stressors are potent medicators of the thermal response to microwave radiation. In: Czerski P. (Ed.), Biologic Effects and Health Hazards of Microwave Radiation. Polish Medical Publishers, Warsaw, Poland.

Justesen, D.R., Riffle, D.W. and Levinson, D.M. (1985) Sensory, motivational, and reinforcing properties of microwaves: an assay of behavioral thermoregulation by mice and rats. In: Monahan J.C. and D'Andrea J.D. (Eds.), Behavioral Effects of Microwave Radiation Absorption. HHS CDRH Publication FDA 85-8238, Rockville, MD, pp. 59–75.

King, N.W., Justesen, D.R. and Clarke, R.L. (1971) Behavioral sensitivity to microwave radiation. Science, 172: 398–401.

Krupp, J.H. (1981) In vivo measurement of radio-frequency radiation absorption. In: Mitchell J.C. (Eds.), Proceedings of a Workshop on the Protection of Personnel against Radiofrequency Electromagnetic Radiation. USAF Aeromedical Review SAM-3-81, School of Aerospace Medicine, Brooks AFB, TX.

Krupp, J.H. (1983) In vivo temperature measurements during the whole-body exposure of *Macaca mulatta* to resonant and non-resonant frequencies. In: Adair E.R. (Ed.), Microwaves and Thermoregulation. Academic Press, New York, NY, pp. 95–107.

Kunz, K. and Luebbers, R. (1993) The Finite Difference Time Domain Method for Electromagnetics. CRC Press, Boca Raton, FL, p. 496.

Lebovitz, R.M. (1983) Pulse modulated and continuous wave microwave radiation yield equivalent changes in operant behavior of rodents. Physiol. Behav., 30: 891–898.

Lee, R.C., Zhang, D. and Hannig, J. (2000) Biophysical injury mechanisms in electrical shock trauma. Annu. Rev. Biomed. Eng., 2: 477–509.

Levinson, D.M., Grove, A., Clarke, R.L. and Justesen, D.R. (1982) Photic cuing of escape by rats from an intense microwave field. Bioelectromagnetics, 3: 105–116.

Levinson, D.M., Justesen, D.R. and Riffle, D.W. (1985) Experimental analysis of aversive behavior: mice and rats in intense microwave fields. In: Monahan J.C. and D'Andrea J.D. (Eds.), Behavioral Effects of Microwave Radiation Absorption. HHS Publication FDA 85-8238, CDRH, Rockville, MD, pp. 36–58.

Lin, J.C. (1978) Microwave Auditory Effects and Applications. Charles C. Thomas, Springfield, IL.

Lin, J.C. (1990) Auditory perception of pulsed microwave radiation. In: Gandhi O.P. (Ed.), Biological Effects and Medical Applications of Electromagnetic Energy. Prentice Hall, Englewood Cliffs, NJ, pp. 277–318.

Lin, J.C., Guy, A.W. and Caldwell, L.R. (1977) Thermographic and behavioral studies of rats in the near field of 918-MHz radiations. IEEE Trans. Microw. Theory Tech. MTT, 25: 833–836.

de Lorge, J.O. (1976) Behavior and Temperature in Rhesus Monkeys Exposed to Low Level Microwave Irradiation. Report NAMRL-1222; Naval Aerospace Medical Research Laboratory, Pensacola, FL: (AD A021769).

de Lorge, J.O. (1979) Operant behavior and colonic temperature of squirrel monkeys during microwave irradiation. Radio Sci., 14: 217–225.

de Lorge, J.O. (1983) The thermal basis for disruption of operant behavior by microwaves in three animal species. In: Adair E.R. (Ed.), Microwaves and Thermoregulation. Academic Press, New York, NY, pp. 379–400.

de Lorge, J.O. (1984) Operant behavior and colonic temperature of *Macaca mulatta* exposed to radio frequency fields at and above resonant frequencies. Bioelectromagnetics, 5: 233–246.

de Lorge, J.O. and Ezell, C.S. (1980) Observing-responses of rats exposed to 1.28- and 5.62-GHz microwaves. Bioelectromagnetics, 1: 183–198.

Lotz, W.G. (1985) Hyperthermia in radiofrequency-exposed rhesus monkeys: a comparison of frequency and orientation effects. Radiat. Res., 102: 59–70.

Lotz, W.G. and Saxton, J.L. (1987) Metabolic and vasomotor responses of rhesus monkeys exposed to 225-MHz radiofrequency energy. Bioelectromagnetics, 8: 73–89.

Lotz, W.G. and Saxton, J.L. (1988) Thermoregulatory responses in the rhesus monkey during exposure at a frequency (255 MHz) near whole-body resonance. In: O'Connor M.E. and Lovely R.H. (Eds.), Electromagnetic Fields and Neurobehavioral Function. Alan R. Liss, Inc., New York, NY, pp. 203–218.

Makrides, L., Heigenhauser, G.J.F. and McCartney, N. (1987) Physical training in young and older healthy subjects. In: Sutton J.R. and Brock R.M. (Eds.), Sports Medicine for the Mature Athlete. Benchmark, Indianapolis, IN, pp. 363–373.

Marr, M.J., de Lorge, J.O. and Olsen, R.G. (1988) Microwaves as reinforcing events in a cold environment. In: O'Connor M.E. and Lovely R.H. (Eds.), Electromagnetic Fields and Neurobehavioral Function. Alan R. Liss, Inc., New York, NY, pp. 219–234.

Mason, P.A., Hurt, W.D., Walters, T.J., D'Andrea, J.A., Gajšek, P., Ryan, K.L., Nelson, D.A., Smith, K.I. and Ziriax, J.M. (2000a) Effects of frequency, permittivity, and voxel size on predicted specific absorption rate values in biological tissue during electromagnetic-field exposure. IEEE Trans. Microw. Theory Tech., 48: 2050–2058.

Mason, P.A., Ziriax, J.M., Hurt, W.D., Walters, T.J., Ryan, K.L., Nelson, D.A. and D'Andrea, J.A. (2000b) Recent advancements in dosimetry measurements and modeling. In: Klauenberg B.J. and Miklavic D. (Eds.), Radio Frequency Radiation Dosimetry. Kluwer Academic Publishers, Dordrecht, The Netherlands, pp. 141–155.

Meltz, M.L. (2003) Radiofrequency exposure and mammalian cell toxicity, genotoxicity, and transformation. Bioelectromagnetics, 24(Suppl 6): S196–S213.

Michaelson, S.M. (1972) Human exposure to nonionizing radiant energy: potential hazards and safety standards. Proc. IEEE, 60: 389–421.

Nagaoka, T., Watanabe, S., Sakurai, K., Kunieda, E., Watanabe, S., Taki, M., Yamanaka, Y. and Kitasato, C.R.L. (2004) Development of realistic high-resolution whole-body voxel models of Japanese adult males and females of average height and weight, and application of models to radio-frequency electromagnetic-field dosimetry. Phys. Med. Biol., 49: 1–15.

NCRP. (1986) Biological effects and exposure criteria for radiofrequency electromagnetic fields. Bethesda: NCRP Report no. 86, National Council on Radiation Protection and Measurements, Bethesda, MD.

NRPB. (2004) Review of the scientific evidence for limiting exposure to electromagnetic fields (0–300 GHz), Documents of the NRPB, vol. 12, no. 3, National Radiological Protection Board, Chilton, Didcot, Oxfordshire, UK.

Olsen, R.G., Griner, T.A. and Prettyman, G.D. (1980) Far-field microwave dosimetry in a rhesus monkey model. Bioelectromagnetics, 1: 149–160.

Osepchuk, J.M. and Petersen, R.C. (2003) Historical review of RF exposure standards and the international committee on electromagnetic safety (ICES). Bioelectromagnetics, 24(Suppl 6): S7–S16.

Phillips, R.D., Hunt, E.L. and King, N.W. (1975) Field measurements, absorbed dose, and biologic dosimetry of microwaves. Ann. N.Y. Acad. Sci., 247: 499–509.

Rawson, R.O. and Quick, K.P. (1970) Evidence of deep-body thermoreceptor response to intra-abdominal heating of the ewe. J. Appl. Physiol., 28: 813–820.

Riu, P.J., Foster, K.R., Blick, D.W. and Adair, E.R. (1997) A thermal model for human thresholds of microwave-evoked warmth sensations. Bioelectromagnetics, 18: 578–583.

Schrot, J. and Hawkins, T.C. (1976) Interaction of microwave frequency and polarization with animal size. In: Biological Effects of Electromagnetic Waves: Selected Papers of the USNC/URSI Annual Meeting, Boulder, CO, October 20–23, 1975. In: Johnson C.C. and Shore M.L. (Eds.), Sponsored by U.S. National Committee of the International Union of Radio Sciences, National Academy of Sciences (Washington, DC). HEW Publication (FDA) 77-8010; Vol. 1, pp. 184–192.

Sharp, J.C., Grove, H.M. and Gandhi, O.P. (1974) Generation of acoustic signals by pulsed microwave energy. IEEE Trans. Microw. Theory Tech., 22: 583–584.

Shlager, K.L. and Schneider, J.B. (1995) A selective survey of the finite-difference time-domain literature. IEEE Antenn. Propag. Mag., 37(4): 39–56.

Shen, K. and Schwartzkroin, P.A. (1988) Effects of temperature alterations on population and cellular activities in hippocampal slices from mature and immature rabbit. Brain Res., 475: 305–316.

Stern, S.L. (1980) Behavioral effects of microwaves. Neurobehav. Toxicol., 2: 49–58.

Stuchly, M.A., Kraszewski, A., Stuchly, S.S., Hartsgrove, G.W. and Spiegel, R.J. (1987) RF energy deposition in a heterogeneous model of man: far-field exposures. IEEE Trans. Biomed. Eng., 34: 951–957.

Taflove, A. (1995) Computational electrodynamics: the finite-difference time-domain method. Artech House, Norwood, MA.

Tell, R.A. and Harlen, F. (1979) A review of selected biological effects and dosimetric data useful for development of radiofrequency safety standards for human exposure. J. Microw. Power, 14: 405–424.

Thomas, J.R., Schrot, J. and Banvard, R.A. (1982) Comparative effects of pulsed and continuous-wave 2.8-GHz microwaves on temporally defined behavior. Bioelectromagnetics, 3: 227–235 [IEEE-824].

Wang, J. and Fujiwara, O. (2000) Dosimetry analysis and safety evaluation of realistic head models for portable telephones. Trans. IEICE, J83-B(5): 720–725.

Webster, M.E.D. (1987) Temperature regulation in children. In: Shiraki K. and Yousef M.K. (Eds.), Man in Stressful Environments: Thermal and Work Physiology. Thomas, Springfield, IL, pp. 35–44.

Yee, K.S. (1966) Numerical solution of initial boundary value problems involving Maxwell's equations in isotropic media. IEEE Trans. Antenn. Propag., 14: 302.

Yousef, M.K. (1987) Thermoregulation in old age: effects of heat. In: Shiraki K. and Yousef M.K. (Eds.), Man in Stressful Environments: Thermal and Work Physiology. Thomas, Springfield, IL, pp. 45–62.

CHAPTER 8

The effect of induced hyperthermia on the immune system

Annette Dieing[1,*], Olaf Ahlers[2], Bert Hildebrandt[3], Thoralf Kerner[2], Ingo Tamm[3], Kurt Possinger[1] and Peter Wust[4]

[1]*Department of Oncology and Hematology, Charité Campus Mitte, University Medicine Berlin, Germany, Schumannstrasse 20/21, D-10117 Berlin, Germany*
[2]*Department of Anesthesiology and Intensive Care Medicine, Charité Medical School, University Medicine Berlin, Campus Virchow Clinic, D-13353 Berlin, Germany*
[3]*Medical Clinic, Department of Hematology and Oncology, Charité Medical School, University Medicine Berlin, Campus Virchow Clinic, D-13353 Berlin, Germany*
[4]*Department of Radiation Medicine, Charité Medical School, University Medicine Berlin, Campus Virchow Clinic, D-13353 Berlin, Germany*

Abstract: Therapeutical hyperthermia has been considered for cancer therapy since William Coley observed tumour remission after induction of fever by bacterial toxins at the end of the 19th century. Because fever is associated with a variety of immunological reactions, it has been suspected, that therapeutical hyperthermia might also activate the immune system in a reproducible manner and thereby positively influence the course of the disease. During the last decade, new insight has been gained regarding the immunological changes taking place during therapeutic hyperthermia. In this chapter, we review the most relevant data known about the effect of hyperthermia on the immune system with special focus on alterations induced by therapeutical whole-body hyperthermia (WBH) in cancer patients.

Keywords: whole-body hyperthermia; apoptosis; lymphocytes; cytokines; heat shock proteins; stress; review

Introduction

The medical subject headings (MeSH) database introduced the term "hyperthermia induced" in 1984 and defined it as "abnormally high temperature intentionally induced in living things regionally or whole body. It is most often induced by radiation (heat waves, infra-red), ultrasound or drugs" (http://www.ncbi.nlm.nih.gov). This definition implies that hyperthermia is a condition provoked by an external source and thus is clearly to be discriminated from other causes of elevated body temperatures, such as fever, malignant hyperthermia or other forms of heat injury like heat stroke. Regarding hyperthermia in this narrower sense, most scientific findings on the effect of external heat application on living organisms have been generated in the scope of cancer research, where preclinical and clinical approaches have been investigated for more than half a decade (reviewed in Hildebrandt et al., 2002; Wust et al., 2002; van der Zee, 2002).

Pioneering radiobiological studies established a synergism between temperature elevation and

*Corresponding author. Tel.: +49-30-450-613458;
Fax: +49-30-450-513952; E-mail: annette.dieing@charite.de

DOI: 10.1016/S0079-6123(06)62008-6

radiation in the early 1970s (Dewey, 1994). The clinical relevance of these findings has been proven in a number of randomised clinical studies on locoregional hyperthermia, with super-agonistic effects observed for the combination of hyperthermia and certain cytotoxic drugs (Wust et al., 2002; van der Zee, 2002). In addition, a number of phase II studies suggest a beneficial effect for whole-body hyperthermia (WBH) as an adjunct to chemotherapy (Hildebrandt et al., 2005). But even if it appears to be proven that local and regional radiofrequency hyperthermia approaches are suitable to enhance the effect of radiotherapy and chemotherapy in a variety of human malignancies, most of the underlying mechanisms are still poorly understood.

Here, we summarise the most relevant data published on the effect of hyperthermia on the cellular immune system with special attention on alterations that have been observed during therapeutical WBH in cancer patients.

Basic effects of hyperthermia

Cytotoxic effects of heat

Early experiments on the cytotoxic effect of *in vitro* hyperthermia on cultured cells revealed a time- and dose-dependant relationship in the temperature range between 41 and 47°C. The thermal doses required to induce hyperthermic cell death varied with factor 10 between different cell types. The slope of the corresponding survival curves showed a typical shoulder, indicating a potential of cells to recover from a thermal insult (reviewed in Dewey, 1994; Dewhirst et al., 2003). Highest heat sensitivity was observed during the mitotic phase where hyperthermia induces microscopically detectable damage of the mitotic apparatus leading to inefficient mitosis and consecutive polyploidy. In contrast, G1-cells exposed to hyperthermia do not exhibit microscopically detectable changes of the mitotic apparatus (termed necrosis) but may undergo a "rapid mode of cell death" immediately after heat exposure (Vidair and Dewey, 1988). Another important finding from these early *in vitro* models was a systematic pattern in the temperature dependency of cell death differentiating temperatures above 43°C. This led to the concept of the "thermal isoeffect dose" (TID) by which a given thermal dose (i.e. temperature elevation for a certain time) can be converted into an "equivalent heating time at 43°C", that has influenced hyperthermia research for the past three decades (Dewey, 1994; Dewhirst et al., 2003). Converting thermometrical data from clinical hyperthermia trials into "equivalent heating minutes at 43°C" (EM 43), significant correlations between this surrogate parameter or similar thermal parameters and clinical response parameters have been shown in various studies on local and regional hyperthermia (but not perfusional techniques or WBH) (Wust et al., 2002; Dewhirst et al., 2003). Therefore, EM 43 seems to be a valuable approach to describe the effectiveness of a clinical heat treatment.

Another groundbreaking observation from the early *in vitro* studies was that hyperthermia not only acts in a cytotoxic way by itself, but also sensitises tumour cells to radiotherapy and various cytostatic drugs at markedly lower temperatures than 43°C ("thermal radiosensitisation" and "thermal chemosensitisation") (reviewed in Dewey, 1989; Hildebrandt et al., 2002; Vujaskovic and Song, 2004). The extent of thermal sensitisation in a given model can be quantified by the quotient of survival fraction of cells treated with radiation/chemotherapy alone and those treated with radiation/chemotherapy at the same dose plus heat ("thermal enhancement ratio", TER). Both thermal radio- and chemosensitisation are reproducible *in vivo* and sufficiently explain that a beneficial clinical effect of radiative locoregional hyperthermia can already be observed at temperatures of 40° to 40.5°C (Myerson et al., 2004).

Although the transferability of the basic mechanisms of hyperthermic cell death on the living organism is generally assumed, the application of therapeutic hyperthermia in conjunction with radio- or chemotherapy to cancer patients is much more complex than in the pre-described preclinical models. Indeed, profound changes concerning the tumour's blood, nutrient and oxygen supply, metabolic situation, cellular signal transduction and immunological pathways as well as pharmacological effects, have been described during or after heat treatment (reviewed in Hildebrandt et al., 2002). However, it is difficult to extract reliable information on the effects of heat from clinical investigations with cancer patients and to separate it from other components of a multimodal treatment scheme (Hildebrandt et al., 2002).

Molecular and cellular effects

A large body of data concerning the cellular and molecular effects of hyperthermia exist, dating back to the 1970s. Hyperthermia has been demonstrated to affect fluidity and stability of cellular membranes and impedes the function of transmembranal transport proteins and cell surface receptors *in vitro* (Coss and Linnemans, 1996; Lepock, 2003). Therefore, it had been assumed that membrane alterations might represent an important target for the induction of heat-induced cell death. In addition, it had been proposed that hyperthermia-induced changes of cytoskeletal organisation (cell shape, mitotic apparatus, intracytoplasmatic membranes such as endoplasmatic reticulum and lysosomes) might correlate with the extent of hyperthermic cell death (reviewed in Lepock, 2003). Borrelli et al. (1986) described for the first time the appearance of "membrane blebbing" in cultured cells exposed to heat, which is a typical feature of programmed cell death (apoptosis).

In the 1960s, heat application was suggested to act similar to radiation by directly damaging nuclear DNA, thereby inducing double-strand breaks. However, it has been shown that heat does not primarily cause severe DNA damage by itself, but rather impedes the repair of radiation-induced cell damage thus boosting radiation-induced DNA fragmentation. Recent data suggest that inhibition of the "base damage repair" system may be the crucial pathogenetic step in hyperthermic radiosensitisation (reviewed in Kampinga et al., 2004).

Generally, hyperthermia exhibits its cell-killing properties by induction of apoptosis, necrosis or cell cycle arrest (reviewed in Hildebrandt et al., 2002), but detailed data on the mode of hyperthermic cell death in the living organism are not yet available. Some *in vivo* data show beneficial as well as disadvantageous effects: Regarding systemic hyperthermia application at moderately elevated temperatures, an animal study using xenotransplanted colon carcinoma revealed that long-term fever-like hyperthermia is suitable to induce significant tumour growth delay due to apoptosis. Analyses of the host tissues revealed a high rate of apoptotic cell death particularly in various lymphatic tissues, a moderately increased apoptosis within the small intestine, but not in any of the remaining organs (Sakaguchi et al., 1995; Yonezawa et al., 1996). In humans, heat-induced apoptosis is suggested to be one of the major pathogenetic mechanism mediating heat-induced developmental defects in foetuses by inducing an irreversible damage to neurogenic cells. A similar mechanism has also been proposed for heat damage of the central nervous system in adults (reviewed in Edwards et al., 2003; Sharma and Hoopes, 2003).

Taken together, data suggest that heat-induced apoptosis represents a central mechanism of heat action that may already be induced when a living organism or tumour-loaded region is subjected to mild to moderate thermal doses. But even if induction of programmed cell death in tumour tissues represents a highly desirable effect in the treatment of human malignancies, the fact that systemic hyperthermia is suitable to cause apoptosis in healthy cells such as lymphocytes raises concerns of potential adverse reactions, particularly in the scope of WBH.

Heat shock proteins

Hyperthermia, as well as various other stress conditions, induces the synthesis of so-called "heat shock proteins" (HSP) which may be mediated by activation of nuclear "heat shock factors" (HSF) within minutes. HSPs represent a heterogenous group of molecular chaperones consisting of at least five subgroups with different molecular mass and varying biologic function that are expressed both constitutively and stress-induced. They are usually divided into small HSPs (molecular mass <40 kDa), and the HSP 60, HSP 70, HSP 90 and HSP 100 protein families. All HSPs are able to unselectively bind to hydrophobic protein sequences liberated by denaturation, thereby preventing irreversible interaction of neighboured proteins ("chaperoning function"). In particular, the HSPs 27 and 70 are able to defend cells against a variety of potentially lethal stimuli and thus are supposed to be "general survival proteins" (Jaattelä, 1999; Jolly and Morimoto, 2000; Kregel, 2002).

HSP synthesis, in general, increases with elevated temperatures. However, above a distinct threshold temperature (typically above 42°C), an inhibition of HSP synthesis occurs, resulting in exponential cell death. Under physiological conditions, HSPs protect cells from potentially lethal heat damage, e.g. by an increase of cellular resistance to apoptosis (Jaattelä, 1999; Jolly and Morimoto, 2000; Kregel, 2002). The major cellular pathways activated by cellular stress and important interactions with HSPs are summarised in Fig. 1. Potential cellular effects of hyperthermia are summarised in Table 1.

Effects of heat on the immune system

Recent advances have highlighted a number of mechanisms by which heat application may interact with the immune system. Some of them are suggesting a similarity between external induction of elevated temperatures and fever, a highly conserved physiological mechanism in the defence against exogenous agents and tumours. The following chapters review important aspects of the complex immunological changes induced by hyperthermia.

In vitro *effects of hyperthermia on lymphocytes*

A number of preclinical studies have investigated the effect of hyperthermia on human lymphocytes *in vitro*. The results are contradictory: Earlier investigations revealed that application of hyperthermia at different temperatures (39°C, 40°C, 42°C) impairs lymphocyte function. Especially natural-killer lymphocyte (NK-cell) activity and cytotoxic T-lymphocyte function have been found to be sensitive to heat (Azocar et al., 1982; Kalland and Dahlquist, 1983; Dinarello et al., 1986; Yang et al., 1992). Another group reported a temperature-dependent effect of hyperthermia: NK-cell activity (NKA) as well as cytotoxic activity of T-lymphocytes (T-cells) were increased at temperatures of 40°C, while incubation at temperatures of 42°C caused decreased activities of both lymphocyte subpopulations as determined in chromium release assays with appropriate target cells (Shen et al., 1994). Again, heat shock proteins seem to play an important role during these processes.

Immunological features of heat shock proteins

Besides their above-mentioned chaperoning function, HSPs are involved in antigen presentation, cross-presentation and tumour immunity. HSPs isolated from cancer tissues have been found to form specific complexes with tumour specific peptides that are internalised into antigen-presenting cells by specific receptors and then presented together with major histocompatibility complex (MHC) class I molecules, thereby inducing a cytotoxic T-cell-activation. It has been shown that HSPs interact with antigen-presenting cells through the CD91 receptor, inducing the re-presentation of chaperoned peptides by MHC molecules and activation of NF-kappa B (Srivastava, 2002). Multhoff and coworkers characterised solid tumour cell lines expressing a stress-inducible form of HSP 70, which mediates MHC-independent lysis. They demonstrated that a cell-surface presentation of HSP 70 may occur constitutively or heat-induced. Thereby, membrane expression of HSP 70 epitopes may represent a target for NK-cells, but also protect tumour cells from radiation damage (Multhoff, 2002; Gehrmann et al., 2005). Because of their unique immunologic features, HSPs are believed to procure more or less specific immunogenic effects induced by hyperthermia and other exogenous stimuli and have attracted particular interest for tumour vaccination strategies (Li and Dewhirst, 2002; Krause et al., 2004). The possible interactions between hyperthermia, heat shock proteins and the immune system are illustrated in Fig. 2.

WBH-induced stress reactions

Induction of HSPs represents only one aspect of WBH-induced stress reaction. Moreover, application of WBH is apparently associated with major physical stress, indicated by marked increases of

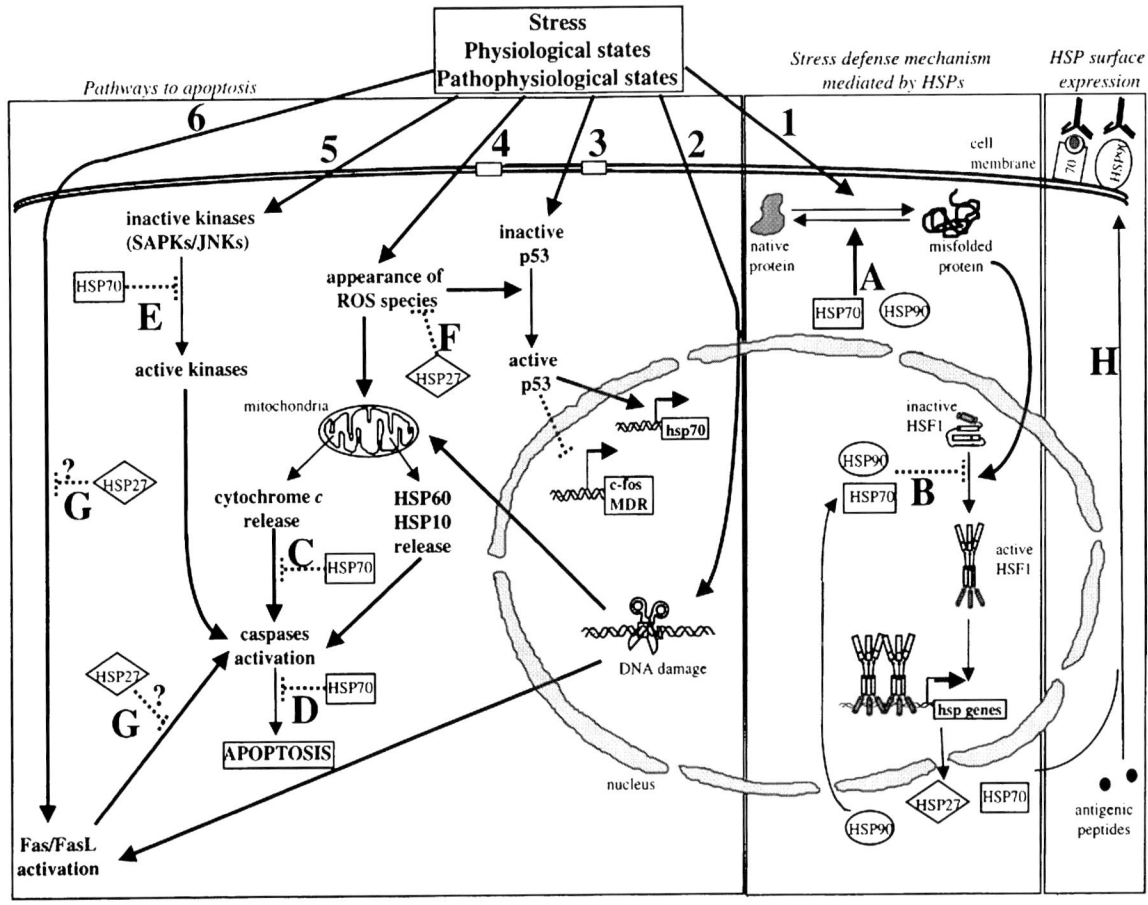

Fig. 1. Interactions between cellular stress pathways and heat shock proteins (adapted with permission from Repasky and Issels, 2002). Major cellular pathways activated by environmental (e.g. heat shock, hyperthermia), physiological and pathophysiological stress (e.g. fever, inflammation): (1) accumulation of misfolded proteins, (2) direct DNA damage (e.g. by UV radiation), (3) activation of p53, (4) increased levels of reactive oxygen species, (5) activation of stress activated protein kinases — SAPKs, jun n-terminal kinases — JNKs, (6) activation of Fas/Fas-L ligand. Pathways by which heat shock proteins (HSPs) interact with stress-induced alterations: (A) HSP assist proteins to acquire a native state, (B) autoregulation of HSP synthesis, (C) blockade of caspase activation by cytochrome C, (D) inhibition of caspase downstream events, (E) inhibition of SAPKs/JNKs activation, (F) inhibition of appearance of ROS species, (G) inhibition of Fas-induced apoptosis, (H) cell surface expression as antigen/antigen-presenting molecule.

heart rate, cardiac output and oxygen consumption (Faithfull et al., 1984; Kerner et al., 1999, 2003). In addition, elevated plasma levels of stress hormones have been described during WBH (Robins et al., 1987a; Kappel et al., 1991; Kearns et al., 1999) and interactions of these stress hormones with the immune system are well known (Hellstrand and Hermodsson, 1989; Madden et al., 1995; Iwakabe et al., 1998; Woiciechowsky et al., 1998; Kohm and Sanders, 2001; Elenkov and Chrousos, 2002). As physical stress (e.g. exercise) physiologically increases body temperature, it is very difficult to distinguish between hyperthermic- and stress-induced effects. Brenner et al. (1997) performed cycle ergometer exercise in a thermo neutral (23°C) or heated (40°C) climatic chamber with resting at the same temperatures as control measurement. Exercise in the thermo neutral surrounding led to significantly increased heart rates and to elevated plasma levels of the stress hormones epinephrine, nor-epinephrine and cortisol. These changes were significantly enhanced by

Table 1. Cellular effects of hyperthermia

Cell membrane, cytoskeleton	Changes in cell shape, "membrane blebbing", stability of plasma membrane, membrane potential, cell surface receptors, transmembrane transport mechanisms, apoptosis
Chromosomes	Impairment of RNA/DNA synthesis, inhibition of DNA-repair mechanisms, "mitotic dysfunction", gene expression, signal transduction
Proteins	Protein synthesis (impaired), misfolding/ denaturation/nuclear aggregation, increased synthesis of heat shock proteins

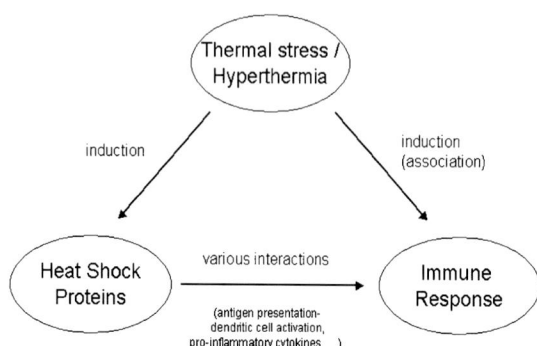

Fig. 2. Possible interactions between hyperthermia and the immune system (adapted with permission from Repasky and Issels, 2002).

additional increase of body temperatures up to 38.1°C during exercise in the heated chamber. None of the parameters changed significantly during resting in 40°C surrounding. Similar results were currently published by Rhind et al. (2004): Cycle ergometer exercise in a cold or hot water bath with rectal temperatures of 37.5 vs. 39.1°C resulted in different elevations of epinephrine and nor-epinephrine. The increase of cortisol during exercise under hot conditions was even abolished in the lower temperature group. Thus additional moderate hyperthermia enhances the secretion of stress hormones induced by physical stress.

The effect of these stress hormones on innate and acquired immunity seems to be different. There is evidence, that the innate immunity represented e.g. by monocytes and macrophages is rather stimulated by stress hormones whereas the acquired cellular immunity, represented by T-lymphocytes, may be transiently suppressed (Galon et al., 2002).

Monocyte stimulation by WBH

The results of investigations performed during clinical trials suggest that WBH at 42°C induces the secretion of pro-inflammatory cytokines such as Interleukin-1 (IL-1), Interleukin-8 (IL-8), and tumour necrosis factor alpha (TNF-α) as well as the pro- and anti-inflammatory cytokine Interleukin-6 (IL-6) through monocytes/macrophages as part of the innate immune system (Robins et al., 1995; Haveman et al., 1996; Katschinski et al., 1999; Atanackovic et al., 2002). Similar effects have been described in patients suffering from heat stroke (Bouchama and Knochel, 2002). In addition, stimulatory effects of WBH on monocyte functions and elevated expression of monocyte markers such as CD14 and CD11b have been reported immediately after induction of moderate WBH in healthy volunteers (Zellner et al., 2002). These results are in line with the hypothesis of stress hormone-induced monocyte activation during WBH. It has been postulated that HSP may also induce secretion of the pro-inflammatory cytokines IL-1, IL-6 and TNF-α by T-cells (Milani et al., 2002). However, it is unclear whether the results of those *in vitro* tests may be due to lipopolysaccarids (LPS) and LPS-associated molecules used during the tests (reviewed in Tsan and Gao, 2004).

Transient impairment of lymphocytes during WBH

Whereas research in the context of extreme WBH at 42°C suggest an activation of innate immunity, the course of lymphocyte controlling cytokines rather indicates a switch to an anti-inflammatory pattern. Plasma levels of Interleukin-10 increased dramatically, while levels of Interleukin-12 (IL-12) and Interferon-gamma (IFN-γ) decreased during and shortly after therapy (Robins et al., 1995; Ahlers et al., 1999; Katschinski et al., 1999). Analyses of lymphocyte subpopulations revealed marked changes during therapeutic WBH: A reversible increase of blood natural killer cells (NK-cells) as well as natural killer T-cells (NKT-cells) and γδ-T-cells — three lymphocyte subpopulations with "innate immune functions" — has been previously observed (Ahlers et al., 1998; Atanackovic et al., 2002). In addition, significant alterations of T-lymphocyte subpopulations could be found during therapeutic WBH: The number of blood $CD4^+$ T-cells decreased, while the number of $CD8^+$ T-cells remained unchanged, resulting in a marked drop in lymphocyte count and T4/T8 ratio. All these changes were reversible within 24 h and could not been found in a control group that received isolated chemotherapy (Ahlers et al., 1998). In addition, therapeutic WBH induced a short period of impaired T-cell proliferation as well as reduced T-cell activity — indicated by reduced serum levels of soluble interleukin-2 receptors (sIL-2R) and reduced numbers of IL-2R positive T-cells (Ahlers et al., 1998; Atanackovic et al., 2002).

Significant alterations of lymphocyte subpopulations can be observed in patients suffering from heat stroke with body core temperatures > 40.1°C when compared with healthy control persons. After heat stroke, percentages of T-lymphocytes as well as the T4/T8 ratio were reduced, while percentages of natural killer (NK) lymphocytes were elevated. These changes resulted in slightly increased numbers of total lymphocytes during heatstroke (Hammami et al., 1998). Declined numbers of T-lymphocytes — without analyses of subpopulations — were also observed in cancer patients undergoing therapeutic WBH of 39° to 40°C (Kraybill et al., 2002). Other authors examined intra-individual immunological changes in healthy volunteers whose core temperatures rose up to 39°C in a water bath. T4-cells and T4/T8 ratio decreased, while NK-cells increased reversibly in these persons. Similarly to heat stroke, these effects resulted in slightly increased numbers of total lymphocytes during moderate WBH (Downing et al., 1988; Kappel et al., 1991; Blazickova et al., 2000). The increase in NK-cell numbers came along with increased NKA during moderate WBH (Downing et al., 1988; Kappel et al., 1991). However, both studies did not calculate the NKA per NK-cell, but measured the total activity of all mononuclear cells. Therefore, the increased total NKA may just be due to the increased numbers of NK-cells (Downing et al., 1988; Kappel et al., 1991). Taken together, moderate (39°C) as well as extreme (42°C) WBH induced marked but transient shifts in lymphocyte subpopulations and -functions: NK-cells, NK-T-cells and γδ-T-cells increased significantly, while T4-cells, total T-cells and T4/T8 ratio decreased. The latter was more pronounced at 42°C, which resulted in a significant decrease of the absolute lymphocyte count at this temperature. T-cell function changed to an anti-inflammatory pattern with increase of IL-10 and decrease of IL-12 and IFN-γ, while per-cell-NKA during WBH has not been measured until now. However, IFN-γ is one of the strongest stimulators of NKA and reduced IFN-γ plasma levels may thus lead to impaired per-cell-NKA (Robertson and Ritz, 1990).

This transient anti-inflammatory phase seems to be followed by a prolonged T-cell activation: Atanackovic et al. (2002) reported increased numbers of the activation marker CD69 on T-cells and increased levels of soluble IL-2R 24 h as well as 48 h after WBH treatment.

The clinical relevance of each of these two phases of lymphocyte response to WBH has to be investigated in further studies. Currently, this phenomenon can only hypothetically be interpreted as an initial, transient compromising of the immune system with decline of T-cells and increase of anti-inflammatory cytokines as an adverse effect of hyperthermal treatment. This seems to be followed by a rather long-term activation of the acquired immune system with activation of T-cells. The latter might contribute to the beneficial effects of therapeutic hyperthermia in cancer patients.

Stress hormones, lymphocyte shift and lymphocyte function

The alterations observed during therapeutic WBH seem to be at least partially due to increased stress hormones accompanying the treatment procedure. Further clues on stress reactions due to this treatment come from the observation that cortisol as well as acute stress and extreme WBH are suitable to induce lymphocyte (i.e. T-lymphocyte) migration out of the blood flow towards skin, bone marrow, lung and gut. This shift might be responsible for the pronounced decrease of T-lymphocytes due to the extreme stress during therapeutic hyperthermia (Dougherty and Frank, 1953; Amaning and Olszewski, 1994; Dhabhar et al., 1996; Pedersen and Hoffman-Goetz, 2000). Shifts from lymphocytes out of the blood flow into lymphoid tissues have also been observed in mice-experiments with long lasting moderate "fever-range" whole-body hyperthermia (FR-WBH). It is postulated, that a decrease in the number of blood lymphocytes and L-selectin-positive lymphocytes is due to an increased property of these cells to migrate into secondary lymphoid tissue, tumour tissues and sites of inflammation.

In addition, FR-WBH has been demonstrated to increase the homing potential of lymphocytes via L-selectin and α4β7 integrin (Burd et al., 1998; Evans et al., 2001; Ostberg et al., 2002; Repasky and Issels, 2002; Shah et al., 2002).

On the other hand, it has been shown that infusion of epinephrine and β2-adrenoceptor agonists may cause mobilisation of lymphocytes from endothelial cells of the blood vessels as well as from lymphatic organs into the blood flow. This effect can be abolished by simultaneous infusion of β2-antagonists (Schedlowski et al., 1996; Benschop et al., 1997; Carlson et al., 1997; Klokker et al., 1997). Density of β2-receptors is high on NK-cells, intermediate on T8-cells and low on T4-cells (Madden et al., 1995). This may lead to different catecholamine-induced mobilisation of lymphocyte subpopulations and may therefore represent one reason for the observed different courses of numbers of blood lymphocyte subpopulations during extreme hyperthermia (Ahlers et al., 1998). In addition, epinephrine decreases T4-cell proliferation (Pedersen and Hoffman-Goetz, 2000) and may thus be responsible for the observed transient impairment of T-cell proliferation described by Atanackovic et al. (2002).

The change towards an anti-inflammatory pattern of lymphocyte controlling cytokines during extreme WBH is also similar to known effects of stress hormones: Cortisol and epinephrine suppress pro-inflammatory plasma levels of IL-12 and IFN-γ as well as expression of IL-12 receptors on lymphocytes, while production of anti-inflammatory IL-10 is increased (Madden et al., 1995;

Iwakabe et al., 1998; Ashwell et al., 2000; Elenkov, 2004). In addition, catecholamines and cortisol have been shown to inhibit the NKA (Madden et al., 1995; Pedersen and Hoffman-Goetz, 2000).

Taken together, activation of both the sympathetic-adrenomedullary system (SAS) and the hypothalamo-pituitary axis (HPA) as two different parts of the stress reaction, leads to marked shifts of lymphocytes between blood and other compartments. In addition, cytokine secretion, mediating lymphocyte function, changes to an anti-inflammatory pattern. Depending on the nature and duration of the stressor, these effects are mostly transient. This is in line with the above-mentioned alterations due to therapeutical hyperthermia.

Lymphocyte apoptosis as part of immune regulation and putative effect of therapeutical hyperthermia

Programmed cell death in lymphocytes is essential for immune regulation. Since alterations of immune responses are induced by hyperthermia, apoptosis might therefore occur after hyperthermia as well. Apoptosis represents a universal death programme that may be induced by a wide variety of stimuli in a relatively uniform way. During therapeutical hyperthermia, a variety of biochemical and immunological functions are altered. Besides direct toxic heat effects as described in the first paragraphs, many other factors could possibly induce apoptosis. The most relevant factors in the context of hyperthermia are therefore discussed in the following section.

Stress-induced apoptosis

As already mentioned above, systemic hyperthermia represents a condition associated with marked stress for both the single cell and the entire organism. The fact that stress might lead to immunosuppression is generally accepted. Recent investigation shows, that apoptosis of lymphocytes as well as of polymorphonuclear neutrophil leukocytes is relevantly involved in this phenomenon (Sasajima et al., 1999; Delogu et al., 2001a, b; Schroeder et al., 2001; Wang et al., 2002; Shi et al., 2003). The decrease of splenocytes due to apoptosis in mice exposed to two 12 h periods of physical stress is as high as 30–40% (Shi et al., 2003). In lymphocytes, the mechanisms involved have been extensively investigated and seem to be critically influenced by opioids. Shi et al. (2003) demonstrated, that endogenous opioids, which are increased in stressful situations, mediate increased expression of Fas-receptor and consecutively increased apoptosis of lymphocytes. On splenocytes mu-opioid-receptor-dependent increase of CD95-expression has been observed by Wang et al. (2002). This is of special interest, as opioids like fentanyl and remifentanil are used to induce analgesic sedation or general anaesthesia during therapeutical hyperthermia, and are also relevantly involved in the temperature regulation in certain circumstances associated with elevated body temperature (Romanovsky and Blatteis, 1998). In humans, the phenomenon of lymphocyte apoptosis due to stress has been reported in several different situations: After surgical trauma and general anaesthesia apoptosis of lymphocytes has been observed. Here, alterations in the mitochondrial energy metabolism seem to be involved in apoptosis induction (Delogu et al., 2001b). In patients with septic disorders, increased apoptosis of lymphocytes was reported as well (Schroeder et al., 2001). In this state of extreme stress and inflammation, certainly many factors may contribute to lymphocyte apoptosis. After major operations, T-cell apoptosis is transient: it appears 6–12 h after onset of stress and disappears 96 h after surgery (Delogu et al., 2001b; Wang et al., 2002; Shi et al., 2003).

In the course of WBH, increased lymphocyte apoptosis can be observed as well (Dieing et al., 2003). The above-described observations, together with the "stress-typical" immunological alterations suggest, that at least part of this phenomenon might be due to the stress caused by this treatment.

Immunologically induced apoptosis

Apoptosis of lymphocytes represents a crucial regulatory mechanism of the immune system, essential for development of immunity, maintenance of peripheral homeostasis and prevention of autoimmunity. Figure 3 gives an overview of the various instances during development; when lymphocytes may undergo programmed cell death. These processes are active in B- and T-cells. In NK-cells, the mechanisms seem to differ; therefore this phenomenon is described separately.

During development, T- and B-lymphocytes are eliminated by apoptosis when basic functions are insufficient, e.g. due to a missing antigen-receptor, or a defective IL-7-receptor with ineffective cytokine stimulation. Later in the maturation, lymphocytes are selected by proper immunologic function. An excessively avid or anergy TCR/MHC interaction leads to negative or failure of positive selection, with consecutive programmed cell death (Rathmell and Thompson, 2002). The mechanisms underlying the elimination of these cells are not entirely clear. However, the induction of nuclear steroid hormone receptor NUR77 might play a role, as it seems to regulate apoptosis through a transcriptional programme and by promoting cytochrome C release (Rathmell and Thompson, 2002). Fas and death receptor signalling through Fas-associated death domain (FADD) and caspase-8 in general are not required for thymocyte negative selection (Strasser and Bouillet, 2003). However, for the negative selection of autoreactive cells Bcl-2-related BH3 protein and receptor-induced apoptosis via TNF play a crucial role (Rathmell and Thompson, 2002; Strasser and Bouillet, 2003).

Homoeostasis of mature lymphocytes is tightly regulated by apoptosis. During early immune response, there is excessive lymphocyte expansion, which needs to be down regulated, when the antigen disappears and inflammation declines. The redundant lymphocytes are eliminated by "activation-induced cell death" (AICD) due to either cytokine withdrawal ("death-by-neglect") or receptor-induced apoptosis (Janeway et al., 2001).

After antigens and inflammatory cytokines are cleared at the end of an immune response, the missing stimulation of the activated lymphocytes leads to reduced glucose metabolism, loss of cellular ATP and mitochondrial dysfunction, and finally to programmed cell death. Death-by-neglect is caspase-independent, but regulated by proteins of the Bcl-2 family (Rathmell and Thompson, 2002).

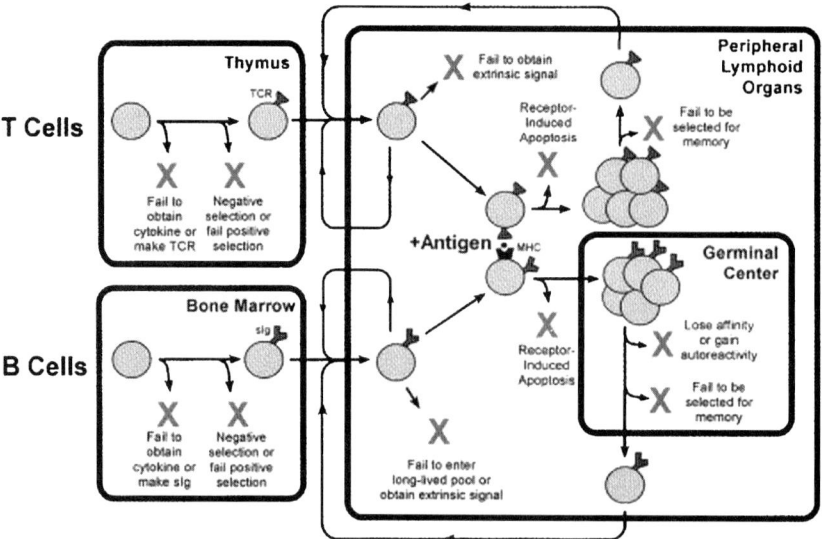

Fig. 3. T- and B-cells undergo apoptosis (red X) in many instances: During development in case of failure to compete for cytokines, failure to properly rearrange an antigen receptor (TCR or sIg), failure to be positively selected, or in case of negative selection. Resting lymphocytes are eliminated if they fail to compete for extrinsic signals and are therefore eliminated from the long-lived recirculating pool. During antigen binding in the periphery, T-cells might undergo receptor-induced apoptosis. B-cells that bind and present antigenic peptides on major histocompatibility complex (MHC) molecules to T-cells can also be eliminated. Loss of affinity for the selecting antigen or acquisition of autoreactivity also leads to programmed cell death. At the end of an immune reaction, lymphocytes that fail to become memory cells after antigen clearance as well undergo apoptotic cell death. (Adapted with permission from Rathmell and Thompson, 2002.)

Receptor-induced apoptosis of activated lymphocytes is mainly regulated by the TNF-receptor family: In early immune response, T-cells upregulate Fas, which acts as costimulator. With continued activation, T-cells as well express Fas-ligand (Fas-L), which leads to induction of apoptosis in the sense of suicide and/or fratricide, e.g. by cells responding to new antigenic challenge. IL-2, necessary for clonal expansion of T-cells, sensitises these cells for apoptosis in the later phase of inflammation (Krueger et al., 2003). Receptor-induced death is initiated by cleavage of the initiator caspase-8 and activation of executor caspases downstream in the cascade such as caspases-3 and -7 (Krueger et al., 2003).

During treatment of cancer patients with extreme WBH at 41.5° to 42°C an increased rate of T-cell-apoptosis can be observed (Fig. 4). This phenomenon is transient and disappears 24 h after treatment (Dieing et al., 2003). However, activation of T-cells, measured by expression of CD69, proliferation activity and serum levels of sIL2-R occur several hours later (Atanackovic et al., 2002). During WBH, however, a short period of immunosuppression dominates (Ahlers et al., 1998; Atanackovic et al., 2002). Thus, it is unlikely, that the observed apoptosis of lymphocytes immediately after WBH is induced by activation of these cells. It might rather be an unwanted side effect, contributing to the transient impairment of the immune system.

In contrast to other T-cells, cytotoxic T-cells increase in number during therapeutical hyperthermia. Therefore cytotoxic T-cells appear to be not affected by the immunosuppressive effect of WBH. It is known, that there is some kind of AICD in these cells as well: Leite-de-Moraes and coworkers observed apoptotic cell death in activated NKT-cells that could not be inhibited by the survival factor IL-7. Fas/Fas-L interaction, as well as activation of caspases are important mechanisms in activation-induced apoptosis of NKT-cells (Leite-de-Moraes et al., 2000). However there is no data on hyperthermia-induced activation or

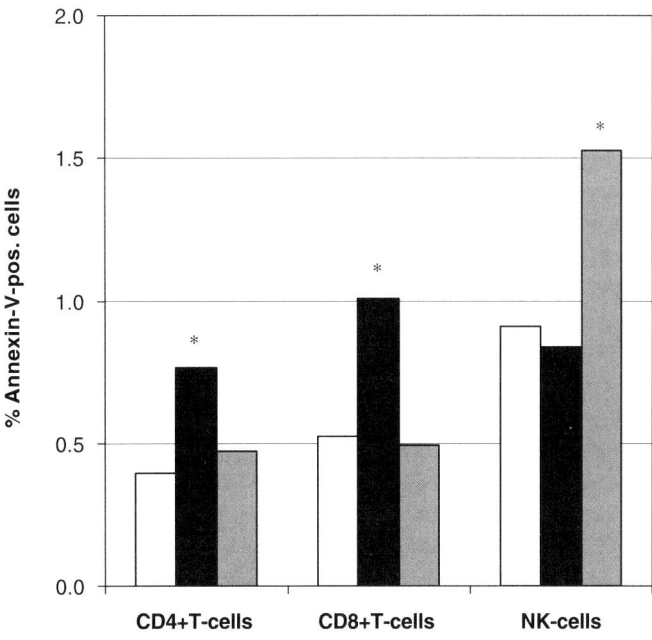

Fig. 4. Apoptosis of T- and NK-cells in per cent of the referring cells before □, during ■ and 24 h after whole-body hyperthermia ▨ in the peripheral blood of patients. *Significant differences compared to the pre-therapeutic value. Notice the delayed apoptosis of NK-cells.

cell death in these cells available and the role of these cells is currently not clear.

In NK-cells, activation induced cell death differs from AICD in T-cells. *In vitro*, NK-cell activation by CD2 or CD16, or ligation of CD94 is followed by induction of apoptosis with detectable DNA-laddering after 2 h (Ortaldo et al., 1995; Ida and Anderson, 1998).

Cell death induced by activation in those cells is independent of TNF-α or Fas-L, as neutralising antibodies against Fas-L or TNF-α cannot inhibit apoptosis. In addition, the caspase-specific peptide inhibitor Z-VADfmk cannot prevent apoptosis, suggesting a caspase-independent pathway (Ida et al., 2003). Instead apoptotic cell death after NK-cell activation seems to involve granzyme B and nitric oxide and requires IL-2-priming (Furuke et al., 1999; Ida et al., 2003). Granzyme B is normally released by NK-cells to induce apoptosis in target cells. NK-cells are protected from apoptosis by proteinase-inhibitor 9 (PI9) that inhibits granzyme B acitivity. However, in IL-2-activated NK-cells, this protecting protein is less induced compared to granzyme B, a constellation that predisposes for apoptotic cell death. Active granzyme B is able to cleave Bid, which in turn promotes mitochondrial dysfunction and thereby induces programmed cell death (Ida et al., 2003).

As the role of NK-cells in tumour defence is well known (reviewed in Brittenden et al., 1996), activation and possible enhancement of cytotoxicity of NK-cells by hyperthermia has been of high interest since the early 1980s. Various studies on the effect of heat on the NK-cell-function were made. These mostly *in vitro* experiments showed a limited potency of heat to enhance activity and cytotoxicity of lymphocytes (Azocar et al., 1982; Kalland and Dahlquist, 1983; Onsrud, 1988; Yang et al., 1992; Shen et al., 1994). A certain role of NK-cells in tumouricidal effects of fever-like hyperthermia has been shown *in vivo* in scid-mice bearing syngeneic breast xenographs: the number of NK-cells increased at the tumour site following 6–8 h 39.8°C hyperthermia, and tumour growth was significantly delayed. This effect could be abrogated by the anti-asialo GM1 antibody, which depletes

NK-cell activity (Burd et al., 1998). One small study showed an increase in cytotoxicity of NKT-cells up to 10% in three patients treated with WBH of 41.8°C for 1 h (Hegewisch-Becker et al., 1998). Although these data are preliminary, and the clinical consequences are unclear, the data could suggest, that the complex, potentially protective interactions of the immune system *in vivo* might allow NK-cell activation at temperatures as high as 41.8°C. The apoptosis rate of NK-cells in patients during therapeutical WBH is differing from the one observed in T-cells. In a study of our group, no increased apoptosis rate in NK-cells, detected by annexin-binding, could be measured during hyperthermia, but it was significantly increased when blood was taken from the same patients the following day (Fig. 4) (Dieing et al., 2003). These data are in line with *in vitro* experiments, where the apoptosis rate of NK-cells increased 24 h after hyperthermal treatment (Dieing et al., 2000). Whether these results indicate delayed cell death after initial activation of NK-cells by the elevated temperature is questionable, since neither enhancement of NK-cell-activity during WBH nor activation-induced cell death under such conditions has been definitely shown yet.

Drug-induced apoptosis

In hyperthermal conditions endogenous substances as steroids or opioids are increased (Robins et al., 1987b; Romanovsky and Blatteis, 1998). When systemic hyperthermia is applied therapeutically, additional drugs such as opioids, steroids and cytostatics are regularly applied.

A series of those drugs are able to induce apoptosis in lymphocytes. One of the earliest and most appealing models to gain insight into the regulation and biochemical mechanisms of apoptosis has been the death of lymphocytes induced by glucocorticoids (GC) (Migliorati et al., 1992; Migita et al., 1997; Planey and Litwack, 2000). In some diseases such as autoimmune disorders, this is part of the intended immune-suppressant therapeutic effect, as well as in cytokine-targeted therapy such as anti-TNF that induces apoptosis of lamina propria T-cells in patients with Crohn's disease (Van Den Brande et al., 2002).

For many other drugs such as cytostatics, opioids or tricyclic antidepressants, apoptosis of healthy lymphocytes is an undesired adverse effect (Xia et al., 1996; Provinciali et al., 1999; Delogu et al., 2001a; Stahnke et al., 2001).

The mechanism by which programmed cell death is induced can be different: Glucocorticoids induce loss of mitochondrial potential and subsequent activation of caspase 9. Mitogen activated protein kinases are very likely to play a role as well (Planey and Litwack, 2000; Frankfurt and Rosen, 2004). After exposure to high-dose steroids, apoptosis occurs within 2 h in lymphocytes *in vivo* and *in vitro* and the effect seems to disappear after 8 h (Migita et al., 1997).

Different mechanisms seem to be active in cytostatic drugs. A number of substances such as cisplatin, doxorubicin, mitomycin, fluorouracil and camptothecin are able to induce or increase CD95-expression and thereby the sensitivity for CD95-induced apoptosis. Doxorubicin and other antineoplastic drugs have been found to induce CD95 ligand, and the CD95 receptor/ligand interaction. Death receptor-independent cleavage of the proximal caspase-8 has also been found in drug-induced apoptosis *in vitro* and *in vivo*. However, experimental caspase-inhibition did not reduce the apoptosis rate, suggesting additional death pathways (Eichhorst et al., 2001; Stahnke et al., 2001).

Taken together, there are several factors that might explain the observed apoptosis induction in lymphocytes during WBH as stress, drugs or immunologic alterations. However, there are differences between the lymphocyte subpopulations. Of special interest is the delayed apoptosis of NK-cells, different to other lymphocyte subsets that might suggest a differentiated immunologic reaction. However, there is not enough data available to closer elucidate such a possible immunologic phenomenon.

Conclusion

Therapeutically utilised schemes of systemic hyperthermia are on the borderline between permanent damage and temporary affection of cellular function as well as enhancement of activation of immunologic cells. Especially NK-cells, for which a special role in tumour defence is known, might be effector cells of a hyperthermia-enhanced immune response. On the other hand, temperatures above 41°C, often used in therapeutical settings frequently impair lymphocytes *in vitro*. However, a series of clinical observations demonstrate a beneficial effect of this therapeutical option for tumour patients. Currently, it remains open, whether enhanced immune response contributes to these positive effects of heat treatment under clinical conditions.

References

Ahlers, O., Boehnke, T., Kerner, T., Deja, M., Keh, D., Löffel, J., Hildebrandt, B., Riess, H., Wust, P., Pappert, D. and Gerlach, H. (1998) Lymphocyte alterations during hyperthermia. Br. J. Anesth., 80: A 301.

Ahlers, O., Boehnke, T., Kerner, T., Deja, M., Keh, D., Löffel, J., Hildebrandt, B., Riess, H., Wust, P., Pappert, D. and Gerlach, H. (1999) Changes in serum cytokine levels during induced whole body hyperthermia. Crit. Care, 3: P 082.

Amaning, E.P. and Olszewski, W.L. (1994) Kinetics of distribution of recirculating lymphocytes during whole body hyperthermia. Arch. Immunol. Ther. Exp. Warsz., 42: 107–113.

Ashwell, J.D., Lu, F.W. and Vacchio, M.S. (2000) Glucocorticoids in T cell development and function. Annu. Rev. Immunol., 18: 309–345.

Atanackovic, D., Nierhaus, A., Neumeier, M., Hossfeld, D.K. and Hegewisch-Becker, S. (2002) 41.8 degrees C whole body hyperthermia as an adjunct to chemotherapy induces prolonged T cell activation in patients with various malignant diseases. Cancer Immunol. Immunother., 51: 603–613.

Azocar, J., Yunis, E.J. and Essex, M. (1982) Sensitivity of human natural killer cells to hyperthermia. Lancet, 1: 16–17.

Benschop, R.J., Schedlowski, M., Wienecke, H., Jacobs, R. and Schmidt, R.E. (1997) Adrenergic control of natural killer cell circulation and adhesion. Brain Behav. Immun., 11: 321–332.

Blazickova, S., Rovensky, J., Koska, J. and Vigas, M. (2000) Effect of hyperthermic water bath on parameters of cellular immunity. Int. J. Clin. Pharmacol. Res., 20: 41–46.

Borrelli, M.J., Calini, W.C., Ransom, B.R. and Dewey, W.C. (1986) Ion-sensitive microelectrode measurements of free intracellular chloride and potassium concentrations in hyperthermia-treated neuroblastoma cells. J. Cell Physiol., 126: 181–190.

Bouchama, A. and Knochel, J.P. (2002) Heat stroke. N. Engl. J. Med., 346: 1978–1988.

Brenner, I.K., Zamecnik, J., Shek, P.N. and Shephard, R.J. (1997) The impact of heat exposure and repeated exercise on circulating stress hormones. Eur. J. Appl. Physiol. Occup. Physiol., 76: 445–454.

Brittenden, J., Heys, S.D., Ross, J. and Eremin, O. (1996) Natural killer cells and cancer. Cancer, 77: 1226–1243.

Burd, R., Dziedzic, T.S., Xu, Y., Caliguiri, M.A., Subjeck, J.R. and Repasky, E.A. (1998) Tumor cell apoptosis, lymphocyte recruitment and tumor vascular changes are induced by low temperature, long-duration (fever-like) whole-body hyperthermia. J. Cell Physiol., 177: 137–147.

Carlson, S.L., Fox, S. and Abell, K.M. (1997) Catecholamine modulation of lymphocyte homing to lymphoid tissues. Brain Behav. Immun., 11: 307–320.

Coss, R.A. and Linnemans, W.A. (1996) The effects of hyperthermia on the cytoskeleton: a review. Int. J. Hyperthermia, 12: 173–196.

Delogu, G., Famularo, G., Moretti, S., De Luca, A., Tellan, G., Antonucci, A., Marandola, M. and Signore, L. (2001a) Interleukin-10 and apoptotic death of circulating lymphocytes in surgical/anesthesia trauma. J. Trauma, 51: 92–97.

Delogu, G., Moretti, S., Famularo, G., Antonucci, A., Signore, L., Marcellini, S., Lo Bosco, L. and De Simone, C. (2001b) Circulating neutrophils exhibit enhanced apoptosis associated with mitochondrial dysfunctions after surgery under general anaesthesia. Acta Anaesthesiol. Scand., 45: 87–94.

Dewey, W.C. (1989) Mechanism of thermal radiosensitization. In: Urano M. and Douple E. (Eds.), Biology of Thermal Potentiation of Radiotherapy. VSP, Utrecht, Tokyo, pp. 1–16.

Dewey, W.C. (1994) Arrhenius relationships from the molecule and cell to the clinic. Int. J. Hyperthermia, 10: 457–483.

Dewhirst, M.W., Viglianti, B.L., Lora-Michiels, M., Hanson, M. and Hoopes, P.J. (2003) Basic principles of thermal dosimetry and thermal thresholds for tissue damage from hyperthermia. Int. J. Hyperthermia, 19: 267–294.

Dhabhar, F.S., Miller, A.H., McEwen, B.S. and Spencer, R.L. (1996) Stress-induced changes in blood leukocyte distribution. Role of adrenal steroid hormones. J. Immunol., 157: 1638–1644.

Dieing, A., Ahlers, O., Kerner, T., Wust, P., Felix, R., Loffel, J., Riess, H. and Hildebrandt, B. (2003) Whole body hyperthermia induces apoptosis in subpopulations of blood lymphocytes. Immunobiology, 207: 265–273.

Dieing, A., Hildebrandt, B., Wust, P. and Riess, H. (2000) Differences in hyperthermia-induced programmed cell death in lymphocyte subpopulations in vitro. Immunobiology, 203: E. 4.

Dinarello, C.A., Dempsey, R.A., Allegretta, M., LoPreste, G., Dainiak, N., Parkinson, D.R. and Mier, J.W. (1986) Inhibitory effects of elevated temperature on human cytokine production and natural killer activity. Cancer Res., 46: 6236–6241.

Dougherty, T.F. and Frank, J.A. (1953) The quantitative and qualitative responses of blood lymphocytes to stress stimuli. J. Lab. Clin. Med., 42: 530–537.

Downing, J.F., Martinez-Valdez, H., Elizondo, R.S., Walker, E.B. and Taylor, M.W. (1988) Hyperthermia in humans enhances interferon-gamma synthesis and alters the peripheral lymphocyte population. J. Interferon Res., 8: 143–150.

Edwards, M.J., Saunders, R.D. and Shiota, K. (2003) Effects of heat on embryos and foetuses. Int. J. Hyperthermia, 19: 295–324.

Eichhorst, S.T., Muerkoster, S., Weigand, M.A. and Krammer, P.H. (2001) The chemotherapeutic drug 5-fluorouracil induces apoptosis in mouse thymocytes in vivo via activation of the CD95(APO-1/Fas) system. Cancer Res., 61: 243–248.

Elenkov, I.J. (2004) Glucocorticoids and the Th1/Th2 balance. Ann. N.Y. Acad. Sci., 1024: 138–146.

Elenkov, I.J. and Chrousos, G.P. (2002) Stress hormones, proinflammatory and antiinflammatory cytokines, and autoimmunity. Ann. N.Y. Acad. Sci., 966: 290–303.

Evans, S.S., Wang, W.C., Bain, M.D., Burd, R., Ostberg, J.R. and Repasky, E.A. (2001) Fever-range hyperthermia dynamically regulates lymphocyte delivery to high endothelial venules. Blood, 97: 2727–2733.

Faithfull, N.S., Reinhold, H.S., Berg, A.P.V.D., Rhoon, G.C.V., Zee, J.V.D. and Wike-Hooley, J.L. (1984) Cardiovascular changes during whole body hyperthermia treatment of advanced malignancy. Eur. J. Appl. Physiol., 53: 274–281.

Frankfurt, O. and Rosen, S.T. (2004) Mechanisms of glucocorticoid-induced apoptosis in hematologic malignancies: updates. Curr. Opin. Oncol., 16: 553–563.

Furuke, K., Burd, P.R., Horvath-Arcidiacono, J.A., Hori, K., Mostowski, H. and Bloom, E.T. (1999) Human NK cells express endothelial nitric oxide synthase, and nitric oxide protects them from activation-induced cell death by regulating expression of TNF-alpha. J. Immunol., 163: 1473–1480.

Galon, J., Franchimont, D., Hiroi, N., Frey, G., Boettner, A., Ehrhart-Bornstein, M., O'Shea, J.J., Chrousos, G.P. and Bornstein, S.R. (2002) Gene profiling reveals unknown enhancing and suppressive actions of glucocorticoids on immune cells. FASEB J., 16: 61–71.

Gehrmann, M., Marienhagen, J., Eichholtz-Wirth, H., Fritz, E., Ellwart, J., Jaattela, M., Zilch, T. and Multhoff, G. (2005) Dual function of membrane-bound heat shock protein 70 (Hsp70), Bag-4, and Hsp40: protection against radiation-induced effects and target structure for natural killer cells. Cell Death Differ., 12: 38–51.

Hammami, M.M., Bouchama, A., Shail, E., Aboul-Enein, H.Y. and Al-Sedairy, S. (1998) Lymphocyte subsets and adhesion molecules expression in heatstroke and heat stress. J. Appl. Physiol., 84(5): 1615–1621.

Haveman, J., Geerdink, A.G. and Rodermond, H.M. (1996) Cytokine production after whole body and localized hyperthermia. Int. J. Hyperthermia, 12: 791–800.

Hegewisch-Becker, S., Nierhaus, A., Panse, J., Wiedenmann, G. and Hossfeld, D.K. (1998) Whole body hyperthermia has a stimulatory effect on the immune cell activity in cancer patients. Ann. Oncol., 9(Suppl 4): 136–137.

Hellstrand, K. and Hermodsson, S. (1989) An immunopharmacological analysis of adrenaline-induced suppression of human natural killer cell cytotoxicity. Int. Arch. Allergy Appl. Immunol., 89: 334–341.

Hildebrandt, B., Hegewisch-Becker, S., Kerner, T., Nierhaus, A., Bakhshandeh-Bath, A., Janni, W., Zumschlinge, R., Sommer, H., Riess, H. and Wust, P. (2005) Current status of radiant whole-body hyperthermia at temperatures >41.5 degrees C and practical guidelines for the treatment of adults. The German 'Interdisciplinary Working Group on Hyperthermia'. Int. J. Hyperthermia, 21: 169–183.

Hildebrandt, B., Wust, P., Ahlers, O., Dieing, A., Sreenivasa, G., Kerner, T., Felix, R. and Riess, H. (2002) The cellular and molecular basis of hyperthermia. Crit. Rev. Oncol./Hematol., 43: 33–56.

Ida, H. and Anderson, P. (1998) Activation-induced NK cell death triggered by CD2 stimulation. Eur. J. Immunol., 28: 1292–1300.

Ida, H., Nakashima, T., Kedersha, N.L., Yamasaki, S., Huang, M., Izumi, Y., Miyashita, T., Origuchi, T., Kawakami, A., Migita, K., Bird, P.I., Anderson, P. and Eguchi, K. (2003) Granzyme B leakage-induced cell death: a new type of activation-induced natural killer cell death. Eur. J. Immunol., 33: 3284–3292.

Iwakabe, K., Shimada, M., Ohta, A., Yahata, T., Ohmi, Y., Habu, S. and Nishimura, T. (1998) The restraint stress drives a shift in Th1/Th2 balance toward Th2-dominant immunity in mice. Immunol. Lett., 62: 39–43.

Jaattelä, M. (1999) Heat shock proteins as cellular lifeguards. Ann. Med., 31: 261–271.

Janeway, C.A., Travers, P., Walport, M. and Shlomchik, M. (2001) Develpoment and survival of lymphocytes. In: *Immunobiology: The Immune System in Health and Disease*. Garland Science, New York.

Jolly, C. and Morimoto, R.I. (2000) Role of the heat shock response and molecular chaperones in oncogenesis and cell death. J. Natl. Cancer Inst., 92: 1564–1572.

Kalland, T. and Dahlquist, I. (1983) Effects of in vitro hyperthermia on human natural killer cells. Cancer Res., 43: 1842–1846.

Kampinga, H.H., Dynlacht, J.R. and Dikomey, E. (2004) Mechanism of radiosensitization by hyperthermia (> or = 43 degrees C) as derived from studies with DNA repair defective mutant cell lines. Int. J. Hyperthermia, 20: 131–139.

Kappel, M., Stadeager, C., Tvede, N., Galbo, H. and Pedersen, B.K. (1991) Effects of in vivo hyperthermia on natural killer cell activity, in vitro proliferative responses and blood mononuclear cell subpopulations. Clin. Exp. Immunol., 84: 175–180.

Katschinski, D., Wiedemann, G.J., Longo, W., d'Oleire, F.R., Spriggs, D. and Robins, H.I. (1999) Whole body hyperthermia cytokine induction: a review, and unifying hypothesis for myeloprotection in the setting of cytotoxic therapy. Cytokine Growth Factor Rev., 10: 93–97.

Kearns, R.J., Ringler, S., Krakowka, S., Tallman, R., Sites, J. and Oglesbee, M.J. (1999) The effects of extracorporal whole body hyperthermia on the functional and phenotypic features of canine peripheral blood mononuclear cells (PBMC). Clin. Exp. Immunol., 116: 188–192.

Kerner, T., Deja, M., Ahlers, O., Loffel, J., Hildebrandt, B., Wust, P., Gerlach, H. and Riess, H. (1999) Whole body hyperthermia: a secure procedure for patients with various malignancies? Intensive Care Med., 25: 959–965.

Kerner, T., Hildebrandt, B., Ahlers, O., Deja, M., Riess, H., Draeger, J., Wust, P. and Gerlach, H. (2003) Anaesthesiological experiences with whole body hyperthermia. Int. J. Hyperthermia, 19: 1–12.

Klokker, M., Secher, N.H., Madsen, P., Pedersen, M. and Pedersen, B.K. (1997) Adrenergic beta1- and beta1+2-receptor blockade suppress the natural killer cell response to head-up tilt in humans. J. Appl. Physiol., 83(5): 1492–1498.

Kohm, A.P. and Sanders, V.M. (2001) Norepinephrine and beta 2-adrenergic receptor stimulation regulate CD4+ T and B lymphocyte function in vitro and in vivo. Pharmacol. Rev., 53: 487–525.

Krause, S.W., Gastpar, R., Andreesen, R., Gross, C., Ullrich, H., Thonigs, G., Pfister, K. and Multhoff, G. (2004) Treatment of colon and lung cancer patients with ex vivo heat shock protein 70-peptide-activated, autologous natural killer cells: a clinical phase I trial. Clin. Cancer Res., 10: 3699–3707.

Kraybill, W.G., Olenki, T., Evans, S.S., Ostberg, J.R., O'Leary, K.A., Gibbs, J.F. and Repasky, E.A. (2002) A phase I study of fever-range whole body hyperthermia (FR-WBH) in patients with advanced solid tumours: correlation with mouse models. Int. J. Hyperthermia, 18: 253–266.

Kregel, K.C. (2002) Heat shock proteins: modifying factors in physiological stress response and acquired thermotolerance. J. Appl. Physiol., 92: 2177–2186.

Krueger, A., Fas, S.C., Baumann, S. and Krammer, P.H. (2003) The role of CD95 in the regulation of peripheral T-cell apoptosis. Immunol. Rev., 193: 58–69.

Leite-de-Moraes, M.C., Herbelin, A., Gouarin, C., Koezuka, Y., Schneider, E. and Dy, M. (2000) Fas/Fas ligand interactions promote activation-induced cell death of NK T lymphocytes. J. Immunol., 165: 4367–4371.

Lepock, J.R. (2003) Cellular effects of hyperthermia: relevance to the minimum dose for thermal damage. Int. J. Hyperthermia, 19: 252–266.

Li, C.Y. and Dewhirst, M.W. (2002) Hyperthermia-regulated immunogene therapy. Int. J. Hyperthermia, 18: 586–596.

Madden, K.S., Sanders, V.M. and Felten, D.L. (1995) Catecholamine influences and sympathetic neural modulation of immune responsiveness. Annu. Rev. Pharmacol. Toxicol., 35: 417–448.

Migita, K., Eguchi, K., Kawabe, Y., Nakamura, T., Shirabe, S., Tsukada, T., Ichinose, Y., Nakamura, H. and Nagataki, S. (1997) Apoptosis induction in human peripheral blood T lymphocytes by high-dose steroid therapy. Transplantation, 63: 583–587.

Migliorati, G., Pagliacci, C., Moraca, R., Crocicchio, F., Nicoletti, I. and Riccardi, C. (1992) Glucocorticoid-induced apoptosis of natural killer cells and cytotoxic T lymphocytes. Pharmacol. Res., 26(Suppl 2): 26–27.

Milani, V., Noessner, E., Ghose, S., Kuppner, M., Ahrens, B., Scharner, A., Gastpar, R. and Issels, R.D. (2002) Heat shock protein 70: role in antigen presentation and immune stimulation. Int. J. Hyperthermia, 18: 563–575.

Multhoff, G. (2002) Activation of natural killer cells by heat shock protein 70. Int. J. Hyperthermia, 18: 576–585.

Myerson, R.J., Roti Roti, J.L., Moros, E.G., Straube, W.L. and Xu, M. (2004) Modelling heat-induced radiosensitization: clinical implications. Int. J. Hyperthermia, 20: 201–212.

Onsrud, M. (1988) Effect of hyperthermia on human natural killer cells. Recent Results Cancer Res., 109: 50–56.

Ortaldo, J.R., Mason, A.T. and O'Shea, J.J. (1995) Receptor-induced death in human natural killer cells: involvement of CD16. J. Exp. Med., 181: 339–344.

Ostberg, J.R., Kaplan, K.C. and Repasky, E.A. (2002) Induction of stress proteins in a panel of mouse tissues by fever-range whole body hyperthermia. Int. J. Hyperthermia, 18: 552–562.

Pedersen, B.K. and Hoffman-Goetz, L. (2000) Exercise and the immune system: regulation, integration, and adaptation. Physiol. Rev., 80: 1055–1081.

Planey, S.L. and Litwack, G. (2000) Glucocorticoid-induced apoptosis in lymphocytes. Biochem. Biophys. Res. Commun., 279: 307–312.

Provinciali, M., Ciavattini, A., Di-Stefano, G., Argentati, K. and Garzetti, G.G. (1999) In vivo amifostine (WR-2721) prevents chemotherapy-induced apoptosis of peripheral blood lymphocytes from cancer patients. Life Sci., 64: 1525–1532.

Rathmell, J.C. and Thompson, C.B. (2002) Pathways of apoptosis in lymphocyte development, homeostasis, and disease. Cell, 109(Suppl): S97–S107.

Repasky, E. and Issels, R. (2002) Physiological consequences of hyperthermia: heat, heat shock proteins and the immune response. Int. J. Hyperthermia, 18: 486–489.

Rhind, S.G., Gannon, G.A., Shephard, R.J., Buguet, A., Shek, P.N. and Radomski, M.W. (2004) Cytokine induction during exertional hyperthermia is abolished by core temperature clamping: neuroendocrine regulatory mechanisms. Int. J. Hyperthermia, 20: 503–516.

Robertson, M.J. and Ritz, J. (1990) Biology and clinical relevance of human natural killer cells. Blood, 76: 2421–2438.

Robins, H.I., Kalin, N.H., Shelton, S.E., Martin, P.A., Shecterle, L.M., Barksdale, C.M., Neville, A.J. and Marshall, J. (1987a) Rise in plasma beta-endorphin, ACTH, and cortisol in cancer patients undergoing whole body hyperthermia. Horm. Metab. Res., 19: 441–443.

Robins, H.I., Kalin, N.H., Shelton, S.E., Shecterle, L.M., Barksdale, C.M., Martin, P.A. and Marshall, J. (1987b) Neuroendocrine changes in patients undergoing whole body hyperthermia. Int. J. Hyperthermia, 3: 99–105.

Robins, H.I., Kutz, M., Wiedemann, G.J., Katschinski, D.M., Paul, D., Grosen, E., Tiggelaar, C.L., Spriggs, D., Gillis, W. and d'Oleire, F. (1995) Cytokine induction by 41.8°C whole body hyperthermia. Cancer Lett., 97: 195–201.

Romanovsky, A.A. and Blatteis, C.M. (1998) Pathophysiology of opioids in hyperthermic states. Prog. Brain Res., 115: 111–127.

Sakaguchi, Y., Stephens, L.C., Makino, M., Kaneko, T., Strebel, F.R., Danhauser, L.L., Jenkins, G.N. and Bull, J.M. (1995) Apoptosis in tumors and normal tissues induced by whole body hyperthermia in rats. Cancer Res., 55: 5459–5464.

Sasajima, K., Inokuchi, K., Onda, M., Miyashita, M., Okawa, K.I., Matsutani, T. and Takubo, K. (1999) Detection of T cell apoptosis after major operations. Eur. J. Surg., 165: 1020–1023.

Schedlowski, M., Hosch, W., Oberbeck, R., Benschop, R.J., Jacobs, R., Raab, H.R. and Schmidt, R.E. (1996) Catecholamines modulate human NK cell circulation and function via spleen-independent beta 2-adrenergic mechanisms. J. Immunol., 156: 93–99.

Schroeder, S., Lindemann, C., Decker, D., Klaschik, S., Hering, R., Putensen, C., Hoeft, A., von Ruecker, A. and Stuber, F. (2001) Increased susceptibility to apoptosis in circulating lymphocytes of critically ill patients. Langenbecks Arch. Surg., 386: 42–46.

Shah, A., Unger, E., Bain, M.D., Bruce, R., Bodkin, J., Ginnetti, J., Wang, W.C., Seon, B., Stewart, C.C. and Evans, S.S. (2002) Cytokine and adhesion molecule expression in primary human endothelial cells stimulated with fever-range hyperthermia. Int. J. Hyperthermia, 18: 534–551.

Sharma, H.S. and Hoopes, P.J. (2003) Hyperthermia induced pathophysiology of the central nervous system. Int. J. Hyperthermia, 19: 325–354.

Shen, R.N., Lu, L., Young, P., Shidnia, H., Hornback, N.B. and Broxmeyer, H.E. (1994) Influence of elevated temperature on natural killer cell activity, lymphokine-activated killer cell activity and lectin-dependent cytotoxicity of human umbilical cord blood and adult blood cells. Int. J. Radiat. Oncol. Biol. Phys., 29: 821–826.

Shi, Y., Devadas, S., Greeneltch, K.M., Yin, D., Allan Mufson, R. and Zhou, J.N. (2003) Stressed to death: implication of lymphocyte apoptosis for psychoneuroimmunology. Brain Behav. Immun., 17(Suppl 1): S18–S26.

Srivastava, P. (2002) Roles of heat-shock proteins in innate and adaptive immunity. Nat. Rev. Immunol., 2: 185–194.

Stahnke, K., Fulda, S., Friesen, C., Strauss, G. and Debatin, K.M. (2001) Activation of apoptosis pathways in peripheral blood lymphocytes by in vivo chemotherapy. Blood, 98: 3066–3073.

Strasser, A. and Bouillet, P. (2003) The control of apoptosis in lymphocyte selection. Immunol. Rev., 193: 82–92.

Tsan, M.F. and Gao, B. (2004) Cytokine function of heat shock proteins. Am. J. Physiol. Cell Physiol., 286: C739–C744.

Van Den Brande, J.M., Peppelenbosch, M.P. and Van Deventer, S.J. (2002) Treating Crohn's disease by inducing T lymphocyte apoptosis. Ann. N.Y. Acad. Sci., 973: 166–180.

Vidair, C. and Dewey, W.C. (1988) Two distinct modes of hyperthermic death. Radiat. Res., 116: 157–171.

Vujaskovic, Z. and Song, C.W. (2004) Physiological mechanisms underlying heat-induced radiosensitization. Int. J. Hyperthermia, 20: 163–174.

Wang, J., Charboneau, R., Barke, R.A., Loh, H.H. and Roy, S. (2002) Mu-opioid receptor mediates chronic restraint stress-induced lymphocyte apoptosis. J. Immunol., 169: 3630–3636.

Woiciechowsky, C., Asadullah, K., Nestler, D., Eberhardt, B., Platzer, C., Schoening, B., Gloeckner, F., Lanksch, W.R., Volk, H.D. and Doecke, W.D. (1998) Sympathetic activation triggers systemic interleukin-10 release in immunodepression induced by brain injury. Nat. Med., 4: 808–813.

Wust, P., Hildebrandt, B., Sreenivasa, G., Rau, B., Gellermann, J., Riess, H., Felix, R. and Schlag, P.M. (2002) Hyperthermia in combined treatment of cancer. Lancet Oncol., 3: 487–497.

Xia, Z., DePierre, J.W. and Nassberger, L. (1996) Dysregulation of bcl-2, c-myc, and Fas expression during tricyclic antidepressant-induced apoptosis in human peripheral lymphocytes. J. Biochem. Toxicol., 11: 203–204.

Yang, H., Lauzon, W. and Lemaire, I. (1992) Effects of hyperthermia on natural killer cells: inhibition of lytic function and microtubule organization. Int. J. Hyperthermia, 8: 87–97.

Yonezawa, M., Otsuka, T., Matsui, N., Tsuji, H., et al. (1996) Hyperthermia induces apoptosis in malignant fibrous histiocytoma cells in vitro. Int. J. Cancer, 66: 347–351.

van der Zee, J. (2002) Heating the patient: a promising approach? Ann. Oncol., 13: 1173–1184.

Zellner, M., Hergovics, N., Roth, E., Jilma, B., Spittler, A. and Oehler, R. (2002) Human monocyte stimulation by experimental whole body hyperthermia. Wien Klin Wochenschr, 114: 102–107.

CHAPTER 9

Cerebral pathophysiology and clinical neurology of hyperthermia in humans

Olaf L. Cremer[1],* and Cor J. Kalkman[2]

[1]*Department of Intensive Care Medicine, University Medical Center, Q04.460, PO Box 85500, 3508 GA Utrecht, The Netherlands*
[2]*Division of Perioperative Care and Emergency Medicine, University Medical Center, E03.511, PO Box 85500, 3508 GA Utrecht, The Netherlands*

Abstract: Deliberate hyperthermia has been used clinically as experimental therapy for neoplastic and infectious diseases. Several case fatalities have occurred with this form of treatment, but most were attributable to systemic complications rather than central nervous system toxicity. Nonetheless, demyelating peripheral neuropathy and neurological symptoms of nausea, delirium, apathy, stupor, and coma have been reported. Temperatures exceeding 40°C cause transient vasoparalysis in humans, resulting in cerebral metabolic uncoupling and loss of pressure-flow autoregulation. These findings may be related to the development of brain edema, intracerebral hemorrhage, and intracranial hypertension observed after prolonged therapeutic hyperthermia. Furthermore, deliberate hyperthermia critically worsens the extent of histopathological damage in animal models of traumatic, ischemic, and hypoxic brain injury. However, it is unknown whether these findings translate to episodes of spontaneous fever in neurologically injured patients. In a clinical setting fever is a strong prognostic marker of a patient's primary degree of neuronal damage, and a causal relation with long-term functional neurological outcome has not been established for most types of brain injury. Furthermore, in the neurosurgical intensive-care unit fever is extremely common whereas antipyretic therapy is only poorly effective. Therefore maintaining strict normothermia may be an impossible goal in many patients. Although there are several physiological arguments for avoiding exogenous hyperthermia in neurologically injured patients, there is no evidence that aggressive attempts at controlling spontaneous fever can improve clinical outcome.

Keywords: therapeutic hyperthermia; pyrexia; brain injury; outcome; cerebral autoregulation; cerebral blood flow; cerebral metabolic rate

Introduction

Problems of hyperthermia and related brain dysfunction have been known since the early periods of civilization. Heat-related deaths and mental illnesses are described in the ancient Indian literature as well as during Biblical times (Sharma and Hoopes, 2003). Hyperthermia denotes a rise in core body temperature above the hypothalamic set point and clinically results from thermoregulatory failure, whereas temperature elevation resulting from intact homeostatic responses is categorized as fever or pyrexia (Simon, 1993). Hyperthermia may

*Corresponding author. Tel.: +31-30-2503261;
Fax: +31-30-2505032; E-mail: o.l.cremer@umcutrecht.nl

evolve in response to direct external causes (fires, burning, and related rescue operations) or to exertion in a hot environment. Hyperthermia can also occur after use of some addictive drugs that cause metabolic activation together with diminished heat dissipation from peripheral vasoconstriction (cocaine and meth-amphetamine). Alternatively, hyperthermia can be deliberately induced for treatment of malignant diseases, or occur accidentally during rewarming after hypothermic cardiopulmonary bypass or treatment of accidental hypothermia ('overshoot') (Robins et al., 1989; Bissonnette et al., 2000; van der Zee, 2002). Compared with spontaneous fever, the rise in body temperature during deliberate or accidental hyperthermia is relatively fast and brief.

The histopathology and pathophysiology of hyperthermic brain injury has been extensively studied in the laboratory, but remains poorly investigated in clinical settings. This chapter focuses on the clinical neurology and cerebral pathophysiology of deliberate hyperthermia in humans. In addition, we critically discuss the relation between spontaneous fever and neurological outcome in patients with central nervous system (CNS) injuries.

Therapeutic hyperthermia in humans

Local, regional, or systemic hyperthermia has been used since ancient times to treat various ailments. Since the 1960s laboratory evidence has generated a scientific rationale for its use in the treatment of neoplastic diseases. Generally, temperatures in excess of 41°C *in vitro* will kill cancer cells exponentially as a function of time (Robins et al., 1989). Compared with local or regional body heating, the potential of whole body hyperthermia is to destroy deep-seated tumors and metastatic disease alone or in combination with radiotherapy and chemotherapy. Temperatures between 40 and 44°C are cytotoxic for cells in an environment with a low pO_2 and low pH. These conditions are found specifically within tumor tissue where blood supply is insufficient (van der Zee, 2002). Recently, whole body hyperthermia has also been explored for treatment of some chronic infectious diseases (Alonso et al., 1994; Ash et al., 1997; Zablow et al., 1997). Whole body hyperthermia has been induced clinically with warm contact media (such as water, water-heated suits or mats, hot air), infrared radiation, or extracorporeal perfusion methods (Bull et al., 1979; Cole et al., 1979; Parks et al., 1979; Robins et al., 1985; Vertrees et al., 1996; Kerner et al., 1999; Vertrees et al., 2000). Generally, heavy sedation or general anesthesia is required with these methods to reduce discomfort and cardiovascular stress.

Clinical effects

In clinical studies of therapeutic hyperthermia, fluid loss, hemodynamic alterations, serum enzyme abnormalities, and other symptoms of variable severity have been described (Table 1). Several case fatalities have occurred with this form of treatment, but most were attributable to systemic complications rather than CNS toxicity (Levin and Blair, 1982; van der Zee et al., 1983; Koga et al., 1985; Alonso et al., 1994). In awake subjects, whole body hyperthermia causes neurological symptoms, including nausea, confusion, disorientation, apathy, delirium, stupor, or coma (Sharma and Hoopes, 2003). The pathogenesis of these abnormalities is likely to be multifactorial, involving hypoxia, hypotension, metabolic derangements, and dehydration (Simon, 1993). Most of these neurologic abnormalities resolve with correction of the underlying disorder and resolution of the hyperthermia. However, from observations in heat stroke it is known that loss of thermoregulatory capacity is associated with a disturbed level of consciousness, brain edema, and a high mortality if the temperature exceeds 40°C (Yaqub and Al Deeb, 1998). Accordingly, during prolonged therapeutic whole body hyperthermia the development of brain edema, intracerebral hemorrhage, intracranial hypertension, and demyelating peripheral neuropathy have been reported (Table 1). These complications are probably related to the temperature-induced morphological changes in axons, nerve cells, glial cells, and vascular endothelium that have been observed in rodents (Sharma and Hoopes, 2003).

Despite the fact that therapeutic hyperthermia has been used for decades, the effects of heat on the structural integrity and function of the human brain are still largely unknown. Clinical observations in selected groups and in patients with cancer who received whole body hyperthermia as treatment suggest that humans can tolerate a body temperature of 41.6° to 42°C for 45 min–8 h, but further studies are necessary (Bynum et al., 1978). In particular, maximum safe exposure time and temperature ('critical thermal maximum'), effects of anesthesia and age in relation to hyperthermia, and the development of thermal tolerance following repeated heat exposure require further investigation in humans.

Physiological effects

Knowledge about the cerebral physiological responses to temperature increase is important for the application of therapeutic hyperthermia, but the influence of temperature on cerebral pressure autoregulation, flow-metabolism coupling, and brain energy consumption is of particular concern to clinicians who encounter fever as a complication in patients recovering from acute neurologic injury. Pressure autoregulation refers to the capacity of the brain to maintain capillary hydrostatic pressure and cerebral blood flow (CBF) approximately constant in spite of variations in cerebral perfusion pressure. Flow-metabolism coupling refers to the ability of the brain to change cerebrovascular resistance in response to regional alterations in demand. If cerebral metabolic demand increases during hyperthermia, it is important to establish whether flow-metabolism coupling remains intact, because injured CNS tissue is potentially at risk for ischemia. If pressure autoregulation becomes impaired during hyperthermia, brain-injured patients may be at particular risk for cerebral hypoperfusion during periods of low blood pressure, whereas they may be prone to the development of vasogenic edema, vascular engorgement, and worsening of intracranial hypertension during periods of high blood pressure.

In the clinical setting of the neurosurgical intensive-care unit (ICU) fever coexists with (and is caused by) an inflammatory response in the injured brain. Thus, it may be impossible to conceptually distinguish the causal effects of either temperature increase or inflammation on brain metabolism and autoregulation. For reviewing the cerebrovascular effects of temperature increase *per se*, therefore, we will strictly focus the discussion on experimental hyperthermia in the non-injured brain.

Cerebral vasomotor responses and autoregulation

In animals, a CBF increase during hyperthermia has generally, but not constantly, been found (Table 2). In dogs, an increase of the rectal temperature to 41.5°C caused cerebral vasoparalysis, resulting in a 7.8%/°C increase of regional CBF as measured by the hydrogen clearance method (Katsumura et al., 1995). In anesthetized swine, CBF measured by microsphere injection increased by 21–24%/°C when temperature was raised to approximately 42°C (Busija et al., 1988; Vertrees et al., 2000). However, Ohmoto et al. (1996) have studied the temporal responses of regional cortical blood flow to varying temperatures in rats. These authors showed that at 41°C cortical blood flow first increases twofold and then subsequently normalizes. At 43°C regional CBF first increases 1.3-fold and subsequently decreases even to below baseline, and at 45°C regional CBF immediately decreases. Other authors have shown that there are also large regional differences within the brain in the response of CBF to heating (Table 2).

In humans, the cerebrovascular effects of hyperthermia are incompletely understood. We recently studied 19 patients with chronic hepatitis C virus infection, who were subjected to experimental therapy with extracorporeal whole body hyperthermia at 41.8°C for 120 min under propofol anesthesia (23 treatment sessions total) (Cremer et al., 2004). During 13 sessions end-tidal carbon dioxide concentrations were allowed to increase during heating, whereas during another 10 sessions end-tidal carbon dioxide was maintained constant. We assessed cerebral pressure-flow autoregulation by static tests using phenylephrine infusion and by assessing the transient hyperemic response in the

Table 1. Clinical experience with therapeutic whole body hyperthermia

Author	Patients	N^a	Heating method	Thermal dose	Neurologic complications	Systemic complications
Pettigrew et al. (1974)	Various neoplasms	227/51	Immersion in hot wax	41.0° to 41.8°C for 4 h	Non-reported	16% Mortality, ventricular fibrillation, myocardial ischemia, liver necrosis, jaundice, disseminated intravascular coagulation, alveolitis, circumoral herpes simplex, pressure necrosis, burns
Cole et al. (1979) Bull et al. (1979), Lees et al. (1980)	Various neoplasms Various neoplasms	?/60 ±500/57	Water blanket Water suit	?°C for 2 h 41.8°C for 4 h	Fatigue, nausea One case fatality (brain herniation), focal cortical hyperactivity, fatigue, nausea, lethargy, neuromuscular irritability, demyelination neuropathy	No mortality, diarrhea, burns Arrhythmias, diarrhea, burns, transiently elevated liver enzymes
Barlogie et al. (1979)	Various neoplasms	4?	Water blanket	42.0°C for 4 h	Nausea, fatigue, seizures, neuropathy	Rhabdomyolysis, electrolyte shifts, disseminated intravascular coagulation
Ostrow et al. (1981)	Various neoplasms	7/7	Water suit	41.8°C for 4 h	Fatigue, nausea and vomiting	Extremity edema, diarrhea, respiratory depression, transiently elevated liver enzymes
Levin and Blair (1982)	Various neoplasms	682/232	Surface heating	41.8°C (for ?h)	Transverse myelitis, coma, disorientation, lethargy, anorexia, peripheral nerve palsies	Case fatalities (due to arrhythmias, disseminated intravascular coagulation, liver failure), pressure necrosis, anemia, circumoral herpes simplex
van der Zee et al. (1983)	Various neoplasms	27/27	Hot air + water mattress	41.8° to 42.0°C for 2 h	Agitation, vomiting	Two case fatalities, arrhythmias, respiratory distress, liver necrosis, muscle damage, electrolyte shifts, diarrhea
Gerad et al. (1984)	Soft tissue sarcoma	35/11	Water suit + blanket	41.8° to 43.0°C for 4 h	Nausea and vomiting, myalgia, transient neuropathy	Edema, diarrhea, burns, arrhythmias, thrombocytopenia, transiently elevated liver enzymes, hyperglycemia, hypophosphatemia, circumoral herpes simplex

Reference	Condition	Treatments/patients[a]	Method	Temperature/duration	Side effects	Complications
Robins et al. (1985)	Various neoplasms	52/12	Radiant heat	41.8°C for 2.5 h	Fatigue	No significant clinical toxicity
Koga et al. (1985)	Gastrointestinal cancer	17/17	Extracorporeal perfusion	41.5°C for 3–5 h	Transient neuropathy	Two case fatalities, worsening hepatorenal syndrome, reversible muscle weakness
Eisler et al. (1985, 1990)	Various neoplasms	116/50	Extracorporeal perfusion	41.8°C for 2–10 h; 42.2°C for 2 h; 42.6°C for 2 h + 15 min at 43°C	Brain edema for hours to days	Cardiac electrical instability, electrolyte shifts, transiently elevated liver enzymes
Adam et al. (1987)	Rectum carcinoma	4/1	?	41.8°C (?)	Severe permanent neuropathy	
Alonso et al. (1994)	HIV-related Kaposi's sarcoma	31/31	Extracorporeal perfusion	42.0°C for 1 h	CNS hemorrhage	Two case fatalities, cardiac arrhythmia, coagulopathy, pressure necrosis
Wiedemann et al. (1994)	Sarcoma and teratoma	49/19	Extracorporeal perfusion	41.8°C for 1 h	Transient neuropathy, no central nervous system toxicity	Severe renal dysfunction (with ifosfamide and carboplatin administration), cardiopulmonary distress, edema, diarrhea, pressure necrosis, circumoral herpes simplex
Pontiggia et al. (1995)	HIV infection	10/10	Infrared radiation	42.0°C for 1 h	None reported	None reported
Kerner et al. (1999, 2003)	Various neoplasms	57/22	Infrared radiation	41.8°C for 1 h	Disorientation and depressed consciousness (up to 6 days)	Transient rise in creatinine, supraventricular tachy-arrhythmias, interstitial lung edema
Kraybill et al. (2002)	Advanced solid tumors	9/9	Infrared radiation	39.5° to 40.0°C for 6 h	Restlessness, frontal headache	Blisters
Cremer et al. (2004)	Chronic hepatitis C	23/19	Extracorporeal perfusion	41.8°C for 2 h	Agitation, permanent peripheral neuropathy	Blisters, pressure necrosis, diarrhea, transiently elevated liver enzymes

[a]Treatments/patients.

Table 2. Effect of induced hyperthermia on cerebral hemodynamics and metabolism

Author	Species	Experimental conditions	Heating method	Target temp. (°C)		Variables measured[a]
Demers et al. (1969)	Dog	Barbiturate anesthesia		37.8→41.5	CBF ≈	CMRO$_2$↑14%/°C
Nemoto and Frankel (1970b)	Dog	Barbiturate anesthesia			CBF ≈	CMRO$_2$↑; CMR$_{gluc}$ ≈
Carlsson et al. (1976)	Rat	Barbiturate anesthesia	Ambient air heating	→42.0	CBF ≈; ↑	CMRO$_2$↑
McCulloch et al. (1982)	Rat	Awake		37.4→40.2		CMR$_{gluc}$ ≈ or ↑ (regional differences)
Busija et al. (1988)	Pig	Newborn piglets	α-Chloralose anesthesia	36.3→41.5	CBF↑19%/°C	CMRO$_2$↑13%/°C
Yamada (1989)	Rabbit	Awake	Hot air	→43	CBF↑20%/°C	Vasoconstriction response preserved
				→44	CBF↑25%/°C	
				→45	CBF↑36%/°C	
Moriyama (1990)	Monkey	Anesthesia	Microwave radiation	37→43	CBF↑10%/°C	Structural damage
				37→45	CBF↑, then ↓↓	
Michenfelder et al. (1991)	Dog	Halothane anesthesia/transient global ischemia	Extracorporeal perfusion	37.0→40.0	CBF↑7%/°C	CMRO$_2$↑8%/°C CMR$_{gluc}$↑27%/°C
Katsumura et al. (1995)	Dog	Barbiturate anesthesia	Extracorporeal perfusion	37.8→41.5	CBF↑8%/°C	ICP↑; autoregulation ↓; no BBB leakage
Ohmoto et al. (1996)	Rat	Isoflurane anesthesia	Microwave radiation	37→41	CBF↑, then ≈	-
				37→43	CBF↑, then ↓	BBB disruption
				37→45	CBF↓↓	BBB disruption
Mickley et al. (1997)	Rat	Ketamine anesthesia	Microwave radiation or hot air	38.0→39.9		CMR$_{gluc}$↑ (all regions)
Vertrees et al. (2000)	Pig	Fentanyl/diazepam anesthesia	Extracorporeal perfusion	35.8→42.4	CBF↑23%/°C	
Heyman et al. (1950)	Human	Awake/neurosyphilis	Bacterial pyrogen	37.2±0.3 → 39.4±0.4	CBF ≈	AVDO$_2$ ≈	CMRO$_2$ ≈
	Human	Awake/dementia paralytica	Bacterial pyrogen	37.1±0.3 → 39.2±0.8	CBF↑14%/°C	AVDO$_2$ ≈ (↓3%/°C)	CMRO$_2$↑11%/°C
Nunneley et al. (2002)	Human	Awake volunteers	Water-heated suit	36.8→38.6			CMR$_{gluc}$↑ and ↓ (regional differences)
Nybo et al. (2002)	Human	Awake volunteers	Exercise induced	37.9→39.5	CBF↓11%/°C (concomitant PaCO$_2$↓)	AVDO$_2$↑21%/°C (concomitant PaCO$_2$↓)	CMRO$_2$↓4%/°C; CMR$_{gluc}$↑6%/°C
Cremer et al. (2004)	Human	Propofol anesthesia/hepatitis C	Extracorporeal perfusion	36.7±0.5 → 41.9±0.1	CBFV↑16%/°C	AVDO$_2$↓9%/°C	CMRO$_2$ ≈ (estimated)

[a] Note: A 50% decrease (division by two) is an effect size of equal magnitude as a 100% increase (multiplying by two).

middle cerebral artery to a 10-s carotid compression and release. Using the latter method, we showed that warming resulted in a gradual impairment of cerebral pressure autoregulation capacity when temperature exceeded approximately 40°C (Fig. 1). This effect remained after multivariate statistical adjustment for changes in $PaCO_2$, mean arterial pressure, and propofol blood concentration. We also found that hyperthermia at 41.8°C resulted in a 1.8-fold increase in middle cerebral artery flow velocity and a simultaneous 1.9-fold decrease in arterial to jugular venous oxygen extraction. Together, these observations suggest that profound systemic hyperthermia >40°C causes a transient vasoparalysis which results in cerebral metabolic uncoupling. At lower temperatures, the cerebrovascular responsiveness remains grossly unaffected, as suggested by Fig. 1 and findings by other authors of an increased index of dynamic autoregulation during a 0.4°C rise of core temperature in awake volunteers taking hot baths (Doering et al., 1999).

The observation of a temperature-induced impairment of cerebral vasoreactivity implies that high fever could potentially aggravate vascular engorgement and intracranial hypertension in patients in the neurosurgical ICU. Although the blood-brain barrier is most likely grossly preserved at temperatures below 42° to 43°C (Katsumura et al., 1995; Ohmoto et al., 1996), it has been suggested that brain edema may occur easily if the arterial blood pressure fluctuates excessively when cerebral autoregulation is absent (van Rhoon and van der Zee, 1983; Rumana et al., 1998; Gupta et al., 2002).

Cerebral metabolism

During hypothermia, there is a well-documented reduction of CBF and metabolic requirement for oxygen by approximately 4.4–6%/°C, i.e., Q_{10} is

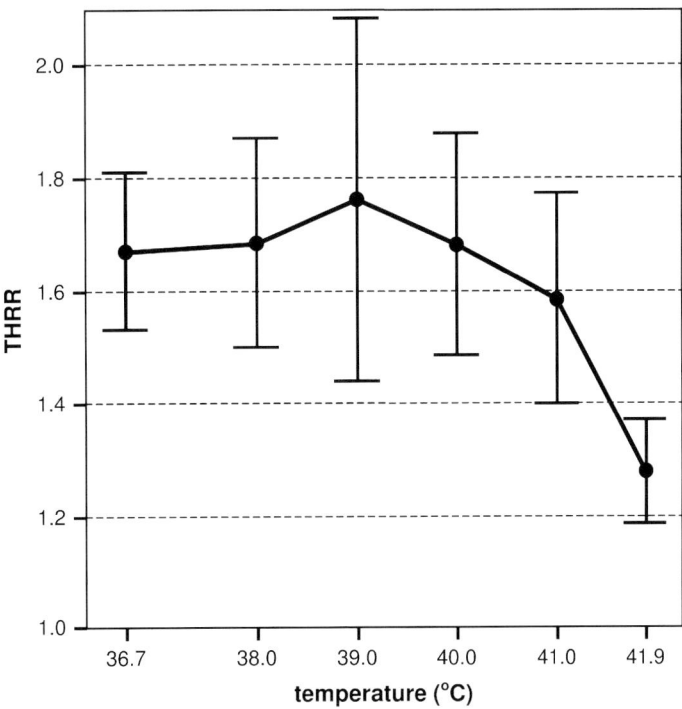

Fig. 1. Cerebral pressure-flow autoregulation during whole body hyperthermia. The transient hyperemic response ratio (THRR) measures the middle cerebral artery blood flow velocity increase in response to a 10-s ipsilateral carotid occlusion. Values near 1 indicate complete absence of a cerebrovascular response, whereas higher values indicate a hyperemic response consistent with normal autoregulation. Values shown are means with 95% confidence intervals.

between 2.0 and 3.0 (Q_{10} is the ratio of the metabolic rates associated with two temperatures that differ by 10°C) (Michenfelder, 1988; van der Linden et al., 1991; Cottrell and Smith, 1994; Walter et al., 2000). It is reasonable to assume that initially the effects of hyperthermia on the brain must be the opposite of those produced by hypothermia, although the relevant temperature range is much smaller, and ultimately the pathophysiology may differ. However, there are relatively few studies that have formally assessed the effect of hyperthermia on CBF and metabolic rate for oxygen or glucose.

Most animal studies indicate that resting cerebral oxygen and glucose consumption appear to increase during hyperthermia, although the magnitude of these metabolic alterations may show considerable regional heterogeneity (Table 2) (McCulloch et al., 1982; Nunneley et al., 2002). In anesthetized newborn piglets, cerebral metabolic rate of oxygen ($CMRO_2$) increased by $16 \pm 6\%/°C$ when the temperature was raised to 42°C (Busija et al., 1988). Michenfelder et al. (1991) showed in a model of sublethal global ischemia at 37 and 40°C in dogs that CBF and $CMRO_2$ both increase by 6%/°C. Similarly, in rats CBF and $CMRO_2$ show an increase of approximately 5%/°C rise in body temperature up to 42°C (Carlsson et al., 1976). Mickley et al. (1997) found that cerebral glucose utilization was increased in all brain regions during induced hyperthermia in rats, whereas Nemoto and Frankel (1970a) found in dogs that during hyperthermia the cerebral metabolic rate for oxygen was increased, but not for glucose.

These reports contrast with studies in humans which indicate that cerebral metabolism does not necessarily increase in all brain regions during induced hyperthermia. Nunneley et al. (2002) used positron emission tomography imaging in 10 awake volunteers wearing a water-heated suit. During a temperature increase from 36.7° to 38.6°C these authors demonstrated increases in cerebral metabolic rate in some brain regions, but decreases in other regions. It is unclear whether these changes should be ascribed to a Q_{10} temperature effect or to changed neuronal activity due to central thermoregulatory responses and changes in somatosensory input in these surface heated awake volunteers.

In our recent study of patients with chronic hepatitis C virus infection who were subjected to experimental treatment with whole body hyperthermia under general anesthesia, we observed a 16%/°C increase in middle cerebral artery blood flow velocity and a simultaneous decrease of cerebral arterial-to-venous oxygen extraction of an approximately equal magnitude (Cremer et al., 2004). Consequently, there appeared to be no increase in the estimated $CMRO_2$ in these patients during profound hyperthermia (Fig. 2). This finding may indicate the absence of an increased demand for oxygen (i.e., metabolic shutdown), but could alternatively also imply the development of a hyperthermia-induced impairment of mitochondrial oxygen metabolism at temperatures exceeding 40°C. In the latter case, if true metabolic demand were high, cerebral lactate production would be expected in most patients during hyperthermia, but this was not observed. As early as 1944, other authors have reported a decline in cerebral oxygen uptake with temperature increase from 40° to 43°C (Field et al., 1944).

Classically, the relationship between temperature and the intrinsic velocity of a chemical reaction (or metabolic rate) is exponential (van't Hoff's law). However, this mathematical model relates only to the kinetics of single chemical reactions, and the pathways by which the brain converts and consumes energy are much more complex. In the hypothermic range the exponential relationship between temperature and $CMRO_2$ is lost between 21 and 18°C, when electrical function ceases (Steen et al., 1983). In a similar way, it can be hypothesized that cortical electrical activity becomes progressively suppressed during profound hyperthermia. This would explain the apparent lack of an increased cerebral oxygen demand during hyperthermia in our study. Indeed, the electroencephalogram in a majority of cases indicated suppression of cortical electrical activity (burst-suppression pattern) during the experimental treatment under propofol anesthesia at 41.8°C (unpublished data).

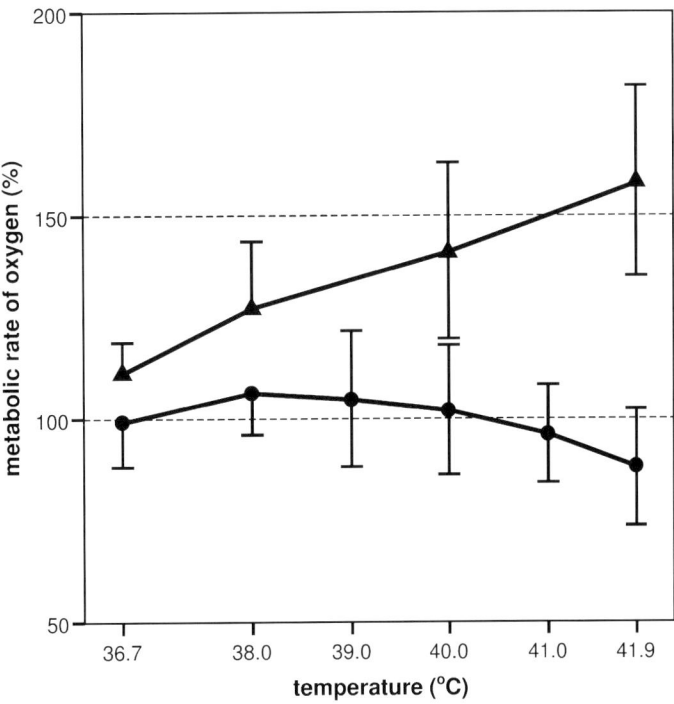

Fig. 2. Oxygen consumption during whole body hyperthermia. Dots show estimated cerebral metabolic rate of oxygen; triangles show systemic metabolic rate of oxygen; both are expressed as a percentage relative to the metabolic rate of oxygen at baseline, before the start of extracorporeal heating. Values shown are means with 95% confidence intervals.

Fever in neurologically injured patients

Induced variations in brain temperature of only 1° to 2°C can critically determine the extent of histopathological injury resulting from transient cerebral ischemia in animal models (Busto et al., 1987; Dietrich, 1992; Wass et al., 1995; Baena et al., 1997). Clinically, elevated body temperature after stroke has been associated with an increased infarct size, death, and poor outcome in survivors (Reith et al., 1996). Similar findings apply for patients who are comatose following traumatic injury or cardiopulmonary resuscitation (Zeiner et al., 2001; Cairns and Andrews, 2002). For these reasons, it is believed that even mild elevations of brain temperature can be detrimental for the hypoxic, ischemic, or injured brain, and it is generally recommended that temperature should be closely monitored and tightly controlled in all patients with neurologic injuries (Ginsberg and Busto, 1998). However, there is currently no evidence that doing so improves clinical outcome.

Epidemiology of fever in the neurosurgical ICU

Fever is extremely frequent in patients admitted to the neurosurgical ICU. Depending on the temperature threshold to define pyrexia, it occurs in at least 50% of neurologically injured patients. Fever is particularly frequent (approximately 73%) after traumatic brain injury (compared with other pathologies), and its incidence appears to be higher in more severe cases and in patients with a prolonged stay (Albrecht et al., 1998; Kilpatrick et al., 2000; Stocchetti et al., 2002). Pyrexia in the ICU is often undertreated (Albrecht et al., 1998). Various antipyretic drugs have been used to control fever, including acetaminophen, indomethacin, diclofenac, barbiturates, and propranolol (Meythaler and

Stinson, 1994; Cormio et al., 2000; Cairns and Andrews, 2002; Dippel et al., 2003). Physical refrigeration has also commonly been employed, using fans, alcohol rubs, ice packs, or gastric lavage with cold saline. However, both techniques are only poorly effective, reducing mean core temperature by only 0.32°C for physical refrigeration alone, 0.58°C for antipyretic drugs alone, and 0.54°C for the combination of the two (Stocchetti et al., 2002). Recently several devices have been introduced designed to rapidly cool patients in the ICU, either via external cooling pads, or via intravenous catheter devices that use counter-current flow of cooled water in combination with increased surface area.

It is important to note that brain temperature commonly exceeds core body temperature. For example, in a series of 20 comatose patients 73% of intracranial temperature measurements were higher than 38°C, but only 58% of pulmonary-artery recordings (Rossi et al., 2001). In observational studies of severely head-injured patients, the difference between brain and core temperature typically ranged from −0.3° to +2.1°C in individuals, depending on characteristics of the patient, on the exact position of the temperature probes, and on changes in physiological variables such as cerebral perfusion pressure and the temperature itself (Henker et al., 1998; Rumana et al., 1998; Rossi et al., 2001). In this regard, it has been noted that the gap between brain and core temperature increases (i.e., brain temperature may be more markedly underestimated) as pyrexia evolves. Considering these findings, brain temperatures >38°C may be so common in the neurosurgical ICU that a temperature above this threshold should probably be considered a 'normal' response to CNS injury.

Inflammation and fever following CNS injury

There is good evidence that the inflammatory response that follows cerebral injury is initiated in the CNS itself. Following traumatic brain injury in adults a classic acute phase response has been reported, including elevations in serum zinc, C-reactive protein, and temperature (Young et al., 1988). In both patients and animal models of head injury, elevated concentrations of interleukins and prostaglandins in the cerebrospinal fluid (DeWitt et al., 1988), expression of endothelial adhesion molecules, and recruitment of neutrophils to contused brain regions have been reported. The presence of these inflammatory mediators in the CNS causes fever. In addition, disruption of the blood-brain barrier after cerebral injury permits systemically released pyrogens to gain increased access to the neural cells that mediate the febrile response (Natale et al., 2000). Subsequently, exposure to hyperthermia can further increase the cerebrovascular permeability, thereby exacerbating cytokine exposure to the brain (Dietrich et al., 1990, 1991). Alternatively, the disruption of hypothalamic temperature regulation by head trauma may cause early hyperthermia more directly.

Fever and neurologic outcome in patients

Despite the fact that many studies have confirmed an association between body temperature and outcome from various neurological conditions, a causal relation has not been established in a clinical setting for most types of brain injury. In contrast to the laboratory, where brain temperature can be rigidly controlled, pyrexia in patients is closely linked to the extent of their neurologic impairment (Reith et al., 1996; Stocchetti et al., 2002). As a consequence, temperature is a prognostic marker and not necessarily a causal factor for the outcome. Furthermore, antipyretic drugs or physical refrigeration are only poorly effective for controlling body temperature (Stocchetti et al., 2002), and it is therefore uncertain whether these therapies can impact clinical outcome in any case. Only intervention trials can provide definite answers to these issues, but because such studies are lacking, we will briefly review observational data to determine the level of clinical evidence for a causal association between fever and poor outcome from various types of brain injury.

Stroke

Early body temperature after stroke is clinically related to initial stroke severity and the presence of intraventricular or subarachnoid blood on the head computed tomography scan (Reith et al., 1996; Schwarz et al., 2000). Recently, the correlation between initial infarct size and the time-course and degree of hyperthermia has been confirmed experimentally (Abraham et al., 2002). This direct relation may confound any etiologic association between fever and poor clinical outcome.

In a group of 260 patients with a hemispheric cerebral infarction, Castillo et al. (1998) found that hyperthermia $>37.5°C$ initiated within the first 24 h from stroke onset, but not afterward, was related to larger infarct volume and higher neurological deficit and dependency at 3 months. Similarly, Wang et al. (2000) found an association between a high admission temperature and 1-year mortality for ischemic stroke (but not for hemorrhagic stroke) which remained after adjustment for other clinical variables of stroke severity. For hemorrhagic stroke the association between admission temperature and mortality was not significant in this study, but others have found the duration of fever to be an independent prognostic factor also in patients following intracerebral hemorrhage (Schwarz et al., 2000). A meta-analysis involving a total of 3790 patients has attempted to quantify the relation between temperature and clinical outcome (Hajat et al., 2000). Seven out of nine studies found pyrexia to be associated with increased mortality or morbidity after stroke, whereas two studies found no association. As a consequence, the combined odds ratio for death following fever was 'only' 1.19 (95% CI: 0.99–1.43). However, this odds ratio does not represent the true independent relation between temperature and outcome, because multivariate adjustment for other markers of stroke severity was not consistently used.

Traumatic brain injury

Compared with (ischemic) stroke, the association between fever and poor neurologic outcome is less well documented for traumatic brain injury. Pyrexia early after head injury is associated with a poor Glasgow Coma Scale on presentation, the presence of diffuse axonal injury, cerebral edema on the initial head computed tomography scan, systolic hypotension, hyperglycemia, and leukocytosis (Cairns and Andrews, 2002). Furthermore, fever can be caused by infection and inflammation (cytokine-mediated) or by hypothalamic dysfunction. As a consequence, it is difficult to determine whether there is a causative relation or simply an association between fever and poor outcome.

In a series of 71 head-injured patients with at least a single episode of pyrexia, the duration of fever was a significant predictor of mortality, but not morbidity, in univariate analyses (Jones et al., 1994). Early fever was associated with a poor Glasgow Coma Scale score on ICU discharge in children in multivariate analyses (Natale et al., 2000). In a study of 110 adults, pyrexia was associated with periods of intracranial hypertension in patients with normal perimesencephalic cisterns, but not in patients with compressed cisterns (Stocchetti et al., 2002). Despite this association, there was no relationship between the neurologic outcome at 6 months and either the presence or duration of fever. This finding shows some resemblance to the results of a large randomized controlled trial of deliberate hypothermia in traumatic brain injury, which showed improved control of intracranial pressure, but no effect on clinical outcome in the head-injured population at large (Clifton et al., 2001).

Non-traumatic subarachnoid hemorrhage

Patients with subarachnoid hemorrhage are at risk for cerebral ischemia due to vasospasm. Hyperthermia may potentially worsen this vasospasm-mediated brain injury. However, blood in the cerebrospinal fluid induces fever in experimental models (Frosini et al., 1999) and temperature is thus a likely marker for the primary severity of the hemorrhage. Nonetheless, fever $>38.3°C$ was associated with poor functional outcome (odds ratio 1.4 per day; 95% CI: 1.11–1.88) in a prospective study of 93 patients with non-traumatic subarachnoid hemorrhage, and this association remained independent of other predictors of outcome (Oliveira-Filho et al., 2001).

Anoxic brain injury

Fever early after successful resuscitation from cardiac arrest is a marker of clinical brain death (Takino and Okada, 1991; Takasu et al., 2001). This association may confound the etiologic (causal) relation between spontaneous fever and poor outcome in this setting. In a prospective study of 151 patients who were admitted after restoration of spontaneous circulation, the maximum attained temperature in the ICU (during the first 48 h) was a strong independent predictor of the maximum attained neurologic performance category of patients (odds ratio 2.26/°C; 95% CI: 1.24–4.12) (Zeiner et al., 2001). However, the endpoint in this study was somewhat peculiar, because the neurologic recovery status was assessed regardless of whether patients eventually died. Furthermore, in their report the authors disregarded the fact that mild hypothermia during the first 4 h after cardiac arrest also predicted severe disability, coma, or a persistent vegetative state, whereas this is the time that the brain may conceptually be most susceptible to secondary ischemic injury. Despite these methodological limitations, it is likely that fever has a causal and detrimental effect on neurological outcome following anoxic brain injury, since two randomized controlled trials have recently provided clinical evidence for the efficacy of deliberate hypothermia for neuroprotection after cardiac arrest (Bernard et al., 2002; Hypothermia after Cardiac Arrest Study Group, 2002).

Cognitive dysfunction after cardiac surgery

It has been suggested that postoperative hyperthermia is also a modulator of neurocognitive dysfunction following cardiac surgery. In a prospective series of 300 patients undergoing coronary artery bypass graft surgery, Grocott et al. (2002) have has addressed this possible relationship. They reported somewhat inconsistent results because the incidence of cognitive dysfunction at 6 weeks was associated with the maximum temperature attained in the postoperative period ($p = 0.05$), but not with the area under the curve for temperatures $>37°C$ ($p = 0.45$).

To address this issue further, van Dijk et al. (2002) at our center recently reanalyzed 198 patients from a randomized controlled trial conducted in The Netherlands, in which the effects of coronary artery bypass graft surgery with (on-pump) and without (off-pump) cardiopulmonary bypass on cognitive outcome was assessed (Octopus Trial). In this trial the on-pump patients, but not the off-pump patients, received an intraoperative 1 mg/kg bolus of dexamethasone to suppress the inflammatory response associated with cardiopulmonary bypass. As a result, 56% of the off-pump patients had a maximum postoperative temperature $>38°C$, compared with only 14% of the on-pump patients ($p<0.001$, unpublished data). However, there was no association between the mean or maximum postoperative temperature (during the first 48 h) and cognitive outcome at either 4 days or 3 or 12 months after surgery, both in univariate or multivariate analyses (personal communication).

Mechanisms of temperature toxicity in experimental CNS injury

In the laboratory it has been well-established that small differences in brain temperature can critically determine the extent of ischemic neuronal injury (Busto et al., 1987; Dietrich, 1992; Wass et al., 1995; Baena et al., 1997). However, several rodent studies that reported increased neuronal damage with hyperthermia may in fact have restored normothermia (38° to 39°C in the rat) from a hypothermic baseline, because anesthesia typically reduces core temperature by at least 1°C. More recently, Noor et al. (2003) studied the effect of true hyperthermia on infarct volume in a focal embolic model of cerebral ischemia in rats. Controlled hyperthermia (brain temperature 39.8°C) significantly increased infarct volume, neurological deficits, and mortality following embolic occlusion of the middle cerebral artery significantly. The same group reported recently that treatment with tissue plasminogen activator significantly reduced infarct volume (preformed clot embolization of the middle cerebral artery) both in normothermic and 38°C hyperthermic rats (Noor et al., 2005).

However, compared with normothermic rats, perfusion deficits in hyperthermic rats were significantly increased at both 3 and 6 h after ischemia, indicating early progression of the ischemic penumbral region to irreversibly damaged tissue. There was also evidence of a more severely disrupted blood-brain barrier, because in the hyperthermic rats Evans blue dye extravasation was increased.

The exact mechanisms by which hypothermia improves and hyperthermia worsens outcome in animal models of brain injury are incompletely understood. The protective effect of profound hypothermia generally has been related to its ability to reduce CBF and metabolic requirements for oxygen and glucose, thus blunting the cascade of secondary injury processes (Dietrich, 1992). At the opposite temperature range, hyperthermia may worsen (or possibly merely accelerate) the secondary cerebral responses to injury. These effects include (1) enhanced release of neurotransmitters, (2) exaggerated oxygen radical production, (3) increased numbers of potentially damaging ischemic depolarizations in the focal ischemic penumbra, (4) impaired recovery of energy metabolism and enhanced inhibition of protein kinases, (5) worsening of cytoskeletal proteolysis, and (6) exacerbation of polymorphonuclear leukocyte infiltration (Ginsberg and Busto, 1998). As a consequence, in rats the oxidative stress that is imposed by deliberate hyperthermia is an important factor in causing breakdown of the blood-brain barrier, brain edema formation, and structural cell damage (Chatzipanteli et al., 2000; Sharma et al., 2000, 2003; Westman et al., 2000). Mitochondrial and plasma membranes appear to be the most temperature-sensitive cellular elements, with irreversible transitions in protein structure or arrangements starting to occur at temperatures higher than 40°C (Dietrich, 1992; Iwagami, 1996; Lepock, 2003).

Most of the knowledge about the effects of temperature variation on the extent of secondary brain injury has been derived from laboratory studies in which animals were subjected to hypothermia or hyperthermia following transient global or focal cerebral ischemia. There are other reviews which discuss these effects in more detail (Dietrich, 1992; Ginsberg and Busto, 1998; Cairns and Andrews, 2002). However, despite a wealth of experimental data it is still unknown whether the studied pathophysiology of deliberate hyperthermia in rats translates to episodes of spontaneous fever in patients.

Conclusions

Hyperthermia is both a cause and a consequence of CNS injury. Excessive heat causes functional neurologic impairment and structural nervous tissue damage, the severity of which depends on the extent and duration of hyperthermia. Thus, prevention of iatrogenic hyperthermia is probably wise in patients who are at risk for or recovering from neurologic injury. In contrast, (mild) fever is extraordinarily common in the neurosurgical ICU and is closely linked to the 'normal' inflammatory response following CNS injury. Although early pyrexia after most types of brain injury is clearly associated with an unfavorable neurologic outcome in patients, it is still uncertain whether fever causes poor outcome or whether processes that result in poor outcome also produce fever. It is conceivable that a spontaneous increase in brain temperature would merely accelerate the rate of cell necrosis and apoptosis, without necessarily aggravating the extent of final neuronal damage. Most antipyretic drugs and routine cooling techniques are only poorly effective in reducing fever. If temperature is to be vigorously controlled within the normothermic range in all patients, invasive refrigeration methods are likely to be required. However, the risk/benefit ratio of such practice has not yet been evaluated. Although there are several physiologic arguments for controlling temperature in neurologically injured patients, there is no sound evidence that aggressive attempts at reducing fever can improve outcome.

Abbreviations

CBF	cerebral blood flow
$CMRO_2$	cerebral metabolic rate of oxygen
CNS	central nervous system
ICU	intensive-care unit

References

Abraham, H., Somogyvari-Vigh, A., Maderdrut, J.L., Vigh, S. and Arimura, A. (2002) Filament size influences temperature changes and brain damage following middle cerebral artery occlusion in rats. Exp. Brain Res., 142: 131–138.

Adam, A.M., Hughes, R.A., Payan, J. and McColl, I. (1987) Peripheral neuropathy and hyperthermia. Lancet, 1: 1270–1271.

Albrecht, R.F., Wass, C.T. and Lanier, W.L. (1998) Occurrence of potentially detrimental temperature alterations in hospitalized patients at risk for brain injury. Mayo Clin. Proc., 73: 629–635.

Alonso, K., Pontiggia, P., Sabato, A., Calvi, G., Curto, F.C., de Bartolomei, E., Nardi, C. and Cereda, P. (1994) Systemic hyperthermia in the treatment of HIV-related disseminated Kaposi's sarcoma. Long-term follow-up of patients treated with low-flow extracorporeal perfusion hyperthermia. Am. J. Clin. Oncol., 17: 353–359.

Ash, S.R., Steinhart, C.R., Curfman, M.F., Gingrich, C.H., Sapir, D.A., Ash, E.L., Fausset, J.M. and Yatvin, M.B. (1997) Extracorporeal whole body hyperthermia treatments for HIV infection and AIDS. ASAIO J., 43: M830–M838.

Baena, R.C., Busto, R., Dietrich, W.D., Globus, M.Y. and Ginsberg, M.D. (1997) Hyperthermia delayed by 24 h aggravates neuronal damage in rat hippocampus following global ischemia. Neurology, 48: 768–773.

Barlogie, B., Corry, P.M., Yip, E., Lippman, L., Johnston, D.A., Khalil, K., Tenczynski, T.F., Reilly, E., Lawson, R., Dosik, G., Rigor, B., Hankenson, R. and Freireich, E.J. (1979) Total-body hyperthermia with and without chemotherapy for advanced human neoplasms. Cancer Res., 39: 1481–1489.

Bernard, S.A., Gray, T.W., Buist, M.D., Jones, B.M., Silvester, W., Gutteridge, G. and Smith, K. (2002) Treatment of comatose survivors of out-of-hospital cardiac arrest with induced hypothermia. N. Engl. J. Med., 346: 557–563.

Bissonnette, B., Holtby, H.M., Davis, A.J., Pua, H., Gilder, F.J. and Black, M. (2000) Cerebral hyperthermia in children after cardiopulmonary bypass. Anesthesiology, 93: 611–618.

Bull, J.M., Lees, D., Schuette, W., Whang-Peng, J., Smith, R., Bynum, G., Atkinson, E.R., Gottdiener, J.S., Gralnick, H.R., Shawker, T.H. and DeVita Jr., V.T. (1979) Whole body hyperthermia: a phase-I trial of a potential adjuvant to chemotherapy. Ann. Intern. Med., 90: 317–323.

Busija, D.W., Leffler, C.W. and Pourcyrous, M. (1988) Hyperthermia increases cerebral metabolic rate and blood flow in neonatal pigs. Am. J. Physiol., 255: H343–H346.

Busto, R., Dietrich, W.D., Globus, M.Y., Valdes, I., Scheinberg, P. and Ginsberg, M.D. (1987) Small differences in intraischemic brain temperature critically determine the extent of ischemic neuronal injury. J. Cereb. Blood Flow Metab., 7: 729–738.

Bynum, G.D., Pandolf, K.B., Schuette, W.H., Goldman, R.F., Lees, D.E., Whang-Peng, J., Atkinson, E.R. and Bull, J.M. (1978) Induced hyperthermia in sedated humans and the concept of critical thermal maximum. Am. J. Physiol., 235: R228–R236.

Cairns, C.J. and Andrews, P.J. (2002) Management of hyperthermia in traumatic brain injury. Curr. Opin. Crit. Care, 8: 106–110.

Carlsson, C., Hagerdal, M. and Siesjo, B.K. (1976) The effect of hyperthermia upon oxygen consumption and blood flow in the cerebral cortex of the rat. Acta Anaesthesiol. Scand., 26: 1001–1006.

Castillo, J., Davalos, A., Marrugat, J. and Noya, M. (1998) Timing for fever-related brain damage in acute ischemic stroke. Stroke, 29: 2455–2460.

Chatzipanteli, K., Alonso, O.F., Kraydieh, S. and Dietrich, W.D. (2000) Importance of posttraumatic hypothermia and hyperthermia on the inflammatory response after fluid percussion brain injury: biochemical and immunocytochemical studies. J. Cereb. Blood Flow Metab., 20: 531–542.

Clifton, G.L., Miller, E.R., Choi, S.C., Levin, H.S., McCauley, S., Smith Jr., K.R., Muizelaar, J.P., Wagner Jr., F.C., Marion, D.W., Luerssen, T.G., Chesnut, R.M. and Schwartz, M. (2001) Lack of effect of induction of hypothermia after acute brain injury. N. Engl. J. Med., 344: 556–563.

Cole, D.R., Pung, J., Kim, Y.D., Berman, R.A. and Cole, D.F. (1979) Systemic thermotherapy (whole body hyperthermia). Int. J. Clin. Pharmacol. Biopharm., 17: 329–333.

Cormio, M., Citerio, G., Spear, S., Fumagalli, R. and Pesenti, A. (2000) Control of fever by continuous, low-dose diclofenac sodium infusion in acute cerebral damage patients. Intensive Care Med., 26: 552–557.

Cottrell, J.E. and Smith, D.S. (1994) Anesthesia and neurosurgery. Mosby-Year Book, St. Louis, MO.

Cremer, O.L., Diephuis, J.C., van Soest, H., Vaessen, P.H., Bruens, M.G., Hennis, P.J. and Kalkman, C.J. (2004) Cerebral oxygen extraction and autoregulation during extracorporeal whole body hyperthermia in humans. Anesthesiology, 100: 1101–1107.

Demers, H.G., Spaich, P. and Usinger, W. (1969) Der Hirnkreislauf bei erhöhter Körper-temperature. Verh. Dtsch. Ges. Kreislaufforschg., 35: 131–140.

DeWitt, D.S., Kong, D.L., Lyeth, B.G., Jenkins, L.W., Hayes, R.L., Wooten, E.D. and Prough, D.S. (1988) Experimental traumatic brain injury elevates brain prostaglandin E2 and thromboxane B2 levels in rats. J. Neurotrauma, 5: 303–313.

Dietrich, W.D. (1992) The importance of brain temperature in cerebral injury. J. Neurotrauma, 9(Suppl 2): S475–S485.

Dietrich, W.D., Busto, R., Halley, M. and Valdes, I. (1990) The importance of brain temperature in alterations of the blood-brain barrier following cerebral ischemia. J. Neuropathol. Exp. Neurol., 49: 486–497.

Dietrich, W.D., Halley, M., Valdes, I. and Busto, R. (1991) Interrelationships between increased vascular permeability and acute neuronal damage following temperature-controlled brain ischemia in rats. Acta Neuropathol. (Berl.), 81: 615–625.

van Dijk, D., Jansen, E.W., Hijman, R., Nierich, A.P., Diephuis, J.C., Moons, K.G., Lahpor, J.R., Borst, C., Keizer, A.M., Nathoe, H.M., Grobbee, D.E., De Jaegere, P.P. and Kalkman, C.J. (2002) Cognitive outcome after off-pump and on-pump coronary artery bypass graft surgery: a randomized trial. JAMA, 287: 1405–1412.

Dippel, D.W., van Breda, E.J., van der Worp, H.B., van Gemert, H.M., Meijer, R.J., Kappelle, L.J. and Koudstaal, P.J. (2003) Effect of paracetamol (acetaminophen) and ibuprofen on body temperature in acute ischemic stroke PISA, a phase II double-blind, randomized, placebo-controlled trial. BMC Cardiovasc. Disord., 3: 2.

Doering, T.J., Aaslid, R., Steuernagel, B., Brix, J., Niederstadt, C., Breull, A., Schneider, B. and Fischer, G.C. (1999) Cerebral autoregulation during whole-body hypothermia and hyperthermia stimulus. Am. J. Phys. Med. Rehabil., 78: 33–38.

Eisler, K., Hipp, R., Gogler, S. and Lange, J. (1990) New clinical aspects of whole body hyperthermia. Adv. Exp. Med. Biol., 267: 393–398.

Eisler, K., Landauer, B., Hipp, R., Kolb, E., Lange, J., Siewert, J.R., Zanker, K. and Blumel, G. (1985) Experiences with therapeutic whole-body hyperthermia. Anaesthesist, 34: 299–303.

Field, J., Fuhrman, F.A. and Martin, A.W. (1944) Effect of temperature on the oxygen consumption of brain tissue. J. Neurophysiol., 7: 117–126.

Frosini, M., Sesti, C., Valoti, M., Palmi, M., Fusi, F., Parente, L. and Sgaragli, G. (1999) Rectal temperature and prostaglandin E2 increase in cerebrospinal fluid of conscious rabbits after intracerebroventricular injection of hemoglobin. Exp. Brain Res., 126: 252–258.

Gerad, H., van Echo, D.A., Whitacre, M., Ashman, M., Helrich, M., Foy, J., Ostrow, S., Wiernik, P.H. and Aisner, J. (1984) Doxorubicin, cyclophosphamide, and whole body hyperthermia for treatment of advanced soft tissue sarcoma. Cancer, 53: 2585–2591.

Ginsberg, M.D. and Busto, R. (1998) Combating hyperthermia in acute stroke: a significant clinical concern. Stroke, 29: 529–534.

Grocott, H.P., Mackensen, G.B., Grigore, A.M., Mathew, J., Reves, J.G., Phillips-Bute, B., Smith, P.K. and Newman, M.F. (2002) Postoperative hyperthermia is associated with cognitive dysfunction after coronary artery bypass graft surgery. Stroke, 33: 537–541.

Gupta, A.K., Al Rawi, P.G., Hutchinson, P.J. and Kirkpatrick, P.J. (2002) Effect of hypothermia on brain tissue oxygenation in patients with severe head injury. Br. J. Anaesth., 88: 188–192.

Hajat, C., Hajat, S. and Sharma, P. (2000) Effects of poststroke pyrexia on stroke outcome: a meta-analysis of studies in patients. Stroke, 31: 410–414.

Henker, R.A., Brown, S.D. and Marion, D.W. (1998) Comparison of brain temperature with bladder and rectal temperatures in adults with severe head injury. Neurosurgery, 42: 1071–1075.

Heyman, A., Patterson Jr., J.L. and Nichols Jr., F.T. (1950) The effects of induced fever on cerebral functions in neurosyphilis. J. Clin. Invest., 29: 1335–1341.

Hypothermia after Cardiac Arrest Study Group. (2002) Mild therapeutic hypothermia to improve the neurologic outcome after cardiac arrest. N. Engl. J. Med., 346: 549–556.

Iwagami, Y. (1996) Changes in the ultrastructure of human cells related to certain biological responses under hyperthermic culture conditions. Hum. Cell, 9: 353–366.

Jones, P.A., Andrews, P.J., Midgley, S., Anderson, S.I., Piper, I.R., Tocher, J.L., Housley, A.M., Corrie, J.A., Slattery, J. and Dearden, N.M. (1994) Measuring the burden of secondary insults in head-injured patients during intensive care. J. Neurosurg. Anesthesiol., 6: 4–14.

Katsumura, H., Kabuto, M., Hosotani, K., Handa, Y., Kobayashi, H. and Kubota, T. (1995) The influence of total body hyperthermia on brain haemodynamics and blood-brain barrier in dogs. Acta Neurochir. (Wien.), 135: 62–69.

Kerner, T., Deja, M., Ahlers, O., Loffel, J., Hildebrandt, B., Wust, P., Gerlach, H. and Riess, H. (1999) Whole body hyperthermia: a secure procedure for patients with various malignancies? Intensive Care Med., 25: 959–965.

Kerner, T., Hildebrandt, B., Ahlers, O., Deja, M., Riess, H., Draeger, J., Wust, P. and Gerlach, H. (2003) Anaesthesiological experiences with whole body hyperthermia. Int. J. Hyperthermia, 19: 1–12.

Kilpatrick, M.M., Lowry, D.W., Firlik, A.D., Yonas, H. and Marion, D.W. (2000) Hyperthermia in the neurosurgical intensive care unit. Neurosurgery, 47: 850–855.

Koga, S., Maeta, M., Shimizu, N., Osaki, Y., Hamazoe, R., Oda, M., Karino, T. and Yamane, T. (1985) Clinical effects of total-body hyperthermia combined with anticancer chemotherapy for far-advanced gastrointestinal cancer. Cancer, 55: 1641–1647.

Kraybill, W.G., Olenki, T., Evans, S.S., Ostberg, J.R., O'Leary, K.A., Gibbs, J.F. and Repasky, E.A. (2002) A phase I study of fever-range whole body hyperthermia (FR-WBH) in patients with advanced solid tumours: correlation with mouse models. Int. J. Hyperthermia, 18: 253–266.

Lees, D.E., Kim, Y.D., Bull, J.M., Whang-Peng, J., Schuette, W., Smith, R. and Macnamara, T.E. (1980) Anesthetic management of whole-body hyperthermia for the treatment of cancer. Anesthesiology, 52: 418–428.

Lepock, J.R. (2003) Cellular effects of hyperthermia: relevance to the minimum dose for thermal damage. Int. J. Hyperthermia, 19: 252–266.

Levin, W. and Blair, R.M. (1982) Pettigrew technique of inducing whole-body hyperthermia. Natl. Cancer Inst. Monogr., 61: 377–379.

van der Linden, J., Priddy, R., Ekroth, R., Lincoln, C., Pugsley, W., Scallan, M. and Tyden, H. (1991) Cerebral perfusion and metabolism during profound hypothermia in children. A study of middle cerebral artery ultrasonic variables and cerebral extraction of oxygen. J. Thorac. Cardiovasc. Surg., 102: 103–114.

McCulloch, J., Savaki, H.E., Jehle, J. and Sokoloff, L. (1982) Local cerebral glucose utilization in hypothermic and hyperthermic rats. J. Neurochem., 39: 255–258.

Meythaler, J.M. and Stinson III, A.M. (1994) Fever of central origin in traumatic brain injury controlled with propranolol. Arch. Phys. Med. Rehabil., 75: 816–818.

Michenfelder, J.D. (1988) Anesthesia and the Brain. Churchill Livingstone, New York, NY.

Michenfelder, J.D., Milde, J.H. and Katusic, Z.S. (1991) Postischemic canine cerebral blood flow is coupled to cerebral metabolic rate. J. Cereb. Blood Flow Metab., 11: 611–616.

Mickley, G.A., Cobb, B.L. and Farrell, S.T. (1997) Brain hyperthermia alters local cerebral glucose utilization: a comparison of hyperthermic agents. Int. J. Hyperthermia, 13: 99–114.

Moriyama, E. (1990) Cerebral blood flow changes during localized hyperthermia. Neurol. Med. Chir. (Tokyo), 30: 923–929.

Natale, J.E., Joseph, J.G., Helfaer, M.A. and Shaffner, D.H. (2000) Early hyperthermia after traumatic brain injury in children: risk factors, influence on length of stay, and effect on short-term neurologic status. Crit. Care Med., 28: 2608–2615.

Nemoto, E.M. and Frankel, H.M. (1970a) Cerebral oxygenation and metabolism during progressive hyperthermia. Am. J. Physiol., 219: 1784–1788.

Nemoto, E.M. and Frankel, H.M. (1970b) Cerebrovascular response during progressive hyperthermia in dogs. Am. J. Physiol., 218: 1060–1064.

Noor, R., Wang, C.X. and Shuaib, A. (2003) Effects of hyperthermia on infarct volume in focal embolic model of cerebral ischemia in rats. Neurosci. Lett., 349: 130–132.

Noor, R., Wang, C.X. and Shuaib, A. (2005) Hyperthermia masks the neuroprotective effects of tissue plaminogen activator. Stroke, 36: 665–669.

Nunneley, S.A., Martin, C.C., Slauson, J.W., Hearon, C.M., Nickerson, L.D. and Mason, P.A. (2002) Changes in regional cerebral metabolism during systemic hyperthermia in humans. J. Appl. Physiol., 92: 846–851.

Nybo, L., Moller, K., Volianitis, S., Nielsen, B. and Secher, N.H. (2002) Effects of hyperthermia on cerebral blood flow and metabolism during prolonged exercise in humans. J. Appl. Physiol., 93: 58–64.

Ohmoto, Y., Fujisawa, H., Ishikawa, T., Koizumi, H., Matsuda, T. and Ito, H. (1996) Sequential changes in cerebral blood flow, early neuropathological consequences and blood-brain barrier disruption following radiofrequency-induced localized hyperthermia in the rat. Int. J. Hyperthermia, 12: 321–334.

Oliveira-Filho, J., Ezzeddine, M.A., Segal, A.Z., Buonanno, F.S., Chang, Y., Ogilvy, C.S., Rordorf, G., Schwamm, L.H., Koroshetz, W.J. and McDonald, C.T. (2001) Fever in subarachnoid hemorrhage: relationship to vasospasm and outcome. Neurology, 56: 1299–1304.

Ostrow, S., Van Echo, D., Whitacre, M., Aisner, J., Simon, R. and Wiernik, P.H. (1981) Physiologic response and toxicity in patients undergoing whole-body hyperthermia for the treatment of cancer. Cancer Treat. Rep., 65: 323–325.

Parks, L.C., Minaberry, D., Smith, D.P. and Neely, W.A. (1979) Treatment of far-advanced bronchogenic carcinoma by extracorporeally induced systemic hyperthermia. J. Thorac. Cardiovasc. Surg., 78: 883–892.

Pettigrew, R.T., Galt, J.M., Ludgate, C.M. and Smith, A.N. (1974) Clinical effects of whole-body hyperthermia in advanced malignancy. Br. Med. J., 4: 679–682.

Pontiggia, P., Bianchi, S.A., Alonso, K. and Santamaria, L. (1995) Whole body hyperthermia associated with beta-carotene supplementation in patients with AIDS. Biomed. Pharmacother., 49: 263–265.

Reith, J., Jorgensen, H.S., Pedersen, P.M., Nakayama, H., Raaschou, H.O., Jeppesen, L.L. and Olsen, T.S. (1996) Body temperature in acute stroke: relation to stroke severity, infarct size, mortality, and outcome. Lancet, 347: 422–425.

van Rhoon, G.C. and van der Zee, J. (1983) Cerebral temperature and epidural pressure during whole body hyperthermia in dogs. Res. Exp. Med. Berl., 183: 47–54.

Robins, H.I., Dennis, W.H., Neville, A.J., Shecterle, L.M., Martin, P.A., Grossman, J., Davis, T.E., Neville, S.R., Gillis, W.K. and Rusy, B.F. (1985) A nontoxic system for 41.8 degrees C whole-body hyperthermia: results of a Phase I study using a radiant heat device. Cancer Res., 45: 3937–3944.

Robins, H.I., Hugander, A. and Cohen, J.D. (1989) Whole body hyperthermia in the treatment of neoplastic disease. Radiol. Clin. North Am., 27: 603–610.

Rossi, S., Zanier, E.R., Mauri, I., Columbo, A. and Stocchetti, N. (2001) Brain temperature, body core temperature, and intracranial pressure in acute cerebral damage. J. Neurol. Neurosurg. Psychiatry, 71: 448–454.

Rumana, C.S., Gopinath, S.P., Uzura, M., Valadka, A.B. and Robertson, C.S. (1998) Brain temperature exceeds systemic temperature in head-injured patients. Crit. Care Med., 26: 562–567.

Schwarz, S., Hafner, K., Aschoff, A. and Schwab, S. (2000) Incidence and prognostic significance of fever following intracerebral hemorrhage. Neurology, 54: 354–361.

Sharma, H.S., Drieu, K., Alm, P. and Westman, J. (2000) Role of nitric oxide in blood–brain barrier permeability, brain edema and cell damage following hyperthermic brain injury. An experimental study using EGB-761 and Gingkolide B pretreatment in the rat. Acta Neurochir. Suppl., 76: 81–86.

Sharma, H.S., Drieu, K. and Westman, J. (2003) Antioxidant compounds EGB-761 and BN-52021 attenuate brain edema formation and hemeoxygenase expression following hyperthermic brain injury in the rat. Acta Neurochir. Suppl., 86: 313–319.

Sharma, H.S. and Hoopes, P.J. (2003) Hyperthermia induced pathophysiology of the central nervous system. Int. J. Hyperthermia, 19: 325–354.

Simon, H.B. (1993) Hyperthermia. N. Engl. J. Med., 329: 483–487.

Steen, P.A., Newberg, L., Milde, J.H. and Michenfelder, J.D. (1983) Hypothermia and barbiturates: individual and combined effects on canine cerebral oxygen consumption. Anesthesiology, 58: 527–532.

Stocchetti, N., Rossi, S., Zanier, E.R., Colombo, A., Beretta, L. and Citerio, G. (2002) Pyrexia in head-injured patients admitted to intensive care. Intensive Care Med., 28: 1555–1562.

Takasu, A., Saitoh, D., Kaneko, N., Sakamoto, T. and Okada, Y. (2001) Hyperthermia: is it an ominous sign after cardiac arrest? Resuscitation, 49: 273–277.

Takino, M. and Okada, Y. (1991) Hyperthermia following cardiopulmonary resuscitation. Intensive Care Med., 17: 419–420.

Vertrees, R.A., Bidani, A., Deyo, D.J., Tao, W. and Zwischenberger, J.B. (2000) Venovenous perfusion-induced systemic hyperthermia: hemodynamics, blood flow, and thermal gradients. Ann. Thorac. Surg., 70: 644–652.

Vertrees, R.A., Tao, W., Pencil, S.D., Sites, J.P., Althoff, D.P. and Zwischenberger, J.B. (1996) Induction of whole body hyperthemia with venovenous perfusion. ASAIO J., 42: 250–254.

Walter, B., Bauer, R., Kuhnen, G., Fritz, H. and Zwiener, U. (2000) Coupling of cerebral blood flow and oxygen metabolism in infant pigs during selective brain hypothermia. J. Cereb. Blood Flow Metab., 20: 1215–1224.

Wang, Y., Lim, L.L., Levi, C., Heller, R.F. and Fisher, J. (2000) Influence of admission body temperature on stroke mortality. Stroke, 31: 404–409.

Wass, C.T., Lanier, W.L., Hofer, R.E., Scheithauer, B.W. and Andrews, A.G. (1995) Temperature changes of $>$ or $=1$ degree C alter functional neurologic outcome and histopathology in a canine model of complete cerebral ischemia. Anesthesiology, 83: 325–335.

Westman, J., Drieu, K. and Sharma, H.S. (2000) Antioxidant compounds EGB-761 and BN-520 21 attenuate heat shock protein (HSP 72 kDa) response, edema and cell changes following hyperthermic brain injury. An experimental study using immunohistochemistry in the rat. Amino Acids, 19: 339–350.

Wiedemann, G.J., d'Oleire, F., Knop, E., Eleftheriadis, S., Bucsky, P., Feddersen, S., Klouche, M., Geisler, J., Mentzel, M. and Schmucker, P. (1994) Ifosfamide and carboplatin combined with 41.8 degrees C whole-body hyperthermia in patients with refractory sarcoma and malignant teratoma. Cancer Res., 54: 5346–5350.

Yamada, N. (1989) The effects of hyperthermia on cerebral blood flow, metabolism and electroencephalogram. No To Shinkei, 41: 205–212.

Yaqub, B. and Al Deeb, S. (1998) Heat strokes: aetiopathogenesis, neurological characteristics, treatment and outcome. J. Neurol. Sci., 156: 144–151.

Young, A.B., Ott, L.G., Beard, D., Dempsey, R.J., Tibbs, P.A. and McClain, C.J. (1988) The acute-phase response of the brain-injured patient. J. Neurosurg., 69: 375–380.

Zablow, A., Shecterle, L.M., Dorian, R., Kelly, T., Fletcher, S., Foreman, M., Myers, R., Holton, M., Sanfilippo, L. and St Cyr, J. (1997) Extracorporeal whole body hyperthermia treatment of HIV patients, a feasibility study. Int. J. Hyperthermia, 13: 577–586.

van der Zee, J. (2002) Heating the patient: a promising approach? Ann. Oncol., 13: 1173–1184.

van der Zee, J., van Rhoon, G.C., Wike-Hooley, J.L., Faithfull, N.S. and Reinhold, H.S. (1983) Whole-body hyperthermia in cancer therapy: a report of a phase I–II study. Eur. J. Cancer Clin. Oncol., 19: 1189–1200.

Zeiner, A., Holzer, M., Sterz, F., Schorkhuber, W., Eisenburger, P., Havel, C., Kliegel, A. and Laggner, A.N. (2001) Hyperthermia after cardiac arrest is associated with an unfavorable neurologic outcome. Arch. Intern. Med., 161: 2007–2012.

SECTION V

Hyperthermia and Brain Pathology

CHAPTER 10

Methods to produce hyperthermia-induced brain dysfunction

Hari Shanker Sharma*

Laboratory of Cerebrovascular Research, Department of Surgical Sciences, Anaesthesiology & Intensive Care Medicine, Uppsala University Hospital, Uppsala University, SE-75185 Uppsala, Sweden

Abstract: The recent increase in the frequency and intensity of killer heat waves across the globe has aroused worldwide medical attention to exploring therapeutic strategies to attenuate heat-related morbidity and/or mortality. Death due to heat-related illnesses often exceeds >50% of heat victims. Those who survive are crippled with lifetime disabilities and exhibit profound cognitive, sensory, and motor dysfunction akin to premature neurodegeneration. Although more than 50% of the world populations are exposed to summer heat waves; our understanding of detailed underlying mechanisms and the suitable therapeutic strategies have still not been worked out. One of the basic reasons behind this is the lack of a reliable experimental model to simulate clinical hyperthermia. This chapter describes a suitable animal model to induce hyperthermia in rats (or mice) comparable to the clinical situation. The model appears to be useful for studying the effects of heat-related illnesses on changes in various organs and systems, including the central nervous system (CNS). Since hyperthermia is often associated with profound brain dysfunction, additional methods to examine some crucial parameters of brain injury, e.g., blood-brain barrier (BBB) breakdown and brain edema formation, are also described.

Keywords: hyperthermia; brain dysfunction; cell injury; blood–brain barrier; brain edema; rectal temperature; anesthetics; rats; behavior; heat stress; heat stroke

Introduction

In recent years the frequency and intensity of heat waves has increased worldwide (CDC, 2001, 2003, 2005, 2006; Naughton et al., 2002; Weisskopf et al., 2002; Davis et al., 2003; Keatinge and Donaldson, 2004; Brucker, 2005; Garssen et al., 2005; Michelozzi et al., 2005; Sharma, 2005a; Simon et al., 2005; Barbieri et al., 2006; Conti et al., 2007; Diaz et al., 2006; Empereur-Bissonnet et al., 2006; Hajat et al., 2007; Tan et al., 2007).

In 2003, the deaths due to heat waves, which exceeded >14,000 lives within less than 20 days in France, were unprecedented (Empereur-Bissonnet et al., 2006; Vandentorren et al., 2006). In 2006 Europe and the United States of America were affected again by major heat waves (CDC, 2006; Empereur-Bissonnet et al., 2006). This resulted in a major shift toward existing health policy by several European and US governments to cope with this new situation (Koppe et al., 2003; DOH, 2004; CDC, 2006; Kovats, 2006).

Although the mechanism of regulation of body temperature in normal individuals is well known (Jendritzky et al., 2000; Journeay et al., 2006; see

*Corresponding author. Tel.:/Fax: +46-18-243899;
E-mail: Sharma@surgsci.uu.se

DOI: 10.1016/S0079-6123(06)62010-4

Cheung; Nybo, in this volume), very little knowledge is available regarding pathological aspects of thermoregulation, e.g., hyperthermia and heat-related mortality and/or its preventive and therapeutic measures (Byard and Riches, 2005; Glazer, 2005; Sucholeiki, 2005). Thus, it is still unclear whether people suffering from various cardiovascular or mental diseases, e.g., psychiatric disorders, depression, hypertension, liver damage, stroke, or diabetes, are at high risk of death during a heat wave or hyperthermia-related illnesses (Bulbena et al., 2006; Kiu et al., 2004; Mastrangelo et al., 2006; Medina-Ramon et al., 2006). Moreover, it is still not well known whether human populations exposed to several therapeutic agents affecting body temperature regulation, such as neuroleptics, anticholinergics, serotonergic, histaminergic (Hadad et al., 2003; Halloran and Bernard, 2004; Gaig et al., 2005; Kwok and Chan, 2005; Sharma et al., 2006c), or psychostimulants, e.g., morphine, methamphetamine, ecstasy, marijuana (Clark et al., 1981; Watson et al., 1993; Kilbourne, 1998; Milroy, 1999; Uemura et al., 2003; Matuszewich and Yamamoto, 2004; Sharma and Ali, 2006), etc., are more susceptible to heat-related illnesses or brain damage.

High environmental heat stress is known to increase the susceptibility of infectious agents, e.g., bacteria or viruses (Gebhardt et al., 2004; Scoville et al., 2004; see also Scadden et al., 2006). Few investigations suggest that virus inoculation following bacterial endotoxin-induced hyperthermia enhances mortality (Franklin, 1999; Kanda et al., 1999; see also Sharma, 2004). This indicates that hyperthermia aggravates several underlying subthreshold diseases. Nanoparticle-induced brain damage is exacerbated following whole-body hyperthermia (WBH), which is in line with this idea (see Chapter 13 in this volume). Thus, additional experimental investigations are needed to determine the effects of environmental heat on healthy and diseased populations in relation to brain function.

Public health measures implemented in Europe after 2003 are centered almost exclusively on heat health warning systems (Bernard and McGeehin, 2004; DOH, 2004; Franklin, 2004; Diaz et al., 2005; CDC, 2006; Diaz, 2006; Empereur-Bissonnet et al., 2006). So far, no concrete plan for scientific investigations leading to reduction in heat-related mortality has been documented. This is largely because of our current understanding (or a lack of it) that heat stress affects only poor elderly people living in urban areas that cannot afford air conditioning, or are homeless (Holstein et al., 2005; O'Neill et al., 2005; Nogueira et al., 2005; Vigotti et al., 2006). Furthermore, it is widely believed that heat illness can be prevented by keeping the patient cool, hydrated, and with adequate salt balance (Knochel, 1983, 1989, 1990; Dickinson, 1994; Faunt et al., 1995; Kark et al., 1996; Rydman et al., 1999; Kaiser et al., 2001; Wexler, 2002; Bricknell and Wright, 2004). So far, the role of brain dysfunction in heat-related mortality or morbidity is largely ignored (Sharma and Hoopes, 2003; Sharma, 2005a). Thus, altered brain function in hyperthermia is still not well examined and requires further investigations.

Heat-related deaths and brain damage

Heat-related illnesses are known since Biblical time (*Judith 8: 2*). However, they are seldom addressed as a medical or health problems in our society (Brahams, 1989; Bricknell, 1994, 1996; Porter, 2000). The magnitude of heat-related death, probably due to an increase in global warming will soon become a huge clinical burden on our health planners (Knochel, 1974, 1983; Kunkel et al., 1996; Luterbacher et al., 2004). The actual number of deaths occurring due to heat-related illness is estimated to be the third largest killer in the world after the cardiovascular and traumatic insults to the CNS (Coris et al., 2004; Sharma, 2005a, b, 2006a).

Approximately 10,000–12,000 deaths are recorded during heat waves when the ambient air temperature ranges between 32° and 34°C (Smoyer et al., 2000; DOH, 2004; Keatinge and Donaldson, 2004; CDC, 2006). Interestingly, high rate of heat-related mortality occurs in cities with relatively high levels of urbanization and high costs of living (Smoyer et al., 2000; Sharma, 2005a, 2006a, b). The scientific reports on heat-related death date back to 1743 when 11,000 persons died in China during one-hot-weather condition in July. Another incidence of heat death was recorded in 1841 in Liverpool, when 33 British soldiers died in 1 day due to hot weather in a ship coming from Muscat to Bushier (Knochel, 1974). Similarly, during 1873 in the "Black Hole of Calcutta", 123 out of 186 British prisoners died in one night (Knochel, 1974; Sharma and Westman, 1998a, b; Sharma, 2005a, b). In the Netherlands, approximately 1000 deaths due to hot weather in 1996 occurred in nursing homes of Rotterdam (Kunst et al., 1993; Garssen et al., 2005). Approximately 700 persons died in 1995 due to hot weather conditions in Chicago during summer months (CDC, 2003).

In spite of the seriousness of this problem, studies related to effects of heat on the central nervous system (CNS) are still largely ignored. Few reports of post-mortem changes in human brain of heat stress victims show profound brain damage in the cerebral cortex, cerebellum, and brain stem (Gauss and Meyer, 1917; Alpers, 1936; Malamud et al., 1946; Austin and Berry, 1956). However, the important components of the brain, such as hippocampus, spinal cord, and vascular endothelium, were not examined at that time. Even after 60 years of this report, new studies on human brain damage in heat stress are still lacking.

Bazille et al. (2005) recently examined three cases of heat-death brains in old ages (63, 74, and 80 years) that had heat stroke during heat waves in France. Neuropathological studies show clear cerebellar atrophy after hyperpyrexia. Pronounced loss of Purkinje cells in the victims was associated with increased expression of heat shock protein (HSP) 70 by Bergmann glia. However, surviving Purkinje cells were negative, indicating that apoptotic mechanisms are not responsible for neuronal death. Degeneration of Purkinje cell axons resulted in myelin pallor of the white matter in the dentate nuclei. However, Ammon's horn and other areas susceptible to hypoxia were spared. These observations further confirm the selective vulnerability of Purkinje cells to heat-induced injury and suggest that CNS is vulnerable during heat stress (Bazille et al., 2005).

Hyperthermia can also induce profound damage to the peripheral nervous system (PNS, see Iwanaga et al., 1996; Sharma and Westman, 1998b). This suggests that individuals with peripheral nerve diseases, e.g., diabetic neuropathy, may be more susceptible to heat-illnesses. Thus, to expand our knowledge in heat-related illnesses, use of new technology and specific molecular markers for neuronal, glial, and myelin pathology at light and electron microscopy are needed in both experimental and in clinical situations. A detailed understanding of molecular mechanisms of the CNS injury will provide new insight to develop potential therapeutic measures to minimize the sufferings of heat victims and to reduce their brain dysfunction.

Heat stress and heat stroke

"Heat Stress" represents the discomfort and physiological strain following physical exercise or daily work for long periods in the hot environment (air temperature $>32°C$) (Anderson et al., 1983; Knochel and Reed, 1994; Bouchama and Knochel, 2002). "Hyperthermia" denotes a rise in core body temperature above the hypothalamic set point ($\approx 37°C$) (Milton, 1994). This is caused by impairment of heat-dissipating mechanisms due to "external" (high environmental temperature) or "internal" (metabolic heat production) factors (Milton, 1994; Blatteis, 1997; see Blatteis in this volume). The internal factors are often influenced by use of drugs and acute and/or chronic diseases (Milton, 1994; Sharma and Westman, 1998a, b).

"Mild" to "Moderate" hyperthermia characterized by a small rise of body temperature (above 37°C not exceeding 40°C, Knochel and Reed, 1994) is associated with "Heat Exhaustion". In this condition, intense thirst, weakness, discomfort, anxiety, dizziness, fainting, and headache are quite common (Milton, 1994; see also Bouchama and Knochel, 2002).

"Heat Stroke" represents severe heat illness when hyperthermia exceeds $>40°C$. This situation is associated with CNS abnormalities, e.g., delirium, convulsions, or coma (Malamud et al., 1946; Sterner, 1990; Knochel and Reed, 1994; Bouchama and Knochel, 2002). Heat stroke can occur either following exposure to environmental heat (classical heat stroke) or following strenuous exercise in hot environment (exertional heat stroke)

(Knochel and Reed, 1994; Bouchama and Knochel, 2002).

Sporadic post-mortem reports of heat stroke victims suggest that CNS is one of the most vulnerable organs (Gauss and Meyer, 1917; Alpers, 1936; Malamud et al., 1946; Austin and Berry, 1956). This is because of the fact that hyperthermia-induced systemic inflammatory response that lead to multi-organ dysfunction is largely dominated by severe encephalopathy (Knochel and Reed, 1994; Bouchama and Knochel, 2002).

Heat treatment of tumors in the brain or in other parts of the body is commonly known as "Therapeutic Hyperthermia". This is also associated with severe adverse reactions in the CNS (Hahn, 1982; Hoopes, 1991; Ryan et al., 1991). "Therapeutic Hyperthermia" induces a relatively fast increase in body temperature compared to hot environment or heat stroke (Bouchama and Knochel, 2002), and thus is used to destroy deep-seated tumors in cancer patients (Hahn, 1982; Hoopes, 1991; Ryan et al., 1991). This method is still effective as one of the potential therapies for cancer treatment in spite of several serious side effects (Hahn, 1982; Hoopes, 1991).

Another cause of heat-related deaths or mental illnesses include fire, burning, and rescue operations (Gauss and Meyer, 1917). In addition, prolonged high fever ($>40°C$ for several h) due to bacterial and/or viral infections is often associated with short- or long-term mental anomalies (Hartman and Major, 1935; Alpers, 1936). Hyperthermia caused by these diverse conditions influences cerebral circulation and metabolism (Siesjö, 1978; Sharma, 1982, 1999, 2004; Sharma and Westman, 1998b, 2000). However, the detailed cellular or molecular mechanisms underlying hyperthermia-induced cell and tissue injuries in the CNS are not well understood. Thus, a suitable animal model is needed to address this important problem seen in a large number of human populations.

Brain hyperthermia

A rise in core body temperature following heat stress is often associated with an increase in brain temperature, probably due to an increase in heat production in the brain and/or altered neuronal metabolic activity (Siesjö, 1978). Altered brain temperature could thus be responsible for impaired neuronal activity and/or brain dysfunction (Siesjö, 1978; Uno et al., 2003; Kiyatkin, 2005).

Regional brain temperatures

Pioneer works of Kiyatkin (see Kiyatkin in this volume) showed that brain temperature varies in different regions (within $0.5°$ to $1.5°C$) and is significantly affected by external environment, such as changes in ambient temperature; normal physiological conditions, e.g., sexual function, other motivated behaviors; and drugs of abuse (for review see Kiyatkin, 2005, 2006 and in this volume).

Most mammalian species including humans exhibit a dorso-ventral temperature gradient in the brain (Kiyatkin, 2005). Thus, the dorsal brain structures are cooler than the ventrally located areas (Andersen and Moser, 1995; Schwab et al., 1997; Horvath et al., 1999; Marota et al., 2000). In "cold" brain regions, most neurons are quiet when at rest whereas, the "warm" brain regions show high neuronal activity even during normal resting states (Kiyatkin and Rebec, 1998; Kiyatkin et al., 2002).

The brain temperature always remained higher than the body or blood temperature in humans (Biddle, 2006). Thus, the cerebral cortex temperature is $0.2°$ to $0.8°C$ higher than the rectal temperature and approximately $1.0°C$ higher than the jugular venous blood temperature (Rumana et al., 1998). The venous blood temperature correlates more strongly with the core temperature than with the brain temperature during hyperthermia (Drust et al., 2005; Kiyatkin, 2005, 2006).

A rapid, specific, and long-lasting alteration in the regional brain temperature occurs during hyperthermia that is faster and greater than the changes in blood temperature (Kiyatkin, 2005; Sharma, 2006a). The striatal temperature changes during hyperthermia are faster than the changes in the cerebellar temperature, which exhibits a considerable delay. However, prolonged changes in cerebellar temperature are seen in several physiological situations compared to striatum (see Kiyatkin, 2005, 2006). A consistent and prolonged increase in brain temperature occurs after elevation of ambient temperature from 23 to 29°C compared to other stressors (Kiyatkin and Brown, 2004, 2005). However, repeated exposure to high environmental heat conditions exhibited a gradual decrease in brain temperature indicating physiological adaptive processes (Kiyatkin and Brown, 2005).

Intense exercise-induced brain hyperthermia (>39.5°C) in humans is normally not associated with fatigue (for details, see Nybo in this volume). However, hyperthermia-induced fatigue is seen in rats when the hypothalamic temperature reaches 40.1° to 42.1°C, compared to body temperature (40° to 40.7°C, 83, see Walters et al., 2000; Watson et al., 2005; Meeusen et al., 2006). This indicates that hypothalamic temperature plays an important role in brain dysfunction, probably due to deep brain heating and concomitant breakdown of the BBB (Watson et al., 2005); development of brain edema, and/or damage to the nerve cells. All these factors together could contribute to fatigue (see Nybo in this volume), a hypothesis that require further investigation.

Psychostimulants and brain temperature

Psychostimulants, such as morphine, cocaine, heroin, amphetamine derivatives, and/or cannabinoids, induce profound hyperthermia by increasing heat production together with the whole body oxygen consumption (Lynch et al., 1990; Ansah et al., 1996; Kiyatkin and Wise, 2002; Green et al., 2003; Pavlov and Epstein, 2003). However, their effects on brain temperature are not well known. Administration of cocaine increases brain temperature (+1.2°C) in nucleus accumbens and in hippocampus (Kiyatkin, 2005). This effect is not correlated with the activation of locomotor behavior (Kiyatkin, 2005; Sharma, 2006a, b). On the other hand, methamphetamine induces a robust increase in brain temperature (+3.4°C) at 23°C, which was further aggravated when the drug was administered at 29°C (Davis et al., 1987; Sharma et al., 2004; 2006c; Sharma and Ali, 2006). This indicates that the magnitude of drug-induced brain hyperthermia is exacerbated at high environmental temperature due to a combination of drug-induced heat production and impairment of heat dissipation (Sharma et al., 2004; Kiyatkin, 2005).

Thus, it appears that increased brain hyperthermia induces neurotoxicity either directly or through disruption of the BBB function. Morphine- and methamphetamine-induced BBB disruption in relation to neurotoxicity seen in our laboratory (Sharma and Ali, 2006; Sharma et al., 2006) supports this idea.

Anesthetics and brain temperatures

Anesthetics induce marked changes in regional brain temperatures (Kiyatkin, 2005). Pentobarbital (50 mg/kg, i.p.) reduces hypothalamic temperature by 4.5°C compared to the body temperature (−3.5°C; see Kiyatkin, 2005; Sharma, 2005b). Urethane (1.5 g/kg, i.p.) lowers hippocampal temperature by 6°C compared to only 4°C decrease in the rectal temperature (Madden and Morrison, 2005). Similar effects of chloralose, chloral hydrate, and halothane anesthesia on brain hypothermia are seen in mammals (Erickson and Lanier, 2003; Zhu et al., 2004). This indicates that anesthetics decrease brain temperature more than the body temperature. Interestingly, body warming during anesthesia maintains largely the core temperature. Only a mild increase in brain temperature occurs during warming because of metabolic inhibition of neuronal activity in anesthesia (Kiyatkin, 2005). Taken together, it appears that the brain is a sensitive organ in hyperthermia and heat-related illnesses. However, studies on brain function in heat-related illnesses are still lacking.

Suitable animal models of heat stress are lacking

Numerous laboratories are engaged in investigations on thermoregulation worldwide. However, studies on heat-related illnesses are carried out by only a few researches using animal models not comparable to the clinical situations of hyperthermia (see below).

The existing methods to induce local or WBH is largely based on the principles of heat treatments of tumors (for details, see Matthew, 1993; Sminia et al., 1994; Gonzalez-Alonso et al., 1999). In these models, excessively high exposure temperatures (40° to 60°C) are used to induce local or WBH resulting in death of a large number of animals (Hubbard et al., 1976; Sharma and Hoopes, 2003). Moreover, these animal models of hyperthermia are not comparable to the local environmental situations to which human populations are normally exposed (Nagashima, 2006; Taylor, 2006).

Another model of hyperthermia is developed to induce heat stroke in rats, rabbits, and other small mammals (Hubbard et al., 1977, 1978; Durkot et al., 1986; Gentile et al., 1996; Lin, 1999; Horowitz, 2002; Leon et al., 2005; Chen et al., 2006; Yan et al., 2006 and this volume). These stroke models also use extreme exposure temperatures (40° to 50°C) for 30 min–2 h duration (Sharma et al., 1998b). Since the skin temperature of laboratory animals varies between 21 and 26°C, exposure to such a high ambient temperature will result in excessive thermal load leading to a quick rise in body temperature to >41°C, causing death in a large number of animals (Sharma, 2004, 2005a; Lin in this volume).

New model of heat-related illnesses and hyperthermia

Keeping the above-mentioned factors in mind, we developed a rat model of mild hyperthermia in which the animals are exposed to 38°C for 4 h. This mild hyperthermia model does not produce heat stroke and is quite comparable to average weather situations during summer months in various parts of the world including Europe and America. The structural and functional changes in brain and spinal cord seen in this model are very similar to that observed in clinical cases (Sharma, 1982, 1999).

Thus, this model could be useful to study hyperthermia-induced brain dysfunction and to explore possible therapeutic measures to treat heat-related illnesses occurring in clinical situations (Sharma et al., 2000a, b, 2003; Sharma and Alm, 2002; Sharma and Westman, 2003). Although we used a rat model, it is quite likely that the experimental set up can also be used to study heat-related disorders in small laboratory animals such as mice, guinea pigs, rabbits, and hamsters (Sharma, 2005a).

Preliminary observations show that the model is also useful to understand the sensitivity of heat-related illnesses in chronic diseases, e.g., diabetes, hypertension, brain tumors, and brain or spinal cord injuries (Sharma, unpublished observations). Based on these observations, it appears that the modalities of heat stress, i.e., exposure temperature and duration, vary considerably in normal animals compared to experimentally diseased rats (Muresanu and Sharma, unpublished observations).

In the following sections, the methods to produce clinical hyperthermia and its consequences in rats are described in detail.

Model of whole-body hyperthermia

Handling of animals prior to heat stress

It is important that animals should be handled with care and are accustomed to pre-heat exposure maneuvers (Sharma, 2005a) for at least 1 week (Sharma and Dey, 1984, 1986a). Handling stress is known to increase body temperature ($\approx \pm 0.5°$ to $1°C$; Sharma, 1982; Sharma and Dey, 1984; see also Verleye and Gillardin, 2004; Kim et al., 2006). In addition, maneuvers regarding measurement of rectal temperature also induce stress in animals (Sharma et al., 1992c; Veening et al., 2004; Fig. 1). Thus, repeated handling is mandatory to minimize stress response prior to heat exposure.

To minimize handling stress (Kosten et al., 2006; Sharma, 2006a), rats or mice should be taken out from the cage gently. Lifting the animals by tail or back skin will induce severe stress in animals, e.g., an increase in heart rate and respiration. Good-quality sterile hand-gloves are required while handling the animals, to avoid accidental bites of the researcher as well as to prevent local infections in animals.

Care should also be taken while weighing the animals. Placing the animals on a weighing pan large enough for easy movement will not induce additional stress. Use of pre-weighed restraint boxes should be avoided as they create additional stress in animals (Sharma and Dey, 1981, 1986b). Weighing animals daily will minimize stress of handling (Sharma, 2005b).

Accurate measurement of the rectal temperature

Use of thermistor probes for high accuracy of temperature measurement is necessary (Eshraghi et al., 2005; see also Sharma, 2005b). The thermistor probes can record temperature with high precision ($\pm 0.01°C$) over the range of $0°$ to $100°C$. These probes are stable over years and can measure temperature changes effectively with $\pm 0.1°C$ accuracy (Sharma and Dey, 1987; Sharma, 2005b). The size of thermistor probe varies depending on the animals used. It is always good to have thinner thermistor probes (o.d. ≈ 3 mm) suitable for rats or mice.

The thermistor probe should be dipped in paraffin oil or glycerin before inserting into the rectum of animals. It is recommended to use liquid paraffin that is prepared from the mineral oil. The liquid paraffin does not contain any stimulant or irritant and exerts no osmotic action (Sharma, 1982; Sharma and Dey, 1986a, b, 1987, 1988). Its main use is as a lubricant and very little of it is absorbed by the body. Thus, its use in rectal temperature measurement is highly recommended because of its innocuous nature. Furthermore, it covers fecal masses and thus prevents absorption of toxins that may interfere with the body temperature measurements (Sharma and Dey, 1984).

The probes should be inserted deeper into the rectum gently to avoid damage of internal organs and internal bleeding or microhemorrhages (Sharma, 1982; Sharma and Dey, 1986a). Insertions approximately 2 cm deep in case of mice and approximately 4 cm deep in case of rats are needed to obtain deep visceral temperature, i.e., liver temperature (Dey and Sharma, 1983). The probes should be held in place at least for 60 s to record stable temperature. Sometimes it is advisable to wrap the tail and the probe with adhesive tape. After recording the temperature, the probes may be taken out gently and cleaned first with cotton followed by mild soap water and then sterilized with 70% ethanol or as directed by the manufacturer for cleaning and maintenance. Care should be taken that animals do not bite the tip of the sensors (Sharma, 2005b).

Heat chamber

One of the frequently used heat chambers is known as biological oxygen demand (BOD) incubator (Sharma and Dey, 1984; Sharma et al., 1992a, b, c, d, 1994, 1997; see also Sharma, 2005b for details). The BOD incubators are made of double-wall steel construction for superior insulation and are equipped with forced air circulation for temperature uniformity inside the whole chamber ($\pm 0.5°C$). These incubators can be used at a temperature range of $+5°$ to $+60°C$ with a precision of $\pm 0.5°C$ (see Table 1).

For heat exposure, the BOD incubator may be set to $38°C$ (or other temperatures as desired). It would be good to check variation in chamber temperatures by placing digital or mercury thermometer at each shelf. The frontal reading of the BOD temperature and the inside thermometer are normally the same ($38 \pm 0.1°C$).

Exposure of rats

Before placing rats or mice inside the incubator, their body weight and core body temperature should be recorded at an interval of 10–15 min (Sharma and Dey, 1986a, 1987). Normally, no more than three rats or mice should be kept per cage as this will help to record the movement and behavioral symptoms properly in individual animals. It would be possible to place two plastic cages at two shelves inside the BOD incubator containing three rats or mice each. Putting more than six rats or mice in BOD during one session may influence minor changes in the blood gases or the incubator temperature variations ($>0.1°C$, Sharma, 1982).

Table 1. Heat stress model in rat (see Sharma, 2005b)

Skin temperature: T_s (°C)	Exposure temperature: T_e (°C)	Heat load T_s–T_e (°C)	Relative humidity (%)	Wind velocity (cm/s)	Exposure duration (h)
24–26	36–38	12–14	47–50	24–26	1–4 (acute)[a]
24–26	34–36	10–12	45–47	20–24	8–9 (1 week)[b]

Source: After Sharma and Dey (1978) and Sharma (1982).
[a] Exposed in a biological oxygen demand (BOD) incubator.
[b] Kept at uncontrolled room temperature during summer.

In any case, introducing new animals into the chamber will lower down the temperature slightly ($\pm 1°C$) for sometime (<10 min). No attempt should be made to correct the internal temperature by increasing the temperature set-up knob on the front panel. The temperature is normally restored within 10–20 min.

Observations during heat exposure

It is easy to observe each rat/mice carefully from the glass window by opening the metal front door of the BOD chamber (Sharma, 2005b). Opening of the front metal door will not affect the internal heat chamber temperature. Normally, movement of the rats in cages will increase (Sharma, 1982, 2004, 2005b; Sharma and Dey, 1986a). An increase in locomotion, exploratory behavior, playing, and sometimes fighting is seen that is mainly due to early reaction of warm stress. The behavioral pattern of rats can be followed at regular intervals (≈ 30 min) or according to the experimental protocol.

Body temperature changes during heat exposure

To record the body temperature rise during heat exposure two procedures can be employed.

(i) At the end of the experiment
Recording of the body temperature can be done at the end of experiment at room temperature. The animals can be taken out one by one at an interval of 10–15 min as they were added into the cage in heat chamber. Normally, it will require not more than 2–3 min to record the body temperature of each rat (Sharma and Dey, 1987). To prevent evaporative heat loss, animals can be wrapped in a cotton towel during the recording of their body temperature after heat exposure. This procedure will avoid additional temperature stress (difference between heat chamber and room temperature). The body temperature can show a variation of $\pm 0.5°C$ with or without the towel. This is particularly important while studying effect of drugs on body temperature changes during heat stress (Sharma, 2005b).

(ii) On-going experiments
To record body temperature during the on-going session of a prolonged experiment, animals should be wrapped loosely with cotton towel immediately after being taken out from the heat chamber. After recording their body temperature, the rats were transferred to the heat chamber again (Sharma and Dey, 1984, 1986a; Sharma, 2005b). The temperature fluctuation (a decrease of $\pm 0.5°C$) in heat chamber while removing and placing the rats will require 5–6 min to stabilize (e.g., $38 \pm 0.1°C$). Thus, an interval of minimum 10 min between two rats is sufficient to avoid fluctuations in heat temperature. A repeat measurement of body temperature can thus be done safely every 30 or 60 min during heat exposure (Sharma and Dey, 1986a).

Heat stress and thermal load

The magnitude of hyperthermia induced in this model of heat stress is based on the amount of thermal overload imposed on the rats (Sharma and

Cervos-Navarro, 1991; Sharma et al., 1991a, b, 1992c). The skin temperature in rats ranges between 24° and 25°C, at the normal room temperature (21°C). When rats are kept in heat chamber at 38°C, a heat load of approximately 13° to 14°C (skin temperature–exposure temperature) is imposed on them (Sharma and Dey, 1986a; Sharma et al., 1998b; Sharma, 2005b; Table 1). This thermal overload activates heat dissipation mechanisms to maintain the normal body temperature (37 ± 0.5°C). However, within 1 h of heat exposure, the rats are no longer able to maintain their body temperature effectively, resulting in hyperthermia (Sharma et al., 1992c).

The total amount of heat load and development of hyperthermia up to 4 h is linear ($r^2 = 0.99$; $P<0.001$; see Fig. 1A; Sharma et al., 2006c, d). Thus, heat stress at 38°C induces hyperthermia >40°C at 4 h (a heat load of ≈ 13°C). On the other hand, for rats exposed to 36°C (a mild heat load of 11° to 12°C), the hyperthermia induced at 4 h is >39°C (see Fig. 1 and Table 1). When the exposure temperature in increased further from 38° to 39°C (≈ 15°C), the animals develop hyperthermia >41°C within 1 or 2 h resulting in death of $>80\%$ animals before 3 h (Sharma et al., 1986, 1998b).

Heat stress without heat stroke

In this model, rats subjected to heat stress do not exhibit symptoms of heat stroke (Sharma et al., 1998a, b; Sharma and Hoopes, 2003). Heat stroke is characterized by an excessively high increase in body temperature, >41°C, with cessation of salivation and urination leading to high mortality rate, $>80\%$. Rats exposed to heat in this model exhibit copious salivation, and urination (see below). Their body temperature normally does not exceed 41°C and the mortality rate is quite low ($<20\%$). These observations suggest that the model is very similar to the clinical situations of heat-related illnesses in humans.

Behavioral observations

Heat exposure activates heat loss mechanisms in animals and leads to symptoms of heat exhaustion

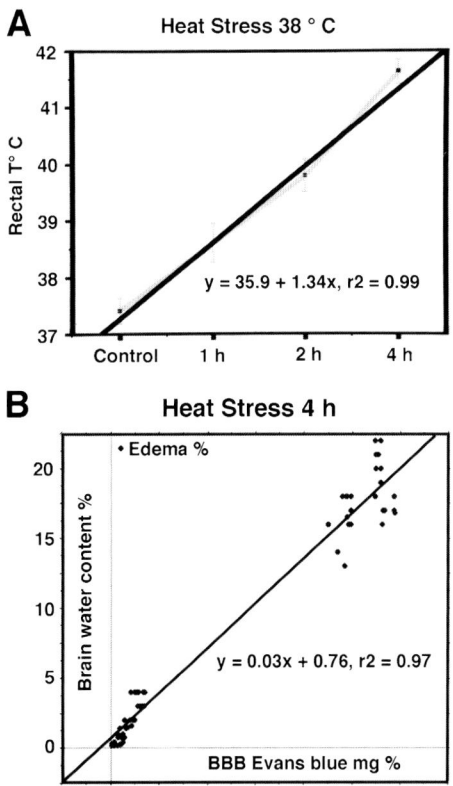

Fig. 1. (A) Whole-body hyperthermia (WBH) in rats. Exposure of rats to heat stress at 38°C in a biological oxygen demand (BOD) incubator resulted in a linear increase in body temperature, which is proportional to the exposure duration seen up to 4 h. (Data from Sharma, 1982; 1999.) This suggests that the model can be used as graded heat load by regulating exposure temperature and duration (for details see text). (B) Strong correlation between breakdown of the blood-brain barrier (BBB) to protein tracer, Evans blue, and development of vasogenic brain edema formation measured as increased brain water content. (Data from Sharma, 1999.) Values at each point represent mean $\pm$ SD of 6–8 rats. (Data from Sharma, 2006a.)

(Table 2). These behavioral parameters can be examined as follows.

(i) Salivation
Rodents lack sweat glands (Gautier, 2000; Whyte and Johnson, 2005; Nagashima, 2006). Thus, the only effective means to lower their body temperature is to spread saliva over the snout and other parts of the body for evaporative heat loss mechanisms (Matthew et al., 1989; Sharma, 2005b).

Table 2. Stress symptoms and physiological variables in control and heat stressed rats (see Sharma, 2005b)

Parameters examined	Control	[a]Heat stress at 38°C in BOD chamber		
		1 h	2 h	4 h
	$n = 8$	$n = 6$	$n = 8$	$n = 12$
		Stress symptoms		
Rectal temperature (°C)	37.42 ± 0.23	38.41 ± 0.32*	39.24 ± 0.21**	41.48 ± 0.23***
Salivation	Nil	+ +	+ + +	+ + + +
Prostration	Nil	Nil	Nil	+ + + +
Gastric hemorrhage	Nil	4 ± 3	8 ± 3	34 ± 8 (microhemorrhages)

Source: Data modified from Sharma (1982) and Sharma and Dey (1986, 1987).
[a]Control rats kept in room temperature (21 ± 1°C); rats exposed in a BOD incubator at 38°C for heat stress (for details see text); nil, absent; + +, mild; + + +, moderate; + + + +, severe.
*, $P<0.05$; **, $P<0.01$; ***, $P<0.001$ Student's unpaired t-test.

The term "salivation" denotes spread of saliva over the snout and also on the other parts of the belly as an effective mechanism for heat dissipation (Sharma and Dey, 1986a). The magnitude and intensity of salivation can be assessed by measuring the spread of saliva over the snout (in mm), which is progressive in nature (see Fig. 1). In extreme conditions, the belly is also wet. However, a wet belly could result due to discharge of urine as well. Thus, spreading of saliva on snout is a more reliable index than the wet belly (Sharma, 1982, 1999).

(ii) Prostration

The state of "heat exhaustion" is reflected by prostration behavior that normally occurs in this model 3 h after heat exposure (see Table 2). At this time, locomotion is considerably reduced and the animals became lethargic (Attia et al., 1983; Sharma and Dey, 1984; Ash et al., 1992). Most of the animals remained confined to the corner of the cage in close association with other animals. Most animals lie flat in supine position and often do not move even after gentle pushing (Damanhouri and Tayeb, 1992; Moran et al., 1999). However, their righting reflexes are not lost. This stage is known as heat-induced "prostration" that can be graded qualitatively (Sharma, 1982, 1999). At this stage the respiratory rate and heartbeats are increased tremendously.

During this stage, care should be taken to clear the snout and nostrils of animals from any obstruction, such as cage walls, corners of the cage, or other small particles from the bedding (Sharma et al., 1998b; Lugo-Amador et al., 2004). Obstruction of breathing will induce death due to asphyxia that is unrelated to heat stress (Sharma et al., 1986).

Loss of body weight after heat exposure

Animals subjected to heat exposure lose body weight considerably because of evaporative water loss and urination (Walters et al., 2000; Cian et al., 2001). These animals normally do not drink water or eat food during heat exposure (Sharma, 1999), as seen in clinical situations (Knochel, 1974). Most marked decrease in body weight of rats is seen after 3–4 h heat exposure (<5–7%), which closely corresponds to the magnitude of hyperthermia (Sharma, 1982). Spread of saliva and evaporative water loss could partially account for such a reduction in body weight after heat stress (Fig. 2).

Thermal nociception

Rats exposed to heat stress do not feel pain but discomfort (Toien and Mercer, 1996; Adelson et al., 1997; Strigo et al., 2000; Sharma, 2005b, 2006a). Thus, mechanical or thermal nociceptive responses of rats subjected to heat stress are not

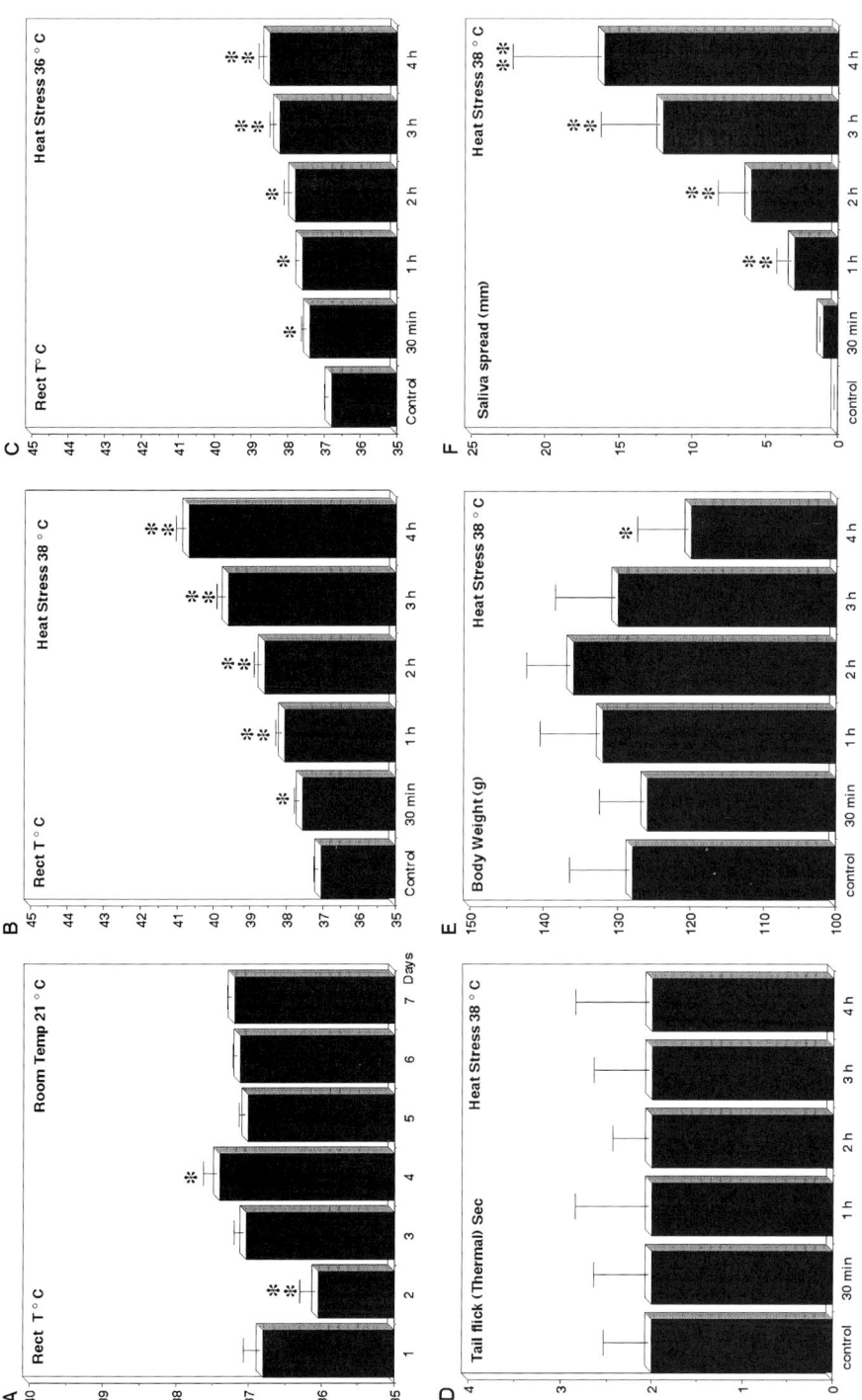

Fig. 2. Changes in body temperature (A–C), pain perception (D), body weight (E), and salivation (F) in control and heat-stressed rats. (A) Handling alone alters the rectal temperature for 1–4 days. The rats get adapted to handling stress from day 5 onwards. A minimum of 7-day handling is necessary to avoid stress in animals. (B) Subjection of rats to heat stress at 38°C in a BOD incubator results in graded hyperthermia that is most severe after 4 h. On the other hand, the magnitude of hyperthermia is considerably less when rats are exposed to heat at 36°C (C). (D) Measurement of tail flick response in heat-exposed rats at 38°C did not show analgesic or hyperalgesic response. This indicates that heat exposure did not influence normal pain pathways. (E) Changes in body temperature during heat exposure of rats at 38°C. Less intake of water and evaporative heat loss through salivation appear to be the important factors in reducing weight of rats exposed to heat stress. A significant reduction in body weight is seen at 4 h heat exposure. (F) Spread of saliva over snout of rats exposed to heat at 38°C. The area of spread (mm) is maximum following 4 h heat exposure in rats. This suggests that rats subjected to 4 h heat stress did not develop stroke. During heat stroke, salivation is stopped (for details see text). *, $P < 0.05$; **, $P < 0.01$; ANOVA followed by Dunnet's test for multiple group comparison from one control group (A–E, from 30 min (F)). Values are mean ± SD of 6–8 rats in each group. Adapted from Sharma (2005b). Reprinted with permission of Current Protocols. (Current Protocols in Toxicology, Suppl. 23, Unit 11.14. 2005) John Wiley & Sons, Inc.

altered (Fig. 2), indicating that thermoreceptors are not damaged in this model of heat exposure. Furthermore, it appears that this model does not influence the animal's sensitivity to pain (Sharma, 1999).

Brain temperature vs. rectal temperature

Rectal temperature closely reflects the brain temperature in several cases of hyperthermia (Dickson et al., 1979; Macy et al., 1985; Mellergard and Nordstrom, 1990; Ko et al., 2001). Thus, when a probe is deeply inserted into the rectum, no difference in brain or rectal temperature is noted following WBH in dogs and pigs (Dickson et al., 1979; Macy et al., 1985). In another study, when dogs with brain tumor were exposed to 42°C for 60 min their liver and brain temperature was only 0.1° and 0.2°C higher than the rectal temperature. However, rise in brain temperature after WBH is not uniform (as above).

Measurement of tympanic membrane temperature using a sensitive thermal probe accurately records cortical brain temperature (± 0.5°C) in rats and mice (Robinson et al., 1967; Hopkins and Fowler, 1998; Mariak et al., 2003; see Zweifler et al., 2004). In our laboratory, tympanic membrane temperature showed a slight but significantly higher hyperthermia (0.5° to 1°C) compared to the rectal temperature at 4 h heat exposure in this model (for details, see Sharma in this volume, Chapter 13). Thus, tympanic membrane temperature in rats could provide useful information about rise in brain temperature. However, a direct recording of brain temperature in various regions is required in this model to further understand these points.

Post-mortem symptoms of heat stress

Microhemorrhages in the stomach

The intensity of heat stress in individual animals can be examined using magnitude of gastric ulceration in animals at post-mortem (Sharma, 1982; Chamberlain, 1993; Cosen-Binker et al., 2004). After a midline incision into the stomach the contents are washed gently under running tap water at room temperature and microhemorrhages or petachae can be seen in mucosal wall of the stomach using a magnifying lens ($\times 3$–5) (Sharma et al., 1992c; Sharma, 2005b). Many microhemorrhages are visible in mucosal wall of the stomach in the fundic and pyloric mucosa, which can be counted manually. Sometimes, the whole mucosal lining is ulcerated. These symptoms represent intense stress caused by heat exposure, as anesthetized animals subject to similar heat exposure do not exhibit gastric ulceration (Sharma, 1982; see below).

Heat stress influences anesthesia

The sensitivity of drugs to the CNS is markedly increased after heat stress (Sharma, 1982, 1999). Thus, a normal dose of anesthetics in heat-stressed rats may induce lethality. To avoid this, the dose of anesthesia in heat stress should be reduced by 30–50% even after 2–3 h heat exposure. For example, if normal animals require 1.5 g/kg, i.p. urethane for surgical-grade anesthesia, the same level of anesthesia is achieved in heat-exposed rats by administering 0.8–1 g/kg, i.p. urethane. This is also true for other anesthetics, e.g., pentobarbital and Equithesin (Sharma, 2005b).

Choice of anesthesia

Several anesthetics can be used in heat stress experiments depending on the local laboratory guidelines and permission from the respective authorities. A brief description of advantages and disadvantages regarding use of particular anesthetics in heat stress is described below.

(i) *Urethane*
   Urethane (ethyl carbamate) is a colorless crystalline substance readily soluble in water (≈ 1 g in 0.5 ml water at room temperature) with a neutral pH. Urethane is a very reliable anesthetic as it induces only mild changes in the cardiovascular and respiratory system and does not influence the sympathetic nervous system activity. For

neurophysiological assessment, urethane is a good choice for anesthesia because it acts on the cortical level and does not depress the respiratory or cardiovascular centers in the brain stem.

The effect of urethane is also long-lasting (>12 h) (Flecknell, 1996). Thus, no maintenance doses of the anesthetic are needed. However, urethane has some carcinogenic effects in animals thus utmost care should be taken for its use in laboratory and exposure to humans. Local safety regulations for the use and disposal of urethane should be followed strictly (Field and Lang, 1988).

(ii) *Pentobarbital*

Surgical-grade anesthesia in rats can be achieved by sodium pentobarbital in a dose of 40–60 mg/kg, i.p. However, this anesthesia induces profound bronchial secretions that may be sometimes fatal (Flecknell, 1996). To avoid bronchial secretions and poor respiration during the experiment, pre-treatment with atropine (0.1 mg/kg, s.c., 20–30 min before induction of anesthesia) can be used (Annis et al., 1976; King et al., 1979), provided this drug does not interfere with the experimental design and protocol. The safety margin with pentobarbital anesthesia is very narrow and the effects of anesthesia are also short lasting (i.e., 60–80 min). Thus, for long-term experiments, maintenance doses are often necessary at a regular interval of 45–60 min.

Pentobarbital anesthesia in heat stress is not recommended, especially when the experiments involve neurophysiological studies, i.e., EEG or evoked potential studies. Heat stress often depresses brain stem reticular function. Thus, use of pentobarbital anesthesia may interfere with the experiments.

(iii) *Equithesin*

A combination of ketamine and a muscle relaxant xylazine are commonly used as anesthetics of choice in various laboratories. However, under this anesthesia several reflexes, e.g., blinking of eyes, swallowing and movements of whiskers, or vibrissa, are present although the animal does not respond to pain stimulus (Deacon and Rawlins, 1996). The duration of anesthesia is short lasting and depends on the dose used. The safety margin is considerably higher than sodium pentobarbital alone (Sonn and Mayevsky, 2006). Thus, choice of a good and safe anesthetic is still elusive. In some cases a mixture of various anesthetics is used to produce a safe anesthesia of varying duration in animals.

One such anesthetic that is widely used in small animals is Equithesin. Equithesin is a mixture of sodium pentobarbital, ketamine, and muscle relaxant (Sharma, 2005c). The anesthesia is very safe in a dose of 3 ml/kg, i.p. in rats. Respiratory depression is seldom seen. However, the main drawback for this anesthesia is its short duration, i.e. 30–40 min (Sharma, 2005b, c). Thus, repeated maintenance doses are necessary to maintain a certain level of anesthesia that is often difficult (Sharma et al., 2006b–d).

The anesthetic works on the brain stem reticular formation; thus, excessive repeated doses may induce cardiovascular and respiratory depression that may lead to death of animals. Since the anesthetic is prepared in a mixture of lipid solvents, i.e., propylene glycol and ethanol. Thus, the anesthesia itself may influence brain function (Gaese and Ostwald, 2001). This aspect should be considered when studying effects of drugs on brain functions in animals after revival from the anesthesia (Sharma, 2005b).

Advantages of this model

One of the main advantages of this model is its close simulation of the clinical conditions of mild-to-moderate hyperthermia, representing "heat stress" and "heat exhaustion". The symptoms of "heat stroke" are largely absent in this model.

Thus, the model represents a physiological response of heat stress reaction in animals. This is further evident from the activation of body's natural defense system, physiological response of heat dissipation, and behavioral alterations. In addition, the exposure temperature is in the physiological range and heat-induced degeneration of thermoreceptors or burning of skin receptors is unlikely in this model.

Disadvantages of this model

In this model, the level of hyperthermia varies according to the individual response of animals to heat exposure. Thus, maintenance of a certain magnitude of hyperthermia in individual animals in this model is not possible. In some animals (<20%) body temperature may increase beyond 41.5°C after 4 h heat exposure resulting in death of animals. Furthermore, animals may die during 12–48 h after heat exposure if hyperthermia exceeds 41°C at 4 h (Sharma et al., 1986; Sharma, 2005b).

Alternative method to produce passive hyperthermia

The other methods to produce hyperthermia in rats and mice under anesthesia include two different protocols.

(i) *Infrared exposure to induce hyperthermia*
In this model, hyperthermia is achieved by exposing the anesthetized animals to infrared lamp (75, 150, or 200 W) at a distance of 15–20 cm (Cox et al., 1975). Care should be taken to avoid excessive heat on skin or localized heating at any point (Whyte et al., 2006). This procedure induces a gradual rise in the body temperature (approximately 1°C within 10–15 min depending on the wattage of the lamp used). The height of lamp can be adjusted to control the rate of rise of the body temperature. When the body temperature reaches the desired level (most commonly 42°C), it can be maintained for 15 or 30 min without lethality. After that the animals may be allowed to revive from anesthesia and returned to the cage. Animals under anesthesia can tolerate the rise of rectal temperature up to 42°C (Hickey et al., 2003; Kenney and Musch, 2004; Gong et al., 2006). The animal may die, if the body temperature reaches >42°C (i.e., 42.5°C). However, in these animals, no stress symptoms can be seen during heat exposure. Microhemorrhages in stomach are also seldom in this model.

This method represents passive heating since physiological and psychological components of heat stress are absent. In addition, the counter mechanisms to cope the stress are seldom activated. However, molecular mechanism of hyperthermia-induced cellular response is activated by this method as evident with the expression of HSPs in the CNS and other tissues (Qian et al., 1998; Westman and Sharma, 1998; Pavlik et al., 2003; Sharma and Westman, 2003, 2004 and Chapters 19 & 20 in this volume; see also Sasara et al., 2004). Histological examination of brains in these animals after 1 or 2 days of survival shows BBB disruption and mild-to-moderate cellular damage (Sharma et al., unpublished observation). Thus, the method can be used to study the adverse effects of hyperthermia on the CNS function (see Fig. 4).

(ii) *Use of heat chambers to induce hyperthermia*
Anesthetized animals can also be kept in BOD or other laboratory incubator for sometime with free airflow facilities at high temperature, i.e., 40° to 45°C, for short duration (30 min–1 h). The rises in body temperature is gradual in anesthetized rats or mice and may reach to the level of the exposure temperature within 1 h. When the body temperature reaches 42°C, this can be maintained for 30 min or 1 h by adjusting the exposure temperature. However, further prolongation of heat exposure will result in death of animals due to excessive heat-induced membrane/enzymatic damage. Thus care should be taken to avoid increase in the body temperature beyond 42°C.

Under anesthesia, the control of thermoregulation is normally non-existing. Thus, if the anesthetized rats or mice are exposed to 38°C incubator temperature, then their body temperature reaches 38°C within 1 or 2 h and is maintained at this level for a long time (Sharma, 1982, 1999; Dey and Sharma, 1983, 1984). For effective heat treatment, animals are exposed to 42°C and this level can be maintained further for sometime or according to experimental design in individual cases.

Advantages and disadvantages of passive hyperthermia

In passive heating models, the death rate is extremely less and a fixed body temperature can be maintained according to the experimental protocol (Sharma, 2005b). Moreover, repeated and long-term effects of heat treatment can be achieved without any further risk of death. However, this method lacks the physiological and psychological effects of heat on the animals and the counter-physiological mechanisms to avoid heat are not activated. Thus, the body response to thermal stress as well as the individual variations among different animals in response to heat treatment is difficult to assess.

Brain dysfunction after heat exposure

One of the most important functions of brain affected by hyperthermia is alteration in the blood-brain barrier (BBB) permeability (Sharma, 1982, 1999; Sharma and Dey, 1986; 1987; Wijsman and Shivers, 1993). Breakdown of the BBB can be visualized using several innocuous dyes, i.e., Trypan blue, bromophenol blue, or Evans blue. Evans blue normally binds to serum proteins when administered into the circulation (Rapoport, 1976; Sharma, 2004). Extravasation of Evans blue dye in the brain can be seen visually and documented using macrophotography. Alternatively, the dye entered into the brain can be measured using colorimetry (Sharma, 1982, 1987).

(i) *BBB permeability to Evans blue*
Evans blue (T-1824), an azo-dye that binds to serum proteins *in vivo*, is commonly used to determine BBB permeability to proteins (Rapoport, 1976; Sharma, 1982). For this purpose, a 2% (w/v) solution of Evans blue in double-distilled water is used (for details, see Sharma, 2005b). After filtration, the Evans blue solution can be kept in a tightly sealed glass bottle at 4°C for approximately 1 year.

Administration

Under Equithesin anesthesia, Evans blue (0.3 ml/100 g body weight) solution is administered into the right jugular vein or in right femoral vein according to standard protocol (Sharma, 1987). This dose of Evans blue in circulation binds with more than 70% of the plasma albumin in rats (Rapoport, 1976). Approximately 1 molecule of Evans blue dye binds to 12 molecules of plasma albumin in circulation. Thus, enough Evans blue should be administered to bind at least 60% of plasma albumin (see Sharma, 2005a for details).

Circulation and washout of dye within intracerebral vessels

The Evans blue dye should remain in the circulation for approximately 5 min. The circulation time for dye (5–15 min) will not influence the permeability. Once the dye enters into the brain tissue, it will bind to the brain tissues and stay there for long time. However, the circulation time of dye ($\approx$10 min) should be maintained fairly constant in a particular group of the experiment.

To assess extravasation of the dye–protein complex into the brain tissue, intracerebrovascular Evans blue dye should be washed out using a brief saline perfusion through heart (90 Torr). For this purpose, a commercial perfusion set or a peristaltic pump should be used. Depending on the size of rats and blood volume (roughly 7 ml/100 g body weight), saline perfusion for 30–45 s is sufficient to washout the remaining Evans blue from the intracerebral blood vessels.

Removal of brain

If the experimental protocol does not permit perfusion with fixative or the fixation of brain *in situ*, then the skull is opened carefully to remove the brain, which is placed in 0.9% saline to examine Evans blue extravasation. Evans blue dye entered into the brain can be measured colorimetrically as described earlier (Sharma, 1982; Sharma and Dey, 1986). However, if the animals are perfused with fixative *in situ*, the brain after removal can be placed into the same fixative at 4°C. The dye entered into the brain will not be diluted or washed out in the fixative if kept at 4°C for a few days.

Examination of brain for dye extravasation

Extravasation of Evans blue dye over the dorsal and ventral surfaces of the brain can be best seen under a white fluorescent lamp using a magnifying lens ($\times 2$–5). The pineal gland always stained deep blue. Extravasation of Evans blue into the cortex may be mild to moderate and may not be as blue as the pineal gland because it lacks the BBB. The cortical surfaces (both dorsal and ventral) may turn light to moderate blue in certain areas (Figs. 3 and 4).

After examination of the blue coloration in dorsal and ventral surfaces, a midline incision can be made to see the inside penetration of the dye, e.g., choroid plexus that are turned deep blue, as the microvessels of the choroid plexus do not contain BBB. Staining of ventricular walls indicates breakdown of the blood-CSF barriers as well. In hyperthermia, mild to moderate blue staining of ventricular walls are clearly seen. However, area postrema in the IVth ventricle always stains deep blue, as it lack the BBB function (Sharma, 2005). The dorsal surface of the hippocampus and caudate nucleus also shows mild staining in heat-stressed rats (see Fig. 3).

Pre- vs. post-heat stress administration of dye

Administration of Evans blue dye in conscious rats either before (through a cannula implanted into the right jugular or femoral vein) or after heat exposure in anesthetized states does not influence the magnitude or intensity of dye extravasation in the brain tissues. Likewise, no difference in pattern or intensity of dye extravasation is seen when Evans blue dye was administered in heat-stressed rats anesthetized with urethane, Equithesin, or pentobarbital. This indicates that the magnitude and intensity of BBB opening to Evans blue dye is not influenced by the timing (before or after heat stress) or state (conscious or anesthetized) of animals in heat stress.

Duration of saline or fixative perfusion

The duration of saline infusion and/or formalin perfusion does not influence the magnitude and intensity of dye extravasation into the brain tissues. Since Evans blue once entered into the brain will bind to brain tissue proteins instantly, washout of the dye by saline or formalin perfusion is quite unlikely. However, it is important to maintain a perfusion pressure at ≈ 90 Torr throughout the procedure. Increased perfusion pressure >100 Torr may rupture the microvessels in the brain causing artifacts of Evans blue staining in the brain or spinal cord tissues (Sharma, 2005b).

Reversibility of the BBB opening in heat stress

To study reversibility of the BBB opening in heat stress, Evans blue dye or any other tracer can be administered in rats at certain time intervals after the termination of heat exposure. Preliminary observations in our laboratory show that opening of the BBB to Evans blue is still present in 4 h heat-stressed rats when the dye was administered after a rest of 2 h (Sharma, 2004). Thus, administration of dye or tracers can be done at different time intervals after the termination of heat stress to examine this question.

(ii) *Brain edema in heat-related illnesses*
Edema of the leptomeninges and an increase in the brain weight by several hundred grams are prominent in the victims of heat-related illnesses (Malamud et al., 1946). Flattening of convolutions and a cerebellar pressure cone along with softening of the brain tissue are the other features seen in clinical cases, which indicate profound volume swelling of heat-injured brains (cf. Malamud et al., 1946; Austin and Berry,

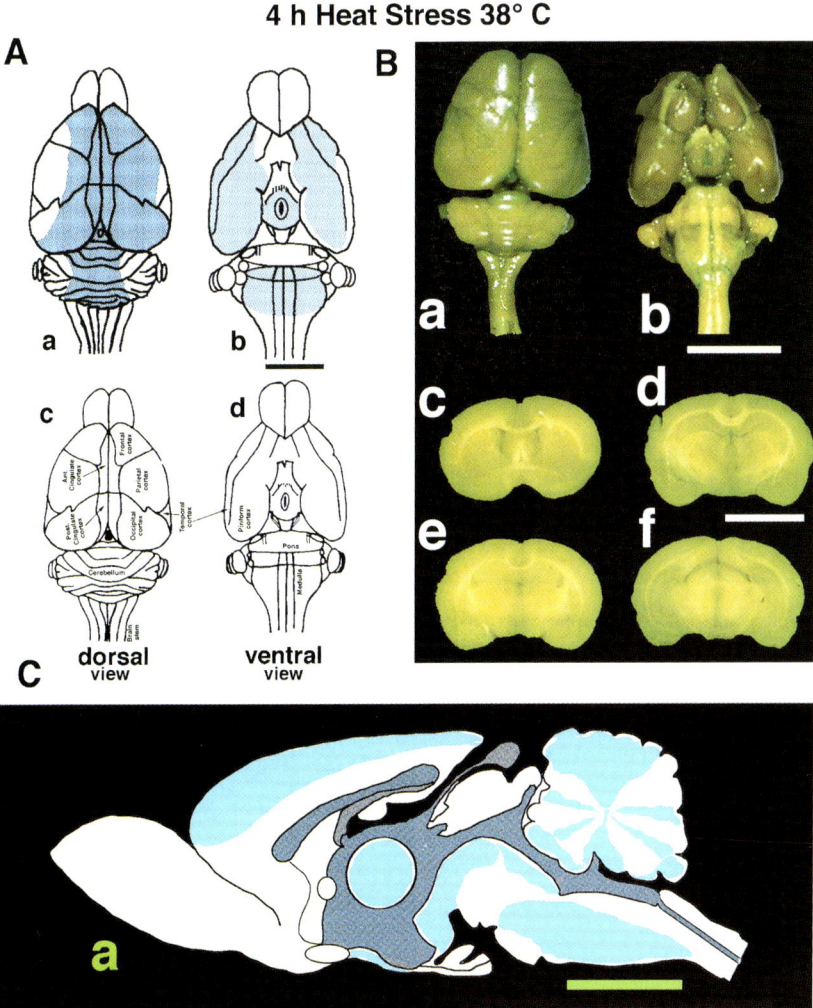

Fig. 3. Evans blue albumin (EBA) extravasation in WBH in rats. (A) Diagrammatic representation of Evans blue leakage on the dorsal (a) and ventral (b) surfaces of the rat brain after 4 h heat stress. Bar = 5 mm. (After Sharma, 2004.) Different regions of the cerebral cortex are divided using hypothetical solid black lines (c, d). Bar = 4 mm. (After Sharma, 2004.) (B) EBA extravasation on the dorsal and ventral surfaces as well as in the deeper parts of the rat brain following WBH. Mild-to-moderate Evans blue staining of somatosensory cortex, piriform cortex, cerebellum, hypothalamus, pons, and brain stem are evident (a, b). Coronal sections of rat brain from four different levels (c–f) showing EBA extravasation in deep brain structures, e.g., caudate putamen (c), hippocampus (d–e), thalamus (d–f), and hypothalamus (d–f). (Co-ordinates for coronal sections from Bregma: $c = +0.10$ to $+0.45$; $d = -3.25$ to -3.90; $e = -4.20$ to -4.60; $f = -5.25$ to -6.65.) Bar = 5 mm. (Modified from Sharma, 2004.) (C) Diagrammatic representation of mid-sagittal section of the rat brain showing extravasation of EBA following WBH (a). Staining of cerebroventricular walls of the lateral ventricles, 4th ventricle, and median eminence is apparent. Bar = 5 mm (from Sharma, 2005b). Reprinted with permission of Current Protocols, (Current Protocols in Toxicology, Suppl. 23, Unit 11.14, 2005) John Wiley & Sons, Inc.

1956). Thus, edema appears to be the most prominent feature of hyperthermic brain injury in clinical cases. However, very little emphasis is given on the functional significance of brain edema formation in heat-related illnesses. Thus, further studies are needed in this direction. This model appears to be useful in measuring heat-related brain edema formation in different regions of the brain and spinal cord.

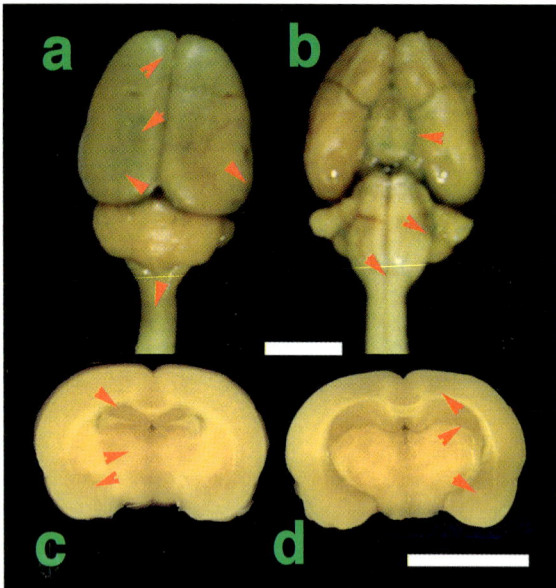

Fig. 4. Extravasation of Evans blue dye in rat brain after infrared lamp heating of Equithesin (3 ml/kg, i.p.) anesthetized rats. In these animals rectal temperature reached 42°C and this was maintained for 15 min. The animals were allowed to survive and the BBB to Evans blue dye was examined 48 h after heat exposure. A mild but marked extravasation of Evans blue dye (arrow heads) is seen on the dorsal (a) and ventral (b) surfaces of the rat brain. Deeper brain structures (c, d) also showed a mild blue staining. The intensity of dye leakage into the brain and the areas involved is considerably different than that seen in animals with WBH (cf. Fig. 2). Co-ordinates for coronal sections from Bregma: $c = -1.10$ to -1.45; $d -3.25$ to -3.90). Bar (a, b) = 5 mm; bar (c, d) = 9 mm.

Leakage of serum proteins within the brain microfluid environment will induce vasogenic brain edema formation (Sharma et al., 1998a, b; see Fig. 1). Edema represents an increase in water content within the extra- or intra cellular compartments of the cerebral components resulting in volume swelling of the brain (Rapoport, 1976; Sharma et al., 1998a, b). Changes in osmotic pressure gradients across the blood-brain interface allow entry of water from the vascular compartment into the brain microfluid environment (for details, see Rapoport, 1976; Sharma et al., 1998b; Sharma, 2006a). The edema fluid spreads over time and depends on the magnitude and severity of the hyperthermia (Lafuente et al., 1990; Sharma and Cervos-Navarro, 1990).

Clinical symptoms of heat-induced brain edema are headache, nausea, vomiting, disturbances of consciousness, and occasionally coma (Rapoport, 1976; Sharma et al., 1998b; Sharma and Hoopes, 2003; Sharma, 2006a). A swollen brain in the closed cranial compartment results in instant death due to compression of vital centers and/or vascular infarction (Sharma et al., 1986, 1998b).

Measurement of brain water content

Brain edema formation in hyperthermia can be examined by measuring water content in different brain regions (Sharma, 2006a). For this purpose, after inspecting Evans blue extravasation either whole brain or selected regions, i.e., cerebral cortex, hippocampus, cerebellum, brain stem, or spinal cord, are weighed immediately on separate pre-weighed filter papers to obtain wet weight of the samples. The brain samples are then placed in a laboratory incubator maintained at 90°C (Sharma, 1982) for 48–72 h. The oven temperature should be maintained well below 95°C to avoid charring of samples. After 48–72 h, the dry weights of samples are recorded on filter paper using same electronic balance. If the last two dry weights are constant then total evaporation of water has occurred (Rapoport, 1976; Sharma and Cervos-Navarro, 1990). The percentage of water content is calculated by the formula: wet weight – dry weight/wet weight × 100 (Sharma, 2005b).

While measuring regional brain water content between experimental and control groups, it is important that the sample sizes (80–300 mg) are as close as possible (± 20–80 mg). In the heat-stressed group, due to profound increase in volume swelling, the dry weight of the sample is considerably reduced compared to the control group. The ideal sample size for brain is over 100 mg, e.g., one hippocampus wet weight ≈ 135 mg in the rat. For spinal cord, sample size below 60 mg may give erroneous results (Sharma et al., 1990).

The whole brain water content (75–76%) is less than the regional brain samples, e.g., cerebral cortex, thalamus, hippocampus, etc., as the gray matter contains more brain water than white matter (Sharma et al., 1998b). Thus, the regional brain water also varies depending on their gray or white matter contents (cerebral cortex, cerebellum, etc., 78–80%; brain stem, spinal cord, 66–72%). Thus, a comparable tissue size and identical regions in control and experimental groups are important for determining regional brain water content (for details, see Sharma, 2005b).

Correlation of brain edema and the BBB permeability

The changes in BBB permeability to protein tracers show a strong correlation with alterations in the brain water content in WBH (Sharma et al., 1998b; Sharma, 1999, 2006a, b). Changes in brain water content and increase in BBB permeability to Evans blue in controls and 4 h heat-stressed rats exhibited a strong correlation ($r^2 = 0.97$; $P < 0.001$; Fig. 1B), indicating a positive role of BBB leakage in brain edema formation (Sharma, 2006b).

Precautions during brain edema measurement

Environmental humidity and airflow are important factors in influencing brain edema measurement of the samples (Rapoport, 1976). Long-time exposure of brain samples to the environment alters wet weight depending on the humidity and airflow in the area where the experiment is conducted. Whether the injured brain is more influenced by humidity and evaporative water loss compared to control group is still not known. To avoid this, the time lag between dissecting out brain regions and recording the wet weight of the sample should be kept to a minimum (<1 min).

The water content can be measured in animals immediately after heat exposure without brain perfusion with saline or formalin. In this situation, removal of blood clots from major blood vessels over the surface of the brain is needed. The content of the blood in microvessels may vary and will affect the wet and dry weights of the samples hence influencing the results. The residual blood volume in brain capillaries in heat-stressed animals could be entirely different compared to normal or drug-treated groups. To avoid this discrepancy, saline-perfused brains may be used for brain edema measurement (Sharma et al., 1998a, b; Sharma, 2004, 2005b, 2006a, b).

Conclusion

The new model of heat-stress described in this chapter appears to induce stress symptoms, hyperthermia, gastric ulcerations, and brain edema formation very similar to the clinical situations. A selective disruption of the BBB to proteins in the regions exhibiting brain edema formation seen in this model further suggest that breakdown of the BBB is instrumental in heat-related brain dysfunction. Thus, the model could be useful to initiate further studies on heat-related illnesses to explore suitable therapeutic measures. It would be interesting to find out whether animals with cardiovascular, endocrine, and metabolic diseases; or following exposure to ultrafine particles (e.g., nanoparticles, engineered, or natural); and/or after psychostimulant treatment (e.g., morphine, cocaine, methamphetamine, ecstasy), are more susceptible to heat-induced brain damage or behavioral dysfunctions. A detailed understanding of the cellular and molecular mechanisms of hyperthermia-induced brain damage in normal and diseased states is very important to explore novel therapeutic strategies to treat human cases effectively in future. This aspect is currently being examined in our laboratory.

Acknowledgments

Thanks are due to reviewers for encouraging the details of the methods for measuring BBB permeability and brain edema formation in relation to heat-stress experiments. Author's research is supported by Swedish Medical Research Council No. 2710; Astra-Zeneca, Mölndal, Sweden, Acure Pharma, Inc., Uppsala, Sweden; Indian Council of Medical Research, New Delhi, India; The University Grants Commission, New Delhi, India; Alexander von Humboldt Foundation, Bonn,

Germany. Secretarial assistance of Aruna Sharma and computer assistance of Suraj Sharma is highly appreciated. This chapter is dedicated to my PhD supervisor, Professor Prasanta Kumar Dey, Department of Physiology, Institute to Medical Sciences, Banaras Hindu University, Varanasi, India, who left this world on August 5, 2006.

References

Adelson, D.W., Wei, J.Y. and Kruger, L. (1997) Warm-sensitive afferent splanchnic C-fiber units in vitro. J. Neurophysiol., 77(6): 2989–3002.

Alpers, B.J. (1936) Hyperthermia due to lesions in the hypothalamus. Arch. Neurol. Psychiatr., 35: 30–42.

Andersen, P. and Moser, E.I. (1995) Brain temperature and hippocampal function. Hippocampus, 5(6): 491–498. Review

Anderson, R.J., Reed, G. and Knochel, J. (1983) Heatstroke. Adv. Intern. Med., 28: 115–140.

Annis, P., Landa, J. and Lichtiger, M. (1976) Effects of atropine on velocity of tracheal mucus in anesthetized patients. Anesthesiology, 44(1): 74–77.

Ansah, T.A., Wade, L.H. and Shockley, D.C. (1996) Changes in locomotor activity, core temperature, and heart rate in response to repeated cocaine administration. Physiol. Behav., 60(5): 1261–1267.

Ash, C.J., Cook, J.R., McMurry, T.A. and Auner, C.R. (1992) The use of rectal temperature to monitor heat stroke. Mo. Med., 89(5): 283–288.

Attia, M., Khogali, M., El-Khatib, G., Mustafa, M.K., Mahmoud, N.A., Eldin, A.N. and Gumaa, K. (1983) Heat stroke: an upward shift of temperature regulation set point at an elevated body temperature. Int. Arch. Occup. Environ. Health, 53(1): 9–17 No abstract available.

Austin, M.G. and Berry, J.W. (1956) Observation on one hundred cases of heatstroke. JAMA, 161: 1525–1529.

Barbieri, A., Pinna, C., Fruggeri, L., Biagioni, E. and Campagna, A. (2006) Heat wave in Italy and hyperthermia syndrome. South Med. J., 99(8): 829–831.

Bazille, C., Megarbane, B., Bensimhon, D., Lavergne-Slove, A., Baglin, A.C., Loirat, P., Woimant, F., Mikol, J. and Gray, F. (2005) Brain damage after heat stroke. J. Neuropathol. Exp. Neurol., 64(11): 970–975.

Bernard, S.M. and McGeehin, M.A. (2004) Municipal heat wave response plans. Am. J. Public Health, 94(9): 1520–1522. Review

Biddle, C. (2006) The neurobiology of the human febrile response. AANA J., 74(2): 145–150. Review

Blatteis, C. (1997) Thermoregulation: recent progress and new frontiers. Ann. N.Y. Acad. Sci., 813: 1–865.

Bouchama, A. and Knochel, J.P. (2002) Heat stroke. N. Engl. J. Med., 346(25): 1978–1988. Review

Brahams, D. (1989) Another British soldier dies from heat illness. Lancet, 2(8673): 1229.

Bricknell, M.C. (1994) Heat illness in Cyprus. J. R. Army Med. Corps, 140(2): 67–69.

Bricknell, M.C. (1996) Heat illness in the army in Cyprus. Occup. Med. (Lond.), 46(4): 304–312.

Bricknell, M.C. and Wright, L.A. (2004) EX SAIF SEREEA II: the field hospital clinical report. J. R. Army Med. Corps, 150(4): 252–255.

Brucker, G. (2005) Vulnerable populations: lessons learnt from the summer 2003 heat waves in Europe. Euro Surveill., 10: 147.

Bulbena, A., Sperry, L. and Cunillera, J. (2006) Psychiatric effects of heat waves. Psychiatr. Serv., 57(10): 1519.

Byard, R.W. and Riches, K.J. (2005) Dehydration and heat-related death: sweat lodge syndrome. Am. J. Forensic Med. Pathol., 26(3): 236–239.

Centers for Disease Control and Prevention (CDC). (2001) Heat-related deaths: Los Angeles County, California, 1999–2000, and United States, 1979–1998. MMWR Morb. Mortal. Wkly. Rep., 50(29): 623–626.

Centers for Disease Control and Prevention (CDC). (2003) Heat-related deaths: Chicago, Illinois, 1996–2001, and United States, 1979–1999. MMWR Morb. Mortal. Wkly. Rep., 52(26): 610–613.

Centers for Disease Control and Prevention (CDC). (2005) Heat-related mortality: Arizona, 1993–2002, and United States, 1979–2002. MMWR Morb. Mortal. Wkly. Rep., 54(25): 628–630.

Centers for Disease Control and Prevention (CDC). (2006) Heat-related deaths: United States, 1999–2003. MMWR Morb. Mortal. Wkly. Rep., 55(29): 796–798.

Chamberlain, C.E. (1993) Acute hemorrhagic gastritis. Gastroenterol. Clin. North Am., 22(4): 843–873. Review

Chen, C.M., Hou, C.C., Cheng, K.C., Tian, R.L., Chang, C.P. and Lin, M.T. (2006) Activated protein C therapy in a rat heat stroke model. Crit. Care Med., 34(7): 1960–1966.

Cian, C., Barraud, P.A., Melin, B. and Raphel, C. (2001) Effects of fluid ingestion on cognitive function after heat stress or exercise-induced dehydration. Int. J. Psychophysiol., 42(3): 243–251.

Clark, W.C., Janal, M.N., Zeidenberg, P. and Nahas, G.G. (1981) Effects of moderate and high doses of marihuana on thermal pain: a sensory decision theory analysis. J. Clin. Pharmacol., 21(8–9 Suppl): 299S–310S.

Conti, S., Masocco, M., Meli, P., Minelli, G., Palummeri, E., Solimini, R., Toccaceli, V. and Vichi, M. (2007) General and specific mortality among the elderly during the 2003 heat wave in Genoa (Italy). Environ. Res., 103(2): 267–274.

Coris, E.E., Ramirez, A.M. and Van Durme, D.J. (2004) Heat illness in athletes: the dangerous combination of heat, humidity and exercise. Sports Med., 34(1): 9–16. Review

Cosen-Binker, L.I., Binker, M.G., Negri, G. and Tiscornia, O. (2004) Influence of stress in acute pancreatitis and correlation with stress-induced gastric ulcer. Pancreatology, 4(5): 470–484.

Cox, B., Green, M.D. and Lomax, P. (1975) Behavioral thermoregulation in the study of drugs affecting body temperature. Pharmacol. Biochem. Behav., 3(6): 1051–1054.

Damanhouri, Z.A. and Tayeb, O.S. (1992) Animal models for heat stroke studies. J. Pharmacol. Toxicol. Methods, 28(3): 119–127. Review

Davis, W.M., Hatoum, H.T. and Waters, I.W. (1987) Toxicity of MDA (3,4-methylenedioxyamphetamine) considered for relevance to hazards of MDMA (Ecstasy) abuse. Alcohol Drug Res., 7(3): 123–134. Review

Davis, R.E., Knappenberger, P.C., Michaels, P.J. and Novicoff, W.M. (2003) Changing heat-related mortality in the United States. Environ. Health Perspect., 111(14): 1712–1718.

Deacon, R.M. and Rawlins, J.N. (1996) Equithesin without chloral hydrate as an anaesthetic for rats. Psychopharmacology (Berl.), 124(3): 288–290.

Department of Health (2004) Heatwave — plan for England — protecting health and reducing harm from extreme heat and heatwaves. London: DoH. www.dh.gov.uk/PublicationsAndStatistics/Publications/PublicationsPolicyAndGuidance/PublicationsPolicyAndGuidanceArticle/fs/en?CONTENT_ID = 4086874&chk = opuHhJ (published 2004, superseded by 2005 edition; last accessed 10 August 2006).

Dey, P.K. and Sharma, H.S. (1983) Ambient temperature and development of traumatic brain oedema in anaesthetized animals. Indian J. Med. Res., 77: 554–563.

Dey, P.K. and Sharma, H.S. (1984) Influence of ambient temperature and drug treatments on brain oedema induced by impact injury on skull in rats. Indian J. Physiol. Pharmacol., 28(3): 177–186.

Diaz, J.H. (2006) Global climate changes, natural disasters, and travel health risks. J. Travel Med., 13(6): 361–372.

Diaz, J., Garcia-Herrera, R., Trigo, R.M., Linares, C., Valente, M.A., De Miguel, J.M. and Hernandez, E. (2006) The impact of the summer 2003 heat wave in Iberia: how should we measure it? Int. J. Biometeorol., 50(3): 159–166.

Diaz, J., Linares, G.C. and Garcia-Herrera, R. (2005) [Impact of extreme temperatures on public health]. Rev Esp Salud Publica., 79(2): 145–157 [Article in Spanish].

Dickinson, J.G. (1994) Heat illness in the services. J. R. Army Med. Corps, 140(1): 7–12.

Dickson, J.A., McKenzie, A. and McLeod, K. (1979) Temperature gradients in pigs during whole-body hyperthermia at 42 degrees C. J. Appl. Physiol., 47(4): 712–717.

Drust, B., Rasmussen, P., Mohr, M., Nielsen, B. and Nybo, L. (2005) Elevations in core and muscle temperature impairs repeated sprint performance. Acta Physiol. Scand., 183(2): 181–190.

Durkot, M.J., Francesconi, R.P. and Hubbard, R.W. (1986) Effect of age, weight, and metabolic rate on endurance, hyperthermia, and heatstroke mortality in a small animal model. Aviat. Space Environ. Med., 57(10 Pt 1): 974–979.

Empereur-Bissonnet, P., Salines, G., Berat, B., Caillere, N. and Josseran, L. (Eds.). (2006) Editorial team. Heatwave in France, July 2006: 112 excess deaths so far attributed to the heat. Euro Surveill., 11(8): E060803.3.

Erickson, K.M. and Lanier, W.L. (2003) Anesthetic technique influences brain temperature, independently of core temperature, during craniotomy in cats. Anesth. Analg., 96(5): 1460–1466.

Eshraghi, A.A., Nehme, O., Polak, M., He, J., Alonso, O.F., Dietrich, W.D., Balkany, T.J. and Van De Water, T.R. (2005) Cochlear temperature correlates with both temporalis muscle and rectal temperatures. Application for testing the otoprotective effect of hypothermia. Acta Otolaryngol., 125(9): 922–928.

Faunt, J.D., Wilkinson, T.J., Aplin, P., Henschke, P., Webb, M. and Penhall, R.K. (1995) The effete in the heat: heat-related hospital presentations during a ten day heat wave. Aust. N.Z. J. Med., 25(2): 117–121.

Field, K.J. and Lang, C.M. (1988) Hazards of urethane (ethyl carbamide): a review of the literature. Lab. Anim., 22(3): 255–262.

Flecknell, P. (1996) Laboratory Animal Anaesthesia. Academic Press, San Diego, CA.

Franklin, C.M. (2004) Lessons from a heat wave. Intensive Care Med., 30(1): 167.

Franklin, Q.J. (1999) Sudden death after typhoid and Japanese encephalitis vaccination in a young male taking pseudoephedrine. Mil. Med., 164(2): 157–159.

Gaese, B.H. and Ostwald, J. (2001) Anesthesia changes frequency tuning of neurons in the rat primary auditory cortex. J. Neurophysiol., 86(2): 1062–1066.

Gaig, C., Marti, M.J., Tolosa, E., Gomez-Choco, M.J. and Amaro, S. (2005) Parkinsonism-hyperpyrexia syndrome not related to antiparkinsonian treatment withdrawal during the 2003 summer heat wave. J. Neurol., 252(9): 1116–1119.

Garssen, J., Harmsen, C. and de Beer, J. (2005) The effect of the summer 2003 heat wave on mortality in the Netherlands. Euro Surveill., 10(7): 165–168.

Gauss, H. and Meyer, K.A. (1917) Heat stroke: report of one hundred and fifty-eight cases from Cook County Hospital, Chicago. Am. J. Med. Sci., 154: 554–564.

Gautier, H. (2000) Body temperature regulation in the rat. J. Therm. Biol., 25(4): 273–279.

Gebhardt, B.M., Kaufman, H.E. and Hill, J.M. (2004) Effect of acyclovir on thermal stress-induced herpesvirus reactivation. Curr. Eye Res., 29(2–3): 137–144.

Gentile, B.J., Szlyk-Modrow, P.C., Durkot, M.J., Krestel, B.A., Sils, I.V., Tartarini, K.A. and Alkhyyat, A.M. (1996) A miniswine model of acute exertional heat exhaustion. Aviat. Space Environ. Med., 67(6): 560–567.

Glazer, J.L. (2005) Management of heatstroke and heat exhaustion. Am. Fam. Physician, 71(11): 2133–2140. Review

Gong, B., Asimakis, G.K., Chen, Z., Albrecht, T.B., Boor, P.J., Pappas, T.C., Bell, B. and Motamedi, M. (2006) Whole-body hyperthermia induces up-regulation of vascular endothelial growth factor accompanied by neovascularization in cardiac tissue. Life Sci., 79(19): 1781–1788.

Gonzalez-Alonso, J., Teller, C., Andersen, S.L., Jensen, F.B., Hyldig, T. and Nielsen, B. (1999) Influence of body temperature on the development of fatigue during prolonged exercise in the heat. J. Appl. Physiol., 86(3): 1032–1039.

Green, A.R., Mechan, A.O., Elliott, J.M., O'Shea, E. and Colado, M.I. (2003) The pharmacology and clinical pharmacology of 3,4-methylenedioxymethamphetamine (MDMA, "ecstasy"). Pharmacol. Rev., 55(3): 463–508. Review

Hadad, E., Weinbroum, A.A. and Ben-Abraham, R. (2003) Drug-induced hyperthermia and muscle rigidity: a practical approach. Eur. J. Emerg. Med., 10(2): 149–154. Review

Hahn, G.M. (1982) Hyperthermia and Cancer. Plenum Press, New York, NY.

Hajat, S., Armstrong, B., Baccini, M., Biggeri, A., Bisanti, L., Russo, A., Paldy, A., Menne, B. and Kosatsky, T. (2006) Impact of high temperatures on mortality: is there an added heat wave effect? Epidemiology, 17(6): 632–638.

Halloran, L.L. and Bernard, D.W. (2004) Management of drug-induced hyperthermia. Curr. Opin. Pediatr., 16(2): 211–215. Review

Hartman, F.W. and Major, R.C. (1935) Pathological changes resulting from accurately controlled artificial fever. Am. J. Clin. Path., 5: 392–410.

Hickey, R.W., Kochanek, P.M., Ferimer, H., Alexander, H.L., Garman, R.H. and Graham, S.H. (2003) Induced hyperthermia exacerbates neurologic neuronal histologic damage after asphyxial cardiac arrest in rats. Crit. Care Med., 31(2): 531–535.

Holstein, J., Canoui-Poitrine, F., Neumann, A., Lepage, E. and Spira, A. (2005) Were less disabled patients the most affected by 2003 heat wave in nursing homes in Paris, France? J. Public Health (Oxf.), 27(4): 359–365.

Hoopes, P.J. (1991) The effects of heat on the nervous system. In: Gutin P.H., Leibel S.A. and Sheline G.E. (Eds.), Radiation Injury to the Nervous System. Raven Press, New York, NY, pp. 407–430.

Hopkins, W.D. and Fowler, L.A. (1998) Lateralized changes in tympanic membrane temperature in relation to different cognitive tasks in chimpanzees (Pan troglodytes). Behav. Neurosci., 112(1): 83–88.

Horowitz, M. (2002) From molecular and cellular to integrative heat defense during exposure to chronic heat. Comp. Biochem. Physiol. A Mol. Integr. Physiol., 131(3): 475–483. Review

Horvath, T.L., Warden, C.H., Hajos, M., Lombardi, A., Goglia, F. and Diano, S. (1999) Brain uncoupling protein 2: uncoupled neuronal mitochondria predict thermal synapses in homeostatic centers. J. Neurosci., 19(23): 10417–10427.

Hubbard, R.W., Bowers, W.D., Matthew, W.T., Curtis, F.C., Criss, R.E., Sheldon, G.M. and Ratteree, J.W. (1977) Rat model of acute heatstroke mortality. J. Appl. Physiol., 42(6): 809–816.

Hubbard, R.W., Matthew, W.T., Criss, R.E., Kelly, C., Sils, I., Mager, M., Bowers, W.D. and Wolfe, D. (1978) Role of physical effort in the etiology of rat heatstroke injury and mortality. J. Appl. Physiol., 45(3): 463–468.

Hubbard, R.W., Matthew, W.T., Linduska, J.D., Curtis, F.C., Bowers, W.D., Leav, I. and Mager, M. (1976) The laboratory rat as a model for hyperthermic syndromes in humans. Am. J. Physiol., 231(4): 1119–1123.

Iwanaga, R., Matsuishi, T., Ohnishi, A., Nakashima, M., Abe, T., Ohtaki, E., Kojima, K., Nagamitsu, S., Ohbu, K. and Kato, H. (1996) Serial magnetic resonance images in a patient with congenital sensory neuropathy with anhidrosis and complications resembling heat stroke. J. Neurol. Sci., 142(1–2): 79–84.

Jendritzky, G., Bucher, K., Laschewski, G. and Walther, H. (2000) Atmospheric heat exchange of the human being, bioclimate assessments, mortality and thermal stress. Int. J. Circumpolar Health, 59(3–4): 222–227.

Journeay, W.S., Carter III, R. and Kenny, G.P. (2006) Thermoregulatory control following dynamic exercise. Aviat. Space Environ. Med., 77(11): 1174–1182.

Kaiser, R., Rubin, C.H., Henderson, A.K., Wolfe, M.I., Kieszak, S., Parrott, C.L. and Adcock, M. (2001) Heat-related death and mental illness during the 1999 Cincinnati heat wave. Am. J. Forensic Med. Pathol., 22(3): 303–307.

Kanda, T., Nakano, M., Yokoyama, T., Hoshino, Y., Okajima, F., Tanaka, T., Saito, Y., Nagai, R. and Kobayashi, I. (1999) Heat stress aggravates viral myocarditis in mice. Life Sci., 64(2): 93–101.

Kark, J.A., Burr, P.Q., Wenger, C.B., Gastaldo, E. and Gardner, J.W. (1996) Exertional heat illness in Marine Corps recruit training. Aviat. Space Environ. Med., 67(4): 354–360.

Keatinge, W.R. and Donaldson, G.C. (2004) The impact of global warming on health and mortality. South Med. J., 97(11): 1093–1099. Review

Kenney, M.J. and Musch, T.I. (2004) Senescence alters blood flow responses to acute heat stress. Am. J. Physiol. Heart Circ. Physiol., 286(4): H1480–H1485.

Kiyatkin, E.A. (2005) Brain hyperthermia as physiological and pathological phenomena. Brain Res. Brain Res. Rev., 50(1): 27–56. Review

Kiyatkin, E.A. (2006) Drug-induced brain hyperthermia. Mechanism and functional implication. Int. J. Neuroprotec. Neuroregen., 2(3): 168–174.

Kiyatkin, E.A. and Brown, P.L. (2004) Modulation of physiological brain hyperthermia by environmental temperature and impaired blood outflow in rats. Physiol. Behav., 83(3): 467–474.

Kiyatkin, E.A. and Brown, P.L. (2005) Brain and body temperature homeostasis during sodium pentobarbital anesthesia with and without body warming in rats. Physiol. Behav., 84(4): 563–570.

Kiyatkin, E.A., Brown, P.L. and Wise, R.A. (2002) Brain temperature fluctuation: a reflection of functional neural activation. Eur. J. Neurosci., 16(1): 164–168.

Kiyatkin, E.A. and Rebec, G.V. (1998) Heterogeneity of ventral tegmental area neurons: single-unit recording and iontophoresis in awake, unrestrained rats. Neuroscience, 85(4): 1285–1309.

Kiyatkin, E.A. and Wise, R.A. (2002) Brain and body hyperthermia associated with heroin self-administration in rats. J. Neurosci., 22(3): 1072–1080.

Kilbourne, E.M. (1998) Cocaine use and death during heat waves. JAMA, 279(22): 1828–1829.

Kim, S.J., Park, S.H., Choi, S.H., Moon, B.H., Lee, K.J., Kang, S.W., Lee, M.S., Choi, S.H., Chun, B.G. and Shin, K.H. (2006) Effects of repeated tianeptine treatment on CRF mRNA expression in non-stressed and chronic mild stress-exposed rats. Neuropharmacology, 50(7): 824–833.

King, M., Engel, L.A. and Macklem, P.T. (1979) Effect of pentobarbital anesthesia on rheology and transport of canine tracheal mucus. J. Appl. Physiol., 46(3): 504–509.

Kiu, A., Horowitz, J.D. and Stewart, S. (2004) Seasonal variation in AF-related admissions to a coronary care unit in a "hot" climate: fact or fiction? J. Cardiovasc. Nurs., 19(2): 138–141.

Knochel, J.P. (1974) Environmental heat illness. An eclectic review. Arch. Intern. Med., 133(5): 841–864.

Knochel, J.P. (1983) Treatment of heat stroke. JAMA, 249(8): 1006–1007.

Knochel, J.P. (1989) Heat stroke and related heat stress disorders. Dis. Mon., 35(5): 301–377. Review

Knochel, J.P. (1990) Catastrophic medical events with exhaustive exercise: "white collar rhabdomyolysis." Kidney Int., 38(4): 709–719. Review

Knochel, J.P. and Reed, G. (1994) Disorders of heat regulation. In: Narins R.G. (Ed.), Maxwell & Kleeman's Clinical Disorders of Fluid and Electrolyte Metabolism (5th ed.). McGraw-Hill, New York, NY, pp. 1549–1590.

Ko, H.K., Flemmer, A., Haberl, C. and Simbruner, G. (2001) Methodological investigation of measuring nasopharyngeal temperature as noninvasive brain temperature analogue in the neonate. Intensive Care Med., 27(4): 736–742.

Koppe, C., Jendritzky, G., Kovats, R.S. and Menne, B. (2003) Heatwaves: impacts and responses. World Health Organization, Copenhagen, Denmark.

Kosten, T.A., Lee, H.J. and Kim, J.J. (2006) Early life stress impairs fear conditioning in adult male and female rats. Brain Res., 1087(1): 142–150.

Kovats, R.S. (2006) Heat waves and health protection. BMJ, 333(7563): 314–315.

Kunkel, K.E., Changnon, S.A., Reinke, B.C. and Arritt, R.W. (1996) The July 1995 heat wave in the Midwest: a climatic perspective and critical weather factors. Bull. Am. Meteorol. Soc., 77(7): 1507–1518.

Kunst, A.E., Looman, C.W. and Mackenbach, J.P. (1993) Outdoor air temperature and mortality in The Netherlands: a time-series analysis. Am. J. Epidemiol., 137(3): 331–341.

Kwok, J.S. and Chan, T.Y. (2005) Recurrent heat-related illnesses during antipsychotic treatment. Ann. Pharmacother., 39(11): 1940–1942.

Lafuente, J.V., Pouschman, E., Cervos-Navarro, J., Sharma, H.S., Schreiner, C. and Korves, M. (1990) Dynamics of tracer distribution in radiation induced brain oedema in rats. Acta Neurochir. Suppl. (Wien.), 51: 375–377.

Leon, L.R., DuBose, D.A. and Mason, C.W. (2005) Heat stress induces a biphasic thermoregulatory response in mice. Am. J. Physiol. Regul. Integr. Comp. Physiol., 288(1): R197–R204.

Lin, M.T. (1999) Pathogenesis of an experimental heatstroke model. Clin. Exp. Pharmacol. Physiol., 26(10): 826–827. Review

Lugo-Amador, N.M., Rothenhaus, T. and Moyer, P. (2004) Heat-related illness. Emerg. Med. Clin. North Am., 22(2): 315–327.

Luterbacher, J., Dietrich, D., Xoplaki, E., Grosjean, M. and Wanner, H. (2004) European seasonal and annual temperature variability, trends, and extremes since 1500. Science, 303: 1499–1503.

Lynch, T.J., Tiseo, P.J. and Adler, M.W. (1990) Morphine-induced pupillary fluctuation: physiological evidence against selective action on the Edinger-Westphal nucleus. J. Ocul. Pharmacol., 6(3): 165–174.

Macy, D.W., Macy, C.A., Scott, R.J., Gillette, E.L. and Speer, J.F. (1985) Physiological studies of whole-body hyperthermia of dogs. Cancer Res., 45(6): 2769–2773.

Madden, C.J. and Morrison, S.F. (2005) Hypoxic activation of arterial chemoreceptors inhibits sympathetic outflow to brown adipose tissue in rats. J. Physiol., 566(Pt 2): 559–573.

Malamud, N., Haymaker, W. and Custer, R.P. (1946) Heat stroke. A clinicopathological study of 125 fatal cases. Mil. Surg., 99: 397–449.

Mariak, Z., White, M.D., Lyson, T. and Lewko, J. (2003) Tympanic temperature reflects intracranial temperature changes in humans. Pflugers Arch., 446(2): 279–284.

Marota, J.J., Mandeville, J.B., Weisskoff, R.M., Moskowitz, M.A., Rosen, B.R. and Kosofsky, B.E. (2000) Cocaine activation discriminates dopaminergic projections by temporal response: an fMRI study in rat. Neuroimage, 11(1): 13–23.

Mastrangelo, G., Hajat, S., Fadda, E., Buja, A., Fedeli, U. and Spolaore, P. (2006) Contrasting patterns of hospital admissions and mortality during heat waves: are deaths from circulatory disease a real excess or an artifact? Med. Hypotheses, 66(5): 1025–1028 [E-pub 2006 January 18].

Matthew, C.B. (1993) Ambient temperature effects on thermoregulation and endurance in anticholinesterase-treated rats. Life Sci., 52(16): 1343–1349.

Matthew, C.B., Hubbard, R.W. and Francesconi, R.P. (1989) Atropine, diazepam, and physostigmine: thermoregulatory effects in the heat-stressed rat. Life Sci., 44(25): 1921–1927.

Matuszewich, L. and Yamamoto, B.K. (2004) Chronic stress augments the long-term and acute effects of methamphetamine. Neuroscience, 124(3): 637–646.

Medina-Ramon, M., Zanobetti, A., Cavanagh, D.P. and Schwartz, J. (2006) Extreme temperatures and mortality: assessing effect modification by personal characteristics and specific cause of death in a multi-city case-only analysis. Environ. Health Perspect., 114(9): 1331–1336.

Meeusen, R., Watson, P., Hasegawa, H., Roelands, B. and Piacentini, M.F. (2006) Central fatigue: the serotonin hypothesis and beyond. Sports Med., 36(10): 881–909.

Mellergard, P. and Nordstrom, C.H. (1990) Epidural temperature and possible intracerebral temperature gradients in man. Br. J. Neurosurg., 4(1): 31–38.

Michelozzi, P., de Donato, F., Bisanti, L., Russo, A., Cadum, E., DeMaria, M., D'Ovidio, M., Costa, G. and Perucci, C.A.

(2005) The impact of the summer 2003 heat waves on mortality in four Italian cities. Euro Surveill., 10(7): 161–165.

Milroy, C.M. (1999) Ten years of 'ecstasy'. J. R. Soc. Med., 92(2): 68–72.

Milton, A.S. (1994) Physiology of Thermoregulation. Birkhauser, Basel, Switzerland, pp. 1–405.

Moran, D.S., Horowitz, M., Meiri, U., Laor, A. and Pandolf, K.B. (1999) The physiological strain index applied to heat-stressed rats. J. Appl. Physiol., 86(3): 895–901.

Nagashima, K. (2006) Central mechanisms for thermoregulation in a hot environment. Ind. Health, 44(3): 359–367. Review

Naughton, M.P., Henderson, A., Mirabelli, M.C., Kaiser, R., Wilhelm, J.L., Kieszak, S.M., Rubin, C.H. and McGeehin, M.A. (2002) Heat-related mortality during a 1999 heat wave in Chicago. Am. J. Prev. Med., 22(4): 221–227.

O'Neill, M.S., Zanobetti, A. and Schwartz, J. (2005) Disparities by race in heat-related mortality in four US cities: the role of air conditioning prevalence. J. Urban Health, 82(2): 191–197.

Nogueira, P.J., Falcao, J.M., Contreiras, M.T., Paixao, E., Brandao, J. and Batista, I. (2005) Mortality in Portugal associated with the heat wave of August 2003: early estimation of effect, using a rapid method. Euro Surveill., 10(7): 150–153.

Pavlik, A., Aneja, I.S., Lexa, J. and Al-Zoabi, B.A. (2003) Identification of cerebral neurons and glial cell types inducing heat shock protein Hsp70 following heat stress in the rat. Brain Res., 973(2): 179–189.

Pavlov, I.F., Epstein, O.I. (2003) Morphine and Antibodies to mu-Opiate Receptors in Ultralow Doses: Effect on Oxygen Consumption. Bull Exp Biol Med. Suppl:137–139.

Porter, A.M. (2000) The death of a British officer-cadet from heat illness. Lancet, 355(9203): 569–571.

Qian, Y.Z., Shipley, J.B., Levasseur, J.E. and Kukreja, R.C. (1998) Dissociation of heat shock proteins expression with ischemic tolerance by whole body hyperthermia in rat heart. J. Mol. Cell Cardiol., 30(6): 1163–1172.

Rapoport, S.I. (1976) Blood-Brain Barrier in Physiology and Medicine. Pergamon Press, New York, NY.

Robinson, S.M., Hutchison, V.H. and Blatt, W.F. (1967) A note on the relationship of tympanic, intraperitoneal, and brain temperatures in the rat. Can. J. Physiol. Pharmacol., 45(2): 355–358.

Rumana, C.S., Gopinath, S.P., Uzura, M., Valadka, A.B. and Robertson, C.S. (1998) Brain temperature exceeds systemic temperature in head-injured patients. Crit. Care Med., 26(3): 562–567.

Ryan, T.P., Hoopes, P.J., Taylor, J.H., Strohbehn, J.W., Roberts, D.W., Double, E.B. and Coughlin, C.T. (1991) Experimental brain hyperthermia: techniques for heat delivery and thermometry. Int. J. Radiat. Oncol. Biol. Phys., 20: 739–750.

Rydman, R.J., Rumoro, D.P., Silva, J.C., Hogan, T.M. and Kampe, L.M. (1999) The rate and risk of heat-related illness in hospital emergency departments during the 1995 Chicago heat disaster. J. Med. Syst., 23(1): 41–56.

Sasara, T., Cizkova, D., Mestril, R., Galik, J., Sugahara, K. and Marsala, M. (2004) Spinal heat shock protein (70) expression: effect of spinal ischemia, hyperthermia (42 degrees C)/hypothermia (27 degrees C), NMDA receptor activation and potassium evoked depolarization on the induction. Neurochem. Int., 44(1): 53–64.

Scadden, D.T., Muse, V.V. and Hasserjian, R.P. (2006) Case records of the Massachusetts General Hospital. Case 30-2006. A 41-year-old man with dyspnea, fever, and lymphadenopathy. N. Engl. J. Med., 355(13): 1358–1368.

Schwab, S., Spranger, M., Aschoff, A., Steiner, T. and Hacke, W. (1997) Brain temperature monitoring and modulation in patients with severe MCA infarction. Neurology, 48(3): 762–767.

Scoville, S.L., Gardner, J.W., Magill, A.J., Potter, R.N. and Kark, J.A. (2004) Nontraumatic deaths during U.S. Armed Forces basic training, 1977–2001. Am. J. Prev. Med., 26(3): 205–212.

Sharma, H.S. (1982) Blood-Brain Barrier in Stress. PhD Thesis, Banaras Hindu University, Varanasi, India, pp. 1–85.

Sharma, H.S. (1987) Effect of captopril (a converting enzyme inhibitor) on blood-brain barrier permeability and cerebral blood flow in normotensive rats. Neuropharmacology, 26(1): 85–92.

Sharma, H.S. (1999) Pathophysiology of blood-brain barrier, brain edema and cell injury following hyperthermia: new role of heat shock protein, nitric oxide and carbon monoxide. An experimental study in the rat using light and electron microscopy. Acta Universitatis Upsaliensis, 830: 1–94.

Sharma, H.S. (2004) Blood-brain and spinal cord barriers in stress. In: Sharma H.S. and Westman J. (Eds.), The Blood-Spinal Cord and Brain Barriers in Health and Disease. Elsevier/Academic Press, San Diego, CA, pp. 231–298.

Sharma, H.S. (2005a) Heat-related deaths are largely due to brain damage. Indian J. Med. Res., 121(5): 621–623.

Sharma, H.S. (2005b) Methods to induce brain hyperthermia. In: Current Protocols in Toxicology (Solicited Contribution), Suppl. 23, Unit 11.14, Wiley, New York, USA, pp. 1–26.

Sharma, H.S. (2005c) Neuroprotective effects of neurotrophins and melanocortins in spinal cord injury: an experimental study in the rat using pharmacological and morphological approaches. Ann. N.Y. Acad. Sci., 1053: 407–421.

Sharma, H.S. (2006a) Hyperthermia induced brain oedema: current status and future perspectives. Indian J. Med. Res., 123(5): 629–652. Review

Sharma, H.S. (2006b) Hyperthermia influences excitatory and inhibitory amino acid neurotransmitters in the central nervous system. An experimental study in the rat using behavioural, biochemical, pharmacological, and morphological approaches. J. Neural Transm., 113(4): 497–519.

Sharma, H.S. and Ali, S.F. (2006) Alterations in blood-brain barrier function by morphine and methamphetamine. Ann. N.Y. Acad. Sci., 1074: 198–224.

Sharma, H.S. and Alm, P. (2002) Nitric oxide synthase inhibitors influence dynorphin A (1–17) immunoreactivity in the rat brain following hyperthermia. Amino Acids, 23(1–3): 247–259.

Sharma, H.S., Alm, P. and Westman, J. (1998a) Nitric oxide and carbon monoxide in the brain pathology of heat stress. Prog. Brain Res., 115: 297–333. Review

Sharma, H.S. and Cervos-Navarro, J. (1990) Brain oedema and cellular changes induced by acute heat stress in young rats. Acta Neurochir. Suppl. (Wien.), 51: 383–386.

Sharma, H.S. and Cervos-Navarro, J. (1991) Role of histamine in pathophysiology of heat stress in rats. Agents Actions Suppl., 33: 97–102.

Sharma, H.S., Cervos-Navarro, J. and Dey, P.K. (1991a) Acute heat exposure causes cellular alteration in cerebral cortex of young rats. Neuroreport, 2(3): 155–158.

Sharma, H.S., Cervos-Navarro, J. and Dey, P.K. (1991b) Rearing at high ambient temperature during later phase of the brain development enhances functional plasticity of the CNS and induces tolerance to heat stress. An experimental study in the conscious normotensive young rats. Brain Dysfunct., 4: 104–124.

Sharma, H. S., Dey, P. K. (1978) Influence of heat and immobilization stressors on the permeability of blood-brain and blood-CSF barriers. Indian J Physiol Pharmacol 22, Supplement II, p. 59–60.

Sharma, H.S. and Dey, P.K. (1981) Impairment of blood-brain barrier (BBB) in rat by immobilization stress: role of serotonin (5-HT). Indian J. Physiol. Pharmacol., 25(2): 111–122.

Sharma, H.S. and Dey, P.K. (1984) Role of 5-HT on increased permeability of blood-brain barrier under heat stress. Indian J. Physiol. Pharmacol., 28(4): 259–267.

Sharma, H.S. and Dey, P.K. (1986a) Probable involvement of 5-hydroxytryptamine in increased permeability of blood-brain barrier under heat stress in young rats. Neuropharmacology, 25(2): 161–167.

Sharma, H.S. and Dey, P.K. (1986b) Influence of long-term immobilization stress on regional blood-brain barrier permeability, cerebral blood flow and 5-HT level in conscious normotensive young rats. J. Neurol. Sci., 72(1): 61–76.

Sharma, H.S. and Dey, P.K. (1987) Influence of long-term acute heat exposure on regional blood-brain barrier permeability, cerebral blood flow and 5-HT level in conscious normotensive young rats. Brain Res., 424(1): 153–162.

Sharma, H.S. and Dey, P.K. (1988) EEG changes following increased blood-brain barrier permeability under long-term immobilization stress in young rats. Neurosci. Res., 5(3): 224–239.

Sharma, H.S., Dey, P.K. and Ashok, K. (1986) Role of circulating 5-HT and lung MAO activity in physiological processes of heat adaptation in conscious young rats. Biomedicine, 6: 31–40.

Sharma, H.S., Drieu, K., Alm, P. and Westman, J. (2000a) Role of nitric oxide in blood-brain barrier permeability, brain edema and cell damage following hyperthermic brain injury. An experimental study using EGB-761 and Gingkolide B pretreatment in the rat. Acta Neurochir. Suppl. (Wein.), 76: 81–86.

Sharma, H.S., Drieu, K. and Westman, J. (2003) Antioxidant compounds EGB-761 and BN-52021 attenuate brain edema formation and hemeoxygenase expression following hyperthermic brain injury in the rat. Acta Neurochir. Suppl. (Wein.), 86: 313–319.

Sharma, H.S., Duncan, J.A. and Johanson, C.E. (2006a) Whole-body hyperthermia in the rat disrupts the blood-cerebrospinal fluid barrier and induces brain edema. Acta Neurochir. Suppl. (Wein.), 96: 426–431.

Sharma, H.S., Gordh, T., Wiklund, L., Mohanty, S. and Sjoquist, P.O. (2006b) Spinal cord injury induced heat shock protein expression is reduced by an antioxidant compound H-290/51. An experimental study using light and electron microscopy in the rat. J. Neural Transm., 113(4): 521–536.

Sharma, H.S. and Hoopes, P.J. (2003) Hyperthermia induced pathophysiology of the central nervous system. Int. J. Hyperthermia, 19(3): 325–354. Review

Sharma, H.S., Kretzschmar, R., Cervos-Navarro, J., Ermisch, A., Ruhle, H.J. and Dey, P.K. (1992a) Age-related pathophysiology of the blood-brain barrier in heat stress. Prog. Brain Res., 91: 189–196.

Sharma, H.S., Lundstedt, T., Boman, A., Lek, P., Seifert, E., Wiklund, L. and Ali, S.F. (2006c) A potent serotonin-modulating compound AP-267 attenuates morphine withdrawal-induced blood-brain barrier dysfunction in rats. Ann. N.Y. Acad. Sci., 1074: 482–496.

Sharma, H.S., Nyberg, F., Cervos-Navarro, J. and Dey, P.K. (1992b) Histamine modulates heat stress-induced changes in blood-brain barrier permeability, cerebral blood flow, brain oedema and serotonin levels: an experimental study in conscious young rats. Neuroscience, 50(2): 445–454.

Sharma, H.S., Olsson, Y. and Dey, P.K. (1990) Early accumulation of serotonin in rat spinal cord subjected to traumatic injury. Relation to edema and blood flow changes. Neuroscience, 36(3): 725–730.

Sharma, H.S., Patnaik, R., Ray, A.K. and Dey, P.K. (2004) Blood-central nervous system barriers in morphine dependence and withdrawal. In: Sharma H.S. and Westman J. (Eds.), The Blood-Spinal Cord and Brain Barriers in Health and Disease. Elsevier Academic Press, San Diego, CA, pp. 299–328.

Sharma, H.S., Sjoquist, P.O., Mohanty, S. and Wiklund, L. (2006d) Post-injury treatment with a new antioxidant compound H-290/51 attenuates spinal cord trauma-induced c-*fos* expression, motor dysfunction, edema formation, and cell injury in the rat. Acta Neurochir. Suppl. (Wein.), 96: 322–328.

Sharma, H.S. and Westman, J. (1998a) Brain function in hot environment — Preface. Progress in Brain Research, Vol. 115. Elsevier, Amsterdam, pp. IX–XIII.

Sharma, H.S. and Westman, J. (1998b) Brain function in hot environment. Progress in Brain Research, Vol. 115. Elsevier, Amsterdam, pp. 1–617.

Sharma, H.S. and Westman, J. (2000) Pathophysiology of hyperthermic brain injury. Current concepts, molecular mechanisms and pharmacological strategies. Research in Legal Medicine Vol. 21 Hyperthermia, Burning and Carbon Monoxide. In: Oehnmichen, M. (Ed.), Lübeck Medical University Publications, Lübeck: Schmidt-Römhild Verlag. Germany, pp. 79–120.

Sharma, H.S. and Westman, J. (2003) Depletion of endogenous serotonin synthesis with p-CPA attenuates upregulation of constitutive isoform of heme oxygenase-2 expression, edema formation and cell injury following a focal trauma to the rat spinal cord. Acta Neurochir. Suppl. (Wein.), 86: 389–394.

Sharma, H.S. and Westman, J. (2004) The heat shock proteins and hemeoxygenase response in central nervous system injuries. In: Sharma H.S. and Westman J. (Eds.), The Blood-Spinal Cord and Brain Barriers in Health and Disease. Elsevier Academic Press, San Diego, CA, pp. 329–360.

Sharma, H.S., Westman, J., Alm, P., Sjoquist, P.O., Cervos-Navarro, J. and Nyberg, F. (1997) Involvement of nitric oxide in the pathophysiology of acute heat stress in the rat. Influence of a new antioxidant compound H-290/51. Ann. N.Y. Acad. Sci., 813: 581–590.

Sharma, H.S., Westman, J. and Nyberg, F. (1998b) Pathophysiology of brain edema and cell changes following hyperthermic brain injury. Prog. Brain Res., 115: 351–412. Review

Sharma, H.S., Westman, J. and Nyberg, F. (2000b) Selective alteration of calcitonin gene related peptide in hyperthermic brain injury. An experimental study in the rat brain using immunohistochemistry. Acta Neurochir. Suppl. (Wein.), 76: 541–545.

Sharma, H.S., Westman, J., Nyberg, F., Cervos-Navarro, J. and Dey, P.K. (1992c) Role of serotonin in heat adaptation: an experimental study in the conscious young rat. Endocr. Regul., 26(3): 133–142.

Sharma, H.S., Westman, J., Nyberg, F., Cervos-Navarro, J. and Dey, P.K. (1994) Role of serotonin and prostaglandins in brain edema induced by heat stress. An experimental study in the young rat. Acta Neurochir. Suppl. (Wien.), 60: 65–70.

Sharma, H.S., Zimmer, C., Westman, J. and Cervos-Navarro, J. (1992d) Acute systemic heat stress increases glial fibrillary acidic protein immunoreactivity in brain: experimental observations in conscious normotensive young rats. Neuroscience, 48(4): 889–901.

Siesjö, B.K. (1978) Brain Energy Metabolism. Wiley, Chichester, New York.

Simon, F., Lopez-Abente, G., Ballester, E., et al. (2005) Mortality in Spain during the heat waves of summer 2003. Euro Surveill., 10: 156–161 [Medline].

Sminia, P., van der Zee, J., Wondergem, J. and Haveman, J. (1994) Effect of hyperthermia on the central nervous system: a review. Int. J. Hyperthermia, 10(1): 1–30. Review

Smoyer, K.E., Rainham, D.G. and Hewko, J.N. (2000) Heat-stress-related mortality in five cities in Southern Ontario: 1980–1996. Int. J. Biometeorol., 44(4): 190–197.

Sonn, J. and Mayevsky, A. (2006) Effects of anesthesia on the responses to cortical spreading depression in the rat brain in vivo. Neurol. Res., 28(2): 206–219.

Sterner, S. (1990) Summer heat illnesses. Conditions that range from mild to fatal. Postgrad. Med., 87(8): 67–70, 73.

Strigo, I.A., Carli, F. and Bushnell, M.C. (2000) Effect of ambient temperature on human pain and temperature perception. Anesthesiology, 92(3): 699–707.

Sucholeiki, R. (2005) Heatstroke. Semin. Neurol., 25(3): 307–314. Review

Tan, J., Zheng, Y., Song, G., Kalkstein, L.S., Kalkstein, A.J. and Tang, X. (2007) Heat wave impacts on mortality in Shanghai, 1998 and 2003. Int. J. Biometeorol., 51(3): 193–200.

Taylor, N.A. (2006) Challenges to temperature regulation when working in hot environments. Ind. Health, 44(3): 331–344. Review

Toien, O. and Mercer, J.B. (1996) Thermosensitivity is reduced during fever induced by *Staphylococcus aureus* cells walls in rabbits. Pflugers Arch., 432(1): 66–74.

Uemura, K., Sorimachi, Y., Yashiki, M. and Yoshida, K. (2003) Two fatal cases involving concurrent use of methamphetamine and morphine. J. Forensic Sci., 48(5): 1179–1181.

Uno, T., Roth, J. and Shibata, M. (2003) Influence of the hypothalamus on the midbrain tonic inhibitory mechanism on metabolic heat production in rats. Brain Res. Bull., 61(2): 129–138.

Vandentorren, S., Bretin, P., Zeghnoun, A., Mandereau-Bruno, L., Croisier, A., Cochet, C., Riberon, J., Siberan, I., Declercq, B. and Ledrans, M. (2006) August 2003 heat wave in France: Risk factors for death of elderly people living at home. Eur. J. Public. Health, 16(6): 583–591.

Veening, J.G., Bouwknecht, J.A., Joosten, H.J., Dederen, P.J., Zethof, T.J., Groenink, L., van der Gugten, J. and Olivier, B. (2004) Stress-induced hyperthermia in the mouse: c-*fos* expression, corticosterone and temperature changes. Prog. Neuropsychopharmacol. Biol. Psychiatry, 28(4): 699–707.

Verleye, M. and Gillardin, J.M. (2004) Effects of etifoxine on stress-induced hyperthermia, freezing behavior and colonic motor activation in rats. Physiol. Behav., 82(5): 891–897.

Vigotti, M.A., Muggeo, V.M. and Cusimano, R. (2006) The effect of birthplace on heat tolerance and mortality in Milan, Italy, 1980–1989. Int. J. Biometeorol., 50(6): 335–341.

Walters, T.J., Ryan, K.L., Tate, L.M. and Mason, P.A. (2000) Exercise in the heat is limited by a critical internal temperature. J. Appl. Physiol., 89(2): 799–806.

Watson, J.D., Ferguson, C., Hinds, C.J., Skinner, R. and Coakley, J.H. (1993) Exertional heat stroke induced by amphetamine analogues. Does dantrolene have a place? Anaesthesia, 48(12): 1057–1060. Review.

Watson, P., Shirreffs, S.M. and Maughan, R.J. (2005) Blood-brain barrier integrity may be threatened by exercise in a warm environment. Am. J. Physiol. Regul. Integr. Comp. Physiol., 288(6): R1689–R1694.

Weisskopf, M.G., Anderson, H.A., Foldy, S., Hanrahan, L.P., Blair, K., Torok, T.J. and Rumm, P.D. (2002) Heat wave morbidity and mortality, Milwaukee, Wis, 1999 vs 1995: an improved response? Am. J. Public Health, 92(5): 830–833.

Westman, J. and Sharma, H.S. (1998) Heat shock protein response in the central nervous system following hyperthermia. Prog. Brain Res., 115: 207–239. Review

Wexler, R.K. (2002) Evaluation and treatment of heat-related illnesses. Am. Fam. Physician, 65(11): 2307–2314. Review

Whyte, D.G., Brennan, T.J. and Johnson, A.K. (2006) Thermoregulatory behavior is disrupted in rats with lesions of the anteroventral third ventricular area (AV3V). Physiol. Behav., 87(3): 493–499.

Whyte, D.G. and Johnson, A.K. (2005) Thermoregulatory role of periventricular tissue surrounding the anteroventral third ventricle (AV3V) during acute heat stress in the rat. Clin. Exp. Pharmacol. Physiol., 32(5–6): 457–461. Review

Wijsman, J.A. and Shivers, R.R. (1993) Heat stress affects blood-brain barrier permeability to horseradish peroxidase in mice. Acta Neuropathol. (Berl.), 86(1): 49–54.

Yan, Y.E., Zhao, Y.Q., Wang, H. and Fan, M. (2006) Pathophysiological factors underlying heatstroke. Med. Hypotheses, 67(3): 609–617.

Zhu, M., Nehra, D., Ackerman, J.H. and Yablonskiy, D.A. (2004) On the role of anesthesia on the body/brain temperature differential in rats. J. Therm. Biol., 29: 599–603.

Zweifler, R.M., Voorhees, M.E., Mahmood, M.A. and Parnell, M. (2004) Rectal temperature reflects tympanic temperature during mild induced hypothermia in nonintubated subjects. J. Neurosurg. Anesthesiol., 16(3): 232–235.

CHAPTER 11

Hyperthermia and central nervous system injury

W. Dalton Dietrich* and Helen M. Bramlett

Department of Neurological Surgery, Miami Project to Cure Paralysis, University of Miami, Miller School of Medicine, 1095 NW 14th Terrace (R-48), Miami, FL 33136, USA

Abstract: Fever is a common occurrence in patients following brain and spinal cord injury (SCI). In intensive care units, large numbers of patients demonstrate febrile periods during the first several days after injury. Over the last several years, experimental studies have reported the detrimental effects of fever in various models of central nervous system (CNS) injury. Small elevations in temperature during or following an insult have been shown to worsen histopathological and behavioral outcome. Thus, the control of fever after brain or SCI may improve outcome if more effective strategies for monitoring and treating hyperthermia were developed. Because of the clinical importance of fever as a potential secondary injury mechanism, mechanisms underlying the detrimental effects of mild hyperthermia after injury have been evaluated. To this end, studies have shown that mild hyperthermia ($>37°C$) can aggravate multiple pathomechanisms, including excitotoxicity, free radical generation, inflammation, apoptosis, and genetic responses to injury. Recent data indicate that gender differences also play a role in the consequences of secondary hyperthermia in animal models of brain injury. The observation that dissociations between brain and body temperature often occur in head-injured patients has again emphasized the importance of controlling temperature fluctuations after injury. Thus, increased emphasis on the ability to monitor CNS temperature and prevent periods of fever has gained increased attention in the clinical literature. Cooling blankets, body vests, and endovascular catheters have been shown to prevent elevations in body temperature in some patient populations. This chapter will summarize evidence regarding hyperthermia and CNS injury.

Keywords: hyperthermia; fever; hypothermia; trauma; ischemia; pathophysiology; inflammation; apoptosis; cell death; blood-brain barrier

Humanity has but three great enemies: fever, famine and war; of these, by far the greatest, by far the most terrible, is fever.

Sir William Osler (Bean, 1968)

Introduction

Fever is a common occurrence in patients with brain and spinal cord injury (SCI) (Sternau et al., 1991; Segatore, 1992; Hayashi et al., 1993; Behr et al., 1997; Albrecht et al., 1998; Henker et al., 1998; Kilpatrick et al., 2000; Natale et al., 2000; Cairns and Andrews, 2002; Thompson et al., 2003a, b; Geffroy et al., 2004; Deogaonkar et al., 2005; Girard, 2005; Johnston et al., 2006; Suz et al., 2006). Temperature recordings, based on measurements of core or cerebral temperature, have shown

*Corresponding author. Tel.: +1 305 243 2297; Fax: 1 305 243 3207; E-mail: ddietrich@miami.edu

that patients following brain or SCI demonstrate elevations in temperature above normothermic levels. Experimentally, findings have also emphasized the importance of temperature on outcome in animal models of cerebral ischemia and trauma (Busto et al., 1987; Globus et al., 1988; Dietrich et al., 1992, 1996b; Thompson et al., 2005). While mild reductions in temperature are neuroprotective, mild elevations in core or central nervous system (CNS) temperature initiated during or following an insult to the brain or spinal cord significantly aggravate outcome (Dietrich et al., 1996a, b; Colbourne et al., 1997). In models of global and focal ischemia, intraischemic elevations in temperature accelerate the progression of neuronal damage as well as aggravate overall histopathological outcome (Busto et al., 1987; Dietrich et al., 1990a, b; Minamisawa et al., 1990; Lundgren et al., 1991). In both global and focal ischemia models, elevations in temperature induced as late as 24 h after the ischemic insult have been shown to increase neuronal vulnerability and infarct volume (Kim et al., 1996; Baena et al., 1997). In the area of traumatic brain and SCI, post-traumatic hyperthermia has also been shown to increase contusion volume, worsen neuronal survival, enhance traumatic axonal injury (TAI), and worsen behavioral outcome (Yu et al., 2001; Dietrich et al., 1996a). Taken together, these clinical and experimental findings emphasize the importance of mild elevations of temperature in the post-injured brain and spinal cord on functional and structural outcomes (Dietrich and Busto, 1997; Manno and Farmer, 2004).

The pathophysiology of brain and SCI is complex and involves multiple injury mechanisms (Tator and Fehlings, 1991; Colbourne et al., 1997; McIntosh et al., 1997; Dirnagl et al., 1999). Recent evidence has emphasized the importance of excitotoxic, apoptotic, and inflammatory cascades in irreversible cell injury and neurological deficits in various injury models (Bramlett and Dietrich, 2004). The fact that temperature has been shown to affect all of these injury cascades emphasizes the importance of temperature elevations in aggravating multiple injury cascades and leading to worse outcomes (Dietrich and Busto, 1997).

Recently, the importance of gender in models of brain and SCI has been emphasized (Hall et al., 1991; Roof et al., 1993; Hurn et al., 1995; Hurn and Maccrae, 2000; Roof and Hall, 2000; Bramlett and Dietrich, 2001). Because clinical studies include both male and female patients, recent experiments have evaluated the importance of gender on both the beneficial effects of mild to moderate hypothermia as well as the detrimental effects of hyperthermia (Suzuki et al., 2003, 2004). This chapter will summarize recent findings concerning the importance of hyperthermia in models of brain and SCI, and will emphasize future directions for this important area of research. The reader is referred to recent review articles and other chapters in this volume concerning therapeutic hyperthermia and hyperthermia-induced pathophysiology of the CNS that relate to this topic (Sharma and Hoopes, 2003).

Global ischemia

The importance of small variations in intraischemic brain temperature was first quantitatively assessed in models of transient forebrain ischemia in rodents (Busto et al., 1987; Chopp et al., 1988; Clifton et al., 1989; Dietrich et al., 1990a, b; Minamisawa et al., 1990; see Table 1). Busto and colleagues (1987) first demonstrated that mild hypothermia (34°C) introduced during a 20 min transient forebrain ischemic insult significantly improved CA1 hippocampal neuron survival 3 days after ischemia. In that study, the effects of small elevations in intraischemic temperature were also noted. Animals in which intraischemic brain temperature was raised to 39°C during the insult had a high mortality rate at 3 days post-ischemia. In a subsequent study, intraischemic hyperthermia was reported to increase the frequency of damaged neurons in the cerebral cortex, CA1 hippocampus, striatum, and thalamus compared to normothermic ischemic animals (Dietrich et al., 1990b). In addition, mild hyperthermia converted some post-ischemic brain regions to frank infarction, a finding that was not seen under normothermic conditions. Finally, intraischemic hyperthermia accelerated the maturation of CA1 pathology compared to normothermic ischemia

Table 1. Hyperthermia in global cerebral ischemia

Reference	Species	Model	Outcome measure
Busto et al. (1987)	Rat	2-VO	Neuronal cell counts, mortality
Chopp et al. (1989)	Cat	Tourniquet	Metabolism
Kuroiwa et al. (1990)	Gerbil	2-VO	Neuronal cell loss
Minamisawa et al. (1990)	Rat	2-VO	Neuronal damage
Dietrich et al. (1990a)	Rat	2-VO	Blood brain barrier
Dietrich et al. (1990b)	Rat	2-VO	Mortality, neuronal damage
Dietrich et al. (1991)	Rat	2-VO	BBB, neuronal damage
Churn et al. (1990)	Gerbil	2-VO	Ca/calmodulin PKII
Colbourne et al. (1993)	Gerbil	2-VO	Spontaneous hyperthermia
Miyazawa et al. (1993)	Rat	4-VO	Neuronal survival, EEG, mortality
Busto et al. (1994)	Rat	2-VO	Protein kinase C
Globus et al. (1995)	Rat	2-VO	Free radicals
Kil et al. (1996)	Rat	2-VO	Free radicals
Enomoto et al. (1996)	Rabbit	Cardiopulmonary bypass	Cerebral oxygen metabolism
Baena et al. (1997)	Rat	2-VO	Neuronal damage
Hickey et al. (2003)	Rat	Cardiac arrest	Neuronal damage
Tomimatsu et al. (2003)	Immature rat	Unilateral CCA occlusion and hypoxia	Caspase-3 activation

Abbreviations: 2-VO, two vessel occlusion; 4-VO, four vessel occlusion; BBB, blood brain barrier; Ca/calmodulin PKII, calcium/calmodulin protein kinase II; EEG, electroencephalogram; CCA, common carotid artery.

(Dietrich, 1999). Under normothermic conditions, CA1 pathology demonstrates a delayed cell death that maturates over a 2–3-day period (Pulsinelli et al., 1980; Kirino, 1982). However, with intraischemic hyperthermia, CA1 pathology was seen as early as 24 h after a cerebral insult. This study emphasized for the first time that in addition to increasing mortality and aggravating overall ischemic pathology, mild intraischemic hyperthermia reduced the therapeutic window for potential therapeutic treatments.

Subsequent studies using this ischemia model showed that intraischemic hyperthermia also significantly aggravated blood-brain barrier (BBB) breakdown (Dietrich et al., 1990a). Thus, in addition to enhancing neuronal pathology, elevated ischemic temperatures also adversely affected the structure and function of the cerebral vasculature. A subsequent study emphasized the importance of these BBB alterations by demonstrating a spatial relationship between sites of protein extravasation (horseradish peroxidase) and acute CA1 neuronal damage after hyperthermic forebrain ischemia (Dietrich et al., 1991). Because post-ischemic hypothermia (30°C) had been recently reported to be neuroprotective after transient global ischemia (Busto et al., 1989a), studies were conducted to determine whether delayed elevations in post-ischemic temperature would also influence hippocampal vulnerability. In one study, a hyperthermic insult initiated 24 h after global ischemia was reported to aggravate CA1 pathology (Baena et al., 1997). Thus, both intraischemic and post-ischemic hyperthermia appears to aggravate outcome under transient ischemic conditions.

In addition to models of transient forebrain ischemia, the potentially detrimental effects of induced hyperthermia in a clinically relevant model of asphyxial cardiac arrest has also been recently investigated (Hickey et al., 2003). At 24 or 48 h after recovery, body temperature either was raised to 40°C or underwent no temperature manipulation. Rats with elevated temperature, 24 h after injury, demonstrated higher mortality rates as well as worse histopathological damage scores compared to rats subjected to asphyxia without induced hyperthermia. This study supported previous investigations regarding delayed post-ischemic hyperthermia and also emphasized the potential detrimental effects of actively rewarming patients resuscitated after cardiac arrest, where active warming may result in hyperthermic overshoots.

Table 2. Hyperthermia in focal cerebral ischemia

Reference	Species	Model	Outcome measure
Chen et al. (1991)	Rat	PMCAO	Infarct volume
Morikawa et al. (1992)	Rat	TMCAO and PMCAO	CBF and infarct volume
Chen et al. (1993)	Rat	TMCAO	Cerebral depolarization, infarct volume
Takagi et al. (1994)	Rat	TMCAO	Glutamate release
Kim et al. (1996)	Rat	TMCAO	Infarct volume
Morimoto et al. (1997)	Rat	TMCAO	Spectrin breakdown
Kim et al. (1998)	Rat	TMCAO	Gene expression

Abbreviations: PMCAO, permanent middle cerebral artery occlusion; TMCAO, transient middle cerebral artery occlusion; CBF, cerebral blood flow.

Focal ischemia

To investigate the pathophysiology and treatment of stroke, many laboratories use rodent models of permanent or transient focal ischemia (Ginsberg and Busto, 1989; see Table 2). Mild elevations in intraischemic temperature have also been shown to affect infarct size in these models of focal ischemia (Chen et al., 1991; Morikawa et al., 1992). Chen and colleagues (1991) first reported the detrimental effects of whole body hyperthermia in a model of permanent middle cerebral artery occlusion (MCAO). In that study, infarct volume was significantly elevated in animals where brain temperature was elevated to 39.5°C compared to hypothermic animals. Subsequent studies from other laboratories showed that hyperthermia aggravates metabolic and histopathological outcomes after transient MCAO in rats (Chen et al., 1991; Morikawa et al., 1992). In a study by Takagi and colleagues (1994), hyperthermia was shown to enhance glutamate release in the ischemic penumbra after MCAO.

Mild spontaneous hyperthermia also occurs in models of global and focal ischemia. In reversible MCAO, Kuluz and colleagues (1993) reported that rectal temperature of awake rats increased to 39.1°C during the 2-h occlusive period. During transient global forebrain ischemia in gerbils, temperature has also been reported to increase (Kato et al., 1991). Importantly, the inhibition of postischemic hyperthermia by halothane treatment was reported to attenuate hippocampal neuronal damage after transient global ischemia (Kuroiwa et al., 1990). Taken together, these experimental studies emphasize the importance of mild hyperthermia in the pathophysiology of both global and focal ischemia, and that measures to prevent temperature elevations may be protective.

Recently, intraischemic hyperthermia has been investigated in a rat model of neonatal hypoxic-ischemic encephalopathy (Tomimatsu et al., 2003). In that study, 7-day-old rats underwent a combination of left common carotid artery ligation and a 15 min hypoxic insult, with body temperatures elevated to 40°C or maintained at normothermia. Hyperthermia was reported to activate caspase-3 activity and reduce microtubule-associated protein-2-positive brain regions. Thus, hyperthermia during a period of hypoxia-ischemia increases the susceptibility of the immature brain to a hypoxic-ischemic insult. Similar to what had been documented in adult animal ischemic studies, hyperthermia during a hypoxic-ischemic insult makes the immature brain extremely susceptible to the insult and that these effects may be mediated in part by the escalation of apoptotic cell death pathways. Clinical data that will be discussed later support these experimental findings showing that fever in acute stroke or after cardiac arrest worsens prognosis (Azzimondi et al., 1995; Hajat et al., 2000; Hickey et al., 2000; Kilpatrick et al., 2000; Kammersgaard et al., 2002).

Traumatic brain injury

Similarities and differences exist between the pathophysiology of cerebral ischemia and traumatic brain injury (TBI) (Bramlett and Dietrich, 2004). Also, mild reductions in post-traumatic core and brain temperature have been reported to be

Table 3. Hyperthermia in traumatic brain injury

Reference	Species	Model	Outcome measure
Dietrich et al. (1996)	Rat	PFP	Pathology, BBB
Whalen et al. (1997)	Rat	CCI	PMNLs and adhesion molecules
Chatzipanteli et al. (2000)	Rat	PFP	PMNLs
Kinoshita et al. (2002b)	Rat	PFP	IL-1β mRNA and protein
Kinoshita et al. (2002a)	Rat	PFP	Hemoglobin
Taylor et al. (2002)	Rat	CCI	Cognition
Suzuki et al. (2004)	Rat	PFP	Neuronal survival and axonal injury
Vitarbo et al. (2004)	Rat	PFP	TNF-α mRNA and protein
Thompson et al. (2005)	Rat	LFP	Inflammatory markers

Abbreviations: PFP, parasagittal fluid percussion; CCI, controlled cortical impact; LFP, lateral fluid percussion; BBB, blood brain barrier; PMNLs, polymorphonuclear leukocytes; IL-1β, interleukin 1-beta; TNF-α, tumor necrosis factor alpha.

neuroprotective and improve behavioral outcome (Clifton et al., 1991; Dietrich et al., 1994; Bramlett et al., 1995 see Table 3). In this regard, experimental studies first demonstrated the beneficial effects of mild hypothermia in models of fluid-percussion (F-P) brain injury (Clifton et al., 1991; Dietrich et al., 1994). Thus, it is not too surprising that mild elevations in brain temperature may also affect outcome in models of TBI (for recent review see Thompson et al., 2003a, b). In one early study, Dietrich and colleagues (1996a) reported that an induced hyperthermic period (39°C/4 h) initiated 24 h after parasagittal moderate F-P brain injury significantly aggravated outcome compared to normothermic animals. In that study, mild hyperthermia increased mortality, significantly increased contusion volume, and also enhanced BBB permeability and inflammatory cell accumulation compared to normothermic rats. In another study, Taylor et al. (2002) investigated the detrimental consequences of post-traumatic hyperthermia on cognitive deficits after TBI. In that study, alcohol consumption after trauma attenuated elevations in temperature and improved behavioral outcome.

These experimental findings emphasized that fever may be detrimental in patients with neurological injuries and that more aggressive control of fever should be undertaken in the intensive care unit. Similar to the stroke literature, this opinion has been supported by recent clinical data (Kilpatrick et al., 2000; Natale et al., 2000; Schwarz et al., 2000; Jiang et al., 2002; Johnston et al., 2006). For example, Kilpatrick and colleagues (2000) showed that febrile episodes occurred in 46.7% of 428 consecutive patients in a neuro-intensive care unit. In that study, febrile episodes, defined as rectal temperature above 38.5°C correlated with length of stay. In another study, rises in temperature seen in severe head injured patients were associated with derangements of intracranial volume homeostasis (Rossi et al., 2001). Brain temperature elevations were concluded to have a significant impact on intracranial pressure (ICP).

Spinal cord injury

The importance of mild hyperthermia in models of cerebral ischemia and trauma has led to concerns about the histopathological and functional impact of moderate to severe hyperthermia in patients with SCI (Holtzclaw, 1992). It is known that fever is a common complication in patients with SCI and that many patients develop infections (i.e., respiratory, urinary tract), accounting for systemic elevations in temperature (Montgomerie, 1997; Sugarman et al., 1982). Thermoregulatory problems, deep venous thrombosis, or fever of unknown etiology are commonly seen in this patient population (Beraldo et al., 1993). In a study by Yu and colleagues (2001), the effects of induced hyperthermia following contusive SCI in rats was evaluated. Animals underwent moderate contusion injury utilizing the NYU Impactor at T10. Beginning 30 min post-trauma, animals' core temperatures were increased to 39.5°C and maintained for a 4-h period. Compared to normothermic (37°C) animals, traumatized rats undergoing a secondary hyperthermic insult demonstrated worse outcomes in both behavioral and histopathological measures. Using the Basso, Beattie, Bresnahan locomotor rating scale (Basso et al., 1995), post-traumatic hyperthermic animals demonstrated significantly less locomotor recovery over the 44 day observation period compared to normothermic SCI rats. In addition, post-traumatic hyperthermia increased the percentage of tissue damage compared to the normothermic group. These data indicate that complications of SCI, including fever and infection, leading to elevations in systemic temperature may contribute to the severity of secondary injury associated with traumatic SCI and significantly affect neurological outcome. Thus, similar to brain injury, it is important that aggressive steps be taken to prevent the onset of fever in SCI patients during the period immediately following injury.

Gender

Recent data have emphasized the importance of gender in clinical and experimental models of cerebral ischemia and trauma (Grosswasser et al., 1998; Farace and Alves, 2000; Bell and Pepping, 2001; Stein, 2001). In studies utilizing focal ischemia, for example, an influence of estrogen, in particular, has been shown to have an effect on outcome (Hurn et al., 1995; Hurn and Maccrae, 2000). Hurn and colleagues (1995) first described an increase in cerebral blood flow and a decrease in hyperemia in female rats treated with 17β estradiol compared to untreated females and males. Alkayed and colleagues (1998) showed that infarct size was reduced in intact females compared to males and ovariectomized females. Gender effects have also been described in models of TBI (Hall et al., 1991; Roof et al., 1993; Bramlett and Dietrich, 2001). For example, Bramlett and Dietrich (2001) reported a significant reduction in contusion volume after F-P injury in intact females vs. males or ovariectomized females. Interestingly, while mild post-traumatic hypothermia was shown to be effective in reducing histopathological damage in male or ovariectomized TBI rats, no significant improvement in outcome was reported in intact female rats with hypothermic treatment (Suzuki et al., 2003).

In the area of hyperthermia, gender effects have also been recently emphasized in the TBI literature (Bramlett and Dietrich, 2004; Suzuki et al., 2004). In one study, male and ovariectomized female rats underwent F-P brain injury followed by a 4-h hyperthermic period. While both female and ovariectomized rats demonstrated increased contusion volumes and frequency of damaged axons as indicated by immunoreactive beta-amyloid precursor protein axonal profiles, the effects of post-traumatic hyperthermia were more pronounced in ovariectomized rats. These findings emphasize that neural hormones, including estrogen and progesterone, may protect not only against primary and secondary injury mechanisms (Stein, 2001), but also against secondary insults, including fever (Fig. 1). These studies underscore the importance of considering gender in both preclinical and clinical studies, while evaluating not only pathophysiological mechanisms of injury but also novel therapeutic treatment strategies.

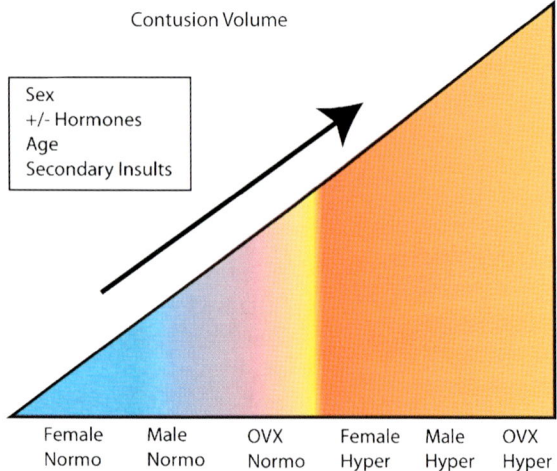

Fig. 1. Schematic drawing of the impact of sex, age, presence, or absence of hormones and secondary insults on contusion volume following traumatic brain injury (TBI). Female normothermic animals have the smallest contusion volume post-TBI. However, as these factors change, for example female becomes male and hyperthermia occurs, then there is an impact on contusion volume. Therefore, it is important to account for these variables when drawing conclusions about the efficacy of treatment strategies and secondary insults in a clinical setting.

Dissociation between brain and core temperature

Although elevations in core temperature are commonly seen in patients with injuries to the brain or spinal cord (Cunha and Tu, 1988; Hayashi et al., 1994), recent clinical studies have demonstrated that brain temperatures can be higher than rectal or bladder temperatures after brain trauma (Mellergard and Nordstrom, 1990, 1991; Sternau et al., 1991; Henker et al., 1998; Rumana, et al., 1998; Hayashi and Dietrich, 2004). In the study by Mellergard and Nordstrom (1991), mean differences between brain temperature and rectal temperature were reported to be 0.10° to 0.76°C in patients with a variety of neurological diagnoses, including head injury, subarachnoid hemorrhage, intracerebral hemorrhage, tumor resection, and hydrocephalus. In a study by Henker and colleagues (1998), brain, rectal, and bladder temperatures were monitored for 5 days after severe TBI, with deep brain microthermistors attached to the ventriculostomy catheter. At virtually all time points, brain temperature was higher than rectal or bladder temperature, with brain temperatures averaging 1°C higher than rectal or bladder temperature. Importantly, differences were greatest when bladder or rectal temperature was elevated. Thus, patients with rectal temperatures of 38° to 39°C were very likely to have brain temperatures of 40° to 41°C.

A similar study was completed by Rumana and colleagues (1998), where brain temperature was frequently 1.1° higher than rectal temperature in severe TBI patients. In that study, the greatest difference between brain and core temperature was observed when cerebral perfusion pressure decreased to less than 50 mmHg, and the smallest differences in brain and core temperature were observed in patients that were treated with high dose barbiturates for control of elevated ICP. Thus, based on animal studies showing a strong association between aggravated outcome and elevated brain temperature, the observation that brain temperature can be higher than core temperature in patients with brain injury emphasizes the importance of controlling temperature in this critically ill patient population.

Rewarming phase

Recent experimental and clinical data have emphasized the importance of the rewarming phase, in terms of hypothermic protection after brain and SCI (Enomoto et al., 1996; Marion et al., 1997; Koizumi and Povlishock, 1998; Nakamura et al., 1999; Clifton et al., 2001; Matsushita et al., 2001; Suehiro et al., 2003; Marion, 2004). In these studies, slow but not rapid rewarming has been shown to provide the most benefit, in terms of histopathological outcome following CNS injury. Clinically, slow rewarming after a period of induced hypothermia is used routinely because of the potential detrimental effects of rebound intracranial hypertension when temperature is increased rapidly (Marion et al., 1997; Clifton et al., 2001). Experimentally, several studies have recently emphasized the importance of slow rewarming in injury models (Koizumi and Povlishock, 1998; Nakamura et al., 1999; Matsushita et al., 2001). For example, in a transient forebrain ischemia study, rapid rewarming failed to provide the neuroprotective effect of hypothermia that was observed with slow rewarming (Nakamura et al., 1999). In that study, rapid rewarming after hypothermia resulted in poor recovery of mean arterial blood pressure and cortical blood flow. These investigators suggested that aggravated uncoupling between blood flow and metabolism through rapid rewarming might reduce the protection of hypothermia. Important to the present discussion, a gradual rewarming protocol after controlled hypothermia has been reported to reduce axonal damage after experimental TBI (Koizumi and Povlishock, 1998). In a study by Matsushita and colleagues (2001), a slow rewarming protocol extended to approximately 2 h after a therapeutic hypothermic period significantly improved histopathological outcome in rats following TBI and secondary hypoxia compared to hypothermic rats that underwent rapid rewarming. In clinical studies, the rewarming phase is critical, in terms of controlling for raised ICP pressure as well as avoiding the potential of temperature overshooting normothermic levels and becoming hyperthermic (Takino and Okada, 1991).

Adverse effects of fever in the clinic

As previously discussed, various experimental studies have reported the detrimental effects of induced hyperthermic periods after brain or SCI. Several retrospective studies have also found a significant association between fever and outcome after intracerebral hemorrhage, subarachnoid hemorrhage, and stroke (Rousseaux et al., 1980; Castillo et al., 1994; Childers et al., 1994; Reith et al., 1996; Hajat et al., 2000; Wang et al., 2000; Oliveira-Filho et al., 2001; Diringer, 2004; Marion, 2004; Deogaonkar et al., 2005; Suz et al., 2006). In a study by Schwarz and colleagues (2000) of 196 patients with spontaneous intracerebral hemorrhage, worse outcomes were observed in those patients who had rectal temperatures of $>37.5°C$ than in those who did not. In another study of patients with subarachnoid hemorrhage, Oliveira-Filho and colleagues (2001) found that patients having rectal temperatures of $>38.5°C$ for more than 2 days during the first week after hemorrhage had poorer outcomes (death, vegetative survival, or severe disability) compared to patients with no fever. More recently, Suz and colleagues (2006) reported a prevalence of fever and infection in a severe pediatric TBI patient population. Both events correlated with a longer length of stay in the pediatric intensive care unit. Several studies have also reported poor outcome in stroke patients with fever (Hindfelt, 1976; Castillo et al., 1994; Azzimondi et al., 1995). In a meta-analysis of 3790 stroke patients by Hajat and colleagues (2000), fever after stroke was associated with a significant increase in neurologic mortality, with a significantly higher incidence of death. Fever in subarachnoid hemorrhage was also reported to be associated with vasospasm and poor outcome independent of hemorrhage severity, pressure, or infection (Oliveira-Filho et al., 2001).

Mechanisms of hyperthermic response after CNS injury

How brain or SCI produces fever remains speculative (Little, 1985; Przelomski et al., 1986; Powers and Scheld, 1996; Schwarz et al., 2000; Thompson et al., 2003a, b, 2005). Elevations in temperature within the first 24 h after brain injury are commonly associated with the "acute phase response" (Young et al., 1988; Moore et al., 2000). Direct damage to thermoregulatory centers in the hypothalamus and brainstem has been shown to cause an increase in CNS temperature (Crompton, 1971; Rudy, 1980; Saper and Breder, 1994). Other studies have indicated that subarachnoid or intraventricular hemorrhage leads to hyperthermia, where blood may mechanically irritate hypothalamic thermoregulatory centers (Oliveira-Filho et al., 2001). It is clear that infections resulting from respiratory, urinary, or unknown etiologies can also lead to an increase in temperature (Dinarello, 1996; Saper, 1998; Grau et al., 1999). In this regard, hyperthermia is commonly experienced in patients in the acute period after brain injury and can be the result of various factors, including those associated with experimental hyperthermia (Kozak et al., 1998). For example, synthesis of heat shock proteins, release of cytokines, cellular responses, and hemodynamic and metabolic alterations have all been shown to be associated with elevations in temperature (for reviews, see Thompson et al., 2003a, b, 2005). Thus, it appears that temperature elevations may be the result of a complex series of metabolic and neurochemical changes that may initially occur as an endogenous protective effect of the injury but potentially result in important secondary injury mechanisms in patients.

Hyperthermic mechanisms of damage

As previously described, the pathophysiology of brain and SCI is multifactorial, involving various injury cascades, leading to both focal and diffuse neuronal injury (McIntosh et al., 1997; Dietrich, 1999; Dirnagl et al., 1999; Morganti-Kossmann et al., 2001). Although the exact mechanism by which hyperthermia may aggravate outcome following brain and SCI is speculative, small and/or delayed increases in temperature have been reported to affect various pathomechanisms felt to be responsible for irreversible neuronal injury (Busto et al., 1989b; Dietrich and Busto, 1997; Ginsberg and

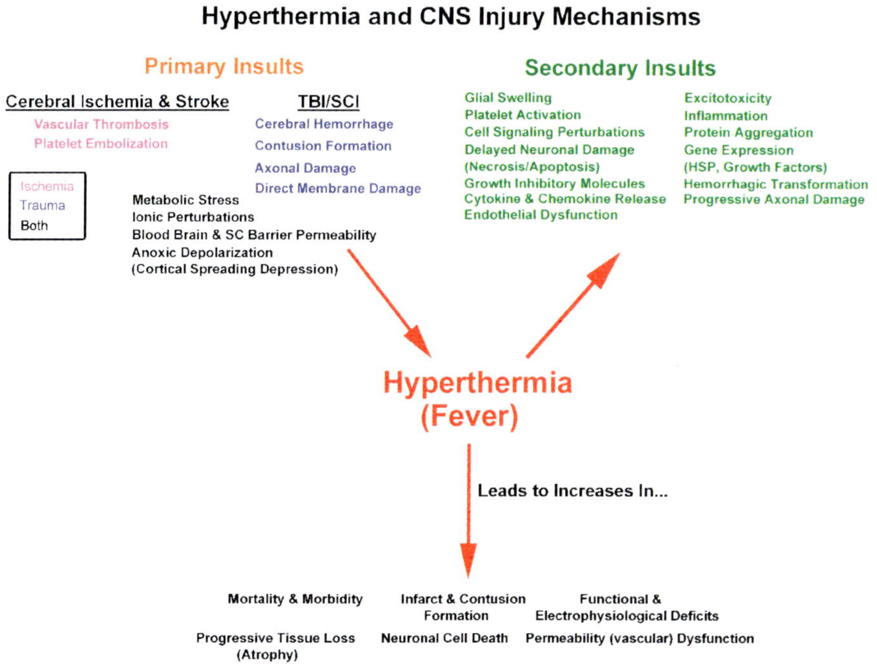

Hyperthermia and CNS Injury Mechanisms

Primary Insults

Cerebral Ischemia & Stroke
- Vascular Thrombosis
- Platelet Embolization

TBI/SCI
- Cerebral Hemorrhage
- Contusion Formation
- Axonal Damage
- Direct Membrane Damage

Ischemia / Trauma / Both
- Metabolic Stress
- Ionic Perturbations
- Blood Brain & SC Barrier Permeability
- Anoxic Depolarization (Cortical Spreading Depression)

Secondary Insults
- Glial Swelling
- Platelet Activation
- Cell Signaling Perturbations
- Delayed Neuronal Damage (Necrosis/Apoptosis)
- Growth Inhibitory Molecules
- Cytokine & Chemokine Release
- Endothelial Dysfunction
- Excitotoxicity
- Inflammation
- Protein Aggregation
- Gene Expression (HSP, Growth Factors)
- Hemorrhagic Transformation
- Progressive Axonal Damage

Hyperthermia (Fever)

Leads to Increases In...
- Mortality & Morbidity
- Infarct & Contusion Formation
- Functional & Electrophysiological Deficits
- Progressive Tissue Loss (Atrophy)
- Neuronal Cell Death
- Permeability (vascular) Dysfunction

Fig. 2. The occurrence of hyperthermia after CNS injury can impact on both primary and secondary injury mechanisms. This in turn leads to an exacerbation of brain damage and an increase in mortality and functional deficits. (Adapted from Bramlett and Dietrich, 2004.)

Busto, 1998; Sharma et al., 2000; Fig. 2). Elevations in temperature have also been shown to aggravate excitotoxicity and free radical generation, as well as to increase metabolism (Nilsson et al., 1975; McCulloch et al., 1982; McDonald et al., 1991; Baiping et al., 1994; Castillo et al., 1999). In a study by Dietrich and colleagues (1990a), intraischemic hyperthermia was shown to significantly alter the blood-brain barrier to protein tracers in a model of transient global ischemia. Also, post-traumatic hyperthermia has been shown to increase inflammatory processes, including elevations in proinflammatory cytokines (Kinoshita et al., 2002a, b; Vitarbo et al., 2004; Thompson et al., 2005), the increased accumulation of polymorphonuclear leukocytes in the post-injured brain and spinal cord (Dietrich et al., 1996a; Whalen et al., 1997; Chatzipanteli et al., 2000) and an increase in reactive astrocytes and interleukin-1β+/ED1+ cells within the hypothalamus after experimental TBI. The hypothalamus is a key center for thermoregulation and damage to this structure can result in hyperthermia.

In the study by Kinoshita and colleagues, (2002b), 3 h of post-traumatic hyperthermia (39°C) was introduced immediately after F-P moderate injury. Compared to normothermia, IL-1β mRNA expression and protein levels were elevated compared to normothermic TBI rats. In another study, post-traumatic hyperthermia had a marked effect on the hemoglobin content of the contralateral (nontraumatized) hemisphere and cerebellum (Kinoshita et al., 2002a). These data supported the detrimental effects of post-traumatic hyperthermia on microvascular function and BBB disruption after CNS injury. Finally, Vitarbo and colleagues (2004) failed to show that post-traumatic hyperthermia altered TNKα expression or levels compared to normothermic conditions. In a more recent investigation, post-traumatic hyperthermia was also associated with periventricular inflammation and worse outcome (Thompson et al., 2005). Thus, it appears that one important mechanism by which post-traumatic hyperthermia can aggravate outcome is by exacerbating inflammatory cascades.

Vascular and inflammatory cascades appear to be extremely sensitive to mild elevations in temperature following CNS injury.

Several lines of evidence also suggest the role of increased neuronal excitotoxicity, in terms of the detrimental effects of hyperthermia (Takagi et al., 1994; Suehiro et al. 1999). In various studies, hyperthermia has been reported to increase neurotransmitter release, accelerate free radical production, increase intracellular glutamate concentrations, and potentiate the sensitivity of neurons to excitotoxic injury (Sternau et al., 1992; Dietrich and Busto, 1997). In one study, Takagi and colleagues (1994) used microdialysis techniques to measure cellular levels of glutamate release in the penumbral cortex of rats after 2 h of normothermic or hyperthermic (39°C) MCAO. In that study, ischemic glutamate release was significantly higher in hyperthermic than normothermic rats, indicating that brain temperature is an important variable in studies of neurotransmitter release. Consistent with the theory of increased excitotoxicity during hyperthermia, other investigations have observed increases in cellular depolarization in the ischemic penumbra surrounding damaged neuronal tissue, increased neural intracellular acidosis, and inhibition of enzymatic protein kinases, which are responsible for synaptic transmission and cytoskeletal function (Chopp et al., 1988; Chen et al., 1993; Morimoto et al., 1997). At the genetic level, elevations in temperature have been shown to enhance the expression of specific immediate early genes such as *hsp*-70 (see Chapters 19, 20 & 21 in this volume), as well as receptor expression associated with glutamate neurotransmission (Kim et al., 1998).

Experimental and human studies have suggested that heat can directly induce nervous system injury (for review, see Sharma and Hoopes, 2003). In these circumstances, increased temperature of a range of 41.6° to 42.0°C for extended periods produces abnormalities in blood-brain barrier, cardiovascular, metabolic, and hemodynamic dysfunction, in some cases leading to an increase in brain water. Thus, mechanisms underlying hyperthermia-induced pathophysiology and secondary hyperthermia following CNS injury are similar in nature. An important point about temperature is that it acts on many pathophysiological mechanisms. Because of this characteristic, elevating or lowering temperature during or following an ischemic insult has the potential to adversely affect many pathophysiological events. Previously undetected periods of mild hyperthermia, especially occurring within the CNS, may therefore have overridden the beneficial effects of pharmacological interventions (Memezawa et al., 1995). This effect could have played a major role in influencing clinical trials that have resulted in failure (Danton and Dietrich, 2004). It is interesting to note that failure of a recent neuroprotective agent in ischemic stroke was potentially attributed to an increase in temperature observed in the treatment group (Enlimomab Acute Stroke Trial Investigators, 2001).

Future clinical trials, where temperature is maintained at normothermic levels or at mild hypothermic levels, may be necessary to allow for neuroprotective drugs to have a beneficial effect on the outcome. Thus, new methods of measuring brain and spinal cord temperature or more sensitive markers of mild hyperthermia must be developed to ensure that these alterations in temperature do not present a barrier to neuroprotective treatments.

Management of hyperthermia

As briefly described, preventing periods of hyperthermia appears to be an important objective, in terms of reducing secondary injury mechanisms in patients following CNS injury (Dietrich et al., 1996b; Dietrich, 1999). Current methods of reducing elevated body temperature are primarily limited to the use of antipyretic agents and external cooling processes (Krieger et al., 2001; Bernard et al., 2002; Greisman and Mackowiak, 2002; Kasner et al., 2002; Diringer et al., 2004; Schulman et al., 2005; Johnston et al., 2006). Thus, novel methods of maintaining normothermia must be developed and tested for safety and efficacy. Kilpatrick and colleagues (2000) reported that anti-pyrexial therapy was provided in 80% of the febrile episodes, either pharmacologically (acetaminophen or ibuprofen) or physically (cooling blankets, axillary ice packs, or alcohol rubs). In a study by Benedek and colleagues (1987), indomethacin was effective against neurogenic hyperthermia. In a randomized, controlled clinical trial, Kasner and colleagues (2002) reported that early acetaminophen treatment resulted in a smaller reduction in core body temperature and was helpful in preventing hyperthermia (37.5°C) in acute stroke patients. Currently, cooling blankets are used in the intensive care units to maintain normothermia as well as to produce mild hypothermia.

The development of more efficient methods of external cooling is also being evaluated, which may more rigorously inhibit periods of temperature elevation (Mayer et al., 2001, 2004). Although some reports have suggested that localized head cooling does not reduce brain temperature (Corbett and Laptook, 1998), recent data suggest that this approach results in improved outcome in cardiac arrest patients (Hachimi-Idrissi et al., 2001). Most recently, the development of endovascular catheters has resulted in a new method of internal cooling (Georgiadis et al., 2001; Marion, 2001; Schmutzhard et al., 2002). Catheters are inserted into specific vessels and have the capacity to cool circulating blood and maintain core temperature at precise temperature levels. These devices also have the ability to regulate temperature relatively quickly, a feature that may be necessary when attempting to reduce temperature spikes that can occur in the early stages of CNS injury.

Marion (2001) reported that endovascular cooling with the CoolGard (Alsius, Irvine, CA) cooling catheter is effective. He showed that head-injured patients whose temperature was controlled with the endovascular device had fever duration of less than one-half of that in the control group. Although no specific complications were reported in this study, the final analysis of the temperature data revealed that there was a 64% reduction in the fever burden in patients with the heat-exchange catheter, compared with control patients. There was also a 61% reduction in the use of cooling blankets, 66% reduction in the use of other physical means of cooling, and a 28% reduction in the use of antipyretic agents in the heat-exchange catheter group. Finally, post hoc analysis also revealed that the fever burden was significantly higher in patients who died than in those who survived (Marion, 2004). Thus, continued development of methods of measuring temperature, both core and CNS, as well as developing devices that help us to maintain normothermia are critical to the continued development of guidelines for the management of brain injured patients (Bullock et al., 1996).

Summary

Temperature is an important factor in most biological processes. Thus, it is not surprising that temperature also plays a major role in the pathophysiological mechanisms of CNS injury. Mild elevations in temperature have been shown in many models of CNS injury to aggravate both histopathological and behavioral outcome. Because hyperthermic periods can affect multiple pathomechanisms, it is important that strategies to inhibit or reduce temperature elevations be developed to improve the management of critically injured CNS patients. It is hoped that, with continued research and study, the detrimental effects of hyperthermia can be reduced or inhibited and that better outcome in our patients will be seen with improved temperature management.

Acknowledgments

This work was supported by NIH grants NS42133, NS43233, and NS30291. The author would like to thank Charlaine Rowlette for expert editorial assistance.

References

Albrecht, R.F., Wass, C. and Lanier, W.L. (1998) Occurrence of potentially detrimental temperature alterations in hospitalized patients at risk for brain injury. Mayo Clinic Proc., 73: 629–635.

Alkayed, N.J., Hartukuni, I., Kimes, A.S., London, E.D., Traystman, R.J. and Hurn, P.D. (1998) Gender-linked brain injury in experimental stroke. Stroke, 29: 159–166.

Azzimondi, G., Bassein, L., Nonino, F., Fiorani, L., Vignatelli, L., Re, G. and D'Allesandro, R. (1995) Fever in acute stroke worsens prognosis: a prospective study. Stroke, 26: 2040–2043.

Baena, R.C., Busto, R., Dietrich, W.D., Globus, M.Y.-T. and Ginsberg, M.D. (1997) Hyperthermia delayed by 24 hours aggravates neuronal damage in rat hippocampus following global ischemia. Neurology, 48: 678–773.

Baiping, L., Xiujuan, T., Hongwei, C., Qiming, X. and Quling, G. (1994) Effect of moderate hypothermia on lipid peroxidation in canine brain tissue after cardiac arrest and resuscitation. Stroke, 25: 147–152.

Basso, D.M., Beattie, M.S. and Bresnahan, J.C. (1995) A sensitive and reliable locomotor rating scale for open field testing in rats. J. Neurotrauma, 12: 1–21.

Bean, W.B. (Ed.). (1968) Sir William Osler: Aphorisms from His Bedside Teachings and Writings. 3rd printing. Charles C. Thomas, Springfield, IL.

Behr, R., Erlingspiel, D. and Becker, A. (1997) Early and longtime modifications of temperature regulation after severe head injury: prognostic implications. Ann. N.Y. Acad. Sci., 813: 722–732.

Bell, K.R. and Pepping, M. (2001) Women and traumatic brain injury. Phys. Med. Rehabil. Clin. N. Am., 12: 169–182.

Benedek, G., Toth-Daru, P., Janaky, J., Hortobagyi, A., Obal Jr., F. and Colner-Sasi, K. (1987) Indomethacin is effective against neurogenic hyperthermia following cranial trauma or brain surgery. Can. J. Neurol. Sci., 14: 145–148.

Beraldo, P.S., Neves, E.G., Alves, C.M., Khan, P., Cirilo, A.C. and Alencar, M.R. (1993) Pyrexia in hospitalised spinal cord injury patients. Paraplegia, 31: 186–191.

Bernard, S.A., Gray, T.W., Buist, M.D., Jones, B.M., Silvester, W., Gutteridge, G. and Smith, K. (2002) Treatment of comatose survivors of out-of-hospital cardiac arrest with induced hypothermia. N. Engl. J. Med., 346: 557–563.

Bramlett, H.M. and Dietrich, W.D. (2001) Neuropathological protection after traumatic brain injury in intact female rats versus males or ovariectomized females. J. Neurotrauma, 18: 891–900.

Bramlett, H.M. and Dietrich, W.D. (2004) Pathophysiology of cerebral ischemia and trauma. Similarities and differences. J. Cereb. Blood Flow Metab., 24: 133–150.

Bramlett, H., Green, J.E., Dietrich, W.D., Busto, R., Globus, M.Y.-T. and Ginsberg, M.D. (1995) Posttraumatic brain hypothermia provides protection from sensorimotor and cognitive behavioral deficits. J. Neurotrauma, 12: 289–298.

Bullock, R., Chesnut, R.M., Clifton, G., Ghajar, J., Marion, D.W., Narayan, R.K., Newell, D.W., Pitts, L.H., Rosner, M.J. and Wilberger, J.W. (1996) Guidelines for the management of severe head injury. Brain Trauma Foundation. Eur. J. Emerg. Med., 3: 109–127.

Busto, R., Dietrich, W.D., Globus, M.Y.-T. and Ginsberg, M.D. (1989a) Post-ischemic moderate hypothermia inhibits CA1 hippocampal ischemic neuronal injury. Neurosci. Lett., 101: 299–304.

Busto, R., Dietrich, W.D., Globus, M.Y.-T., Valdes, I., Scheinberg, P. and Ginsberg, M.D. (1987) Small differences in intraischemic brain temperature critically determine the extent of ischemic neuronal injury. J. Cereb. Blood Flow Metab., 7: 729–738.

Busto, R., Globus, M.Y.-T., Dietrich, W.D., Martinez, E., Valdes, I. and Ginsberg, M.D. (1989b) Effect of mild hypothermia on ischemia-induced release of neurotransmitters and free fatty acids in rat brain. Stroke, 20: 904–910.

Busto, R., Globus, M.Y.-T., Neary, J.T. and Ginsberg, M.D. (1994) Regional alterations of protein kinase C activity following transient cerebral ischemia: effects of intraischemic brain temperature modulation. J. Neurochem., 63: 1095–1103.

Cairns, C.J.S. and Andrews, P.J.D. (2002) Management of hyperthermia in traumatic brain injury. Curr. Opin. Crit. Care, 8: 106–110.

Castillo, J., Davalos, A. and Noya, M. (1999) Aggravation of acute ischemic stroke by hyperthermia is related to an excitotoxic mechanism. Cerebrovasc. Dis., 9: 22–27.

Castillo, J., Martinez, F., Leira, R., et al. (1994) Mortality and morbidity of acute cerebral infarction related to temperature and basal analytical parameters. Cerebrovasc. Dis., 4: 56–71.

Chatzipanteli, K., Alonso, O.F., Kraydieh, S. and Dietrich, W.D. (2000) Importance of posttraumatic hypothermia and hyperthermia on the inflammatory response after fluid percussion brain injury: biochemical and immunocytochemical studies. J. Cereb. Blood Flow Metab., 20: 531–542.

Chen, Q., Chopp, M., Bodzin, G. and Chen, H. (1993) Temperature modulations of cerebral depolarization during focal cerebral ischemia in rats: correlation with ischemic injury. J. Cereb. Blood Flow Metab., 13: 389–394.

Chen, H., Chopp, M. and Welch, K.M.A. (1991) Effect of mild hyperthermia on the ischemic infarct volume after middle cerebral artery occlusion in the rat. Neurology, 41: 1133–1135.

Childers, M.K., Rupright, J. and Smith, D.W. (1994) Posttraumatic hyperthermia in acute brain injury rehabilitation. Brain Inj., 8: 335–343.

Chopp, M., Knight, R., Tidwell, C.D., Helpern, J.A., Brown, E. and Welch, K.M.A. (1989) The metabolic effects of mild hypothermia on global cerebral ischemia and recirculation in the cat: comparison to normothermia and hyperthermia. J. Cereb. Blood Flow Metab., 9: 141–148.

Chopp, M., Welch, K.M.A., Tidwell, C.D., Knight, R. and Helpern, J.A. (1988) Effect of mild hyperthermia on recovery of metabolic function after global cerebral ischemia in cats. Stroke, 19: 1521–1525.

Churn, S.B., Taft, W.C., Billingsley, M.S., Bair, R.E. and De Lorenzo, R.J. (1990) Temperature modulation of ischemic neuronal death and inhibition of calcium/calmodulin-dependent protein kinase II in gerbils. Stroke, 21: 1715–1721.

Clifton, G.L., Jiang, J.Y., Lyeth, B.G., Jenkins, L.W., Hamm, R.J. and Hayes, R.L. (1991) Marked protection by moderate hypothermia after experimental traumatic brain injury. J. Cereb. Blood Flow Metab., 11: 114–121.

Clifton, G.L., Miller, E.R., Choi, S.C., Levin, H.S., McCauley, S., Smith Jr., K.R., Muizelaar, J.P., Wagner Jr., F.C., Marion, D.W., Luerssen, T.G., Chesnut, R.M. and Schwartz, M. (2001) Lack of effect of induction of hypothermia after acute brain injury. N. Engl. J. Med., 344: 556–563.

Clifton, G.L., Taft, W.C., Blair, R.E., Choi, S.C. and DeLorenzo, R.J. (1989) Conditions for pharmacologic evaluation in the gerbil model of forebrain ischemia. Stroke, 20: 1545–1552.

Colbourne, F., Nurse, S.M. and Corbett, D. (1993) Spontaneous postischemic hyperthermia is not required for severe CA1 ischemic damage in gerbils. Brain Res., 623: 1–5.

Colbourne, F., Sutherland, G. and Corbett, D. (1997) Postischemic hypothermia. A critical appraisal with implications for clinical treatment. Mol. Neurobiol., 14: 171–201.

Corbett, R.J.T. and Laptook, A.R. (1998) Failure of localized head cooling to reduce brain temperature in adult humans. Neuro. Rep., 9: 2721–2725.

Crompton, M.R. (1971) Hypothalamic lesions following closed head injury. Brain, 94: 165–172.

Cunha, B.A. and Tu, R.P. (1988) Fever in the neurosurgical patient. Heart & Lung: J. Acute Crit. Care, 17: 608–611.

Danton, G.H. and Dietrich, W.D. (2004) The search for neuroprotective strategies in stroke. Am. J. Neuroradiol., 25: 181–194.

Deogaonkar, A., De Georgia, M., Bae, C., Abou-Chebl, A. and Andrefsky, J. (2005) Fever is associated with third ventricular shift after intracerebral hemorrhage: pathophysiologic implications. Neurol. India, 53: 202–206.

Dietrich, W.D. (1992) The importance of brain temperature in cerebral injury. J. Neurotrauma, 9: 374–384.

Dietrich, W.D. (1999) Temperature and ischemic stroke. In: Fisher M. and Bogousslavsky J. (Eds.), Current Review of Cerebrovascular Disease. Butterworth Heinemann, Boston, MA, pp. 43–52.

Dietrich, W.D., Alonso, O., Busto, R., Globus, M.Y.-T. and Ginsberg, M.D. (1994) Post-traumatic brain hypothermia reduces histopathological damage following contusive brain injury in the rat. Acta Neuropathol., 87: 250–258.

Dietrich, W.D., Alonso, O., Halley, M. and Busto, R. (1996a) Delayed post-traumatic brain hyperthermia worsens outcome after fluid percussion brain injury: a light and electron microscopic study in rats. Neurosurgery, 38: 533–541.

Dietrich, W.D. and Busto, R. (1997) Hyperthermia and brain ischemia. In: Welch K.M.A., Caplan L.R., Reis D.J., Siesjo B.K. and Weir B. (Eds.), Primer on Cerebrovascular Diseases. Academic Press, San Diego, CA, pp. 163–165.

Dietrich, W.D., Busto, R., Globus, M.Y.-T. and Ginsberg, M.D. (1996b) Brain damage and temperature: cellular and molecular mechanisms. In: Siesjo B. and Wieloch T. (Eds.), Annals of Neurology: Cellular and Molecular Mechanisms, Vol. 71. Lippincott-Raven Publishers, Philadelphia, pp. 177–198.

Dietrich, W.D., Busto, R., Halley, M. and Valdes, I. (1990a) The importance of brain temperature in alterations of the blood-brain barrier following cerebral ischemia. J. Neuropathol. Exp. Neurol., 49: 486–497.

Dietrich, W.D., Busto, R.D., Valdes, I., et al. (1990b) Effects of normothermic versus mild hyperthermic forebrain ischemia in rats. Stroke, 21: 1318–1325.

Dietrich, W.D., Halley, M., Valdes, I. and Busto, R. (1991) Interrelationships between increased vascular permeability and acute neuronal damage following temperature-controlled brain ischemia in rats. Acta Neuro., 81: 615–625.

Dinarello, C.A. (1996) Thermoregulation and the pathogenesis of fever. Infect. Dis. Clin. North Am., 10: 433–449.

Diringer, M.N. for the Neurocritical Care Fever Reduction Trial Group (2004) Treatment of fever in the neurologic intensive care unit with a catheter-based heat exchange system. Crit. Care Med., 32: 559–564.

Diringer, M.N., Reaven, N.L., Funk, S.E., et al. (2004) Elevated body temperature independently contributes to increased length of stay in neuro ICU patients. Crit. Care Med., 32: 1489–1495.

Dirnagl, U., Iadecola, C. and Moskowitz, M.A. (1999) Pathobiology of ischaemic stroke: an integrated view. Trends Neurosci., 22: 391–397.

Enlimomab Acute Stroke Trial Investigators. (2001) Use of anti-ICAM-1 therapy in ischemic stroke. Results of the Enlimomab acute stroke trial. Neurology, 57: 1428–1434.

Enomoto, S., Hindman, B.J., Dexter, F., Smith, T. and Cutkomp, J. (1996) Rapid rewarming causes an increase in the cerebral metabolic rate for oxygen that is temporarily unmatched by cerebral blood flow. A study during cardiopulmonary bypass in rabbits. Anesthesiology, 84: 1392–1400.

Farace, E. and Alves, W.M. (2000) Do women fare worse: a metaanalysis of gender differences in traumatic brain injury outcome. J. Neurosurg., 94: 862–864.

Geffroy, A., Bronchard, R., Merckx, P., Seince, P.-F., Faillot, T., Albaladejo, P. and Marty, J. (2004) Severe traumatic head injury in adults: Which patients are at risk of early hyperthermia? Intensive Care Med., 30: 785–790.

Georgiadis, D., Schwarz, S., Kollmar, R. and Schwab, S. (2001) Endovascular cooling for moderate hypothermia in patients with acute stroke. First results of a novel approach. Stroke, 32: 2550–2553.

Ginsberg, M.D. and Busto, R. (1989) Rodent models of cerebral ischemia. Stroke, 20: 1627–1642.

Ginsberg, M.D. and Busto, R. (1998) Combating hyperthermia in acute stroke: a significant clinical concern. Stroke, 29: 529–534.

Girard, F. (2005) Managing head injured patients. Curr. Opin. Anaesthesiol., 18: 471–476.

Globus, M.Y., Busto, R., Dietrich, W.D., Martinez, E., Valdes, I. and Ginsberg, M.D. (1988) Effect of ischemia on the in vivo release of striatal dopamine, glutamate, and GABA studied by intracerebral microdialysis. J. Neurochem., 51: 1455–1464.

Globus, M.Y., Busto, R., Lin, B., Schnippering, H. and Ginsberg, M.D. (1995) Detection of free radical activity during transient global ischaemia and recirculation: effects of intraischemic brain temperature modulation. J. Neurochem., 65: 1250–1256.

Grau, A.J., Buggle, F., Schnitzler, P., Spiel, M., Lichy, C. and Hacke, W. (1999) Fever and infection early after ischemic stroke. J. Neurol. Sci., 171: 115–120.

Greisman, L.A. and Mackowiak, P.A. (2002) Fever: beneficial and detrimental effects of antipyretics. Curr. Opin. Infect. Dis., 15: 241–245.

Grosswasser, Z., Cohen, M. and Keren, O. (1998) Female TBI patients recover better than males. Brain Inj., 12: 805–808.

Hachimi-Idrissi, S., Corne, L., Ebinger, G., Michotte, Y. and Huyghens, L. (2001) Mild hypothermia induced by a helmet device: a clinical feasibility study. Resuscitation, 51: 275–281.

Hajat, C., Hajat, S. and Sharma, P. (2000) Effects of poststroke pyrexia on stroke outcome: a meta-analysis of studies in patients. Stroke, 31: 410–414.

Hall, E.D., Pazara, K.E. and Linseman, K.L. (1991) Sex differences in postischemic neuronal necrosis in gerbils. J. Cereb. Blood Flow Metab., 11: 292–298.

Hayashi, N. and Dietrich, W.D. (Eds.). (2004) Brain Hypothermia Treatment. Springer, Tokyo.

Hayashi, N., Hirayama, T. and Ohata, M. (1993) The computed cerebral hypothermia management techniques to the critical head injury patients. Adv. Neurotrauma Res., 5: 61–64.

Hayashi, N., Hirayama, T. and Utagawa, A. (1994) The cerebral thermo-pooling and hypothermia treatment of critical head injury patients. In: Nagai H. (Ed.), Intracranial Pressure IX. Springer, Berlin, pp. 589–599.

Henker, R.A., Brown, S.D. and Marion, D.W. (1998) Comparison of brain temperature with bladder and rectal temperature in adults with severe head injury. Neurosurgery, 42: 1071–1075.

Hickey, R.W., Kochanek, P.M., Ferimer, H., Alexander, H.L., Garman, R.H. and Graham, S.H. (2003) Induced hyperthermia exacerbates neurologic neuronal histologic damage after asphyxial cardiac arrest in rats. Crit. Care Med., 31: 531–535.

Hickey, R.W., Kochanek, P.M., Ferimer, H., et al. (2000) Hypothermia and hyperthermia in children after resuscitation from cardiac arrest. Pediatrics, 106: 118–122.

Hindfelt, B. (1976) The prognostic significance of subfebrility and fever in ischaemic cerebral infarction. Acta Neurol. Scand., 53: 72–79.

Holtzclaw, B. (1992) The febrile response in critical care: state of the science. Heart Lung, 21: 482–500.

Hurn, P.D., Littleton-Kearney, M.T., Kirsch, J.R., Dharmarajan, A.M. and Traystman, R.J. (1995) Postischemic cerebral blood flow recovery in the female: effect of 17β-estradiol. J. Cereb. Blood Flow Metab., 15: 666–672.

Hurn, P.D. and Maccrae, I.M. (2000) Estrogen as a neuroprotectant in stroke. J. Cereb. Blood Flow Metab., 20: 631–652.

Jiang, J.Y., Gao, G.Y., Li, W.P., Yu, M.K. and Zhu, C. (2002) Early indicators of prognosis in 846 cases of severe traumatic brain injury. J. Neurotrauma, 19: 869–874.

Johnston, N.J., King, A.T., Protheroe, R. and Childs, C. (2006) Body temperature management after severe traumatic brain injury: methods and protocols used in the United Kingdom and Ireland. Resuscitation, 70: 254–262.

Kammersgaard, L.P., Jorgensen, H.S., Rungby, J.A., Reith, J., Nakayama, H., Weber, U.J., Houth, J. and Olsen, T.S. (2002) Admission body temperature predicts long-term mortality after acute stroke. The Copenhagen Stroke Study. Stroke, 33: 1759–1762.

Kasner, S.E., Wein, T., Piriyawat, P., Villar-Cordova, C.E., Chalela, J.A., Krieger, D.W., Morgenstern, L.B., Kimmel, S.E. and Grotta, J.C. (2002) Acetaminophen for altering body temperature in acute stroke. A randomized clinical trial. Stroke, 33: 130–135.

Kato, H., Araki, T. and Kogure, K. (1991) Postischemic spontaneous hyperthermia is not a major aggravating factor for neuronal damage following repeated brief cerebral ischemia in the gerbil. Neurosci. Lett., 126: 21–24.

Kil, H.Y., Zhang, J. and Piantadosi, C.A. (1996) Brain temperature alters hydroxyl radical production during cerebral ischemia/reperfusion in rats. J. Cereb. Blood Flow Metab., 16: 100–106.

Kilpatrick, M.M., Lowry, D.W., Firlik, A.D., Yonas, H. and Marion, D.W. (2000) Hyperthermia in the neurosurgical intensive care unit. Neurosurgery, 47: 850–856.

Kim, Y., Busto, R., Dietrich, W.D., et al. (1996) Delayed postischemic hyperthermia in awake rats worsens the histopathological outcome of transient focal ischemia. Stroke, 27: 2274–2280.

Kim, Y., Truettner, J., Zhao, W., Busto, R. and Ginsberg, M.D. (1998) The influence of delayed postischemic hyperthermia following transient focal ischemia: alterations of gene expression. J. Neurol. Sci., 159: 1–10.

Kinoshita, K., Chatzipanteli, K., Alonso, O.F., Howard, M. and Dietrich, W.D. (2002a) The effect of brain temperature on hemoglobin extravasation after traumatic brain injury. J. Neurosurg., 97: 945–953.

Kinoshita, K., Chatzipanteli, K., Vitarbo, E., Truettner, J.S., Alonso, O.F. and Dietrich, W.D. (2002b) Interleukin-1beta messenger ribonucleic acid and protein levels after fluid-percussion brain injury in rats: importance of injury severity and brain temperature. Neurosurgery, 51: 195–203.

Kirino, T. (1982) Delayed neuronal cell death in the gerbil hippocampus following ischemia. Brain Res., 239: 57–69.

Koizumi, H. and Povlishock, J.T. (1998) Posttraumatic hypothermia in the treatment of axonal damage in an animal model of traumatic axonal injury. J. Neurosurg., 89: 303–309.

Kozak, W., Kluger, M.J., Soszynski, D., Conn, C.A., Rudolph, K., Leon, L.R. and Zheng, H. (1998) IL-6 and IL-1 beta in fever. Studies using cytokine-deficient (knockout) mice. Ann. N.Y. Acad. Sci., 856: 33–47.

Krieger, D.W., DeGeorgia, M.A., Abou-Chgebl, A., Andrefsky, J.C., Sila, C.A., Katzan, I.L., Mayberg, M.R. and Furlan, A.J. (2001) Cooling for acute ischemic brain damage (Cool Aid): an open pilot study of induced hypothermia in acute ischemic stroke. Stroke, 32: 1847–1854.

Kuluz, J.W., Gregory, G.A., Han, Y., Dietrich, W.D. and Schleien, C.L. (1993) Fructose-1,6-biphosphate reduces infarct volume after reversible middle cerebral artery occlusion in rats. Stroke, 24: 1576–1583.

Kuroiwa, T., Bonnekoh, P. and Hossman, K.-A. (1990) Prevention of postischemic hyperthermia prevents ischemic injury of CA1 neurons in gerbils. J. Cereb. Blood Flow Metab., 10: 550–556.

Little, R.A. (1985) Heat production after injury. Br. Med. Bull., 41: 226–231.

Lundgren, J., Smith, M.-L. and Siesjo, B.K. (1991) Influence of moderate hypothermia on ischemic brain damage incurred under hyperthermic conditions. Exp. Brain Res., 84: 91–101.

Manno, E.M. and Farmer, J.C. (2004) Acute brain injury: if hypothermia is good, then is hyperthermia bad? Crit. Care Med., 32: 1611–1612.

Marion, D.W. (2001) Therapeutic moderate hypothermia and fever. Curr. Pharm. Des., 7: 1533–1536.

Marion, D.W. (2004) Breakthroughs in resuscitation "Therapeutic hypothermia, from hibernation to resuscitation." Controlled normothermia in neurologic intensive care. Crit. Care Med., 32: S43–S45.

Marion, D.W., Penrod, L.E., Kelsey, S.F., et al. (1997) Treatment of traumatic brain injury with moderate hypothermia. N. Engl. J. Med., 336: 540–546.

Matsushita, Y., Bramlett, H., Alonso, O. and Dietrich, W.D. (2001) Post-traumatic hypothermia is neuroprotective in a model of traumatic brain injury complicated by secondary hypoxic insult. Crit. Care Med., 29: 2060–2066.

Mayer, S.A., Commichau, C., Scarmeas, N., Presciutti, M., Bates, J. and Copeland, D. (2001) Clinical trial of an air-circulating cooling blanket for fever control in critically ill neurologic patients. Neurology, 56: 292–298.

Mayer, S.A., Kowalski, R.G., Prescuitti, M., Ostapkovich, N.D., McGann, E., Fitzsimmons, B.-F., Yavagal, D.R., Du, Y.E., Naidech, A.M., Janjua, N.A., Claassen, J., Kreiter, K.T., Parra, A. and Commichau, C. (2004) Neurologic critical care: clinical trial of a novel surface cooling system for fever control in neurocritical care patients. Crit. Care Med., 32: 2508–2515.

McCulloch, J., Savaki, H.E., Jehle, J. and Sokoloff, L. (1982) Local cerebral glucose utilization in hypothermic and hyperthermic rats. J. Neurochem., 39: 255–258.

McDonald, J.W., Chen, C.K., Trescher, W.H. and Johnston, M.V. (1991) The severity of excitotoxic brain injury is dependent on brain temperature in immature rat. Neurosci. Lett., 126: 83–86.

McIntosh, T.K., Saatman, K.E. and Raghupathi, R. (1997) Calcium and the pathogenesis of traumatic CNS injury: cellular and molecular mechanisms. Neuroscientist, 3: 169–175.

Mellergard, P. and Nordstrom, C.-H. (1990) Epidural temperature and possible intracerebral temperature gradients in man. Br. J. Neurosurg., 4: 31–38.

Mellergard, P. and Nordstrom, C.-H. (1991) Intracerebral temperature in neurosurgical patients. Neurosurgery, 28: 709–713.

Memezawa, H., Zhao, O., Smith, M.L., et al. (1995) Hyperthermia nullifies the ameliorating effect of dizocilpine maleate (MK-801) in focal cerebral ischemia. Brain Res., 23: 48–52.

Minamisawa, H., Smith, M.L. and Siesjo, B.K. (1990) The effect of mild hyperthermia and hypothermia on brain damage following 5, 10 and 15 minutes of forebrain ischemia. Ann. Neurol., 28: 26–33.

Miyazawa, T., Bonnekoh, P., Widmann, R. and Hossmann, K.-A. (1993) Heating of the brain to maintain normothermia during ischemia aggravates brain injury in the rat. Acta Neuropathol., 85: 488–494.

Montgomerie, J.Z. (1997) Infections in patients with spinal cord injuries. Clin. Infect. Dis., 25: 1285–1292.

Moore, T.H., Osteen, T.L., Chatziioannou, T.F., Hovda, D.A. and Cherry, T.R. (2000) Quantitative assessment of longitudinal metabolic changes in vivo after traumatic brain injury in the adult rat using FDG-microPET. J. Cereb. Blood Flow Metab., 20: 1492–1501.

Morganti-Kossmann, M.C., Rancan, M., Otto, V.I., Stahel, P.F. and Kossmann, T. (2001) Role of cerebral inflammation after traumatic brain injury: a revisited concept. Shock, 16: 165–177.

Morikawa, E., Ginsberg, M.D., Dietrich, W.D., Duncan, R.C., Kraydieh, S., Globus, M.Y.-T. and Busto, R. (1992) The significance of brain temperature in focal ischemia: histopathological consequences of middle cerebral artery occlusion in the rat. J. Cereb. Blood Flow Metab., 12: 380–389.

Morimoto, T., Ginsberg, M.D., Dietrich, W.D., et al. (1997) Hyperthermia enhances spectrin breakdown in transient focal cerebral ischemia. Brain Res., 746: 43–51.

Nakamura, T., Miyamoto, O., Yamagami, S.-I., Hayashida, Y., Itano, T. and Nagao, S. (1999) Influence of rewarming conditions after hypothermia in gerbils with transient forebrain ischemia. J. Neurosurg., 91: 114–120.

Natale, J.E., Joseph, J.G., Helfaer, M.A. and Schaffner, D.H. (2000) Early hyperthermia after traumatic brain injury in children: risk factors, influence on length of stay, and effect on short-term neurologic status. Crit. Care Med., 28: 1608–1615.

Nilsson, L., Kogure, K. and Busto, R. (1975) Effect of hypothermia and hyperthermia on brain energy metabolism. Acta Anaesthesiol. Scand., 19: 199–205.

Oliveira-Filho, J., Ezzeddine, M.A., Segal, A.Z., et al. (2001) Fever in subarachnoid hemorrhage: relationship to vasospasm and outcome. Neurology, 56: 1299–1304.

Powers, J.H. and Scheld, W.M. (1996) Fever in neurologic diseases. Infect. Dis. Clinics North Am., 10: 45–66.

Przelomski, M.M., Roth, R.M., Gkleckman, A. and Marcus, E.M. (1986) Fever in the wake of a stroke. Neurology, 36: 427–429.

Pulsinelli, W.A., Brierley, J.B. and Plum, F. (1980) Temporal profile of neuronal damage in a model of transient forebrain ischemia. Ann. Neurol., 11: 491–498.

Reith, J., Jorgensen, H.S., Pedersen, P.M., Nakayama, H., Raaschou, H.O., Jeppesen, L.L. and Olsen, T.S. (1996) Body temperature in acute stroke: relation to stroke severity, infarct size, mortality, and outcome. Lancet, 347: 422–425.

Roof, R.L., Duvdevani, R. and Stein, D.G. (1993) Gender influences outcome of brain injury: progesterone plays a protective role. Brain Res., 607: 333–336.

Roof, R.L. and Hall, E.D. (2000) Gender differences in acute CNS trauma and stroke: neuroprotective effects of estrogen and progesterone. J. Neurotrauma, 17: 367–388.

Rossi, S., Roncati Zanier, E., Mauri, I., Columbo, A. and Stocchetti, N. (2001) Brain temperature, body core temperature, and intracranial pressure in acute cerebral damage. J. Neurol. Neurosurg. Psychiat., 71: 448–454.

Rousseaux, P., Scherpereel, B., Bernard, M.H., Graftieaux, J.P. and Guyot, J.F. (1980) Fever and cerebral vasospasm in ruptured intracranial aneurysms. Surg. Neurol., 14: 459–465.

Rudy, T.A. (1980) Pathogenesis of fever associated with cerebral trauma and intracranial hemorrhage. In: Cox B., et al. (Eds.), Thermoregulatory Mechanisms and Therapeutic Implications 4th International Symposium on the Pharmacology of Thermoregulation. Karger, Basel, Switzerland, pp. 75–81.

Rumana, C.S., Gopinath, S.P., Uzura, J., Valadka, A.B. and Robertson, C.S. (1998) Brain temperature exceeds systemic temperature in head-injured patients. Crit. Care Med., 26: 562–567.

Saper, C.B. (1998) Neurobiological basis of fever. Ann. N.Y. Acad. Sci., 856: 90–94.

Saper, C.B. and Breder, C.D. (1994) The neurologic basis of fever. N. Engl. J. Med., 330: 1880–1886.

Schmutzhard, E., Engelhardt, K., Beer, R., et al. (2002) Safety and efficacy of a novel intravascular cooling device to control body temperature in neurologic intensive care patients: a prospective pilot study. Crit. Care Med., 30: 2481–2488.

Schulman, C.I., Namias, N., Doherty, J., Manning, R.J., Li, P., Alhaddad, A., Lasko, D., Amortegui, J., Dy, D.J., Dlugasch, L., Baracco, G. and Cohn, S.M. (2005) The effect of antipyretic therapy upon outcomes in critically ill patients: a randomized, prospective study. Surg. Infect. (Larchmt.), 6: 369–375.

Schwarz, S., Hafner, K., Aschof, A. and Schwab, S. (2000) Incidence and prognostic significance of fever following intracerebral hemorrhage. Neurology, 54: 354–361.

Segatore, M. (1992) Fever after traumatic brain injury. J. Neurosci. Nurs., 24: 104–109.

Sharma, H.S., Drieu, K., Alm, P. and Westman, J. (2000) Role of nitric oxide in blood-brain barrier permeability, brain edema and cell damage following hyperthermic brain injury. An experimental study using EGB-761 and Gingkolide B pretreatment in the rat. Acta Neurochir., 76(Suppl): 81–86.

Sharma, H.S. and Hoopes, P.J. (2003) Hyperthermia induced pathophysiology in the central nervous system. Int. J. Hyperthermia, 19: 325–354.

Stein, D.G. (2001) Brain damage, sex hormones and recovery: a new role for progesterone and estrogen? TINS, 24: 386–391.

Sternau, L.L., Globus, M.Y.-T., Dietrich, W.D., Martinez, E., Busto, R. and Ginsberg, M.D. (1992) Ischemia-induced neurotransmitter release: effects of mild intraischemic hyperthermia. In: Globus M.Y.-T. and Dietrich W.D. (Eds.), The Role of Neurotransmitters in Brain injury. Plenum, New York, pp. 33–38.

Sternau, L., Thompson, C., Dietrich, W.D., Busto, R., Globus, M.Y.-T. and Ginsberg, M.D. (1991) Intracranial temperature — observations in the human brain. J. Cereb. Blood Flow Metab., 11: S123.

Suehiro, E., Fujisawa, H., Ito, H., Ishikawa, T. and Maeikawa, T. (1999) Brain temperature modifies glutamate neurotoxicity in vivo. J. Neurotrauma, 16: 285–297.

Suehiro, E., Ueda, Y., Wei, E.P., Kontos, H.A. and Povlishock, J.T. (2003) Posttraumatic hypothermia followed by slow rewarming protects the cerebral microcirculation. J. Neurotrauma, 20: 381–390.

Sugarman, B., Brown, D. and Muscher, D. (1982) Fever and infection in spinal cord injury patients. JAMA, 248: 66–70.

Suz, P., Vavilala, M.S., Souter, M., Muangman, S. and Lam, A.M. (2006) Clinical features of fever associated with poor outcome in severe pediatric traumatic brain injury. J. Neurosurg. Anesthesiol., 18: 5–10.

Suzuki, T., Bramlett, H.M. and Dietrich, W.D. (2003) The importance of gender on the beneficial effects of posttraumatic hypothermia. Exp. Neurol., 184: 1017–1026.

Suzuki, T., Bramlett, H.M., Ruenes, G. and Dietrich, W.D. (2004) The effects of early post-traumatic hyperthermia in female and ovariectomized rats. J. Neurotrauma, 21: 842–853.

Takagi, K., Ginsberg, M.D., Globus, M.Y., Martinez, E. and Busto, R. (1994) Effect of hyperthermia on glutamate release in ischemic penumbra after middle cerebral artery occlusion in rats. Am. J. Physiol., 267: H1770–H1776.

Takino, M. and Okada, Y. (1991) Hyperthermia following cardiopulmonary resuscitation. Intensive Care Med., 17: 419–420.

Tator, C.H. and Fehlings, M.G. (1991) Review of the secondary injury theory of acute spinal cord trauma with emphasis on vascular mechanisms. J. Neurosurg., 75: 15–26.

Taylor, A.N., Romeo, H.E., Beylin, A.V., Tio, D.L., Rahman, S.U. and Hovda, D.A. (2002) Alcohol consumption in traumatic brain injury: attenuation of TBI-induced hyperthermia and neurocognitive deficits. J. Neurotrauma, 19: 1597–1608.

Thompson, H.J., Hoover, R.C., Tkacs, N.C., Saatman, K.E. and McIntosh, T.K. (2005) Development of posttraumatic hyperthermia after traumatic brain injury in rats is associated with increased periventricular inflammation. J. Cereb. Blood Flow Metab., 25: 163–176.

Thompson, H.J., Pinto-Martin, J. and Bullock, M.R. (2003a) Neurogenic fever after traumatic brain injury: an epidemiological study. J. Neurol. Neurosurg. Psychiatr., 74: 614–619.

Thompson, H.J., Tkacs, N.C., Saatman, K.E., Raghupathi, R. and McIntosh, T.K. (2003b) Hyperthermia following traumatic brain injury: a critical evaluation. Neurobiol. Dis., 12: 163–173.

Tomimatsu, T., Fukuda, H., Kanagawa, T., Mu, J., Kanzaki, T. and Murata, M.D. (2003) Effects of hyperthermia on hypoxic-ischemic brain damage in the immature rat: its influence on caspase-3-like protease. Am. J. Obstet. Gynecol., 88: 768–773.

Vitarbo, E.A., Chatzipanteli, K., Kinoshita, K., Truettner, J.S., Alonso, O.F. and Dietrich, W.D. (2004) Tumor necrosis factor α and protein levels after fluid percussion injury in rats: the effect of injury severity and brain temperature. Neurosurgery, 55: 416–425.

Wang, Y., Lim, L.L., Levi, C., Heller, R.F. and Fisher, J. (2000) Influence of admission body temperature on stroke mortality. Stroke, 31: 404–409.

Whalen, M.J., Carlos, T.M., Clark, R.S.B., Marion, D.W., DeKosky, S.T., Heineman, S., Schiding, J.K., Memarzadeh, F. and Kochanek, P.M. (1997) The effect of brain temperature on acute inflammation after traumatic brain injury in rats. J. Neurotrauma, 14: 561–572.

Young, A.B., Ott, L.G., Beard, D., Dempsey, R.J., Tibbs, P.A. and Mcclain, C.J. (1988) The acute-phase response of the brain-injured patient. J. Neurosurg., 69: 375–380.

Yu, C.-G., Jagid, J., Ruenes, G., Dietrich, W.D., Marcillo, A.E. and Yezierski, R.P. (2001) Detrimental effects of systemic hyperthermia on locomotor function and histopathological outcome after traumatic spinal cord injury in the rat. Neurosurgery, 49: 152–159.

CHAPTER 12

Physiological and pathological brain hyperthermia

Eugene A. Kiyatkin*

Cellular Neurobiology Branch, National Institute on Drug Abuse — Intramural Research Program, National Institutes of Health, DHHS, 5500 Nathan Shock Drive, Baltimore, MD 21224, USA

Abstract: While brain temperature is usually considered a stable, tightly regulated parameter, recent animal research revealed relatively large and rapid brain temperature fluctuations (~3°C) during various forms of naturally occurring physiological and behavioral activities. This work demonstrates that physiological brain hyperthermia has an intra-brain origin, resulting from enhanced neural metabolism and increased intra-brain heat production, and discusses its possible mechanisms and functional consequences. This work also shows that brain hyperthermia may also be induced by various drugs of abuse. While each individual drug (i.e., heroin, cocaine, meth-amphetamine, MDMA) has its own, dose-dependent effects on brain and body temperatures, these effects are strongly modulated by the individual's activity state and environmental conditions, showing dramatic alterations during the development of drug-taking behavior. While brain temperatures may also increase due to environmental overheating and diminished heat dissipation from the brain, adverse environmental conditions and physiological activation strongly potentiate thermal effects of psychomotor stimulant drugs, resulting in dangerous brain overheating. Since hyperthermia exacerbates drug-induced toxicity and is destructive to neural cells and brain functions, use of these drugs under conditions that restrict heat loss may pose a significant health risk, resulting in both acute life-threatening complications and chronic destructive CNS changes. We argue that brain temperature is an important physiological parameter, affecting various neural functions, and show the potential of brain temperature monitoring for studying alterations in metabolic neural activity under physiological and pathological conditions. Finally, we discuss brain temperature as a factor affecting various neuronal and neurochemical evaluations made in different animal preparations (*in vitro* slices, general anesthesia, awake, freely moving conditions) and consider a possible contribution of temperature fluctuations to behavior-related and drug-induced alterations in neuronal and neurochemical parameters.

Keywords: brain temperature; metabolism; cerebral blood flow; hyperthermia; arousal; sexual behavior; physical exercise; psychomotor stimulants; hot and humid environment; neural damage; neurotoxicity

Introduction

By developing sophisticated mechanisms that maintain stabile internal temperatures, living organisms actively function within a relatively wide range of environmental temperatures. Humans, for example, maintain a body temperature of ~37°C whether they enter cold water or a sauna with air temperature near water's boiling point. This high stability of internal temperatures is due to the ability to sense changes in environmental temperatures and to adjust the organism's heat production and heat loss to the external environment. While various effector

---

*Corresponding author. Tel.: +1 (410) 550-5551; Fax: +1 (410) 550-5553; E-mail: ekiyatki@intra.nida.nih.gov

systems are involved in thermoregulation, the brain plays a crucial role in both sensing environmental temperatures and maintaining temperature homeostasis. Following bacterial or viral invasion or inflammation, the fine adjustment between heat production and heat loss is actively dis-regulated and body temperature transiently increases $\sim 2°$ to $3°C$. This hyperthermia (fever) is significant for the organism's fight against these potentially dangerous challenges (Schmidt-Nielsen, 1997), and the brain is crucial for the development of a fever response (Schiltz and Sawchenko, 2003; Jansky and Vybiral, 2004).

The idea of thermoregulation as a centrally organized adjustment between heat production and heat loss, however, becomes much more obscure and controversial with respect to temperature homeostasis of the brain itself. While the first direct measurements of brain temperature were performed in 1870 (Schiff, 1870 cited by James, 1892), long before neuronal activity was first recorded, brain temperature remains a generally unknown homeostatic parameter. It is surprising because brain temperature is essentially entwined with brain metabolism and cerebral circulation — two hot topics of modern neuroscience. Although the brain represents only 2% of the human body mass, it accounts for $\sim 20\%$ of the organism's total oxygen consumption, requiring several orders of magnitude more energy than other cells. The average power consumption of a single neuron at resting conditions is ~ 0.5–4.0 nW, or 200–2500 times more than that of the average body cell (1.6 pW). Since all energy used for neural metabolism is finally transformed into heat (Siesjo, 1978), intense heat production appears to be an essential feature of brain metabolism. While under normal, quiet resting conditions intra-brain heat production is balanced by heat dissipation from the brain, under several physiological conditions, this temperature balance becomes shifted and brain temperatures either increase above (hyperthermia) or decrease below (hypothermia) their "normal" values. Brain temperature balance can also be changed by various pharmacological drugs, which affect brain metabolism and alter heat dissipation to the external environment. Therefore, brain hyperthermia and hypothermia may also occur following pharmacological challenges. Finally, brain hyperthermia may result from extreme environmental overheating and a diminished ability to dissipate metabolic heat to the external environment, or occur during various pathological conditions associated with brain damage and/or impaired heat dissipation from the brain.

Since most physical and chemical processes governing neuronal activity are temperature-dependent, naturally occurring fluctuations of brain temperatures may have important functional consequences, affecting various neural functions. Since brain cells are very sensitive to heat damage, high brain temperatures may have both direct destructive effects on neural cells and brain functions as well as potentiate neural damage associated with CNS pathology or neurotoxic drugs.

This work will demonstrate that brain temperature is a fluctuating homeostatic parameter and that brain hyperthermia is a normal phenomenon occurring under various physiological and behavioral conditions. We will also present data on brain hyperthermia induced by various pharmacological drugs that cause metabolic brain activation and impair normal heat dissipation. Our focus will be on addictive drugs — substances that are self-administered by humans for the purpose of altered psycho-emotional states. We will demonstrate that hyperthermic effects of amphetamine-like psychostimulant drugs are greatly potentiated during physiological activation or by environmental conditions that restrict heat dissipation to the environment. Finally, we will consider the functional implications of brain hyperthermia as a factor that affects various neural functions under physiological conditions and induces, or potentiates, neural damage.

Brain hyperthermia as a physiological phenomenon

While it is usually believed that brain temperature is a stable, tightly regulated homeostatic parameter, direct measurements of brain temperature in awake animals revealed relatively large increases (2–3°C) following exposure to various environmental challenges and during different behaviors (Abrams and Hammel, 1964; Delgado and Hanai,

1966; McElligott and Melzack, 1967; Blumberg et al., 1987; Moser et al., 1993). Significant changes in brain temperature also occurred during spontaneous fluctuations in activity state, particularly in the transition from sleep to wakefulness (Serota, 1939; Abrams and Hammel, 1964; Delgado and Hanai, 1966). These temperature fluctuations correlated with the biological significance of environmental challenges, spontaneous and stimuli-induced changes in EEG, and had some structural specificity with respect to the modality of sensory stimuli. Although all these findings point to metabolic neural activity as a cause of brain temperature fluctuations, temperature changes were generally correlative in different brain structures and typically associated with similar changes in body core temperature, suggesting a possible heat arrival to the brain from the body. The only way to clarify this issue is to determine the temperature gradient between the arterial blood arriving to the brain and brain tissue.

The source of brain temperature fluctuations

To determine the source of physiological brain hyperthermia, temperatures were measured simultaneously in several brain structures (dorsal and ventral striatum, and cerebellum) and in arterial blood supply in rats at stable environmental temperatures (23°C) (Kiyatkin et al., 2002). Because of specifics of temperature measurement in blood flow, we placed our thermocouple in the tip of a thin polyethylene catheter chronically implanted into abdominal aorta via *a. caudalis*. While the carotid artery may appear to be a better choice, it is quite difficult to record temperature in the carotid artery without a full block or significant decrease in blood flow, thus affecting the measured parameter. By contrast, our probe in the abdominal aorta did not produce an evident decrease in blood flow and measurements were made in the body core — the area with the highest temperature. These measurements were made in freely moving rats exposed to a number of physiologically relevant arousing stimuli ranging from simple sensory to clearly aversive ones (placement in the cage, tail-pinch, social interaction with either a male or female companion, sound).

This study provided several important findings. First, each brain structure had its own basal temperature (ventral stratum: 37.57°C; dorsal striatum: 37.20°C; cerebellum: 37.34°C), which was significantly ($p<0.001$) higher than that in arterial blood (36.62°C). Second, all stimuli induced rapid and relatively long temperature elevations in each brain structure and arterial blood, greatly exceeding the duration of stimulation. The temperature elevation was minimal and shortest (~0.1°C for 6–7 min) after a 20-s sound and maximal and most prolonged after the rat was transferred from its home to experimental cage (~1.8°C for 2–4 h). Three-minute social interaction between a recorded male and either male or female companion and a 3-min tail-pinch (see Fig. 1) were all accompanied by moderate temperature elevations (~1.0°C for 20–30 min). Third, changes in each brain structure were significantly faster and stronger in amplitude than those in arterial blood. For example, temperature increase in both striatal compartments following tail-pinch, social interaction with female and sound presentation became significant from 10 to 12 s after stimulus onset, but in arterial blood these latencies were 41, 32, and 46 s, respectively. Fourth, despite a generally correlative time-course, temperature changes had structural specificity. Both striatal divisions showed quite similar changes, but temperature increases in cerebellum were more delayed and prolonged following each stimulus. Finally, brain temperature increases induced by environmental challenges showed changes within repeated tests. While the elevations following tail-pinch and social interaction remained relatively stable over five repeated daily sessions, responses to sound showed a clear habituation with a complete disappearance of the thermal response by the fifth session. The initial temperature increase associated with environmental change was relatively stable over repeated sessions, but on each subsequent day temperatures decreased more quickly to lower baselines, enhancing the relative response magnitude. Thus, during repeated habituation to the same experimental environment, rats showed a trend of decreasing brain temperatures, which was parallel to a decrease in general motor activity.

Given that the blood supply to the brain was cooler than the brain itself, and that brain

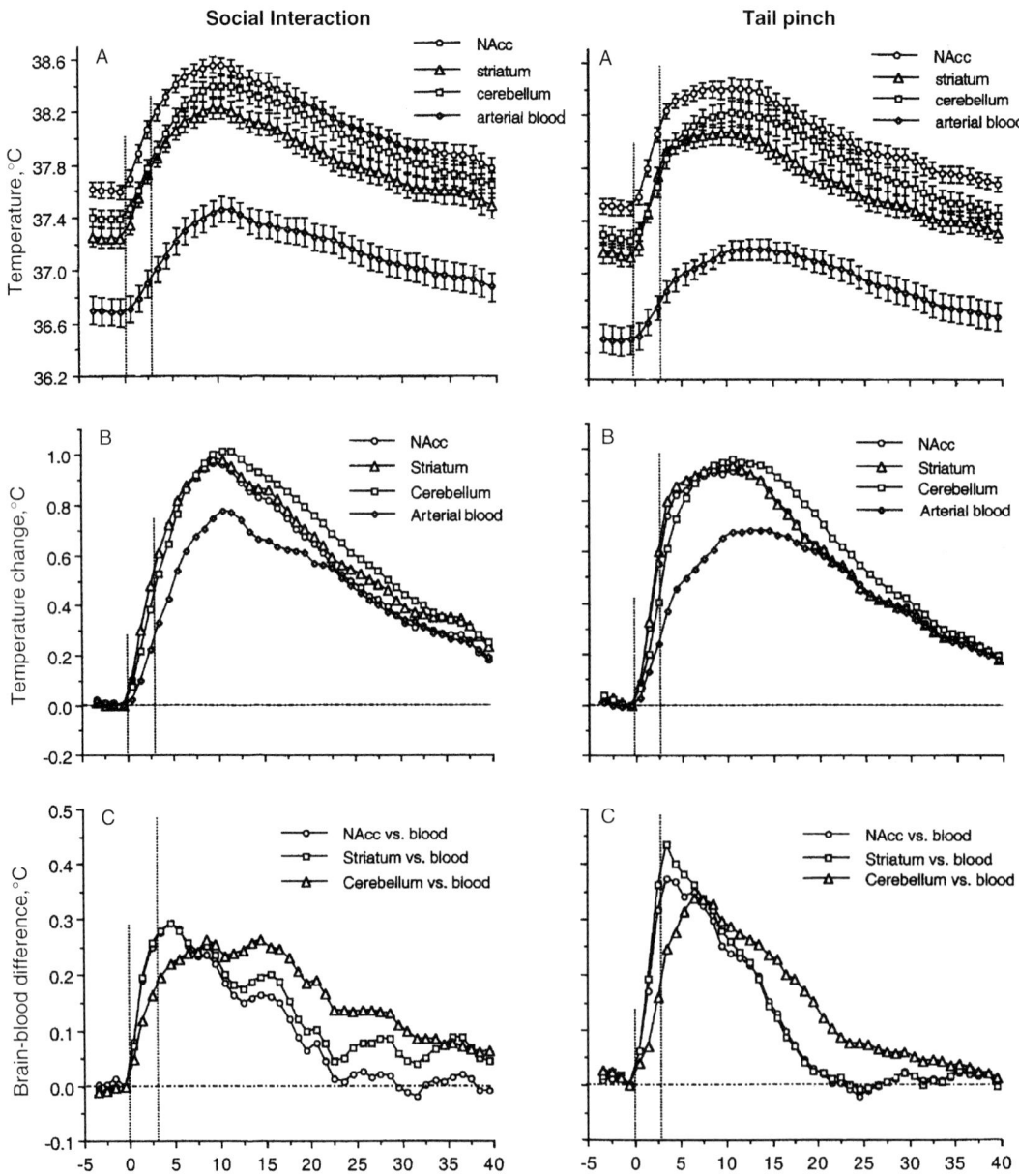

Fig. 1. Changes in brain (NAcc, striatum, cerebellum) and arterial blood temperatures in male rats during 3-min social interaction with a female (left) and 3-min tail-pinch (right). (A) Shows absolute temperatures, (B) shows relative temperatures with respect to pre-stimulus baseline, and (C) shows temperature differentials between individual brain structures and arterial blood.

temperatures rose more quickly and to a larger extent than arterial blood temperature in response to all challenges, intra-brain heat production appears to be the primary cause of functional brain hyperthermia. While arterial blood temperature also gradually increases in response to all challenges, brain-blood temperature differentials grew consistently during behavioral activation, showing an apparent increase in intra-brain heat production. Therefore, it seems that under physiological conditions blood circulation removes heat from, rather than delivers heat to, the brain. Tail-pinch similar to that used in our study induced an almost twofold increase in striatal blood flow (Fellows et al., 1993), evidence that such a mechanism exists. Similar to brain temperature, this increase was rapid and greatly exceeded the duration of stimulation. Phasic, large increases in striatal blood flow (80–120%) were also found during grooming and eating (Fellows et al., 1993), activities consistently associated with brain temperature increases (Kiyatkin and Wise, 2001). Hence, brain circulation is a significant factor in the re-distribution of locally released heat within brain tissue and in its removal from the brain, thus contributing to brain temperature fluctuations occurring under behavioral conditions.

The brain hyperthermia we observed was caused by a variety of stimuli. With the exception of sound, which caused only a momentary arousal response or no visible effect at all, behavioral activation always accompanied brain hyperthermia. The present data, therefore, indicate that brain hyperthermia is not a response specific to stressful stimuli or events, unless, as has been proposed by Selye (1979), any stimulus successful in eliciting an arousal response is taken to be stressful to some degree. The "stress-induced" or "emotional" hyperthermia described both in animals (see Moltz, 1993 for review) and humans (Briese, 1995), and based largely on recording of body core or rectal temperature, may best be viewed as a consequence of a more general phenomenon of "arousal-related" brain hyperthermia. In turn, arousal can be viewed not only as an electrophysiological phenomenon, but also as a generalized metabolic neural activation and functional brain hyperthermia.

Brain temperature as a factor affecting neural functions

Although heat release is an obvious "by-product" of metabolic activity, the changes in brain temperature it triggers may play an important integrative role, involving and uniting numerous central neurons within the brain. Although this is obviously not the main, nor the most efficient type of inter-neuronal communication, it may have adaptive significance. Since most physical and chemical processes governing neural activity are temperature-dependent, changes in brain temperature not only reflect metabolic neural activity but also affect numerous neural functions. For example, dopamine uptake is known to double with a 3°C increase in temperature (Xie et al., 2000), a range easily achieved in the brain under conditions of physiological activation. Since such a temperature increase affects the activity of ionic channels (particularly Na^+ and Ca^{++}; Rosen, 1996, 2001), an increase in uptake should be compensated for by increased dopamine release. By increasing both release and uptake, it appears that brain hyperthermia makes neurotransmission more efficient and neural functions more effective at reaching behavioral goals.

Although numerous data suggest that increased brain metabolism is accompanied by increased brain circulation, and changes in local cerebral blood flow are widely used as a measure of functional brain activation, the relationships between brain metabolism and cerebral blood flow are complex and currently poorly understood (see Mintun et al., 2001; Raichle, 2003 for review). Though a discussion of these relationships is beyond the scope of the present review, it appears that metabolism-related changes in brain temperature may play an important role in coupling and adjusting local circulation to meet the demands of enhanced metabolism. While temperature is usually omitted from equations relating metabolism and blood flow (Yablonskiy et al., 2000), direct relations between temperature and blood flow have been well established in peripheral tissues. An increase in local temperature was accompanied by linear and strong blood flow increases in skin (Ryan et al., 1997; Charkoudian, 2003), muscular tissue (Oobu, 1993), the intestine (Nagata et al., 2000), and the liver (Nakajima et al., 1992). This relationship has also been observed in brain tissue, as shown in monkeys (Moriyama, 1990), rats (Uda and Tanaka, 1990), and humans (Nybo et al., 2002). Increases in local brain temperature resulting from increased neural metabolism can, therefore, be understood as one of the factors that increase local blood flow. This factor may explain at least partially why the increases in cerebral blood flow observed during functional brain activation typically exceed metabolic activity of brain tissue (see Mintun et al., 2001 for review). Because of this mechanism, the brain is able to increase blood flow more, and in advance of, actual metabolic demands, thus providing a crucial advantage for successful goal-directed behavior and the organism's best adaptation to potential energetic demands. By increasing blood flow over its current demand, more oxygen and nutrients are delivered to the areas of potential demand and more potentially dangerous metabolic heat is removed from intensively working brain tissue.

Within-brain and brain-body temperature homeostasis under stable environmental conditions

While basal temperatures both in various brain structures and peripheral sites (temporal muscle, body core, skin) were different, it is not surprising that they correlated since each was recorded from the same physical body. The strength of correlation, however, was different. Figure 2 shows the relationships between basal temperatures in body core, skin, and two brain structures (medial preoptic area of the hypothalamus or MPAH and hippocampus or Hippo) evaluated in intact freely moving rats at normal ambient temperatures (23°C). Animals were habituated to the recording environment for several daily sessions and the values were obtained during quiet resting conditions.

While it is reported that body temperature evaluated in humans based on rectal measurements is typically 0.2° to 0.8°C lower than brain temperature (Mariak et al., 1998, 1999, 2000; Rumana et al., 1998), we found that in quietly resting rats core body temperature (37.31 ± 0.59°C) is virtually identical to MPAH temperature (37.34 ± 0.60°C) and significantly higher than that in Hippo (36.58 ± 0.53°C) (A). Temperatures in both brain structures tightly correlated ($r = 0.934$; $p < 0.001$) and both tightly correlated with body core temperatures ($r = 0.931$ and 0.934 for MPAH-body and Hippo-body, respectively, both $p < 0.001$). Skin has the lowest temperature in an organism and its correlation with core body ($r = 0.690$) and brain temperatures (MPAH: $r = 0.464$; Hippo: $r = 0.666$) was minimal. Interestingly, regression lines for MPAH-body and Hippo-body (A) were generally parallel to the line of no effect (hatched line), suggesting that increases in brain temperature are accompanied by linear increases in body temperatures. In contrast, at higher body temperatures (hyperthermia), the body-skin difference became larger, obviously reflecting increased heat loss from skin surfaces. At lower body temperatures (hypothermia), the differences became smaller, reflecting diminished heat dissipation from skin surfaces. While the correlation was weaker, similar direct relationships were found between temperatures in specific brain sites and skin (B).

Within the brain, each structure has its own temperature: more ventrally located structures (ventral tegmental area of midbrain, or VTA, 37.30°C; MPAH, 37.34°C; nucleus accumbens or NAcc, 37.26°C) were consistently warmer than more dorsally located structures (Hippo, 36.58°C; dorsal striatum, 36.74°C). While a dorso-ventral temperature gradient was described in both humans (Schwab et al., 1997; Mariak et al., 2000) and animals (Serota and Gerard, 1938; Horvath et al., 1999), its mechanisms remain unclear. One point of view suggests that higher temperatures in more ventrally located structures reflect their further location from the colder environment and more heating by, supposedly, warm blood from the body. Another point relates these differences with metabolic activity of different brain structures, particularly with expression of brain uncoupling proteins that regulate uncoupling in mitochondria and local heat production (Horvath et al., 1999). These temperature differences may also reflect known differences in electrical activity of dorsally and ventrally located structures in the number of spontaneously active and silent neurons and their discharge characteristics.

Brain hyperthermia and rapid temperature fluctuations during natural motivated behavior

In developing brain thermorecording, we sought a tool that could assess alterations in neural activity in different brain structures during the development and performance of goal-directed behavior. This

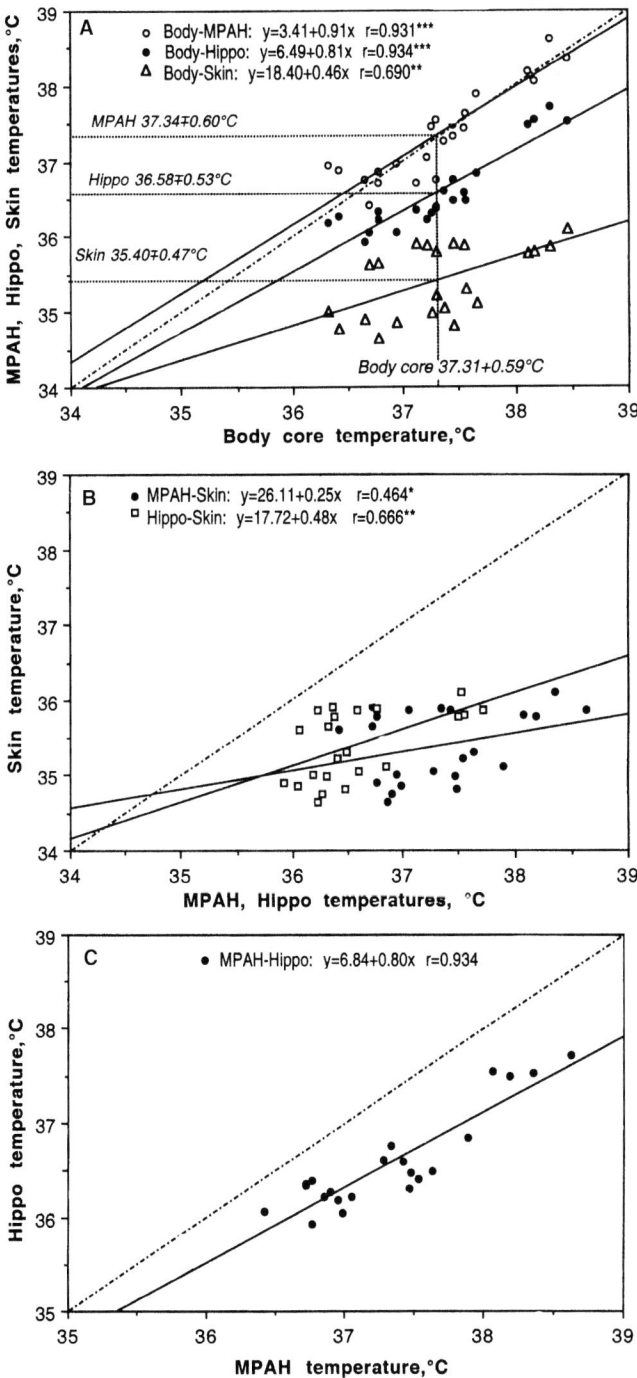

Fig. 2. Relationships between basal temperatures recorded from different brain (MPAH, Hippo) and body (body core, skin) sites in awake, unrestrained rats under quiet resting conditions.

approach has been applied to sexual behavior, perhaps the most important type of social interaction and the most coordinated, energy-consuming form of goal-directed behavior (Kiyatkin and Mitchum, 2003). We developed a simple protocol, which included exposure of a sexually experienced male to a receptive female with subsequent unrestricted copulatory behavior. Recordings were taken from three brain sites: MPAH — a structure implicated in central regulation of both sexual functions and thermoregulation, the NAcc — a structure crucial for any motivated behavior, and Hippo — a structure usually not implicated in the central organization of sexual behavior. Body was represented by a recording site in *musculus temporalis*, a non-locomotor head muscle that has a blood supply similar to that of the brain.

Robust tonic increases and phasic fluctuations in brain and muscle temperature occurred in male rats during sexual interaction (Fig. 3). With respect to baseline, temperatures rapidly increased (+1.5° to 2.0°C), when the female was placed into a neighboring compartment, and peaked (38° to 39°C) when full interaction was allowed. As the male repeatedly mounted and achieved intromission with the female, temperature further increased, peaking at the first ejaculation (+0.4°C), but remaining relatively stable at each subsequent ejaculation point. Temperature abruptly dropped after complete cessation of sexual activity, but increased once again after the female was removed from the cage.

While the time-course of temperature fluctuations was generally similar in each of the four recording sites, changes in brain temperature were larger than those in the muscle, resulting in significant increases in brain-muscle differentials (Fig. 3B). These increases were stronger and almost identical in MPAH and NAcc and lower in Hippo. The maximal increase in brain-muscle differentials occurred during sexual arousal and at the first ejaculation; each subsequent ejaculation was associated with slightly weaker increases and smaller changes in brain-muscle differentials. Similar to the introduction of the female, removal of the female was associated with a rise in brain-muscle differentials.

During rapid time-course analysis, it was found that the arousal-related temperature increase is significantly quicker and larger in each of the three brain structures than in the muscle (Figs. 4A and B). Among brain structures, the NAcc showed the shortest onset latencies (10–20 s) and most rapid acceleration of the initial temperature increase. Muscle temperatures showed the longest latencies and minimal increase. The NAcc- and MPAH-muscle temperature difference peaked at ~5 min (~0.35°C; see Fig. 4B) and increased again when the rats began to interact freely. There was a tight correlation between temperature at the moment of arousing stimulation and magnitude of temperature elevation associated with arousal (NAcc: $r = -0.899$). When the rat was at quiet rest and temperatures were low, the arousal-related temperature increase was strong (1.6° to 2.2°C), while it was progressively smaller (0.6° to 0.8°C) when the rat was active and basal temperatures were higher.

In contrast to monotonic temperature elevation during exposure to sexually arousing stimuli, biphasic temperature fluctuations occurred during copulatory behavior (Figs. 4C and D). The male copulatory cycle consisted of a chain of repeated mounts and intromissions with the final intromission followed by ejaculation. As can be seen in Fig. 4C, temperature gradually increased during repeated mounts and intromissions, peaked within 1–3 min after ejaculation (zero time), and abruptly decreased to a lower point (10–12 min), from which temperature increased again during the next copulatory cycle. There were structural differences, with maximal and almost identical temperature fluctuations in the NAcc and MPAH, smaller in the Hippo, and minimal in the muscle (see Fig. 4D). While NAcc- and MPAH-muscle differentials gradually increased during copulatory behavior, peaked immediately preceding the ejaculation, and rapidly decreased for the next 5 min, the Hippo-muscle differences were much smaller.

This study revealed that sexual behavior is accompanied by tonic temperature increase and phasic temperature fluctuations tightly associated with key behavioral events. The strong increase in brain temperature occurring after a male's exposure to sexually relevant sensory stimuli is an obvious manifestation of sexual arousal — a generalized neural activation that is an essential prelude to

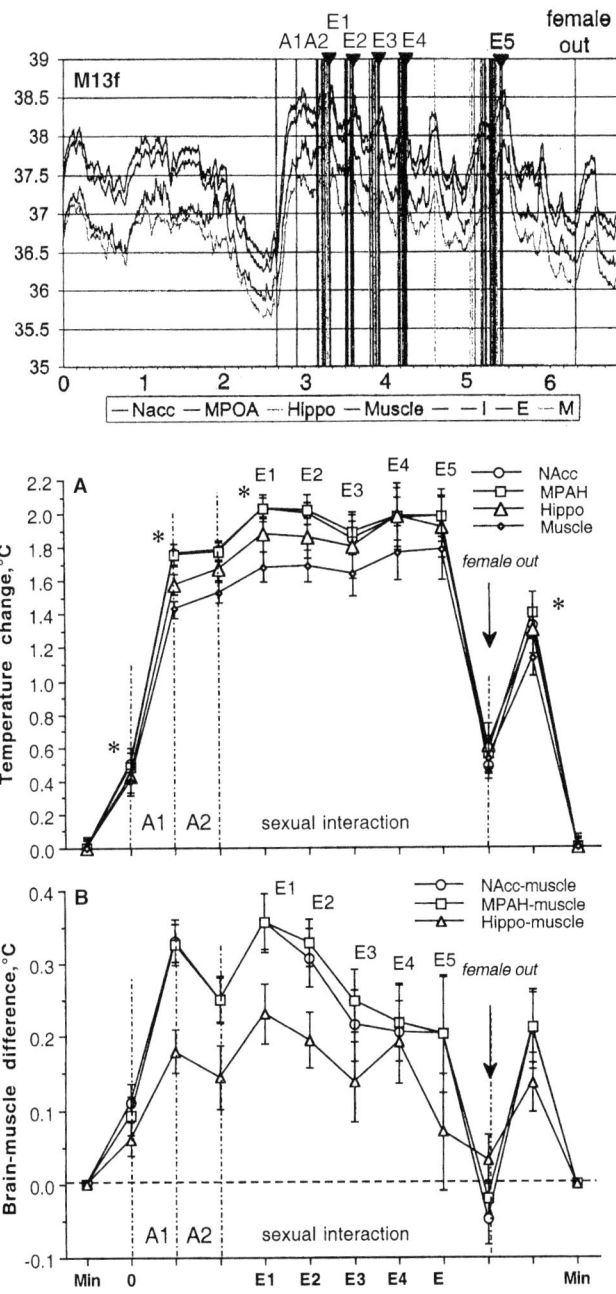

Fig. 3. Original example (top) and mean changes in brain (MPAH, NAcc, Hippo) and muscle temperatures in male rats during the session of sexual interaction with female. A, arousal; E, ejaculation.

copulatory behavior. This increase was more rapid and stronger than that in the muscle, suggesting the brain as its source, and it was maintained when the rat initiated copulatory behavior. Copulatory behavior was associated with biphasic temperature fluctuations, which were superimposed on tonic temperature increase. Importantly, brain temperature increase began for ~90 s before the rat made

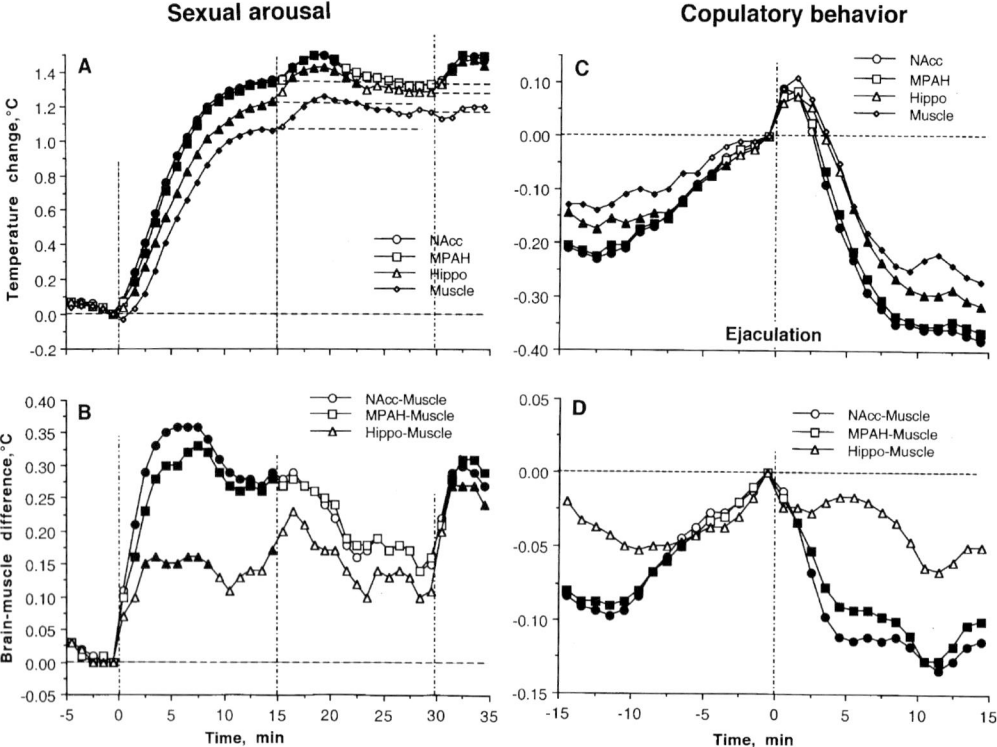

Fig. 4. Rapid-time course analysis of temperature changes associated with sexual arousal and copulatory behavior in male rats. (A) and (C) show relative temperature changes with respect to onset of arousing stimulation (A) and ejaculation (C), (B) and (D) show brain-muscle temperature differentials with respect to the same reference points. Filled symbols indicate values significantly different from baseline.

his first mount or intromission, suggesting neural activation as a factor that causes behavior. This activation gradually progressed during copulatory behavior, peaked at the point of ejaculation, and rapidly and abruptly decreased after ejaculation. Such a pattern of neural activation suggested by our temperature measurements agree well with physiological measurements in humans (Bartlett, 1956; Nemec et al., 1976; Bohlen et al., 1984; Exton et al., 2000; Stein, 2000), neuronal recordings during sexual behavior (Oomura et al., 1988; Shimura et al., 1994), and the four-phase pattern of sexual arousal or physiological excitation studied in humans (Masters and Johnson, 1966). If extremely sharp temperature acceleration followed by an abrupt transient temperature fall seen in association with the male's ejaculation reflects the pattern of neural activity, it explains why the male orgasm is very sharp, short, and followed by a refractory period and general somnolence, which can occur not only because of previous energy expenditure but because of a global decrease in brain activity.

Although tightly correlated with individual events of sexual behavior, these temperature fluctuations may reflect some general alterations in neural activity associated with any cyclic goal-directed behavior. Rapid brain hyperthermia seen during sexually arousing stimulation (sexual arousal) also occurred after different environmental challenges ranging from simple sensory to traditional stress-ogenic stimuli (see previous section). Thus, widespread neural activation (arousal) appears to be an essential factor in the organization of adaptive behavioral responses to quite different changes in the environmental continuum. Triggering some gender-specific consummatory mechanisms of sexual behavior, neural activation associated with sexual arousal may be generally identical to other forms of "specific" arousal, particularly those induced by either food or food-related sensory stimuli in hungry animals (Kiyatkin and Gratton, 1994) and by drug exposure or drug-related sensory stimuli in drug-experienced individuals (Kiyatkin and Wise, 2002; Kiyatkin and Brown, 2003). While arousal-related neural activation is an essential pre-stage of successful copulatory behavior, it progresses and phasically accelerates during repeated mounts and intromissions, sharply peaking at the ejaculation point. Independent of whether this activation is viewed as anticipatory or associated with preparation and acquisition of behavioral activity, this change also appears to be common to other types of motivated behavior since gradual temperature increases were consistently seen in both heroin- and cocaine-experienced rats before the first drug self-injection of a session (Kiyatkin and Wise, 2002; Kiyatkin and Brown, 2003). Similarly, an abrupt cessation of neural activation after ejaculation (viewed as the event of sexual reward) may be an essential feature of any type of reward common to various goal-directed behaviors. Although different in time-course, qualitatively similar biphasic fluctuations have been found during operant feeding and drug-taking behavior at the level of various parameters (i.e., single-unit impulse activity, temperature, arterial blood pressure) both in animals (Kiyatkin and Stein, 1993; Peoples and West, 1996; Kiyatkin and Rebec, 2001; Kiyatkin and Wise, 2002) and humans (Battig et al., 1993; Hasenfratz et al., 1993). The mechanisms underlying behavior-related neural activation and its reward-associated cessation are obviously different in each case (ejaculation, consumption of earned food, rewarding action of self-injected drug in the brain), but these converging data suggest that this dramatic neural alteration plays a crucial role in reinforcement — a key mechanism in regulating any goal-directed behavior.

Brain and body temperature homeostasis during physical activity under conditions of environmental warming and impairment of venous blood outflow

While enhanced intra-brain heat production associated with functional brain activation is accompanied by a compensatory increase in heat dissipation to the external environment, limiting brain hyperthermia within physiological limits, robust brain hyperthermia may also develop during several situations due to a diminished ability to remove metabolic heat from the brain.

It is known that intense physical activity is associated with robust energy consumption and significant heat production in both animals and humans. Oxygen consumption in humans may increase up to 10-fold in the transition between quiet resting conditions and intense running (Schmidt-Nielsen, 1997). This increase corresponds to whole-body metabolic heat production, which increases from ~ 1 W/kg at rest to ~ 10 W/kg during heavy exercise (Donaldson et al., 2003). Enhanced heat production is typically compensated by enhanced heat loss, resulting in a relatively stable body temperature. The high effectiveness of heat loss mechanisms in humans depends on a well-developed ability to sweat and on a dynamic range of blood flow rates in the skin, which can increase from ~ 0.2 to 0.5 l/min in thermally neutral conditions to 7–8 l/min under maximally tolerable heat stress (Rowell, 1983). Sweat rates under these conditions may reach up to 2.0 l/h, providing a potential evaporative rate of heat loss in excess of 1 kW (or ~ 14 W/kg), i.e., more than maximal potential heat production. These compensatory mechanisms, however, become progressively less effective in hot, humid conditions, resulting in progressive heat accumulation in the organism and robust hyperthermia. For example, body temperatures measured at the end of a marathon run on a warm day were found to be as high as 40°C and cases of fatigue during marathon running were associated with even higher temperatures (Cheuvront and Haymes, 2001). The ability to dissipate metabolic heat appears to be crucial for the development of hyperthermia. While 90-min of intense cycling in experienced cyclists at normal environmental temperatures increased body temperature less than 1°C, 2.0° to 2.5°C increases were found when participants cycled in water-impermeable suits that restricted heat loss via skin surfaces (Nybo et al., 2002).

Although robust body hyperthermia associated with intense physical activity under conditions that restrict heat loss is a known phenomenon (Schaefer, 1979), the impact of these conditions on the brain is a matter of intense speculation and controversy (Cabanac, 1998) because of very limited data with direct measurement of brain temperature. Since the direct recording of brain temperature in humans is usually impossible because of ethical considerations, tympanic temperature has been used as an indirect measure of brain temperature (Brinnel et al., 1987; Cabanac et al., 1987). From tympanic temperature measurements during exercise with and without active cooling of the head, it was concluded that brain temperatures during intense body heat production remain lower than body temperatures, suggesting selective brain cooling as a mechanism preventing brain overheating under extreme body hyperthermia. While tympanic temperature was long considered to be a valid index of brain temperature, recent applications of more sophisticated physiological techniques called the existence of a brain cooling mechanism in humans into question (Nybo and Nielson, 2001; Nybo et al., 2002). In these experiments, human volunteers were equipped with two thermosensor probes placed in arterial blood entering the brain (carotid artery) and venous blood exiting the brain (internal jugular vein); venous blood temperature should be equal or very close to brain temperature. Using this approach, it was found that venous blood is ~0.3°C warmer than arterial blood at rest and this positive venous–arterial difference remains during intense physical exercise, suggesting that brain temperature is higher than arterial blood temperature even under intense physical activity. When cycling was done in water-impermeable suits, the increases were more dramatic (arterial from 36.75° to 39.30°C, venous from 37.15° to 39.5°C with individual changes up to 40.4°C), but again the difference remained positive. After termination of exercise, arterial blood temperature rapidly dropped, while venous temperature decreased more slowly and the venous–arterial difference reached 0.9°C before a slow return toward the ~0.3°C baseline. Simultaneous measurements of tympanic temperatures, moreover, revealed that their values are consistently lower than, and independent of, arterial and venous temperatures.

While diminished heat outflow from the brain appears to be the primary cause of intra-brain heat accumulation during physical exercise, this activity is also associated with increased brain metabolism (Ide and Secher, 2000; Ide et al., 2000) and consequently intra-brain thermogenesis. Taking into account uptake of oxygen, glucose, and lactate, an almost a twofold increase in global brain metabolism was reported during 10-min intense cycling at normal environmental temperatures. Although cerebral blood flow increases during physical exercise and this increase is in excess of the increases in global cerebral metabolic activity (Ide et al., 2000; Nybo and Nielson, 2001; Nybo et al., 2002), cerebral blood flow gradually decreases during maximal exercise at hyperthermic conditions because of a hyperventilation-induced decrease in CO_2 pressure (Nybo and Nielson, 2001). In contrast to more intense heat removal from the brain by blood flow during physical exercise under normal conditions, under hyperthermic conditions compromised cerebral blood outflow is an additional and powerful factor restricting heat dissipation from the brain and determining intra-brain heat accumulation.

Although robust brain hyperthermia (~39.5°C) did occur during intense cycling in harsh environmental conditions (Nybo et al., 2002), such activity did not result in clear fatigue. Hypothalamic temperatures associated with forced exercise-induced fatigue in rats were larger (40.1° to 42.1°C; Walters et al., 2000) than those found in humans during self-motivated exercise (39.7° to 40.3°C; Nielsen et al., 2001). While a pathological condition by itself, fatigue developed during intense exercise may protect the brain against further overheating.

To investigate how environmental temperatures and impairment of venous blood outflow affect basal brain and body temperatures as well as their responses to environmental challenges we employed two procedures. First, rats were tested at warm environmental temperature (29°C), which corresponds to normothermic conditions (Romanovsky et al., 2002), when heat production is balanced with heat loss at minimal energy expenditure. Second, tests were performed in rats with chronic bilateral occlusion of the jugular veins (Kiyatkin and Brown, 2004b). Since under intact conditions more than 90% of venous blood is removed from the brain via jugular veins, chronic bilateral occlusion of these passages served as a valid model for impaired blood outflow.

Compared to normal, 23°C environmental temperatures, rats exposed to 29°C had slightly larger basal temperatures in muscle (36.36±0.39 vs. 36.04±0.42°C; $p<0.01$) and Hippo (36.39±0.35 vs. 36.03±0.47°C; $p<0.05$) with no differences in NAcc (36.89±0.39 vs. 36.82±0.59°C). Both groups had a virtually identical, strong correlation between brain and muscle temperatures. Animals housed at warmer temperatures, however, showed slightly attenuated but more prolonged brain hyperthermic responses to environmental challenges than those seen in rats housed at 23°C (Fig. 5).

Animals with chronically occluded jugular veins tested 1 week after surgery did not have evident differences from control animals in their somatic status and behavioral activity, and had virtually similar brain and muscle temperatures. These animals, however, showed lower correlations in basal temperatures between NAcc and Hippo ($r = 0.589$ vs. 0.868 in control), larger brain-muscle differentials, and much weaker correlation between brain and muscle temperatures (muscle–NAcc, $r = 0.330$, muscle–Hippo, $r = 0.318$ vs. $r = 0.701$ and 0.922 in control). When exposed to environmental challenges (novelty, tail-pinch, social interaction with female), animals with chronically occluded jugular veins showed strongly diminished temperature responses with very profound increases in brain-muscle differentials (Fig. 5).

Brain temperature as a factor inducing or potentiating neuronal damage

Similar to systemic hyperthermia (fever), which is adaptive within some limits but may rise to

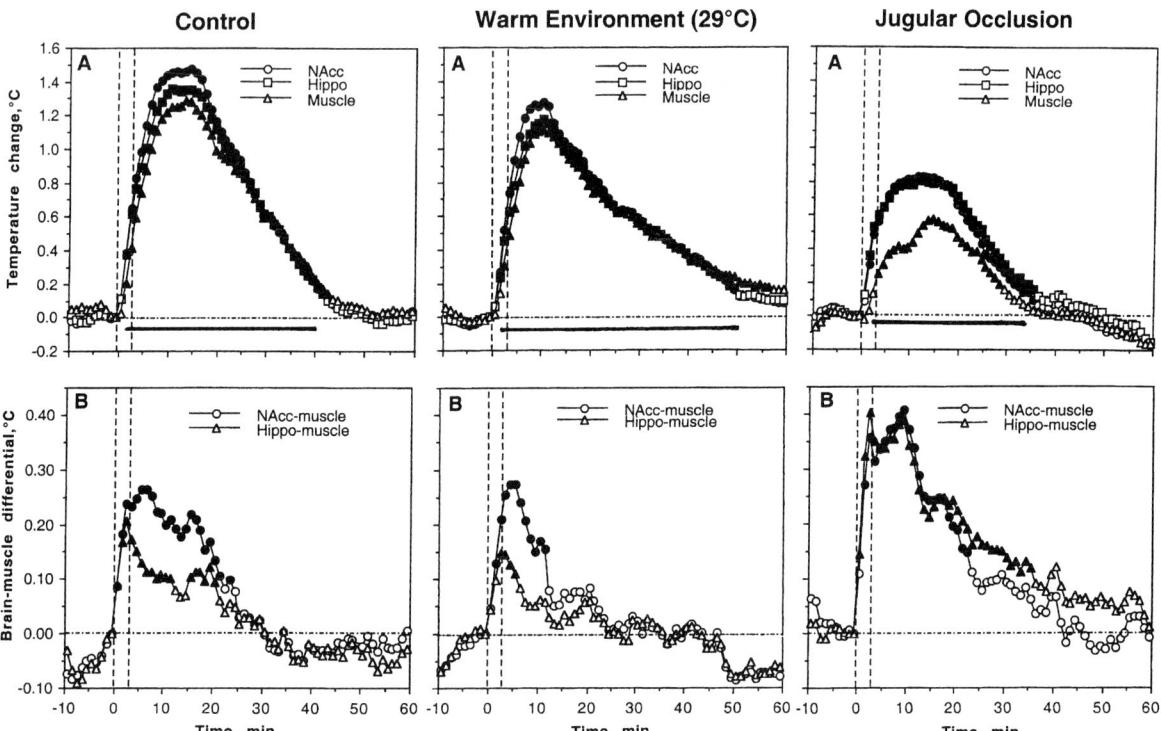

Fig. 5. Changes in brain and muscle temperature in a male rat induced by 3-min social interaction with female under three conditions: in intact animals at 23°C (control, left panel), in intact animals at 29°C (middle panel), in animals with chronically occluded jugular veins at 23°C (right panel). Filled symbols indicate values significantly different from baseline.

pathological levels, it is quite difficult to draw the line between physiological and pathological brain hyperthermia. It is important, however, to realize that temperature increase above some limit has a direct destructive action on cells, which will grow exponentially with slight increases above this limit. For example, 79% of bovine endothelial cells survived in culture at 37°C incubation temperature, but survival dropped to 9.0 and 0.2% at 41 and 43°C, respectively (Lin et al., 1991). In another model, normal pulmonary fibroblasts completely stopped proliferation at 41.0°C (Iwagami, 1996). Finally, in lung cells, morphologically verified onset of hyperthermic cell death occurred between 40 and 41.5°C (Lepock et al., 1983). Although all tissues are sensitive to hyperthermia, the brain and testicles are among the most sensitive organs (Dewhirst et al., 2003). The most temperature-sensitive cellular elements of neural cells are mitochondrial and plasma membranes, in which irreversible transitions in protein structure or arrangements begin to occur at temperatures higher than 40°C (Lepock et al., 1983; Iwagami, 1996; Lepock, 2003). Thus, 40.0°C may be considered the threshold of pathological brain hyperthermia.

Although all cells are affected by high temperature, hyperthermia-induced destruction of endothelial cells of the brain and spinal cord, and leakage of serum proteins across the brain-blood barrier are the two important factors determining brain edema (Sharma and Hoopes, 2003), the most dangerous acute complication of pathological brain hyperthermia (Kalant, 2001; Dewhirst et al., 2003). While heat-induced damage occurs in different brain structures, the damage is usually stronger within the edematous areas of the brain, suggesting edematous swelling as an important cofactor of heat-induced damage of brain tissue (Sharma et al., 1998). Heat-induced brain injury is not limited to neural cells and includes glial cells and cerebral microvessels. Temperature-induced destruction occurs within the whole brain, but some brain areas are especially vulnerable to damage. For example, heat-induced increase in blood–brain barrier permeability was maximal in cingulated and occipital cortex and cerebellum (where edema was also maximal), but less in hypothalamus and thalamus. In contrast, maximal axonal damage, detected by myelin basic protein staining, was prominent in brain stem reticular formation, pons, medulla, and the spinal cord (Sharma et al., 1998). Excessive hyperthermia also results in expression of heat shock proteins (Welsh, 1992), which is a reliable index of thermal injury in the form of denaturated protein (Lepock, 2003). These proteins, however, appear to play an important role as endogenous neuroprotectors by binding to partially folded and misfolded proteins, thus preventing their irreversible denaturation (Bouchama and Knochel, 2002). Since these proteins are also expressed during hypoxia, ischemia, heat trauma, neurodegenerative diseases, and epilepsy (Reynolds and Allen, 2003), they are obviously a non-specific marker of neural damage and an important part of endogenous neuroprotective mechanisms activated by all these pathological conditions.

Although high temperature *per se* may be a factor in cellular damage, more often it is a potentiating factor. An increased temperature strongly increases neural damage induced by experimental hypoxia, ischemia, and cerebral trauma in animals, while hypothermia has a neuroprotective action (see Maier and Steinberg, 2003; Miyazawa et al., 2003; Olsen et al., 2003 for review). For example, hyperthermia strongly potentiates the cytotoxic effects of reactive oxygen species *in vitro* (Lin et al., 1991) and glutamate-induced neurotoxicity (Suehiro et al., 1999). While prevention of fever and mild hypothermia may be important therapeutic tools to minimize the extent and severity of neural damage associated with these pathological conditions (see Maier and Steinberg, 2003 for review), it is unknown how brain temperature is changed during these situations. Without this crucial information, it is quite difficult to establish a role for brain temperature as a factor contributing to neurotoxicity. The changes in temperature, however, may be an important factor for the neuroprotective action of various pharmacological drugs. For example, barbiturates, which are effective neuroprotective drugs (see Maier and Steinberg, 2003 for review), strongly inhibit brain metabolism (Crane et al., 1978; Michenfelder, 1988) and robustly decrease brain temperature (Kiyatkin and Brown, 2005).

Pharmacological brain hyperthermia

All addictive drugs induce behavioral, autonomic, and psycho-emotional stimulation (Wise and Bozarth, 1987; Wise, 2002), which may reflect metabolic neural activation, a presumed common feature of all addictive drugs. Opiates at low doses, for example, increase whole-body oxygen consumption and heat production (Lynch et al., 1990; Pavlov and Epstein, 2003), parameters that point to metabolic activation, while they decrease these parameters at higher doses (Endoh et al., 1999). Cocaine used in drug-naive animals also increases cerebral oxygen consumption (Robinson et al., 2000), body temperatures (Ansah et al., 1996), and cerebral blood flow (Marota et al., 2000; Robinson et al., 2000; Howell et al., 2002), although preferential decreases in cerebral blood flow were reported in experienced drug users expecting drug administration (Gollib et al., 1998: Li et al., 2000). Metabolic activation is also typical to amphetamine-like substances (i.e., amphetamine, meth-amphetamine (METH), and 3,4-methylenedioxymethamphetamine ("ecstasy" or MDMA)) (Green et al., 2003), nicotine (Perkins et al., 1996; Jessen et al., 2003), cannabinoids (Cota et al., 2002), and ketamine (Langsjo et al., 2003). Humans abuse all these drugs.

Our use of brain thermorecording for studying addictive drugs was motivated by two reasons. First, monitoring brain and body temperatures during passive administration of various addictive drugs and their self-administration allows for the study of their central effects and provides important clues on the development and regulation of drug-taking behavior. This approach was used with respect to heroin (Kiyatkin and Wise, 2002) and cocaine (Kiyatkin and Brown, 2002, 2004a). Second, since amphetamine-like psychomotor stimulants are known to induce robust hyperthermia, which is a major symptom of drug intoxication, monitoring brain temperature was used for studying the factors and mechanisms that determine hyperthermia and the adverse health effects of these drugs.

Brain hyperthermia induced by METH and MDMA and its state and environmental modulation

METH and related amphetamine-like compounds (i.e., MDMA or ecstasy) are popular drugs of abuse that can cause serious health problems ranging from acute toxicity and mortality to brain damage with chronic use (Davidson et al., 2001; Kalant, 2001; Rawson et al., 2002; Green et al., 2003). Since these substances induce abnormal release of various endogenous transmitters, including catecholamines and glutamate (Stephens and Yamamoto, 1989; Ohmori et al., 1996; Seiden and Sabol, 1996), some toxic products of their metabolism (i.e., nitric oxide, catechol-quinones, peroxynitrite, and arachodonic acid) are usually considered primary contributors to neural cell damage via oxidative stress (Spina and Cohen, 1989; Lipton and Rosenberg, 1994; Kuhn and Geddes, 2000; Cadet et al., 2001). These drugs are also known to cause hyperthermia in both humans (Kalant, 2001) and rodents (Sandoval et al., 2000; Mechan et al., 2001); this action appears to contribute to both the acute mortality and neurotoxicity induced by these drugs. Hyperthermia potentiates dopamine and tyrosine-hydroxylase depletion, astrocytosis, and oxidative stress (Omar et al., 1987; Lin et al., 1991), whereas hypothermia protects against these effects (Bowyer et al., 1993, 1994; Miller and O'Callaghan, 1994).

Although it is known that both METH and MDMA induce body hyperthermia (Sandoval et al., 2000; Mechan et al., 2001), associated changes in brain temperature and their relationships to body hyperthermia have not been thoroughly examined. Since amphetamine-like drugs are typically consumed in social situations (see Kalant, 2001 for review), which are accompanied by brain and body hyperthermia, it is important to know how the effects of these drugs will be modulated under activated conditions. These drugs have vasoconstrictive action (Pederson and Blessing, 2001) and they are often used in warm, humid, and crowded environments, which diminish heat dissipation to the environment. Therefore, we studied the effects of METH and MDMA on brain and body temperature in quiet resting conditions at normal environmental temperatures (23°C), under activated conditions (during social interaction with female), and at elevated ambient temperatures (29°C) (Brown et al., 2003; Brown and Kiyatkin, 2004, 2005). Since rats with chronically occluded jugular veins had significant changes in brain and body temperature responses to natural environmental challenges (Kiyatkin and Brown, 2004b), this animal preparation was used to assess the role of venous brain outflow in mediating the hyperthermic effects of MDMA.

As shown in Fig. 6, METH (9 mg/kg, sc) used under quiet resting conditions at 23°C ambient temperatures (left side) induced a robust hyperthermic response (~3.4°C), which was significantly stronger in brain sites (NAcc, Hippo) than in the muscle. Importantly, brain-muscle differentials increased more strongly and for a much longer period of time than those induced by natural stimuli. Therefore, increased body temperature, the most dangerous symptom of METH intoxication, is a consequence of excessive metabolic brain activation induced by the drug. Since this activation exceeds the physiological limits in both amplitude and duration of temperature response, it may be viewed as "pathological." When METH was administered during social interaction (right side), both the amplitude and, especially, the duration of hyperthermia were larger than those it caused in quiet conditions. The potentiating effect was especially strong with respect to brain temperatures.

A similar pattern of state–drug interaction was also typical of MDMA (Fig. 7). In this case, the overall increase in temperature induced by MDMA at the same dose (9 mg/kg, sc) under quiet resting conditions was weaker than that induced by METH (~0.9°C) and had a delayed onset (left panel). Similar to METH, brain-muscle differentials were elevated for more than 3 h, suggesting sustained metabolic activation and continuous intra-brain heat production as a primary cause of brain hyperthermia. When the drug was injected during social interaction, the increase was significantly stronger and continued for more than 5 h (center panel). Potentiation of drug action was also evident in brain-muscle differentials, which remained elevated for a longer time after drug administration under activated conditions. Brain hyperthermia induced by MDMA was greatly potentiated in animals with chronic occlusion of jugular veins. Brain-muscle differentials increased in this case much greater and for a longer time than in control (B).

The effects of both METH and MDMA were greatly increased when the drugs were used at warm environmental temperatures (29°C) due to a combination of drug-induced heat production and drug-induced impairment of heat dissipation (Fig. 8). In this case, brain temperatures gradually increased from the moment of drug administration, most tested rats (four of six with METH and five of six with MDMA) developed a malignant hyperthermic state and died within 7 h. This mortality rate of 83% for 9 mg/kg MDMA seems quite high considering that the commonly accepted LD_{50} in the rat is 49 mg/kg (Davis et al., 1978). It is also quite high for METH, which never induced lethality at this dose at normal environmental conditions.

The present data underscore the importance of both brain state and environmental conditions in determining the adverse effects of METH and MDMA. This drug–state interaction is important with respect to both acute adverse drug effects and the slow neurotoxic action typical of chronic drug use over extended periods of time. Since a wide range of activated conditions are associated with brain hyperthermia and these drugs themselves induce metabolic brain activation and diminish heat dissipation to the external environment, drug use under these activated conditions will result in stronger effects. Under these conditions, body and brain heat production is increased (Nybo et al., 2002), adaptive increases in cerebral blood flow are slowed (Nybo and Nielson, 2001) (because of

---

Fig. 6. Changes in brain and muscle temperature induced by meth-amphetamine (9 mg/kg, sc) administered under quiet resting conditions and after 3-min social interaction with female. The first hatched vertical line (left panel) indicates the moment of drug administration in quietly resting rats and two vertical lines (right panel) indicate the start of social interaction and drug administration (+30 min), respectively. A shows absolute temperature changes, (B) shows relative temperature changes, (C) shows brain-muscle differentials, and (D) shows differences in area under curve analyzed as a time above 2°C. [* Indicates significant increase for activated vs. quiet resting conditions ($p<0.05$).] Filled symbols indicate values significantly different from baseline.

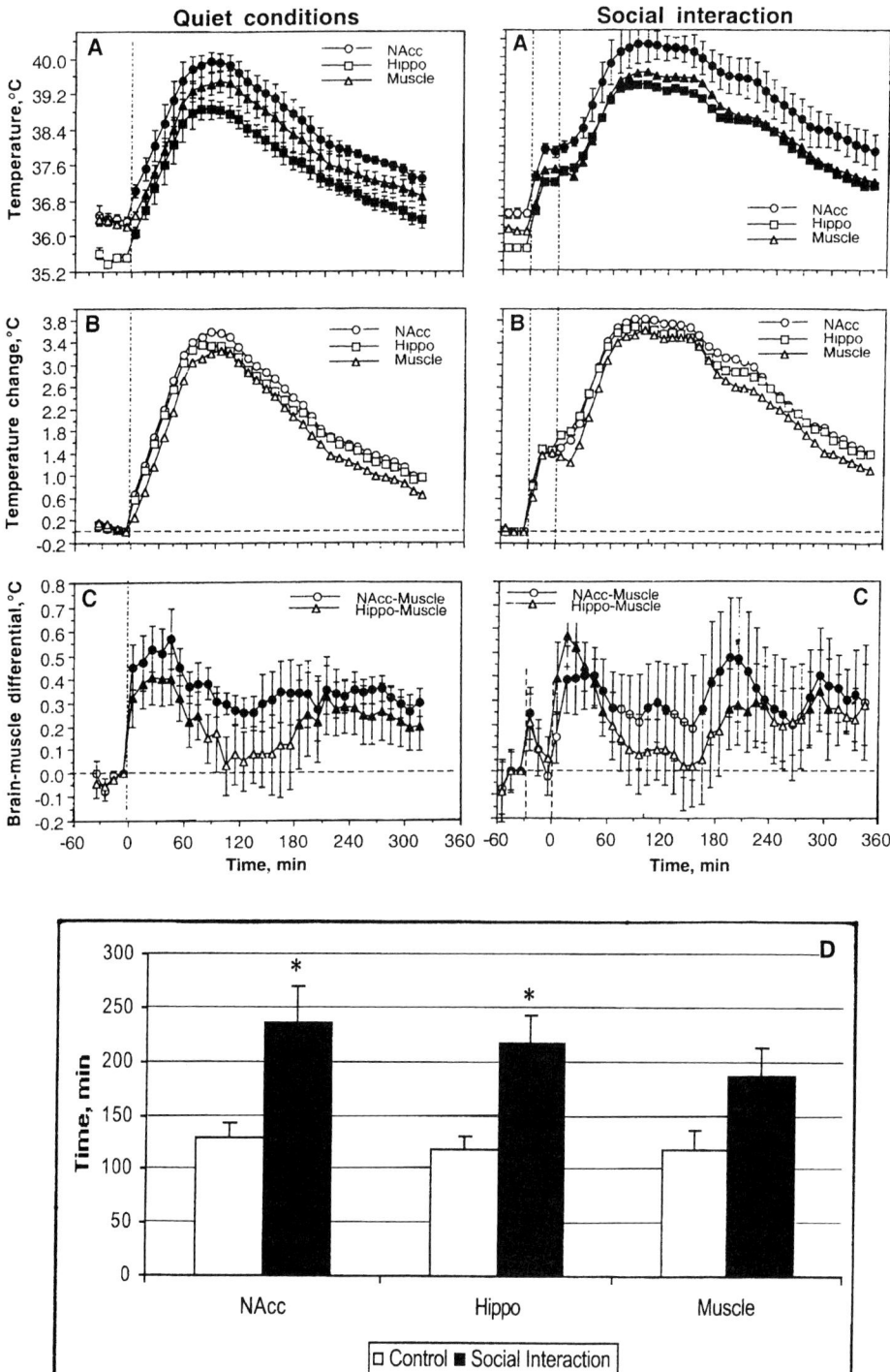

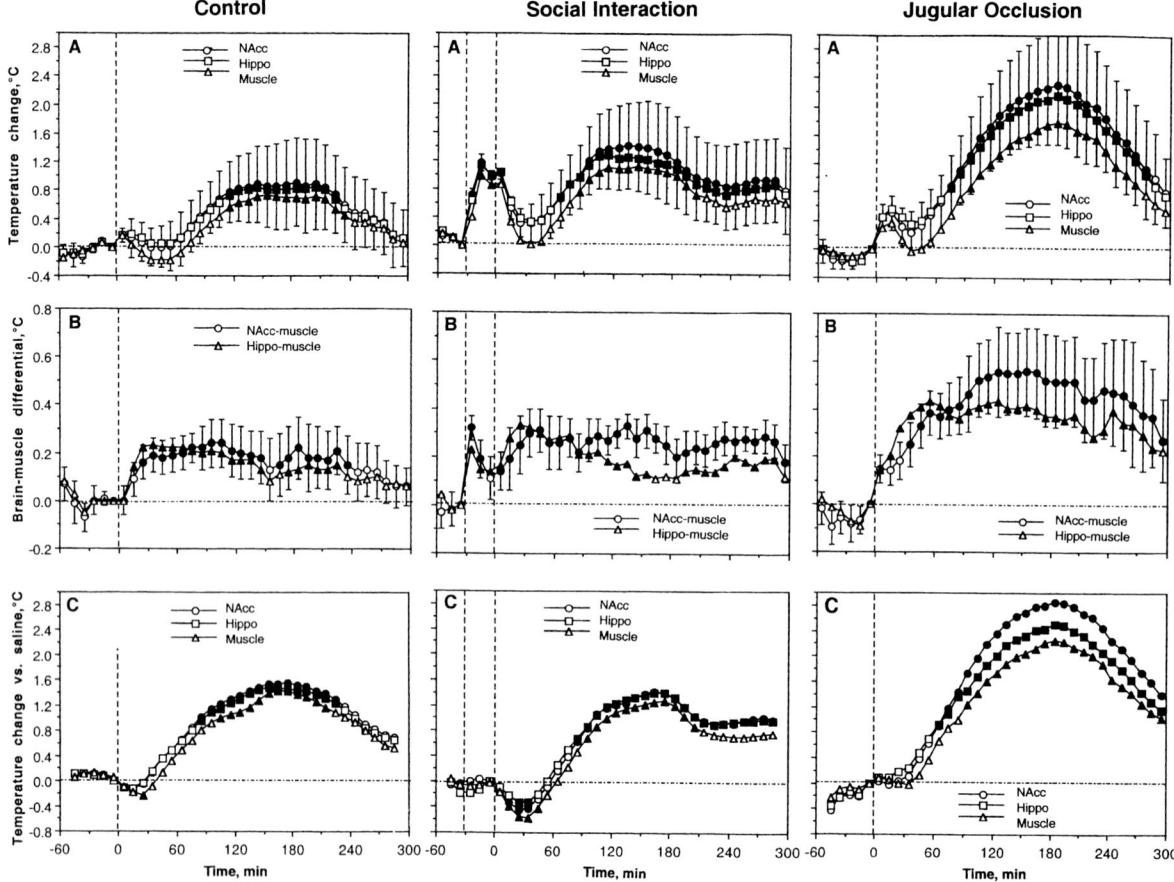

Fig. 7. Changes in brain and muscle temperature induced by MDMA (9 mg/kg, sc) administered in three conditions: in intact animals under quiet conditions at 23°C (control, left panel), in intact animals under activated conditions at 23°C (30 min after the start of 3-min social interaction with a female; central panel), and in animals with chronically occluded jugular veins at 23°C (right panel). (A) Shows relative temperature change with respect to baseline, (B) shows relative change in brain-muscle differentials, and (C) shows relative temperature change vs. saline. Filled symbols indicate values significantly different from baseline.

hyperventilation-induced decrease in arterial CO_2 tension), and peripheral heat loss is strongly impaired. These effects are strongly potentiated in environments that diminish an organism's ability to dissipate heat, further contributing to malignant hyperthermia and fatality. Therefore, it is a combination of specific activity state and environmental conditions coupled with individual drug predisposition that makes amphetamine-like substances especially dangerous. While pathological hyperthermia is a dangerous symptom of acute intoxication induced by amphetamine-like drugs, it is also a factor inducing irreversible damage of mitochondria and cell death (Iwagami, 1996; Willis et al., 2000; Lepock, 2003). Since the neurotoxic effects of amphetamine-like drugs are temperature-dependent (Gordon et al., 1991; Bowyer et al., 1993, 1994; Alberts and Sonsalla, 1995; Miller and O'Callaghan, 1994), brain hyperthermia is also a significant factor potentiating neurotoxicity — another dangerous complication of chronic use of amphetamine-like substances.

Brain temperature as a factor affecting neuronal and neurochemical evaluations

Naturally occurring and drug-induced fluctuations in brain temperature also affect neuronal and

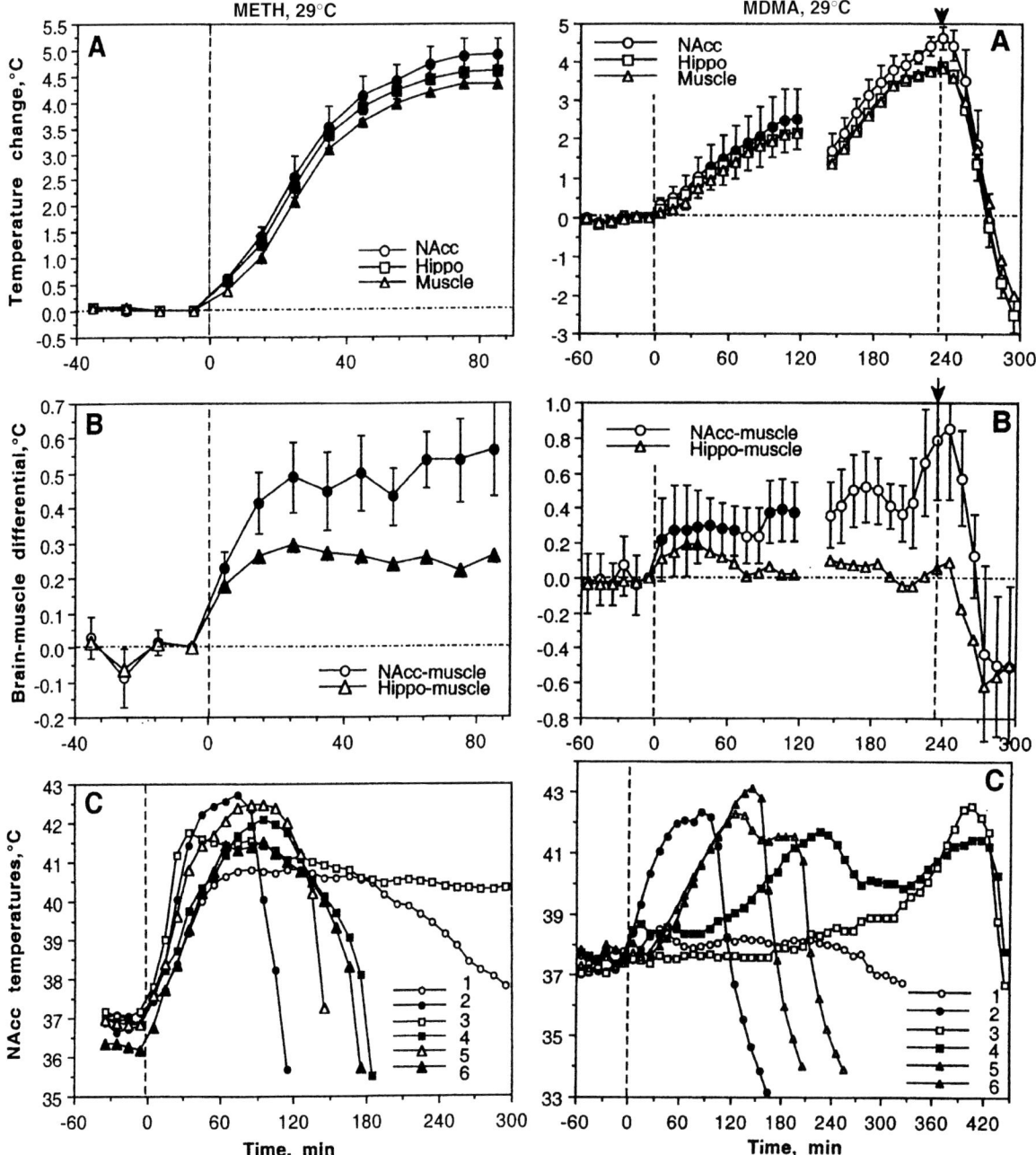

Fig. 8. Changes in brain and muscle temperature induced by meth-amphetamine (9 mg/kg, sc) and MDMA (9 mg/kg, sc) administered to rats at warm environmental temperatures (29°C). (A) Shows mean (±SEM) temperature change relative to baseline, (B) shows mean (±SEM) changes in brain-muscle differentials relative to baseline, and (C) shows changes in NAcc temperature in each individual animal. Filled symbols indicate values significantly different from baseline. The first vertical line shows the moment of drug administration and the second vertical line shows the moment of death.

neurochemical measurements made in behaving animals. While the change in electrical activity of single neurons is usually viewed as a consequence of synaptic stimulation, temperature appears to be an additional factor affecting this parameter. It is quite difficult to determine the contribution of this factor for *in vivo* recordings, but *in vitro* data suggest that many CNS cells are highly temperature-sensitive, showing robust changes in discharge rate and synaptic responses following relatively small changes in slice temperature. For example, 22% of medial thalamic neurons were found to be highly temperature-sensitive (Travis et al., 1995), showing a positive thermal coefficient ($>0.8\,\text{imp/s/}^\circ\text{C}$) similar to the classic warm-sensitive neurons of the anterior and posterior hypothalamus (Boulant, 1998). High temperature sensitivity has been also shown on many other central neurons, including those located in cortex (Lee et al., 2004; Volgushev et al., 2004), Hippo (Thompson et al., 1985), medulla (Tryba and Ramirez, 2004), superchiasmatic nucleus of the hypothalamus (Ruby and Heller, 1996), and midbrain dopamine-containing neurons (Guatteo et al., 2005). In the latter study, it has been found that the discharge rate of substantia nigra dopamine neurons increased with warming within the physiological range (34° to 39°C: $Q10 = 2.7$) and dramatically decreased with cooling below physiological range (34° to 29°C: $Q10 = 7.7$).

While behavior-related and drug-induced fluctuations in brain temperature should affect various parameters of neuronal activity in freely moving animals, temperature appears to be important for interpreting neuronal evaluations in anesthetized animals and *in vitro* slice preparations maintained at cold, room temperatures ($\sim 22^\circ\text{C}$ or $\sim 15^\circ\text{C}$ below normal baseline). Most general anesthetic drugs drastically reduce brain metabolism, inducing robust brain and body hypothermia (Crane et al., 1978; Michenfelder, 1988; Wass et al., 1998; Zhu et al., 2004). Recently, we showed that brain temperature falls up to 4.0° to 4.5°C below normal, quiet resting baseline after a single administration of sodium pentobarbital at a typical anesthetic dose (50 mg/kg, i.p.) (Kiyatkin and Brown, 2005). Importantly, the decrease was stronger in brain sites than in muscle, suggesting primary metabolic brain inhibition, and it could not be fully compensated by the efficient, feedback-controlled body warming. While body warming was able to almost fully eliminate body hypothermia, it amplified brain-body temperature differentials, making the metabolically inhibited brain relatively cooler than the artificially warmed body. External body warming during anesthesia may also invert the normal direction of heat exchange between the brain and arterial blood, thus affecting the various metabolic and blood flow evaluations made in the brain of anesthetized animals. In addition to robust hypothermia, brain temperature in unwarmed animals is highly unstable, showing phasic decreases and subsequent increases during the development of and awakening from general anesthesia. For these reasons, work conducted with anesthetized animal preparations as well as in cold-maintained brain slices can produce an obscured, distorted view of the active brain (Micheva and Smith, 2005).

Temperature appears to be an important factor affecting the results of *in vivo* microdialysis and electrochemical measurements made in behaving animals. It is well known that both the diffusion of neuroactive substances via semi-permeable membranes, the basis of *in vivo* microdialysis, and generation of oxidative currents by applied potentials, the basis of *in vivo* electrochemistry, are temperature-dependent. Therefore, if brain temperature goes up or down during dialysis, more or less substance, respectively, will diffuse from the brain tissue to dialysate, determining erroneous increases or decreases in concentration of the measured substance. In addition to its effect on diffusion via dialysis membrane, naturally occurring and drug-induced changes in brain temperature will affect the dynamics of neurotransmission because of temperature-dependence of both the release and reuptake of neuroactive substances as well as their diffusion from releasing points to neuroreceptors (Xie et al., 2000), additionally distorting the precision of neurochemical measurements. Finally, all these factors will determine the differences in neurochemical evaluations between awake, freely moving and anesthetized animal preparations.

Conclusions

The present work demonstrates that brain temperature in awake, freely moving animals in a temperature-stable environment is not stable, showing profound fluctuations under conditions of physiological activation. While mean temperature in the MPAH, a structure heavily implicated in the central regulation of body temperature, under quiet resting conditions in rats was ~37.3°C, it decreased below 36°C during sleep and transiently peaked well above 39°C (up to 39.9°C in individual animals) during sexual behavior at ejaculation. Despite evident but subtle differences, similar state- and activity-related fluctuations also occurred in other brain structures. This range, exceeding 3°C, therefore, can be viewed as fluctuations in normal brain temperatures and brain hyperthermia may be viewed as a part of normal brain functioning, not an index of disease. Brain hyperthermia within 3°C was also induced by various addictive drugs (i.e., heroin, cocaine, METH, MDMA) at doses comparable or equivalent to those used by humans; this hyperthermia results from both enhanced brain metabolism and diminished heat dissipation to the external environment because of peripheral vasoconstriction. At higher doses, amphetamine-like psychomotor stimulants induced prolonged hyperthermia up to 40° to 41°C. While tolerated by animals, this increase clearly exceeds the normal range and may be viewed as pathological hyperthermia. These drug effects are dramatically modulated by relatively weak increases in ambient temperature, resulting in fatality of most tested animals.

Although humans have more efficient mechanisms for heat dissipation than rats, this drug–environment interaction is important for understanding potential health hazards of psychomotor stimulant drugs, which are typically used during activated conditions in hot, humid environments. Thus, the acute adverse effects of psychomotor stimulants are not purely a function of the drug and its dose, but a combination of drug, associated activity, specific environmental conditions, and individual predisposition. These drug-independent factors may also play an important role in the development of drug-induced neurotoxicity, a latent but dangerous complication of chronic use of psychomotor stimulants. Although undetected, this latent drug-related neurotoxicity, being summated with age-related degeneration and environmental neurotoxic influences, can result in earlier development and stronger progression of various neurodegenerative diseases.

The data presented above point toward metabolic neural activation accompanied by intra-brain heat production as the primary cause of functional brain hyperthermia and cerebral circulation as the primary means to dissipate metabolic heat to the body and then to the external environment. While intra-brain heat production is an obvious by-product of cerebral metabolism, functional hyperthermia affects various neural functions ranging from the activity of single ionic channels to such global functions as transmitter release and uptake. Therefore, naturally occurring fluctuations in brain temperature should be considered as a communicative or integrative factor, involving and uniting numerous central neurons within the brain. Although this type of interneuronal communication is obviously not the main, nor the most efficient, it has an important adaptive significance.

Abbreviations

Hippo	hippocampus
MDMA	3,4-methylenedioxymethamphetamine ("ecstasy")
METH	meth-amphetamine
MPAH	medial preoptic area of the hypothalamus
NAcc	nucleus accumbens
VTA	ventral tegmental area of midbrain

Acknowledgments

The research discussed in this review was supported by the Intramural Research Program of the NIH, NIDA. I would like to express my deep gratitude to P. Leon Brown, a collaborator in most experiments described in this article, for multiple discussions and valuable help in clearly formulating the ideas discussed in this manuscript.

References

Abrams, R. and Hammel, H.T. (1964) Hypothalamic temperature in unanesthetized albino rats during feeding and sleeping. Am. J. Physiol., 206: 641–646.

Alberts, D.S. and Sonsalla, P.K. (1995) Methamphetamine-induced hyperthermia and dopaminergic neurotoxicity in mice: pharmacological profile of protective and nonprotective agents. J. Pharmacol. Exp. Ther., 275: 1104–1114.

Ansah, T.A., Wade, L.H. and Shockley, D.C. (1996) Changes in locomotor activity, core temperature, and heart rate in response to repeated cocaine administration. Physiol. Behav., 60: 1261–1267.

Bartlett, J.R. (1956) Physiological responses during coitus. J. Appl. Physiol., 9: 472–496.

Battig, K., Jacober, A. and Hasenfratz, M. (1993) Cigarette smoking related variation of heat rate and physical activity with ad libitum smoking under field conditions. Psychopharmacology (Berl.), 110: 371–373.

Blumberg, M.S., Mannella, J.A. and Moltz, H. (1987) Hypothalamic temperature and deep body temperature during copulation in the male rat. Physiol. Behav., 39: 367–370.

Bohlen, J.G., Held, J.P., Sanderson, M.O. and Patterson, R.P. (1984) Heart rate, rate-pressure product, and oxygen uptake during four sexual activities. Arch. Intern. Med., 144: 1745–1748.

Bouchama, A. and Knochel, J.P. (2002) Heat stroke. N. Engl. J. Med., 346: 1978–1988.

Boulant, J.A. (1998) Hypothalamic neurons. Mechanisms of sensitivity to temperature. Ann. N.Y. Acad. Sci., 856: 108–115.

Bowyer, J.F., Davies, D.L., Schmued, L., Broening, H.W., Newport, G.D., Slikker, W. and Holson, R.R. (1994) Further studies of the role of hyperthermia in methamphetamine neurotoxicity. J. Pharmacol. Exp. Ther., 268: 1571–1580.

Bowyer, J.F., Gough, B., Slikker, W., Lipe, G.W., Newport, G.D. and Holson, R.R. (1993) Effects of a cold environment or age on methamphetamine-induced dopamine release in the caudate putamen of female rats. Pharmacol. Biochem. Behav., 44: 87–98.

Briese, E. (1995) Emotional hyperthermia and performance in humans. Physiol. Behav., 58: 615–618.

Brinnel, H., Nagasaka, T. and Cabanac, M. (1987) Enhanced brain protection during passive hyperthermia in humans. Eur. J. Appl. Physiol. Occup. Physiol., 56: 540–545.

Brown, P.L. and Kiyatkin, E.A. (2004) Brain hyperthermia induced by MDMA (ecstasy): modulation by environmental conditions. Eur. J. Neurosci., 20: 51–58.

Brown, P.L. and Kiyatkin, E.A. (2005) Fatal intra-brain heat accumulation induced by meth-amphetamine at normothermic conditions in rats. Int. J. Neuroprotec. Neuroregen., 1: 86–90.

Brown, P.L., Wise, R.A. and Kiyatkin, E.A. (2003) Brain hyperthermia is induced by methamphetamine and exacerbated by social interaction. J. Neurosci., 23: 3924–3929.

Cabanac, M. (1998) Selective brain cooling and thermoregulatory set-point. J. Basic Clin. Physiol. Pharmacol., 9: 3–17.

Cabanac, M., Germain, M. and Brinnel, H. (1987) Tympanic temperatures during hemiface cooling. Eur. J. Appl. Physiol., 56: 534–539.

Cadet, J.L., Thiriet, N. and Jayanthi, S. (2001) Involvement of free radicals in MDMA-induced neurotoxicity in mice. Ann. Med. Intern., 152(Suppl 3): IS57–IS59.

Charkoudian, N. (2003) Skin blood flow in adult human thermoregulation: how it works, when it does not, and why. Mayo Clin. Proc., 78: 603–612.

Cheuvront, S.N. and Haymes, E.M. (2001) Thermoregulation and marathon running: biological and environmental influences. Sports Med., 31: 743–762.

Cota, D., Marsicano, G., Tschop, M. and Glubler, Y. (2002) The endogenous cannabinoid system affects energy balance via central orexigenic drive and peripheral lipogenesis. J. Clin. Invest., 112: 423–431.

Crane, P., Braun, L., Cornford, E., Cremer, J., Glass, J. and Oldendorf, M. (1978) Dose-dependent reduction of glucose utilization by pentobarbital in rat brain. Stroke, 9: 12–18.

Davidson, C., Gow, A.J., Lee, T.H. and Ellinwood, E.H. (2001) Methamphetamine neurotoxicity: necrotic and apoptotic mechanisms and relevance to human abuse and treatment. Brain Res. Rev., 36: 1–22.

Davis, W.M., Hatoum, H.T. and Walters, I.W. (1978) Toxicity of MDA (2,4-methylenedioxyamphetamine) considered for relevance to hazards of MDMA (Ecstasy) abuse. Alcohol Drug Res., 7: 123–134.

Delgado, J.M.R. and Hanai, T. (1966) Intracerebral temperatures in free-moving cats. Am. J. Physiol., 211: 755–769.

Dewhirst, M.W., Viglianti, B.L., Lora-Michiels, M., Hanson, M. and Hoopes, P.J. (2003) Basic principles of thermal dosimetry and thermal thresholds for tissue damage from hyperthermia. Int. J. Hyperthermia, 19: 267–294.

Donaldson, G.C., Keatinge, W.R. and Saunders, R.D. (2003) Cardiovascular responses to heat stress and their adverse consequences in healthy and vulnerable human populations. Int. J. Hyperthermia, 19: 225–235.

Endoh, H., Taga, K., Yamakura, T., Sato, K., Watanabe, I., Fukuda, S. and Shimoji, K. (1999) Effects of naloxone and morphine on acute hypoxic survival in mice. Crit. Care Med., 27: 1923–1933.

Exton, N.G., Truong, T.C., Exton, M.S., Wingenfeld, S.A., Leygraf, N., Saller, B., Hartman, U. and Schedlowski, M. (2000) Neuroendocrine response to film-induced sexual arousal in men and woman. Psychoendocrinology, 25: 187–199.

Fellows, L.K., Boutelle, M.G. and Fillenz, M. (1993) Physiological stimulation increases nonoxidative glucose metabolism in the brain of the freely moving rat. J. Neurochem., 60: 1258–1263.

Gollib, R.L., Breiter, H.C., Kantor, H., Kennedy, D., Gastfriend, D. and Mathew, R.T. (1998) Cocaine decreases cortical cerebellar blood flow but does not obscure regional activation in functional magnetic resonance imaging in human subjects. J. Cereb. Blood Flow Metab., 18: 724–734.

Gordon, C.J., Watkinson, W.O., O'Callaghan, J.P. and Miller, D.B. (1991) Effects of 3,4-methylenedioxymethamphetamine on autonomic thermoregulatory responses of the rat. Pharmacol. Biochem. Behav., 38: 339–344.

Green, A.R., Mechan, A.O., Elliott, J.M., O'Shea, E. and Colado, M.I. (2003) The pharmacology and clinical pharmacology of 3,4-methylenedioxymethamphetamine (MDMA, "Ecstasy"). Pharmacol. Rev., 55: 463–508.

Guatteo, E., Chung, K.K.H., Bowala, T.K., Bernardi, G., Mercuri, N.R. and Lipski, J. (2005) Temperature sensitivity of dopaminergic neurons of the substantia nigra pars compacta: involvement of TRP channels. J. Neurophysiol., 94: 3069–3080.

Hasenfratz, M., Jacober, A. and Battig, K. (1993) Smoking-related subjective and physiological changes: pre- to post-puff and pre to post-cigarette. Pharmacol. Biochem. Behav., 46: 527–534.

Horvath, T.L., Warden, C.H., Hajos, M., Lombardi, A., Goglia, F. and Diano, S. (1999) Brain uncoupling protein 2: uncoupled neuronal mitochondria predict thermal synapses in homeostatic centers. J. Neurosci., 19: 10417–10427.

Howell, L.L., Hoffman, J.N., Votaw, J.R., Landrum, A.M., Wilcox, K.M. and Lindsey, K.P. (2002) Cocaine-induced brain activation determined by positron emission tomography neuroimaging in conscious rhesus monkeys. Psychopharmacology, 159: 154–160.

Ide, K., Schmalbruch, I.K., Quistorff, B., Horn, A. and Secher, N.H. (2000) Lactate, glucose, and oxygen uptake in human brain during recovery from maximal exercise. J. Physiol., 522: 159–164.

Ide, K. and Secher, N.H. (2000) Cerebral blood flow and metabolism during exercise. Prog. Neurobiol., 61: 397–414.

Iwagami, Y. (1996) Changes in ultrastructure of human cell related to certain biological responses under hyperthermic culture conditions. Human Cell, 9: 353–366.

James, W. (1892) Psychology: Briefer Course. Henry Holt, New York, NY.

Jansky, L. and Vybiral, S. (2004) Thermal homeostasis in systemic inflammation: modulation of neural mechanisms. Front. Biosci., 9: 3068–3084.

Jessen, A.B., Toubro, S. and Astrup, A. (2003) Effect of chewing gum containing nicotine and caffeine on energy expenditure and substrate utilization in men. Am. J. Clin. Nutr., 77: 1442–1447.

Kalant, H. (2001) The pharmacology and toxicology of "ecstasy" (MDMA) and related drugs. Can. Med. Assoc. J., 165: 917–928.

Kiyatkin, E.A. and Brown, P.L. (2003) Fluctuations in neural activity during cocaine self-administration: clues provided by brain thermorecording. Neuroscience, 116: 525–538.

Kiyatkin, E.A. and Brown, P.L. (2004a) Brain temperature fluctuations during passive vs. active cocaine administration: clues for understanding the pharmacological determination of drug-taking behavior. Brain Res., 1005: 101–116.

Kiyatkin, E.A. and Brown, P.L. (2004b) Modulation of physiological brain hyperthermia by the environmental temperature and impaired blood outflow in rats. Physiol. Behav., 83: 467–474.

Kiyatkin, E.A. and Brown, P.L. (2005) Brain and body temperature homeostasis during sodium pentobarbital anesthesia with and without body warming in rats. Physiol. Behav., 84: 563–570.

Kiyatkin, E.A., Brown, P.L. and Wise, R.A. (2002) Brain temperature fluctuation: a reflection of functional neural activation. Eur. J. Neurosci., 16: 164–168.

Kiyatkin, E.A. and Gratton, A. (1994) Electrochemical monitoring of extracellular dopamine in nucleus accumbens of rats lever-pressing for food. Brain Res., 652: 225–234.

Kiyatkin, E.A. and Mitchum, R. (2003) Fluctuations in brain temperatures during sexual behavior in male rats: an approach for evaluating neural activity underlying motivated behavior. Neuroscience, 119: 1169–1183.

Kiyatkin, E.A. and Rebec, G.V. (2001) Impulse activity of ventral tegmental area neurons during heroin self-administration in rats. Neuroscience, 102: 565–580.

Kiyatkin, E.A. and Stein, E.A. (1993) Behavior-associated changes in blood pressure during heroin self-administration. Pharmacol. Biochem. Behav., 46: 561–567.

Kiyatkin, E.A. and Wise, R.A. (2001) Striatal hyperthermia associated with arousal: intracranial thermorecordings in behaving rats. Brain Res., 918: 141–152.

Kiyatkin, E.A. and Wise, R.A. (2002) Brain and body hyperthermia associated with heroin self-administration in rats. J. Neurosci., 22: 1072–1080.

Kuhn, D.M. and Geddes, T.J. (2000) Molecular footprints of neurotoxic amphetamine action. Ann. N.Y. Acad. Sci., 914: 92–103.

Langsjo, J.W., Kaisti, K.K., Aalto, S., Hinkka, S., Aantaa, R., Oikenen, V., Supila, H., Kurki, T., Silvanto, M. and Scheinin, H. (2003) Effects of subanesthetic doses of ketamine on regional cerebral blood flow, oxygen consumption, and blood volume in humans. Anesthesiology, 99: 614–623.

Lee, J.C.F., Callaway, J.C. and Foehring, R.C. (2004) The effects of temperature on calcium transients and Ca^{2+}-dependent afterhyperpolarizations in neocortical pyramidal neurons. J. Physiol., 93: 2012–2020.

Lepock, J.R. (2003) Cellular effects of hyperthermia: relevance to the minimum dose for thermal damage. Int. J. Hyperthermia, 19: 252–266.

Lepock, J.R., Cheng, K.-H., Al-Qysi, H. and Kruuv, J. (1983) Thermotropic lipid and protein transitions in Chinese hamster lung cell membranes: relationship to hyperthermic cell killing. Can. J. Biochem. Cell Biol., 61: 421–427.

Li, S.J., Biswal, B., Risinger, R., Rainey, C., Cho, J.H., Salmeron, B.J. and Stein, E.A. (2000) Cocaine administration decreases functional connectivity in human primary visual and motor cortex as detected by functional MRI. Magn. Reson. Med., 43: 43–51.

Lin, P.S., Quamo, S., Ho, K.C. and Gladding, J. (1991) Hyperthermia enhances the cytotoxic effects of reactive oxygen species to Chinese hamster cells and bovine endothelial cells in vitro. Radiat. Med., 126: 43–51.

Lipton, S.A. and Rosenberg, P.A. (1994) Excitatory amino acids as a final common pathway for neurologic disorders. N. Engl. J. Med., 330: 613–622.

Lynch, T.J., Adler, M.W. and Eisenstein, T.K. (1990) Comparison of the mechanisms on interleukin-1 and morphine-induced hyperthermia in the rat. Ann. N.Y. Acad. Sci., 594: 469–471.

Maier, C.M. and Steinberg, G.K. (2003) Hypothermia and Cerebral Ischemia. Humana Press, New York, NY.

Mariak, Z., Jadeszko, M., Lewko, J., Lebkowski, W. and Lyson, T. (1998) No specific brain protection against thermal stress in fever. Acta Neurochir. (Wien), 140: 585–590.

Mariak, Z., Lebkowski, W., Lyson, T., Lewko, J. and Piekarski, P. (1999) Brain temperature during craniotomy in general anesthesia. Neurol. Neurochir. Pol., 33: 1325–1327.

Mariak, Z., Lyson, T., Peikarski, P., Lewko, J., Jadeszko, M. and Szydlik, P. (2000) Brain temperature in patients with central nervous system lesions. Neurol. Neurosurg. Pol., 34: 509–522.

Marota, J.J., Mendeville, J.B., Weisskoff, R.M., Moskowitz, M.A., Rosen, B.R. and Kosovsky, B.E. (2000) Cocaine activation discriminates dopaminergic projections by temporal response; an fMRI study in rats. Neuroimage, 11: 13–23.

Masters, W.H. and Johnson, V.E. (1966) Human Sexual Response. Little, Brown and Co., Boston, MA.

McElligott, J.C. and Melzack, R. (1967) Localized thermal changes evoked in the brain by visual and auditory stimulation. Exp. Neurol., 17: 293–312.

Mechan, A.O., O'Shea, E., Elliot, J.M., Colado, M.I. and Green, A.R. (2001) A neurotoxic dose of 3,4-methylenedioxymethamphetamine (MDMA; ecstasy) to rats results in a long-term deficit in thermoregulation. Psychopharmacology, 155: 413–418.

Michenfelder, J. (1988) Anesthesia and the Brain: Clinical, Functional, and Vascular Correlates. Churchill Livingstone, New York, NY.

Micheva, K.D. and Smith, S.J. (2005) Strong effects of subphysiological temperature on the function and plasticity of mammalian presynaptic terminals. J. Neurosci., 17: 7481–7488.

Miller, D.B. and O'Callaghan, J.P. (1994) Environment-, drug- and stress-induced alterations in body temperature affect the neurotoxicity of substituted amphetamines in the C57BL/6J mouse. J. Pharmacol. Exp. Ther., 270: 752–760.

Mintun, M.A., Lundstrom, B.N., Snyder, A.Z., Vlasenko, A.G., Shulman, G.L. and Raichle, M.E. (2001) Blood flow and oxygen delivery to human brain during functional activity: theoretical modeling and experimental data. Proc. Natl. Acad. Sci., 98: 6859–6864.

Miyazawa, T., Tamara, A., Fukui, S. and Hossmann, K.A. (2003) Effects of mild hypothermia on focal cerebral ischemia: review of experimental studies. Neurol. Res., 25: 457–464.

Moltz, H. (1993) Fever: causes and consequences. Neurosci. Biobehav. Rev., 17: 237–269.

Moriyama, E. (1990) Cerebral blood flow changes during localized hyperthermia. Neurol. Med. Chir. (Tokyo), 30: 923–929.

Moser, E., Mathesen, I. and Andersen, P. (1993) Association between brain temperature and dentate field potentials in exploring and swimming rats. Science, 259: 1324–1326.

Nagata, Y., Katayama, K., Manivel, C.J. and Song, C.W. (2000) Changes in blood flow in locally heated intestine of rats. Int. J. Hyperthermia, 16: 159–170.

Nakajima, T., Rhee, J.G., Song, C.W. and Onoyama, Y. (1992) Effect of a second heating on rat liver blood flow. Int. J. Hyperthermia, 8: 679–687.

Nemec, E.D., Mansfiled, L. and Kennedy, J.W. (1976) Heart rate and blood pressure responses during sexual activity in normal males. Am. Heart J., 92: 274–277.

Nielsen, B., Hyldig, T., Bidsrup, F., Gonzalez-Alonso, J. and Christoffersen, G.R. (2001) Brain activity and fatigue during prolonged exercise in the heat. Pflugers Arch., 442: 41–48.

Nybo, L. and Nielson, N. (2001) Middle cerebral artery blood velocity is reduced with hyperthermia during prolonged exercise in humans. J. Physiol., 534: 279–286.

Nybo, L., Secher, N.H. and Nielson, B. (2002) Inadequate heat release from the human brain during prolonged exercise with hyperthermia. J. Physiol., 545: 697–704.

Ohmori, T., Abekawa, T. and Koyama, T. (1996) The role of glutamate in behavioral and neurotoxic effects of methamphetamine. Neurochem. Int., 29: 301–307.

Olsen, T.S., Weber, U.J. and Kammersgaard, L.P. (2003) Therapeutic hypothermia for acute stroke. Lancet Neurol., 2: 410–416.

Oobu, K. (1993) Experimental studies on the effect of heating on blood flow in the tongue of golden hamsters. Fukuoka Igaku Zasshi, 84: 497–511.

Oomura, Y., Aou, S., Koyama, Y., Fujita, I. and Yoshimatsu, H. (1988) Central control of sexual behavior. Brain Res. Bull., 20: 863–870.

Omar, R.A., Yano, S. and Kikkawa, T. (1987) Antioxydant enzymes and survival of normal and simian 40-transformed embryo cells after hyperthermia. Cancer Res., 47: 3473–3476.

Pavlov, I.F. and Epstein, O.I. (2003) Morphine and antibodies to mu-opiate receptors in ultra-low doses: effect on oxygen consumption. Bull. Exp. Biol. Med., 135–136(Suppl. 1): 137–139.

Pederson, N.R. and Blessing, W.W. (2001) Cutaneous vasoconstriction contributes to hyperthermia induced by 3,4-methylenedioxymethamphetamine (ecstasy) on conscious rabbits. J. Neurosci., 21: 8648–8654.

Peoples, L.L. and West, M.O. (1996) Phasic firing of single neurons in the rat nucleus accumbens correlated with the timing of intravenous cocaine self-administration. J. Neurosci., 16: 3459–3473.

Perkins, K.A., Sexton, J.E. and DiMarco, A. (1996) Acute thermogenic effects of nicotine and alcohol in healthy male and female smokers. Physiol. Behav., 60: 305–309.

Raichle, M.E. (2003) Functional brain imaging and human brain functions. J. Neurosci., 23: 3959–3962.

Rawson, R.A., Anglin, M.D. and Ling, W. (2002) Will the methamphetamine problem go away? J. Addict. Dis., 21: 5–19.

Reynolds, L.P. and Allen, G.V. (2003) A review of heat shock protein induction following cerebellar injury. Cerebellum, 2: 171–177.

Robinson, R., Iida, H., O'Brien, T.P., Pane, M.A., Trystman, R.J. and Gleason, C.A. (2000) Comparison of cerebellar effects of intravenous cocaine injection in fetal, newborn, and adult sheep. Am. J. Physiol., 279: H1–H6.

Romanovsky, A.A., Ivanov, A.I. and Shimansky, Y.P. (2002) Ambient temperature for experiments in rats: a new method for determining the zone of thermal neutrality. J. Appl. Physiol., 92: 2667–2679.

Rosen, A.D. (1996) Temperature modulation of calcium channel function in GH3 cells. Am. J. Physiol., 271: C863–C868.

Rosen, A.D. (2001) Nonlinear temperature modulation of sodium channel kinetics in GH3 cells. Biochem. Biophys. Acta, 1511: 391–396.

Rowell, L.B. (1983) Cardiovascular aspects of human thermoregulation. Circ. Res., 52: 367–376.

Ruby, N.F. and Heller, H.C. (1996) Temperature sensitivity of the suprachiasmatic nucleus of ground squirrels and rats in vitro. J. Biol. Rhythms, 11: 126–136.

Rumana, C.S., Gopinath, S.P., Uzura, M., Valadka, A.B. and Robertson, C.S. (1998) Brain temperatures exceeds systemic temperatures in head-injured patients. Clin. Care Med., 26: 562–567.

Ryan, K.L., Taylor, W.F. and Bishop, V.S. (1997) Arterial baroreflex modulation of heat-induced vasodilation in the rabbit ear. J. Appl. Physiol., 83: 2091–2097.

Sandoval, V., Hanson, G.R. and Fleckenstein, A.E. (2000) Methamphetamine decreases mouse striatal dopamine transporter activity: roles of hyperthermia and dopamine. Eur. J. Pharmacol., 409: 265–271.

Schaefer, C.F. (1979) Possible teratogenic hyperthermia and marathon running. JAMA, 241: 1892.

Schmidt-Nielsen, K. (1997) Animal Physiology: Adaptation and Environment (5th Ed.). Cambridge Univ. Press, Cambridge, UK.

Schiltz, J.C. and Sawchenko, P.E. (2003) Signaling the brain in systemic inflammation: the role of perivascular cells. Front. Biosci., 8: S1321–S1329.

Schwab, S., Spranger, M., Aschoff, A., Steiner, T. and Hacke, W. (1997) Brain temperature monitoring and modulation in patients with severe MCA infarction. Neurology, 48: 762–767.

Seiden, L.S. and Sabol, K.E. (1996) Methamphetamine and methylenedioxymethamphetamine neurotoxicity: possible mechanisms of cell destruction. NIDA Res. Monogr., 163: 251–276.

Selye, H. (1979) Stress without Distress. New American Library, New York, NY.

Serota, H.M. (1939) Temperature changes in the cortex and hypothalamus during sleep. J. Neurophysiol., 2: 42–47.

Serota, H.M. and Gerard, R.W. (1938) Localized thermal changes in cat's brain. J. Neurophysiol., 1: 115–124.

Siesjo, B. (1978) Brain Energy Metabolism. Wiley, New York, NY.

Sharma, H.S., Alm, P. and Westman, P.J. (1998) Nitric oxide and carbon monoxide in the pathophysiology of brain functions in heat stress. Prog. Brain Res., 115: 297–333.

Sharma, H.S. and Hoopes, P.J. (2003) Hyperthermia-induced pathophysiology of the central nervous system. Int. J. Hyperthermia, 19: 325–354.

Shimura, T., Yamamoto, T. and Shimokochi, M. (1994) The medial preoptic area is involved in both sexual arousal and performance in male rats: reevaluation of neuron activity in freely moving animals. Brain Res., 640: 215–222.

Spina, M.B. and Cohen, G. (1989) Dopamine turnover and glutathione oxidation: implications for Parkinson disease. Proc. Natl. Acad. Sci. U.S.A., 86: 1398–1400.

Stephens, S.E. and Yamamoto, B.K. (1989) Methamphetamine-induced neurotoxicity: roles for glutamate and dopamine efflux. Synapse, 17: 203–209.

Stein, R.A. (2000) Cardiovascular response to sexual activity. Am. J. Cardiol., 86: 27F–29F.

Suehiro, E., Fujisawa, H., Ito, H., Ishikawa, T. and Maekawa, T. (1999) Brain temperature modifies glutamate neurotoxicity in vivo. J. Neurotrauma, 16: 285–297.

Thompson, S.M., Musakawa, L.M. and Rince, D.A. (1985) Temperature dependence of intrinsic membrane properties and synaptic potentials in hippocampal CA1 neurons in vitro. J. Neurosci., 5: 817–824.

Travis, K.A., Bockholt, H.J., Zardetto-Smith, A.M. and Johnson, A.K. (1995) In vitro thermorensitivity of the midline thalamus. Brain Res., 686: 17–22.

Tryba, A.K. and Ramirez, J.-M. (2004) Hyperthermia modulates respiratory pacemaker bursting properties. J. Neurophysiol., 92: 2844–2852.

Uda, M. and Tanaka, Y. (1990) Arterial blood flow changes after hyperthermia on normal liver, normal brain, and normal small intestine. Gan. No. Rinsho., 36: 2362–2366.

Volgushev, M., Kudryashov, I., Chistiakova, M., Mukovski, M., Niesmann, J. and Eysel, U.T. (2004) Probability of transmitter release at neocortical synapses at different temperatures. J. Neurophysiol., 92: 212–220.

Walters, T.J., Rynan, K.L., Tate, L.M. and Mason, P.A. (2000) Exercise in the heat is limited by a critical internal temperature. J. Appl. Physiol., 89: 799–806.

Wass, C., Cable, D., Schaff, H. and Lanier, W. (1998) Anesthetic technique influences brain temperature during cardiopulmonary bypass in dogs. Ann. Thorac. Surg., 65: 454–460.

Welsh, W.J. (1992) Mammalian stress response: cell physiology, structure/function of stress proteins, and implications for medicine and disease. Physiol. Rev., 72: 1063–1081.

Willis, W.T., Jackman, M.R., Bizeau, M.E., Pagliassotti, M.J. and Hazel, J.R. (2000) Hyperthermia impairs liver mitochondrial functions. Am. J. Physiol., 278: R1240–R1246.

Wise, R.A. (2002) Brain reward circuitry: insights from unsensed incentives. Neuron, 36: 229–240.

Wise, R.A. and Bozarth, M.A. (1987) Psychomotor stimulant theory of addiction. Psychol. Rev., 96: 469–492.

Xie, T., McGann, U.D., Kim, S., Yuan, J. and Ricaurte, G.A. (2000) Effect of temperature on dopamine transporter function and intracellular accumulation of methamphetamine: implications for methamphetamine-induced dopaminergic neurotoxicity. J. Neurosci., 20: 7838–7845.

Yablonskiy, D.A., Ackerman, J.H. and Raichle, M.E. (2000) Coupling between changes in human brain temperature and oxidative metabolism during prolonged visual stimulation. Proc. Natl. Acad. Sci., 97: 7603–7608.

Zhu, M., Nehra, D., Ackerman, J.H. and Yablonskiy, D.A. (2004) On the role of anesthesia on the body/brain temperature differential in rats. J. Thermal Biol., 29: 599–603.

CHAPTER 13

Nanoparticles aggravate heat stress induced cognitive deficits, blood–brain barrier disruption, edema formation and brain pathology

Hari Shanker Sharma* and Aruna Sharma

Laboratory of Cerebrovascular Research, Department of Surgical Sciences, Anesthesiology and Intensive Care Medicine, University Hospital, Uppsala University, SE-75185 Uppsala, Sweden

Abstract: Our knowledge regarding the influence of nanoparticles on brain function *in vivo* during normal or hyperthermic conditions is still lacking. Few reports indicate that when nanoparticles enter into the central nervous system (CNS) they may induce neurotoxicity. On the other hand, nanoparticle-induced drug delivery to the brain enhances neurorepair processes. Thus, it is likely that the inclusion of nanoparticles in body fluid compartments alters the normal brain function and/or its response to additional stress, *e.g.*, hyperthermia. New data from our laboratory show that nanoparticles derived from metals (*e.g.*, Cu, Ag or Al, ≈ 50–60 nm) are capable of inducing brain dysfunction in normal animals and aggravating the brain pathology caused by whole-body hyperthermia (WBH). Thus, normal animals treated with nanoparticles (for 1 week) exhibited mild cognitive impairment and cellular alterations in the brain. Subjection of these nanoparticle-treated rats to WBH resulted in profound cognitive and motor deficits, exacerbation of blood–brain barrier (BBB) disruption, edema formation and brain pathology compared with naive animals. These novel observations suggest that nanoparticles enhance brain pathology and cognitive dysfunction in hyperthermia. The possible mechanisms of nanoparticle-induced exacerbation of brain damage in WBH and its functional significance in relation to our current knowledge are discussed in this review.

Keywords: nanoparticles; silver; copper; aluminum; cognitive function; body temperature; heat stress; blood–brain barrier; brain edema; cerebral blood flow; hyperthermia; neurotoxicity

Introduction

Nanoparticles and their potential applications in biomedicine generated enormous scientific interest recently (Euliss et al., 2006; Farokhzad and Langer, 2006). Particles derived from metals, semiconductor materials or semi-solid or soft substances, which are in the size range of 1–200 nm, are known as "Nanoparticles" (Lanone and Boczkowski, 2006; Wagner et al., 2006). Liposomes are one of the examples of semi-solid or soft nanoparticles (Fang, 2006). The properties of materials alter when their sizes reach the nanoscale. Thus, copper (Cu) nanoparticles below 50 nm sizes are extremely hard materials that do not exhibit the same malleability and ductility as bulk Cu (Zhang et al., 2003). It appears that nanoparticles derived from metals could behave differently within the biological systems. However, the influence of nanoparticles on

*Corresponding author. Tel.: +46 18 243899;
Fax: +46 18 243899; E-mail: Sharma@surgsci.uu.se

brain function in normal or in pathological conditions is not known. It is still unclear whether the use of nanoparticles in biomedicine or their clinical application is safe in nature.

Nanoparticles made from semiconductor materials are referred to as quantum dots (Mertens et al., 2006). These nanoscale particles are now commonly used in biomedical applications as drug carriers or as imaging agents (Fang, 2006; Sharma et al., 2006a; Williams, 2006). Liposome nanoparticles are currently being used for delivery of anticancer drugs and/or vaccines (see Fang, 2006). There are reasons to believe that in the coming years, numerous engineered nanoscale products have widespread applications in our society causing significant exposure of human populations to these nanomaterials (He et al., 2006; Moghimi, 2006). Thus, the potential adverse effects of nanoparticles, if any, on human health require detailed investigation (see Borm et al., 2006). Few *in vitro* studies suggest that brain cells are important targets of nanomaterials (Koziara et al., 2006). However, *in vivo* effects of nanoparticles on brain function are still unexplored.

Absorption, distribution, metabolism, excretion and toxicity of nanoparticles are largely dependent on their physicochemical properties and the surrounding environmental conditions (Lam et al., 2006; Teeguarden et al., 2007). Thus, the use of nanoparticles in drug targeting and *in vivo* biomedical imaging could significantly help in improving health care. However, their possible risks to human health with special regard to neurotoxicity require further investigation. Keeping these views into consideration, there is an urgent need to expand our knowledge about the potentially harmful side effects of these nanomaterials in the central nervous system (CNS), particularly *in vivo* situations.

Nanoparticles may enter into the body fluid environments through inhalation (Borm et al., 2006; Lam et al., 2006; Mills et al., 2006). The inhaled nanoparticles from the ambient air are incorporated into a variety of non-neural cells through endocytosis (Kim et al., 2006a, b) and may reside there for weeks to months (Dunning et al., 2004). The potential risks and the cellular toxicity imposed by these nanoparticles *per se* are still unknown (see Xia et al., 2006). In addition, it is still uncertain whether external and internal stressful situations, such as environmental conditions and disease processes, respectively, further alter the effects of these intracellular nanoparticles *in vivo*.

Recent observations show that the distribution of quantum dots in breast tissues is significantly altered following hyperthermia induced by chemotherapy (Minet et al., 2004). Thus, after chemotherapy the nanoparticles can enter into the cell bodies of the breast tissues and enhance cell membrane damage causing aggravation of cell and tissue injuries (Minet et al., 2004). This observation suggests that hyperthermia caused by heat stress, exercise in hot environment or psychostimulant drugs influences the distribution and/or effects of nanoparticles in the CNS.

This review, largely based on our own investigations, is focused on the influence of nanoparticles on brain function in normal and in hyperthermic condition with special regard to the altered cognitive function, behavioral changes, blood–brain barrier (BBB) dysfunction, edema formation and brain pathology in an *in vivo* rat model.

Nanoparticles and neurotoxicity: an emerging new concept?

Although the *in vivo* effects of nanoparticles on neurotoxicity are still lacking, few recent reports suggest that the ultrafine particles from metals induce cellular toxicity both *in vivo* and *in vitro* experiments (see below).

The nanoparticles have higher inflammatory potential to the exposed cells and tissues, compared with larger particles of the same material because of their small sizes (10–100 nm). Thus, when titanium dioxide (TiO_2) particles in the range of 20 nm were applied intratracheally into rats or mice, an intense inflammatory neutrophil response was observed in the lung compared with TiO_2 in the range of 250 nm size at identical doses (Oberdörster, 1996; Oberdörster et al., 2000, 2005a, b; see Brown et al., 2001).

Toxicological effects of nanoparticles are primarily due to their particle chemistry, especially the surface chemistry in addition to their particle size (Oberdörster, 1996). Thus, exposure of rats and humans to polytetrafluoroethylene (PTFE) fume (normally within the size range of 18 nm) induces high acute toxicity and mortality (Cavagna et al., 1961; Coleman et al., 1968; Griffith et al., 1973; National Toxicology Program, 1997). However, the gas phase alone is not toxic, as aging of the PTFE fume particles for 3 min, which increases their particle size to >100 nm, results in a considerable loss of toxicity (Johnston et al., 2000). Changes in particle surface chemistry and not just the larger particle size during aging contribute to the loss of toxic effects (Johnston et al., 1996).

The engineered nanomaterials occurring in different shapes also induce toxicity depending on their dose, dimension, durability and distribution that is species specific (Zhu et al., 2006). Administration of carbon nanotubes (2–50 nm) within the trachea causes acute inflammatory effects in the lungs in mice (0.3–1.3 mg/kg; Shvedova et al., 2005) but not in rats (1–5 mg/kg; Warheit et al., 2004). This suggests that toxic effects of nanoparticles are dependent on their biodistribution and the species used. Whether the acute nanotoxicity observed in mammalian species is due to direct effects of nanoparticles *per se*, or these particles induced alterations in cellular and molecular functions, is still unclear.

One of the first *in vivo* studies using carbon nanotubes shows that the large doses of instilled aggregated nanotubes are responsible for obstruction of the airways leading to death (Warheit et al., 2004). The carbon nanotubes induce profound oxidative stress *in vitro*, as evidenced by the formation of free radicals, accumulation of peroxidative products and depletion of cellular antioxidants (Shvedova et al., 2004; Manna et al., 2005; Fenoglio et al., 2006). This indicates that nanoparticle-induced cellular and molecular reactions in the biological system are important factors in inducing cell and tissue injuries. However, relatively high doses used in *in vitro* studies make it difficult to compare with the findings obtained in *in vivo* investigations.

Thus, it is extremely important to develop animal models to simulate human exposure to engineered nanoparticles to examine their neurotoxic effects *in vivo* in relation to their physicochemical characteristics, aggregation states and concentration (number, mass, surface area), a subject currently being examined in our laboratory.

Translocation of nanoparticles in the biological system

Nanoparticles normally enter into the body fluid compartments through the respiratory tracts (Brooking et al., 2001; Bennett, 2002). Depending on their sizes, nanoparticles are translocated readily to other tissues and organs using different mechanisms. By transcytosis, nanoparticles cross epithelia of the respiratory tract and enter into the interstitium and then the blood stream directly or through lymphatics (Bennett, 2002). Once nanoparticles enter into the blood stream, they are distributed throughout the body (Florence et al., 1995). Nanoparticles may also gain access into the CNS through their uptake by sensory nerve endings in airway epithelia, followed by axonal translocation to the brain and spinal cord (see Gao et al., 2006). However, further studies are needed to establish these pathways, as the CNS is strictly protected by the BBB.

The BBB does not allow particles or ions to enter into the brain fluid microenvironment even within the size range of 10–12 Å (Xu and Ling, 1994; see Sharma, 2004). In rats and mice, lanthanum (molecular diameter 12 Å) when introduced into the circulation does not enter into the brain fluid compartments (see Sharma, 2004a). The electron-dense tracer is stopped at the tight junctions present between the endothelial cells by the luminal surface of the endothelial cell membranes (see Fig. 1; Pettersson et al., 1990; Sharma, 2000, 2005a). This suggests that the entry of nanoparticles into the brain fluid microenvironment is prevented by the cerebral endothelium constituting the BBB. Since transcytosis or pinocytosis is almost non-existent in the CNS microvessels, uptake of nanoparticles through similar routes in the brain is also unlikely (see Rapoport, 1976). However, it is quite possible that disease processes or stressful situations facilitate the entry of nanoparticles into the CNS fluid microenvironment due to a compromised BBB function.

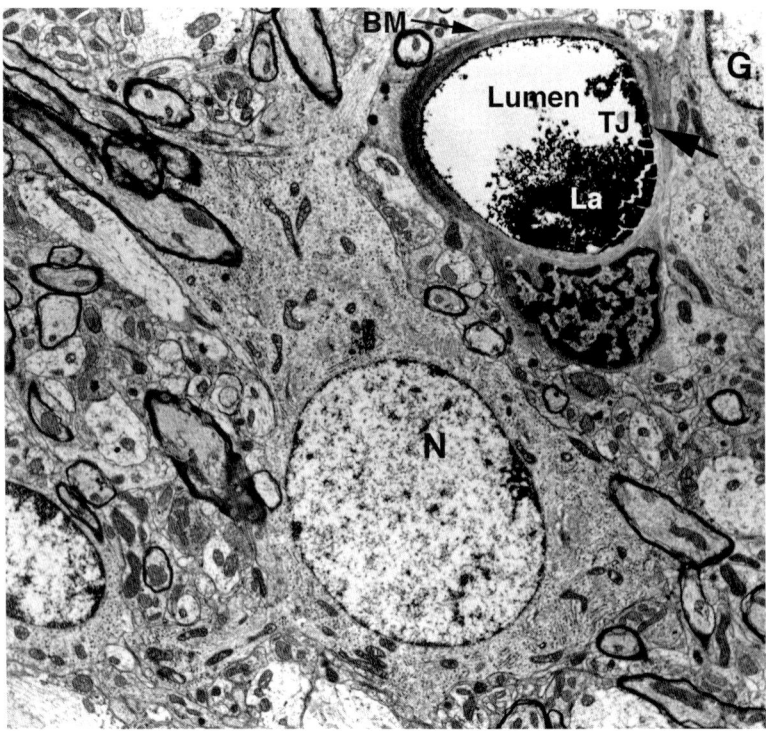

Fig. 1. Blood-spinal cord barrier (BSCB) to lanthanum in rats. Low-power electron micrograph of the rat spinal cord ventral horn showing one microvessel and its surroundings. The electron-dense tracer lanthanum (La) when added into the fixative in situ is localized into the lumen. The passage of lanthanum across the spinal cord endothelial cells is stopped at the tight junction (arrow). A thick basement membrane (BM, 10–12 nm) around the spinal cord endothelial cell (arrow), the glial cell (G) and the surrounding astrocytic processes (Asp) and the nerve cell (N) are clearly seen. Exudation of lanthanum across the endothelial cell membrane or its infiltration within the cell cytoplasm and/or in the basal lamina is absent (for details see text). Bar = 1 μm. Adapted with permission from Sharma (2005a).

Translocation of nanoparticles in neurons

Translocation of solid nanoparticles from the respiratory tract into the neuronal axons was first described ∼65 years ago (Howe and Bodian, 1940; Bodian and Howe, 1941a, b). However, nanoparticles uptake in brain and its consequences on the CNS with regard to neurotoxicity has received very little attention till today.

The nasal and tracheobronchial regions comprise sensory nerve endings of the olfactory and the trigeminus nerves forming an intricate network around this region (see Oberdörster et al., 2005a). Polio virus (≈ 30 nm), administered intranasally into chimpanzees or rhesus monkeys, enter into the CNS through the olfactory nerve and olfactory bulbs (Howe and Bodian 1940; Bodian and Howe, 1941a, b). Since the nasal olfactory mucosa and olfactory bulb are in close proximity (De Lorenzo, 1957), neural transport of virus requires very little time. The transport velocity of the virus in the axoplasm is estimated to be ∼2.5 mm/h (Bodian and Howe, 1941b).

De Lorenzo (1970) further confirmed these observations in squirrel monkeys using intranasally instilled silver-coated colloidal gold particles (≈ 50 nm). In this experiment, translocation of gold particles anterogradely in the axons of the olfactory nerves to the olfactory bulbs took only few hours. Interestingly, the nanoparticles in the olfactory bulb are seen in the mitochondria and not in cytoplasm, indicating a potential adverse effect of these particles on cells through production of reactive oxygen species (ROS).

Recently, inhalation of solid manganese nanoparticles (≈ 30 nm; manganese dioxide MnO_2) in rats showed a 3.5-fold increase in Mn in the olfactory bulb after a 12-day exposure schedule compared with only a twofold increase of Mn in the lung (see Oberdörster et al., 2005a, b). When one nostril is occluded during a 6-h exposure, the accumulation of Mn was seen only in the olfactory bulb of the open nostril (Elder et al., 2006). These observations suggest that nanoparticles in the air can enter into the CNS during accidental or prolonged environmental or occupational exposure to humans.

Although the olfactory mucosa of the human nose is only 5% of the total nasal mucosal surface in rats, translocation of 20 nm particles is 2–10 times higher in the human olfactory bulb than in rats (Oberdörster et al., 2005b). Thus, the translocated nanoparticles in humans can enter into the deeper brain structures in short exposure time (Gianutsos et al., 1997). This idea is supported by the fact that human meningitis and herpes viruses reach trigeminal neurons through olfactory routes rapidly and can trigger outbreaks (Terasaki et al., 1997; Kennedy and Chaudhuri, 2002). Transport of nanoparticles from the trachea to the ganglion nodosum in the neck area in the guinea pig through the vagal system also occurs quite rapidly (Hunter and Undem, 1999). Taken together, it appears that nanoparticles in the ambient air are translocated to the autonomic nervous system *via* circulation or to the CNS through sensory nerves in the respiratory tract, and induce serious cardiovascular or neurological effects (Utell et al., 2002).

A significant increase of tumor necrosis factor-α in brain and a decrease in dopaminergic neurons in mice caused by solid nanoparticles from the air are in line with this idea (Campbell et al., 2005; Veronesi et al., 2005). This confirms that nanoparticles within the environment can influence brain function. Profound inflammatory and neurodegenerative changes in the olfactory mucosa in dogs from the heavily polluted areas in Mexico City, compared with nonexistent degenerative changes seen in animals living in a less polluted area, further support this hypothesis (Calderon-Garcidueñas et al., 2002).

It is likely that nanoparticles from the ambient air, when they reach the CNS microenvironment, initiate a series of cellular and molecular reactions leading to slowly developing neurodegenerative diseases. Although the unique biokinetic properties of nanoparticles, *e.g.*, cellular endocytosis, transcytosis, neuronal and circulatory translocation or distribution, are desirable for medical therapeutic or diagnostic applications, nanoparticle-induced toxicity is another important aspect that requires further investigation. An enhanced or facilitated drug delivery to the CNS with the use of nanoparticles may therefore be considered with caution as their translocation to specific cell types or to subcellular structures in the brain may cause neurotoxicity at a later stage.

Presence of nanoparticles in the body fluid microenvironment or in the CNS is likely to alter the biological response of the organisms to additional external and internal stressful situations, *e.g.*, extreme heat or cold and cardiovascular or endocrine malfunction. These aspects may be considered for the future use of nanoparticles in biomedicine.

Nanoparticles and blood–brain barrier function

Brain is a privileged organ that is protected by the BBB, as mentioned above (see Rapoport, 1976; Sharma, 2004). The BBB resides mainly within the endothelial cells of the cerebral capillaries (Sharma, 2004) that are connected with tight junctions (Fig. 2). In addition, the cerebral endothelium does not normally possess vesicles for macromolecular transport (see Fig. 2) compared with non-cerebral capillaries (Fig. 2). Thus, the permeability properties of the BBB are very similar to that of an extended plasma membrane (see Rapoport, 1976; Sharma, 2004).

Moreover, the endothelial cells of the cerebral capillaries are surrounded by astrocytic end feet ($\sim 85\%$) and possess a thick basement membrane (Pries and Kuebler, 2006). On the other hand, the non-cerebral capillaries have thin basement membrane and do not possess any astrocytic end feet around them (see Sharma, 2004; Fig. 2). The intimate anatomical relationship between astrocytes, cerebral capillaries and basement membrane suggests their active role in BBB function (Sharma, 2004). However, a definite role of basal lamina

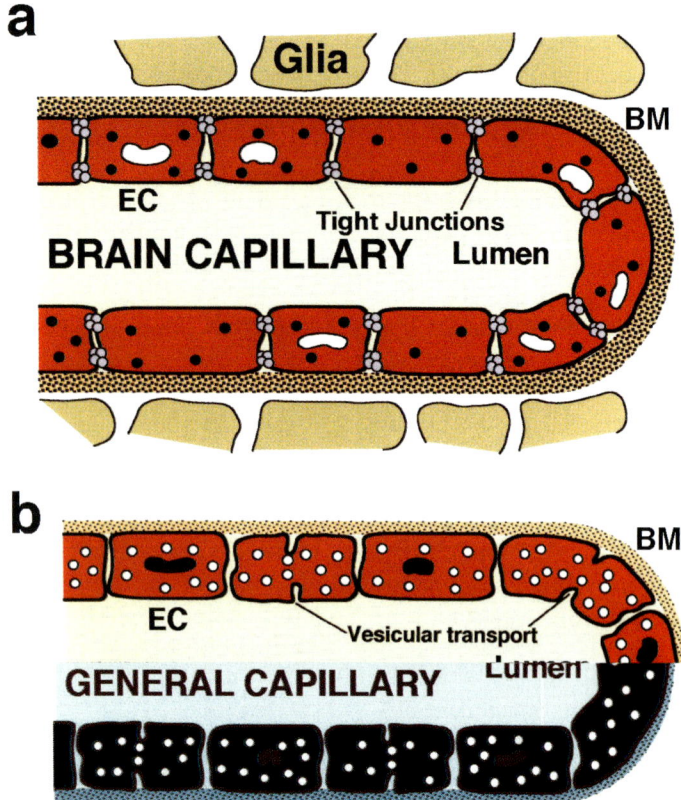

Fig. 2. Diagrammatic representation of the blood–brain barrier (BBB). The cerebral capillaries (a) are different from non-cerebral capillaries (b). The endothelial cells (EC) of cerebral capillaries are connected with tight junctions and do not contain microvesicles for vesicular transport compared with the non-cerebral capillary. The endothelial cells (EC) in the brain are surrounded by thick basement membrane (BM) and glial cells (G). The permeability of the BBB is thus very similar to that of an extended plasma membrane under normal conditions (modified after Sharma, 1999).

and/or astrocytic end feet in influencing BBB permeability in healthy and diseased tissues still requires further investigation.

Nanoparticles, when they reach into circulation, may influence endothelial cell membrane toxicity and/or disrupt the tight junctions. In addition, nanoparticles may stimulate vesicular transport to gain access into the CNS microenvironment as well. However, these new aspects of translocation of nanoparticles from blood to brain require further research. Alternatively, nanoparticles induce oxidative stress and generate free radicals that could disrupt the endothelial cell membrane causing BBB dysfunction. Our laboratory has initiated detailed investigation on the influence of nanoparticles on the structure and function of the BBB in normal animals and their modification by several environmental factors and disease processes with regard to neurotoxicity or neurorepair.

Facilitation of drug transport by nanoparticles across the BBB

The recent advancement in nanotechnology resulted in production of polymer nanoparticles that can be used as carriers to transport the entrapped or adsorbed drugs across the BBB. Kreuter et al. (1995) demonstrated for the first time that polysorbate 80-coated polybutylcyanoacrylate (PBCA) nanoparticles can deliver the peptide "dalargin" into CNS to induce its analgesic effects. Later, several workers were able to successfully deliver high amount of various drugs into the brain using this colloidal carrier (Schroder and Sabel, 1996; Alyautdin et al., 1997, see Friese et al., 2000; and Olivier, 2005). This indicates that drug delivery to the brain can be enhanced using nanoparticles.

Nanoparticles without the surfactant coating are largely engulfed by mononuclear phagocyte system and thus are unable to reach the brain in desirable quantities. Thus, no drug is detected in CSF or brain tissues when uncoated nanoparticles are administered (Gao and Jiang, 2006). Based on these results, it appears that long-circulating PEGylated polycyanoacrylate nanoparticles can offer an alternative as they cross the BBB in a significant amount (Calvo et al., 2001). Endocytosis within the brain endothelial cells is likely to be the preferred route for transfer of coated nanoparticles into the brain.

The mechanisms of enhanced transport of surface-coated nanoparticles across the BBB are not fully known. However, it is assumed that surface-coated nanoparticles mimic LDL and thus cross the BBB through receptor-mediated mechanisms (Sharma, 2004; see Tiwari and Amiji, 2006). The polysorbate 80 behaves like an anchor between nanoparticles and the apolipoprotein, especially ApoE (Liu et al., 2006). Nanoparticles combined with the apolipoprotein act like LDL, and thus LDL receptor-mediated transcytosis enhances the drug delivery together with nanoparticles across the BBB (see Liu et al., 2006). For this purpose, nanoparticles with size similar to that of LDL (in the range of 20–25 nm) are most effective in drug delivery across the BBB (Kreuter, 2004; Zheng et al., 2005). The smaller nanoparticles degrade more rapidly after their translocation into the endothelial cells causing a rapid release of the drug into the brain. Thus, the size of nanoparticles is an important factor in drug delivery to the brain. However, polysorbate 80 also had its toxic and hemolytic effects that restrict its use in drug delivery to the brain.

Surface charge of nanoparticles and the BBB

Apart from the physical properties of the endothelial cell membranes of brain capillaries and size of the nanoparticles, electrostatic charges play important roles in transport of nanomaterials across the BBB regardless of endocytosis (Hagenbuch and Meier, 2003). Thus, cationic charged molecules occupy anionic areas at the BBB endothelium (Nagy et al., 1983) and increase the endothelial cell permeability probably by disrupting the junctions (Nagy et al., 1981; Hardebo and Kahrstrom, 1985, see below). In *in vitro* studies, the cationized nanoparticles have higher brain translocation compared with anionic or neutral nanoparticles (Fenart et al., 1999). Thus, both the size and the charge of colloidal drug carriers are important factors in determining drug or nanoparticle delivery across the BBB or in brain parenchyma (Sahagun et al., 1990). However, there are little *in vivo* data regarding brain permeability of cationized nanoparticles.

Lockman et al. (2004) recently showed that the nanoparticles with either neutral or low concentrations of anionic surface coating charges do not disrupt the BBB integrity. On the other hand, high anionic and/or cationic charged nanoparticles induce breakdown of the BBB function within <60 s in a dose-dependent manner. This BBB disruption by low concentration of nanoparticles (25 μg/kg) in rats is reversible in nature (Saija et al., 1997).

Koziara et al. (2003) demonstrated that administration of negatively charged nanoparticles in a dose range of 100–200 μg per animal resulted in its accumulation in the cellular matrix without apparent neurotoxicity. However, only free fraction of nanoparticles in formulations is related to BBB opening in presence of anionic nanoparticles (see Koziara et al., 2006). Thus, further studies are needed to find out the optimal concentration of negatively charged nanoparticles that can be achieved in brain without apparent neurotoxicity.

The BBB disruption by cationic nanoparticles may be due to opening of inter-endothelial routes, *i.e.*, widening of tight junctions without their damage (Nagy et al., 1983). The structural changes in tight junction may occur without any decrease in the high electrical resistance of cerebral endothelium. An accumulation of small-sized particles into brain in these experiments suggests that the endothelial cell membrane permeability is not altered (Hart et al., 1987). Increased permeability of [^{14}C] sucrose (that normally does not enter into the brain) across the BBB after exposure of cerebral endothelium to the cationic nanoparticles is in line with this idea (Lockman et al., 2004). Interestingly, in these experiments, accumulation of [^{14}C] sucrose is widespread in several brain regions (Lockman et al., 2004).

Taken together, these observations suggest that since cationic nanoparticles are neurotoxic, neutral and low concentrations of anionic nanoparticles can be used as colloidal drug carriers for enhanced delivery to the brain safely. Thus, surface charges of nanoparticles should be taken into consideration apart from their sizes while targeting the brain for drug delivery in future. It would be interesting to find out whether a combination of nanoparticles having different surface charges can prolong the BBB opening up to 2–3 h. This would facilitate enhanced drug delivery for a long time using colloidal formulations safely and effectively in future.

Nanoparticles, blood–brain barrier and neurotoxicity

Works of Kim et al. (2006a) show that intraperitoneal administration of nanoparticles reach almost all organs in mice in a time-dependent manner. Most of the nanoparticles are taken up by the liver and then redistributed to the other organs, *e.g.*, spleen, lungs, heart and kidney. This study further shows that nanoparticles (<50 nm) are also present in the brain and testes after bypassing the blood–brain and the blood-testis barriers, respectively, without inducing any apparent toxicity (Kim et al., 2006a).

The tight junctions that constitute the BBB have a gap of only 4–6 nm (Kniesel and Wolburg, 2000). Thus, the nanoparticles are likely to pass through the endothelial cell membrane rather than *via* inter-endothelial junctions. Since nanoparticles are lipid insoluble and endothelial cells lack pinocytosis, their penetration into the brain through circumventricular organs lacking a tight barrier is also quite likely (Ambruosi et al., 2005). Using this route, nanoparticles may gain access into the brain without affecting the BBB permeability. However, further studies are needed to clarify these issues.

Nanoparticles disrupt the normal BBB function in vivo

Research from our laboratory shows that intravenous (30 mg/kg), intraperitoneal (50 mg/kg) or intracerebral (20 μg in 10 μl) administration of Ag, Cu or Al nanoparticles ($\approx$ 50–60 nm) open the BBB to Evans blue albumin in rats and mice in a highly selective and specific manner (Sharma et al., 2006b). Leakage of Evans blue dye was observed largely in the ventral surface of the brain and in the proximal frontal cortex 24 h after nanoparticle administration. The dorsal surfaces of cerebellum and the thoracic spinal cord also exhibited mild-to-moderate Evans blue staining. The ventral surface of the spinal cord is mildly stained with Evans blue that was seen predominantly in the thoracic and lumbar segments of the spinal cord (Sharma and Sharma, unpublished observations). However, BBB to Evans blue dye was least affected following intraperitoneal administration of nanoparticles. Interestingly, Al nanoparticles exerted only a minimal effect on the BBB function compared to the Ag and Cu nanoparticles at identical doses (Sharma et al., 2006a).

Opening of the BBB with nanoparticles in our experiments was associated with neurotoxicity in mice and rats in several areas of the brain (results not shown). These neurotoxic effects were most pronounced following intravenous and intracerebral administration of Ag nanoparticles followed by Cu and Al nanoparticles (Sharma HS, unpublished observations). In general, the nerve cell, glial cell and myelin changes were most prominent in the brain or spinal cord areas exhibiting BBB disruption (Sharma HS, unpublished observation). This indicates that breakdown of the BBB by nanoparticles is instrumental in neurotoxicity. A lack of BBB leakage and neurotoxicity in animals following intraperitoneal administration of nanoparticles is in line with this idea (Sharma et al., 2006b).

These observations clearly suggest that the route of administration of nanoparticles and their chemistry are important factors in inducing BBB dysfunction and brain pathology. However, further studies using dose- and size-related effects of nanoparticles are needed to understand their effects on neurotoxicity *in vivo*.

Nanoparticles induce oxidative stress and formation of free radicals

There are reasons to believe that oxidative stress and formation of free radicals by nanoparticles is somehow responsible for cell membrane damage leading to BBB disruption and brain pathology. Nanoparticles, *e.g.*, cobalt, carbon tubes, quantum dots and ultrafine particles (20–80 nm), induce production of ROS, especially following concomitant exposure to light, ultraviolet or transition metals (Brown et al., 2000, 2001; Li et al., 2003; Derfus et al., 2004; Joo et al., 2004; Oberdörster, 2004a; Sayes et al., 2004; Shvedova et al., 2004, 2005). Since oxidative stress alone is able to induce brain pathology (Sharma et al., 2003a, b, 2006b), it is likely that nanoparticles induce ROS activation and play further crucial roles in BBB disruption and neurotoxicity.

Nanoparticles of various sizes and different chemical compositions influence mitochondria to stimulate ROS overproduction and interfere with antioxidant defense mechanism (De Lorenzo, 1970; Gopinath et al., 1978; Foley et al., 2002; Li et al., 2003; Rodoslav et al., 2003). Furthermore, photo excitation; release of free electrons; creating redox active intermediates, particularly *via* metabolism of cytochrome P450s and/or inflammatory responses that release oxygen radicals through macrophages could also contribute to ROS production by nanoparticles (see Sharma et al., 2006a). However, the detailed mechanisms of nanoparticle-induced ROS production is not yet well understood.

Nanoparticles enhance tumor temperature after thermotherapy

Recent advancement in nanotechnology resulted in novel therapeutic strategies for the treatment of cancer by enhancing tumor temperatures using magnetic nanoparticles (see Jordan et al., 2006). The idea of "Thermotherapy" using nanoparticles for tumor treatment is based on controlled heating of intratumorally administered magnetic nanoparticles (see Plotkin et al., 2006). Local administration of magnetic nanoparticles within the tumor results in a 3° to 4°C higher temperature compared with tumors without the nanoparticles. With magnetic nanoparticles a selectively high tumor temperature can be achieved for effectively killing the cancer cells. Interestingly, nanoparticle-induced high temperature in tumors also limits proliferation by inducing localized necrosis around the peritumoral zones (see Plotkin et al., 2006).

Animal experiments show that two sessions of thermotherapy following a single intratumoral injection of aminosilane-coated iron oxide nanoparticles, but not dextran-coated particles, result in a 4.5-fold prolongation in survival. Morphological studies revealed large necrotic areas close to nanoparticle deposits. A decrease in proliferation rate and profound reactive astrogliosis near the tumor site was also observed. These observations point out that the effects of localized interstitial thermotherapy are enhanced by magnetic nanoparticles leading to an effective treatment for malignant brain tumors.

Tanaka et al. (2005a, b) further demonstrate that magnetic nanoparticle-induced hyperthermia enhances antitumor immunity. This effect is potentiated by adjuvant therapy including cytokines. Since dendritic cells (DCs) are potent antigen-presenting cells, they could play important roles in regulating immune responses in cancer. Thus, therapeutic effects of hyperthermia enhanced by magnetic nanoparticles in combination with DC immunotherapy appear to be a promising tool for future cancer therapy. Magnetite cationic liposome nanoparticles (MCLN), which have a positive surface charge and generate heat in an alternating magnetic field (AMF), were employed in a mouse model of EL4 T-lymphoma (Tanaka et al., 2005a). The MCLN were administered into an EL4 nodule in C57BL/6 mice and subjected to AMF for 30 min resulting in tumor temperature up to 45°C, which was maintained by regulating the magnetic field intensity. This hyperthermia treatment was repeated twice at 24 h intervals followed by injection of immature DCs directly into the EL4 nodule. The enhanced heat treatment using MCLN resulted in complete regression of tumors in six out of eight mice compared with one out of eight mice without nanoparticles. These observations suggest that enhanced heat treatment with nanoparticles provides novel therapeutic strategies for even advanced malignancies in clinical situations in the future. Optimal treatment schedules with other potential combination strategies using different magnetic nanoparticles are thus needed to find out an effective antitumor therapy in clinical settings.

Our investigations on the influence of nanoparticles on brain dysfunction

This section is based on our observations on the *in vivo* effects of nanoparticles derived from metals on the brain functions in normal and in heat-stressed rats. The salient new findings show aggravation of neurotoxicity in rats subjected to WBH compared with naïve animals (see below).

Methodological consideration

We examined the effect of three nanoparticles derived from metals, *e.g.*, Al, Ag and Cu (50–60 nm sizes), on brain function in rats kept at normal room temperature or subjected to WBH. The results are compared with heat-stressed rats without receiving any nanoparticles.

All animal experiments described in this review were conducted according to National Institute of Health (NIH), United States Government guidelines for care, handling and maintenance of animals and approval by Local Institutional Ethics Committee for Animal Care and Research.

Body temperature and handling of animals

Experiments were carried out on male Sprague Dawley rats (weighing between 150 and 200 g; age 12–15 weeks) housed at normal room temperature ($21\pm1°C$) with 12 h light and 12 h dark schedule. Food and tap water were provided *ad libitum*. The animals were maintained for 1 week in the laboratory to adapt to the local environment and handling (see Sharma, 2006a; Sharma, Chapter 10 in this volume). During the maintenance period, each animal was taken out from the cage daily to record body weight, rectal and tympanic membrane temperatures using clinical digital thermometer or thermistor probes (Sharma, 2006a).

Studies show that the tympanic membrane temperature is only $<0.5°C$ lower than the brain temperature and thus represents the hypothalamic changes in the temperature (Sharma et al., 1998a, b). On the other hand, temperatures recorded in other peripheral parts of the body, *e.g.*, skin temperature or muscle temperatures, can be influenced by the ambient air temperature (Sharma, 1982). The deep visceral temperature, *i.e.*, core body temperature, can be recorded using a thermistor probe inserted deep into the rectum (~ 4–6 cm, see Sharma, 1999). Normally the core temperature is $\sim <1°C$ from the average brain temperature (see Benzinger, 1969).

After 4 days of handling, recording of body or tympanic membrane temperatures did not show stress-induced fluctuations (see Sharma, 2006a; Sharma, Chapter 10 in this volume).

Selection of nanoparticles from metals

We have used three commonly occurring metal nanoparticles in the environment to which humans are normally exposed in daily life or during work-related activities.

(a) Cu nanoparticles (Cu $\approx$ 50 nm)
The natural sources of Cu exposure to humans and the environment are windblown dust, volcanoes, decaying vegetation, forest fires and sea spray. Other emission sources for industrialized nations include smelters, iron foundries, power stations and municipal incinerators. Agricultural use of Cu products accounts for 2% of Cu released into the soil. The average background concentrations of Cu in the air in rural areas normally range between 5 and 50 ng/m^3. This concentration is sufficient enough in the environment to affect human exposure (WHO, Geneva, 1998).
In addition, Cu is widely used in cooking utensils and water distribution systems, as well as fertilizers, bactericides, fungicides, algaecides and antifouling paints. Cu is also utilized in several animal feed additives and growth promoters, as well as for disease control in livestock and poultry. Cu is

employed in production of wood preservatives, electroplating, azo-dye manufacture, petroleum refining and as a mordant for textile dyes and in petroleum refining as well. (Source: International Programme on Chemical Safety. Environmental Health Criteria 200; Copper, WHO, Geneva, 1998.)

(b) Al nanoparticles (Al ≈ 60 nm)

Al is not an essential element for mammals (Venugopal and Luckey, 1978). However, acid rain increased the availability of aluminum to biological systems (Goyer, 1991). Al metal is used as a structural material in the construction, automotive and aircraft industries, in the production of metal alloys and in the electrical industry in power lines, insulated cables and wiring. Other uses of aluminum metal include cooking utensils, decorations, fencing, highway signs, cans, food packaging, foil and dental crowns and dentures (Agency for Toxic Substances and Disease Registry, ATSDR, 2006). The fine Al powder is used as flashlight in photography, in explosives, fireworks and aluminum paints as well as for absorbing occluded gases in the manufacture of steel (Agency for Toxic Substances and Disease Registry, ATSDR, 2006, Toxicological Profile for Aluminum. U.S. Department of Health and Human Services. Public Health Service).

(c) Silver nanoparticles (Ag ≈ 50 nm)

Silver is a rare metal that occurs naturally in the earth's crust and is released into the environment from several industrial sources (Agency for Toxic Substances and Disease Registry, ATSDR, 1990). Human exposure to silver may occur orally, dermally or by inhalation. Silver and silver compounds are mainly used in photographic materials, electroplating, electrical conductors, dental alloys, solder and brazing alloys, paints, jewelry, coins and mirror production. Silver is also used as an antibacterial agent and in water purification (Nordberg and Gerhardsson, 1988).

Thus, it appears that these commonly occurring nanoparticles may influence human health and/or brain function under normal and altered environmental conditions.

Treatment with nanoparticles

The metal nanoparticles (Cu, Al and Ag, ≈ 50–60 nm in size) were suspended in Tween 80 and administered separately in three different groups of rats intraperitoneally at a dose of 50 mg/kg (weight by volume) once daily for 7 days. The animals were observed for behavioral changes at every 24 h interval and their body weight, rectal and tympanic membrane temperatures were recorded daily. In addition, in some animals the mean arterial blood pressure (MABP), blood gases, heart rate and respiration were also monitored (see Sharma, 1999, 2004; see Table 1). Furthermore, several behavioral and functional parameters such as cognitive dysfunction, blood flow, BBB permeability and brain pathology were also determined (see below).

Effect of nanoparticles in normal animals

Normal animals treated with nanoparticles did not exhibit gross abnormal behavior. However, physiological and behavioral functions show minor modifications 1 week after nanoparticle administration (see Tables 1 and 2).

Body temperature and physiological parameters

Treatment with nanoparticles slightly increased the rectal and/or tympanic membrane temperatures (0.1° to 0.25°C) compared with saline administration (Table 1). On the other hand, there was a significant decrease in body weights of animals (7–13 g) in nanoparticle-treated rats compared with the saline controls (Table 1).

A significant increase in respiratory cycle (7–10 cycles/min) and heart rate (30–40 beats/min) was observed in nanoparticle-treated rats (Table 1). In these animals, MABP showed a significant increase (12–19 Torr) after 1 week of nanoparticles administration (Table 1). A slight but significant decrease in the arterial PaO_2 (2–2.5 Torr) and a

Table 1. Effect of nanoparticles on physiological variables in normal rats and animals subjected to 4 h whole body hyperthermia (WBH) at 38° C. For details see text

Experiment type	n	Temperature (°C) Rectal	Temperature (°C) Tympanic membrane	Body (Wt/g)	MABP (Torr)	Arterial pH	PaO$_2$ (Torr)	PaCO$_2$ (Torr)	Heart Rate (beats/min)	Respiration (cycles/min)
A. Normal animals										
Saline	6	37.41±0.45	38.40±0.12	255±12	120±8	7.40±0.02	81.34±0.12	35.56±0.21	315±12	60±4
Cu	8	37.67±0.23	38.68±0.21	242±8*	132±12*	7.30±0.02	79.34±0.89	37.34±0.33*	345±8*	68±6*
Ag	8	37.58±0.23	38.49±0.32	248±10*	134±15*	7.20±0.08	78.56±0.22*	36.54±0.34*	351±7	70±8*
Al	7	37.56±0.12	38.48±0.32	246±12*	139±8*	7.33±0.08	78.89±0.89*	36.34±0.23	355±12*	67±4*
B. Heat Stressed Animals (38°C for 4h)										
Saline	5	40.48±0.15^{aa}	42.46±0.14^{aa}	248±8^{a}	90±6^{aa}	7.30±0.06^{a}	83.24±0.56^{a}	34.48±0.52	425±11^{aa}	74±7^{a}
Cu	9	41.67±0.13^{bb}	43.38±0.21^{b}	242±8^{b}	82±8^{b}	7.33±0.06*	82.44±0.21*	33.16±0.34^{b*}	428±11*	70±8*
Ag	7	41.27±0.19^{b}	43.48±0.24^{bb}	238±7^{b}	70±6^{bb*}	7.21±0.04^{b}	83.18±0.14*	32.43±0.66^{b*}	418±12*	68±6^{b*}
Al	8	40.78±0.24*	42.89±0.11**	234±7^{b*}	74±4^{b*}	7.28±0.06*	82.36±0.26^{b*}	33.64±0.18*	435±9^{b*}	68±6^{b*}

The nanoparticles were suspended in Tween 80 and administered separately intraperitoneally in rats daily once in a dose of 50 mg/kg (weight/volume). Values are Mean±SD from 5 to 9 animals at each data point. * = $P<0.05$; ** = $P<0.01$ compared from saline in normal animals, ^{a} = $P<0.05$; ^{aa} = $P<0.01$, compared from 4 h saline treatment in normal vs. heat stressed. ^{b} = $P<0.05$, ^{bb} = $P<0.01$ compared from saline treated heat stressed animals. ANOVA followed by Dunnett's test for multiple group comparison from one control group (Data from Sharma HS, unpublished observations).

Table 2. Effect of nanoparticles on cognitive, motor function and heat-stress symptoms in normal rats and animals subjected to 4 h whole body hyperthermia (WBH) at 38° C. For details see text

Experiment type	n	Cognitive function			Motor function				Salivation saliva spread	Prostration grade	Micro-hemorrhage in stomach (spots, no.)
		Rota Rod (180 s)	Grid-walking total steps (60 s)	Placement error (60 s) %	Inclined plane angle° (5 s)	Foot-print analysis			Snout (mm)		
						Hind-feet distance (mm)	Stride length (mm)				
A. Normal animals											
Saline	6	120±8	40±6	nil	60±4	50±5	130±6		nil	nil	nil
Cu	8	118±10	35±9	2±1	55±8	45±8	120±12		nil	nil	2±4
Ag	8	110±12	32±9	4±3	52±6*	42±11	118±8*		nil	nil	8±4
Al	8	116±9	36±8	3±4	56±8	46±12	124±16		nil	nil	2±3
B. Heat Stressed Animals (38 C for 4 h)											
Saline	8	80±4**	20±5**	35±6**	35±6**	85±6**	75±8**		22±6**	+++	48±12#**
Cu	8	65±12a	18±6	38±8	38±8	78±8a	80±12		18±6	+++	55±8#
Ag	8	68±6a	16±4a	40±6	30±5*	80±8a	72±6		20±6	++++	60±12#a
Al	8	74±6	18±5	36±7	36±8	81±5	72±9		20±8	+++	36±8#

= many microhaemorrhages seen. * = $P<0.05$; ** = $P<0.01$, χ^2 test, compared from saline treated control; a = Student's unpaired t-test compared from saline treated heat stressed group.
The nanoparticles were suspended in Tween 80 and administered separately intraperitoneally in rats daily once in a dose of 50 mg/kg (weight/volume). Data are Mean±SD from 5 to 8 animals at each data point. Nil = absent. Data from Sharma HS unpublished observations.

mild increase in the arterial $PaCO_2$ (1–1.5 Torr) are also seen. However, the arterial pH was not affected by nanoparticle treatment (Table 1).

Thus, chronic treatment with nanoparticles affects important physiological variables in animals. It appears that Cu and Ag nanoparticles have strong effects on body temperature and physiological variables compared with Al nanoparticle.

Cognitive, sensory, motor functions and thermal stress symptoms

Influence of nanoparticles on cognitive and motor functions was examined on well-trained rats on the eighth day in a blinded fashion (see Sharma, 2006a).

Cognitive functions

(a) *The Rota-Rod performance.* The Rota-Rod treadmill is used to assess cognitive functions and fatigue in rats using a fixed speed of 16 rotations per minute (RPM). Rats not falling off the Rota-Rod for 2 min were considered normal during a 3 min session (see Sharma, 2006a).

(b) *Grid walking.* This test is used to determine changes in locomotor behavior, gait and overall walking skill using an elevated level (30°) of stainless steel grid with a mesh size of 30 mm (Sharma, 2006a). The total number of paired steps, *i.e.*, placement of both forelimbs, was estimated during 1 min period. The number of misplaced limbs error, *viz.*, the forelimbs fell through the grid, was also recorded (see Sharma, 2006a).

Motor functions

(a) *Inclined plane test.* The motor function disturbances were determined by placing rats on inclined plane test using an angle in such a way that the animal could stay on the plane for 5 s without falling (Sharma, 2005b, 2006a).

(b) *Foot-print analysis.* The gait and motor disturbances were determined using foot-print analysis obtained on a paper coated with bromophenol blue dissolved in acetone (see Sharma, 2006a). The distance (mm) between hind paws was measured from the base of the central pads. The stride length (mm) was assessed by measurement of the distances between hind paws in two consecutive steps (Sharma, 2005b, c, 2006a).

Administration of nanoparticles did not influence cognitive functions significantly. However, mild but significant alterations in motor behavior were observed (Table 2). A significant reduction in inclined plane angle ($-8°$) as well as in stride length (12 mm) in Ag nanoparticle-treated rats was seen (Table 2). A mild reduction of inclined plane angle (4–5°) and stride length (6–10 mm) was also seen in animals treated with Cu and Al nanoparticles. These observations suggest that nanoparticle treatment in normal animals could influence motor function.

Thermal stress symptoms

(a) *Salivation.* Salivation is a well-known measure of heat dissipation in rodents caused by heat stress (see Damas, 1994; Sharma 2005a, b; Whyte et al., 2006; Sharma, Chapter 10 in this volume). The spread of saliva over the snout is directly proportional to the rise in body temperature (Sharma, 2005b, c, 2006a, b). The presence and absence of such salivation was recorded in animals subjected to nanoparticle treatment.

(b) *Prostration.* A rise in body temperature results in sluggish movement and alteration in the locomotor behavior caused by stress associated with hyperthermia (see Ajalat, 1960; Harata, 1970; Galina et al., 1983; Sharma, 2005b). This behavior can be graded into four stages (1 = normal locomotion; 2 = severe reduction in locomotion, but no support flattening posture; 3 = animals lay prostrate often in the cage and normally do not move for some times; 4 = animal lay prostrate in cage and often do not move even after gentle pushing) (for details see Sharma, 1982, 1999, 2004, 2005a, b; see Table 2).

(c) *Microhemorrhages in stomach wall.* All kinds of stressors are known to induce gastric ulceration in the stomach (see Evangelista, 2006; Lou et al., 2006). The intensity of gastric ulcerations depends on the magnitude and intensity of the stressor. Previously, we have shown that heat stress-induced hyperthermia causes massive gastric ulcerations (see Sharma, 1982, 2005a, b) in the mucosal wall of stomach, which were examined at post-mortem under a magnifying lens (Sharma, 1999).

Administration of nanoparticles did not influence thermal stress symptoms, *e.g.*, salivation or prostration (see Table 2). However, mild incidences of gastric ulceration were noted in animals in nanoparticle-treated group (see Table 2). This indicates that administration of nanoparticles may cause mild stress symptoms leading to gastric ulceration.

Blood–brain barrier permeability

The influence of nanoparticles on the BBB function was examined in animals using two different kinds of *in vivo* protein tracers, *i.e.*, Evans blue and $^{[131]}$Iodine that bound to endogenous serum albumin (Sharma, 2005a, see Sharma, Chapter 10 in this volume). These tracers were administered into the right femoral vein and allowed to circulate for 5–10 min. After intravascular tracers were washed out with 0.9% saline through heart, different brain and spinal cord regions were dissected out and leakage of tracers was measured (see Sharma, 1987).

Administration of nanoparticles slightly but significantly increased the extravasation of Evans blue and radioiodine in the brain and spinal cord (Table 3). Extravasation of radioiodine tracer was more pronounced compared with Evans blue leakage in the brain or spinal cord (Table 3). It appears that Ag and Cu nanoparticles are more potent in causing breakdown of the BBB compared with Al (Table 3).

Regional BBB permeability

A selective increase in regional BBB permeability is noted in animals subjected to chronic nanoparticle administration (Table 4). Thus, extravasation of radioiodine was most pronounced in the cerebellum (217–233%) followed by hippocampus (85–108%), brain stem (78–111%), cortex (76–100%) and thalamus and hypothalamus (25–36%, see Table 4). In general, Ag was the most potent nanoparticle inducing extensive leakage of radiotracer in most brain areas. On the other hand, Al was the least effective in opening the BBB to radiotracer in all the brain regions, except cerebellum (Table 4).

These observations are the first to demonstrate that chronic administration of nanoparticles, derived from metals, is capable of inducing BBB breakdown in rats. This increase in BBB dysfunction caused by nanoparticles is dependent on the type of metal used and is highly specific in different brain regions. The possible functional significance of such a selective increase in BBB permeability by metal nanoparticles is unclear.

Brain and spinal cord blood flow

To understand the influence of nanoparticles on cerebral ischemia, brain and spinal cord blood flow was measured using $^{[125]}$Iodine-labeled carbonized microspheres (see Sharma, 1987). Animals treated with nanoparticles exhibited a significant reduction in the brain and spinal cord blood flow compared with controls. This effect is most pronounced in Cu nanoparticle-treated groups (Table 3). However, the magnitude of flow reduction alone is not sufficient enough to induce BBB breakdown (see Sharma, 2004).

These novel observations suggest that chronic administration of nanoparticles alter microcirculation in the brain and spinal cord. Whether the nanoparticles exert a direct effect on the cerebral microvessels to impair blood flow in the CNS or through some neurochemical mediators is not known.

Brain edema formation

Breakdown of BBB to large molecules, *e.g.*, protein tracers, is often associated with formation of vasogenic edema in the brain and spinal cord (Sharma et al., 1998a, 2006a, b, c; Sharma, 2004, 2005a, b, c, d). We examined brain and spinal cord water content to evaluate edema formation in animals receiving nanoparticles.

Table 3. Effect of nanoparticles on BBB permeability, cerebral blood flow and brain edema in normal rats and animals subjected to 4 h whole body hyperthermia (WBH) at 38° C. For details see text

Experiment type	n	BBB permeability		BSCB Permeability		Blood flow (ml/g/min)		Edema formation		Volume Swelling	
		Evans blue (mg %)	[131]Iodine (%)	Evans blue (mg %)	[131]Iodine (%)	Brain	Spinal cord	Brain water (%)	Spinal cord water (%)	% Brain	% Spinal cord
A. Normal animals											
Saline	6	0.28 ± 0.08	0.36 ± 0.06	0.22 ± 0.04	0.30 ± 0.05	1.16 ± 0.06	0.94 ± 0.04	74.23 ± 0.86	65.25 ± 0.21	nil	nil
Cu	8	$0.56 \pm 0.13^{**}$	$0.68 \pm 0.11^{**}$	$0.67 \pm 0.08^{**}$	$0.76 \pm 0.10^{**}$	$0.98 \pm 0.08^{*}$	$0.84 \pm 0.09^{*}$	$75.14 \pm 0.13^{*}$	$66.16 \pm 0.08^{*}$	3.53	2.61
Ag	8	$0.64 \pm 0.12^{**}$	$0.74 \pm 0.11^{**}$	$0.64 \pm 0.08^{**}$	$0.75 \pm 0.05^{**}$	$0.96 \pm 0.10^{*}$	$0.88 \pm 0.06^{*}$	$75.04 \pm 0.08^{*}$	$66.08 \pm 0.11^{*}$	3.14	2.38
Al	8	$0.46 \pm 0.08^{*}$	$0.58 \pm 0.06^{*}$	$0.58 \pm 0.06^{*}$	$0.64 \pm 0.08^{*}$	$1.04 \pm 0.06^{*}$	$0.88 \pm 0.08^{*}$	$74.84 \pm 0.13^{*}$	$65.75 \pm 0.31^{*}$	2.36	1.43
B. Heat Stressed Animals (38 °C for 4 h)											
Saline	5	2.06 ± 0.25^{aa}	2.86 ± 0.24^{aa}	1.21 ± 0.21^{aa}	1.46 ± 0.22^{aa}	0.76 ± 0.08^{aa}	0.70 ± 0.06^{a}	81.34 ± 0.62^{aa}	68.23 ± 0.21^{aa}	27.59	8.57
Cu	6	2.68 ± 0.33^{b}	3.17 ± 0.26^{b}	2.10 ± 0.8^{b}	2.83 ± 0.08^{b}	0.70 ± 0.06^{b}	$0.65 \pm 0.07^{b*}$	$83.12 \pm 0.34^{b*}$	$68.76 \pm 0.43^{*}$	34.49	10.01
Ag	8	2.64 ± 0.33^{b}	3.14 ± 0.12^{bb}	1.87 ± 0.6^{b}	$2.08 \pm 0.07^{bb*}$	0.74 ± 0.08^{b}	$0.68 \pm 0.06^{*}$	$82.32 \pm 0.26^{b*}$	$68.56 \pm 0.28^{*}$	31.39	9.52
Al	8	$2.13 \pm 0.36^{*}$	$3.08 \pm 0.21^{b*}$	$1.46 \pm 0.32^{b*}$	$1.89 \pm 0.42^{b*}$	$0.72 \pm 0.06^{*}$	$0.68 \pm 0.10^{b*}$	$82.16 \pm 0.24^{*}$	$68.34 \pm 0.21^{b*}$	30.77	8.89

The nanoparticles were suspended in Tween 80 and administered separately intraperitoneally in rats daily once in a dose of 50 mg/kg (weight/volume).
Values are Mean ± SD from 5 to 9 animals at each data point. * = $P < 0.05$; ** = $P < 0.01$ compared from saline in normal animals, ^{a} = $P < 0.05$; ^{aa} = $P < 0.01$, compared from 4 h saline treatment in normal vs. heat stressed. ^{b} = $P < 0.05$, ^{bb} = $P < 0.01$ compared from saline treated heat stressed animals. ANOVA followed by Dunnett's test for multiple group comparison from one control group (Data from Sharma HS, unpublished observations).

Table 4. Effect of nanoparticles on regional BBB permeability and regional brain edema in normal rats and animals subjected to 4 h whole body hyperthermia (WBH) at 38°C. For details see text

Experiment type	BBB permeability [131]Iodine %				Regional brain edema (water content %)			
	Cerebral cortex	Hippocampus + caudate N	Cerebellum	Thalamus + hypothalamus	Brain Stem	Cerebral cortex	Hippocampus + caudate N	Cerebellum	Thalamus + hypothalamus	Brain Stem

Experiment type	Cerebral cortex	Hippocampus + caudate N	Cerebellum	Thalamus + hypothalamus	Brain Stem	Cerebral cortex	Hippocampus + caudate N	Cerebellum	Thalamus + hypothalamus	Brain Stem
A. Normal animals ($n = 6$ to 8)										
Saline	0.34 ± 0.06	0.26 ± 0.04	0.12 ± 0.04	0.56 ± 0.07	0.18 ± 0.04	72.34 ± 0.21	80.38 ± 0.21	78.32 ± 0.12	76.33 ± 0.43	66.35 ± 0.21
Cu	$0.62 \pm 0.11^{**}$ (+82%)	$0.48 \pm 0.08^{**}$ (+85%)	$0.38 \pm 0.05^{**}$ (+217%)	$0.71 \pm 0.10^{**}$ (+27%)	$0.38 \pm 0.06^{*}$ (+111%)	$74.38 \pm 0.19^{*}$ (+2.82%)	$81.34 \pm 0.23^{*}$ (+1.19%)	$80.33 \pm 0.14^{*}$ (+2.56%)	$78.18 \pm 0.23^{*}$ (+2.42%)	$67.34 \pm 0.12^{*}$ (+1.49%)
Ag	$0.68 \pm 0.14^{**}$ (+100%)	$0.54 \pm 0.06^{**}$ (+108%)	$0.40 \pm 0.06^{**}$ (+233%)	$0.76 \pm 0.08^{**}$ (+36%)	$0.36 \pm 0.07^{*}$ (+100%)	$74.21 \pm 0.16^{*}$ (+2.58)	$81.65 \pm 0.27^{*}$ (+1.57%)	$80.67 \pm 0.34^{*}$ (+3.00%)	$78.23 \pm 0.13^{**}$ (+2.48%)	$67.67 \pm 0.21^{*}$ (+1.98%)
Al	$0.60 \pm 0.09^{*}$ (+76%)	$0.50 \pm 0.08^{**}$ (+92%)	$0.41 \pm 0.03^{**}$ (+242%)	$0.70 \pm 0.06^{**}$ (+25%)	$0.32 \pm 0.08^{**}$ (+78%)	$73.67 \pm 0.23^{*}$ (+1.83%)	$81.75 \pm 0.43^{*}$ (+1.70%)	$80.56 \pm 0.28^{*}$ (+2.86%)	$77.65 \pm 0.21^{*}$ (+1.72%)	$67.04 \pm 0.14^{*}$ (+1.04%)
B. Heat Stressed Animals (38°C for 4 h, $n = 6$ to 8)										
Saline	1.86 ± 0.15^{aa} (+447%)	2.06 ± 0.28^{aa} (+692%)	2.01 ± 0.11^{aa} (+1575%)	2.56 ± 0.24^{aa} (+357%)	0.76 ± 0.08^{aa} (+322%)	$78.34 \pm 0.12^{**}$ (+8.29%)	82.34 ± 0.12^{aa} (+2.43%)	81.43 ± 0.08^{aa} (+3.97%)	$78.78 \pm 0.21^{**}$ (+3.21%)	$70.48 \pm 0.08^{**}$ (+6.22%)
Cu	2.48 ± 0.13^{b} (+612%)	2.87 ± 0.16^{b} (+1004%)	2.20 ± 0.18^{b} (+1733%)	2.87 ± 0.18^{b} (+412%)	1.45 ± 0.06^{b} (+706%)	$79.54 \pm 0.18^{**}$ (+9.95%)	$83.87 \pm 0.18^{b*}$ (+4.34%)	$82.04 \pm 0.13^{*}$ (+4.74%)	$79.56 \pm 0.14^{**}$ (+4.23%)	$71.56 \pm 0.22^{**}$ (+7.85%)
Ag	2.34 ± 0.14^{b} (+588%)	3.01 ± 0.21^{bb} (+1058%)	2.37 ± 0.16^{b} (+1875%)	$2.76 \pm 0.17^{bb*}$ (+393%)	1.70 ± 0.06^{b} (+844%)	$79.16 \pm 0.23^{*}$ (+9.43%)	$83.02 \pm 0.16^{b*}$ (+3.28%)	$81.76 \pm 0.12^{*}$ (+4.39%)	$79.65 \pm 0.08^{*}$ (+4.34%)	$71.48 \pm 0.06^{**}$ (+7.73%)
Al	$2.13 \pm 0.12^{*}$ (+526%)	$2.78 \pm 0.14^{b*}$ (+969%)	$2.16 \pm 0.13^{b*}$ (+1700%)	$2.89 \pm 0.12^{b*}$ (+416%)	$0.96 \pm 0.08^{*}$ (+433%)	$78.97 \pm 0.21^{*}$ (+9.165%)	$82.96 \pm 0.13^{*}$ (+3.20%)	$81.84 \pm 0.17^{b*}$ (+4.49%)	79.23 ± 0.11 (+3.79%)	$71.52 \pm 0.21^{*}$ (+7.79%)

The nanoparticles were suspended in Tween 80 and administered separately intraperitoneally in rats daily once in a dose of 50 mg/kg (weight/volume). Values are Mean ± SD from 5 to 9 animals at each data point. $^{*} = P < 0.05$; $^{**} = P < 0.01$ compared from saline in normal animals, $^{a} = P < 0.05$; $^{aa} = P < 0.01$, compared from 4 h saline treatment in normal vs. heat stressed. $^{b} = P < 0.05$, $^{bb} = P < 0.01$ compared from saline treated heat stressed animals, ANOVA followed by Dunnett's test for multiple group comparison from one control group (Data from Sharma HS, unpublished observations).

Our results are the first to show that chronic administration of metal nanoparticles induces brain and spinal cord edema formation (0.5–1%; Table 3). Analysis of volume swelling (%f) based on alteration in the water content (∼1% increase in water content represents more than 3% volume swelling) revealed 2–4% increase in %f in brain and ∼1–3% increase in %f in the spinal cord (Table 3). The Ag nanoparticles exert most pronounced effects on edema formation compared with Cu and Al (Table 3). This suggests that breakdown of the BBB caused by nanoparticles is associated with vasogenic edema formation.

Profound increase in volume swelling of the brain and spinal cord in nanoparticle-treated group is likely to precipitate neurological consequences and neurodegenerative changes. Thus, ingestion or inhalation of nanoparticles from the environment for long periods may induce brain dysfunction, probably by modifying the BBB function leading to neurodegeneration.

Cellular changes in the brain and spinal cord

Disruption of BBB allows entry of several toxic substances that are normally excluded from the CNS (Sharma, 2004) resulting in alterations in cellular structure and function (see Sharma et al., 1998a, b; Sharma, 2006a, b). Using morphological approaches, we examined changes in neuronal, glial, axonal and endothelial cells in animal treated with nanoparticles.

Selective morphological alterations in the neural and non-neural cells in different brain and spinal cord regions were observed in animals treated with nanoparticles (Sharma HS, unpublished observations). Distortion of nerve cells in different brain regions is common in nanoparticle-treated group. Nissl staining showed condensed cytoplasm and chromatolysis in many nerve cells (Table 5). Damage to nuclear membranes and dense karyoplasm with eccentric nucleoli are common (Table 5). A general sponginess and edema are frequent in brain and spinal cord regions showing BBB leakage. These intracellular alterations are very similar to those observed in neurodegeneration.

The non-neural cells are also affected by nanoparticle treatment. Thus, light and electron microscopic studies demonstrate profound changes in astrocytes and endothelial cells in the brain (results not shown). Upregulation of glial fibrillary acidic protein (GFAP) immunoreactivity, a specific marker of astrocytes, is observed in many brain and spinal cord regions after nanoparticle treatment (Table 5). In most cases prominent gliosis around the perivascular regions is a frequent finding (results not shown).

Infiltration of lanthanum within the endothelial cell cytoplasm as well as in the vesicular profiles is frequent in nanoparticle-treated rats (Table 5). Structural alterations in the endothelial cells especially at the endothelial–glial interface are also prominent (result not shown).

Degradation of myelin is evident in several brain areas in nanoparticle-treated animals as seen by Luxol Fast Blue staining and myelin basic protein (MBP) immunoreactivity. Distortion and/or loss of myelin are present in brain and spinal cord regions showing BBB disruption and edema formation (Table 5).

These structural changes are most pronounced in animals that received Ag nanoparticle followed by Cu and Al (see Table 5). Taken together, our observations provide the first experimental evidence for nanoparticle-induced brain pathology *in vivo*.

Effect of nanoparticles on heat-stressed animals

In general, WBH in nanoparticle-treated rats worsened functional and pathological outcomes. This effect is most pronounced in Cu and Ag nanoparticle-treated rats compared with Al. However, Al nanoparticle-treated rats also exhibited considerably higher pathological and functional disturbances following WBH compared with saline-treated group.

Body temperature changes

Nanoparticle-treated rats showed a higher increase in rectal (+3.3° to +4.2°C) and tympanic membrane (+4.4° to +5°C) temperatures following WBH compared with saline-treated rats (body

Table 5. Effect of nanoparticles on morphological changes in the neurons, astrocytes, myelin and endothelial cells in normal rats and animals subjected to 4 h whole body hyperthermia (WBH) at 38°C. For details see text. Semiquantitative grading was done by two independent workers in a blinded fashion and was averaged in the final score

Experiment type	Neuronal reaction					Glial reaction		Axonal reaction		Lanthanum extravasation	
	Distortion cell shape	Chromatolysis + degeneration	Nuclear damage	Eccentric + nucleolus	Sponginess	Reactive gliosis	Perivascular + gliosis	Loss of myelin	Vesiculation of myelin	Endothelial cell cytoplasm	Basal lamina
A. Normal animals ($n = 6$ to 8)											
Saline	nil	nil	nil	nil	nil	+/−	+/−	nil	+/−	nil	nil
Cu	++	+/−	+	+/−	+	++	++	++	+	++	+/−
Ag	+++	+	+	+	+	++	++	+++	++	++	+/−
Al	+	++	+	++	+	++	+++	+	+±	+	+/−
B. Heat Stressed Animals (38°C for 4h, $n = 6$ to 8)											
Saline	++++	++++	+++	+++	+++	+++++	+++	+++	+++	+++	+++
Cu	+++++	++++	+++++	+++++	++++	+++++	+++++	+++++	+++++	++++	++++
Ag	++++	++++	++++	++++	++++	+++++	+++++	+++++	+++++	++++	+++
Al	++++	++++	+++	+++	+++	++++	++++	++++	+++	++++	+++

nil = absent; **+++** = severe; **++++** = moderate; **++** = mild; **+** = present; **+/−** = possible.

The nanoparticles were suspended in Tween 80 and administered separately intraperitoneally in rats daily once in a dose of 50 mg/kg (weight/volume). Light microscopy and immunohistochemistry was used to asses neuronal, glial and myelin damage. Ultrastructural changes were used to determine eccentric nucleolus, perivascular gliosis, myelin vesiculation and lanthanum extravasation. Lanthanum as dark electron dense particle can easily be seen within endothelial cell cytoplasm or in the basal lamina as dark particulate deposits under transmission electron microscope, for details see Chapter 10 and 15 in this volume (Data from Sharma HS, unpublished observations).

T +3.0°C; tympanic membrane T +4°C). The effect was most marked in Cu and Ag (body T +0.8° to 1.3°C; tympanic membrane T +0.9° to 1°C) treated nanoparticles compared with Al (body T +0.3°C; tympanic membrane T +0.4°C) (see Table 1B). This suggests that nanoparticles are able to enhance WBH-induced hyperthermia depending on their chemical composition.

Changes in body weight

There is also enhanced reduction in body weight in nanoparticle-treated rats after WBH (Table 1). The most pronounced change is seen in Al-treated nanoparticles (−22 g) followed by Ag (−18 g) and Cu (−13 g; see Table 1). These observations suggest that perception of heat stress and activation of evaporative heat loss are significantly increased in nanomaterial-treated animals following WBH.

Cardiovascular and respiratory functions

Al nanoparticle treatment significantly increased the heart rate (10 beats/min) following WBH compared with saline treatment (Table 1). On the other hand, treatment with Cu and Al nanoparticles did not influence WBH-induced changes in heart rate (Table 1). Mild-to-moderate decrease in respiration is seen in nanoparticle-treated animals after WBH. This effect is most pronounced in Ag- and Al-treated rats (see Table 1).

Nanoparticle-treated animals exhibited more pronounced decrease in MABP following WBH (Table 1). This effect is most marked in Ag nanoparticle-treated group followed by Al and Cu. However, only mild-to-moderate changes in PaO_2 or $PaCO_2$ in nanoparticle-treated stressed rats are seen following WBH (Table 1). The arterial pH did not differ between saline- or nanoparticle-treated group after WBH (Table 1). These observations suggest that treatment with nanoparticles influences cardiovascular and respiratory functions in hyperthermia.

Cognitive, sensory, motor functions and thermal stress symptoms

More pronounced disturbances in cognitive and motor functions are seen in nanoparticle-treated heat-stressed rats compared with saline-treated animals (Table 2). Cu-treated nanoparticles exhibited more pronounced decline in Rota-Rod performance after WBH compared to Ag and Al treatment. On the other hand, Ag nanoparticle-treated rats showed most marked decrease in the number of steps taken during a grid walking session after WBH than Cu and Al treatment (Table 2). However, no significant difference in placement error was seen in saline- or nanoparticle-treated animals after WBH (Table 2).

A marked decrease in the angle of inclined plane in Ag-treated rats is seen after WBH (Table 2). The distance between hind feet is also decreased in Ag-treated stressed rats. However, no differences in stride length are observed in Ag- or saline-treated rats after WBH (Table 2). Interestingly, there were no significant differences in inclined plane test or foot print analysis in Cu- or Al-treated rats after WBH (Table 2). These observations suggest that disturbances in the cognitive and motor functions caused by WBH in animals are markedly influenced by nanoparticles depending on their chemistry.

Thermal stress symptoms

Interestingly, the thermal stress symptoms, *e.g.*, salivation and prostration, are not significantly different following WBH in saline- or nanoparticle-treated groups (Table 2). However, a significant increase in WBH-induced gastric ulceration is seen in Ag-treated rats followed by Cu-treated animals. In contrast, Al nanoparticles attenuated the WBH-induced gastric ulceration in rats (Table 2).

These observations support the idea that nanoparticles, depending on their chemical composition, influence a variety of biological functions following WBH, the probable mechanism of which is unclear and requires further investigation.

Blood–brain and spinal cord permeability

Approximately 3–30% increase in the BBB and ~16–75% increase in blood–spinal cord barrier (BSCB) permeability to protein tracers are seen in nanoparticle-treated animals after WBH (Table 3). The effect is most pronounced in Cu-treated nanoparticles followed by Ag and Al treatment (Table 3). The magnitude of BBB disruption after WBH is most marked in the spinal cord compared with the brain in nanoparticle-treated animals. This suggests that nanoparticles are likely to produce severe damage to the spinal cord resulting in sensory-motor dysfunction. The mechanism behind a selective vulnerability of the spinal cord in nanoparticle-treated heat-stressed rats compared with the brain is not known.

Regional BBB permeability

The magnitude of radiotracer extravasation varies in different brain regions following WBH in nanoparticle-treated rats (Table 4). Most marked increase in regional BBB permeability is seen in Ag-treated group followed by Cu- and Al-treated animals. The highest increase in regional BBB to radioiodine was seen in brain stem ($\approx 400\%$) followed by hippocampus and cerebellum ($\approx 300\%$), cerebral cortex (200%) and thalamus and hypothalamus ($\approx 150\%$, See Table 4). These observations indicate that nanoparticles are able to influence regional BBB function in a selective manner depending on their chemistry.

Brain and spinal cord blood flow

The magnitude of reduction in brain and spinal cord blood flow is more pronounced in nanoparticle-treated rats after WBH (Table 3). This effect is most evident in Cu nanoparticle-treated animals followed by Ag and Al. Interestingly, spinal cord did not exhibit greater reduction in blood flow compared to the brain in nanoparticle-treated heat-stressed rats. This suggests that the reduction in blood flow is unrelated to breakdown of the BBB or BCSB function in WBH.

Brain edema formation

Development of brain edema in nanoparticle-treated heat-stressed animals is significantly higher compared to saline-treated rats (Table 3). The brain water is increased by 0.8–2%, and the spinal cord water shows an elevation by 0.2–0.5% in nanoparticle-treated rats after WBH. This effect is most pronounced in Cu- and Ag-treated animals compared with Al-treated group (Table 3). These observations are in line with the idea that nanoparticles, depending on their chemical properties, are able to influence brain edema formation in WBH. A less swelling of spinal cord compared with the brain in spite of high tracer extravasation in the cord is unclear. Greater intracellular accumulation of water in astrocytes and endothelial cells in the gray matter compared with the white matter in WBH could possibly account for this discrepancy. Alternatively, time course of development of edema or its resolution in spinal cord may be different due to differences in tissue compliance or tissue pressure (Rapoport, 1976).

Regional brain edema formation

A higher accumulation of water in nanoparticle-treated rats after WBH is seen in several brain regions compared with saline-treated group (Table 4). The most pronounced increase in regional brain water is seen in Cu-treated nanoparticles followed by Ag and Al (Table 4). Cerebral cortex and hippocampus showed higher increase in brain water in Cu-treated group followed by cerebellum, brain stem and thalamus and hypothalamus (Table 4). On the contrary, Ag- and Al-treated rats exhibited only a mild increase in regional brain edema formation. These observations suggest that the chemical composition of nanoparticles plays an important role in the development of regional brain edema in WBH.

Cellular changes in the brain and spinal cord

Exacerbation of neuronal, glial, axonal and endothelial cell damages are seen in nanoparticle-treated rats after WBH (see Table 5). The most marked effects of WBH-induced aggravation of cellular injuries are seen in the case of Ag nanoparticle followed by Cu and Al.

Neuronal damage

Large number of distorted neurons are present in nanoparticle-treated stressed rats. The magnitude

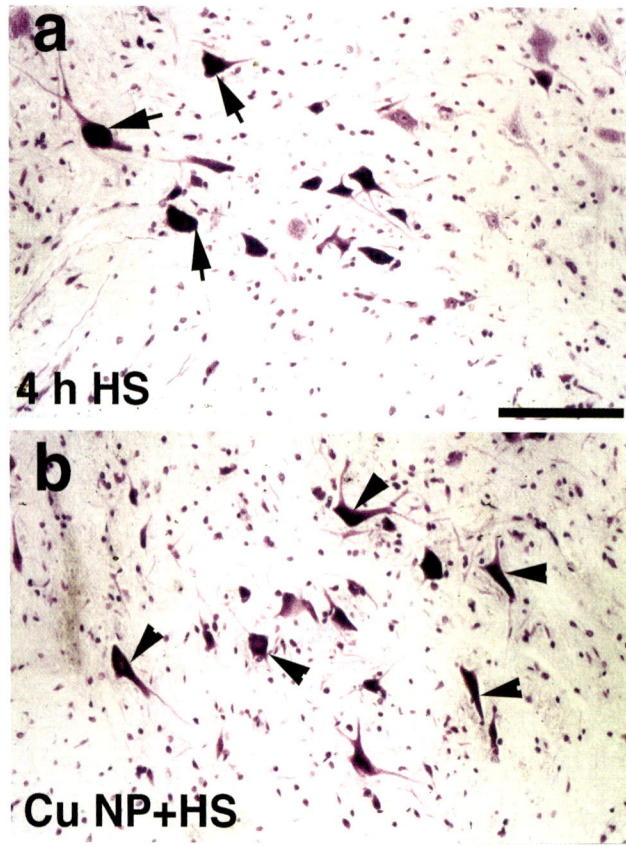

Fig. 3. Neuronal damage in heat stress (a) and its exacerbation with Cu nanoparticles (b). Many nerve cells (arrows) are dark and distorted in heat-stressed rat (a), and sign of sponginess and edema are clearly seen. These cell changes appear to be more prominent in rat that received Cu nanoparticle treatment (for 1 week) before heat stress (b). The number of dark and distorted nerve cells (arrow heads) is much more pronounced in Cu nanoparticle-treated rats after heat exposure (b) and saline-treated rat (a). Bar = 100 μm.

and intensity of neuronal damage, e.g., chromatolysis, degeneration, distortion of cell nucleus and nucleolus, is more in Ag- and Cu-treated animals (Fig. 3). A greater expansion, sponginess and edema are common in these rats compared with saline treatment (Fig. 3).

Glial and axonal cell injuries

Massive upregulation of GFAP and perivascular edema are apparent in Ag- and Cu-treated animals following WBH (Table 5). The magnitude and intensity of GFAP immunostaining is most pronounced in Ag and Cu nanoparticle-treated animals compared with Al. Loss of myelinated fibers and myelin vesiculation is also prominent in Ag- and Cu-treated animals (Table 5).

Endothelial cell membrane disruption

At the ultrastructural level, the endothelial cells of cerebral microvessels showed widespread exudation of lanthanum in Ag and Cu nanoparticle-treated rats after WBH compared with saline treatment (Fig. 4). Diffusion of lanthanum into the adjacent neuropil is a common finding in nanoparticle-treated rats after WBH, whereas in saline-treated rats, the electron-dense tracer is largely confined to the basal lamina (Fig. 4). These observations are the first to show that the magnitude and intensity of

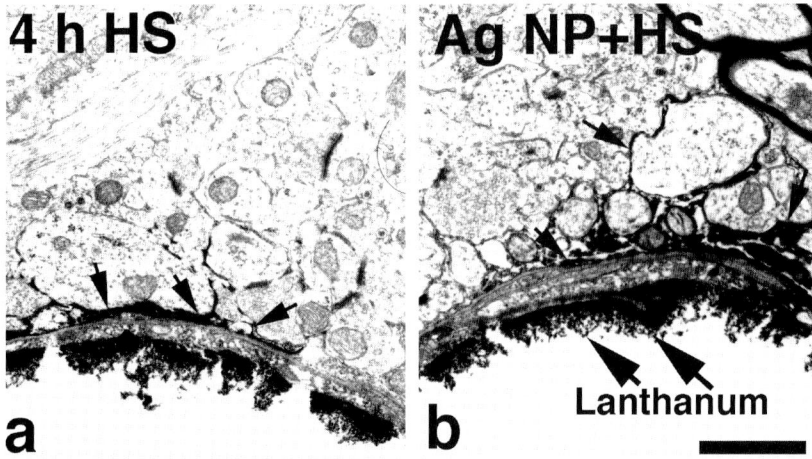

Fig. 4. Blood-spinal cord barrier (BSCB) to electron-dense tracer lanthanum after 4 h heat stress in saline-treated (a) and in Ag nanoparticle-treated (1 week) rat (b). High-power electron micrograph showing presence of lanthanum in the basal lamina (arrows) of one spinal cord microvessel from the ventral horn of the T9 segment (a). Deposition of lanthanum within the basal lamina and in extracellular space is clearly seen in the neuropil (a). In Ag nanoparticle-treated stressed rat, the magnitude and intensity of lanthanum extravasation across the BSCB and in the neuropil is much more pronounced (b) Thus, the endothelial cell cytoplasm is heavily infiltrated in nanoparticle-treated heat-stressed rat (b). The extent of penetration of the tracer in the extracellular compartment of the spinal cord neuropil is much deeper within 4 h (b) compared with saline-treated stressed rats for the same duration (a). Perivascular edema and myelin degeneration are prominent in the neuropil of heat-stressed rats (a, b). However, the tight junctions are intact to lanthanum (b). Bar = a, b = 600 nm.

endothelial cell membrane dysfunction to lanthanum (molecular diameter 12 Å) following WBH is significantly increased in nanoparticle-treated group compared with saline treatment.

However, it is still unclear whether nanoparticles (50–60 nm) may have entered into the brain parenchyma after disruption of the BBB during WBH to induce greater neurotoxicity. To confirm this idea, localization of nanoparticles in brain and spinal cord is needed, a subject that is currently being investigated in our laboratory.

Possible mechanisms of nanoparticle-induced exacerbation of brain damage

New data presented in this review clearly demonstrate that when nanoparticles enter the body fluid compartments they are able to influence the sensory, motor and cognitive functions and induce BBB disruption leading to the development of brain pathology. Another most important finding of this investigation is that subjection of nanoparticle-treated animals to additional stress, *e.g.*, WBH, exacerbates the brain pathology and sensory motor disturbances.

These findings have immense strategic significance with regard to defense planning and military exercise, particularly in hot environments in various parts of the world. It appears that both military and non-military personnels, when exposed to nanoparticles from the ambient air at home or abroad, are more vulnerable to additional environmental stress. Further research is necessary to find out suitable therapeutic strategies to attenuate nanoparticle-induced exacerbation of brain pathology in stressful situations.

Epidemiological studies in humans show that exposure of airborne particles (< 10 μm) enhances asthma attack, cardiovascular dysfunctions, hospital admissions as well as morbidity and mortality (Chang et al., 2005; Namdeo and Bell, 2005). The mass of particles, their size, number, the reactive surface area and bulk chemistry, and their clearance from biological tissues could all play important roles in inducing cellular toxicity (BéruBé et al., 2006). The equivalent masses of small particles are more inflammatory than larger sized particles of similar composition in lungs (Osier and Oberdörster, 1997). Instillation of diesel exhaust particles and Cabosil into the trachea increases vascular permeability, edema formation and cell damage in pulmonary tissues (see Evans et al., 2006). Our observations in WBH further suggest that nanoparticles, depending on their chemistry, are able to increase BBB permeability to large molecules, which could be instrumental in edema formation and neurotoxicity.

The possible mechanisms by which nanoparticles aggravate hyperthermia-induced brain damage are unclear. Carbon nanoparticles, depending on their sizes, aggravate lung inflammation caused by bacterial endotoxin. Thus, smaller particles (14 nm) induce most prominent damage in the lung tissues compared to large particles (≈ 50 nm; Inoue et al., 2006). This exacerbation of inflammatory changes by nanoparticles is mediated through enhanced local expression of proinflammatory cytokines and induction of oxidative stress (Inoue et al., 2006). In addition, nanoparticles also induce coagulatory disturbance that could play important roles in aggravating lung inflammation (Inoue et al., 2006). It seems quite likely that similar mechanisms are operating in WBH-induced aggravation of brain damage by nanoparticles.

Several environmental nanoparticles induce profound oxidative stress leading to aggravation of lung tissue injury (Yanagisawa et al., 2003). Since heat stress alone induces massive oxidative stress and lipid peroxidation (Sharma et al., 2003a, b, 2006b; Sharma, 2004), exacerbation of brain damage after WBH in nanoparticle-treated animals may, at least in part, be mediated through enhanced production of oxidative stress and generation of free radicals. Our study provides new evidences showing that apart from sizes, the chemistry of nanoparticles is also important in exacerbation of neurotoxicity in WBH.

Nanoparticles alone induce mild brain damage in normal animals probably due to induction of oxidative stress by them. Thus, additional exposure of nanoparticle-treated rats to WBH could reflect the synergistic effects of two inflammatory agents (heat stress and nanoparticles) resulting in potentiation of the pathological outcome.

Whether nanoparticles facilitate coagulatory disturbance leading to inflammation of CNS or other organs is unclear from this study. It appears that endothelial cell membrane damage by WBH allows large amounts of smaller nanoparticles to pass easily into the circulation. Once these nanoparticles enter into the brain fluid microenvironment, they may generate local oxidative stress and free radical production (MacNee and Donaldson, 2000). This in turn enhances brain inflammation and cellular damage. Furthermore, coagulatory disturbances in cerebral circulation have additional impact on brain injury. In WBH, synergistic effects of increased oxidative stress and coagulatory disturbances in the brain by nanoparticles exacerbate neurotoxicity. However, further studies are needed to confirm this hypothesis.

Alternatively, nanoparticles within the body fluid microenvironment are more sensitive to heat. Based on their thermodynamic properties, each nanoparticle can induce different thermal effects in the surrounding tissues. Thus, WBH in nanoparticle-treated animals induces a greater rise in local body and brain temperatures. Obviously a greater increase in tissue temperature aggravates brain damage. A selective exacerbation of brain damage by different chemical composition of the nanoparticles in WBH further supports this idea.

Increased cognitive and motor dysfunction after WBH in nanoparticle-treated rats correlates well with the development of brain edema formation and cell injuries. This suggests that WBH-induced brain damage is crucial in inducing motor and cognitive disorders. A reduction in brain edema and cell injury with naloxone treatment in WBH together with cognitive and motor dysfunction in rats (Sharma, 2006a; Sharma in this volume, Chapter 10) further supports this hypothesis (see Chapter 15 in this volume).

Conclusion

In conclusion, the novel data presented in this review indicate that nanoparticles depending on their chemical composition can exacerbate brain pathology in WBH. This effect is most pronounced with Ag and Cu nanoparticles. It appears that increased local oxidative stress and/or tissue temperature in nanoparticle-treated rats after WBH play important roles. Further studies using potent antioxidant compounds in nanoparticle-treated normal and heat-stressed animals are needed to clarify these points.

Acknowledgments

This investigation is partially supported by the Air Force Office of Scientific Research (London), Air Force Material Command, USAF, under grant number FA8655-05-1-3065. The U.S. Government is authorized to reproduce and distribute reprints for Government purpose notwithstanding any copyright notation thereon. The views and conclusions contained herein are those of the authors and should not be interpreted as necessarily representing the official policies or endorsements, either expressed or implied, of the Air Force Office of Scientific Research or the U.S. Government. We express sincere gratitude to several laboratories where a part of the work is done or some data is recorded and evaluated. The help from Drs. SK Patnaik and R Patnaik (Varanasi, India) and D Muresanu (Cluj, Romania) in extending necessary laboratory facilities and experimental support for heat stress is greatly acknowledged.

References

Agency for Toxic Substances and Disease Registry (ATSDR). (1990) Toxicological profile for silver. Atlanta, GA: U.S. Department of Health and Human Services, Public Health Service. http://www.atsdr.cdc.gov/toxprofiles/tp146.html#book mark12

Agency for Toxic Substances and Disease Registry (ATSDR). (2006) Toxicological profile for aluminum. Draft for Public Comment September 2006. http://www.atsdr.cdc.gov/toxprofiles/tp22.html#bookmark13

Ajalat, M.P. (1960) Recovery from heat prostration and body temperature of 109°F. Calif. Med., 92(May): 350.

Alyautdin, R.N., Petrov, V.E., Langer, K., Berthold, A., Kharkevich, D.A. and Kreuter, J. (1997) Delivery of loperamide across the blood-brain barrier with polysorbate 80-coated polybutylcyanoacrylate nanoparticles. Pharm. Res., 14(3): 325–328.

Ambruosi, A., Yamamoto, H. and Kreuter, J. (2005) Body distribution of polysorbate-80 and doxorubicin-loaded [^{14}C]poly(butyl cyanoacrylate) nanoparticles after i.v. administration in rats. J. Drug Target, 13(10): 535–542.

Bennett, W.D. (2002) Rapid translocation of nanoparticles from the lung to the bloodstream? Am. J. Respir. Crit. Care Med., 165(12): 1671–1672; author reply 1672.

Benzinger, T.H. (1969) Heat regulation: homeostasis of central temperature in man. Physiol. Rev., 49(4): 671–759. Review

BéruBé, K.A., Jones, T.P., Moreno, T., Sexton, K., Balharry, D., Hicks, M., Merolla, L. and Mossman, B.T. (2006) Characterisation of airborne particulate matter and related mechanisms of toxicity. In: Ayres J., Maynard R. and Richards R. (Eds.), Air Pollution Reviews, Vol. 3. Imperial College Press, London, pp. 69–109.

Bodian, D. and Howe, H.A. (1941a) Experimental studies on intraneural spread of poliomyelitis virus. Bull. Johns Hopkins Hosp., 69: 248–267.

Bodian, D. and Howe, H.A. (1941b) The rate of progression of poliomyelitis virus in nerves. Bull. Johns Hopkins Hosp., 69: 79–85.

Borm, P.J., Robbins, D., Haubold, S., Kuhlbusch, T., Fissan, H., Donaldson, K., Schins, R., Stone, V., Kreyling, W., Lademann, J., Krutmann, J., Warheit, D. and Oberdorster, E. (2006) The potential risks of nanomaterials: a review carried out for ECETOC. Part. Fibre Toxicol., 3(Aug. 14): 11.

Brooking, J., Davis, S.S. and Illum, L. (2001) Transport of nanoparticles across the rat nasal mucosa. J. Drug Target, 9(4): 267–279.

Brown, D.M., Stone, V., Findlay, P., MacNee, W. and Donaldson, K. (2000) Increased inflammation and intracellular calcium caused by ultrafine carbon black is independent of transition metals or other soluble components. Occup. Environ. Med., 57: 685–691.

Brown, D.M., Wilson, M.R., MacNee, W., Stone, V. and Donaldson, K. (2001) Size-dependent proinflammatory effects of ultrafine polystyrene particles: a role for surface area and oxidative stress in the enhanced activity of ultrafines. Toxicol. Appl. Pharmacol., 175: 191–199.

Calderon-Garcidueñas, L., Azzarelli, B., Acuna, H., Garcia, R., Gambling, T.M., Osnaya, N., et al. (2002) Air pollution and brain damage. Toxicol. Pathol., 30(3): 373–389.

Calvo, P., Gouritin, B., Chacun, H., Desmaele, D., D'Angelo, J., Noel, J.P., Georgin, D., Fattal, E., Andreux, J.P. and Couvreur, P. (2001) Long-circulating PEGylated polycyanoacrylate nanoparticles as new drug carrier for brain delivery. Pharm. Res., 18(8): 1157–1166.

Campbell, A., Oldham, M., Becaria, A., Bondy, S.C., Meacher, D., Sioutas, C., et al. (2005) Particulate matter in polluted air may increase biomarkers of inflammation in mouse brain. Neurotoxicology, 26: 133–140.

Cavagna, G., Finulli, M. and Vigliani, E.C. (1961) Experimental study on the pathogenesis of fevers caused by the inhalation of Teflon (polytetrafluoroethylene) fumes. Article in Italian. Med. Lav., 52(Apr.): 251–261.

Chang, C.C., Tsai, S.S., Ho, S.C. and Yang, C.Y. (2005) Air pollution and hospital admissions for cardiovascular disease in Taipei, Taiwan. Environ. Res., 98: 114–119.

Coleman, W.E., Scheel, L.D., Kupel, R.E. and Larkin, R.L. (1968) The identification of toxic compounds in the pyrolysis products of polytetrafluoroethylene (PTFE). Am. Ind. Hyg. Assoc. J., 29(1): 33–40.

Copper, WHO, Geneva. (1998) EHC 200. http://www.inchem.org/documents/ehc/ehc/ehc200.htm

Damas, J. (1994) Involvement of the kallikrein-kinin system in the salivary secretion elicited in rats by heat stress. Braz. J. Med. Biol. Res., 27(8): 2013–2020.

Derfus, A.M., Chan, W.C.W. and Bhatia, S.N. (2004) Probing the cytotoxicity of semiconductor quantum dots. Nano Lett., 4(1): 11–18.

De Lorenzo, A.J. (1957) Electron microscopic observations of the olfactory mucosa and olfactory nerve. J. Biophys. Biochem. Cytol., 3: 839–850.

De Lorenzo, A.J. (1970) The olfactory neuron and the blood-brain barrier. In: Wolstenholme G. and Knight J. (Eds.), Taste and Smell in Vertebrates. Churchill, London, pp. 151–176.

Dunning, M.D., Lakatos, A., Loizou, L., Kettunen, M., ffrench-Constant, C., Brindle, K.M. and Franklin, R.J. (2004) Superparamagnetic iron oxide-labeled Schwann cells and olfactory ensheathing cells can be traced in vivo by magnetic resonance imaging and retain functional properties after transplantation into the CNS. J. Neurosci., 24(44): 9799–9810.

Elder, A., Gelein, R., Silva, V., Feikert, T., Opanashuk, L., Carter, J., Potter, R., Maynard, A., Ito, Y., Finkelstein, J. and Oberdorster, G. (2006) Translocation of inhaled ultrafine manganese oxide particles to the central nervous system. Environ. Health Perspect., 114(8): 1172–1178.

Euliss, L.E., Dupont, J.A., Gratton, S. and Desimone, J. (2006) Imparting size, shape, and composition control of materials for nanomedicine. Chem. Soc. Rev., 35(11): 1095–1104.

Evangelista, S. (2006) Role of sensory neurons in restitution and healing of gastric ulcers. Curr. Pharm. Des., 12(23): 2977–2984. Review

Evans, S.A., Al-Mosawi, A., Adams, R.A. and Berube, K.A. (2006) Inflammation, edema, and peripheral blood changes in lung-compromised rats after instillation with combustion-derived and manufactured nanoparticles. Exp. Lung. Res., 32(8): 363–378.

Fang, J.Y. (2006) Nano- or submicron-sized liposomes as carriers for drug delivery. Chang Gung. Med. J., 29(4): 358–362. Review

Farokhzad, O.C. and Langer, R. (2006) Nanomedicine: developing smarter therapeutic and diagnostic modalities. Adv. Drug Deliv. Rev., 58(14): 1456–1459.

Fenart, L., Casanova, A., Dehouck, B., Duhem, C., Slupek, S., Cecchelli, R. and Betbeder, D. (1999) Evaluation of effect of charge and lipid coating on ability of 60-nm nanoparticles to cross an in vitro model of the blood-brain barrier. J. Pharmacol. Exp. Ther., 291(3): 1017–1022.

Fenoglio, I., Tomatis, M., Lison, D., Muller, J., Fonseca, A., Nagy, J.B. and Fubini, B. (2006) Reactivity of carbon nanotubes: free radical generation or scavenging activity? Free Radic. Biol. Med., 40(7): 1227–1233.

Florence, A.T., Hillery, A.M., Hussain, N. and Jani, P.U. (1995) Factors affecting the oral uptake and translocation of polystyrene nanoparticles: histological and analytical evidence. J. Drug Target, 3(1): 65–70.

Foley, S., Crowley, C., Smaihi, M., Bonfils, C., Erlanger, B.F., Seta, P., et al. (2002) Cellular localisation of a water-soluble fullerene derivative. Biochem. Biophys. Res. Commun., 294: 116–119.

Friese, A., Seiller, E., Quack, G., Lorenz, B. and Kreuter, J. (2000) Increase of the duration of the anticonvulsive activity of a novel NMDA receptor antagonist using poly(butylcyanoacrylate) nanoparticles as a parenteral controlled release system. Eur. J. Pharm. Biopharm., 49(2): 103–109.

Galina, Z.H., Sutherland, C.J. and Amit, Z. (1983) Effects of heat-stress on behavior and the pituitary adrenal axis in rats. Pharmacol. Biochem. Behav., 19(2): 251–256.

Gao, K. and Jiang, X. (2006) Influence of particle size on transport of methotrexate across blood brain barrier by polysorbate 80-coated polybutylcyanoacrylate nanoparticles. Int. J. Pharm., 310(1–2): 213–219.

Gao, X., Tao, W., Lu, W., Zhang, Q., Zhang, Y., Jiang, X. and Fu, S. (2006) Lectin-conjugated PEG-PLA nanoparticles: preparation and brain delivery after intranasal administration. Biomaterials, 27(18): 3482–3490.

Gianutsos, G., Morrow, G.R. and Morris, J.B. (1997) Accumulation of manganese in rat brain following intranasal administration. Fundam. Appl. Toxicol., 37: 102–105.

Gopinath, P.G., Gopinath, G. and Kumar, A. (1978) Target site of intranasally sprayed substances and their transport across the nasal mucosa: a new insight into the intranasal route of drug delivery. Curr. Ther. Res., 23(5): 596–607.

Goyer, R.A. (1991) Toxic Effects of Metals. In: Amdur M.O., Doull J. and Klaassen C.D. (Eds.), Casarett and Doull's Toxicology. The Basic Science of Poisons, 4th ed. Permagon Press, pp. 662–663.

Griffith, F.D., Stephens, S.S. and Tayfun, F.O. (1973) Exposure of Japanese quail and parakeets to the pyrolysis products of fry pans coated with Teflon and common cooking oils. Am. Ind. Hyg. Assoc. J., 34(4): 176–178.

Hagenbuch, B. and Meier, P.J. (2003) The superfamily of organic anion transporting polypeptides. Biochim. Biophys. Acta, 1609(1): 1–18. Review

Harata, I. (1970) Heat exhaustion and prostration. Pa Med., 73(4): 58–60.

Hardebo, J.E. and Kahrstrom, J. (1985) Endothelial negative surface charge areas and blood–brain barrier function. Acta Physiol. Scand., 125(3): 495–499.

Hart, M.N., VanDyk, L.F., Moore, S.A., Shasby, D.M. and Cancilla, P.A. (1987) Differential opening of the brain endothelial barrier following neutralization of the endothelial luminal anionic charge in vitro. J. Neuropathol. Exp. Neurol., 46(2): 141–153.

He, W., Yong, T., Ma, Z.W., Inai, R., Teo, W.E. and Ramakrishna, S. (2006) Biodegradable polymer nanofiber mesh to maintain functions of endothelial cells. Tissue Eng., 12(9): 2457–2466.

Howe, H.A. and Bodian, D. (1940) Portals of entry of poliomyelitis virus in the chimpanzee. Proc. Soc. Exp. Biol. Med., 43: 718–721.

Hunter, D.D. and Undem, B.J. (1999) Identification and substance P content of vagal afferent neurons innervating the epithelium of the guinea pig trachea. Am. J. Respir. Crit. Care Med., 159: 1943–1948.

Inoue, K., Takano, H., Yanagisawa, R., Hirano, S., Sakurai, M., Shimada, A. and Yoshikawa, T. (2006) Effects of airway exposure to nanoparticles on lung inflammation induced by bacterial endotoxin in mice. Environ. Health Perspect., 114(9): 1325–1330.

Johnston, C.J., Finkelstein, J.N., Gelein, R., Baggs, R. and Oberdorster, G. (1996) Characterization of the early pulmonary inflammatory response associated with PTFE fume exposure. Toxicol. Appl. Pharmacol., 140(1): 154–163.

Johnston, C.J., Finkelstein, J.N., Mercer, P., Corson, N., Gelein, R. and Oberdorster, G. (2000) Pulmonary effects induced by ultrafine PTFE particles. Toxicol. Appl. Pharmacol., 168(3): 208–215.

Joo, S.H., Feitz, A.J. and Waite, T.D. (2004) Oxidative degradation of the carbothioate herbicide, molinate, using nanoscale zerovalent iron. Environ. Sci. Technol., 38: 2242–2247.

Jordan, A., Scholz, R., Maier-Hauff, K., van Landeghem, F.K., Waldoefner, N., Teichgraeber, U., Pinkernelle, J., Bruhn, H., Neumann, F., Thiesen, B., von Deimling, A. and Felix, R. (2006) The effect of thermotherapy using magnetic nanoparticles on rat malignant glioma. J. Neurooncol., 78(1): 7–14.

Kennedy, P. and Chaudhuri, A. (2002) Herpes simplex encephalitis. J. Neurol. Neurosurg. Psychiatry, 73: 237–238.

Kim, J.S., Yoon, T.J., Yu, K.N., Kim, B.G., Park, S.J., Kim, H.W., Lee, K.H., Park, S.B., Lee, J.K. and Cho, M.H. (2006a) Toxicity and tissue distribution of magnetic nanoparticles in mice. Toxicol. Sci., 89(1): 338–347.

Kim, J.S., Yoon, T.J., Yu, K.N., Noh, M.S., Woo, M., Kim, B.G., Lee, K.H., Sohn, B.H., Park, S.B., Lee, J.K. and Cho, M.H. (2006b) Cellular uptake of magnetic nanoparticle is mediated through energy dependent endocytosis in A549 cells. J. Vet. Sci., 7(4): 321–326.

Kniesel, U. and Wolburg, H. (2000) Tight junctions of the blood-brain barrier. Cell Mol. Neurobiol., 20(1): 57–76. Review

Koziara, J.M., Lockman, P.R., Allen, D.D. and Mumper, R.J. (2003) In situ blood-brain barrier transport of nanoparticles. Pharm. Res., 20(11): 1772–1778.

Koziara, J.M., Lockman, P.R., Allen, D.D. and Mumper, R.J. (2006) The blood-brain barrier and brain drug delivery. J. Nanosci. Nanotechnol., 6(9–10): 2712–2735.

Kreuter, J. (2004) Influence of the surface properties on nanoparticle-mediated transport of drugs to the brain. J. Nanosci. Nanotechnol., 4(5): 484–488. Review

Kreuter, J., Alyautdin, R.N., Kharkevich, D.A. and Ivanov, A.A. (1995) Passage of peptides through the blood-brain barrier with colloidal polymer particles (nanoparticles). Brain Res., 674(1): 171–174.

Lam, C.W., James, J.T., McCluskey, R., Arepalli, S. and Hunter, R.L. (2006) A review of carbon nanotube toxicity and assessment of potential occupational and environmental health risks. Crit. Rev. Toxicol., 36(3): 189–217. Review

Lanone, S. and Boczkowski, J. (2006) Biomedical applications and potential health risks of nanomaterials: molecular mechanisms. Curr. Mol. Med., 6(6): 651–663.

Li, N., Sioutas, C., Cho, A., Schmitz, D., Misra, C., Sempf, J., et al. (2003) Ultrafine particulate pollutants induce oxidative stress and mitochondrial damage. Environ. Health Perspect., 111: 455–460.

Liu, G., Men, P., Harris, P.L., Rolston, R.K., Perry, G. and Smith, M.A. (2006) Nanoparticle iron chelators: a new therapeutic approach in Alzheimer disease and other neurologic disorders associated with trace metal imbalance. Neurosci. Lett., 406(3): 189–193.

Lockman, P.R., Koziara, J.M., Mumper, R.J. and Allen, D.D. (2004) Nanoparticle surface charges alter blood-brain barrier integrity and permeability. J. Drug Target, 12(9–10): 635–641.

Lou, L.X., Geng, B., Yu, F., Zhang, J., Pan, C.S., Chen, L., Qi, Y.F., Ke, Y., Wang, X. and Tang, C.S. (2006) Endoplasmic reticulum stress response is involved in the pathogenesis of stress induced gastric lesions in rats. Life Sci., 79(19): 1856–1864.

MacNee, W. and Donaldson, K. (2000) How can ultrafine particles be responsible for increased mortality? Monaldi Arch. Chest Dis., 55(2): 135–139. Review.

Manna, S.K., Sarkar, S., Barr, J., Wise, K., Barrera, E.V., Jejelowo, O., Rice-Ficht, A.C. and Ramesh, G.T. (2005) Single-walled carbon nanotube induces oxidative stress and activates nuclear transcription factor-kappaB in human keratinocytes. Nano Lett., 5(9): 1676–1684.

Mertens, H., Biteen, J.S., Atwater, H.A. and Polman, A. (2006) Polarization-selective plasmon-enhanced silicon quantum-dot luminescence. Nano Lett., 6(11): 2622–2625.

Mills, N.L., Amin, N., Robinson, S.D., Anand, A., Davies, J., Patel, D., de la Fuente, J.M., Cassee, F.R., Boon, N.A., Macnee, W., Millar, A.M., Donaldson, K. and Newby, D.E. (2006) Do inhaled carbon nanoparticles translocate directly into the circulation in humans? Am. J. Respir. Crit. Care Med., 173(4), 426–431.

Minet, O., Dressler, C. and Beuthan, J. (2004) Heat stress induced redistribution of fluorescent quantum dots in breast tumor cells. J. Fluoresc., 14(3): 241–247.

Moghimi, S.M. (2006) Recent developments in polymeric nanoparticle engineering and their applications in experimental and clinical oncology. Anticancer Agents Med. Chem., 6(6): 553–556. Review

Nagy, Z., Peters, H. and Huttner, I. (1981) Endothelial surface charge: blood-brain barrier opening to horseradish peroxidase induced by the polycation protamin sulfate. Acta Neuropathol. Suppl. (Berl.), 7: 7–9.

Nagy, Z., Peters, H. and Huttner, I. (1983) Charge-related alterations of the cerebral endothelium. Lab. Invest., 49(6): 662–671.

Namdeo, A. and Bell, M.C. (2005) Characteristics and health implications of fine and coarse particulates at roadside, urban background and rural sites in UK. Environ. Int., 31: 565–573.

National Toxicology Program. (1997) NTP toxicology and carcinogenesis studies of tetrafluoroethylene (CAS No. 116-14-3) in F344 rats and B6C3F1 mice (inhalation studies). Natl. Toxicol. Program Tech. Rep. Ser., 450(Apr.): 1–321.

Nordberg, G.F. and Gerhardsson, L. (1988) Silver. In: Seiler H.G. and Sigel H. (Eds.), Handbook on Toxicity of Inorganic Compounds. Marcel Dekker, Inc., New York, pp. 619–623.

Oberdörster, E. (2004a) Manufactured nanomaterials (fullerenes, C60) induce oxidative stress in brain of juvenile largemouth bass. Environ. Health Perspect., 112: 1058–1062.

Oberdörster, G. (1996) Significance of particle parameters in the evaluation of exposure-dose-response relationships of inhaled particles. Inhal. Toxicol., 8(Suppl): 73–89. Review

Oberdörster, G., Finkelstein, J.N., Johnston, C., Gelein, R., Cox, C., Baggs, R. and Elder, A.C. (2000) Acute pulmonary effects of ultrafine particles in rats and mice. Res. Rep. Health Eff. Inst., Aug. (96): 5–74; discussion 75–86.

Oberdörster, G., Maynard, A., Donaldson, K., Castranova, V., Fitzpatrick, J., Ausman, K., Carter, J., Karn, B., Kreyling, W., Lai, D., Olin, S., Monteiro-Riviere, N., Warheit, D., Yang, H. and ILSI Research Foundation/Risk Science Institute Nanomaterial Toxicity Screening Working Group. (2005a) Principles for characterizing the potential human health effects from exposure to nanomaterials: elements of a screening strategy. Part. Fibre Toxicol., 2(6): 8.

Oberdörster, G., Oberdorster, E. and Oberdorster, J. (2005b) Nanotoxicology: an emerging discipline evolving from studies of ultrafine particles. Environ. Health Perspect., 113(7): 823–839. Review

Olivier, J.C. (2005) Drug transport to brain with targeted nanoparticles. NeuroRx, 2(1): 108–119. Review

Osier, M. and Oberdörster, G. (1997) Intratracheal inhalation versus intratracheal instillation: differences in particulate effects. Fundam. Appl. Toxicol., 40: 220–227.

Pettersson, C.A., Sharma, H.S. and Olsson, Y. (1990) Vascular permeability of spinal nerve roots. A study in the rat with Evans blue and lanthanum as tracers. Acta Neuropathol. (Berl.), 81(2): 148–154.

Plotkin, M., Gneveckow, U., Meier-Hauff, K., Amthauer, H., Feussner, A., Denecke, T., Gutberlet, M., Jordan, A., Felix, R. and Wust, P. (2006) 18F-FET PET for planning of thermotherapy using magnetic nanoparticles in recurrent glioblastoma. Int. J. Hyperthermia, 22(4): 319–325.

Pries, A.R. and Kuebler, W.M. (2006) Normal endothelium. Handb. Exp. Pharmacol., (176 Pt 1): 1–40.

Rapoport, S.I. (1976) Blood-Brain Barrier in Physiology and Medicine. Raven Press, New York.

Rodoslav, S., Laibin, L., Eisenberg, A. and Dusica, M. (2003) Micellar nanocontainers distribute to defined cytoplasmic organelles. Science, 300: 615–618.

Sahagun, G., Moore, S.A. and Hart, M.N. (1990) Permeability of neutral vs. anionic dextrans in cultured brain microvascular endothelium. Am. J. Physiol., 259(1 Pt 2): H162–H166.

Saija, A., Princi, P., Trombetta, D., Lanza, M. and De Pasquale, A. (1997) Changes in the permeability of the blood-brain barrier following sodium dodecyl sulphate administration in the rat. Exp. Brain Res., 115(3): 546–551.

Sayes, C., Fortner, J., Guo, W., Lyon, D., Boyd, A.M., Ausman, K.D., et al. (2004) The differential cytotoxicity of water-soluble fullerenes. Nano Lett., 4: 1881–1887.

Schroder, U. and Sabel, B.A. (1996) Nanoparticles, a drug carrier system to pass the blood-brain barrier, permit central analgesic effects of i.v. dalargin injections. Brain Res., 710(1–2): 121–124.

Sharma, H.S. (1982) Blood-Brain Barrier in Stress. PhD Thesis, Banaras Hindu University, Varanasi, India, pp. 1–85.

Sharma, H.S. (1987) Effect of captopril (a converting enzyme inhibitor) on blood-brain barrier permeability and cerebral blood flow in normotensive rats. Neuropharmacology, 26(1): 85–92.

Sharma, H.S. (1999) Pathophysiology of blood-brain barrier, brain edema and cell injury following hyperthermia: new role of heat shock protein, nitric oxide and carbon monoxide. An experimental study in the rat using light and electron microscopy. Acta Universitatis Upsaliensis, 830: 1–94.

Sharma, H.S. (2000) A bradykinin BK2 receptor antagonist HOE-140 attenuates blood-spinal cord barrier permeability following a focal trauma to the rat spinal cord. An experimental study using Evans blue, ^{131}I-sodium and lanthanum tracers. Acta Neurochir. Suppl. (Wien.), 76: 159–163.

Sharma, H.S. (2004) Pathophysiology of the blood-spinal cord barrier in traumatic injury. In: Sharma H.S. and Westman J. (Eds.), The Blood-Spinal Cord and Brain Barriers in Health and Disease. Elsevier Academic Press, San Diego, CA, pp. 437–518.

Sharma, H.S. (2005a) Pathophysiology of blood-spinal cord barrier in traumatic injury and repair. Curr. Pharm. Des., 11(11): 1353–1389. Review

Sharma, H.S. (2005b) Methods to induce brain hyperthermia. In: Current Protocols in Toxicology (Solicited Contribution), Suppl 23, Unit 11.14. Wiley, New York, USA, pp. 1–26.

Sharma, H.S. (2005c) Selective neuronal vulnerability, blood-brain barrier disruption and heat shock protein expression in stress induced neurodegeneration. Invited review. In: Sarbadhikari S.N. (Ed.), Depression and Dementia: Progress in Brain Research, Clinical Applications and Future Trends. Nova Science Publishers, Inc., New York, pp. 97–152.

Sharma, H.S. (Ed.). (2005d) Heat-related deaths are largely due to brain damage. Editorial. Indian J. Med. Res., 121(5): 621–623.

Sharma, H.S. (2006a) Hyperthermia influences excitatory and inhibitory amino acid neurotransmitters in the central nervous system. An experimental study in the rat using behavioural, biochemical, pharmacological, and morphological approaches. J. Neural Transm., 113(4): 497–519.

Sharma, H.S. (2006b) Hyperthermia induced brain oedema: current status and future perspectives. Indian J. Med. Res., 123(5): 629–652. Review

Sharma, H.S., Ali, S.F., Schlager, J. and Hussain, S. (2006a) Effect of nanoparticles on the blood-brain barrier. Int. J. Neuroprotec. Neuroregen., 2(3): 78.

Sharma, H.S., Alm, P. and Westman, J. (1998a) Nitric oxide and carbon monoxide in the brain pathology of heat stress. Prog. Brain Res., 115: 297–333. Review

Sharma, H.S., Drieu, K. and Westman, J. (2003a) Antioxidant compounds EGB-761 and BN-52021 attenuate brain edema formation and hemeoxygenase expression following hyperthermic brain injury in the rat. Acta Neurochir. Suppl. (Wien.), 86: 313–319.

Sharma, H.S., Gordh, T., Wiklund, L., Mohanty, S. and Sjoquist, P.O. (2006b) Spinal cord injury induced heat shock protein expression is reduced by an antioxidant compound H-290/51. An experimental study using light and electron microscopy in the rat. J. Neural Transm., 113(4): 521–536.

Sharma, H.S., Sjoquist, P.O. and Alm, P. (2003b) A new antioxidant compound H-290151 attenuates spinal cord injury induced expression of constitutive and inducible isoforms of nitric oxide synthase and edema formation in the rat. Acta Neurochir. Suppl. (Wien.), 86: 415–420.

Sharma, H.S., Westman, J. and Nyberg, F. (1998b) Pathophysiology of brain edema and cell changes following hyperthermic brain injury. Prog. Brain Res., 115: 351–412. Review

Sharma, P., Brown, S., Walter, G., Santra, S. and Moudgil, B. (2006c) Nanoparticles for bioimaging. Adv. Colloid Interface Sci., 123–126(Nov. 16): 471–485.

Shvedova, A.A., Kisin, E.R., Mercer, R., Murray, A.R., Johnson, V.J., Potapovich, A.I., Tyurina, Y.Y., Gorelik, O., Arepalli, S., Schwegler-Berry, D., Hubbs, A.F., Antonini, J., Evans, D.E., Ku, B.K., Ramsey, D., Maynard, A., Kagan, V.E., Castranova, V. and Baron, P. (2005) Unusual inflammatory and fibrogenic pulmonary responses to single-walled carbon nanotubes in mice. Am. J. Physiol. Lung Cell Mol. Physiol., 289(5): L698–L708.

Shvedova, A.A., Kisin, E.R., Murray, A., Kommineni, C., Vallyathan, V. and Castranova, V. (2004) Pro/antioxidant status in murine skin following topical exposure to cumene hydroperoxide throughout the ontogeny of skin cancer. Biochemistry (Mosc.), 69(1): 23–31.

Tanaka, K., Ito, A., Kobayashi, T., Kawamura, T., Shimada, S., Matsumoto, K., Saida, T. and Honda, H. (2005a) Heat immunotherapy using magnetic nanoparticles and dendritic cells for T-lymphoma. J. Biosci. Bioeng., 100(1): 112–115.

Tanaka, K., Ito, A., Kobayashi, T., Kawamura, T., Shimada, S., Matsumoto, K., Saida, T. and Honda, H. (2005b) Intratumoral injection of immature dendritic cells enhances antitumor effect of hyperthermia using magnetic nanoparticles. Int. J. Cancer, 116(4): 624–633.

Teeguarden, J.G., Hinderliter, P.M., Orr, G., Thrall, B.D. and Pounds, J.G. (2007) Particokinetics in vitro: dosimetry considerations for in vitro nanoparticle toxicity assessments. Toxicol. Sci., 95(2): 300–312.

Terasaki, S., Kameyama, T. and Yamamoto, S. (1997) A case of zoster in the 2nd and 3rd branches of the trigeminal nerve associated with simultaneous herpes labialis infection — a case report. Kurume Med. J., 44(1): 61–66 [PubMed].

Tiwari, S.B. and Amiji, M.M. (2006) A review of nanocarrier-based CNS delivery systems. Curr. Drug. Deliv., 3(2): 219–232. Review

Utell, M., Frampton, M., Zareba, W., Devlin, R. and Cascio, W. (2002) Cardiovascular effects associated with air pollution: potential mechanisms and methods of testing. Inhal. Toxicol., 14: 1231–1247 [PubMed].

Venugopal, B. and Luckey, T.D. (1978) Metal Toxicity in Mammals, Vol. 2, New York, Plenum Press, pp. 104–112.

Veronesi, B., Makwana, O., Pooler, M. and Chen, L.C. (2005) Effects of subchronic exposures to concentrated ambient particles. VII. Degeneration of dopaminergic neurons in Apo $E^{+/-}$ mice. Inhal. Toxicol., 17(4–5): 235–241.

Wagner, V., Dullaart, A., Bock, A.K. and Zweck, A. (2006) The emerging nanomedicine landscape. Nat. Biotechnol., 24(10): 1211–1217.

Warheit, D.B., Laurence, B.R., Reed, K.L., Roach, D.H., Reynolds, G.A. and Webb, T.R. (2004) Comparative pulmonary toxicity assessment of single-wall carbon nanotubes in rats. Toxicol. Sci., 77(1): 117–125.

Whyte, D.G., Brennan, T.J. and Johnson, A.K. (2006) Thermoregulatory behavior is disrupted in rats with lesions of the anteroventral third ventricular area (AV3V). Physiol. Behav., 87(3): 493–499.

Williams, D. (2006) Quantum dots in medical technology. Med. Device Technol., 17(4): 8–9.

Xia, T., Kovochich, M., Brant, J., Hotze, M., Sempf, J., Oberley, T., Sioutas, C., Yeh, J.I., Wiesner, M.R. and Nel, A.E. (2006) Comparison of the abilities of ambient and manufactured nanoparticles to induce cellular toxicity according to an oxidative stress paradigm. Nano Lett., 6(8): 1794–1807.

Xu, J. and Ling, E.A. (1994) Studies of the ultrastructure and permeability of the blood-brain barrier in the developing corpus callosum in postnatal rat brain using electron dense tracers. J. Anat., 184(Pt 2): 227–237.

Yanagisawa, R., Takano, H., Inoue, K., Ichinose, T., Sadakane, K., Yoshino, S., Yamaki, K., Kumagai, Y., Uchiyama, K., Yoshikawa, T. and Morita, M. (2003) Enhancement of acute lung injury related to bacterial endotoxin by components of diesel exhaust particles. Thorax, 58(7): 605–612.

Zhang, W.W., Cao, Q.Q., Xie, J.L., Ren, X.M., Lu, C.S., Zhou, Y., Yao, Y.G. and Meng, Q.J. (2003) Structural, morphological, and magnetic study of nanocrystalline cobalt-nickel-copper particles. J. Colloid Interface Sci., 257(2): 237–243.

Zheng, G., Chen, J., Li, H. and Glickson, J.D. (2005) Rerouting lipoprotein nanoparticles to selected alternate receptors for the targeted delivery of cancer diagnostic and therapeutic agents. Proc. Natl. Acad. Sci. U.S.A., 102(49): 17757–17762.

Zhu, S., Oberdorster, E. and Haasch, M.L. (2006) Toxicity of an engineered nanoparticle (fullerene, C60) in two aquatic species, Daphnia and fathead minnow. Mar. Environ. Res., 62 Suppl(Suppl): S5–S9.

SECTION VI

Neurochemicals and Hyperthermia

CHAPTER 14

Neuropeptides in hyperthermia

Fred Nyberg* and Mathias Hallberg

Department of Pharmaceutical Biosciences, Division of Biological Research on Drug Dependence, Uppsala University, S-751 24 Uppsala, Sweden

Abstract: Brain damage as a result of hyperthermia or heat-stress has been the focus of attention in many areas of neuroscience in recent years. Heat-induced alterations in structural components of the central nervous system (CNS) will obviously also influence the relevant transmitter systems, which may be involved in a variety of different behaviors. Indeed, many studies have indicated that excitatory amino acids, and monoaminergic and peptidergic systems are affected during hyperthermia. This chapter will address past and current research on various neuropeptides that have been implicated in the consequences of hyperthermia and various other heat disorders. However, considering the large and even increasing number of identified neuroactive peptides, it is necessary to limit this chapter to a few peptides or peptide systems, which have received particular attention in relation to hyperthermia. Among these are the opioid peptides, the tachykinins, calcitonin gene-related peptide (CGRP), and peptides belonging to the angiotensin system. Most of these neuropeptides are not only affected by hyperthermia and abnormal alterations in the body temperature but also are involved in the endogenous mechanisms of regulating body temperature. This review does not endeavor to fully cover the field but it does aim to give the reader an idea of how various neuropeptides may be involved in the control of body heat and how peptidergic systems are affected during various thermal changes, including both immediate and long-term consequences.

Keywords: neuropeptides; opiod peptides; dynorphin; substance P; calcitonin gene-related peptides (CGRP); angiotensin; heat stress; hyperthermia

Introduction

Reports dealing with studies on the effects of stress on neuropeptides are abundant in the scientific literature, particularly in connection with stress associated with pain, trauma, endocrine disorders, and psychiatric diseases. Studies on peptides and peptide systems during stress that has emerged as a result of hyperthermia are less abundant and are confined to a limited number of neuropeptide systems. Consequently, this review has been confined to certain key peptides, which have attracted the most attention among researchers in the field. The chapter provides a brief review of the location of these compounds in the body, their production, and how they interact with receptors on their target cells to produce the relevant response. A number of studies on individual peptide systems during various conditions of heat stress are also reviewed.

Hyperthermia and heat stress

Heat-related disorders may be categorized into several subgroups, including heat cramps, heat

*Corresponding author. Tel.: +46-18-471-4166; Fax: +46-18-501920; E-mail: Fred.Nyberg@farmbio.uu.se

stress, heat exhaustion, and heat stroke (Sharma et al., 1998; Koko et al., 2004; Yeo, 2004; Ahmed, 2005; Park et al., 2005). A number of signs accompanying the milder forms of hyperthermia, such as muscle spasms, weakness, fatigue, nausea, vomiting, etc., are well documented. Treatment for heat cramps, heat stress, and heat exhaustion may require removal of the patient to a cool place, rest, and the administration of oral fluids; in more severe cases, the addition of intravenous fluids and electrolytes may be necessary. Heat stroke, the most serious of the heat-related disorders, can lead to collapse and may arise in young people during vigorous exercise as a result of neurological changes due to organic and metabolic disturbances; however, in elderly people, similar neurological changes may arise during climatic heat waves. Heat stress is a serious clinical condition associated with dysfunctions in the CNS caused by heat-induced changes in various brain circuits. Hyperthermia is believed to result from failure of the body's normal thermoregulation devices. A breakdown of the thermoregulatory homeostasis leading to increased body temperature may arise from an uncontrolled rise in heat production, failure of heat dissipation systems, extreme environmental heat, or in some cases a hypothalamic malfunction. Prolonged hyperthermia usually results in damage to brain cells. The underlying mechanism causing this cell damage has so far not been well characterized. Brain edema and microhemorrhage or CNS cell injury, which are often seen as a result of hyperthermia in both clinical and experimental observations, appear to be related to the breakdown of the blood-brain-barrier (BBB) following hyperthermia.

Studies of neuropeptides under stressful conditions

The molecular mechanisms of hyperthermia in relation to effects on the CNS transmitters have been extensively investigated. A large number of studies have focused on the excitatory amino acids (Le Greves et al., 1997; Chang et al., 2004), monoaminergic systems (Sharma et al., 1992b, 1994; Kao and Lin, 1996; Yang and Lin, 1999; McGugan, 2001; Lieberman et al., 2005), and many endogenous pyrogenic compounds (e.g., Sharma et al., 1992a; Suganuma et al., 2002; Watanabe et al., 2004; Romanovsky et al., 2005). Several studies directed to the role of various neuropeptides in thermogenesis can also be found in the scientific literature. In humans, studies on the neuropeptide systems during stress are associated with limitations such as restricted availability of human CNS tissue. The activity of neuroactive peptides has, however, been studied by measuring their levels in various body fluids and also by investigating the response to drugs that act on neuropeptide receptors. More extensive studies have been carried out using various animal models of stress. In these models, behavioral studies have been combined with various morphological and neurochemical approaches. In order to understand the impact of these methods, it is necessary to recall the fundamentals of peptide biosynthesis, processing, and action through the receptors. The rest of this chapter includes a summary of currently available methods for investigating activity in the various neuropeptide systems. However, the main focus is on results emerging from studies of the involvement of endogenous opioids and related peptides, tachykinins and related peptides, and peptides belonging to the angiotensin system.

Neuropeptide biosynthesis, processing, and inactivation

Among all the bioactive compounds that are known to exert significant influence on the CNS, the neuroactive peptides have received special interest. These peptides differ in many ways (e.g., in size and biosynthesis/inactivation mechanisms) from the classical neurotransmitters such as monoamines and excitatory amino acids. Immunohistochemistry and modern molecular biological techniques have confirmed the presence of a variety of neuropeptides in the CNS. Of these, many have been discovered and characterized only in the last few years. Like most other neurotransmitters or neuromodulators, the neuropeptides are released from the nerve terminals of specific neurons; they induce their effects by acting on pre- or post-synaptically located cell membrane receptors. Peptides produced within the CNS are synthesized as large inactive protein precursors (prepropeptides) within the cell bodies of the particular peptide neuron. However, neuropeptide precursors are also formed in peripheral tissues. The active peptides are formed from their precursors following a number of proteolytic cleavages and modification steps. Details of these processes are reviewed elsewhere (von Eggelkraut-Gottanka and Beck-Sickinger, 2004; Hallberg et al., 2005). Following release and interaction with its receptor, the active neuropeptide is inactivated by enzymatic degradation. In some cases, the released active peptide is converted to a fragment with some retained or altered biological activity.

Prepropeptide and precursor processing

Biosynthesis of the neuropeptide precursor protein occurs at the ribosomes located on the endoplasmic reticulum in the cell body of the peptidergic neuron. A general feature of neuropeptide precursors (which are usually 100–250 amino acid residues in length) is that they contain an N-terminal signal sequence adjacent to a variable stretch of unknown function, which is followed by a peptide-containing region including several copies of active peptide units. The pathways for releasing the bioactive peptides include a sequence of enzymatic steps. The first cleavage is the removal of the signal peptide, which is followed by core glycosylation and glucose trimming to allow suitable folding of the propeptide. Further modifications of the sugar content may then occur within the Golgi compartment, in which the initial release of intermediate-sized peptides also occurs. Additional modification, such as sulfation (Bundgaard et al., 1997; Bonetto et al., 1999), phosphorylation, amidation (Katopodis et al., 1991), or truncation, may also occur at this stage.

The release of smaller-sized precursor fragments containing the active units is accomplished by proteases acting at paired basic amino acid residues, or sometimes at single arginine sites (Foulon et al., 1996; Canaff et al., 1999; Seidah and Chretien, 1999). The completion of proteolyses, including amino- and carboxy-terminal trimming as well as terminal modifications such as acetylation (Jornvall, 1975; Loh, 1987) and amidation (Eipper et al., 1987), takes place in the secretory granules. These granules, which contain the active peptide, are stored near the nerve endings and, following activation of the relevant neuron, they are fused with the plasma membrane and the peptide is released. The structures of the various peptides discussed in this article are given in Tables 1–3. A number of proteolytic enzymes involved in neuropeptide precursor processing have been

Table 1. Opioid and related peptides

Peptide	Amino acid sequence	Preferred receptor
Leu-enkephalin	H-Tyr-Gly-Gly-Phe-Leu-OH	DOP
Met-enkephalin	H-Tyr-Gly-Gly-Phe-Met-OH	DOP
Dynorphin A	H-Tyr-Gly-Gly-Phe-Leu-Arg-Arg-Ile-Arg-Pro-Lys-Leu-Lys-Trp-Asp-Asn-Gln-OH	KOP
Dynorphin B	H-Tyr-Gly-Gly-Phe-Leu-Arg-Arg-Gln-Phe-Lys-Val-Val-Thr-OH	KOP
β-Endorphin (human)	H-Tyr-Gly-Gly-Phe-Met-Thr-Ser-Glu-Lys-Ser-Gln-Thr-Pro-Leu-Val-Thr-Leu-Phe-Lys-Asn-Ala-Ile-Ile-Lys-Asn-Ala-Tyr-Lys-Lys-Gly-Glu-OH	MOP
Nociceptin/orphanin FQ	H-Phe-Gly-Gly-Phe-Thr-Gly-Ala-Arg-Lys-Ser-Ala-Arg-Lys-Leu-Ala-Asn-Gln-OH	ORL1
Endomorphin-1	H-Tyr-Pro-Trp-Phe-NH$_2$	MOP
Endomorphin-2	H-Tyr-Pro-Phe-Phe-NH$_2$	MOP
β-Casomorphin-7 (bovine)	H-Tyr-Pro-Phe-Pro-Gly-Pro-Ile-OH	MOP
β-Casomorphin-7 (human)	H-Tyr-Pro-Phe-Val-Glu-Pro-Ile-Pro-OH	MOP
Hemorphin-4 (human)	H-Leu-Val-Val-Tyr-Pro-Trp-Thr-OH	MOP
Hemorphin-7 (human)	H-Tyr-Pro-Trp-Thr-Gln-Arg-Phe-OH	MOP

Table 2. Calcitonin gene-related peptide, tachykinins and related peptides

Peptide	Amino acid sequence	Preferred receptor
α-CGRP (human)	N-Ala-Cys-Asp-Thr-Ala-Thr-Cys-Val-Thr-His-Arg-Leu-Ala-Gly-Leu-Leu-Ser-Arg-Ser-Gly-Gly-Val-Val-Lys-Asn-Asn-Phe-Val-Pro-Thr-Asn-Val-Gly-Ser-Lys-Ala-Phe-NH$_2$	CGRP-1
β-CGRP (human)	H-Ala-Cys-Asn-Thr-Ala-Thr-Cys-Val-Thr-His-Arg-Leu-Ala-Gly-Leu-Leu-Ser-Arg-Ser-Gly-Gly-Met-Val-Lys-Ser-Asn-Phe-Val-Pro-Thr-Asn-Val-Gly-Ser-Lys-Ala-Phe-NH$_2$	CGRP-1
CGRP(8-37) (human)	H-Val-Thr-His-Arg-Leu-Ala-Gly-Leu-Leu-Ser-Arg-Ser-Gly-Gly-Met-Val-Lys-Ser-Asn-Phe-Val-Pro-Thr-Asn-Val-Gly-Ser-Lys-Ala-Phe-NH$_2$	
Substance P	H-Arg-Pro-Lys-Pro-Gln-Gln-Phe-Phe-Gly-Leu-Met-NH$_2$	NK-1
Substance P (1–7)	H-Arg-Pro-Lys-Pro-Gln-Gln-Phe-OH	
Neurokinin A	H-His-Lys-Thr-Asp-Ser-Phe-Val-Gly-Leu-Met-NH$_2$	NK-2
Neurokinin B	H-Asp-Met-His-Asp-Phe-Phe-Val-Gly-Leu-Met-NH$_2$	NK-3

Table 3. Angiotensin II and related peptides

Peptide	Amino acid sequence	Preferred receptor
Angiotensin I	H-Asp-Arg-Val-Tyr-Ile-His-Pro-Phe-His-Leu-OH	
Angiotensin II	H-Asp-Arg-Val-Tyr-Ile-His-Pro-Phe-OH	AT1/AT2
Angiotensin III	H-Arg-Val-Tyr-Ile-His-Pro-Phe-OH	AT3
Angiotensin IV	H-Val-Tyr-Ile-His-Pro-Phe-OH	AT4 (IRAP)
Angiotensin(1–7)	H-Asp-Arg-Val-Tyr-Ile-His-Pro-OH	AT(1–7)?

characterized (Foulon et al., 1996; Hook et al., 1996; Canaff et al., 1999; Seidah and Chretien, 1999). These include a family of convertases, which cleave the propeptides at selected sites composed of single or paired basic amino acids. The exopeptidase enzyme group also participates in modification reactions (Fricker and Snyder, 1983); these remove single basic residues and are involved in acetylation (Loh, 1987) and amidation (Eipper et al., 1987) reactions. All the steps involved in neuropeptide processing are to some extent represented in the pathways known to be involved in the formation of the peptides discussed in this chapter.

Opioid peptides

Three genetically distinct opioid peptide (OP) precursor proteins, known as pro-opiomelanocortin (POMC), proenkephalin (ProENK), and prodynorphin (ProDYN), have been described to date (Cooper et al., 1996). Each precursor protein contains several active units, which are released during processing. β-Endorphin, and acetylated or truncated forms thereof, is thought to be released from POMC (Table 1). ProENK may give rise to at least seven enkephalin-containing peptides, while proDYN includes dynorphin A, dynorphin B, and α/β-neo-endorphin within its structure. Pro-orphanin is a more recently discovered precursor protein, which gives rise to the opioid-related peptide orphanin or nociceptin (Meunier et al., 1995; Reinscheid et al., 1995). Under some conditions, orphanin/nociceptin appears to have opioid effects, while in other circumstances it exerts antiopioid effects.

OPs may also be generated via an alternative pathway involving a group of peptides containing an N-terminal Tyr-Pro sequence. These peptides, named the atypical endogenous opioids (Table 1), are formed by partial hydrolysis of structural proteins. The N-terminal dipeptide Tyr-Pro is essential for both opioid activity and their comparatively high metabolic stability. Peptides in this family include the β-casomorphins, which can be generated from the milk protein β-casein (Brantl et al., 1981; Nyberg et al., 1989), and the hemorphins derived from the blood protein hemoglobin (Brantl et al., 1986; Nyberg et al., 1997). Although these peptides are found endogenously in reasonably high amounts, their agonistic actions on opioid receptors are relatively weak. In contrast, two recently discovered OPs, endomorphin-1 and endomorphin-2 (Zadina et al., 1997; Horvath, 2000), which have some sequence homology with the hemorphins and the β-casomorphins, respectively, exhibit very high potency and selectivity for a subgroup of opioid receptors, the so-called μ-opioid peptide (MOP) receptors. However, the precursor protein from which the endomorphins are released has not yet been identified.

Furthermore, recent studies have revealed that under certain conditions the precursor ProDYN is not completely processed but remains in a partially processed form ("big dynorphin"). The "big dynorphin" molecule, a peptide that consists of dynorphin A and dynorphin B, has recently been identified and characterized by Bakalkin and co-workers (Tan-No et al., 2002). This peptide induced nociceptive behavior mediated by activation of the N-methyl-D-aspartyl (NMDA) receptor ion-channel complex after acting on the NMDA receptor NR2B subunit in the mouse spinal cord. Moreover, three novel ProDYNs expressed from spliced or truncated ProDYN transcripts are still processed into dynorphins or translated into N-terminally truncated proteins despite lacking a central segment (Nikoshkov et al., 2005). One truncated ProDYN, which was found in the cell nucleus, is thought to be involved in a non-opioid function for this entity. The complexity of ProDYN expression and diversity of its protein products is believed to be relevant to the differential plasticity in adaptive responses seen with the ProDYN system (Nikoshkov et al., 2005).

Tachykinins and other neuropeptides

At least three separate precursor proteins have been described for the tachykinin family of neuropeptides, namely the α-, β-, and γ-preprotachykinins (Hokfelt et al., 1994). These are thought to give rise to several neuroactive products sharing a common C-terminal sequence ending with an amide (Table 2). The most studied tachykinin peptide is substance P (SP). SP is an undecapeptide with an affinity for the neurokinin receptors, preferentially the NK1 receptor. It is found in high concentrations in the primary afferent neurons of the spinal cord (Hokfelt et al., 1994). Calcitonin gene-related peptide (CGRP) is a peptide composed of 37 amino acids produced by alternative splicing of the calcitonin gene (Table 2). The actual gene product is processed in a typical pattern involving amidation of the C-terminal residue. Studies have suggested that CGRP may be involved in a large number of biological actions. The peptide produces a large number of effects in both the CNS and the periphery (e.g., the cardiovascular system). In the CNS, CGRP is co-localized with SP and it has been suggested that both peptides have a role in pain processing. In the spinal cord, CGRP is believed to regulate the expression of NK1 receptors via a pathway involving activation of the transcription factor cAMP response element binding protein (CREB). CGRP may induce its effects by acting on two types of G-protein-coupled receptors, CGRP-1 and CGRP-2. Angiotensin II (Table 3), a vasoconstrictor, is derived from the renin-angiotensin system (RAS) and is originally released from a precursor protein of 20 kDa. The effects of angiotensin II are mediated through the AT1 and AT2 receptors, mainly known for their involvement in cardiovascular function.

Neuropeptide conversion

Among the various enzymatic steps involved in the process of neuropeptide generation is the limited hydrolysis of neuroactive peptides, which leads to the formation of fragments with retained or different biological activities (Hallberg and Nyberg, 2003). This kind of hydrolysis is effected by relatively specific endoproteases, denoted convertases. Several neuropeptide convertases capable of releasing bioactive fragments from their substrate peptides have been identified in various CNS tissues (Devi and Goldstein, 1986; Camargo et al., 1987; Skidgel et al., 1987; Silberring and Nyberg, 1989; Silberring et al., 1992) and also in the cerebrospinal fluid (CSF) (Persson et al., 1992, 1995; Nyberg, 2004). For instance, the OP dynorphin A is converted to Leu-enkephalin-Arg and subsequently to Leu-enkephalin (Persson et al., 1995). While both these fragments retain their biological activity, their receptor activation profile differs from that of the parent peptide, dynorphin A. Similarly, SP is converted by several endoproteases, including neutral endopeptidase (NEP) and substance P endopeptidase (SPE), to a bioactive N-terminal heptapeptide fragment SP(1–7) (Zhou et al., 2001; Michael-Titus et al., 2002). The heptapeptide retains some effects but opposes others ascribed to the undecapeptide (Zhou et al., 1998, 2000). It is suggested that SPE through this conversion (Fig. 1) modulates the response of SP (Hallberg and Nyberg, 2003). The bioactive angiotensin II, known to bind to and stimulate the AT-1 and AT-2 receptors, is converted to angiotensin IV (i.e., angiotensin 3–8) with preference for the AT-4 receptor, or to angiotensin (1–7), which is not recognized by any of these receptors. Both angiotensin IV and angiotensin (1–7) are biologically active. For example, angiotensin (1–7) retains some of the effects associated with angiotensin II but counteracts others. CGRP also undergoes proteolytic conversion to biologically active fragments. For example, release of the first seven N-terminal amino acids from CGRP leads to the formation of CGRP(8–37), a fragment with properties antagonistic to those of the parent molecule (Wimalawansa, 1996). This peptide has high affinity for the CGRP-1 receptor but low antagonistic activity on the CGRP-2 receptor. The N-terminal of CGRP is usually associated with agonistic properties while the C-terminal portion appears to be related to an antagonistic function. Finally, the more recently discovered neuropeptide nociceptin may also be converted to fragments with retained biological activity. The N-terminal heptapeptide nociceptin (1–7), and the extended N-terminal fragments nociceptin (1–9) and (1–13) modulate nociceptin-induced scratching, biting, and licking in mice (Sakurada et al., 2000). Thus, it seems that some effects of nociceptin can be counteracted or antagonized by its bioactive fragments. However, some actions of nociceptin are retained in its fragments and this may indicate that some effects attributed to nociceptin may be due to the release of its fragments. For instance, since nociceptin produces both antinociception and hyperalgesia, it has been suggested that its antinociceptive activity arises after the conversion of the original peptide. In fact, this speculation was supported by studies showing that i.c.v. injection of the nociceptin fragments (1–7) and (1–11) elicited antinociception without giving rise to hyperalgesia (Rossi et al., 1997).

Inactivation of neuropeptides

The inactivation of neuroactive peptides mainly involves enzymes hydrolyzing the peptides via an exopeptidase action. These enzymes include several aminopeptidases (Hersh, 1985; Taylor, 1993), carboxypeptidases (Fricker and Snyder, 1983), metallo-endopeptidases (Roques et al., 1993), and dipeptidyl aminopeptidases (Yaron and Naider, 1993). A general route for neuropeptide inactivation is accomplished by sequential degradation by an aminopeptidase. For instance, enkephalins lose their opioid activity after removal of the N-terminal tyrosine residue, and the SP action associated with the N-terminal sequence is abolished by the action of a dipeptidyl aminopeptidase

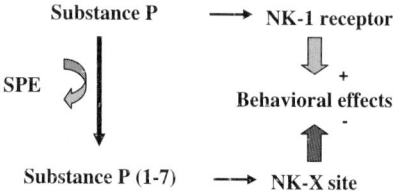

Fig. 1. Substance P endopeptidase (SPE) is shown to convert SP to its bioactive N-terminal fragment SP(1–7). The heptapeptide has been demonstrated to modulate or counteract SP-induced effects or behavior in several animal models (see Hallberg and Nyberg, 2003; Wiktelius et al., 2006).

named dipeptidase IV or DPIV (Demuth et al., 1993). However, other enzymatic pathways for neuropeptide degradation have also been described. One typical example involves degradation of enkephalins by NEP (Roques et al., 1993). Another is typified by inactivation of bradykinin, as a result of hydrolysis by angiotensin-converting enzyme (Skidgel et al., 1987).

Neuropeptide receptors

Following release, the neuroactive peptides bind to and activate specific receptors located pre- or post-synaptically on the surface of the target cells. The neuropeptide receptors, which are cellular components imbedded in the plasma membrane, belong to the family of G-protein-coupled receptors (Gudermann et al., 1996; Berthold and Bartfai, 1997). The primary structures of these receptors contain an extracellular NH_2-terminal sequence, seven putative α-helical transmembrane domains, and an intracellular COOH terminal domain of variable length. The G-protein-coupled receptor affects their function mainly through the activation of three distinct effector systems: (1) the adenylate cyclase/cAMP system, (2) the phospholipase C/inositol phosphate system, and (3) the regulation of ion channels. Neuropeptide effects on the adenylate cyclase system result in raising or lowering the concentration of cAMP within the cell. cAMP in turn regulates the activity of various protein kinases, which control the cell function in many different ways. Activation of the phospholipase C/inositol phosphate system causes an increase of intracellular calcium, which may lead to several events including contraction, secretion, enzyme activation, and membrane hyperpolarization. The neuropeptide action via its G-protein-coupled receptor directly linked to ion channels may result in such effects as membrane excitability and transmitter release. Repeated agonist stimulation of the G-protein-coupled receptor may cause desensitization. This is believed to be due to a reduced interaction between the receptor and the G-protein as a result of phosphorylation. Following phosphorylation, the receptors are reactivated by phosphatases or internalized by endocytosis subsequent to degradation or recycling (Koenig and Edwardson, 1997). Agonist-induced internalization is thought to be involved in the mechanism of opiate tolerance (Lefkowitz, 1998; Chaturvedi, 2003; Bohn et al., 2004).

Direct G-protein-mediated effects, through activation of opioid receptors, include activation of an inwardly rectifying potassium channel, inhibition of voltage-operated calcium channels and inhibition of adenylate cyclase (Grudt and Williams, 1993). These effects are involved in, for example, the ability of OPs to modulate the activity of pain transmitting pathways (Pasternak, 1993; Yaksh, 1997). At present, three distinct opioid receptors have been cloned and characterized, μ, δ, and κ. According to the IUPHAR nomenclature, these receptors are now named the OP receptors MOP, DOP, and KOP, respectively (Table 1). In some cases the OP receptors have also been shown to dimerize or oligomerize among themselves and each other, which may well influence their functional properties. OP receptors have also been shown to undergo homo- and hetero-dimerization, a process with potential implications for opioid physiology and pharmacology (Bohn et al., 2004; Wang et al., 2005).

The endogenous ligands for the MOP receptor are β-endorphin and, in particular, the endomorphins (Zadina et al., 1997; Horvath, 2000). The atypical endogenous OPs β-casomorphins and hemorphins are also selective for the MOP receptor (Nyberg et al., 1997). The most selective endogenous ligands for the DOP receptor are the enkephalins, whereas the KOP receptor recognizes the ProDYN-derived OPs as its endogenous ligands. The opioid-related peptide orphanin/nociceptin appears to act on the so-called ORL1 receptor (Meunier et al., 1995; Reinscheid et al., 1995).

Three tachykinin receptor subtypes are known to exist (Table 2): NK1, NK2, and NK3. SP exerts its effects through the tachykinin receptor NK1 (Table 2) via stimulation of adenylate cyclase as well as the inositol phosphate system (Maggio, 1988; Maggi et al., 1993; Datar et al., 2004). Stimulation of the intracellular signal systems for the NK1 receptor appears to be involved in the effects of SP on smooth muscle and pain transmission

(Datar et al., 2004). Neurokinin A (NKA), acting on NK2 receptors, produces slow excitatory synaptic potentials in the dorsal horn neurons. Neurokinin B (NKB) activates NK3 receptors, which are mainly located in the brain (Maggi et al., 1993; Datar et al., 2004).

The CGRP receptors have been classified as CGRP-1 and CGRP-2 subtypes, mainly depending on their affinity for the antagonist CGRP(8–37). The inhibitory effect of the CGRP(8–37) fragment on the activity of wide-dynamic-range neurons in the spinal dorsal horn of rats was recently shown (Yan and Yu, 2004) to be attenuated by the MOP receptor antagonist naloxone and also by application of the KOP receptor antagonist nor-binaltorphimine or the MOP receptor antagonist β-funaltrexamine, but not by the DOP-receptor antagonist naltrindole. It was suggested that this result indicated that the KOP and MOP receptors may be involved in the modulation of CGRP(8–37)-induced antinociception in the dorsal horn of the spinal cord in rats (Yan and Yu, 2004).

At least three distinct receptor sites have been identified and characterized for angiotensin. The AT1 and AT2 receptors are considered to be G-protein-coupled receptors (Zhuo et al., 1998). The AT3, AT4 and AT(1–7) sites, recognized by some fragments of angiotensin II (see Table 3), have been identified by binding assays.

Measurements of neuropeptides in heat stress

Most of the studies investigating the function of neuropeptides in heat stress have used techniques allowing the assessment of peptide levels in tissues and various body fluids. Other studies are based on investigation of peptide receptors or genes encoding the various peptides or their receptors. Studies of neuropeptides in healthy humans or those with pathological conditions are limited by, among other things, the restricted availability of human CNS tissue. Further, it is not ethically possible to use surgical techniques for experimental modeling, as is possible for animal experiments. Therefore, most studies on neuropeptides in relation to CNS disorders that have been carried out in the human CNS have been restricted to brain imaging and analysis of peptide levels in CSF and plasma. In the area of heat stress, most studies involving neuropeptides are based on assessment of their levels in plasma. Most procedures developed for quantification of plasma concentrations of neuropeptides have used immunological techniques. These include enzyme-linked immunosorbent assay (ELISA) and, in particular, radioimmunoassay (RIA). RIA is usually used for measuring OPs in plasma. Over the years, this technique has proven to be very sensitive and reproducible (Nyberg, 2004). It can detect peptides at levels down to some fmole/ml plasma and allows screening of a large number of samples within a relatively short time interval. A major difficulty with the RIA, however, is that the antibodies used may cross-react with molecules of similar structure to the antigen; furthermore, salt and proteins present in the sample may also interfere with the assay. To overcome this problem, many RIA procedures include a preextraction step using minicolumns with reversed-phase silica gels or with an anion exchanger. An additional disadvantage of the RIA method is that, because of the use of different antibodies and different procedures for preseparation of the CSF samples, outcomes can significantly vary between different laboratories. Nevertheless, RIA procedures have been a valuable tool for improving our understanding of the roles of various peptides in hyperthermia.

Neuropeptides in hyperthermia

Opioid peptides

Previous studies have suggested that the OPs may have a role in the regulation of body temperature; these compounds have thus been studied with respect to the response pattern to various kinds of heat stress (Nemeroff et al., 1979; Widdowson et al., 1983; Handler et al., 1992, 1995; Vescovi and Coiro, 1993; Sharma et al., 1997; Xin et al., 1997; Patel et al., 2002; Sharma and Alm, 2002; Kraemer et al., 2003). Soon after their discovery, the involvement of the endogenous opioids in hyperthermia became an area of interest. The antinociceptive and hypothermic effects of intracisternal OPs and morphine were evaluated in a mouse model more than 25 years ago (Nemeroff et al., 1979). It was found that β-endorphin exerted significant antinociceptive and hypothermic effects at a higher potency than Leu- and Met-enkephalin. In a subsequent study, the ability of OPs, and derivatives thereof, to influence rectal temperature after injection into the periaqueductal gray (PAG) region of the brain was investigated using a rat model (Widdowson et al., 1983). In a more recent study applying microdialysis techniques in a rat model, it was demonstrated that MOP receptor agonists elicit hyperthermia following administration into both the preoptic anterior hypothalamus and the PAG region (Xin et al., 1997). Moreover, it was also shown in this study that KOP receptor agonists induced hypothermia following microdialysis delivery into the PAG region. These data support the hypothesis that, at least in the rat, the hyperthermic response to opioids is mediated by the MOP receptor and the hypothermic response is mediated by the KOP receptor (Fig. 2). Thus, there seems to exist a tonic balance between the MOP and KOP receptors that serves as a homeostatic mechanism for maintaining the body temperature, as has been recently suggested (Chen et al., 2005).

Dynorphin A(1–17) immunoreactivity in the rat brain in response to hyperthermia was examined using a biological oxygen demand (BOD) incubator, where animals were subjected to 38°C for 4h (Sharma and Alm, 2002). The results indicated a marked upregulation of dynorphin immunoreactivity in several brain regions, including the cerebral cortex, hippocampus, cerebellum, and brain stem. Interestingly, preadministration of the potent nitric oxide synthase (NOS) inhibitors L-NAME or L-NMMA significantly attenuated dynorphin A(1–17) immunoreactivity in the brain. These drugs also reduced hyperthermia-induced BBB permeability, brain edema formation, and cell injury. The authors (Sharma and Alm, 2002) suggested that the hyperthermia produced in their model increases activity in the ProDYN system in several brain areas, which may contribute to heat stress and elicit damage; they also suggested that the mechanism for this effect involves nitric oxide (NO).

Animal studies also revealed that pyrogenic compounds such as interleukin (IL)-1β and prostaglandin E2, known to elicit fever following i.c.v. injection in the rat, also induce upregulation of β-endorphin immunoreactivity in the preoptic anterior hypothalamus (Tsai et al., 2003). However, the hyperthermic effect of these compounds was significantly reduced by pretreatment of buprenorphine, a MOP receptor blocker. This observation was thought to indicate that the actual pyrogens may enhance β-endorphin release in the hypothalamus and trigger fever, which can be attenuated by buprenorphine, in this case acting as an opioid receptor antagonist (Tsai et al., 2003).

Furthermore, studies have revealed that the involvement of β-endorphin in temperature regulation may be essential for the specific heat-stroke response in humans. A recent study (Kraemer et al., 2003) examined the response patterns of plasma levels of β-endorphin at baseline, during a 7-day exercise-heat acclimatization period (HA; 90 min per day at 40°C), and during a 6-h exercise-heat tolerance test (HTT; 6h at 40°C) performed before and after HA. The study included individuals previously diagnosed as heat stroke patients and a group of matched controls. No differences between the two groups were observed at rest, prior to HTT. The study showed, however, that β-endorphin levels in the patient group were significantly higher than resting values following 6h of HTT and were significantly higher than those in control subjects. During HA tests, a significant

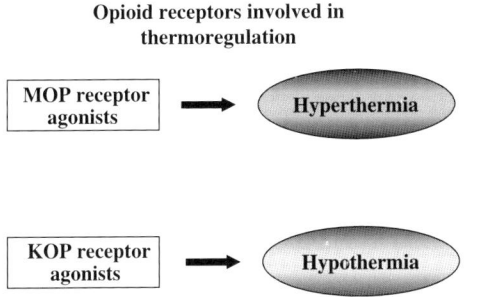

Fig. 2. The relationship between opioid receptors MOP and KOP in thermoregulation. The hyperthermic response to opioids is mediated by the MOP receptor and the hypothermic response is mediated by the KOP receptor.

enhancement in plasma concentrations of β-endorphin was recorded in the patients before exercise on day 1 and day 7, but not on day 4. Also, a significant increase was seen after exercise in both groups on all days of HA. However, no differences in β-endorphin levels were detected between the groups on days 4 and 7. The authors concluded that the OP response in individuals with a history of heat stroke was greater than in control subjects during the initial 6 h of HTT, probably as a result of decreased physical training among the patients. The disappearance of this difference after 7 days of HA, occurring as a result of reduced β-endorphin responses in patients, was suggested to reflect an adaptation of the pituitary function (Kraemer et al., 2003).

The role of endogenous opioids in response to hyperthermia has also been studied in drug or alcohol addicts. In this regard, β-endorphin was studied rather than enkephalins, because of the above-mentioned involvement in adaptation to heat in the CNS. However, it appears from the literature that the response of β-endorphin to the thermal stress of a sauna is absent in heroin, cocaine, or alcohol addicts (Vescovi et al., 1992). It was suggested that the opioid response to thermal stress, and thus the adaptation of the body to heat, is disrupted by the long-term stimulation of hypothalamic opioid neurotransmission that is produced directly or indirectly by alcohol, drugs or cocaine. Moreover, since chronic alcohol drinking is known to cause profound alterations in hypothalamic-pituitary function, the opioid response to heat was also examined in chronic alcoholics. Circulating β-endorphin and Met-enkephalin and even cardiovascular (blood pressure) changes in response to hyperthermic stress (sauna at 90°C for 30 min) were investigated in healthy men and in male alcoholics after 5 weeks of abstinence (Vescovi et al., 1997). It was found that alcoholic subjects had significantly lower increments in systolic blood pressure and lower β-endorphin increases in response to sauna than normal controls. In contrast, sauna-induced hyperthermia did not significantly alter the plasma levels of Met-enkephalin in either normal controls or chronic alcoholics. From this study, it was concluded that impairment of the adaptive response to heat stress affects male alcoholics even after a few weeks of abstinence from alcohol (Vescovi et al., 1997).

The opioid-related peptide nociceptin has also been studied regarding its role in hyperthermia using animal models. A recent study was designed to investigate the effects of nociceptin on body temperature and to explore whether the mechanism behind the alterations in thermogenesis induced by this peptide involved the opioid system (Chen et al., 2001). It was shown that i.c.v. injection of nociceptin at high doses produced hypothermia in adult rats. It was further observed that nociceptin reduced morphine-induced hyperthermia. Neither the OP receptor antagonist naloxone nor the KOP receptor antagonist nor-BNI affected the hypothermia induced by i.c.v. injection of nociceptin. Finally, an antisense oligo against the ORL1 receptor significantly decreased the hypothermia induced by i.c.v. injection of nociceptin. From these results the authors suggested (Chen et al., 2001) that the hypothermia elicited by i.c.v. injection of nociceptin at high doses is at least partially mediated by the ORL1 receptor, independently or downstream of OP receptors in the rat brain. The authors further suggested that nociceptin may act as a physiological antagonist to reduce morphine-induced hyperthermia.

The tachykinins

The involvement of tachykinins, CGRP, and their receptors in hyperthermia has been suggested from several independent studies. These peptides are located in close proximity to each other in many areas in the CNS, and both CGRP and SP are released following carrageenan-induced inflammation and hyperthermia (Garry and Hargreaves, 1992). Three hours after carrageenan injection, rat hindpaws exhibited hyperalgesia, edema, and hyperthermia. Spontaneous and capsaicin-evoked release of CGRP and SP were significantly increased following inflammation (Garry and Hargreaves, 1992). The change in body temperature after central administration of SP to rats has also been examined (Yakimova and Ovtcharov, 1988). Preimplanted cannules in the preoptic anterior part of the hypothalamus in normal rats allowed demonstration of an SP-elicited rise in temperature, which was significantly lower in adrenalectomized (AX) animals. When the animals were pretreated with the $GABA_A$ receptor agonist muscimol at doses not affecting the temperature, the SP-induced rise in temperature was significantly lower than that without muscimol in the non-AX rats. It was concluded from this study that a link between the GABA-ergic and the SP-ergic systems and the role of the suprarenal gland might exist for SP-induced hyperthermia (Yakimova and Ovtcharov, 1988).

Another link between the SP system and hyperthermia seems to involve the NO system. Studies confirming an interaction between NO and SP in neurogenic inflammatory responses have been carried out by measuring the change in the degree of Evans blue leakage and NO levels in perfusates from the subcutaneous space in the rat following noxious heat stimulation (Yonehara and Yoshimura, 2000). Furthermore, the effects of drugs affecting NOS were examined. In these studies, it was found that noxious heat stimulation elicited an increase in NO production, as recorded by the assessment of nitrite/nitrate (NO_x), into the perfusate concomitant with plasma extravasation. Chronic administration of the NOS inhibitor L-NAME significantly suppressed the increase in Evans blue extravasation induced by heat stimulation, whereas acute administration of L- and D-NAME had no significant effect. Heat-induced release of NO was also significantly suppressed by chronic, but not acute, pretreatment with L-NAME. The addition of SP to the perfusate caused a remarkable increase in the release of NO_x into the perfusate. Intra-arterial injection of the NK-1 receptor antagonist RP67580 (but not the NK-2 receptor antagonist SR48968) on the perfused side significantly attenuated the increases in Evans blue leakage and release of NO_x during heat stimulation. It was suggested by the authors of this study (Yonehara and Yoshimura, 2000) that heat-induced SP release from the peripheral endings of small-diameter afferent fibers causes NO generation mediated through the NK-1 receptor, and that NO enhances the inflammatory responses.

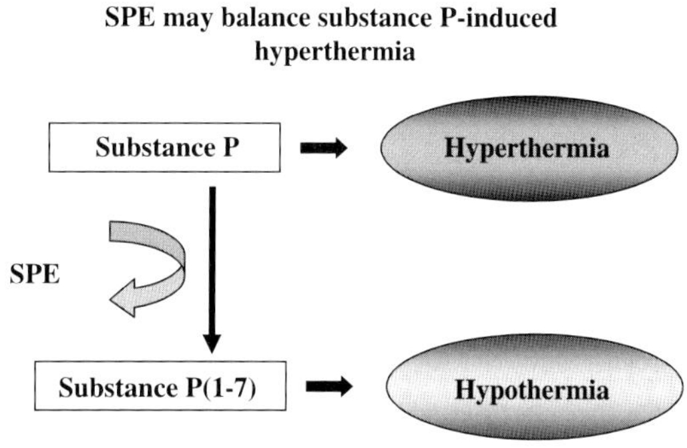

Fig. 3. The activity SPE is increased in the rat brain during heat stress (Karlsson et al., 2006) and may thereby reduce the action of SP and increase the level of the modulatory N-terminal fragment, SP(1–7). This is suggested to reflect an endogenous mechanism to counteract the expression of heat stress in the animal.

Additional studies on the role of tachykinins in hyperthermia have indicated that although neurokinins may be important for the early response to thermal injury, they do not appear to play an important role in the continual edema response (Waller et al., 1997). Thus, peptides like the tachykinins, as well as other vasoactive mediators, appear to have a role in the acute plasma extravasation observed after thermal injury, but not in the ongoing inflammatory injury.

Moreover, the observed increase in the SP regulatory enzyme SPE (Karlsson et al., 2006) seen in heat stressed rats may reflect a compensatory effect to counteract the action of SP. That is, the SPE product SP(1–7), which has an antiinflammatory activity (Wiktelius et al., 2006) may reduce the SP-induced hyperthermia (Fig. 3).

Calcitonin gene-related peptide

As mentioned above, CGRP, along with the tachykinins, has been attributed a role in hyperthermia. The peptide is released during carrageenan-induced hyperthermia and increased levels of CGRP have also been observed with other modes of heat stimuli. It is known that heat stress protects the myocardium against ischemia-reperfusion injury. CGRP has been shown to play an important role in the mediation of ischemic preconditioning, and a recent study has examined whether this peptide is involved in early or delayed preconditioning induced by retrograde hyperthermic perfusion *in vitro* or by whole-body hyperthermia *in vivo* (Song et al., 1999). It was shown that retrograde hyperthermic perfusion (42°C) for 5 min improved the recovery of cardiac function and elevated the level of CGRP in the coronary effluent. The fragment CGRP(8–37), a selective CGRP receptor antagonist, abolished the heat-stress induced cardioprotection. Furthermore, pretreatment with capsaicin, a compound previously shown to deplete the transmitter content in sensory neurons, abolished both the cardioprotection and the increased release of CGRP. Hyperthermia (42°C for 15 min) in intact animals elicited an enhancement in the plasma concentration of CGRP and early or delayed protection was shown in the hearts obtained from the animals subjected to whole-body hyperthermia both at 10 min and 48 h before the experiments. The early or delayed protection induced by heat stress in these animals was also abolished by pretreatment with capsaicin. The authors of this study concluded that, in the rat, CGRP is involved in both early and delayed cardioprotection induced by heat stress. In a study examining the role of CGRP in NO-mediated myocardial delayed preconditioning induced by heat stress, it was observed that hyperthermia significantly improved the recovery of cardiac protection, reduced the release of creatine kinase, and increased plasma concentrations of CGRP (Tan et al., 2001). Pretreatment with the NOS inhibitor L-NAME or capsaicin depleted the protective effects and attenuated the hyperthermia-induced increased concentration of CGRP. Thus, it seems that, at least in the rat, NO is involved in heat stress-induced cardioprotection and that the beneficial effects of NO involve the action of CGRP. It has further been shown that the cardioprotection afforded by CGRP-mediated preconditioning is due to inhibition of cardiac tumor necrosis factor-alpha (TNFα) production (Li and Peng, 2002).

A study on the expression of CGRP concentrations in the rat brain subjected to hyperthermic injury was examined in a comparatively recent study (Sharma et al., 2000). Thus, applying immunohistochemical techniques, the CGRP content was assessed in various brain regions in control animals and in rats subjected to heat stress. As a result of exposure to hyperthermic stress in a BOD incubator at 38°C for 4 h, a marked alteration in the expression of CGRP was observed. Thus, this mode of heat stress profoundly increased CGRP immunoreactivity in brain areas including the cerebral cortex, hippocampus, cerebellum, medulla, and spinal cord whereas, in other regions such as the brain stem and pons, the CGRP activity was decreased. It was concluded that heat stress influenced the levels of CGRP immunoreactivity in several brain areas previously suggested to be associated with the molecular mechanisms underlying hyperthermic stress.

Angiotensin and related peptides

Angiotensin II is a bioactive peptide that, in addition to its presence in peripheral cells and organs, is also widely distributed within the CNS. It is well recognized for its important roles in blood-pressure and body-fluid regulation but has recently also been reported to be involved in normal thermoregulation and fever (Watanabe et al., 2004). Angiotensin II acts as a hypothermia-inducing agent and lowers body temperature when it is centrally or systemically administered. However, studies have shown that, in its endogenous function, angiotensin II is involved both in heat-reducing responses in a hot environment and in thermogenesis in the cold. It has therefore been suggested that angiotensin II has a function in maintaining body temperature at the set-point. The peptide appears to mediate or modulate stress-induced fever through the AT1 receptor (Watanabe et al., 2004). Furthermore, as shown by these authors, at the final step in pyrogen-induced fever, angiotensin II facilitates the fever induced by PGE2 through its action on the AT2 receptor, while the first step in fever elicited by pyrogenic cytokines (e.g., IL-1β), as a response to lipopolysaccharides (LPS), seems to involve angiotensin II stimulation of its AT1 receptor. On the other hand, as it has been suggested that LPS at high doses may induce hypothermia in rodents through an interaction with TNFα, Watanabe et al. (2004)

have suggested that angiotensin II, which contributes to the LPS-induced production of cytokines such as IL-1β, is also involved in TNF production leading to LPS-induced hypothermia. They concluded that angiotensin II acting through its AT1 and AT2 receptors gives rise to a number of contributions to normal thermoregulation and fever, as well as the hypothermia in systemic inflammation.

These investigators also examined whether i.c.v. injection of angiotensin II may contribute to LPS-induced fever and to IL-1β production in the rat brain (Shimizu et al., 2004). Results indicated that LPS may induce dose-dependent fever and increases in the concentrations of IL-1β in several areas of the brain, such as the hypothalamus, hippocampus, and cerebellum. The observed effects were significantly inhibited by i.c.v. administration of either an ACE inhibitor or an AT1 receptor antagonist. In contrast, they found that the ACE inhibitor had no effect on the IL-1β (i.c.v.)-induced fever, whereas the effect was enhanced by the AT1 receptor antagonist. These observations were thought to suggest that, in the rat brain, angiotensin II may be involved in the LPS-induced production of IL-1β through stimulation of the AT1 receptor, leading to the fever induced by the presence of LPS (Shimizu et al., 2004). An interaction between the NO system and AT1/AT2 receptors in thermoregulation has also been suggested (Schwimmer et al., 2004). Furthermore, a review article dealing with LPS-induced hyperthermia and the role of PGs, particularly PGE2 and its receptors (Romanovsky et al., 2005) has recently been published. Although there are many PGE2 receptors, EP3 is suggested to be a primary "fever receptor".

Conclusions

It is well established that the inflammatory signaling and thermoeffector pathways involved in fever and hyperthermia are modulated by neuropeptides as well a number of other compounds involved in neuronal transmission. There are a large number of neuroactive peptides and peptide hormones that have a putative role in the molecular mechanisms underlying hyperthermia. Over recent years, the roles of some compounds have become better understood and in some cases even revised. Current progress in peptide research using new technology will undoubtedly lead to new discoveries about neuropeptide functions in hyperthermia and this may in turn provide a basis for the development of new pharmacological tools for treating various types of disorders related to heat stress and hyperthermia. This includes the discovery of new targets for antipyretic therapy but also the development of drugs for the treatment of damage caused by heat exposure. However, it should be noted that most studies on neuropeptides in relation to hyperthermia have been carried out using animal models, with an obvious lack of studies investigating these compounds in humans.

Acknowledgments

This study was supported by the Swedish Medical Research Council (grant no. 9459).

References

Ahmed, R.G. (2005) Heat stress induced histopathology and pathophysiology of the central nervous system. Int. J. Dev. Neurosci., 23: 549–557.

Berthold, M. and Bartfai, T. (1997) Modes of peptide binding in G protein-coupled receptors. Neurochem. Res., 22: 1023–1031.

Bohn, L.M., Dykstra, L.A., Lefkowitz, R.J., Caron, M.G. and Barak, L.S. (2004) Relative opioid efficacy is determined by the complements of the G protein-coupled receptor desensitization machinery. Mol. Pharmacol., 66: 106–112.

Bonetto, V., Jornvall, H., Andersson, M., Renlund, S., Mutt, V. and Sillard, R. (1999) Isolation and characterization of sulphated and nonsulphated forms of cholecystokinin-58 and their action on gallbladder contraction. Eur. J. Biochem., 264: 336–340.

Brantl, V., Gramsch, C., Lottspeich, F., Mertz, R., Jaeger, K.H. and Herz, A. (1986) Novel opioid peptides derived from hemoglobin: hemorphins. Eur. J. Pharmacol., 125: 309–310.

Brantl, V., Teschemacher, H., Blasig, J., Henschen, A. and Lottspeich, F. (1981) Opioid activities of beta-casomorphins. Life Sci., 28: 1903–1909.

Bundgaard, J.R., Vuust, J. and Rehfeld, J.F. (1997) New consensus features for tyrosine O-sulfation determined by mutational analysis. J. Biol. Chem., 272: 21700–21705.

Camargo, A.C., Oliveira, E.B., Toffoletto, O., Metters, K.M. and Rossier, J. (1987) Brain endo-oligopeptidase A: a

putative enkephalin converting enzyme. J. Neurochem., 48: 1258–1263.
Canaff, L., Bennett, H.P. and Hendy, G.N. (1999) Peptide hormone precursor processing: getting sorted? Mol. Cell Endocrinol. 156: 1–6.
Chang, C.K., Chiu, W.T., Chang, C.P. and Lin, M.T. (2004) Effect of hypervolaemic haemodilution on cerebral glutamate, glycerol, lactate and free radicals in heatstroke rats. Clin. Sci. (Lond.), 106: 501–509.
Chaturvedi, K. (2003) Opioid peptides, opioid receptors and mechanism of down regulation. Indian J. Exp. Biol., 41: 5–13.
Chen, X., McClatchy, D.B., Geller, E.B., Liu-Chen, L., Tallarida, R.J. and Adler, M.W. (2001) Possible mechanism of hypothermia induced by intracerebroventricular injection of orphanin FQ/nociceptin. Brain Res., 904: 252–258.
Chen, X., McClatchy, D.B., Geller, E.B., Tallarida, R.J. and Adler, M.W. (2005) The dynamic relationship between mu and kappa opioid receptors in body temperature regulation. Life Sci., 78: 329–333.
Cooper, J.R., Bloom, F.E. and Roth, R.H. (1996) Biochemical basis of neuropharmacology. Oxford University Press, New York, NY.
Datar, P., Srivastava, S., Coutinho, E. and Govil, G. (2004) Substance P: structure, function, and therapeutics. Curr. Top. Med. Chem., 4: 75–103.
Demuth, H.U., Schlenzig, D., Schierhorn, A., Grosche, G., Chapot-Chartier, M.P. and Gripon, J.C. (1993) Design of (omega-N-(O-acyl)hydroxy amid) aminodicarboxylic acid pyrrolidides as potent inhibitors of proline-specific peptidases. FEBS Lett., 320: 23–27.
Devi, L. and Goldstein, A. (1986) Conversion of leumorphin (dynorphin B-29) to dynorphin B and dynorphin B-14 by thiol protease activity. J. Neurochem., 47: 154–157.
von Eggelkraut-Gottanka, R. and Beck-Sickinger, A.G. (2004) Biosynthesis of peptide hormones derived from precursor sequences. Curr. Med. Chem., 11: 2651–2665.
Eipper, B.A., Park, L.P., Dickerson, I.M., Keutmann, H.T., Thiele, E.A., Rodriguez, H., Schofield, P.R. and Mains, R.E. (1987) Structure of the precursor to an enzyme mediating COOH-terminal amidation in peptide biosynthesis. Mol. Endocrinol., 1: 777–790.
Foulon, T., Cadel, S., Chesneau, V., Draoui, M., Prat, A. and Cohen, P. (1996) Two novel metallopeptidases with a specificity for basic residues: functional properties, structure and cellular distribution. Ann. N.Y. Acad. Sci., 780: 106–120.
Fricker, L.D. and Snyder, S.H. (1983) Purification and characterization of enkephalin convertase, an enkephalin-synthesizing carboxypeptidase. J. Biol. Chem., 258: 10950–10955.
Garry, M.G. and Hargreaves, K.M. (1992) Enhanced release of immunoreactive CGRP and substance P from spinal dorsal horn slices occurs during carrageenan inflammation. Brain Res., 582: 139–142.
Grudt, T.J. and Williams, J.T. (1993) kappa-opioid receptors also increase potassium conductance. Proc. Natl. Acad. Sci. U.S.A., 90: 11429–11432.

Gudermann, T., Kalkbrenner, F. and Schultz, G. (1996) Diversity and selectivity of receptor-G protein interaction. Annu. Rev. Pharmacol. Toxicol., 36: 429–459.
Hallberg, M., Le Greves, P. and Nyberg, F. (2005) Neuropeptide processing. In: Lendeckel U. and Hooper N.M. (Eds.), Proteases in Biology and Disease. Proteases in the Brain. Springer Science, New York, NY, pp. 203–234.
Hallberg, M. and Nyberg, F. (2003) Neuropeptide conversion to bioactive fragments: an important pathway in neuromodulation. Curr. Protein Pept. Sci., 4: 31–44.
Handler, C.M., Geller, E.B. and Adler, M.W. (1992) Effect of mu-, kappa-, and delta-selective opioid agonists on thermoregulation in the rat. Pharmacol. Biochem. Behav., 43: 1209–1216.
Handler, C.M., Mondgock, D.J., Zhao, S.F., Geller, E.B. and Adler, M.W. (1995) Interaction between opioid agonists and neurotensin on thermoregulation in the rat. I. Body temperature. J. Pharmacol. Exp. Ther., 274: 284–292.
Hersh, L.B. (1985) Characterization of membrane-bound aminopeptidases from rat brain: identification of the enkephalin-degrading aminopeptidase. J. Neurochem., 44: 1427–1435.
Hokfelt, T., Castel, M.-N., Morino, P., Zhang, X. and Dagerlind, A. (1994) General overview of neuropeptides. Raven Press, New York, NY.
Hook, V.Y., Schiller, M.R., Azaryan, A.V. and Tezapsidis, N. (1996) Proenkephalin-processing enzymes in chromaffin granules: model for neuropeptide biosynthesis. Ann. N.Y. Acad. Sci., 780: 121–133.
Horvath, G. (2000) Endomorphin-1 and endomorphin-2: pharmacology of the selective endogenous mu-opioid receptor agonists. Pharmacol. Ther., 88: 437–463.
Jornvall, H. (1975) Acetylation of protein N-terminal amino groups structural observations on alpha-amino acetylated proteins. J. Theor. Biol., 55: 1–12.
Kao, T.Y. and Lin, M.T. (1996) Brain serotonin depletion attenuates heatstroke-induced cerebral ischemia and cell death in rats. J. Appl. Physiol., 80: 680–684.
Karlsson, K., Sharma, H. and Nyberg, F. (2006) Chromatographic characterization of substance P endopeptidase in the rat brain reveals affected enzyme activity following heat stress. Biomed. Chromatogr., 20: 77–82.
Katopodis, A.G., Ping, D.S., Smith, C.E. and May, S.W. (1991) Functional and structural characterization of peptidylamidoglycolate lyase, the enzyme catalyzing the second step in peptide amidation. Biochemistry, 30: 6189–6194.
Koenig, J.A. and Edwardson, J.M. (1997) Endocytosis and recycling of G protein-coupled receptors. Trends Pharmacol. Sci., 18: 276–287.
Koko, V., Djordjeviae, J., Cvijiae, G. and Davidoviae, V. (2004) Effect of acute heat stress on rat adrenal glands: a morphological and stereological study. J. Exp. Biol., 207: 4225–4230.
Kraemer, W.J., Armstrong, L.E. and Watson, G. (2003) The effects of exertional heatstroke and exercise-heat acclimation on plasma beta-endorphin concentrations. Aviat. Space Environ. Med., 74: 758–762.

Lefkowitz, R.J. (1998) G protein-coupled receptors. III. New roles for receptor kinases and beta-arrestins in receptor signaling and desensitization. J. Biol. Chem., 273: 18677–18680.

Le Greves, P., Sharma, H.S., Westman, J., Alm, P. and Nyberg, F. (1997) Acute heat stress induces edema and nitric oxide synthase upregulation and down-regulates mRNA levels of the NMDAR1, NMDAR2A and NMDAR2B subunits in the rat hippocampus. Acta Neurochir. Suppl., 70: 275–278.

Li, Y.J. and Peng, J. (2002) The cardioprotection of calcitonin gene-related peptide-mediated preconditioning. Eur. J. Pharmacol., 442: 173–177.

Lieberman, H.R., Georgelis, J.H., Maher, T.J. and Yeghiayan, S.K. (2005) Tyrosine prevents effects of hyperthermia on behavior and increases norepinephrine. Physiol. Behav., 84: 33–38.

Loh, Y.P. (1987) Peptide precursor processing enzymes within secretory vesicles. Ann. N.Y. Acad. Sci., 493: 292–307.

Maggi, C.A., Patacchini, R., Rovero, P. and Giachetti, A. (1993) Tachykinin receptors and tachykinin receptor antagonists. J. Auton. Pharmacol., 13: 23–93.

Maggio, J.E. (1988) Tachykinins. Annu. Rev. Neurosci., 11: 13–28.

McGugan, E.A. (2001) Hyperpyrexia in the emergency department. Emerg. Med. (Fremantle), 13: 116–120.

Meunier, J.C., Mollereau, C., Toll, L., Suaudeau, C., Moisand, C., Alvinerie, P., Butour, J.L., Guillemot, J.C., Ferrara, P., Monsarrat, B., et al. (1995) Isolation and structure of the endogenous agonist of opioid receptor-like ORL1 receptor. Nature, 377: 532–535.

Michael-Titus, A.T., Fernandes, K., Setty, H. and Whelpton, R. (2002) In vivo metabolism and clearance of substance P and co-expressed tachykinins in rat striatum. Neuroscience, 110: 277–286.

Nemeroff, C.B., Osbahr III, A.J., Manberg, P.J., Ervin, G.N. and Prange Jr., A.J. (1979) Alterations in nociception and body temperature after intracisternal administration of neurotensin, beta-endorphin, other endogenous peptides, and morphine. Proc. Natl. Acad. Sci. U.S.A., 76: 5368–5371.

Nikoshkov, A., Hurd, Y.L., Yakovleva, T., Bazov, I., Marinova, Z., Cebers, G., Pasikova, N., Gharibyan, A., Terenius, L. and Bakalkin, G. (2005) Prodynorphin transcripts and proteins differentially expressed and regulated in the adult human brain. FASEB J., 19: 1543–1545.

Nyberg, F. (2004) Opioid peptides in cerebrospinal fluid-methods for analysis and their significance in the clinical perspective. Front. Biosci., 9: 3510–3525.

Nyberg, F., Lieberman, H., Lindstrom, L.H., Lyrenas, S., Koch, G. and Terenius, L. (1989) Immunoreactive beta-casomorphin-8 in cerebrospinal fluid from pregnant and lactating women: correlation with plasma levels. J. Clin. Endocrinol. Metab., 68: 283–289.

Nyberg, F., Sanderson, K. and Glamsta, E.L. (1997) The hemorphins: a new class of opioid peptides derived from the blood protein hemoglobin. Biopolymers, 43: 147–156.

Park, H.G., Han, S.I., Oh, S.Y. and Kang, H.S. (2005) Cellular responses to mild heat stress. Cell. Mol. Life Sci., 62: 10–23.

Pasternak, G.W. (1993) Pharmacological mechanisms of opioid analgesics. Clin. Neuropharmacol., 16: 1–18.

Patel, H.H., Hsu, A. and Gross, G.J. (2002) Attenuation of heat shock-induced cardioprotection by treatment with the opiate receptor antagonist naloxone. Am. J. Physiol. Heart Circ. Physiol., 282: H2011–H2017.

Persson, S., Le Greves, P., Thornwall, M., Eriksson, U., Silberring, J. and Nyberg, F. (1995) Neuropeptide converting and processing enzymes in the spinal cord and cerebrospinal fluid. Prog. Brain Res., 104: 111–130.

Persson, S., Post, C., Holmdahl, R. and Nyberg, F. (1992) Decreased neuropeptide-converting enzyme activities in cerebrospinal fluid during acute but not chronic phases of collagen induced arthritis in rats. Brain Res., 581: 273–282.

Reinscheid, R.K., Nothacker, H.P., Bourson, A., Ardati, A., Henningsen, R.A., Bunzow, J.R., Grandy, D.K., Langen, H., Monsma Jr., F.J. and Civelli, O. (1995) Orphanin FQ: a neuropeptide that activates an opioidlike G protein-coupled receptor. Science, 270: 792–794.

Romanovsky, A., Almeida, M.C., Aronoff, D.M., Ivanov, A.I., Konsman, J.P., Steiner, A.A. and Turek, V.F. (2005) Fever and hypothermia in systemic inflammation: recent discoveries and revisions. Front. Biosci., 10: 2193–2216.

Roques, B.P., Noble, F., Dauge, V., Fournie-Zaluski, M.C. and Beaumont, A. (1993) Neutral endopeptidase 24.11: structure, inhibition, and experimental and clinical pharmacology. Pharmacol. Rev., 45: 87–146.

Rossi, G.C., Leventhal, L., Bolan, E. and Pasternak, G.W. (1997) Pharmacological characterization of orphanin FQ/nociceptin and its fragments. J. Pharmacol. Exp. Ther., 282: 858–865.

Sakurada, T., Sakurada, S., Katsuyama, S., Hayashi, T., Sakurada, C., Tan-No, K., Johansson, H., Sandin, J. and Terenius, L. (2000) Evidence that N-terminal fragments of nociceptin modulate nociceptin-induced scratching, biting and licking in mice. Neurosci. Lett., 279: 61–64.

Schwimmer, H., Gerstberger, R. and Horowitz, M. (2004) Nitric oxide and angiotensin II: neuromodulation of thermoregulation during combined heat and hypohydration stress. Brain Res., 1006: 177–189.

Seidah, N.G. and Chretien, M. (1999) Proprotein and prohormone convertases: a family of subtilases generating diverse bioactive polypeptides. Brain Res., 848: 45–62.

Sharma, H.S. and Alm, P. (2002) Nitric oxide synthase inhibitors influence dynorphin A (1–17) immunoreactivity in the rat brain following hyperthermia. Amino Acids, 23: 247–259.

Sharma, H.S., Nyberg, F., Cervos-Navarro, J. and Dey, P.K. (1992a) Histamine modulates heat stress-induced changes in blood-brain barrier permeability, cerebral blood flow, brain oedema and serotonin levels: an experimental study in conscious young rats. Neuroscience, 50: 445–454.

Sharma, H.S., Westman, J., Cervos-Navarro, J., Dey, P.K. and Nyberg, F. (1997) Opioid receptor antagonists attenuate heat stress-induced reduction in cerebral blood flow, increased blood-brain barrier permeability, vasogenic edema and cell changes in the rat. Ann. N.Y. Acad. Sci., 813: 559–571.

Sharma, H.S., Westman, J. and Nyberg, F. (1998) Pathophysiology of brain edema and cell changes following hyperthermic brain injury. Prog. Brain Res., 115: 351–412.

Sharma, H.S., Westman, J. and Nyberg, F. (2000) Selective alteration of calcitonin gene related peptide in hyperthermic brain injury. An experimental study in the rat brain using immunohistochemistry. Acta Neurochir. Suppl., 76: 541–545.

Sharma, H.S., Westman, J., Nyberg, F., Cervos-Navarro, J. and Dey, P.K. (1992b) Role of serotonin in heat adaptation: an experimental study in the conscious young rat. Endocr. Regul., 26: 133–142.

Sharma, H.S., Westman, J., Nyberg, F., Cervos-Navarro, J. and Dey, P.K. (1994) Role of serotonin and prostaglandins in brain edema induced by heat stress. An experimental study in the young rat. Acta Neurochir. Suppl. (Wien.), 60: 65–70.

Shimizu, H., Miyoshi, M., Matsumoto, K., Goto, O., Imoto, T. and Watanabe, T. (2004) The effect of central injection of angiotensin-converting enzyme inhibitor and the angiotensin type 1 receptor antagonist on the induction by lipopolysaccharide of fever and brain interleukin-1beta response in rats. J. Pharmacol. Exp. Ther., 308: 865–873.

Silberring, J., Castello, M.E. and Nyberg, F. (1992) Characterization of dynorphin A-converting enzyme in human spinal cord. An endoprotease related to a distinct conversion pathway for the opioid heptadecapeptide? J. Biol. Chem. 267: 21324–21328.

Silberring, J. and Nyberg, F. (1989) A novel bovine spinal cord endoprotease with high specificity for dynorphin B. J. Biol. Chem., 264: 11082–11086.

Skidgel, R.A., Defendini, R. and Erdos, E.G. (1987) Angiotensin I converting enzyme and its role in neuropeptide metabolism. In: Turner A.J. (Ed.), Neuropeptides and their Peptidases. Ellis Horwood, Chichester, England, pp. 165–182.

Song, Q.J., Li, Y.J. and Deng, H.W. (1999) Early and delayed cardioprotection by heat stress is mediated by calcitonin gene-related peptide. Naunyn Schmiedebergs Arch. Pharmacol., 359: 477–483.

Suganuma, T., Irie, K., Fujii, E., Yoshioka, T. and Muraki, T. (2002) Effect of heat stress on lipopolysaccharide-induced vascular permeability change in mice. J. Pharmacol. Exp. Ther., 303: 656–663.

Tan, B., He, S.Y., Deng, H.W. and Li, Y.J. (2001) Role of calcitonin gene-related peptide in nitric oxide-mediated myocardial delayed preconditioning induced by head stress. Acta Pharmacol. Sin., 22: 851–856.

Tan-No, K., Esashi, A., Nakagawasai, O., Niijima, F., Tadano, T., Sakurada, C., Sakurada, T., Bakalkin, G., Terenius, L. and Kisara, K. (2002) Intrathecally administered big dynorphin, a prodynorphin-derived peptide, produces nociceptive behavior through an N-methyl-D-aspartate receptor mechanism. Brain Res., 952: 7–14.

Taylor, A. (1993) Aminopeptidases: structure and function. FASEB J., 7: 290–298.

Tsai, S.M., Lin, M.T., Wang, J.J. and Huang, W.T. (2003) Pyrogens enhance beta-endorphin release in hypothalamus and trigger fever that can be attenuated by buprenorphine. J. Pharmacol. Sci., 93: 155–162.

Vescovi, P.P. and Coiro, V. (1993) Hyperthermia and endorphins. Biomed. Pharmacother., 47: 301–304.

Vescovi, P.P., Coiro, V., Volpi, R., Giannini, A. and Passeri, M. (1992) Hyperthermia in sauna is unable to increase the plasma levels of ACTH/cortisol, beta-endorphin and prolactin in cocaine addicts. J. Endocrinol. Invest., 15: 671–675.

Vescovi, P.P., DiGennaro, C. and Coiro, V. (1997) Hormonal (ACTH, cortisol, beta-endorphin, and met-enkephalin) and cardiovascular responses to hyperthermic stress in chronic alcoholics. Alcohol Clin. Exp. Res., 21: 1195–1198.

Waller, J., Siney, L., Hoult, J.R. and Brain, S.D. (1997) A study of neurokinins and other oedema-inducing mediators and mechanisms in thermal injury. Clin. Exp. Pharmacol. Physiol., 24: 861–863.

Wang, D., Sun, X., Bohn, L.M. and Sadee, W. (2005) Opioid receptor homo- and heterodimerization in living cells by quantitative bioluminescence resonance energy transfer. Mol. Pharmacol., 67: 2173–2184.

Watanabe, T., Miyoshi, M. and Imoto, T. (2004) Angiotensin II: its effects on fever and hypothermia in systemic inflammation. Front. Biosci., 9: 438–447.

Widdowson, P.S., Griffiths, E.C. and Slater, P. (1983) Body temperature effects of opioids administered into the periaqueductal grey area of rat brain. Regul. Pept., 7: 259–267.

Wiktelius, D., Khalil, Z. and Nyberg, F. (2006) Modulation of peripheral inflammation by the substance P N-terminal metabolite substance P1–7. Peptides, 27: 1490–1497.

Wimalawansa, S.J. (1996) Calcitonin gene-related peptide and its receptors: molecular genetics, physiology, pathophysiology, and therapeutic potentials. Endocr. Rev., 17: 533–585.

Xin, L., Geller, E.B. and Adler, M.W. (1997) Body temperature and analgesic effects of selective mu and kappa opioid receptor agonists microdialyzed into rat brain. J. Pharmacol. Exp. Ther., 281: 499–507.

Yakimova, K. and Ovtcharov, R. (1988) Effect of the body temperature upon central administration of substance P: the role of adrenalectomy and GABA-activation. Acta Physiol. Pharmacol. (Bulg.), 14: 65–70.

Yaksh, T.L. (1997) Pharmacology and mechanisms of opioid analgesic activity. Acta Anaesthesiol. Scand., 41: 94–111.

Yan, Y. and Yu, L.C. (2004) Involvement of opioid receptors in the CGRP8-37-induced inhibition of the activity of wide-dynamic-range neurons in the spinal dorsal horn of rats. J. Neurosci. Res., 77: 148–152.

Yang, Y.L. and Lin, M.T. (1999) Heat shock protein expression protects against cerebral ischemia and monoamine overload in rat heatstroke. Am. J. Physiol., 276: H1961–H1967.

Yaron, A. and Naider, F. (1993) Proline-dependent structural and biological properties of peptides and proteins. Crit. Rev. Biochem. Mol. Biol., 28: 31–81.

Yeo, T.P. (2004) Heat stroke: a comprehensive review. AACN Clin. Issues, 15: 280–293.

Yonehara, N. and Yoshimura, M. (2000) Interaction between nitric oxide and substance P on heat-induced inflammation in rat paw. Neurosci. Res., 36: 35–43.

Zadina, J.E., Hackler, L., Ge, L.J. and Kastin, A.J. (1997) A potent and selective endogenous agonist for the mu-opiate receptor. Nature, 386: 499–502.

Zhou, Q., Karlsson, K., Liu, Z., Johansson, P., Le Greves, M., Kiuru, A. and Nyberg, F. (2001) Substance P endopeptidase-like activity is altered in various regions of the rat central nervous system during morphine tolerance and withdrawal. Neuropharmacology, 41: 246–253.

Zhou, Q., Le Greves, P., Ragnar, F. and Nyberg, F. (2000) Intracerebroventricular injection of the N-terminal substance P fragment SP(1–7) regulates the expression of the N-methyl-D-aspartate receptor NR1, NR2A and NR2B subunit mRNAs in the rat brain. Neurosci. Lett., 291: 109–112.

Zhou, Q., Liu, Z., Ray, A., Huang, W., Karlsson, K. and Nyberg, F. (1998) Alteration in the brain content of substance P(1–7) during withdrawal in morphine-dependent rats. Neuropharmacology, 37: 1545–1552.

Zhuo, J., Moeller, I., Jenkins, T., Chai, S.Y., Allen, A.M., Ohishi, M. and Mendelsohn, F.A. (1998) Mapping tissue angiotensin-converting enzyme and angiotensin AT1, AT2 and AT4 receptors. J. Hypertens., 16: 2027–2037.

CHAPTER 15

Interaction between amino acid neurotransmitters and opioid receptors in hyperthermia-induced brain pathology

Hari Shanker Sharma*

Abstract: This review is focused on the possible interaction between amino acid neurotransmitters and opioid receptors in hyperthermia-induced brain dysfunction. A balance between excitatory and inhibitory amino acids appears to be necessary for normal brain function. Increased excitotoxicity and a decrease in inhibitory amino acid neurotransmission in hyperthermia are associated with brain pathology and cognitive impairment. This is supported by recent data from our laboratory that show a marked increase in glutamate and aspartate and a decrease in GABA and glycine in several brain areas following heat stress at the time of brain pathology. Blockade of multiple opioid receptors with naloxone restored the heat stress-induced decline in GABA and glycine and thwarted the elevation of glutamate and aspartate in the CNS. In naloxone-treated stressed animals, cognitive dysfunction and brain pathology are largely absent. Taken together, these new findings suggest that an intricate balance between excitatory and inhibitory amino acids is important for brain function in heat stress. In addition, opioid receptors play neuromodulatory roles in amino acid neurotransmission in hyperthermia.

Keywords: hyperthermia; brain injury; naloxone; blood-brain barrier permeability; brain edema formation; cell injury; glutamate; GABA; glycine; aspartate; amino acids; neurotransmitters

Introduction

Hyperthermia associated with heat stroke is a life-threatening illness, particularly when the body temperature reaches beyond 40°C (Sharma, 1999, 2005a, b, c, 2006a, b). Severe hyperthermia induces central nervous system (CNS) dysfunction, such as delirium, convulsion and coma (see Bouchama and Knochel, 2002). The death rate is often higher than 50% of heat stroke victims, despite lowering of the body temperature by cooling or therapeutic intervention (see Sharma and Hoopes, 2003). Those who survive exhibit permanent neurological deficit (see Bouchama and Knochel, 2002 for details). Thus, further studies are needed to understand the possible mechanisms of heat stress-induced CNS dysfunction.

The CNS is very rich in amino acids and opioid peptides (Nyberg et al., 1995; Singewald and Philippu, 1998; Chavkin, 2000; Sprenger et al., 2005; Wang and Yang, 2005; Heja et al., 2006; Sharma, 2006a). It appears that a balance between excitatory and inhibitory amino acids is necessary for normal brain function (Aguilar et al., 2005; Ohnuma et al., 2005). Since opioids modulate neurotransmission either through classical μ-, δ- or κ-opioid receptors and/or through NMDA/AMPA receptors in the CNS (Hughes and Woodruff, 1992; Kozela and Popik, 2002), it is likely that

*Corresponding author. Tel.:/Fax: +4618-243899;
E-mail: Sharma@surgsci.uu.se

their interaction with other neurotransmitters plays an important role in hyperthermia. Thus, new studies directed to understand the role of various neurotransmitters and/or their interaction with several neuromodulators and their receptors in hyperthermia are needed to expand our knowledge on the molecular mechanisms of brain pathology in heat stress.

This review is based on our investigations on whole-body hyperthermia (WBH)-induced alterations in excitatory amino acids, glutamate and aspartate, and inhibitory amino acids, GABA and glycine in the CNS. Furthermore, the role of opioids in neuromodulation of amino acid neurotransmitters in hyperthermia based on new data from our laboratory is also discussed.

Problems of hyperthermia among human populations

Heat stress-induced hyperthermia poses severe clinical problems in the world, particularly during the summer season (see Sharma and Hoopes, 2003; Sharma, 2005a, b, 2006a, b). The magnitude of heat-related deaths among human populations has reached an alarming level, probably due to global warming, resulting in increased frequency and intensity of heat waves worldwide (Keatinge and Donaldson, 2004; Poumadere et al., 2005; see Sharma, 1982, 1999, 2004a; Sharma and Westman, 1998, 2004; Sharma and Hoopes, 2003; Stott et al., 2004).

Heat waves during summer are responsible for a large number of deaths in the United States of America, Canada, Europe, Asia and Australia (Sharma, 2005a, 2006a, see Sharma, Chapter 10 in this volume). Approximately 8–10,000 deaths are recorded worldwide during heat waves when the air temperature ranges between 32° and 34°C (Buechley et al., 1972; Smoyer, 1998; Rydman et al., 1999; Smoyer et al., 2000; Dessai, 2003; Pirard et al., 2005; Centers for Disease Control and Prevention (CDC), 2006). Thus, the heat-related deaths in fact are the highest compared to mortality caused by other major natural disasters.

Interestingly, heat-related mortality occurred in cities with relatively high levels of urbanization and high costs of living, e.g., in Canada (Smoyer et al., 2000). However, it is still unclear whether populations suffering from acute or chronic cardiovascular, endocrine or metabolic disease are more vulnerable to heat-related deaths.

Amino acid neurotransmitters in thermoregulation

The CNS contains high amounts of amino acid neurotransmitters (see Chavkin, 2000). However, their role in heat stress-induced brain pathology or neurological deficits is still not well known (see Dascombe, 1985; Wischmeyer, 2002). Taurine and γ-amino butyric acid (GABA) are present in high concentrations in various hypothalamic nuclei regulating thermoregulatory mechanisms (see Kerwin and Pycock, 1979; Bligh, 1981; DeFeudis, 1984; Keil et al., 1994). Frontal cerebral cortex that controls heat production is rich in aspartate and glutamate (see Monda et al., 1998; Madden and Morrison, 2003). These excitatory amino acids are released in high quantities in the extracellular fluid following prostaglandin E1 (PGE1)-induced hyperthermia (Monda et al., 1998). Intracerebroventricular (icv) administration of excitatory amino acids, aspartate and glutamate induces hyperthermia in rats (Bligh, 1981; Cremades and Penafiel, 1982), and central or systemic administration of GABA or taurine induces dose-dependent hypothermia in rodents (Serrano et al., 1985; Sgaragli and Palmi, 1985). These early observations suggest that amino acid neurotransmitters are involved in the physiological mechanisms of temperature regulation.

Amino acid neurotransmitters in pathophysiology of hyperthermia

There are reasons to believe that amino acid neurotransmitters are also involved in the pathological mechanisms of hyperthermia. This idea is supported by the fact that a localized brain heating causes glutamate release, the magnitude of which depends on the intensity and severity of hyperthermia (see Sharma, 2006). However, glutamate-induced neurotoxicity in hyperthermia can only be seen when the brain temperature reaches beyond 43°C but not at 41°C. These findings suggest that glutamate release in hyperthermia induces neurotoxicity.

Apart from excitatory amino acids, the inhibitory amino acids are also released in hyperthermia (see Frosini et al., 2004, 2006, and Chapter 22 in this volume). Thus, fever induced by intracisternal injection of pyrogens increases taurine and GABA concentration in the frontal cerebral cortex (Keil et al., 1994; Frosini et al., 2000a, b, and in this volume). An increase in GABA depresses axonal conduction and thereby contributes to the secondary axonal damage. This alteration in the GABA and glycine depends on the magnitude and intensity of hyperthermia. However, studies on interaction between excitatory and inhibitory amino acids in hyperthermia and their roles in brain pathology are still lacking.

Opioid peptides and their interaction with neurotransmitters

High concentrations of opioid peptides are present in various parts of the CNS (Nyberg et al., 1995). The opioid neurotransmission is important in memory function, sleep, stress, thermoregulation, behaviour and drug addiction (Björklund et al., 1990; Nyberg et al., 1995). Opioid peptides are known neuromodulators and thus could influence the function of other neurotransmitters, such as serotonin, prostaglandins, histamine and/or amino acids (see Nyberg et al., 1995 and Chapter 14 in this volume). Most of the opioid peptides are often co-localized with a large number of other classical neurotransmitters (Hökfelt et al., 1978, 1987). It is quite likely that modulation of opioid receptors influences the function of other neurotransmitters in the CNS.

One of the important opioid peptides is dynorphin that is involved in brain and spinal cord pathology following ischaemia, trauma and hyperthermia (Sharma et al., 1993a, 1995; Thornwall et al., 1997; Sharma and Alm, 2002; Sharma, 2004b; 2006). The dynorphins represent one of the three major groups of opioid peptides and are present in very high quantity in the CNS (Sharma et al., 1997b, c, 1992c; for details see Fallon and Ciofi, 1990). In some neurons, dynorphins co-exist with enkephalins (Hökfelt et al., 1978). A widespread localization of opioids in several brain regions suggests that the peptidergic neurotransmission is important in CNS functions (Guthrie and Basbaum, 1984; Herrera-Marschitz et al., 1984; Sasek and Elde, 1986). This idea is supported by the fact that the physiological and behavioural characteristics of the peptide correlate well with their anatomical distributions (Smith and Lee, 1988).

The opioids are involved in the central mechanisms of analgesia (Basbaum and Fields, 1984), self-stimulation, reward mechanisms, stress response and drug abuse (Watson et al., 1989; Akil et al., 1984; Herrera-Marschitz et al., 1984). The opioid and non-opioid peptides participate in cardiovascular depression at the levels of hypothalamus and medulla (Gautret and Schmitt, 1985; Punnen and Sapru, 1986), brain trauma, stroke (Baskin et al., 1985) and spinal cord injuries (Faden, 1993; Sharma et al., 1995).

Opioids and hyperthermia

Although the opioid peptides are involved in the central mechanisms of temperature regulation (Lee, 1984), respiratory depression (Woo et al., 1983) and neuroendocrine functions of hypothalamus and pituitary (Akil et al., 1984; Karlsson et al., 2006; Przewlocki et al., 1987), their role in hyperthermia-induced brain pathology is still not well known. Previous works from our laboratory show that blockade of opioid receptors using naloxone and naltrexone significantly attenuated heat stress-induced blood-brain barrier (BBB) dysfunction, edema formation and brain pathology (Sharma et al., 1997b, c; see Sharma, 2004a, b). This indicates a potential role of opioids in heat-induced neurotoxicity.

Marked upregulation of dynorphin occurs in several brain and spinal cord areas showing cell injury in heat stress (Sharma and Alm, 2002). This upregulation of dynorphin and cell damage are considerably reduced by pretreatment with potent nitric oxide synthase (NOS) inhibitors (Sharma and Alm, 2002). These observations suggest that dynorphin is involved in heat-induced neurotoxicity and interaction between opioids, and nitric oxide play important roles in hyperthermia-induced brain damage (see Sharma and Alm, 2002).

However, dynorphin apart from its interaction with nitric oxide can stimulate glutamate release in the CNS through N-methyl-D-aspartate (NMDA) receptors (Tang et al., 2000; Woods et al., 2006). Blockade of NMDA receptors *in vivo* with dynorphin antiserum strongly supports this idea (Hauser et al., 2001). In fact, dynorphin antiserum is more potent in reducing allodynia-induced pain than the classical NMDA receptor antagonist MK-801 (Nichols et al., 1997; Kuzmin et al., 2006). These findings point out an important interaction between dynorphin and glutamate in the CNS neurotoxicity.

Activation of glutamate receptors induces generation of free radicals and formation of nitric oxide (Azbill et al., 1997). Furthermore, blockade of dynorphin upregulation in hyperthermia with drugs inhibiting nitric oxide is capable of inducing neuroprotection (Sharma and Alm, 2002). Thus, it is quite likely that opioids interact with several neurotransmitters and modulators to regulate CNS function in healthy and diseased tissues. However, additional studies are needed to understand the interaction between opioids and amino acid neurotransmitters in hyperthermia.

Animal models of hyperthermia

The existing animal models of hyperthermia do not simulate the clinical situations of behavioural and neurological deficits (Sharma, 1982, 1999, 2004a, 2005c, 2006a, b; Sharma et al., 1992a, b, 1998a, b). A suitable animal model of hyperthermia is thus essential to explore possible therapeutic measures and to develop new strategies to reduce heat-related morbidity or mortality. We developed a new mode of heat stress in rats (for details see Sharma, Chapter 10 in this volume) that induces behavioral alterations, brain dysfunction and disturbances in cognitive functions very similar to those observed in clinical situations (Sharma, 1982, 2004a; Sharma et al., 1998a, b; Sharma and Alm, 2002).

In this model, animals are subjected to heat stress in a biological oxygen demand (BOD) incubator maintained at 38°C for 30 min, 1, 2, 3 and 4 h without showing stroke symptoms (Sharma, 1982, 2005c). The relative humidity (45–47%) and wind velocity (20–26 cm/s) were kept fairly constant. These experiments were conducted between 8:00 and 9:00 AM to avoid circadian variation. All experiments were carried out according to the care and guidelines of National Institute of Health, USA (Sharma, 1982) and approved by the ethics committee of Uppsala University, Uppsala, Sweden and Banaras Hindu University, Varanasi, India (Sharma et al., 1998b; Sharma, Chapter 10 in this volume).

Using this model, our laboratory demonstrated for the first time that heat stress without heat stroke increases BBB permeability to protein tracers (see Sharma, 1982; Sharma and Dey, 1984, 1986, 1987). This effect appears to be caused by several neurochemical mediators, such as serotonin (Sharma et al., al., 1994), prostaglandins (Sharma et al., 1997b) and opioid peptides (Sharma et al., 1997c). This indicates that hyperthermia-induced brain damage is complex.

Alterations in amino acid neurotransmitters in hyperthermia

We measured two excitatory amino acids, glutamate and aspartate, and two inhibitory amino acids, GABA and glycine, in the cerebral cortex, hippocampus, thalamus, hypothalamus, brain stem and spinal cord using a high-performance liquid chromatography (HPLC)-fluorometric system employing o-phthalaldehyde method (Fosse et al., 1986; Sharma et al., 2006a; see Zhang et al., 2005 for details). The fluorescence was measured at an excitatory filter of 340 nm and an emission filter of 460 nm. The amino acid content was quantified from standard solution and expressed as micromoles per gram (μmol/g) wet tissue (see Sharma et al., 1995; Sharma, 2006a).

(a) *Excitatory amino acids glutamate and aspartate*
Subjection of rats to heat stress profoundly altered the excitatory amino acid levels in

the CNS. The glutamate and aspartate are elevated in most of the brain regions at 4 h after heat stress, except in cerebral cortex (Fig. 1). Increased levels of aspartate and glutamate are seen in the cerebral cortex (2–4-fold) following 1 and 2 h of heat exposure, which returned to normal level at 4 h (Fig. 1), whereas a significant increase in

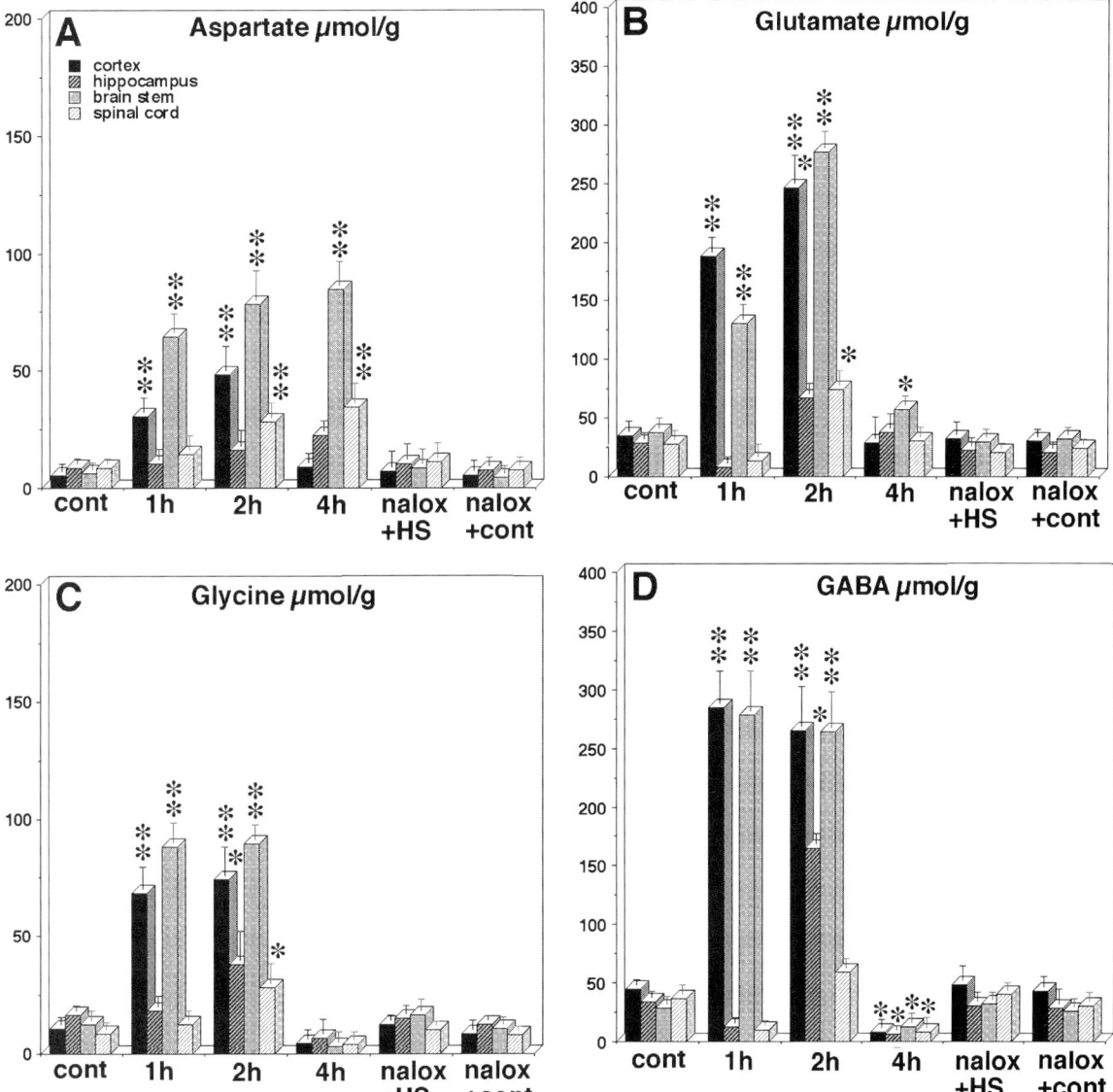

Fig. 1. Heat stress-induced alterations in excitatory amino acids, aspartate (A) and glutamate (B), and inhibitory amino acids, glycine (C) and GABA (D), in the brain and spinal cord compared to controls. Widespread alterations in amino acid content in the brain and spinal cord were noted following heat stress compared to the control group. Pretreatment with naloxone at high doses (10 mg) thwarted the alterations in the amino acids. The drug treatment alone did not influence basal levels of the amino acid concentration. $*P<0.05$; $**P<0.001$, ANOVA followed by Dunnett's test for multiple group comparison from one control group. Data modified from Sharma (2004a, 2006a).)

these amino acids is observed in hippocampus at 4 h heat exposure only (Fig. 1). On the other hand, the brain stem showed a progressive increase in aspartate and glutamate up to 1 and 2 h heat exposure. At 4 h, glutamate level reduced significantly although the aspartate levels continue to remain high. Spinal cord showed a significant elevation of the excitatory amino acids at 2 and 4 h heat stress (Fig. 1). These observations show that excitatory amino acid levels are increased in different brain regions following heat exposure-induced hyperthermia.

(b) *Inhibitory amino acids GABA and glycine*
The inhibitory amino acids also showed profound redistribution in several brain regions during heat stress. An early rise in GABA and glycine levels is seen after 2 h heat stress, which declined in several brain regions at 4 h heat exposure. After 1 h heat exposure, glycine and GABA showed a 4–6-fold increase in the cerebral cortex and in brain stem, which continued up to 2 h. At the end of 4 h heat stress, the glycine level reached normal; however; the GABA level declined below the control value (Fig. 3). An elevation in glycine and GABA is observed at 2 h heat exposure in hippocampus. However, glycine level returned to normal at 4 h heat stress, but GABA levels continued to decline further. In the spinal cord, glycine increased significantly 2 h after heat stress and returned to normal level at 4 h. On the other hand, no significant changes in spinal GABA levels were observed at 1 or 2 h heat exposure; a significant decrease in the spinal cord GABA is seen after 4 h heat exposure (Fig. 1). This suggests that inhibitory amino acid levels are largely declined following hyperthermia caused by heat exposure.

Amino acid neurotransmitters and cell injury in hyperthermia

A decrease in inhibitory amino acids and an increase in excitatory amino acid neurotransmitters at 4 h heat stress are associated with brain pathology (Sharma, 2006a). The detailed mechanisms by which amino acid neurotransmitters induce cell death in heat stress are still unclear. It appears that glutamate-induced programmed cell death through activation of neuronal glutamate receptors and a Ca^{2+} influx mediated through ionotropic receptors plays important roles (Babot et al., 2005; see Erreger et al., 2004; Waring, 2005; Chavez et al., 2006; Wang et al., 2006).

Excessive release of amino acids and intrasynaptic accumulation of glutamate in ischaemia causes sodium and calcium influx resulting in an increase in intracellular water and cell swelling (Choi, 1990; Unterberg et al., 2004). Accumulation of Ca^{2+} activates synthesis of NOS causing increased production of NO leading to cell death (Brann et al., 1997; Bredt, 1999; Boullerne and Benjamins, 2006; Ishii et al., 2006). It appears that similar mechanisms are involved in amino acid-induced cell death in hyperthermia. Neuroprotection induced by inhibitors of NOS in hyperthermia is in agreement with this hypothesis (Sharma et al., 1997a, 1998a; Sharma and Alm, 2004).

Induction of oxidative stress by glutamate could also be another important factor in causing cell death in heat stress (see Rossler et al., 2004; Kern and Jones, 2006). An imbalance between oxidants and antioxidants in the CNS following hyperthermia could be responsible for cell injury. Neuroprotective effects induced by pretreatment with an antioxidant compound H-290/51 in hyperthermia is in line with this hypothesis (Sharma et al., 1997a; Alm et al., 1998, 2000; Sharma, 2004a). However, it remains to be seen whether H-290/51 can influence amino acid neurotransmitters in the CNS of hyperthermic rats, a feature currently being examined in our laboratory.

Naloxone treatment influences amino acid neurotransmitters in hyperthermia

Pretreatment with naloxone (10 mg) significantly attenuated the rise in glutamate or aspartate levels in different brain regions following heat stress. Interestingly, the decrease in GABA or glycine levels in heat stress was also thwarted by naloxone treatment (Fig. 1). This effect of naloxone was absent in animals that received low doses of the compound (1 mg or 5 mg, results not shown). These observations suggest that a modulation of opioid receptors influences amino acid neurotransmission in heat stress.

The mechanism by which naloxone influences amino acid neurotransmitters in hyperthermia is not known. High doses of naloxone inhibit κ-opioid receptors, whereas low doses of the compound preferentially antagonize either μ- or δ-receptors (Sharma et al., 1997b, c; Sharma, 2004a). It seems likely that opioid antagonist naloxone attenuates hyperthermia-induced cellular stress response (Olsson et al., 1995). A reduction in cellular or systemic stress attenuates BBB dysfunction and consequently edema formation and/or neuronal injury (Sharma et al., 1991a, b; McCormack et al., 2005). Alternatively, naloxone can influence Ca^{2+} permeability at the membrane level (Di Sole et al., 2001; Minoia and Sciorsci, 2001; Irnaten et al., 2003) by interfering with opening of neurotransmitter-gated ion-channels. Neuroprotection achieved by L-type Ca^{2+} channel blocker nimodipine in heat stress is in line with this idea (Sharma and Cervós-Navarro, 1990a, b). Alterations in Ca^{2+} channel also influences amino acid neurotransmitter release and accumulation (Choi et al., 1988). These observations indicate a novel functional interrelationship between opioids and amino acid neurotransmitters in hyperthermia.

Brain pathology in hyperthermia

To understand the neuroprotective role of naloxone in amino acid-induced brain pathology in hyperthermia, cell and tissue injuries were examined using standard morphological techniques (Sharma, 1999). After heat stress, the animals were perfused with modified Somogyi fixative (containing 4% paraformaldehyde in 0.1 M phosphate buffer, pH 7.0 with 2.5% picric acid) *in situ* and the identical brain regions used for amino acid measurement were dissected out. The tissue pieces were embedded in paraffin or epon for light or electron microscopy, respectively (see Sharma et al., 1998b).

(a) *Light microscopy*

About 3 μm thick paraffin sections were cut from identical brain regions and stained with haematoxylin and eosin or Nissl (Sharma et al., 1998a, b). Some sections were immunostained for glial fibrillary acidic protein (GFAP) and myelin basic protein (MBP) (see Sharma et al., 1993b; Sharma, 2003, 2006b).

Profound nerve cell, glial cell reaction and myelin vesiculation were seen following 4 h heat exposure in several brain regions (Table 1). There were many dark and distorted nerve cells in the cerebral cortex of 4 h heat-stressed rats (Fig. 2b). Loss of nerve cells, sponginess and edema were evident at this time (Figs. 2a, b). Selective cell loss and nerve cell distortion were seen in the CA 4 region of the hippocampus (Fig. 5). The dentate gyrus also exhibited profound nerve cell damage (Fig. 2). Degradation of MBP immunostaining is prominent in the brain stem of heat-exposed rats (Figs. 2f, h). Edematous swelling and loss of myelinated nerve fibres were also observed (Fig. 2).

(b) *Electron microscopy*

Representative tissue samples from brain and spinal cord were used for transmission electron microscopy (Sharma, 1999). About 1 μm thick Epon sections were cut and stained with toloudine blue for high-resolution light microscopy. The desired portion of the blocks were then trimmed out and ultrathin sections were cut using diamond knife on a LKB Ultramicrotome (Sweden). The ultrathin sections were collected on a mesh copper grid and counterstained with lead acetate and uranyl citrate before examination under a Phillips 400 transmission electron microscope (Sharma et al., 1998b).

The ultrastructural studies confirmed nerve cell, glial cell and myelin damage in heat stress (Fig. 3). Thus, dark and distorted nerve cells are common in the cerebral cortex and in the brain stem regions in 4 h heat-stressed animals (Fig. 3A.a, b). Synaptic damage, edema, membrane disruption and myelin vesiculation are frequent at this time (Fig. 3). In

Table 1. Brain edema and cell injury following heat stress and its modification with naloxone pretreatment

Type of exp.	Regional brain water content %			CNS damage/distortion			
	Cortex	Hippocampus	Brain stem	Spinal cord	Nerve cells	Glial cell	Myelin
Control ($n = 5$)	76.86±0.23	78.42±0.21	68.56±0.23	64.35±0.21	Nil	Nil	Nil/?
Naloxone 10 mg/kg, i.p. ($n = 5$)	76.04±0.08	77.82±0.11	67.67±0.37	64.47±0.10	Nil	Nil	Nil/?
1 h Heat stress ($n = 6$)	76.38±0.28	78.56±0.23	68.77±0.43	64.85±0.33	Nil/?	Nil/?	Nil/+
2 h Heat stress ($n = 6$)	76.78±0.44	78.67±0.42	68.76±0.56	65.78±0.54	Nil/+	Nil/+	Nil/+
4 h Heat stress ($n = 6$)	80.54±0.23***	81.34±0.23***	73.24±0.19***	67.34±0.14***	+ + + +	+ + + +	+ + + +
Naloxone 10 mg + 4 h heat stress ($n = 6$)	77.45±0.18**,†	79.48±0.11**,†	69.38±0.45**,†	65.67±0.21**,†	+ +	+ +	+ +
Naloxone 5 mg + 4 h heat stress ($n = 6$)	80.23±0.56***	80.76±0.45***	72±0.33***	66.86±0.34	+ + + +	+ + +?	+ + +?
Naloxone 1 mg + 4 h heat stress ($n = 6$)	81.76±0.34‡	82.34±0.22‡	74.65±0.26‡	67.67±0.13‡	+ + + +	+ + + +	+ + +

Notes: Naloxone was given (1 mg, 5 mg or 10 mg/kg, i.p.) 30 min before the onset of heat stress; regional brain water content was measured in identical sample sizes used to determine amino acid neurotransmitters in various experimental groups; values are mean ± SD, Student's unpaired t-test; nil = absent, + = occasional, + + = mild, + + + + = severe, ? = unclear. **$P<0.01$, compared from control; ***$P<0.001$, compared from control; †$P<0.001$, compared from 4 h heat stress; ‡$P<0.05$, compared from naloxone 10 mg + heat stress (data after Sharma, 1999, 2006a).

many nerve cells, damage to cell nucleus is also seen. Irregular shape of nuclear membrane and loss of nucleolus are common findings (Fig. 3A). Collapse of microvessels, perivascular edema and cell membrane damage are observed in rats at 4 h heat exposure (Fig. 3A.c).

Distortion of endothelial cell membrane and leakage of lanthanum tracer in several brain regions are apparent (Fig. 3B.c–g). There is leakage of lanthanum across the endothelial cell membrane without widening of the tight junctions (Fig. 3B.c, d). Lanthanum is stopped at the tight junction (Fig. 3B.d, h); however, the underlying basal lamina of the microvessel exhibited lanthanum extravasation (Fig. 3B.d). Several endothelial cells exhibited lanthanum infiltration within the cell cytoplasm (Fig. 3B.d–h). In few cases, lanthanum was seen in microvesicular profiles within the endothelial cell cytoplasm (Fig. 3B.e–h).

Influence of naloxone on structural changes in hyperthermia

High doses of naloxone (10 mg) markedly attenuated the nerve cell damage, glial cell injury and myelin vesiculation after 4 h heat exposure (Table 1). In drug-treated animals, several neurons with normal dendrites are present in the cortex after heat stress. Many nerve cells are normal in appearance, and signs of sponginess and edema are less frequent (Fig. 2a). On the other hand, low doses of naloxone (1 mg and 5 mg) did not influence cell damage in heat stress (results not shown). This indicates that naloxone at high doses is capable of inducing neuroprotection in hyperthermia.

At the ultrastructural level, the neuropil is less distorted in naloxone-treated (10 mg) stressed rats (Fig. 3B, b). Signs of collapsed microvessels, perivascular edema, cell membrane damage and myelin vesiculation are less evident (Sharma, unpublished observations). The cerebral endothelial cells did not show leakage of lanthanum across the BBB (results not shown). Damage to glial cells, synaptic membrane, nerve cell nucleus and nuclear membrane are less pronounced. On the other hand, lower doses of naloxone (1 and 5 mg) were ineffective in reducing hyperthermia-induced brain damage at the ultrastructural level (results not shown).

These observations suggest that attenuation in excitatory amino acids (glutamate and aspartate) and augmentation of inhibitory amino acids (GABA and glycine) in the CNS of heat-stressed

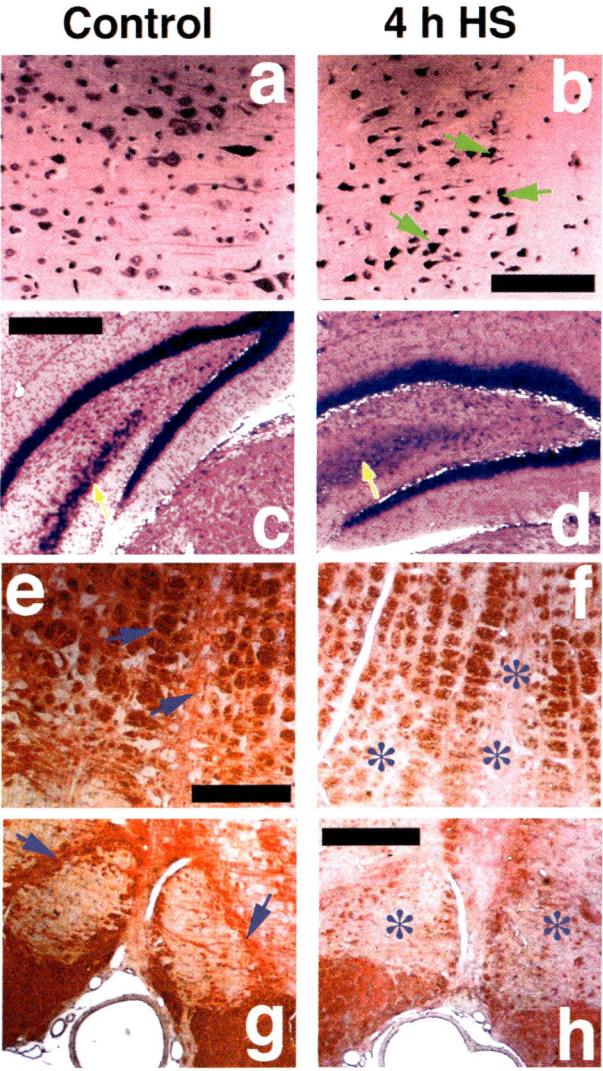

Fig. 2. Low-power light micrograph showing structural changes in the cerebral cortex (a,b), hippocampus (c,d), brain stem reticular formation (e,f) and pons-medulla (g,h) regions following 4 h heat exposure in rats compared to the control group. Damage to several nerve cells (arrows), edema and sponginess are apparent in the cerebral cortex of heat-exposed rats (b) compared to controls (a). In the hippocampus, most pronounced cell loss and cell death are evident in the CA4 region (arrow) (d) compared to controls (c). Loss of myelinated fibres (*) as seen by immunostaining of myelin basic protein (MBP) is most pronounced in the brain stem reticular formation following heat stress (f) compared to controls (arrows, e). Similar damage to myelin (*) is seen in the pons medulla region of heat-exposed rats (h) compared to normal groups (arrows, g). Bars: a,b = 50 μm; c,d = 80 μm; e,f = 30 μm; g,h = 35 μm. Data modified from Sharma et al. (1998) and Sharma (1999, 2004a, 2006a).

animals by naloxone is responsible for neuroprotection. These observations clearly show that modulation of opioids influence amino acid neurotransmitters in the brain in heat stress leading to neuroprotection. Pharmacological blockade of various amino acid neurotransmitter receptors in heat stress is needed to further clarify this point.

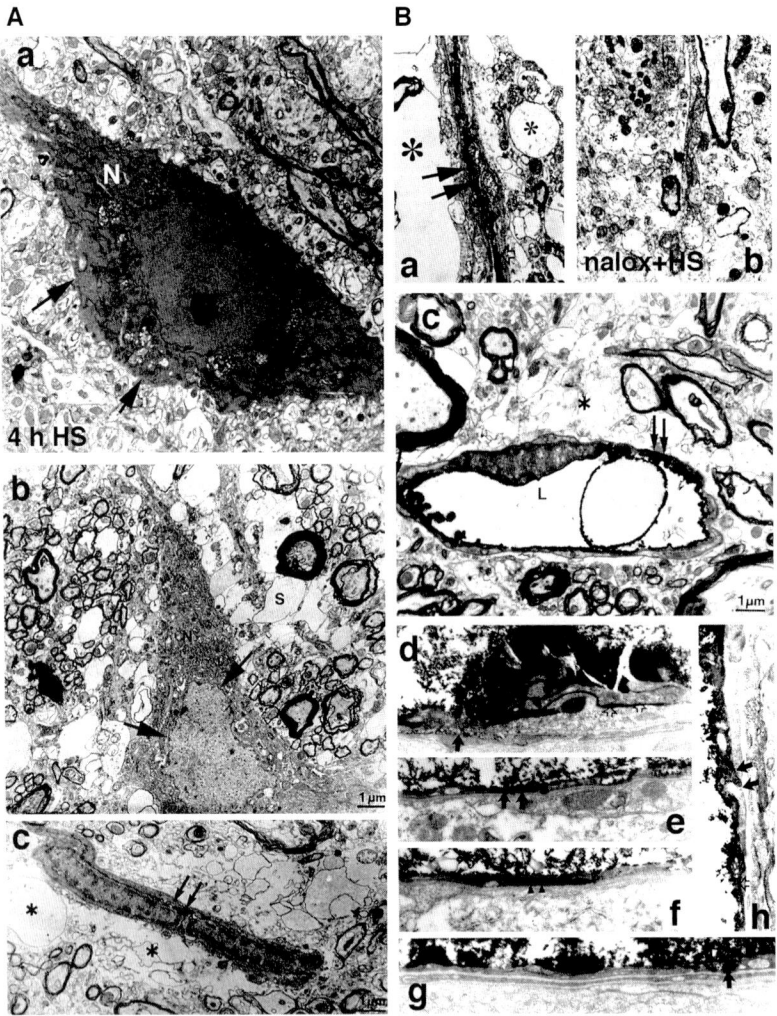

Fig. 3. Low-power electron micrograph showing ultrastructural changes in the brain and their modification with naloxone. One dark nerve cell (N) containing electron-dense cytoplasm (arrows) and karyoplasm is seen in the parietal cortex after heat stress (A.a). Eccentric nucleolus, perineuronal edema and damage to neuropil are evident. In deep thalamic region, a dark nerve cell (arrows) with condensed cytoplasm is visible (A.b). In this nerve cell, irregular shape of nuclear membrane and loss of nucleolus are apparent. Synaptic damage (s), myelin vesiculation and edema are common in the neuropil. In the brain stem region, one completely collapsed microvessel (arrows) is seen. Perivascular edema (*), membrane vacuolation and damage to neuropil are common findings (A.c). Pretreatment with naloxone at high dose (10 mg) reduced the structural changes in the cerebral cortex (B.b) compared to untreated rats (B.a). Several microvessels in various brain regions show extravasation of lanthanum across the blood-brain barrier (BBB) in heat-stressed animals (B.c–g). In most cases, lanthanum infiltration is seen across a part of the endothelial cell (arrows), whereas other regions of the microvessel are completely normal. Lanthanum can also be seen within the luminal side (L). High-power electron micrograph across one tight junction of the endothelial cell reveals that lanthanum is largely present within the endothelial cleft and is stopped at the tight junction complex (B.d, arrows). Lanthanum infiltrates within the endothelial cell cytoplasm (arrows) in certain parts of the endothelial cell membrane (B.e.f, arrows). The remaining part of the endothelial cell membrane is impermeable to the tracer. In some microvessels, lanthanum is seen within microvesicular profiles located in the endothelial cell cytoplasm (B.g), whereas in others, one endothelial cell becomes completely infiltrated with lanthanum around the areas covering the tight junction, although the tight junction remains intact (B.h). Bars: A.a = 1 μm; b,c = 1.5 μm; B.a,b = 2 μm; c = 1 μm; d = 0.4 μm; e–h = 0.6 μn. Data modified from Sharma et al. (1998a) and Sharma (1999, 2004a, 2006a).

Regional blood-brain barrier permeability in hyperthermia

To understand the role of amino acids on BBB dysfunction in hyperthermia, regional BBB permeability was measured using Evans blue albumin (EBA) and ^{131}Iodine tracers (see Sharma and Dey, 1986). The tracers were administered together into the right femoral vein immediately after termination of heat exposure under Equithesin anaesthesia. Extravasation of the tracers in various brain areas was analysed according to the standard protocol (see Sharma, 1987).

Subjection of rats to 4 h heat stress induced extravasation of Evans blue and radioiodine tracers in several brain regions (Fig. 4). The blue staining is seen in eight regions, viz., cingulate cortex, occipital cortex, parietal cortex, cerebellum, temporal cortex, frontal cortex, hypothalamus and thalamus. Mild-to-moderate staining of the ventricular walls was also observed, whereas extravasation of radioiodine was seen in all the 14 regions examined, including hippocampus, caudate nucleus, superior and inferior colliculi, pons and medulla (Fig. 4A).

Subjection of rats to shorter duration of heat stress, i.e., 1 or 2 h, did not induce BBB disruption to either tracer (results not shown). These observations suggest that amino acid neurotransmitters contribute to regional BBB disturbances.

A significant higher permeability of radioiodine in the brain compared to Evans blue in heat-stressed rats may be due to a difference in the molecular size of the tracers. Alternatively, difference in the proteins to which these tracers bind *in vivo* also influences their distribution (Mayhan and Heistad, 1985; Sharma et al., 1990).

Influence of naloxone on regional BBB permeability in hyperthermia

That the amino acid neurotransmitters are capable of inducing BBB disruption in various brain areas is further strengthened by naloxone experiments. Pretreatment with naloxone (10 mg) significantly reduced hyperthermia-induced regional BBB breakdown to Evans blue and radiotracers. However, low doses of naloxone (5 and 1 mg) were not effective in reducing regional BBB permeability in heat stress significantly (Fig. 4A). These observations are in line with the idea that blockade of opioid receptors attenuates BBB disruption in heat stress (Sharma et al., 1997a, c).

Pretreatment with naloxone attenuated the increase in excitatory amino acids, arrested the decline in inhibitory amino acids and thus reduced the BBB disruption. This indicates that amino acid neurotransmitters somehow contribute to BBB disruption. The mechanism by which naloxone influences amino acid neurotransmitters and/or BBB function is not known. Since naloxone influences Ca^{2+} permeability at the membrane level (see Minoia and Sciorsci, 2001), it is likely that blockade of opioid receptors induces amino acid neurotransmitter release and/or their accumulation (Choi et al., 1988).

Regional cerebral blood flow in hyperthermia

To understand the role of local ischaemia in the breakdown of regional BBB permeability in hyperthermia, regional cerebral blood flow (CBF) was measured using carbonized microspheres (15 ± 0.6 o.d.) labelled with ^{125}I (Sharma and Dey, 1986). The tracer microspheres ($\sim$72 000–85 000) were injected into the left cardiac vertical through an indwelling polythene cannula over 20–30 s (Sharma and Dey, 1986). The reference samples from the femoral artery were withdrawn (at the rate of 0.8 ml/min) starting from 30 s before infusion and continued up to 90 s after infusion. The brains were removed and divided into 14 identical anatomical regions (as above) and counted for radioactivity. At the end of the experiment, whole blood radioactivity was also determined. The CBF was calculated using formula: CBF (ml/g/min) = $C_B \times RBF \div C_R$, where C_B is cpm/g brain, RBF reference blood flow (rate of withdrawal of blood samples from reference artery, 0.8 ml/min) and C_R total counts in the reference blood samples (for details see Sharma, 1987).

Using this technique, the regional CBF showed a decline in all the 14 regions at the end of 4 h heat stress (cortical regions −38 to −53%; subcortical

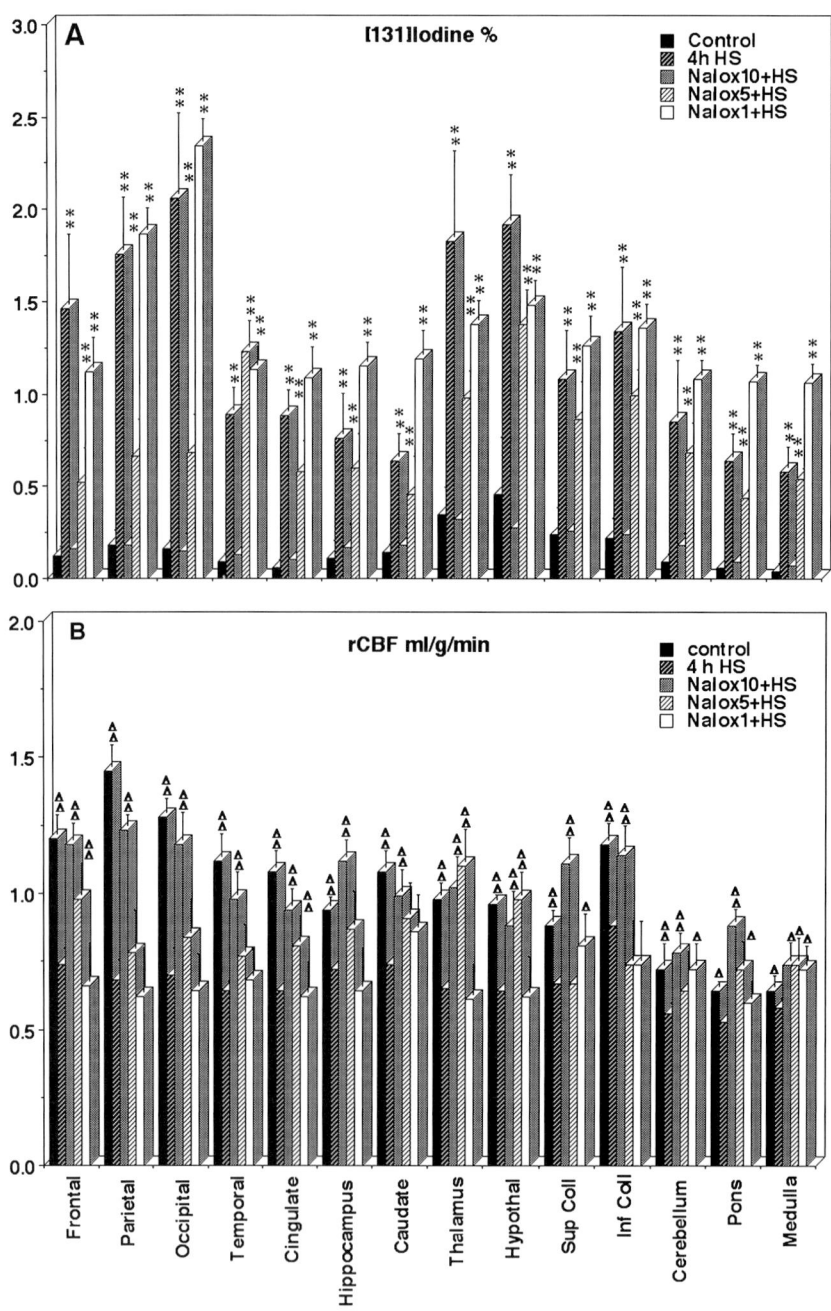

Fig. 4. Regional blood-brain barrier (rBBB) permeability and regional cerebral blood flow (rCBF) in rats subjected to 4 h heat stress and their modification with naloxone pretreatment. A significant increase in the rBBB permeability to radioiodine (A) and a marked decrease in the rCBF (B) following 4 h heat stress were observed in all the 14 brain regions compared to the control group. Pretreatment with naloxone at high dose (10 mg) significantly attenuated rBBB and rCBF changes (A, B). Lower doses of the compound (1 and 5 mg) were much less effective on alterations in the rBBB or rCBF in heat stress. Data modified from Sharma et al.(1997, 1998) and Sharma (2004). Each column represents mean ± SD of 6–8 rats. *$P<0.05$; **$P<0.001$, ANOVA followed by Dunnett's test for multiple group comparison from one control group. $\Delta = P<0.05$; $\Delta\Delta = P<0.01$ compared from 4 h heat stress. Data modified from Sharma (2004a, 2006a).

region −23 to −31%; cerebellum and brain stem −15 to −22%, see Fig. 4B). However, the severity of flow reduction did not coincide with the magnitude of increased BBB permeability. On the other hand, rats subjected to 1 or 2 h heat stress did not show significant changes in the CBF (results not shown). These observations suggest that a decline in regional CBF alone is not sufficient to impair BBB function. This decline in regional CBF could be due to the effects of various vasoactive neurotransmitters in the brain, e.g., serotonin, prostaglandins, opioids and histamine, following 4 h heat stress (see Sharma et al., 1992b, d, e; Sharma, 1999, 2004a).

Influence of naloxone on regional CBF in hyperthermia

It appears that opioids influence vasomotor tones of cerebral microvessels in hyperthermia (Sharma et al., 1997a–c). This is evident from the finding that pretreatment with a high dose of naloxone (10 mg) significantly prevented the regional CBF decline, whereas lower doses of the compound (5 and 1 mg) were ineffective (results not shown).

Whether opioids directly influence regional microcirculation in hyperthermia is unclear from this investigation. However, it is likely that blockade of opioid receptors also influences the function of NMDA receptor and Ca^{2+}-gated ion channel (Nyberg et al., 1995; see Sharma, 2006a). Obviously, an interaction between opioids and other vasoactive compounds in brain could be responsible for alterations in the regional microcirculation during hyperthermia.

Regional brain edema

Extravasation of serum proteins and regional ischaemia could influence the transport of water from vascular to the cerebral compartment (Sharma et al., 1998b; Sharma, 2006b). To understand the contribution of regional extravasation of protein tracers in inducing brain edema formation following hyperthermia, the regional brain water content was examined (see Sharma, Chapter 10 in this volume). Immediately after heat exposure, the brain was removed and the desired parts were dissected out. The samples were weighed to record wet weight and then placed in an oven at 90°C for 48 h to obtain their dry weights. The water content was calculated from the differences between dry and wet weights of the sample (Sharma and Cervós-Navarro, 1990a, b).

The regions rich in white matter, e.g., brain stem and spinal cord, show much less water compared to the areas rich in grey matter, viz., cerebral cortex and hippocampus (see Sharma et al., 1998b). Subjection of animals to 4 h heat stress resulted in a significant increase in the regional brain water content (Table 1). The most pronounced increase in the brain water was seen in the brain stem region (∼5%) followed by cerebral cortex (3%), hippocampus and spinal cord (∼3% each). On the other hand, subjection of rats to 1 or 2 h heat stress did not result in increased water content in any brain region (Table 1). These observations suggest that breakdown of the regional BBB to proteins is related with vasogenic edema formation.

Influence of naloxone on regional brain edema in hyperthermia

That the vasogenic edema formation is related with regional BBB disturbances is further supported by our investigations using pharmacological blockade opioid receptors with naloxone. Treatment with naloxone (10 mg) attenuated edema formation along with regional BBB dysfunction (see Sharma et al., 1997a, b; Sharma, 2006a), a feature not seen with low doses of the compound (Table 1). This suggests that high doses of naloxone are needed to thwart regional brain edema formation in hyperthermia. It is quite likely that blockade of opioid receptors may interfere with release of various vasoactive substances including amino acid neurotransmitters and alter the function of several ion-gated channels leading to attenuation in the regional BBB permeability. Obviously, a lack of BBB disruption to protein could be instrumental in attenuating brain edema formation.

Cognitive and motor function deficits in hyperthermia

The magnitude and intensity of brain damage seen in various important anatomical regions, e.g., cerebellum, hippocampus, thalamus, hypothalamus, cerebral cortex and spinal cord, suggest severe memory impairment and deficits in cognitive and motor functions in hyperthermia (see Chapter 10 in this volume).

We analysed motor and cognitive functions in the rats following heat stress using standard behavioural tests in well-trained animals (see Sharma, 2006a). Normal animals were trained for cognitive or motor function tests before the experiment for a minimum of two sessions daily for 4 days and all the behavioural analyses were conducted in a blinded fashion (Sharma, 2005c).

Cognitive functions

(a) *Rota-Rod performances*

The Rota-Rod treadmill is used to assess motor functions and fatigue in rats. A fixed speed of 16 rotations per minute (RPM) on Rota-Rod equipment was used with manual counting. Rats were given training to Rota-Rod at this setting for 10 min every day twice (at an interval of 4 h) for 4 days prior to the experiment. Rats not falling off the Rota-Rod for 2 min were considered normal during a 3 min session (Sharma, 2006a).

A significant decline in Rota-Rod performance was seen in rats following 3-h stress, which was progressive in nature. The duration of stay on the Rota-Rod was reduced at 3-h and 4-h of heat exposure compared to the control group (Fig. 5D).

(b) *Grid walking*

To determine changes in locomotor behaviour, gait and overall walking skill, an elevated level (30°) of stainless steel grid was used with a mesh size of 30 mm. Rats were trained for 1 min twice everyday for 4 days prior to the experiments. The animals were placed on the grid for 1 min and the total number of paired steps, i.e., placement of both forelimbs, was estimated. During this period, the number of misplaced limbs error, viz., the forelimbs fell through the grid, was recorded. The total number of errors for each forelimb was also evaluated manually (see Sharma, 2006a).

Rats subjected to 4 h heat exposure were unable to walk normally during a grid walking session. The number of steps taken during a 60 s grid walking session was significantly reduced in heat-stressed rats at 4 h (Fig. 5E). These animals exhibited greater placement errors of hind legs on the grid following 3 and 4 h after heat exposure (Fig. 5F).

Motor functions

(a) *Inclined plane test*

The motor disturbances in the rat after heat stress, if any, were determined using the inclined plane test. Each rat was trained on inclined plane test using an angle in such a way that the animal could stay on the plane for 5 s without falling. Each rat was trained twice daily for 4 days before the experiments (Sharma, 2006a).

In general, significant motor deficit in animals is seen during late periods of heat exposure (see below). Thus, a significant decrease in the angle of the inclined plane was observed in animals subjected to 3 and 4 h stress compared to the control group (Fig. 5A).

(b) *Footprint analysis*

Footprint analysis is a good measure of gait and motor disturbances. For this purpose, the hind paws were wetted and the animals were allowed to walk on a paper coated with bromophenol blue dissolved in acetone. When rats walk on the coated paper, the hind paw imprints were used to assess motor function behaviour (see Sharma, 2005c, 2006a). The distance (millimetre) between hind paws was measured from the base of the central pads. The stride length (millimetre) is measured by the distance between hind paws in two consecutive steps.

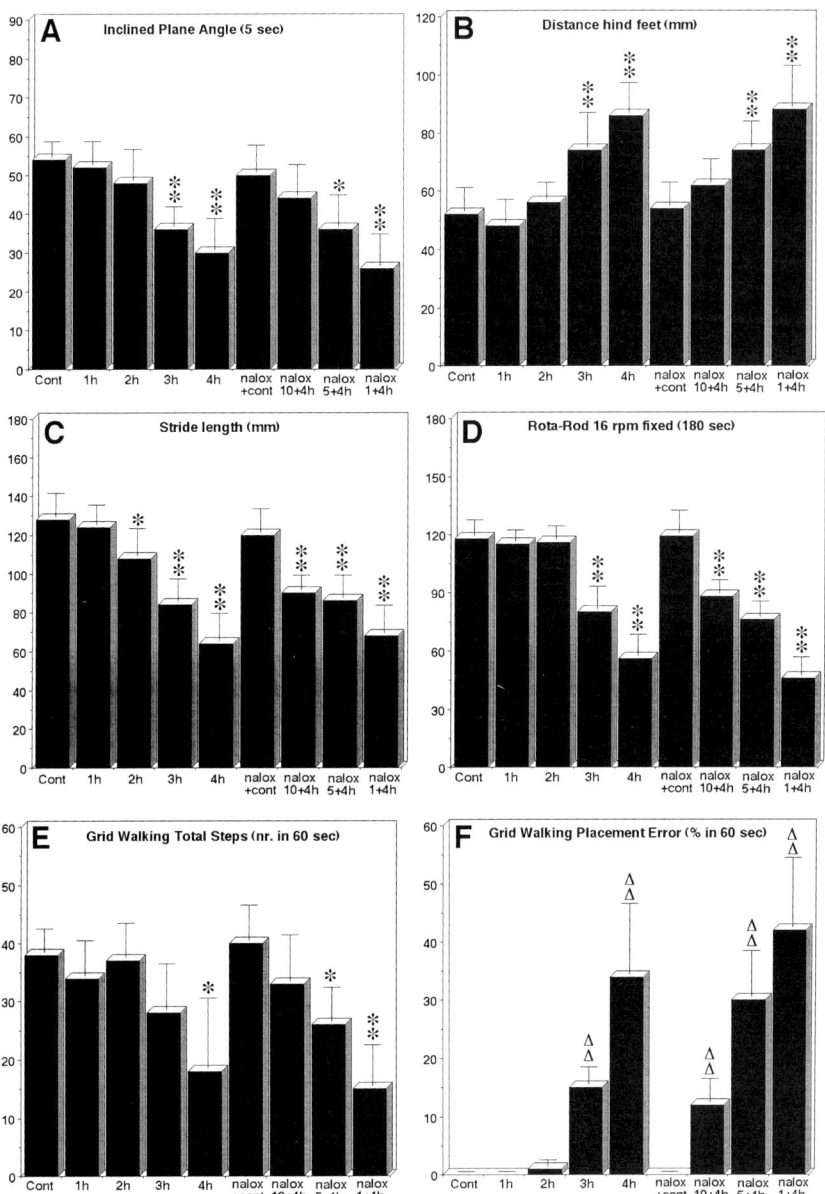

Fig. 5. Changes in motor and cognitive functions in rats after heat stress and their modification with naloxone pretreatment. A significant decrease in the angle of inclined plane test reflecting alterations in motor functions was noted in rats after 3 h after heat exposure, which was progressive up to 4 h (A). Marked decline in cognitive function was also seen in heat-stressed animals when they were subjected to Rota-Rod performance (B). At 3 h heat stress, rats could not stand on Rota-Rod for more than 75 s, which was subsequently reduced to 60 s at 4 h (B). Measurement of gait and walking pattern revealed deficit in the placement of legs (c) and the stride length (D) from 3 h and onwards. Heat-stressed rats also showed significant deficit in grid walking and placement errors compared to controls (E, F). A decrease in the number of steps taken in a grid walking test for 60 s was evident in rats subjected to 4 h heat stress (E). However, placement errors of hind feet can be seen as early as 3 h after heat exposure (F). Pretreatment with naloxone at high doses improved the motor and cognitive functions in rats after 4 h heat stress (A–F). However, low doses of naloxone (1 and 5 mg) did not alter these deficits significantly. Data at each column represent mean ± SD of 5–6 rats. *$P<0.05$; **$P<0.01$, ANOVA followed by Dunnett's test for multiple group comparison from one control group. $\Delta\Delta = P<0.01$, χ^2-test from the control group. Data after permission from Sharma (2006a).

The transverse distance between the hind feet, a measure of disturbed gait, was increased significantly at 4 h (Fig. 5B). On the other hand, the stride length calculated as longitudinal distance between hind feet during stepping showed a significant decrease from 2 h following heat exposure, which was progressive in nature (Fig. 5C).

Influence of naloxone on cognitive and motor deficits in hyperthermia

Pretreatment with naloxone at high dose (10 mg) significantly improved the cognitive functions in animals subjected to 4 h heat stress (Fig. 5D–F) and exhibited marked improvement on the Rota-Rod performance and grid walking with minimum placement errors. This beneficial effect of the compound was not seen when lower doses were used (Fig. 5).

Blockade of opioid receptor with high dose of naloxone also improved motor functions in heat-stressed rats at 4 h (Fig. 5), whereas low doses of the compound (naloxone 1 and 5 mg) did not alter heat-induced motor impairment (Fig. 5).

These observations strongly indicate that hyperthermia-induced brain damage is somehow responsible for impairment of the cognitive and motor function in heat-stressed animals.

Stress symptoms and physiological variables in hyperthermia

Heat exposure induces profound hyperthermia and other stress symptoms in clinical situations. We examined changes in rectal temperature, behavioural salivation and prostration as major clinical signs of heat stress in our model. Animals subjected to heat stress do not suffer pain (see Sharma, 2005c). This was confirmed by assessing pain threshold using thermal pain stimulus after heat exposure (see Sharma, 2005c). Loss of body fluids due to spread of saliva or urination was examined by recording their body weight before and after the exposure (Sharma, 2005c). Changes in mean arterial blood pressure (MABP), arterial pH and blood gases were also analysed (see Sharma, 1987).

Rats subjected to heat stress at 38°C showed a progressive hyperthermia that was most marked at the end of 4 h (see Sharma, 2006a; Chapter 10 in this volume). On the other hand, rats exposed to 38°C for 30 min, 1 or 2 h showed only mild increase in the body temperature (Sharma, 2006a). The rise in body temperature appears to be linear in the model and positively correlates with the duration of heat exposure (for details see Sharma, Chapter 10 in this volume).

Activation of heat loss mechanisms was evident during stress as rats showed profuse salivation (Table 2), the magnitude and intensity of which was related to the duration of heat exposure. Saliva spread over snout is increased by 10-fold at 4 h compared to 30 min heat exposure (Sharma, 2005b). A significant decrease in body weight at the end of 4 h heat exposure (Table 2) confirms the evaporative water loss from the animals after heat stress (see Sharma, 2006a).

A mild but significant hypotension is observed at the end of 4 h in heat-stressed rats. At this time PaO_2 showed a slight decrease and $PaCO_2$ is increased mildly (Table 2). Short duration of heat exposure did not alter blood gases or arterial pH significantly (Table 2).

Influence of naloxone on stress symptoms and physiological variable in hyperthermia

Blockade of opioid receptors prior to heat stress with naloxone (10 mg) considerably reduced the development of hyperthermia (Table 2), whereas low doses of the compound (1 and 5 mg) were ineffective in lowering the body temperature after heat exposure. Loss of body weight, salivation and prostration were also significantly reduced with high doses of naloxone (Sharma, 2006a). The magnitude of hypotension is considerably reduced with naloxone (10 mg) without affecting the blood gases or the arterial pH (Table 2).

These observations suggest that although hypotension and altered blood gases may somehow contribute to minor alterations in the regional microcirculation, a slight reduction in hypotension in naloxone-treated animals is in line with this hypothesis. A minor but significant decrease in

Table 2. Stress symptoms and physiological variables in control and heat-stressed rats

Parameters examined	Control ($n = 8$)	Heat stress at 38°C in BOD chamber		
		1 h ($n = 6$)	2 h ($n = 8$)	4 h ($n = 12$)
A. Stress symptoms				
Rectal T°C	37.42 ± 0.23	$38.41 \pm 0.32^*$	$39.24 \pm 0.21^{**}$	$41.48 \pm 0.23^{***}$
Salivation	Nil	++	+++	++++
Prostration	Nil	Nil	Nil	++++
Gastric haemorrhage	Nil	4 ± 3	8 ± 3	34 ± 8 (microhaemorrhages)
B. Physiological variables				
MABP torr	101 ± 6	94 ± 8	$124 \pm 8^{**}$	$76 \pm 4^{**}$
Arterial pH	7.38 ± 0.04	7.36 ± 0.03	7.33 ± 0.10	7.34 ± 0.08
PaCO$_2$ torr	33.46 ± 1.04	33.56 ± 0.76	34.13 ± 0.24	$32.12 \pm 0.11^*$
PaO$_2$ torr	78.24 ± 1.22	79.12 ± 0.54	79.34 ± 0.26	$82.14 \pm 0.23^{**}$

$^* P < 0.05$; $^{**} = P < 0.01$; $^{***} = P < 0.001$, Student's unpaired t-test (data modified after Sharma, 1982; Sharma and Dey, 1986, 1987; Sharma et al., 1998).

hyperthermia may somehow reduce the magnitude and intensity of cellular stress and/or release of neurochemicals in the brain (Sharma, 1999). However, minor alterations in stress symptoms and the physiological variables following heat stress may not contribute directly to the BBB dysfunction, brain edema and cell injury (Sharma, 2006). Additional investigations are needed to further understand these points.

Probable mechanisms of amino acid and opioid interaction in hyperthermia

Our new data show that both excitatory and inhibitory amino acids participate in the pathophysiology of hyperthermia-induced brain damage. Whether brain damage caused by hyperthermia is through specific glutamate/aspartate receptors is still unclear. A decline in inhibitory influence at the time of increased excitotoxicity in heat stress is likely to contribute to brain damage. Obviously, a balance between excitatory and inhibitory amino acids for hyperthermia is crucial for cell death or cell survival (see Artemowicz et al., 2004).

Blockade of cell death by high dose of naloxone suggests that opioids are involved in the pathophysiology of hyperthermic brain injuries (Sharma et al., 1997a–c). Both excitatory and inhibitory effects of opioid neurotransmission have been described in the brain and spinal cord (North, 1993). The excitatory effects of opioids are usually antagonized by naloxone, whereas the inhibitory effects are not affected (Duggan and Fleetwood-Walker, 1993). Several electrophysiological effects of dynorphin in the CNS are blocked by naloxone, probably through κ-opioid receptors (Duggan and Fleetwood-Walker, 1993). This idea is supported by the fact that a selective blockade of κ-opioid receptor with nor-binaltrophimine (nor-BNI) completely blocks the inhibitory effects of dynorphin in a concentration that has no effect on μ- or δ-opioid receptors (Smith and Lee, 1988).

Previous studies from our laboratory suggest that high doses of naloxone (10 mg/kg) attenuate depression in spinal cord evoked potential (SCEP) after trauma (Winkler et al., 1994, 2002, 2003; Sharma and Winkler, 2002), which was not observed with low doses of the compound (1 and 5 mg) (Sharma, 2004a, b). Probably, high doses of naloxone are capable of antagonizing κ-opioid receptor that is needed to have a beneficial effect on the spinal cord pathophysiology following trauma.

Our findings in heat stress indicate that opioid peptides are also involved in the pathophysiology of hyperthermic brain injury. Since lower doses of naloxone are not effective, it appears that the compound at high doses can block the κ-opioid receptors either alone or in combination with μ- and δ-receptors (see Sharma et al., 1997b, c). Naloxone at high doses inhibits κ-opioid receptors, whereas at low doses, these compounds preferentially antagonize only μ- or δ-receptors (Nyberg et al., 1995). Thus, it is quite likely that blockade of κ-opioid receptors by naloxone at high doses in hyperthermia also attenuates brain pathology.

The physiological mechanisms underlying naloxone-induced neuroprotection in hyperthermia could be related to reduction in the tracer transport across the cerebral microvessels. Although the presence of opioid receptors on cerebral microvessels is not yet fully clarified, our observations indicate an active presence of multiple opioid receptors on the cerebral microvasculature, a feature that requires additional investigation.

The molecular mechanisms by which opioids reduce the extravasation of tracer transport across the BBB are still unclear. It may be that the endogenous opioids influence the function of other neurochemicals, *e.g.*, amino acid neurotransmitters, their receptors or related ion-gated channels. A less number of vesicular profiles loaded with lanthanum in naloxone-treated heat-stressed rats further suggest that opioids can influence transendothelial transport rather than inter-endothelial passage (Sharma et al., 1990, 1998a, b, Sharma, 2004a, b), i.e., through opening of tight junctions (Sharma, unpublished observation). Obviously, a reduction in vascular permeability is primarily responsible for neuroprotection.

Alterations in the BBB function and edema formation appear to be major factors in inducing abnormal brain function in heat stress. A close parallelism between BBB disturbances, edema formation and behavioural deficits supports this idea. Disturbances in the fluid microenvironment of the brain after BBB disruption are likely to contribute to functional deficits, edema formation and cell injury. A reduction in heat stress-induced BBB disruption, edema formation and motor or cognitive dysfunction in naloxone-pretreated animals supports this hypothesis. A reduction in the regional CBF following heat stress is unrelated with BBB disruption, as the magnitude of CBF reduction does not correlate with the intensity of BBB permeability in any brain region.

Systemic hypotension caused by hyperthermia alone or changes in the arterial pH or blood gases does not contribute to BBB disruption. Activation of hypothalamic–pituitary–adrenal axis and/or related compensatory mechanisms is likely to alter these physiological variables in hyperthermia (Harri and Kuusela, 1986; Zelena et al., 1999; Sharma, 2004a, b).

Conclusion

In conclusion, data presented in this chapter suggest that a balance between excitatory and inhibitory amino acids is crucial for hyperthermia-induced cell injury or cell survival. Furthermore, blockade of opioid receptors significantly altered amino acid neurotransmission and induced neuroprotection in hyperthermic rats. This indicates that an interaction between opioid receptors and amino acid neurotransmitters in hyperthermia is important for brain function. The BBB disruption appears to be instrumental in hyperthermia-induced cognitive brain dysfunction, edema formation and brain damage. However, additional studies are needed to explore the role of amino acid neurotransmitter receptors in hyperthermia-induced brain dysfunction.

Acknowledgements

Author's original research presented in this review is supported by grants from Swedish Medical Research council no. 2710, Göran Gustafsson Foundation, Sweden; Alexander Humboldt Foundation, Bonn, Germany; University Grants Commission, New Delhi, India; Indian Council of Medical Research, New Delhi, India. Thanks to reviewers for providing important inputs on opioid-induced neuroprotection in relation to hyperthermia. Thanks are due to our collaborators for extending laboratory facilities for heat stress experiments and/or providing some data presented in this review. Assistance from Drs. F. Fonnum (Oslo, Norway); A. Ermisch, V. Bigl (Leipzig, Germany); J. Cervós-Navarro (Berlin, Germany); P. K. Dey (Varanasi, India); W. Slikker, S. F. Ali (NCTR/US-FDA) is also acknowledged. The secretarial assistance of Aruna Sharma and Angela Ludwig and graphic assistance of Suraj Sharma are highly appreciated.

References

Aguilar, E., Tena-Sempere, M. and Pinilla, L. (2005) Role of excitatory amino acids in the control of growth hormone secretion. Endocrine, 28(3): 295–302. Review

Akil, H., Watson, S.J., Young, E., Lewis, M.E., Khachaturian, H. and Walker, J.M. (1984) Endogenous opioids: biology and functions. Ann. Rev. Neurosci., 7: 223–232.

Alm, P., Sharma, H.S., Hedlund, S., Sjoquist, P.O. and Westman, J. (1998) Nitric oxide in the pathophysiology of hyperthermic brain injury. Influence of a new anti-oxidant compound H-290/51. A pharmacological study using immunohistochemistry in the rat. Amino Acids, 14(1–3): 95–103.

Alm, P., Sharma, H.S., Sjoquist, P.O. and Westman, J. (2000) A new antioxidant compound H-290/51 attenuates nitric oxide synthase and heme oxygenase expression following hyperthermic brain injury. An experimental study using immunohistochemistry in the rat. Amino Acids, 19(1): 383–394.

Artemowicz, B., Sotowiej, E. and Sobaniec, W. (2004) The distribution of the oxidants-antioxidants balance in epileptic children. Int. J. Neuroprotec. Neuroregen., 1(1): 27–31.

Azbill, R.D., Mu, X., Bruce-Keller, A.J., Mattson, M.P. and Springer, J.E. (1997) Impaired mitochondrial function, oxidative stress and altered antioxidant enzyme activities following traumatic spinal cord injury. Brain Res., 765: 283–290.

Babot, Z., Cristofol, R. and Sunol, C. (2005) Excitotoxic death induced by released glutamate in depolarized primary cultures of mouse cerebellar granule cells is dependent on GABA-A receptors and niflumic acid-sensitive chloride channels. Eur. J. Neurosci., 21(1): 103–112.

Basbaum, A.I. and Fields, W.L. (1984) Endogenous pain control system: brainstem spinal pathway and endorphin circuitry. Ann. Rev. Neurosci., 7: 309–327.

Baskin, D.S., Kuroda, H., Hosobuchi, Y. and Lee, N.M. (1985) Treatment of stroke with opiate antagonists — effects of exogenous antagonists and dynorphin 1–13. Neuropeptides, 5: 307–312.

Björklund, A., Hökfelt, T. and Kuhar, M.I. (1990) Neuropeptides in the CNS, Part II. Handbook Chem. Neuroanat., 9: 1–538.

Bligh, J. (1981) Amino acids as central synaptic transmitters or modulators in mammalian thermoregulation. Fed. Proc., 40(13): 2746–2749.

Bouchama, A. and Knochel, J.P. (2002) Heat stroke. N. Engl. J. Med., 346(25): 1978–1988.

Boullerne, A.I. and Benjamins, J.A. (2006) Nitric oxide synthase expression and nitric oxide toxicity in oligodendrocytes. Antioxid. Redox. Signal., 8(5–6): 967–980.

Brann, D.W., Bhat, G.K., Lamar, C.A. and Mahesh, V.B. (1997) Gaseous transmitters and neuroendocrine regulation. Neuroendocrinology, 65(6): 385–395. Review

Bredt, D.S. (1999) Endogenous nitric oxide synthesis: biological functions and pathophysiology. Free Radic. Res., 31(6): 577–596. Review

Buechley, R.W., Van Bruggen, J. and Truppi, L.E. (1972) Heat island equals death island? Environ. Res., 5(1): 85–92.

Centers for Disease Control and Prevention (CDC). (2006) Heat-related deaths — United States, 1999–2003. MMWR Morb. Mortal. Wkly. Rep., 55(29): 796–798.

Chavez, A.E., Singer, J.H. and Diamond, J.S. (2006) Fast neurotransmitter release triggered by Ca influx through AMPA-type glutamate receptors. Nature, 443(7112): 705–708.

Chavkin, C. (2000) Dynorphins are endogenous opioid peptides released from granule cells to act neurohumorly and inhibit excitatory neurotransmission in the hippocampus. Prog. Brain Res., 125: 363–367. Review

Choi, D.W. (1990) Methods for antagonizing glutamate neurotoxicity. Cerebrovasc. Brain Metab. Rev., 2(2): 105–147. Review

Choi, D.W., Koh, J.Y. and Peters, S. (1988) Pharmacology of glutamate neurotoxicity in cortical cell culture: attenuation by NMDA antagonists. J. Neurosci., 8(1): 185–196.

Cremades, A. and Penafiel, R. (1982) Hyperthermia and brain neurotransmitter amino acid levels in infant rats. Gen. Pharmacol., 13(4): 347–350.

Dascombe, M.J. (1985) The pharmacology of fever. Prog. Neurobiol., 25(4): 327–373. Review

DeFeudis, F.V. (1984) Involvement of GABA and other inhibitory amino acids in thermoregulation. Gen. Pharmacol., 15(6): 445–447. Review

Dessai, S. (2003) Heat stress and mortality in Lisbon Part II. An assessment of the potential impacts of climate change. Int. J. Biometeorol., 48(1): 37–44.

Di Sole, F., Guerra, L., Bagorda, A., Reshkin, S.J., Albrizio, M., Minoia, P. and Casavola, V. (2001) Naloxone inhibits A6 cell Na(+)/H(+) exchange by activating protein kinase C via the mobilization of intracellular calcium. Exp. Nephrol., 9(5): 341–348.

Duggan, A.W. and Fleetwood-Walker, S.M. (1993) Opioids and sensory processing in the central nervous system. Handbook Exp. Pharmacol., 104(II): 731–771.

Erreger, K., Chen, P.E., Wyllie, D.J. and Traynelis, S.F. (2004) Glutamate receptor gating. Crit. Rev. Neurobiol., 16(3): 187–224. Review

Faden, A.I. (1993) Role of endogenous opioids and opioid receptors in central nervous system injury. Handbook Exp. Pharmacol., 104(Part I): 325–341.

Fallon, J.H. and Ciofi, P. (1990) Dynorphin-containing neurons. Handbook Chem. Neuroanat., 9: 1–286.

Fosse, V.M., Kolstad, J. and Fonnum, F. (1986) A bioluminescence method for the measurement of L-glutamate: applications to the study of changes in the release of L-glutamate from lateral geniculate nucleus and superior colliculus after visual cortex ablation in rats. J. Neurochem., 47(2): 340–349.

Frosini, M., Ricci, L., Saponara, S., Palmi, M., Valoti, M. and Sgaragli, G. (2006) GABA-mediated effects of some taurine derivatives injected i.c.v. on rabbit rectal temperature and gross motor behavior. Amino Acids, 30(3): 233–242.

Frosini, M., Sesti, C., Palmi, M., Valoti, M., Fusi, F., Mantovani, P., Bianchi, L., Della Corte, L. and Sgaragli, G. (2000)

The possible role of taurine and GABA as endogenous cryogens in the rabbit: changes in CSF levels in heat-stress. Adv. Exp. Med. Biol., 483: 335–344.

Frosini, M., Sesti, C., Saponara, S., Donati, A., Palmi, M., Valoti, M., Machetti, F. and Sgaragli, G. (2000) Effects of taurine and some structurally related analogues on the central mechanism of thermoregulation: a structure-activity relationship study. Adv. Exp. Med. Biol., 483: 273–282.

Frosini, M., Valoti, M. and Sgaragli, G. (2004) Changes in rectal temperature and ECoG spectral power of sensorimotor cortex elicited in conscious rabbits by i.c.v. injection of GABA, GABA(A) and GABA(B) agonists and antagonists. Br. J. Pharmacol., 141(1): 152–162.

Gautret, B. and Schmitt, H. (1985) Central and peripheral sites for cardiovascular actions of dynorphin-(1–13) in rats. Eur. J. Pharmacol., 111: 263–271.

Guthrie, J. and Basbaum, A.I. (1984) Colocalization of immunoreactive proenkephalin and prodynorphin products in medullary neurons of the rat. Neuropeptides, 4(6): 437–445.

Harri, M. and Kuusela, P., (1986) Is swimming exercise or cold exposure for rats? Acta Physiol. Scand., 126(2): 189–197.

Hauser, K.F., Knapp, P.E. and Turbek, C.S. (2001) Structure-activity analysis of dynorphin A toxicity in spinal cord neurons: Intrinsic neurotoxicity of dynorphin A and its carboxyl-terminal, nonopioid metabolites. Exp. Neurol., 168: 78–87.

Heja, L., Karacs, K. and Kardos, J. (2006) Role for GABA and Glu plasma membrane transporters in the interplay of inhibitory and excitatory neurotransmission. Curr. Top Med. Chem., 6(10): 989–995. Review

Herrera-Marschitz, M., Hökfelt, T., Ungerstedt, V., Terenius, L. and Goldstein, M. (1984) Effect of intranigral injections of dynorphin, dynorphin fragments and neoendorphin in rotational behaviour in the rat. Eur. J. Pharmacol., 102: 213–217.

Hökfelt, T., Elde, R., Goldstein, M., Nilsson, G., Pernow, B., Terenius, L., Ganten, D., Jeffocate, S.L., Rechfeld, J. and Said, S. (1978) Distribution of peptide containing neurons. In: Lipton M.A., DiMascio A. and Killam K.F. (Eds.), Psychopharmacology: A Generation of Progress. Raven Press, New York, pp. 39–66.

Hökfelt, T., Johansson, O., Holets, V., Meister, B. and Melander, T. (1987) Distribution of neuropeptides with special reference to their coexistence with classical neurotransmitters. In: Meltzer H.Y. (Ed.), Psychopharmacology: The Third Generation of Progress. Raven Press, New York, pp. 401–416.

Hughes, J. and Woodruff, G.N. (1992) Neuropeptides. Function and clinical applications. Arzneimittelforschung, 42(2A): 250–255. Review

Irnaten, M., Aicher, S.A., Wang, J., Venkatesan, P., Evans, C., Baxi, S. and Mendelowitz, D. (2003) Mu-opioid receptors are located postsynaptically and endomorphin-1 inhibits voltage-gated calcium currents in premotor cardiac parasympathetic neurons in the rat nucleus ambiguous. Neuroscience, 116(2): 573–582.

Ishii, H., Shibuya, K., Ohta, Y., Mukai, H., Uchino, S., Takata, N., Rose, J.A. and Kawato, S. (2006) Enhancement of nitric oxide production by association of nitric oxide synthase with N-methyl-D-aspartate receptors via postsynaptic density 95 in genetically engineered Chinese hamster ovary cells: real-time fluorescence imaging using nitric oxide sensitive dye. J. Neurochem., 96(6): 1531–1539.

Karlsson, K., Sharma, H. and Nyberg, F. (2006) Chromatographic characterization of substance P endopeptidase in the rat brain reveals affected enzyme activity following heat stress. Biomed. Chromatogr., 20: 77–82.

Keatinge, W.R. and Donaldson, G.C. (2004) The impact of global warming on health and mortality. South Med. J., 97(11): 1093–1099. Review

Keil, R., Gerstberger, R. and Simon, E. (1994) Hypothalamic thermal stimulation modulates vasopressin release in hyperosmotically stimulated rabbits. Am. J. Physiol., 267(4 Pt 2): R1089–R1097.

Kern, J.K. and Jones, A.M. (2006) Evidence of toxicity, oxidative stress, and neuronal insult in autism. J. Toxicol. Environ. Health B Crit. Rev., 9(6): 485–499. Review

Kerwin, R.W. and Pycock, C.J. (1979) Role of taurine as a possible transmitter in the thermoregulatory pathways of the rat. J. Pharm. Pharmacol., 31(7): 466–470.

Kozela, E. and Popik, P. (2002) The effects of NMDA receptor antagonists on acute morphine antinociception in mice. Amino Acids, 23(1–3): 163–168. Review

Kuzmin, A., Madjid, N., Terenius, L., Ogren, S.O. and Bakalkin, G. (2006) Big dynorphin, a prodynorphin-derived peptide produces NMDA receptor-mediated effects on memory, anxiolytic-like and locomotor behavior in mice. Neuropsychopharmacology, 31(9): 1928–1937.

Lee, N.M. (1984) The role of dynorphin in narcotic tolerance mechanisms. Nat. Inst. Drug Abuse Res. Mon. Ser., 54: 162–166.

Madden, C.J. and Morrison, S.F. (2003) Excitatory amino acid receptor activation in the raphe pallidus area mediates prostaglandin-evoked thermogenesis. Neuroscience, 122(1): 5–15.

Mayhan, W.G. and Heistad, D.D. (1985) Permeability of blood-brain barrier to various sized molecules. Am. J. Physiol., 248(5 Pt 2): H712–H718.

McCormack, A.L., Atienza, J.G., Johnston, L.C., Andersen, J.K., Vu, S. and Di Monte, D.A. (2005) Role of oxidative stress in paraquat-induced dopaminergic cell degeneration. J. Neurochem., 93(4): 1030–1037.

Minoia, P. and Sciorsci, R.L. (2001) Metabolic control through L calcium channel, PKC and opioid receptors modulation by an association of naloxone and calcium salts. Curr. Drug Targets Immune Endocr. Metabol. Disord., 1(2): 131–137. Review

Monda, M., Viggiano, A., Sullo, A. and De Luca, V. (1998) Aspartic and glutamic acids increase in the frontal cortex during prostaglandin E1 hyperthermia. Neuroscience, 83(4): 1239–1243.

Nichols, M.L., Lopez, Y., Ossipov, M.H., Bian, D. and Porreca, F. (1997) Enhancement of the antiallodynic and antinociceptive efficacy of spinal morphin by antisera to dynorphin A(1–13) or MK-801 in nerve-ligation model of peripheral neuropathy. Pain, 69: 317–322.

North, R.A. (1993) Opioid actions on membrane ion channels. Handbook Exp. Pharmacol., 104(II): 773–797.

Nyberg, F., Sharma, H.S. and Wissenfeld-Hallin, Z. (1995) Neuropeptides in the Spinal Cord, Progress in Brain Research, Volume 104. Elsevier, Amsterdam, pp. 1–430.

Ohnuma, T., Suzuki, T. and Arai, H. (2005) Hypothesis: minimal changes in neural transmission in schizophrenia: decreased glutamatergic and GABAergic functions in the prefrontal cortex. Prog. Neuropsychopharmacol. Biol. Psychiatry, 29(6): 889–894. Review

Olsson, Y., Sharma, H.S., Nyberg, F. and Westman, J. (1995) The opioid receptor antagonist naloxone influences the pathophysiology of spinal cord injury. Prog. Brain Res., 104: 381–399. Review

Pirard, P., Vandentorren, S., Pascal, M., Laaidi, K., Le Tertre, A., Cassadou, S. and Ledrans, M. (2005) Summary of the mortality impact assessment of the 2003 heat wave in France. Euro Surveill., 10(7): 153–156.

Poumadere, M., Mays, C., Le Mer, S. and Blong, R. (2005) The 2003 heat wave in France: dangerous climate change here and now. Risk Anal., 25(6): 1483–1494.

Przewlocki, R., Lason, W., Hollt, V., Silberring, J. and Herz, A. (1987) The influence of chronic stress on multiple opioid peptide systems in the rat: pronounced effects upon dynorphin in spinal cord. Brain Res., 413(2): 213–219.

Punnen, S. and Sapru, H.N. (1986) Cardiovascular responses to medullary micro-injections of opiate agonists in urethane-anaesthetized rats. J. Cardiovasc. Pharmacol., 8: 950–958.

Rossler, O.G., Bauer, I., Chung, H.Y. and Thiel, G. (2004) Glutamate-induced cell death of immortalized murine hippocampal neurons: neuroprotective activity of heme oxygenase-1, heat shock protein 70, and sodium selenite. Neurosci. Lett., 362(3): 253–257.

Rydman, R.J., Rumoro, D.P., Silva, J.C., Hogan, T.M. and Kampe, L.M. (1999) The rate and risk of heat-related illness in hospital emergency departments during the 1995 Chicago heat disaster. J. Med. Syst., 23(1): 41–56.

Sasek, C.A. and Elde, R.P. (1986) Coexistence of enkephalin and dynorphin immunoreactivities in neurons in the dorsal gray commissure of the sixth lumbar and first sacral spinal cord segments in rat. Brain Res., 381(1): 8–14.

Serrano, J.S., Minano, F.J., Sancibrian, M. and Duran, J.A. (1985) Involvement of bicuculline-insensitive receptors in the hypothermic effect of GABA and its agonists. Gen. Pharmacol., 16(5): 505–508.

Sgaragli, G.P. and Palmi, M. (1985) The role and mechanism of action of taurine in mammalian thermoregulation. Prog. Clin. Biol. Res., 179: 343–357. Review

Sharma, H.S. (1982) Blood-Brain Barrier in Stress, PhD Thesis, Banaras Hindu University, Varanasi, India, pp. 1–85.

Sharma, H.S. (1987) Effect of captopril (a converting enzyme inhibitor) on blood-brain barrier permeability and cerebral blood flow in normotensive rats. Neuropharmacology, 26(1): 85–92.

Sharma, H.S. (1999) Pathophysiology of blood-brain barrier, brain edema and cell injury following hyperthermia: New role of heat shock protein, nitric oxide and carbon monoxide. An experimental study in the rat using light and electron microscopy. Acta Universitatis Upsaliensis, 830: 1–94.

Sharma, H.S. (2003) Neurotrophic factors attenuate microvascular permeability disturbances and axonal injury following trauma to the rat spinal cord. Acta Neurochir. Suppl., 86: 383–388.

Sharma, H.S. (2004a) Blood-brain and spinal cord barriers in stress. In: Sharma H.S. and Westman J. (Eds.), The Blood-Spinal Cord and Brain Barriers in Health and Disease. Elsevier Academic Press, San Diego, CA, pp. 231–298.

Sharma, H.S. (2004b) Pathophysiology of the blood-spinal cord barrier in traumatic injury. In: Sharma H.S. and Westman J. (Eds.), The Blood-Spinal Cord and Brain Barriers in Health and Disease. Elsevier Academic Press, San Diego, CA, pp. 437–518.

Sharma, H.S. (Ed.). (2005a) Heat-related deaths are largely due to brain damage. Editorial. Indian J. Med. Res., 121(5): 621–623.

Sharma, H.S. (2005b) Selective neuronal vulnerability, blood-brain barrier disruption and heat shock protein expression in stress induced neurodegeneration. Invited review. In: Sarbadhikari S.N. (Ed.), Depression and Dementia, Progress in Brain Research, Clinical Applications and Future Trends. Nova Science Publishers, Inc., New York, pp. 97–152.

Sharma, H.S. (2005c) Methods to induce brain hyperthermia, In: Current Protocols in Toxicology (Solicited Contribution), Suppl. 23, Unit 11.14, pp. 1–26.

Sharma, H.S. (2006a) Hyperthermia influences excitatory and inhibitory amino acid neurotransmitters in the central nervous system. An experimental study in the rat using behavioural, biochemical, pharmacological, and morphological approaches. J. Neural. Transm., 113(4): 497–519.

Sharma, H.S. (2006b) Hyperthermia induced brain oedema: current status and future perspectives. Indian J. Med. Res., 123(5): 629–652. Review

Sharma, H.S. and Alm, P. (2002) Nitric oxide synthase inhibitors influence dynorphin A (1–17) immunoreactivity in the rat brain following hyperthermia. Amino Acids, 23(1–3): 247–259.

Sharma, H.S. and Alm, P. (2004) Role of nitric oxide on the blood-brain and the spinal cord barriers. In: Sharma H.S. and Westman J. (Eds.), The Blood-Spinal Cord and Brain Barriers in Health and Disease. Elsevier Academic Press, San Diego, pp. 191–230.

Sharma, H.S., Alm, P. and Westman, J. (1998a) Nitric oxide and carbon monoxide in the brain pathology of heat stress. Prog. Brain Res., 115: 297–333. Review

Sharma, H.S. and Cervós-Navarro, J. (1990) Brain oedema and cellular changes induced by acute heat stress in young rats. Acta Neurochir. Suppl. (Wien.), 51: 383–386.

Sharma, H.S. and Cervós-Navarro, J. (1990) Nimodipine improves cerebral blood flow and reduces brain edema, cellular damage and blood-brain barrier permeability following heat

Sharma, H.S., Cervos-Navarro, J. and Dey, P.K. (1991a) Acute heat exposure causes cellular alteration in cerebral cortex of young rats. Neuroreport, 2(3): 155–158.

Sharma, H.S., Cervós-Navarro, J. and Dey, P.K. (1991b) Rearing at high ambient temperature during later phase of the brain development enhances functional plasticity of the CNS and induces tolerance to heat stress. An experimental study in the conscious normotensive young rats. Brain Dysfunct., 4: 104–124.

Sharma, H.S. and Dey, P.K. (1984) Role of 5-HT on increased permeability of blood-brain barrier under heat stress. Indian J. Physiol. Pharmacol., 28(4): 259–267.

Sharma, H.S. and Dey, P.K. (1986) Probable involvement of 5-hydroxytryptamine in increased permeability of blood-brain barrier under heat stress in young rats. Neuropharmacology, 25(2): 161–167.

Sharma, H.S. and Dey, P.K. (1987) Influence of long-term acute heat exposure on regional blood-brain barrier permeability, cerebral blood flow and 5-HT level in conscious normotensive young rats. Brain Res., 424(1): 153–162.

Sharma, H.S. and Hoopes, P.J. (2003) Hyperthermia induced pathophysiology of the central nervous system. Int. J. Hypertherm., 19: 325–354.

Sharma, H.S., Kretzschmar, R., Cervos-Navarro, J., Ermisch, A., Ruhle, H.J. and Dey, P.K. (1992a) Age-related pathophysiology of the blood-brain barrier in heat stress. Prog. Brain Res., 91: 189–196.

Sharma, H.S., Nyberg, F., Cervos-Navarro, J. and Dey, P.K. (1992b) Histamine modulates heat stress-induced changes in blood-brain barrier permeability, cerebral blood flow, brain oedema and serotonin levels: an experimental study in conscious young rats. Neuroscience, 50(2): 445–454.

Sharma, H.S., Nyberg, F., Gordh, T. and Alm, P. (2006) Topical application of dynorphin A (1–17) antibodies attenuates neuronal nitric oxide synthase up-regulation, edema formation, and cell injury following focal trauma to the rat spinal cord. Acta Neurochir. Suppl., 96: 309–315.

Sharma, H.S., Nyberg, F. and Olsson, Y. (1992c) Dynorphin A content in the rat brain and spinal cord after a localized trauma to the spinal cord and its modification with p-chlorophenylalanine. An experimental study using radioimmunoassay technique. Neurosci. Res., 14(3): 195–203.

Sharma, H.S., Nyberg, F., Thornwall, M. and Olsson, Y. (1993a) Met-enkephalin-Arg6-Phe7 in spinal cord and brain following traumatic injury to the spinal cord: influence of p-chlorophenylalanine. An experimental study in the rat using radioimmunoassay technique. Neuropharmacology, 32(7): 711–717.

Sharma, H.S., Olsson, Y. and Cervos-Navarro, J. (1993b) Early perifocal cell changes and edema in traumatic injury of the spinal cord are reduced by indomethacin, an inhibitor of prostaglandin synthesis. Experimental study in the rat. Acta Neuropathol. (Berl.), 85(2): 145–153.

Sharma, H.S., Olsson, Y. and Dey, P.K. (1990) Blood-brain barrier permeability and cerebral blood flow following elevation of circulating serotonin level in the anaesthetized rats. Brain Res., 517: 215–223.

Sharma, H.S., Olsson, Y. and Nyberg, F. (1995) Influence of dynorphin A antibodies on the formation of edema and cell changes in spinal cord trauma. Prog. Brain Res., 104: 401–416. Review

Sharma, H.S. and Westman, J. (1998) Brain Functions in Hot Environment, Progress in Brain Research, Volume 115. Elsevier, Amsterdam, pp. 1–516.

Sharma, H.S. and Westman, J. (2004) The Blood-Spinal Cord and Brain Barriers in Health and Disease. Academic Press, San Diego, CA, pp. 1–617 (Release date: Nov. 9, 2003).

Sharma, H.S., Westman, J., Alm, P., Sjoquist, P.O., Cervos-Navarro, J. and Nyberg, F. (1997a) Involvement of nitric oxide in the pathophysiology of acute heat stress in the rat. Influence of a new antioxidant compound H-290/51. Ann. N.Y. Acad. Sci., 813: 581–590.

Sharma, H.S., Westman, J., Cervos-Navarro, J., Dey, P.K. and Nyberg, F. (1997b) Opioid receptor antagonists attenuate heat stress-induced reduction in cerebral blood flow, increased blood-brain barrier permeability, vasogenic edema and cell changes in the rat. Ann. N.Y. Acad. Sci., 813: 559–571.

Sharma, H.S., Westman, J., Cervos-Navarro, J. and Nyberg, F. (1997c) Role of neurochemicals in brain edema and cell changes following hyperthermic brain injury in the rat. Acta Neurochir. Suppl., 70: 269–274.

Sharma, H.S., Westman, J. and Nyberg, F. (1998b) Pathophysiology of brain edema and cell changes following hyperthermic brain injury. Prog. Brain Res., 115: 351–412. Review

Sharma, H.S., Westman, J., Nyberg, F., Cervos-Navarro, J. and Dey, P.K. (1992d) Role of serotonin in heat adaptation: an experimental study in the conscious young rat. Endocr. Regul., 26(3): 133–142.

Sharma, H.S., Westman, J., Nyberg, F., Cervos-Navarro, J. and Dey, P.K. (1994) Role of serotonin and prostaglandins in brain edema induced by heat stress. An experimental study in the young rat. Acta Neurochir. Suppl. (Wien.), 60: 65–70.

Sharma, H.S. and Winkler, T. (2002) Assessment of spinal cord pathology following trauma using early changes in the spinal cord evoked potentials: a pharmacological and morphological study in the rat. Muscle Nerve Suppl., 11: S83–S91.

Sharma, H.S., Zimmer, C., Westman, J. and Cervos-Navarro, J. (1992e) Acute systemic heat stress increases glial fibrillary acidic protein immunoreactivity in brain: experimental observations in conscious normotensive young rats. Neuroscience, 48(4): 889–901.

Singewald, N. and Philippu, A. (1998) Release of neurotransmitters in the locus coeruleus. Prog. Neurobiol., 56(2): 237–267. Review

Smith, A.P. and Lee, N.M. (1988) Pharmacology of dynorphin. Annu. Rev. Pharmacol. Toxicol., 28: 123–140. Review

Smoyer, K.E. (1998) A comparative analysis of heat waves and associated mortality in St. Louis, Missouri — 1980 and 1995. Int. J. Biometeorol., 42(1): 44–50.

Smoyer, K.E., Rainham, D.G. and Hewko, J.N. (2000) Heat-stress-related mortality in five cities in Southern Ontario: 1980–1996. Int. J. Biometeorol., 44(4): 190–197.

Sprenger, T., Berthele, A., Platzer, S., Boecker, H. and Tolle, T.R. (2005) What to learn from in vivo opioidergic brain imaging? Eur. J. Pain, 9(2): 117–121. Review.

Stott, P.A., Stone, D.A. and Allen, M.R. (2004) Human contribution to the European heatwave of 2003. Nature, 432(7017): 610–614.

Tang, Q., Lynch, R.M., Porreca, F. and Lai, J. (2000) Dynorphin A elicits an increase in intracellular calcium in cultured neurons via a non-opioid, non-NMDA mechanims. J. Neurophysiol., 83: 2610–2615.

Thornwall, M., Sharma, H.S., Gordh, T., Sjoquist, P.O. and Nyberg, F. (1997) Substance P endopeptidase activity in the rat spinal cord following injury: influence of the new anti-oxidant compound H 290/51. Acta Neurochir. Suppl., 70: 212–215.

Wang, B.W., Liao, W.N., Chang, C.T. and Wang, S.J. (2006) Facilitation of glutamate release by nicotine involves the activation of a Ca^{2+}/calmodulin signaling pathway in rat prefrontal cortex nerve terminals. Synapse, 59(8): 491–501.

Wang, S.J. and Yang, T.T. (2005) Role of central glutamatergic neurotransmission in the pathogenesis of psychiatric and behavioral disorders. Drug News Perspect., 18(9): 561–566. Review

Waring, P. (2005) Redox active calcium ion channels and cell death. Arch. Biochem. Biophys., 434(1): 33–42. Review

Watson, S.J., Trujillo, K.A., Herman, J.P. and Akil, H. (1989) Neuroanatomical and neurochemical substrates of drug-seeking behaviour: overview and future directions. In: Goldstein A. (Ed.), Molecular and Cellular Aspects of the Drug Addictions. Springer-Verlag, New York, pp. 29–38.

Winkler, T., Sharma, H.S., Gordh, T., Badgaiyan, R.D., Stalberg, E. and Westman, J. (2002) Topical application of dynorphin A (1–17) antiserum attenuates trauma induced alterations in spinal cord evoked potentials, microvascular permeability disturbances, edema formation and cell injury: an experimental study in the rat using electrophysiological and morphological approaches. Amino Acids, 23(1–3): 273–281.

Winkler, T., Sharma, H.S., Stalberg, E., Badgaiyan, R.D., Gordh, T. and Westman, J. (2003) An L-type calcium channel blocker, nimodipine influences trauma induced spinal cord conduction and axonal injury in the rat. Acta Neurochir. Suppl., 86: 425–432.

Winkler, T., Sharma, H.S., Stalberg, E., Olsson, Y. and Nyberg, F. (1994) Opioid receptors influence spinal cord electrical activity and edema formation following spinal cord injury: experimental observations using naloxone in the rat. Neurosci. Res., 21(1): 91–101.

Wischmeyer, P.E. (2002) Glutamine and heat shock protein expression. Nutrition, 18(3): 225–228. Review

Woods, A.S., Kaminski, R., Oz, M., Wang, Y., Hauser, K., Goody, R., Wang, H.Y., Jackson, S.N., Zeitz, P., Zeitz, K.P., Zolkowska, D., Schepers, R., Nold, M., Danielson, J., Graslund, A., Vukojevic, V., Bakalkin, G., Basbaum, A. and Shippenberg, T. (2006) Decoy peptides that bind dynorphin noncovalently prevent NMDA receptor-mediated neurotoxicity. J Proteome Res., 5(4): 1017–1023.

Woo, S.K., Tulunay, F.C., Loh, H.H. and Lee, N.M. (1983) Effects of dynorphin-(1–13)and related peptides on respiratory rate and morphine-induced respiratory rate depression. Eur. J. Pharmacol., 86: 117–123.

Unterberg, A.W., Stover, J., Kress, B. and Kiening, K.L. (2004) Edema and brain trauma. Neuroscience, 129(4): 1021–1029. Review

Zelena, D., Haller, J., Halasz, J. and Makara, G.B. (1999) Social stress of variable intensity: physiological and behavioral consequences. Brain Res. Bull., 48(3): 297–302.

Zhang, S., Takeda, Y., Hagioka, S., Takata, K., Aoe, H., Nakatsuka, H., Yokoyama, M. and Morita, K. (2005) Measurement of GABA and glutamate in vivo levels with high sensitivity and frequency. Brain Res. Protoc., 14(2): 61–66.

SECTION VII

Hyperthermia and Gene Expression

CHAPTER 16

Exertional heat illness and human gene expression

Larry A. Sonna[1,*], Michael N. Sawka[2] and Craig M. Lilly[3]

[1]*Division of Pulmonary and Critical Care Medicine, University of Maryland School of Medicine, Baltimore, MD 21201, USA*
[2]*Thermal and Mountain Medicine Division, US Army Research Institute of Environmental Medicine, Natick, MA 01760, USA*
[3]*University of Massachusetts School of Medicine, Worcester, MA, USA*

Abstract: Microarray analysis of gene expression at the level of RNA has generated new insights into the relationship between cellular responses to acute heat shock *in vitro*, exercise, and exertional heat illness. Here we discuss the systemic physiology of exertional hyperthermia and exertional heat illness, and compare the results of several recent microarray studies performed *in vitro* on human cells subjected to heat shock and *in vivo* on samples obtained from subjects performing exercise or suffering from exertional heat injury. From these comparisons, a concept of overlapping component responses emerges. Namely, some of the gene expression changes observed in peripheral blood mononuclear cells during exertional heat injury can be accounted for by normal cellular responses to heat, exercise, or both; others appear to be specific to the disease state itself. If confirmed in future studies, these component responses might provide a better understanding of adaptive and pathological responses to exercise and exercise-induced hyperthermia, help find new ways of identifying individuals at risk for exertional heat illness, and perhaps even help find rational molecular targets for therapeutic intervention.

Keywords: gene expression; exercise; heat injury; heat stroke; genomics; microarrays; peripheral blood mononuclear cells; heat shock proteins; cellular stress response

Introduction

Humans respond to environmental challenges at many levels, including behavioral, systemic, cellular, and molecular. Improvements in our ability to identify molecular alterations have led to new mechanistic insights into the effects of heat and cold on cellular and systemic function. For example, it has long been known that hyperthermia can produce changes in cellular gene expression, both *in vivo* and *in vitro*, and that some of these changes, such as increased expression of heat shock proteins (HSPs), are related to beneficial responses such as the development of thermotolerance (Lindquist, 1986; Parsell and Lindquist, 1993). More recent findings, including studies of gene expression on a large scale using microarrays, demonstrate that the gene expression response of human cells to thermal stress also includes many genes that are not traditionally designated as HSPs but that are nonetheless involved in functional pathways likely to be of physiological importance (Dinh et al. 2001; Sonna et al., 2002a, b).

It has been recognized that gene expression responses can readily be identified in human

*Corresponding author. Tel.: +1 (410) 328-8141; Fax: +1 (410) 328-8087; E-mail: larry_sonna@yahoo.com

peripheral blood mononuclear cells (PBMCs), both *in vitro* (e.g., Sonna et al., 2002b) and *in vivo* (e.g., Sonna et al., 2004). This clinically accessible human cell type exhibits many of the responses to thermal stress that occur in other cell types traditionally used to investigate the effects of heat shock. As will be discussed in this chapter, it is increasingly evident that there are important differences in the PBMC gene expression responses that occur in response to normal physiological stresses, such as physical exercise, and those that occur during pathophysiological states, such as exertional heat illness. These differences have the potential to provide novel insights into the molecular mechanisms of exertional heat illnesses and are starting to implicate pathways not formerly associated with these diseases. We believe that a better understanding of how these pathways interact with those already known to be associated with both normal and pathophysiological responses to heat will provide the conceptual basis for the next generation of advances in the diagnosis and treatment of exertional heat illness.

This chapter will review recent insights gained from the application of microarray technologies to the study of gene expression responses to heat shock *in vitro*, and both exercise and exertional heat illness *in vivo*. We will briefly discuss the systemic physiology of exertional hyperthermia and exertional heat illness; review insights gained from *in vitro* models of human heat stress; and finally, compare the results of microarray studies of human responses to acute physical exercise and exertional heat injury.

Physiology of exertional hyperthermia and exertional heat illness

Physical exercise involves skeletal muscle contraction, which causes metabolic heat production that must be dissipated into the surrounding environment (Sawka et al., 1996; Sawka and Pandolf, 2001). During exercise, core temperature rises as a result of differences between metabolic heat production and heat dissipation. In humans, most heat dissipation occurs via transfer of heat to the surface through increases in skin blood flow and then heat loss by convection and sweat evaporation. If heat production exceeds heat dissipation, core temperature rises until dissipation matches production and a new steady state is achieved.

Compensable vs. uncompensable heat stress

Heat stress is said to be *compensable* when the mechanisms that dissipate heat are able to increase enough to quantitatively match those which produce it, resulting in an elevated steady-state core temperature (Table 1). Heat loads that exceed the capacity of heat dissipation mechanisms are said to be *uncompensable* and lead to progressive elevations of core temperature until the point of exhaustion is reached. It is important to note that the terms "compensable" and "uncompensable" refer to the core temperature response to a heat load; as will be discussed later, both compensable and uncompensable heat stress are capable of producing heat illnesses.

Table 1. Compensable vs. uncompensable heat stress

	Compensable heat stress	Uncompensable heat stress
Definition	Heat losses match heat production	Heat production exceeds heat losses
Example	Well-acclimatized and well-hydrated individual exercising in light clothing in hot weather	Individual in heavy industrial protective clothing, doing heavy work in a hot environment
Core temperature	Rises to a steady-state that is proportional to exercise intensity	Continues to rise until exhaustion
	Sustained core temperatures of $>40°C$ are possible	Exhaustion often occurs at core temperatures $<39°C$
Common exercise-limiting factor(s)	Hydration	Diversion of cardiac output to skin
	Fitness	
	Energy depletion	

The outcome of a heat stress depends on the magnitude of the heat load applied, the organism's ability to dissipate heat, and the individual's capacity to function at a higher body temperature. The environmental conditions that produce the two types of heat stress differ in important ways, as do the physiological and pathophysiological consequences thereof. During compensable heat stress, the biophysics of heat exchange permits sufficient surface heat loss by evaporative, conductive, and radiant loss mechanisms to maintain a steady-state elevation in core temperature. A typical example of this is a physically fit, well-acclimatized individual who is wearing light clothing and who is exercising under conditions of moderate heat but low humidity. Under these conditions, the core temperature achieved is roughly proportional to the exercise intensity. Core temperatures that are substantially above normal can be achieved and sustained for relatively long durations, until factors such as dehydration and energy depletion make further exertion impossible. For well-acclimatized, fit individuals subjected to compensable heat stress, function can be sustained even with core temperatures of $>40\,°C$.

By contrast, during uncompensable heat stress, the biophysics of heat exchange does not allow adequate heat dissipation and core temperature continually rises during the period of exposure. An example of a setting in which uncompensable heat stress occurs is intense exercise under hot and humid conditions by an individual who is wearing heavy protective clothing. Under these conditions, skin temperature typically rises very quickly due to inadequate evaporative cooling, which in turn produces a high cutaneous blood flow at the expense of flow to other tissues. This in turn leads to diversion of blood flow from viscera and is clinically manifest as exhaustion and collapse that often occurs before the high core temperatures that are typical of compensable heat stress have been reached. For individuals exposed to uncompensable heat stress, core temperatures $<39\,°C$ at the time of exhaustion or collapse can be common.

Physiological consequences of heat dissipation responses under both compensable and uncompensable conditions include ongoing loss of body water in the form of sweat and diversion of blood flow from viscera to active muscles and skin (Sawka et al., 1996; Sawka and Pandolf, 2001). This combination of reduction in circulating blood volume and redistribution of blood flow away from vital organs can have important secondary consequences. For example, ischemia of the splanchnic circulation may compromise intestinal mucosal integrity, allowing translocation of bacteria and their immuno-stimulatory products, endotoxemia, activation of innate immune systems, and the production of oxidative-nitrosative intermediates that can cause tissue damage. In addition to these perfusion-mediated problems, excessively high tissue temperatures ($>41\,°C$, $105.8\,°F$) can produce tissue injury directly when protective cellular mechanisms fail. There is considerable variation in the extent to which individuals can endure these consequences, and this physiological reserve can be affected by a number of identifiable factors such as degree of hydration, heat acclimatization, and aerobic fitness. Less is known about the cellular mechanisms that account for this variability in sensitivity or threshold at which exhaustion occurs. However, even in low-risk individuals, the physiological demands generated by the need for increased heat dissipation cannot be endured indefinitely. Coupled with elevated core temperature, they lead to circulatory insufficiency, cellular dysfunction, and organ injury.

The exertional heat illnesses: heat exhaustion, exertional heat injury, and exertional heat stroke

Compensable or not, any heat stress that continues beyond the individual's physiological limits of tolerance can produce illness within a spectrum of clinically related, overlapping syndromes: heat exhaustion (HE), exertional heat injury (EHI), and exertional heat stroke (EHS) (Gardner and Kark, 2001). Collectively these can be referred to as "exertional heat illnesses". Although there are no universally accepted definitions of these conditions, HE is typically defined as simple inability to continue exercise in the context of hyperthermia. EHI is characterized by evidence of end-organ injury such as elevations in serum liver enzymes but without neurological impairment beyond transient

Table 2. Examples of risk factors for heat illness

Protective against heat illness	Increased risk of heat illness
Low adiposity	High adiposity
High cardiovascular fitness	Low cardiovascular fitness
Adequate hydration	Dehydration
Prior heat acclimatization	Lack of prior heat acclimatization
	Impaired perspiration
	Antecedent febrile illness
	Previous history of heat illness
	Prior-day prodromal illness or substantial heat load
Low-intensity exercise	High-intensity exercise
Low ambient temperature and humidity	High ambient temperature and humidity
Clothing with low insulating capacity	Highly insulating or water-impermeable clothing

confusion. EHS involves significant neurological signs and symptoms, such as persistent confusion, delirium, or coma. These three conditions are best thought of as a continuum rather than separate disease processes, and it is not unusual for individuals with heat stroke to exhibit evidence of other organ injury or dysfunction.

The onset, type, and severity of heat illness are all influenced by a number of contributing factors. Common recognized clinical risk factors for the development of exertional heat illnesses are listed in Table 2. As noted, both compensable and uncompensable heat stress can produce exertional heat illness. In general terms,[1] compensable heat stress, which tends to produce marked and prolonged elevations of core temperature, is more likely to cause severe EHI and heat stroke than is uncompensable heat stress. By contrast, uncompensable heat stress is more commonly associated with HE and less severe EHI.

Historically, exertional heat illnesses were thought to occur predominantly in high risk persons such as those of low physical fitness, high body fat, unacclimatized to heat, or dehydrated (Shibolet et al., 1976; Epstein et al., 1999). It is now increasingly appreciated that exertional heat illnesses also occur in fit individuals and at lower exertional levels and heat loads than were previously considered a risk for exertional heat illness.

For example, several studies have reported that many cases of exertional heat illness occur under relatively temperate conditions (Kark et al., 1996; Epstein et al., 1999; Gardner and Kark, 2001). Furthermore, the duration of exercise that precipitates decompensation need not be prolonged (<1 h in 43% of cases in one series (Epstein et al., 1999)).

There has been speculation in the literature that for some heat injury/stroke victims, a previous heat injury or illness might augment the hyperthermia of exercise, thus inducing heat illness under environmental conditions that would normally be tolerated (Montain et al., 2000). For others, antecedent viral infection or febrile illness might limit the ability of cells or tissues to adapt to heat stress and thus make them more susceptible to injury (Hasday and Singh, 2000; Sonna et al., 2004). As many as 16–18% of individuals who develop exertional heat illness also report having suffered from a non-specific prodromal illness in the days leading up to the acute decompensation (Shibolet et al., 1967; Epstein et al., 1999). Prior-day exposure to heat stress has also been identified as a risk factor for development of exertional heat illness (Kark et al., 1996). Regardless, the incidence of exertional heat illness among individuals otherwise thought to be at low risk suggests the possibility that there may be additional, as of yet unidentified, risk factors for the development of exertional heat illness.

In summary, the risk of exertional heat illness is influenced by the magnitude and duration of hyperthermia (which in turn is determined by the balance between heat load, and dissipative

---

[1]This assumes that all other factors are comparable and that the individual's rise in core temperature is halted promptly on collapse (e.g., by moving to temperate conditions and achieving adequate heat dissipation).

capacity), and by the ability of cells, tissues, and organ systems to function at increased temperature. Risk factors for the development of exertional heat illness can be gleaned from epidemiological studies; some of these likely exert effects systemically (such as the insulating effects of high body fat and the limiting effects of dehydration on cardiac output). Others, such as antecedent illness, seem likely to affect the ability of cells, tissues, and organs to resist injury and continue functioning at high temperature. Thus, a comprehensive understanding of physiological and pathophysiological responses to heat must combine knowledge of systemic, organ, and tissue responses, with insights into the heat-related pathways that govern cell function.

Cellular responses to heat

Exposure to heat exerts direct effects on the constituent molecules that comprise cells. It also triggers cellular responses designed to cope with those effects, restore homeostasis, and render the cell tolerant to subsequent heat stress. These cellular responses, referred to collectively as "heat shock responses", are ubiquitous and highly conserved in eukaryotes (Lindquist, 1986; Parsell and Lindquist, 1993; Katschinski, 2004).

The effects of heat on cells include (Fig. 1): denaturation and misaggregation of proteins (Lindquist, 1986); activation of transcription factors such as heat shock factor 1 (HSF1) (Morimoto, 1998) that can lead to increased expression of stress proteins; an otherwise generalized disruption of transcription (Lindquist, 1986), RNA processing (Lindquist, 1986; Bond, 1988; Yost and Lindquist, 1991), and translation (Panniers, 1994); acceleration of enzymatic reactions (per the Arrhenius equation); changes in the activities of many signal phosphatases and kinases (Dubois and Bensaude, 1993; Han et al., 2000, 2001; Obata et al., 2000; Dorion and Landry, 2002); increased

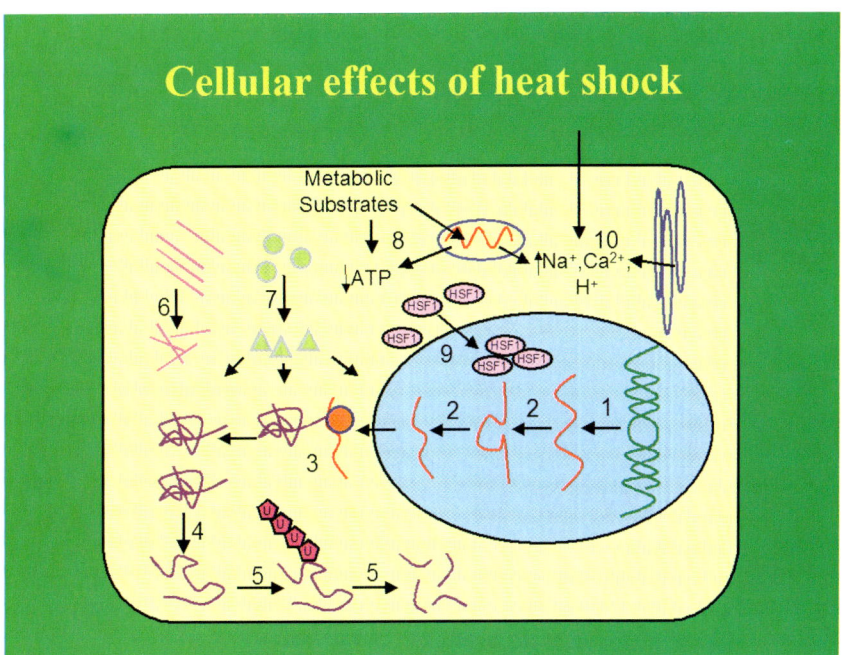

Fig. 1. Cellular effects of acute heat shock. These include effects on (1) transcription, (2) RNA processing, (3) translation, (4) protein conformation, (5) protein degradation, (6) cytoskeletal organization, (7) activities of signal transduction proteins, including stress kinases, (8) metabolism and biosynthesis, sometimes resulting in decreased cellular ATP, (9) cellular distribution and activities of critical proteins, such as transcription factor HSF-1, and (10) changes in membrane permeability leading to changes in intracellular ion concentrations.

protein degradation via the proteasomal and lysosomal pathways (Parag et al., 1987; Xu et al., 1997; Mathew and Morimoto, 1998); cell-cycle arrest (Helmbrecht et al., 2000; Kuhl and Rensing, 2000); and activation of pro- or anti-apoptotic pathways (Punyiczki and Fesus, 1998; Creagh et al., 2000; Beere, 2004). Heat can also alter membrane permeability (Weitzel et al., 1987; Gaffin et al., 1997; Skrandies et al., 1997; Koratich and Gaffin, 1999) to produce increases in intracellular sodium, hydrogen ion, and calcium concentrations, disrupt cytoskeletal components (Han et al., 2000; Dorion and Landry, 2002), and decrease intracellular stores of ATP (Findly et al., 1983; Weitzel et al., 1987).

As will be discussed in detail, the cellular response to acute heat stress itself appears to comprise at least three components. The first involves changes in the functional activities of previously translated proteins. These include activation of constitutively expressed transcription factors such as HSF1 (Morimoto, 1998), and of components of stress- and mitogen-activated protein kinase pathways (Dubois and Bensaude, 1993). The second component involves altered expression levels of proteins that comprise an acute homeostatic response. Two of the best-characterized protein families in this group are the HSPs, many of which re-fold denatured proteins (Lindquist, 1986), and the ubiquitins, which are ligated to proteins that cannot be re-folded, thus targeting them for degradation via the proteasome system (Parsell and Lindquist, 1993). Proteins involved in redox control may be part of this acute homeostatic response. The early gene expression response to heat shock also appears to involve changes in expression of molecules such as transcription factors that likely influence subsequent gene expression (Sonna et al., 2002a). The third and least understood component involves changes in expression of proteins that appear to be involved in restoring cellular function, directing cellular remodeling, and determining cellular fate after heat shock, such as regulatory proteins, proteins involved in cell-cycle control, structural proteins, and proteins involved in pro- and anti-apoptotic pathways (Sonna et al., 2002a, b). In at least some cell lines, this component also includes a number of molecules involved in cell signaling (Sonna et al., 2002a).

At the cellular level, the ultimate outcome of given heat stress appears to occur in gradations, with survival and adaptation occurring if conditions permit and if not, apoptosis in preference to necrosis (Creagh et al., 2000; Dorion and Landry, 2002). The factors that determine cellular fate after heat shock are still an active area of investigation, but it is clear that antecedent history is one of the major determinants. For example, cells that have previously been subjected to heat shock and are thus expressing high levels of HSPs can often tolerate subsequent stresses would be lethal to unconditioned cells (Lindquist, 1986; Parsell and Lindquist, 1993). In contrast, prior exposure to stimuli such as LPS, TNF-alpha, interferons, or cytokines can sensitize cells to heat shock-induced apoptosis (Buchman et al., 1993; Abello and Buchman, 1994).

We propose that a better understanding of the balance between pathways that support adaptation and survival and those that steer cells toward apoptosis and/or necrosis is critical to the advancement of our knowledge of heat-related illness. It is well accepted that non-lethal thermal stress activates several cellular adaptation pathways, which in turn contribute to the ability of the organism to withstand subsequent and increasingly severe thermal stresses (Kregel, 2002). We propose that, conversely, individuals who are predisposed to heat related illness by virtue of an antecedent episode may be prone to failure of adaptation and activation of apoptotic and cell death pathways at heat loads that would ordinarily be tolerated. If this hypothesis is confirmed, then the factors that influence this balance may hold the key to successful therapeutic manipulation that will increase systemic performance during thermal stress.

Heat shock responses *in vitro*

Experimentally, the best-characterized response to heat shock is increased expression of several families of HSPs (Lindquist, 1986; Parsell and Lindquist, 1993; Katschinski, 2004). In models involving tissue culture or primary isolates of cells, a very strong heat shock response can often be elicited by exposing cells to a temperature of 5° to 6°C above the normal culture temperature for 20–60 min, followed by normothermic recovery for varying periods of time. While changes in HSP expression are often detectable during the period of hyperthermia itself, the maximal expression of HSPs commonly occurs several hours into the normothermic recovery period (e.g., Sonna et al., 2002a, b). Changes in expression have also been reported in many genes belonging to functional categories other than the 'classic' HSPs (Sonna et al., 2002a).

One of the most important mechanisms whereby heat shock alters gene expression involves activation of transcription factors, of which the best characterized is HSF1 (Morimoto, 1998; Pirkkala et al., 2001). This ubiquitous transcription factor is expressed constitutively in cytoplasmic multiprotein complexes that include HSPs and co-chaperones. When denatured protein residues accumulate in the cytoplasm as a result of heat or other protein-denaturing stresses, the HSF1 monomers are released and appear shortly thereafter in the nucleus in a trimeric, transcriptionally active form that is capable of inducing transcription of HSPs and other heat shock-responsive genes (Morimoto, 1998). Additionally, some HSPs have specialized features that permit their expression under hyperthermic conditions that globally disrupt gene expression. For example, some members of the HSP 70 family lack introns (Lindquist, 1986) and therefore can be expressed even at temperatures that disrupt RNA splicing.

Heat shock also causes changes in the activities of a number of stress-activated signal transduction pathways. These include mitogen-activated protein kinases, stress-activated protein kinases such as jun-N-terminal kinases (JNK), p38, and extracellular signal-regulated kinases (ERK) (Dubois and Bensaude, 1993; Han et al., 2000, 2001; Obata et al., 2000; Dorion and Landry, 2002). These changes modulate cellular responses to heat stress and may influence cell fate after heat shock (i.e., survival and adaptation vs. apoptosis).

HSPs serve a variety of functional roles. Many HSPs (best typified by the HSP 70 family of proteins) are chaperonins, enzymes whose primary function is to re-fold denatured proteins into a native conformation (Lindquist, 1986; Georgopoulos and Welch, 1993). Some, like members of the HSP 90 family, are also involved in the normal processing of regulatory proteins such as steroid receptors (Georgopoulos and Welch, 1993). Others are involved in regulation of cellular redox state and cellular signaling (HSP 32, better known as heme oxygenase-1 (Otterbein and Choi, 2000)), and still others are involved in targeting for degradation of proteins (the ubiquitins) (Parsell and Lindquist, 1993).

As noted, the heat shock response also involves genes other than those traditionally designated as HSPs (Sonna et al., 2002a). These include components of transcription factors (such as jun) that may produce downstream changes in gene expression. They also include several genes that likely have substantial effects on cellular function after heat shock, such as cell-cycle proteins p53 and p21 (WAF-1), signal transduction molecules such as DUSP1, and molecules involved in redox control such as Cu, Zn superoxide dismutase (Sonna et al., 2002a).

Microarray studies of the heat shock response in vitro

DNA microarray studies have confirmed the hypothesis that responses to heat shock produces extensive changes in gene expression, at least with respect to mRNA. Two recent examples of the application of DNA microarrays to the study of heat stress responses involved the study of retinal pigment epithelial cells that had survived laser burns (Dinh et al., 2001) and human PBMCs subjected to heat shock *in vitro* (43°C for 20 min followed by recovery at 37°C for 2 h and 40 min) (Sonna et al., 2002b). In both studies, extensive changes were noted in expression both of HSPs and of genes not traditionally considered to be HSPs. Changes occurred both in pathways previously known from biochemical and physiological studies to be involved in the human response to heat shock as well as in pathways whose role in the heat shock response is less well defined (Sonna et al., 2002b). Given the overlap between the changes observed in PBMCs with those typically found in other cell lines in which heat shock responses have been studied, the identification of genes involved in cell-cycle control, gene expression, and pro- and anti-apoptotic pathways provides novel opportunities for understanding how heat stress could affect tissue function and integrated physiological responses without the need for obtaining human cells by highly invasive methods.

The heat shock response in animal models

The responses of animals exposed to hyperthermic stress has been studied extensively, and the reader is referred to excellent reviews on this topic (Moseley, 1997; Kregel, 2002) and recent work in a baboon model (Bouchama et al., 2005), mice (Leon et al., 2005, 2006), and rat (Sharma, 2006). A wide variety of tissues exhibit increased expression of HSPs in response to hyperthermic stress, and the increased expression of these proteins correlates with the development of thermotolerance. Other factors such as desensitization of the p38 and JNK pathways may also be involved in the acquisition of thermotolerance (Dorion and Landry, 2002). Expression of HSPs may also be involved in increased tolerance to non-thermal stresses such as oxidative stress and ischemia-reperfusion (Kregel, 2002) as well as in acclimation to heat (the ability to perform increasing work at high temperatures (Moseley, 1997)). The expression of HSPs within cells can therefore confer benefit at the cellular, tissue, organ, and whole organism levels. These models have the potential not only to link cellular changes to integrated physiological responses but also to ascertain the extent to which pathways other than those traditionally associated with thermal stress are involved.

Heat shock responses and moderate (febrile-range) hyperthermia

Adaptive responses induced by exercise hyperthermia may be associated with heat acclimatization (Sawka et al., 1996), the acquisition of thermotolerance, or both, as evidenced by observation that exercise-trained rats have reduced mortality when exposed to severe heat stress (Fruth and Gisolfi, 1983). Although high temperatures will elicit a maximal heat shock response (42° to 45°C for most human cell lines), recent studies also show that heat shock responses can be produced at much lower temperatures, including temperatures commonly achieved in the febrile range (38.0° to 39.5°C) (Hasday and Singh, 2000; Park et al., 2005). Temperatures in this range are commonly achieved during acute physical exercise, which raises the possibility that acute physical exercise *in vivo* might cause some of the same gene expression changes that are observed after more severe heat stress. This is conceptually important because, by comparing changes in gene expression that occur during acute physical exercise to changes in expression that occur during heat illness, it may be possible to distinguish adaptive from pathological pathways and mechanisms.

Effect of acute physical exercise on gene expression in humans

Cell types used to study gene expression responses to exercise

It is generally accepted that exercise training produces changes in gene expression in cell types such as skeletal muscle (Fluck and Hoppeler, 2003; Goldspink, 2003). Recently it has been recognized that even a single bout of exercise can produce acute changes in gene expression in skeletal muscle, at least in younger men (Jozsi et al., 2000; Mahoney et al., 2004, 2005; Bickel et al., 2005). That even a single period of exercise can induce such changes can be explained by the observation that exercising muscle is subject to a variety of stimuli that are known to alter gene expression, including elevated temperatures, hormones, neuronal activation (leading to membrane depolarization and increases in intracellular calcium), tissue hypoxia, acidosis, and mechanical deformation (Fluck and Hoppeler, 2003). Some of these factors are likely to affect not only skeletal muscle itself, but also any cells that traverse skeletal muscle capillary beds during exercise (such as circulating leukocytes). Furthermore, the exercising muscle itself releases factors that are known to influence gene expression in other cell types. For example, several studies have reported exercise-induced release of the anti-inflammatory cytokine IL-6 (for a recent review, see Petersen and Pedersen, 2005), which in turn can induce expression of both the IL-1 receptor antagonist (also elevated in exercise, see below) and IL-10, and inhibit production of TNF-alpha (Petersen and Pedersen, 2005).

Skeletal muscle biopsy can, therefore, provide important insights into human gene expression changes that occur with exercise and hyperthermia. However, the technique is invasive, and skeletal muscle gene expression changes during exercise only yield information about the effects of exercise on that single tissue type, albeit one that is undeniably central to our understanding of the molecular physiology of exercise in humans. An understanding of normal and abnormal effects of exercise and hyperthermia on other tissue types (such as liver, bowel, neural tissue, etc.) requires animal models, the application of invasive biopsy techniques to human volunteers, or the use of representative samples of cells such as those that circulate through these tissues.

Fortunately, PBMCs, which can be readily obtained from humans, are an accessible and potentially informative cell type with which to study non-muscle responses to heat stress and exercise. PBMCs express many genes and are responsive to a wide variety of stimuli *in vitro* and *in vivo*, including heat stress. As inflammatory cells, they are centrally involved in many important systemic inflammatory responses under both physiological and pathophysiological conditions. As circulating cells, they are exposed both to systemic signals and to local signals present in perfused tissues during exercise, including temperature, pH, oxygen tension, cytokines, mechanical stresses, oxidative stress, and other local tissue factors. Indeed, there is increasing evidence that gene expression in circulating PBMCs is influenced by disease processes in isolated organs, such as pulmonary hypertension (Bull et al., 2004). Finally, studies have demonstrated convincingly that PBMCs generate a gene expression response *in vivo* to thermal stress that has remarkable fidelity to the HSP response of many other cell types (Ryan et al., 1991; Fehrenbach et al., 2000a, b, 2001; Schneider et al., 2002).

An important limitation to the use of unsorted PBMCs is that they represent a heterogeneous cell population. Measured changes in expression can, therefore, result from either actual changes in transcript level within individual cells or from changes in the distribution of PBMC subpopulations produced by a given stimulus. Although this can be of considerable utility in its own right (e.g., a stimulus that markedly alters the ratio of CD4 to CD8 lymphocytes would likely produce measurable changes in the gene expression profile of peripherally obtained PBMCs that could be used for diagnostic purposes), it is an inherent limitation of the use of PBMCs that must be kept in mind when evaluating the gene expression literature.

Microarray studies of exercise-induced gene expression changes in PBMCs

Two important microarray studies have recently shed light on the responses of PBMCs to acute physical exercise.

Connolly et al. (2004) studied the effect of acute physical exercise on PBMC gene expression in 15 moderately fit (VO$_2$ peak, 35–45 ml/min/kg) young men aged 18–30. The exercise protocol consisted of cycling at ~80% of peak VO$_2$ for 30 min. They did not report body temperature measurements, but such exercise intensities would be expected to increase core temperatures to ~39°C (Sawka et al., 1996). Samples were obtained for microarray analysis before exercise, immediately after exercise, and after 60 min of post-exercise recovery. RNA samples were analyzed on a single-dye platform containing 22,283 sequences (Affymetrix HU133A). Moderately strict criteria were used to call a difference in expression statistically significant (estimated false positives $\leq 1\%$).

In this study, the subjects demonstrated a rise in serum lactate and growth hormone that peaked at the end of exercise, as well as a rise in IL-6 and IL-1ra that peaked at the end of the recovery period. An increase in circulating leukocytes was noted at the end of exercise in all three major lineages (granulocytes, lymphocytes, and monocytes), which returned to baseline values at the recovery time point. The array data showed a substantial, time-dependent gene expression response to exercise involving several hundred genes. Interestingly, increases in gene expression dominated the expression pattern at the end of exercise, whereas both increases and decreases in gene expression were evident at the recovery time point. In all, these authors reported that 311 genes demonstrated significant expression changes during exercise (post- vs. pre-exercise) and 292 genes demonstrated expression changes after recovery as compared to the pre-exercise baseline (recovery vs. pre-exercise). A total of 552 genes showed differences in expression between the post-exercise and the recovery time points.

The genes affected by exercise (at one or both of the time points studied) included many "classic" HSPs, some well-known non-specific stress proteins (such as DUSP1), and several inflammatory modulators. Changes in IL-6 mRNA were not detected, consistent with prior literature concluding that skeletal muscle, not PBMCs, is the predominant source of the circulating IL-6 produced during exercise (Petersen and Pedersen, 2005). Additionally, many of the affected genes are involved in cell growth, proliferation, and differentiation as well as in transcription and signal transduction.

The breadth of the gene expression response to acute physical exercise was similar to observations made in heat shock microarray studies of PBMCs *in vitro* (Sonna et al., 2002b) and prompted us to perform a more detailed comparison of the specific genes affected by the two stresses. In the published supplement to their report, Connolly et al. (2004) listed 433 specific sequences that were affected by exercise in one or more of their paired comparisons. From this list, 345 corresponding sequences were identified on the arrays used to perform the previous *in vitro* heat shock experiments (the U95A)[2] (Sonna et al., 2002b). We queried this list to identify genes that displayed a statistically significant change in expression as a result of *in vitro* heat shock. To maximize methodological comparability with the data of Connolly et al., we performed our query without the post-hoc filters used in the previous *in vitro* study (Sonna et al., 2002b), such as expression calls and arbitrary fold-change cutoffs.

Table 3 provides a comparison of gene expression from acute physical exercise and *in vitro* heat shock. Of the 345 sequences affected by exercise for which cross-platform comparisons could be made, 184 sequences were different from baseline at the end of exercise (168 increased and 16 decreased) and 104 were different from baseline after recovery (61 increased and 43 decreased). Approximately 1/5 of these genes showed similar changes after heat shock *in vitro* (Table 3). This overlap is somewhat larger than that found (11–14%) in a control comparison of the effects of acute physical exercise in PBMCs to the effects of hypoxia on a different cell line (exposure of HepG2 hepatocytes in culture to 1% oxygen for 24 h (Sonna et al., 2003)). Furthermore, the degree of overlap increased substantially when the analysis was limited to genes considered to be of high interest by Connolly et al., that were further classified as "stress proteins and HSPs". Of 15 sequences for which cross-platform comparisons could be made (representing 11 genes of this class)[2], all showed increased expression at one or both of the time points after exercise as compared to baseline and of these, 12 (80%) were also increased significantly by heat shock *in vitro*. The three sequences that showed responses after heat shock that differed from acute physical exercise corresponded to dual specificity phosphatase 5, HIF1A (both unaffected by heat shock *in vitro*) and HIF-1 responsive RTP801 (decreased by heat shock).

Among the "high interest" genes identified by Connolly et al., the degree of overlap between the effects of exercise and the *in vitro* PBMC heat shock responses reported by Sonna et al. (2002b) was also greater for stress proteins and HSPs than for other categories. For inflammatory response genes, the overlap with heat shock responses was 17 and 31% at end-exercise and recovery, respectively. For genes classified as growth and

---

[2] In some cases, direct correspondences were redundant (meaning, more than one sequence could be found on one array that corresponded to the targeted sequence on the other). In others, correspondence was indirect (i.e., corresponding sequences were identified on the two platforms that targeted the same gene though not necessarily not the same target sequence).

Table 3. Overlap between the effects on gene expression of acute exercise *in vivo* (Connolly et al., 2004; Zieker et al., 2005) and heat shock *in vitro* (Sonna et al., 2002b)

Study	Time point	Category	Direction of change		
			Increased	Decreased	Either
Connolly et al. (PBMCs)	Post-exercise	Total sequences affected[a]	168 (100%)	16 (100%)	184 (100%)
		Sequences similarly affected by heat shock	34 (20%)	7 (44%)	41 (22%)
		Sequences differently affected by heat shock[b]	134 (80%)	9 (56%)	143 (78%)
	Recovery	Total sequences affected[a]	61 (100%)	43 (100%)	104 (100%)
		Sequences similarly affected by heat shock	15 (25%)	6 (14%)	21 (20%)
		Sequences differently affected by heat shock[b]	46 (75%)	37 (86%)	83 (80%)
Zieker et al. (Whole blood)	Post-exercise	Total sequences affected[a,*]	13 (100%)	22 (100%)	35 (100%)
		Sequences similarly affected by heat shock	4 (31%)	8 (36%)	12 (34%)
		Sequences differently affected by heat shock[b]	9 (69%)	14 (64%)	23 (66%)

[a] Relative to pre-exercise baseline.
[b] Includes sequences not significantly affected by heat shock and sequences that showed a statistically significant change of opposite direction.
*Includes all differentially-expressed sequences regardless of Bonferroni-adjusted P value.

transcription factors, the overlap was 19 and 8%. Specific examples of genes that display overlap in the responses to heat shock and acute physical exercise are listed in Table 4.

In summary, a portion of the *in vivo* PBMC gene expression response to physical exercise appears to involve changes that are also produced by heat shock *in vitro*, particularly among stress proteins and HSPs. These observations are consistent with the hypothesis that physical exercise is a complex stimulus in which the accompanying hyperthermia might account for at least some of the observed changes in PBMC gene expression.

Zieker et al. (2005) studied trained runners (peak VO_2 not reported but average weekly distance run was 52.2 km (32.4 miles) and all had been training for at least 2 years), aged 28–58, who were competing in a half-marathon on a cool and humid day (ambient temperature 1°C or 33.8°F) on hilly terrain. The average race time was 105 min (range 77–139 min). Samples were drawn for microarray analysis before, immediately after, and 24 h after the race. In this study, RNA was extracted from whole blood, not simply PBMCs. Samples were analyzed using a custom dual-dye spotted array platform containing 277 probes primarily focused on genes involved in inflammatory responses. Findings of interest were confirmed by real-time PCR. As in the study by Connolly et al. (2004), an increase in circulating leukocyte count had occurred at the end of exercise that was principally attributable to an increase granulocytes. The number of monocytes also increased but there was a decrease in lymphocytes. These changes returned to baseline by the recovery time point.

The microarray array data demonstrated that ~10% of the genes (29 sequences representing 28 genes) showed expression differences after the race as compared to baseline. The magnitudes of the changes reported were generally small, with only three genes showing increases of twofold or greater (selectin L, thioredoxin, IL 8 receptor alpha) and only four showing decreases of twofold or greater (CD81, ICAM2, Chemokine receptor 1 and CD2). Furthermore, although all 29 sequences showed changes that could be considered statistically significant by unadjusted P values, only seven met the very stringent criteria for statistical significance that the authors applied to minimize false positive reports (Bonferroni adjustment).[3] Importantly, the

---

[3] The affected sequences corresponded to: selectin L (upregulated), CD81, CD244, integrin alpha X, glutathione S-transferase M3, ICAM2, and chemokine receptor 1 (all downregulated). To account for multiple comparisons, the authors used a factor of 345 to adjust P values (Zieker, personal communication). Accordingly, we included all 29 genes reported by the authors as having Bonferroni-adjusted $P<1$ (i.e., unadjusted $P<0.003$) in our comparative analysis. This broader list of 29 genes included the IL-1 receptor antagonist, which had a Bonferroni-adjusted P value of 0.93 on the microarray but which the authors found on real-time PCR to be upregulated in this study, in keeping with observations made by others.

Table 4. Examples of genes that are similarly affected in PBMCs by acute physical exercise and heat shock

Stress response genes and HSPs		Inflammatory and immune response genes	
Increased		Increased	
DUSP1	Dual-specificity phosphatase 1	CSF3R	Colony stimulating factor 3 receptor (granulocyte)
DUSP2	Dual-specificity phosphatase 2	NCR3	Lymphocyte antigen 117
DUSP3	Dual-specificity phosphatase 3	PTPN22	Protein tyrosine phosphatase, non-receptor type 22 (lymphoid)
HSPB1	Heat shock 27 kDa protein 1	TNFSF14	Tumor necrosis factor (ligand) superfamily, member 14
DNAJA1	DnaJ (Hsp 40) homolog, subfamily A, member 1	XCL2	Chemokine (C-motif) ligand 2
DNAJB1	DnaJ (Hsp 40) homolog, subfamily B, member 1	Decreased	
HSPA1A	Heat shock 70 kDa protein 1A	IGHM	Immunoglobulin heavy constant mu
HSPA1B	Heat shock 70 kDa protein 1B	NCF1	Neutrophil cytosolic factor 1 (47 kDa, chronic granulomatous disease, autosomal 1)
HSPCA	Heat shock 90 kDa protein 1, alpha	TNFRSF12	Tumor necrosis factor receptor superfamily, member 12 (translocating chain association membrane protein)
HSPH1	Heat shock protein 105		
STIP1	Stress-induced phosphoprotein 1 (HSP 70/HSP 90 organizing protein)		
Growth factors and transcription		Metabolism and biosynthesis	
Increased		Increased	
EGR1	Early growth response 1	BPGM	2,3-bisphosphoglycerate mutase
NCOA1	Nuclear receptor co-activator 1	NDUFB7	NADH dehydrogenase (ubiquinone) 1 beta subcomplex, 7, 18 kDa
NR4A2	Nuclear receptor subfamily 4, group A, member 2	Other	
RUNX3	Runt-related transcription factor 3	Increased	
S100A9	S100 calcium binding protein A9 (calgranulin B)	MADH7	MAD, mothers against decapentaplegic homolog 7 (*Drosophila*)
TCF8	Transcription factor 8		
TIEG	TGFB inducible early growth response	SNTB2	Syntrophin, beta 2 (dystrophinassociated protein A1, 59 kDa, basic component 2)
YES1	v-yes-1 Yamaguchi sarcoma viral oncogene homolog 1	STK39	Serine threonine kinase 39 (STE20/SPS1 homolog, yeast)

authors confirmed several of the observed gene expression changes by real-time PCR, including some that would have been excluded based on their Bonferroni adjustment, and furthermore, carefully demonstrated that some of the observed changes were quantitatively attributable to changes in cell type distribution as determined by cell surface markers. The gene expression profile had returned to baseline by the 24-h recovery time point, as no significant differences in gene expression from pre-exercise baseline were detected at that time point.

A substantial fraction of the 29 genes affected by exercise in whole blood in the study by Zieker et al. were similarly affected by heat shock of PBMCs *in vitro*. A cross-platform comparison analogous to the one described above for the Connolly study identified 35 sequences on the array used for *in vitro* heat shock (Sonna et al., 2002b) that corresponded to genes on the platform used by Zieker et al. Of these, about one third were similarly affected by *in vitro* heat shock and *in vivo* exercise (Table 3). This is comparable in magnitude to the previously noted overlap between heat shock in vitro and the effects of exercise on the expression of "high interest" inflammatory response genes in the Connolly study (17–31%, see above).

The two exercise studies used markedly different array platforms, and we identified only eight genes affected by exercise in the Zieker study for which comparisons could be made between studies. The changes observed were more commonly discordant than concordant (i.e., increased in one study but decreased in the other or affected in the Zieker study but not in the Connolly study). The only concordance between the two datasets occurred between the immediate post-exercise sample obtained by Zieker et al. and the recovery time point obtained 1 h after the race by Connolly et al., time points that both occurred about the same time after the initiation of exercise (105 min in the Zieker study, 90 min in the Connolly study). At this time point, three genes showed concordant responses, namely, heat shock 27 kDa protein-1 (HSPB1) (increased in both studies), CD14 (increased in both studies), and interleukin 2 receptor beta (decreased in both studies). One important difference between the two studies was the ambient temperature at which subjects exercised. Whereas the Connolly study was performed in a laboratory setting at room temperature, the Zieker study was performed outdoors on a cool and humid day (ambient temperature 1°C). Also, the Connolly study examined gene expression in PBMCs, whereas the Zieker study used whole blood.

Despite the differences in methodology and findings between these two studies, the following general inferences can be made. First, acute physical exercise in fit males can produce changes in gene expression that are easily detectable in peripheral blood and circulating PBMCs using DNA microarrays and other technologies. Second, the changes in gene expression appear to be strongly time-dependent. Third, some of these changes in gene expression are likely due to exercise-induced changes in the distribution of PBMC subtypes whereas other changes cannot be explained solely by this phenomenon. Fourth, some of the PBMC gene expression changes include molecules that are inflammatory modulators, but not IL-6, which is consistent with the hypothesis that cells other than PBMCs (such as skeletal muscle) are the source of circulating IL-6 during exercise. Fifth, expression of at least some HSPs appears to be increased by exercise. Sixth, the data by Connolly et al. suggest that a large part of the PBMC response to exercise involves genes other than stress proteins and inflammatory modulators.

Some of the differences between in vitro and in vivo responses to heat stress can be accounted for by established mechanisms. For example, the IL1 receptor antagonist (IL1RN) was increased by exercise (e.g., the real-time PCR data in the Zieker study) but unaffected (1 sequence) or decreased (1 sequence) by heat shock in vitro. It has been postulated that the increased expression of IL1RN that occurs during exercise is induced by IL6 released by tissues such as skeletal muscle (Petersen and Pedersen, 2005). By contrast, increased expression of IL6 is not thought to occur in PBMCs during exercise (Connolly et al., 2004; Zieker et al., 2005) and in vitro, heat shock actually decreased IL6 mRNA levels in PBMCs (geometric mean expression ratio 0.37, 95% CI 0.16–0.85) (Sonna et al., 2002b). Thus, the discrepancies between the in vitro and in vivo IL1RN expression responses can be accounted for by the differences in availability of IL6.

Another informative difference between in vitro and in vivo responses is exemplified by CD14. Expression of CD14 mRNA was increased by exercise in vivo in both studies but was strongly decreased by heat shock in vitro. Interestingly, CD14 is a receptor for both bacterial LPS and apoptotic cells (Gupta et al., 1996; Devitt et al., 1998), and there is evidence in the literature that heat stress in vivo can lead to entry of LPS into the systemic circulation, putatively originating from the bowel (Hall et al., 2001; Lambert et al., 2002; Lambert, 2004). As with IL1RN, the discrepancy suggests the presence of a stimulus other than heat that modulates CD14 expression in vivo, and indeed, the observed change in CD14 in the Zieker study can be well accounted for by the observed changes in cell type distribution (Zieker et al., 2005).

The comparison of expression responses of PBMCs in vitro and in vivo suggest that exercise is a complex stimulus in which systemic and local factors can modify or override the effects of hyperthermia on PBMC gene expression that would be expected based on the *in vitro* data. Responses that are similar between the two experimental systems can be accounted for, at least in part, by postulating a direct effect of hyperthermia on gene expression. By contrast, responses that differ between the two systems can lead to novel mechanistic hypotheses and insights, such as a search for novel mediators of responses to exercise, or can reaffirm the mechanistic importance of modifiers that have previously been identified (such as IL6). The concept that these complex responses can be dissected into component parts is perhaps the more important one, as it provides the intellectual basis for studies that compare gene expression responses that are beneficial (such as those involved in training and acclimatization) to those that are harmful (such as ones that contribute to dysfunction in EHI). Comparison of the component responses of beneficial and pathological responses may help identify candidate pathways for experimental and therapeutic manipulation.

Gene expression changes in PBMCs caused by exertional heat injury

Given the observation that PBMCs exhibit large gene expression responses to heat shock *in vitro* and to acute bouts of exercise *in vivo*, it is not surprising that extensive changes have been reported in these cells in the context of EHI. Comparisons of the PBMC expression responses to different stressors are informative, in that they can help identify candidate genes that might serve as markers of EHI as well as others that might be involved mechanistically in the disease processes.

Gene expression responses in EHI: similarity and differences to acute physical exercise

The effects of EHI on gene expression by PBMCs were recently reported in a study of Marine Corps recruits (Sonna et al., 2004). Samples were obtained from four subjects who presented to the medical clinic for emergency treatment of exertional heat injury. Samples were obtained at the time of presentation, 3 h after active cooling, and at a 24–48 h follow-up visit. The subjects had documented hyperthermia with evidence of organ injury (such as elevations in serum liver enzymes and/or creatine kinase), and had experienced prodromal symptoms suggestive of viral illness in the days preceding the onset of EHI. None met neurological criteria for the diagnosis of heat stroke. Control samples were obtained from three recruits several days before and several days after an intense field training exercise in hot weather. Gene expression analysis was performed on pooled samples using single-dye oligonucleotide arrays (Affymetrix U95Av2).

The results showed that the subjects experiencing EHI exhibited a time-dependent change in PBMC gene expression. The expression response was large, with 361 sequences showing increased expression at one or more of the time points studied and 331 showing decreased expression (the total expression response was slightly smaller than the 692 predicted by summing the two numbers, as some sequences showed increases at some time points and decreases at others). Many of the gene expression changes could be accounted for by heat shock responses previously documented in PBMCs *in vitro*, most notably among the HSPs. Others, however, could not be readily explained by the *in vitro* data. Importantly, and unlike the *in vitro* responses of PBMCs, about one fourth of the sequences most highly upregulated in EHI corresponded to interferon-inducible genes that are induced by interferons -alpha, -gamma, or both.

It is informative to compare the results of the Marine Corps recruit study to the changes reported by Connolly et al. (2004) (Tables 5A–C). Among the sequences whose expression was increased by exertional heat injury, approximately 1/6 were also increased by acute physical exercise, and many of these were also increased by heat shock *in vitro*. However, few of the highly upregulated EHI genes that were comparably affected by physical exercise were also interferon-inducible (Table 5A). By contrast, sequences that were highly upregulated by EHI but not by acute physical exercise commonly included interferon-inducible genes (Table 5B). A concept of component responses therefore emerges. Namely, EHI and exercise share an expression response that can be ascribed to thermal stress *per se*, and the expression signature of EHI includes a component response that is functionally related to interferons but that is not present in acute physical exercise or in *in vitro* models of heat shock.

The dissection of gene signature responses into distinct components has important implications for improving the application of gene expression studies to the study of physiological states, as linking component responses to well-described functional outcomes allows correlations of molecular changes with the clinical and biochemical features of the disease. For example, increases in circulating interferons alpha and gamma do not appear to be part of the normal response to exhaustive exercise (Suzuki et al., 2002). In one case, a study of 16 elite marathon runners (race times less than 2 h and 38 min) found no significant differences in circulating interferons alpha and gamma levels before and after a marathon; by contrast, these subjects had marked elevations of IL-6 and IL-1 receptor antagonist, as expected (Suzuki et al., 2000). Likewise, neither the Connolly study nor the Zieker study reported an increase in mRNA encoding interferons alpha or gamma. By contrast, in the EHI study, a sequence corresponding to interferon gamma was significantly and strongly increased in EHI but was excluded from the final list of affected genes because of the strict post-hoc filter criteria used in that study (specifically, expression calls). Thus, these comparisons provide support for the hypothesis that interferons might play a role in the pathophysiology of some cases of human exertional heat injury.

Among sequences that exhibited decreased expression, little overlap occurred between EHI and acute physical exercise. Indeed, the number of overlapping downregulated sequences was far smaller than the number of overlapping upregulated sequences, representing only approximately 2% of all genes that were significantly decreased by EHI (as compared to ~1/6 of upregulated sequences, see above). The observation that the patterns of genes downregulation differ to such a great extent between the two conditions suggests the possibility that they might be used as biological markers to help distinguish between the two states.

These differences notwithstanding, there was nonetheless some overlap observed in the sequences that were downregulated by EHI and those downregulated by acute physical exercise. For example, both EHI and acute physical exercise resulted in decreased expression of the transcription factors *jun* and *myc*, which are known to have broad downstream effects. These shared changes in expression might represent normal responses to exercise that also occur in EHI, and might even be physiologically important. Interestingly, in tissue culture, decreased expression of *myc* has been shown to be important for cell survival after heat shock (Wennborg et al., 1995).

Because of the apparent interferon-inducible response observed among the upregulated genes in the EHI study, we re-examined the list of immune function genes that were downregulated by EHI. Interestingly, about half of these 43 downregulated sequences corresponded to genes that are normally expressed by T-cells (particularly activated and cytotoxic T-cells) and/or natural killer (NK) cells. Corresponding sequences in the Connolly study were found for 12 of these 43 sequences and of these, only 3 (2 of which were T-cell/NK cell related sequences) showed decreased expression at one of the time points examined; the remainder were upregulated. This observation gives rise to the hypothesis that EHI may produce changes in PBMC cell type distribution (specifically, of T-cells and/or NK cells) that are not part of the normal response to exercise.

Finally, several sequences were identified that were increased in response to exercise but that failed to show a significant expression response to EHI (Table 5C). This included genes that have broad effects on cellular function, such as NR4A2, CREM, EGR1, DUSP2, and DUSP3. Some of these were similarly affected by heat shock *in vitro*. One possible explanation for the finding of genes that are upregulated by heat shock *in vitro* and by acute physical exercise *in vivo* but not by EHI is the presence of signals during EHI that antagonize components of PBMC hyperthermia-induced responses. The identification of genes that distinguish normal exercise responses from those that characterize heat injury provides a rational basis for studies that explore their role as part of

Table 5. Gene expression responses of exercise, exertional heat injury, and heat shock

Sequence	Name	Interferon-responsive gene?	Expression ratio (EHI/control)			Effect of acute exercise		Effect of heat shock in vitro
			At presentation	After cooling	At follow-up	End-exercise/pre	Recovery/pre	

A: Examples of genes similarly affected by exercise and exertional heat injury

Stress response genes and HSPs

Increased

HSPA1B/A	Heat shock 70 kDa protein 1B; HSP 70-2		32	18.7	1.1 (NS)	Increased	Increased	Up
HSPA1A	Heat shock 70 kDa protein 1A; HSP 70-1		31.4	22.3	1.2	NS or Increased	Increased	Up
HSPB1	Heat shock 27 kDa protein 1; HSP 28; HSP 27-1		9.6	32.4	1.1 (NS)	NS	Increased	Up
DNSJB1	DnaJ (Hsp 40) homolog, subfamily B, member 1; Hsp 40		8.4	1.5 (NS)	0.71 (NS)	Increased	NS	Up
HSPCA	Heat shock 90 kDa protein 1, alpha; HSP 90-1, alpha; HSP 90A		5.8	6.8	1.1 (NS)	NS	Increased	Up
DNAJA1	DnaJ (Hsp 40) homolog, subfamily A, member 1; HSPF4		3.5	1.4 (NS)	1.2 (NS)	Increased	NS	Up

Immune function

Increased

GZMB	Granzyme B (granzyme 2, cytotoxic T-lymphocyte-associated serine esterase 1)		5.6	1.8 (NS)	2.3	Increased	NS	Down
KIR3DL1	Killer cell immunoglobulin-like receptor, three domains, long cytoplasmic tail, 2		3.5	1.6 (NS)	0.58 (NS)	Increased	NS	NS
CTLA1	Similar to granzyme B (granzyme 2, cytotoxic T lymphocyte-associated serine esterase 1) (H. sapiens)		3.1	0.85 (NS)	2.0	Increased	NS	NS

Decreased

IGHM	Immunoglobulin heavy constant mu		0.34	0.44	0.35	Decreased	NS	Down

Transcription

Decreased

MYC	v-myc myelocytomatosis viral oncogene homolog (avian)		0.19	0.50	0.58 (NS)	Decreased	NS	NS
JUN	v-jun sarcoma virus 17 oncogene homolog (avian)		0.80	0.33	0.54	NS	Decreased	Up

Other							
Increased							
PLSCR1	Phospholipid scramblase 1	4.5	5.3	4.3	NS	Increased	NS
BCL2A1	BCL2-related protein A1	2.7	4.7	2	NS	Increased	NS
CCR1	Chemokine (C–C motif) receptor 1	1.9	4.5	3.3	NS	Increased	Down
S100A8	S100 calcium binding protein A8 (calgranulin A)	1.6	3.3	2.9	NS	Increased	NS
S100A9	S100 calcium binding protein A9 (calgranulin B)	1.7	3.4	2.6	NS	Increased	Up
S100A12	S100 calcium binding protein A12 (calgranulin C)	1.2 (NS)	3.3	1.8 (NS)	NS	Increased	NS
Decreased							
TBC1D4	TBC1 domain family, member 4	0.073	0.32	0.48 (NS)	Decreased	NS	NS

Sequence	Name	Interferon-responsive gene?	Expression ratio (EHI/Control)			Effect of heat shock *in vitro*
			At presentation	After cooling	At follow-up	

B: Examples of genes affected by exertional heat injury but not by exercise

Upregulated sequences

Stress response genes and HSPs

HSPA6	HSP 70B'		24.7	3.6	3.8	Up
HSPA6	HSP 70B'		19.8	3.8	2.3	Up
SERPINH1	Serine (or cysteine) proteinase inhibitor, clade H (heat shock protein 47), member 1, (collagen binding protein 1); HSP 47; colligin-1; colligin 2; SERPINH2		6.6	7.4	0.66 (NS)	Up

Immune function

RSAD2	Radical S-adenosyl methionine domain containing 2' cig5; Viperin	Yes	9.9	19.8	20.0	NS
IFIT3/IFIT4	Interferon-induced protein with tetratricopeptide repeats 3; interferon-induced protein 60; IFI 60; interferon-induced protein with tetratricopeptide repeats 4	Yes	9.1	15.6	19.4	Down
IFIT2	Interferon-induced protein with tetratricopeptide repeats; 2G10P2; cig42; IFI-54; GARG-39; ISG-54K	Yes	6.8	8.5	9.2	Down
OASL	2–5'-oligoadenylate synthetase-like; thyroid hormone receptor interactor 14; TRIP14; p59OASL	Yes	6.1	9.5	8.5	NS
IFI27	Interferon alpha-inducible protein 27; p27; ISG12	Yes	2.8	39.8	20.7	NS

338

Table 5 (continued)

Sequence	Name	Interferon-responsive gene?	Expression ratio (EHI/Control)			Effect of heat shock in vitro
			At presentation	After cooling	At follow-up	
FCGR1A	Fc fragment of IgG, high affinity Ia, receptor for (CD64; CD64; Fc-gamma receptor I A1		3.4	10.9	8.6	NS
C3AR1	Complement component 3a receptor 1		3.3	8.6	3.0	Down
Other						
HISTH2H2AA	Histone 2, H2aa; H2AFO; H2A histone family, member O		10.7	22.3	17.6	NS
IFI44L	Interferon-induced protein 44-like; C1orf29; chromosome 1 open reading frame 29	Possibly	6.2	7.8	5.7	NS
SLC15A1	Solute carrier family 15 (oligopeptide transporter), member 1; peptide transporter HPEPT1; PEPT1	No[a]	5.8	1.0 (NS)	4.5	NS
PPP2R1B	Protein phosphatase 2 (formerly 2A), regulatory subunit A (PR 65), beta isoform		2.3	8.2	12.2	NS
Downregulated sequences						
Cell growth, proliferation, and differentiation						
GOS2	Putative lymphocyte G_0/G_1 switch gene		0.033	1.4 (NS)	0.018	Down
HIPK3	Homeodomain interacting protein kinase 3		0.26	0.34	0.66 (NS)	NS
JAG1	Jagged 1 (Alagille syndrome)		0.75 (NS)	0.10	1.9	NS
Immune function						
FCER1A	Fc fragment of IgE, high affinity I, receptor for; alpha polypeptide		0.039	0.13	0.21	NS
CD96	CD96 antigen; TACTILE; T-cell activated increased late expression		0.11	0.12	0.041	Down
MAL	mal, T-cell differentiation protein		0.14	0.29	0.22	Up
Signal transduction						
IBTK	Inhibitor of Burton's tyrosine kinase		0.25	0.62 (NS)	0.43	Down
ARHH	Ras homolog gene family, member H		0.54 (NS)	0.19	0.48	Down
Transcription						
DRAP1	DR1-associated protein 1 (negative cofactor 2 alpha); Dr1-associated corepressor, mRNA sequence		0.13	0.39	0.15	NS

Sequence	Name	Interferon-responsive gene?			
ZNF85	Zinc finger protein 85	0.89 (NS)	0.077	0.61 (NS)	NS
HMGA1	High mobility group AT-hook 1	0.57 (NS)	0.21	0.45	NS
Miscellaneous					
MS4A1	Membrane-spanning 4-domains, subfamily A, member 1; CD20 antigen	0.090	0.37	0.52 (NS)	NS
NUFIP1	Nuclear fragile X mental retardation protein interacting protein 1	0.13	0.39	0.087	Up
AHSA2[b]	Activator of heat shock 90kDa protein ATPase homolog 2 (yeast)	0.26	0.31	0.56	Up
AGL	Amylo-1, 6-glucosidase, 4-alpha-glucanotransferase (glycogen debranching enzyme, glycogen storage disease type III)	0.39	0.20	0.34	NS
GAD1	Glutamate decarboxylase 1 (brain, 67 kDa)	0.40 (NS)	0.088	0.44 (NS)	NS
KIAA0982	KIAA0982 protein	0.61	0.16	0.69	Up
CENTD1	Centaurin, delta 1	0.80 (NS)	0.21	0.40	NS

Sequence	Name	Interferon-responsive gene?	Effect of acute exercise		Effect of heat shock *in vitro*
			End-exercise/pre	Recovery/pre	

C: Examples of genes affected by exercise but not by exertional heat injury

Upregulated sequences

Growth factors and transcription

NR4A2	Nuclear receptor subfamily 4, group A, member 2		Increased	NS	Up
RGS1	Regulator of G-protein signaling-1		Increased	NS	Down
EREG	Epiregulin		Increased	NS	NS
CREM	cAMP responsive element modulator		Increased	Increased	NS
SAP30	Sin3-associated peptide, 30 kDa		NS	Increased	NS
EGR1	Early growth response 1		NS	Increased	Up
TLE3	Transducin-like enhancer of split 3 (E(sp1) homolog, *Drosophila*)		NS	Increased	NS
TGFB1	Transforming growth factor, beta-induced, 69 kDa		NS	Increased	NS

Immune function

KIR2DL2	Killer cell immunoglobulin-like receptor, two domains, long cytoplasmic tail, 2		Increased	NS	NS
CPM	Carboxypeptidase M		NS	Increased	NS
CSF3R	Colony stimulating factor 3 receptor (granulocyte)		NS	Increased	Up

Stress response genes and HSPs

DUSP2	Dual specificity phosphatase 2		Increased	NS	Up
DUSP3	Dual specificity phosphatase 3		NS	Increased	Up

Table 5 (continued)

Sequence	Name	Interferon-responsive gene?	Effect of acute exercise		Effect of heat shock in vitro
			End-exercise/pre	Recovery/pre	
Downregulated sequences					
LFNG	Lunatic fringe homolog (Drosophila)		Decreased	NS	NS
MEN1	Multiple endocrine neoplasia I		Decreased	NS	NS
TNFRSF12	Tumor necrosis factor receptor superfamily, member 12 (translocating chain association membrane protein)		Decreased	NS	Down
CD22	CD22 antigen		Decreased	Decreased	NS
NELL2	NEL-like 2 (chicken)		Decreased	NS	Up

Source: Data are from Connolly et al. (2004), Sonna et al. (2004), and Sonna et al. (2002b).
[a]The transport activity of this protein, however, is increased by IFN-gamma in some cell types (Buyse et al., 2003).
[b]Previously identified as KIAA0570/KIAA0729 based on UniGene clusters.

normal compensatory and even beneficial effects of exercise.

The following picture emerges from these comparative analyses of published studies is as follows. First, the gene expression signature of EHI can be subdivided into various identifiable components that appear to have mechanistic implications. One component includes genes that are similarly affected by heat shock *in vitro*, some of which are also part of the normal responses to exercise. These genes are good candidates for study as potential targets of heat-activated transcription factors, such as HSF-1. A second component includes genes that are similarly affected by exercise and EHI but that do not appear to be strictly dependent on hyperthermia for their change in expression. Changes in expression of these genes as a result of exercise would thus appear more likely to depend on factors other than heat *per se*. A third component involves genes that seem limited to EHI. This component includes several interferon-inducible genes that are upregulated in EHI as well as several genes that appear to be involved in T- and NK-cell function that are downregulated in EHI. Additional, better comparisons involving additional environmental exposures and time-dependent changes are likely to lead to the identification of additional informative component responses.

Additional conclusions that can be drawn from the comparative analysis presented include the following. First, the overlap between EHI and normal exercise responses may be much greater for sequences that show increases in expression than for sequences that show decreases in expression. These differences in the patterns of downregulation might, therefore, eventually prove useful as markers for injury. Second, finding genes that are affected by acute physical exercise but not EHI suggests that some of the changes characteristic of EHI might represent a failure to alter the expression of important genes that play a compensatory or protective role in normal exercise. Last, and most importantly, some of the differences in expression detected in these studies may be attributable not to changes in genes expression within a particular cell type, but rather to differences in the composition of circulating PBMCs (i.e., a change in the relative distributions of PBMC subtypes). For example, some of the downregulated genes might be accounted for by selective loss of certain subsets of T- and NK-cells in the peripheral circulation.

Inflammatory mediators, lymphocyte function, and exertional heat illness

The Marine Corps recruit study of gene expression responses to EHI does not permit a firm distinction to be made between changes that contribute to the pathophysiology EHI, those which are consequences of EHI (and as such, markers of the condition), and those which are epiphenomena. As commonly occurs in clinical studies, the Marine Corps recruit study did have methodological limitations that must be kept in mind when attempting to generalize the findings. These limitations include a low number of subjects with varying degrees of severity of EHI, use of pooled samples, and use of samples from control subjects that were drawn at different times from the index cases and at rest, rather than shortly after exhaustive exercise. There also may have been an artifact of population sampling given the small size of the study.

These limitations notwithstanding, two lines of evidence support the findings of the Marine Corps recruit study that interferons and other pro-inflammatory proteins might play a mechanistic role in some cases of EHI. The first is a series of *in vitro* studies that have reported that the response of cultured cells to heat shock is substantially altered by previous exposure history. For example, pre-exposure of porcine endothelial cells to LPS- or to TNF-alpha leads to a dose-dependent increase in cell death by apoptosis after heat shock (Buchman et al., 1993). A similar phenomenon was observed in a transformed murine cell line after exposure to interferon gamma (Abello and Buchman, 1994). These outcomes contrast with the normal response of these cells to heat shock in the absence of pre-exposure, which is survival and adaptation. The *in vitro* data thus suggest that delivery of a heat shock in the context of a pre-existing pro-inflammatory stimulus can lead to cell death by apoptosis, rather than survival and recovery.

The mechanism by which pro-inflammatory stimuli, such as LPS and TNF-alpha, can lead to decreased cellular survival after heat shock might involve transcription factor NF-kappa-B (DeMeester et al., 2001). Normally, NF-kappa-B exists in an inactivated state in the cytoplasm, bound to the inhibitor I-kappa-B (Malhotra and Wong, 2002). Activation of NF-kappa-B by pro-inflammatory stimuli involves phosphorylation of this inhibitor by I-kappa-B kinase (IKK), which in turn to the ubiquitinylation and proteasome-mediated degradation of I-kappa-B, translocation of NF-kappa-B to the nucleus, and activation of NF-kappaB — dependent transcription (Curry et al., 1999; Shanley et al., 2000; Yoo et al., 2000). Heat shock inhibits the activation of NF-kappa-B through a variety of mechanisms, of which the most important is believed to be inhibition of IKK activity, which in turn prevents phosphorylation of I-kappa-B and its dissociation from NF-kappa-B (Wong et al., 1997; Scarim et al., 1998; Curry et al., 1999; Shanley et al., 2000; Yoo et al., 2000). Increases in I-kappa-B expression have also been observed as a result of heat shock (Wong et al., 1999; Pritts et al., 2000). In the presence of pre-existing pro-inflammatory stimuli, which activate NF-kappa-B, the subsequent delivery of a heat shock appears to be sufficient to shift cellular fate toward apoptosis (DeMeester et al., 2001).

As noted previously, although exertional heat illness can result from a single exposure to heat stress that overwhelms available compensatory responses, epidemiological data suggest that for some individuals, exertional heat illness may occur as the result of a "two-hit" process. These subjects present with exertional heat illness at ambient temperatures and exercise intensities that would normally be tolerated (Kark et al., 1996; Epstein et al., 1999; Gardner and Kark, 2001), commonly report feeling ill in the days preceding exertional heat illness (Shibolet et al., 1967; Epstein et al., 1999), and frequently report prior day exposure to heat (Kark et al., 1996). It thus seems plausible that some precedent pro-inflammatory exposure (previous heat injury or viral illness) could make a person more susceptible to exertional heat illness during a subsequent unremarkable exercise-heat exposure. The initial exposure might act to augment the hyperthermia of exercise, make tissue more susceptible to injury for a given heat stress, or both.

That interferons could provide such a first hit stimulus is supported by a case series in which circulating levels of interferon gamma were measured in patients who presented to the emergency department with heat stroke (Bouchama et al., 1993). Of 10 patients for which interferon levels were available, half showed elevated levels of circulating interferon gamma that generally diminished after treatment. As noted (previous section), increased expression of interferons alpha and gamma are not generally considered a normal feature of acute physical exercise. Thus, the observation of increases in circulating interferon gamma in heat stroke suggests that, at least in some cases of human heat illness, interferons might play a role in the pathophysiology of the disease. Although the observational data cannot distinguish between a causal role for interferon gamma and the possibility that it is merely a consequence of heat illness, when coupled with the observed responses of animal cells *in vitro*, it seems at least plausible to suggest that a pre-existing stressor that increases circulating levels of interferon gamma (or alpha) might presensitize cells and tissues to heat stress in a manner that produces morbidity under heat loads that might otherwise be well tolerated. Further observational and experimental works are needed to test this hypothesis.

Finally, in addition to increases in the expression of genes that respond to inflammatory mediators, the Marine Corps recruit study identified a component of altered T- and NK-cell gene expression in subjects with EHI. This finding suggests that EHI may involve protein-driven changes that involve cellular immunity. Understanding the intersection between thermal stress and immune function may thus expand and redirect our thinking about thermal biology.

Summary

Gene expression signature data to date shows that EHI produces time-dependent changes in expression that are easily detectable in PBMCs. These changes can be dissected into component parts that include effects resulting for thermal stress, exercise, and those that likely reflect alterations in immune function. Cross-platform comparisons such as those outlined in this chapter have significant limitations that do not prevent the identification of major component responses, but do limit the confidence with which many individual sequences can be implicated.

Exertional heat illnesses occur when heat load exceeds heat dissipation beyond the point that the individual can physiologically tolerate. Thermal biology research cannot alter that fact that every individual has a heat load at which function is no longer possible. It can, however, define the mechanisms that allow for more efficient heat dissipation, those that allow more effective adaptation to a given heat load, and those that alter the point at which an individual becomes sick. Conceptually, cellular responses to a heat load can be dichotomized into survival and adaptation, which typically permit continued effective function at a new (and higher) thermal steady state, and dysfunction, which tend toward cell death (by apoptosis if possible, necrosis if not). The factors that determine which pathways will predominate and the mediators on which they depend are now being elucidated, in part by comparison of pathway-specific components of gene expression signature and associations with the clinical and biochemical profiles of heat related injury.

Acknowledgments and disclaimer

The views, opinions, and/or findings contained in this report are those of the authors and should not be construed as an official Department of the Army position, or decision, unless so designated by other official documentation. Approved for public release; distribution unlimited.

The authors wish to thank Ms. Pamela M. Paradee and SPC Erik B. Lloyd for bibliographic assistance. We also wish to thank Derek Zieker, Elvira Fehrenbach, and Jeffrey Hasday for their review and comments.

This work was funded in part by NIH grants RO1 HL-080073 and HL 072114 and funding by the Department of Army Medical Research and Material Command.

References

Abello, P.A. and Buchman, T.G. (1994) Heat shock-induced cell death in murine microvascular endothelial cells depends on priming with tumor necrosis factor-alpha or interferon-gamma. Shock, 2: 320–323.

Beere, H.M. (2004) "The stress of dying": the role of heat shock proteins in the regulation of apoptosis. J. Cell. Sci., 117: 2641–2651.

Bickel, C.S., Slade, J., Mahoney, E., Haddad, F., Dudley, G.A. and Adams, G.R. (2005) Time course of molecular responses of human skeletal muscle to acute bouts of resistance exercise. J. Appl. Physiol., 98: 482–488.

Bond, U. (1988) Heat shock but not other stress inducers leads to the disruption of a sub-set of snRNPs and inhibition of in vitro splicing in HeLa cells. EMBO J., 7: 3509–3518.

Bouchama, A., Roberts, G., Al Mohanna, F., El Sayed, R., Lach, B., Chollet-Martin, S., Ollivier, V., Al Baradei, R., Loualich, A., Nakeeb, S., Eldali, A. and de Prost, D. (2005) Inflammatory, hemostatic, and clinical changes in a baboon experimental model for heatstroke. J. Appl. Physiol., 98: 697–705.

Bouchama, A., al Sedairy, S., Siddiqui, S., Shail, E. and Rezeig, M. (1993) Elevated pyrogenic cytokines in heatstroke. Chest, 104: 1498–1502.

Buchman, T.G., Abello, P.A., Smith, E.H. and Bulkley, G.B. (1993) Induction of heat shock response leads to apoptosis in endothelial cells previously exposed to endotoxin. Am. J. Physiol., 265: H165–H170.

Bull, T.M., Coldren, C.D., Moore, M., Sotto-Santiago, S.M., Pham, D.V., Nana-Sinkam, S.P., Voelkel, N.F. and Geraci, M.W. (2004) Gene microarray analysis of peripheral blood cells in pulmonary arterial hypertension. Am. J. Respir. Crit. Care Med., 170: 911–919.

Buyse, M., Charrier, L., Sitaraman, S., Gewirtz, A. and Merlin, D. (2003) Interferon-gamma increases hPepT1-mediated uptake of di-tripeptides including the bacterial tripeptide fMLP in polarized intestinal epithelia. Am. J. Pathol., 163: 1969–1977.

Connolly, P.H., Caiozzo, V.J., Zaldivar, F., Nemet, D., Larson, J., Hung, S.P., Heck, J.D., Hatfield, G.W. and Cooper, D.M. (2004) Effects of exercise on gene expression in human peripheral blood mononuclear cells. J. Appl. Physiol., 97: 1461–1469.

Creagh, E.M., Sheehan, D. and Cotter, T.G. (2000) Heat shock proteins: modulators of apoptosis in tumour cells. Leukemia, 14: 1161–1173.

Curry, H.A., Clemens, R.A., Shah, S., Bradbury, C.M., Botero, A., Goswami, P. and Gius, D. (1999) Heat shock inhibits radiation-induced activation of NF-kappaB via inhibition of I-kappaB kinase. J. Biol. Chem., 274: 23061–23067.

DeMeester, S.L., Buchman, T.G. and Cobb, J.P. (2001) The heat shock paradox: does NF-kappaB determine cell fate? FASEB J., 15: 270–274.

Devitt, A., Moffatt, O.D., Raykundalia, C., Capra, J.D., Simmons, D.L. and Gregory, C.D. (1998) Human CD14 mediates recognition and phagocytosis of apoptotic cells. Nature, 392: 505–509.

Dinh, H.K., Zhao, B., Schuschereba, S.T., Merrill, G. and Bowman, P.D. (2001) Gene expression profiling of the response to thermal injury in human cells. Physiol. Genom., 7: 3–13.

Dorion, S. and Landry, J. (2002) Activation of the mitogen-activated protein kinase pathways by heat shock. Cell Stress Chaperones, 7: 200–206.

Dubois, M.F. and Bensaude, O. (1993) MAP kinase activation during heat shock in quiescent and exponentially growing mammalian cells. FEBS Lett., 324: 191–195.

Epstein, Y., Moran, D.S., Shapiro, Y., Sohar, E. and Shemer, J. (1999) Exertional heat stroke: a case series. Med. Sci. Sports Exerc., 31: 224–228.

Fehrenbach, E., Niess, A.M., Schlotz, E., Passek, F., Dickhuth, H.H. and Northoff, H. (2000a) Transcriptional and translational regulation of heat shock proteins in leukocytes of endurance runners. J. Appl. Physiol., 89: 704–710.

Fehrenbach, E., Niess, A.M., Veith, R., Dickhuth, H.H. and Northoff, H. (2001) Changes of HSP72-expression in leukocytes are associated with adaptation to exercise under conditions of high environmental temperature. J. Leukoc. Biol., 69: 747–754.

Fehrenbach, E., Passek, F., Niess, A.M., Pohla, H., Weinstock, C., Dickhuth, H.H. and Northoff, H. (2000b) HSP expression in human leukocytes is modulated by endurance exercise. Med. Sci. Sports Exerc., 32: 592–600.

Findly, R.C., Gillies, R.J. and Shulman, R.G. (1983) In vivo phosphorus-31 nuclear magnetic resonance reveals lowered ATP during heat shock of Tetrahymena. Science, 219: 1223–1225.

Fluck, M. and Hoppeler, H. (2003) Molecular basis of skeletal muscle plasticity: from gene to form and function. Rev. Physiol. Biochem. Pharmacol., 146: 159–216.

Fruth, J.M. and Gisolfi, C.V. (1983) Work-heat tolerance in endurance-trained rats. J. Appl. Physiol., 54: 249–253.

Gaffin, S.L., Koratich, M. and Hubbard, R.W. (1997) The effect of hyperthermia on intracellular sodium concentrations of isolated human cells. Ann. N.Y. Acad. Sci., 813: 637–639.

Gardner, J.W. and Kark, J.A. (2001) Clinical diagnosis, management, and surveillance of exertional heat illness. In: Pandolf K.B. and Burr R.E. (Eds.), Medical Aspects of Harsh Environments, Vol. 1. Office of the Surgeon General, Department of the Army, USA/Borden Institute, Washington, DC, pp. 231–279.

Georgopoulos, C. and Welch, W.J. (1993) Role of the major heat shock proteins as molecular chaperones. Ann. Rev. Cell Biol., 9: 601–634.

Goldspink, G. (2003) Gene expression in muscle in response to exercise. J. Muscle Res. Cell. Motil., 24: 121–126.

Gupta, D., Kirkland, T.N., Viriyakosol, S. and Dziarski, R. (1996) CD14 is a cell-activating receptor for bacterial peptidoglycan. J. Biol. Chem., 271: 23310–23316.

Hall, D.M., Buettner, G.R., Oberley, L.W., Xu, L., Matthes, R.D. and Gisolfi, C.V. (2001) Mechanisms of circulatory and intestinal barrier dysfunction during whole body hyperthermia. Am. J. Physiol. Heart Circ. Physiol., 280: H509–H521.

Han, S.I., Ha, K.S., Kang, K.I., Kim, H.D. and Kang, H.S. (2000) Heat shock-induced actin polymerization, SAPK/JNK activation, and heat-shock protein expression are mediated by genistein-sensitive tyrosine kinase(s) in K562 cells. Cell Biol. Int., 24: 447–457.

Han, S.I., Oh, S.Y., Woo, S.H., Kim, K.H., Kim, J.H., Kim, H.D. and Kang, H.S. (2001) Implication of a small GTPase Rac1 in the activation of c-Jun N-terminal kinase and heat shock factor in response to heat shock. J. Biol. Chem., 276: 1889–1895.

Hasday, J.D. and Singh, I.S. (2000) Fever and the heat shock response: distinct, partially overlapping processes. Cell Stress Chaperones, 5: 471–480.

Helmbrecht, K., Zeise, E. and Rensing, L. (2000) Chaperones in cell cycle regulation and mitogenic signal transduction: a review. Cell Prolif., 33: 341–365.

Jozsi, A.C., Dupont-Versteegden, E.E., Taylor-Jones, J.M., Evans, W.J., Trappe, T.A., Campbell, W.W. and Peterson, C.A. (2000) Aged human muscle demonstrates an altered gene expression profile consistent with an impaired response to exercise. Mech. Ageing Dev., 120: 45–56.

Kark, J.A., Burr, P.Q., Wenger, C.B., Gastaldo, E. and Gardner, J.W. (1996) Exertional heat illness in Marine Corps recruit training. Aviat. Space Environ. Med., 67: 354–360.

Katschinski, D.M. (2004) On heat and cells and proteins. News Physiol. Sci., 19: 11–15.

Koratich, M. and Gaffin, S.L. (1999) Mechanisms of calcium transport in human endothelial cells subjected to hyperthermia. J. Therm. Biol., 24: 245–249.

Kregel, K.C. (2002) Heat shock proteins: modifying factors in physiological stress responses and acquired thermotolerance. J. Appl. Physiol., 92: 2177–2186.

Kuhl, N.M. and Rensing, L. (2000) Heat shock effects on cell cycle progression. Cell Mol. Life Sci., 57: 450–463.

Lambert, G.P. (2004) Role of gastrointestinal permeability in exertional heatstroke. Exerc. Sport. Sci. Rev., 32: 185–190.

Lambert, G.P., Gisolfi, C.V., Berg, D.J., Moseley, P.L., Oberley, L.W. and Kregel, K.C. (2002) Selected contribution: hyperthermia-induced intestinal permeability and the role of oxidative and nitrosative stress. J. Appl. Physiol., 92: 1750–1761.

Leon, L.R., Blaha, M.D. and DuBose, D.A. (2006) Time course of cytokine, corticosterone, and tissue injury responses in mice during heat strain recovery. J. Appl. Physiol., 100: 1400–1409.

Leon, L.R., DuBose, D.A. and Mason, C.W. (2005) Heat stress induces a biphasic thermoregulatory response in mice. Am. J. Physiol. Regul. Integr. Comp. Physiol., 288: R197–R204.

Lindquist, S. (1986) The heat-shock response. Ann. Rev. Biochem., 55: 1151–1191.

Mahoney, D.J., Carey, K., Fu, M.H., Snow, R., Cameron-Smith, D., Parise, G. and Tarnopolsky, M.A. (2004) Real-time RT-PCR analysis of housekeeping genes in human skeletal muscle following acute exercise. Physiol. Genom., 18: 226–231.

Mahoney, D.J., Parise, G., Melov, S., Safdar, A. and Tarnopolsky, M.A. (2005) Analysis of global mRNA expression in human skeletal muscle during recovery from endurance exercise. FASEB J., 19: 1498–1500.

Malhotra, V. and Wong, H.R. (2002) Interactions between the heat shock response and the nuclear factor-kappa B signaling pathway. Crit. Care Med., 30: S89–S95.

Mathew, A. and Morimoto, R.I. (1998) Role of the heat-shock response in the life and death of proteins. Ann. N.Y. Acad. Sci., 851: 99–111.

Montain, S.J., Latzka, W.A. and Sawka, M.N. (2000) Impact of muscle injury and accompanying inflammatory response on thermoregulation during exercise in the heat. J. Appl. Physiol., 89: 1123–1130.

Morimoto, R.I. (1998) Regulation of the heat shock transcriptional response: cross talk between a family of heat shock factors, molecular chaperones, and negative regulators. Genes Dev., 12: 3788–3796.

Moseley, P.L. (1997) Heat shock proteins and heat adaptation of the whole organism. J. Appl. Physiol., 83: 1413–1417.

Obata, T., Brown, G.E. and Yaffe, M.B. (2000) MAP kinase pathways activated by stress: the p38 MAPK pathway. Crit. Care Med., 28: N67–N77.

Otterbein, L.E. and Choi, A.M. (2000) Heme oxygenase: colors of defense against cellular stress. Am. J. Physiol. Lung Cell. Mol. Physiol., 279: L1029–L1037.

Panniers, R. (1994) Translational control during heat shock. Biochimie, 76: 737–747.

Parag, H.A., Raboy, B. and Kulka, R.G. (1987) Effect of heat shock on protein degradation in mammalian cells: involvement of the ubiquitin system. EMBO J., 6: 55–61.

Park, H.G., Han, S.I., Oh, S.Y. and Kang, H.S. (2005) Cellular responses to mild heat stress. Cell Mol. Life Sci., 62: 10–23.

Parsell, D.A. and Lindquist, S. (1993) The function of heat-shock proteins in stress tolerance: degradation and reactivation of damaged proteins. Ann. Rev. Genet., 27: 437–496.

Petersen, A.M. and Pedersen, B.K. (2005) The anti-inflammatory effect of exercise. J. Appl. Physiol., 98: 1154–1162.

Pirkkala, L., Nykanen, P. and Sistonen, L. (2001) Roles of the heat shock transcription factors in regulation of the heat shock response and beyond. FASEB J., 15: 1118–1131.

Pritts, T.A., Wang, Q., Sun, X., Moon, M.R., Fischer, D.R., Fischer, J.E., Wong, H.R. and Hasselgren, P.O. (2000) Induction of the stress response in vivo decreases nuclear factor-kappa B activity in jejunal mucosa of endotoxemic mice. Arch. Surg., 135: 860–866.

Punyiczki, M. and Fesus, L. (1998) Heat shock and apoptosis. The two defense systems of the organism may have overlapping molecular elements. Ann N.Y. Acad. Sci., 851: 67–74.

Ryan, A.J., Gisolfi, C.V. and Moseley, P.L. (1991) Synthesis of 70 K stress protein by human leukocytes: effect of exercise in the heat. J. Appl. Physiol., 70: 466–471.

Sawka, M.N. and Pandolf, K.B. (2001) Physical exercise in hot climates: Physiology, performance, and biomedical issues. In: Pandolf K.B. and Burr R.E. (Eds.), Medical Aspects of Harsh Environments, Vol. 1, Office of the Surgeon General, Department of the Army, USA/Borden Institute, Washington, DC, pp. 87–133.

Sawka, M.N., Wenger, C.B. and Pandolf, K.B. (1996) Thermoregulatory responses to acute exercise: heat stress and heat acclimation. In: Blatteis C.M. and Fregley M.J. (Eds.), Handbook of Physiology. Section 4: Environmental Physiology. Oxford University Press for the American Physiological Society, New York, NY, pp. 157–186.

Scarim, A.L., Heitmeier, M.R. and Corbett, J.A. (1998) Heat shock inhibits cytokine-induced nitric oxide synthase expression by rat and human islets. Endocrinology, 139: 5050–5057.

Schneider, E.M., Niess, A.M., Lorenz, I., Northoff, H. and Fehrenbach, E. (2002) Inducible hsp70 expression analysis after heat and physical exercise: transcriptional, protein expression, and subcellular localization. Ann. N.Y. Acad. Sci., 973: 8–12.

Shanley, T.P., Ryan, M.A., Eaves-Pyles, T. and Wong, H.R. (2000) Heat shock inhibits phosphorylation of I-kappaBalpha. Shock, 14: 447–450.

Sharma, H.S. (2006) Hyperthermia influences excitatory and inhibitory amino acid neurotransmitters in the central nervous system. An experimental study in the rat using behavioural, biochemical, pharmacological, and morphological approaches. J. Neural Transm., 113: 497–519.

Shibolet, S., Coll, R., Gilat, T. and Sohar, E. (1967) Heatstroke: its clinical picture and mechanism in 36 cases. Q. J. Med., 36: 525–548.

Shibolet, S., Lancaster, M.C. and Danon, Y. (1976) Heat stroke: a review. Aviat. Space Environ. Med., 47: 280–301.

Skrandies, S., Bremer, B., Pilatus, U., Mayer, A., Neuhaus-Steinmetz, U. and Rensing, L. (1997) Heat shock- and ethanol-induced ionic changes in C6 rat glioma cells determined by NMR and fluorescence spectroscopy. Brain Res., 746: 220–230.

Sonna, L.A., Cullivan, M.L., Sheldon, H.K., Pratt, R.E. and Lilly, C.M. (2003) Effect of hypoxia on gene expression by human hepatocytes (HepG2). Physiol. Genom., 12: 195–207.

Sonna, L.A., Fujita, J., Gaffin, S.L. and Lilly, C.M. (2002a) Invited review: effects of heat and cold stress on mammalian gene expression. J. Appl. Physiol., 92: 1725–1742.

Sonna, L.A., Gaffin, S.L., Pratt, R.E., Cullivan, M.L., Angel, K.C. and Lilly, C.M. (2002b) Effect of acute heat shock on gene expression by human peripheral blood mononuclear cells. J. Appl. Physiol., 92: 2208–2220.

Sonna, L.A., Wenger, C.B., Flinn, S., Sheldon, H.K., Sawka, M.N. and Lilly, C.M. (2004) Exertional heat injury and gene expression changes: a DNA microarray analysis study. J. Appl. Physiol., 96: 1943–1953.

Suzuki, K., Nakaji, S., Yamada, M., Totsuka, M., Sato, K. and Sugawara, K. (2002) Systemic inflammatory response to exhaustive exercise. Cytokine kinetics. Exerc. Immunol. Rev., 8: 6–48.

Suzuki, K., Yamada, M., Kurakake, S., Okamura, N., Yamaya, K., Liu, Q., Kudoh, S., Kowatari, K., Nakaji, S. and Sugawara, K. (2000) Circulating cytokines and hormones with immunosuppressive but neutrophil-priming potentials rise after endurance exercise in humans. Eur. J. Appl. Physiol., 81: 281–287.

Weitzel, G., Pilatus, U. and Rensing, L. (1987) The cytoplasmic pH, ATP content and total protein synthesis rate during heat-shock protein inducing treatments in yeast. Exp. Cell. Res., 170: 64–79.

Wennborg, A., Classon, M., Klein, G. and von Gabain, A. (1995) Downregulation of c-myc expression after heat shock in human B-cell lines is independent of 5′ mRNA sequences. Biol. Chem. Hoppe-Seyler, 376: 671–680.

Wong, H.R., Ryan, M.A., Menendez, I.Y. and Wispe, J.R. (1999) Heat shock activates the I-kappaBalpha promoter and increases I-kappaBalpha mRNA expression. Cell Stress Chaperones, 4: 1–7.

Wong, H.R., Ryan, M. and Wispe, J.R. (1997) The heat shock response inhibits inducible nitric oxide synthase gene expression by blocking I kappa-B degradation and NF-kappa B nuclear translocation. Biochem. Biophys. Res. Commun., 231: 257–263.

Xu, C., Fracella, F., Richter-Landsberg, C. and Rensing, L. (1997) Stress response of lysosomal cysteine proteinases in rat C6 glioma cells. Comp. Biochem. Physiol. B Biochem. Mol. Biol., 117: 169–178.

Yoo, C.G., Lee, S., Lee, C.T., Kim, Y.W., Han, S.K. and Shim, Y.S. (2000) Anti-inflammatory effect of heat shock protein induction is related to stabilization of I kappa B alpha through preventing I kappa B kinase activation in respiratory epithelial cells. J. Immunol., 164: 5416–5423.

Yost, H.J. and Lindquist, S. (1991) Heat shock proteins affect RNA processing during the heat shock response of *Saccharomyces cerevisiae*. Mol. Cell. Biol., 11: 1062–1068.

Zieker, D., Fehrenbach, E., Dietzsch, J., Fliegner, J., Weidmann, M., Nieselt, K., Gebicke-Haerter, P., Spanagel, R., Simon, P., Niess, A.M. and Northoff, H. (2005) cDNA-microarray analysis reveals novel candidate genes expressed in human peripheral blood following exhaustive exercise. Physiol. Genom., 23: 287–294.

CHAPTER 17

Cellular mechanisms of neuronal damage from hyperthermia

Michael G. White[1,3], Luminita E. Luca[1,3], Doris Nonner[1], Osama Saleh[1], Bingren Hu[2,3], Ellen F. Barrett[1,3] and John N. Barrett[1,3,*]

[1]*Department of Physiology and Biophysics, Miller School of Medicine, University of Miami, Miami, FL 33136, USA*
[2]*Department of Neurology, Miller School of Medicine, University of Miami, Miami, FL 33136, USA*
[3]*Neuroscience Program, Miller School of Medicine, University of Miami, Miami, FL 33136, USA*

Abstract: Hyperthermia can cause brain damage and also exacerbate the brain damage produced by stroke and amphetamines. The developing brain is especially sensitive to hyperthermia. The severity of, and mechanisms underlying, hyperthermia-induced neuronal death depend on both temperature and duration of exposure. Severe hyperthermia can produce necrotic neuronal death. For a window of less severe heat stresses, cultured neurons exhibit a delayed death with apoptotic characteristics including cytochrome c release and caspase activation. Little is known about mechanisms of hyperthermia-induced damage upstream of these late apoptotic effects. This chapter considers several possible upstream mechanisms, drawing on both *in vivo* and *in vitro* studies of the nervous system and other tissues. Hyperthermia-induced damage in some non-neuronal cells includes endoplasmic reticular stress due to denaturing of nascent polypeptide chains, as well as nuclear and cytoskeletal damage. Evidence is presented that hyperthermia produces mitochondrial damage, including depolarization, in cultured mammalian neurons.

Keywords: hyperthermia; heat stress; neuron; apoptosis; caspase; mitochondria; heat shock; ischemia

Introduction

Hyperthermia can produce or contribute to neuronal injury under multiple environmental and clinical conditions. Recently hyperthermia has received attention due to high profile incidents involving athletes, amphetamine use, soldiers in hot environments, and the increased incidence of severe heat waves. High ambient temperatures, especially when accompanied by intense physical activity and dehydration, can lead to the rapid onset hyperthermia of heat stroke. Heat stroke occurs when the brain reaches temperatures above 105–106°F (40.5° to 41°C). At this temperature, the ability of the hypothalamus to coordinate thermoregulation becomes compromised and there can be a further increase in brain temperature, low blood pressure, an increase in intracerebral pressure, monoamine overload, and multi-organ dysfunction (Simon, 1993; Lee-Chiong and Stitt, 1995; Yang and Lin, 1999; Bouchama and Knochel, 2002; Chang et al., 2004, see Chapters 10 and 15 in this volume). An individual who survives these complications may experience persisting deficits including attention and memory loss, personality changes, and brain lesions resulting in global dementia (Romero et al., 2000; Worfolk, 2000).

*Corresponding author. Tel.: +1 (305) 243-6357; Fax: +1 (305) 243-5931; E-mail: jbarrett@med.miami.edu

Certain drugs can induce or exacerbate hyperthermia-induced brain damage. For example, the combination of amphetamines, physical activity, and a warm environment causes brain damage, due in part due to an amphetamine-induced increase in brain metabolism (e.g., Brown and Kiyatkin, 2004; Sanchez et al., 2004). Neuroleptics and certain anesthetics can produce neuroleptic malignant syndrome (malignant hyperthermia, Halliday, 2003, see Chapter 6 in this volume), which can produce brain damage (Park et al., 2004).

Even hyperthermia too mild to produce heat stroke can produce long-lasting neurological deficits (Chia and Teo, 2003) and pathological changes in the brain (reviewed by Sharma and Hoopes, 2003). Fever caused by infection or left unchecked for medical purposes (Styrt and Sugarman, 1990), may potentially injure the brain. Likewise, hyperthermia is used to increase the effectiveness of some anti-cancer strategies (Westermann et al., 2003; Ma et al., 2004) and to treat certain brain cancers (Stea et al., 1992), but carries the risk of neuronal injury (Narayan et al., 2004). A better knowledge of the mechanisms by which hyperthermia damages the brain should be helpful in devising therapies to minimize this brain damage.

The developing nervous system is especially vulnerable to hyperthemia

Hyperthermia during pregnancy is teratogenic, inducing a number of developmental defects including embryonic death, growth retardation, and mental retardation. Exposure to temperatures of 41°C or above can damage the embryonic brain (Edwards, 1998; Paula-Lopes and Hansen, 2002). Developmental events thought to be especially sensitive to hyperthermia include neuronal migration and proliferation of neuronal progenitors. For example, the death of guinea pig embryos subjected to heat in utero was associated with abnormal cardiac development, attributed to heat-induced defects in migration of neural crest cells (Upfold et al., 1991). Hinoue et al. (2001) found that just 12 min of exposure to 43°C during embryonic development could produce neuronal apoptosis, reducing the thickness of the cortical gray matter. In guinea pigs, a 1 h heat exposure during early neurogenesis increased the incidence of microencephaly (Edwards et al., 2003). Mouse embryos exposed for 12 min to 42°C or for 7–10 min to 43°C exhibited anterior neural tube defects (exencephaly, anencephaly, and cranial neural tube defects with facial cleft, Shiota, 1988). Two teratogenic windows were described in guinea pigs exposed to maternal hyperthermia: embryonic day 13 (E13), which corresponds to closure of the neural groove and anterior neuropore, and E21, corresponding to early stages of cortical plate formation. Heat stress at E13 produced a high incidence of neural tube defects (e.g., exencephaly) associated with microphthalmia and scoliosis/kyphosis (Cawdell-Smith et al., 1992). After birth the cerebellum remains susceptible to heat stress during the granule cell proliferative period (Khan and Brown, 2002).

Epidemiological data strongly suggest an association of fever in pregnant women with neural tube defects in their offspring. Human embryos exposed to hyperthermia during the equivalent period of neural groove closure (E23–E25) showed neural tube defects similar to those seen in the guinea pigs mentioned above (Smith et al., 1992). Mothers of children with spina bifida had a higher incidence of fever during pregnancy than mothers of unaffected children (Layde et al., 1980). Milunsky et al. (1992) found that exposure to hot tubs, saunas, or fever during the first trimester of pregnancy was correlated with increased risk of neural tube defects. In human embryos, exencephaly (defect resulting in anencephaly) is considered a specific neural tube defect induced by fever during early pregnancy (Shiota, 1982).

The mechanisms by which these neural tube defects are produced include abnormal apoptosis. Fever increases apoptosis in dividing cells of the testis and thymus, and also in neocortex and tectum of embryonic day 17 rats, suggesting that actively-dividing populations are highly susceptible to hyperthermia-induced apoptosis (Khan and Brown, 2002). Mitotic activity was inhibited in embryonic tissues for several hours after hyperthermia, followed by a burst of mitosis, suggesting that neural tube defects seen in mouse embryos are the result of cell death and cessation of cell proliferation (Shiota, 1988).

Buckiova and Brown (1999) assessed the effect of in utero hyperthermia on expression of several transcription factors that act as developmental control genes. Neural tube defects induced by hyperthermia in rat embryos correlated with decreased expression of the Krox20 gene, normally expressed in hindbrain, whereas the expression pattern of other studied genes (Otx2, Emx2, and hoxb1) was unaltered. The defects might be a result of disrupted signaling from the midbrain/hindbrain boundary.

Some aspects of the age-dependence of neuronal vulnerability to damage from hyperthermia can be reproduced in cell culture. Figure 1 plots the viability of septal neurons from embryonic day 15 rats that were cultured for 5 or 21 days and then subjected to a heat stress (43°C for 2 h). The younger cells died more rapidly than the older cells. Glia resisted the heat stress better than neurons. Co-culture with glia for 5 days did not improve neuronal survival.

Hyperthermia exacerbates post-ischemic neuronal death

Mild temperature elevations which by themselves are not sufficient to produce neuronal death can exacerbate the delayed neuronal death that follows an ischemic insult. For example, Castillo et al. (1998) demonstrated that 61% of patients with hemispheric cerebral infarction exhibited hyperthermia (>37.5°C), and that the mortality rate in hyperthermic patients was 15-fold greater than in normothermic patients. Hyperthermia initiated within the first 24 h correlated with a larger infarct volume and a greater neurological deficit assessed at 3 months, regardless of whether or not the hyperthermia arose from infectious or non-infectious (e.g., hypothalamic damage) causes. Similar findings have been documented in a variety of animal models of brain ischemia accompanied by hyperthermia (e.g., Busto et al., 1987; Minamisawa et al., 1990; Chen et al., 1991; Meden et al., 1994). The detrimental effect of post-ischemic hyperthermia was evident with brain temperatures of 40°C (Kim et al., 1996; Baena et al., 1997). Cooling the brain below 37°C afforded protection in animal ischemia models, suggesting that there is a graded relationship between post-ischemic brain damage and brain temperature (reviewed by Ginsberg et al., 1992; also Maher and Hachinski, 1993; Barone et al., 1997). Because of these findings, anti-pyretic treatments are routinely undertaken in cases of post-ischemic hyperthermia, but clinical trials testing methods for cooling the post-ischemic brain have to date yielded inconclusive results (reviewed by Zaremba, 2004). Physically reducing the body temperature to 32° to 33°C is associated with multiple complications including cardiac arrhythmia, hypotension, a decrease in blood platelets, and increased risk of pneumonia (Georgiadis et al., 2001; Schwab et al., 2001). Milder hypothermia (35° to 35.5°C) precluded the need for sedation but still required intensive care and pethidine to control shivering (Kammersgaard et al., 2000).

Recent pharmacological experiments and clinical trials have focused on controlling elevated body temperature during stroke using high doses of acetaminophen. However, results have been disappointing in achieving hypothermia robust enough for clinical impact, suggesting that alternative pharmacologic approaches should be explored (Dippel et al., 2001; Kasner et al., 2002). Other experiments have focused on identifying mechanisms by which hyperthermia may compound ischemic conditions during stroke. These approaches have yielded evidence that body temperature can complicate ischemia-induced energy stress (Chopp et al., 1988; Chen et al., 1993; Mies et al., 1993), excitotoxic glutamate release (Globus et al., 1991; Takagi et al., 1994), free radical production (Globus et al., 1995; Kil et al., 1996), alterations in blood-brain barrier permeability (Dietrich et al., 1990, 1991), and cytoskeletal integrity (Churn et al., 1990; Eguchi et al., 1997; Morimoto et al., 1997). Understanding how heat stress triggers these secondary stroke events and their role in determining neuronal survival and infarct size during stroke may lead to improved therapies.

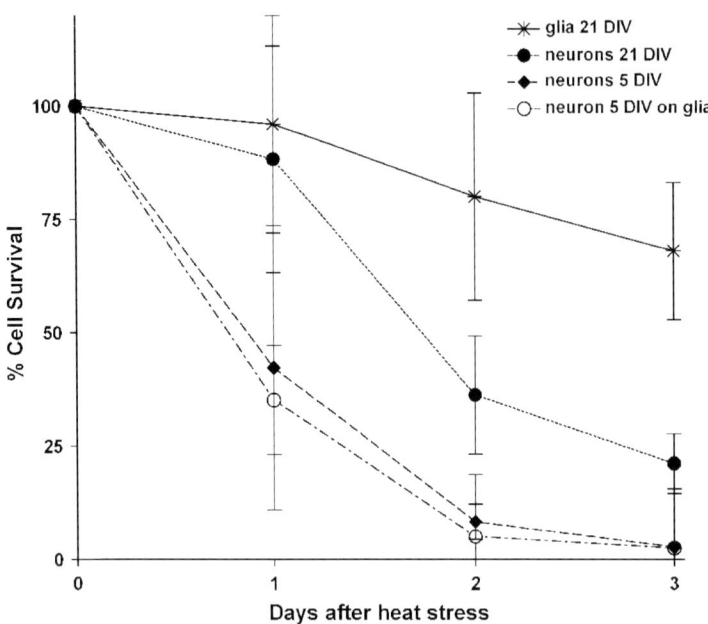

Fig. 1. Delayed death of rat septal cells following hyperthermia, showing that older neurons are more resistant than younger neurons, and that glia are more resistant than neurons. Sister cultures prepared from embryonic day 15 rats were maintained for 5 or 21 days and then stressed at 43°C for 2 h. Surviving cells were counted at the indicated times following stress termination. Young neurons grown on 4-week-old glial cultures did not become more resistant to the heat stress. Each point is the mean ± SEM of cell counts from at least 14 cultures.

Cellular mechanisms of neuronal damage from hyperthermia

The degree of cellular damage depends on both the duration and the intensity of the heat stress. Over a range of temperatures (42° to 50°C) the degree of cellular damage increases with the duration of hyperthermia in an approximately logarithmic manner for many cell types (Harmon et al., 1990). A 1.5–2 h exposure to 43°C was required to kill cultured neurons, whereas at a higher temperature (45°C) a much shorter duration (30 min) was sufficient (compare White et al., 2003 with Vogel et al., 1997). Very high temperatures produce cell swelling and necrotic death that become obvious during or shortly after the heat stress. More moderate hyperthermia does not produce obvious signs of neuronal damage during the stress, but neurons die over the following days (as in Fig. 2).

After stressing cultured primary striatal neurons for 2 h at 43°C, the first light microscopic sign of damage is blebbing of processes, visible ~24 h following stress termination. Condensed nuclei, probably indicative of apoptosis (although sometimes appearing in necrotic cells), appear by 24–36 h. By 48–60 h, most neurons are dead (Fig. 4).

Hyperthermia has also been shown to produce a delayed, perhaps apoptotic, death in cortical neurons (Vogel et al., 1997), cerebellar granule neurons (Lowenstein et al., 1991), dorsal root ganglion neurons (Uney et al., 1993), and septal neurons (Fig. 1).

Evidence for caspase activation in heat-stressed neurons

The line between necrotic and apoptotic death is hard to define for some cellular stresses (Kanduc et al., 2002). Some of the commonly used apoptotic markers such as TUNEL or annexin V staining may not clearly distinguish between these modes of death. Apoptosis often involves caspase activation. Caspases are cysteine-dependent aspartate-specific

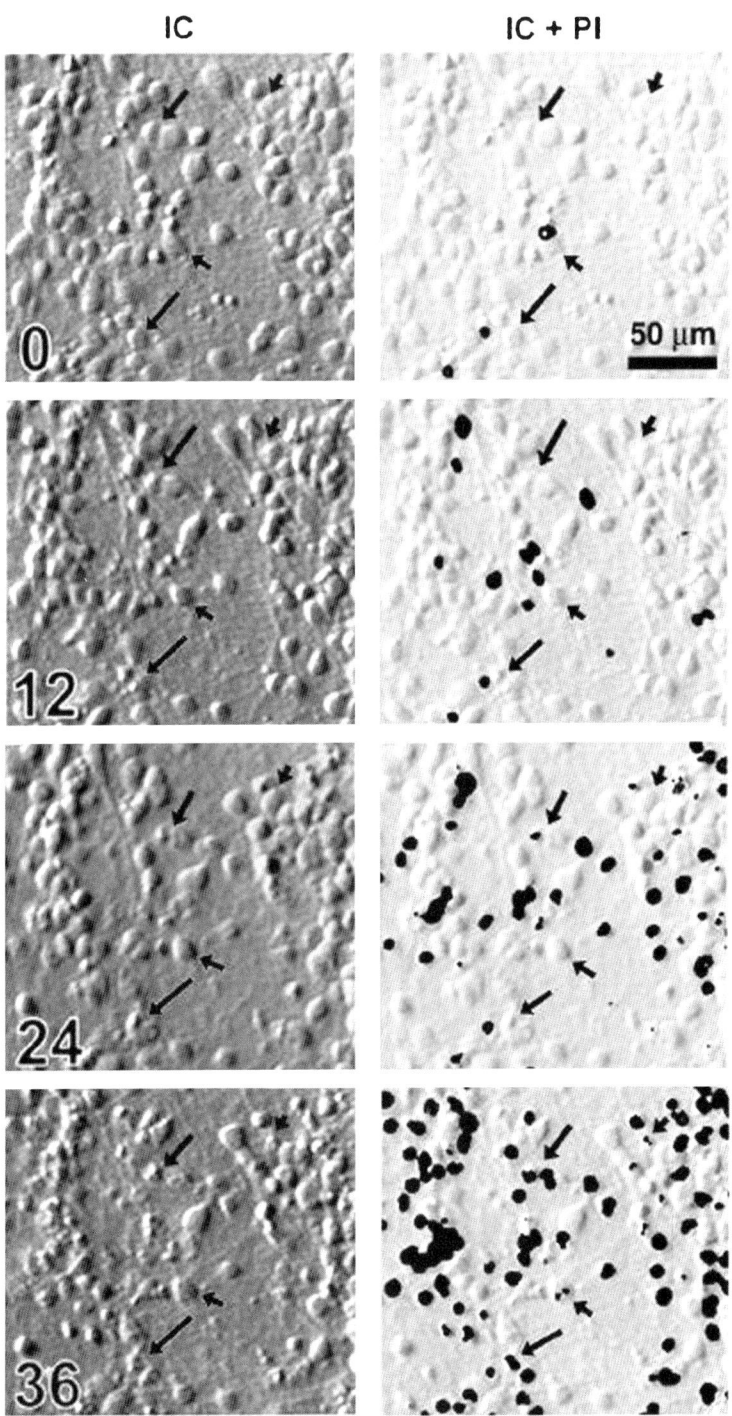

Fig. 2. Delayed death of rat striatal neurons following a 2 h 43°C stress. The left column shows a field of cells imaged with interference contrast (IC) immediately after stress termination (0 h, top) and at 12, 24, and 36 h thereafter. Paired images in the right column show the same field overlaid with staining for propidium iodide (PI, black), which marks the nuclei of dead/dying cells. By 24 h many cells exhibit signs of apoptosis including shrunken somata and condensed and fragmented nuclei. Arrows identify four cells that developed these apoptotic changes. Cells were maintained 8 days in culture prior to the stress.

proteases that exist as pro-caspase zymogens until they are activated by proteolytic cleavage. Upstream initiator caspases (1, 2, 8, 9, 12) are involved in activating apoptotic pathways. Particular caspases are often associated with a particular stress or particular stressed organelles (e.g., caspase-12 is associated with endoplamic reticulum (ER) stress). Both upstream and downstream executioner caspases (3, 6, 7) cleave cellular substrates, resulting in cell death. The executioner caspase-3 is activated in neurons of embryonic mice subjected to hyperthermia in utero (Umpierre et al., 2001). Caspase activation follows heat stress in neuroblastoma cells (Bijur et al., 2000; Bijur and Jope, 2001) and in primary rat neurons in culture (White et al., 2003). For example, in cultured striatal neurons maintained 1–2 weeks in vitro, caspase-3 and caspase-7 are activated with a delay following heat stress for 1.5–2 h at 43°C (Fig. 3A). Caspase-3 activation correlated closely with the time frame and stress conditions needed to achieve delayed death. In addition, the level of caspase-3 activation by the heat stress decreased with increasing resistance of septal neurons to heat with time in culture (Fig. 3B).

The general caspase inhibitors Z-Val-Ala-Asp(OMe)-fluoromethylketone (zVAD-fmk) and quinolyl-valyl-O-methyl-aspartyl-[2,6-difluoro phenoxyl-methyl ketone (qVD-OPH) prolonged neuronal survival by ~1 day (Fig. 4). zVAD-fmk and qVD-OPH also reduced the level of caspase-3 activation, but not as effectively as they improved survival of heat-stressed neurons (Fig. 3C). In addition, Ac-Asp-Glu-Val-Asp-aldehyde (DEVD-CHO) inhibited caspase-3 more effectively than the general caspase inhibitors in both test tube and in vitro assays (Fig. 3D), but appeared to be less effective in prolonging post-hyperthermic neuronal survival. Therefore, the general caspase inhibitors probably protect by inhibiting an upstream caspase in addition to caspase-3.

These upstream caspases may include caspase-9, since cytochrome c is released from mitochondria following the heat stress (White et al., 2003). Cytochrome c is also released from brain mitochondria in heated murine embryos (Mirkes and Little, 2000). Release of cytochrome c from the mitochondrial intermembrane space induces APAF-1 to associate with procaspase-9, forming the apoptosome, which cleaves procaspase-9 into its activated state, leading to subsequent activation of caspase-3 and cell death (reviewed by Wang, 2001; Fiskum, 2004). Consistent with this pathway, Western blots showed the presence of cleaved caspase-9 (Little and Mirkes, 2002; and our preliminary data). It is less clear whether activation of another upstream caspase preceded activation of caspase-9. Caspase-8 is a good candidate for an upstream caspase, since heat stress contributes to sensitizing Jurkat and HeLa cells to activation of apoptosis by the Fas ligand (Tran et al., 2003). In these cells heat stress leads to activation of caspase-8 by reducing the amount of the caspase-8 inhibitory protein FLIP. Whether a similar mechanism operates in heat-stressed neurons is not yet known, although some neurons do express Fas ligand receptors (e.g., Jayanthi et al., 2005).

The above-mentioned pan-caspase inhibitors protected neuronal cultures more robustly against a 1.5–2 h heat stress than against a staurosporine stress (White et al., 2003). Death following exposure to staurosporine, a non-selective kinase inhibitor, involves both caspase activation and a non-caspase dependent mechanism involving apoptosis-inducing factor (AIF, Susin et al., 1999, 2000; Cregan et al., 2002; AIF reviewed by Candé et al., 2002). However the heat-stressed neurons were not permanently protected by caspase inhibitors; neurons usually died 3–5 days after termination of the hyperthermia. This suggests that parallel and/or caspase-independent pathways, like those utilizing AIF, may also be involved in hyperthermia-induced death.

Consistent with this idea, some heat stresses produce a delayed neuronal death that does not exhibit classical apoptotic characteristics. For example, when the duration of the 43°C heat stress was increased from 2 to 4 h, ~20% of the neurons were dead immediately after the stress. Most of the remaining neurons died after a delay of 1–2 days, but this delayed death occurred without detectable caspase-3 activation (diamonds in Fig. 3A), and was not postponed by pan-caspase inhibitors. Studies on non-neuronal cells have also documented a transition from death involving caspase activation to caspase-independent death as the severity of the heat

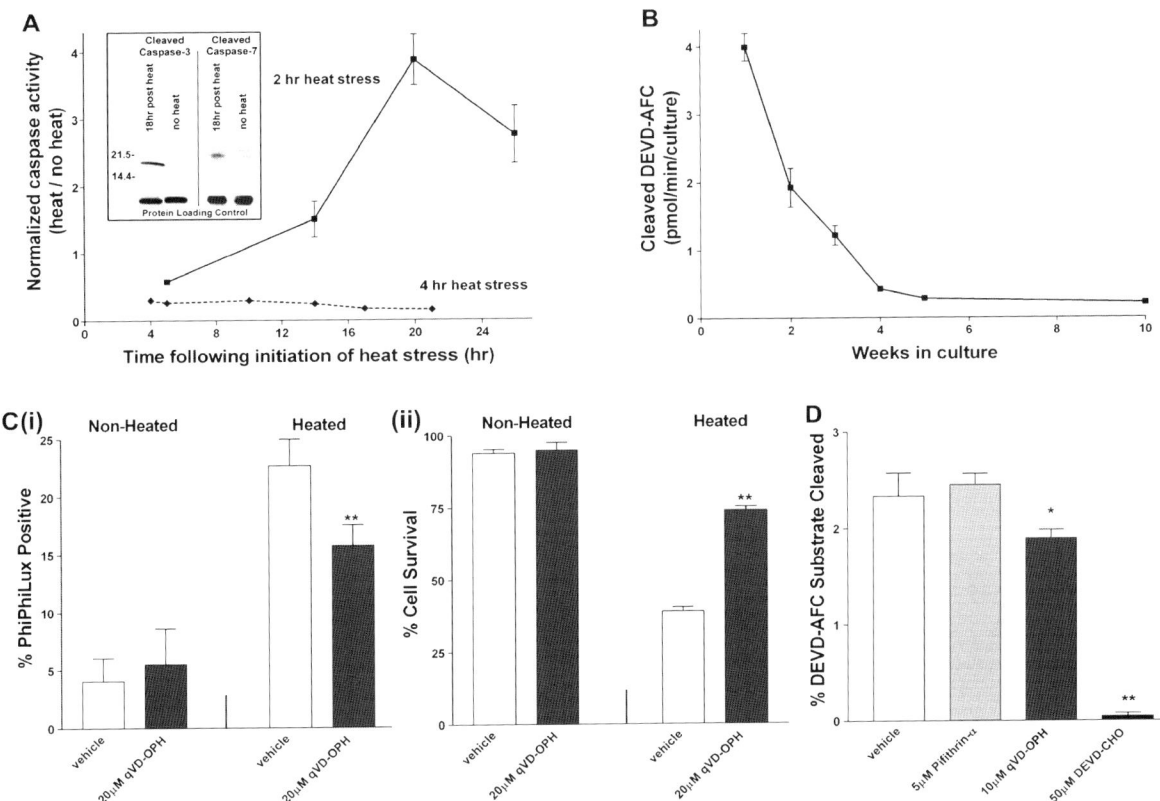

Fig. 3. Caspase-3 activation following a 2 h 43°C stress. (A) Caspase-3 activity increases with a delay following a 2 h stress, but does not increase following a 4 h stress. Activity was measured using a caspase-3 substrate zDEVD-AFC that becomes fluorescent following cleavage. Inset: Western blots show the appearance of cleaved caspase-3 and -7 by 18 h post-stress. Lower bands are loading controls; numbers indicate molecular weights. (B) Caspase-3 activity assayed 18 h following a 2 h heat stress is higher in younger than in older cultures. (C(i)) qVD-OPH reduces the number of heated cells that stain with the membrane-permeable caspase-3 substrate PhiPhiLux, which becomes fluorescent following cleavage. qVD-OPH was added 20 min post-stress, and PhiPhiLux was added 17 h post-stress, with fluorescent cells assayed 5 h thereafter. (C(ii)) qVD-OPH increased survival of heated cells assayed in the same cultures 45 h post-stress. (D) Caspase-3 activity measured using zDEVD-AFC 18.5 h after a 2 h heat stress is reduced by the pan-caspase inhibitor qVD-OPH and by the caspase-3 inhibitor DEVD-CHO, but not by the p53 inhibitor pifithrin-α. Inhibitors and vehicle were added 1 h post-stress. Vehicle included 0.5–1% DMSO. Data in (B) came from septal cultures; the remaining data came from striatal cultures maintained 8 (A, C, D) or 16 (inset in A) days *in vitro*. Data are expressed as means ± SEM; $n = 10$–48 in A, B, and D; $n = 4$ in (C). $*p<0.05$ and $**p<0.001$ compared with corresponding vehicle controls, one-way ANOVA with Student–Newman–Keuls test. Additional experimental details in White et al. (2003).

stress was increased (Harmon et al., 1990; O'Neill et al., 1998; Moriyama-Gonda et al., 2002; Dunn et al., 2004). Thus there is a window of heat stress severity that produces apoptotic death.

Excitotoxic contribution to hyperthermia-induced neuronal death

Stress-induced increases in cytosolic $[Ca^{2+}]$ increase transmitter release from nerve terminals, and there are elevated levels of the neurotransmitters glutamate and glycine in the cerebrospinal fluid of hyperthermic stroke patients (Castillo et al., 1999). Excitotoxicity is thought to contribute to the damage associated with hyperthermia in children (Toth et al., 1998) and is involved in hyperthermic damage following traumatic brain injury (Thompson et al., 2003). Febrile hyperthermia in children can cause seizures, and this combination produces brain damage that probably

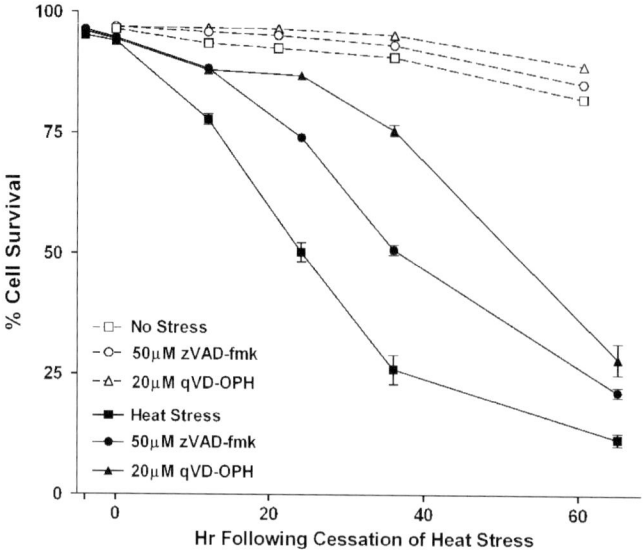

Fig. 4. Pan-caspase inhibitors postpone death following a 2 h 43°C heat stress. zVAD-fmk or qVD-OPH was added to heated (filled symbols, solid lines) and unheated (open symbols, dashed lines) sister striatal cultures at the end of the heat stress. The percentage of surviving cells was measured pre-stress (−4 h), and at the indicated times post-stress. Caspase inhibitors significantly increased survival of all stressed cells between 12 and 65 h post-stress. qVD-OPH was effective at lower concentrations than zVAD-fmk. The survival assay counted cells stained with propidium iodide, and is described in detail in White et al. (2003). Mean ± SEM, $n = 11$–18 culture wells for each point.

involves excitotoxicity (Yager et al., 2004). Excessive glutamate release or impaired glutamate reuptake can cause large increases in intracellular $[Ca^{2+}]$ that may activate nitric oxide synthetase and increase the levels of reactive oxygen species (ROS, see later section). The resulting oxidative DNA damage may lead to overactivation of poly(ADP-ribose) polymerase (PARP) and apoptotic death (Berger, 1985; Carson et al., 1986; Skaper, 2003; Yu et al., 2003; Cregan et al., 2004).

However, young cultured striatal and septal neurons (9 days *in vitro*) were not protected from hyperthermia by drugs that block action potentials (tetrodotoxin and saxitoxin, which block voltage-dependent Na^+ channels) or by drugs that block ionotropic glutamate receptor channels (CNQX to block AMPA/kainate receptors, MK801 to block NMDA receptor channels; White et al., 2003). Thus hyperthermia can kill cultured neurons even in the absence of excitotoxicity. Damaging effects of excitotoxicity probably increase as neurons mature and express more glutamate receptors, and *in vivo*, where neurons are more closely packed than they are *in vitro*.

Thus it is likely that heat stress damages neurons by both excitotoxic and non-excitotoxic mechanisms. The relative importance of these mechanisms may depend on the developmental maturity of the neurons.

Hyperthermia-induced activation of death mechanisms upstream of caspase activation

Although hyperthermia-induced death has been investigated in many non-neuronal cells, the mechanisms that initiate death pathways upstream of caspase-3 in neurons have not been extensively explored. Potential candidates include the hyperthermia-induced damage mechanisms reported in non-neuronal cells such as those involving damage to mitochondria, ER, nucleus, and/or cytoskeleton. Some of these pathways are diagrammed in Fig. 5 and discussed below.

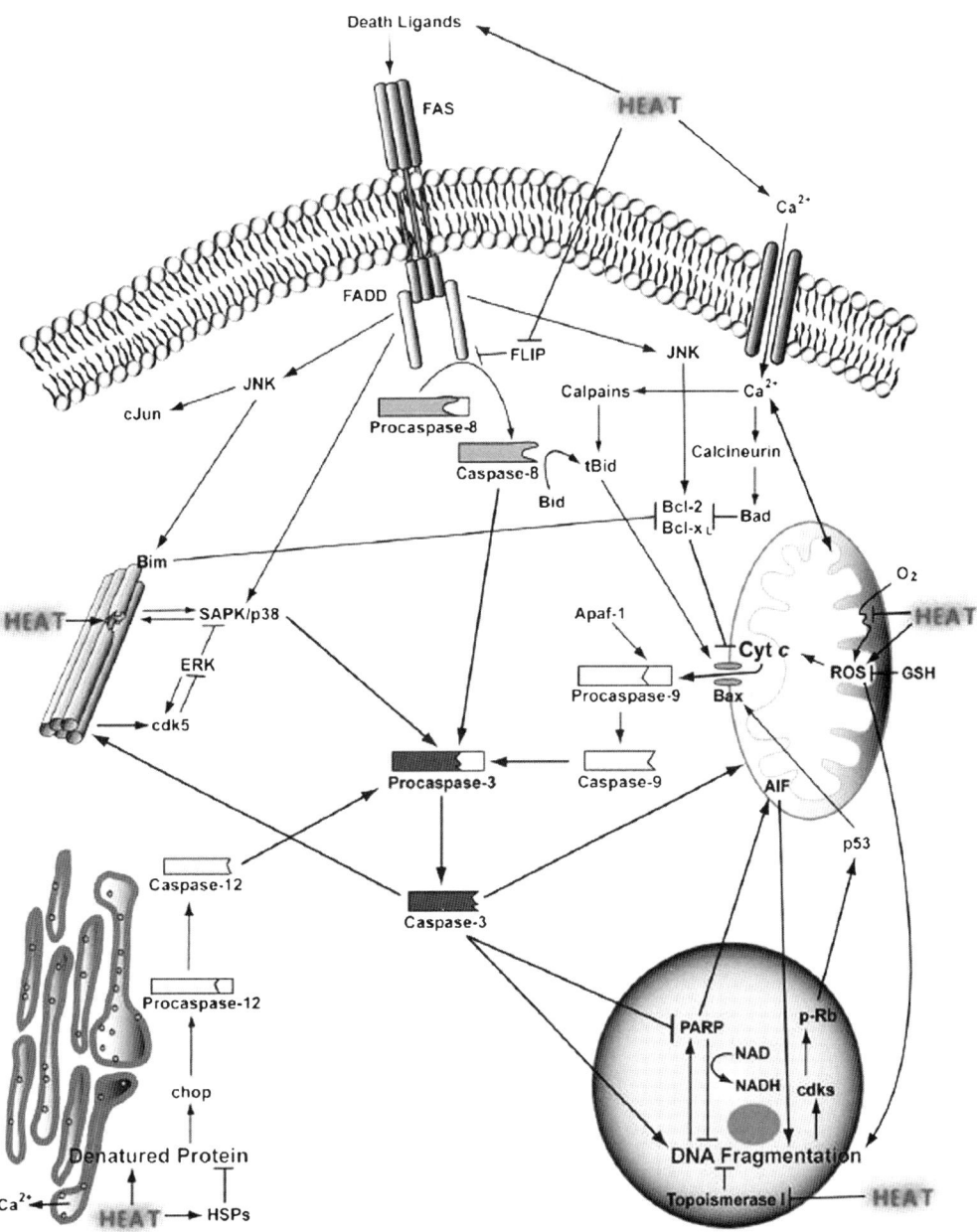

Fig. 5. Possible sites of cellular disruption and activation of apoptotic pathways by heat stress, as described in the text. This simplified diagram omits some interactions among apoptotic pathways. GSH, glutathione.

Mitochondrial damage contributes to hyperthermia-induced neuronal death

Mitochondria often play a critical role in apoptosis (Green and Reed, 1998; Desagher and Martinou, 2000), and may be involved in hyperthermia-induced death as a site of initial dysfunction and/or as an amplification step for apoptosis. As mentioned above, cytochrome c is released from neuronal mitochondria following a heat stress. One mechanism proposed for this release involves cleavage of BID (a member of the BH3- only subset of the Bcl-2 family of proteins) by the Ca^{2+}-activated protease calpain (Takano et al., 2005) or by caspase-8 (Li et al., 1998). The truncated form of BID (tBID) then causes insertion of Bcl-2-associated X protein (Bax) into the mitochondrial outer membrane, forming channels that allow release of apoptotic proteins such as cytochrome c from the mitochondrial intermembrane space.

Hyperthermia might also damage mitochondria directly, since our preliminary data reveal early signs of mitochondrial dysfunction during a heat stress. Figure 6 indicates that cultured neurons exposed to 43°C for 1.5–2 h show a partial depolarization of the mitochondrial membrane potential during the heat stress. Following return to physiological temperatures there is a partial recovery from this initial depolarization, followed by a later, irreversible depolarization. One possible mechanism for the initial mitochondrial depolarization is a hyperthermia-induced increase in the permeability of the mitochondrial inner membrane (Nichols et al., 1980; Willis et al., 2000). Hyperthermia induces mitochondrial uncoupling in rat cardiomyocytes (Qian et al., 2004) and mud clams (Abele et al., 2002).

Reduced mitochondrial respiration might also contribute to the observed mitochondrial depolarization. Moderate hyperthermia enhances mitochondrial respiratory activity in rat cardiac cells within 25 min (Sammut et al., 2001; Sammut and Harrison, 2003), but more severe heat stress impairs oxidative phosphorylation in isolated liver mitochondria (Willis et al., 2000; Qian et al., 2004). Consistent with this scenario, our preliminary data indicate that exposure to 43°C increased oxygen consumption in cultured striatal neurons during the first 20–30 min, but more prolonged exposure to 43°C produced a steady decline in oxygen consumption (not shown). Temperature-dependent partial inactivation of complex I (NADH-ubiquinone oxidoreductase) of the electron transport chain becomes evident at temperatures as low as 37°C and increases with increasing temperature (Grivennikova et al., 2001). This inactivation is rapidly reversed when the temperature falls back into the normal range, consistent with our finding of transient partial recovery of the mitochondrial potential in cultured neurons when the heat stress was terminated. Electron transport chain activity could also be reduced by dilution of the electron acceptor cytochrome c following its release into the cytoplasm after disruption of the outer mitochondrial membrane, as described above.

Mitochondria can also be depolarized by opening of the mitochondrial permeability transition pore (MPTP), either in a relatively low conductance, reversible mode or in a high conductance, irreversible mode. It is unlikely that irreversible opening of the mitochondrial transition pore contributes to the early, partially reversible depolarization of cultured neurons during hyperthermia, but it might contribute to the progressive, irreversible depolarization that develops following termination of the heat stress. MPTP opening may be modulated by electron flux through complex I (Fontaine et al., 1998; Fontaine and Bernardi, 1999; Chauvin et al., 2001).

Mitochondrial depolarization interferes with two important mitochondrial functions, ATP production and calcium sequestration. Our preliminary measurements of ATP in cultured neurons show that they maintain $81.2 \pm 8.2\%$ ($n = 10$–12) of their preheated ATP levels immediately following a 2 h, 43°C stress that produces delayed neuronal death. ATP levels during the hyperthermia may be sustained by glycolysis in these young neurons. Consistent with these *in vitro* findings, Busto et al.'s (1987) measurements of energy substrates following combined heat stress and ischemia *in vivo* showed that brain temperature during ischemia did not affect energy metabolite levels measured at the conclusion of ischemia: the ischemia-induced depletion of brain ATP, phosphocreatine, glucose and glycogen, and the elevation of lactate, were not altered by brain temperature during ischemia over the range examined. Thus factors besides ATP depletion appear to be more critical in mediating the damaging effects of hyperthermia following ischemia.

Hyperthermia-induced disruption of Ca^{2+} transport across mitochondrial, ER, or plasma membranes would be expected to increase cytosolic $[Ca^{2+}]$. Heat stress indeed increases intracellular $[Ca^{2+}]$ in cardiomyocytes (Everett et al., 2001) and cultured neurons (Greffrath et al., 2001; and our preliminary data). Elevation of cytosolic $[Ca^{2+}]$

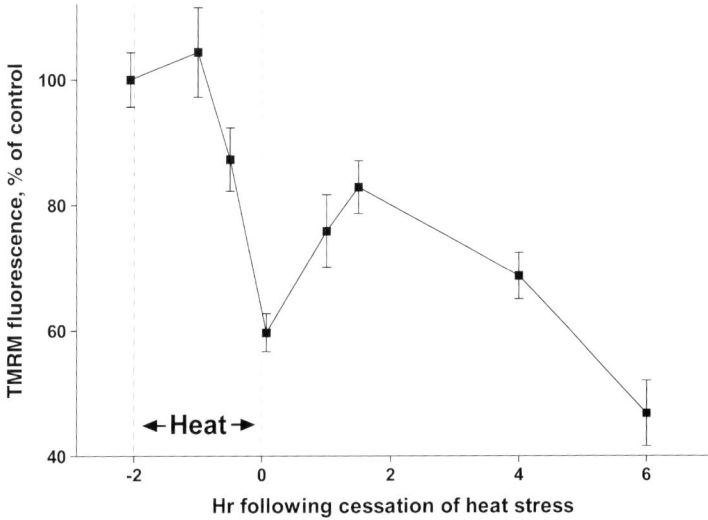

Fig. 6. Changes in mitochondrial membrane potential in striatal cultures during and following a 2 h 43°C stress. Tetramethyl rhodamine methyl ester (TMRM) is a positively charged Nernstian potentiometric fluorescent dye that accumulates in mitochondria due to the negative-inside potential across their inner membrane. A decrease in TMRM fluorescence indicates depolarization. Cells partially depolarized during the heat stress, partially repolarized within 1–1.5 h following a return to physiological temperatures, and then began irreversibly depolarizing by 4–6 h post-stress. This figure combines data from multiple experiments. A similar time course of changes was measured when the experiment was repeated in isotonic (145 mM) KCl to depolarize the plasma membrane potential (not shown). Method: Cultures were incubated 30 min in 1 μM TMRM, then washed with TMRM-free medium and lysed with 50% DMSO in distilled water. The remaining TMRM was measured with a fluorescence plate reader. Values are expressed as percentage of pre-stress control TMRM fluorescence; mean ± SEM, $n = 12$–36.

would contribute to the excitotoxic damage described previously, and would also activate Ca^{2+}-dependent proteases such as the calpains. Calpains may play a role in germ cell apoptosis (Somwaru et al., 2004). They cleave and thereby activate pro-apoptotic proteins such as Bax (Hwang et al., 2003) and Bid (Mandic et al., 2002), and also cleave the cyclin-dependent kinase 5 (cdk5) regulator p35 (Nath et al., 2000, see later section).

Do reactive oxygen species (ROS) contribute to hyperthermic stress?

Hyperthermia-induced disruptions of mitochondrial oxidative phosphorylation and coupling may promote apoptosis not only by aggravating an energy stress, but also by increasing ROS production. Several studies have detected an increase in cellular ROS during heat stress (e.g., Ikeda et al., 1999; Abele et al., 2002; Venkataraman et al., 2004). Elevated levels of free radicals and ROS can in turn exacerbate mitochondrial dysfunction, including inactivation of complex I (Bautista et al., 2000; Riobó et al., 2001).

One ROS whose production is increased by mitochondrial damage is superoxide, which is converted to hydrogen peroxide. Other radicals (e.g., •OH) can also be produced if the catalase and glutathione peroxidase ROS defense systems become overwhelmed. Free radicals can induce DNA damage that leads to apoptosis. However, it is not yet clear whether increased ROS production contributes significantly to hyperthermia-induced neuronal damage. Our preliminary data show a $31.9 \pm 21.1\%$ ($n = 6$) decrease in glutathione levels 1 h following heat stress in cultured neurons, suggesting a reduced ability to scavenge free radicals. However, preliminary data also indicate that gamma-glutamate cysteine, a precursor for reduced glutathione, does not protect these neurons from a heat stress. In addition there was no significant protection by free radical scavengers and anti-oxidants such as N-aceytl-cysteine and 1%

Table 1. Drugs tested for protection against heat stress

	Drugs (μM)	Percentage change from non-treated heat-stressed control	Drug/no drug ratio
Caspase inhibitors	qVD-OPH (20)***	190.48 ± 38.48	2.90
	zVAD-fmk (100)***	179.61 ± 36.14	2.80
PARP inhibitor	DPQ (50)	21.07 ± 38.22	1.21
Protein synthesis inhibitor	Cycloheximide (20 μg/ml)	5.37 ± 5.07	1.05
p38 inhibitor	SB203580 (0.6)	−3.96 ± 5.13	0.96
Anti-oxidants	N-acetyl cysteine (10)	9.35 ± 5.38	1.07
	PBN (500)	2.85 ± 6.43	1.02
	MnTBAP (100)	−3.28 ± 12.73	0.97
	DMSO (1%)	−1.68 ± 6.11	0.98

Note: Pharmacologic agents were added to culture wells at various concentrations before or after the heat stress. The ability of the drugs to affect survival was evaluated from 24 to 52 h following the cessation of the stress as described in White et al. (2003). Cultures treated with pan-caspase inhibitors had statistically better saving compared with non-treated cultures. No significant difference in survival was seen in cultures treated with free radical scavengers or inhibitors of PARP, p38, or protein synthesis. Results shown use drug concentrations that gave maximal protection from heat stress.
***$p < 0.0005$, unpaired two-tailed t-test.

dimethylsulfoxide (DMSO), by the spin trap reagent PBN, by the superoxide dismutase mimetic and peroxynitrite scavenger MnTBAP (Table 1), or by reducing the partial pressure of oxygen during the heat stress (unpublished observation).

Protein misfolding may contribute to hyperthermia-induced neuronal death

Certain temperature elevations produce cellular damage only when present for long times, suggesting that heat-induced damage cumulates over time. One candidate for this cumulative damage is unfolding or misfolding of proteins. Many proteins begin to be denatured even at temperatures below 37°C (Rees and Robertson, 2001), but at this normal body temperature the rate at which they are renatured or replaced balances this slow denaturing rate. Perhaps the threshold temperature for hyperthermia-induced cell damage is the temperature at which the rate of denaturing processes exceeds the rate of renaturing or replacement of some heat-labile proteins. In the nervous system, this threshold temperature is likely to depend on the type of neuron, its stage of development, and its expression of chaperone proteins (see later section), as well as on other stresses that might be present. Thus rather than an invariant threshold for damage, the relationship between hyperthermia and damage is likely to depend on factors that influence the balance between denaturing and repair processes.

As diagrammed in Fig. 7, heat-induced protein denaturation can contribute to cell death by inactivating functionally important proteins, by overwhelming normal mechanisms for protein degradation (proteosomes, lysosomes), and/or by toxic properties acquired by abnormal denatured proteins and/or abnormal protein aggregates. Newly synthesized proteins and proteins in the process of being synthesized on ribosomes are thought to be especially sensitive to damage from hyperthermia. Since the ER is involved in synthesis, post-translational modification, oligomerization, and folding of proteins (see review by Helenius et al., 1992), it would be expected to be especially vulnerable to heat-induced denaturation and misfolding of proteins. Indeed, hyperthermia has been shown to produce ER stress by denaturing newly synthesized proteins in some cell types (Ghibelli et al., 1992). Under these stress conditions the unfolded protein response (UPR) can be triggered in the ER, causing transcriptional upregulation of several proteins including chaperone proteins. Accumulation of denatured and misfolded proteins in the ER can induce an apoptotic pathway involving CHOP/GADD153 and activation of caspase-12 (van der Sanden et al., 2003). More severe ER stress suppresses protein translocation, which may also induce apoptotic mechanisms (Ron, 2002).

Blocking protein synthesis before and during heat stress prevents this accumulation of proteins in the ER and robustly saves some heat-stressed

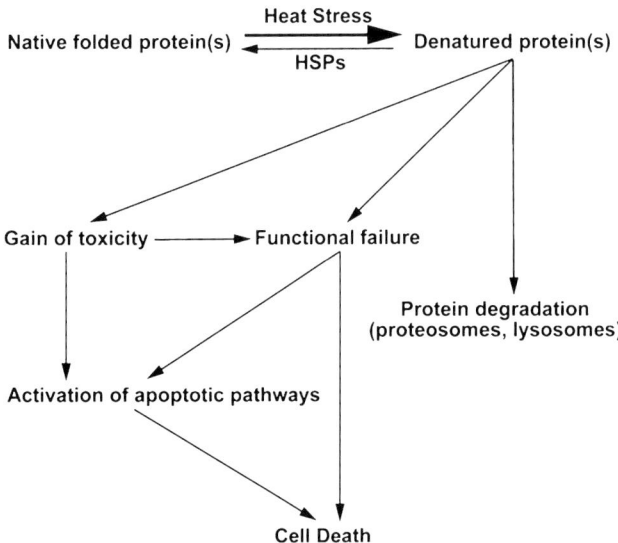

Fig. 7. Possible mechanisms of cellular damage from denatured proteins. Heat stress increases denaturation of both nascent proteins on the ribosomes and mature proteins.

cells, including fibroblasts (Lee and Dewey, 1986, 1987; Lee et al., 1990, 1991; Michels et al., 2000). However, the protein synthesis inhibitors puromycin and cycloheximide failed to improve survival rates for heat-stressed cultured neurons (Table 1). Cycloheximide added following the heat stress also failed to protect, suggesting that for these neurons, hyperthermia-induced death does not require protein synthesis.

Accumulation of aggregates of abnormal proteins is thought to contribute to neuronal damage in several chronic diseases of the nervous system including Alzheimer's, Parkinson's and Huntington's diseases, and amyotrophic lateral sclerosis (reviewed by Muchowski and Wacker, 2005). However, in these chronic diseases the accumulation of protein aggregates occurs over a much slower time course than any aggregate accumulation induced by hyperthermia and/or energy stresses.

Molecular chaperone and protein recycling systems normally protect against hyperthermia

Hyperthermia can increase expression of heat shock proteins (HSPs), which help protect cells from heat stress by acting as chaperones that stabilize properly folded protein conformations and help refold denatured proteins. Many of these HSPs act together in complexes. For example, HSP10 and HSP60 form a complex that helps stabilize proteins within mitochondria. The HSP40 and HSP70 families work both independently and together to help prevent misfolding of nascent polypeptide chains on cytoplasmic and endoplasmic reticulum-associated ribosomes (reviewed in Borges and Ramos, 2005). HSP70 and HSP40 family members working together use ATP to provide the necessary energy for refolding proteins (Kelley, 1998; Fan et al., 2003). The expression of particular HSPs in the nervous system is heterogeneous. For example, HSP70 family members such as HSP12A and HSP12B are expressed more in particular neuronal populations (e.g., Pongrac et al., 2004). The histogram in Fig. 8 shows that HSP27, HSP60, and HSP70 were detected by immunohistochemistry in some but not all stressed septal neurons grown in culture for 4 weeks. These HSPs were hard to detect in younger cultures.

HSP70 overexpression protects dorsal root ganglion neurons from heat stress (Uney et al., 1993), and protects hippocampal neurons from ischemic

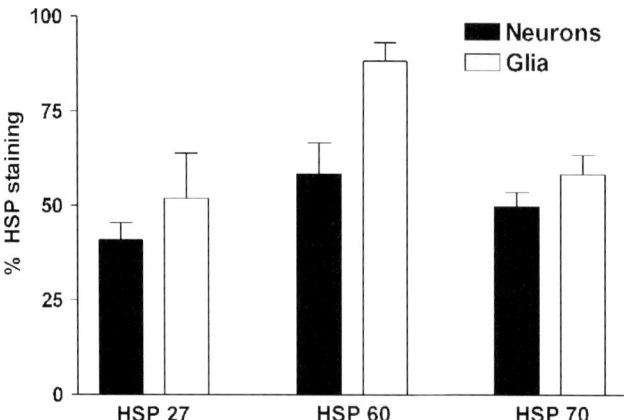

Fig. 8. Heat shock protein (HSP) expression assayed in 4-week-old septal cultures 6 days after a 2 h 43°C stress. Counts of cells stained immunohistochemically for HSP27, HSP60, or HSP70 are expressed as a percentage of the total number of surviving neurons (NeuN-positive) and astrocytes (GFAP-positive). Only about half the neurons expressed detectable levels of any one of these HSPs. Costaining (not shown) indicated that individual neurons or glia often showed detectable levels of one, but not all, of these HSPs.

brain damage (Giffard et al., 2004). HSPs also protect against a variety of other neuronal stresses (Magrane et al., 2004; Dong et al., 2005).

Possible contribution of Fas death receptors to hyperthermia-induced damage

Perhaps the most intensively studied mechanisms of apoptosis are those triggered by stimulating the Fas death receptor. Fas-dependent apoptosis is involved in heat-induced apoptosis in testicular germ cells (Yin et al., 2002). The Fas receptor, a member of the tumor necrosis factor receptor family, recruits the FADD adapter molecule that binds procaspase-8, forming the death-inducing signaling complex (DISC). While associated with the DISC, procaspase-8 and/or procaspase-10 is cleaved into its activated form, initiating a proteolytic cascade resulting in caspase-3 activation and death (reviewed by Gupta, 2003).

Fas receptor-mediated death is often referred to as an extrinsic mechanism of apoptosis since binding of an extracellular ligand to the Fas receptor can initiate the apoptotic cascade. However, heat-induced apoptotic mechanisms in neuronal cultures do not appear to be triggered by any stable factor secreted into the medium, since addition of medium conditioned by heat-stressed neurons did not induce damage in non-stressed cultures. Non-stressed cells receiving heat-stressed cell conditioned media had no significant increase in survival ($10.9 \pm 15.0\%$, $n = 12$). Likewise, replacing the medium of heated cultures with medium from non-heated cultures does not reduce heat-induced damage. Heat-stressed cells receiving non-stressed cell conditioned media had no significant change in survival ($15.0 \pm 8.8\%$, $n = 12$).

As mentioned above, heat stress downregulates FLIP, which interferes with the interaction between caspase-8 and FADD, thus sensitizing cells to Fas receptor-mediated apoptosis (Tran et al., 2003). Perhaps levels of Fas receptor agonist in neuronal cultures are sufficient to induce apoptosis in neurons that have undergone heat stress. Alternatively the Fas receptor ligand may not be required for initiating the caspase-8 pathway. In Jurkat and cancer cell lines caspase-8 can also be activated by cytotoxic drugs in a manner independent of the Fas/FADD receptor complex (Ferrari et al., 1998; Ferreira et al., 2000).

Possible contribution of DNA damage to hyperthermia-induced death

ROS (see above) can produce DNA damage and strand breaks. In addition, topoisomerase I, an enzyme involved in DNA repair, is heat-sensitive and its inactivation could prevent correction of

DNA damage (Ciavarra et al., 1992). DNA damage can lead to apoptosis by several mechanisms including overactivation of PARP (see above), an enzyme that uses NAD as a substrate and is activated by binding to DNA breaks. Depletion of NAD by PARP can lead to apoptosis (Berger, 1985; Skaper, 2003). PARP activation in neurons can also induce the release of apoptotic inducing factor (AIF) from mitochondria, resulting in a non-caspase mediated apoptotic death (Cheung et al., 2005). Nevertheless, our preliminary data suggest that the PARP inhibitor DPQ (3,4-dihydro-5-[4(1-piperidinyl)butoxy]-1(2H)-isoquinolinone), which protects some cells from DNA damage-induced apoptosis (Takahashi et al., 1999; Yu et al., 2002; Meli et al., 2004; Wang et al., 2004; Zong et al., 2004), does not significantly protect cultured neurons from heat stress (Table 1).

The tumor suppressor p53 can lead to cell cycle arrest (allowing repair of DNA), but can also induce apoptosis via Bax translocation to the mitochondrial membrane (Cregan et al., 1999). P53 participates in heat-induced apoptosis in glioblastoma cells (Ohnishi et al., 1996, 1998) and fibroblasts (Matsumoto et al., 1997). The p53 inhibitor pifithrin-α protects several cell types from p53-mediated apoptosis (Komarov et al., 1999; Culmsee et al., 2001; Toillon et al., 2002; Dagher, 2004). However, our preliminary data indicate that pifithrin-α does not enhance survival or reduce caspase activation in cultured neurons following heat stress (Fig. 3D).

Some forms of apoptotic death in neurons, including those initiated by DNA damage, involve activation of the cell cycle (Padmanabhan et al., 1999; Park et al., 2000). Activated cyclin dependent kinases (cdk) 2, 4, or 6 can phosphorylate the tumor suppressor retinoblastoma protein (Rb) that controls cell cycle entry at the G_1-restriction point and can lead to apoptosis in some circumstances (Herwig and Strauss, 1997). However, our preliminary data indicate that neither inhibitors of cdks (indirubin-3′-monoxime, butyrolactone-1, Tat-LFG, fascaplysin, see Kitagawa et al., 1993; Chen et al., 1999; Knockaert et al., 2002), nor overexpression of non-phosphorylatable Rb can substantially protect cultured neurons from heat stress (not shown).

Possible contribution of cytoskeletal damage to hyperthermia-induced death

Hyperthermia can affect the cytoskeleton (Wachsberger and Coss, 1990; Coss and Linnemans, 1996; Morimoto et al., 1997), and as mentioned above one of the first signs of hyperthermia-induced damage is blebbing of neuronal processes (axon and dendrites, see White et al., 2003). Disruption of the cytoskeleton can lead to apoptosis via activation of SAPK2/p38 mitogen-activated protein (MAP) kinase (Garcia et al., 2002). Thus cytoskeletal damage is a possible mechanism underlying heat-induced apoptosis.

Bim, a BH3-only domain member of the Bcl-2 family of proteins, is attached to the microtubule-associated dynein motor complex via dynein light chain-1. Bim and the related protein Bmf are thought to serve as monitors of cytoskeletal integrity (Bouillet et al., 1999; Puthalakath et al., 2001; Chen et al., 2002; Fukazawa et al., 2004). During stress, Bim can be phosphorylated by c-jun N-terminal kinase (JNK), releasing Bim from the dynein motor complex (Lei and Davis, 2003; Putcha et al., 2003) and allowing Bim to translocate to mitochondria (Puthalakath et al., 1999; Andoniou et al., 2004). Bim also associates with and antagonizes the anti-apoptotic proteins Bcl-2 and Bcl-x_L (Puthalakath et al., 1999; Terradillos et al., 2002; Morishima et al., 2004).

Furthermore, the cdk5/p35 complex is involved in regulating the actin cytoskeleton in neuronal processes (Nikolic et al., 1998), and cdk5 promotes apoptosis in heated astrocytoma cells (Gao et al., 2001). Therefore, cdk5 could participate in heat stress-induced neuronal apoptosis initiated by damage to the cytoskeleton.

Possible involvement of MAP and JNK kinase pathways in hyperthermia-induced neuronal damage

Several intracellular pathways might be triggered by heat stress and contribute to activating apoptotic pathways. Pathways involving MAP kinases such as SAPK2/p38 and JNK (both mentioned above) and extracellular signal-regulated kinase (ERK) can be activated by heat stress (Dorion and Landry, 2002), and can be either pro- or anti-apoptotic under different circumstances (Chang and Karin, 2001).

As mentioned above, SAPK2/p38 is activated by cytoskeletal disruption. In addition, SAPK2/p38 induction by heat can be triggered by glutathione S-transferase Mu1-1 (GSTM1-1) separation from apoptosis signal-regulating kinase-1 (ASK-1, Dorion et al., 2002). Dorion and Landry (2002) conjectured that heat-induced release of hydrophobic compounds such as sphingosine and ceramide may act to titrate out GSTM1-1, thus freeing ASK-1 for activation. They also speculated that mammalian cells might have a general heat sensor similar to the transmembrane protein Wsc1-Hcs77 which acts as a heat shock and cell wall mechanosensor in yeast (Philip and Levin, 2001). In this scenario, heat-induced changes in membrane fluidity could be monitored by a protein similar to Wsc1-Hsc77 and could lead to SAPK2/p38 mediated apoptosis. Another possible mechanism involves capsaicin, which induces phosphorylation of p38 (Sweitzer et al., 2004; Mizushima et al., 2005) and binds to the vanilloid TRPV1 ion channel receptor, which is activated at elevated temperatures (Caterina et al., 1997). TRPV1 channel RNA is found in the brain (Mezey et al., 2000; Roberts et al., 2004) and thus may play a role in p38 activation-induced apoptosis in neurons. However, our preliminary data suggest that pharmacological inhibitors of p38 (SB239063 and SB203580) do not protect cultured neurons from heat stress (Table 1).

Prevention of JNK dephosphorylation due to direct inhibition of JNK phosphatases by heat stress may trigger JNK activation-induced apoptosis (Meriin et al., 1999). Activation of the JNK pathway has also been linked to DNA damage in neurons (Ghahremani et al., 2002) and to ER stress via recruitment of TNF receptor-associated factor 2 (TRAF2) by the IRE1-α stress sensor protein (Urano et al., 2000). In cardiac myocyte reoxygenation studies, electron transport-coupled calcium influx and mitochondrial-generated ROS provide the initial stimulation of JNK via proline-rich tyrosine kinase 2 (Pyk2, Dougherty et al., 2004). Thus heat itself and/or hyperthermia-induced DNA damage, ER stress, and/or mitochondrial dysfunction might lead to JNK activation. Despite this, our preliminary data suggest that a JNK inhibitor, SB600125, does not protect cultured neurons from heat stress (not shown).

Activation of ERK in NIH-3T3 fibroblasts by hyperthermia produces agonist-independent phosphorylation and activation of the epidermal growth factor receptor (Lin et al., 1997; Ng and Bogoyevitch, 2000). ERK is also known to be activated following cytoskeletal disruption (Irigoyen et al., 1997), DNA damage (Tang et al., 2002), and oxidative stress (Houle et al., 2003). However, ERK's role in heat-induced neuronal apoptosis is undetermined.

Additionally, the glycogen synthase kinase 3 (GSK-3) apoptotic pathway has been implicated in heat stress in neuroblastoma cells (Bijur et al., 2000). However, the GSK-3 inhibitors valproate and SB415286 did not protect neuronal cultures from heat stress (Fig. 9).

Mechanisms underlying the synergistic damaging effects of combined hyperthermia and ischemia

As mentioned previously, hyperthermia greatly increases brain damage following stroke/ischemia and hypothermia reduces stroke/ischemia-induced brain damage in animal models. The mechanisms for heat-induced neuronal damage considered above suggest several possible explanations for the deadly synergy between ischemia and hyperthermia. Like heat, ischemia can denature proteins and lead to protein aggregation within neurons in animal models (Hu et al., 2000, 2001; Ouyang and Hu, 2001). Renaturing of proteins by cellular chaperone systems requires ATP (Hartl and Hayer-Hartl, 2002; Fan et al., 2003). Thus protein denaturation may contribute to the synergistic damaging effects when an energy stress is combined with hyperthermia. Synergistic interactions causing proteosome overload and/or damage to lysosomes are also possible.

Both hyperthermia and ischemia damage mitochondria. Ischemia produces mitochondrial damage involving calcium overload, free radical damage, calpain-dependent damage (Blomgren et al., 2001) and opening of the MPTP (Siesjo et al., 1999). Mitochondria loaded with calcium during ischemia might release that Ca^{2+} into the

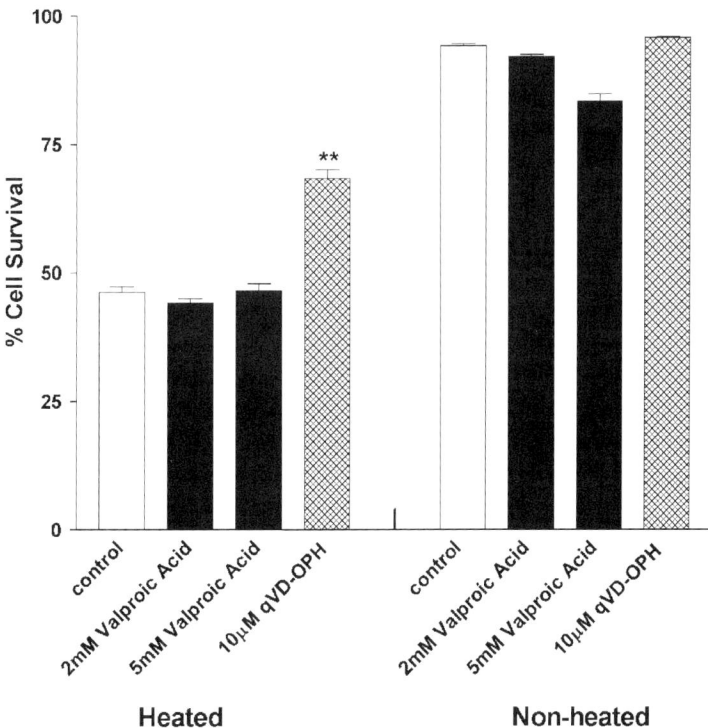

Fig. 9. Valproate, an inhibitor of glycogen synthase kinase (GSK)-3, does not enhance survival of striatal neurons assayed 47 h following cessation of a 2 h 43°C stress. Survival was determined as described in White et al. (2003). The pan-caspase inhibitor qVD-OPH did enhance survival. Another GSK-3 inhibitor, SB415286, also failed to protect (not shown). Mean ± SEM, $n = 6$–12 culture wells. **$p < 0.001$. Stress administered after 8 days $in\ vitro$.

cytosol because of heat stress-induced mitochondrial depolarization, thus activating calpain-like proteases.

Conclusions

Both $in\ vivo$ and $in\ vitro$ studies demonstrate that neurons are especially vulnerable to heat stress. Damage depends on neuronal maturity, on the degree and the duration of the temperature elevation, and on whether or not other stresses are also present. Temperature elevations that by themselves cause no damage can become damaging in combination with other stresses involving disrupted energy metabolism (ischemia, hypoglycemia) or in conditions where the natural defense mechanisms against heat stress are compromised. Stresses involving DNA damage or abnormal accumulation and aggregation of proteins would also be expected to increase hyperthermia-induced damage. Work summarized in this chapter suggests that denatured proteins and mitochondrial damage contribute to heat-induced neuronal death. Much more work remains to be done to clarify these (and perhaps other) mechanisms, with the eventual goal of therapeutic treatments to mitigate hyperthermia-induced neuronal damage.

Abbreviations

AIF	apoptosis inducing factor
AMPA	alpha-amino-3-hydroxy-5-methyl-4-isoxazole-4-propionate
APAF-1	apoptotic protease activating factor 1
ASK-1	apoptosis signal-regulating kinase-1
ATP	adenosine triphosphate

Bad	Bcl-x$_L$/Bcl-2 associated death promoter
Bax	Bcl-2-associated X protein
Bcl-2	B-cell lymphoma protein 2
BH3	Bcl-2 homology 3
Bid	BH3 interacting domain death agonist
Bmf	Bcl-2 modifying factor
cdk	cyclin-dependent kinase
CNQX	6-cyano-7-nitroquinoxaline-2,3-dione disodium
DMSO	dimethylsulfoxide
DEVD-CHO	Ac-Asp-Glu-Val-Asp-aldehyde
DISC	death-inducing signaling complex
ER	endoplasmic reticulum
GSTM1-1	glutathione S-transferase Mu1-1
ERK	extracellular signal-regulated kinase
FADD	Fas-associating protein with death domain
FLIP	FLICE-inhibitory proteins
GFAP	glial fibrillary acidic protein
GSH	glutathione
GSK-3	glycogen synthase kinase 3
HSP	heat shock protein
IRE1-α	inositol-requiring ER-to-nucleus signal kinase 1 alpha
JNK	c-jun N-terminal kinase
MAP kinase	mitogen activated protein kinase
MK801	(5S,10R)-(+)-5-methyl-10,11-dihydro-5H-dibenzo[a,d]-cyclohepten-5,10-imine maleate
MnTBAP	manganese(III) tetrakis (4-benzoic acid) porphyrin
MPTP	mitochondrial permeability transition pore
NAD	nicotinamide adenine dinucleotide (oxidized)
NADH	nicotinamide adenine dinucleotide (reduced)
NeuN	neuron-specific nuclear protein
NMDA	N-methyl-D-aspartate
PARP	poly(ADP-ribose) polymerase
PBN	N-tert-butyl-α-phenylnitrone
Pyk2	proline-rich tyrosine kinase 2
qVD-OPH	quinolyl-valyl-O-methyl-aspartyl-[2,6-difluorophenoxy]l-methyl ketone
Rb	retinoblastoma protein
ROS	reactive oxygen species
SAPK	stress-activated protein kinase
tBid	truncated Bid
TMRM	tetramethyl rhodamine methyl ester
TRAF2	TNF-receptor-associated factor 2
TRPV1	transient receptor potential channel type V1
TUNEL	terminal deoxynucleotidyl transferase biotin-dUTP nick end labeling
UPR	unfolded protein response
zDEVD-AFC	carbobenzoxy-Asp-Glu-Val-Asp-7-amino-4-trifluoromethylcoumarin
zVAD-fmk	Z-Val-Ala-Asp(OMe)-fluoromethyl ketone

Acknowledgments

We thank Jessica Kos and Chris Granville for their help in collecting and analyzing data and Dr. Laurie Stepanek for evaluation of the manuscript. This work was supported by NIH grants NS 12404 and NS 12207 and a grant from the American Heart Association (Florida). MGW was supported by NIH 2T32 NS 07044. LEL was supported by fellowships from the American Heart Association and the University of Miami Center on Aging. MGW and LEL were Lois Pope Leaders in Furthering Education Fellows.

References

Abele, D., Heise, K., Portner, H.O. and Puntarulo, S. (2002) Temperature-dependence of mitochondrial function and production of reactive oxygen species in the intertidal mud clam *Mya arenaria*. J. Exp. Biol., 205: 1831–1841.

Andoniou, C.E., Andrews, D.M., Manzur, M., Ricciardi-Castagnoli, P. and Degli-Esposti, M.A. (2004) A novel checkpoint in the Bcl-2-regulated apoptotic pathway revealed by murine cytomegalovirus infection of dendritic cells. J. Cell Biol., 166: 827–837.

Baena, R.C., Busto, R., Dietrich, W.D., Globus, M.Y. and Ginsberg, M.D. (1997) Hyperthermia delayed by 24 hours aggravates neuronal damage in rat hippocampus following global ischemia. Neurology, 48: 768–773.

Barone, F.C., Feuerstein, G.Z. and White, R.F. (1997) Brain cooling during transient focal ischemia provides complete neuroprotection. Neurosci. Biobehav. Rev., 21: 31–44.

Bautista, J., Corpas, R., Ramos, R., Cremades, O., Gutierrez, J.F. and Alegre, S. (2000) Brain mitochondrial complex I inactivation by oxidative modification. Biochem. Biophys. Res. Commun., 275: 890–894.

Berger, N.A. (1985) Poly(ADP-ribose) in the cellular response to DNA damage. Radiat. Res., 101: 4–15.

Bijur, G.N., De Sarno, P. and Jope, R.S. (2000) Glycogen synthase kinase-3beta facilitates staurosporine- and heat shock-induced apoptosis. Protection by lithium. J. Biol. Chem., 275: 7583–7590.

Bijur, G.N. and Jope, R.S. (2001) Proapoptotic stimuli induce nuclear accumulation of glycogen synthase kinase-3 beta. J. Biol. Chem., 276: 37436–37442.

Blomgren, K., Zhu, C., Wang, X., Karlsson, J.O., Leverin, A.L., Bahr, B.A., Mallard, C. and Hagberg, H. (2001) Synergistic activation of caspase-3 by m-calpain after neonatal hypoxia-ischemia: a mechanism of "pathological apoptosis"? J. Biol. Chem., 276: 10191–10198.

Borges, J.C. and Ramos, C.H. (2005) Protein folding assisted by chaperones. Protein Pept. Lett., 12: 257–261.

Bouchama, A. and Knochel, J.P. (2002) Heat stroke. N. Engl. J. Med., 346: 1978–1988.

Bouillet, P., Metcalf, D., Huang, D.C., Tarlinton, D.M., Kay, T.W., Kontgen, F., Adams, J.M. and Strasser, A. (1999) Proapoptotic Bcl-2 relative Bim required for certain apoptotic responses, leukocyte homeostasis, and to preclude autoimmunity. Science, 286: 1735–1738.

Brown, P.L. and Kiyatkin, E.A. (2004) Brain hyperthermia induced by MDMA (ecstasy): modulation by environmental conditions. Eur. J. Neurosci., 20: 51–58.

Buckiova, D. and Brown, N.A. (1999) Mechanism of hyperthermia effects on CNS development: rostral gene expression domains remain, despite severe head truncation; and the hindbrain/otocyst relationship is altered. Teratology, 59: 139–147.

Busto, R., Dietrich, W.D., Globus, M.Y., Valdes, I., Scheinberg, P. and Ginsberg, M.D. (1987) Small differences in intraischemic brain temperature critically determine the extent of ischemic neuronal injury. J. Cereb. Blood Flow Metab., 7: 729–738.

Candé, C., Cecconi, F., Dessen, P. and Kroemer, G. (2002) Apoptosis-inducing factor (AIF): key to the conserved caspase-independent pathways of cell death? J. Cell Sci., 115: 4727–4734.

Carson, D.A., Seto, S., Wasson, D.B. and Carrera, C.J. (1986) DNA strand breaks, NAD metabolism, and programmed cell death. Exp. Cell Res., 164: 273–281.

Castillo, J., Davalos, A., Marrugat, J. and Noya, M. (1998) Timing for fever-related brain damage in acute ischemic stroke. Stroke, 29: 2455–2460.

Castillo, J., Davalos, A. and Noya, M. (1999) Aggravation of acute ischemic stroke by hyperthermia is related to an excitotoxic mechanism. Cerebrovasc. Dis., 9: 22–27.

Caterina, M.J., Schumacher, M.A., Tominaga, M., Rosen, T.A., Levine, J.D. and Julius, D. (1997) The capsaicin receptor: a heat-activated ion channel in the pain pathway. Nature, 389: 816–824.

Cawdell-Smith, J., Upfold, J., Edwards, M. and Smith, M. (1992) Neural tube and other developmental anomalies in the guinea pig following maternal hyperthermia during early neural tube development. Teratog. Carcinog. Mutagen, 12(1): 1–9.

Chang, C.P., Lee, C.C., Chen, S.H. and Lin, M.T. (2004) Aminoguanidine protects against intracranial hypertension and cerebral ischemic injury in experimental heatstroke. J. Pharmacol. Sci., 95: 56–64.

Chang, L. and Karin, M. (2001) Mammalian MAP kinase signalling cascades. Nature, 410: 37–40.

Chauvin, C., De Oliveira, F., Ronot, X., Mousseau, M., Leverve, X. and Fontaine, E. (2001) Rotenone inhibits the mitochondrial permeability transition-induced cell death in U937 and KB cells. J. Biol. Chem., 276: 41394–41398.

Chen, D., Wang, M., Zhou, S. and Zhou, Q. (2002) HIV-1 Tat targets microtubules to induce apoptosis, a process promoted by the pro-apoptotic Bcl-2 relative Bim. EMBO J., 21: 6801–6810.

Chen, H., Chopp, M. and Welch, K.M. (1991) Effect of mild hyperthermia on the ischemic infarct volume after middle cerebral artery occlusion in the rat. Neurology, 41: 1133–1135.

Chen, Q., Chopp, M., Bodzin, G. and Chen, H. (1993) Temperature modulation of cerebral depolarization during focal cerebral ischemia in rats: correlation with ischemic injury. J. Cereb. Blood Flow Metab., 13: 389–394.

Chen, Y.N., Sharma, S.K., Ramsey, T.M., Jiang, L., Martin, M.S., Baker, K., Adams, P.D., Bair, K.W. and Kaelin Jr., W.G. (1999) Selective killing of transformed cells by cyclin/cyclin-dependent kinase 2 antagonists. Proc. Natl. Acad. Sci. U.S.A., 96: 4325–4329.

Cheung, E.C., Melanson-Drapeau, L., Cregan, S.P., Vanderluit, J.L., Ferguson, K.L., McIntosh, W.C., Park, D.S., Bennett, S.A. and Slack, R.S. (2005) Apoptosis-inducing factor is a key factor in neuronal cell death propagated by BAX-dependent and BAX-independent mechanisms. J. Neurosci., 25: 1324–1334.

Chia, S.E. and Teo, K.J. (2003) Prognosis of adult men with heat exhaustion with regard to postural stability and neurobehavioral effects: a 6-month follow-up study. Neurotoxicol. Teratol., 25: 503–508.

Chopp, M., Welch, K.M., Tidwell, C.D., Knight, R. and Helpern, J.A. (1988) Effect of mild hyperthermia on recovery of metabolic function after global cerebral ischemia in cats. Stroke, 19: 1521–1525.

Churn, S.B., Taft, W.C., Billingsley, M.S., Blair, R.E. and DeLorenzo, R.J. (1990) Temperature modulation of ischemic neuronal death and inhibition of calcium/calmodulin-dependent protein kinase II in gerbils. Stroke, 21: 1715–1721.

Ciavarra, R.P., Duvall, W. and Castora, F.J. (1992) Induction of thermotolerance in T cells protects nuclear DNA topoisomerase I from heat stress. Biochem. Biophys. Res. Commun., 186: 166–172.

Coss, R.A. and Linnemans, W.A. (1996) The effects of hyperthermia on the cytoskeleton: a review. Int. J. Hyperthermia, 12: 173–196.

Cregan, S.P., Dawson, V.L. and Slack, R.S. (2004) Role of AIF in caspase-dependent and caspase-independent cell death. Oncogene, 23: 2785–2796.

Cregan, S.P., Fortin, A., MacLaurin, J.G., Callaghan, S.M., Cecconi, F., Yu, S.W., Dawson, T.M., Dawson, V.L., Park, D.S., Kroemer, G. and Slack, R.S. (2002) Apoptosis-inducing factor is involved in the regulation of caspase-independent neuronal cell death. J. Cell Biol., 158: 507–517.

Cregan, S.P., MacLaurin, J.G., Craig, C.G., Robertson, G.S., Nicholson, D.W., Park, D.S. and Slack, R.S. (1999) Bax-dependent caspase-3 activation is a key determinant in p53-induced apoptosis in neurons. J. Neurosci., 19: 7860–7869.

Culmsee, C., Zhu, X., Yu, Q.S., Chan, S.L., Camandola, S., Guo, Z., Greig, N.H. and Mattson, M.P. (2001) A synthetic inhibitor of p53 protects neurons against death induced by ischemic and excitotoxic insults, and amyloid beta-peptide. J. Neurochem., 77: 220–228.

Dagher, P.C. (2004) Apoptosis in ischemic renal injury: roles of GTP depletion and p53. Kidney Int., 66: 506–509.

Desagher, S. and Martinou, J.C. (2000) Mitochondria as the central control point of apoptosis. Trends Cell Biol., 10: 369–377.

Dietrich, W.D., Busto, R., Halley, M. and Valdes, I. (1990) The importance of brain temperature in alterations of the blood-brain barrier following cerebral ischemia. J. Neuropathol. Exp. Neurol., 49: 486–497.

Dietrich, W.D., Halley, M., Valdes, I. and Busto, R. (1991) Interrelationships between increased vascular permeability and acute neuronal damage following temperature-controlled brain ischemia in rats. Acta Neuropathol. (Berl.), 81: 615–625.

Dippel, D.W., van Breda, E.J., van Gemert, H.M., van der Worp, H.B., Meijer, R.J., Kappelle, L.J. and Koudstaal, P.J. (2001) Effect of paracetamol (acetaminophen) on body temperature in acute ischemic stroke: a double-blind, randomized phase II clinical trial. Stroke, 32: 1607–1612.

Dong, Z., Wolfer, D.P., Lipp, H.P. and Bueler, H. (2005) Hsp70 gene transfer by adeno-associated virus inhibits MPTP-induced nigrostriatal degeneration in the mouse model of Parkinson disease. Mol. Ther., 11: 80–88.

Dorion, S., Lambert, H. and Landry, J. (2002) Activation of the p38 signaling pathway by heat shock involves the dissociation of glutathione S-transferase Mu from Ask1. J. Biol. Chem., 277: 30792–30797.

Dorion, S. and Landry, J. (2002) Activation of the mitogen-activated protein kinase pathways by heat shock. Cell Stress Chaperones, 7: 200–206.

Dougherty, C.J., Kubasiak, L.A., Frazier, D.P., Li, H., Xiong, W.C., Bishopric, N.H. and Webster, K.A. (2004) Mitochondrial signals initiate the activation of c-Jun N-terminal kinase (JNK) by hypoxia-reoxygenation. FASEB J., 18: 1060–1070.

Dunn, S.R., Thomason, J.C., Le Tissier, M.D. and Bythell, J.C. (2004) Heat stress induces different forms of cell death in sea anemones and their endosymbiotic algae depending on temperature and duration. Cell Death Differ., 11: 1213–1222.

Edwards, M.J. (1998) Apoptosis, the heat shock response, hyperthermia, birth defects, disease and cancer. Where are the common links? Cell Stress Chaperones, 3: 213–220.

Edwards, M.J., Saunders, R.D. and Shiota, K. (2003) Effects of heat on embryos and foetuses. Int. J. Hyperthermia, 19: 295–324.

Eguchi, Y., Yamashita, K., Iwamoto, T. and Ito, H. (1997) Effects of brain temperature on calmodulin and microtubule-associated protein 2 immunoreactivity in the gerbil hippocampus following transient forebrain ischemia. J. Neurotrauma, 14: 109–118.

Everett, T.H.t., Nath, S., Lynch III, C., Beach, J.M., Whayne, J.G. and Haines, D.E. (2001) Role of calcium in acute hyperthermic myocardial injury. J. Cardiovasc. Electrophysiol., 12: 563–569.

Fan, C.Y., Lee, S. and Cyr, D.M. (2003) Mechanisms for regulation of Hsp70 function by Hsp40. Cell Stress Chaperones, 8: 309–316.

Ferrari, D., Stepczynska, A., Los, M., Wesselborg, S. and Schulze-Osthoff, K. (1998) Differential regulation and ATP requirement for caspase-8 and caspase-3 activation during CD95- and anticancer drug-induced apoptosis. J. Exp. Med., 188: 979–984.

Ferreira, C.G., Span, S.W., Peters, G.J., Kruyt, F.A. and Giaccone, G. (2000) Chemotherapy triggers apoptosis in a caspase-8-dependent and mitochondria-controlled manner in the non-small cell lung cancer cell line NCI-H460. Cancer Res., 60: 7133–7141.

Fiskum, G. (2004) Mechanisms of neuronal death and neuroprotection. J. Neurosurg. Anesthesiol., 16: 108–110.

Fontaine, E. and Bernardi, P. (1999) Progress on the mitochondrial permeability transition pore: regulation by complex I and ubiquinone analogs. J. Bioenerg. Biomembr., 31: 335–345.

Fontaine, E., Eriksson, O., Ichas, F. and Bernardi, P. (1998) Regulation of the permeability transition pore in skeletal muscle mitochondria. Modulation By electron flow through the respiratory chain complex i. J. Biol. Chem., 273: 12662–12668.

Fukazawa, H., Noguchi, K., Masumi, A., Murakami, Y. and Uehara, Y. (2004) BimEL is an important determinant for induction of anoikis sensitivity by mitogen-activated protein/extracellular signal-regulated kinase kinase inhibitors. Mol. Cancer Ther., 3: 1281–1288.

Gao, C., Negash, S., Wang, H.S., Ledee, D., Guo, H., Russell, P. and Zelenka, P. (2001) Cdk5 mediates changes in morphology and promotes apoptosis of astrocytoma cells in response to heat shock. J. Cell Sci., 114: 1145–1153.

Garcia, J.G., Wang, P., Schaphorst, K.L., Becker, P.M., Borbiev, T., Liu, F., Birukova, A., Jacobs, K., Bogatcheva, N. and Verin, A.D. (2002) Critical involvement of p38 MAP kinase in pertussis toxin-induced cytoskeletal reorganization and lung permeability. FASEB J., 16: 1064–1076.

Georgiadis, D., Schwarz, S., Kollmar, R. and Schwab, S. (2001) Endovascular cooling for moderate hypothermia in patients with acute stroke: first results of a novel approach. Stroke, 32: 2550–2553.

Ghahremani, M.H., Keramaris, E., Shree, T., Xia, Z., Davis, R.J., Flavell, R., Slack, R.S. and Park, D.S. (2002) Interaction of the c-Jun/JNK pathway and cyclin-dependent kinases in death of embryonic cortical neurons evoked by DNA damage. J. Biol. Chem., 277: 35586–35596.

Ghibelli, L., Nosseri, C., Oliverio, S., Piacentini, M. and Autuori, F. (1992) Cycloheximide can rescue heat-shocked L cells from death by blocking stress-induced apoptosis. Exp. Cell Res., 201: 436–443.

Giffard, R.G., Xu, L., Zhao, H., Carrico, W., Ouyang, Y., Qiao, Y., Sapolsky, R., Steinberg, G., Hu, B. and Yenari, M.A. (2004) Chaperones, protein aggregation, and brain protection from hypoxic/ischemic injury. J. Exp. Biol., 207: 3213–3220.

Ginsberg, M.D., Sternau, L.L., Globus, M.Y., Dietrich, W.D. and Busto, R. (1992) Therapeutic modulation of brain temperature: relevance to ischemic brain injury. Cerebrovasc. Brain Metab. Rev., 4: 189–225.

Globus, M.Y., Busto, R., Lin, B., Schnippering, H. and Ginsberg, M.D. (1995) Detection of free radical activity during transient global ischemia and recirculation: effects of intraischemic brain temperature modulation. J. Neurochem., 65: 1250–1256.

Globus, M.Y., Ginsberg, M.D. and Busto, R. (1991) Excitotoxic index: a biochemical marker of selective vulnerability. Neurosci. Lett., 127: 39–42.

Green, D.R. and Reed, J.C. (1998) Mitochondria and apoptosis. Science, 281: 1309–1312.

Greffrath, W., Kirschstein, T., Nawrath, H. and Treede, R. (2001) Changes in cytosolic calcium in response to noxious heat and their relationship to vanilloid receptors in rat dorsal root ganglion neurons. Neuroscience, 104: 539–550.

Grivennikova, V.G., Kapustin, A.N. and Vinogradov, A.D. (2001) Catalytic activity of NADH-ubiquinone oxidoreductase (complex I) in intact mitochondria: evidence for the slow active/inactive transition. J. Biol. Chem., 276: 9038–9044.

Gupta, S. (2003) Molecular signaling in death receptor and mitochondrial pathways of apoptosis (review). Int. J. Oncol., 22: 15–20.

Halliday, N.J. (2003) Malignant hyperthermia. J. Craniofac. Surg., 14: 800–802.

Harmon, B.V., Corder, A.M., Collins, R.J., Gobe, G.C., Allen, J., Allan, D.J. and Kerr, J.F. (1990) Cell death induced in a murine mastocytoma by 42–47 degrees C heating in vitro: evidence that the form of death changes from apoptosis to necrosis above a critical heat load. Int. J. Radiat. Biol., 58: 845–858.

Hartl, F.U. and Hayer-Hartl, M. (2002) Molecular chaperones in the cytosol: from nascent chain to folded protein. Science, 295: 1852–1858.

Helenius, A., Marquardt, T. and Braakman, I. (1992) The endoplasmic reticulum as a protein-folding compartment. Trends Cell Biol., 2: 227–231.

Herwig, S. and Strauss, M. (1997) The retinoblastoma protein: a master regulator of cell cycle, differentiation and apoptosis. Eur. J. Biochem., 246: 581–601.

Hinoue, A., Fushiki, S., Nishimura, Y. and Shiota, K. (2001) In utero exposure to brief hyperthermia interferes with the production and migration of neocortical neurons and induces apoptotic neuronal death in the fetal mouse brain. Brain Res. Dev. Brain Res., 132: 59–67.

Houle, F., Rousseau, S., Morrice, N., Luc, M., Mongrain, S., Turner, C.E., Tanaka, S., Moreau, P. and Huot, J. (2003) Extracellular signal-regulated kinase mediates phosphorylation of tropomyosin-1 to promote cytoskeleton remodeling in response to oxidative stress: impact on membrane blebbing. Mol. Biol. Cell, 14: 1418–1432.

Hu, B.R., Janelidze, S., Ginsberg, M.D., Busto, R., Perez-Pinzon, M., Sick, T.J., Siesjo, B.K. and Liu, C.L. (2001) Protein aggregation after focal brain ischemia and reperfusion. J. Cereb. Blood Flow Metab., 21: 865–875.

Hu, B.R., Martone, M.E., Jones, Y.Z. and Liu, C.L. (2000) Protein aggregation after transient cerebral ischemia. J. Neurosci., 20: 3191–3199.

Hwang, S.Y., Paik, S., Park, S.H., Kim, H.S., Lee, I.S., Kim, S.P., Baek, W.K., Suh, M.H., Kwon, T.K., Park, J.W., Park, J.B., Lee, J.J. and Suh, S.I. (2003) N-phenethyl-2-phenylacetamide isolated from Xenorhabdus nematophilus induces apoptosis through caspase activation and calpain-mediated Bax cleavage in U937 cells. Int. J. Oncol., 22: 151–157.

Ikeda, M., Kodama, H., Fukuda, J., Shimizu, Y., Murata, M., Kumagai, J. and Tanaka, T. (1999) Role of radical oxygen species in rat testicular germ cell apoptosis induced by heat stress. Biol. Reprod., 61: 393–399.

Irigoyen, J.P., Besser, D. and Nagamine, Y. (1997) Cytoskeleton reorganization induces the urokinase-type plasminogen activator gene via the Ras/extracellular signal-regulated kinase (ERK) signaling pathway. J. Biol. Chem., 272: 1904–1909.

Jayanthi, S., Deng, X., Ladenheim, B., McCoy, M.T., Cluster, A., Cai, N.S. and Cadet, J.L. (2005) Calcineurin/NFAT-induced up-regulation of the Fas ligand/Fas death pathway is involved in methamphetamine-induced neuronal apoptosis. Proc. Natl. Acad. Sci. U.S.A., 102: 868–873.

Kammersgaard, L.P., Rasmussen, B.H., Jorgensen, H.S., Reith, J., Weber, U. and Olsen, T.S. (2000) Feasibility and safety of inducing modest hypothermia in awake patients with acute stroke through surface cooling: a case-control study: the Copenhagen Stroke Study. Stroke, 31: 2251–2256.

Kanduc, D., Mittelman, A., Serpico, R., Sinigaglia, E., Sinha, A.A., Natale, C., Santacroce, R., Di Corcia, M.G., Lucchese, A., Dini, L., Pani, P., Santacroce, S., Simone, S., Bucci, R. and Farber, E. (2002) Cell death: apoptosis versus necrosis (review). Int. J. Oncol., 21: 165–170.

Kasner, S.E., Wein, T., Piriyawat, P., Villar-Cordova, C.E., Chalela, J.A., Krieger, D.W., Morgenstern, L.B., Kimmel, S.E. and Grotta, J.C. (2002) Acetaminophen for altering body temperature in acute stroke: a randomized clinical trial. Stroke, 33: 130–134.

Kelley, W.L. (1998) The J-domain family and the recruitment of chaperone power. Trends Biochem. Sci., 23: 222–227.

Khan, V.R. and Brown, I.R. (2002) The effect of hyperthermia on the induction of cell death in brain, testis, and thymus of the adult and developing rat. Cell Stress Chaperones, 7: 73–90.

Kil, H.Y., Zhang, J. and Piantadosi, C.A. (1996) Brain temperature alters hydroxyl radical production during cerebral ischemia/reperfusion in rats. J. Cereb. Blood Flow Metab., 16: 100–106.

Kim, Y., Busto, R., Dietrich, W.D., Kraydieh, S. and Ginsberg, M.D. (1996) Delayed postischemic hyperthermia in awake rats worsens the histopathological outcome of transient focal cerebral ischemia. Stroke, 27: 2274–2280 discussion 2281.

Kitagawa, M., Okabe, T., Ogino, H., Matsumoto, H., Suzuki-Takahashi, I., Kokubo, T., Higashi, H., Saitoh, S., Taya, Y., Yasuda, H., et al. (1993) Butyrolactone I, a selective inhibitor of cdk2 and cdc2 kinase. Oncogene, 8: 2425–2432.

Knockaert, M., Greengard, P. and Meijer, L. (2002) Pharmacological inhibitors of cyclin-dependent kinases. Trends Pharmacol. Sci., 23: 417–425.

Komarov, P.G., Komarova, E.A., Kondratov, R.V., Christov-Tselkov, K., Coon, J.S., Chernov, M.V. and Gudkov, A.V. (1999) A chemical inhibitor of p53 that protects mice from the side effects of cancer therapy. Science, 285: 1733–1737.

Layde, P.M., Edmonds, L.D. and Erickson, J.D. (1980) Maternal fever and neural tube defects. Teratology, 21: 105–108.

Lee, Y.J., Armour, E.P., Corry, P.M. and Dewey, W.C. (1990) Mechanism of drug-induced heat resistance: the role of protein degradation? Int. J. Hyperthermia, 6: 591–595.

Lee, Y.J. and Dewey, W.C. (1986) Protection of Chinese hamster ovary cells from hyperthermic killing by cycloheximide or puromycin. Radiat. Res., 106: 98–110.

Lee, Y.J. and Dewey, W.C. (1987) Effect of cycloheximide or puromycin on induction of thermotolerance by heat in Chinese hamster ovary cells: dose fractionation at 45.5 degrees C1. Cancer Res., 47: 5960–5966.

Lee, Y.J., Hou, Z.Z., Kim, D., al-Saadi, A. and Corry, P.M. (1991) Inhibition of protein synthesis and heat protection: histidinol-resistant mutant cell lines. J. Cell Physiol., 149: 396–402.

Lee-Chiong Jr., T.L. and Stitt, J.T. (1995) Disorders of temperature regulation. Compr. Ther., 21: 697–704.

Lei, K. and Davis, R.J. (2003) JNK phosphorylation of Bim-related members of the Bcl2 family induces Bax-dependent apoptosis. Proc. Natl. Acad. Sci. U.S.A., 100: 2432–2437.

Li, H., Zhu, H., Xu, C.J. and Yuan, J. (1998) Cleavage of BID by caspase 8 mediates the mitochondrial damage in the Fas pathway of apoptosis. Cell, 94: 491–501.

Lin, R.Z., Hu, Z.W., Chin, J.H. and Hoffman, B.B. (1997) Heat shock activates c-Src tyrosine kinases and phosphatidylinositol 3-kinase in NIH3T3 fibroblasts. J. Biol. Chem., 272: 31196–31202.

Little, S.A. and Mirkes, P.E. (2002) Teratogen-induced activation of caspase-9 and the mitochondrial apoptotic pathway in early postimplantation mouse embryos. Toxicol. Appl. Pharmacol., 181: 142–151.

Lowenstein, D.H., Chan, P.H. and Miles, M.F. (1991) The stress protein response in cultured neurons: characterization and evidence for a protective role in excitotoxicity. Neuron, 7: 1053–1060.

Ma, N., Szmitko, P., Brade, A., Chu, I., Lo, A., Woodgett, J., Klamut, H. and Liu, F.F. (2004) Kinase-dead PKB gene therapy combined with hyperthermia for human breast cancer. Cancer Gene Ther., 11: 52–60.

Magrane, J., Smith, R.C., Walsh, K. and Querfurth, H.W. (2004) Heat shock protein 70 participates in the neuroprotective response to intracellularly expressed beta-amyloid in neurons. J. Neurosci., 24: 1700–1706.

Maher, J. and Hachinski, V. (1993) Hypothermia as a potential treatment for cerebral ischemia. Cerebrovasc. Brain Metab. Rev., 5: 277–300.

Mandic, A., Viktorsson, K., Strandberg, L., Heiden, T., Hansson, J., Linder, S. and Shoshan, M.C. (2002) Calpain-mediated Bid cleavage and calpain-independent Bak modulation: two separate pathways in cisplatin-induced apoptosis. Mol. Cell Biol., 22: 3003–3013.

Matsumoto, H., Takahashi, A., Wang, X., Ohnishi, K. and Ohnishi, T. (1997) Transfection of p53-knockout mouse fibroblasts with wild-type p53 increases the thermosensitivity and stimulates apoptosis induced by heat stress. Int. J. Radiat. Oncol. Biol. Phys., 38: 1089–1095.

Meden, P., Overgaard, K., Pedersen, H. and Boysen, G. (1994) The influence of body temperature on infarct volume and thrombolytic therapy in a rat embolic stroke model. Brain Res., 647: 131–138.

Meli, E., Pangallo, M., Picca, R., Baronti, R., Moroni, F. and Pellegrini-Giampietro, D.E. (2004) Differential role of poly-(ADP-ribose) polymerase-1in apoptotic and necrotic neuronal death induced by mild or intense NMDA exposure in vitro. Mol. Cell Neurosci., 25: 172–180.

Meriin, A.B., Yaglom, J.A., Gabai, V.L., Zon, L., Ganiatsas, S., Mosser, D.D., Zon, L. and Sherman, M.Y. (1999) Protein-damaging stresses activate c-Jun N-terminal kinase via inhibition of its dephosphorylation: a novel pathway controlled by HSP72. Mol. Cell Biol., 19: 2547–2555.

Mezey, E., Toth, Z.E., Cortright, D.N., Arzubi, M.K., Krause, J.E., Elde, R., Guo, A., Blumberg, P.M. and Szallasi, A. (2000) Distribution of mRNA for vanilloid receptor subtype 1 (VR1), and VR1-like immunoreactivity, in the central nervous system of the rat and human. Proc. Natl. Acad. Sci. U.S.A., 97: 3655–3660.

Michels, A.A., Kanon, B., Konings, A.W., Bensaude, O. and Kampinga, H.H. (2000) Cycloheximide- and puromycin-induced heat resistance: different effects on cytoplasmic and nuclear luciferases. Cell Stress Chaperones, 5: 181–187.

Mies, G., Iijima, T. and Hossmann, K.A. (1993) Correlation between peri-infarct DC shifts and ischaemic neuronal damage in rat. Neuroreport, 4: 709–711.

Milunsky, A., Ulcickas, M., Rothman, K.J., Willett, W., Jick, S.S. and Jick, H. (1992) Maternal heat exposure and neural tube defects. JAMA, 268: 882–885.

Minamisawa, H., Smith, M.L. and Siesjo, B.K. (1990) The effect of mild hyperthermia and hypothermia on brain damage following 5, 10, and 15 min of forebrain ischemia. Ann. Neurol., 28: 26–33.

Mirkes, P.E. and Little, S.A. (2000) Cytochrome c release from mitochondria of early postimplantation murine embryos exposed to 4-hydroperoxycyclophosphamide, heat shock, and staurosporine. Toxicol. Appl. Pharmacol., 162: 197–206.

Mizushima, T., Obata, K., Yamanaka, H., Dai, Y., Fukuoka, T., Tokunaga, A., Mashimo, T. and Noguchi, K. (2005) Activation of p38 MAPK in primary afferent neurons by noxious stimulation and its involvement in the development of thermal hyperalgesia. Pain, 113: 51–60.

Morimoto, T., Ginsberg, M.D., Dietrich, W.D. and Zhao, W. (1997) Hyperthermia enhances spectrin breakdown in transient focal cerebral ischemia. Brain Res., 746: 43–51.

Morishima, N., Nakanishi, K., Tsuchiya, K., Shibata, T. and Seiwa, E. (2004) Translocation of Bim to the endoplasmic reticulum (ER) mediates ER stress signaling for activation of caspase-12 during ER stress-induced apoptosis. J. Biol. Chem., 279: 50375–50381.

Moriyama-Gonda, N., Igawa, M., Shiina, H., Urakami, S., Shigeno, K. and Terashima, M. (2002) Modulation of heat-induced cell death in PC-3 prostate cancer cells by the antioxidant inhibitor diethyldithiocarbamate. BJU Int., 90: 317–325.

Muchowski, P.J. and Wacker, J.L. (2005) Modulation of neurodegeneration by molecular chaperones. Nat. Rev. Neurosci., 6: 11–22.

Narayan, P., Crocker, I., Elder, E. and Olson, J.J. (2004) Safety and efficacy of concurrent interstitial radiation and hyperthermia in the treatment of progressive malignant brain tumors. Oncol. Rep., 11: 97–103.

Nath, R., Davis, M., Probert, A.W., Kupina, N.C., Ren, X., Schielke, G.P. and Wang, K.K. (2000) Processing of cdk5 activator p35 to its truncated form (p25) by calpain in acutely injured neuronal cells. Biochem. Biophys. Res. Commun., 274: 16–21.

O'Neill, K.L., Fairbairn, D.W., Smith, M.J. and Poe, B.S. (1998) Critical parameters influencing hyperthermia-induced apoptosis in human lymphoid cell lines. Apoptosis, 3: 369–375.

Ng, D.C. and Bogoyevitch, M.A. (2000) The mechanism of heat shock activation of ERK mitogen-activated protein kinases in the interleukin 3-dependent ProB cell line BaF3. J. Biol. Chem., 275: 40856–40866.

Nichols, J.W., Hill, M.W., Bangham, A.D. and Deamer, D.W. (1980) Measurement of net proton-hydroxyl permeability of large unilamellar liposomes with the fluorescent pH probe, 9-aminoacridine. Biochim. Biophys. Acta, 596: 393–403.

Nikolic, M., Chou, M.M., Lu, W., Mayer, B.J. and Tsai, L.H. (1998) The p35/Cdk5 kinase is a neuron-specific Rac effector that inhibits Pak1 activity. Nature, 395: 194–198.

Ohnishi, T., Wang, X., Ohnishi, K., Matsumoto, H. and Takahashi, A. (1996) p53-dependent induction of WAF1 by heat treatment in human glioblastoma cells. J. Biol. Chem., 271: 14510–14513.

Ohnishi, K., Wang, X., Takahashi, A. and Ohnishi, T. (1998) Contribution of protein kinase C to p53-dependent WAF1 induction pathway after heat treatment in human glioblastoma cell lines. Exp. Cell Res., 238: 399–406.

Ouyang, Y.B. and Hu, B.R. (2001) Protein ubiquitination in rat brain following hypoglycemic coma. Neurosci. Lett., 298: 159–162.

Padmanabhan, J., Park, D.S., Greene, L.A. and Shelanski, M.L. (1999) Role of cell cycle regulatory proteins in cerebellar granule neuron apoptosis. J. Neurosci., 19: 8747–8756.

Park, D.S., Morris, E.J., Bremner, R., Keramaris, E., Padmanabhan, J., Rosenbaum, M., Shelanski, M.L., Geller, H.M. and Greene, L.A. (2000) Involvement of retinoblastoma family members and E2F/DP complexes in the death of neurons evoked by DNA damage. J. Neurosci., 20: 3104–3114.

Park, J.W., Choi, Y.B., Park, S.K., Kim, Y.I. and Lee, K.S. (2004) Magnetic resonance imaging reveals selective vulnerability of the cerebellum and basal ganglia in malignant hyperthermia. Arch. Neurol., 61: 1462–1463.

Paula-Lopes, F.F. and Hansen, P.J. (2002) Heat shock-induced apoptosis in preimplantation bovine embryos is a developmentally regulated phenomenon. Biol. Reprod., 66: 1169–1177.

Philip, B. and Levin, D.E. (2001) Wsc1 and Mid2 are cell surface sensors for cell wall integrity signaling that act through Rom2, a guanine nucleotide exchange factor for Rho1. Mol. Cell Biol., 21: 271–280.

Pongrac, J.L., Middleton, F.A., Peng, L., Lewis, D.A., Levitt, P. and Mirnics, K. (2004) Heat shock protein 12A shows reduced expression in the prefrontal cortex of subjects with schizophrenia. Biol. Psychiatry, 56: 943–950.

Putcha, G.V., Le, S., Frank, S., Besirli, C.G., Clark, K., Chu, B., Alix, S., Youle, R.J., LaMarche, A., Maroney, A.C. and Johnson Jr., E.M. (2003) JNK-mediated BIM phosphorylation potentiates BAX-dependent apoptosis. Neuron, 38: 899–914.

Puthalakath, H., Huang, D.C., O'Reilly, L.A., King, S.M. and Strasser, A. (1999) The proapoptotic activity of the Bcl-2 family member Bim is regulated by interaction with the dynein motor complex. Mol. Cell, 3: 287–296.

Puthalakath, H., Villunger, A., O'Reilly, L.A., Beaumont, J.G., Coultas, L., Cheney, R.E., Huang, D.C. and Strasser, A. (2001) Bmf: a proapoptotic BH3-only protein regulated by interaction with the myosin V actin motor complex, activated by anoikis. Science, 293: 1829–1832.

Qian, L., Song, X., Ren, H., Gong, J. and Cheng, S. (2004) Mitochondrial mechanism of heat stress-induced injury in rat cardiomyocyte. Cell Stress Chaperones, 9: 281–293.

Rees, D.C. and Robertson, A.D. (2001) Some thermodynamic implications for the thermostability of proteins. Protein Sci., 10: 1187–1194.

Riobó, N.A., Clementi, E., Melani, M., Boveris, A., Cadenas, E., Moncada, S. and Poderoso, J.J. (2001) Nitric oxide inhibits mitochondrial NADH:ubiquinone reductase activity through peroxynitrite formation. Biochem. J., 359: 139–145.

Roberts, J.C., Davis, J.B. and Benham, C.D. (2004) [3H]Resiniferatoxin autoradiography in the CNS of wild-type and TRPV1 null mice defines TRPV1 (VR-1) protein distribution. Brain Res., 995: 176–183.

Romero, J.J., Clement, P.F. and Belden, C. (2000) Neuropsychological sequelae of heat stroke: report of three cases and discussion. Mil. Med., 165: 500–503.

Ron, D. (2002) Translational control in the endoplasmic reticulum stress response. J. Clin. Invest., 110: 1383–1388.

Sammut, I.A. and Harrison, J.C. (2003) Cardiac mitochondrial complex activity is enhanced by heat shock proteins. Clin. Exp. Pharmacol. Physiol., 30: 110–115.

Sammut, I.A., Jayakumar, J., Latif, N., Rothery, S., Severs, N.J., Smolenski, R.T., Bates, T.E. and Yacoub, M.H. (2001) Heat stress contributes to the enhancement of cardiac mitochondrial complex activity. Am. J. Pathol., 158: 1821–1831.

Sanchez, V., O'Shea, E., Saadat, K.S., Elliott, J.M., Colado, M.I. and Green, A.R. (2004) Effect of repeated ('binge') dosing of MDMA to rats housed at normal and high temperature on neurotoxic damage to cerebral 5-HT and dopamine neurones. J. Psychopharmacol., 18: 412–416.

van der Sanden, M.H., Houweling, M., van Golde, L.M. and Vaandrager, A.B. (2003) Inhibition of phosphatidylcholine synthesis induces expression of the endoplasmic reticulum stress and apoptosis-related protein CCAAT/enhancer-binding protein-homologous protein (CHOP/GADD153). Biochem. J., 369: 643–650.

Schwab, S., Georgiadis, D., Berrouschot, J., Schellinger, P.D., Graffagnino, C. and Mayer, S.A. (2001) Feasibility and safety of moderate hypothermia after massive hemispheric infarction. Stroke, 32: 2033–2035.

Sharma, H.S. and Hoopes, P.J. (2003) Hyperthermia induced pathophysiology of the central nervous system. Int. J. Hyperthermia, 19: 325–354.

Shiota, K. (1982) Neural tube defects and maternal hyperthermia in early pregnancy: epidemiology in a human embryo population. Am. J. Med. Genet., 12: 281–288.

Shiota, K. (1988) Induction of neural tube defects and skeletal malformations in mice following brief hyperthermia in utero. Biol. Neonate, 53: 86–97.

Siesjo, B.K., Hu, B. and Kristian, T. (1999) Is the cell death pathway triggered by the mitochondrion or the endoplasmic reticulum? J. Cereb. Blood Flow Metab., 19: 19–26.

Simon, H.B. (1993) Hyperthermia. N. Engl. J. Med., 329: 483–487.

Skaper, S.D. (2003) Poly(ADP-ribose) polymerase-1 in acute neuronal death and inflammation: a strategy for neuroprotection. Ann. N.Y. Acad. Sci., 993: 217–228 discussion 287–218.

Smith, M.S., Upfold, J.B., Edwards, M.J., Shiota, K. and Cawdell-Smith, J. (1992) The induction of neural tube defects by maternal hyperthermia: a comparison of the guinea-pig and human. Neuropathol. Appl. Neurobiol., 18: 71–80.

Somwaru, L., Li, S., Doglio, L., Goldberg, E. and Zirkin, B.R. (2004) Heat-induced apoptosis of mouse meiotic cells is suppressed by ectopic expression of testis-specific calpastatin. J. Androl., 25: 506–513.

Stea, B., Kittelson, J., Cassady, J.R., Hamilton, A., Guthkelch, N., Lulu, B., Obbens, E., Rossman, K., Shapiro, W., Shetter, A., et al. (1992) Treatment of malignant gliomas with interstitial irradiation and hyperthermia. Int. J. Radiat. Oncol. Biol. Phys., 24: 657–667.

Styrt, B. and Sugarman, B. (1990) Antipyresis and fever. Arch. Intern. Med., 150: 1589–1597.

Susin, S.A., Daugas, E., Ravagnan, L., Samejima, K., Zamzami, N., Loeffler, M., Costantini, P., Ferri, K.F., Irinopoulou, T., Prevost, M.C., Brothers, G., Mak, T.W., Penninger, J., Earnshaw, W.C. and Kroemer, G. (2000) Two distinct pathways leading to nuclear apoptosis. J. Exp. Med., 192: 571–580.

Susin, S.A., Lorenzo, H.K., Zamzami, N., Marzo, I., Snow, B.E., Brothers, G.M., Mangion, J., Jacotot, E., Costantini, P., Loeffler, M., Larochette, N., Goodlett, D.R., Aebersold, R., Siderovski, D.P., Penninger, J.M. and Kroemer, G. (1999) Molecular characterization of mitochondrial apoptosis-inducing factor. Nature, 397: 441–446.

Sweitzer, S.M., Peters, M.C., Ma, J.Y., Kerr, I., Mangadu, R., Chakravarty, S., Dugar, S., Medicherla, S., Protter, A.A. and Yeomans, D.C. (2004) Peripheral and central p38 MAPK mediates capsaicin-induced hyperalgesia. Pain, 111: 278–285.

Takagi, K., Ginsberg, M.D., Globus, M.Y., Martinez, E. and Busto, R. (1994) Effect of hyperthermia on glutamate release in ischemic penumbra after middle cerebral artery occlusion in rats. Am. J. Physiol., 267: H1770–H1776.

Takahashi, K., Pieper, A.A., Croul, S.E., Zhang, J., Snyder, S.H. and Greenberg, J.H. (1999) Post-treatment with an inhibitor of poly(ADP-ribose) polymerase attenuates cerebral damage in focal ischemia. Brain Res., 829: 46–54.

Takano, J., Tomioka, M., Tsubuki, S., Higuchi, M., Iwata, N., Itohara, S., Maki, M. and Saido, T.C. (2005) Calpain mediates excitotoxic DNA fragmentation via mitochondrial pathways in adult brains: evidence from calpastatin-mutant mice. J. Biol. Chem., 280: 16175–16184.

Tang, D., Wu, D., Hirao, A., Lahti, J.M., Liu, L., Mazza, B., Kidd, V.J., Mak, T.W. and Ingram, A.J. (2002) ERK activation mediates cell cycle arrest and apoptosis after DNA damage independently of p53. J. Biol. Chem., 277: 12710–12717.

Terradillos, O., Montessuit, S., Huang, D.C. and Martinou, J.C. (2002) Direct addition of BimL to mitochondria does not lead to cytochrome c release. FEBS Lett., 522: 29–34.

Thompson, H.J., Tkacs, N.C., Saatman, K.E., Raghupathi, R. and McIntosh, T.K. (2003) Hyperthermia following traumatic brain injury: a critical evaluation. Neurobiol. Dis., 12: 163–173.

Toillon, R.A., Chopin, V., Jouy, N., Fauquette, W., Boilly, B. and Le Bourhis, X. (2002) Normal breast epithelial cells induce p53-dependent apoptosis and p53-independent cell cycle arrest of breast cancer cells. Breast Cancer Res. Treat., 71: 269–280.

Toth, Z., Yan, X.X., Haftoglou, S., Ribak, C.E. and Baram, T.Z. (1998) Seizure-induced neuronal injury: vulnerability to febrile seizures in an immature rat model. J. Neurosci., 18: 4285–4294.

Tran, S.E., Meinander, A., Holmstrom, T.H., Rivero-Muller, A., Heiskanen, K.M., Linnau, E.K., Courtney, M.J., Mosser, D.D., Sistonen, L. and Eriksson, J.E. (2003) Heat stress downregulates FLIP and sensitizes cells to Fas receptor-mediated apoptosis. Cell Death Differ., 10: 1137–1147.

Umpierre, C.C., Little, S.A. and Mirkes, P.E. (2001) Co-localization of active caspase-3 and DNA fragmentation (TUNEL) in normal and hyperthermia-induced abnormal mouse development. Teratology, 63: 134–143.

Uney, J.B., Kew, J.N., Staley, K., Tyers, P. and Sofroniew, M.V. (1993) Transfection-mediated expression of human Hsp70i protects rat dorsal root ganglian neurones and glia from severe heat stress. FEBS Lett., 334: 313–316.

Upfold, J.B., Smith, M.S. and Edwards, M.J. (1991) Interference with neural crest migration by maternal hyperthermia as a cause of embryonic death due to heart failure. Med. Hypotheses, 35: 244–246.

Urano, F., Wang, X., Bertolotti, A., Zhang, Y., Chung, P., Harding, H.P. and Ron, D. (2000) Coupling of stress in the ER to activation of JNK protein kinases by transmembrane protein kinase IRE1. Science, 287: 664–666.

Venkataraman, S., Wagner, B.A., Jiang, X., Wang, H.P., Schafer, F.Q., Ritchie, J.M., Patrick, B.C., Oberley, L.W. and Buettner, G.R. (2004) Overexpression of manganese superoxide dismutase promotes the survival of prostate cancer cells exposed to hyperthermia. Free Radic. Res., 38: 1119–1132.

Vogel, P., Dux, E. and Wiessner, C. (1997) Evidence of apoptosis in primary neuronal cultures after heat shock. Brain Res., 764: 205–213.

Wachsberger, P.R. and Coss, R.A. (1990) Effects of hyperthermia on the cytoskeleton and cell survival in G_1 and S phase in Chinese hamster ovary cells. Int. J. Hyperthermia, 6: 67–85.

Wang, H., Yu, S.W., Koh, D.W., Lew, J., Coombs, C., Bowers, W., Federoff, H.J., Poirier, G.G., Dawson, T.M. and Dawson, V.L. (2004) Apoptosis-inducing factor substitutes for caspase executioners in NMDA-triggered excitotoxic neuronal death. J. Neurosci., 24: 10963–10973.

Wang, X. (2001) The expanding role of mitochondria in apoptosis. Genes Dev., 15: 2922–2933.

Westermann, A.M., Wiedemann, G.J., Jager, E., Jager, D., Katschinski, D.M., Knuth, A., Vorde Sive Vording, P.Z., Van Dijk, J.D., Finet, J., Neumann, A., Longo, W., Bakhshandeh, A., Tiggelaar, C.L., Gillis, W., Bailey, H., Peters, S.O. and Robins, H.I. (2003) A Systemic Hyperthermia Oncologic Working Group trial. Ifosfamide, carboplatin, and etoposide combined with 41.8 degrees C whole-body hyperthermia for metastatic soft tissue sarcoma. Oncology, 64: 312–321.

White, M.G., Emery, M., Nonner, D. and Barrett, J.N. (2003) Caspase activation contributes to delayed death of heat-stressed striatal neurons. J. Neurochem., 87: 958–968.

Willis, W.T., Jackman, M.R., Bizeau, M.E., Pagliassotti, M.J. and Hazel, J.R. (2000) Hyperthermia impairs liver mitochondrial function in vitro. Am. J. Physiol. Regul. Integr. Comp. Physiol., 278: R1240–R1246.

Worfolk, J.B. (2000) Heat waves: their impact on the health of elders. Geriatr. Nurs., 21: 70–77.

Yager, J.Y., Armstrong, E.A., Jaharus, C., Saucier, D.M. and Wirrell, E.C. (2004) Preventing hyperthermia decreases brain damage following neonatal hypoxic-ischemic seizures. Brain Res., 1011: 48–57.

Yang, Y.L. and Lin, M.T. (1999) Heat shock protein expression protects against cerebral ischemia and monoamine overload in rat heatstroke. Am. J. Physiol., 276: H1961–H1967.

Yin, Y., Stahl, B.C., DeWolf, W.C. and Morgentaler, A. (2002) P53 and Fas are sequential mechanisms of testicular germ cell apoptosis. J. Androl., 23: 64–70.

Yu, S.W., Wang, H., Dawson, T.M. and Dawson, V.L. (2003) Poly(ADP-ribose) polymerase-1 and apoptosis inducing factor in neurotoxicity. Neurobiol. Dis., 14: 303–317.

Yu, S.W., Wang, H., Poitras, M.F., Coombs, C., Bowers, W.J., Federoff, H.J., Poirier, G.G., Dawson, T.M. and Dawson, V.L. (2002) Mediation of poly(ADP-ribose) polymerase-1-dependent cell death by apoptosis-inducing factor. Science, 297: 259–263.

Zaremba, J. (2004) Hyperthermia in ischemic stroke. Med. Sci. Monit., 10: RA148–RA153.

Zong, W.X., Ditsworth, D., Bauer, D.E., Wang, Z.Q. and Thompson, C.B. (2004) Alkylating DNA damage stimulates a regulated form of necrotic cell death. Genes Dev., 18: 1272–1282.

CHAPTER 18

Heat acclimation and cross-tolerance against novel stressors: genomic–physiological linkage

Michal Horowitz*

Laboratory of Environmental Physiology, The Hebrew University, POB 12272, Jerusalem 91120, Israel

Abstract: Heat acclimation (AC) is a "within lifetime" reversible phenotypic adaptation, enhancing thermotolerance and heat endurance via a transition to "efficient" cellular performance when acclimatory homeostasis is reached. An inseparable outcome of AC is the development of cross-tolerance (C-T) against novel stressors. This chapter focuses on central plasticity and the molecular–physiological linkage of acclimatory and C-T responses. A drop in temperature thresholds (T-Tsh) for activation of heat-dissipation mechanisms and an elevated T-Tsh for thermal injury development imply autonomic nervous system (ANS) and cytoprotective network involvement in these processes. During acclimation, the changes in T-Tsh for heat dissipation are biphasic. Initially T-Tsh drops, signifying the early autonomic response, and is associated with perturbed peripheral effector cellular performance. Pre-acclimation values return when acclimatory homeostasis is achieved. The changes in the ANS suggest that acclimatory plasticity involves molecular and cellular changes. These changes are manifested by the activation of central peripheral molecular networks and post-translational modifications. Sympathetic induction of elevated HSP 72 reservoirs, with faster heat shock response, is only one example of this. The global genomic response, detected using gene-chips and cluster analyses imply upregulation of genes encoding ion channels, pumps, and transporters (markers for neuronal excitability) in the hypothalamus at the onset of AC and down regulation of metabotrophic genes upon long term AC. Peripherally, the transcriptional program indicates a two-tier defense strategy. The immediate transient response is associated with the maintenance of DNA and cellular integrity. The sustained response correlates with long-lasting cytoprotective-signaling networks. C-T is recorded against cerebral hypoxia, hyperoxia, and traumatic brain injury. Using the highly developed ischemic/reperfused heart model as a baseline, it is evident that C-T stems via protective shared pathways developed with AC. These comprise constitutive elevation of HIF 1α and associated target pathways, HSPs, anti-apoptosis, and antioxidative pathways. Collectively the master regulators of AC and C-T are still enigmatic; however, cutting-edge investigative techniques, using a broad molecular approach, challenge current ideas, and the data accumulated will pinpoint novel pathways and provide new perspectives.

Keywords: heat acclimation; cross-tolerance; HSP 72 (heat shock protein 72 kDa); cytoprotection; autonomic acclimation plasticity; brain stem evoked response (ABR); hypothalamic renin–angiotensin system (Hypothalamic RAS); global genomic response; heat acclimation–oxygen deprivation cross tolerance (or heat acclimation–hypoxia cross tolerance); HIF-1; hypoxia; ischemia/reperfusion; oxygen toxicity; traumatic brain injury; heat acclimation–hypohydration interference; iNOS

---

*Corresponding author. Tel.: +972 2 6757588;
Fax: +972 3 6439736; E-mail: horowitz@cc.huji.ac.il

Introduction

Heat acclimation is a "within lifetime" phenotypic adaptation involving adjustments at all levels of body organization to enhance thermotolerance and heat endurance. The reprogramming of gene expression and post-transcriptional mechanisms are essential components of acclimatory mechanisms. An inseparable outcome of these acclimatory processes is that an adjustment to one environmental stressor can, in addition to the primary adaptation, augment the degree of adjustment to additional stressors. Such cross-reinforcement raises the possibility of inducing adaptation to a given stressor without prior exposure to it, a process defined as cross-tolerance (exaptation). Some stressors override or interfere with acclimation. To date, our knowledge about the thermal plasticity leading to heat acclimation and cross-tolerance or interference, particularly in the brain is very limited. This chapter addresses these questions and reviews the current knowledge on this topic. Our comprehensive integrative model of heat acclimation (Horowitz, 2002) provides a solid basis for the interpretation of the dynamics of heat acclimation.

Heat acclimation: the general concept

Heat acclimation is a conserved phenotypic adaptive response to a prolonged transfer to higher ambient temperatures that not only confers protection against acute heat stress but also delays thermal injury (Horowitz, 2001, 2002). In terms of integrative physiological mechanisms, heat acclimation is characterized by decreased heat production, core temperature, and heart rate. Concomitantly, the capacities of the vascular and the evaporative cooling systems for heat dissipation are augmented (Horowitz, 2001, 2002). A drop in temperature thresholds for the activation of heat-dissipation mechanisms and an elevated temperature threshold for the development of thermal injury imply the involvement of the autonomic nervous system and cytoprotective networks, respectively, in these processes (Horowitz et al., 1983, 1999, 2004; Schwimmer et al., 2004).

In order to understand the concept of, and the mechanisms leading to heat acclimation, i.e., the kinetics of the process, namely, the changes occurring during the course of heat acclimation must be examined. Heat acclimation is a biphasic process involving cross-talk between the peripheral effector organs and central, autonomic control. In homeotherms, with constant body temperature generally independent of the surrounding temperature, an apparent acclimated state (e.g., reduced heart rate, enhanced evaporative cooling, and increased thermal endurance) emerges shortly (2–5 days) after exposure to acclimating conditions. The development of the new acclimatory homeostasis, however, is a prolonged process (3–4 weeks). Effector organ-autonomic cross-talk at each specific acclimatory phase phenotype is achieved via different mechanisms. At the onset of heat acclimation (Phase I — Short Term Heat Acclimation (STHA)), an increased excitability of the autonomic nervous system compensates for the impaired cellular performance resulting from impaired signaling pathways. This phase is dominated by changes in the cell membrane, leading to a desensitization of G-protein-coupled receptors (Kloog et al., 1985), their associated signaling pathways including Ca^{2+} signals (Kaspler and Horowitz, 2001), and target sensitivity. When acclimatory homeostasis has been attained (Phase II — Long Term Heat Acclimation (LTHA)), metabolic alterations improve cellular function leading to enhanced efficiency — namely an increased effector-organ-output/excitation-signal ratio, suggesting decreased neural excitability (Horowitz, 1998; Kaspler and Horowitz, 2001). During the second acclimatory phase, enhanced integrative physiological mechanisms expand the dynamic thermoregulatory range (Horowitz, 2002). Conceptually, the process of heat acclimation (in homeotherms) represents a transition from an early transient, "inefficient" to very "efficient" cellular performance when acclimatory homeostasis has been reached (Horowitz, 2003). Figure 1 illustrates the time course of heat acclimation in the rat acclimation model.

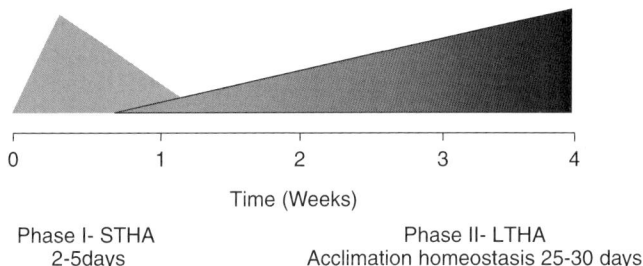

Fig. 1. Phases in heat acclimation. Heat acclimation (continuous exposure to 34°C) is a bi-phasic process, comprising a transient excitatory phase (Phase I) during which the autonomic nervous system plays a major regulatory role. When acclimatory homeostasis has been attained (Phase II), metabolic alterations improve cellular function leading to enhanced efficiency. STHA — Short Term Heat Acclimation (Phase I); LTHA — Long Term Heat Acclimation (Phase II).

Heat acclimation and plasticity of the autonomic responses

A lesson from central temperature thresholds for activation of peripheral thermoregulatory effectors

The temperature threshold (T-Tsh) for activating thermoregulatory effectors plays a pivotal role in the process of acclimation. The T-Tsh changes following chronic exposure to different ambient temperatures (e.g., Horowitz et al., 1983; Shido et al., 1995, 1999; Yamazaki and Hamasaki, 2003), to light/dark cycling, and to seasonality (Nakamura and Okamura, 1998). Laboratory thermal manipulations can selectively affect the T-Tsh of individual thermoregulatory control loops (Shido et al., 1999). The adjustability of the T-Tsh, in response to chronic sensory inputs from the environment, implies that long-term cellular processes are an underlying mechanism of neuronal plasticity.

In studying the dynamics of heat acclimation, we have demonstrated that the variations in T-Tsh for the major heat-dissipation mechanisms during the progression of acclimation occur in a biphasic manner, time-correlated with the changes recorded for the peripheral effectors, but in reciprocal directions (Horowitz et al., 1983; Horowitz and Meiri, 1985a; Schwimmer et al., 2004). Namely, the marked initial T-Tsh drop, signifying the early autonomic response (during STHA), is associated with perturbed peripheral effector cellular performance (Horowitz et al., 1983; Kloog et al., 1985), and pre-acclimation values return when acclimatory homeostasis (LTHA) is achieved (Horowitz et al., 1983; Kloog et al., 1985; Horowitz, 2001, 2002). Given that our concept of acclimation is one of orchestrated interactions among peripheral central feedback loops, we hypothesize that neuromodulation occurs during the process of acclimation. In other words, the inherent intrinsic thresholds of the various thermoregulatory loops are modulated by a variety of signals, not necessarily thermal, that are evoked by the deviation from homeostasis (e.g., rectal temperature, body fluid osmolarity and volume, and hormonal status).

Physiological evidence for changes in central sensitivity

Bligh (1973) hypothesized that heat acclimation involves changes in the central controller of temperature regulation. Nevertheless, the physiological evidence for central adaptations occurring during acclimation is limited. In guinea pigs, tonic activity of neurons in the nucleus raphe magnus and subcoeruleus regions changes after cold and warm adaptations in response to reciprocal ambient stimulation (Hinckel and Schroder-Rosenstock, 1982; Hinckel and Perschel, 1987). In rats, Pierau's group (Pierau et al., 1998) showed that heat acclimation leads to a considerable decrease in the number of warm-sensitive neurons with a diminished number having very low temperature coefficient (0.6–0.8 imp/s/°C) compared to matched populations of neurons in hypothalami of cold-acclimated rats. An additional clear effect of warm acclimation (in contrast to cold or no adaptation) was the ability to convert a larger number of insensitive to warm sensitive neurons by mediators, e.g., bombesin (Pierau et al., 1998). Based on the existing neuronal models, that the central temperature controller relies on three sets of neurons with unique thermal sensitivities (Hammel, 1968; Boulant, 1998, 2006), the reported data imply that the altered availability of these neurons differing in their sensitivity or tonic activity is an integral part of acclimation-mediated adaptive responses. Teleologically, the altered ratio between warm-sensitive/cold-sensitive neurons reflects changes in the activation of thermoregulatory effectors during particular acclimatory states. In contrast, the study of neuronal activity, when acclimatory homeostasis has been achieved (Attias et al., 1988), used different approaches to investigate changes in brainstem activity during the critical phases of the heat acclimation regimen. Measuring the auditory nerve-brainstem evoked responses (ABR) during a 2-month acclimation regimen, the authors demonstrated that following STHA, the time intervals between successive waves (reflecting compound action potentials of the brainstem auditory pathway) are prolonged. After 60 days of acclimation, the initial delay in neural activity was followed by shorter time intervals accompanied by significant increases in the compound action potential of the auditory nerve. The shorter time intervals reflect more rapid activity and enhanced firing synchronization between the axons contributing to the compound amplitude. Upon hyperthermia, the latencies of ABR parameters in LTHA further decrease. Whether such effects are due to axonal and/or synaptic changes is unclear. However, conceptually, this finding is consistent with Hinckel's and Pierau's (Hinckel and Perschel, 1987; Pierau et al., 1998). Evidence for acclimatory changes at the biochemical level were reported by Christman and Gisolfi (1980, 1985). They showed that prior acclimation leads to enhanced hypothalamic sensitivity to norepinephrine, a consensus neurotransmitter in temperature regulation.

Chronic central acclimatory changes: the need for biochemical and molecular experimental paradigms

The phenomena described above do not provide an explanation for the underlying mechanisms of acclimatory neuronal plasticity. Currently, neuronal electrophysiological activity under situations of chronic stress is believed to depend on structural flexibility and molecular changes; therefore, new concepts of the core processes leading to acclimation plasticity may develop (Armstrong and Stoppani, 2002).

Collectively, the evidence implies that long-term accommodation to a changing environment involves functional neuronal remodeling. Transcriptional plasticity during long-term processes in the brain has been documented during the response to environmental enrichment, namely, a different surrounding every day (Gagne et al., 1998), hypoxia (Xia and Haddad, 1999), and learning and memory (Hess et al., 1997; Rattiner et al., 2004). Each of these factors caused long-term changes in behavior and physiological mechanisms that evolved, at least in part, from alterations at the transcription level. We advanced the hypothesis that (at least) one mode of acclimatory plasticity involves changes at the cellular and molecular level and used two approaches to assess this hypothesis: (1) studying the effects of acclimation on neuromodulators, and (2) studying the global genomic response of the hypothalamus and a peripheral organ during the course of heat acclimation.

Neuromodulation — a lesson from the hypothalamic renin–angiotensin system (RAS)

Evidence for a link between angiotensin and thermoregulation has been provided by immunohistochemical and physiological studies. Cell bodies, immunopositive for angiotensin II (Ang II), are found in hypothalamic nuclei related to thermoregulatory circuits on both sides of the blood-brain barrier (Lind et al., 1985), with Ang II receptors of the AT_1 subtype highly expressed (Lenkei et al., 1997). Ang II was also identified as a contributor to numerous central thermoregulatory processes, including thermally induced water loss, enhanced cutaneous evaporative heat loss via central stimulation (Wilson and Fregley, 1985), modulation of vasopressin release into the circulation in response to heat load (Kregel et al., 1994), and sympathetically-mediated splanchnic vasomotor responses (Kregel et al., 1994). Such adjustments depend on the rapid responsiveness of the autonomic systems controlling the thermoregulatory loops. Additionally, Ang II can induce overexpression of transcription factors in brain nuclei associated with the central control of blood pressure and salt and water homeostasis. The onset of the effects of these peptides is delayed by the synthesis of new proteins along the targeted signaling pathways (clearly, the synthesis of new proteins takes more time than the release of neurotransmitters or effector peptides) (Blume et al., 2002). We found profound differences in the sensitivity of various thermoregulatory signaling cascades, including angiotensinergic pathways during STHA and LTHA, e.g., angiotensin activation patterns with two components, immediate and delayed. (Horowitz et al., 1999). Hence, we further hypothesized that the angiotensinergic system mirrors one mode of the neuromodulatory plasticity of thermoregulatory mechanisms during heat acclimation. This system involves the interplay between two antagonistic Ang II signaling pathways, that of AT1 and AT2 receptors. Both receptors play integratory roles in body homeostasis. Likewise, AT2 is involved in chronic adaptive processes, partly via nitric oxide (NO) mediation (Gohlke et al., 1998). The modulatory role of Ang II was unraveled by its disruption following the intracerebroventricular administration of losartan (an AT1 receptor blocker). Ang II seems to have opposing effects on vasodilatation and salivation thresholds namely, decreased salivation and increased vasodilatation thresholds. Additionally, there are profound differences in the sensitivity of AT1 signaling during STHA compared to LTHA (Horowitz et al., 1999; Schwimmer et al., 2004). The responsiveness of post-synaptic receptors depends on both the density and the sensitivity of the receptor protein. Hence, our first step in understanding the plasticity of acclimation via Ang II modulation was to study the level of active Ang II receptor fractions (membrane receptors) and their cellular reserves. Heat acclimation did not change the density of the membrane AT1 receptor protein but rather led to a significantly higher (AT1 membrane/AT1 cytosolic receptor) ratio in LTHA, suggesting the maintenance of AT1 receptor levels by cellular trafficking. In contrast, membrane AT2 receptors were significantly upregulated following LTHA, at the expense of their cytosolic reserves. Pharmacological studies *in vivo*, using specific AT receptor blockade, confirmed that whereas AT1 receptors decrease the evaporative-cooling Tsh and, in turn, the core temperature (Tre), AT2 receptors elevate body temperature (the latter effect has also been documented during fever (Watanabe et al., 1999)). We suggest that Ang II receptors provide the fine-tuning to a "rudimentary" regulatory pathway, abolishing extreme deviations in integrative physiological responses. Fine tuning works in concert with other mechanisms, prolonging utilization of the intracellular water reservoir for evaporative cooling and, in turn, enhancing heat endurance.

We conclude that one mode of acclimatory plasticity to promote body temperature homeostasis is mediated via the interaction of neuromodulators with reciprocal effects. In the case of the hypothalamic RAS, the delicate balance between AT1 and AT2 receptors leads to a predominance of AT2 signaling upon LTHA. Receptor translocation plays a pivotal role in adjusting the levels of the active AT receptors. Yet, transcriptional changes occurring during the initial phase of heat acclimation, where mRNA transcripts are upregulated by 50 and 6 orders of magnitude respectively, (Schwimmer et al., 2004), are essential processes for the establishment of adequate AT receptor reserves. Such a transient change is in agreement with the hypothesis that upon transfer to new environmental conditions, increased transcription persists until protein levels reach the appropriate new steady-state levels (Maloyan et al., 1999; Gasch et al., 2000).

Heat acclimation — what do genes and their products tell us?

Our knowledge, particularly in homeotherms, of genomic responses associated with heat acclimation is sparse. Accruing sporadic data on the changes in cellular constitutive proteins suggested to us that the induction of the heat-acclimated phenotype involves the transcription of genes encoding constitutive proteins and stress-inducible molecules (e.g., Eynan et al., 2000, 2002; Maloyan et al., 1999, 2005; Schulte, 2001). For example, alterations in Ca^{2+} regulatory and contractile proteins in various taxa and glycolytic enzymes have been the targets of several investigations (Goldspink et al., 1992; Schulte, 2001; Eynan et al., 2002). Among heat-inducible genes, those encoding the heat shock protein (HSP) 72 kDa and HSP 90 kDa have been the most extensively studied e.g., in mammalian species (Moseley, 1997; Maloyan et al., 1999; Maloyan and Horowitz, 2002). Recently, broad-scale genomic response studies using gene-chip analyses (Horowitz et al., 2004; Schwimmer et al., 2006) expanded our knowledge in this respect.

Heat shock proteins

HSP 72 kDa and HSP 90 kDa confer cytoprotection via chaperoning the correct folding of other proteins or the degradation of abnormal proteins, or, alternatively, via their facilitatory interaction with the molecular signaling of cytoprotective pathways (Morimoto and Santoro, 1998). Heat acclimation has several effects on HSPs namely, (1) hastening of the transcriptional response (Horowitz et al., 1997; Maloyan et al., 1999), and (2) increasing the cellular reserves of the inducible species of HSP 72 in the heat acclimated phenotype. It has been demonstrated that HSP 72 protein is elevated in dwellers of hot environments compared to their matched, normothermic counterparts (Lyashko et al., 1994). Taken together, this implies that in the acclimated phenotype, cytoprotection can be accomplished without de novo HSP synthesis. This above-mentioned buildup of HSP cellular reserves is also fascinating. Due to the fact that heat acclimation in mammals does not induce extreme hyperthermia, mediators other than severely elevated body temperature must be able to increase *hsp* gene transcription. The failure to build up HSP 72 cellular reserves during acclimation in the presence of adrenergic blockade led us to postulate that the excitable sympathetic system, primarily during STHA, is a likely mediator of the transcriptional process. This was confirmed in heat acclimating rats with adrenergic blockade (Maloyan and Horowitz, 2002). These rats demonstrated elevated salivation T-Tsh and markedly decreased salivation and heat endurance (Fig. 2; Shlaier, Y., Eli-Berchoer, B. and Horowitz, M., unpublished observations). Taken together, the results provide causal evidence between augmented HSP 72 reserves and delayed thermal injury, implying that the HSP defense pathway is an integral part of the heat acclimation repertoire. Elevated HSP 72 was detected in hearts (Maloyan et al., 1999), brains (Oppenheim et al., 1996; Arieli et al., 2003), salivary glands, and the evaporative cooling organs of rats (Robinson, Marmary and Horowitz, unpublished observation). This HSP response confers a high degree of protection to the controller and the heat dissipating effectors. The finding that prior heat shock (brief exposure to sublethal temperatures) induces the upregulation of HSP species and affords thermotolerance to Drosophila neuronal synapses, and extends their performance at higher than normal temperatures (e.g., Karunanithi et al., 1999, 2002) further emphasizes the importance of the acclimation-mediated elevation in HSP reserves in delaying thermal injury and maintaining acclimatory homeostasis.

Global genomic responses to heat acclimation

In view of the large number of inducible stress genes responding during acute heat stress (Sonna et al., 2002), we postulated that the induction of HSP genes alone is unlikely to be sufficient to confer heat-acclimation-induced thermotolerance to an organism. Furthermore, HSP 72 by itself can target other genes (e.g., Choi et al., 2005). We, thus hypothesized that heat acclimation involves the activation of more than one molecular cytoprotective network and additional genes, including those not currently among the consensus of the heat-inducible set of genes. Additionally, along with the concept of *heat acclimation*heat acclimation *mediated functional remodeling of neuronal*

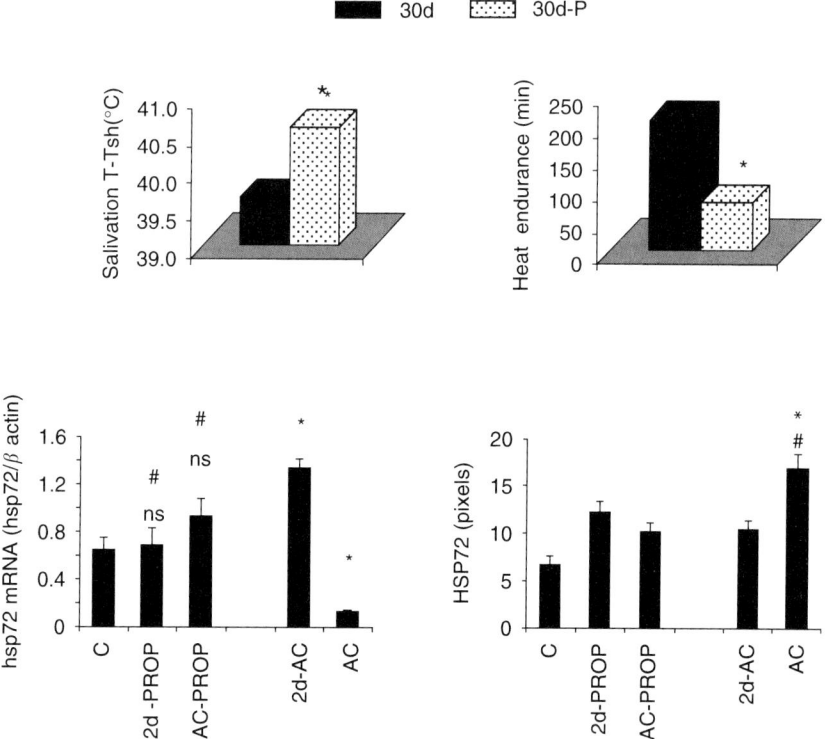

Fig. 2. There is causal evidence between augmented HSP 72 reserves and delayed thermal injury, implying that the HSP defense pathway is an integral part of heat acclimation. Upper panel: Heat acclimation (34°C) for 1 mo, in the presence of adrenergic blockade (propranolol), resulted in elevated salivation T-Tsh (left) and decreased heat endurance (at 40°C, right). (Unpublished data; $n = 5$ per group; Shlaier, Y., Eli-Berchoer, B., Horowitz, M.) Bottom: Heat acclimation for 1 mo, in the presence of adrenergic blockade as above, resulted in decreased hsp 72 transcription on STHA (left) and in turn, greater HSP 72 expression following LTHA (right). For comparison: Heat acclimation leads to intense hsp 72 transcription on STHA leading to the build up of large HSP 72 cellular reserves upon LTHA. *Significant difference from untreated C rats, #significant difference between AC to AC-PROP rats ($p < 0.05$). C = control, AC = heat acclimated, PROP = propranolol, 2d = 2 days. (Adapted from Maloyan et al., 1999; Maloyan and Horowitz, 2002; Shlaier, Eli-Berchoer and Horowitz, unpublished.)

circuits, the storage and processing of information that requires de novo protein synthesis (Armstrong and Stoppani, 2002) is required. Stimuli that cause increased excitability and lead to receptor upregulation/desensitization (e.g., Kloog et al., 1985), and change transmitter release and morphology, all occur during acclimation, and effect transcriptional activation. Hence, a broad-scale genomic approach could be used to expand our knowledge of gene families that have not yet been associated with heat acclimation. Among homeotherms, the limited data available are derived from gene-chip analyses and studies of targeted genes that change under stressful environmental conditions.

Global genomic responses — a lesson from genechip analyses

To achieve a broad scale genomic approach we used cDNA Atlas arrays of genes representing homeostatic responses and stress-associated genes (Horowitz et al., 2004; Schwimmer et al., 2006) respectively, using RNA from rat hypothalamic and heart tissues sampled during the course of heat acclimation. This method allowed us to study the dynamics of acclimation as mirrored by changes in the expression of the responding genes. Due to the resemblance we found between the major inducible cytoprotective networks in the brain and heart, we could take an integrative approach.

Clustering all visible genes during the acclimation period, based on their mutual expression behavior, provided a dynamic profile of the global genomic response throughout the acclimation process. Temporal variations found in the expression of the different clusters suggested a continuum of changes in gene activation/silencing. Whereas several clusters of genes were phase specific, others showed a persistent response. Among the stress responders (Horowitz et al., 2004), four different gene profiles were delineated: (1) genes for which acclimation influenced only their basal transcript level; (2) genes for which acclimation only affected their response to heat stress; (3) genes for which acclimation affected both their basal steady-state level and their response to heat stress; and (4) genes that responded independently to heat acclimation and heat stress. Notable were those genes assigned to category (1), representing transient upregulation in the heart (Horowitz et al., 2004) and up/down regulation in the brain (Schwimmer et al., 2006); and (http://www.ncbi.nlm.nih.gov/geo/ GEO Series accession number genbank:GSE2890), peaking at STHA and then silencing or resuming preacclimation levels following LTHA. Most genes assigned to this category are associated with maintaining DNA integrity, e.g., topoisomerase II and ERCC1, both engaged in double-stranded DNA repair and DNA synthesis (Gaffney et al., 2000; Kultz and Chakravarty, 2001; Zhang et al., 2003), and with free-radical scavenging (e.g., cytochrome b reductase, gluthathion s transferase). Collectively, this genomic response confirms that the onset of heat acclimation is a stressful phase, in which the organism encounters the release of cytochrome P450 and reactive oxygen species that can lead to DNA damage (Kultz and Chakravarty, 2001; Zhang et al., 2003). Both damaged DNA and damaged proteins (Kultz and Chakravarty, 2001) are likely inducers of the transient cellular stress response via the activation of target genes and perhaps switching on long-lasting acclimatory responses. At least in the heart, neither of the transiently induced genes respond to acute heat stress, suggesting that during the acclimation process sustained heat strain (such as that imposed at the initial phase of heat acclimation) is required for their induction.

Stress-associated genes classified under additional categories (2–4) include those that respond to heat stress (following LHTA) by a more profound transcriptional activation than that found in the non-acclimated phenotype. A high percentage of these expressed genes are stress-response regulators and effectors, and their activation possibly enhances a heat-acclimation-specific cytoprotective signaling network as a coping mechanism against the perturbations in the acclimatory homeostasis achieved following LTHA. Overall, these genes are assigned to three major cytoprotective networks: (1) HSPs, (2) antiapoptotic, and (3) antioxidative (Horowitz et al., 2004, and http://www.ncbi.nlm.nih.gov/geo/ GEO Series accession number genbank:GSE2890).

In summary, the delineated transcriptional program of the expressed stress genes indicates a two-tier defense strategy. Whereas the immediate transient response is associated with the maintenance of DNA and cellular integrity during the strain developed at the onset of acclimation, the sustained response correlates with adaptive, long-lasting cytoprotective signaling networks.

Considering the acclimated phenotype and acclimation dynamics, one should question whether the hypothalamus responds with a genomic expression program of genes linked with physiological integrative functions to match the stress loads at each acclimation phase. The clustering and functional analyses of the expression profile of a battery of genes representing various central regulatory functions of body homeostasis demonstrated, similar to the stress-response profile, a bi-phasic acclimation profile which, indeed, explains the molecular basis underlying our integrative physiological model of heat acclimation (Schwimmer et al., 2006). The initial phase of heat acclimation (Phase I — STHA), characterized by a transient upregulation or down regulation of genes encoding voltage-gated ion channels, ion pumps, or transporters, as well as hormone/transmitter receptors and cellular messengers, collectively points to enhanced membrane depolarization, and in turn, a release of neurotransmitters and neuronal excitability. Concomitantly, a transient down regulation among the groups of genes participating in intracellular protein trafficking, metabolism, or phosphorylation processes, e.g., Janus kinases phosphodiesterase (JANKa, genbank:U13396, genbank:D28508), implies a perturbation in cellular maintenance. Following LTHA (Phase II), most genes returned to pre-acclimation expression levels. Genes encoding hormones and neuropeptides, linked with metabolic rate and food intake, were down regulated in both STHA and LTHA. Such a temporal genomic profile is consistent with our previously established integrative acclimation model, demonstrating enhanced autonomic excitability during STHA and its resumption upon LTHA (Horowitz and Meiri, 1985a), Aldolase A (Canete-Soler et al., 2005), and myelin basic protein (Korolainen et al., 2006). Collectively, the data now at hand allow us to construct an integrative model showing a physiologic–genomic linkage during the course of heat acclimation (Fig. 3).

The cross-tolerance and interference phenomena

Adolph, in his chapter on "Perspectives of adaptation: some general properties" (Adolph, 1964), infers the phenomenon of cross-tolerance (although not regarding heat acclimation); however the underlying cellular mechanism was not described until recently, when molecular studies were performed (Horowitz et al., 1997, 2004; Maloyan et al., 1999, 2005; Horowitz, 2002; Maloyan and Horowitz, 2002; Arieli et al., 2003). We now have evidence that heat acclimation induces cross-tolerance against several environmental stressors associated with altered oxygen supplementation (hypoxia, ischemia/reperfusion, hyperoxia (Levi et al., 1993; Arieli et al., 2003)), ionizing radiation (Robinson, Horowitz and Marmary, in preparation), and several other traumatic situations (noise, traumatic injury (Shohami et al., 1994; Paz et al., 2004)). In contrast to the beneficial effects of cross-tolerance, there are environmental stressors that can interfere with, or alternatively, override heat acclimation. For example, when hypohydration stress was superimposed on acclimated animals undergoing heat stress, the benefits of heat acclimation on heat endurance were marked (Horowitz and Meiri, 1985b; Schwimmer et al., 2006, 2004).

We hypothesize that cross-tolerance or interference emerges from alterations in the capacity or responsiveness of molecular signaling pathways shared by the primary and the secondary stressors (heat tolerance, PO_2 or hypohydration, respectively). Our well established heat acclimation model (Fig. 2) allows us to test this hypothesis. Two approaches will be discussed below: (1) comparison of the pathways that are alerted during heat acclimated to those responding to the novel stressor, and (2) examination of the rate of response of these pathways to the novel stressor. The novel environmental stressors for which data are available include oxygen deprivation, hyperoxia, noxious trauma, and hypohydration.

Heat acclimation — oxygen deprivation cross-tolerance

Oxygen deprivation results in several well characterized negative effects. Some of the pathways/mechanisms leading to these effects are tissue specific, whereas others are shared. The brain and the heart are among the most susceptible organs to impaired oxygen delivery. Therefore, the discovery of the protective role of ischemic/hypoxic/thermal preconditioning against cerebral and cardiac ischemia enhanced our understanding of the signaling pathways involved. Preconditioning is an inducible molecular response to stress, where protection is conveyed by prior sublethal exposures to the particular stress or to another stress (Xi et al., 2001), providing that there is cross-tolerance between the two stresses. This preconditioning effect is considered to be part of a rapid adaptive mechanism induced by many physiological and pharmacological stressors, and shared by numerous cells and organs. The preconditioning effect is of limited duration and represents the cellular attempt to protect structure and function from stress-induced damage. Classical preconditioning is immediate and lasts 1–2 h, involving the activation of adenosine receptors and mitochondrial K ATP dependent channels. A second window of protection emerges 24–48 h later and lasts for ~2 days (Sharp et al., 2004; Hausenloy et al., 2005), in heart and brain, respectively. While the immediate

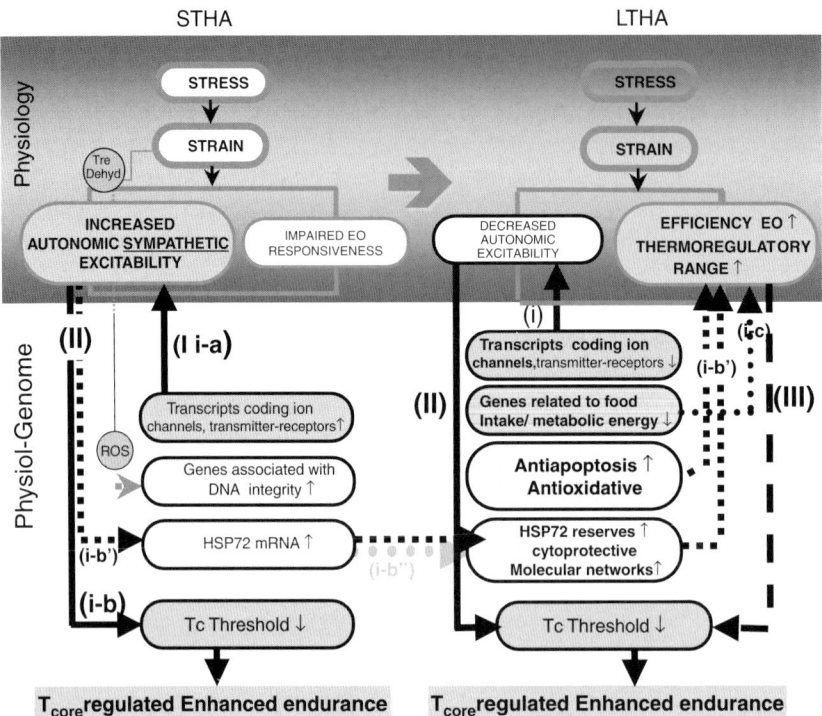

Fig. 3. Conceptual model of heat acclimation: *Molecular–Physiological linkage*. Heat acclimation is a bi-phasic process; at the onset of acclimation (STHA — 2d at 34°C, 30–40% R.H. — left cascade) increased autonomic excitability activates transient defense mechanisms to alleviate the initial strain. Upon long-term heat acclimation (LTHA — 30d at 34°C, 30–40% R.H. — right cascade) acclimatory homeostasis has been achieved. The upper area of the model (STHA and LTHA cascades) illustrates integrative physiological responses whereas the lower area represents the molecular mechanisms underlying the integrative responses. Black lines, (continuous and broken) indicate identified connections. Light gray broken lines indicate hypothetical pathways. STHA: [I i–a] exhibits hypothalamic molecular responses associated with stressful conditions, leading to increased excitability and in turn, altered T-Tsh for activation of heat-dissipation effectors (pathway i–b). Concomitantly, increased sympathetic activity leads to increased *hsp* 72 mRNA in peripheral and central organs (i–b) and in turn, greater HSP 72 reserves upon LTHA. The stress response enhances transient activation of genes linked to maintenance of DNA integrity. LTHA: Decreased T-Tsh is accompanied by enhanced activation of cytoprotective networks, thus delaying the Tsh for thermal injury (pathways (i–b)). Collectively, the leftward and the rightward shifts in salivation of T-Tsh and thermal injury Tsh, respectively, expand the dynamic thermoregulatory temperature range. Down regulation of genes involved with metabolism and food intake (i–c) decrease heat production and possibly integrate with additional cellular mechanisms leading to the LTHA-enhanced-efficiency (pathway III). Pathway i–b (light gray pathway) denotes hypothetical pathway representing an orchestrated mechanism leading to upregulation of cytoprotective molecules. These mechanisms await investigation. (This model was compiled from data in Horowitz et al., 1983, 2004; Horowitz and Meiri, 1985a; Horowitz, 1998, 2002; Maloyan et al., 1999; Maloyan and Horowitz, 2002; Schwimmer et al., 2006.)

response stems from the activation of adenosine and salvage kinases, which were studied in detail in the heart (Hausenloy et al., 2005), the second window of protection requires new RNA and protein synthesis and activates a host of pathways including HSPs and HIF-1 targeted pathways (e.g., glycolytic enzymes, glucose transporters, iNOS cascade, erythropoietin). Mitochondrial K ATP that down regulates the mitochondrial transition permeability pores, is shared by all organs exhibiting preconditioning effects (Sharp et al., 2004). Organ specific pathways also contribute. Because the late window of protection allows time for transcriptional and translational processes, and because many pharmacologic agents can mimic the effects of ischemia in inducing the late phase of preconditioning, this phenomenon might be exploited for therapeutic intervention.

Heat acclimation mediated ischemic and hypoxic cross-tolerance differs from both the classical and the 2nd window of protection because it lasts for weeks, rather than days. The phenomenon that heat acclimation reinforces endurance during oxygen shortage was initially discovered by Hiestand et al. (1955) who observed that heat acclimated mice can survive drowning. This protection was attributed to decreased thyroxine levels occurring upon heat acclimation. A focused extensive study on heat acclimation–oxygen deprivation cross tolerance is currently continuing in our laboratory. Early studies conducted on the hearts of heat acclimated rats, were mostly at integrative levels. Hemodynamic recovery of heat acclimated hearts from either ischemic/reperfusion or hypoxic insults is markedly enhanced compared to non-acclimated hearts. ^{32}P and ^{13}C NMR studies provided solid evidence that enhanced anaerobic ATP generation, at a slower rate (favoring intracellular pH regulation), and decreased ATP utilization during global ischemia, were essential to the ischemic protection achieved (Levi et al., 1993; Eynan et al., 2002).

In order to gain insights into the master regulators of the cellular mechanisms involved with this cross-tolerance response, we were guided by (1) the finding that heat acclimation *per se* enhances the HSP cytoprotective networks and (2) the fact that enhanced glycolysis, greater glycogen stores, and elevated levels of the glucose transporter (GLUT1) found following heat acclimation are all fingerprints of HIF-1α — the master regulator of oxygen homeostasis mediated cascades (Semenza, 2004). We showed that, similarly to the effects of heat stress (see Heat shock proteins, this chapter), ischemic insult in heat acclimated animals induces faster *hsp* 72 transcription, to maintain the HSP 72 stores (Maloyan and Horowitz, unpublished). Concomitantly, we provided substantial evidence that in rats and mice, HIF-1α expression increased significantly, under normoxic conditions in both heart and brain tissues following long-term heat acclimation. The functional role of HIF-1α in heat acclimation was confirmed by characterization of the HIF-1α activation profile, namely its binding capacity to the hypoxia response element and subsequent increased transcription of target genes such as vascular endothelial growth factor (VEGF), following heat acclimation with superimposed ischemia (Maloyan et al., 2005). *Epo* gene and *EpoR* (Erythropoietin and Erythropoietin receptor, respectively) transcripts were upregulated in response to acclimation and this transcription level was maintained upon exposure to the additional stress. *In accordance with our concept of heat-acclimation-mediated cytoprotection* cytoprotection *via larger cytoprotective protein reserves*, the markedly enhanced levels of HIF-1α and EpoR, which mediates the protective function of erythropoietin via targeting metabolic and anti-apoptotic cascades, ensures that the heat acclimated ischemic heart is protected and thus provides a potential molecular basis for cross-tolerance-mediated mechanisms. It is noteworthy that heat acclimation leads to upregulation of *Epo* transcript in the kidney, the major site of its production in our body, implying that acclimation has global effects (Maloyan et al., 2005). The larger HSP 72 reserves in the brains of heat acclimated rats (Oppenheim et al., 1996; Arieli et al., 2003) and higher basal stores of nuclear HIF-1α and EpoR in heat acclimated mice brains (Shein et al., 2005) confirm this conclusion.

Taken together, we can conclude that enhanced reserves and the faster response of cytoprotective components are important hallmarks of heat-acclimation-mediated cross-tolerance. Given that both HSPs and HIF-1α are documented as putative factors in preconditioning it is likely that heat acclimation mediated cross-tolerance utilizes many of the consensus salvage pathways of preconditioning. However, considering that the duration of the heat acclimation/cross-tolerance cytoprotective time scale is profoundly longer (than preconditioning), perhaps due to different responsiveness of the numerous cytoprotective networks, the mechanisms involved require further investigation.

HSP 72 and HIF-1α play putative protective role in neuroprotection achieved by altitude adaptation (Chavez et al., 2000; Ramaglia and Buck, 2004). It is thus reasonable to believe that protective pathways induced by these master regulators are also associated with heat acclimation induced neuroprotection. Accordingly, data accumulated in our laboratory (Jacobi, A., Stern Bach, Y. and Horowitz, M., unpublished observations) confirm heat acclimation mediated neuroprotection during whole body hypoxia (14.5% PO_2, 15 min). Protection manifested as improved outcome in cognitive tests. A new piece in our cross-tolerance mechanism puzzle was the decrease in NMDA receptor levels in heat acclimated brains (hippocampus and frontal cortex). This decrease occurred coincidentally with a change in the receptor subunit ratio, favoring neuroprotection via decreased Ca^{2+} influx (Erreger et al., 2005; Shin et al., 2005) during the hypoxic insult (Jacobi, A., Stern Bach, Y. and Horowitz, M., unpublished observations).

Interestingly many of the adaptive protective mechanisms emerging following heat acclimation are similar to those characterizing anoxic tolerant species (Lutz and Nilsson, 1997). The brain of the freshwater turtle *Trachemys scripta* can withstand anoxia for days at room temperature. Elevated basal HSP 72 level, enhanced HIF-1α mediated pathways, greater glycogen stores, and changes in the NMDA receptors as well as altered activity of the K ATP dependent channels are among the constitutive features that predispose to tolerance to anoxia (Pek-Scott and Lutz, 1998; Ramaglia and Buck, 2004; Milton and Lutz, 2005; Gorr et al., 2006). The similarity in these constitutive adaptations to brain hypoxia in fish, reptiles, and mammals suggest that this adaptation is conserved throughout evolution. It is therefore likely that heat and altitude "within lifetime" acclimations recapitulate evolutionary adaptation.

Ischemia/reperfusion and cross-tolerance — a lesson from the global genomic response

We propose that cross-tolerance emerges from an enhanced capacity or responsiveness of molecular signaling shared by adaptation to the primary stress (heat tolerance) and the secondary stress (oxygen shortage) and its consequences. This hypothesis was tested in a battery of stress-associated genes in the heart. Gene-cluster analysis showed a divergence between genes responding to heat stress and those responding to ischemia/reperfusion insult. A noteworthy finding was also the earlier activation threshold of e.g., HSP 70 transcript to the insult (Fig. 4). Yet, in accordance with our

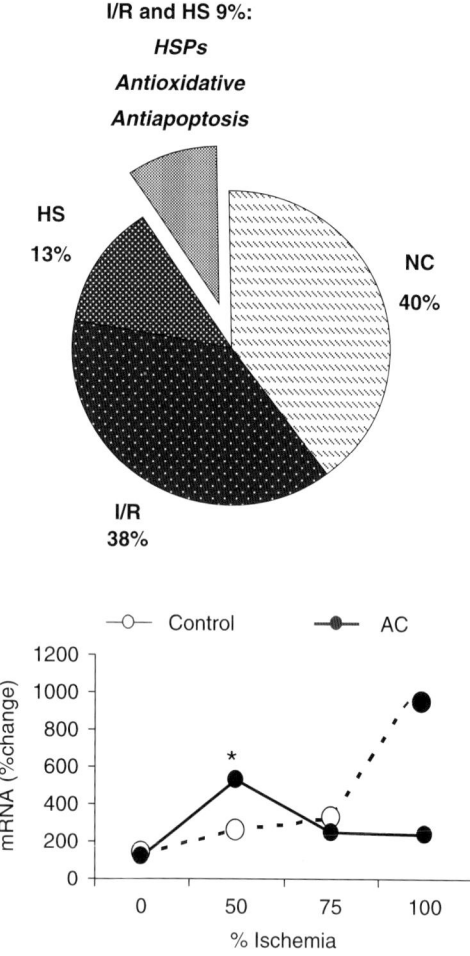

Fig. 4. Upper panel: Cluster analysis of stress-associated genes from hearts subjected to ischemia/reperfusion (I/R) insult or from heat-stressed rats. The percentage of genes responding to I/R and/or heat stress is shown. Among the "shared" genes, 60% showed an enhanced response to heat stress and/or to I/R insult, but only after heat acclimation. The functional categories to which the genes are assigned are listed. Bottom: Percent change of *hsp* 72 steady state mRNA level in heart tissue during progressively graded ischemia. The molecular response of the heat acclimated hearts was markedly faster than that of the non-acclimated controls. (Adapted from Horowitz et al., 2004.)

hypothesis that "shared signaling cascades" are the underlying mechanism of cross-tolerance, there were several genes with enhanced responses to heat stress and to I/R insult after heat acclimation only. The cellular functions associated with these genes represent (1) anti-apoptosis and the maintenance of protein integrity networks (e.g., decreased bcl-2

(BAD), JNK$_{1\&2}$, and flavin containing monooxygenase (FMO)) and upregulation of DNAJ2; and (2) maintenance of DNA and chromatin integrity (e.g., upregulation of DNA mismatch/repair protein, down regulation of MPI (M-phase inducer phosphatase 2, which inhibits damaged DNA signaling following genotoxic stress) (Horowitz et al., 2004)). We hypothesize that the profound change in the expression of these genes reinforces inherent coordinated cardioprotective mechanisms and contributes to the cytoprotective aspects of the cross-tolerance phenomenon. The discussion on the role of HIF-1α and targeted pathways (Heat acclimation — oxygen deprivation cross-tolerance), which is highly expressed during both heat stress and ischemic reperfusion in hearts and brains of heat acclimated animals, further strengthens the postulate of alerted shared pathways and recapitulation of evolutionary adaptation.

Heat acclimation cross-tolerance against central nervous system oxygen toxicity

Hyperbaric/hyperoxic adverse conditions leading to oxygen toxicity may be encountered in artificial/abnormal environmental conditions such as during diving with an oxygen-enriched mixture or using a closed-circuit apparatus during hyperbaric medical treatments, or alternatively, following catastrophic accidents while at sea or underground (Dean et al., 2003). Under these adverse conditions O_2 toxicity may result from oxidative stress coupled with a failure of the antioxidant defenses to overcome excessive production of reactive oxygen species. This O_2 toxicity impairs in ionic conductance, and synaptic transmission, and causes DNA damage etc. (Dean et al., 2003). Arieli et al. (2003) demonstrated that heat acclimation confers protection against oxygen toxicity in the brain by prolonging the threshold of neurotoxicity. This protection continues for 3 weeks of de-acclimation, and positively correlates with the ability to maintain HSP 72 levels (Fig. 5). The same authors found that the antioxidant CuZnSOD, is also elevated by acclimation. In contrast to HSP 72, CuZnSOD remains elevated during de-acclimation (Eynan et al., 2005). The reported time scale for de-acclimation

agrees with that reported for the heart (Cohen et al., 2001). The causal relationship between HSP 72 and oxygen-toxicity neuroprotection implies that this chaperon plays an integrative role in neuroprotection induced by heat acclimation. HSP 72 promotes neuroprotection via multimodal effects, leading to the preservation of protein structure, RNA and DNA integrity, and reduced NO levels. The role played by inducible HSP 70 on antioxidant levels has also been documented (Ohtsuki et al., 1992; Choi et al., 2005). When examining the genomic profile of the heat acclimated "cytoprotected" phenotype (Horowitz et al., 2004 and http://www.ncbi.nlm.nih.gov/geo/ GEO Series accession number genbank:GSE2890), we can conclude that heat acclimation hyperoxic protection reinforcement fits with our basic prediction of shared pathways as the underlying cytoprotective mechanism of cross-tolerance.

Heat acclimation cross-tolerance against closed head injury (CHI)

An additional aspect of the cross-tolerance mechanism is its involvement in the attenuation of the damage induced by closed head injury (CHI) and the acceleration of clinical recovery. The manifestation of the integrative profile of this cross-tolerance effect is summarized in Table 1. Biochemical evidence, indicative of genomic responses, demonstrates the upregulation of low molecular weight antioxidants (LMWA) together with improved clinical recovery after subjection of heat acclimated rats to CHI compared to decreased LMWA and delayed recovery in non-acclimated rats (Beit-Yannai et al., 1997). This suggests that the antioxidative pathways have enhanced capacity upon heat acclimation. Considering the importance of HIF-1α (Bergeron et al., 2000) and erythropoietin (a target of HIF-1α transcription factor) in neuro- and cardioprotection (Bernaudin et al., 1999; Cai et al., 2003, respectively), together with the finding that HIF-1α is an integral component of heat acclimation/and cross-tolerance against oxygen deprivation repertoire (Maloyan et al., 2005; Bromberg and Horowitz, unpublished). Shein et al. (2005) tested the role of HIF-1α-erythropoietin axis in the

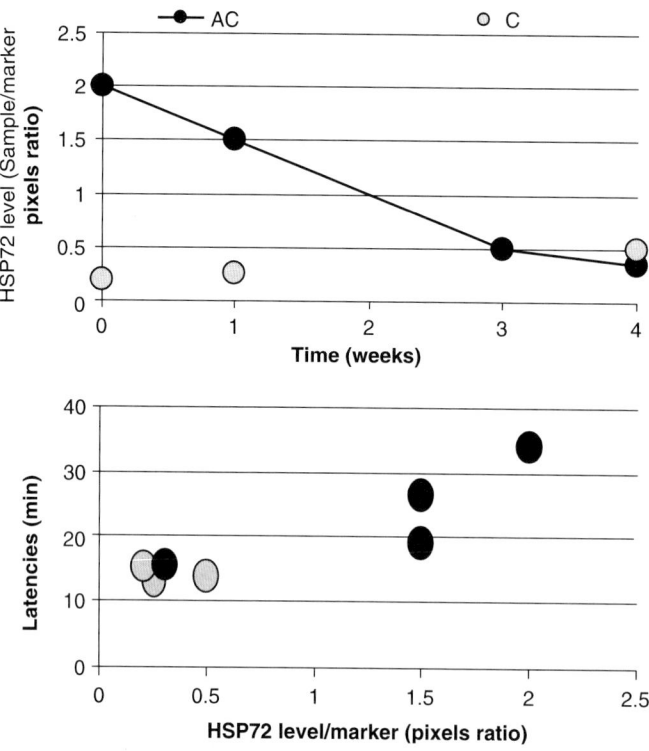

Fig. 5. Heat acclimation confers protection against oxygen toxicity in the brain by prolonging the threshold of neurotoxicity. There is causal evidence between HSP 72 levels and the protection observed. Upper panel: Brain HSP 72 levels during 4 weeks of de-acclimation. Bottom: Latency to CNS oxygen toxicity vs. HSP 72 levels during the course of de-acclimation. Black circles: heat acclimation; gray circles: non-acclimated. (Adapted from Arieli et al., 2003.)

recovery of heat acclimated mice from CHI. Together with improved functional recovery of the heat acclimated mice vs. the non-acclimated, increased expression of nuclear HIF-1α and EpoR were detected providing causal evidence that HIF-1α is involved in the heat acclimation cross-tolerance phenomenon.

Does cross-tolerance depend on alerted shared signaling pathways — conclusion

Predisposition to improved coping with novel stressors following heat acclimation, mirrored by genomic responses, has been studied for oxygen deprivation, hyperoxia and CHI. Despite the sparse data available, and the inconsistency in approaches used and pathways studied by the different authors, it is apparent that (1) the protective networks enhanced by heat acclimation are exploited by the novel stressors and (2) the same mediators render protection against opposing stressors (e.g., hyperoxia vs. O_2 deprivation). Enhancement is dual; an upregulated constitutive component, as well as faster molecular responses such as those seen in the HSP system in the heart (Maloyan et al., 1999) and the upregulation of LMWA in the brain and the heart following CHI (Beit-Yannai et al., 1997). Given that evolvement of the acclimated phenotype depends on (1) a continuum of changes in gene expression and (2) up/down regulation in gene and protein expression when acclimatory homeostasis is achieved, it is likely that (with the exception of the common "master regulators") organ specific functional remodeling contributes to the cross-tolerance

Table 1. Evidence for cross-tolerance between heat acclimation and closed head injury (CHI)

Measure	Rats						Mice				Reference
	Con		Heat-AC				Con		Heat-AC		
Time post trauma (h)											
Time	0	48	0	48			0	1	0	1	
Neurological severity score; scale: 0–20; 0 = maximal recovery	17.5	9.0	16.6	4.0*							Shohami et al. (1994)
Edema formation (%H_2O)		84.0		81.5*				83		81.5*	Shohami et al. (1994), Shein et al. (2005)
BBB disruption; Evans blue (ng/g tissue)		315.0		50.0*							Shohami et al. (1994)
Cognitive test (min to explore new object; sham vs. post trauma)							59 vs. 49		72 vs. 51*		Shein et al. (2005)

*Significant difference from non-acclimated at the marked time point ($p < 0.05$).

Table 2. Heat acclimation mediated cross-tolerance — evidence of exploitation of enhanced-altered cytoprotective pathways by novel stressors

	Heat acclimation				
	Cytoprotection				Metabolic
	HSPs	Antiapoptosis	Antioxidation		HIF-1
Heat acclimation	↑	↑	↑		↑
Heat stress	↑	↑	↑		↑
Oxygen deprivation	↑	↑	↑		↑
Hyperoxia	↑		↑		
CHI	↑		↑		↑

Note: In each row, the cytoprotective networks exploited by the novel stressor are marked. Empty cells denote "unavailable." Data were compiled from Oppenheim et al. (1996), Beit-Yannai et al. (1997), Arieli et al. (2003), Horowitz et al. (2004), Maloyan et al. (2005), and Shein et al. (2005).

mechanism (as evident from the NMDA receptors). Table 2 enlists the "shared" pathways in the cross-tolerance set up.

Hypohydration masks the beneficial effects of heat acclimation

Thermal acclimation depends on the regulation of shifts in body fluid during heat loads. Hypohydration affects the sensitivity to thermal stress. Under conditions of hypohydration, the T-Tsh for evaporative cooling is markedly elevated as is body temperature. There also is increased reliance on the physical pathways for heat dissipation to regulate body temperature. Hypohydration affects sensitivity to thermal stress differently at each acclimation phase, depending on fluid compartmentalization dynamics (Horowitz and Samueloff, 1979; Horowitz, 2001). For example, our previous findings (Horowitz, 2001) revealed that heat acclimation leads to a greater exploitation of the intracellular water compartment for evaporative cooling. This conserves circulating blood volume, so that it can be used for heat dissipation. Yet, following hypohydration, the similarity in heat endurance of LTHA rats to that of non-acclimated animals suggests that hypohydration abrogates the effect of heat acclimation (Fig. 6). Physiological studies, based on T-Tsh data infer that hypohydration desensitizes the central receptors of neuromodulatory pathways (Horowitz et al., 1999; Schwimmer et al., 2004). However, the phenomenon of receptor desensitization alone does not adequately explain the underlying cellular mechanisms that interfere with the benefits of heat acclimation. Using a global genomic approach, it is evident that in the hypothalamus, hypohydration upregulates a multitude of genes, and most of the upregulated functional groups comprise genes encoding ion channels, transporters, transmitters, hormones linked with cell volume regulation, neuronal excitability, and immune system proteins (Fig. 6 and Schwimmer et al., 2006). This upregulated gene profile is similar for non-acclimated and LTHA hypothalami, and resembles that of the "impaired" STHA euhydrated state. However, physiologically there was a discrepancy between T-Tsh for evaporative cooling and thermal endurance observed for STHA (decreased) and hypohydration (elevated) even though the gene profile was similar (Schwimmer et al., 2006). The two states also differed in the number of down regulated genes; however, we cannot provide a purely genomic explanation of the interference noted. Water loss in acclimated-hypohydrated heat-stressed rats is significantly higher than that of control rats (Schwimmer et al., 2006). Considering that the salivary glands of heat-acclimated rats (the evaporating cooling effector of this species) increase their output/stimulus ratio (Horowitz, 1998, 2002) when acclimated, we hypothesize that the disruption of acclimatory thermal endurance in the hypohydrated state is due to a failure in adjusting the secretion volume at the glandular level rather than to a central failure.

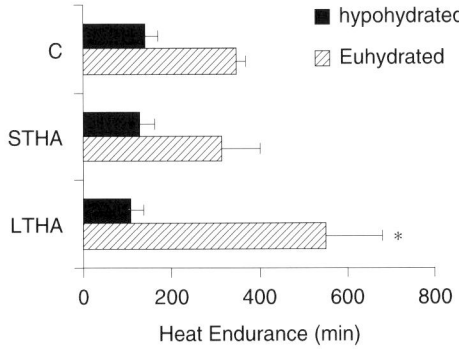

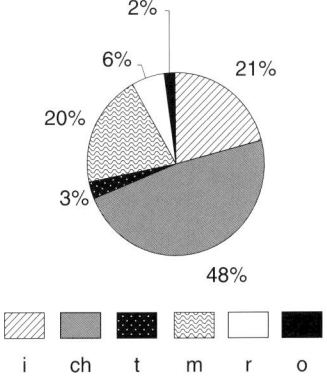

Fig. 6. Upper panel: Heat endurance upon subjection to heat stress at 39°C. Heat-acclimated rats improve heat endurance by Δ 125% compared to non-acclimated rats. Water deprivation, resulting in 10% loss of body weight, abrogates this beneficial effect at each acclimatory phase. C: controls non-acclimated, STHA: short term heat acclimation, LTHA: long term heat acclimation. Bottom: Global genomic response to hypohydration (10% loss of body weight) in LTHA rats. Distribution of genes showing significant upregulation in their expression levels (⩾1.5 fold) according to functional categories. i: immune response, ch: ion channels, transporters, pumps, t: trafficking, m: metabolism, r: receptors, o-kinases and proteases. (Data were compiled from Schwimmer et al., 2006. For detailed list of genes see http://www.ncbi.nlm.nih.gov/geo/

Summary

This review, based on the link between early physiological and recent genomic studies, summarizes the progression of our concepts regarding biochemical neuromodulation, and the plasticity of central and peripheral mechanisms leading to heat acclimation homeostasis. A novel approach in the field of heat acclimation in mammalian species is the employment of gene-chip technology and molecular analyses of identified physiological features. Many of the processes are still enigmatic; however, the data accumulated provide direction for further detailed studies. This approach has already revealed novel acclimatory pathways such as the exploitation of HIF-1 by a variety of physiological normoxic adaptive mechanisms requiring both metabolic changes and enhancement of cytoprotective networks, thus offering a genomic explanation to both thermal endurance and the cross-tolerance mechanisms.

Acknowledgment

The research of the author of this chapter was supported over the years by the ISF, Israel Science Foundation, founded by the Israel Academy of Sciences and Humanities. Studies in collaboration with R. Gerstberger were supported by the German Israeli Foundation Research Grants to M. Horowitz and R. Gerstberger. His contribution is highly appreciated. I am also deeply indebted to Sharon Robinson from The Faculty of Dental Medicine, The Hebrew University, for valuable comments.

References

Adolph, E. (1964) Perspectives of adaptation: some general properties. In: Dill D.B., Adolph E.F. and Wilber C.G. (Eds.), Handbook of Physiology: Adaptation to the Environment. Am. Physiol. Soc., Washington D.C., pp. 27–36.

Arieli, Y., Eynan, M., Gancz, H., Arieli, R. and Kashi, Y. (2003) Heat acclimation prolongs the time to central nervous system oxygen toxicity in the rat: possible involvement of HSP72. Brain Res., 962: 15–20.

Armstrong, L. and Stoppani, J. (2002) Central nervous system control of heat acclimation adaptations: an emerging paradigm. Rev. Neurosci., 13: 271–285.

Attias, J., Gold, S., Sohmer, H., Schmuel, M., Horowitz, M. and Shahar, A. (1988) Effects of long-term heat exposure on the auditory nerve-brainstem evoked response. J. Therm. Biol., 13: 175–177.

Beit-Yannai, E., Kohen, R., Horowitz, M., Trembovler, V. and Shohami, E. (1997) Changes of biological reducing activity in rat brain following closed head injury: a cyclic voltammetry study in normal and heat-acclimated rats. J. Cereb. Blood Flow Metab., 17: 273–279.

Bergeron, M., Gidday, J., Yu, A., Semenza, G., Ferriero, D. and Sharp, F.R. (2000) Role of hypoxia-inducible factor-1 in hypoxia-induced ischemic tolerance in neonatal rat brain. Ann. Neurol., 48: 285–296.

Bernaudin, M., Marti, H.H., Roussel, S., Divoux, D., Nouvelot, A., MacKenzie, E.T. and Petit, E. (1999) A potential role for erythropoietin in focal permanent cerebral ischemia in mice. J. Cereb. Blood Flow Metab., 19: 643–651.

Bligh, J. (1973) Temperature Regulation in Mammals and Other Vertebrates. North-Holland, Amsterdam.

Blume, A., Neumann, C., Dorenkamp, M., Culman, J. and Unger, T. (2002) Involvement of adrenoceptors in the angiotensin II-induced expression of inducible transcription factors in the rat forebrain and hypothalamus. Neuropharmacology, 42: 281–288.

Boulant, J.A. (1998) Cellular mechanisms of temperature sensitivity in hypothalamic neurons. Prog. Brain Res., 115: 3–8.

Boulant, J.A. (2006) Neuronal basis of Hammel's model for set-point thermoregulation. J. Appl. Physiol., 100: 1347–1354.

Cai, Z., Manalo, D.J., Wei, G., Rodriguez, E.R., Fox-Talbot, K., Lu, H., Zweier, J.L. and Semenza, G.L. (2003) Hearts from rodents exposed to intermittent hypoxia or erythropoietin are protected against ischemia-reperfusion injury. Circulation, 108: 79–85.

Canete-Soler, R., Reddy, K.S., Tolan, D.R. and Zhai, J. (2005) Aldolases A and C are ribonucleolytic components of a neuronal complex that regulates the stability of the light-neurofilament mRNA. J. Neurosci., 25: 4353–4364.

Chavez, J.C., Agani, F., Pichiule, P. and LaManna, J.C. (2000) Expression of hypoxia-inducible factor-1alpha in the brain of rats during chronic hypoxia. J. Appl. Physiol., 89: 1937–1942.

Choi, S., Park, K., Lee, H., Park, M.S., Lee, J., Park, K., Kim, M., Lee, S., Seo, J. and Yoon, B. (2005) Expression of Cu/Zn SOD protein is suppressed in hsp 70.1 knockout mice. J. Biochem. Mol. Biol., 38: 111–114.

Christman, J. and Gisolfi, C. (1980) Effects of repeated heat exposure on hypothalamic sensitivity to norepinephrine. J. Appl. Physiol., 49: 942–945.

Christman, J. and Gisolfi, C. (1985) Heat acclimation: role of norepinephrine in the anterior hypothalamus. J. Appl. Physiol., 58: 1923–1928.

Cohen, O., Stern, M. and Horowitz, M. (2001) Heat acclimation improves cardiac contractility and ischemic tolerance. Is acclimation memorized? J. Mol. Cell. Cardiol., 33: A22.

Dean, J.B., Mulkey, D.K., Garcia III, A.J., Putnam, R.W. and Henderson III, R.A. (2003) Neuronal sensitivity to hyperoxia, hypercapnia, and inert gases at hyperbaric pressures. J. Appl. Physiol., 95: 883–909.

Erreger, K., Dravid, S.M., Banke, T.G., Wyllie, D.J.A. and Traynelis, S.F. (2005) Subunit-specific gating controls rat NR1/NR2A and NR1/NR2B NMDA channel kinetics and synaptic signalling profiles. J. Physiol. (Lond.), 563: 345–358.

Eynan, M., Arieli, Y., Ertracht, O., Gancz, H., Arieli, R. and Kashi, Y. (2005) Protection against CNS oxygen toxicity induced by heat acclimation involves increased levels of HSP72 and CuZnSOD. Proc. IUPS, San-Diego (abst.).

Eynan, M., Gross, C., Hasin, Y., Palmon, A. and Horowitz, M. (2000) Changes in cardiac mechanics with heat acclimation: adrenergic signaling and SR-Ca regulatory proteins. Am. J. Physiol. Regul. Integr. Comp. Physiol., 279: R77–R85.

Eynan, M., Knubuvetz, T., Meiri, U., Navon, G., Gerstenblith, G., Bromberg, Z., Hasin, Y. and Horowitz, M. (2002) Heat acclimation-induced elevated glycogen, glycolysis, and low thyroxine improve heart ischemic tolerance. J. Appl. Physiol., 93: 2095–2104.

Gaffney, D., Lundquist, M., Warters, R. and Rowley, R. (2000) Effects of modifying topoisomerase II levels on cellular recovery from radiation damage. Radiat. Res., 154: 461–466.

Gagne, J., Gelinas, S., Martinoli, M., Foster, T., Ohayon, M., Thompson, R., Baudry, M. and Massicotte, G. (1998) AMPA receptor properties in adult rat hippocampus following environmental enrichment. Brain Res. Mol. Brain Res., 799: 16–25.

Gasch, A.P., Spellman, P.T., Kao, C.M., Carmel-Harel, O., Eisen, M.B., Storz, G., Botstein, D. and Brown, P.O. (2000) Genomic expression programs in the response of yeast cells to environmental changes. Mol. Biol. Cell, 11: 4241–4257.

Gohlke, P., Pees, C. and Unger, T. (1998) AT2 receptors stimulation increases aortic cyclic GMP in SHRSP by a kinin-dependent mechanism. Hypertension, 81: 349–355.

Goldspink, G., Turay, L., Hansen, E., Ennion, G. and Gerlach, G. (1992) Switch in fish myosin genes induced by environment temperature in muscle of the carp. Symp. Soc. Exp. Biol., 46: 139–149.

Gorr, T., Gassmann, M. and Wappner, P. (2006) Sensing and responding to hypoxia via HIF in model invertebrates. J. Insect. Physiol., 52: 349–364.

Hammel, H.T. (1968) Regulation of internal body temperature. Annu. Rev. Physiol., 30: 641–710.

Hausenloy, D., Tsang, A. and Yellon, D. (2005) The reperfusion injury salvage kinase pathway: a common target for both ischemic preconditioning and postconditioning. Trends Cardiovasc. Med., 15: 69–75.

Hess, U., Gall, C., Granger, R. and Lynch, G. (1997) Differential patterns of c-fos mRNA expression in amygdale during successive stages of odor discrimination learning. Learn. Mem., 4: 262–283.

Hiestand, W., Stemler, F. and Jasper, R. (1955) Increased anoxic resistance resulting from short period heat adaptation. Proc. Soc. Exp. Biol. Med., 88: 94–95.

Hinckel, P. and Perschel, W.T. (1987) Influence of cold and warm acclimation on neuronal responses in the lower brain stem. Can. J. Physiol. Pharmacol., 65: 1281–1289.

Hinckel, P. and Schroder-Rosenstock, K. (1982) Central thermal adaptation of lower brain stem units in the guinea-pig. Pflugers Arch., 395: 344–346.

Horowitz, M. (1998) Do cellular heat acclimatory responses modulate central thermoregulatory activity? NIPS, 13: 218–225.

Horowitz, M. (2001) Heat acclimation: phenotypic plasticity and cues to underlying mechanisms. J. Thermal. Biol., 26: 357–363.

Horowitz, M. (2002) From molecular and cellular to integrative heat defense during exposure to chronic heat. Comp. Biochem. Physiol. A Mol. Integr. Physiol., 131: 475–483.

Horowitz, M. (2003) Matching the heart to heat-induced circulatory load: heat-acclimatory responses. News Physiol. Sci., 18: 215–221.

Horowitz, M., Argov, D. and Mizrahi, R. (1983) Interrelationships between heat acclimation and salivary cooling mechanism in conscious rats. Comp. Biochem. Physiol. A Mol. Integr. Physiol., 74: 945–949.

Horowitz, M., Eli-Berchoer, L., Wapinski, I., Friedman, N. and Kodesh, E. (2004) Stress-related genomic responses during the course of heat acclimation and its association with ischemic-reperfusion cross-tolerance. J. Appl. Physiol., 97: 1496–1507.

Horowitz, M., Kaspler, P., Simon, E. and Gerstberger, R. (1999) Heat acclimation and hypohydration: involvement of central angiotensin II receptors in thermoregulation. Am. J. Physiol. Regul. Integr. Comp. Physiol., 277: R47–R55.

Horowitz, M., Maloyan, A. and Shlaier, J. (1997) HSP 70 kDa dynamics in animals undergoing heat stress superimposed on heat acclimation. Ann. N.Y. Acad. Sci., 813: 617–619.

Horowitz, M. and Meiri, U. (1985a) Altered responsiveness to parasympathetic activation of submaxillary salivary gland in the heat acclimated rat. Comp. Biochem. Physiol., 80A: 57–60.

Horowitz, M. and Meiri, U. (1985b) Thermoregulatory activity in the rat: effects of hypohydration, hypovolemia and hypertonicity and their interaction with short-term heat acclimation. Comp. Biochem. Physiol. A Mol. Integr. Physiol., 82: 577–582.

Horowitz, M. and Samueloff, S. (1979) Plasma water shifts during thermal dehydration. J. Appl. Physiol., 47: 738–744.

Karunanithi, S., Barclay, J., Brown, I., Robertson, R. and Atwood, H. (2002) Enhancement of presynaptic performance in transgenic Drosophila overexpressing heat shock protein Hsp70. Synapse, 44: 8–14.

Karunanithi, S., Barclay, J., Robertson, R., Brown, I.R. and Atwood, H.L. (1999) Neuroprotection at Drosophila synapses conferred by prior heat shock. J. Neurosci. Series, 19: 4360–4369.

Kaspler, P. and Horowitz, M. (2001) Heat acclimation and heat stress have different effects on cholinergic-induced calcium mobilization. Am. J. Physiol. Regul. Integr. Comp. Physiol., 280: R1688–R1696.

Kloog, Y., Horowitz, M., Meiri, U., Galron, R. and Avron, A. (1985) Regulation of submaxillary gland muscarinic receptors during heat acclimation. Biochimica. Biophysica. Acta, 845: 428–435.

Korolainen, M.A., Goldsteins, G., Nyman, T.A., Alafuzoff, I., Koistinaho, J. and Pirttila, T. (2006) Oxidative modification of proteins in the frontal cortex of Alzheimer's disease brain. Neurobiol. Aging, 27: 42–53.

Kregel, K., Strauss, H. and Ungar, T. (1994) Modulation of autonomic nervous system adjustments to heat stress by central AngII receptor antagonism. Am. J. Physiol. Regul. Integr. Comp. Physiol., 267: R1089–R1097.

Kultz, D. and Chakravarty, D. (2001) Maintenance of genomic integrity in mammalian kidney cells exposed to hyperosmotic stress. Comp. Biochem. Physiol. A, 130: 421–428.

Lenkei, Z., Palkovits, M., Corvol, P. and Llorence-Cortes, C. (1997) Expression of angiotensin type-1 (AT1) and type-2 (AT2) receptor mRNA in the adult rat brain: a functional neuroanatomical review. Front. Neuroendocrinol., 18: 383–439.

Levi, E., Vivi, A., Hasin, Y., Tassini, M., Navon, G. and Horowitz, M. (1993) Heat acclimation improves cardiac mechanics and metabolic performance during ischemia and reperfusion. J. Appl. Physiol., 75: 833–839.

Lind, W., Swanson, L. and Ganten, D. (1985) Organization of angiotensin II immunoreactive cells and fibers in the rat central nervous system. Neuroendocrinology, 40: 2–24.

Lutz, P.L. and Nilsson, G.E. (1997) Contrasting strategies for anoxic brain survival — glycolysis up or down. J. Exp. Biol., 200: 411–419.

Lyashko, V.N., Vikulova, V.K., Chernicov, V.G., Ivanov, V.I., Ulmasov, K.A., Zatsepina, O.G. and Evgen'ev, M.B. (1994) Comparison of the heat shock response in ethnically and ecologically different human populations. PNAS, 91: 12492–12495.

Maloyan, A., Eli-Berchoer, L., Semenza, G.L., Gerstenblith, G., Stern, M.D. and Horowitz, M. (2005) HIF-1{alpha}-targeted pathways are activated by heat acclimation and contribute to acclimation-ischemic cross-tolerance in the heart. Physiol. Genomics, 23: 79–88.

Maloyan, A. and Horowitz, M. (2002) Beta -adrenergic signaling and thyroid hormones affect HSP72 expression during heat acclimation. J. Appl. Physiol., 93: 107–115.

Maloyan, A., Palmon, A. and Horowitz, M. (1999) Heat acclimation increases the basal HSP72 level and alters its production dynamics during heat stress. Am. J. Physiol. Regul. Integr. Comp. Physiol., 276: R1506–R1515.

Milton, S. and Lutz, P. (2005) Adenosine and ATP-sensitive potassium channels modulate dopamine release in the anoxic turtle (Trachemys scripta) striatum. Am. J. Physiol. Regul. Integr. Comp. Physiol., 289: R77–R83.

Morimoto, R.I. and Santoro, M.G. (1998) Stress-inducible responses and heat shock proteins: new pharmacologic targets for cytoprotection. Nat. Biotech., 16: 833–838.

Moseley, P.L. (1997) Heat shock proteins and heat adaptation of the whole organism. J. Appl. Physiol., 83: 1413–1417.

Nakamura, Y. and Okamura, K. (1998) Seasonal variation of sweating responses under identical heat stress. Appl. Human Sci., 17: 167–172.

Ohtsuki, T., Matsumoto, M., Kuwabara, K., Kitagawa, K., Suzuki, K., Taniguchi, N. and Kamada, T. (1992) Influence of oxidative stress on induced tolerance to ischemia in gerbil hippocampal neurons. Brain Res., 599: 246–252.

Oppenheim, A., Beit-Yannai, E., Horowitz, M. and Shohami, E. (1996) Production of heat shock protein-72 in rat brain after closed head injury: study in heat acclimated and non-acclimated (Abstract). Isr. J. Med. Sci., 32: S38.

Paz, Z., Freeman, S., Horowitz, M. and Sohmer, H. (2004) Prior heat acclimation confers protection against noise-induced hearing loss. Audiol. Neurootol., 9: 363–369.

Pek-Scott, M. and Lutz, P.L. (1998) ATP-sensitive K^+ channel activation provides transient protection to the anoxic turtle brain. Am. J. Physiol. Regul. Integr. Comp. Physiol., 275: R2023–R2027.

Pierau, F., Sann, H., Yakimova, K. and Haug, P. (1998) Plasticity of hypothalamic temperature-sensitive neurons. Prog. Brain Res., 115: 63–84. Review

Ramaglia, V. and Buck, L.T. (2004) Time-dependent expression of heat shock proteins 70 and 90 in tissues of the anoxic western painted turtle. J. Exp. Biol., 207: 3775–3784.

Rattiner, L., Davis, M. and Ressler, K. (2004) Differential regulation of brain-derived neurotrophic factor transcripts during the consolidation of fear learning. Learn. Mem., 11: 727–731.

Schulte, P. (2001) Environmental adaptations as windows on molecular evolution. Comp. Biochem. Physiol. B Biochem. Mol. Biol., 128: 597–611.

Schwimmer, H., Eli-Berchoer, L. and Horowitz, M. (2006) Acclimation-phase specificity of gene expression during the course of heat acclimation and superimposed hypohydration in the rat hypothalamus. J. Appl. Physiol., 100(6): 1992–2003.

Schwimmer, H., Gerstberger, R. and Horowitz, M. (2004) Heat acclimation affects the neuromodulatory role of AngII and nitric oxide during combined heat and hypohydration stress. Brain Res. Mol. Brain Res., 130: 95–108.

Semenza, G.L. (2004) Hydroxylation of HIF-1: oxygen sensing at the molecular level. Physiology, 19: 176–182.

Sharp, F., Ran, R., Lu, A., Tang, Y., Strauss, K., Glass, T., Ardizzone, T. and Bernaudin, M. (2004) Hypoxic preconditioning protects against ischemic brain injury. NeuroRx, 1: 26–35.

Shein, N.A., Horowitz, M., Alexandrovich, A.G., Tsenter, J. and Shohami, E. (2005) Heat acclimation increases hypoxia-inducible factor 1[alpha] and erythropoietin receptor expression: implication for neuroprotection after closed head injury in mice. 25: 1456–1465.

Shido, O., Sakurada, S., Sugimoto, N. and Nagasaka, T. (1995) Shifts of thermoeffector thresholds in heat-acclimated rats. J. Physiol., 483(Pt 2): 491–497.

Shido, O., Sugimoto, N., Tanabe, M. and Sakurada, S. (1999) Core temperature and sweating onset in humans acclimated to heat given at a fixed daily time. Am. J. Physiol. Regul. Integr. Comp. Physiol., 276: R1095–R1101.

Shin, D.S.-H., Wilkie, M.P., Pamenter, M.E. and Buck, L.T. (2005) Calcium and protein phosphatase 1/2A attenuate N-methyl-D-aspartate receptor activity in the anoxic turtle cortex. Comp. Biochem. Physiol. A Mol. Integr. Physiol., 142: 50–57.

Shohami, E., Novikov, M. and Horowitz, M. (1994) Long term exposure to heat reduces edema formation after closed head injury in the rat. Acta Neurochir. Suppl. (Wien.), 60: 443–445.

Sonna, L.A., Fujita, J., Gaffin, S.L. and Lilly, C.M. (2002) Molecular biology of thermoregulation: invited review: effects of heat and cold stress on mammalian gene expression. J. Appl. Physiol., 92: 1725–1742.

Watanabe, T., Hashimoto, M., Okuyama, S., Inagami, T. and Nakamura, S. (1999) Effects of targeted disruption of the mouse angiotensin II type 2 receptors gene on stress-induced hyperthermia. J. Physiol., 515: 881–885.

Wilson, K. and Fregley, M. (1985) Factors affecting angiotensin II-induced hypothermia in rats. Peptides, 6: 695–701.

Xi, L., Tekin, D., Bhargava, P. and Kukreja, R. (2001) Whole body hyperthermia and preconditioning of the heart: basic concepts, complexity, and potential mechanisms. Int. J. Hyperthermia, 17: 439–455.

Xia, Y. and Haddad, G. (1999) Effect of prolonged O_2 deprivation on Na^+ channels: differential regulation in adult versus fetal rat brain. Neuroscience, 94: 1231–1241.

Yamazaki, F. and Hamasaki, K. (2003) Heat acclimation increases skin vasodilation and sweating but not cardiac baroreflex responses in heat-stressed humans. J. Appl. Physiol., 95: 1567–1574.

Zhang, H.J., Xu, L., Drake, V.J., Xie, L., Oberley, L.W. and Kregel, K.C. (2003) Heat-induced liver injury in old rats is associated with exaggerated oxidative stress and altered transcription factor activation. FASEB J., 03-0139fje.

SECTION VIII

Heat Shock Proteins in Hyperthermia

CHAPTER 19

Heat shock protein expression in brain: a protective role spanning intrinsic thermal resistance and defense against neurotropic viruses

Matthew A. Buccellato[1], Thomas Carsillo[2], Zachary Traylor[1] and Michael Oglesbee[1,2,*]

[1]Department of Veterinary Biosciences, The Ohio State University, 1925 Coffey Rd., Columbus, OH 43210, USA
[2]Department of Molecular Virology, Immunology and Medical Genetics, The Ohio State University, 333 West 10th Avenue, Columbus, OH 43210, USA

Abstract: Heat shock proteins (HSPs) play an important role in the maintenance of cellular homeostasis, particularly in response to stressful conditions that adversely affect normal cellular structure and function, such as hyperthermia. A remarkable intrinsic resistance of brain to hyperthermia reflects protection mediated by constitutive and induced expression of HSPs in both neurons and glia. Induced expression underlies the phenomenon of hyperthermic pre-reconditioning, where transient, low-intensity heating induces HSPs that protect brain from subsequent insult, reflecting the prolonged half-life of HSPs. The expression and activity of HSPs that is characteristic of nervous tissue plays a role not just in the maintenance and defense of cellular viability, but also in the preservation of neuron-specific luxury functions, particularly those that support synaptic activity. In response to hyperthermia, HSPs mediate preservation or rapid recovery of synaptic function up to the point where damage in other organ systems becomes evident and life threatening. Given the ability of HSPs to enhance gene expression by neurotropic viruses, the constitutive and inducible HSP expression profiles would seem to place nervous tissues at risk. However, we present evidence that the virus–HSP relationship can promote viral clearance in animals capable of mounting effective virus-specific cell-mediated immune responses, potentially reflecting HSP-dependent increases in viral antigenic burden, immune adjuvant effects and cross-presentation of viral antigen. Thus, the protective functions of HSPs span the well-characterized intracellular roles as chaperones to those that may directly or indirectly promote immune function.

Keywords: hyperthermia; brain; synapse; hsp72 protein; heat shock proteins

Introduction

A recent study involving whole body hyperthermia in dogs provides one of the first in-depth examinations of the intrinsic resistance of the brain to heat stress (Oglesbee et al., 2002a). In this case, the controlled induction of hyperthermia by heating blood (either directly or indirectly via peritoneal lavage) was used to analyze the effects of a thermal dosage that falls just short of the threshold at which systemic effects on hemostasis are realized (Oglesbee et al., 1999, Fig. 1). Beyond this threshold, the effects of hyperthermia on the mammalian central nervous system become complicated by the onset of cerebral ischemia, edema and

*Corresponding author. Tel.: +1-614-292-9672; Fax: +1-614-292-6473; E-mail: oglesbee.1@osu.edu

DOI: 10.1016/S0079-6123(06)62019-0

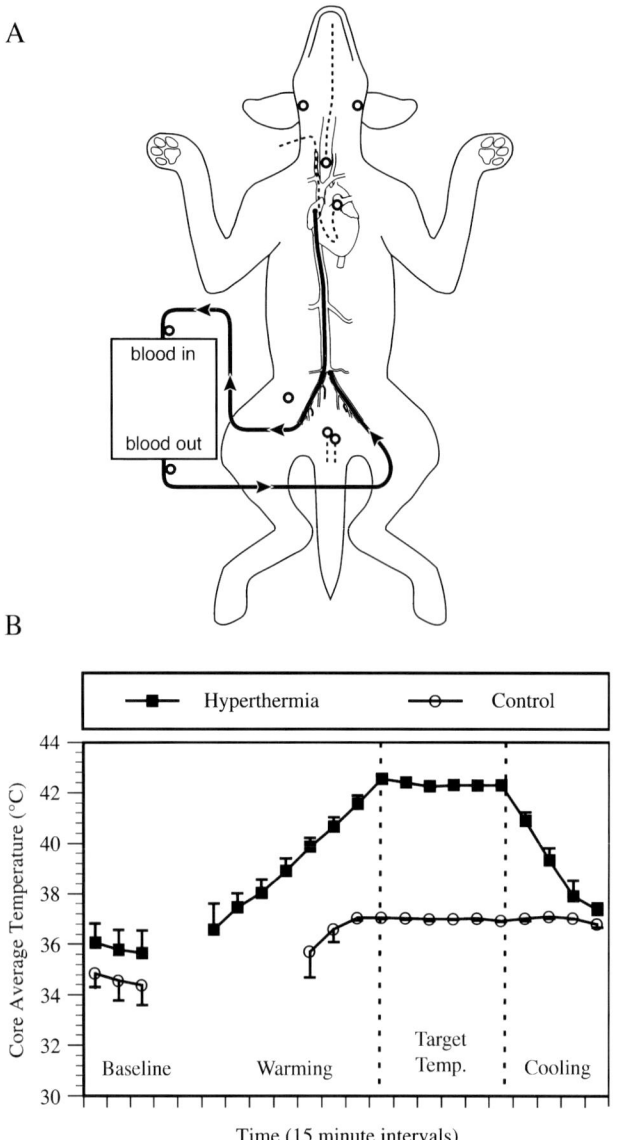

Fig. 1. Extracorporeal blood heating used to measure the physiologic response to hyperthermia in the canine brain. (A) The extracorporeal blood circulation system allows for the application of a controlled thermal dose to the brain. In this schematic, the subject is shown in dorsal recumbency. Vascular access is gained at the left and right femoral veins, and the flow circuit between the pump/heating apparatus and the body is indicated by arrows. Open circles indicate the placement of thermisters (from rostral to caudal: tympanic membranes (2), esophagus, pulmonary artery, subcutaneous, bladder, deep rectum). Two thermisters also monitored temperature of blood entering and exiting the heating device, with global temperature monitoring used to regulate core body temperature. (B) Average core temperature +/−SD of a dog exposed to extracorporeal whole body hyperthermia and a euthermic extracorporeal circulation control dog. Panel (A) reprinted with permission from Alterations in hemostasis associated with hyperthermia in a canine model, Diehl et al. (2000). Panel (B) reprinted with permission from Veterinary Immunology and Immunopathology, Oglesbee et al. (1999).

hemorrhage, although there are species-specific differences in the precise thermal dose at which such lesions occur (Sminia et al., 1994). For dogs, the systemic heat stress threshold is surpassed when the average body temperature is maintained at 42.5°C for longer than 90 min. Beyond this range, the indicators of fulminant disseminated intravascular coagulopathy (DIC) become evident, characterized by severe thrombocytopenia, hypofibrinogenemia, elevated expression of fibrin degradation products and prolonged clotting times. Serum biochemistries and histopathological findings indicate that hepatocellular function defines the maximally tolerated thermal dose, and it is hepatocellular injury that triggers the onset of DIC (Diehl et al., 2000). This coagulopathy leads to characteristic pathological consequences that culminate in multiorgan dysfunction and death, with the initial evidence of ischemic damage in brains of dogs being selective neuronal necrosis, particularly in cerebellar Purkinje cells.

The intrinsic thermostability of the brain was determined by analyzing its structural, biochemical and functional characteristics after being subjected to sub-threshold hyperthermia. There were little to no permanent, observable alterations in neurologic function, as measured by level of consciousness, motor function, brainstem reflexes and brainstem auditory evoked potential (BAEP) recordings. Although changes in BAEP recordings are reported to occur during heating (Takahashi et al., 1991), these findings suggest that such alterations are transient. Likewise, there was a lack of structural and biochemical changes in the brain that would indicate tissue injury. Of the parameters measured, the only alteration was increased expression of the inducible 72 kDa heat shock protein (HSP) (Fig. 2), a well-documented indicator of cellular stress response induction (Mizzen and Welch, 1988).

The stress response is well conserved throughout evolution, and the genetic and protein constituents share significant homology in both prokaryotes and eukaryotes (reviewed in Snoeckx et al., 2001). In the first described example of a cellular reaction to an applied stress, Ritossa noted a characteristic "puffing" in the genes of *Drosophila* salivary gland cells exposed to transient elevations in temperature that was shown to be accompanied by the elevated expression of a specific set of proteins (Ritossa, 1962; Tissieres et al., 1974). Further characterization of these and other related proteins has shown that this class of proteins, initially termed the HSPs, is in fact comprised of an array of molecules involved in the maintenance of normal cellular homeostasis as well as the response to and recovery from all manner of stressful stimuli. The HSPs are divided into families based upon molecular mass, such as the well-characterized 90 kDa, 70 kDa and small molecular mass groups, each of which has members that are expressed constitutively and/or in response to stress. Constitutively expressed HSPs are most often associated with functions required for normal cell growth and metabolism, including the regulation of protein maturation, folding and intracellular trafficking (reviewed in Oglesbee et al., 2002b), while the inducible HSPs respond to conditions that promote protein denaturation and dysregulation of cellular metabolic pathways, a list of which includes (but is not limited to) thermal stress, oxidative injury, ischemia, chemical or heavy metal toxicity and viral infection (reviewed in Oglesbee et al., 2002b; Sharma and Westman, 2004). In the brain, as in other organs, induction of the cellular stress response results in a state of heightened resistance to a variety of insults in the post-stress interval. Exposure to transient hyperthermia is perhaps the best-known means of inducing this resistance phenotype, a treatment known as hyperthermic pre-conditioning (Yager and Asselin, 1999; Kwong et al., 2003).

In the central nervous system, there are distinct expression patterns for specific HSP isoforms, and in some cases there appears to be considerable functional overlap between the proteins involved. Studies have examined the effect of hyperthermia on the induction of a variety of HSP families within the brain. A large portion of this work focuses upon the constitutive and inducible 70 kDa family members under such conditions (i.e., hsc70 and hsp72, respectively), largely because elevated levels of hsp72 are a primary mediator of hyperthermic pre-conditioning, with temporal elevation of hsp72 levels corresponding to the window of cellular protection (Kwong et al., 2003). Recently, other HSPs have also been shown to be important

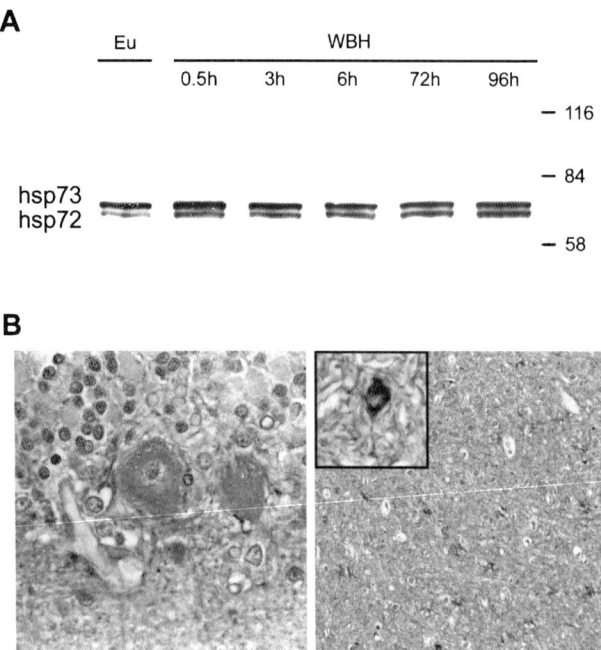

Fig. 2. Hsp72 expression in the canine brain following the induction of whole body hyperthermia by extracorporeal blood heating. (A) Western blot analysis of total protein from cerebellum using a monoclonal antibody recognizing a common epitope of hsc70 (denoted in this figure as hsp73) and hsp72. Hsp72 is elevated over time following hyperthermia in comparison with levels in euthermic control tissue (Eu), while hsc70 shows no evidence of induction. Note the prominent basal expression of hsp72 in control tissue. (B) Immunohistochemical staining of hsp72 in cerebellum 24 h after whole body hyperthermia using an hsp72-specific monoclonal antibody. In the cortex (left panel), positive hsp72 staining was localized to both the nucleus and the cytoplasm of Purkinje cells and the nuclei of neurons and glial cells in the internal granular layer. In white matter regions (right panel), hsp72 expression is localized to cells with oligodendroglial morphologies (inset). Reprinted with permission from Neuroscience, Oglesbee et al. (2002a).

components in the integrated response of the brain to thermally induced changes in cellular homeostasis. The maintenance of normal clinical neurologic status and electrophysiologic activity in dogs indicates that the responses of the central nervous system to hyperthermia include mechanisms by which synaptic activity is preserved in addition to those that promote cell viability, at least to a threshold thermal dosage where CNS injury occurs as a secondary event related to loss of function in other organs. The participation of HSPs in the preservation of synaptic function in the face of cellular stress is supported by work in many *in vivo* and *in vitro* model systems (Dawson-Scully and Meldrum, 1998; Bechtold and Brown, 2000; Bechtold et al., 2000; Karunanithi et al., 2002; Kelty et al., 2002; Oglesbee et al., 2002a; Lin et al., 2004). This review will focus on the potential role of constitutive and induced expression of 70 kDa, 90 kDa and low molecular mass HSPs in preservation of cell viability and synaptic performance in the CNS in the face of hyperthermia. Special consideration will also be given to the occurrence and probable significance of extracellular release of HSPs, specifically hsp72.

The protective roles of hsp72 will also be examined in the realm of infectious disease. Several neurotropic viruses respond to hyperthermic preconditioning of host cells by increasing viral gene expression and cytopathic effect. We will present evidence that supports a model describing how this virus–cell relationship can paradoxically promote viral clearance, reflecting the HSP-dependent increases in viral antigenic burden and the immunostimulatory roles of extracellular HSPs. From this perspective, HSPs can be viewed collectively as both supporting normal neurologic function during episodes of physiological hyperthermia (i.e., fever), and as components of the febrile response that ultimately provides protection against microbial invasion.

Constitutive expression of heat shock proteins in the brain

HSP70 Family

Members of the 70 kDa HSP family (HSP70) are one of the most abundantly expressed classes of proteins in mammalian cells (Welch and Mizzen, 1988). They are involved in protein folding, the assembly of multimeric protein complexes and the trafficking of proteins between cellular compartments (Beckmann et al., 1990; Ngosuwan et al., 2003). Thus they are important for guiding and assisting protein precursors/subunits to their fully active structural state and preventing inappropriate aggregation. These actions are mediated by a conserved binding pocket, which recognizes a broad array of protein substrates through the transient interaction with exposed hydrophobic domains, targeting linear sequences of eight to nine amino acids (Fourie et al., 1994; Rudiger et al., 1997). The reversible nature of the affinity of these HSPs for their targets is mediated by the cyclic binding and hydrolysis of adenosine triphosphate (ATP) and release of adenosine diphosphate (ADP) by a nucleotide-binding domain within the molecule (McCarty et al., 1995). For HSP70 bound to ATP, the peptide binding pocket is open, a conformation which promotes rapid association/dissociation with ligand (Misselwitz et al., 1998). ATP hydrolysis to ADP causes closure of the binding pocket, thereby locking HSP70 onto the bound peptide. ADP–ATP nucleotide exchange allows for the repetitive cycle of binding and release. Thus by binding exposed hydrophobic domains and preventing the formation of insoluble multimeric complexes, HSPs provide nascent or denatured proteins with the opportunity to attain their normal structure and function. The chaperone functions of HSP70 involve more than just protein folding. HSP70 supports conformational changes that attend transmembrane protein transport (Pilon and Schekman, 1999), assembly and disassembly of multimeric protein complexes (Barouch et al., 1994), and trafficking of binding partners into cell compartments such as the nucleus (Oglesbee and Krakowka, 1993).

The constitutively expressed 73 kDa isoform (hsc70) carries out its functions by chaperoning newly formed peptides during protein synthesis and maturation, and by determining the assembly state of multimeric protein complexes within the cell. The highly inducible 72 kDa HSP (hsp72, also known as hsp70) is typically expressed under varying conditions of cellular stress, including heat shock (Li et al., 1992), ischemia (Chopp et al., 1991) and toxin or heavy metal exposure (Kato et al., 1997), and is responsible for the rescue of denatured cytoplasmic proteins, allowing them to refold and return to their normal form and function. Studies demonstrate that HSP70 isoforms exhibit species-specific differences in both constitutive and induced expression in the brain. Since these isoforms exhibit functional similarities that reflect their high amino acid sequence homology (Brown et al., 1993), the presence of one may compensate for the absence of the other in any particular cell population, and their complimentary nature allows both to contribute to the intrinsic thermal resistance of the brain.

HSP70 also recognize small patches of surface hydrophobicity on native proteins, resulting in altered activity of the substrate, a function known as HSP-mediated activity control (Gamer et al., 1996). Substrate activity may either be enhanced or diminished as a consequence of HSP70 binding, reflecting changes in conformation that influence interaction with binding partners or rate of degradation. As general examples of this capacity, HSP70 modulates cell cycle progression and induction of apoptosis by binding p53 (Zylicz et al., 2001), kinases of the mitogen activated signal cascade (Song et al., 2001), or apoptosis inducing factor (Ravagnan et al., 2001). As well, interactions with native proteins underlie some of the functions of hsc70/hsp72 in the maintenance and protection of synaptic signaling (Chamberlain and Burgoyne, 1997; Bronk et al., 2001; Morgan et al., 2001).

Hsc70

Hsc70, being the constitutively expressed member of the HSP70 family in the cytosol of most cells, is particularly prevalent in the brain. In fact, cellular levels of basal hsc70 expression in the brain exceeds that of most other organs (Bechtold et al., 2000; Morrison et al., 2000; Belay and Brown, 2006). The reasons for such high basal levels have not been fully determined, but specific expression patterns may provide clues as to the requisite functions of this molecule in nervous tissue. A study of constitutive HSP70 expression (including both hsc70 and hsp72) during development of the rat CNS revealed that peak levels occur during the first 2–4 months of age, and its decline to typical adult levels correlates with the onset of physiologic maturity (Bodega et al., 2002).

Immunohistochemical studies show hsc70 in both the cell bodies and dendrites/perikaryal extensions of neurons in the cortex, hippocampus and cerebellum. Electron microscopy and fractionation studies have been used to demonstrate hsc70 in and around synapses, in both pre- and post-synaptic compartments (Suzuki et al., 1999; Bechtold et al., 2000; Moon et al., 2001; Bodega et al., 2002). Fractionation studies also show that hsc70 and other constitutively expressed HSPs are localized to lipid rafts in the rat brain (Chen et al., 2005b). Lipid rafts are membrane domains enriched in cholesterol and sphingolipids (reviewed in Simons and Toomre, 2000), and in neurons are proposed to be a critical component of the synapse (reviewed in Suzuki, 2002). While the specific roles of hsc70 in synaptic transmission have not been completely elucidated, several functions of hsc70 either have been shown or would be predicted to support regulation of signaling between neurons. Its activity as a protein chaperone directing the proper folding of nascent polypeptides would be required during local protein production, which is necessary for the maintenance of synaptic plasticity (Martin, 2004). In addition to hsc70, a wide variety of mRNA transcripts and the protein complexes that are components of the cellular translational machinery have been demonstrated in post-synaptic densities (Tian et al., 1999; Asaki et al., 2003). In this context, the role of hsc70 as a chaperone for locally produced proteins is underscored by its co-localization with hsp40 at post-synaptic sites (Suzuki et al., 1999). The association with hsp40 boosts the ATPase activity of hsc70, regulating the peptide binding and release cycle that underlies its chaperone function (Minami et al., 1996).

Distinct from its function as a protein-folding chaperone, hsc70 is known to be operative in synaptic vesicle cycling, the basis for the rapid release and recovery of neurotransmitters by pre-synaptic neurons (reviewed in Ohtsuka and Suzuki, 2000). Two other recognized hsc70 co-chaperones, cysteine string protein (CSP) and auxilin, are present in synaptic fractions and appear to mediate the separable activities of hsc70 in vesicular exo- and endocytosis, illustrating how specific collaborations between hsc70 and its partners determine its functional activity. CSP and hsc70 are required for normal, Ca^{2+}-dependent vesicular exocytosis and are components of a complex containing hsp90 that regulates recycling of Rab3A, a small G protein that mediates membrane fusion in the pre-synaptic terminal (Dawson-Scully et al., 2000; Bronk et al., 2001; Sakisaka et al., 2002; Chen et al., 2005a). Alternatively, hsc70 binding to auxilin is necessary for uncoating of clathrin-coated endocytotic vesicles in pre-synaptic nerve terminals. Inhibition of hsc70 interactions with auxilin results in a breakdown of synaptic vesicle trafficking, which eventually brings normal synaptic signaling to a halt (Morgan et al., 2001).

Hsp72

A number of studies report no constitutive expression of hsp72 mRNA or protein in any region (i.e., forebrain, cerebellum or brainstem) of the adult rat brain in the absence of a clearly defined stressor (Khan and Brown, 2002). Occasionally, low level constitutive protein expression is described, although often only slightly above background (Ostberg et al., 2002; Belay and Brown, 2003). It should be noted that the finding of low or no basal expression of hsp72 is in contrast to the significant, regional hsp72 levels described in both unstressed rabbits and dogs (Manzerra et al., 1997;

Oglesbee et al., 2002a, Fig. 2) and in a group of elderly human patients that died acutely without evidence of agonal stress or anti-mortem fever (Pardue et al., 2007). Interestingly, constitutive hsp72 expression has been identified in post-synaptic fractions in both the cerebral cortex and cerebellum in tight association with synaptic elements (Moon et al., 2001). This example underscores the potential for functional overlap between the two cytosolic HSP70 isoforms in regulation of synaptic function in the broader sense, given that hsp72 may bind and be activated by both hsp40 and CSP in the same manner discussed previously for hsc70 (Chamberlain and Burgoyne, 1997; Ohtsuka and Hata, 2000).

Hsp90

Also widely expressed in the brain is the constitutive member of the 90 kDa family, hsp90 (Quraishi et al., 1996; D'Souza and Brown, 1998). Like hsc70, it is well recognized as a chaperone involved in normal protein maturation, although hsp90 has a narrower range of substrates that includes regulatory kinases, steroid hormone receptors and myosin. The hsp90 molecule is also involved in protein sorting to various organelles, including the proteosome (reviewed in Young et al., 2004). Although studies examining functions of hsp90 specific to the brain are few, those available do provide additional examples of the participation of HSPs in activities critical to normal CNS activity. Relatively equivalent expression in the unstressed rat brain has been demonstrated in both cerebrum and cerebellum and, as with hsc70, its presence was noted in synaptosomes, specifically the pre-synaptic fraction (Bechtold et al., 2000). As stated previously, this localization has been linked to the requirement of hsp90, in complex with hsc70 and CSP, for efficient neurotransmitter release by the pre-synaptic neuron (Sakisaka et al., 2002; Gerges et al., 2004). A post-synaptic function, the regulation of continuous AMPA-type glutamate receptor cycling, has also been determined (Gerges et al., 2004). Such constant cycling is proposed to account for the maintenance of synaptic glutamate receptor expression profiles in the face of receptor turnover, a phenomenon that may be the molecular basis for memory retention (Malinow et al., 2000).

Low molecular weight heat shock proteins

The small HSPs (Hsp27 and Hsp32) have a restricted pattern of constitutive expression in the brain. Little to no demonstrable protein is present in the rat hippocampus, cerebellum or cerebral cortex, although expression of hsp27 in the brain stem is reported to be more robust (Bechtold and Brown, 2000, 2003; Krueger-Naug et al., 2000). The latter finding is compatible with developmental studies in the mouse, where constitutive expression of hsp25 (the murine homologue of rat hsp27) is noted in a number of cranial nerves and their nuclei (i.e., cranial nerves V, VII, X and XII) just after birth and their levels remain constant into adulthood (Armstrong et al., 2001). The functional significance of this constitutive expression has yet to be elucidated.

Hyperthermic induction of heat shock proteins in the brain

Hsp72

A common method of inducing a hyperthermic state in the central nervous system in various experimental settings has been the use of externally applied whole body hyperthermia, either via extended exposure to elevated ambient temperature or the use of heating pads, although drug-induced hyperthermia, specifically treatment with lysergic acid diethylamide (LSD), is used for the study of HSP expression in the brain in rabbits (Manzerra and Brown, 1990). The exact thermal dosage that is applied (i.e., the duration and degree to which measured core temperatures are elevated) can vary and this may account for minor differences in specific expression patterns that have been described within a particular animal species, in addition to contributing to the interspecies differences that have been noted. Because the method of heating can influence the precise thermal dose to the brain (and thus the hsp72 expression profile), monitored heating of the blood was chosen as an approach for induced hyperthermia in dogs (Oglesbee et al., 2002a). This method resembles that used to deliver therapeutic doses of radiant energy in the treatment of human cancers, and is amenable to use in dogs, whose large size in comparison to other common laboratory animal species allows for constant, real-time monitoring of an array of specific physiologic parameters during the hyperthermic interval. When combined with concomitant fluid therapy and mechanical ventilation, which is correct for the systemic effects of hyperthermia that may secondarily influence the central nervous system (i.e., reduced mean arterial pressure and respiratory alkalosis), this form of heating can be successfully used to measure the effects of sublethal hyperthermia without the interference of other heat-related physiologic responses.

Regardless of the specific method by which a thermal dosage is delivered, it has been well established that hsp72 is one of the main induced HSPs in the mammalian brain. As discussed previously, basal hsp72 expression in the brain is species-dependent and often difficult to document. However, even mild hyperthermia within the physiological range of most febrile responses (i.e., up to a 3.5°C rise in body temperature for at least 60 min) results in measurable increases in hsp72 protein expression within 3–5 h after the onset of temperature elevation, with the most striking expression in the cerebellum (Bechtold et al., 2000; Morrison et al., 2000; Khan and Brown, 2002; Oglesbee et al., 2002a; Maroni et al., 2003). The onset of stress-related transcriptional activity is regulated by cytosolic heat shock factors (HSF), particularly the HSF1 isoform in mammals (reviewed in Kiang and Tsokos, 1998). Under normal conditions, HSF is bound to HSPs in the cytosol. Following the onset of cellular stress, HSF1 is released and then activated by serine/threonine phosphorylation. This promotes homotrimerization, which is followed by translocation to the nucleus, where the HSF1 complex associates with heat shock elements (HSE), DNA binding regions that promote transcription of stress-responsive genes, like that of hsp72 (Schiaffonati et al., 2001). The role of HSF1 in hsp72 induction is supported by the finding that hsp72 mRNA and protein expression patterns mimic those noted for HSF1 expression and DNA binding, which is more prominent in the cerebellum than in the cerebral cortex or hippocampus (Morrison et al., 2000; Maroni et al., 2003). There is a similar hsp72 expression gradient in dog brain following hyperthermia, with the greatest induction in the cerebellum, then decreasing sequentially from the hippocampus to the hypothalamus and frontal cortex, and finally becoming unapparent in the myelencephalon (Oglesbee et al., 2002a).

The basis for regional differences in *in vivo* expression of hsp72 following heat shock beyond differences in HSF1 expression is less clear. Studies of hippocampal neurons provide evidence of regional regulatory mechanisms of hsp72 expression. In the rat, hyperthermic treatment (core temperature of 42° to 42.5°C) showed significant elevations in hsp72-specific mRNA levels in neurons in the pyramidal and dentate granule cell layers, whereas protein expression was transient and of relatively low intensity when compared to that of local glial cells (Krueger et al., 1999). The authors suggest that such findings demonstrate post-transcriptional control of hsp72 expression in these neuronal populations, although a specific mechanism has yet to be delineated. Should this be the case, there appears to be a threshold beyond which such control is relaxed. The effect of increased duration and intensity of hyperthermic treatment on hsp72 expression can be documented when comparing this study to one in which a longer exposure to higher elevations in ambient temperature (43.5°C for 50 min.) was performed on hippocampal explant cultures. Here, neurons in the pyramidal layer and dentate gyrus demonstrate quite a robust pattern of hsp72 protein expression, on par with adjacent glial cell populations, and remain elevated at least 5 days post-shock. Regional differences in hsp72 expression following heat shock may also reflect differences in the need for chaperone function during hyperthermia, where there is overlap in the contributions of hsp72 and hsc70, the latter being abundant in neurons (Suzuki et al., 1999) and active in cells responding to hyperthermic stress (Brown et al., 1993; Ellis et al., 2000; Cvoro and Matic, 2002). This view is supported by recent findings showing prominent basal hsc70 expression in neuronal populations that do not appear to express elevated levels of inducible hsp72, yet remain protected from apoptosis following hyperthermia (Belay and Brown, 2006).

In the cerebellum of adult rats, the use of light microscopy and immunocytochemistry detected robust expression only in the neurons of the granular cell layer following a physiologic increase in body temperature, while other neuronal populations, such as Purkinje cells, seemed nonresponsive to heat stress (Bechtold and Brown, 2000). While there was no evidence of neuronal apoptosis in adult rats for this degree of CNS hyperthermia (Khan and Brown, 2002), a more profound, supraphysiologic elevation in temperature showed a correlation between hsp72 expression in neurons and resistance to apoptosis (Belay and Brown, 2003). Indeed, in a primary culture system, the induction of hsp72 by hyperthermic preconditioning was shown to prevent repolarization-induced apoptosis in granular layer neurons, confirming its protective effects in this cell population (Chen et al., 2004). In neonates (postnatal day 7), hyperthermia-induced apoptosis in the cerebellum was confined to the external granular cell layer, which at this point in development is comprised of actively dividing cells (Belay and Brown, 2003). Increased basal hsp72 expression was noted in mature, non-dividing cells but not in the external granular cell layer, indicating that the hsp72 expression pattern is inversely related to the susceptibility of heat shocked cells to apoptosis, reinforcing the general view that HSP expression protects against apoptotic cell death (reviewed in Sreedhar and Csermely, 2004).

Our *in vitro* studies demonstrate a direct role for hsp72 in protecting neural cells against hyperthermic insult. Cell lines that were stably transfected with an hsp72 expression plasmid showed enhanced survival and colony forming ability following an otherwise lethal hyperthermic treatment (Fig. 3). These experiments further demonstrate that constitutive high-level expression of hsp72 does not perturb normal cellular metabolism, perhaps reflecting the cell's ability to buffer the activity provided by overexpressed HSP. For example, hsp72 overexpression does not alter the chemical induction of neuroblastoma cell differentiation by lithium chloride treatment, as assessed by both morphologic and biochemical parameters (Fig. 4). In both control and hsp72 overexpressing cell populations, the degree of differentiation and the time to maximal effect was the same. Antisense oligonucleotide treatment was used to show that hsp72 in particular was responsible for the hyperthermia-related protection of CA1 neurons from the lethal effects of glutamate treatment (Sato and Matsuki, 2002). Such findings extend previous work documenting the protective effect of hsp72 on hippocampal neuron survival in culture (Sato et al., 1996) to a form of injury specific to the central nervous system (i.e., glutamate toxicity).

Beyond the global ability of hsp72 to protect cells from the stress-related loss of maintenance functions, we see a role for hsp72 in preserving synaptic activity, a particularly important luxury function of neurons. Following a physiologic heating event, hsp72 expression was markedly up-regulated in both pre- and post-synaptic compartments derived from separate cerebellar and forebrain extracts, and was determined to be an integral component of the post-synaptic density, similar to that of its constitutive isoform described previously (Bechtold et al., 2000). These findings are mirrored by the rapid appearance of hsp72 in rat brain lipid raft extracts following a similar hyperthermic treatment (Chen et al., 2005b). Studies using hippocampal neurons have also demonstrated the importance of such expression on the maintenance of synaptic function in injured neuronal populations. Expression of hsp72 in pyramidal layer neurons following hyperthermia has recently been shown to have protective effects upon synaptic plasticity. In a rat hippocampal explant model, hyperthermic treatment eliminated the suppression of long-term potentiation in CA1 neurons induced by scopolamine. Involvement of hsp72 in this effect is supported by a similar time course of both hsp72 induction and retention of normal synaptic function in the face of drug-related muscarinic receptor blockade (Lin et al., 2004). Further evidence of the protective effect on synaptic transmission is provided in work using *Drosophila* that overexpresses native hsp72 in response to heat shock (i.e., more than normal reactive response levels) (Karunanithi et al., 2002).

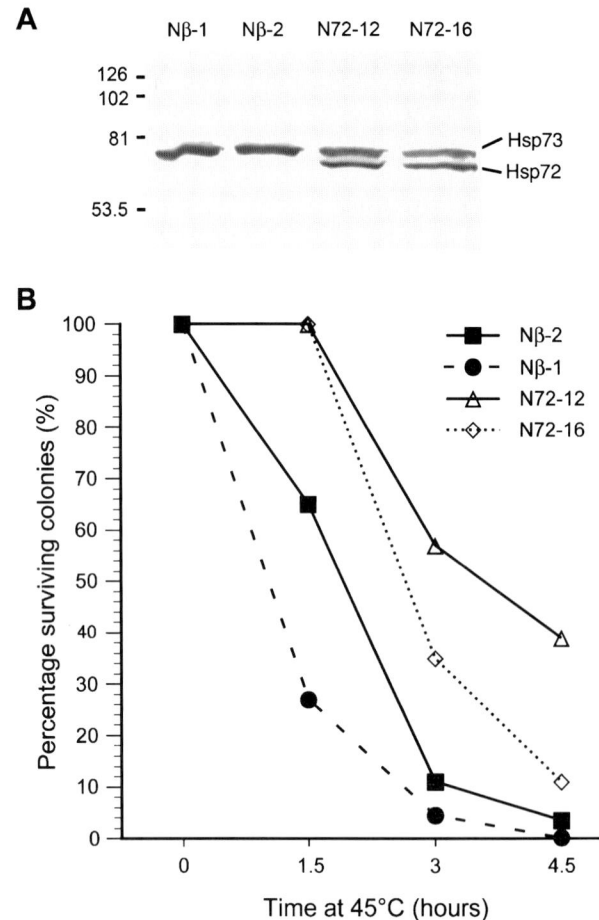

Fig. 3. Constitutive expression of hsp72 in murine neuroblastoma cells mediates cellular tolerance to subsequent heat stress. (A) Western blot analysis of total protein using a monoclonal antibody that recognizes a common epitope of hsc70 (denoted in this figure as hsp73) and hsp72. Murine neuroblastoma cell lines (N72-12, N72-16) are stably transfected with a plasmid that directs constitutive expression of hsp72 via the beta actin promoter. Vector transfected control cell lines (Nβ-1, Nβ-2) do not express hsp72, characteristic of murine cell lines. All cells express the normal constitutive HSP70 isoform (hsc70). (B) The cell lines expressing hsp72 constitutively from (A) show increased colony forming ability following severe heat shock compared with vector transfected control lines.

This increased expression promotes synaptic thermotolerance by enhancing pre-synaptic neuron performance, allowing for maintenance of signaling at elevated temperature, similar to results seen in another insect-based system (Dawson-Scully and Meldrum, 1998). The effect of heat shock on maintenance of synaptic transmission appears to act via the modulation of pre-synaptic calcium levels at the nerve terminal and involves components of the cellular cytoskeleton, although the specific interactions involved have yet to be elucidated (Barclay and Robertson, 2003; Klose et al., 2004). Such findings have been extended in a mammalian system where, in the mouse medullary pre-Botzinger complex, thermal pre-conditioning showed an analogous preservative affect on synaptic activity in the face of subsequent hyperthermic challenge, again apparently through a specific modulation of pre-synaptic neurotransmitter release (Kelty et al., 2002).

There are only a few studies that focus upon the *in vivo* expression of hsp72 in glial cells following hyperthermia relative to the number of reports on neuronal expression. In rabbit, the effect of brief,

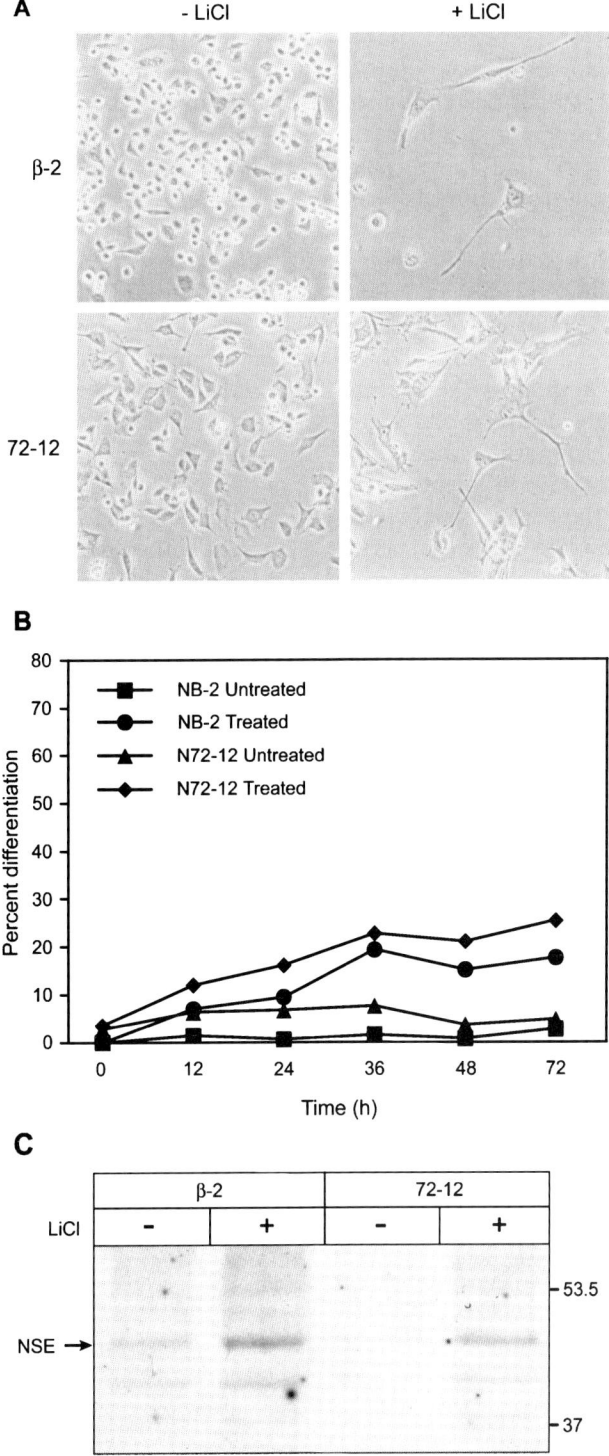

Fig. 4. Lithium chloride (LiCl) treatment induced neuronal differentiation in a murine neuroblastoma cell line that constitutively expresses hsp72 (72-12) and in a vector-transfected control cell line (β-2). (A) Morphologic differentiation in both cell lines was expressed as the development of axon-like processes greater than or equal to two cell diameters. (B) The number of cells categorized as differentiated based on morphologic criteria was increased to a similar degree in both cell lines following LiCl treatment. (C) Biochemical confirmation of cellular differentiation was based on the expression of neuron specific enolase (NSE) in lysates from LiCl treated cells, shown here by Western blot analysis.

fever-range hyperthermia on cell-specific production of hsp72 messenger RNA has been examined in glia-rich regions of the forebrain, including the axon fiber tracts of the fimbria and corpus callosum, cortical layer 1 and the hippocampal fissure (Foster and Brown, 1997). The most abundant hsp72 mRNA levels were noted in all regional oligodendrocyte populations, while microglial transcript production varied depending on the anatomic region examined, ranging from 50% induction in cortical layer 1–90% in the fimbria. Astrocytes were not observed to express hsp72 transcripts in any region, supporting the finding that different glial cell types have different thresholds for the induction of the hsp72 response following heat shock. This pattern of hsp72-specific RNA expression in glial cells in the hyperthermic rabbit closely mirrors glial hsp72 protein localization in dogs (Oglesbee et al., 2002a) and rats. In the latter, induction of hyperthermia in the physiologic range (41.5°C) resulted in prominent, early hsp72 protein induction in oligodendrocytes within the white matter, scattered expression in microglia and little to no observable protein in astrocytes or neurons (Pavlik et al., 2003, see Chapter 20 in this volume). The strong oligodendroglial up-regulation in hsp72 expression matches that seen in cultured rat oligodendrocytes following more intense hyperthermic treatment (Goldbaum and Richter-Landsberg, 2001). Such induction was shown to be protective only against subsequent hyperthermic episodes, but not against insults promoting cellular oxidative damage.

A growing number of studies support the hypothesis that HSP70 (both constitutive hsc70 and induced hsp72) produced by glial cells may be transported to neurons and protect them from subsequent insult. An *in vitro* study shows that cultured glioma cells release hsp72 into the extracellular milieu under normal, non-stressed conditions, and that such expression is increased following heat shock (Guzhova et al., 2001). This result is consistent with reports in invertebrates, which demonstrate transport of hsp72 from glial cells to neurons (Tytell et al., 1986; Sheller et al., 1995, 1998). Furthermore, exogenous HSP70 may be taken up by neurons and prevent cell death related to heat shock or the induction of apoptosis.

In a differentiated neuroblastoma cell culture sensitive to the induction of cell death/apoptosis by heat shock or staurosporine, rapid uptake of a mixture of hsc70/hsp72 applied to the cells resulted in a phenotype that is much more resistant to the cytotoxic effects of either treatment (Guzhova et al., 2001). In another example of the protective effect of exogenous hsp72 on neuron function, the incubation of a mouse medullary slice preparation in a solution containing hsp72 prior to heat shock resulted in the preservation of synaptic transmission in comparison to non-supplemented control samples, mimicking the effects of hyperthermic pre-conditioning (Kelty et al., 2002). Similarly, pre-treatment of slice preparations of rat olfactory cortex with hsp72 protects glutamatergic synaptic transmission from the excitotoxic effects of glutamate supplementation (Mokrushin et al., 2005). A preliminary number of *in vivo* studies in mammals, in addition to those in invertebrates mentioned above, have so far confirmed the ability of extracellular HSP70 supplementation to protect neurons against injurious stimuli. Labeled-hsc70/hsp72 injected intravitreally was found to be rapidly and widely distributed in neurons within the retina and protected photoreceptors from light-induced damage (Yu et al., 2001), mirroring the beneficial effect of hyperthermic pre-conditioning on retinal survival following a deleterious level of light exposure (Barbe et al., 1988). A similar application of either mixed or individual HSP70 proteins increased the survival of damaged motor or sensory neurons, respectively, in the murine lumbar spinal cord following sciatic nerve axotomy (Houenou et al., 1996; Tidwell et al., 2004).

Although *in vitro* studies demonstrate the capacity for neural cells to release hsp72 to the extracellular environment, there is a conspicuous absence of reports describing cerebrospinal fluid (CSF) levels of hsp72. Some cell types, such as hepatocytes, can release extracellular hsp72 in response to stress even in the absence of necrosis or apoptosis, resulting in elevated serum levels of hsp72 (Febbraio et al., 2002; Fleshner et al., 2004). Here, extracellular hsp72 invokes pro-inflammatory responses and may be viewed as an acute phase reactant (Milani et al., 2002; Oglesbee et al., 2002b; Campisi and Fleshner, 2003; Campisi et al., 2003).

Our studies on whole body hyperthermia in the dog reveal that CSF levels of hsp72, determined using a commercially available ELISA (StressGen Biotechnologies), are not significantly elevated despite the prominent neuronal and glial induction that is detected by both immunohistochemistry and Western blot analysis of brain total protein (Oglesbee et al., 2002a; Fig. 2). Dogs in which whole body hyperthermia was induced by peritoneal lavage did not show statistically significant changes from a pre-treatment value of 1.1 ± 0.5 (SD)–1.6 ± 1.1 ng/ml at time zero (i.e., the end of the hyperthermic treatment) and 2.1 ± 2.2 ng/ml at 24 h post-treatment ($n = 6$ animals). In contrast, serum levels increased from a baseline of $0.6 \pm .8$–6.1 ± 5.0 ng/ml at time zero and 4.9 ± 2.6 ng/ml at 24 h post-treatment. Control animals received peritoneal lavage without heating and their serum and CSF hsp72 levels were within range of pre-treatment values. Sample analysis from dogs receiving their thermal dosage from heated blood (extracorporal circulation or ECC whole body hyperthermia) revealed CSF levels of hsp72 that were comparable to the peritoneal lavage whole body hyperthermia treatment groups. Serum levels in control animals were within the reference range reported above, whereas serum levels in the hyperthermia group were much higher than in the peritoneal lavage groups. At 24 h post-treatment, serum levels of hsp72 were 166.3 ± 163.0 ng/ml ($n = 4$), declining to 10.6 ± 9.8 ng/ml at 3 d post-treatment ($n = 2$) and returning to baseline by 8 d (0.8 ± 0.9 ng/ml, $n = 4$).

On the other hand, results from a separate body of ongoing work indicates that extracellular expression of hsp72 in the CNS may relate to the nature of the stress imparted to the tissue, as we have demonstrated significant elevations in CSF levels of hsp72 in dogs following ischemia/reperfusion injury of the spinal cord (unpublished observation). In this case, a transient aortic cross-clamp results in poliomyelomalacia affecting the lumbosacral spinal cord. Preliminary data show that CSF levels of hsp72 can rise to approximately 35 ng/ml, whereas serum levels of hsp72 are not affected. Elevated CSF levels of hsp72 are correlated to increased hsp72 expression in ependymal cells in addition to parenchymal glia and neurons.

Based upon these two lines of experimentation, it appears that release of hsp72 into the CSF is restricted to events associated with tissue injury, where hsp72 may both act as a danger signal and passively impart injury tolerance to neurons. In the absence of tissue injury, there is an appropriate lack of such signaling, such that physiological episodes of hyperthermia do not, in and of themselves, precipitate inflammatory responses. Moreover, it is apparent that CSF and serum levels of hsp72 may change independently as a function of tissue source (i.e., brain versus liver).

Hsp90

While the inducible expression of hsp90 in the brain following hyperthermia has been examined, there is little evidence that protein levels change significantly (Quraishi and Brown, 1995; Quraishi et al., 1996; Bechtold et al., 2000). This may reflect the relatively high basal levels in comparison to other organs, such as the kidney, where induction following hyperthermia is robust (Quraishi and Brown, 1995). Although hsp90 does play a role in protecting neurons from thermal stress (Mailhos et al., 1994; Wyatt et al., 1996), further studies describing its neuron-specific functions are restricted to its activity in the unstressed state.

Low molecular weight heat shock proteins

As opposed to the relatively low constitutive level of expression of small HSPs in the CNS, in particular hsp27 and hsp32, hyperthermia-induced expression appears to be robust and has a particular effect on the maintenance of synaptic activity, with a pattern distinct from that of hsp72. Adult rats subjected to a 41.5°C heat shock for 1 h results in elevated expression of these low mass HSPs in glial cells (Bergmann glia) in the cerebellum by 15 h post-treatment, but not in neurons. By electron microscopy (EM), hsp27 and 32 are localized to synaptic sites, particularly post-synaptic neurons and peri-synaptic glia. The authors suggest that these HSPs help preserve synaptic function and raise the possibility of secreted HSPs as mediators of recovery from injury (Bechtold and

Brown, 2000). Further work identified a similar pattern of hsp27 and 32 expression in glial cells (astrocytes and activated microglia) in the hippocampus, particularly in areas where a high number of synaptic connections are made (Bechtold and Brown, 2003). In a separate study, transient (15 min) 42° to 42.5°C heat shock was used to examine induced hsp27 expression throughout the brain. Western blot analysis showed high basal expression in brainstem, very weak expression in control cerebellum and hippocampus and none in cerebrum. Heat shock resulted in elevated expression in all regions, in similar proportions as the constitutively expressed molecule when measured at 24 h. Expression was strongly induced in glial cells, mostly astrocytes, in all regions studied by immunohistochemistry. Expression begins at 1.5–3 h, peaks at 24 h and remains present in some cells up to 6 days post-shock. Cerebellar Bergmann glia showed a pattern of hsp27 expression similar to that reported above. Constitutive expression was patchy in ependyma of lateral, third and fourth ventricles, while there was strong induction in all ependyma and in nearby parenchymal cells. Maximal expression is observed at 24 h, being largely maintained up to 6 days post-treatment. Cells of the choroid plexus showed a similar pattern. Various neuronal populations within a number of regions (hippocampus, circumventricular organs, hypothalamus and dorsal vagal complex) also express elevated hsp27 levels after heat shock (Krueger-Naug et al., 2000). As with hsp90, the functional significance of these expression patterns in relation to preservation of synaptic function has yet to be described.

Hyperthermic induction of heat shock proteins and viral neurovirulence

The febrile state is a potent inducer of hsp72 expression (Li et al., 1992) and is a characteristic host response to infection by numerous microbial pathogens. At the same time, hyperthermic pre-conditioning has been shown to enhance gene expression of several neurotropic viruses, including human T lymphotropic virus type 1 (HTLV-1) (Andrews et al., 1995, 1998), herpes simplex virus type 1 (Halford et al., 1996), measles virus (MV) (Vasconcelos et al., 1998a) and canine distemper virus (CDV) (Oglesbee et al., 1993). The mechanism by which stress-conditioning enhances viral gene expression is best characterized for the latter two examples, which are closely related members of the morbillivirus genus. Here, we have shown that hsp72 binds the viral core particle, specifically the nucleocapsid protein that packages the viral RNA genome (Zhang et al., 2002, 2005). Viral transcription and genome replication are enhanced through hsp72-dependent modifications of this ribonucleoprotein template, resulting in enhanced viral protein expression and virus-induced cytopathic effect (Vasconcelos et al., 1998a, b). Results are identical to infection of cells pre-conditioned with transient hyperthermia or constitutively overexpressing an hsp72 transgene, including the murine neuroblastoma cells illustrated in Fig. 3 (Carsillo et al., 2006). The question thus arises as to whether the febrile response and increased hsp72 expression in brain is detrimental or protective to the host in the face of such pathogens.

Using a mouse model of MV CNS infection, we have shown that the induction of hsp72 following transient hyperthermia of neonatal Balb/c mice is correlated to an increased immune-mediated clearance of the Edmonston strain of the measles virus (Ed-MV) from the brain following intracranial inoculation (Oglesbee et al., 2002b; Carsillo et al., 2004). Whole body hyperthermia was induced by placing animals in a heated chamber, and induction of hsp72 was demonstrated by Western blot analysis of brain total protein. Viral challenge groups included mice inoculated at the time of elevated hsp72 expression in brain and a control (non-heated) group. Forty seven percent of the non-heated animals supported persistent neuronal infection at 21 d post-infection, based upon real time RT-PCR analysis of brain viral RNA burden, in contrast to only 5% in the pre-conditioned animals (i.e., 95% clearance). The temporal onset and progression of clearance was associated with the development of virus-specific cell mediated immune responses known to be responsible for viral clearance from brain.

A model to explain the enhanced viral clearance in preconditioned tissues involves hsp72-dependent increases in viral antigenic burden and the extracellular release of hsp72 from virus-infected cells, where hsp72 can serve as an adjuvant and a mediator of antigenic cross-presentation, with both properties serving to enhance virus-specific cell mediated adaptive immune responses (reviewed in Oglesbee et al., 2002b). Key to viral persistence in multiple systems is the restricted antigen expression that confounds the immune-recognition of virus infected cells, particularly in the central nervous system (Liebert and Finke, 1995). In this regard, hsp72-mediated increases in viral gene expression serve to unmask these viral reservoirs, thereby promoting clearance at the expense of the initially infected cells. HSP release into the extracellular environment has been shown to occur following cell necrosis, but not after apoptosis, suggesting this release is restricted to times when cell death is uncontrolled (Basu et al., 2000). Thus, hsp72-dependent increases in MV cytopathic effect, a reflection of increased viral gene expression, are likely associated with the extracellular release of HSPs. Stimulatory effects of extracellular HSPs on the immune response include induction of inflammatory cytokine secretion (INFα, Il-1β, IL-12 and GM-CSF), NO production, chemokine production (MCP-1, MIP-2, RANTES) and dendritic cell maturation with translocation of NFκB to the nucleus (reviewed in Srivastava, 2002). The extracellular HSPs would also include those associated with viral antigenic peptides. Peptides derived from a number of viruses, including Vesicular Stomatitis Virus (Suto and Srivastava, 1995), Herpes-Simplex Virus 2 (Navaratnam et al., 2001), Simian Virus 40 (Blachere et al., 1993) and Hepatitis B Virus (Meng et al., 2001), contain residues of defined antigenicity that bind HSPs. HSP/antigen complexes have been shown to be potent immunogens against a wide variety of viral, bacterial and neoplastic diseases, including UV-induced tumors in C3H mice (Janetzki et al., 1998) and *M. tuberculosis* and LCMV in Balb/c mice (Zugel et al., 2001). It is known that antigen presenting cells recognize and bind HSP/antigen complexes via the CD91 membrane receptor, resulting in re-presentation of HSP-associated antigens by either major histocompatability complex (MHC) type I or type II molecules (Binder et al., 2000; Basu et al., 2001). Some components of the cytoplasmic processing mechanism for hsp72-associated antigens have been identified, including HSP interaction with the transporter-associated with antigen processing (Basu et al., 2001). Cross-presentation is particularly attractive in the CNS in that it allows viral antigen produced in a neuron to prime immune responses in uninfected antigen presenting cells. CSF would serve as the probable vehicle of antigen transport to regional lymph nodes via nerve sheaths, with priming of immune responses occurring in those regional lymphoid organs.

The hsp72-mediated enhancement of viral gene expression that promotes viral clearance would also appear to place nervous tissues at inherent risk in a host that is immunocompromised. For MV, the resistance or susceptibility of inbred mice to viral encephalitis correlates with the strain-specific expression of MHC haplotype that in turn dictates the efficiency of MV-specific adaptive T cell responses mediating clearance (Niewiesk et al., 1993). Balb/c mice, which express the H-2^d haplotype, are inherently more resistant to MV-induced encephalitis and can generate a protective cytotoxic T lymphocyte (CTL) response in the face of intracerebral virus inoculation. It follows that the increase in hsp72 expression in the CNS following hyperthermia in this strain thus leads to more rapid clearance of the virus, given the ability of hsp72 to act as an endogenous adjuvant for bound antigens and promote cross-priming of CTL responses. In contrast, mice expressing the H-2^b haplotype (e.g., C57BL/6) are more susceptible to MV-induced encephalitis due to deficient virus-specific CTL responses (Niewiesk et al., 1993).

We have recently generated transgenic C57BL/6 mice that constitutively overexpress hsp72 in neurons. In these animals, intracerebral inoculation with the non-rodent-adapted Ed-MV strain results not in more rapid clearance, as in the heat-shocked Balb/c mice, but in widely disseminated infection and a six-fold increase in mortality. Death is associated with increased brain viral RNA burdens that are almost two orders of magnitude greater than that of wild type C57BL/6 mice infected with the same virus. The outcome mimics the characteristic pathological features of measles virus inclusion body encephalitis (MIBE), a fulminate infection of the CNS of immune compromised humans that can be caused by either wild type or vaccine strains of MV. When the same hsp72 construct is expressed in H-2^d (resistant) mice, viral RNA burdens are significantly reduced relative to inoculated non-transgenic control mice and mortality is not observed. These results support the importance of H-2 haplotype, and thus the capacity for a robust antiviral immune response, in determining whether elevated hsp72 levels are host protective or a host determinant that promotes viral neurovirulence.

A recombinant infectious MV variant has also been generated that exhibits an attenuated response to hsp72-dependent increases in viral transcription (Zhang et al., 2002), and this virus fails to exhibit enhanced mortality in the transgenic hsp72 overexpressing H-2^b mice relative to the wild type controls. Although less neurovirulent, this viral variant would also be more likely to persist in the face of a competent immune response despite elevations in hsp72. Thus we are beginning to see that the fever-induced expression of hsp72 in the CNS may have disparate effects on the outcome of infection in the brain that reflects the nature of the interaction between hsp72 and the inciting agent as well as the ability of the immune system to respond to the biological effect of such an interaction.

Conclusions

Synaptic activity might be considered a dispensable luxury function, yet to lose it would constitute central nervous system failure. Profiles of HSP expression in the brain appear to have evolved in such a way as to preserve this function in addition to promoting cell survival in the face of hyperthermia and other stimuli that otherwise result in a loss of cellular homeostasis. So far, the HSP70 family provides the best example of this protective effect and the disparate expression patterns of hsc70 and hsp72 reflect the functional overlap between the two molecules and emphasize their complimentary roles. Here, hsp72 may again play a protective role by promoting clearance of neurotropic viruses. Our model for this clearance involves the hsp72-mediated increases in antigenic burden and the extracellular release of HSPs, a phenomenon documented in the CNS but restricted to instances where cell injury is involved. Hsp72-mediated increases in viral gene expression may drive the extracellular release of HSPs by promoting viral cytopathic effect, and both would enhance virus-specific adaptive immune responses leading to clearance. There appears to be a liability associated with this protective mechanism, in that enhanced neurovirulence can be the outcome in the face of an immune deficiency, and that viral variants can emerge that are capable of establishing persistent infection by becoming non-hsp72 responsive.

Abbreviations

ADP	adenosine diphosphate
ATP	adenosine triphosphate
BAEP	brainstem auditory evoked potential
CDV	canine distemper virus
CSF	cerebrospinal fluid
CSP	cysteine string protein
CTL	cytotoxic T lymphocyte
DIC	disseminated intravascular coagulopathy
Ed-MV	Edmonston strain of the measles virus
EM	electron microscopy
hsc70	constitutive isoform of the 70-kDa heat shock family
HSE	heat shock element
HSF	heat shock factor
HSP	heat shock protein
HSP70	70-kDa heat shock protein family
hsp72	inducible isoform of the 70-kDa heat shock protein family
hsp90	isoform of the 90-kDa heat shock protein family
hsp27, hsp32	isoforms of the small molecular weight heat shock protein family
kDa	kilodalton
LSD	lysergic acid diethylamide
MHC	major histocompatability complex
MIBE	measles inclusion body encephalitis
mRNA	messenger ribonucleic acid
MV	measles virus

Acknowledgments

The authors wish to thank Dr. Steve Krakowka for his help in review of this manuscript and Dr. Hamdy Awad for sharing data relevant to the spinal cord ischemia-reperfusion injury model. Our work is supported by funds from the National Institute of Neurological Disorders and Stroke (R01 NS31693).

References

Andrews, J.M., Oglesbee, M. and Lairmore, M.D. (1998) The effect of the cellular stress response on human T-lymphotropic virus type I envelope protein expression. J. Gen. Virol., 79(Pt 12): 2905–2908.

Andrews, J.M., Oglesbee, M.J., Trevino, A.V., Guyot, D.J., Newbound, G.C. and Lairmore, M.D. (1995) Enhanced human T-cell lymphotropic virus type I expression following induction of the cellular stress response. Virology, 208: 816–820.

Armstrong, C.L., Krueger-Naug, A.M., Currie, R.W. and Hawkes, R. (2001) Constitutive expression of heat shock protein HSP25 in the central nervous system of the developing and adult mouse. J. Comp. Neurol., 434: 262–274.

Asaki, C., Usuda, N., Nakazawa, A., Kametani, K. and Suzuki, T. (2003) Localization of translational components at the ultramicroscopic level at postsynaptic sites of the rat brain. Brain Res., 972: 168–176.

Barbe, M.F., Tytell, M., Gower, D.J. and Welch, W.J. (1988) Hyperthermia protects against light damage in the rat retina. Science, 241: 1817–1820.

Barclay, J.W. and Robertson, R.M. (2003) Role for calcium in heat shock-mediated synaptic thermoprotection in *Drosophila* larvae. J. Neurobiol., 56: 360–371.

Barouch, W., Prasad, K., Greene, L.E. and Eisenberg, E. (1994) ATPase activity associated with the uncoating of clathrin baskets by Hsp70. J. Biol. Chem., 269: 28563–28568.

Basu, S., Binder, R.J., Ramalingam, T. and Srivastava, P.K. (2001) CD91 is a common receptor for heat shock proteins gp96, hsp90, hsp70, and calreticulin. Immunity, 14: 303–313.

Basu, S., Binder, R.J., Suto, R., Anderson, K.M. and Srivastava, P.K. (2000) Necrotic but not apoptotic cell death releases heat shock proteins, which deliver a partial maturation signal to dendritic cells and activate the NF-kappa B pathway. Int. Immunol., 12: 1539–1546.

Bechtold, D.A. and Brown, I.R. (2000) Heat shock proteins Hsp27 and Hsp32 localize to synaptic sites in the rat cerebellum following hyperthermia. Brain Res. Mol. Brain Res., 75: 309–320.

Bechtold, D.A. and Brown, I.R. (2003) Induction of Hsp27 and Hsp32 stress proteins and vimentin in glial cells of the rat hippocampus following hyperthermia. Neurochem. Res., 28: 1163–1173.

Bechtold, D.A., Rush, S.J. and Brown, I.R. (2000) Localization of the heat-shock protein Hsp70 to the synapse following hyperthermic stress in the brain. J. Neurochem., 74: 641–646.

Beckmann, R.P., Mizzen, L.E. and Welch, W.J. (1990) Interaction of Hsp 70 with newly synthesized proteins: implications for protein folding and assembly. Science, 248: 850–854.

Belay, H.T. and Brown, I.R. (2006) Cell death and expression of heat-shock protein Hsc70 in the hyperthermic rat brain. J. Neurochem., 97(Suppl. 1): 116–119.

Belay, H.T. and Brown, R. (2003) Spatial analysis of cell death and Hsp70 induction in brain, thymus, and bone marrow of the hyperthermic rat. Cell Stress Chaperones, 8: 395–404.

Binder, R.J., Han, D.K. and Srivastava, P.K. (2000) CD91: a receptor for heat shock protein gp96. Nat. Immunol., 1: 151–155.

Blachere, N.E., Udono, H., Janetzki, S., Li, Z., Heike, M. and Srivastava, P.K. (1993) Heat shock protein vaccines against cancer. J. Immunother., 14: 352–356.

Bodega, G., Hernandez, C., Suarez, I., Martin, M. and Fernandez, B. (2002) HSP70 constitutive expression in rat central nervous system from postnatal development to maturity. J. Histochem. Cytochem., 50: 1161–1168.

Bronk, P., Wenniger, J.J., Dawson-Scully, K., Guo, X., Hong, S., Atwood, H.L. and Zinsmaier, K.E. (2001) *Drosophila* Hsc70-4 is critical for neurotransmitter exocytosis in vivo. Neuron, 30: 475–488.

Brown, C.R., Martin, R.L., Hansen, W.J., Beckmann, R.P. and Welch, W.J. (1993) The constitutive and stress inducible forms of hsp 70 exhibit functional similarities and interact with one another in an ATP-dependent fashion. J. Cell Biol., 120: 1101–1112.

Campisi, J. and Fleshner, M. (2003) Role of extracellular HSP72 in acute stress-induced potentiation of innate immunity in active rats. J. Appl. Physiol., 94: 43–52.

Campisi, J., Leem, T.H. and Fleshner, M. (2003) Stress-induced extracellular Hsp72 is a functionally significant danger signal to the immune system. Cell Stress Chaperones, 8: 272–286.

Carsillo, T., Carsillo, M., Niewiesk, S., Vasconcelos, D. and Oglesbee, M. (2004) Hyperthermic pre-conditioning promotes measles virus clearance from brain in a mouse model of persistent infection. Brain Res., 1004: 73–82.

Carsillo, T., Zhang, X., Vasconcelos, D., Niewiesk, S. and Oglesbee, M. (2006) A single codon in the nucleocapsid protein C terminus contributes to in vitro and in vivo fitness of Edmonston measles virus. J. Virol., 80: 2904–2912.

Chamberlain, L.H. and Burgoyne, R.D. (1997) Activation of the ATPase activity of heat-shock proteins Hsc70/Hsp70 by cysteine-string protein. Biochem. J., 322(Pt 3): 853–858.

Chen, C.Y., Sakisaka, T. and Balch, W.E. (2005a) Use of Hsp90 inhibitors to disrupt GDI-dependent Rab recycling. Methods Enzymol., 403: 339–347.

Chen, L.J., Su, X.W., Qiu, P.X., Huang, Y.J. and Yan, G.M. (2004) Thermal preconditioning protected cerebellar granule neurons of rats by modulating HSP70 expression. Acta Pharmacol. Sin., 25: 458–461.

Chen, S., Bawa, D., Besshoh, S., Gurd, J.W. and Brown, I.R. (2005b) Association of heat shock proteins and neuronal membrane components with lipid rafts from the rat brain. J. Neurosci. Res., 81: 522–529.

Chopp, M., Li, Y., Dereski, M.O., Levine, S.R., Yoshida, Y. and Garcia, J.H. (1991) Neuronal injury and expression of 72-kDa heat-shock protein after forebrain ischemia in the rat. Acta Neuropathol. (Berl.), 83: 66–71.

Cvoro, A. and Matic, G. (2002) Hyperthermic stress stimulates the association of both constitutive and inducible isoforms of 70 kDa heat shock protein with rat liver glucocorticoid receptor. Int. J. Biochem. Cell Biol., 34: 279–285.

Dawson-Scully, K., Bronk, P., Atwood, H.L. and Zinsmaier, K.E. (2000) Cysteine-string protein increases the calcium sensitivity of neurotransmitter exocytosis in Drosophila. J. Neurosci., 20: 6039–6047.

Dawson-Scully, K. and Meldrum, R.R. (1998) Heat shock protects synaptic transmission in flight motor circuitry of locusts. Neuroreport, 9: 2589–2593.

Diehl, K.A., Crawford, E., Shinko, P.D., Tallman Jr., R.D. and Oglesbee, M.J. (2000) Alterations in hemostasis associated with hyperthermia in a canine model. Am. J. Hematol., 64: 262–270.

Ellis, S., Killender, M. and Anderson, R.L. (2000) Heat-induced alterations in the localization of HSP72 and HSP73 as measured by indirect immunohistochemistry and immunogold electron microscopy. J. Histochem. Cytochem., 48: 321–332.

Febbraio, M.A., Ott, P., Nielsen, H.B., Steensberg, A., Keller, C., Krustrup, P., Secher, N.H. and Pederson, B.K. (2002) Exercise induces hepatosplanchnic release of heat shock protein 72 in humans. J. Physiol., 544(Pt3): 957–962.

Fleshner, M., Campisi, J., Amiri, L. and Diamond, D.M. (2004) Cat exposure induces both intra- and extracellular Hsp72: the role of adrenal hormones. Psychoneuroendocrinology, 29: 1142–1152.

Foster, J.A. and Brown, I.R. (1997) Differential induction of heat shock mRNA in oligodendrocytes, microglia, and astrocytes following hyperthermia. Brain Res. Mol. Brain Res., 45: 207–218.

Fourie, A.M., Sambrook, J.F. and Gething, M.J. (1994) Common and divergent peptide binding specificities of hsp70 molecular chaperones. J. Biol. Chem., 269: 30470–30478.

Gamer, J., Multhaup, G., Tomoyasu, T., McCarty, J.S., Rudiger, S., Schonfeld, H.J., Schirra, C., Bujard, H. and Bukau, B. (1996) A cycle of binding and release of the DnaK, DnaJ and GrpE chaperones regulates activity of the Escherichia coli heat shock transcription factor sigma32. EMBO J., 15: 607–617.

Gerges, N.Z., Tran, I.C., Backos, D.S., Harrell, J.M., Chinkers, M., Pratt, W.B. and Esteban, J.A. (2004) Independent functions of hsp90 in neurotransmitter release and in the continuous synaptic cycling of AMPA receptors. J. Neurosci., 24: 4758–4766.

Goldbaum, O. and Richter-Landsberg, C. (2001) Stress proteins in oligodendrocytes: differential effects of heat shock and oxidative stress. J. Neurochem., 78: 1233–1242.

Guzhova, I., Kislyakova, K., Moskaliova, O., Fridlanskaya, I., Tytell, M., Cheetham, M. and Margulis, B. (2001) In vitro studies show that Hsp70 can be released by glia and that exogenous Hsp70 can enhance neuronal stress tolerance. Brain Res., 914: 66–73.

Halford, W.P., Gebhardt, B.M. and Carr, D.J. (1996) Mechanisms of herpes simplex virus type 1 reactivation. J. Virol., 70: 5051–5060.

Houenou, L.J., Li, L., Lei, M., Kent, C.R. and Tytell, M. (1996) Exogenous heat shock cognate protein Hsc 70 prevents axotomy-induced death of spinal sensory neurons. Cell Stress Chaperones, 1: 161–166.

Janetzki, S., Blachere, N.E. and Srivastava, P.K. (1998) Generation of tumor-specific cytotoxic T lymphocytes and memory T cells by immunization with tumor-derived heat shock protein gp96. J. Immunother., 21: 269–276.

Karunanithi, S., Barclay, J.W., Brown, I.R., Robertson, R.M. and Atwood, H.L. (2002) Enhancement of presynaptic performance in transgenic Drosophila overexpressing heat shock protein Hsp70. Synapse, 44: 8–14.

Kato, K., Yamanaka, K., Nakano, M., Hasegawa, A., Oku, N. and Okada, S. (1997) Cell-nuclear accumulation of 72-kDa stress protein induced by dimethylated arsenics. Biol. Pharm. Bull., 20: 364–369.

Kelty, J.D., Noseworthy, P.A., Feder, M.E., Robertson, R.M. and Ramirez, J.M. (2002) Thermal preconditioning and heat-shock protein 72 preserve synaptic transmission during thermal stress. J. Neurosci., 22: RC193.

Khan, V.R. and Brown, I.R. (2002) The effect of hyperthermia on the induction of cell death in brain, testis, and thymus of the adult and developing rat. Cell Stress Chaperones, 7: 73–90.

Kiang, J.G. and Tsokos, G.C. (1998) Heat shock protein 70 kDa: molecular biology, biochemistry, and physiology. Pharmacol. Ther., 80: 183–201.

Klose, M.K., Armstrong, G. and Robertson, R.M. (2004) A role for the cytoskeleton in heat-shock-mediated thermoprotection of locust neuromuscular junctions. J. Neurobiol., 60: 453–462.

Krueger, A.M., Armstrong, J.N., Plumier, J., Robertson, H.A. and Currie, R.W. (1999) Cell specific expression of Hsp70 in neurons and glia of the rat hippocampus after hyperthermia and kainic acid-induced seizure activity. Brain Res. Mol. Brain Res., 71: 265–278.

Krueger-Naug, A.M., Hopkins, D.A., Armstrong, J.N., Plumier, J.C. and Currie, R.W. (2000) Hyperthermic induction of the 27-kDa heat shock protein (Hsp27) in neuroglia and neurons of the rat central nervous system. J. Comp. Neurol., 428: 495–510.

Kwong, J.M., Lam, T.T. and Caprioli, J. (2003) Hyperthermic pre-conditioning protects retinal neurons from N-methyl-D-aspartate (NMDA)-induced apoptosis in rat. Brain Res., 970: 119–130.

Li, Y., Chopp, M., Yoshida, Y. and Levine, S.R. (1992) Distribution of 72-kDa heat-shock protein in rat brain after hyperthermia. Acta Neuropathol. (Berl.), 84: 94–99.

Liebert, U.G. and Finke, D. (1995) Measles virus infections in rodents. Curr. Top. Microbiol. Immunol., 191: 149–166.

Lin, Y.W., Yang, H.W., Min, M.Y. and Chiu, T.H. (2004) Heat-shock pretreatment prevents suppression of long-term potentiation induced by scopolamine in rat hippocampal CA1 synapses. Brain Res., 999: 222–226.

Mailhos, C., Howard, M.K. and Latchman, D.S. (1994) Heat shock proteins hsp90 and hsp70 protect neuronal cells from thermal stress but not from programmed cell death. J. Neurochem., 63: 1787–1795.

Malinow, R., Mainen, Z.F. and Hayashi, Y. (2000) LTP mechanisms: from silence to four-lane traffic. Curr. Opin. Neurobiol., 10: 352–357.

Manzerra, P. and Brown, I.R. (1990) Time course of induction of a heat shock gene (hsp70) in the rabbit cerebellum after LSD in vivo: involvement of drug-induced hyperthermia. Neurochem. Res., 15: 53–59.

Manzerra, P., Rush, S.J. and Brown, I.R. (1997) Tissue-specific differences in heat shock protein hsc70 and hsp70 in the control and hyperthermic rabbit. J. Cell Physiol., 170: 130–137.

Maroni, P., Bendinelli, P., Tiberio, L., Rovetta, F., Piccoletti, R. and Schiaffonati, L. (2003) In vivo heat-shock response in the brain: signalling pathway and transcription factor activation. Brain Res. Mol. Brain Res., 119: 90–99.

Martin, K.C. (2004) Local protein synthesis during axon guidance and synaptic plasticity. Curr. Opin. Neurobiol., 14: 305–310.

McCarty, J.S., Buchberger, A., Reinstein, J. and Bukau, B. (1995) The role of ATP in the functional cycle of the DnaK chaperone system. J. Mol. Biol., 249: 126–137.

Meng, S.D., Gao, T., Gao, G.F. and Tien, P. (2001) HBV-specific peptide associated with heat-shock protein gp96. Lancet, 357: 528–529.

Milani, V., Noessner, E., Ghose, S., Kuppner, M., Ahrens, B., Scharner, A., Gastpar, R. and Issels, R.D. (2002) Heat shock protein 70: role in antigen presentation and immune stimulation. Int. J. Hyperthermia, 18: 563–575.

Minami, Y., Hohfeld, J., Ohtsuka, K. and Hartl, F.U. (1996) Regulation of the heat-shock protein 70 reaction cycle by the mammalian DnaJ homolog, Hsp40. J. Biol. Chem., 271: 19617–19624.

Misselwitz, B., Staeck, O. and Rapoport, T.A. (1998) J proteins catalytically activate Hsp70 molecules to trap a wide range of peptide sequences. Mol. Cell, 2: 593–603.

Mizzen, L.A. and Welch, W.J. (1988) Characterization of the thermotolerant cell. I. Effects on protein synthesis activity and the regulation of heat-shock protein 70 expression. J. Cell Biol., 106: 1105–1116.

Mokrushin, A.A., Pavlinova, L.I. and Plekhanov, A.Y. (2005) Heat shock protein HSP70 increases the resistance of cortical cells to glutamate excitotoxicity. Bull. Exp. Biol. Med., 140: 1–5.

Moon, I.S., Park, I.S., Schenker, L.T., Kennedy, M.B., Moon, J.I. and Jin, I. (2001) Presence of both constitutive and inducible forms of heat shock protein 70 in the cerebral cortex and hippocampal synapses. Cereb. Cortex, 11: 238–248.

Morgan, J.R., Prasad, K., Jin, S., Augustine, G.J. and Lafer, E.M. (2001) Uncoating of clathrin-coated vesicles in presynaptic terminals: roles for Hsc70 and auxilin. Neuron, 32: 289–300.

Morrison, A.J., Rush, S.J. and Brown, I.R. (2000) Heat shock transcription factors and the hsp70 induction response in brain and kidney of the hyperthermic rat during postnatal development. J. Neurochem., 75: 363–372.

Navaratnam, M., Deshpande, M.S., Hariharan, M.J., Zatechka Jr., D.S. and Srikumaran, S. (2001) Heat shock protein-peptide complexes elicit cytotoxic T-lymphocyte and antibody responses specific for bovine herpesvirus 1. Vaccine, 19: 1425–1434.

Ngosuwan, J., Wang, N.M., Fung, K.L. and Chirico, W.J. (2003) Roles of cytosolic Hsp70 and Hsp40 molecular chaperones in post-translational translocation of presecretory proteins into the endoplasmic reticulum. J. Biol. Chem., 278(9): 7034–7042.

Niewiesk, S., Brinckmann, U., Bankamp, B., Sirak, S., Liebert, U.G. and ter Meulen, V. (1993) Susceptibility to measles virus-induced encephalitis in mice correlates with impaired antigen presentation to cytotoxic T lymphocytes. J. Virol., 67: 75–81.

Oglesbee, M. and Krakowka, S. (1993) Cellular stress response induces selective intranuclear trafficking and accumulation of morbillivirus major core protein. Lab. Invest., 68: 109–117.

Oglesbee, M.J., Alldinger, S., Vasconcelos, D., Diehl, K.A., Shinko, P.D., Baumgartner, W., Tallman, R. and Podell, M. (2002a) Intrinsic thermal resistance of the canine brain. Neuroscience, 113: 55–64.

Oglesbee, M.J., Diehl, K., Crawford, E., Kearns, R. and Krakowka, S. (1999) Whole body hyperthermia: effects upon canine immune and hemostatic functions. Vet. Immunol. Immunopathol., 69: 185–199.

Oglesbee, M.J., Kenney, H., Kenney, T. and Krakowka, S. (1993) Enhanced production of morbillivirus gene-specific RNAs following induction of the cellular stress response in stable persistent infection. Virology, 192: 556–567.

Oglesbee, M.J., Pratt, M. and Carsillo, T. (2002b) Role for heat shock proteins in the immune response to measles virus infection. Viral Immunol., 15: 399–416.

Ohtsuka, K. and Hata, M. (2000) Molecular chaperone function of mammalian Hsp70 and Hsp40: a review. Int. J. Hyperthermia, 16: 231–245.

Ohtsuka, K. and Suzuki, T. (2000) Roles of molecular chaperones in the nervous system. Brain Res. Bull., 53: 141–146.

Ostberg, J.R., Kaplan, K.C. and Repasky, E.A. (2002) Induction of stress proteins in a panel of mouse tissues by fever-range whole body hyperthermia. Int. J. Hyperthermia, 18: 552–562.

Pardue, S., Wang, S., Miller, M.M. and Morrison-Bogorad, M. (2007) Elevated levels of inducible heat shock 70 proteins in human brain. Neurobiol. Aging, 28(2): 314–324.

Pavlik, A., Aneja, I.S., Lexa, J. and Al Zoabi, B.A. (2003) Identification of cerebral neurons and glial cell types inducing heat shock protein Hsp70 following heat stress in the rat. Brain Res., 973: 179–189.

Pilon, M. and Schekman, R. (1999) Protein translocation: how Hsp70 pulls it off. Cell, 97: 679–682.

Quraishi, H. and Brown, I.R. (1995) Expression of heat shock protein 90 (hsp90) in neural and nonneural tissues of the control and hyperthermic rabbit. Exp. Cell Res., 219: 358–363.

Quraishi, H., Rush, S.J. and Brown, I.R. (1996) Expression of mRNA species encoding heat shock protein 90 (hsp90) in control and hyperthermic rabbit brain. J. Neurosci. Res., 43: 335–345.

Ravagnan, L., Gurbuxani, S., Susin, S.A., Maisse, C., Daugas, E., Zamzami, N., Mak, T., Jaattela, M., Penninger, J.M., Garrido, C. and Kroemer, G. (2001) Heat-shock protein 70 antagonizes apoptosis-inducing factor. Nat. Cell Biol., 3: 839–843.

Ritossa, F. (1962) A new puffing pattern induced by temperature shock and DNP in *Drosophila*. Experientia, 18: 571–573.

Rudiger, S., Germeroth, L., Schneider-Mergener, J. and Bukau, B. (1997) Substrate specificity of the DnaK chaperone determined by screening cellulose-bound peptide libraries. EMBO J., 16: 1501–1507.

Sakisaka, T., Meerlo, T., Matteson, J., Plutner, H. and Balch, W.E. (2002) Rab-alphaGDI activity is regulated by a Hsp90 chaperone complex. EMBO J., 21: 6125–6135.

Sato, K. and Matsuki, N. (2002) A 72 kDa heat shock protein is protective against the selective vulnerability of CA1 neurons and is essential for the tolerance exhibited by CA3 neurons in the hippocampus. Neuroscience, 109: 745–756.

Sato, K., Saito, H. and Matsuki, N. (1996) HSP70 is essential to the neuroprotective effect of heat-shock. Brain Res., 740: 117–123.

Schiaffonati, L., Maroni, P., Bendinelli, P., Tiberio, L. and Piccoletti, R. (2001) Hyperthermia induces gene expression of heat shock protein 70 and phosphorylation of mitogen activated protein kinases in the rat cerebellum. Neurosci. Lett., 312: 75–78.

Sharma, H.S. and Westman, J. (2004) The heat shock proteins and hemeoxygenase response in central nervous system injuries. In: Sharma, H.S. and Westman, J. (Eds.), Blood-Spinal Cord and Brain Barriers in Health and Disease, xvi; 17, Elsevier, Amsterdam, pp. 329–360.

Sheller, R.A., Smyers, M.E., Grossfeld, R.M., Ballinger, M.L. and Bittner, G.D. (1998) Heat-shock proteins in axoplasm: high constitutive levels and transfer of inducible isoforms from glia. J. Comp. Neurol., 396: 1–11.

Sheller, R.A., Tytell, M., Smyers, M. and Bittner, G.D. (1995) Glia-to-axon communication: enrichment of glial proteins transferred to the squid giant axon. J. Neurosci. Res., 41: 324–334.

Simons, K. and Toomre, D. (2000) Lipid rafts and signal transduction. Nat. Rev. Mol. Cell Biol., 1: 31–39.

Sminia, P., van der, Z.J., Wondergem, J. and Haveman, J. (1994) Effect of hyperthermia on the central nervous system: a review. Int. J. Hyperthermia, 10: 1–30.

Snoeckx, L.H., Cornelussen, R.N., Van Nieuwenhoven, F.A., Reneman, R.S. and Van Der Vusse, G.J. (2001) Heat shock proteins and cardiovascular pathophysiology. Physiol. Rev., 81: 1461–1497.

Song, J., Takeda, M. and Morimoto, R.I. (2001) Bag1-Hsp70 mediates a physiological stress signalling pathway that regulates Raf-1/ERK and cell growth. Nat. Cell Biol., 3: 276–282.

D'Souza, S.M. and Brown, I.R. (1998) Constitutive expression of heat shock proteins Hsp90, Hsc70, Hsp70 and Hsp60 in neural and non-neural tissues of the rat during postnatal development. Cell Stress Chaperones, 3: 188–199.

Sreedhar, A.S. and Csermely, P. (2004) Heat shock proteins in the regulation of apoptosis: new strategies in tumor therapy: a comprehensive review. Pharmacol. Ther., 101: 227–257.

Srivastava, P. (2002) Interaction of heat shock proteins with peptides and antigen presenting cells: chaperoning of the innate and adaptive immune responses. Annu. Rev. Immunol., 20: 395–425.

Suto, R. and Srivastava, P.K. (1995) A mechanism for the specific immunogenicity of heat shock protein-chaperoned peptides. Science, 269: 1585–1588.

Suzuki, T. (2002) Lipid rafts at postsynaptic sites: distribution, function and linkage to postsynaptic density. Neurosci. Res., 44: 1–9.

Suzuki, T., Usuda, N., Murata, S., Nakazawa, A., Ohtsuka, K. and Takagi, H. (1999) Presence of molecular chaperones, heat shock cognate (Hsc) 70 and heat shock proteins (Hsp) 40, in the postsynaptic structures of rat brain. Brain Res., 816: 99–110.

Takahashi, H., Tanaka, R., Sekihara, Y. and Hondo, H. (1991) Auditory brainstem response during systemic hyperthermia. Int. J. Hyperthermia, 7: 613–620.

Tian, Q.B., Nakayama, K., Okano, A. and Suzuki, T. (1999) Identification of mRNAs localizing in the postsynaptic region. Brain Res. Mol. Brain Res., 72: 147–157.

Tidwell, J.L., Houenou, L.J. and Tytell, M. (2004) Administration of Hsp70 in vivo inhibits motor and sensory neuron degeneration. Cell Stress Chaperones, 9: 88–98.

Tissieres, A., Mitchell, H.K. and Tracy, U.M. (1974) Protein synthesis in salivary glands of *Drosophila melanogaster*: relation to chromosome puffs. J. Mol. Biol., 84: 389–398.

Tytell, M., Greenberg, S.G. and Lasek, R.J. (1986) Heat shock-like protein is transferred from glia to axon. Brain Res., 363: 161–164.

Vasconcelos, D., Norrby, E. and Oglesbee, M. (1998a) The cellular stress response increases measles virus-induced cytopathic effect. J. Gen. Virol., 79(Pt 7): 1769–1773.

Vasconcelos, D.Y., Cai, X.H. and Oglesbee, M.J. (1998b) Constitutive overexpression of the major inducible 70 kDa heat shock protein mediates large plaque formation by measles virus. J. Gen. Virol., 79(Pt 9): 2239–2247.

Welch, W.J. and Mizzen, L.A. (1988) Characterization of the thermotolerant cell. II. Effects on the intracellular distribution of heat-shock protein 70, intermediate filaments, and small nuclear ribonucleoprotein complexes. J. Cell Biol., 106: 1117–1130.

Wyatt, S., Mailhos, C. and Latchman, D.S. (1996) Trigeminal ganglion neurons are protected by the heat shock proteins hsp70 and hsp90 from thermal stress but not from programmed cell death following nerve growth factor withdrawal. Brain Res. Mol. Brain Res., 39: 52–56.

Yager, J.Y. and Asselin, J. (1999) The effect of pre hypoxic-ischemic (HI) hypo and hyperthermia on brain damage in the immature rat. Brain Res. Dev. Brain Res., 117: 139–143.

Young, J.C., Agashe, V.R., Siegers, K. and Hartl, F.U. (2004) Pathways of chaperone-mediated protein folding in the cytosol. Nat. Rev. Mol. Cell Biol., 5: 781–791.

Yu, Q., Kent, C.R. and Tytell, M. (2001) Retinal uptake of intravitreally injected Hsc/Hsp70 and its effect on susceptibility to light damage. Mol. Vis., 7: 48–56.

Zhang, X., Bourhis, J.M., Longhi, S., Carsillo, T., Buccellato, M., Morin, B., Canard, B. and Oglesbee, M. (2005) Hsp72 recognizes a P binding motif in the measles virus N protein C-terminus. Virology, 337: 162–174.

Zhang, X., Glendening, C., Linke, H., Parks, C.L., Brooks, C., Udem, S.A. and Oglesbee, M. (2002) Identification and characterization of a regulatory domain on the carboxyl terminus of the measles virus nucleocapsid protein. J. Virol., 76: 8737–8746.

Zugel, U., Sponaas, A.M., Neckermann, J., Schoel, B. and Kaufmann, S.H. (2001) gp96-peptide vaccination of mice against intracellular bacteria. Infect. Immun., 69: 4164–4167.

Zylicz, M., King, F.W. and Wawrzynow, A. (2001) Hsp70 interactions with the p53 tumour suppressor protein. EMBO J., 20: 4634–4638.

CHAPTER 20

Cerebral neurons and glial cell types inducing heat shock protein Hsp70 following heat stress in the rat

Alfred Pavlik* and Inderjeet S. Aneja

Department of Physiology, Faculty of Medicine, Kuwait University, P.O. Box 24923, Safat 13110, Kuwait

Abstract: In this chapter, the distribution of Hsp70 in brain cell types following whole body hyperthermia is reviewed. The prevalence of Hsp70 expression in oligodendrocytes, microglia, and vascular cells in this type of stress contrasts with scarcity of Hsp70 induction in astrocytes and most neurons of the hyperthermic brain. However, a similarity between hyperthermic- and arsenite-induced brain patterns of Hsp70 expression supports the view that denaturation of specific proteins plays a major role in the selectivity of glial/vascular expression also during hyperthermia *in vivo*. The mechanism of neuronal Hsp70 non-responsiveness in heat stress despite their ability to use Hsc70 in a partial heat stress response remains to be elucidated.

Keywords: Hsp70; Hsc70; hyperthermia; neuron; oligodendrocyte; microglia; capillary; immunohistochemistry

Introduction

Since the discovery of heat shock proteins (Hsps) in *Drosophila* (Tissieres et al., 1974), avian, and mammalian cells (Kelley and Schlesinger, 1978), we have witnessed an impressive development of a new research domain of cellular stress response (for reviews, see Welch, 1992; Morimoto, 1993; Hightower and Hendershot, 1997; Mogk and Bukau, 2004). Studies of Hsps function in normal cells have brought about their recognition as molecular chaperones that bind to unfolded polypeptides and assist their correct folding (Ellis and van der Vries, 1991; Young et al., 2004). Hsps also chaperoned proteins being transported between cellular compartments or destined for disposal (Barral et al., 2004). They were implicated in the protection and stabilization of receptors and signaling proteins (Pratt, 1993; Wegele et al., 2004). In cells under stress, they rescued denatured, even aggregated proteins (for reviews, see Sharp et al., 1999; Sherman and Goldberg, 2001; Lee et al., 2004).

Neurodegeneration is often accompanied by buildup of abnormal proteins in affected neurons and glial cells, which mobilize their cellular stress response to handle the load of misfolded proteins. Recently, Hsps role in counteracting an apoptotic process was firmly established (Sharp et al., 1999; Sherman and Goldberg, 2001; Takayama et al., 2003).

Several excellent reviews were written on Hsps in the brain cells (Brown, 1994; Brown and Sharp, 1995; Westman and Sharma, 1998; Sharp et al., 2001). We would like to focus, therefore, on two major Hsps of HSP70 family: the constitutive

*Corresponding author. Tel.: +965 5312300 ext. 6369; Fax: +965 5338937; E-mail: alfred@hsc.edu.kw

Hsc70 and inducible Hsp70 and their *in vivo* expression in brain cell types following heat stress.

Neuroglia, but not neurons, express Hsp70 following heat stress *in vivo*

Increase of brain temperature to the range of 41° to 42°C is considered a very serious clinical situation. When prolonged it may result in heat stroke with frequent fatalities (Sminia et al., 1994, see Chapters 10 and 15 in this volume). The same outcome may follow an experimental hyperthermia in animals if heat exposure extends for longer time, e.g., more than 1 h. Clinical and experimental studies of heat stroke claimed that its symptoms indicated a serious brain dysfunction and they were supported by pathophysiological and histopathological findings including neuronal death (Sharma et al., 1991, 1998, 1994; Lin, 1997; Yang et al., 1998; Sharma and Hoopes, 2003). However, if animals were exposed to a short-term hyperthermia of the same range, the signs of brain pathology were missing (Li et al., 1992; McCabe and Simon, 1993; Brown, 1994; Matsumi et al., 1994; Takahashi et al., 1999; Lee et al., 2000). Evidently, short-term hyperthermia of the brain is only a stressful but physiological state, which resolves without permanent damage of brain cells (Brown, 1994; Brown and Sharp, 1999).

One of the most significant indicators of heat stress response in mammalian cells *in vitro* is Hsp70 expression. So it came as a surprise that the majority of most important brain cell type — the neuron — did not show this sort of stress response *in vivo*. Contrary to expectation, the expression of Hsp70 in hyperthermic brain was localized to neuroglial cells (Brown et al., 1982; Sprang and Brown, 1987; Blake et al., 1990; for reviews, see Brown, 1990, 1994; Sharp et al., 2001). Such a selective expression pattern of Hsp70 in brain cells was completely different from other types of stress. In ischemia and neurotoxic stress, the pattern was just opposite: expression of Hsp70 occurred mainly in neurons and rarely in glial cells (Vass et al., 1988; Nowak, 1991; Nowak and Jacewicz, 1994; Armstrong et al., 1996; Krueger et al., 1999). Combined neuronal/glial expression patterns were reported after brain trauma and inflammation (see Brown, 1990, 1994; Sharp et al., 1999).

Oligodendrocytes are major producers of Hsp70 in the hyperthermic brain

Availability of selective mRNA probes and monoclonal antibodies for Hsp70 enabled to follow the process of its mRNA and protein appearance after heat stress in the brain (Sprang and Brown, 1987; Marini et al., 1990; Miller et al., 1991; Li et al., 1992; Manzerra et al., 1993; McCabe and Simon, 1993). These pioneering in situ hybridization/immunohistochemical studies of Hsp70 mRNA/protein distribution in the brain agreed on the general pattern of cellular localization. Glial cells, identified only by morphological criteria as oligodendrocytes, produced both Hsp70 mRNA and protein in the forebrain. In the cerebellum, however, granule neurons were the major source of Hsp70.

The typical pattern of Hsp70 induction appeared between 60 and 90 min following the end of hyperthermic period (Manzerra and Brown, 1996; Krueger et al., 1999; Pavlik et al., 2003). After heat stress (between 41 and 42°C for 15–60 min), brain sections became punctuated with small dots corresponding to glial cells both in the white and gray matter, together with staining of the cerebral vessels. Later, the glial dots were more marked with sharper contours and prevalent localization of induced Hsp70 in glial nuclei. Hsp70 immunostaining was accompanied with a low background. In our sections from normothermic control rats or immunohistochemical controls, it showed few lightly stained vessels, meningeal, ependymal, and choroid plexus cells (Fig. 1).

Practically identical pattern was reported in in situ hybridization studies on hyperthermic expression of Hsp70 mRNA. As expected, mRNA labeling appeared sooner than protein (Brown and Rush, 1990; Miller et al., 1991; Manzerra and Brown, 1992b). In the cerebellum, besides the granule neurons, oligodendrocytes and Bergmann glia were expressing Hsp70 (Manzerra et al., 1993). Induction of Hsp70 was reported also in oligodendrocytes of spinal cord tracts following

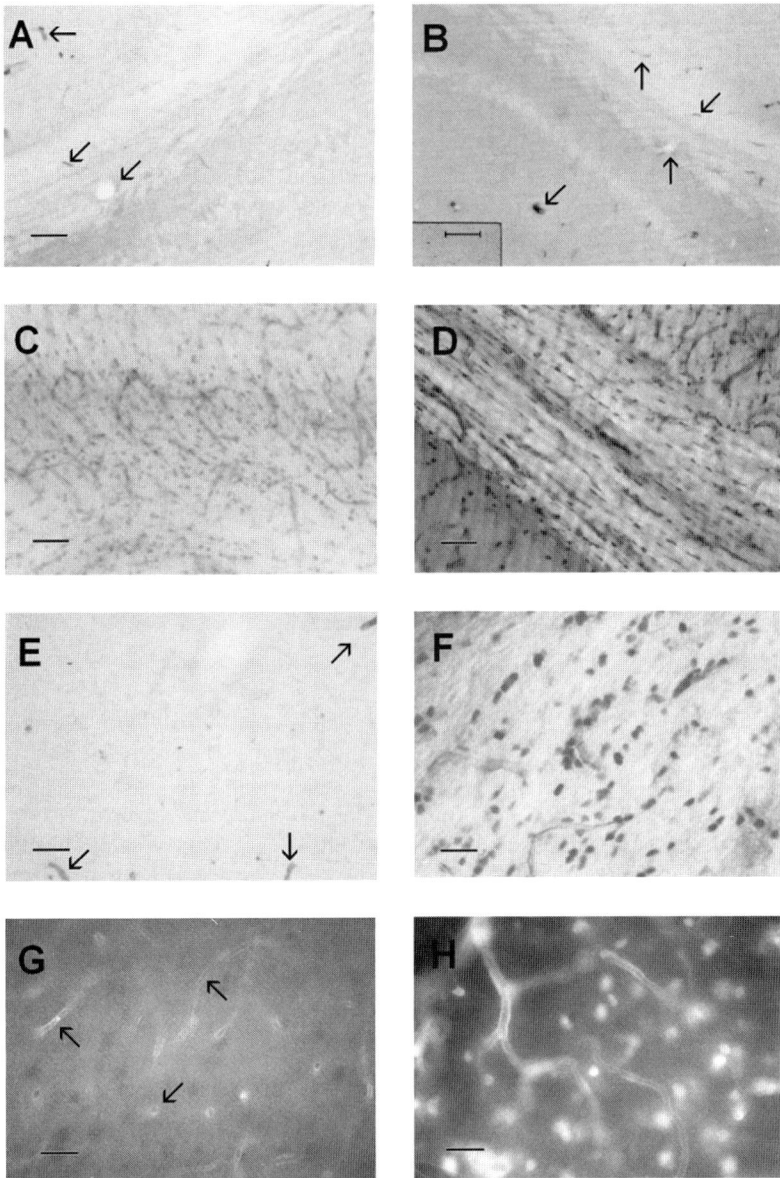

Fig. 1. Hsp70 immunostaining in the control and hyperthermic forebrain. Hsp70 staining (DAB) of the vibratome section from an anesthetized control rat (A) and the rat sacrificed 30 min post-hyperthermia (B) showed only low background and a few stained vessels (arrows). Presumably glial cells and vessels are stained with anti-Hsp70 antibody at 1.5 (C) and 4 h (D) following hyperthermia in the corpus callosum. E, an immunohistochemical control (Hsp70 antibody was replaced by non-immune ascites) with a few vessels (arrows) is compared to Hsp70 staining of presumably glial cells and vessels in the internal capsule (F). The fluorescent detection of Hsp70 with Alexa Fluor 488-tagged secondary antibodies weakly labeled some vessels (arrows) in the thalamus of the non-heated control rat (G). In the heated rat, the cells and vessels of the same region are shown to induce a distinct Hsp70 staining at 4 h post-hyperthermia (H). Calibration bar: A, B — 100 μm; C, D — 50 μm; E–H — 25 μm. (Reprinted from Pavlik et al., 2003 with permission from Elsevier.)

hyperthermic exposure of the animal (Manzerra and Brown, 1992a, 1996; Sasara et al., 2004).

Foster et al. (1995) first reported confirmation of oligodendrocyte as the cell expressing Hsp70 mRNA by using lysergic acid diethylamide-induced hyperthermia in the rabbit. They also identified microglia as another significant producer of Hsp70 mRNA in hyperthermic rabbit brain (Foster and Brown, 1997). In their study, glial fibrillary acidic protein (GFAP)-labeled astrocytes did not show any co-localization with Hsp70 mRNA in heat-stressed rabbit brain. We attempted identification of Hsp70 protein inducing oligodendrocytes with oligodendrocyte marker lectin GS II from *Griffonia simplicifolia* (Supler et al., 1994) as in Foster and Brown (1997). In Fig. 2, the co-localization of GS II lectin and Hsp70 protein was presented for the first time showing nearly complete coincidence in oligodendrocytes of cerebral white matter (Pavlik et al., 2003). However, using a reliable microglial marker OX-42 (monoclonal antibody against CD 11b) in the rat, we found substantially less co-localization with Hsp70 indicating that some Hsp70 mRNA might not be translated in heat-stressed microglia (Fig. 3C). Finally, we confirmed the negative finding on astrocyte expression of Hsp70 since less than 5% of GFAP-positive astrocytes co-localized with Hsp70-stained cells (Figs. 3A, B).

We believe that gray matter oligodendrocytes express Hsp70 in a similar way as the interfascicular oligodendrocytes of the white matter (Fig. 4). However, GS II lectin was not an ideal marker for gray matter oligodendrocytes, which was also suggested by Supler et al. (1994). Some of these oligodendrocytes were recognized earlier by Rio del Hortega as perivascular or perineuronal satellites (see Szuchet, 1995). Cerebral satellites seem to be the least studied glial cell types, perhaps because of difficulty to re-create mutual relationship between satellites and their neurons in cell culture. They may behave like other oligodendrocytes since they were active in normal myelination as well as remyelination processes (Ludwin, 1979, 1984). However, perineuronal satellites were reported to remove detritus due to toxic chemicals or lipofuscin from the cytoplasm of attached neurons (Sturrock, 1987; Cavanagh et al., 1990).

Perivascular satellites are supposed to present antigens to the immune system and might be one of antigen presenting cell subgroups in the brain (see Angelov et al., 1998).

Some 30 years ago we reported an autoradiographic study of newly synthesized proteins in brain slices incubated with radioactive leucine (Pavlik, 1974; Pavlik and Jakoubek, 1976). These proteins were localized mainly in vessels, glial cells, and in nuclei of perineuronal satellites (Fig. 5). At the same time, Tissieres et al. (1974) observed the hyperthermic induction of Hsps for the first time. Then, it was found that most of these newly synthesized proteins in brain slices was Hsp70 (SP 71) induced in glial and vascular cells by trauma of cutting brain tissue (White, 1980). So it took some time to bring the evidence that individual types of brain cells possess selective *in vivo* ways to respond to heat stress, and most probably to other types of stress as well.

Nuclear translocation of Hsp70

Nuclear translocation of some Hsps including Hsp70 as a part of heat stress response is also a well-known phenomenon existing in the brain cells (Welch and Feramisco, 1984; Marini et al., 1990; Manzerra and Brown, 1996; Xu et al., 1998; Pavlik et al., 2003). These reports agreed that Hsp70 was swiftly translocated to nuclei and/or nucleoli but after some time it was relocated back to cytoplasm. Cytoplasmic relocation of Hsp70 in oligodendrocytes was reported to occur by 5 h post-hyperthermia in the rabbit brain (Manzerra and Brown, 1996). We studied nuclear translocation of Hsp70 and its relocation in glial cells of the white matter at 2, 4, 8, 24, 48, and 72 h after hyperthermia. The results are presented in Fig. 6. At 2 h, 40% of Hsp70-positive cells in the corpus callosum showed prevalence of nuclear translocation. Most glial cells kept Hsp70 localized in their nuclei for 8 h following hyperthermia. Significant relocation was seen only by 24 h post-hyperthermia. Nuclear localization then continued to decrease and the specific staining of Hsp70 nearly disappeared by 72 h (not shown) corresponding to long half-life of Hsp70 protein

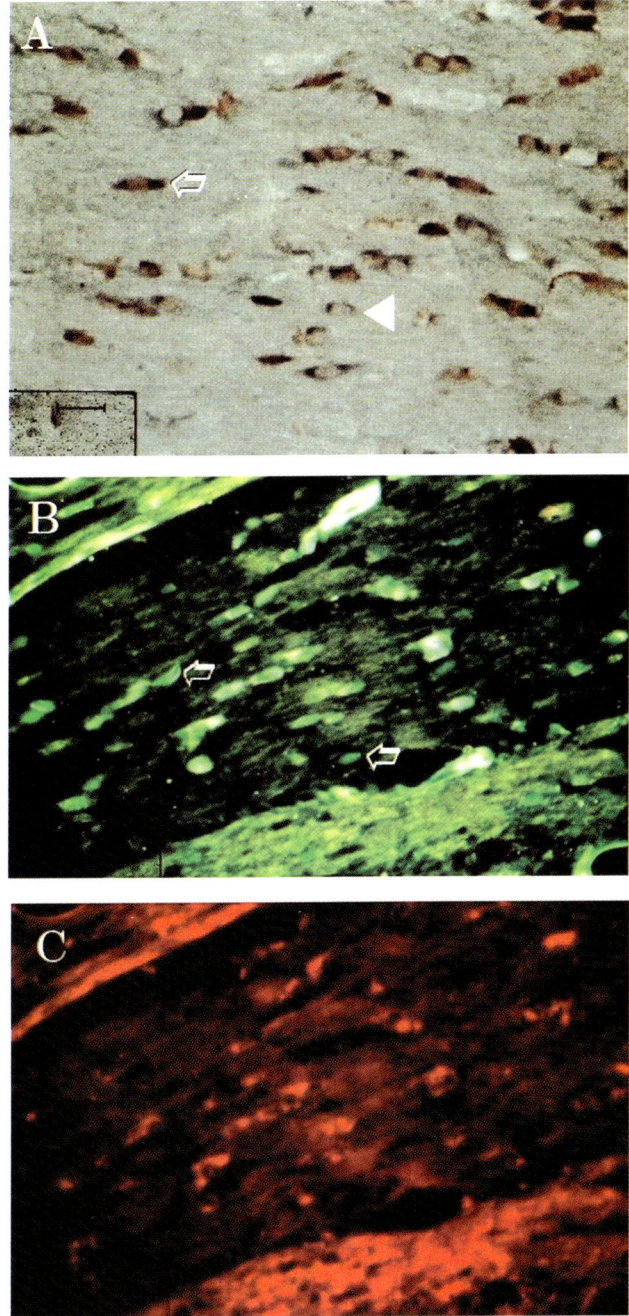

Fig. 2. Double immunohistochemistry of oligodendrocytes in the hyperthermic forebrain. Hsp70 (NovaRed) and GS II lectin (DAB-Nickel) label interfascicular oligodendrocytes in the corpus callosum. Typical Hsp70-negative GS II-positive oligodendrocyte is indicated with arrowhead, double-labeled cell with arrow (A). In paraffin section, Hsp70 is visualized with green Alexa Fluor 488 (B) and GS II lectin with red Alexa Fluor 546 (C). The arrows in (B) indicate Hsp70-positive cells that are not seen in (C). Calibration bar: A–C — 25 μm. (Reprinted from Pavlik et al., 2003 with permission from Elsevier.)

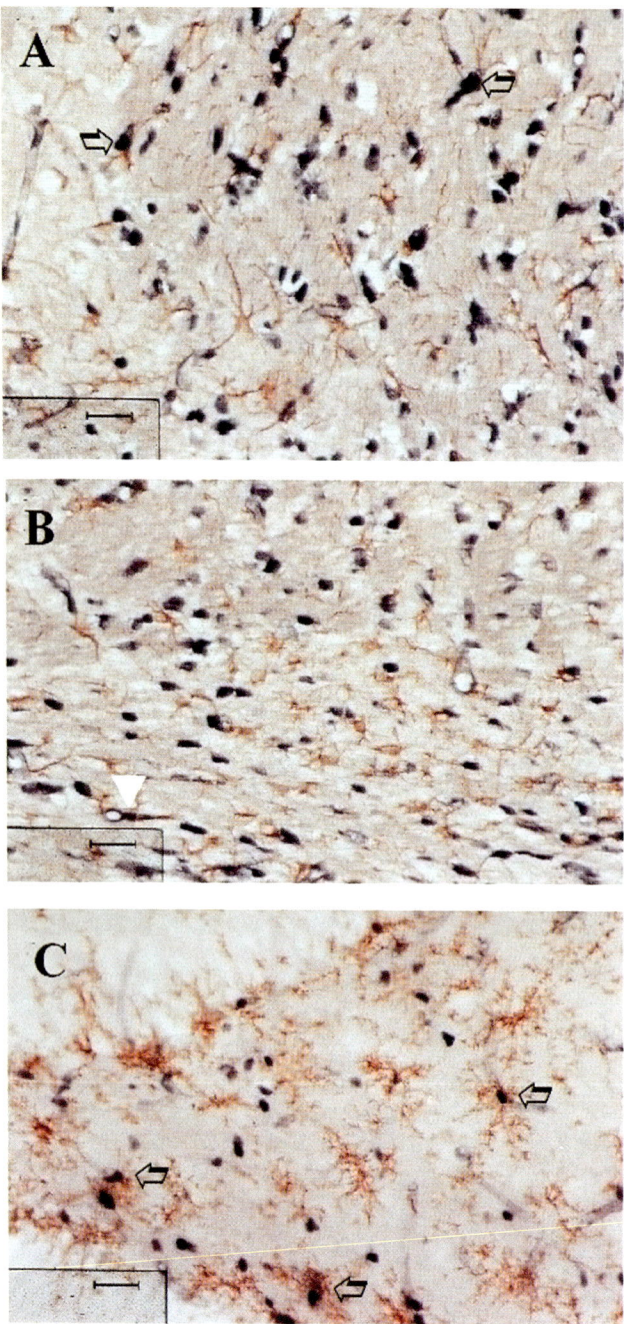

Fig. 3. Double immunohistochemistry of astrocytes and microglia in the hyperthermic forebrain. GFAP-positive astrocytes (NovaRed) in the internal capsule (A) and corpus callosum (B) rarely coincide with Hsp70 staining (DAB-Nickel, arrows and arrowhead). Only three double-labeled microglia (arrows) illustrate low occurrence of co-localization of Hsp70-positive (DAB-Nickel) and OX-42-positive (NovaRed) cells in the hippocampal hilus (C). Calibration bar: A–C — 25 μm. (Reprinted from Pavlik et al., 2003 with permission from Elsevier.)

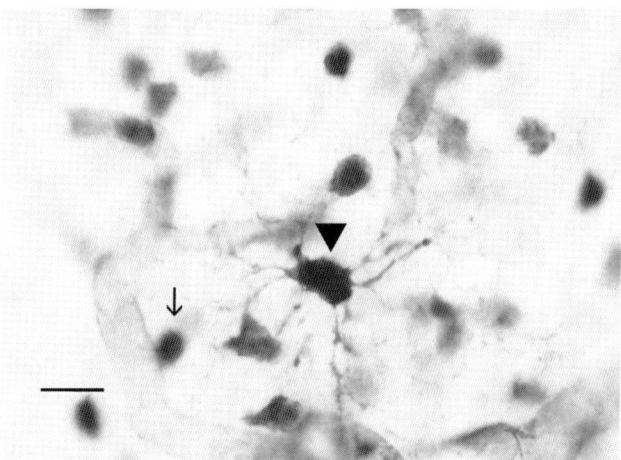

Fig. 4. Glial and vascular induction of Hsp70 in the gray matter. Perivascular satellite (arrow), presumably type I oligodendrocyte (arrowhead) and capillaries are positively stained for Hsp70 (SG) in the thalamic nuclei 4 h after hyperthermia. Calibration bar: 10 μm. (Reprinted from Pavlik et al., 2003 with permission from Elsevier.)

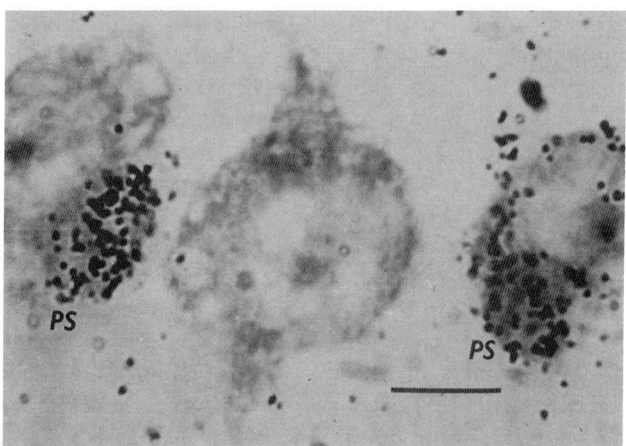

Fig. 5. Autoradiograph of the first slice (type II) incubated with [^{3}H]leucine for 30 min. The surplus of grain density over perineuronal satellites (PS) is demonstrated in comparison with their neurons. The labeled satellites are rather sunken into the neuronal bodies indicating the neuronophagia in progress. Exposed for one month. Thionin stain; calibration bar = 10 μm. (Reprinted from Pavlik and Jakoubek, 1976 with permission from Elsevier.)

(Pepper et al., 1998). Signs of nuclear presence of Hsp70 by 24 h after hyperthermia were also seen in illustrations of other reports (Marini et al., 1990; Li et al., 1992; McCabe and Simon, 1993; Krueger et al., 1999). Protection of sensitive nuclear/nucleolar targets by Hsp70 may be indispensable for the coping with the heat stress and for cell survival. The same may be true for its timely cytoplasmic relocation (Xu et al., 1998).

Neuroglial cells in culture

Hsps expression in glial cells in culture was recently reviewed by Sharp et al. (2001). Induction of Hsp70 following heat stress was found in astrocytes, oligodendrocytes, and microglia (Marini et al., 1990; Nishimura et al., 1991; Satoh et al., 1992; Satoh and Kim, 1994; Nishimura and Dwyer, 1996). In glial cultures from prenatal/neonatal brain,

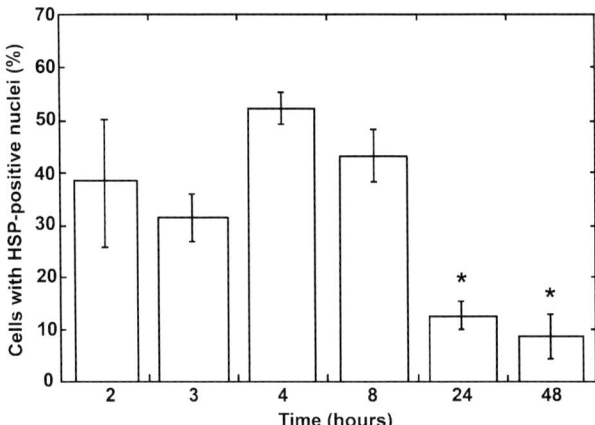

Fig. 6. Nuclear translocation of Hsp70 into glial nuclei of the corpus callosum. Hsp70-immunopositive nuclei of glial cells in the corpus callosum were counted at indicated time intervals following the end of hyperthermia. The marked nuclear translocation of Hsp70 was already apparent at 2 h interval and its cytoplasmic relocation occurred between 8 and 24 h following hyperthermia. For each interval, 100 of Hsp70 stained cells in each brain section were evaluated as showing prevalence of nuclear or cytoplasmic staining. Percentages of average values were plotted with SEM. The bars labeled with asterisks were significantly different at $P \leqslant 0.05$.

Hsp70 production in astrocytes and microglia was superior to oligodendrocytes. However, astrocytes derived from adult human brain were unable to induce Hsp70 despite being heated for 1.5 h at 43°C (Freedman et al., 1992; D'Souza et al., 1994) resembling astrocytes in adult rat brain (Foster and Brown, 1997; Pavlik et al., 2003). Astrocytes may contain some other Hsps expressed constitutively, e.g., Hsc70, or they may express other Hsps than Hsp70. Hsp27 and alpha B-crystallin, members of small HSP family, were induced in cultured astrocytes by heat stress (Satoh and Kim, 1995; Brzyska et al., 1998). Following hyperthermia *in vivo*, Hsp27 and Hsp70 were localized to astrocytic processes around synapses and in synaptic fractions, respectively (Bechtold and Brown, 2000; Bechtold et al., 2000). Hsp27 was predominantly expressed in astrocytes by non-heat types of stress as well (see Sharp et al., 2001).

Preparation of primary cell cultures or brain slices from the adult brain was apparently accompanied by traumatic stress of tissue dissociation, which was followed by substantial decrease of protein synthesis rate, abnormal distribution of newly synthesized protein, and Hsps induction (Blomstrand and Hamberger, 1970; Tiplady and Rose, 1971; Dunlop et al., 1974; Pavlik, 1974; Pavlik and Jakoubek, 1976; White, 1980; White and Currie, 1982). On the contrary, neither inhibition of protein synthesis nor Hsps induction was seen in brain slices of early postnatal brain (Pavlik, 1974; Pavlik and Jakoubek, 1978; White and Currie, 1982). Culturing neural cells may, therefore, result in recovery or loss of ability to express some Hsps according to the cell type. Consequently, it is rather difficult to compare neuroglial/neuronal stress responses *in vitro* and *in vivo*.

Proteotoxic stress by sodium arsenite mimics the cellular distribution of Hsp70 following hyperthermia

Heat stress response is most probably caused by increasing amount of unfolded and/or misfolded/denatured protein in cells (Gosslau et al., 2001). Sodium arsenite, which causes swift protein denaturation and stress response, was used to induce cellular stress *in vitro* and *in vivo* (Lee and Hahn, 1988). However, systemic administration of sodium arsenite (5 mg/kg, i.p.) induced Hsp70 only in vascular cells of the brain (Pavlik, unpublished data) and, therefore, we attempted to administer sodium arsenite intraventricularly. Following slow infusion of 5 nmol of sodium arsenite into lateral ventricle of urethane-anesthetized rat, we observed

Hsp70 induction in the white and gray matter surrounding the lateral and third ventricles (Fig. 7A). Most of Hsp70-positive cells were oligodendrocytes distributed in the corpus callosum, striatum, thalamus, and hypothalamus. Surprisingly, the same groups of neurons were stained for Hsp70 in the hypothalamus as we had earlier reported (n. habenulae medialis (Fig. 7B), n. paraventricularis, and n. dorsomedialis). Hsp70 induction in the neurons of the striatum, thalamus, or hippocampus was not observed. However, very strong immunostaining was present in ependymal cells lining the ventricles (Fig. 7A), in some vessels close to ventricular surface, and in tanycytes (Fig. 7C). Their processes between the wall of the third ventricle and vessels were filled with Hsp70. Hsp70 was missing in GFAP-positive astrocytes and only rarely seen in OX-42 labeled microglia (not shown). This distribution pattern was practically identical with that we have seen following hyperthermia.

Sodium arsenite is denaturing the cytosolic proteins, which are also denatured by hyperthermia (Lee and Hahn, 1988; Freeman et al., 1995). This similarity implied that in oligodendrocytes and other brain cells expressing Hsp70 after heat stress, the primary target was a population of cell proteins which was either very sensitive to denaturation or might be unprotected by constitutive chaperones. Alternatively, these protein molecules when denatured were very effective in inducing activation of heat shock factor 1 (HSF1), the critical process which might be accelerated by a plethora of cell-specific co-activators like protein kinases, phospholipases, phosphatases, etc. Evidently, the cell type-specific regulation of heat stress response in the brain needs further studies.

Cerebral vessels increase Hsp70 staining following heat stress

Heat stress induced expression of Hsp70 in cerebral vascular cells is mentioned usually as an additional finding to glial/neuronal distribution (Blake et al., 1990; Brown, 1990, 1994; Marini et al., 1990; Li et al., 1992). Hyperthermia has been reported to induce Hsp70 in systemic arteries also (Amrani et al., 1993; Udelsman et al., 1994).

We found some cerebral vessels weakly stained in sections of various types of controls (Figs. 1A, E, G). More vessels became positive for Hsp70 by 1.5 h and later intervals following hyperthermia (Figs. 1C, D, F, H). Capillaries stained less for Hsp70 than glial cells. Endothelial nuclei were distinctly darker than cytoplasm (Fig. 4). Actually, an increased Hsp70 staining of endothelial nuclei is another indication of vascular cell stress. There was significant staining of cells in larger vessels like penetrating branches of cortical arteries (not shown). Since vascular cells might take up some immunoglobulins during heat stress as a consequence of blood-brain barrier opening, we checked this possibility with goat or rabbit biotinylated secondary antibodies against rat immunoglobulins. However, only background level of staining was seen even in 4 h posthyperthermic sections like in controls (Pavlik et al., 2003) (Figs. 6 and 7).

It was reported that estradiol increased Hsp70 protein in brain arteries of both male and female rats (Lu et al., 2002). Physiological levels of circulating estrogens may have enhanced induction of Hsp70 in brain vessels of hyperthermic female rats. However, gender difference in hyperthermic Hsp70 induction was not observed (Pavlik, unpublished data).

Microvessels of brain slices incubated with radioactive amino acids were actually among the first brain cells found to induce Hsp70 named at that time as stress protein 1 (SP-1) (White, 1980; White and Currie, 1982). We have seen many vessels of brain slices to contain considerable amount of newly synthesized protein as compared to brain vessels after systemic administration of the tracer (Pavlik, 1974; Pavlik and Jakoubek, 1976). We also reported the developmental dependence of such abnormal cellular distribution of newly synthesized protein in brain slices since it appeared only after 14 days of postnatal life (Pavlik and Jakoubek, 1978).

There are several other brain cell types, which induce Hsp70 following hyperthermia or may even show a constitutive Hsp70 expression. Epithelial (ependymal) cells covering choroid plexus contained constitutively expressed Hsp70 in the rabbit

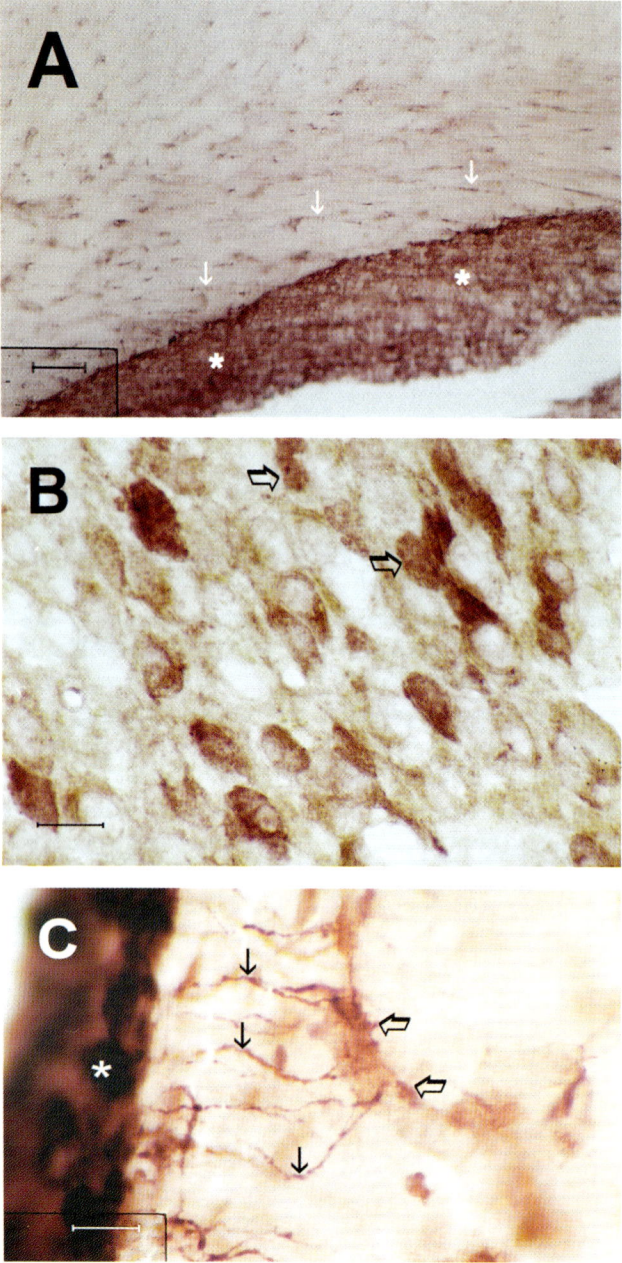

Fig. 7. Hsp70 induced by intraventricular arsenite in the periventricular areas of the rat brain. (A) Induction of Hsp70 at 4 h following administration of 5 nanomoles of sodium arsenite. Immunoreactive cells in the corpus callosum were apparent interfascicular oligodendrocytes containing Hsp70 mainly in their chain-forming nuclei (arrows). Marked Hsp70 immunostaining (NovaRed) was seen in the ependymal cells covering the wall of the lateral ventricle (asterisks). Hsp70 labeling in the corpus callosum decreased with distance from the ventricle. (B) Hsp70 was present in many neurons of the nucleus of medial habenula, mainly in the perikaryal cytoplasm, and less often in the neuronal nuclei (arrows). (C) Vascular processes of tanycytes (black arrows) filled with Hsp70 were visible in the vicinity of the third ventricle wall (asterisk). The draining vessel (empty arrows) showed Hsp70 staining in its wall and nuclei. Calibration bars: A — 100 μm, B, C — 25 μm.

brain, which was transferred to their nuclei following drug-induced hyperthermia (Manzerra and Brown, 1996). Similarly, some ependymal cells lining the ventricles and some meningeal cells also showed presence of constitutively expressed Hsp70 in control rat brain (Pavlik et al., 2003). Constitutive expression of Hsp70 in these supporting cells may be responsible for positive Hsp70 immunoblots from the brain regions containing choroid plexus, ventricular ependyma, or meningeal coverings in control rabbits (Manzerra et al., 1993; Manzerra and Brown, 1996).

Neuronal Hsp70 non-responsiveness to heat stress

Heat-induced Hsp70 response in neurons was limited to a few brain regions (Sprang and Brown, 1987; Blake et al., 1990; Li et al., 1992; Pavlik et al., 2003). They included both magnocellular and parvocellular neurons in some hypothalamic nuclei (medial habenula; paraventricular; supraoptic; suprachiasmatic; dorsomedial). In addition, subnuclei of amygdaloidal complex were also expressing Hsp70 after hyperthermia. Hsp70 was distributed in nuclei, perikarya, as well as in the processes of these neurons (Pavlik et al., 2003). It was suggested that these neurons might be activated as a part of heat-induced neuroendocrine stress response (Blake et al., 1990; Li et al., 1992). However, increasing plasma osmolarity by 20% failed to induce Hsp70 expression in magnocellular neuroendocrine neurons (Pavlik et al., 2003). These neurons did not enhance vasopressin immunoreactivity in the paraventricular nucleus of heat-stroked rats despite the marked activation of c-Fos expression (Tsay et al., 1999).

In the cerebellum, a massive expression of Hsp70 mRNA after heat stress was originally observed in granule neurons of the internal granule cell layer (Sprang and Brown, 1987). Later, Hsp70 mRNA appeared in Purkinje cells by 10 h after hyperthermia but it was not translated into Hsp70 protein (Manzerra et al., 1993). Similarly, expression of Hsp70 mRNA was also reported in hippocampal neurons and dentate gyrus granule cells (Blake et al., 1990; Krueger et al., 1999). However, the translation of heat-induced message has evidently been dampened, since only weak Hsp70 immunoreactivity was observed in dentate gyrus (Li et al., 1992; Krueger et al., 1999). Others did not see any staining of granule neurons in the dentate gyrus (McCabe and Simon, 1993; Manzerra and Brown, 1996; Pavlik et al., 2003). It cannot be excluded that this difference might be caused by higher temperature used in the experiments of Li et al. (1992) and Krueger et al. (1999). Finally, decreased degradation of Hsp70 mRNA following heat stress (Pepper et al., 1998) might increase their levels and explain cases of their posthyperthermic detection in some neurons (Blake et al., 1990; Manzerra et al., 1993; Krueger et al., 1999).

Hsp70 immunoreactivity was reported to appear in cortical neurons and hippocampal pyramidal cells following prolonged heat exposure of animals (Sharma et al., 1991, 1992; Lin, 1997; see Sharma et al., 1998). Longer-lasting hyperthermia may have led to heat-stroke, which is complicated with brain edema, intracranial hypertension, systemic hypotension, and brain hypoxia/ischemia (Sharma et al., 1998; Yang et al., 1998). Widespread neuronal Hsp70 expression in such a situation may actually indicate that the transition point between physiological hyperthermia and pathological heat-stroke has already been reached.

Constitutive expression of Hsp70 mRNA was observed in cortical and hippocampal neurons of the rabbit (Foster and Brown, 1996). However, it seems that either such basal level of Hsp70 message was not translated into protein or the amount of translated protein was below the detection limits of the immunohistochemical methods used (Manzerra and Brown, 1996; Krueger et al., 1999; Pavlik et al., 2003).

Other two constitutively expressed Hsps — Hsc70 and Hsp90 — attain high levels in the mammalian brain neurons (Manzerra et al., 1993; Quraishi et al., 1996; Foster and Brown, 1996). Hsc70 was mainly detected in large cerebral neurons, cerebellar Purkinje cells, deep cerebellar neurons, and motoneurons of the spinal cord (Manzerra and Brown, 1992a; Foster and Brown, 1996). Since extensive homology exists between Hsc70 and Hsp70, abundant Hsc70 may be used in heat-stressed neurons instead of Hsp70 to buffer elevation of unfolded or denatured proteins. Therefore, neurons may not need to activate Hsp70 expression (Brown, 1990, 1994; Brown and Sharp, 1999). Hsc70 was reported to translocate to neuronal nuclei if the intensity of hyperthermia was increased (Manzerra and Brown, 1996). Such a partial heat-shock response of Hsc70 in neurons may protect nuclear components similarly as Hsp70 does in nuclei of other cell types. Furthermore, high neuronal levels of Hsp90 may be able to keep HSF1 inactivated by Hsp90 complex and to prevent initiation of genomic heat-stress response (Zou et al., 1998). It was suggested that some neurons might not contain HSF1 but HSF2, which did not support heat expression of Hsp70 (Marcuccilli et al., 1996). However, neuronal unresponsiveness of Hsp70 expression to heat stress is exceptional because many other types of stress do induce Hsp70 in brain neurons.

Expression of Hsp70 after hyperthermia is considered as an important condition for inducing thermotolerance or preconditioning (see Chapter 19 in this volume). Brain neurons may be made thermotolerant by preceding heat stress to following serious heat shock (see Kregel, 2002). Since neither Hsp70 nor constitutive Hsc70 are induced in them by heat stress *in vivo*, these Hsps could hardly be the mediators of thermotolerance in brain neurons. We may speculate that glial and/or vascular cells, which actually induce most of Hsp70 in the brain, confer thermotolerance or even preconditioning to heat- as well as non-heat types of injurious stress. However, a possibility of involvement of other neuronal Hsps in thermotolerance/preconditioning phenomena cannot be excluded. Nevertheless, the induction of Hsp70 offers protective advantage to responsive cells making them thermotolerant or preconditioned to more severe stress. The question why so many neurons are devoid of heat stress induced Hsp70 expression would deserve further elucidation.

Abbreviations

DAB	diaminobenzidine
c-Fos	immediate-early protein c-Fos
GFAP	glial fibrillary acidic protein
GS II	lectin II from *Griffonia simplicifolia*
Hsc70	heat shock cognate protein 70 (constitutive)
HSF1 & 2	heat shock factor 1 & 2
Hsp27	heat shock protein 27
HSP70	family of heat shock proteins of 70 kDa molecular weight
Hsp70	heat shock protein 70 (inducible)
Hsp90	heat shock protein 90
Hsps	heat shock proteins
NovaRed	chromagen
OX-42	monoclonal antibody against CD 11b
SG	chromagen
SP-1	stress protein 1 (Hsp70)

Acknowledgments

We thank Dr. J. Lexa and G. Alexander for their help. The study was supported by grants MY 031 and MY 033 from Research Administration of Kuwait University.

References

Amrani, M., O'Shea, J., Corbett, J., Dunn, M.J., Tadjkarimi, S., Theodoropoulos, S., Pepper, J. and Yacoub, M.H. (1993) Role of catalase and heat shock protein on recovery of cardiac endothelial cells and mechanical function after ischemia. Cardioscience, 4: 193–198.

Angelov, D.N., Walther, M., Streppel, M., Gemtinas-Lichius, O. and Neiss, W.F. (1998) The cerebral perivascular cells. Adv. Anat. Embryol. Cell Biol., 147: 1–87.

Armstrong, J.N., Plumier, J.C., Robertson, H.A. and Currie, R.W. (1996) Cell specific expression of Hsp70 in neurons and glia of the rat hippocampus after hyperthermia and kainic acid-induced seizure activity. Mol. Brain Res., 71: 265–278.

Barral, J.M., Broadley, S.A., Schaffar, G. and Hartl, F.U. (2004) Roles of molecular chaperones in protein misfolding disease. Semin. Cell Dev. Biol., 15: 17–29.

Bechtold, D.A. and Brown, I.R. (2000) Heat shock proteins hsp27 and hsp32 localize to synaptic sites in the rat cerebellum following hyperthermia. Mol. Brain Res., 75: 309–320.

Bechtold, D.A., Rush, S.J. and Brown, I.R. (2000) Localization of heat-shock protein Hsp70 to the synapse following hyperthermic stress in the brain. J. Neurochem., 74: 641–646.

Blake, M.J., Nowak Jr., T.S. and Holbrook, N.J. (1990) In vivo hyperthermia induces expression of HSP70 mRNA in brain regions controlling the neuroendocrine response to stress. Mol. Brain Res., 8: 89–92.

Blomstrand, C. and Hamberger, V. (1970) Amino-acid incorporation in vitro into protein of neuronal and glial cell-enriched fractions. J. Neurochem., 17: 1187–1195.

Brown, I.R. (1990) Induction of heat shock (stress) genes in the mammalian brain by hyperthermia and other traumatic events: a current perspective. J. Neurosci. Res., 27: 247–255.

Brown, I.R. (1994) Induction of heat shock genes in the mammalian brain by hyperthermia and tissue injury. In: Mayer R.J. and Brown I.R. (Eds.), Heat Shock Proteins in the Nervous System. Academic Press, London, pp. 31–54.

Brown, I.R., Cosgrove, J.W. and Clark, B.D. (1982) Physiologically relevant increases in body temperature induce the synthesis of a heat shock protein in mammalian brain and other organs. In: Schlesinger M.J., Ashburner M. and Tissieres A. (Eds.), Heat Shock from Bacteria to Man. Cold Spring Harbor Laboratory, New York, pp. 361–368.

Brown, I.R. and Rush, S.J. (1990) Expression of heat shock genes (hsp70) in the mammalian brain: distinguishing constitutively expressed and hyperthermia inducible mRNA species. J. Neurosci. Res., 25: 14–19.

Brown, I.R. and Sharp, F. (1999) The cellular stress gene response in brain. In: Latchman D.S. (Ed.), Stress Proteins: Handbook of Experimental Pharmacology. Springer, Heidelberg, pp. 243–263.

Brzyska, M., Stege, G.J., Renkawek, K. and Bosman, G.J. (1998) Heat shock, but not the reactive state per se, induces increased expression of small stress proteins hsp25 and alpha B-crystallin in glial cells in vitro. Neuroreport, 9: 1549–1552.

Cavanagh, J.B., Nolan, C.C. and Brown, A.W. (1990) Glial cell intrusions actively remove detritus due to toxic chemicals from within nerve cells. Neurotoxicology, 11: 1–12.

D'Souza, S.D., Antel, J.P. and Freedman, M.S. (1994) Cytokine induction of heat shock protein expression in human oligodendrocytes: an interleukin-1-mediated mechanism. J. Neuroimmunol., 50: 17–24.

Dunlop, D.S., van Elden, W. and Lajtha, A. (1974) Measurements of rates of protein synthesis in rat brain slices. J. Neurochem., 22: 821–830.

Ellis, R.J. and van der Vries, S.M. (1991) Molecular chaperones. Annu. Rev. Biochem., 60: 321–347.

Foster, J. and Brown, I. (1996) Basal expression of stress-inducible hsp70 mRNA detected in hippocampal and cortical neurons of normal rabbit brain. Brain Res., 724: 73–83.

Foster, J.A. and Brown, I.R. (1997) Differential induction of heat shock mRNA in oligodendrocytes, microglia, and astrocytes following hyperthermia. Mol. Brain Res., 45: 207–218.

Foster, J.A., Rush, S.J. and Brown, I.R. (1995) Localization of constitutive and hyperthermia-inducible heat shock mRNAs (hsc70 and hsp70) in the rabbit cerebellum and brainstem by non-radioactive in situ hybridization. J. Neurosci. Res., 41: 603–612.

Freedman, M.S., Buu, N.N., Ruijs, T.C.J., Williams, K. and Antel, J.P. (1992) Differential expression of heat shock proteins by human glial cells. J. Neuroimmunol., 41: 231–238.

Freeman, M.L., Borrelli, M.J., Syed, K., Senistera, G., Stafford, D.M. and Lepock, J.R. (1995) Characterization of a signal generated by oxidation of protein thiols that activates the heat shock transcription factor. J. Cell Physiol., 164: 356–366.

Gosslau, A., Ruoff, P., Mohsenzadeh, S., Hobohm, U. and Rensing, L. (2001) Heat shock and oxidative stress-induced exposure of hydrophobic protein domains as common signal in the induction of hsp68. J. Biol. Chem., 276: 1814–1821.

Hightower, L.E. and Hendershot, L.M. (1997) Molecular chaperones and the heat shock response at Cold Spring Harbor. Cell Stress Chaperones, 2: 1–11.

Kelley, P.M. and Schlesinger, M.J. (1978) The effects of amino acid analogs and heat shock gene expression in chicken embryo fibroblasts. Cell, 19: 1277–1286.

Kregel, K.C. (2002) Heat shock proteins: modifying factors in physiological stress responses and acquired thermotolerance. J. Appl. Physiol., 92: 2177–2186.

Krueger, A.M.R., Armstrong, J.N., Plumier, J.C., Robertson, H.A. and Currie, R.W. (1999) Cell specific expression of Hsp70 in neurons and glia of the rat hippocampus after hyperthermia and kainic acid-induced seizure activity. Mol. Brain Res., 71: 265–278.

Lee, K.J. and Hahn, G.M. (1988) Abnormal proteins as the trigger for induction of stress responses: heat, diamide, and sodium arsenite. J. Cell Physiol., 136: 411–420.

Lee, S.Y., Lee, S.H., Akuta, K., Uda, M. and Song, C.W. (2000) Acute histological effects of interstitial hyperthermia on normal rat brain. Int. J. Hyperthermia, 16: 73–83.

Lee, S., Sowa, M.E., Choi, J.M. and Tsai, F.T. (2004) The ClpB/Hsp104 molecular chaperone: a protein disaggregating machine. J. Struct. Biol., 146: 99–105.

Li, Y., Chopp, M., Yoshida, Y. and Levine, S.R. (1992) Distribution of 72-kDa heat-shock protein in rat brain after hyperthermia. Acta Neuropathol., 84: 94–99.

Lin, M.T. (1997) Heat-stroke induced cerebral ischemia and neuronal damage. Ann. N.Y. Acad. Sci. (U.S.A.), 813: 572–580.

Lu, A., Clark, R.Q., Reilly, M., Nee, A. and Sharp, F.R. (2002) 17-Beta-estradiol induces heat shock proteins in brain arteries and potentiates ischemic heat shock protein induction in glia and neurons. J. Cereb. Blood Flow Metab., 22: 183–195.

Ludwin, S.K. (1979) The perineuronal satellite oligodendrocyte: a role in remyelination. Acta Neuropathol. (Berl.), 47: 49–53.

Ludwin, S.K. (1984) The function of perineuronal satellite oligodendrocytes: an immunohistochemical study. Neuropathol. Appl. Neurobiol., 10: 143–149.

Manzerra, P. and Brown, I.R. (1992a) Distribution of the constitutive- and hyperthermia-inducible heat shock mRNA species (hsp70) in the Purkinje layer of the rabbit cerebellum. Neurochem. Res., 17: 559–564.

Manzerra, P. and Brown, I.R. (1992b) Expression of heat shock genes (hsp70) in the rabbit spinal cord: Localization of constitutive and hyperthermia-inducible mRNA species. J. Neurosci. Res., 31: 606–615.

Manzerra, P. and Brown, I.R. (1996) The neuronal stress response: nuclear translocation of heat shock proteins as an indicator of hyperthermic stress. Exp. Cell Res., 229: 35–47.

Manzerra, P., Rush, S.J. and Brown, I.R. (1993) Temporal and spatial distribution of heat shock mRNA and protein (hsp70) in the rabbit cerebellum in response to hyperthermia. J. Neurosci. Res., 36: 480–490.

Marcuccilli, C.J., Mathur, S.K., Morimoto, R.I. and Miller, R.J. (1996) Regulatory differences in the stress response of hippocampal neurons and glial cells after heat shock. J. Neurosci., 16: 478–485.

Marini, A.M., Kozuka, M., Lipsky, R.H. and Nowak Jr., T.S. (1990) 70-kilodalton heat shock protein induction in cerebellar astrocytes and cerebellar granule cells in vitro: comparison with immunocytochemical localization after hyperthermia in vivo. J. Neurochem., 54: 1509–1516.

Matsumi, N., Matsumoto, K., Mishima, N., Moriyama, E., Furuta, T., Nishimoto, A. and Taguchi, K. (1994) Thermal damage threshold of brain tissue: histological study of heated normal monkey brains. Neurol. Med. Chir. (Tokyo), 34: 209–215.

McCabe, T. and Simon, R.P. (1993) Hyperthermia induces 72 kDa heat shock protein expression in rat brain in non-neuronal cells. Neurosci. Lett., 159: 163–165.

Miller, E.K., Raese, J.D. and Morrison-Bogorad, M. (1991) Expression of heat shock protein 70 and heat shock cognate 70 messenger RNAs in rat cortex and cerebellum after heat shock or amphetamine treatment. J. Neurochem., 56: 2060–2071.

Mogk, A. and Bukau, B. (2004) Molecular chaperones: structure of protein disaggregase. Curr. Biol., 14: R78–R80.

Morimoto, R.I. (1993) Cells in stress: transcriptional activation of heat shock genes. Science, 259: 1409–1410.

Nishimura, R.N. and Dwyer, B.E. (1996) Evidence for different mechanisms of induction of HSP 70i: a comparison of cultured cortical neurons with astrocytes. Mol. Brain Res., 36: 227–239.

Nishimura, R.N., Dwyer, B.E., Vinters, H.V., DeVellis, J. and Cole, R. (1991) Heat shock in cultured neurons and astrocytes: correlation of ultrastructure and heat shock protein synthesis. Neuropathol. Appl. Neurobiol., 17: 139–147.

Nowak Jr., T.S. (1991) Localization of 70 kDa stress protein mRNA induction in gerbil brain after ischemia. J. Cereb. Blood Flow Metab., 11: 432–439.

Nowak Jr., T.S. and Jacewicz, M. (1994) The heat shock/stress response in focal cerebral ischemia. Brain Pathol., 4: 67–76.

Pavlik, A. (1974) Biochemical and autoradiographic aspects of protein synthesis in brain slices. PhD Thesis, Czechoslovak Academy of Sciences, Prague.

Pavlik, A., Aneja, I.S., Lexa, J. and Al-Zoabi, B.A. (2003) Identification of cerebral neurons and glial cell types inducing heat shock protein Hsp70 following heat stress in the rat. Brain Res., 973: 179–189.

Pavlik, A. and Jakoubek, B. (1976) Distribution of protein bound radioactivity in brain slices of the adult rat incubated with labeled leucine. Brain Res., 101: 113–128.

Pavlik, A. and Jakoubek, B. (1978) Developmental changes of protein bound radioactivity distribution in rat brain slices incubated with labeled leucine. Brain Res., 154: 95–104.

Pepper, A., Grimbergen, C.A., Spaan, J.A.E., Souren, J.E.M. and van Wijk, R. (1998) A mathematical model of the hsp70 regulation in the cell. Int. J. Hyperthermia, 14: 97–124.

Pratt, W.B. (1993) The role of heat shock proteins in regulating the function, folding, and trafficking of the glucocorticoid receptors. J. Biol. Chem., 268: 21455–21458.

Quraishi, H., Rush, S.J. and Brown, I.R. (1996) Expression of mRNA species encoding heat shock protein 90 (hsp90) in control and hyperthermic rabbit brain. J. Neurosci. Res., 43: 335–345.

Sasara, T., Cizkova, D., Mestril, R., Galik, J., Sugahara, K. and Marsala, M. (2004) Spinal heat shock protein (70) expression: effect of spinal ischemia, hyperthermia (42°C)/hypothermia (27°C), NMDA receptor activation and potassium evoked depolarization on the induction. Neurochem. Int., 44: 53–64.

Satoh, J.I. and Kim, S.U. (1994) HSP72 induction by heat stress in human neurons and glial cells in culture. Brain Res., 653: 243–250.

Satoh, J.I. and Kim, S.U. (1995) Differential expression of heat shock protein HSP27 in human neurons and glial cells in culture. J. Neurosci. Res., 41: 805–818.

Satoh, J.I., Yamamura, T., Kunishita, T. and Tabira, T. (1992) Heterogeneous induction of 72-kDa heat shock protein (HSP72) in cultured mouse oligodendrocytes and astrocytes. Brain Res., 73: 37–43.

Sharma, H.S., Cervos-Navarro, J. and Dey, P.K. (1991) Acute heat exposure causes cellular alteration in cerebral cortex of young rats. Neuroreport, 2: 155–158.

Sharma, H.S. and Hoopes, P.J. (2003) Hyperthermia induced pathophysiology of the central nervous system. Int. J. Hyperthermia, 19: 325–354.

Sharma, H.S., Westman, J., Cervos-Navarro, J. and Gosztonyi, G. (1992) Acute systemic heat exposure increases heat shock protein (HSP-70 kDa) immunoreactivity in the brain and spinal cord of young rats. Clin. Neuropathol., 11: 174–175.

Sharma, H.S., Westman, J. and Nyberg, F. (1998) Pathophysiology of brain edema and cell changes following hyperthermic brain injury. In: Sharma H.S. and Westman J. (Eds.), Brain Function in Hot Environment, Progress in Brain Research, Vol. 115. Elsevier, Amsterdam, pp. 351–412.

Sharma, H.S., Westman, J., Nyberg, F., Zimmer, C., Cervos-Navarro, J. and Dey, P.K. (1994) Selective vulnerability of rat hippocampus in heat stress. In: Milton A.S. (Ed.), Temperature Regulation (Adv. Pharmacol. Sci.). Birkhauser, Basel, pp. 267–272.

Sharp, F.R., Bernaudin, M., Bartels, M. and Wagner, K.R. (2001) Glial expression of heat shock proteins (HSPs) and oxygen-regulated proteins (ORPs). In: Lopez B.C. and Nieto-Sampedro M. (Eds.), Progress in Brain Research, Vol. 132. Elsevier, Amsterdam, pp. 427–440.

Sharp, F.R., Massa, S.M. and Swanson, R.A. (1999) Heat-shock protein protection. Trends Neurosci., 22: 97–99.

Sherman, M.Y. and Goldberg, A. (2001) Cellular defenses against unfolded proteins: a cell biologist thinks about neurodegenerative diseases. Neuron, 20: 15–32.

Sminia, P., van der Zee, J., Wondergem, J. and Haveman, J. (1994) Effect of hyperthermia on the central nervous system: a review. Int. J. Hyperthermia, 10: 1–30.

Sprang, G.K. and Brown, I.R. (1987) Selective induction of a heat shock gene in fiber tracts and cerebellar neurons of the rabbit brain detected by in situ hybridization. Mol. Brain Res., 3: 89–93.

Sturrock, R.R. (1987) A morphological study of the neostriatum of aged mice with particular reference to neuroglia. J. Hirnforsch., 28: 505–515.

Supler, M.L., Semple-Rowland, S.L. and Streit, W.J. (1994) Oligodendrocytes produce low molecular weight glycoproteins containing N-acetyl-D-glucosamine in their Golgi apparatus. Glia, 10: 193–201.

Szuchet, S. (1995) Morphology and ultrastructure of oligodendrocytes and their functional implication. In: Kettenmann H. and Ransom B.R. (Eds.), Neuroglia. Oxford University Press, New York, pp. 23–43.

Takahashi, S., Tanaka, R., Watanabe, M., Takahashi, H., Kakinuma, K., Suda, T., Yamada, M. and Takahashi, H. (1999) Effects of whole body hyperthermia on the canine central nervous system. Int. J. Hyperthermia, 15: 203–216.

Takayama, S., Reed, J.C. and Homma, S. (2003) Heat-shock proteins as regulators of apoptosis. Oncogene, 22: 9041–9047.

Tiplady, B. and Rose, S.P.R. (1971) Amino-acid incorporation into protein in neuronal cell body and neuropil fractions in vitro. J. Neurochem., 18: 549–558.

Tissieres, A., Mitchell, H.K. and Tracy, V.M. (1974) Protein synthesis in salivary glands of Drosophila melanogaster: relation to chromosomal puffs. J. Mol. Biol., 84: 389–398.

Tsay, H.J., Li, H.Y., Lin, C.H., Yang, Y.L., Yeh, J.Y. and Lin, M.T. (1999) Heatstroke induces c-fos expression in the rat hypothalamus. Neurosci. Lett., 262: 41–44.

Udelsman, R., Blake, M.J., Stagg, C.A. and Holbrook, N.J. (1994) Endocrine control of stress-induced heat shock protein 70 expression in vivo. Surgery, 115: 611–616.

Vass, K., Welch, W.J. and Nowak Jr., T.S. (1988) Localization of 70 kDa stress protein induction in gerbil brain after ischemia. Acta Neuropathol., 77: 126–135.

Wegele, H., Muller, L. and Buchner, J. (2004) Hsp70 and Hsp90—a relay team for protein folding. Rev. Physiol. Biochem. Pharmacol., 151: 1–44.

Welch, W.J. (1992) Mammalian stress response: cell physiology, structure/function of stress proteins, and implications for medicine and disease. Physiol. Rev., 72: 1063–1081.

Welch, W.J. and Feramisco, J.R. (1984) Nuclear and nucleolar localization of the 72,000-dalton heat shock protein in heat-shocked mammalian cells. J. Biol. Chem., 259: 4501–4513.

Westman, J. and Sharma, H.S. (1998) Heat shock protein response in the central nervous system following hyperthermia. In: Sharma H.S. and Westman J. (Eds.), Brain Function in Hot Environment, Progress in Brain Research, Vol. 115. Elsevier, Amsterdam, pp. 207–239.

White, F.P. (1980) Protein synthesis in rat telencephalon slices: high amount of newly synthesized protein found in association with brain capillaries. Neuroscience, 5: 173–178.

White, F.P. and Currie, R.W. (1982) A mammalian response to trauma: the synthesis of a 71-kDa protein. In: Schlesinger M.J., Ashburner M. and Tissieres A. (Eds.), Heat shock from bacteria to man. Cold Spring Harbor Laboratory, New York, pp. 379–388.

Xu, M., Wright, W.D., Higashikubo, R. and Roti, R.J. (1998) Intracellular distribution of hsp70 during long duration moderate hyperthermia. Int. J. Hyperthermia, 14: 211–225.

Yang, Y.L., Lu, K.T., Tsay, H.J., Lin, C.H. and Lin, M.T. (1998) Heat shock protein expression protects against death following exposure to heatstroke in rats. Neurosci. Lett., 252: 9–12.

Young, J.C., Agshe, V.R., Siegers, K. and Hartl, F.U. (2004) Pathways of chaperone-mediated protein folding in cytosol. Nat. Rev. Mol. Cell. Biol., 5: 781–791.

Zou, J., Guo, Y., Guettouche, T., Smith, D.F. and Voellmy, R. (1998) Repression of heat shock transcription factor HSF1 activation by HSP90 (HSP90 complex) that forms a stress-sensitive complex with HSF1. Cell, 94: 471–480.

CHAPTER 21

Heat shock proteins and the heat shock response during hyperthermia and its modulation by altered physiological conditions

Michal Horowitz[*] and Sharon D.M. Robinson

Laboratory of Environmental Physiology, Faculty of Dental Medicine, The Hebrew University, POB 12272, Jerusalem 91120, Israel

Abstract: The fundamental functions of heat shock proteins (HSPs) are molecular chaperoning and cellular repair. There is little literature on the association between the numerous functions of HSPs and systemic integrative responses, particularly those controlled by the central nervous system. This chapter focuses on the role played by members of the HSP70 superfamily, universally recognized as cytoprotectants during heat stress, within the physiological context of hyperthermia and with its superimposition on situations of chronic stress. In the nucleus tractus solitarius, HSP70 levels enhance the sensitivity of sympathetic and parasympathetic arms of the autonomic nervous system to attenuate heat stroke-induced cerebral ischemia and hypotension. Chronic stressors that alter the heat shock response may affect the physiological profile during hyperthermic conditions. Upon aging, significantly lower HSP70 production is noted in the ventral paraventricular and lateral magnocellular nuclei. Likewise, results from cultured cells suggest that the age-related decline in HSP70 expression is constitutive and is due to decreased binding of the heat shock factor 1 (HSF-1) to the heat shock element (HSE) and diminished HSP70 transcription. These changes may be associated with decreased thermotolerance upon aging, although HSP70 production in response to other stressors is not affected. Heat acclimation (AC), in contrast, increases tissue reserves of HSP70 and accelerates the heat shock response. AC protects epithelial integrity, vascular reactivity and interactions with cellular signaling networks, enhancing protection and delaying thermal injury. The link between HSP70 and the immune system is discussed with respect to exercise. Exercise enhances the immune response via production of HSP72 in central and peripheral structures. At least in part, the effects of HSP72 in the brain are mediated via eHSP72-circulating HSPs providing a "danger signal" to activate the immune response. In summary, HSPs are primarily cytoprotective components, the physiological situations described in this chapter infer their pivotal role in central control of integrative systems.

Keywords: HSP70; eHSP72; hyperthermia; heat stroke; cytoprotection; aging; heat acclimation; exercise.

Introduction

Animals exposed to heat, as well as to numerous other stressors overexpress a set of highly conserved proteins in a hierarchical manner, the heat

[*]Corresponding author. Tel.: +972-2-6757588; Fax: +972-3-6439736; E-mail: horowitz@cc.huji.ac.il

shock proteins (HSPs). The fundamental functions of these proteins are molecular chaperoning and cellular repair. Hence, their accumulation protects tissues but can also be considered a stress biomarker. There are abundant publications regarding their cellular mode of action and molecular signaling. Less literature is devoted to the link between the many functions of the HSPs and systemic integrative responses, particularly those controlled by the central nervous system (CNS). Because of the excellent reviews available, much of the basic cellular mechanisms will not be covered here. In this chapter we will focus on the role played by HSPs within the *in vivo* physiological context of hyperthermia *per se* and with its superimposition on situations of chronic stress. Both hyperthermia and chronic stress situations are known to modulate the HSP response and, in turn, physiological responses. Hence, these models pinpoint the HSP-physiological linkage. Inducible HSP72, the most responsive HSP to heat stress is the central player in our discussion.

Heat shock proteins: an introductory overview

HSPs are among the functional class of molecular chaperones, these unrelated protein families (Dietz and Somero, 1992; Maloyan et al., 1999; Baek et al., 2000) assist in the correct non-covalent assembly of other polypeptide containing structures *in vivo*, but are not components of these assembled structures while they are performing their biological functions (Rico et al., 1999). The chaperones recognize and selectively bind non-native proteins under physiological and stress conditions, inhibit incorrect interactions and the formation of non-functional structures (Buchner, 1996; Calderwood et al., 1996; Bouchama and Knochel, 2002). HSPs are ubiquitous in nature; they are present in cells under normal conditions and their accumulation increases, transiently, after a wide variety of insults (Moseley, 1996). Chronic insults, such as long-term/continuous exposure to heat, result in a constitutive elevation of cellular HSP reserves (Dietz and Somero, 1992; Maloyan et al., 1999), in the redistribution of HSP isoforms (Norris and Hightower, 2002) and also in an alteration of their response rate to other stresses (Maloyan et al., 1999).

HSPs are grouped into families based on molecular weight and amino acid sequence (Feige and Polla, 1994; Westman and Sharma, 1998), structure and function (Park et al., 2000). In mammalian cells the five major HSP families are HSP 110, 90, 70, 60 and 27 (Freeman et al., 1999). Each HSP family has defined functions in different cellular compartments (Feige and Polla, 1994). Among the plethora of HSPs, the members of HSP70 family are universally recognized for their cytoprotective function (Volloch and Rits, 1999). However, during severe stresses protection is also conferred by other HSPs such as HSP90 and HSP25 in a hierarchical manner (Lindquist and Kim, 1996; Baek et al., 2000).

The protective role of HSPs is, at least in part, related to enhancing the cell's ability to cope with denatured proteins, to prevent their aggregation and to facilitate renaturation of these proteins or their degradation (Feige and Polla, 1994). Therefore, the stress response can be viewed as an amplification of preexisting chaperone functions (Calderwood et al., 1996).

Collectively, the cellular salvage stress-response is thought to have two components, one which protects in the short term — up to 60 min, not requiring the synthesis of specific proteins, but activating defined salvage pathways involving the adenosine receptor, various kinases and mitochondrial $_K$ATP-dependent channels (Hausenloy and Yellon, 2006). It is known as the first window of protection, or classical preconditioning. The second, delayed response is a longer lasting phase and involves the preferential synthesis of HSPs (Calderwood et al., 1996) and other proteins. Because of this sequence of events, short/acute sublethal stress preconditions cells to cope better with subsequent, more severe stresses, administered during the delayed phase of protection.

The stressors inducing HSPs are denatured proteins produced by exposure to heat shock, ischemia, heavy metals and toxins (Moseley, 1996). Stresses such as heat, ethanol, arsenite, transition metals, release from anoxia and mutagens also induce HSP production (Baek et al., 2000). The type and severity of the stress determine the HSP response and the outcome for the cell — survival or death (Morimoto and Santoro, 1998).

The cytoprotective role of HSPs has been attributed largely to the HSP70 family, in particular to HSP70 (Baek et al., 2000). Data show that the HSP70 family prevents protein damage, restores function to damaged proteins and prevents cellular destruction (Mestril et al., 1996; Mosser et al., 1997; Weiss et al., 2002) by interference with programmed cell death (apoptosis), thus providing cells with time to repair damage (Volloch and Rits, 1999). In addition to these functions, evidence indicates that HSPs have roles in both normal processes of the immune system and in specific immune responses (Rico et al., 1999). For example, macrophages that accumulated HSPs showed transcriptional inhibition and decreased secretion of the inflammatory cytokines TNF-α and IL-1 (Moseley, 1998). Using whole animals, those that underwent a conditioning heat stress sufficient to cause HSP70 accumulation had decreased circulating TNF-α after endotoxin exposure. Additionally, the cellular accumulation of HSP renders cells resistant to the cytotoxic effects of TNF-α, and TNF-α and IL-1 upregulate the HSPs (Moseley, 1997). Other aspects of the stress proteins include their ability to activate the cellular immune response. This is achieved at least in part by the expression of HSP72 on the cell surface (especially noted on tumor cells and virally infected cells) and by the development of stress protein-specific peptide immune complexes (Moseley, 2000; Radons and Multhoff, 2005). Accruing data imply HSP interaction with integrative physiological responses, although the mechanisms involved are not yet understood (Li et al., 2001; Kelty et al., 2002). Another important feature of HSPs is their ability to confer tolerance to one stress via the exposure to another (Moseley, 1996; Horowitz et al., 2004). Collectively, the biological responses to stress (described above) are conserved across species from bacteria to humans (Calderwood et al., 1996).

Heat shock factor 1 (HSF1) is a transcription factor essential to heat shock gene transcription. There are at least two suggested methods by which HSF1 "senses" stress, one involves a conformational change in HSF1 during stress and the second involves a cytoplasmic inhibitory protein such as HSP70 (Calderwood et al., 1996), or misfolded proteins. In simple terms, when the cell is stressed, the HSF1 monomer is released from the (HSP70–HSF1) complex, trimerizes in the cytoplasm, is then phosphorylated, migrates to the nucleus (Feige and Polla, 1994; Calderwood et al., 1996). In the nucleus it binds to the heat shock response element (HSE) found in the promoters of all heat shock genes initiating gene statement, thereby increasing HSP synthesis (Calderwood et al., 1996; Ding et al., 1997a, b, 1998). Recently, more details have been discovered. Shamovsky et al. (2006) reported that an RNA translation factor (eEF1A) regulates the activity of a transcriptional activator essential to the mammalian heat shock response. HSR1 and eEF1A are present in all cells; eEF1A is a translational elongation factor, whereas the exact role of HSR1 is unknown. Upon heat shock, two events lead to the trimerization of HSF1. Initially HSF1 dissociates from a complex comprising Hsp90, p23 and immunophilin, thereby freeing the monomers to trimerize by associating with eE1F1 and HSR1. The homotrimer then binds to the HSE and activates transcription (Shamovsky et al., 2006).

Heat stroke and hyperthermia

The pathophysiology of heat stroke

Homeotherms control their body temperature within a very narrow range. Despite the fact that the thermoregulatory system competes with other homeostatic drives for shared effectors, it operates at a very high gain. Therefore, when body temperature increases, clinical failure (heat illness) may rapidly develop (Hales et al., 1996; Horowitz and Hales, 1998). Heat illness represents a continuum of disorders ranging in severity from temporary and mild to fatal, depending not only on hyperthermia *per se* but also on a variety of "host factors". These "host factors" affect the magnitude and rate of rise of body temperature by interfering with temperature regulation or other related homeostatic control systems (Hales et al., 1996). Numerous factors lead to the development of the heat stroke syndrome, including the hyperthermic level at which heat stroke develops. Nevertheless, the final cascade of pathophysiological consequences is similar in all cases. It is conceivable that a critical factor in the development of heat stroke is the inability to sustain circulatory function (Hales et al., 1996; Horowitz and Hales, 1998).

The thermoregulatory system relies on adjustments in cardiovascular activity to transport heat to the body surface. Concomitantly, however, the cardiovascular system must perfuse all tissues at a level/rate adequate to meet their metabolic demands; for this, arterial pressure must be regulated. During mild heat stress, body surface blood flow is elevated without impinging on the blood supply to other tissues, via an increase and/or redistribution of the cardiac output, namely diverting blood flow from non-vital tissues, e.g. the splanchnic area and probably also the kidneys, muscle and fat to the skin. As the heat load increases, skin blood flow is further augmented and profuse sweating is seen. This increases peripheral blood volume. Concomitantly, the decrease in total blood volume (due to enhanced evaporation), causes a depletion of central blood volume. This reduces central venous pressure, which, presumably, via low-pressure baroreceptors in the cardiopulmonary regions, diminishes skin blood flow and volume as well as evaporative cooling. Collectively, these changes may provoke the development of hyperthermia. Maintenance of blood pressure becomes threatened. The demands for body temperature control may conflict with the requirements for blood pressure regulation. This conflict in homeostatic drives is the dangerous factor, ultimately leading to physiological breakdown and the development of the heat stroke syndrome (Hales et al., 1996; Horowitz and Hales, 1998). The cardiovascular needs eventually take priority over body temperature control. Severe body hyperthermia *per se* is accompanied by detrimental cellular/molecular changes that lead to the potentially fatal cascade of events including acidosis, cerebral hypoxia, cellular fatigue, neural dysfunction and decreased blood pressure (Hales et al., 1996; Horowitz and Hales, 1998).

Extensive studies on experimental animal models as well as case reports on humans suffering from heat stroke or post-mortem examinations, substantiated causal relations between elevated circulating plasma cytokines (Leon et al., 2006), brain cytokines and monoamines (Kao et al., 1994; Lin et al., 1995), suggesting a role for these substances in the pathophysiology of this syndrome (Lin, 1999; Bouchama and Knochel, 2002).

Does HSP72 attenuate the pathophysiology of hyperthermia and heat stroke?

Bazille et al. (2005) described that in heat stroke victims, severe diffuse loss of Purkinje cells was associated with HSP70 expression by Bergmann glia, DNA internucleosomal breakages in the dentate nuclei and centromedian nuclei of the thalamus and degeneration of cerebellar efferent pathways. Pavlik et al. (2003), Westman and Sharma (1998) and Yang et al. (1998), provided substantial evidence (using rodent models) of a marked elevation in the inducible form of HSP70 in CNS tissues, in response to heat stress and hyperthermia. Upregulation of HSP70 was demonstrated in canine heat stroke model as well (Oglesbee et al., 2002, see Chapter 19 in this volume). Furthermore, the relative resistance to heat stroke in this model was linked to the high levels of HSPs in the brains of this species. The fundamental role of this ubiquitous protein is to protect nascent and unfolded proteins. Therefore, this protein is also considered as a reliable stress biomarker. Additionally, prior overexpression of HSP70 (e.g., via preconditioning) affords protection against tissues damage (Reshef et al., 1996; Hausenloy and Yellon, 2006). Similarly, a large number of reports provide evidence that thermal, chemical and hypoxic preconditioning, leading to the overexpression of HSP70, delay detrimental heat stroke induced cerebral ischemia and hypotension. For example, Yang et al. (1998) using the anesthetized rat heat stroke model discovered that prior acute heat shock or chemical preconditioning (both these stresses elevate HSP72) confer significant protection against heat stroke-induced cerebral ischemia, neuronal damage and fatal arterial hypotension. This protection lasted for approximately 2 days and terminated with the return of HS72 to its preconditioning level. These data suggest that the presence of HSP72 increases survival in the rat model of heat stroke.

The neuroprotective role of HSP70 has been established in many stressful situations (Welch, 2001; Yenari, 2002). However, it is intriguing to find out whether HSPs have functions in addition to their chaperoning abilities, or alternatively, mediate physiological outcomes via cellular safeguarding. The rational for this question is the debate regarding the role of HSPs in cardioprotection during the second window of protection (SWOP). For example, in the rat heart, ischemic preconditioning, in contrast to thermal preconditioning, does not induce the classical SWOP despite enhanced expression of HSP72k, suggesting that SWOP and heat stress, have distinct mechanisms of protection that may not be exclusively related to HSP72 expression (Qian et al., 1999).

In the same vein, Li et al. (2001) questioned whether HSP70 potentiates the baroreflex response by acting on the nucleus tractus solitarius (NTS), the recipient "relay station" of baroreceptor afferent fibers. These authors demonstrated that HSP70 levels in the NTS enhanced the sensitivity of both sympathetic and parasympathetic arms of the autonomic nervous system and thus attenuated heat stroke-induced cerebral ischemia and hypotension. Microinjection of anti-HSP antibodies to the NTS prior to thermal preconditioning abolished this effect. The same authors observed that a threshold level of HSP expression in the NTS is required for the onset of the baroreflex response. These observations are consistent with the intriguing finding that thermal preconditioning and increased HSP72 preserved synaptic transmission during thermal stress (Kelty et al., 2002). This was seen in medullary slices from neonatal mice, containing the neural network that generates respiratory rhythm. They observed attenuation of the frequency increases of GABAergic, glutaminergic and glycinergic miniature post synaptic currents compared with mice without HSP72 overexpression. Due to the fact that HSP is taken up by neurons it can interact with various intracellular proteins to regulate transmitter release (Jiang et al., 2000; Ohtsuka and Suzuki, 2000). Figure 1 summarizes the experimental evidence supporting the

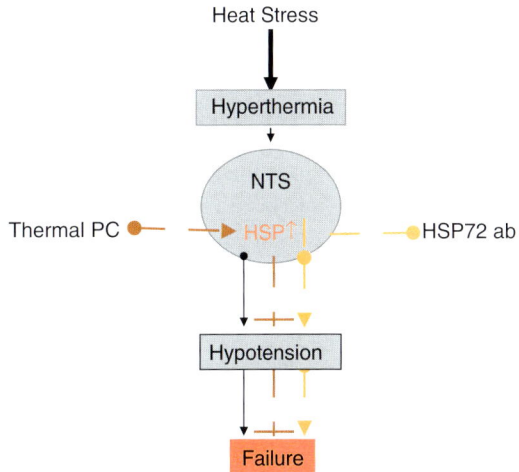

Fig. 1. Experimental evidence supporting the link between augmented HSP72 reserves in nucleus tractus solitarius (NTS) and protection against heat stroke associated hypotension and circulatory failure leading to irreversible heat stroke. Thermal PC, thermal preconditioning; HSP72 ab, antibodies against HSP72. Adapted from Li et al. (2001).

link between HSP72 and the NTS-mediated control of blood pressure during hyperthermia and heat stroke.

We can conclude that HSP72 protects using its multiple functions by direct interaction with autonomic neuronal networks. The mode of this interaction is not completely understood. There are indications, however, that molecular pathways associated with synaptic transmission are involved. We hypothesize that maintenance of protein integrity plays a pivotal role in this process. It is noteworthy that neurons show cell specificity and regulatory differences in their HSP70 gene response (Kaarniranta et al., 2002). Interestingly, chronic stressors, leading to altered HSPs responses, may affect the physiological profile during hyperthermic conditions. We will discuss these physiological aspects using heat acclimation and aging as models. Both these altered physiological states show seemingly similar sluggishness in the profile of increasing body temperature during severe hyperthermia. Likewise both demonstrate sustained low thyroxine levels (Horowitz et al., 1986) that impact on HSPs (Maloyan and Horowitz, 2002).

Aging impedes coping with hyperthermia: is central HSP70 important?

A lesson from whole body responses

In vivo hyperthermia induces cell damage in the brains of both the newborn and the adult. It also activates cell survival pathways. However, these pathways are more vulnerable to heat shock in the young (Schiaffonati et al., 2001). Aging is associated with a loss of ability at the molecular, cellular and whole organism levels to cope with environmental stressors (Blake et al., 1990; Kregel et al., 1990; Papaconstantinou, 1994). The mechanisms underlying these alterations seemingly include modifications in transcriptional regulation of stress-inducible genes (Papaconstantinou, 1994), inappropriate production of inflammatory cytokines (Ershler, 1993), blunted synthesis of protective acute-phase proteins such as HSPs (Blake et al., 1990; Kregel et al., 1990; Kregel and Moseley, 1996) and, at an integrated level, multi-organ system failure (Kregel et al., 1990; Papaconstantinou, 1994). In aged human populations, this is demonstrated by the increase in heat-related morbidity and mortality rates reported during periodic heat waves in urban areas (Semenza et al., 1996; Hall et al., 2000).

Due to their central role in maintaining cellular homeostasis, HSPs and the heat shock response during hyperthermia provide us with a good model to understand some key aspects of cellular cytoprotection during aging. Overall, during environmental heating, the ability to generate protective intracellular stress proteins is blunted in old compared with young rats (Kregel et al., 1995; Kregel, 2002). Additionally, Kregel et al. (1990) discovered that older rats are significantly less thermotolerant and have higher mortality rates when exposed to repeated heat challenges. Linking thermal tolerance to HSP72 levels, Blake et al. (1991) postulated that the age-related impairment in the induction of heat-shock proteins, the normal physiological response to heat stress (Currie and White, 1981) may contribute to the reduction in thermotolerance noted with age. Mapping the ability of various organs to produce HSPs, it was confirmed that with increased heating, the brain, lung and skin generate HSP72 transcript in a graded manner but there was less HSP70 expression in tissues of older rats (Blake et al., 1991). Walters et al. (2001) demonstrated that the HSP response in the brain is site specific, and that depending on the brain region and the sampling time post heat stress, aged rats displayed significantly greater, lesser or similar increases in this protein compared with young rats, but the majority of the brain loci studied demonstrated a delayed heat shock response. Significantly lower HSP70 production was noted in the ventral paraventricular and lateral magnocellular nuclei. Given the recent findings of Li et al. (2001) on the pivotal role played by HSP72 levels in the NTS in the initiation of the baroreflex response to delay heat stroke-mediated hypotension and cerebral ischemia (mentioned above in "Heat shock proteins: an introductory overview"), we can hypothesize that the attenuated heat shock response in hypothalamic nuclei controlling homeostatic drives may be linked with the decreased thermotolerance seen in the aged.

Using similar experimental setups Blake et al. (1991) and Walters et al. (2001) found that the decrease in HSP70 expression (both transcript and protein) with aging correlated with an attenuated increase in colonic temperature (T_c) of the old rats subjected to heat stress, suggesting that the attenuated T_c rise rather than impaired regulation of HSP70 gene expression is age dependent. Decreased heat production during exposure to high ambient temperatures (Parmacek et al., 1979), leading to sluggish elevation in T_c have been observed by various authors (e.g., Meiri et al., 1991). Therefore this factor should be taken into consideration when interpreting the responses of aged individuals to hyperthermia.

A lesson from cellular responses

Evidence from studies on livers and hepatocytes of young and senescent rats confirm the above conclusions. Kregel and Moseley (1996) examined the pattern of accumulation of HSP72 in the livers of mature (12-month-old) and senescent (24-month-old) rats after either passive or exertional heat stress. Rats from both age groups increased T_c to 41°C. Liver HSP72 accumulation was increased in mature (vs. control) rats by 192 and 292% for sedentary and exercising groups respectively. In contrast, the senescent rats demonstrated no significant increase in inducible HSP72 with heating but a large increase with exercise (232%) was seen. These data suggest that the blunted HSP response to heating observed with aging is not necessarily due to an inability to produce inducible HSP72. Investigating age changes in hepatocytes, Heydari et al. (1993) and Wu et al. (1993) examined the ability of hepatocytes isolated from young/adult (4–7-month-old) and old (22–28-month-old) rats to express HSP70 after acute heat shock. They found that the induction of HSP70 synthesis and mRNA levels by heat shock was 40–50% lower in the hepatocytes isolated from the old rats compared with those from the young rats. In a more recent study Heydari et al. (2000) provided evidence for impaired thermostability of the DNA binding activity of HSF1 with age in a cell-free system as well as in isolated hepatocytes. These studies demonstrated that the age-related decline in HSP70 expression is constitutive and is due to a decline in binding of the HSF to the HSE and decreased HSP70 transcription.

Interestingly, caloric restriction reversed this age-related decline in the induction of HSP70 expression. Similarly Hall et al. (2000) investigated the influence of caloric restriction on stress tolerance in old rats exposed to an environmental heating protocol on 2 consecutive days, demonstrating that all the old caloric restricted rats survived the stress, compared with only 50% of the control-fed old rats. The tissue specificity of caloric restriction was demonstrated by Pahlavani et al. (1996). They applied a similar protocol to lymphocytes from young and old rats and showed that the percentage viability (an index of thermosensitivity) as well as the induction of HSP70 protein and mRNA of the lymphocytes isolated from old rats was lower than that in the young rats. Food restriction did not reverse this result. Rao et al. (1999) also demonstrated age-related attenuation of HSPs 105, 90, 70, 60, 47, 40, 27 and 16 in lymphocytes.

There is a discrepancy in the conclusions drawn from whole animal studies vs. those obtained from the direct heating of cells. While the latter model provides conclusive evidence of the diminished capacity of the HSR at the cellular level, whole body heating/exercise experiments show attenuated HSR but not necessarily decreased capacity. It should be noted that there are differences in the tissues examined and in the models used. Likewise, the whole animal model is a setup in which humoral as well as neural factors exert their effect. HSPs are but one cell signaling pathway and impairments in thermotolerance or thermal preconditioning result from many signals along numerous cascades (Gray et al., 2000; Honma et al., 2003).

Table 1 summarizes systemic thermoregulatory responses, central heat shock response and thermotolerance with aging.

Heat acclimation modulates the heat shock response during hyperthermia

The process of heat acclimation enhances the organism's ability to perform in the heat due to improvements in heat dissipating mechanisms. There is much evidence of the role of HSPs in this (thermal) adaptation. HSP70 levels correlate with environmental temperatures, the higher the ambient temperature of a particular environmental niche, the greater the amount of constitutive HSP70 found during non-stress/physiological conditions. In humans, skin fibroblasts obtained from Turkmen living in the Central Asian desert and from European Russians living in temperate climates revealed differences in HSP70 after a 42.5°C heat shock, with greater increases in inducible HSP70 in desert-originated fibroblasts (Lyashko et al., 1994). Heat acclimation increases HSP reserves in many tissues, suggesting that preexisting elevation in HSP70 might allow the animal to continue to perform work under acute temperature elevations, without de novo synthesis of additional HSPs. Given the multiple roles played by HSPs we suggest that HSPs are essential to thermotolerance but also protect epithelial integrity (Moseley, 1997; Dokladny et al., 2006), vascular reactivity (Itoh et al., 1994) and interact with cellular signaling networks (above in this chapter).

Table 1. Thermoregulatory responses, central heat shock response and thermotolerance with aging

Heat stress	Ex vivo	In vivo	
Thermoregulation			
Central autonomic responses		↓	
Heat dissipation effectors		↓	
Heat production		↓	
T_b rise		↓	
Thermal tolerance		↓	
Heat shock response (HSR)			
Central HSP70 level	No compromise with aging	↓	Decreased central HSP may stem from sluggish heating rate rather than decreased cellular capacity
Peripheral HSP70 level, caloric restriction "preconditioning"	No compromise with aging	↓	Attenuated HSR may contribute to decreased thermal tolerance

Source: Data were compiled from Blake et al. (1990, 1991); Meiri et al. (1991); Heydari et al. (1993, 2000); Kregel et al. (1995); Kregel and Moseley (1996); Walters et al. (1998, 2001); Kenney et al. (2000); Kregel (2002).

Characterization of HSP70 and the HSR upon acclimation, in the context of the global thermoregulatory response, can be approached by studying the mRNA encoding HSP70 and the *HSP70* gene product profile during subjection to progressively increasing environmental HS. Under these conditions, the invocation of the HSR is due to the cumulative heat strain induced by the rate of heating and the hyperthermic level of the animals. Heat strain and hyperthermic levels depend on the physiological mechanisms for heat dissipation.

Analysis of the differences in m*HSP*/HSP during stress helps pinpoint the thermal physiological parameters involved in the invocation of the HSR. Likewise, the correlation between the thermoregulatory span (i.e., the period during which body temperature is regulated) and the injury temperature threshold of acclimated and control rats, and the HSP basal level/production rate (during stress), allows evaluation of the contributions of the HSPs vs. the physiological features to the welfare of the acclimated vs. the non-acclimated animals during heat stress.

Thermoregulatory capacity during hyperthermia is demonstrated by the T_c profile during exposure to heat stress. Interestingly, heat stress at 41°C accelerates the heating rate of the acclimated rats; whereas heat stress at 43°C, in contrast, profoundly attenuates heating in this group, leading to decreased heat strain. In non-acclimated/control animals, elevation of the ambient temperature resulted in a rise in T_c and heat strain. Considering these responses, some general conclusions regarding the HSR-physiological linkage can be drawn. Time to peak HSP72 transcript level and HSP72 accumulation are faster following acclimation with the elevation of environmental strain. However, with increased environmental stress the enhanced physiological capacity of the acclimated animals allows them to attenuate the rise in body temperature and, in turn, heat strain and HSP production (Fig. 2).

Although there is a strong basis for the role of HSP70 in physiological modulation of the respiratory center (Kelty et al., 2002) and the NTS controlling blood pressure (Li et al., 2001; Ouyang et al., 2006) under acute stress no information is available regarding these integrative functions in acclimated animals.

Physical activity alters brain HSP and cytokine responses to immunological challenges

Physical activity and brain vs. plasma HSPs

HSP70 induction has been shown to take place in the CNS tissues not only in response to heat stress, but in response to a number of non-thermal stressors such as ischemia and hypoxia and hypoxia (Latchman, 2004). Exercise at room temperature, however, is not one of these stimuli (Walters et al.,

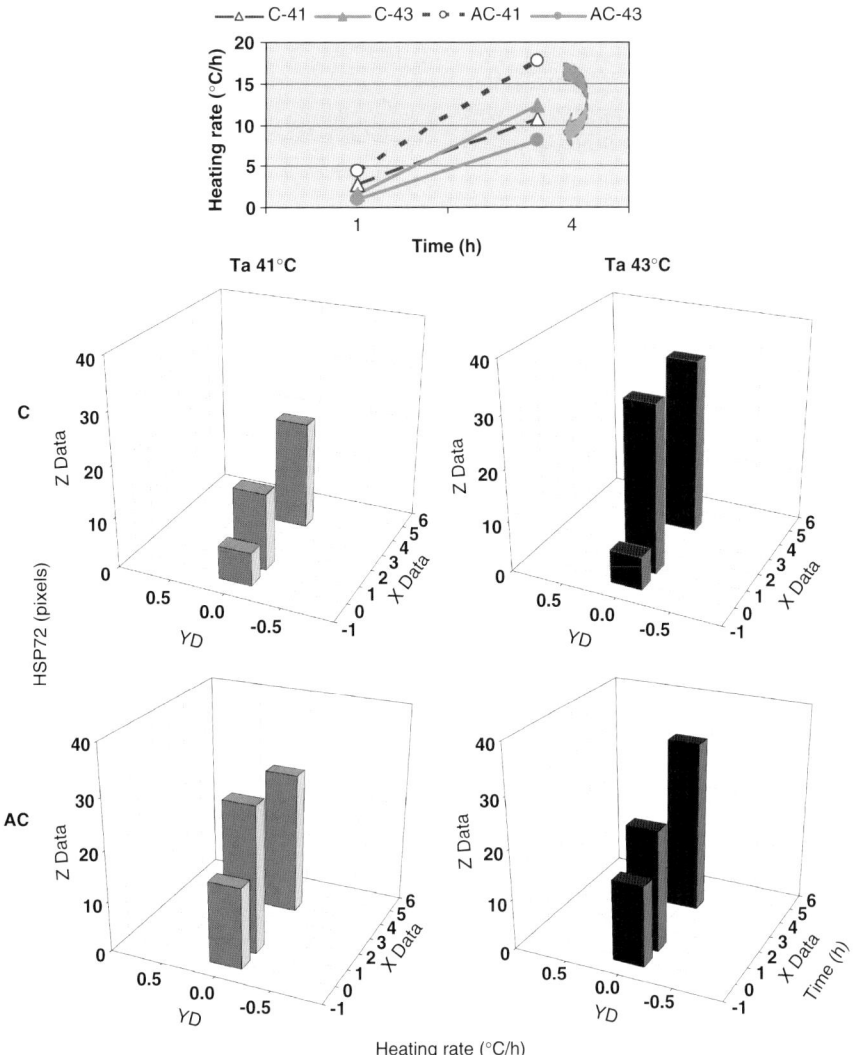

Fig. 2. Heat acclimation modulates the heat shock response during hyperthermia. Acclimation elevates basal HSP72 reserves and accelerates the heat shock response. During recovery from a 2-h exposure to 41°C at room temperature, peak HSP72 levels in acclimated rats (AC) are seen after 1 h vs. 4 h in the non-acclimated (C) rats (left panel C vs. AV). With increased strain, caused by exposure to 43°C, the AC rats show an attenuated heating rate and in turn, HSP72 production while the C rats demonstrate enhanced heating rates and HSP72 synthesis (right panels-red bars vs. left panels-black bars). Inset: Heating rates (T_c°C/h) of C and AC rats during 2-h exposure to 41°C (black lines) and 43°C (red lines) are depicted. It is notable that the AC rats demonstrate a profound decrease in heating rate at 43°C. Adapted from Maloyan et al. (1999).

1998; Nickerson et al., 2005). However, during exercise, the heat production in muscles and peripheral tissues is adequate to upregulate HSP70. On the other hand, elevations in ambient temperature increase HSP70 in brain regions coordinating the neuroendocrine response to stress. Pavlik et al. (2003) in a detailed immunohistological study provided solid evidence that following heat stress, HSP70 appears mainly in glial and vascular cells. Higher numbers of HSP70-positive cells were detected in the white matter and diencephalic region than in the cerebral cortex. HSP70 was localized in many oligodendrocytes, in some vessels. Microglia showed less HSP70 than the previous two cell types. In contrast, only a few HSP70-stained cells were positive for the astrocyte marker GFAP. In addition to glial/vascular HSP70 staining, neuronal HSP70 induction was observed only in discrete regions including the paraventricular, supraoptic, suprachiasmatic and other hypothalamic nuclei, and in the amygdale, areas associated with homeostatic functions. The phenomenon of differential regulation was also validated when the regulation of HSP70 genes during heat shock was examined in several primary neuron cultures (Kaarniranta et al., 2002). HSP70 response is regulated at the translational level in Neuro-2a neuroblastoma cells, whereas the IMR-32 neuroblastoma cells respond to stress by classical HSF1-driven transcriptional regulatory mechanisms. Rat primary hippocampal neurons showed a lack of HSF1 and did not induce the HSP70 gene at all, indicating that different neuronal cells have unique HSP70 gene expression patterns, and this fact may explain the differing sensitivities of specific neuron types to stress (Kaarniranta et al., 2002).

Given the above findings, and the observations that exercise (1) protects immune cells from oxidative mediated apoptosis (Avula et al., 2001), (2) prevents stress-induced immunosuppression (Moraska and Fleshner, 2001), (3) enhances natural killer cell cytotoxicity protection in response to immune challenge (Jonsdottir et al., 2000) and that exhaustive exercise causes a greater and faster production of HSP72 in a number of central and peripheral structures (Campisi et al., 2003a, b) the hypothesis that, at least in part, brain HSP72 effects are mediated via external HSP72 sources and enhanced HSP72 — immune system interactions (Nickerson et al., 2005) is promoted. It seems that the recently discovered eHSP72 (circulating HSPs) may fulfill this function.

It is now recognized that glial cells, B cells and peripheral blood mononuclear cells (PBMC) release eHSP72 by hormone receptor mediated exocytosis into the blood stream and that eHSP72 levels are elevated within 10–25 min in response to various stressors (Nickerson et al., 2005). Febbraio et al. (2002) demonstrated that splanchnic tissues release HSP72 during exercise and that this release is partially responsible for the elevated systemic HSP72 concentration during exercise in humans.

Physical activity affects brain responses to immunological challenges

Given that physical activity increases eHSP72, it is reasonable to believe that prior exercise facilitates the immune response (this is analogous to the changes seen following thermal or ischemic/hypoxic preconditioning, or heat acclimation). Nickerson et al. (2005) showed that prior free wheel exercise training at room temperature enhances brain HSP72 generation in response to immunological challenge (*E. coli* administration) profoundly more than in sedentary rats, together with a robust increase in IL-β, apparently contributing to faster bacterial clearance and reduced sickness. Interestingly, Nickerson et al. (2005) also demonstrated parallel enhancement of the proinflamatory cytokine IL-β and protective intracellular HSP72. This phenomenon was previously elucidated in experiments where other stressors were applied (Campisi and Fleshner, 2003; Campisi et al., 2003b; Fleshner et al., 2004; Fleshner and Johnson, 2005), suggesting that eHSP72 provides the "danger signal" to potentiate the immune response. There is evidence that TLR2 and/or TLR4 act as cell surface receptors for eHSP72, these receptors transduce the inflammatory signals to innate immune cells (for details, see Campisi et al., 2003a; Fleshner and Johnson, 2005; Johnson and Fleshner, 2006) and thus prepare the immune system for possible, subsequent pathogenic challenges and for bacterial killing via nitric oxide and or cytokines (Campisi et al., 2003a). The pathways and mechanisms involved and the specific physiological context of these proteins in relation to their origin are detailed in Wang et al. (2005) and Johnson and Fleshner (2006).

Summary

In this chapter we discussed the effects of thermal and exercise stress on central HSP70 accumulation and function as manifested by the response of the whole body in an integrated manner. We brought evidence of cell specificity in the appearance and accumulation of HSPs, however, their effect on homeostatic drives is unequivocal. Chronic situations affect the HSP profile. There is attenuation in their level and their response rate is sluggish during aging. In contrast, following heat acclimation (which has other similarities with aging), basal HSP level is augmented and their response rate following heat stress, is accelerated compared with non-acclimated animals. During extreme heat stress, the physiological mechanisms for heat dissipation of the acclimated animals have a greater impact on heat endurance than the chaperoning actions of the HSPs. Exercise, which does not enhance levels of tissue HSPs increases eHSP production to potentiate innate immunity.

Interestingly, despite of the plethora of publications on HSPs, reviews on the link between their molecular/cellular functions and integrative systemic responses are not abundant.

References

Avula, C.P.R., Muthukumar, A.R., Zaman, K., McCarter, R. and Fernandes, G. (2001) Inhibitory effects of voluntary wheel exercise on apoptosis in splenic lymphocyte subsets of C57BL/6 mice. J. Appl. Physiol., 91: 2546–2552.

Baek, S.H., Min, J.N., Park, E.M., Han, M.Y., Lee, Y.S., Lee, Y.J. and Park, Y.M. (2000) Role of small HSP25 in radio-resistance and glutathione-redox cycle. J. Cell. Physiol., 183: 100–107.

Bazille, C., Megarbane, B., Bensimhon, D., Lavergne-Slove, A., Baglin, A.C., Loirat, P., Woimant, F., Mikol, J. and Gray, F. (2005) Brain damage after heat stroke. J. Neuropathol. Exp. Neurol., 64: 970–975.

Blake, M.J., Fargnoli, J., Gershon, D. and Holbrook, N.J. (1991) Concomitant decline in heat-induced hyperthermia and HSP70 mRNA expression in aged rats. Am. J. Physiol., 260(4 Pt 2): R663–R667.

Blake, M.J., Gershon, D., Fargnoli, J. and Holbrook, N.J. (1990) Discordant expression of heat shock protein mRNAs in tissues of heat-stressed rats. J. Biol. Chem., 265: 15275–15279.

Bouchama, A. and Knochel, J.P. (2002) Heat stroke. N. Engl. J. Med., 346: 1978–1988.

Buchner, J. (1996) Supervising the fold: functional principles of molecular chaperones. FASEB J., 10: 10–18.

Calderwood, S., Freeman, M., Laszlo, A., Li, G., Subjeck, J. and Weber, L. (1996) The stress response: a radiation study section workshop. Radiat. Res., 145: 107–117.

Campisi, J. and Fleshner, M. (2003) Role of extracellular HSP72 in acute stress-induced potentiation of innate immunity in active rats. J. Appl. Physiol., 94: 43–52.

Campisi, J., Leem, T.H. and Fleshner, M. (2003a) Stress-induced extracellular Hsp72 is a functionally significant danger signal to the immune system. Cell Stress Chaperones, 8: 272–286.

Campisi, J., Leem, T.H., Greenwood, B.N., Hansen, M.K., Moraska, A., Higgins, K., Smith, T.P. and Fleshner, M. (2003b) Habitual physical activity facilitates stress-induced HSP72 induction in brain, peripheral, and immune tissues. Am. J. Physiol. Regul. Integr. Comp. Physiol., 284: R520–R530.

Currie, R.W. and White, F.P. (1981) Trauma-induced protein in rat tissues: a physiological role for a "heat shock" protein? Science, 214: 72–73.

Dietz, T.J. and Somero, G.N. (1992) The threshold induction temperature of the 90-kDa heat shock protein is subject to acclimatization in eurythermal goby fishes (genus *Gillichthys*). PNAS, 89: 3389–3393.

Ding, X.Z., Tsokos, G.C. and Kiang, J.G. (1997a) Heat shock factor-1 protein in heat shock factor-1 gene-transfected human epidermoid A431 cells requires phosphorylation before inducing heat shock protein-70 production. J. Clin. Invest., 99: 136–143.

Ding, X.Z., Tsokos, G.C. and Kiang, J.G. (1998) Overexpression of HSP-70 inhibits the phosphorylation of HSF1 by activating protein phosphatase and inhibiting protein kinase C activity. FASEB J., 12: 451–459.

Ding, X.Z., Tsokos, G.C., Smallridge, R.C. and Kiang, J.G. (1997b) Heat shock gene-expression in HSP70 and HSF1 gene-transfected human epidermoid A-431 cells. Mol. Cell Biochem., 167: 145–152.

Dokladny, K., Moseley, P.L. and Ma, T.Y. (2006) Physiologically relevant increase in temperature causes an increase in intestinal epithelial tight junction permeability. Am. J. Physiol. Gastrointest. Liver Physiol., 290: G204–G212.

Ershler, W.B. (1993) Interleukin-6: a cytokine for gerontologists. J. Am. Geriatr. Soc., 41: 176–181.

Febbraio, M.A., Ott, P., Nielsen, H.B., Steensberg, A., Keller, C., Krustrup, P., Secher, N.H. and Pedersen, B.K. (2002) Exercise induces hepatosplanchnic release of heat shock protein 72 in humans. J. Physiol. (Lond.), 544: 957–962.

Feige, U. and Polla, B.S. (1994) Hsp70: a multi-gene, multi-structure, multi-function family with potential clinical applications. Experientia, 50: 979–986.

Fleshner, M., Campisi, J., Amiri, L. and Diamond, D.M. (2004) Cat exposure induces both intra- and extracellular Hsp72: the role of adrenal hormones. Psychoneuroendocrinology, 29: 1142–1152.

Fleshner, M. and Johnson, J.D. (2005) Endogenous extracellular heat shock protein 72: releasing signal(s) and function. Int. J. Hyperthermia, 21: 457–471.

Freeman, M.L., Borrelli, M.J., Meredith, M.J. and Lepock, J.R. (1999) On the path to the heat shock response: destabilization and formation of partially folded protein intermediates, a consequence of protein thiol modification. Free Radic. Biol. Med., 26: 737–745.

Gray, C.C., Amrani, M., Smolenski, R.T., Taylor, G.L. and Yacoub, M.H. (2000) Age dependence of heat stress mediated cardioprotection. Ann. Thorac. Surg., 70: 621–626.

Hales, J.R.S., Hubbard, R.W. and Gaffin, S.L. (1996) Limitations of heat tolerance. In: Fregly M.J. and Blatteis C.M. (Eds.), Handbook of Physiology, Section 4, Environmental Physiology, Vol. 1. (Chap 15). Oxford University Press, Oxford, pp. 285–359.

Hall, D.M., Oberley, T.D., Moseley, P.M., Buettner, G.R., Oberley, L.W., Weindruch, R. and Kregel, K.C. (2000) Caloric restriction improves thermotolerance and reduces hyperthermia-induced cellular damage in old rats. FASEB J., 14: 78–86.

Hausenloy, D.J. and Yellon, D.M. (2006) Survival kinases in ischemic preconditioning and postconditioning. Cardiovasc. Res., 70: 240–253.

Heydari, A.R., Wu, B., Takahashi, R., Strong, R. and Richardson, A. (1993) Expression of heat shock protein 70 is altered by age and diet at the level of transcription. Mol. Cell. Biol., 13: 2909–2918.

Heydari, A.R., You, S., Takahashi, R., Gutsmann-Conrad, A., Sarge, K.D. and Richardson, A. (2000) Age-related alterations in the activation of heat shock transcription factor 1 in rat hepatocytes. Exp. Cell Res., 256: 83–93.

Honma, Y., Tani, M., Yamamura, K., Takayama, M. and Hasegawa, H. (2003) Preconditioning with heat shock further improved functional recovery in young adult but not in middle-aged rat hearts. Exp. Gerontol., 38: 299–306.

Horowitz, M., Eli-Berchoer, L., Wapinski, I., Friedman, N. and Kodesh, E. (2004) Stress-related genomic responses during the course of heat acclimation and its association with ischemic-reperfusion cross-tolerance. J. Appl. Physiol., 97: 1496–1507.

Horowitz, M. and Hales, J.R.S. (1998) Pathophysiology of hyperthermia. In: Blatteis C.M. (Ed.), Physiology and Pathophysiology of Temperature Regulation, Chapter 12. World Scientific Publishing Co., Singapore, pp. 229–245.

Horowitz, M., Peyser, Y.M. and Muhlrad, A. (1986) Alterations in cardiac myosin isoenzymes distribution as an adaptation to chronic environmental heat stress in the rat. J. Mol. Cell Cardiol., 18: 511–515.

Itoh, Y.H., Aihara, M., Oguri, T. and Miyata, N. (1994) Thermotolerance and heat shock protein (HSP 70) in vascular endothelial cells. Nippon Igaku Hoshasen Gakkai Zasshi, 54: 1187–1189.

Jiang, R., Gao, B., Prasad, K., Greene, L.E. and Eisenberg, E. (2000) Hsc70 chaperones clathrin and primes it to interact with vesicle membranes. J. Biol. Chem., 275: 8439–8447.

Johnson, J.D. and Fleshner, M. (2006) Releasing signals, secretory pathways, and immune function of endogenous extracellular heat shock protein 72. J. Leukoc. Biol., 79: 425–434.

Jonsdottir, I.H., Hellstrand, K., Thoren, P. and Hoffmann, P. (2000) Enhancement of natural immunity seen after voluntary exercise in rats. Role of central opioid receptors. Life Sci., 66: 1231–1239.

Kaarniranta, K., Oksala, N., Karjalainen, H.M., Suuronen, T., Sistonen, L., Helminen, H.J., Salminen, A. and Lammi, M.J. (2002) Neuronal cells show regulatory differences in the hsp70 gene response. Brain Res. Mol. Brain Res., 101: 136–140.

Kao, T.Y., Chio, C.C. and Lin, M.T. (1994) Hypothalamic dopamine release and local cerebral blood flow during onset of heatstroke in rats. Stroke, 25: 2483–2486.

Kelty, J.D., Noseworthy, P.A., Feder, M.E., Robertson, R.M. and Ramirez, J.-M. (2002) Thermal preconditioning and heat-shock protein 72 preserve synaptic transmission during thermal stress. J. Neurosci., 22: 193RC.

Kenney, M.J., Pickar, J.G., Weiss, M.L., Saindon, C.S. and Fels, R.J. (2000) Effects of midbrain and spinal cord transections on sympathetic nerve responses to heating. Am. J. Physiol. Regul. Integr. Comp. Physiol., 278: R1329–R1338.

Kregel, K.C. (2002) Molecular biology of thermoregulation: invited review: heat shock proteins: modifying factors in physiological stress responses and acquired thermotolerance. J. Appl. Physiol., 92: 2177–2186.

Kregel, K.C. and Moseley, P.L. (1996) Differential effects of exercise and heat stress on liver HSP70 accumulation with aging. J. Appl. Physiol., 80: 547–551.

Kregel, K.C., Moseley, P.L., Skidmore, R., Gutierrez, J.A. and Guerriero Jr., V. (1995) HSP70 accumulation in tissues of heat-stressed rats is blunted with advancing age. J. Appl. Physiol., 79: 1673–1678.

Kregel, K.C., Tipton, C.M. and Seals, D.R. (1990) Thermal adjustments to nonexertional heat stress in mature and senescent Fischer 344 rats. J. Appl. Physiol., 68: 1337–1342.

Latchman, D.S. (2004) Protective effect of heat shock proteins in the nervous system. Curr. Neurovasc. Res., 1: 21–27.

Leon, L.R., Blaha, M.D. and DuBose, D.A. (2006) Time course of cytokine, corticosterone, and tissue injury responses in mice during heat strain recovery. J. Appl. Physiol., 100: 1400–1409.

Li, P.-L., Chao, Y.-M., Chan, S.H.H. and Chan, J.Y.H. (2001) Potentiation of baroreceptor reflex response by heat shock protein 70 in nucleus tractus solitarii confers cardiovascular protection during heatstroke. Circulation, 103: 2114–2119.

Lin, M.T. (1999) Pathogenesis of an experimental heatstroke model. Clin. Exp. Pharmacol. Physiol., 26: 826–827.

Lin, M.T., Kao, T.Y., Chio, C.C. and Jin, Y.T. (1995) Dopamine depletion protects striatal neurons from heatstroke-induced ischemia and cell death in rats. Am. J. Physiol. Heart Circ. Physiol., 269: H487–H490.

Lindquist, S. and Kim, G. (1996) Heat-shock protein 104 expression is sufficient for thermotolerance in yeast. Proc. Natl. Acad. Sci. U.S.A., 93: 5301.

Lyashko, V.N., Vikulova, V.K., Chernicov, V.G., Ivanov, V.I., Ulmasov, K.A., Zatsepina, O.G. and Evgen'ev, M.B. (1994) Comparison of the heat shock response in ethnically and ecologically different human populations. PNAS, 91: 12492–12495.

Maloyan, A. and Horowitz, M. (2002) beta-Adrenergic signaling and thyroid hormones affect HSP72 expression during heat acclimation. J. Appl. Physiol., 93(1): 107–115.

Maloyan, A., Palmon, A. and Horowitz, M. (1999) Heat acclimation increases the basal HSP72 level and alters its production dynamics during heat stress. Am. J. Physiol. Regul. Integr. Comp. Physiol., 276: R1506–R1515.

Meiri, U., Shochina, M. and Horowitz, M. (1991) Heat acclimated hypohydrated rats: age varying vasomotor and plasma volume response to heat stress. J. Therm. Biol., 16: 241–247.

Mestril, R., Giordano, F.J., Conde, A.G. and Dillmann, W.H. (1996) Adenovirus-mediated gene transfer of a heat shock protein 70 (hsp 70i) protects against simulated ischemia. J. Mol. Cell Cardiol., 28: 2351–2358.

Moraska, A. and Fleshner, M. (2001) Voluntary physical activity prevents stress-induced behavioral depression and anti-KLH antibody suppression. Am. J. Physiol. Regul. Integr. Comp. Physiol., 281: R484–R489.

Morimoto, R.I. and Santoro, M.G. (1998) Stress-inducible responses and heat shock proteins: new pharmacological targets for cytoprotection. Nat. Biotechnol., 16: 833–838.

Moseley, P. (2000) Stress proteins and the immune response. Immunopharmacology, 48: 299–302.

Moseley, P.L. (1996) Heat shock proteins: a broader perspective. J. Lab. Clin. Med., 128: 233–234.

Moseley, P.L. (1997) Heat shock proteins and heat adaptation of the whole organism. J. Appl. Physiol., 83: 1413–1417.

Moseley, P.L. (1998) Heat shock proteins and the inflammatory response. Ann. N.Y. Acad. Sci., 856: 206–213.

Mosser, D., Caron, A., Bourget, L., Denis-Larose, C. and Massie, B. (1997) Role of human heat shock protein hsp70 in protection against stress-induced apoptosis. Mol. Cell Biol., 17: 5317–5327.

Nickerson, M., Elphick, G.F., Campisi, J., Greenwood, B.N. and Fleshner, M. (2005) Physical activity alters the brain Hsp72 and IL-1{beta} responses to peripheral E. coli challenge. Am. J. Physiol. Regul. Integr. Comp. Physiol., 289: R1665–R1674.

Norris, C.E. and Hightower, L.E. (2002) Discovery of two distinct small heat shock protein (HSP) families in the desert fish Poeciliopsis. Prog. Mol. Subcell. Biol., 28: 19–35.

Oglesbee, M.J., Alldinger, S., Vasconcelos, D., Diehl, K.A., Shinko, P.D., Baumgartner, W., Tallman, R. and Podell, M. (2002) Intrinsic thermal resistance of the canine brain. Neuroscience, 113: 55–64.

Ohtsuka, K. and Suzuki, T. (2000) Roles of molecular chaperones in the nervous system. Brain Res. Bull., 53: 141–146.

Ouyang, Y.B., Xu, L.J., Sun, Y.J. and Giffard, R.G. (2006) Overexpression of inducible heat shock protein 70 and its mutants in astrocytes is associated with maintenance of mitochondrial physiology during glucose deprivation stress. Cell Stress Chaperones, 11: 180–186.

Pahlavani, M.A., Harris, M.D., Moore, S.A. and Richardson, A. (1996) Expression of heat shock protein 70 in rat spleen lymphocytes is affected by age but not by food restriction. J. Nutr., 126: 2069–2075.

Papaconstantinou, J. (1994) Unifying model of the programmed (intrinsic) and stochastic (extrinsic) theories of aging. The stress response genes, signal transduction-redox pathways and aging. Ann. N.Y. Acad. Sci., 719: 195–211.

Park, S.H., Lee, S.J., Chung, H.Y., Kim, T.H., Cho, C.K., Yoo, S.Y. and Lee, Y.S. (2000) Inducible heat-shock protein 70 is involved in the radioadaptive response. Radiat. Res., 153: 318–326.

Parmacek, M.S., Fox, J.H., Harrison, W.H., Garron, D.C. and Swenie, D. (1979) Effect of aging on brain respiration and carbohydrate metabolism of CBF1 mice. Gerontology, 25: 185–191.

Pavlik, A., Aneja, I.S., Lexa, J. and Al-Zoabi, B.A. (2003) Identification of cerebral neurons and glial cell types inducing heat shock protein Hsp70 following heat stress in the rat. Brain Res., 973(2): 179–189.

Qian, Y.-Z., Bernardo, N.L., Nayeem, M.A., Chelliah, J. and Kukreja, R.C. (1999) Induction of 72-kDa heat shock protein does not produce second window of ischemic preconditioning in rat heart. Am. J. Physiol. Heart Circ. Physiol., 276: H224–H234.

Radons, J. and Multhoff, G. (2005) Immunostimulatory functions of membrane-bound and exported heat shock protein 70. Exerc. Immunol. Rev., 11: 17–33.

Rao, D.V., Watson, K. and Jones, G.L. (1999) Age-related attenuation in the expression of the major heat shock proteins in human peripheral lymphocytes. Mech. Ageing Dev., 107: 105–118.

Reshef, A., Sperling, O. and Zoref-Shani, E. (1996) Preconditioning of primary rat neuronal cultures against ischemic injury: characterization of the 'time window of protection'. Brain Res., 741: 252–257.

Rico, A.I., Angel, S.O., Alonso, C. and Requena, J.M. (1999) Immunostimulatory properties of the Leishmania infantum heat shock proteins HSP70 and HSP83. Mol. Immunol., 36: 1131–1139.

Schiaffonati, L., Maroni, P., Bendinelli, P., Tiberio, L. and Piccoletti, R. (2001) Hyperthermia induces gene expression of heat shock protein 70 and phosphorylation of mitogen activated protein kinases in the rat cerebellum. Neurosci. Lett., 312: 75–78.

Semenza, J.C., Rubin, C.H., Falter, K.H., Selanikio, J.D., Flanders, W.D., Howe, H.L. and Wilhelm, J.L. (1996) Heat-related deaths during the July 1995 heat wave in Chicago. N. Engl. J. Med., 335: 84–90.

Shamovsky, I., Ivannikov, M., Kandel, E.S., Gershon, D. and Nudler, E. (2006) RNA-mediated response to heat shock in mammalian cells. Nature, 440: 556–560.

Volloch, V. and Rits, S. (1999) A natural extracellular factor that induces HSP72, inhibits apoptosis and restores stress resistance in aged human cells. Exp. Cell Res., 253: 483–492.

Walters, T.J., Ryan, K.L. and Mason, P.A. (2001) Regional distribution of Hsp70 in the CNS of young and old food-restricted rats following hyperthermia. Brain Res. Bull., 55: 367–374.

Walters, T.J., Ryan, K.L., Tehrany, M.R., Jones, M.B., Paulus, L.A. and Mason, P.A. (1998) HSP70 expression in the CNS in response to exercise and heat stress in rats. J. Appl. Physiol., 84: 1269–1277.

Wang, Y., Whittall, T., McGowan, E., Younson, J., Kelly, C., Bergmeier, L.A., Singh, M. and Lehner, T. (2005) Identification of stimulating and inhibitory epitopes within the heat shock protein 70 molecule that modulate cytokine production and maturation of dendritic cells. J. Immunol., 174: 3306–3316.

Weiss, Y.G., Maloyan, A., Tazelaar, J., Raj, N. and Deutschman, C.S. (2002) Adenoviral transfer of HSP-70 into pulmonary epithelium ameliorates experimental acute respiratory distress syndrome. J. Clin. Invest., 110: 801–806.

Welch, W.J. (2001) Heat shock proteins as biomarkers for stroke and trauma. Am. J. Med., 111: 669–670.

Westman, J. and Sharma, H.S. (1998) Heat shock protein response in the central nervous system following hyperthermia. Prog. Brain Res., 1215: 207–239.

Wu, B., Gu, M.J., Heydari, A.R. and Richardson, A. (1993) The effect of age on the synthesis of two heat shock proteins in the hsp70 family. J. Gerontol., 48: B50–B56.

Yang, Y.-L., Lu, K.-T., Tsay, H.-J., Lin, C.-H. and Lin, M.-T. (1998) Heat shock protein expression protects against death following exposure to heatstroke in rats. Neurosci. Lett., 252: 9–12.

Yenari, M.A. (2002) Heat shock proteins and neuroprotection. Adv. Exp. Med. Biol., 513: 281–299.

SECTION IX

Hyperthermia and Cerebrospinal Fluid

CHAPTER 22

Changes in CSF composition during heat stress and fever in conscious rabbits

Maria Frosini*

Dipartimento di Scienze Biomediche, Sezione di Farmacologia, Fisiologia e Tossicologia Università di Siena, Polo Scientifico di S. Miniato viale A. Moro 2, lotto C 53100 Siena, Italy

Abstract: Elevation of brain temperature after stroke can lead to severe brain injury and even a moderate hyperthermia correlates with increased nervous damage. The role of endogenous cryogens in the pathways that down-regulate body temperature are of overwhelming interest in view of their effectiveness in protecting brain from such damage. The aim of the present work was to study whether heat stress (HS) or fever generates brain homeostatic responses aimed at counteracting the resulting rise in body temperature. Conscious rabbits, with cannulas chronically implanted in the *cisterna magna* and lateral ventricle, underwent HS (50 min, 40°C) or were injected with 25 ng of endogenous pyrogen IL-1β, while cerebrospinal fluid (CSF) levels of amino acids involved in central mechanisms of thermoregulation like taurine, GABA, aspartate and glutamate were monitored. The concentrations of some CSF cations (Na^+, K^+, Mg^{2+} and Ca^{2+}) were also determined in view of their purported role (sodium and calcium in particular) in establishing the thermal set point within the hypothalamus. Results show that during HS-induced hyperthermia, CSF taurine and GABA levels were significantly increased. On the contrary, IL-1β caused an increase in CSF taurine and, concurrently, a decrease in CSF GABA. Aspartate and glutamate did not change in both conditions. Furthermore, among CSF cations, only calcium and sodium underwent changes. In particular, calcium content increased both in HS- and febrile-animals, while CSF sodium decreased significantly only under IL-1β-injected treatment. In conclusion, GABA and taurine contribute as endogenous cryogens in a different fashion to the central mechanisms, which regulate dissipation of body heat in hyperthermia or heat production in fever, possibly in coordination with extracellular calcium and sodium.

Keywords: heat stress; fever; IL-1β; taurine; GABA; cerebrospinal fluid (or CSF) calcium concentration; cerebrospinal fluid (or CSF) sodium concentration

Introduction

Hyperthermia is a common issue in neurological intensive care unit and, independently of its etiology, is associated with a poor neurological outcome in stroke, spontaneous subarachnoid hemorrhage and traumatic brain injury (Geffroy et al., 2004). The role of endogenous cryogens in the pathways that down-regulate body temperature are of overwhelming interest in view of their possible effectiveness in protecting the brain from hyperthermia-induced damage. GABA and taurine have received increased attention with regard to their ability to lower body temperature in mammals. Taurine, in fact, when injected intracerebroventricularly (i.c.v.),

*Corresponding author. Tel.: +39 0577 234441; Fax: +39 0577 234446; E-mail: frosinim@unisi.it

DOI: 10.1016/S0079-6123(06)62022-0

induces a dose-related hypothermia accompanied by depression of gross motor behavior and peripheral vasodilation (Sgaragli et al., 1981, Frosini et al., 2003b), whereas the taurine antagonist 6-aminomethyl-3-methyl-4H-1,2,4-benzothiadiazine-1,1-dioxide (TAG) increases core temperature (Frosini et al., 2003b). Central or systemic injection of GABA, $GABA_A$ or $GABA_B$ receptors agonists diminishes core temperature, whereas the injection of antagonists of either class of receptors induces hyperthermia (Sancibrian et al., 1991; Frosini et al., 2004).

In the present work, recent results concerning the possibility that hyperthermia or fever generate brain homeostatic responses aimed at counteracting the resulting rise in body temperature that involves these two amino acids are presented and discussed. In addition, since aspartate and glutamate are excitatory neurotransmitters in the cerebral cortex, the area of the CNS that participate in the control of heat production (Monda et al., 1998), their possible involvement was investigated.

The contribution of some cations (Na^+, K^+, Mg^{2+} and Ca^{2+}) was also considered in view of their purported importance (sodium and calcium in particular) in the establishment of the thermal set point within the hypothalamus (Myers and Veale, 1970). The study was carried out on conscious rabbits implanted with cannulas in the *cisterna magna* (for withdrawal of cerebrospinal fluid (CSF) samples) and lateral ventricle (for intracerebroventricular injection), according to the methods described by Frosini et al. (2000, 2004). It was assumed that changes in CSF amino acid and cation concentrations reflected modifications in the brain extracellular milieu.

Experimental models for induction of fever and hyperthermia

Fever is a regulated elevation in the preoptic temperature set point in response to blood pyrogenic cytokines, where heat production and heat loss are still balanced and the temperature does not exceed the upper limit that in humans is ~42°C (Saper and Breder, 1994). In contrast, such balance is missed in hyperthermia in which there is no resetting of the set point and the thermoregulatory mechanisms are overwhelmed by excessive heat gain subsequent to insufficient heat dissipation (Saper and Breder, 1994). Despite their physiologic differences, hyperthermia and fever cannot be differentiated clinically on the basis of the height of the temperature or its pattern (Simon, 1993). The experimental procedure widely used for inducing fever in many animal species is the injection of pyrogenic, proinflammatory cytokines such as Interleukin-1β (IL-1β) (for a review, see Luheshi, 1998; Rothwell and Luheshi, 2000). Fever induced by central administration of IL-1β is almost indistinguishable from that induced by intravenous *E. coli* endotoxin and was inhibited by antipyretics (Palmi et al., 1992, 1994). Consequently, in the present study, fever was caused by injecting i.c.v. 25 ng of human recombinant IL-1β.

Hyperthermia can be induced by the injection of several drugs such as amphetamines, atropine, cocaine and calcium antagonists (Palmi and Sgaragli, 1989; Damanhouri and Tayeb, 1992), but it is accompanied by other drugs-mediated side effects that may interfere with the regular process of the increase of body temperature (Damanhouri and Tayeb, 1992) and, for this reason, it is not a recommended approach to induce hyperthermia. Exposure to high ambient temperature (heat stress, HS) causes an increase in body temperature by interfering with heat-dissipating mechanisms and is considered a suitable, simple method. Prolonged exposure to high ambient temperature, however, can lead to heat stroke that in rabbit is characterized by deep hyperthermia with rectal temperatures ≥43°C, loss of sensation, decreased muscle tone, unconsciousness and coma (Shih et al., 1984). It has been reported that cerebral ischemia is the main cause for the onset of the heat stroke syndrome (Lin and Lin, 1992). The mean exposure time at 40°C ambient temperature for the onset of heat stroke was found to be, in the rabbit, ~90 min (Lin and Lin, 1992). However, we have demonstrated that, in the rabbit, the exposure to 40°C for 50 min is a period sufficient to elicit a significant increase in rectal temperature without inducing heat stroke (Frosini et al., 2000). This experimental protocol can be successfully employed to investigate the effects of HS-induced

Table 1. Summary of the effects elicited by i.c.v. injection of IL-1β or heat stress on RT in conscious rabbits

Treatment	Basal RT	ΔRT_{max} at (min)	ΔRT_{mean} (125–375 min)
Vehicle ($n = 6$) or control ($n = 6$)	39.0 ± 0.1	0.5 ± 0.06 (375)	0.3 ± 0.04
Heat stress ($n = 5$)[a]	39.0 ± 0.1	3.1 ± 0.22 (225)	1.42 ± 0.31
IL-1β ($n = 4$)	38.8 ± 0.1	1.4 ± 0.25 (250)	0.9 ± 0.16

Note: Data are reported as mean ± s.e.mean. Basal Rectal Temperature (RT) refers to the mean of RT values (°C) recorded during 100 min interval period at neutral temperature ($T_a = 20°C$). Heat stress (HS) was induced by exposing the rabbits at 40°C for 50 min. IL-1β was injected i.c.v. at the dose of 25 ng and vehicle consisted in pyrogen-free water. ΔRT_{max} represents the peak of hyperthermia or fever at the corresponding time (min). ΔRT_{mean} is the mean of ΔRT changes recorded during 275 min interval after the induction of HS of fever.
[a] From Frosini et al. (2000), with permission.

hyperthermia in conscious rabbits. A summary of the effects elicited by i.c.v. injection of IL-1β or HS on rectal temperature in conscious rabbits are described in Table 1.

Effect of heat-stress-induced hyperthermia on CSF taurine, GABA, aspartate and glutamate contents

The aim of the present work was to ascertain whether in HS are still operating homeostatic responses intended at counteracting the resulting rise in body temperature, with the involvement of GABA and taurine as endogenous cryogens. There is strong evidence, in fact, that these two amino acids participate in the neuronal network that down-regulates body temperature (DeFeudis, 1984; Pierau et al., 1997; Frosini et al., 2000, 2003a, 2004). Moreover, recent data from this laboratory have demonstrated the existence of a specific, central taurinergic pathway in the rabbit, which runs in parallel with GABA-ergic system(s) (Frosini et al., 2003b).

During HS brain taurine and GABA metabolism are modified so that an increase in their CSF concentrations is induced (Fig. 1). Interestingly, Sharma (2006) reported that in a rat model of HS, a 4–6-fold increase in cerebral cortex and brain stem glycine and GABA contents was observed. CSF taurine and GABA rise, however, cannot be ascribed to blood contamination resulting from a transient opening of the blood-CSF barrier, as exhaustively discussed in a previous paper (Frosini et al., 2000). It is conceivable that the increased output of taurine and GABA from the brain into CSF is aimed at counteracting the hyperthermia promoted by exposure to heat. This becomes evident when the time-course for HS-induced changes in rectal temperature (RT) and that of CSF taurine and GABA contents are compared, since RT decreased as soon as taurine and GABA CSF contents rose (Frosini et al., 2000). The increase of CSF taurine and GABA in HS rabbits, however, can be important not only in view of the ability of these amino acids to counteract the rise in body temperature, but also for their neuroprotective properties. In HS, in fact, hyperthermia is accompanied by cerebral vascular congestion and edema, giving rise to intracranial hypertension thus contributing to a reduction in cerebral perfusion pressure, responsible for cerebral ischemia and neuronal damage, which is common features in heat stroke (Lin, 1997; Hales and Sakurada, 1998). The increase of GABA-ergic transmission, which leads to increased CSF GABA content, can antagonize excessive glutamatergic excitation that occurs during brain ischemia concurrently promoting hypothermia that is "*per se*" neuroprotective, as shown in experimental as well as clinical settings (for a review, see Schwartz-Bloom and Sah, 2001). Moreover, also taurine has been demonstrated to be a neuroprotective agent, possibly acting through several different mechanisms that involve inhibition of neurotransmitter release and normalization of intracellular osmolality (O'Byrne and Tipton, 2000; Pessina et al., 2000; Hayes et al., 2001; Shuaib, 2003). Thus, the increased output of taurine and GABA during HS can be envisaged not only as an homeostatic response aimed at counteracting the rise in body temperature, but also as a potential neuroprotective strategy to prevent neuronal injury, especially when the progression of HS to heat stroke occurs.

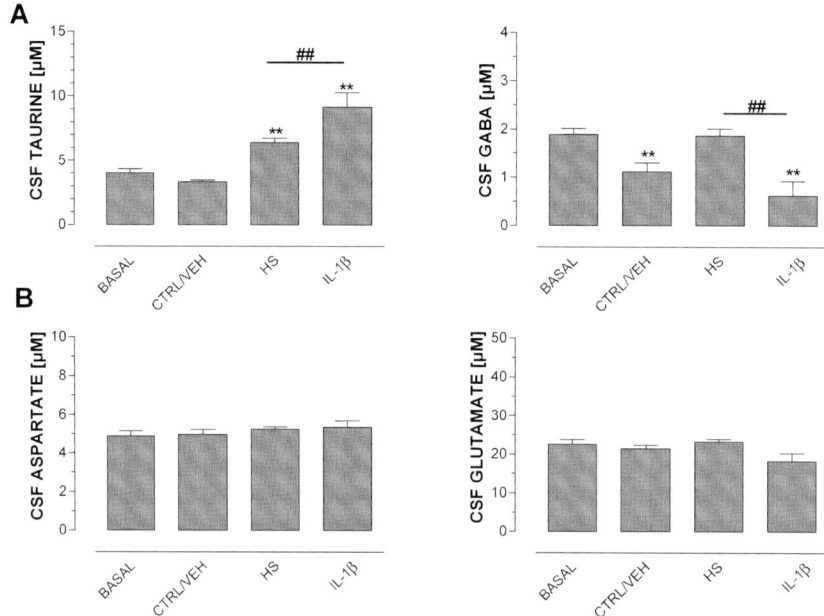

Fig. 1. Effects of heat stress (HS) or i.c.v. injection of IL-1β (25 ng) on cerebrospinal fluid (CSF) taurine, GABA (panel A), aspartate and glutamate (panel B) contents. *Basal:* CSF samples collected before HS or i.c.v. injection. *Vehicle/control:* CSF samples of vehicle-rabbits (i.e., animals injected i.c.v. with pyrogen-free water, $n = 6$) or controls (animals kept at ambient temperature of 20°C, $n = 6$). *HS*: CSF samples of HS rabbits (animal exposed for 50 min at 40°C + 125 min at 20°C, $n = 5$). *IL-1β:* CSF samples of rabbits (injected i.c.v. with 25 ng of IL-1β, $n = 5$). Amino acid CSF content was determined by reversed-phase HPLC with o-phthalaldehyde pre-column derivatization (Bianchi et al., 1999). Values are reported as mean ± S.E.M. The statistical significance of differences between data relative to HS or IL-1β treatment vs. data observed during the same period in control- or vehicle-rabbits, respectively, was checked by analysis of variance (ANOVA). $*P<0.05$, $**P<0.01$. $^{\#\#}P<0.01$. Data relative to HS are adapted with permission from Frosini et al. (2000).

In the present study, HS-induced hyperthermia was not accompanied by modifications in CSF glutamate and aspartate contents (Fig. 1). Relatively little is known about the role of aspartate and glutamate in *in vivo* models of HS or heat stroke. Cremades and Peñafiel (1982) observed that exposure of infant rats (7- or 14-day-old) to elevated T_a (40°C for 90 min) induced hyperthermia accompanied by a significant increase and a significant decrease in glutamate and aspartate levels in brain tissues, respectively. On the contrary, in adult rats (21-day-old), HS induced a greater rise in body temperature, which was associated to slight changes in brain content of these amino acids. The authors concluded that the different outcome to heat exposure in infant or adult rats could be related to different maturation of the blood–brain barrier at the two age groups. Adachi et al. (1995) observed in rat that localized brain hyperthermia (41°C) does not cause a change of glutamate extracellular levels, while either moderate (43°C) or severe (45°C) hyperthermia increases glutamate extracellular concentrations toward neurotoxic levels. Sharma (2006) reported that the exposure of rats to 4 h HS at 38°C resulted in a marked increase in glutamate and aspartate in some brain regions. These observations suggest that glutamate-mediated excitotoxicity might play an important role in hyperthermia-induced cellular injury to CNS, and that a balance between excitatory and inhibitory amino acid is crucial for hyperthermia-induced cell injury or cell survival (Sharma, 2006). However, since in the present study hyperthermia was not accompanied by modifications in CSF glutamate and aspartate contents, we can conclude that in this experimental model of HS, these two amino acids do not play a role.

Effect of IL-1-induced fever on CSF taurine, GABA, aspartate and glutamate contents

As reported in Fig. 1, fever caused an increase in CSF taurine, but, at variance with HS, it elicited a significant decrease of CSF GABA content. The central point for fever development is the up-regulation of the thermostatic set point in the prooptic area of the hypothalamus (POAH). Recently, the intracellular mechanisms at the POAH for fever production have begun to be clarified. IL-1β, in fact, decreases the firing rate of warm-sensitive (warm-activated) neurons, while increasing the firing rate of cold-sensitive (cold-activated) neurons, with no influence at all on the tonic activity of temperature-insensitive neurons (Shibata and Blatteis, 1991; Xin and Blatteis, 1992). However, fever-inducing substances appear to modulate not only the firing rate, but also the thermosensitivity of thermosensitive neurons. In fact, a shift of the neuronal thermal thresholds of thermosensitive neurones to high temperatures has been described during IL-1β-induced fever (Vasilenko et al., 2000); the threshold of warm- and cold-sensitive POAH neurons has been suggested to drive both the shift and the long-lasting maintenance at high values of body temperature. The present results indicate that GABA might be involved in such mechanisms. IL-1β affects many neurotransmitter systems in the brain, including GABA (De Simoni and Imeri, 1998). Some studies, in fact, have reported that IL-1β enhances (Miller et al., 1991; Yu and Shinnick-Gallagher, 1994; Luk et al., 1999) or depresses (Pringle et al., 1996; Wang et al., 2000) $GABA_A$ receptor function. IL-1β, either directly or via NO$^•$ generation, reduces GABAergic input on parvocellular neurones in the paraventricular nucleus of the rat hypothalamus, thus eliciting a depolarization of neuronal cells through disinhibition. These effects were blocked by bicuculline, suggesting the $GABA_A$ receptor involvement (Ferri and Ferguson, 2003). Feleder et al. (1998), however, showed that IL-1β reduced GABA and taurine release from the preoptic/mediobasal hypothalamic area of the rat. In the light of these observations, a cell-specific action of IL-1β has been proposed (Wang et al., 2000). IL-1β modified brain GABA metabolism so that a significant decrease in CSF concentrations of this amino acid is induced: this observation suggests that, in order to reset the hypothalamic set point toward higher value(s), a down-regulation of brain GABA metabolisms is required. However, it is interesting to outline the specific involvement of GABA but not taurine in this mechanism. CSF taurine, in fact, increased after i.c.v. injection of IL-1β, thus suggesting a different role of this amino acid during fever. Taurine, for example, can play a protective role against fever-associated events such as excitatory influences or electrolyte alterations. I.c.v. injection of taurine in rabbits, in fact, was shown to reduce the febrile response caused by *Salmonella typhosa* endotoxin, leukocytic pyrogen or prostaglandin E_2 (Lipton and Ticknor, 1979; Sgaragli et al., 1981).

In the present study IL-1β did not modify CSF aspartate and glutamate. At variance with this observation, fever induced by injection of PGE_1, PGE_2 or *E. coli* is accompanied by an increase of aspartate and glutamate levels in CSF or in microdialysate samples collected from the cortex of the rat (Perry et al., 1993; Malmberg et al., 1995; Monda et al., 1998), thus suggesting that these amino acids might be involved in the pathogenesis of fever in this animal species.

Effect of heat-stress-induced hyperthermia on CSF calcium, sodium, potassium and magnesium contents

HS is able to modify CSF composition by increasing calcium content (Fig. 2). Previous studies from this laboratory showed that when thermoregulation is set toward the promotion of heat dissipation, as it happens when animals develop hypothermia following i.c.v. injection of taurine, there is a significant and long-lasting increase in CSF calcium content (Sgaragli et al., 1994). Conversely, when thermoregulation is set toward the promotion of heat dissipation on exposure to high T_a, brain calcium metabolism changes give rise to an increase of CSF calcium concentration, which is accompanied by the increase in CSF taurine, as it was also observed in the present study. This effect was specific since CSF concentrations of sodium, magnesium and potassium were not modified (Fig. 2).

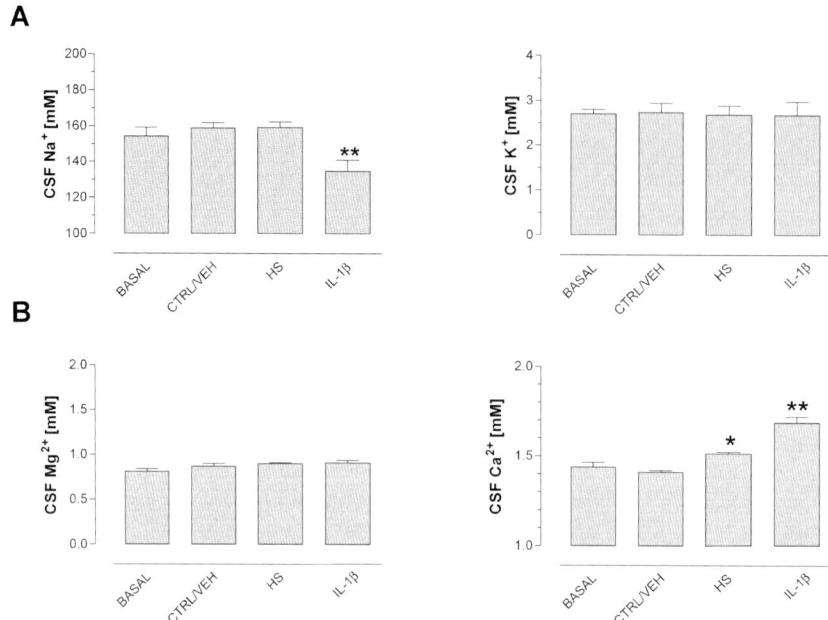

Fig. 2. Effects of heat stress (HS) or i.c.v. injection of IL-1β (25 ng) on cerebrospinal fluid (CSF) sodium, potassium (panel A), magnesium and calcium (panel B) contents. *Basal:* CSF samples collected before HS or i.c.v. injection. *Vehicle/control:* CSF samples of vehicle-rabbits (i.e., animals injected i.c.v. with pyrogen-free water, $n = 6$) or controls (animals kept at ambient temperature of 20°C, $n = 6$). *HS*: CSF samples of HS-rabbits (animal exposed for 50 min at 40°C + 125 min at 20°C, $n = 5$). *IL-1β:* CSF samples of rabbits (injected i.c.v. with 25 ng of IL-1β, $n = 5$). Cation CSF content was determined by an HPLC-conductimetric detection method according to Frosini et al. (1993). Values are reported as mean ± S.E.M. The statistical significance of differences between data relative to HS or IL-1β treatment vs. data observed during the same period in control- or vehicle-rabbits, respectively, was checked by analysis of variance (ANOVA). *$P<0.05$, **$P<0.01$. ##$P<0.01$. Data relative to HS are adapted with permission from Frosini et al. (2000).

We can hypothesize that during hyperthermia, the rise in CSF calcium concentration may determine the activation of thermoeffectors, as advocated by previous studies (Myers and Veale, 1970). In this activation, taurine might have an important role. This hypothesis, although attractive, needs further experimental data to demonstrate the effective cause–effect relationship between calcium and taurine increase in CSF on one side, and the activation of thermoeffectors on the other side, thus excluding a mere temporal parallelism of the two phenomena.

Effect of IL-1-induced fever on CSF calcium, sodium, potassium and magnesium contents

IL-1β-promoted fever was accompanied by CSF calcium content increase, while CSF sodium decreased significantly. Interestingly, Stanton et al (1986) observed that IL-1β induces a marked and prolonged increase in intracellular sodium via stimulation of the Na+/H+ exchanger and a transient fall in total intracellular calcium in 70Z/3 cells.

Previous work, from this laboratory, on the mechanisms of fever production has shown how the increase in CSF calcium was linearly and positively correlated to the degree of temperature gain (Palmi et al., 1992, 1994) and that other pyrogens in addiction to IL-1β, caused CSF calcium increase; acetylsalicylic acid and dexamethasone antagonized this effect as well as the increase in body temperature (Palmi et al., 1992, 1994). The blood-brain barrier allows constant levels of calcium within the brain and CSF to be maintained, even during conditions of plasma hyper- and hypocalcemia (Murphy et al., 1986, 1989). This condition of ionic balance may, however, be perturbed during fever; consequently, IL-1β might enhance the access of blood calcium to brain, thereby contributing to the rise in CSF. This hypothesis, however, seems unlikely since it has been demonstrated that this cytokine stimulates calcium release from striatal brain slices (Palmi et al., 1994), and by cultured hypothalamic GABA-ergic neurones (De et al., 2002) by a specific receptor-mediated mechanism. Furthermore, the NO$^{\bullet}$/cGMP-signaling pathway is part of the mechanisms transducing IL-1β-evoked calcium mobilization from inositol-(1,4,5)-trisphosphate-sensitive calcium stores in rat striatal brain slices as well as in human astrocytoma U-373 MG cells (Meini et al., 2000; Palmi and Meini, 2002). All together these findings indicate that the extra amount of calcium in CSF is pumped out by nervous cells, thus corroborating its role in thermoregulation (Myers and Veale, 1970; Palmi and Sgaragli, 1989). It has been proposed, in fact, that the set point of body temperature in mammals coincides with the concentration ratio of calcium and sodium in the extracellular fluid within the posterior hypothalamus. The calcium/sodium ratio seems to act by modulating the firing frequency of a distinct population of cells, thus determining the "set point" for body temperature (Jones et al., 1980).

Concluding remarks

The aim of the present work was to study whether HS or fever generate brain homeostatic responses aimed at counteracting the resulting rise in body temperature. GABA and taurine contribute as endogenous cryogens in a different fashion to the central mechanisms, which regulate dissipation of body heat in hyperthermia or heat production in fever, possibly in coordination with extracellular calcium and sodium, as summarized in Fig. 3. The pharmacological manipulation of GABA-ergic and/or taurinergic system is susceptible of therapeutic exploitation in brain diseases such as stroke, spontaneous subarachnoid hemorrhage and traumatic brain injury in which hyperthermia is associated with a poor neurological outcome; this is the premise for the development of new drugs, able

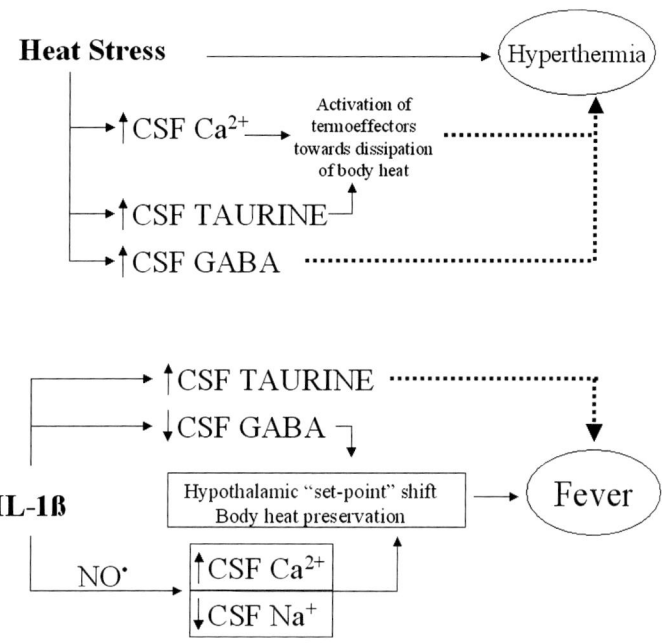

Fig. 3. Possible relationships in mammals between cerebrospinal fluid (CSF) GABA, taurine, calcium and sodium contents and heat stress- or IL-1β-induced hyperthermia and fever, respectively. Solid line: activation. Dotted line: inhibition or counteraction. ↓: decrease. ↑: increase.

to induce hypothermia rapidly and safely, to be used in the emergency care medicine.

Acknowledgments

This work has been performed with the financial support of the University of Siena, Piano di Ateneo per la Ricerca (PAR). The valuable contribution of Loria Bianchi and Beatrice Gorelli in HPLC determinations is gratefully acknowledged. The author thanks Prof. Giampietro Sgaragli for the critical reading of the manuscript, and the several, fruitful discussions on this topic.

References

Adachi, H., Fujisawa, H., Maekawa, T., Yamashita, T. and Ito, H. (1995) Changes in extracellular glutamate concentrations in the rat cortex following localised hyperthermia. Int. J. Hyperthermia, 11: 587–599.

Bianchi, L., Della Corte, L. and Tipton, K.F. (1999) Simultaneous determination of basal and evoked output levels of aspartate, glutamate, taurine and 4-aminobutyric acid during microdialysis and from superfused brain slices. J. Chromatogr. B Biomed. Sci. Appl., 723: 47–59.

Cremades, A. and Peñafiel, R. (1982) Hyperthermia and brain neurotransmitter amino acid levels in infant rats. Gen. Pharmacol., 13: 347–350.

Damanhouri, Z.A. and Tayeb, O.S. (1992) Animal models for heat stroke studies. J. Pharmacol. Toxicol. Methods, 28: 119–127.

De, A., Churchill, L., Obal Jr., F., Simasko, S.M. and Krueger, J.M. (2002) GHRH and IL1beta increase cytoplasmic Ca(2+) levels in cultured hypothalamic GABAergic neurons. Brain Res., 949: 209–212.

De Simoni, M.G. and Imeri, L. (1998) Cytokine-neurotransmitter interactions in the brain. Biol. Signals Recept., 7: 33–44.

DeFeudis, F.V. (1984) Involvement of GABA and other inhibitory amino acids in thermoregulation. Gen. Pharmacol., 15: 445–447.

Feleder, C., Refojo, D., Nacht, S. and Moguilevsky, J.A. (1998) Interleukin-1 stimulates hypothalamic inhibitory amino acid neurotransmitter release. Neuroimmunomodulation, 5: 1–4.

Ferri, C.C. and Ferguson, A.V. (2003) Interleukin-1 beta depolarizes paraventricular nucleus parvocellular neurones. J. Neuroendocrinol., 15: 126–133.

Frosini, M., Gorelli, B., Matteini, M., Palmi, M., Valoti, M. and Sgaragli, G.P. (1993) HPLC determination of inorganic cation levels in CSF and plasma of conscious rabbits. J. Pharmacol. Toxicol. Methods, 29: 99–104.

Frosini, M., Sesti, C., Dragoni, S., Valoti, M., Palmi, M., Dixon, H.B., Machetti, F. and Sgaragli, G.P. (2003a) Interactions of taurine and structurally related analogues with the GABAergic system and taurine binding sites of rabbit brain. Br. J. Pharmacol., 138: 1163–1171.

Frosini, M., Sesti, C., Palmi, M., Valoti, M., Fusi, F., Mantovani, P., Bianchi, L., Della Corte, L. and Sgaragli, G.P. (2000) Heat-stress-induced hyperthermia alters CSF osmolality and composition in conscious rabbits. Am. J. Physiol. Regul. Integr. Comp. Physiol., 279: R2095–R2103.

Frosini, M., Sesti, C., Saponara, S., Ricci, L., Valoti, M., Palmi, M., Machetti, F. and Sgaragli, G.P. (2003b) A specific taurine recognition site in the rabbit brain is responsible for taurine effects on thermoregulation. Br. J. Pharmacol., 139: 487–494.

Frosini, M., Valoti, M. and Sgaragli, G.P. (2004) Changes in rectal temperature and ECoG spectral power of sensorimotor cortex elicited in conscious rabbits by i.c.v. injection of GABA, GABA(A) and GABA(B) agonists and antagonists. Br. J. Pharmacol., 141: 152–162.

Geffroy, A., Bronchard, R., Merckx, P., Seince, P.F., Faillot, T., Albaladejo, P. and Marty, J. (2004) Severe traumatic head injury in adults: which patients are at risk of early hyperthermia? Intensive Care Med., 30: 785–790.

Hales, J.R. and Sakurada, S. (1998) Heat tolerance. A role for fever? Ann. N.Y. Acad. Sci., 856: 188–205.

Hayes, J., Tipton, K.F., Bianchi, L. and Della Corte, L. (2001) Complexities in the neurotoxic actions of 6-hydroxydopamine in relation to the cytoprotective properties of taurine. Brain Res. Bull., 55: 239–245.

Jones, D.L., Veale, W.L. and Cooper, K.E. (1980) Alterations in body temperature elicited by intrahypothalamic administration of tetrodotoxin, ouabain and A23187 ionophore in the conscious cat. Brain Res. Bull., 5: 75–80.

Lin, M.T. (1997) Heatstroke-induced cerebral ischemia and neuronal damage. Involvement of cytokines and monoamines. Ann. N.Y. Acad. Sci., 813: 572–580.

Lin, M.T. and Lin, S.Z. (1992) Cerebral ischemia is the main cause for the onset of heat stroke syndrome in rabbits. Experientia, 48: 225–227.

Lipton, J.M. and Ticknor, C.B. (1979) Central effect of taurine and its analogues on fever caused by intravenous leukocytic pyrogen in the rabbit. J. Physiol. (Lond.), 287: 535–543.

Luheshi, G.N. (1998) Cytokines and fever. Mechanisms and sites of action. Ann. N.Y. Acad. Sci., 856: 83–89.

Luk, W.P., Zhang, Y., White, T.D., Lue, F.A., Wu, C., Jiang, C.G., Zhang, L. and Moldofsky, H. (1999) Adenosine: a mediator of interleukin-1beta-induced hippocampal synaptic inhibition. J. Neurosci., 19: 4238–4244.

Malmberg, A.B., Hamberger, A. and Hedner, T. (1995) Effects of prostaglandin E_2 and capsaicin on behaviour and cerebrospinal fluid amino acids concentrations of unanesthetised rats: a microdialysis study. J. Neurochem., 65: 2185–2193.

Meini, A., Benocci, A., Frosini, M., Sgaragli, G., Pessina, G., Aldinucci, C., Youmbi, G.T. and Palmi, M. (2000) Nitric oxide modulation of interleukin-1[beta]-evoked intracellular Ca^{2+} release in human astrocytoma U-373 MG cells and brain striatal slices. J. Neurosci., 20: 8980–8986.

Miller, L.G., Galpern, W.R., Dunlap, K., Dinarello, C.A. and Turner, T.J. (1991) Interleukin-1 augments gamma-aminobutyric acidA receptor function in brain. Mol. Pharmacol., 39: 105–108.

Monda, M., Viaggiano, A., Sullo, A. and De Luca, V. (1998) Aspartic and glutamic acids increase in the frontal cortex during prostaglandin E_1 hyperthermia. Neuroscience, 83(4): 1239–1243.

Murphy, V.A., Smith, Q.R. and Rapoport, S.I. (1986) Homeostasis of brain and cerebrospinal fluid calcium concentrations during chronic hypo- and hypercalcemia. J. Neurochem., 47: 1735–1741.

Murphy, V.A., Smith, Q.R. and Rapoport, S.I. (1989) Uptake and concentrations of calcium in rat choroid plexus during chronic hypo- and hypercalcemia. Brain Res., 484: 65–70.

Myers, R.D. and Veale, W.L. (1970) Body temperature: possible ionic mechanism in the hypothalamus controlling the set-point. Science, 170: 95–97.

O'Byrne, M.B. and Tipton, K.F. (2000) Taurine-induced attenuation of MPP+ neurotoxicity in vitro: a possible role for the GABA(A) subclass of GABA receptors. J. Neurochem., 74: 2087–2093.

Palmi, M., Frosini, M., Becherucci, C., Sgaragli, G.P. and Parente, L. (1994) Increase of extracellular brain calcium involved in interleukin-1 beta-induced pyresis in the rabbit: antagonism by dexamethasone. Br. J. Pharmacol., 112: 449–452.

Palmi, M., Frosini, M. and Sgaragli, G.P. (1992) Calcium changes in rabbit CSF during endotoxin, IL-1 beta, and PGE2 fever. Pharmacol. Biochem. Behav., 43: 1253–1262.

Palmi, M. and Meini, A. (2002) Role of the nitric oxide/cyclic GMP/Ca^{2+} signaling pathway in the pyrogenic effect of interleukin-1beta. Mol. Neurobiol., 25: 133–147.

Palmi, M. and Sgaragli, G.P. (1989) Hyperthermia induced in rabbits by organic calcium antagonists. Pharmacol. Biochem. Behav., 34: 325–330.

Perry, V.L., Young, R.S.K., Aquila, W.J. and During, M.J. (1993) Effect of experimental *Escherichia coli* meningitis on concentrations of excitatory and inhibitory amino acids in the rabbit brain: in vivo microdialysis study. Pediatr. Res., 34(2): 187–191.

Pessina, F., Matteucci, G., Esposito, L., Gorelli, B., Valoti, M. and Sgaragli, G.P. (2000) Protection of intrinsic nerves of guinea-pig detrusor strips against anoxia/glucopenia and reperfusion injury by taurine. Adv. Exp. Med. Biol., 483: 325–333.

Pierau, F.K., Yakimova, K.S., Sann, H. and Schmid, H.A. (1997) Specific action of GABAB ligands on the temperature sensitivity of hypothalamic neurons. Ann. N.Y. Acad. Sci., 813: 146–155.

Pringle, A.K., Gardner, C.R. and Walker, R.J. (1996) Reduction of cerebellar GABAA responses by interleukin-1 (IL-1) through an indomethacin insensitive mechanism. Neuropharmacology, 35: 147–152.

Rothwell, N.J. and Luheshi, G.N. (2000) Interleukin 1 in the brain: biology, pathology and therapeutic target. Trends Neurosci., 23: 618–625.

Sancibrian, M., Serrano, J.S. and Minano, F.J. (1991) Opioid and prostaglandin mechanisms involved in the effects of GABAergic drugs on body temperature. Gen. Pharmacol., 22: 259–262.

Saper, C.B. and Breder, C.D. (1994) The neurologic basis of fever. N. Engl. J. Med., 330: 1880–1886.

Schwartz-Bloom, R.D. and Sah, R. (2001) Gamma-aminobutyric acid(A) neurotransmission and cerebral ischemia. J. Neurochem., 77: 353–371.

Sgaragli, G., Carla, V., Magnani, M. and Galli, A. (1981) Hypothermia induced in rabbits by intracerebroventricular taurine: specificity and relationships with central serotonin (5-HT) systems. J. Pharmacol. Exp. Ther., 219: 778–785.

Sgaragli, G.P., Frosini, M., Palmi, M., Bianchi, L. and Della Corte, L. (1994) Calcium and taurine interaction in mammalian brain metabolism. Adv. Exp. Med. Biol., 359: 299–308.

Sharma, H.S. (2006) Hyperthermia influences excitatory and inhibitory amino acid neurotransmitters in the central nervous system. An experimental study in the rat using behavioural, biochemical, pharmacological and morphological approaches. J. Neural Transm., 113: 497–519.

Shibata, M. and Blatteis, C.M. (1991) Differential effects of cytokines on thermosensitive neurons in guinea pig preoptic area slices. Am. J. Physiol., 261: R1096–R1103.

Shih, C.-J., Lin, M.-T. and Tsai, S.-H. (1984) Experimental study on the pathogenesis of heat stroke. J. Neurosurg., 60: 1246–1252.

Shuaib, A. (2003) The role of taurine in cerebral ischemia: studies in transient forebrain ischemia and embolic focal ischemia in rodents. Adv. Exp. Med. Biol., 526: 421–431.

Simon, H.B. (1993) Hyperthermia. N. Engl. J. Med., 329: 483–487.

Stanton, T.H., Maynard, M. and Bomsztyk, K. (1986) Effect of interleukin-1 on intracellular concentration of sodium, calcium, and potassium in 70Z/3 cells. J. Biol. Chem., 261: 5699–5701.

Vasilenko, V.Y., Petruchuk, T.A., Gourine, V.N. and Pierau, F.K. (2000) Interleukin-1beta reduces temperature sensitivity but elevates thermal thresholds in different populations of warm-sensitive hypothalamic neurons in rat brain slices. Neurosci. Lett., 292: 207–210.

Wang, S., Cheng, Q., Malik, S. and Yang, J. (2000) Interleukin-1beta inhibits gamma-aminobutyric acid type A (GABA(A)) receptor current in cultured hippocampal neurons. J. Pharmacol. Exp. Ther., 292: 497–504.

Xin, L. and Blatteis, C.M. (1992) Blockade by interleukin-1 receptor antagonist of IL-1 beta-induced neuronal activity in guinea pig preoptic area slices. Brain Res., 569: 348–352.

Yu, B. and Shinnick-Gallagher, P. (1994) Interleukin-1 beta inhibits synaptic transmission and induces membrane hyperpolarization in amygdala neurons. J. Pharmacol. Exp. Ther., 271: 590–600.

CHAPTER 23

Blood–cerebrospinal fluid barrier in hyperthermia

Hari Shanker Sharma[1] and Conrad Earl Johanson[2],*

[1]*Laboratory of Cerebrovascular Research, Institute of Surgical Sciences, Department of Anaesthesiology and Intensive Care, University Hospital, Uppsala University, SE-75185 Uppsala, Sweden*
[2]*Department of Neurosurgery, Brown Medical School, Rhode Island Hospital, 593 Eddy Street, Providence, RI 02903, USA*

Abstract: The blood–CSF barrier (BCSFB) in choroid plexus works with the blood–brain barrier (BBB) in cerebral capillaries to stabilize the fluid environment of neurons. Dysfunction of either transport interface, i.e., BCSFB or BBB, causes augmented fluxes of ions, water and proteins into the CNS. These barrier disruptions lead to problems with edema and other compromised homeostatic mechanisms. Hyperthermic effects on BCSFB permeability and transport are not as well known as for BBB. However, it is becoming increasingly appreciated that elevated prostaglandin synthesis from fever/heat activation of cyclooxygenases (COXs) in the BCSFB promotes water and ion transfer from plasma to the ventricles; this harmful fluid movement into the CSF-brain interior can be attenuated by agents that inhibit the COXs. Moreover, new functional data from our laboratory animal model indicate that the BCSFB (choroidal epithelium) and the CSF-bordering ependymal cells are vulnerable to whole body hyperthermia (WBH). This is evidenced from the fact that rats subjected to 4 h of heat stress (38°C) showed a significant increase in the translocation of Evans blue and [131]Iodine from plasma to cisternal CSF, and manifested blue staining of the dorsal surface of the hippocampus and caudate nucleus. Degeneration of choroidal epithelial cells and underlying ependyma, a dilated ventricular space and damage to the underlying neuropil were frequent. A disrupted BCSFB is associated with a marked increase in edema formation in the hippocampus, caudate nucleus, thalamus and hypothalamus. Taken together, these findings suggest that the breaching of the BCSFB in hyperthermia significantly contributes to cell and tissue injuries in the CNS.

Keywords: heat stress; whole body hyperthermia; fever; blood–cerebrospinal fluid barrier; choroid plexus; brain edema; neurodegeneration; Evans blue; [131]Iodine transport into CSF; prostaglandins; cyclooxygenases; blood–brain barrier; pyrogens

An overview of the CNS barriers or transport interfaces

Optimal neuronal functioning would be impossible without exquisitely regulated transport of water-soluble substances into and out of the brain and cerebrospinal fluid (CSF). Transport regulation occurs at multiple interfaces, each highly specialized to carry out specific functions essential for brain metabolic and fluid homeostasis (Spector and Johanson, 1989; Jones et al., 1992; Saunders et al., 1999; Segal 2000; Chodobski and Szmydynger-Chodobska, 2001; Ghersi-Egea et al., 2001; Preston, 2001; Johanson, 2003; Angelow et al., 2004; Redzic et al., 2005; Hawkins and Davis, 2005).

*Corresponding author. Tel.: +1 (401) 444 8739; Fax: +1 (401) 444 8727; E-mail: Conrad_Johanson@Brown.edu

Transport interfaces, sometimes called 'barriers', can be broadly categorized as endothelial or epithelial in nature. Figure 1 depicts examples of interfaces between the blood and various regions of the central nervous system (CNS). In general, the endothelial interfaces regulate 'solute traffic' between the cerebrovascular system and underlying neuronal networks, whereas the epithelial interfaces control movement of substances between blood and CSF.

The term 'barrier' for certain endothelial and epithelial interfaces has been used for over a century. The concept of barrier implies restriction to the diffusion of hydrophilic molecules as small as urea with a molecular weight of 60 (Johanson and Woodbury, 1978). Most interfaces between blood and the vast majority of brain and cord regions (Sharma, 2005a, b, c) are true 'barriers', i.e., collectively known as the blood–brain barrier (BBB). The choroid plexuses also offer restriction to molecular diffusion (Parandoosh and Johanson, 1982), and so are called the blood–CSF barrier (BCSFB). However, other regions allow relatively unrestricted diffusion across their interfaces. Such non-barrier regions lacking tight junctions include the capillaries of the circumventricular organs (CVOs) as well as the ependymal lining that separates the brain interstitial fluid (ISF) from the large cavity CSF (Johanson, 2003). Permeable interfaces such as the ependyma and pia-glia contain the less restrictive gap junctions between cells rather than the 'tighter' intercellular occlusive junctions of the BBB and BCSFB.

What is the anatomical substrate for the 'barriers'? Tight junctions or *zonulae occludentes* (Brightman and Reese, 1969; Begley and Brightman, 2003) between the cerebral endothelial cells and choroid epithelial cells restrict free diffusion, thereby causing most water-soluble solutes to be sieved (screened) at these transport interfaces. Tight junctions are 'welds' that completely encircle the cells at the interfaces between blood and brain or CSF. Diffusing molecules are thus impeded, partially or fully (depending on size), by the tight junctions. Because diffusion is thus thwarted, the CNS is normally greatly protected against potentially harmful molecules in the bloodstream such as cytokines (Banks et al., 1995) and drug metabolites (Johanson et al., 2005a, b).

The presence of tight junctions between the endothelial and epithelial cells make the permeability characteristics of these barriers similar to those of an extended plasma membrane. Thus, the permeability properties of BBB or BCSFB are largely regulated through the transcellular route depending on the physicochemical properties of the cell membrane (Rapoport, 1976). Under normal conditions, lipid-soluble solutes can easily penetrate the plasma membrane at the BBB or BCSFB interface, whereas lipid insoluble non-electrolytes and proteins are severely restricted at the barrier sites. The BBB is a relatively strict regulatory

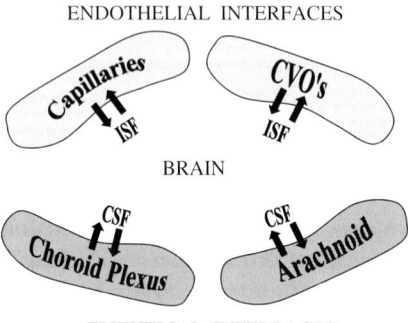

Fig. 1. Highly-idealized schema for several transport interfaces that encase the brain and CSF compartments: endothelial and epithelial interfaces that serve as barriers for the regulated entry and removal of substances from the brain and CSF. The endothelium of most brain capillaries has tight junctions between cells and is therefore relatively impervious to most water-soluble molecules. On the other hand, the capillaries in the circumventricular organs (CVOs), such as the subfornical organ in the 3rd ventricle region (Johanson, 2003), are permeable and thus allow fluid-regulating peptides such as angiotensin II to diffuse from blood to the parenchymal neuroendocrine cells within the CVO. Another major group of barrier-type cells are the choroidal and arachnoidal epithelium of the blood–CSF barrier. Although there is secretory (arrows directed into the CSF-CNS) and reabsorptive transport (arrows directed out of the CSF-CNS) at both of the blood–CSF barriers, the secretory phenomena are prominent at the choroid plexus whereas reabsorptive solute/water movements are the main function of the arachnoid membrane. This review emphasizes ion, water and protein fluxes at the choroid plexus and cerebral capillary interfaces in normothermia and hyperthermia. Little is known about the effects of hyperthermia and heat stress on the CVOs and the arachnoidal epithelium. ISF = interstitial fluid.

interface between blood and the CNS. However, the BCBSFB appears to be less stringent because of a special need for facilitated diffusion, aqueous secretion and extensive active ion transport across the blood–CSF interfaces. Accordingly, at the BCSFB interface, intravascular substances enter into the CSF at faster rates compared to the BBB site (Davson, 1967).

In certain diseases and other perturbations of the brain (Palm et al., 1995; Lu et al., 2004), the tight junctions and basal lamina are damaged, thereby reducing the barrier protection (Zipser et al., 2006). As the permeability of the BBB is increased (Brown and Davis, 2002), there can be disrupted homeostasis that normally ensures stability of the fluid surrounding the neurons. Consequently there are compelling reasons for studying the breakdown and possible repair of the BBB (Johanson et al., 2005).

A comparative analysis of the blood–CSF vs. blood–brain barriers

In order to better understand the initiation and progression of many brain disorders, it is essential to compare the functional characteristics of the BCSFB vs. BBB. There are striking differences between these major transport interfaces. This has implications for the various mechanisms of homeostatic disruption and their pharmacological remediation. Of additional significance in CNS disease onset are the respective locations of the choroid plexuses (circumscribed to the ventricular interior of the brain) and the brain capillaries (diffusely distributed throughout the parenchyma). Subsequent discussion deals with four distinguishing functional features of the blood–CSF vs. blood–brain interfaces: (i) the permeability of the tight junctions, (ii) the nature of the active transporters in the external limiting membranes, (iii) the expression of specific peptides and proteins, and (iv) the fluid-generating capacity of the constituent cells of the barrier. By understanding the fundamental differences between the choroidal epithelium and the cerebral endothelium, it should be possible to gain a fuller appreciation of CNS region-specific physiology as well as pathology.

Barrier permeability

Tight junctions are interposed between the epithelial cells of choroid plexus and the endothelial cells of cerebral capillaries, but they differ quantitatively in their degree of 'leakiness'. Molecular tracers such as microperoxidase (20 kDa; 2 nm diameter), horseradish peroxidase (40 kDa; 6 nm diameter) and Evans blue-albumin (68 kDa) administered into the vascular compartment permeate choroidal capillaries and fill the connective tissue stroma. These tracers then easily diffuse into lateral intercellular spaces between the choroidal epithelial cells up to the apical tight junctions (near the CSF), where their passage into the ventricles is thwarted.

However, the choroid plexus epithelial lining is relatively less tight (i.e., not as restrictive to diffusion) compared to the BBB. Thus the electrical resistance of choroidal epithelium is only 150–175 $\Omega\,cm^2$ (Strazielle and Ghersi-Egea, 1999; Shu et al., 2002) compared to 8000 $\Omega\,cm^2$ for the cerebral endothelium (Smith and Rapoport, 1986; Butt and Jones, 1992). This fits with findings that the influx rate constants for mannitol (0.18 kDa) and inulin (5 kDa) for their respective entry into CSF are 1.4 and 0.5 $\mu l/g/min$ (Wilson et al., 1984; Smith and Rapoport, 1986), compared to 1.2 and 0.05 $\mu l/g/min$ for cerebral cortex (Smith and Rapoport, 1986; Masada et al., 2001). Conversion of these influx rate constants into corresponding permeability-area products (ml/min) indicates a 100-fold greater permeability for inulin across the BCSFB compared to the BBB (Rapoport, 1976).

Furthermore, an ultrastructural study of cat choroid plexus with lanthanum also supports the idea of a more permeable tight junction in the BCSFB vs. the BBB. Thus, perfusion of buffered lanthanum chloride solutions into the internal carotid artery resulted in deposits of the electron dense tracer in the lumen of the choroidal blood vessels, perivascular space and along the length of the lateral spaces between adjacent choroidal epithelial cells (Castel et al., 1974). In this study, lanthanum was also occasionally present in the ventricular lumen. Conversely, when these investigators administered lanthanum into the cerebral ventricles (i.e., on the CSF side of the plexus), the tracer passed paracellularly through the tight junctions between the epithelial cells but was not seen in any vesicular profiles within the cell cytoplasm. On the basis of these observations Castel et al. (1974) concluded that the tight junctions between choroid plexus epithelial cells are leaky. On the other hand, tight junctions between brain endothelial cells stopped the penetration of lanthanum (1.2 nm diameter) at the tight junction (Brightman and Reese, 1969; Rapoport, 1976).

Ultrastructural differences in the tight junctions of the BCSFB (choroid plexus) compared to BBB (cerebral capillaries) have functional implications (Fig. 2). Choroidal tight junctions are less restrictive to protein diffusion than the BBB counterparts. Thus there is a slow leak of plasma proteins transchoroidally from blood to CSF. This protein flux across the BCSFB in healthy mammals leads to a certain albeit low concentration of serum albumin and IgG in the CSF. Damage to the BCSFB, induced experimentally or observed in particular disease states, results in enhanced penetration of these proteins into the ventricles. This leads to a greater CSF albumin/IgG concentration ratio, compared to that for plasma (serum). Using this approach, investigators have made assessments of the integrity of the BCSFB, based on alterations in the relative distribution of albumin and/or IgG between blood and CSF. In experimental models such as multiple sclerosis (MS) or hyperthermia, with accompanying BCSFB

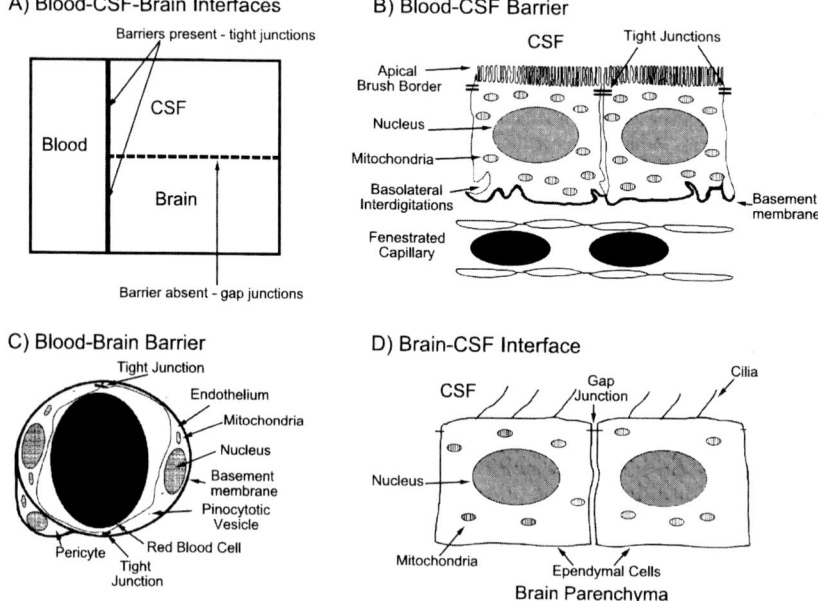

Fig. 2. Tight junctions vs. gap junctions, and the compartments which they separate: (A) Overview of the three main compartments in the CNS: blood, CSF and brain. Due to the restrictive effects of tight junctions in the BBB and BCSFB, Evans blue dye placed in the blood normally stains neither the brain nor the CSF system. However, blue dye administered into the CSF readily stains the brain because the gap junctions at the CSF-brain interface allow even large molecules to pass through. (B) The anatomical substrate for the blood–CSF barrier in the choroid plexus is the tight junction, which melds the apical poles of adjacent epithelial cells. In health, the paracellular diffusion of water-soluble solutes through the lateral spaces (between cells) is at least partially restricted. Tight junction occlusive ability can be reduced in some disorders or disease states. In contrast to the brain, the capillaries in the choroid plexus are permeable to all molecules. (C) The main locus of the BBB is the tight junction between endothelial cells. It has a high electrical resistance and a low hydraulic conductivity. Most of the capillaries in the brain and spinal cord have tight junctions. (D) The ependymal cell layer separates the large-cavity CSF from the interstitial fluid of the brain. The ependyma is a monolayer of cells, most of which allow free exchange of solutes between ventricular CSF and the brain extracellular space. Gap junctions, unlike tight junctions, do not fully wrap around the ependymal cells. It is appropriate to speak of the brain–CSF interface, *not* the brain–CSF barrier (reproduced with permission from Smith, Johansson and Keep, 2004).

breakdown, there is enhanced uptake of blood-borne macromolecules by CSF. Another issue is whether the tight junctions of BCSFB vs. BBB are differentially susceptible to insults from ischemia, pathogen infection and fever.

Active transport systems at the barriers

There are fundamental differences in the active transporters for organic solutes and ions at the BCSFB vs. BBB. The cerebral endothelial cells are specialized to extensively transport amino acids, glucose and free fatty acids, serially across their luminal *and* abluminal membranes into the brain. By comparison, the choroid plexus transports these same organic solutes across the basolateral membrane into its epithelial compartment but *not* to any appreciably further extent, i.e., across the apical membrane into the CSF compartment. Significantly, the choroidal epithelium is 'streamlined' for a high-capacity active transport of Na and Cl into the CSF. Conversely, the cerebral capillaries do not transport Na and Cl as extensively from blood into the brain ISF (Smith et al., 1982). This difference in ion transport capacity at the barriers can be exemplified by carbonic anhydrase-sensitive transport of Na and Cl (Murphy and Johanson, 1989). Acetazolamide substantially reduces (by 40–50%) the transport of ^{22}Na and ^{36}Cl from blood to plexus to CSF; but this carbonic anhydrase inhibitor does not alter the low-capacity transport of these ions from blood to brain.

The choroid plexus also uniquely transports certain micronutrients into the CNS via the CSF. A prime example is the transport of vitamin C. The active Na-ascorbate cotransporter is present at the BCSFB but not at the BBB (Spector and Johanson, 2006). Thus, ascorbate gains access to the neurons by way of the plexus-CSF-brain nexus, not via the cerebral capillary beds. Clearly the differences in transport by the choroidal epithelium *vs.* the brain endothelium are qualitative as well as quantitative.

The cerebral capillaries therefore have the main responsibility of providing neurons with energy substrate and amino acids for protein synthesis (Smith, 1991). On the other hand, the choroid plexuses furnish a large amount of fluid (ions plus water) as well as vitamins and growth factors to the CNS interior (Johanson et al., 2000). Accordingly, due to the distinctive transporters at the BCSFB *vs.* BBB, imbalances in compromised transport at the respective interfaces in various disorders would be expected to differentially affect brain status. This is why it is critical in disease modeling, *e.g.*, MS or hyperthermia, to analyze both the BBB (Pan and Kastin, 2004) and BCSFB, in order to have a more complete picture of the pathophysiological disruption and compensation.

The 'differential dynamics' in the transfer of solutes and water across the barriers, BCSFB vs. BBB, is a salient consideration in experimental modeling. Tracers such as radioactive ions (*e.g.*, ^{22}Na and ^{36}Cl) administered parenterally appear quickly (within minutes) in the CSF and can be conveniently sampled from the cisterna magna (Knuckey et al., 1991; Johanson et al., 1994). Furthermore, following disruption of the BCSFB, blood-borne tracers such as ^{131}RISHA, appear rapidly in cisternal CSF where their activity reflects transport across the choroid plexus (Murphy and Johanson, 1985). Therefore, CSF analyses, such as those indicated above as well as those for albumin/IgG quotients, provide a fair assessment of the functional status (permeability as well as transport capacity) of the BCSFB after experimental treatment or in disease.

Gene expression by cells at the barriers

The epithelial cells of the plexus express many peptides and growth factors involved in neuroendocrine regulation of CNS fluid balance. The consequences of heat stress include barrier disruption as well as an imbalance of CSF formation and osmolality. In the CSF system, the sympathomimetic amines, angiotensin II and AVP work in concert (Nilsson et al., 1992; Saavedra et al., 2006) to help in adaptive responses to stressful states. Moreover, basic fibroblast growth factor (FGF2) is thought to work in tandem with arginine vasopressin (AVP) to effect CSF fluid homeostasis (Szmydynger-Chodobska et al., 2002) in states involving dehydration (Johanson et al., 2001). AVP, like FGF2, is synthesized by choroid plexus

(Matthews et al., 1993; Chodobski et al., 1997). Following osmolality disturbances (Zemo and McCabe, 2001), such as which occur in the dehydration of whole body hyperthermia (WBH) when water is not available to the organism, AVP is up-regulated in the plexus and presumably released into CSF. An analysis of CSF concentrations of AVP in a series of febrile children supports the hypothesis of synchronous regulation of CSF osmolality and AVP levels (Kiviranta et al., 1996). It is interesting that perfusion of exogenous AVP suppresses the hyperthermia induced by PGE2 infused into a lateral ventricle (Ruwe et al., 1985). In order to help control brain edema and intracranial pressure during hyperthermia, the well-known ability of AVP to reduce (or antagonize changes in) CSF production by the plexus likely counters the vasopressinergic-modulated enhanced uptake of water across the BBB. Overall, CSF AVP exerts multiple effects (Yang and Zhi, 1994) in blunting perturbations, or in attempting to restore water and ion homeostasis following increased fluid and protein uptake by the CSF-brain system during heat stress.

CSF formation

The primary function of the choroid plexus is to manufacture CSF. Approximately 70% of CSF is produced by secretion at the choroid plexuses located in the floor of the lateral ventricles and on the rooves of third and fourth ventricles. The CSF production rate in rats is $\sim 3\,\mu l/min$. This represents a CSF turnover or renewal rate of $\sim 0.4\%/min$. Secretion of CSF is a temperature-sensitive process, with the rate of fluid formation increasing with brain temperature elevation. Thus, as the cellular metabolic rate is generally enhanced in hyperthermia, one would expect fluid formation by the choroidal epithelium to be augmented in fever or heat stress. In addition, the cyclooxygenase-generated prostaglandins (PGs) that are elevated in hyperthermia promote vasodilation and fluid turnover. The ability of indomethacin to inhibit cyclooxygenases, reduce intracranial pressure (Forderreuther and Straube, 2000) and modulate CSF formation (Schalk et al., 1992), suggests usefulness of COX inhibitors to minimize the deleterious effects of hyperthermia on fluid imbalance in the CNS.

PGE2, a major prostanoid in the reaction to heat stress, increases fluid turnover at the BCSFB (Deng, 1986). PGE2, as the 'final mediator' in fever, is a pivotal PG in the exacerbation of hyperthermia with its deleterious effects on brain function. The CSF concentration of PGE2 rises in hyperthermia. Intraventricularly administered PGE2 is pyrogenic (Coceani et al., 1988). PGE2 injected into the CSF of normothermic rats greatly enhances the transport of ^{36}Cl from blood to CSF, thus pointing to increased production of CSF (Deng, 1986). Thus, pharmacological strategies to prevent the buildup of PGE2 in CSF might lessen the severity of fluid retention in the CNS as well as fever development. This is suggested by the observation that attenuated fever in pregnant rats is associated with decreased syntheses of PGE2 and brain cyclooxygenase-2 (COX-2) (Imai-Matsumura et al., 2002). Conversely, COX-1 and COX-2, which catalyze PG formation, are both expressed in choroid plexus (Maslinska et al., 1999) and up-regulated there by LPS-induced fever (Vellucci and Parrott, 1998). Cyclooxygenase inhibition may thus be a strategy for curtailing the barrier-generated prostanoids that increase temperature and edema secondary to heat stress. Moreover, in fever initiation, one of the principal PG carriers, PGT, is up-regulated in the cerebral endothelium as well as choroidal epithelium (Kis et al., 2006). Thus, regulation of PG transport across the CNS barriers by PGT may be yet another mechanism for reducing the untoward effects of hyperthermia on brain fluid homeostasis.

Altered CSF composition: how much is due to a disrupted blood–CSF barrier?

It is instructive to analyze the CSF composition of solutes in chronic illnesses and other disorders such as hyperthermia that damages barrier functions in humans. Changes in CSF solute concentration can reflect altered BCSFB function as well as changes in brain metabolism. Sometimes it is difficult to differentiate why CSF concentrations are altered in disease. However, because the choroid plexus has a dominant effect on the composition of the CSF, and a high turnover rate (every 6 h or so), it is important to consider the role of the BCSFB in diseases of the CNS. Levine (1987) has emphasized the need for more extensive neuropathological attention to the BCSFB in explaining brain pathophysiology. Consequently, this points to the need for more frequent sampling of choroid plexus and CSF in patient studies as well as for animal modeling.

There are several factors that can cause significant changes in CSF composition. Because most of the CSF emanates from the plexus, and because the bulk flow of the secreted CSF through the ventriculo-subarachnoid spaces is an order of magnitude greater than that of the brain ISF (Johanson et al., 2005b), it is likely that CSF composition is determined usually by phenomena at the BCSF interface: synthesis/secretion (*e.g.*, transthyretin) from the choroidal epithelium (Chen et al., 2005) as well as transchoroidal transport (*e.g.*, prolactin) from the blood (Smith et al., 2004). Another factor affecting CSF composition is the secretory/reabsorptive activity by other CSF-bordering cells: the ependyma, pia-glia and arachnoid. However, the profile of secreted substances by the ependymo-meningeal type of cells is quite similar to that of the choroidal epithelium. A third factor affecting CSF composition is the transport of substances from neurons and glia into the brain ISF, which readily communicates (by relatively unrestricted diffusion) with the large-cavity CSF. In some cases of severe pathology, this 'brain metabolism' factor can lead to significant changes in CSF neurochemistry. Microdialysis in the animal brain can shed light on ISF concentrations, whereas blood-borne radioisotope tracers (described below for hyperthermia) sampled from the CSF can provide insight on molecular/ionic fluxes across the BCSFB.

Several studies of CSF composition are discussed below. Both human and experimental data are treated. In some cases they point to a compromised BCSFB, but they mainly prompt new experimental approaches by illustrating the challenge in interpreting steady-state CSF compositional data in various disorders.

An elevated level of substance P in the CSF of patients with posttraumatic stress disorder (PTSD) and/or major depression was shown by Geracioti et al. (2006). Substance P concentrations in patient's CSF increased by 169% and 91% of baseline levels at 10 and 70 min, respectively, after the start of a traumatic videotape compared to a neutral videotape. These observations in relation to findings by other workers indicate that elevated substance P concentrations in the CSF are involved in major depression and PTSD. A release of substance P from the CNS and/or transient breakdown of the BCSFB could be responsible for an elevation of CSF substance P concentrations during and after the symptom-provoking stimuli.

A significant increase in protein S100B in CSF occurs in an animal model of depression (Busnello et al., 2006). Repeated electroconvulsive shock (ECS) treatment in rats induced depression-like behavior increases in the S100B level 6 h after shock. Lactate levels in the CSF increased instantly (Busnello et al., 2006). No increase in CSF neuron specific enolase (NSE) was seen in this model. It appears that a significant increase in CSF proteins and lactate may therefore be due to altered BCSFB function.

A *decrease* in CSF levels of various proteins or peptides may also indicate a dysfunctional BCSFB. Measurement of one lectin, tetranectin (TN), a glycoprotein that plays a prominent role in tissue remodeling, was done by Stoevring et al. (2006) in the CSF of patients with MS. The authors found a significant decrease of TN concentrations in CSF of MS patients compared to the control group. This decrease of TN significantly correlated with the CSF albumin concentration. Likewise, Rasmusson et al. (2006) described reduced CSF levels of allopregnanolone and GABA in women with PTSD. A low level of CSF allopregnanolone in premenopausal women with PTSD indicates an imbalance in inhibitory versus excitatory neurotransmission, resulting in increased depressive symptoms.

Altered CSF composition is also seen in schizophrenia and other metal illnesses. Recently Huang et al. (2006) measured CSF proteins and peptides in 179 cases comprising 58 schizophrenia patients, 16 depressed cases, 5 subjects with obsessive-compulsive disorder, 10 patients with Alzheimer disease and 90 controls. Their findings showed highly selective alterations in proteins/peptide concentrations in mental illnesses compared to healthy subjects. At the onset of schizophrenia, there was an up-regulation of a 40-amino acid vascular growth factor (VGF)-derived peptide and the down-regulation of transthyretin at ~4 kDa. There was also a peptide cluster at ~6.8 to 7.3 kDa. However, only mild changes in the CSF were seen in these proteins in other kinds of mental illness. Overall, such observations indicate a wide range of changes in the CSF composition of proteins and peptides in mental diseases. Taken together, it appears that the CSF composition is a sensitive indicator of specific neurological and mental illnesses. In some cases the altered CSF content may reflect BCSFB dysfunction that could impair brain function.

Because hyperthermia can be life threatening, it is imperative to gain a better understanding of how the CSF compartment responds to hyperthermia, heat stress and fever episodes. Despite the serious clinical implications of rising body temperature, unfortunately only a few studies have examined CSF composition in hyperthermia induced by various agents. Dascombe and Milton (1976) examined the concentration of adenosine 3′, 5′-monophosphate (cyclic AMP) in CSF of un-anaesthetized cats following hyperthermia (+2.5°C) caused by heat exposure (44° to 45°C, for 3.5 h) or *Shigella dysenteriae*- (2 and 20 μg/kg, i.v.) induced fever. The investigators noted a dose-related increase in cyclic AMP concentration in CSF during pyrogen-induced fever; in a related manner, intraventricularly administered cAMP increased the ^{36}Cl transport from blood to CSF in rats, a response that was consistent with the observed increase in CSF formation rate (Deng and Johanson, 1992). Interestingly, no alteration in cyclic AMP levels in CSF was seen following non-pyrogenic heat exposure in the investigation by Dascombe and Milton (1976). The activation of the cyclooxygenases in some febrile disorders can be inferred from the finding that the PG synthase inhibitor, paracetamol (75 mg/kg, i.p.), injected before fever onset suppressed the increase in body temperature and CSF cyclic AMP response to pyrogen (Dascombe and Milton, 1976).

A 5–6-fold increase in PGE levels in the CSF of rabbits following systemic administration of bacterial endotoxin was observed by Bernheim et al. (1980). This elevation in PGE levels in CSF correlates well with the magnitude of hyperthermia induced by pyrogens (Bernheim et al., 1980). On the other hand, no increase in PGE level was seen in the CSF following heat exposure: induced hyperthermia (+0.5°C) or local heating of the hypothalamus (Bernheim et al., 1980). These observations suggest that CSF is a sensitive indicator of brain function and can perceive changes in the body temperature caused by external or internal heat stimuli, and/or by bacterial pyrogens.

An altered Na^+/Ca^{2+} ratio in the CSF of an avian species, *Gallus domesticus*, following hyperthermia induced by heat exposure was described by Maki et al. (1990). At a thermoneutral ambient temperature (28°C), the CSF Na^+/Ca^{2+} in hens was 61.7. Subjection of hens to heat stress (39°C) altered the Na^+/Ca^{2+} to 59.4 at 30 min; 62.6 at 3 h and 52.4 at 10 h (Maki et al., 1990). These observations suggest that heat stress stimulates Na^+-transport mechanisms in the CSF. Frosini et al. (2000a) showed profound alterations in CSF concentrations of taurine, GABA, aspartate and glutamate in conscious rabbits following heat exposure at high ambient temperature (40°C for 50 min; for details, see Frosini, Chapter 22 in this volume). Significant increases in CSF osmolality, protein, Ca^{2+}, taurine and GABA levels were observed (Corte and Sgaragli, 2000). These changes correlated well with the rise in body temperature (Frosini et al., 2000b). An increased level of CSF protein usually indicates breakdown of the BCSFB function in hyperthermia. However, neither BBB nor BCSFB permeability was measured in these experiments.

Altered CSF composition is also seen in clinical cases of hyperthermia. Thus, Iwanaga et al. (1996) reported a significant elevation of CSF myelin basic protein (MBP) in a 9-month-old febrile child (+3.5°C) on the 16th day of hospital admission. The patient had a persistent high fever, developed a second generalized tonic clonic convulsion and became comatose. On the 20th hospital day, magnetic resonance imaging (MRI) showed bilateral symmetrical paracentral hypo-intensity of the white matter with occipital hypo-intensity. These

MRI and CSF findings suggest a loss of water from the cerebral cortices and deep white matter indicating brain damage due to high fever. The alterations in several CSF enzymes and the brain damage in patients suffering heat stroke (Kew et al., 1967) are in line with this idea. Thus, there are reasons to think that hyperthermia caused by heat exposure or heat stroke has the capacity to markedly alter CSF protein composition. However, definitive breakdown of the BCSFB in heat-related illnesses has still not been unequivocally demonstrated in patients and therefore requires additional investigation in animal models of hyperthermia.

This review is focused on the breakdown of the BCSFB in hyperthermia in relation to brain pathology based on our own investigations. New data clearly show that BCSFB dysfunction plays an important role in heat stress-induced brain edema formation and cell injury.

Why study the effects of heat stress on the 'barriers' in the CNS?

Many stressors, *e.g.*, trauma, ischemia and virulent pathogens, open up the BBB and expose the neuronal networks to plasma-derived proteins, peptides, growth factors and cytokines. Such an invasion of blood-borne solutes wreaks havoc with a host of CNS functions. Heat stress adversely affects humans by many different causes; yet ironically, hyperthermia analyses of the brain (a very vulnerable system to 'heat loading') have not been adequately carried out.

Hyperthermia and heat-related illnesses cause large numbers of death (~800–2000 cases) during summer months in the United States of America and in Europe (Bouchama and Knochel, 2002; Davis et al., 2003; Belmin and Golmard, 2005; Centers for Disease Control and Prevention, 2005; Conti et al., 2005. For details, see Sharma, Chapter 10, in this volume). Significantly, heat-related death far exceeds in number that of any other natural calamity such as floods, cyclones or hurricanes (Gauss and Meyer, 1917; Malamud et al., 1946; Kaiser et al., 2001; Moore et al., 2002; Sweeney, 2002; Scoville et al., 2004; Stott et al., 2004; Belmin and Golmard, 2005; Bouchama et al., 2005). Recently, the intensity of heat stroke-induced deaths is increasing with global warming. There is a worldwide upsurge in the frequency and intensity of heat waves (Sharma, 1982, 2005; Sharma and Dey, 1986, 1987; Kaiser et al., 2001; Bouchama and Knochel, 2002).

Heat stress and associated heat stroke are life-threatening illnesses in which body temperature increases above 40°C causing severe CNS dysfunction, *e.g.*, delirium, convulsion and coma (Bouchama and Knochel, 2002; Belmin and Golmard, 2005; Bouchama et al., 2005). More than 50% of heat stroke victims die within a short time despite lowering of the body temperature and therapeutic intervention (Bouchama and Knochel, 2002; Davis et al., 2003; Centers for Disease Control and Prevention, 2005; Conti et al., 2005). Those who survive heat stroke often show permanent neurological deficit (Bouchama and Knochel, 2002; Moore et al., 2002; Bouchama, 2004; Scoville et al., 2004).

Interestingly, WBH is commonly used as an adjunct to cytotoxic therapy for cancer treatment (Thrall et al., 1986; Katschinski et al., 1999; Sharma, 2005; Haveman et al., 2005; Hildebrandt et al., 2005). Recently, it has been recognized that if WBH is combined with cytotoxic therapy during cancer treatment, it causes increased DNA adduct formation, inhibition of DNA repair, increased drug permeation and decreased resistance to DNA damaging agents (Robins et al., 2003). New clinical and experimental results show that WBH enhances cytotoxic ionizing radiation and chemotherapy (Dewhirst et al., 2003; Robins et al., 2003; Haveman et al., 2005; Hildebrandt et al., 2005; Sharma, 2005). There are reasons to believe that WBH-induced severe side effects, including altered brain function, are probably due to breakdown of BBB function (Sharma, 1982, 2004, 2005).

Hyperthermia-induced damage to choroid plexus and neighboring CSF–brain regions

Although greater attention has been focused on the permeability properties of the BBB in hyperthermia (Sharma and Dey, 1986, 1987; Sharma,

1999, 2004, 2005; Sharma and Hoopes, 2003), there are reasons to think that the BCSFB also plays an important role in hyperthermia-induced brain dysfunctions.

Previous experiments in our laboratory on WBH suggest that alterations in the brain fluid microenvironment following heat stress are responsible for hyperthermia-induced brain damage (Sharma et al., 1991a, b, 1997, 1998, 2003). However, studies on alterations in the BCSFB in heat stress are still lacking. The BCSFB maintains the composition of CSF (Rapoport, 1976; Bradbury, 1979; Johanson et al., 2005). Thus, massive breakdown of the BCSFB will adversely influence the CNS structure and function.

This section is based on our own investigations demonstrating that heat stress-induced hyperthermia is able to alter BCSFB permeability to exogenously-administered tracers; that the choroid plexus and ependymal cells exhibit morphological alterations in hyperthermia; and that BCSFB breakdown has implications for edema in periventricular regions of brain. Before discussing hyperthermia effects on the choroid plexus-CSF-brain system, we will briefly review the salient aspects of the model being used for WBH.

We developed the WBH model in rats to simulate heat stress situations seen in clinical conditions (see Sharma, Chapter 10 in this volume). In this model, rats were exposed to WBH in a biological oxygen demand (BOD) incubator (relative humidity 45–50%; wind velocity 18–25 cm/s) maintained at 38°C for 1–4 h (Sharma et al., 1997, 1998, 2003; Sharma, 1999). The experiments were conducted according to NIH guidelines for care of animals (Sharma et al., 1986, 1997, 2003). These experimental conditions were approved by the Ethics Committee of Uppsala University.

The BCSFB was examined *in vivo* using Evans blue (2%; 0.3 ml/100 g) and ^{131}Iodine (10 µCi/100 g) as tracers (Sharma et al., 1990; Sharma and Westman, 1998; Sharma, 1999, 2004, 2006; Sharma and Hoopes, 2003). The tracers were administered into the right femoral artery and allowed to circulate for 5 min. At the end of the experiment an approximately 100 µl CSF sample, without blood contamination, was drawn from the cisterna magna. The animals were then perfused with 0.9% saline through the heart and the brain dissected out. Extravasation of Evans blue dye was visually examined in the walls of the lateral, 3rd and 4th ventricles. Then various parts of the brain were dissected out, weighed and counted for radioactivity in a Gamma counter (energy window of 500–800 KeV) (Sharma, 1999). Before perfusion, a sample of whole blood was withdrawn from the left ventricle by cardiac puncture and the radioactivity determined as above (Sharma et al., 1998, 2003; Sharma, 1999). Tracer extravasation into CSF and periventricular brain regions was expressed as percentage increase over the whole blood radioactivity (Sharma, 1999). In some regions, the Evans blue dye that entered into the brain was also measured colorimetrically (Sharma et al., 1998, 2003; Sharma, 1999, 2004).

There was substantial evidence for severe rupture of the BCSFB. Rats subjected to 4 h of WBH at 38°C in a BOD incubator exhibited profound alterations in the BCSFB permeability to Evans blue and radioiodine tracers. The choroid plexus had deep blue staining. Mild to moderate blue staining of the walls in the lateral, 3rd and 4th cerebroventricles were noted. Moreover the dorsal surface of the hippocampus and caudate nucleus showed moderate staining. Measurement of Evans blue dye in selected brain regions, *e.g.*, hippocampus, caudate nucleus, mid-thalamus (massa intermedia), hypothalamus, dorsal surface of the brain stem and the ventral surface of the cerebellum showed a marked increase from the control group (Table 1). A significant increase in Evans blue and ^{131}Iodine tracers was also observed in the CSF samples obtained from cisterna magna following 4-h WBH compared to the control group (Table 1).

Interesting differences in the response to WBH, as a function of the *duration* of the heat stress, were noted. Thus, blue staining of the ventricular walls and/or surface of the structures within the cerebral ventricles was absent in animals subjected to only 1- or 2-h heat stress. At these earlier times, no significant increase in Evans blue or radioiodine tracer was noted in various brain regions and/or CSF samples (Table 1).

These new findings show a marked increase in the BCSFB permeability to Evans blue and radio-iodine tracer following 4 h of WBH in rats. This

Table 1. Shows changes in blood–cerebrospinal fluid barrier (BCSFB), brain edema and structural changes in 4 h heat-stressed rats. Rats were subjected to whole body hyperthermia (WBH) at 38°C in a biological oxygen demand (BOD) incubator for 4 h [for details, see text].

Parameters measured	n	Control	Heat stress 38°C in a biological oxygen demand incubator		
			1 h	2 h	4 h
BCSFB permeability[#]	5				
¹³¹Iodine %					
Whole brain		0.35 ± 0.06	0.33 ± 0.08	0.42 ± 0.08	$1.88 \pm 0.24^{**}$
Cisternal CSF		0.18 ± 0.04	0.12 ± 0.11	0.16 ± 0.12	$0.76 \pm 0.12^{**}$
Hippocampus		0.42 ± 0.12	0.47 ± 0.11	nd	$0.84 \pm 0.23^{**}$
Caudate nucleus		0.28 ± 0.08	0.32 ± 0.14	nd	$0.93 \pm 0.12^{**}$
Cerebellum		0.13 ± 0.08	0.16 ± 0.08	nd	$0.65 \pm 0.10^{**}$
Thalamus		0.48 ± 0.12	0.46 ± 0.08	nd	$0.89 \pm 0.14^{**}$
Hypothalamus		0.54 ± 0.21	nd	nd	$0.87 \pm 0.23^{**}$
Brain stem		0.18 ± 0.08	nd	nd	$0.34 \pm 0.14^{*}$
Water content[#]	5				
Whole brain %		76.12 ± 0.18	76.04 ± 0.13	76.4 ± 0.14	$80.18 \pm 0.24^{**}$
Hippocampus		78.43 ± 0.23	78.11 ± 0.21	78.21 ± 0.34	$81.56 \pm 0.34^{**}$
Caudate nucleus		77.43 ± 0.24	77.34 ± 0.32	nd	$81.48 \pm 0.54^{**}$
Cerebellum		74.43 ± 0.21	74.33 ± 0.32	nd	$79.34 \pm 0.23^{**}$
Thalamus		75.21 ± 0.22	75.12 ± 0.33	nd	$78.56 \pm 0.23^{**}$
Hypothalamus		74.54 ± 0.12	nd	nd	$76.45 \pm 0.23^{**}$
Brain stem		68.54 ± 0.12	nd	nd	$69.78 \pm 0.12^{**}$
Structural changes	5				
Neuronal damage		Nil	Nil	Nil	+ + + +
Glial cell injury		Nil	Nil	+/−?	+ + + +
Myelin damage		Nil	Nil	+/−?	+ + + +

Note: Values are mean ± SD of five rats in each group. # = tissue sample size (135–180 mg); CSF sample = 50–80 μl; +/−? = uncertain; + + + + = severe cell damage; nil = absent; nd = not done. Data after with permission from Sharma et al., 2006.
*$P<0.05$ (compared from control); ANOVA followed by Dunnett's test from one control.
**$P<0.01$ (compared from control); ANOVA followed by Dunnett's test from one control.

indicates that heat stress-induced hyperthermia (>40°C) was capable of disrupting the BCSFB to large molecule tracers. The duration/magnitude of exposure to hyperthermia appears to determine the degree of BCSFB disruption (or not) in heat stress. This was supported by the additional finding that the shorter heat exposure of the animals to 1- or 2-h hyperthermia at <39°C did not impair the BCSFB function to proteins (Sharma et al., 2006).

Alterations in the composition of CSF and/or its osmolality are known to occur following lipopolysaccharide (LPS)-induced fever or heat stress in rabbits (Frosini et al., 2000a, b). WBH is known to increase plasma viscosity and probably alters the plasma tonicity (Thrall et al., 1986; Bouchama and Knochel, 2002; Bouchama, 2004). Consequently, a possibility exists that hyperosmolality of plasma and/or CSF following WBH are important factors in disrupting the BCSFB (Johanson et al., 1974). Osmolality changes in WBH, and their effects on CNS barriers (Fig. 1), deserve further attention.

It is unlikely that the simple heating *per se* of animals during the 4 h of WBH directly caused BCSFB leakage (Sharma and Cervós-Navarro, 1990; Sharma and Hoopes, 2003; Sharma, 2005). This conclusion is based on the finding that when anesthetized animals were subjected to 4 h of WBH, no disruption of the BCSFB was observed (Sharma and Johanson, unpublished observation). Thus, there are reasons to believe that WBH-induced alterations in the plasma and CSF composition play important roles in BCSFB disruption.

Brain edema formation was evaluated by the water content of the samples, either in the whole brain or in the corresponding brain regions used for radiotracer measurement (as above) (Sharma, Chapter 10 in this volume). Measurements of the respective water contents of corresponding brain regions showing leakage of Evans blue or radiotracers (*e.g.*, hippocampus, caudate nucleus, mid-thalamus (massa intermedia), hypothalamus, dorsal surface of the brain stem and the ventral surface of the cerebellum) demonstrated significant increases in water content after 4-h WBH (Table 1). However, consistent with the tracer analyses in the 1–2-h experiments (described above), the rats subjected to 1- or 2-h heat exposure did not have any increases in brain water content compared to corresponding regions for the control group (Table 1).

These observations indicate that breakdown of the BCSFB and BBB in WBH plays an important role in brain edema formation. An increase in Evans blue and radioiodine concentrations in the CSF from rats subjected to 4 h of WBH support the idea of a breakdown of the BCSFB. Evans blue or radioiodine injected into the circulation binds to endogenous serum proteins (Rapoport, 1976). Approximately one molecule of Evans blue binds to 12 molecules of serum albumin *in vivo* (Rapoport, 1976; Sharma, 1999). Thus, Evans blue uptake by CSF indicates leakage of the serum protein-dye complex across the choroid plexus (Sharma, 1999). Permeation of serum proteins into the CSF compartment would alter the osmotic gradient across the choroid plexus epithelium and the tight junctions, resulting in enhanced transport of water and other solutes from the choroidal vascular compartment. This would lead to edema formation (Rapoport, 1976; Reulen et al., 1978; Sharma, 1999). However, it is unclear whether changes in the intracranial pressure, CSF bulk flow and hydrostatic pressure gradients are contributing to accelerated protein access to brain parenchyma in heat stress. This important topic requires further investigation.

An increase in water content of various intracerebral structures following WBH is in line with this hypothesis. Since WBH is known to disrupt BBB in these areas as well (Sharma and Dey, 1987; Sharma and Cervós-Navarro, 1990), it is likely that BCSFB breakdown further aggravates regional brain edema formation (from BBB damage) due to the ventricle-to-brain percolation of CSF rich in serum proteins and other osmotically active solutes. Accumulation of serum proteins in the brain extracellular fluid likely initiates a series of cellular and molecular events leading to neuronal injury and death (Sharma, 2005). The damaged nerve cells, occurrence of sponginess, vacuolation and edema in many brain areas following 4 h of WBH (Sharma, 2005) are in line with this idea.

That the breakdown of the BCSFB contributes to edema formation and cell injury in WBH is further supported by the fact that the short duration of heat exposure (1 or 2 h) is not associated either with leakage of tracers into CSF or an increase in water content and cell damage. These observations strongly suggest that both the magnitude and duration of WBH determines BCSFB damage and brain pathology.

We examined structural changes in the brain and choroid plexus following hyperthermia using light and electron microscopy (Sharma, 1999). Although the CSF tracer data implicate a damaged BCSFB in WBH, this needs to be corroborated by direct assessments of choroid plexus specimens. For this purpose, rats under anaesthesia were perfused transcardially with 4% paraformaldehyde in 0.1 M phosphate buffer (pH 7.4) preceded with a brief saline rinse at the end of heat exposure (Sharma et al., 1991a, b; Sharma and Hoopes, 2003). The animals were wrapped in aluminum foil and kept at 4°C overnight. Next day, the brain and spinal cord were dissected out and small pieces were embedded in paraffin. Approximately 3 µm thick sections were cut and stained with hematoxylin and eosin, or Nissl, and examined under a Leica bright field microscope for neurodegenerative changes (Sharma and Dey, 1986). Small tissue pieces from hippocampus and cerebral cortex were post-fixed in osmium tetraoxide and embedded for transmission electron microscopy (see Sharma, 1999). For semi-quantitative assessments of cell injury, rough scores of 0 (no damage), or 1 (least damage) to 4 (maximum damage) were assigned in a blinded fashion (Sharma, 2005).

Morphological analysis showed degeneration of choroid epithelial cells and the nearby ependyma in rats subjected to 4-h WBH (Fig. 3). The ventricular

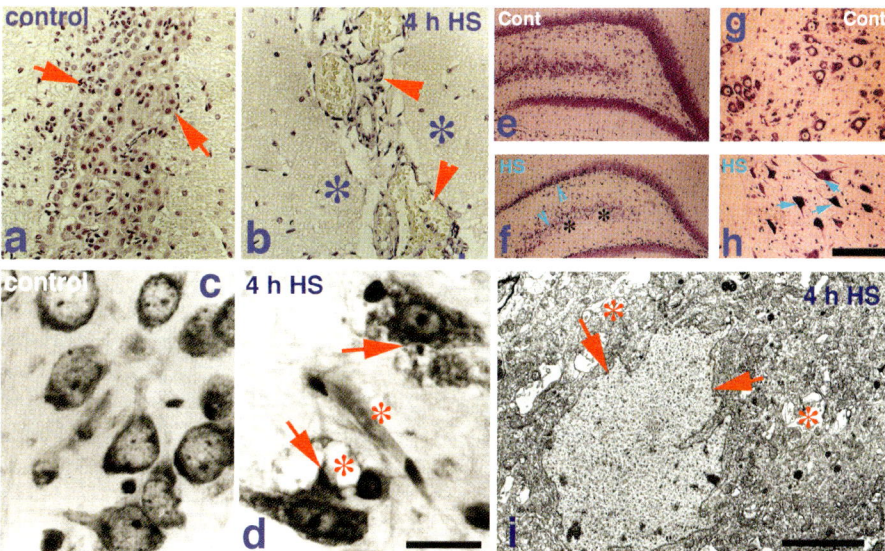

Fig. 3. Shows structural changes in choroidal epithelium, hippocampus and cerebral cortex following 4-h whole body hyperthermia at 38°C [for details see text]. Degeneration of choroidal epithelium (arrowheads) in the lateral ventricle of a 4-h heat stressed rat (b) is clearly seen compared to normal animal (a). Damage of nerve cells (arrows) and edema (*) is quite evident in the cerebral cortex (d), hippocampus CA4 region (f) and brain stem (h) in heat-stressed rat compared to controls (c,e,g). Low power electron micrograph of a neuron in hippocampus area CA4 of a 4-h heat-stressed animal (i). Increased density of cytoplasm, cytoplasmic vacuoles and swollen mitochondria are evident. The nuclear membrane shows a high degree of folding (arrows, i). Dark and condensed karyoplasm (without nucleolus) is clearly visible. Vacuolation and edema (*) in the neuropil is apparent. Bars: a,b,e,f = 100 µm, c,d = 30 µm, g,h = 20 µm, i = 500 nm (reproduced with permission from Sharma et al., 2006).

space appeared to be dilated and the underlying neuropil showed neurodegenerative changes. Neuronal damage and edema in the hippocampus, cerebral cortex, thalamus, hypothalamus and brain stem were common in 4-h heat stressed rats (Figs. 3–5). Ultrastructural studies of the CA4 subfield of hippocampus in conscious animals subjected to 4-h WBH showed neuronal changes affecting both perikarya and surrounding neuropil (see Figs. 4 and 5). Thus, the CA4 region in hippocampus showed complete collapse of microvessels (Fig. 4), damage of synapse, astrocytes, myelin vesiculation and edema (Fig. 4). Another region of hippocampus exhibited leakage of lanthanum tracer across the microvessels suggesting BBB disruption (Fig. 5). In these areas, perivascular edema, membrane damage and distortion of nerve and glial cells were prominent (Figs. 4 and 5). In general, the nerve cell somata demonstrated an increased density of the cytoplasm containing numerous vacuoles and swollen mitochondria (Fig. 4). The cell nucleus showed an increased folding of the nuclear membrane (Fig. 3). The neuropil surrounding the neuronal perikarya had pronounced vacuolation and swollen mitochondria in dendrites, astroglial cells and other structures (Figs. 3 and 4). The myelin sheaths were swollen and their periodic structure lost. Vacuolation and membrane damage were quite frequent in this region (Figs. 4 and 5). Microvessels were mainly collapsed and perivascular edema was common (Fig. 4). In some vessels extravasation of lanthanum occurred across the endothelium, and in some cases, lanthanum was present in the basement membrane. However, the tight junctions were mainly intact (Fig. 4). Similar changes are also seen in the cerebellum and brain stem (Fig. 5).

On the other hand, rats subjected to the shorter 1- or 2-h WBH did not show such structural changes (Sharma, unpublished observations). Thus the ultrastructural data were consistent with the functional data, both temporally and spatially.

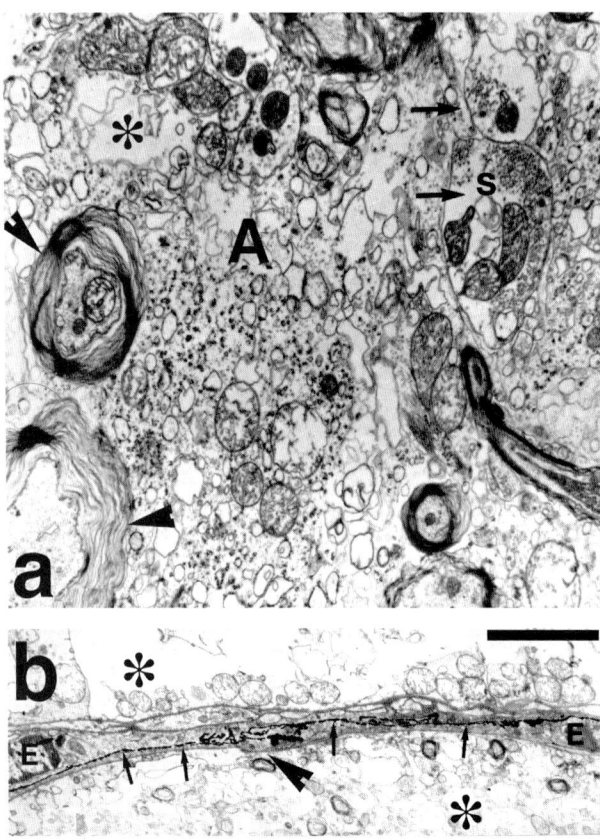

Fig. 4. High power electron micrograph of a 4-h heat-stressed animal showing neuropil in CA4 subfield of hippocampus (a). An astrocyte (A) with swollen mitochondria and cytoplasmic vacuoles is clearly visible. Arrows point to synaptic terminals, which are also damaged. Note the enlarged myelin sheath (arrowheads). Vacuolation and edema (*) are frequent in the neuropil. (b) High power electron micrograph of a vessel from CA4 subfield of hippocampus in a 4-h heat-stressed animal. Totally collapsed capillary is recognizable by the lanthanum particles adhering to the endothelium (E) (arrows). An infiltration of lanthanum across endothelium is clearly visible. However, the tight junction is intact (arrowhead). The perivascular space demonstrates edema (*) and swollen mitochondria. Bar = 300 nm.

The microvessels supplying choroid plexus are leaky (Rapoport, 1976; Bradbury, 1979). Therefore the choroid epithelial cells are in direct contact with the hyperosmolar blood plasma (Rapoport, 1976). Exposure of the choroid basolateral epithelial membrane to this osmotic stress in WBH could result in increased membrane damage. Furthermore, CSF hyperosmolality also affects the tight junction permeability of the choroidal epithelium from the ependymal side (Johanson et al., 1974). The structural changes seen in choroid epithelium and adjacent ependyma fit this hypothesis. However, to further confirm these points, ultrastructural investigations of choroidal epithelium and its tight junctions are needed for better understanding of the damage inflicted by WBH.

Our observations further show that leakiness of the BCSFB is associated with marked cellular changes in several brain regions near the cerebroventricles. Thus, substantial cell damage is seen in the hippocampus, caudate nucleus, thalamus, hypothalamus, cerebellum and brain stem. Collectively, these observations indicate that alterations in the BCSFB in WBH are somehow contributing to neurodegenerative changes.

Overall, the internal consistency of the data among the water content, tracer uptake and ultrastructural observations makes a compelling

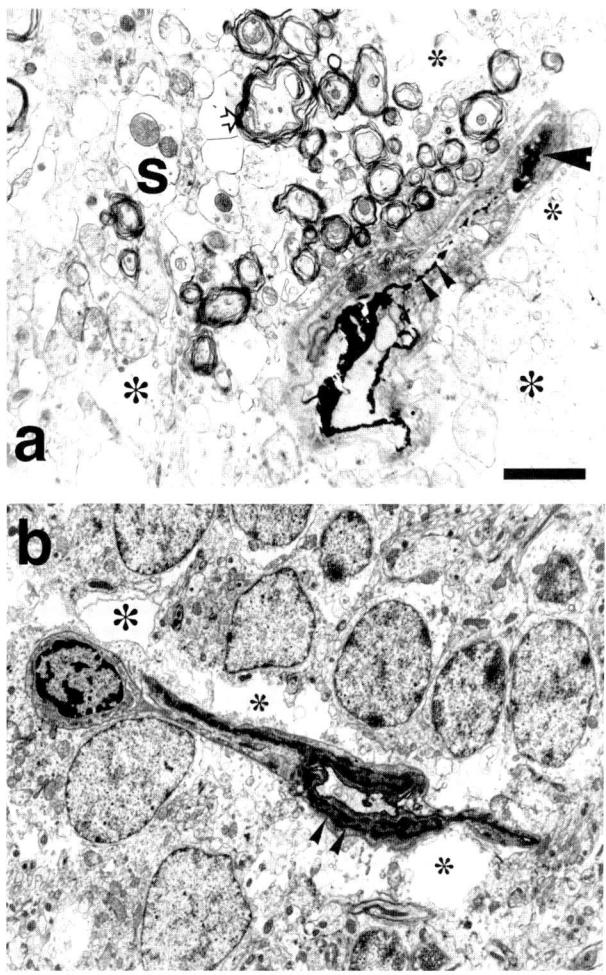

Fig. 5. Low power electron micrograph showing leakage of lanthanum across one cerebral microvessel in the brain stem (a) and in cerebellum (b) of a heat stressed rat. Perivascular edema (*) and damage of surrounding neuropil is clearly evident. Myelin vesiculation (blank arrow), synaptic damage (S) swollen mitochondria are very common in the brain stem of heat stressed rat. Bar = 1 μm. Data (b) modified after Sharma, 2004a.

argument that WBH disrupts the CNS barriers, allows penetration of water and solutes into the CSF-brain nexus and eventually causes damage or death to the neuronal elements. In conclusion, new observations from our laboratory suggest that hyperthermia induced by WBH is capable of breaking down the BCSFB and contributing to widespread cell injury in the CNS. It would be interesting to see whether neuroprotective drugs are able to attenuate BCSFB damage in WBH, and if this in turn could reduce neuronal impairment in heat stress disorders.

Neuropathological *vs.* neurotherapeutic roles of the choroid plexus–CSF system: future studies

Recent findings in our laboratory suggest that CSF is a conduit for several neurohormones and is capable of transporting many peptides and growth factors in various disease processes (Johanson et al., 2005). Thus, it is likely that CSF can play an active role in neurodegeneration (Wang et al., 2005). However, the CSF circulation also represents a pharmacological or therapeutic opportunity for promoting neuroregeneration

(Spector and Johanson, 2006) and/or neuroprotection (Johanson et al., 2005). To further explore potential therapeutic strategies involving the CSF microenvironment (Johanson et al., 2004), the administration of hormones, growth factors or neuropeptides (Smith et al., 2004) into the cerebroventricular spaces should be done in WBH, an approach currently being examined in our laboratory.

Acknowledgments

Authors' research is supported by grants from Swedish Medical Research Council (2710, HSS); Astra-Zeneca, Mölndal (HSS), Sweden; National Institute of Health (NS27601, CEJ), USA; Alexander von Humboldt Foundation (HSS), Germany; The University Grants Commission (HSS), New Delhi, India; and The Indian Council of Medical Research (HSS), New Delhi, India. The expert technical assistance of Kärstin Flink, Kerstin Rystedt, Franziska Drum and Katherin Kern; the secretarial assistance of Aruna Sharma; and the graphical and editing contributions of Nancy and Julie Johanson are all greatly appreciated.

References

Angelow, S., Zeni, P. and Galla, H.J. (2004) Usefulness and limitation of primary cultured porcine choroid plexus epithelial cells as an in vitro model to study drug transport at the blood–CSF barrier. Adv. Drug Deliv. Rev., 56(12): 1859–1873.
Banks, W.A., Kastin, A.J. and Broadwell, R.D. (1995) Passage of cytokines across the blood–brain barrier. Neuroimmunomodulation, 2(4): 241–248.
Begley, D.J. and Brightman, M.W. (2003) Structural and functional aspects of the blood–brain barrier. Prog. Drug Res., 61: 39–78.
Belmin, J. and Golmard, J.L. (2005) Mortality related to the heatwave in 2003 in France: forecasted or over the top? Presse. Med., 34(9): 627–628.
Bernheim, H.A., Gilbert, T.M. and Stitt, J.T. (1980) Prostaglandin E levels in third ventricular cerebrospinal fluid of rabbits during fever and changes in body temperature. J. Physiol., 301: 69–78.
Bouchama, A. (2004) The 2003 European heat wave. Intensive Care Med., 30(1): 1–3.
Bouchama, A. and Knochel, J.P. (2002) Heat stroke. N. Engl. J. Med., 346(25): 1978–1988.
Bouchama, A., Roberts, G., Al Mohanna, F., El-Sayed, R., Lach, B., Chollet-Martin, S., Ollivier, V., Al Baradei, R., Loualich, A., Nakeeb, S., Eldali, A. and de Prost, D. (2005) Inflammatory, hemostatic, and clinical changes in a baboon experimental model for heatstroke. J. Appl. Physiol., 98(2): 697–705.
Bradbury, M.W.B. (1979) Blood–brain barrier. In: Physiology and Medicine. Raven Press, New York, pp. 1–316.
Brightman, M.W. and Reese, T.S. (1969) Junctions between intimately apposed cell membranes in the vertebrate brain. J. Cell Biol., 40: 648–677.
Brown, R.C. and Davis, T.P. (2002) Calcium modulation of adherens and tight junction function: a potential mechanism for blood–brain barrier disruption after stroke. Stroke, 33(6): 1706–1711.
Busnello, J.V., Leke, R., Oses, J.P., Feier, G., Bruch, R., Quevedo, J., Kapczinski, F., Souza, D.O. and Cruz Portela, L.V. (2006) Acute and chronic electroconvulsive shock in rats: effects on peripheral markers of neuronal injury and glial activity. Life Sci., 78(26): 3013–3017.
Butt, A.M. and Jones, H.C. (1992) Effect of histamine and antagonists on electrical resistance across the blood–brain barrier in rat brain-surface microvessels. Brain Res., 569: 100–105.
Castel, M., Sahar, A. and Erlij, D. (1974) The movement of lanthanum across diffusion barriers in the choroid plexus of the cat. Brain Res., 67(1): 178–184.
Centers for Disease Control and Prevention (CDC). (2005) Heat-related mortality — Arizona, (1993–2002), and United States, 1979–2002. Morb. Mortal. Wkly. Rep., 54(25): 628–630.
Chen, R.L., Athauda, S.B., Kassem, N.A., Zhang, Y., Segal, M.B. and Preston, J.E. (2005) Decrease of transthyretin synthesis at the blood–cerebrospinal fluid barrier of old sheep. J. Gerontol. A Biol. Sci. Med. Sci., 60(7): 852–858.
Chodobski, A., Loh, Y.P., Corsetti, S., Szmydynger-Chodobska, J., Johanson, C.E., Lim, Y.P. and Monfils, P.R. (1997) The presence of arginine vasopressin and its mRNA in rat choroid plexus epithelium. Brain Res. Mol. Brain Res., 48(1): 67–72.
Chodobski, A. and Szmydynger-Chodobska, J. (2001) Choroid plexus: target for polypeptides and site of their synthesis. Microsc. Res. Tech., 52(1): 65–82.
Coceani, F., Lees, J. and Bishai, I. (1988) Further evidence implicating prostaglandin E2 in the genesis of pyrogen fever. Am. J. Physiol., 254(3 Pt 2): R463–R469.
Conti, S., Meli, P., Minelli, G., Solimini, R., Toccaceli, V., Vichi, M., Beltrano, C. and Perini, L. (2005) Epidemiologic study of mortality during the Summer 2003 heat wave in Italy. Environ. Res., 98(3): 390–399.
Corte, L. and Sgaragli, G. (2000) The possible role of taurine and GABA as endogenous cryogens in the rabbit: changes in CSF levels in heat-stress. Adv. Exp. Med. Biol., 483: 335–344.
Dascombe, M.J. and Milton, A.S. (1976) Cyclic adenosine 3′,5′-monophosphate in cerebrospinal fluid during thermoregulation and fever. J. Physiol., 263(3): 441–463.

Davis, R.E., Knappenberger, P.C., Michaels, P.J. and Novicoff, W.M. (2003) Changing heat-related mortality in the United States. Environ. Health Perspect., 111(14): 1712–1718.

Davson, H. (1967) Physiology of the Cerebrospinal Fluid. Churchill, London.

Deng, Q.S. (1986) Drug Modification of Chloride Transport in the Choroid Plexus-Cerebrospinal Fluid System of the Rat. PhD Thesis, University of Utah, pp. 1–55.

Deng, Q.S. and Johanson, C.E. (1992) Cyclic AMP alteration of chloride transport into the choroid plexus-cerebrospinal fluid system. Neurosci. Lett., 143(1–2): 146–150.

Dewhirst, M.W., Viglianti, B.L., Lora-Michiels, M., Hanson, M. and Hoopes, P.J. (2003) Basic principles of thermal dosimetry and thermal thresholds for tissue damage from hyperthermia. Int. J. Hyperthermia, 19(3): 267–294.

Forderreuther, S. and Straube, A. (2000) Indomethacin reduces CSF pressure in intracranial hypertension. Neurology, 55(7): 1043–1045.

Frosini, M., Sesti, C., Palmi, M., Valoti, M., Fusi, F., Mantovani, P., Bianchi, L., Della Corte, L. and Sgaragli, G. (2000a) The possible role of taurine and GABA as endogenous cryogens in the rabbit: changes in CSF levels in heat-stress. Adv. Exp. Med. Biol., 483: 335–344.

Frosini, M., Sesti, C., Palmi, M., Valoti, M., Fusi, F., Mantovani, P., Bianchi, L., Della Corte, L. and Sgaragli, G. (2000b) Heat-stress-induced hyperthermia alters CSF osmolality and composition in conscious rabbits. Am. J. Physiol. Regul. Integr. Comp. Physiol., 279(6): R2095–R2103.

Gauss, H. and Meyer, K.A. (1917) Heat stroke: report of one hundred and fifty-eight cases from Cook County Hospital, Chicago. Am. J. Med. Sci., 154: 554–564.

Geracioti Jr., T.D., Carpenter, L.L., Owens, M.J., Baker, D.G., Ekhator, N.N., Horn, P.S., Strawn, J.R., Sanacora, G., Kinkead, B., Price, L.H. and Nemeroff, C.B. (2006) Elevated cerebrospinal fluid substance P concentrations in posttraumatic stress disorder and major depression. Am. J. Psychiatry, 163(4): 637–643.

Ghersi-Egea, J.F., Strazielle, N., Murat, A., Edwards, J. and Belin, M.F. (2001) Are blood–brain interfaces efficient in protecting the brain from reactive molecules? Adv. Exp. Med. Biol., 500: 359–364.

Haveman, J., Sminia, P., Wondergem, J., van der Zee, J. and Hulshof, M.C. (2005) Effects of hyperthermia on the central nervous system: what was learnt from animal studies? Int. J. Hyperthermia, 21(5): 473–487.

Hawkins, B.T. and Davis, T.P. (2005) The blood–brain barrier/neurovascular unit in health and disease. Pharmacol. Rev., 57(2): 173–185.

Hildebrandt, B., Hegewisch-Becker, S., Kerner, T., Nierhaus, A., Bakhshandeh-Bath, A., Janni, W., Zumschlinge, R., Sommer, H., Riess, H., Wust, P. and The German Interdisciplinary Working Group on Hyperthermia. (2005) Current status of radiant whole-body hyperthermia at temperatures >41.5°C and practical guidelines for the treatment of adults. The German 'Interdisciplinary Working Group on Hyperthermia'. Int. J. Hyperthermia, 21(2): 169–183.

Huang, J.T., Leweke, F.M., Oxley, D., Wang, L., Harris, N., Koethe, D., Gerth, C.W., Nolden, B.M., Gross, S., Schreiber, D., Reed, B. and Bahn, S. (2006) Disease biomarkers in cerebrospinal fluid of patients with first-onset psychosis. PLoS Med., 3(11): e428.

Imai-Matsumura, K., Matsumura, K., Terao, A. and Watanabe, Y. (2002) Attenuated fever in pregnant rats is associated with blunted syntheses of brain cyclooxygenase-2 and PGE2. Am. J. Physiol. Regul. Integr. Comp. Physiol., 283(6): R1346–R1353.

Iwanaga, R., Matsuishi, T., Ohnishi, A., Nakashima, M., Abe, T., Ohtaki, E., Kojima, K., Nagamitsu, S., Ohbu, K. and Kato, H. (1996) Serial magnetic resonance images in a patient with congenital sensory neuropathy with anhidrosis and complications resembling heat stroke. J. Neurol. Sci., 142(1–2): 79–84.

Johanson, C. (2003) The choroid plexus-CSF nexus: gateway to the brain. In: Michael Conn P. (Ed.), Neuroscience in Medicine. Humana Press, Totowa: NJ, pp. 165–195.

Johanson, C., Duncan, J., Baird, A., Stopa, E. and McMillan, P. (2005a) Choroid plexus: a key player in neuroprotection and neuroregeneration. Int. J. Neuroprotec. Neuroregen., 1(2): 77–85.

Johanson, C.E., Duncan, J.A., Stopa, E.G. and Baird, A. (2005b) Enhanced prospects for drug delivery and brain targeting by the choroid plexus-CSF route. Pharm. Res., 22(7): 1011–1037.

Johanson, C.E., Foltz, F.M. and Thompson, A.M. (1974) The clearance of urea and sucrose from isotonic and hypertonic fluids perfused through the ventriculo-cisternal system. Exp. Brain Res., 20(1): 18–31.

Johanson, C.E., Gonzalez, A.M. and Stopa, E.G. (2001) Water-imbalance-induced expression of FGF-2 in fluid-regulatory centers: choroid plexus and neurohypophysis. Eur. J. Pediatr. Surg., 11(Suppl 1): S37–S38.

Johanson, C.E., Palm, D.E., Dyas, M.L. and Knuckey, N.W. (1994) Microdialysis analysis of effects of loop diuretics and acetazolamide on chloride transport from blood to CSF. Brain Res., 641(1): 121–126.

Johanson, C.E., Palm, D.E., McMillan, P.N., Stopa, E.G., Doberstein, C.E. and Duncan, J.A. (2004) Volume transmission-mediated protective impact of choroid plexus-CSF growth factors on forebrain ischemic injury. In: Sharma H.S. (Ed.), Blood–Spinal Cord and Brain Barriers in Health and Disease. Academic Press, San Diego, CA, pp. 361–384.

Johanson, C.E., Palm, D.E., Primiano, M.J., McMillan, P.N., Chan, P., Knuckey, N.W. and Stopa, E.G. (2000) Choroid plexus recovery after transient forebrain ischemia: role of growth factors and other repair mechanisms. Cell Mol. Neurobiol., 20(2): 197–216.

Johanson, C.E. and Woodbury, D.M. (1978) Uptake of [^{14}C]urea by the in vivo choroid plexus–cerebrospinal fluid-brain system: Identification of sites of molecular sieving. J. Physiol., 275: 167–176.

Jones, H.C., Keep, R.F. and Butt, A.M. (1992) The development of ion regulation at the blood–brain barrier. Prog. Brain Res., 91: 123–131.

Kaiser, R., Rubin, C.H., Henderson, A.K., Wolfe, M.I., Kieszak, S., Parrott, C.L. and Adcock, M. (2001) Heat-related death and mental illness during the 1999 Cincinnati heat wave. Am. J. Forensic Med. Pathol., 22(3): 303–307.

Katschinski, D.M., Wiedemann, G.J., Longo, W., d'Oleire, F.R., Spriggs, D. and Robins, H.I. (1999) Whole body hyperthermia cytokine induction: a review, and unifying hypothesis for myeloprotection in the setting of cytotoxic therapy. Cytokine Growth Factor Rev., 10(2): 93–97.

Kew, M.C., Bersohn, I., Peter, J., Wyndham, C.H. and Seftel, H.C. (1967) Preliminary observations on the serum and cerebrospinal fluid enzymes in heatstroke. S. Afr. Med. J., 41(21): 530–532.

Kis, B., Isse, T., Snipes, J.A., Chen, L., Yamashita, H., Ueta, Y. and Busija, D.W. (2006) Effects of LPS stimulation on the expression of prostaglandin carriers in the cells of the blood–brain and blood–cerebrospinal fluid barriers. J. Appl. Physiol., 100(4): 1392–1399.

Kiviranta, T., Tuomisto, L., Jolkkonen, J. and Airaksinen, E.M. (1996) Vasopressin in the cerebrospinal fluid of febrile children with or without seizures. Brain Dev., 18(2): 110–113.

Knuckey, N.W., Fowler, A.G., Johanson, C.E., Nashold, J.R. and Epstein, M.H. (1991) Cisterna magna microdialysis of ^{22}Na to evaluate ion transport and cerebrospinal fluid dynamics. J. Neurosurg., 74(6): 965–971.

Levine, S. (1987) Choroid plexus: target for systemic disease and pathway to the brain. Lab. Invest., 56(3): 231–233.

Lu, T.S., Chen, H.W., Huang, M.H., Wang, S.J. and Yang, R.C. (2004) Heat shock treatment protects osmotic stress-induced dysfunction of the blood–brain barrier through preservation of tight junction proteins. Cell Stress Chaperones, 9(4): 369–377.

Maki, A.A., Beck, M.M., Gleaves, E.W., DeShazer, J.A. and Eskridge, K.M. (1990) CSF ion composition and manipulation during thermoregulation in an avian species, *Gallus domesticus*. Comp. Biochem. Physiol. A, 96(1): 135–140.

Malamud, N., Haymaker, W. and Custer, R.P. (1946) Heat stroke. A clinicopathological study of 125 fatal cases. Milit. Surg., 99: 397–449.

Masada, T., Hua, Y., Xi, G., Ennis, S.R. and Keep, R.F. (2001) Attenuation of ischemic brain edema and cerebrovascular injury following ischemic preconditioning in the rat. J. Cereb. Blood Flow Metab., 21: 22–33.

Maslinska, D., Kaliszek, A., Opertowska, J., Toborowicz, J., Deregowski, K. and Szukiewicz, D. (1999) Constitutive expression of cyclooxygenase-2 (COX-2) in developing brain. A. Choroid plexus in human fetuses. Folia Neuropathol., 37(4): 287–291.

Matthews, S.G., Parrott, R.F. and Sirinathsinghji, D.J. (1993) Distribution and cellular localization of vasopressin mRNA in the ovine brain, pituitary and pineal glands. Neuropeptides, 25(1): 11–17.

Moore, R., Mallonee, S., Sabogal, R.I., Zanardi, L., Redd, J. and Malone, J. (2002) Heat-related deaths — four states, July–August 2001, and United States, 1979–1999. JAMA, 288(8): 950–951.

Murphy, V.A. and Johanson, C.E. (1985) Adrenergic-induced enhancement of brain barrier system permeability to small nonelectrolytes: choroid plexus versus cerebral capillaries. J. Cereb. Blood Flow Metab., 5(3): 401–412.

Murphy, V.A. and Johanson, C.E. (1989) Acidosis, acetazolamide, and amiloride: effects on ^{22}Na transfer across the blood–brain and blood–CSF barriers. J. Neurochem., 52(4): 1058–1063.

Nilsson, C., Lindvall-Axelsson, M. and Owman, C. (1992) Neuroendocrine regulatory mechanisms in the choroid plexus-cerebrospinal fluid system. Brain Res. Brain Res. Rev., 17(2): 109–138.

Palm, D., Knuckey, N., Guglielmo, M., Watson, P., Primiano, M. and Johanson, C. (1995) Choroid plexus electrolytes and ultrastructure following transient forebrain ischemia. Am. J. Physiol., 269(1 Pt 2): R73–R79.

Pan, W. and Kastin, A.J. (2004) Why study transport of peptides and proteins at the neurovascular interface. Brain Res. Brain Res. Rev., 46(1): 32–43.

Parandoosh, Z. and Johanson, C.E. (1982) Ontogeny of blood–brain barrier permeability to, and cerebrospinal fluid sink action on [^{14}C]urea. Am. J. Physiol., 243(3): R400–R407.

Preston, J.E. (2001) Ageing choroid plexus-cerebrospinal fluid system. Microsc. Res. Tech., 52(1): 31–37.

Rapoport, S.I. (1976) The Blood–Brain Barrier in Physiology and Medicine. Raven Press, New York, pp. 1–380.

Rasmusson, A.M., Pinna, G., Paliwal, P., Weisman, D., Gottschalk, C., Charney, D., Krystal, J. and Guidotti, A. (2006) Decreased cerebrospinal fluid allopregnanolone levels in women with posttraumatic stress disorder. Biol. Psychiatry, 60(7): 704–713.

Redzic, Z.B., Preston, J.E., Duncan, J.A., Chodobski, A. and Szmydynger-Chodobska, J. (2005) The choroid plexus-cerebrospinal fluid system: from development to aging. Curr. Top. Dev. Biol., 71: 1–52.

Reulen, H.J., Tsuyumu, M., Tack, A., Fenske, A.R. and Prioleau, G.R. (1978) Clearance of edema fluid into cerebrospinal fluid. A mechanism for resolution of vasogenic brain edema. J. Neurosurg., 48(5): 754–764.

Robins, H.I., Peterson, C.G. and Mehta, M.P. (2003) Combined modality treatment for central nervous system malignancies. Semin. Oncol., 30(4 Suppl 9): 11–22.

Ruwe, W.D., Naylor, A.M. and Veale, W.L. (1985) Perfusion of vasopressin within the rat brain suppresses prostaglandin E-hyperthermia. Brain Res., 338(2): 219–224.

Saavedra, J.M., Benicky, J. and Zhou, J. (2006) Angiotensin II: multitasking in the brain. J. Hypertens. Suppl., 24(1): S131–S137.

Saunders, N.R., Habgood, M.D. and Dziegielewska, K.M. (1999) Barrier mechanisms in the brain, I. Adult brain. Clin. Exp. Pharmacol. Physiol., 26(1): 11–19.

Schalk, K.A., Faraci, F.M. and Heistad, D.D. (1992) Effect of endothelin on production of cerebrospinal fluid in rabbits. Stroke, 23(4): 560–563.

Scoville, S.L., Gardner, J.W., Magill, A.J., Potter, R.N. and Kark, J.A. (2004) Nontraumatic deaths during U.S. Armed

Forces basic training, 1977–2001. Am. J. Prev. Med., 26(3): 205–212.

Segal, M.B. (2000) The choroid plexuses and the barriers between the blood and the cerebrospinal fluid. Cell Mol. Neurobiol., 20(2): 183–196.

Sharma, H.S. (1982) Blood–Brain Barrier in Stress, PhD Thesis, Banaras Hindu University, Varanasi, India, pp. 1–85.

Sharma, H.S. (1999) Pathophysiology of blood–brain barrier, brain edema and cell injury following hyperthermia: new role of heat shock protein, nitric oxide and carbon monoxide. An experimental study in the rat using light and electron microscopy. Acta Universitatis Upsaliensis, 830: 1–94.

Sharma, H.S. (2004) Blood–brain and spinal cord barriers in stress. In: Sharma H.S. and Westman J. (Eds.), The Blood–Spinal Cord and Brain Barriers in Health and Disease. Elsevier Academic Press, San Diego, CA, pp. 231–298.

Sharma, H.S. (2005a) Heat-related deaths are largely due to brain damage. Indian J. Med. Res., 121(5): 621–623.

Sharma, H.S. (2005b) Methods to induce brain hyperthermia. In: Costa E. (Ed.), Current Protocols in Toxicology. Suppl 23, Unit 11.14, Wiley, New York, USA, pp. 1–26.

Sharma, H.S. (2005c) Pathophysiology of blood–spinal cord barrier in traumatic injury and repair. Curr. Pharm. Des., 11(11): 1353–1389.

Sharma, H.S. (2006) Hyperthermia influences excitatory and inhibitory amino acid neurotransmitters in the central nervous system. An experimental study in the rat using behavioural, biochemical, pharmacological, and morphological approaches. J. Neural Transm., 113(4): 497–519.

Sharma, H.S. and Cervós-Navarro, J. (1990) Brain oedema and cellular changes induced by acute heat stress in young rats. Acta Neurochir. Suppl. (Wien.), 51: 383–386.

Sharma, H.S., Cervós-Navarro, J. and Dey, P.K. (1991a) Acute heat exposure causes cellular alteration in cerebral cortex of young rats. Neuroreport, 2: 155–158.

Sharma, H.S., Cervós-Navarro, J. and Dey, P.K. (1991b) Rearing at high ambient temperature during later phase of the brain development enhances functional plasticity of the CNS and induces tolerance to heat stress. An experimental study in the conscious normotensive young rats. Brain Dysfunct., 4: 104–124.

Sharma, H.S. and Dey, P.K. (1986) Probable involvement of 5-hydroxytryptamine in increased permeability of blood–brain barrier under heat stress. Neuropharmacology, 25: 161–167.

Sharma, H.S. and Dey, P.K. (1987) Influence of long-term acute heat exposure on regional blood–brain barrier permeability, cerebral blood flow and 5-HT level in conscious normotensive young rats. Brain Res., 424: 153–162.

Sharma, H.S. and Hoopes, P.J. (2003) Hyperthermia induced pathophysiology of the central nervous system. Int. J. Hyperthermia, 19(3): 325–354.

Sharma, H.S. and Westman, J. (1998) Brain Functions in Hot Environment. Progress in Brain Research, Vol. 115. Elsevier, Amsterdam, pp. 1–516.

Sharma, H.S., Dey, P.K., Ashok Kumar (1986) Role of circulating 5-HT and lung MAO activity in physiological processes of heat adaptation in conscious young rats. Biomedicine 6: 31–40.

Sharma, H.S., Drieu, K. and Westman, J. (2003) Antioxidant compounds EGB-761 and BN-52021 attenuate brain edema formation and hemeoxygenase expression following hyperthermic brain injury in the rat. Acta Neurochir. Suppl. (Wien.), 86: 313–319.

Sharma, H.S., Duncan, J.A. and Johanson, C.E. (2006) Whole-body hyperthermia in the rat disrupts the blood–cerebrospinal fluid barrier and induces brain edema. Acta Neurochir. Suppl. (Wien.), 96: 426–431.

Sharma, H.S., Olsson, Y. and Dey, P.K. (1990) Changes in blood–brain barrier and cerebral blood flow following elevation of circulating serotonin level in anesthetized rats. Brain Res., 517(1–2): 215–223.

Sharma, H.S., Westman, J., Cervós-Navarro, J. and Nyberg, F. (1997) Role of neurochemicals in brain edema and cell changes following hyperthermic brain injury in the rat. Acta Neurochir. Suppl. (Wien.), 70: 269–274.

Sharma, H.S., Westman, J. and Nyberg, F. (1998) Pathophysiology of brain edema and cell changes following hyperthermic brain injury. In: Sharma H.S. and Westman J. (Eds.), Brain Functions in Hot Environment. Progress in Brain Research, Vol. 115. Elsevier, Amsterdam, pp. 351–412.

Shu, C., Shen, H., Keep, R.F. and Smith, D.E. (2002) Role of PEPT2 in peptide/mimetic trafficking at the blood–CSF barrier: studies in rat choroid plexus epithelial cells in primary culture. J. Pharmacol. Exp. Ther., 301: 820–829.

Smith, Q.R. (1991) The blood–brain barrier and the regulation of amino acid uptake and availability to brain. Adv. Exp. Med. Biol., 291: 55–71.

Smith, D.E., Johanson, C.E. and Keep, R.F. (2004) Peptide and peptide analog transport systems at the blood–CSF barrier. Adv. Drug Deliv. Rev., 56(12): 1765–1791.

Smith, Q.R. and Rapoport, S.I. (1986) Cerebrovascular permeability coefficients to sodium, potassium, and chloride. J. Neurochem., 46: 1732–1742.

Smith, Q.R., Woodbury, D.M. and Johanson, C.E. (1982) Kinetic analysis of [^{36}Cl]-, [^{22}Na]- and [^{3}H]mannitol uptake into the in vivo choroid plexus-cerebrospinal fluid brain system: ontogeny of the blood brain and blood–CSF barriers. Brain Res., 255(2): 181–198.

Spector, R. and Johanson, C. (1989) The mammalian choroid plexus. Sci. Am., 261(5): 68–74.

Spector, R. and Johanson, C. (2006) Micronutrient and urate transport in choroid plexus and kidney: implications for drug therapy. Pharm. Res., 23(11): 2515–2524.

Stoevring, B., Jaliashvili, I., Thougaard, A.V., Ensinger, C., Hogdall, C.K., Rasmussen, L.S., Sellebjerg, F. and Christiansen, M. (2006) Tetranectin in cerebrospinal fluid of patients with multiple sclerosis. Scand. J. Clin. Lab. Invest., 66(7): 577–584.

Stott, P.A., Stone, D.A. and Allen, M.R. (2004) Human contribution to the European heatwave of 2003. Nature, 432(7017): 559–560.

Strazielle, N. and Ghersi-Egea, J.F. (1999) Demonstration of a coupled metabolism-efflux process at the choroid plexus as a mechanism of protection toward xenobiotics. J. Neurosci., 19: 6275–6289.

Sweeney, K.G. (2002) Heat-related deaths. J. Insur. Med., 34(2): 114–119.

Szmydynger-Chodobska, J., Chun, Z.G., Johanson, C.E. and Chodobski, A. (2002) Distribution of fibroblast growth factor receptors and their co-localization with vasopressin in the choroid plexus epithelium. Neuroreport, 13(2): 257–259.

Thrall, D.E., Page, R.L., Dewhirst, M.W., Meyer, R.E., Hoopes, P.J. and Kornegay, J.N. (1986) Temperature measurements in normal and tumor tissue of dogs undergoing whole body hyperthermia. Cancer Res., 46(12 Pt 1): 6229–6235.

Vellucci, S.V. and Parrott, R.F. (1998) Expression of mRNAs for vasopressin, oxytocin and corticotrophin releasing hormone in the hypothalamus, and of cyclooxygenases-1 and -2 in the cerebral vasculature, of endotoxin-challenged pigs. Neuropeptides, 32(5): 439–446.

Wang, Y.F., Gwathmey, J.K., Zhang, G., Soriano, S.G., He, S. and Wang, Y. (2005) Cerebrospinal fluid may mediate CNS ischemic injury. Cerebrospinal Fluid Res., 2: 7.

Wilson, J.F., Anderson, S., Snook, G. and Llewellyn, K.D. (1984) Quantification of the permeability of the blood–CSF barrier to alpha-MSH in the rat. Peptides, 5: 681–685.

Yang, Y. and Zhi, D. (1994) Effects of acupuncture hypothermia and its relationship to changes of AVP contents in the plasma and CSF in the rabbits. Zhen Ci Yan Jiu, 19(2): 56–59.

Zemo, D.A. and McCabe, J.T. (2001) Salt-loading increases vasopressin and vasopressin 1b receptor mRNA in the hypothalamus and choroid plexus. Neuropeptides, 35(3–4): 181–188.

Zipser, B.D., Johanson, C.E., Gonzalez, L., Berzin, T.M., Tavares, R., Hulette, C.M., Vitek, M.P., Hovanesian, V. and Stopa, E.G. (2006) Microvascular injury and blood–brain barrier leakage in Alzheimer's disease. Neurobiol. Aging [Epub ahead of print].

SECTION X

Heat Stroke and Hyperthermia

CHAPTER 24

Heat stroke and cytokines

Lisa R. Leon*

US Army Research Institute of Environmental Medicine, Thermal and Mountain Medicine Division, Natick, MA 01760-5007, USA

Abstract: Heat stroke is a life-threatening illness that affects all segments of society, including the young, aged, sick, and healthy. The recent high death toll in France (Dorozynski, 2003) and the death of high-profile athletes has increased public awareness of the adverse effects of heat injury. However, the etiology of the long-term consequences of this syndrome remains poorly understood such that preventive/treatment strategies are needed to mitigate its debilitating effects. Cytokines are important modulators of the acute phase response (APR) to stress, infection, and inflammation. Current data implicating cytokines in heat stroke responses are mainly from correlation studies showing elevated plasma levels in heat stroke patients and experimental animal models. Correlation data fall far short of revealing the mechanisms of cytokine actions such that additional research to determine the role of these endogenous substances in the heat stroke syndrome is required. Furthermore, cytokine determinations have occurred mainly at end-stage heat stroke, such that the role of these substances in progression and long-term recovery is poorly understood. Despite several studies implicating cytokines in heat stroke pathophysiology, few studies have examined the protective effect(s) of cytokine antagonism on the morbidity and mortality of heat stroke. This is particularly surprising since heat stroke responses resemble those observed in the endotoxemic syndrome, for which a role for endogenous cytokines has been strongly implicated. The implication of cytokines as mediators of endotoxemia and the presence of circulating endotoxin in heat stroke patients suggests that much knowledge can be gained from applying our current understanding of endotoxemic pathophysiology to the study of heat stroke. Heat shock proteins (HSPs) are highly conserved proteins that function as molecular chaperones for denatured proteins and reciprocally modulate cytokine production in response to stressful stimuli. HSPs have been shown repeatedly to confer protection in heat stroke and injury models. Interactions between HSPs and cytokines have received considerable attention in the literature within the last decade such that a complex pathway of interactions between cytokines, HSPs, and endotoxin is thought to be occurring *in vivo* in the orchestration of the APR to heat injury. These data suggest that much of the pathophysiologic changes observed with heat stroke are not a consequence of heat exposure, *per se*, but are representative of interactions among these three (and presumably additional) components of the innate immune response. This chapter will provide an overview of current knowledge regarding cytokine, HSP, and endotoxin interactions in heat stroke pathophysiology. Insight is provided into the potential therapeutic benefit of cytokine neutralization for mitigation of heat stroke morbidity and mortality based on our current understanding of their role in this syndrome.

Keywords: heat stroke; heat stress; heat injury; hyperthermia; hypothermia; interleukin; tumor necrosis factor

*Corresponding author. Tel.: +1 508 233 4862; Fax: +1 508 233 5298; E-mail: lisa.r.leon@us.army.mil

Table 1. The heat illness continuum

Term	Definition
Heat cramps	Intermittent cramping pain in muscles subjected to strenuous activity; normal T_c; may occur in cold environment
Heat exhaustion (heat prostration; heat collapse)	Heat illness due to salt or water depletion resulting from strenuous physical exercise or prolonged exposure to a hot environment; T_c may or may not be elevated; decreased cardiac output
Heat stroke	Life-threatening illness characterized by $T_c 41°C$ and CNS abnormalities (delirium, fainting, seizures, and coma) resulting from prolonged exposure to a hot environment (classic) or strenuous physical exercise (exertional)

The heat illness continuum

The heat illness syndrome is typically depicted as a series of discrete events, characterized by pathophysiologic responses that increase in severity as one moves from the mildly innocuous condition of heat cramps to heat exhaustion and heat stroke (Table 1; Petersdorf, 1994). Heat cramps are the most benign condition, precipitated by strenuous muscle activity and profuse sweating that results in a loss of electrolytes. Spasms of skeletal muscles in the extremities may be sporadic, but painful in this condition (Wexler, 2002). Heat cramps typically occur following exercise in the cold and are not associated with elevated environmental temperature (Petersdorf, 1994). Heat exhaustion (also referred to as heat prostration or heat collapse) is the most common heat syndrome, which results from water or salt depletion in a hot environment. This mild to moderate illness is associated with an inability to maintain adequate cardiac output resulting in elevation of core temperature and potential for collapse. The use of diuretics and other medications may predispose individuals to heat exhaustion (Petersdorf, 1994). Heat stroke is the most serious condition resulting from prolonged exposure to a hot environment. The clinical definition of heat stroke includes core temperature in excess of 41.0°C, hot, dry flushed skin, and central nervous system (CNS) dysfunction (Petersdorf, 1994). While removal from the heat and rapid cooling are essential for heat stroke survival, a variety of complications ensuing after heat exposure make the choice of treatment modalities difficult, thus enhancing the probability of permanent neurological damage in survivors (Malamud et al., 1946; Dematte et al., 1998).

The absence or presence of an exertional component during heat exposure allows further categorization of heat stroke into its classic (i.e., passive) or exertional form. Classic heat stroke results from passive exposure to a hot environment and is typically observed in immunocompromised and aging populations, which show enhanced mortality during heat waves (Dematte et al., 1998; Naughton et al., 2002; Dorozynski, 2003). Pre-existing conditions, such as mental illness, alcoholism, or drug use (e.g., diuretics, anticholinergics) can predispose individuals to classic heat stroke (Levine, 1969; Naughton et al., 2002). Conversely, exertional heat stroke typically occurs in healthy, young individuals undergoing strenuous physical activity in hot environments. Athletes and soldiers represent two high-risk populations for this form of heat injury, although heat acclimatization of these populations can reduce risk (Coris et al., 2004). Exertional heat stroke is a particularly complicated heat syndrome to study since it is difficult to dissociate the direct effects of strenuous physical activity from that imposed by exposure to a hot environment.

Epidemiology of heat wave mortalities

In 1999, Chicago experienced its second deadliest heat wave in a decade in which 80 mortalities were reported (Naughton et al., 2002). These deaths occurred despite extensive programs to educate and provide interventions to high-risk populations, such as the elderly. A 20-fold higher rate of heat-related mortality was reported in persons >75 years of age compared to younger populations (Naughton et al., 2002). Naughton et al. (2002) reported several social factors predisposing individuals to heat mortality, including living alone, an inability or unwillingness to leave one's home, residing on the top floor of buildings (heat rises) and an annual income of <$10,000/year (Naughton et al., 2002). A working air conditioner was shown to be the strongest protective factor, while fan cooling did not afford protection (Naughton et al., 2002). The availability and *use* of air conditioning units may be directly related to socioeconomic status, since these units are energetically expensive to run and in many cases are nonfunctional (Dematte et al., 1998; Naughton et al., 2002). Ineffective cooling prolongs exposure time to elevated ambient temperatures that increases the risk of mortality or permanent neurological damage in heat stroke survivors (Hart et al., 1982). The unwillingness of the elderly to leave their residences and the presence of higher indoor temperatures in the absence of air conditioning units (or refusal to use those units) magnifies the intensity of heat exposure. Vandentorren et al. (2004) suggest that the high death toll (~15,000) in the 2003 France heat wave may be partly explained by culture, as air conditioning units are typically not used in residences, retirement homes, or hospitals in that country. It is assumed that the large population (~10,000) of people aged 100 and over in France, in combination with a lack of air conditioning explains the high mortality rate in that country compared to other European regions that experienced the same heat wave (Dorozynski, 2003). As the average human lifespan increases, the development of education and intervention strategies will become increasingly important to prevent an increase in heat stroke deaths in the aged.

Urban structure also impacts heat mortality rates. Inhabitants of urban dwellings are exposed to greater intensity and longer duration of heat exposure since concrete structures do not effectively dissipate heat as nighttime temperatures decrease (Clarke, 1972). Landsberg (1970) reports city temperatures ~0.5° to 1.0°C warmer than rural areas; these "urban heat islands" represent the warmest areas which are typically in the center of the city and whose magnitude is dependent on city size. This relates to recent findings from Stott et al. (2004) demonstrating that activities that increase the production of greenhouse gases will double the risk for extreme climate fluctuations. Thus, as man's activities increase in urban centers it is not surprising that more heat deaths are reported, and expected in these areas (Shattuck and Hilferty, 1932, 1933).

Most heat waves occur across 3 or more days with the majority of hospitalizations and deaths occurring within 24 h of the onset of the event (Ramlow and Kuller, 1990; Kark et al., 1996; Dematte et al., 1998; Naughton et al., 2002). Thus, the duration of heat exposure and a lack of heat acclimatization increase the risk of heat stroke during the initial days of a heat wave. Several additional pre-disposing factors have been identified as increasing an individual's susceptibility to heat stroke mortality. Austin and Berry (1956) examined 100 cases of heat stroke and found cardiovascular illness in 84% of patients. Similarly, Levine (1969) found heat stroke mortalities in the elderly (200 cases) to be associated with arteriosclerotic heart disease (72%) and hypertension (12%). Animal studies have also shown a reduction in thermotolerance in spontaneously hypertensive rats (Wright et al., 1977). Heat strain imposes large cardiovascular demands on the body as blood flow is shunted from core organs to the skin to dissipate excess heat to the environment. A cardiac deficiency (e.g., congestive heart failure) impedes heat dissipation and an inability to maintain cardiac output during prolonged heat exposure leads to circulatory collapse and death. The high death rate (>600) in the Chicago 1995 heat wave was associated with hypertension, alcohol abuse, diuretic medications, aspirin, and pre-existing infections in the elderly population (Dematte et al., 1998). Alcohol depresses vasomotor reflexes and stimulates metabolism, resulting in increased heat gain and heat production. Thus, the high death toll due to excessive heat *per se* may be small compared to those caused by the aggravation in severity of a pre-existing condition. In addition, concurrent infections may predispose to heat stroke as individuals are immunocompromised prior to heat exposure (Knochel, 1989; Sonna et al., 2004).

Heat stroke causes changes in immune status such as disturbances in leukocyte distribution, production of cytokines, and bacterial translocation following gut ischemia that may predispose to early infection (Bouchama et al., 1992, 1993; Hall et al., 2001). The interaction of all of these events in heat stroke mortality only serves to complicate the etiology of this syndrome, thus limiting the success of currently established medical interventions.

Although soldiers and athletes represent young, healthy populations that do not have pre-existing physical ailments due to aging, significant hospitalization and death rates from exertional heat stroke have been recorded. Perhaps the most notorious example is the 1967 Six-Day War between Israel and Egypt in which 20,000 Egyptian soldiers suffered heat stroke deaths (Hubbard et al., 1982). Between 1911 and 1926, the US Navy reported 2049 cases of heat stroke deaths (Wakefield and Hall, 1927). Currently, the US Army hospitalizes 25–70 soldiers per 100,000 due to heat injury per year and the incidence of heat stroke hospitalizations has increased almost 10-fold over the past 20 years (Carter et al., 2005). During peacetime exercises, Malamud et al. (1946) noted that ~25% of fatal heat stroke cases occurred in military recruits that had been in training camp for less than ~2 weeks. The majority of cases occurred during the hottest summer months (typically July) and was likely a consequence of intense exercise with a lack of heat acclimatization. Obesity was also a significant factor (Malamud et al., 1946). Given the increasing incidence of obesity in the U.S., this may be a particular concern for future military populations. Carter et al. (2005) examined heat illness hospitalization rates and deaths for the US Army from 1980 to 2002 and reported the highest incidence in those recruits enlisted for <12 months, with recruits from northern, cold climate states at higher risk than those from southern, warm climate states.

Protective clothing may be a significant pre-disposing factor to heat stroke deaths in athletic and military populations. Under normal clothing conditions, sufficient heat exchange between the skin surface and the environment can occur to regulate core temperature within a narrow range, thus supporting thermal homeostasis. Protective clothing, which may consist of multiple layers and often encapsulates the head (a site of significant heat exchange; Rasch et al., 1991), forms an insulative layer of air between the skin and the environment, thus impeding heat exchange. Fifty-one cases of exertional heat illness were observed in military trainees in San Antonio, Texas, during participation in a 5.8 mile march in full battle dress uniform and boots (Smalley et al., 2003). Athletic uniforms also limit evaporative and convective heat loss during strenuous activity. Similar to that observed with heat waves and military training exercises, a lack of acclimatization to the uniform and high environmental temperatures results in the majority of heat stroke cases observed on the second or third day of football practice in hot weather (Graber et al., 1971; Roberts, 2004). To avoid heat exposure, exercise regimens are typically scheduled for the early morning hours, when it is cooler; guidelines for adequate hydration are also implemented in athletic and military populations (Armstrong et al., 1996; Departments of Army and Air Force, 2003). Paradoxically, the use of nutritional supplements and implementation of fluid replacement guidelines may result in longer duration of exposure to elevated temperatures, potentially increasing heat stroke incidence (Carter et al., 2005). A lack of heat acclimatization as well as increased emphasis on high intensity, long-duration exercise regimens, which adds a metabolic heat load to the body, has likely increased heat stroke hospitalization rates in military and athletic populations in the past two decades (Carter et al., 2005). To further complicate the issue, exercise has been shown to induce similar changes in core temperature (values as high as 41.9°C have been noted in marathon runners; Maron et al., 1977) and immune parameters (Cross et al., 1996; Nielsen and Pedersen, 1997) as heat exposure such that teasing apart the influence of these two factors in exertional heat stroke is made more difficult.

Systemic responses to heat stroke

Heat stroke is clinically defined as core temperature in excess of 41.0°C, hot, flushed dry skin and CNS dysfunction, such as delirium, coma, and seizures (Petersdorf, 1994). However, a myriad of thermoregulatory, hemodynamic, and organ abnormalities are manifest in heat illness beyond those immediately recognized by the clinical definition of this syndrome (Table 2). The implication of cytokines as regulators/modulators of several of the responses listed in Table 2 suggests that neutralization of one or more of the actions of these endogenous mediators may be beneficial in the mitigation of heat stroke morbidity and mortality.

Despite the myriad of complications associated with heat illness, an elevation of core temperature above 41.0°C (often referred to as fever or hyperpyrexia) is the most widely recognized symptom of this syndrome. Core temperature is extremely labile to environmental and physiological perturbations and is simple to measure, thus providing rapid and powerful information regarding homeostatic balance of the individual. The basis for a specific core temperature cut-off value of 41°C as a heat stroke criterion is not readily apparent, but may reflect an attempt to dissociate the degree of hyperthermia observed in heat illness from that of infection, in which fevers rarely exceed 41°C (Dubois, 1949).

Core temperature varies dramatically in heat stroke patients with ranges of 41° to 42°C commonly observed and values as high as ~47°C reported (Bouchama et al., 1991, 1993; Chang, 1993; Hammami et al., 1997; Hashim et al., 1997; Lu et al., 2004; Sonna et al., 2004). Large variability in core temperature values may be due to several factors, including (1) differences in the time of clinical presentation such that patients' temperature is obtained at varying stages of heat stroke progression and treatment (i.e., hyperthermia vs. cooling), (2) individual differences in the critical thermal maximum (CTM) associated with heat stroke mortality, and (3) the site of the temperature measurement. CTM is defined as the minimum core temperature that is lethal to an organism (Cowles and Bogert, 1944; Hutchison, 1961). As previously discussed predisposing factors such as medications, infection, and cardiovascular disease may enhance susceptibility to heat stroke, which would presumably manifest as a lower CTM prior to collapse. Attempts to define the CTM in several species indicate wide species and

Table 2. Complications associated with heat stroke

Thermoregulatory
 T_c 41°C (hyperpyrexia)
 Hot, dry skin
 Fever (long-term symptom)
 Hypothermia (animal studies only)

Hemodynamic
 Dehydration/hemoconcentration
 Disseminated intravascular coagulation (DIC)
 Elevated C-reactive protein (CRP)
 Endotoxemia
 Hyperglycemia/hypoglycemia
 Hypotension/cardiovascular abnormalities
 Increased cortisol/corticosterone
 Increased pro- and anti-inflammatory cytokines
 Lactic acidosis
 Leukocytosis
 Rhabdomyolysis

Tissue injury
 Adrenal ischemia
 Cardiac focal necrosis
 Cerebral edema/ischemia
 Increased intestinal permeability
 Liver necrosis (long-term symptom)
 Lung edema
 Renal failure
 Spleen necrosis
 Multi-organ system failure (ultimate cause of mortality)

inter-individual variability. Austin and Berry (1956) report core temperature values ranging from 38.5 to 44.0°C in heat stroke patients with 10% of the reported mortalities occurring below 41.1°C. Thus, many patients do not meet the clinical core temperature criterion of heat stroke.

Obviously, it is not ethical or desirable to experimentally determine the CTM of humans. Therefore, animal models of heat stroke have been developed to more precisely determine the CTM in various species in order to study the mechanisms of thermoregulatory control during heat exposure. Adolph (1947) determined the CTM in cats (~43.5°C), dogs (41.7°C), and rats (42.5°C), proposing differences in tissue susceptibility as responsible for variability in CTM between species (tissue injury was not measured in this study). Wide ranges of CTM are reported in monkeys (~44.5°C; Gathiram et al., 1987a), dogs (37.7° to 41.1°C; Drobatz and Macintire, 1996), sheep (43.7° to 44.0°C; Hales et al., 1987), rats (40.4° to 45.4°C; Ohara et al., 1975; Hubbard et al., 1976; Wright et al., 1977; DuBose et al., 1983a; Lord et al., 1984), mice (42.7°C and ~44° to 45°C; Wright, 1976; Leon et al., 2005), and salamanders (~33°C; Hutchison and Murphy, 1985). Malamud et al. (1946) proposed direct thermal injury to the thermoregulatory centers of the brain as the primary mechanism of heat stroke mortality, despite an inability to detect thermal injury to the hypothalamus at autopsy of 125 fatal cases of heat stroke. The ability to induce core temperatures of 41.6° to 42.0°C in humans (within the reported CTM range) with no adverse clinical effects illustrates the inability to rely on a specific CTM for injury predictions (Bynum et al., 1978). Similarly, rectal temperatures of 41.9°C have been recorded in competitive runners showing no adverse clinical signs of heat injury, suggesting that this level of elevated core temperature is tolerable in humans (Maron et al., 1977). Of course, runners may be acclimatized to elevated core temperatures due to repeated exposure during intensive training regimens.

Inconsistency in CTM values may be due as much to individual and species variability in thermal resistance as to methodological differences between studies. One of the most dramatic differences between studies is the ambient temperature used to induce heat stroke. The ambient temperature used in animal studies ranges from 38.6 to 59.4°C, making comparisons between studies difficult (Adolph, 1947; Ohara et al., 1975; Wright, 1976; Hubbard et al., 1977; Wright et al., 1977; DuBose et al., 1983a; Gathiram et al., 1987a; Leon et al., 2005). The physiological relevance of 59.4°C is questionable since this ambient temperature is not routinely encountered in nature (Adolph, 1947). Similarly, the majority of these studies exposed animals to preheated environmental chambers (Adolph, 1947; Ohara et al., 1975; Wright, 1976; Hubbard et al., 1977; Wright et al., 1977; DuBose et al., 1983a; Gathiram et al., 1987a; Heidemann et al., 2000), which represents a heat "shock" rather than heat "stress" paradigm. *In vitro* studies have also used heat shock paradigms (e.g., 42° to 43°C water bath exposure for 1 h) to examine responses in different cell types (D'Souza et al., 1994; Watanabe et al., 1997, 1998). Heat stroke severity is influenced by the rate of heating such that rapid exposure to a pre-heated chamber may not provide sufficient lag time for the thermoregulatory system to sense and respond to a dramatic shift in ambient temperature before core temperature reaches lethal levels (Hutchison, 1961; Flanagan et al., 1995). Thus, the physiological relevance of the heat shock paradigm is questionable.

Core temperature values of heat stroke patients will differ depending on the site of measurement. In humans, esophageal temperature is the most accurate and responsive to changes in blood temperature, although instrumentation may not be feasible in severely injured, unresponsive patients. Rectal temperature has a slower response rate and gives slightly higher readings than esophageal temperature (Bynum et al., 1978). Oral temperature is rapidly measured, but may provide inaccurate (low) readings due to hyperventilation in the heat stroke patient (Cole, 1983). The recent development of remote core temperature sensing by radiotelemetry is a powerful technique that is applicable to both human and animal studies. In humans, core temperature may vary as the pill is swallowed and travels through the gastrointestinal (GI) tract. In animal models, the use of radiotelemetry has significantly improved experimental design by permitting an assessment of rapid and long-term core temperature changes induced by heat exposure of conscious, freely moving animals (Leon et al., 2005). It is anticipated that these advances in physiological monitoring will significantly improve our ability to model heat stroke responses in experimental test species.

Thermoregulatory consequences of heat stroke

While several characteristics of the thermoregulatory response during progression to heat stroke collapse are well defined, the core temperature response observed during recovery has received less attention. This is rather surprising since the magnitude, duration, and direction of core temperature changes displayed during recovery may provide information regarding severity and etiology of the initial heat insult. In experimental animals, hypothermia is the

Table 3. Incidence of hypothermia during heat stroke recovery

Species	Maximum T_c	Recovery T_a	Hypothermia	Authors
Cat	NS	NS	33.5, 35.0°C	Adolph (1947)
Guinea pig	NS	NS	37°C	Adolph (1947)
	IPH; 43.9°C	21° to 23°C	T_h; 34°C	Romanovsky and Blatteis (1996)
Mouse	42°C	10, 25°C	32°C	Wright (1976)
	38.7° to 41.9°C	24°C	29° to 35°C	Wilkinson et al. (1988)
	42.4, 42.7, 43.0°C	25°C	29° to 31°C	Leon et al. (2005)
Rat	41.5° to 42.8°C	Heat pad 39° to 40°C	35° to 36°C	Lord et al. (1984)
Salamander	33.7°C	Thermal gradient	10°C below controls	Hutchison and Murphy (1985)

Abbreviations: IPH — intraperitoneal heating; T_a — ambient temperature; T_c — core temperature; T_h — hypothalamic temperature; NS — not specified.

predominant heat stress recovery response (Table 3). Heat-induced hypothermia is the term used to define the seemingly paradoxical decrease of core temperature below baseline levels (Romanovsky and Blatteis, 1996). Note that the depth of hypothermia varies widely between studies, which may be species-specific or a result of nonconformity between experimental designs. Nevertheless, regardless of the experimental conditions, hypothermia of > 1.0°C is commonly observed in both mammals and poikilotherms (e.g., salamanders; animals without effective autonomic temperature regulation; IUPS, 2001). In mice, hypothermia is quite profound such that core temperature may be regulated only a few degrees above ambient temperature (Wright, 1976; Wilkinson et al., 1988; Leon et al., 2005). The depth of hypothermia is also affected by the ambient temperature to which animals are exposed during recovery. As shown in Fig. 1, the hypothermic response of mice during heat stress recovery is significantly blunted during exposure to a thermal gradient, which allows the behavioral selection of a wide range of ambient temperatures, compared to that observed during housing at a constant temperature of 25°C (cool ambient temperature for mice). The implication of these findings to heat stroke recovery responses in humans is currently unknown, although rapid cooling (which would be facilitated at lower ambient temperatures) is the most effective therapy for recovery.

In mice, the depth (~1.0° to 5.0°C) and duration (~1–24 h) of hypothermia is directly related to the severity of the heat insult, suggesting that core temperature responses displayed during recovery may serve as sensitive biomarkers of injury (Wilkinson et al., 1988; Leon et al., 2005). This is also apparent in comparisons of individual responses to heat stress; those animals experiencing the longest duration of heat exposure typically show the most profound heat-induced hypothermic response during recovery (Fig. 2). Exposure to a recovery ambient temperature that prevents hypothermia development enhances heat-induced intestinal damage and significantly decreases survival in mice (Wilkinson et al., 1988; Leon et al., 2005).

The connection between hypothermia and tissue injury suggests that cooling of heat stroke patients to a hypothermic level (i.e., core temperature <37°C) may be beneficial for the prevention of tissue injury. Further support for this contention is provided by the use of induced hypothermia, in which core temperature is physically decreased using cooling blankets or other methods, as a protective measure during cardiopulmonary bypass surgery and as treatment for cerebral ischemia and stroke (Marion et al., 1997; Dietrich and Kuluz, 2003). The realization that hypothermic treatment would be more efficacious if regulated, rather than forced reductions in core temperature were implemented suggests that further studies are required to determine the regulated nature of hypothermia under injurious conditions (Gordon, 2001). If cytokines are regulators/modulators of heat-induced hypothermia or their production is influenced by hypothermia, as previously described for

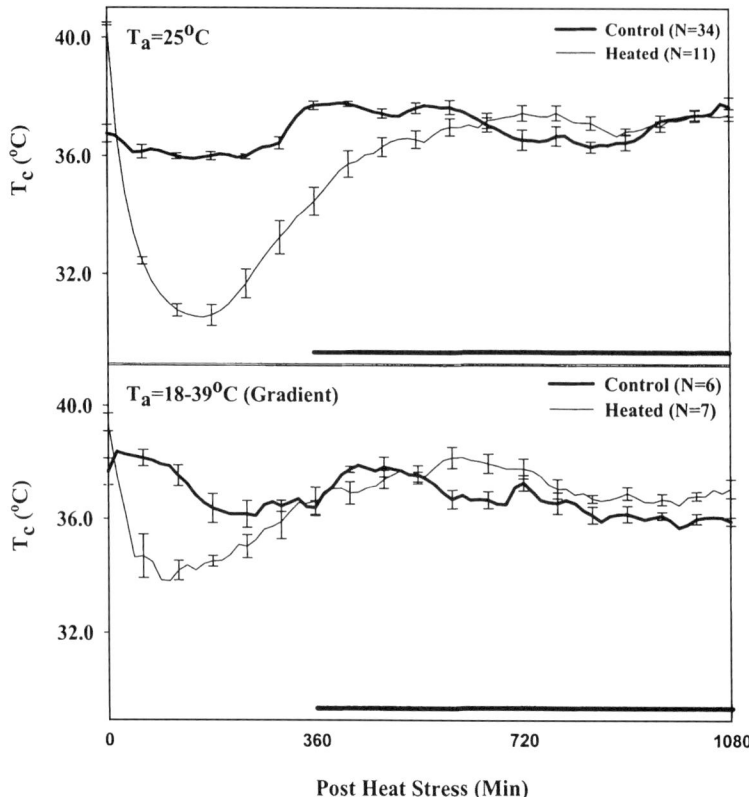

Fig. 1. Twenty-four hour core temperature ($\pm 0.1^\circ$C) response of male C57BL/6J mice during heat stress recovery at a constant ambient temperature (T_a) of $25 \pm 2^\circ$C (top) or in a thermal gradient that allows the behavioral selection of a wide range of ambient temperatures (18° to 39°C; bottom). Mice were heat stressed at an ambient temperature of $39.5 \pm 0.5^\circ$C, in the absence of food and water, until their core temperature reached 42.4°C (data not shown; Leon et al., 2005). Control mice were not exposed to heat stress. Immediately following heat exposure, mice were placed at 25°C or in the thermal gradient for recovery observations. Note that the hypothermic response displayed within the first ~360 min of recovery is significantly more pronounced at 25°C than in the thermal gradient. Black horizontal bar represents lights-off period in a 12:12 h L:D cycle. Sample sizes are indicated in parentheses. Core temperature was collected at 1 min intervals in conscious, freely moving mice using intraperitoneally implanted radiotelemetry devices (Data Sciences International, St. Paul, MN).

bacterial infection, they may represent one class of substances that could be targeted to induce hypothermia in a regulated fashion and minimize tissue injury in heat stroke patients (Arons et al., 1999; Fairchild et al., 2004; Leon et al., 2006).

In humans, anecdotal evidence suggests that fever is a symptom of heat stroke, persisting for 7–14 days following clinical presentation in some patients (Malamud et al., 1946; Austin and Berry, 1956; Attia et al., 1983). The occurrence of this thermoregulatory response late in the heat stroke syndrome suggests that fever, which is a tightly controlled physiologic response to stress, is more directly related to the complications ensuing after heat exposure, than to the initial heat insult. Although "fever" is reported immediately upon admission in many reports, it is likely that this represents the hyperthermic response to direct heat exposure rather than a true fever. Fever is defined as a regulated increase in the hypothalamic thermal setpoint and is observed in response to several stimuli including bacterial infection, stress, and tissue inflammation (Kluger, 1991). Fever results from the coordinated action of behavioral and physiological mechanisms that increase heat production and decrease heat loss to raise core

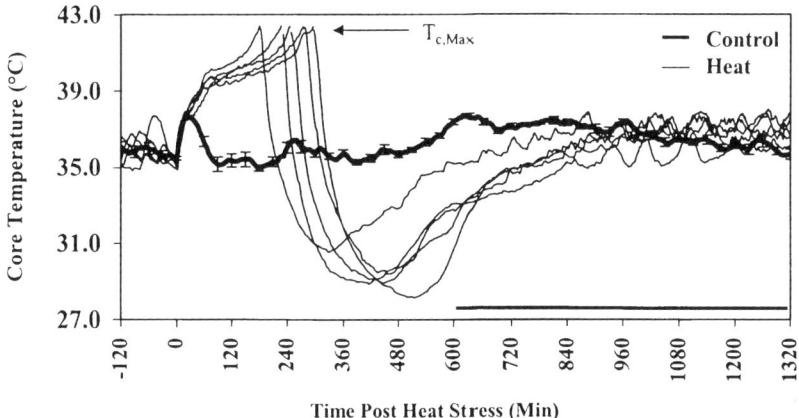

Fig. 2. Core temperature response ($\pm 0.1°C$) of male C57BL/6J mice during heat exposure (from time 0 to $T_{c,Max}$) and recovery at a constant ambient temperature of $25 \pm 2°C$ (from $T_{c,Max}$ to 1320 min). $T_{c,Max} = 42.4°C$. All mice served as their own controls. During the control condition, mice remained in their original cage location in the animal room and were not exposed to heat stress. Details of the heat stress protocol have been described elsewhere (Leon et al., 2005). Note that those mice experiencing the longest duration of heat exposure displayed the lowest depth and duration of hypothermia. Black horizontal bar represents lights-off period in a 12:12 h L:D cycle. Core temperature was collected at 1 min intervals in conscious, freely moving mice using intraperitoneally implanted radiotelemetry devices (Data Sciences International, St. Paul, MN).

temperature to a new elevated level. The presence of tissue injury and endotoxemia (both described in more detail below) in heat stroke suggests that the persistence of fever beyond the initial day of heat exposure (and clinical admission) may be a result of a systemic inflammatory response. This may also account for fever not being widely recognized as a heat stroke recovery response in animal studies. Due to reliance on rectal probes, restraint, and/or anesthesia for core temperature measurements, thermoregulatory responses to heat stress have typically not been examined across multiple circadian cycles in animal models (Lin et al., 1994, 1997; Romanovsky and Blatteis, 1996; Bouchama et al., 2005). The advent of radiotelemetry has alleviated this experimental limitation and shown that a "fever-like" core temperature elevation is observed 24–36 h following heat exposure in mice (Leon et al., 2005). Interestingly, mice show virtually identical increases in the fever-like core temperature response ($\sim 1.0°$ to $1.5°C$) irrespective of heat severity (Leon et al., 2005). This is in direct contrast to hypothermia, which is directly related to heat severity in mice, as previously described (Wilkinson, et al., 1988; Leon et al., 2005), indicating that fever is not a reliable biomarker of heat severity. Interestingly, mice that do not recover from hypothermia and develop fever, succumb to heat stroke (Leon et al., 2005). It is unclear if this is indicative of a protective function of fever or a debilitating effect of prolonged hypothermia in this syndrome.

Hemodynamic changes

Significant hemodynamic alterations occur in heat stroke patients (Table 2). Evaporative cooling (sweating in humans, salivary spreading in rodents) is a primary mechanism of core cooling under heat stress conditions. Prolonged heat exposure can induce significant dehydration and hemoconcentration. Disturbances in plasma glucose homeostasis (i.e., hyper- or hypoglycemia) are also common in heat stroke and may be directly related to thermal injury of the liver (Bouchama et al., 1996). Phosphoenolpyruvate carboxykinase (PEPCK) is a key regulatory enzyme of the hepatic gluconeogenic pathway — alterations in PEPCK regulation following thermal injury to the liver have been hypothesized as a mechanism of heat-induced hypoglycemia, although this has not been experimentally confirmed (Paidas et al., 2002). Previous reports showing an ability of dehydration and hypoglycemia to induce hypothermia in small rodents suggest that these pathophysiologic changes may represent two, of perhaps several, physiological stimuli driving the development of heat-induced hypothermia in rodents (Buchanan et al., 1991; Ibuka and Fukumura, 1997); the implications of this for the human condition is unknown. Immune dysfunction is common with disturbances in the distribution of several peripheral lymphocyte subpopulations (Bouchama et al., 1992; DuBose et al., 2003). The degree of hyperthermia is directly correlated with increased lymphocytes and T suppressor-cytotoxic cells (Bouchama et al., 1992). Changes in regional blood flow, increased catecholamine and cortisol release, and direct effects of exercise, cytokines, and endotoxin are proposed mechanisms of increases in these cell types with heat stroke (Levine, 1969; Maron et al., 1977; Haynes and Fauci, 1978; Al-Hadramy, 1989; Bouchama et al., 1991, 1992; Kappel et al., 1991; Frisina et al., 1994; Cross et al., 1996; Nielsen and Pedersen, 1997; Mitchell et al., 2002; Dubose et al., 2003).

Disseminated intravascular coagulation (DIC)

Disseminated intravascular coagulation (DIC) is a systemic intravascular disorder resulting from uncontrolled activation of coagulation and/or impairment of fibrinolysis or anticoagulation (Fig. 3; Bouchama et al., 1996; Bouchama and Knochel, 2002). Studies of the sepsis syndrome have provided the majority of data regarding DIC pathway components that are affected by cytokines. Cytokine modulation of DIC is supported by several lines of evidence including: (1) increased plasma levels of interleukin (IL)-1β, IL-6, and tumor necrosis factor (TNF) in patients with DIC — with high IL-6 correlating with organ failure (Wada et al., 1991a, b, 1993), (2) alteration of coagulation following IL-1, IL-2, IL-6, IL-8, IL-10, IL-12, or TNF injection (van der Poll et al., 1990; Baars et al., 1992; Pradier et al., 1993; Jansen et al., 1995; Stouthard et al., 1996; Neumann et al., 1997), and (3) attenuation of coagulation following injection of cytokine neutralizing antibodies (Paleolog et al., 1994; van der Poll et al., 1994; Schmid et al., 1995).

There are several different components of the coagulation, fibrinolytic, and anticoagulation pathways that have been shown to be affected by cytokines and endotoxin in the sepsis syndrome (Fig. 3). Tissue factor (TF) is a cell surface receptor expressed by monocytes and vascular endothelial cells; increased expression of TF results in initiation of the coagulation cascade via the extrinsic pathway (Fig. 3). TF expression is induced following exposure to blood; thus major trauma, burns, hereditary vascular or chronic inflammatory disorders increase an individual's susceptibility to DIC via increased TF expression (Baker, 1989; Bakhshi and Arya, 2003). The expression of TF is regulated by endotoxin, as well as pro- and anti-inflammatory cytokines with TNF, IL-1α, IL-1β, IL-6, IL-8, leukemia inhibitory factor, interferon (IFN)γ, and monocyte chemoattractant protein (MCP)-1 stimulating TF and TGFβ, IL-4, IL-10, and IL-13 inhibiting its expression (Herbert et al., 1992; Pradier et al., 1993; Schwager and Jungi, 1994; Del Prete et al., 1995; Neumann et al., 1997). In addition to affecting the extrinsic pathway of coagulation, endotoxin and cytokines modulate the fibrinolytic and anticoagulation pathways. The fibrinolytic pathway is a natural anticoagulant pathway that is important for host protection against excessive clotting. Data from human and animal experiments suggest that cytokines mediate lipopolysaccharide (LPS)-induced increases in plasminogen activator inhibitor (PAI)-1 in several organs, including liver, kidney, and lung (Fig. 3B; Sawdey and Loskutoff, 1991). PAI-1 is a negative inhibitor of tissue plasminogen activator (tPA), an essential enzyme involved in fibrin clot degradation (Fig. 3B). High PAI-1 levels have been shown to precede DIC incidence and correlate with poor outcome; PAI-1 knockout mice are resistant to LPS-induced kidney thrombosis (Sawdey and Loskutoff, 1991; Yamamoto and Loskutoff, 1996). Finally, cytokines can also modulate the protein C–protein S anticoagulation pathway at several levels, preventing proteolytic cleavage of several factors (Va and VIIIa) involved in the coagulation pathway (Fig. 3C; Redl et al., 1995; Yamamoto et al., 1999). The reader is referred to reviews that discuss the intricacies of cytokine modulation of these pathways with DIC (Levi, 2001; van der Poll et al., 2001).

A. Coagulation Pathway

INTRINSIC PATHWAY

Damaged Surface
↓
Kininogen/Kallikrein
XII ⟶ XIIa
　XI ⟶ XIa
　　IX ⟶ IXa
　　　↓ (VIIIa)
　X ⟶ Xa

EXTRINSIC PATHWAY

Injury/Inflammation
↓
VIIa ← VII
↓   Tissue Factor
X ⟶ Xa
　　+ ← LPS, IL-1, TNF

↓ (Va)
Prothrombin (II) ⟶ Thrombin (IIa)
　　Fibrinogen (I) ⟶ Fibrin (Ia)
　　　　↓ (XIIIa)
　　　　Clot

B. Fibrinolytic Pathway

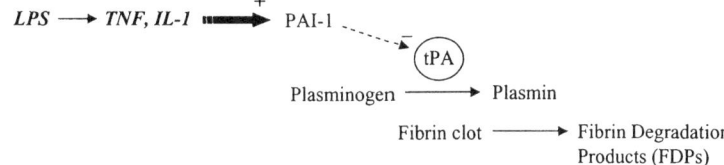

LPS ⟶ TNF, IL-1 ⟹+ PAI-1 ⊣ (tPA)
Plasminogen ⟶ Plasmin
Fibrin clot ⟶ Fibrin Degradation Products (FDPs)

C. Anticoagulant Pathway

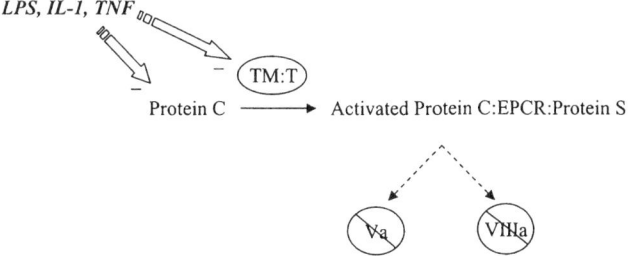

LPS, IL-1, TNF ⟶ − (TM:T)
Protein C ⟶ Activated Protein C:EPCR:Protein S
　　　　↓   ↓
　　　(Va) (VIIIa)

Fig. 3. Coagulative (A), fibrinolytic (B), and anticoagulative (C) pathways involved in disseminated intravascular coagulation. The coagulation cascade generates a fibrin clot through intrinsic and/or extrinsic pathways. LPS, IL-1, and TNF have been shown to elicit tissue factor formation. This leads to formation of factor Va–Xa and VIIIa–IXa complexes, which result in the generation of thrombin and clot formation. The fibrinolytic pathway is activated by LPS via the production of TNF and IL-1. Cytokine activation of plasminogen activator inhibitor (PAI)-1 results in inhibition of tissue plasminogen activator (tPA), which impairs fibrinolysis. LPS, IL-1, and TNF also affect the anticoagulation pathway through downregulation of thrombomodulin (TM) expression, which ultimately prevents the inactivation of factors Va and VIIIa, thus prolonging clot formation.

Tissue injury responses

Heat-induced tissue damage may be extensive, as injury to the liver, kidney, spleen, heart, lung, small intestine, brain, and skeletal muscle (rhabdomyolysis) is commonly observed in human and animal studies (Malamud et al., 1946; Graber et al., 1971; Bouchama et al., 1996, 2005; Dematte et al., 1998;

Lu et al., 2004). The severity of heat stroke is purported to be primarily related to the extent of damage incurred in the CNS, liver, and kidneys (Kew et al., 1967, 1970). The extent of peripheral organ damage is most readily assessed by analysis of serum enzyme levels, such as creatine phosphokinase (CPK; skeletal muscle), uric acid (kidney), alanine aminotransferase (ALT; liver), and aspartate amino transferase (AST; liver). However, due to plasma levels of these enzymes being altered by heat as well as exhaustive exercise (in the absence of heat stroke), a differential diagnosis of heat stroke is not always possible based on these measures alone.

Tissue histopathology studies have provided detailed analysis of the extent of thermal damage to several organs in heat stroke cases. GI barrier dysfunction is a common complication of heat exposure (Moseley et al., 1994; Lambert et al., 2002, 2004). Prolonged heat exposure induces a reduction in splanchnic blood flow as a greater proportion of cardiac output is shunted to the skin to facilitate heat dissipation. Resultant GI ischemia (with subsequent cytokine and free-radical production) contributes to barrier disruption, which is commonly observed as dilation of the central lacteals of intestinal villi (Figs. 4A, B; Hall et al., 2001; Lambert et al., 2002). Renal failure is a common complication of heat stroke, as characterized by glomerular ischemia and hemorrhages (Malamud et al., 1946; Chao et al., 1981; Bouchama et al., 1996). Tubular necrosis is also common in human and animal heat stroke models (Figs. 4C, D); protein clumping in tubular epithelial cells of the kidney is thought to be a consequence of direct thermal injury, rhabdomyolysis, or DIC (Kew et al., 1967; Graber et al., 1971; Raju et al., 1973; Chao et al., 1981; Kilbourne et al., 1982; Lu et al., 2004; Carter et al., 2005). Cytoplasmic protein clumping in the spleen is thought to be a direct result of hyperthermia as the organs are essentially "cooked and coagulated" (Figs. 4E, F; Chao et al., 1981). Fatty change is observed in the liver of exertional heat stroke cases and may be a consequence of enhanced breakdown of fats induced by hyperthermia and/or an inability of the liver to mobilize the fat (Chao et al., 1981). Typically, liver injury is seen in long-term survivors, suggesting that it may be a consequence of the inflammatory response that ensues during recovery, rather than representing an acute (immediate) response to hyperthermia (Malamud et al., 1946). This has also been observed in a mouse model of passive heat exposure in which liver damage was absent through 24 h of recovery, but appeared ~72 h after the initial heat insult (Fig. 5). The presence of circulating endotoxin in some heat stroke patients is thought to be due to leakage following GI barrier disruption, but may also be related to liver damage since this organ is one of the major sites of endotoxin clearance (Bradfield, 1974; Nolan, 1981). Similarly, renal failure has been suggested as a potential mechanism for increased cytokine (i.e., soluble TNF receptor) concentrations, as cytokine clearance is a reported function of this organ (Hammami et al., 1997). Ultimately, multi-organ system dysfunction is the cause of heat stroke mortality and is revealed at autopsy as edema and micro-hemorrhages in several organs of the periphery as well as specific brain regions (Malamud et al., 1946; Chao et al., 1981).

Cytokines and heat stroke

> Although shock undoubtedly plays a significant role in the course of heat stroke...it is regarded as a secondary manifestation and therefore non-specific and unessential to the fundamental pathogenesis of the disorder.
> Malamud et al., 1946

The heat illness syndrome is a continuum of increasing severity that is inclusive not only of the conditions incurred during direct heat exposure, but also those pathophysiologic responses that manifest during long-term recovery. As such, multi-organ system failure is now thought to result from heat cytotoxicity in combination with a subsequent systemic inflammatory response syndrome (SIRS) of the host to tissue injury (Bouchama and Knochel, 2002). Based on this realization, Bouchama and Knochel (2002) proposed a new definition of heat stroke as "a form of hyperthermia associated with a systemic inflammatory response leading to a syndrome of multi-organ dysfunction in which encephalopathy

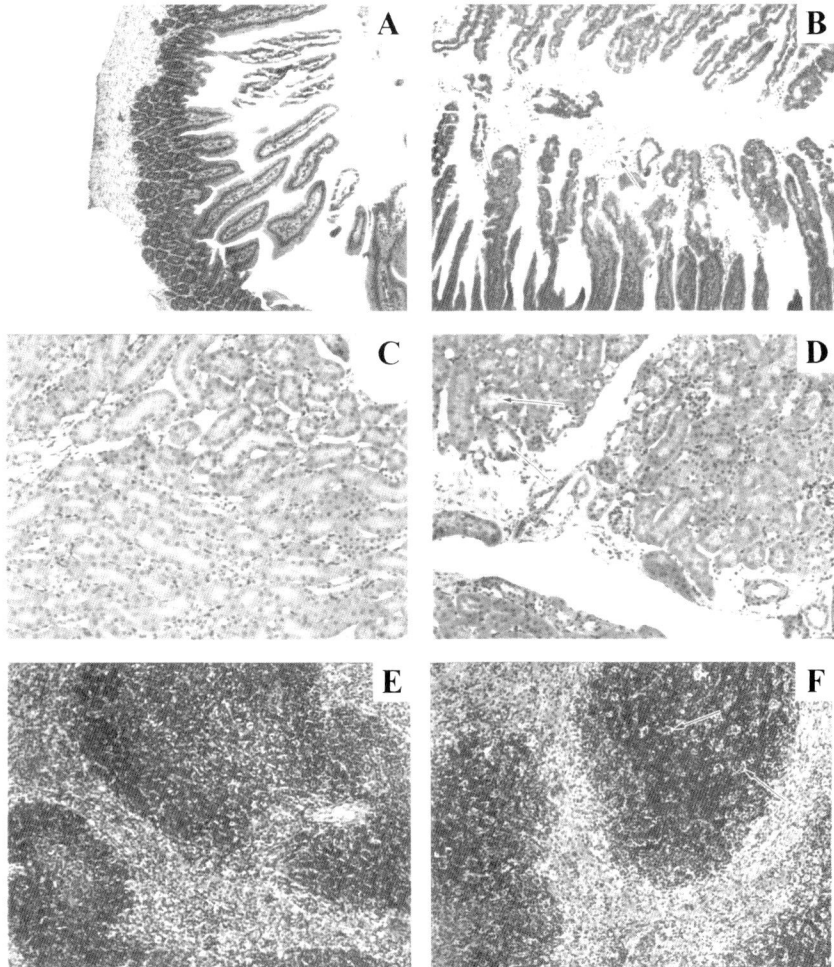

Fig. 4. Representative photomicrographs of small intestine (top), kidney (middle), and spleen (bottom) obtained from male C57BL/6J mice exposed to a control (A, C, E) or heat stress condition (B, D, F). Tissues are from male C57BL/6J mice heat stressed to a maximum core temperature of 42.7°C and allowed to recover until hypothermia, at which time tissues were collected (Leon et al., 2005). Tissues were stained with hematoxylin and eosin for microscopic evaluation at magnification 200 ×. Arrows indicate identified tissue lesions, which are described in the text.

predominates." Note the lack of inclusion of a specific core temperature value in this definition, which may be reflective of the wide variability of core temperature responses observed in heat stroke cases.

Endogenous cytokines may be important mediators of SIRS in heat stroke patients. Cytokines are intercellular chemical messengers released by a variety of cell types, including macrophages, T and B cells, endothelial cells, and astrocytes (Kelker et al., 1985; Chensue et al., 1989; Van Dam et al., 1992; Sharif et al., 1993; Malyak et al., 1994). Their defining characteristics include a general lack of constitutive production and pleiotropy or redundancy of actions. This latter feature has important implications since it is rare that a cytokine is released in the absence of other endogenous substances that may influence its action. Furthermore, antagonism of the physiological actions of one cytokine may be compensated by overlapping properties of a different, perhaps related cytokine. Different classes of cytokines have redundant

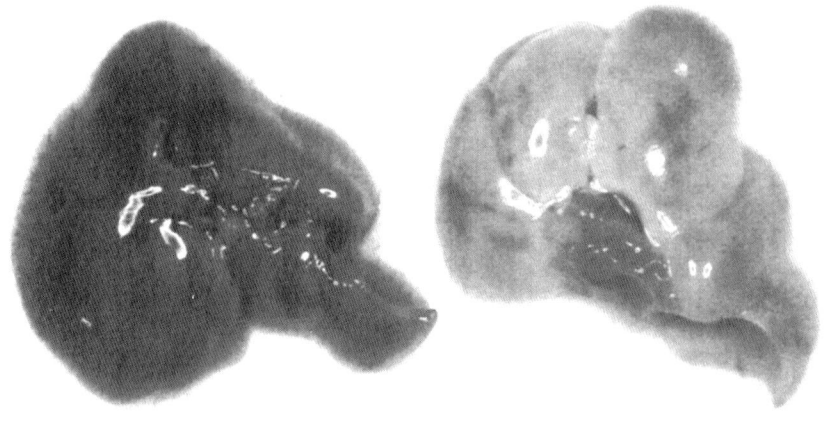

Nonheated Control **Heated Nonsurvivor**

Fig. 5. Photograph showing the "fatty liver" change observed in mice at ~72 h following heat stress. Livers from a nonheated control (left) and heat stressed nonsurvivor (right) are shown.

effects on specific cell types and combinations of cytokines can be synergistic or antagonistic, depending on the targeted cell types and the combination of cytokines that are present (the cytokine "milieu"). Several pro- and anti-inflammatory cytokines are elevated concomitantly in the circulation of heat stroke patients. However, due to an examination at end-stage heat stroke or at clinical presentation (often after cooling has occurred), our understanding of changes in the balance of pro- and anti-inflammatory cytokines over long-term progression of this syndrome remains poorly understood. Although attempts have been made to classify particular cytokines as more harmful than others in the morbidity and mortality of human heat stroke (e.g., IL-6), these efforts have been hindered by a lack of data beyond that provided by correlation studies.

Cytokine-inducing stimuli include bacterial and viral infection (Patel et al., 1994; Drexler, 1995), psychological stress (Maes et al., 2000; Oka et al., 2001), heat stress or whole body hyperthermia (WBH; Neville and Sauder, 1988; Bouchama et al., 1991, 1993, 2005; Lin et al., 1994; Haveman et al., 1996), exercise (Camus et al., 1997; Moldoveanu et al., 2001; Nieman et al., 2001; Suzuki et al., 2003), and other cytokines (Content et al., 1985; Neta et al., 1992). Note that heat stroke patients may be exposed to several of these stimuli concomitantly, complicating etiology of their condition. Heat exposure influences and/or induces a variety of physiological responses that are known to be modulated by endogenous cytokines, including fever (Kluger, 1991; Leon et al., 2005), hypothermia (Romanovsky and Blatteis, 1996; Leon, 2004, 2005), increased gut permeability (Moseley et al., 1994; Hall et al., 2001; Oshima et al., 2001; Desai et al., 2002; Lambert et al., 2002; Wang and Hasselgren, 2002), activation of the hypothalamic-pituitary-adrenal (HPA) axis (e.g., glucocorticoid release; Berkenbosch et al., 1987), and hypotension (Lin et al., 1997).

In order to clearly delineate a role for an endogenous cytokine in heat stroke, a series of criteria need to be experimentally satisfied (adapted from Kluger, 1991). While these criteria were originally described for characterization of endogenous cytokines in fever, they can be similarly applied to the study of heat stroke pathophysiologic responses. Briefly, a cytokine should elicit the expected response(s) when injected or infused (*Criterion 1*), its endogenous release/production should be temporally correlated with the appearance of heat stroke symptoms (*Criterion 2*), and antagonism of a cytokine's action or production should inhibit or eliminate the physiological symptom (*Criterion 3*). While the first two criteria provide evidence in favor of a role of the targeted protein in a response, efficacy provided by protein neutralization is the most stringent evidence in favor of that endogenous substance playing a role in the response of interest. Surprisingly few studies have examined the effectiveness of cytokine antagonism in heat stroke morbidity and mortality (Table 4). The reasons for this may be several-fold. While the injection of antibodies or protein inhibitors appears to be a straightforward method, there are several technical difficulties inherent in drug application studies (for a review, see Leon, 2005). In some cases commercial availability of an antagonist is lacking such that an experiment using this traditional technique is not feasible. It is anticipated that the increased availability of gene knockout models and more sophisticated, specific drug agents (e.g., small interfering RNAs (siRNAs)) will alleviate this experimental limitation for future studies. In other cases, the inherent properties of the cytokine and its receptors make study difficult. While in most cases effective neutralization of a cytokine can be achieved using an antibody or soluble receptor, there are instances when these substances paradoxically increase the endogenous action of the cytokine. For example, IL-6 antibodies as well as IL-6:soluble IL-6 receptor (sIL-6R) complexes have been shown to enhance, rather than limit, several endogenous actions of IL-6 (May et al., 1993; Schobitz et al., 1995; Peters et al., 1996).

Evidence supporting a role for cytokines in heat stroke pathophysiology includes: (1) increased circulating levels of cytokines in patients and experimental animals at end-stage heat stroke or in response to WBH, (2) beneficial effects of IL-1 antagonism on rat heat stroke survival (the effectiveness of antagonism of other cytokines has not been reported and IL-1 antagonism in other species has not been tested), and (3) the induction of heat stroke symptoms following cytokine injection in experimental animal models. Indirect evidence for a role of endogenous cytokines in heat stroke is suggested by the association of endotoxemia with heat stroke, the known role of cytokines in the endotoxemic syndrome, and reciprocal cytokine and heat shock protein (HSP) interactions that have been demonstrated *in vitro* in response to endotoxin treatment with or without heat stress. Each line of research will be discussed in more detail below.

Table 4. Current data supporting a role for cytokines in heat stroke

	Criterion 1: injection effects	*Criterion 2*: increased production	*Criterion 3*: neutralization
IL-1α	NT	✓	✓
IL-1β	✓	✓	✓
IL-1ra	NT	✓	NT
IL-1R1	NT	NT	✓
IL-2	NT	✓	NT
IL-2R	NT	✓	NT
IL-4	NT	✓	NT
IL-6	✓	✓	✓
sIL-6R	NT	✓	NT
IL-8	NT	✓	NT
IL-10	✓	✓	NT
IL-12p40	NT	✓	NT
IL-12p70	NT	NT	NT
IFNα	NT	✓	NT
IFNγ	NT	✓	NT
TNFα	✓	✓	NT
TNFβ	NT	NT	NT
sTNFR p55	NT	✓	NT
sTNFR p75	NT	✓	NT
TNFR p55/p75	NT	NT	✓
MIP-1α	NT	✓	NT
MCP-1	NT	✓	NT
GCSF	NT	✓	NT

Source: Adapted and used with permission from Kluger, 1991.
Note: *Criterion 1* — injected/infused cytokine induces heat stroke morbidity/mortality; *Criterion 2* — cytokine gene expression/release/production is temporally correlated with heat stroke morbidity/mortality; *Criterion 3* — cytokine neutralization inhibits or eliminates the heat stroke morbidity/mortality, NT — not tested.

Human heat stroke: correlation studies

Due to the ethical concerns of exposing human subjects to thermal extremes, the study of heat stroke responses is limited to clinical and field studies of those patients presenting with the syndrome. The annual Muslim pilgrimage to Makkah (the Hajj) has provided a rich source of data regarding cytokine changes during exertional heat stroke. While heat stroke during the Hajj has been described as the classical form (Bouchama et al., 1991, 1993), it is more representative of a mixed type since exertional features are clearly present. This is a consequence of the variety of rituals associated with this event, many of which require significant physical exertion by the participant. As such, heat stroke etiology in this population is expected to be multi-factorial.

Located in the hot, arid desert environment of Saudi Arabia, the Hajj takes place in the extreme weather months of May to September, when temperatures range from 38 to 50°C (Khogali, 1983). These weather conditions combined with physical exertion (first day consists of a 3.5 km jog), heavy clothing that is traditional to the region (limited heat dissipation), lack of sleep (due to a rigorous schedule), and an aged population (~50 years, which is an advanced age for this region) predisposes many individuals to heat injury. The effect of clothing is a particular concern for Muslim women who are required to wear more clothing that covers a larger surface area of the body and is darker in color than that worn by men (Hashim et al., 1997). The life-threatening effects of heat stroke under these conditions are the major concern, but heat exhaustion with water or salt depletion is also prevalent. It is likely that the high occurrence of this milder form of heat illness is a result of many of the religious participants being from neighboring regions in which acclimatization to the warmer ambient conditions has occurred. However, medical conditions such as diabetes, cardiovascular abnormalities or communicable, parasitic diseases are common and predispose to heat stroke in this population (Khogali, 1983). As exemplified in the 1980s when ~2 million people participated in the Hajj, overcrowding and congestion also impose large demands on sanitation services, raising health concerns. Unfortunately, it is expected that advances in modern technologies (e.g., more rapid transport to the area) will introduce additional factors (e.g., lack of acclimatization, increased congestion, air pollution) to this complex situation.

In the late 1980s to early 1990s, several studies were conducted during the Hajj to characterize peripheral cytokine and immune disturbances in heat stroke cases. Several studies at the Hajj determined circulating cytokine concentrations at the time of clinical presentation and following cooling therapy. At the time of admission, elevations in circulating concentrations of IL-1α, IL-1β, IL-1 receptor antagonist (IL-1ra), IL-6, soluble IL-6 receptor (sIL-6R), IL-10, interferon (IFN)γ, TNFα, and soluble TNF receptors (sTNFR60 and sTNFR80) are observed (Bouchama, et al., 1991, 1993, 2000; Hammami et al., 1997; Hashim et al., 1997). In some cases, only 30–40% of patients show increased concentration of a particular cytokine (e.g., IL-1β and IL-10; Bouchama et al., 1993, 2000), whereas other cytokines, such as IL-6, are often significantly elevated in 100% of patients (Bouchama et al., 1993). IL-6 shows the highest correlation with mortality and neurologic symptoms, thus implicating it as a potential therapeutic target for heat stroke prevention/treatment strategies (Bouchama et al., 1993; Hammami et al., 1997; Hashim et al., 1997). Although attempts to correlate IL-6 with core temperature at admission have been unsuccessful, this may be related to variability in presentation times (i.e., wide core temperature range), as previously described. However, sIL-6R, sTNFR60, and sTNFR80 concentrations show a direct correlation with post-cooling in one study (Hammami et al., 1997).

An important aspect of cytokine analysis in heat stroke research is to understand the relationship of endogenous levels of a cytokine to its soluble receptor (or natural antagonists), such as exists for TNF and IL-6. It is currently unclear if antagonism of endogenous cytokine levels by their soluble receptors is directly responsible for cooling in heat stroke patients. In the study by Hammami et al. (1997), TNFα and β concentrations were undetectable at the time of clinical admission, whereas sTNFR60 and sTNFR80 concentrations were significantly elevated above controls. The inability to detect circulating TNF concentrations may be the result of localized production of the cytokine (not detectable in serum samples) or the neutralizing activity of the sTNFRs, the latter of which may interfere with assay detection of the cytokine. Interestingly, heat stroke survivors had higher sTNFR concentrations than non-survivors, suggesting a potential detrimental effect of TNF in this syndrome; unfortunately, the small sample size ($N = 3$) in this study precludes a definitive conclusion as to the role of these receptors and endogenous TNF in human heat stroke mortality.

IL-6 has been correlated with heat stroke mortality and shows reciprocal changes with respect to its sIL-6R from time of clinical admission to post-cooling (Hammami et al., 1997). A definitive role for the sIL-6R in heat stroke pathophysiology is currently unrecognized, but two scenarios have been postulated (Hammami et al., 1997). The first scenario suggests that in the presence of high IL-6, the formation of IL-6:sIL-6R complexes may potentiate the effects of endogenous IL-6. A potentiating effect of sIL-6R on IL-6-induced responses has been previously demonstrated. In rats, the intracerebroventricular (i.c.v.) injection of sIL-6R augments and prolongs the fever and motor activity suppressing effect of IL-6 injection (Schobitz et al., 1995). These effects are mediated following its integration into cell membranes and subsequent association with the signal transducing gp130 molecule. The net effect of sIL-6R injection is an increase in the total concentration of available IL-6 signaling receptors on any particular cell type (Schobitz et al., 1995). The alternative scenario postulates that in the presence of low concentrations of IL-6 (such as occurs during cooling), sIL-6R may compete with cellular bound IL-6R for the low bioavailability of IL-6, resulting in reduced signaling. It is unclear if either of these scenarios is operating *in vivo* in the heat stroke syndrome. Future studies to examine the relationship of circulating cytokines and their soluble receptors (IL-6:sIL-R, TNF:sTNFR) or receptor antagonists (IL-1:IL-1ra) in heat stroke pathophysiology are warranted to test these hypotheses.

There is controversy in the literature regarding the most efficacious cooling method of heat stroke patients (for a review, see Hadad et al., 2004). The studies conducted at the Hajj employed evaporative cooling techniques (the most common clinical cooling therapy), thus providing insight into the effect of rapid core temperature reductions on cytokine production and clearance. Not surprisingly, rapid cooling rates are correlated with survival, cooling rates differ widely between heat stroke patients, and core temperature varies widely following cooling therapy (<38.0° to 39.4°C), which unfortunately corresponds to the time of sample collection in some studies (Hammami et al., 1997; Hashim et al., 1997). An analysis of cytokine levels post-cooling indicates increases in sIL-6R, sTNFR80 and decreases in IL-1, IL-10, and TNFα (Bouchama et al., 1991, 2000; Hammami et al., 1997; Hashim et al., 1997). Again, high IL-6 levels are associated with mortality, as sustainment of high IL-6 levels post-cooling correlates with non-survival (Hashim et al., 1997).

The correlation of high IFNγ levels with poor prognosis in non-heat related illness has stimulated investigations into a potential role of this cytokine in the morbidity and mortality of heat stroke. Bouchama et al. (1993) showed increased serum IFNγ levels in >50% of patients presenting to the clinic with exertional heat stroke prior to cooling therapy. Elevated circulating levels of IL-1β and IL-6 were also observed, although these samples were obtained in a different set of patients than those in which IFNγ was detected. Whereas IL-6 levels tend to be highest in non-survivors, IFNγ levels do not correlate as strongly with heat stroke morbidity/mortality (Bouchama et al., 1993). Sonna et al. (2004) hypothesized prodromal viral illness as the stimulus for increased expression of interferon-inducible (IFI) genes observed in peripheral blood mononuclear cells (PBMC) from military recruits that collapsed from exertional heat stroke during basic training. It is difficult to speculate from this study as to the incidence of the combination of viral illness with heat stroke in this population due to the small number of screened subjects ($N=4$). However, it does provide food-for-thought regarding the appropriateness of pre-screening individuals for illness and/or elevated cytokine levels prior to heat exposure as a potential protective measure against injury or death.

In military recruits with exertional heat stroke, plasma levels of IFNγ are elevated concomitantly with IL-1β, IL-6, TNFα, IL-2 receptor (IL-2R), and IL-8; elevated IFNγ, IL-6, and IL-2R levels correlate with morbidity in this study (Lu et al., 2004). Differences between studies in the correlation of IFNγ with mortality (e.g., Bouchama et al., 1993; Lu et al., 2004) may be due to several factors. First, military recruits in the study by Lu et al. (2004) participated in a well-defined exercise regimen whereas the specific rituals experienced by participants of the Hajj (Bouchama et al., 1993) are not as well documented for each exertional heat stroke case. Second, circulating cytokine levels in heat stroke patients were compared to normal controls in one study (Bouchama et al., 2000) and to exercising controls that did not experience heat stroke symptoms in the other (Lu et al., 2004). A complicating factor in many studies of this nature is the control population that is used for comparison of cytokine responses to the heat stroke condition. For those studies conducted at the Hajj, it is a logistical impossibility to obtain plasma samples from a group of individuals that represent an appropriate control population. In many cases, comparisons to a control population are not provided; rather, pre- (time of clinical admission) and post-cooling (8 or 24 h after cooling) values are directly compared to one another with, in most cases, further characterization provided between survivors and non-survivors (Bouchama et al., 1991; Hashim et al., 1997). Other studies have compared heat stroke values to those observed in a control "normal" population (typically undefined, but presumably at rest; Bouchama et al., 1993) or, more appropriately, to individuals participating in the same event with heat exposure, but in the absence of heat stroke (Bouchama et al., 2000; Lu et al., 2004). A comparison of heat stroke values to both heat exposed and normothermic controls in the resting and exercise condition is the most appropriate scientific design, but is achieved in few studies. Studies on military populations are typically more controlled, but also present with inherent difficulties since non-exercising controls are typically not available at the time of heat stroke collapse. As such, these difficulties speak to the importance of animal experimentation for an understanding of the complex etiology of exertional heat stroke since the multitude of factors inherent in this syndrome can be more easily controlled in the laboratory setting. However, as will be described in more detail below, animal models of passive, rather than exertional heat stroke (the latter being more representative of the human population) are typically used to explore cytokine responses *in vivo*.

Several attempts have been made to correlate cytokine changes with different aspects of the heat stroke syndrome, such as hyperthermia, immune disturbances, and heat severity. It is typically difficult to find a strong correlation between cytokine levels and core temperature in human heat stroke cases, due to differences in clinical treatment strategies and presentation times. For example, weak correlations between reported core temperature and cytokine values are common as patients are subjected to different cooling regimens and durations (Bouchama et al., 1991, 1993, 2000; Hammami et al., 1997; Hashim et al., 1997; Sonna et al., 2004). However, in one study, the ability to cool patients from 40° to 38°C was dependent on serum IL-1β (Chang, 1993). This is the only report implicating endogenous IL-1β in the control of core temperature responses in heat stroke patients. This seems rather surprising since several of the cytokines implicated in heat stroke pathophysiology are known regulators of core temperature in health and disease. On the other hand, the inability to correlate circulating cytokine concentrations with heat-induced core temperature changes may be due to tissue concentrations being more important than circulating levels in the mediation of these responses.

Several studies have attempted to correlate changes in circulating cytokine concentrations with different aspects of the heat-induced acute phase response (APR). IL-6 modulates the APR to infection/inflammation (such as that induced by endotoxin) and is known to stimulate hepatic synthesis of C-reactive protein (CRP, a sensitive marker of inflammation; Heinrich et al., 1990). Hashim et al. (1997) reported elevated CRP levels in survivors and non-survivors of exertional heat stroke, but were unable to correlate these changes with elevations in IL-6. Elevated ALT levels suggested potential hepatic dysfunction in these patients, but did not correlate with CRP. LPS is a cell wall component of gram-negative bacteria that increases in concentration in human heat stroke cases, although this response does not show a direct correlation with circulating levels of IL-1 or TNFα (Bouchama et al., 1991). The two hypotheses postulated to account for these findings include having missed the time of peak cytokine levels in some patients as presentation times varied so widely, and the greater importance of tissue vs. circulating cytokine concentrations in these analyses (Bouchama et al., 1991). Although several additional pathophysiological responses have been observed, including hyper- and hypoglycemia, lactic acidosis, and leukocytosis, potential correlations between cytokine levels and these responses have not been examined. Attempts to correlate elevated cytokine levels with one another have also provided contradictory results. IL-1 and TNFα levels did not correlate with one another in one study, whereas significant correlations between IL-1β, TNFα, and IL-6 have been observed (Chang, 1993). The reason(s) for these discrepant results is not immediately evident.

Animal models of heat stroke

While several animal models have been developed for the study of pathophysiologic responses to heat stroke (Adolph, 1947; Hubbard et al., 1976, 1977; Wright, 1976; Wright et al., 1977; Dubose et al., 1983a; Lord et al., 1984; Gathiram et al., 1987a; Wilkinson et al., 1988; Romanovsky and Blatteis, 1996; Hall et al., 2001), few studies have examined the role of cytokines. Elevated circulating concentrations of IL-1, IL-6, IL-8, IL-10, TNF, and granulocyte colony stimulating factor have been reported with localized or WBH in primates (Bouchama et al., 2005), rabbits (Lin et al., 1994), mice (Neville and Sauder, 1988), and rats (Chiu et al., 1995, 1996; Haveman et al., 1996; Lin et al., 1997; Liu et al., 2000). As shown in human heat stroke, IL-6 is correlated with heat stroke severity in a primate model of passive heat stroke (Bouchama et al., 2005).

Unfortunately, most animal studies have determined cytokine concentrations at maximum heat stress, thus ignoring changes in the production of these mediators throughout exposure *and recovery*. Furthermore, cytokine determinations have typically been performed on plasma/serum samples — the relation of these measurements to cytokine changes occurring at the tissue level (i.e., potential site(s) of heat injury) are unknown. Recently, my laboratory characterized the plasma and tissue (liver) profile of 11 cytokines (IL-1α, IL-1β, IL-2, IL-4, IL-6, IL-10, IL-12p40, IL-12p70, IFNγ, MIP-1α, TNFα) at maximum heat stress and throughout 24 h of recovery in a mouse model of passive heat exposure. The goal of this experiment was to correlate cytokine changes with the profound thermoregulatory responses (hypothermia and fever) observed during recovery. As shown in Fig. 6, only plasma IL-12p40 was elevated at maximum core temperature (CTM of 42.7°C in this model; Leon et al., 2005, 2006). On the other hand, hypothermia was associated with significantly elevated concentrations of IL-1β, IL-6, and IL-10; it is unknown if these cytokines are regulating this core temperature response or are a consequence of low core temperature, since hypothermia has been shown to exacerbate cytokine production in other models (Arons et al., 1999; Fairchild et al., 2004). Heat-induced hypothermia was also associated with increased levels of IL-1α in the liver, despite an inability to detect this cytokine in the plasma (Fig. 7); as previously described, these data suggest a potential dissociation between cytokine levels measured in the blood and those directly related to organ (dys)function. Interestingly, at 24 h after the start of heat exposure (~0900–1000 h the following day), mice showed a fever-like core temperature response (~1.0° to 1.5°C above controls) that correlated with significantly increased levels of IL-6, a known endogenous pyrogen (Fig. 6; Leon et al., 2006). At present, it is unclear if IL-6 is modulating this core temperature response, which again speaks to the importance of investigating the effectiveness of cytokine antagonism/neutralization in altering core temperature responses and mortality in the heat syndrome. Importantly, similarities in cytokine responses between passive (animal) and exertional (human) cases of heat stroke are supportive of the appropriateness of animal models to address this issue, although exertional animal models need to be developed.

IL-1 and heat stroke

Lin et al. (1994) reported increased plasma and hypothalamic levels of IL-1β in heat stroked rabbits. IL-1 levels in other brain areas, including the medulla oblongata, cortex, and spinal cord, were unaffected. The anterior hypothalamus is thought

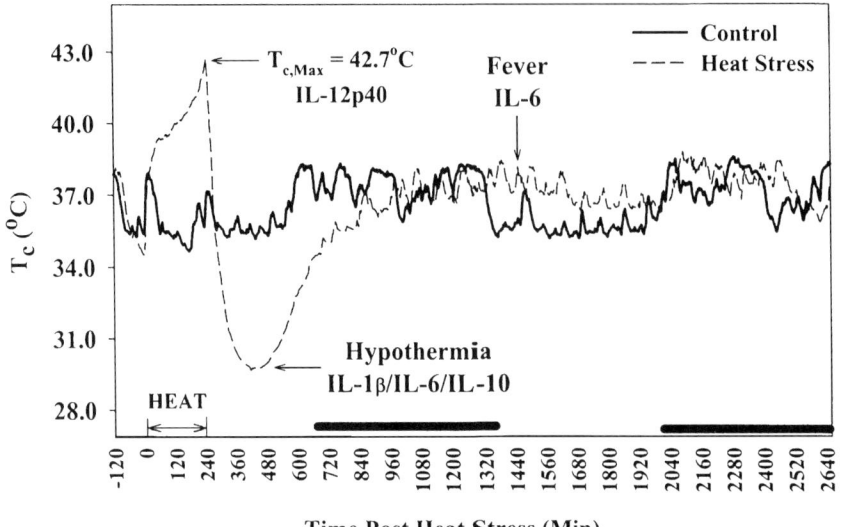

Fig. 6. Summary of plasma cytokine changes observed in male C57BL/6J mice during 48 h of heat stress recovery. Details of the heat stress protocol and cytokine findings are described elsewhere (Leon et al., 2005, 2006). Mice were heat stressed to a maximum core temperature of 42.7°C and then allowed to recover at an ambient temperature of 25 ± 2°C until sample collection. Mice were sacrificed at each time point for tissue collection ($N = 8$–11 mice/group). Core temperature response is depicted for one representative control and heat stressed mouse. Note that IL-12p40 was the only cytokine significantly elevated at $T_{c,Max}$ (42.7°C). The largest cytokine changes were observed at hypothermia, where IL-1β, IL-6, and IL-10 were significantly elevated. IL-6 was the only cytokine that remained elevated 24 h following heat stress, which is the time point at which mice displayed a fever-like (~1.0° to 1.5°C) core temperature elevation. Note that IL-1α, IL-2, IL-4, IL-12p70, IFNγ, MIP-1α, and TNFα were not detectable in the plasma at any time point. Black horizontal bars represent lights-off period in a 12:12 h L:D cycle. Core temperature was collected at 1 min intervals in conscious, freely moving mice using intraperitoneally implanted radiotelemetry devices (Data Sciences International, St. Paul, MN). Cytokine determinations were performed on duplicate samples using the FlowMetrix™ System (Luminex, Austin, TX), which permits the simultaneous quantitation of multiple cytokines (Leon et al., 2006).

to be the main integration site of thermoregulatory responses. While systemic injection of the IL-1ra (a naturally occurring antagonist of endogenous IL-1 actions) attenuated the rectal temperature response to heat stroke, the effectiveness of this treatment following administration directly into the hypothalamus was not tested (Lin et al., 1994). In addition, it is unclear if attenuation of the hyperthermic response was an indirect response due to a reduction in cardiovascular strain or vice versa; thus, the specific role of endogenous IL-1β (and other cytokines) on thermoregulatory control mechanisms during heat exposure and/or recovery remains unknown. Unfortunately, control levels of IL-1β were also elevated in this study in response to anesthesia. Anesthesia is commonly used in animal experimentation and represents a confounding factor due to its known effects on thermoregulatory control processes and cytokine production (Stoen and Sessler, 1990; Hanagata et al., 1995; Brix-Christensen et al., 1998). It is also unclear if increased heat-induced brain levels of IL-1β were due to local production by hypothalamic tissue or systemic production and infiltration into hypothalamic tissue after IL-1β had traversed the blood brain barrier (BBB; Lin et al., 1994).

Increased hypothalamic IL-1β levels have been implicated in heat-induced cerebral ischemia in animal models (Lin et al., 1994). A connection between heat stroke, IL-1β and central concentrations of dopamine (DA) and serotonin (5-HT) may account for this response. Sharma et al. (1994) demonstrated a protective effect of the selective 5-HT$_2$-receptor antagonist ketanserin as cerebral edema was reduced following a 4 h heat exposure in the rat. Selective brain depletion of 5-HT following i.c.v. administration of

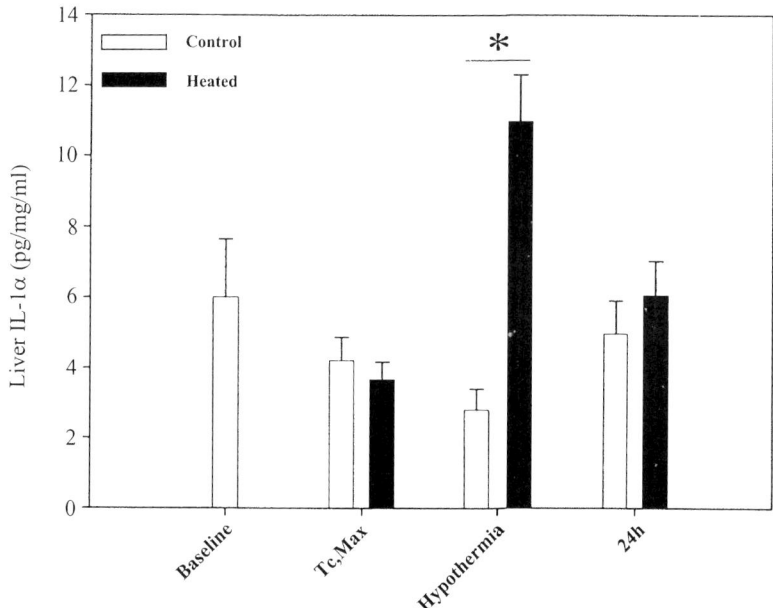

Fig. 7. Liver IL-1α levels in control and heat stressed male C57BL/6J mice. Mice were heat stressed to a maximum core temperature of 42.7°C and allowed to recover at an ambient temperature of 25 ± 2°C until sample collection. Mice were sacrificed at each time point for tissue collection ($N = 4$–7 mice/group). IL-1α was the only cytokine that was detectable at an increased level in the liver of heat stressed mice compared to controls. IL-1α was significantly elevated in heat stressed mice at hypothermia; note that increased plasma levels of IL-1α were not detectable at any time point in heat stressed mice. Cytokine levels were normalized to tissue protein concentration (~50 mg/ml of sample). Cytokine determinations were performed on duplicate samples using the FlowMetrix System (Luminex, Austin, TX), which permits the simultaneous quantitation of multiple cytokines. *Indicates significance at $P < 0.05$.

5,7-dihydroxytryptamine (5,7-DHT) also showed protection in heat-stroked rats, as indicated by significant reductions in arterial hypotension, cerebral ischemia, hypothalamic 5-HT accumulation, cerebral neuronal damage, and prolonged heat stroke survival time (Kao and Lin, 1996). The mechanism of 5-HT effects in the induction of cerebral edema is thought to be due to a reduction in mean arterial pressure (MAP) and an increase in intracerebral pressure, which inhibits cerebral flow resulting in neuronal injury (Kao and Lin, 1996). Similarly, intravenous (i.v.) administration of the IL-1ra prolonged survival time and attenuated arterial hypotension and hypothalamic 5-HT accumulation (Chiu et al., 1995). The ability of intrahypothalamically injected IL-1β to induce 5-HT production has also been demonstrated in a rat model (Shintani et al., 1993), and IL-1 and 5-HT have been implicated in the regulation of BBB permeability (Sharma and Dey, 1986, 1987).

Together, these studies support a central action of 5-HT and IL-1 in the mediation of heat injury and cerebral ischemia. The potential role of other endogenous cytokines in these 5-HT-mediated events is currently unknown.

Increased striatal DA concentrations are correlated with cerebral ischemia in the rat (Kao et al., 1994a, b; Chiu et al., 1996) and destruction of the nigrostriatal dopaminergic system following i.c.v. injection with 6-hydroxydopamine (6-OHDA) increases striatal blood flow, reduces ischemic damage, and prolongs survival time in rats (Lin et al., 1995). A connection between IL-1 and DA is suggested by the correlation between prolonged heat stroke survival and attenuation of striatal DA release in rats treated with the IL-1ra (Chiu et al., 1996); the ability of IL-1β to increase hypothalamic levels of DA has also been reported (Shintani et al., 1993). DA also appears to modulate the HPA axis response to heat-induced dehydration, as the D_2

receptor antagonist haloperidol attenuates cortisol release in humans (Hennig et al., 1995). Glucocorticoid hormones (corticosterone in rodents and cortisol in humans) are produced by the adrenal gland and represent an integral component of the HPA axis. In addition to the mobilization of energy stores during stress, glucocorticoids inhibit cytokine production in response to stress (Waage et al., 1990; Barber et al., 1993; Di Santo et al., 1996). Classical heat stress has been shown to induce cortisol secretion with heat stroke patients showing up to a fivefold elevation of plasma concentrations (Collins et al., 1969; Follenius et al., 1982; Laatikanen et al., 1988; Al-Harthi et al., 1990). Glucocorticoids are commonly used prophylactically, or their production is inhibited by adrenalectomy (ADX), to examine changes in cytokine and core temperature responses to environmental stimuli in experimental animals. Liu et al. (2000) treated rats with the synthetic glucocorticoid dexamethasone and showed an attenuation of heat stroke-induced hypotension, cerebral ischemia, neuronal damage, and prolongation to time of death; the protective effect of dexamethasone treatment was not due to a reduction in heat-induced core temperature. In the same study, the removal of endogenous glucocorticoids by ADX increased susceptibility of rats to heat stroke — an effect that was reversed by dexamethasone replacement therapy (Liu et al., 2000). Unfortunately, ADX rats were not provided replacement baseline levels of glucocorticoids in this study, which may have altered the observed responses. For example, the exacerbated arterial hypotension observed in ADX rats may have resulted from the absence of the normal permissive action of glucocorticoids on the cardiovascular response to catecholamines and other hormones that are released by heat exposure (Liu et al., 2000). However, the effects of glucocorticoids were correlated with their ability to decrease plasma IL-1β levels at the onset of heat stroke, implicating cytokine antagonism as the mechanism of glucocorticoid protective effects.

Cytokine injection induces heat stroke symptoms

Systemic injection of IL-1 induces many heat stroke symptoms including hypotension, decreased systemic vascular resistance, depressed myocardial function, and tissue necrosis (for review, see Dinarello, 1991). Prostanoid synthesis has been proposed as a mechanism of many IL-1 induced effects. In rabbits, cyclooxygenase inhibitors mitigate the hypotensive effects to IL-1 alone or in combination with TNF (Okusawa et al., 1988). As previously described, the ability of the IL-1ra to reduce many of the hemodynamic effects of IL-1 has directly implicated this cytokine in the mediation of these effects in heat stroke (Chiu et al., 1995, 1996; Lin et al., 1997). IL-1 acts synergistically with other cytokines, such as IL-6 and TNF, in the mediation of hemodynamic shock and tissue damage (Dinarello, 1991). The peripheral injection of IL-1β, IL-6, IL-10, and TNF induces many of the pathophysiological responses observed in human and animal heat stroke studies, including hypothermia, fever, HPA activation, increased vascular permeability, hemodynamic shock, and death (Kluger, 1991; Oshima et al., 2001; Leon, 2002, 2004; Kuwagata et al., 2003; Wieczorek et al., 2005).

Cytokine-mediated heat stroke responses: results from gene knockout models

As previously discussed, the study of cytokine-mediated heat stroke responses has been limited by technical difficulties, such as a lack of commercial availability of antibodies and/or antagonists. The recent advent of gene knockout technology provides a technique by which the efficacy of cytokine neutralization can be examined *in vivo*. Gene knockout mice are genetically engineered to lack a functional gene in every tissue of the body and essentially function as "chronic protein neutralization systems" (Sigmund, 1993). There is wide commercial availability of cytokine and cytokine receptor knockout mice, many of which have been used for the study of infectious and inflammatory responses. While the development of functional redundancy is an important concern in physiological research (i.e., the redundant/pleiotropic properties of cytokines may allow developmental redundancy to compensate for a missing gene's action), these models also provide several methodological advantages over traditional techniques (reviewed in Leon, 2005).

Although high IL-6 levels are typically associated with mortality in heat stroke patients and animal models (Hashim et al., 1997; Bouchama et al., 2005), protective effects of IL-6 neutralization have not been experimentally examined. Recent data from my laboratory demonstrate *decreased* heat stress survival in IL-6 knockout mice compared to wild-type controls (33 vs. 100% survival, respectively; $P = 0.025$) with the maximum decrease occurring within 24 h of heat stress recovery (Fig. 8). These data are suggestive of a dual role for endogenous IL-6 in heat stroke responses with exacerbated (pathophysiological) levels being detrimental (as suggested by correlation studies; this has not been experimentally verified), but baseline (permissive) levels required for heat stroke survival (as suggested by results of Fig. 8). Similar results are reported for IL-6 knockout mice in the endotoxemic syndrome (Leon et al., 1998). As shown in Fig. 8B, preliminary data from TNF p55/p75 receptor (TNFR) knockout mice are equally contradictory to the findings from correlation studies. Although high TNF levels are implicated as harmful in the heat stroke condition, TNFR knockout mice (mice that produce endogenous TNF, but are unable to respond to the cytokine due to an absence of the signaling receptors) showed a tendency toward decreased survival compared to wild-type controls (40 vs. 100% survival, respectively; $P = 0.10$; note small sample size). Although future studies using traditional methods of cytokine neutralization are important to verify the findings, these data suggest that the results obtained from correlation studies may have been overstated with regards to the deleterious role of endogenous cytokines in the heat stroke syndrome.

GI permeability and endotoxemia with heat stroke

The primary cardiovascular response to heat exposure is an increase in skin blood flow to promote heat loss and reduce the rate of heat gain from the environment. Increased skin blood flow is accompanied by a fall in splanchnic blood flow as a compensatory mechanism to maintain blood pressure. A severe reduction of intestinal blood flow results in GI ischemia and increased vascular permeability — the latter response facilitates the leakage of endogenous bacteria (and its toxic cell wall component, LPS) into the systemic circulation, resulting in endotoxemia. While under normal conditions the liver serves as an adequate clearance organ for endotoxin, thermal injury to this organ may compromise its clearance function, permitting endotoxin leakage into the portal circulation. It has been postulated that increased cytokine expression is a direct consequence of endotoxin leakage under heat stroke conditions (Fig. 9).

GI permeability and heat stroke

Physical and physiological disruptions of GI barrier function facilitate luminal translocation of intestinal contents to the blood. Endothelial barrier dysfunction occurs in response to a variety of stimuli including endotoxemia (Unno et al., 1997), trauma/hemorrhage (Roumen et al., 1993), strenuous exercise (Brock-Utne et al., 1988; Pals et al., 1997), and hyperthermia/heat stroke (Wijsman and Shivers, 1993; Moseley et al., 1994; Hall et al., 2001). With heat exposure, increased endothelial permeability is commonly observed in the GI tract and the BBB (Wijsman and Shivers, 1993; Moseley et al., 1994; Lambert et al., 2002, 2004). A clinical indicator of GI barrier disruption is detectable levels of endotoxin in the portal or systemic circulation.

Compensatory splanchnic vascular responses to heat strain cause a reduction in intestinal blood flow and the promotion of intestinal oxidative and nitrosative stress, which compromises GI permeability (Hall et al., 2001; Lambert, 2004). Nitric oxide (NO; otherwise known as endothelium-derived relaxing factor) is one of several free radicals whose production increases with hyperthermia. Increased NO concentrations are detectable in peripheral splanchnic vascular beds of hyperthermic animals and are thought to increase tight junction permeability (Hall et al., 1994). It is thought that constitutive NO synthesis provides a protective function

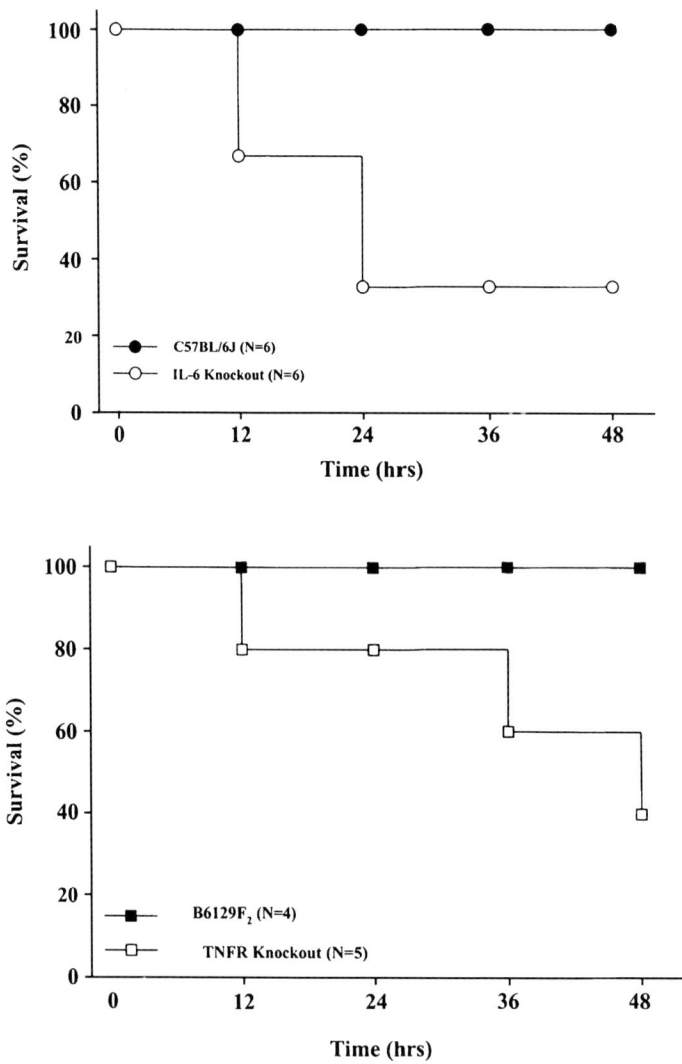

Fig. 8. Forty-eight hour survival curves of heat stressed male IL-6 (top) and TNF p55/p75 receptor (TNFR) knockout mice (bottom) compared to their respective controls. Details of the heat stress protocol have been described in detail elsewhere (Leon et al., 2005). Mice were heat stressed to a maximum core temperature of 42.7°C and allowed to recover, undisturbed, for 48 h at an ambient temperature of 25 ± 2°C. Survival was assessed visually and through inspection of radiotelemetry measurements every 12 h. Note that IL-6 and TNFR knockout mice showed significantly decreased survival rates compared to controls, despite high serum IL-6 and TNF levels being correlated with heat stroke mortality in human and animal models (see text for details).

under hyperthermic conditions by buffering increases in splanchnic vasoconstrictor activity and cellular stress (Hall et al., 2001). Increased intestinal barrier permeability occurs following cytokine-induced stimulation of NOS II enzymatic activity, which increases cellular NO flux above constitutive levels; the resultant overproduction of NO inhibits splanchnic resistance, leading to hypotension and circulatory collapse with heat stroke. Findings of lowered heat tolerance and increased venous portal endotoxin levels in heat stressed rats treated with the nitric oxide synthase inhibitor, L-NAME, support this hypothesis. Increased microvascular permeability has also been reported in cats, under non-heat stress conditions, following L-NAME treatment (Kubes and Granger, 1992). An increase in

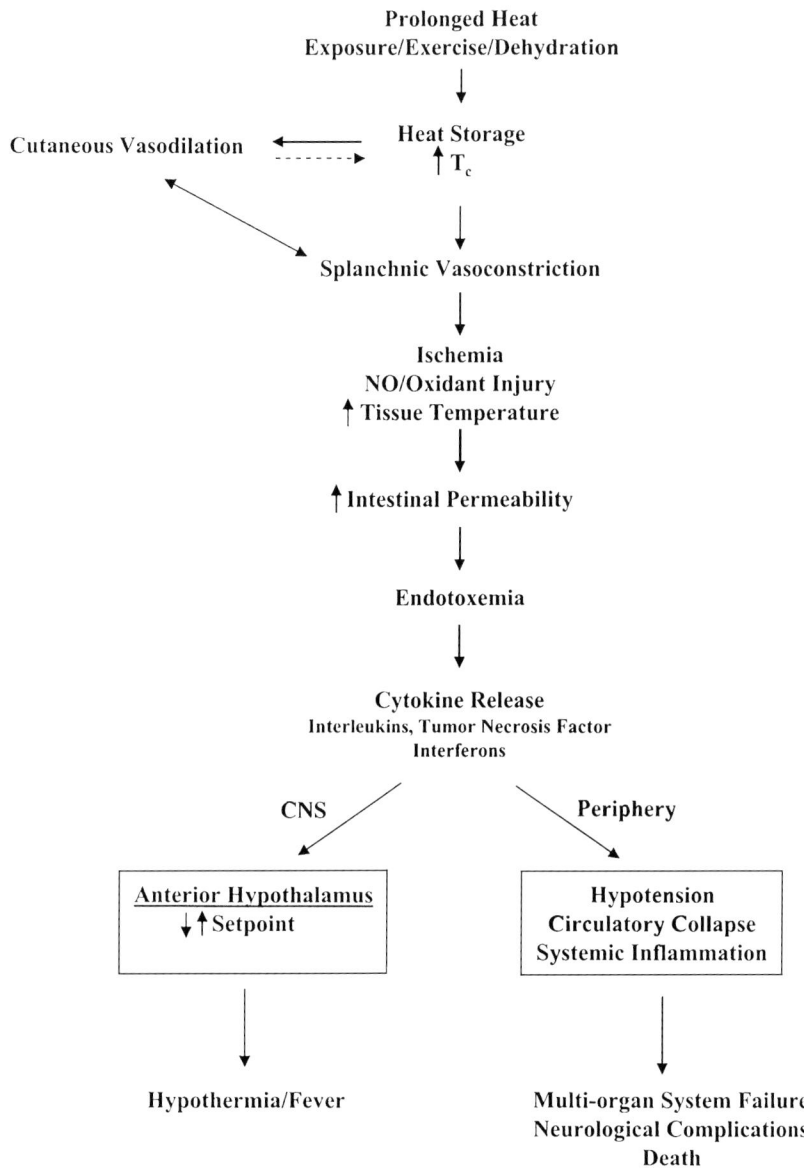

Fig. 9. Effect of prolonged heat exposure on changes in blood flow, intestinal permeability, and cytokine release. Heat exposure induces a rise in core temperature (T_c), which stimulates an increase in cutaneous blood flow to facilitate heat dissipation to the environment. Splanchnic vasoconstriction represents a compensatory response that is aimed at facilitating cutaneous vasodilation. The resultant ischemia of the intestines causes oxidative and nitrosative stress, which results in an increase in gut permeability. As intestinal membrane permeability is increased, endotoxin leakage is enhanced and stimulates cytokine (e.g., IL-1, IL-6, IL-10) production. Increased cytokine production signals to the periphery and CNS, which may induce changes in core temperature (fever, hypothermia) and multi-organ system failure (the ultimate cause of heat stroke mortality).

leukocyte adhesion is the proposed mechanism of L-NAME effects, as NO is a known modulator of leukocyte adherence (Kubes et al., 1991; Kubes and Granger, 1992).

Inhibition of leukocyte recruitment is also a proposed mechanism for the protective effects of IL-10 on alterations of vascular permeability with endotoxemia. Hickey et al. (1998) reported greater leukocyte recruitment and vascular permeability of IL-10 knockout mice compared to wild-type controls at 4 h following peripheral injection of LPS. The combined blockade of E- and P-selectin reduced leukocyte rolling to control levels in IL-10 knockout mice, suggesting a direct effect of IL-10 on the expression of these adhesion molecules (Hickey et al., 1998). IL-10 is a Th2 type cytokine that has potent inhibitory actions on the production of pro-inflammatory cytokines, such as IFNγ, IL-1, IL-6, and TNF. In a model of delayed-type hypersensitivity in mice, IL-10 treatment significantly inhibited footpad swelling induced by injection of Th1 clones and reduced local footpad levels of IL-2, IL-6, IFNγ, and TNF (Li et al., 1994). IL-10 is a known inhibitor of contact hypersensitivity — Berg et al. (1995) examined the extent of ear swelling in response to croton oil (a skin irritant) and showed twice the amount of swelling in IL-10 knockout mice compared to their wild-type controls. Swelling persisted through 48 h in the knockout mice, whereas it showed ∼50% abatement at this time point in wild-type mice. Treatment with an anti-TNF antibody reduced tissue necrosis in IL-10 knockout mice and eliminated focal necrosis in wild-type mice; thus, negative regulation of TNF production is a proposed permeability control mechanism mediated by IL-10 (Berg et al., 1995).

Additional cytokine interactions in the control of intestinal permeability include IL-10 modulation of IFNγ and IL-6 production. In a cecal ligation and puncture (CLP) model of sepsis, wild-type mice showed significantly increased IL-6 levels in the ileal mucosa which correlated with increased intestinal permeability (Wang et al., 2001a). In IL-6 knockout mice, CLP was without effect on intestinal permeability, presumably due to ∼20-fold higher mucosal level of IL-10 in these animals. The occurrence of increased intestinal permeability in wild-type mice, despite increased IL-10 levels (albeit below the level observed in the knockout animals) is suggestive of the importance of understanding the ratio of IL-6 to IL-10 levels in the mediation of this response in sepsis (Wang et al., 2001a). Finally, IFN-γ has been implicated in endothelial barrier dysfunction due to its ability to enhance expression of endothelial cell adhesion molecules — this effect being mediated via the stimulated production of other cytokines (TNFα, IL-1β) and free oxygen radicals (e.g., NO; Ruszczak et al., 1990). IL-10 pretreatment has been shown to prevent IFN-γ-induced increases in vascular permeability in human umbilical vein endothelial cells (Oshima et al., 2001).

Endotoxemia and heat stroke

In exertional heat stroke patients, TNFα, IL-1α, and endotoxin were detectable at clinical admission (core temperature ∼42.1°C), and showed a significant decrease following the completion of cooling (Bouchama et al., 1991). Changes in LPS concentrations occurred independently from cytokine production and the decrease in core temperature with cooling. High endotoxin levels were detected in a young athlete (core temperature 40.6°C) on the second day of football practice and may have been related to hemorrhagic necrosis of the liver (Graber et al., 1971). Conversely, Chung et al. (1999) were unable to measure plasma endotoxin in former heat stroke patients or controls that were exposed to a 60 min heat stress (core temperature <39.5°C). Presumably, this negative response was related to the relatively low core temperature increase induced by the 1 h heat exposure; in primates, elevations in circulating endotoxin were detectable as core temperature approached 41.5°C and then showed a precipitous increase starting at ∼43.0°C (Gathiram et al., 1987a). It is noteworthy that splanchnic blood flow shows an initial decrease at core temperature of 40°C (Hall et al., 2001) and liver damage is typically detectable at core temperatures of ∼42° to 43°C (Kew et al., 1970; Bowers et al., 1978; Chao et al., 1981).

Several methods have been used to explore the role of endotoxin in heat stroke responses. Bynum et al. (1979) demonstrated increased survival to experimental heat stroke in the dog following a reduction of gut flora by antibiotics — 18 h survival increased more than three-fold, as long as treatment was provided prior to heat exposure. Similarly, the rise in core temperature and incidence of endotoxemia in rabbits was reduced following oral antibiotics (Butkow et al., 1984). In primates, 24 h pretreatment with anti-LPS hyper-immune serum returned plasma LPS levels to baseline and reversed mortality (Gathiram et al., 1987b). However, this protective effect was inhibited following heating of the animals to a higher final core temperature, indicating that hyperthermia alone may cause irreversible organ damage and death (Gathiram et al., 1987b); interestingly, mortality was observed following only ~0.3°C further increase in core temperature (43.5 vs. 43.8°C). Dubose et al. (1983a) reported similar findings in endotoxin-tolerant rats which were protected from heat stroke mortality under moderate heat stress conditions — this protective effect dissipated once the accumulated thermal load (calculated as thermal area) exceeded a threshold value of >60°C/min (referred to as severe heat stress). Interestingly, an increase in endotoxin sensitivity following zymosan (complement antagonist) injection had no effect on thermotolerance (Dubose et al., 1983a). Furthermore, increased endotoxin concentrations were undetectable in plasma or tissues of heat stressed rats in this study. The authors concluded from these findings that the protective effect of endotoxin tolerance may have been unrelated to heat-induced endotoxemia *per se*, but rather a generalized protective response against the shock-like (SIRS) syndrome induced by heat — similar to that observed in response to hemorrhage or trauma (Dubose, et al., 1983a). A subsequent study by this same laboratory suggested that the protective effect of endotoxin tolerance was related to enhanced stimulation of the reticuloendothelial (RES) system, a major route of endotoxin clearance (Dubose et al., 1983b). As such, RES stimulation reduced and RES blockade increased mortality of heat stressed rats (Dubose et al., 1983b).

While the results from the aforementioned studies support a role for multiple cytokine interactions in the maintenance of epithelial barrier integrity, an examination into their role in the vascular permeability changes that occur in the heat stroke syndrome has not yet been performed.

Heat shock proteins

Heat shock proteins are phylogenetically conserved proteins that function as molecular chaperones to prevent the misfolding and aggregation of proteins under stressful conditions (Lindquist and Craig, 1988; Fink, 1999; Jaattela, 1999; Hartl and Hayer-Hartl, 2002). Based on these functions, HSPs have traditionally been regarded as intracellular proteins; however, the presence of HSPs in the circulation of normal individuals and increased expression in response to a variety of stressful stimuli has stimulated interest in their extracellular functions (Lindquist and Craig, 1988; Jaattela, 1999; Pockley et al., 1999; Basu et al., 2000). HSPs are found in all organisms that have been examined, from bacteria to humans, and are thought to have a major role in providing cytoprotection in the face of exposure to environmental (heavy metals, heat stress), physiological (cell differentiation, protein translation), and pathological (infections, ischemia/reperfusion) stimuli (Lindquist and Craig, 1988; Jaattela, 1999; Kregel, 2002). A decrease in HSP expression with aging and increased expression under pathophysiological conditions such as hypertension and atherosclerosis suggest that these molecules may be important biomarkers of stress susceptibility (Pockley et al., 2000; Rea et al., 2001; Xu, 2002).

HSPs range in size from 27,000 to 110,000 Da and are grouped into families according to their molecular mass, cellular localization, and function. HSP 27 is a constitutively expressed protein that resides in the cell cytoplasm, but undergoes increased expression and translocation to the cell nucleus in response to heat exposure (Arrigo et al., 1988). The function of HSP 27 is thought to reside in its ability to stabilize cytoskeletal protein organization under stressful conditions (Lavoie et al., 1993). The HSP 60 family consists of mitochondrial and cytosolic members that function as molecular chaperones to facilitate protein folding (Bukau and Horwich, 1998). It has been suggested that HSPs of the 60-kDa family function as danger signals to the innate immune system, inducing the release of several cytokines implicated in chronic inflammatory conditions, as well as atherosclerosis (Kaufmann, 1990; Nomoto and Yoshikai, 1991). The ability of HSPs to interact with pattern recognition receptors, such as the LPS receptors known as CD14 and Toll, to induce cytokine release suggests that interactions between HSPs and the host immune system may be an important first line of defense against infection/inflammation. The HSP 70 family consists of constitutive cytosolic HSP 73 (also known as HSC 70), stress-inducible cytosolic HSP 72, endoplasmic reticulum Bip (also known as Grp78), and mitochondrial HSP 70 (Lindquist and Craig, 1988; Fink, 1999; Hartl and Hayer-Hartl, 2002). Proteins of the HSP 70 family have been extensively studied for their protective function(s) against a variety of stressful insults, including thermal stress (Yang et al., 1998; Li et al., 2001; Kelty et al., 2002), ischemia/reperfusion (Marber et al., 1995; Stojadinovic et al., 1995; Rajdev et al, 2000), tissue injury (Brown et al., 1989), metabolic stress such as glucose deprivation (Williams et al., 1993), and sepsis (Hotchkiss et al., 1993; Villar et al., 1994). HSP 70s are also known to function in concert with other molecular chaperones, such as HSP 90 and HSP 110. The HSP 90 family consists of cytosolic HSP 90 and glucose regulation protein (GRP) 96, the latter of which is upregulated in the ER in response to glucose starvation (Kabakov et al., 1990; Morita et al., 2000). HSP 90 has been shown to play a role in glucocorticoid receptor (GR) functioning, through facilitation of the folding of the receptor's hormone binding domain, receptor intracellular trafficking, and stabilization of the receptor against proteolytic degradation (Pratt et al., 1999). HSP 70 is thought to facilitate GR–HSP 90 heterocomplex formation and has also been shown to participate in HSP 70–HSP 90–LPS interactions (Hutchison et al., 1994). HSP 110 is a molecular chaperone that has been strongly implicated in anti-tumor immune responses. Increased HSP 110 expression has been observed in hepatocellular carcinoma and human colorectal cancer (Hwang et al., 2003; Gotoh et al., 2004) and has been hypothesized to function as a "danger signal" by stimulating antigen presenting cells (e.g., dendritic cells) to release pro-inflammatory cytokines, thus alerting the immune system to tumors (Manjili et al., 2005).

HSPs and protein folding

In addition to their protective functions in stressful environments, HSP 70 family members also interact transiently with a variety of cellular proteins to facilitate natural protein folding and maturation during normal growth and physiological functioning. Beckmann et al. (1990) suggested that most proteins interact with HSP 72 and HSP 73 during synthesis (emergence from the ribosome) as a mechanism of maintaining a stable conformation during the translation process; upon release of the HSPs, each newly synthesized protein folds into its final conformational state. Presumably, the affinity of nascent protein–HSP interactions is altered as the protein folds into its mature conformation and neighboring peptide domains interact with one another, thus releasing HSP in an ATP-dependent process (Beckmann et al., 1990). There are several lines of evidence to suggest that HSP 70s exist in an equilibrium state between free and substrate-bound forms; that is, as free HSP is reduced following binding to newly synthesized proteins, additional HSP synthesis occurs. When the protein folds into its mature conformation and releases HSP 70, the free pool is increased once again and new HSP synthesis is halted. Evidence to support this type of regulation includes: (1) HSP 70 synthesis is increased in direct proportion to heat stress severity (and protein denaturation); (2) HSP 70 synthesis following a second stressor is influenced by the amount of pre-existing HSP 70 induced by the initial stressor; (3) HSP 70 synthesis is activated following the microinjection of denatured proteins into frog oocytes — this response may be directly related to the binding of HSP 70 to the newly injected denatured proteins, which subsequently reduces the free pool of HSP 70; (4) an immediate induction of HSP 70 synthesis occurs following the injection of anti-HSP 70 antibodies; and (5) the injection of puromycin, which stops protein translation prior to the completion of synthesis, induces increased HSP 70 synthesis (Ananthan et al., 1986; Mizzen and Welch, 1988; Beckmann et al., 1990). Taken together, these results suggest that HSP 70s transiently function as molecular chaperones to facilitate proper folding of newly synthesized proteins.

Alterations in HSP expression

HSPs were originally discovered in *Drosophila melanogaster*, in which puffs appeared on the giant chromosomes of the salivary glands in response to heat exposure (Ritossa, 1962). Tissières et al. (1974) later identified novel protein synthesis in the salivary glands of *Drosophila* that was related to the chromosomal puffs induced by heat exposure. It was subsequently hypothesized that the denaturation of mature proteins inside the cell following stress exposure was the triggering event for increased protein synthesis (Hightower, 1980). As previously described, there are now abundant data to support this hypothesis.

Numerous studies have examined the effect of stressful stimuli on the time course of HSP gene expression in a variety of cell and tissue types. Schena et al. (1996) examined the heat shock response in human T cells (43°C for 4 h) and heart tissue using cDNA microarray technology and noted a significant increase in HSP 90 expression. In human PBMCs, maximal expression of intracellular HSP 70 was observed between 4 and 6 h after heat shock (43°C for 20 min; Sonna et al., 2002). Increased HSP gene expression (HSP 10, 20, 40, 60, 70, 90, and 110) has also been observed in response to exertional heat stroke and hypoxia in PBMC and human hepatocytes, respectively (Sonna et al., 2003, 2004, see Chapter 16 in this volume). *In vivo* approaches have also been successful in demonstrating increased HSP gene expression in response to heat and other stressors. Blake et al. (1990) showed that the ambient temperature to which rats were exposed, as well as the duration of exposure to a given temperature were factors that determined the magnitude of HSP 70 mRNA induction in brain, lung, and skin. Whether core body temperature was elevated by exposure to high temperature for a short duration or to a lower temperature for a longer duration, there was a direct correlation between maximum core temperature attained and the level of HSP gene expression (Blake et al., 1990). Flanagan et al. (1995) directly tested the effect of heating rate on the tissue-specific HSP 70 response in rats and showed that a high rate of passive heating (0.175°C/min) induced greater HSP 70 expression in the liver, small intestine, and kidney than a low rate of heating (0.05°C/min). The kinetics of HSP 70 induction has also been shown to differ by organ. In rats, brain, lung, and skin showed the most rapid induction (∼1 h), whereas HSP 70 was maximally expressed in the liver at 6 h after heat exposure (Blake et al., 1990). Liver induction of HSP 70 has also been observed in rats 24 h following a 5–10 min exposure to ambient temperature of 40°C (Kluger et al., 1997). The time course of increased HSP 70 expression correlated with reduced TNF production and enhanced LPS-induced fever in heat stressed rats, supporting a role for HSP 70 in the modulation of core temperature responses (through an alteration of cytokine production) during infection. TNF has been implicated as an endogenous antipyretic, such that the inhibition of its production by HSPs would be expected to enhance fever responses (Kluger et al., 1997).

In the rat brain, the distribution of HSP 72 has been localized to neurons located in the dentate gyrus, medial habenula, hypothalamus, granular layer of the cerebellum, glia, endothelial cells of the arterioles, choroid plexus, and the olfactory area following *in vivo* hyperthermia that induced a core temperature elevation of 41.5°C (Li et al., 1992a). Increased HSP 72 expression in the hypothalamus and hippocampus suggests a direct connection between thermal injury and alteration of thermoregulatory control mechanisms, which are thought to reside in those brain areas. Focal cerebral ischemia had a similar effect on neuronal HSP 72 expression in the rat brain (Li et al., 1992b). In human fetal astrocytes, HSP 27 is constitutively expressed and shows increased phosphorylation following a 30 min exposure to heat shock, cytokines, or growth factors (Satoh and Kim, 1995). Importantly, it is now known that a "heat shock" paradigm is not needed to evoke the expression of HSPs in mammals. Heat acclimation for several weeks to a hot but not lethal environment is sufficient to elicit significant HSP expression. For example, rats maintained for 4 weeks at an ambient temperature of 34°C undergo a 175% increase in HSP 72 levels in cardiac tissue (Maloyan et al., 1999). The degree to which core or peripheral tissue temperatures must increase to elicit the HSP response is not clear. One might expect the threshold temperature for HSP induction to correlate with the temperatures in which a particular species normally lives (reviewed in

Feder and Hofmann, 1999). Although one would expect a significant elevation in core temperature of the rat when maintained continuously at 34°C, core temperature is commonly not measured in such studies (Maloyan et al., 1999).

Trauma is also sufficient to induce HSP 70 expression in rat brain (Brown et al., 1989; Gower et al., 1989). Thus, in addition to serving as a biomarker of thermal injury, HSP expression is useful for the identification of reactive cells that respond to trauma in the presumed absence of a core temperature change (Brown et al., 1989). Exercise is another potential stimulator of HSP induction, although the mechanisms that lead to the heat shock response are not fully understood. The etiology of exercise-induced effects on homeostasis has been attributed to increased core temperature, oxidative stress, accumulation of lactic acid, alteration in calcium homeostasis, and glucose deprivation (Salo et al., 1991; Kilgore et al., 1998; Clarkson and Sayers, 1999). Moderate intensity exercise induced HSP 70 expression in locomotor muscles of the rat hindlimb, despite maintenance of core temperature at baseline levels (Skidmore et al., 1995); however, muscle temperature was not measured so it is difficult to determine if a complete dissociation between exercise and heat effects was achieved in this study. Walters et al. (1998) measured exercise-induced brain mRNA and protein levels of HSP 70 in which brain temperature was maintained at baseline levels; a change in HSP expression was undetectable in seven fore- and midbrain regions or three hindbrain regions of the rat, suggesting that exercise alone was not sufficient for HSP induction in their model.

Thermotolerance and injury protection

Acquired thermotolerance is the term used to describe the non-inheritable, transient resistance to lethal heat stress that is acquired following an initial, short exposure to a non-lethal heat treatment. Several studies have demonstrated a temporal relationship between the development of thermotolerance and HSP expression, accumulation, and degradation (Landry et al., 1982; Li and Werb, 1982; Subjeck et al., 1982). Of the several types of HSPs that are synthesized in response to heat exposure, HSP 70 concentrations show the best correlation with thermotolerance. Li et al. (1991) used rat fibroblasts expressing different levels of a cloned HSP 70-encoding human gene to show that the higher the level of expressed HSP 70, the greater the thermal resistance of cells to a heat shock treatment of 45°C for 90 min. Na arsenite, hypoxia, and ethanol also induced tolerance to a subsequent heat exposure in Chinese hamster fibroblasts, which correlated with increased synthesis of HSP 70, 87, and 97 (Li and Werb, 1982). Thus, stressors in addition to heat can confer protection to subsequent thermal damage. Additional evidence in support of a role of HSP 70 in thermotolerance is provided by studies examining the effect of competitive inhibition or over-expression of HSP function. Johnston and Kucey (1988) showed that a 90% reduction of endogenous HSP 70 levels, achieved following the introduction of a dominant negative mutant HSP 70 gene into Chinese hamster ovary cells, increased thermosensitivity as illustrated by a more rapid time course of cell death and reduced colony formation compared to controls. In a similar manner, the introduction of affinity-purified monoclonal antibodies to HSP 70 into rat fibroblasts rendered those cells incapable of surviving a 30 min heat shock at 45°C, a response that correlated with a loss of cell membrane integrity (Riabowol et al., 1988). As an important control, heat-denatured HSP 70 antibodies as well as control antibody injections had no effect on cell survival, indicating that antigen:antibody complexes were not responsible for the observed change in thermosensitivity. Further analysis revealed that increasing dilutions of injected HSP 70 antibodies resulted in an increase in thermal cytotoxicity, indicating that the level of thermotolerance was directly dependent on the concentration of HSP 70. Finally, cells microinjected with the HSP 70 antibody showed low or absent levels of nuclear staining for the protein, suggesting that nuclear translocation (and presumed effects on gene transcription) was necessary for the thermotolerant effects observed by prior heat treatment in control cells.

Due to a reduction of ATP levels during ischemia, protein translation is halted; the protective effect of HSP 70 under these conditions is thought to reside in their ability to assist in the refolding of denatured proteins following commencement of reperfusion. An in situ study of the hearts of transgenic mice overexpressing rat HSP 70 protein showed a 40% reduction in the zone of infarction, which coincided with a two-fold increase in contractile function following ischemic injury (Marber et al., 1995). Unfortunately, this study was conducted in a non-functional, buffer perfused heart, making applicability to the *in vivo* condition questionable. However, Hutter et al. (1996) confirmed this result in transgenic mice overexpressing HSP 72 in brain, cardiac, and skeletal muscle by showing significantly decreased infarct size following a 30 min left coronary artery occlusion. Similar results have been reported in response to cerebral infarction (Rajdev et al., 2000). Cross tolerance, in which induction of HSPs from one stress agent confers protection against a different subsequent insult, provided myocardial protection against ischemia (Hutter et al., 1994; Stojadinovic et al., 1995), phospholipase A_2- and sepsis-induced acute lung injury (Villar et al., 1993, 1994), and heat stroke and endotoxin mortality (Ryan et al., 1992, 1994; Hotchkiss et al., 1993; Lappas et al., 1994; Yang et al., 1998). Thus, the term "thermotolerance" is a bit of a misnomer since tolerance can be acquired by pretreatment with agents other than heat, such as Na arsenite, as long as those treatments are able to induce HSP synthesis. Furthermore, increased HSP 70 expression can protect against a variety of stressors in additional to thermal toxicity.

Heat shock proteins and cytokines

HSP gene expression is mediated primarily at the level of gene transcription by a family of heat shock transcription factors (HSF) that interact with a regulatory element, known as the heat shock element (HSE), in the promoter region of genes. The major stress responsive HSF in mammalian cells is HSF-1, which is constitutively present in a non-DNA binding state, but is rapidly transformed to a nuclear form following exposure to heat or other stresses. Subsequent to nuclear translocation, HSF-1 binds to heat shock elements (HSEs) to regulate gene transcription (Voellmy, 1994; Wu, 1995). Under stressful conditions, gene promoter activity is significantly inhibited. A hypothesis to account for this response is that the inhibition of transcription serves to limit the accumulation of nascent or denatured proteins until more favorable environmental conditions are restored.

Mechanism of HSP protection: inhibition of cytokine transcription

The inhibition of IL-1, IL-6, and TNFα production provides protection against the morbidity and mortality of endotoxin exposure in several experimental models (Silva et al., 1990; Lundblad et al., 1995; Luheshi et al., 1996; Leon et al., 1998). HSF-1 is activated by febrile-range temperatures and the interaction of HSF-1 with HSEs in the promoter region of cytokine genes is a potential mechanism by which HSPs can inhibit cytokine gene transcription and confer protection against endotoxin and other infectious/inflammatory stimuli. Several studies have shown a direct effect of stress-induced HSP production on cytokine levels. In gene-transfected human PBMCs, overexpression of HSP 70 significantly reduced LPS-induced (*Brucella melitensis*) production of TNFα, IL-1β, IL-10, and IL-12, an effect that was reversible following treatment with antisense HSP 70 treatment (Ding et al., 2001). Others have shown opposite effects on pro- and anti-inflammatory cytokines, with a decrease in LPS-induced IL-12 (pro-inflammatory) levels and increase in IL-10 (anti-inflammatory) levels following heat exposure (Wang et al., 2001b). Differences in cytokine profiles may have been related to the time of cytokine measurement following heat shock treatment (4h vs. 24h), which complicates inter-study comparisons in most cases.

Interestingly, HSP 70 expression has been shown consistently to be without effect on IL-6 levels (Ensor et al., 1994, 1995; Ding et al., 2001; Wang et al., 2001b). The maintenance of IL-6 levels may be an indirect mechanism of HSP-induced inhibition of IL-1 and TNF, since IL-6 is a negative regulator of these cytokines (a proposed anti-inflammatory effect of IL-6; Ding et al., 2001). The downregulation of IL-1 and TNF production has been demonstrated in a variety of heat shock models. Heat shock treatment, administered at several different time points prior to LPS stimulation was effective in down-regulating TNFα and IL-1β production in human and rat macrophage, Kupffer and glomerular cells (Velasco et al., 1991; Fouqueray et al., 1992; Snyder et al., 1992; Ensor et al., 1994). Not surprisingly, the protective effect of heat treatment was dependent on the dose of LPS — the higher the LPS dose, the less effective the heat treatment or the higher temperature required to confer protection (Ensor et al., 1994). Importantly, a time course of heat-induced HSP expression was temporally correlated with cytokine inhibition in most studies — thus, it is not heat *per se*, but stressor-induced HSP production (i.e., heat, Na arsenite, etc.) that is mediating cytokine inhibition. In a human monocytic cell line, heat shock (42.5°C for 30 min) induced HSP 70 mRNA and a consequent inhibition of LPS-induced IL-1β and TNFα transcription (Xie et al., 2002). This heat shock effect was mediated through direct interaction of HSF-1 with the nuclear factor of IL-6 (NF-IL6, also known as a CCAAT enhancer binding protein or C/EBPβ), which is a direct regulator of IL-1β transcription (Xie et al., 2002). That is, HSF-1 bound to NF-IL6 to inhibit its ability to interact with the IL-1β promoter (Xie et al., 2002). Transcriptional control of TNFα has also been demonstrated — overexpression of HSF-1 or exposure to febrile range temperatures reduced TNFα promoter activity in macrophages (Singh et al., 2000). The relevance of transcriptional regulation of cytokine repression to the *in vivo* condition is demonstrated in HSF-1 knockout mice, which show increased mortality and an exaggerated TNFα response after endotoxic challenge (Xiao et al., 1999).

Aging and heat stroke

At the time of the writing of this chapter, Hurricane Katrina ravaged the Gulf Coast of the United States and, due to the destructive forces of the storm on electrical power supplies, caused significant mortalities of the elderly as air conditioning units were rendered inoperable in the face of rising local temperatures (as high as 110°F were reported; *NY Times*, Sept. 19, 2005). Even if the implementation of cooling or hydration strategies could have been implemented in the aftermath of this destructive storm, the presence of pre-existing illnesses in the elderly population of this region (many of the elderly resided in nursing homes) may have rendered them unresponsive to medical treatment. Clearly, a more thorough understanding of age-associated alterations in heat-induced responses is required for proper diagnosis, prevention, and treatment of this unique population.

Hyperthermia, fever, HSP expression, and cytokine production are all hallmarks of heat exposure that may be altered in aged individuals, an effect that may account for increased heat stroke morbidity and mortality in this population. It is expected that during the life of an organism there is an accumulation of protein damage resulting from continual oxidant/free radical activity. As stress tolerance deteriorates with aging, due to a variety of compromised mechanisms, an organism becomes less able to mitigate the adverse effects of protein denaturation with a resultant increase in stress morbidity and mortality (Holbrook and Udelsman, 1994; Lee et al., 1996). Alzheimer's disease has been postulated to be a consequence of decreased HSP function, which results in increased deposition of abnormally folded proteins (Morrison-Bogorad et al., 1995). In *Drosophila*, the original organism in which HSP were discovered, lifespan is extended by heat shock treatment or following the addition of HSP 70 gene copies, suggesting that an increase in protein chaperonin activity provides protection during aging (Khazaeli et al., 1997; Tatar et al., 1997). As suggested by Kim (2003), the identification of individuals with high HSP antibody titers might be useful in the identification of those particularly susceptible to heat stroke (or other stressors).

Several studies have reported altered heat stress responses with aging. In rats, aging was associated with a significant reduction in liver HSP expression following passive heat exposure (Kregel and Moseley, 1996). However, the ability of aged rats to show a similar HSP 70 response as mature rats to an exertional heat load suggested that the response to passive heating was not due to a global inability to express HSP; rather, a reduction in the threshold for HSP stimulation appeared to have changed during the aging process (Kregel and Moseley, 1996). Similarly, Fargnoli et al. (1990) showed that a global reduction in protein synthesis was not responsible for decreased HSP induction in aged lung fibroblasts. Overall protein synthesis patterns were similar between old and young cultures before and after heat stress (including HSP 27), but HSP 70 was significantly reduced with aging. One of the confounding variables in many studies is the lower core temperature increase induced in aged vs. mature rats during heat exposure. For example, aged rats showed a significant reduction in HSP 70 mRNA levels with heating, but it is unclear in many studies if this is a consequence of a lower heating rate, lower starting core temperature, and/or lower final core temperature in aged compared to young rats (Blake et al., 1991).

Several studies have reported impairment of baroreceptor reflex modulation (Stauss et al., 1997), lower sweating rate and longer onset to sweating (Inoue et al., 1999), and diminished renal and splanchnic sympathetic nerve discharge in aged organisms (Kenney and Fels, 2002). Minson et al. (1998) demonstrated that older men relied on a higher percentage of their cardiac chronotropic reserve compared to young men during heat exposure. This finding may have particular relevance to those individuals experiencing a heat wave with a pre-existing condition, such as coronary artery disease (Minson et al., 1998). One would expect a greater risk for a cardiac event under those conditions, as a higher heart rate would be required to maintain cardiac function in a hot environment. Additional factors to consider with aging is that the time course of HSP expression may differ in young and old rats, such that analysis at only time point may not be adequate to determine alterations in expression profiles in this population (Kregel and Moseley, 1996). Differences in body size are also important to consider. Walters et al. (2001) maintained constant body mass differences and heating profiles of aged and young rats to assess effect of heat on the time course of HSP 70 expression in the CNS — they demonstrated a time-dependent and regionally specific alteration of HSP 70 expression in aged rats.

An elevation in core temperature, whether due to an external heat load or internal generation of fever, has been shown to induce HSP synthesis *in vivo*. As described previously in this chapter, hyperthermia, hypothermia, and fever are common thermoregulatory responses to heat exposure that may be regulated by endogenous HSP and/or cytokine interactions in the host. Several studies have shown decreased fever responses in aged rats in response to IL-1β, IL-6, endotoxin, and prostaglandin (the final mediator of fever) injections (Norman et al., 1988; Miller et al., 1995; Satinoff et al., 1999; Krabbe et al., 2001; Buchanan et al., 2003; Peloso et al., 2003). In some cases, cytokine profiles in response to endotoxin were altered in the aged. For example, aging was associated with a prolonged fever and higher TNFα:IL-10 ratio following endotoxin injection in humans (Krabbe et al., 2001). Thus, an exaggerated pro-inflammatory cytokine response was sustained in these individuals, suggestive of an initial hyperreactivity and delayed termination of the APR in this population (Krabbe et al., 2001). Altered mechanisms of fever development in the aged may range from reduced sensitivity of endogenous pyrogen receptors, altered cytokine expression, changes in BBB permeability that reduces cytokine entrance into the brain to initiate an increase in the thermal setpoint, or changes in heat production capabilities (Chorinchath et al., 1996; Tateda et al., 1996; McLay et al., 2000; Buchanan et al., 2003; Peloso et al., 2003). Interestingly, access to warm ambient temperatures facilitates the development of endotoxin-induced fever in aged rats (Peloso et al., 2003). Thus, as shown with mice during heat stroke recovery, the ambient temperature that an organism experiences during manifestation of SIRS can have a profound impact on thermoregulatory responses and recovery. These effects in the aged may have profound consequences on heat stroke mortality rates in that the absence of hyperthermia or fever in the aged may prevent proper heat stroke diagnosis, and altered cytokine and thermoregulatory (hypothermia/fever) responses may exacerbate and/or prolong heat-induced SIRS.

Conclusion

As the average lifespan of the human population and the incidence of global warming increases, a rise in the incidence of heat stroke mortality may be anticipated. Current health care strategies for the prevention and/or treatment of heat stroke focus on the implementation of hydration guidelines, acclimatization protocols, and rapid cooling therapies. However, despite successful implementation of these techniques, significant adverse consequences of prolonged heat exposure continue to be realized. The fact that many heat stroke survivors incur permanent neurological damage despite cooling therapy suggests that current knowledge of the mechanisms responsible for the adverse consequences of this syndrome are not fully understood. Perhaps the most important outcome to-date of the study of heat stroke pathophysiology has been the realization that it is a "syndrome" that encompasses not only the responses elicited during direct heat exposure, but also those that ensue during long-term recovery. Given the current implication of cytokines in the pathophysiology of heat stroke and improvements in experimental techniques (e.g., radiotelemetry, transgenic technologies, stem cell therapy, etc.), it is anticipated that rapid advancements will be made in the near future in our understanding of the role of cytokines, and other physiological mediators, in the morbidity and mortality of this syndrome.

Abbreviations

5-HT	serotonin
5,7-DHT	5,7-dihydroxytryptamine
6-OHDA	6-hydroxydopamine
ADX	adrenalectomy
ALT	alanine aminotransferase
APR	acute phase response
AST	aspartate amino transferase
BBB	blood brain barrier
CLP	cecal ligation and puncture
CNS	central nervous system
CPK	creatine phosphokinase
CRP	C-reactive protein
CTM	critical thermal maximum
DA	dopamine
DIC	disseminated intravascular coagulation
GI	gastrointestinal
HPA	hypothalamic-pituitary-adrenal axis
HSP	heat shock protein
ICP	intracranial pressure
IFI	interferon inducible
IFN	interferon
IL	interleukin
IL-1ra	interleukin-1 receptor antagonist
IL-2R	interleukin-2 receptor
IPH	intraperitoneal heating
LPS	lipopolysaccharide
MAP	mean arterial pressure
MCP-1	monocyte chemoattractant protein-1
PAI-1	plasminogen activator inhibitor
PBMC	peripheral blood mononuclear cells
sIL-6R	soluble interleukin-6 receptor
siRNA	small interfering RNA
SIRS	systemic inflammatory response syndrome
sTNFR	soluble tumor necrosis factor receptor
T_a	ambient temperature
T_c	core temperature
$T_{c,Max}$	maximum core temperature
T_{es}	esophageal temperature
TF	tissue factor
T_h	hypothalamic temperature
TNF	tumor necrosis factor
tPA	tissue plasminogen activator
T_{re}	rectal temperature
WBH	whole body hyperthermia

Acknowledgments

The data provided in this chapter were collected through the tireless efforts of several individuals in my laboratory, including M.D. Blaha, J.D. Castor, B.C. Nephew, and L.D. Walker. The opinions or assertions contained herein are the private views of the author(s) and are not to be construed as official or reflecting the views of the Army or the Department of Defense. In conducting the research described in this report, the investigators adhered to the "Guide for Care and Use of Laboratory Animals" as prepared by the Committee on Care and Use of Laboratory Animals of the Institute of Laboratory Animal Resources, National Research Council. Any citations of commercial organizations and trade names in this report do not constitute an official Department of the Army endorsement of approval of the products or services of these organizations.

References

Adolph, E.F. (1947) Tolerance to heat and dehydration in several species of mammals. Am. J. Physiol., 151: 564–575.
Al-Hadramy, M.S. (1989) Catecholamines in heat stroke. Mil. Med., 1554: 263–264.
Al-Harthi, S.S., Karrar, O., Al-Mashhadani, S.A. and Saddique, A.A. (1990) Metabolite and hormonal profiles in heat stroke patients in Mecca Pilgrimage. Intern. Med., 228: 343–346.
Ananthan, J., Goldberg, A.L. and Voellmy, R. (1986) Abnormal proteins serve as eukaryotic stress signals and trigger the activation of heat shock genes. Science, 232: 522–524.
Armstrong, L.E., Epstein, Y., Greenleaf, J.E., Haymes, E.M., Hubbard, R.W., Roberts, W.O. and Thompson, P.D. (1996) American College of Sports Medicine position stand: heat and cold illnesses during distance running. Med. Sci. Sports Exerc., 28: i–x.
Arons, M.M., Wheeler, A.P., Bernard, G.R., Christman, B.W., Russell, J.A., Schein, R., Summer, W.R., Steinberg, D.P., Fulkerson, W., Wright, P., Dupont, W.D. and Swindell, B.B. (1999) Effects of ibuprofen on the physiology and survival of hypothermic sepsis: ibuprofen in sepsis study group. Crit. Care Med., 27: 699–707.
Arrigo, A.-P., Suhan, J.P. and Welch, W.J. (1988) Dynamic changes in the structure and intracellular locale of the mammalian low-molecular-weight heat shock protein. Mol. Cell Biol., 8: 5059–5071.
Attia, M., Khogali, M., El-Khatib, G., Mustafa, M.K.E., Mahmoud, M.A., Eldin, A.N. and Gumaa, K. (1983) Heatstroke: an upward shift of temperature regulation set point at an elevated body temperature. Int. Arch. Occup. Environ. Health, 53: 9–17.
Austin, M.G. and Berry, J.W. (1956) Observations on one hundred cases of heatstroke. JAMA, 161: 1525–1529.
Baars, J.W., de Boer, J.P., Wagstaff, J., Roem, D., Eerenberg-Belmer, A.J., Nauta, J., Pinedo, H.M. and Hack, C.E. (1992) Interleukin-2 induces activation of coagulation and fibrinolysis: resemblance to the changes seen during experimental endotoxaemia. Br. J. Haematol., 82: 295–301.

Baker, W.F. (1989) Clinical aspects of disseminated intravascular coagulation: a clinician's point of view. Semin. Thromb. Hemostas., 15: 1–57.
Bakhshi, S. and Arya, L.S. (2003) Etiopathophysiology of disseminated intravascular coagulation. J. Assoc. Physicians India, 51: 796–800.
Barber, A.E., Coyle, S.M., Marano, M.A., Fischer, E., Calvano, S.E., Fong, Y., Moldawer, L.L. and Lowry, S.F. (1993) Glucocorticoid therapy alters hormonal and cytokine responses to endotoxin in man. J. Immunol., 150: 1999–2006.
Basu, S., Binder, R.J., Ramalingam, T. and Srivastava, P.K. (2000) Necrotic but not apoptotic cell death releases heat shock proteins, which deliver a partial maturation signal to dendritic cells and activate the NF-kB pathway. Int. Immunol., 12: 1539–1546.
Beckmann, R.P., Mizzen, L.A. and Welch, W.J. (1990) Interaction of Hsp70 with newly synthesized proteins: implications for protein folding and assembly. Science, 248: 850–854.
Berg, D.J., Leach, M.W., Kuhn, R., Rajewsky, K., Muller, W., Davidson, N.J. and Rennick, D. (1995) Interleukin 10 but not interleukin 4 is a natural suppressant of cutaneous inflammatory responses. J. Exp. Med., 182: 99–108.
Berkenbosch, F., Van Oers, J., Del Rey, A., Tilders, F. and Besedovsky, H.O. (1987) Corticotropin-releasing factor-producing neurons in the rat activated by interleukin-1. Science, 238: 524–526.
Blake, M.J., Fargnoli, J., Gershon, D. and Holbrook, N.J. (1991) Concomitant decline in heat-induced hyperthermia and HSP70 mRNA expression in aged rats. Am. J. Physiol., 260: R663–R667.
Blake, M.J., Gershon, D., Fargnoli, J. and Holbrook, N.J. (1990) Discordant expression of heat shock protein mRNAs in tissues of heat-stressed rats. J. Biol. Chem., 265: 15275–15279.
Bouchama, A., Bridey, F., Hammami, M.M., Lacombe, C., al-Shail, E., al-Ohali, Y., Combe, F., al-Sedairy, S. and de Prost, D. (1996) Activation of coagulation and fibrinolysis in heatstroke. Thromb. Haemost., 76: 909–915.
Bouchama, A., Hammami, M.M., Al Shail, E. and De Vol, E. (2000) Differential effects of in vitro and in vivo hyperthermia on the production of interleukin-10. Intensive Care Med., 26: 1646–1651.
Bouchama, A., al Hussein, K., Adra, C., Rezeig, M., al Shail, E. and al Sedairy, S. (1992) Distribution of peripheral blood leukocytes in acute heatstroke. J. Appl. Physiol., 73: 405–409.
Bouchama, A. and Knochel, J.P. (2002) Heat stroke. N. Engl. J. Med., 346: 1978–1988.
Bouchama, A., Parhar, R.S., el-Yazigi, A., Sheth, K. and al-Sedairy, S. (1991) Endotoxemia and release of tumor necrosis factor and interleukin 1 alpha in acute heatstroke. J. Appl. Physiol., 70: 2640–2644.
Bouchama, A., Roberts, G., Al Mohanna, F., El-Sayed, R., Lach, B., Chollet-Martin, S., Ollivier, V., Al Baradei, R., Loualich, A., Nakeeb, S., Eldali, A. and de Prost, D. (2005) Inflammatory, hemostatic, and clinical changes in a

baboon experimental model for heatstroke. J. Appl. Physiol., 98: 697–705.

Bouchama, A., al-Sedairy, S., Siddiqui, S., Shail, E. and Rezeig, M. (1993) Elevated pyrogenic cytokines in heatstroke. Chest, 104: 1498–1502.

Bowers, W.D., Hubbard, R.W., Leav, I., Dawn, R., Conlon, M., Hamlet, M.P., Mager, M. and Brandt, P. (1978) Alterations of rat liver subsequent to heat overload. Arch. Pathol. Med., 102: 154–157.

Bradfield, J.W. (1974) Control of spillover: the importance of Kupffer-cell function in clinical medicine. Lancet, 2: 883–886.

Brix-Christensen, V., Tonnesen, E., Sorensen, I.J., Bilfinger, T.V., Sanchez, R.G. and Stefano, G.B. (1998) Effects of anaesthesia based on high versus low doses of opioids on the cytokine and acute-phase protein responses in patients undergoing cardiac surgery. Acta Anaesthesiol. Scand., 42: 63–70.

Brock-Utne, J.G., Gaffin, S.L., Wells, M.T., Gathiram, P., Sohar, E., James, M.F., Morrell, D.F. and Norman, R.J. (1988) Endotoxemia in exhausted runners after a long-distance race. S. Afr. Med. J., 73: 533–536.

Brown, I.R., Rush, S. and Ivy, G.O. (1989) Induction of a heat shock gene at the site of tissue injury in the rat brain. Neuron, 2: 1559–1564.

Buchanan, J.B., Peloso, E. and Satinoff, E. (2003) Thermoregulatory and metabolic changes during fever in young and old rats. Am. J. Physiol., 285: R1165–R1169.

Buchanan, T.A., Cane, P., Eng, C.C., Sipos, G.F. and Lee, C. (1991) Hypothermia is critical for survival during prolonged insulin-induced hypoglycemia in rats. Metabolism, 40: 330–334.

Bukau, B. and Horwich, A.L. (1998) The Hsp70 and hsp60 chaperone machines. Cell, 92: 351–366.

Butkow, N., Mitchell, D., Laburn, H. and Kenedi, E. (1984) Heat stroke and endotoxaemia in rabbits. In: Hales J.R.S. (Ed.), Thermal Biology. Raven Press, New York, pp. 511–514.

Bynum, G., Brown, J., Dubose, D., Marsili, M., Leav, I., Pistole, T.G., Hamlet, M., LeMaire, M. and Caleb, B. (1979) Increased survival in experimental dog heatstroke after reduction of gut flora. Aviat. Space Environ. Med., 50: 816–819.

Bynum, G.D., Pandolf, K.B., Schuette, W.H., Goldman, R.F., Lees, D.E., Whang-Peng, J., Atkinson, E.R. and Bull, J.M. (1978) Induced hyperthermia in sedated humans and the concept of critical thermal maximum. Am. J. Physiol., 235: R228–R236.

Camus, G., Poortmans, J., Nys, M., Deby-Dupont, G., Duchateau, J., Deby, C. and Lamy, M. (1997) Mild endotoxemia and the inflammatory response induced by a marathon race. Clin. Sci. (Lond.), 92: 415–422.

Carter, R., Cheuvront, S.N., Williams, J.O., Kolka, M.A., Stephenson, L.A., Amoroso, P.J. and Sawka, M.N. (2005) Epidemiology of hospitalizations and deaths from heat illness in soldiers from 1980 through 2002. Med. Sci. Sports Exerc., 37: 1328–1334.

Chang, D.M. (1993) The role of cytokines in heat stroke. Immunol. Invest., 22: 553–561.

Chao, T.C., Sinniah, R. and Pakiam, J.E. (1981) Acute heat stroke deaths. Pathology, 13: 145–156.

Chensue, S.W., Shmyr-Forsch, C., Otterness, I.G. and Kunkel, S.L. (1989) The beta form is the dominant interleukin 1 released by murine peritoneal macrophages. Biochem. Biophys. Res. Commun., 160: 404–408.

Chiu, W.T., Kao, T.Y. and Lin, M.T. (1995) Interleukin-1 receptor antagonist increases survival in rat heatstroke by reducing hypothalamic serotonin release. Neurosci. Lett., 202: 33–36.

Chiu, W.T., Kao, T.Y. and Lin, M.T. (1996) Increased survival in experimental rat heatstroke by continuous perfusion of interleukin-1 receptor antagonist. Neurosci. Res., 24: 159–163.

Chorinchath, B.B., Kong, L.Y., Mao, L. and McCallum, R.E. (1996) Age-associated differences in TNF-alpha and nitric oxide production in endotoxic mice. J. Immunol., 156: 1525–1530.

Chung, N.K., Shabbir, M. and Lim, C.L. (1999) Cytokine levels in patients with previous heatstroke under heat stress. Mil. Med., 164: 306–310.

Clarke, J.F. (1972) Some effects of the urban structure on heat mortality. Environ. Res., 5: 92–104.

Clarkson, P.M. and Sayers, S.P. (1999) Etiology of exercise-induced muscle damage. Can. J. Appl. Physiol., 24: 234–248.

Cole, R.D. (1983) Heat stroke during training with nuclear, biological, and chemical protective clothing: case report. Mil. Med., 148: 624–625.

Collins, K.J., Few, J.D., Forward, T.J. and Giec, L.A. (1969) Stimulation of adrenal glucocorticoid secretion in man by raising body temperature. J. Physiol., 202: 645–660.

Content, J., De Wit, L., Poupart, P., Opdenakker, G., Van Damme, J. and Billiau, A. (1985) Induction of a 26-kDa-protein mRNA in human cells treated with an interleukin-1-related, leukocyte-derived factor. Eur. J. Biochem., 152: 253–257.

Coris, E.E., Ramirez, A.M. and Van Durme, D.J. (2004) Heat illness in athletes: the dangerous combination of heat, humidity and exercise. Sports Med., 34: 9–16.

Cowles, R.B. and Bogert, C.M. (1944) A preliminary study of the thermal requirements of desert reptiles. Bull. Am. Mus. Nat. Hist., 83: 265–296.

Cross, M.C., Radomski, M.W., Vanhelder, W.P., Rhind, S.G. and Shepard, R.J. (1996) Endurance exercise with and without a thermal clamp: effects on leukocytes and leukocyte subsets. J. Appl. Physiol., 81: 822–829.

van Dam, A.-M., Brouns, M., Louisse, S. and Berkenbosch, F. (1992) Appearance of interleukin-1 in macrophages and in ramified microglia in the brain of endotoxin-treated rats: a pathway for the induction of non-specific symptoms of sickness? Brain Res., 588: 291–296.

Del Prete, G., De Carli, M., Lammel, R.M., D'Elios, M.M., Daniel, K.C., Giusti, B., Abbate, R. and Romagnani, S. (1995) Th1 and Th2 T-helper cells exert opposite regulatory effects on procoagulant activity and tissue factor production by human monocytes. Blood, 86: 250–257.

Dematte, J.E., O'Mara, K., Suescher, J., Whitney, C.G., Forsythe, S., McNamee, T., Adiga, R.B. and Ndukwu, I.M. (1998) Near-fatal heat stroke during the 1995 heat wave in Chicago. Ann. Intern. Med., 129: 173–181.

Departments of Army and Air Force. (2003) Heat stress control and heat casualty management. Tech. Bull. Med., 507/Air Force Pamphlet, pp. 48–152.

Desai, T.R., Leeper, N.J., Hynes, K.L. and Gewertz, B.L. (2002) Interleukin-6 causes endothelial barrier dysfunction via the protein kinase C pathway. J. Surg. Res., 104: 118–123.

Dietrich, W.D. and Kuluz, J.W. (2003) New research in the field of stroke: therapeutic hypothermia after cardiac arrest. Stroke, 34: 1051–1053.

Dinarello, C.A. (1991) Interleukin-1 and interleukin-1 antagonism. Blood, 77: 1627–1652.

Ding, X.Z., Fernandez-Prada, C.M., Bhattacharjee, A.K. and Hoover, D.L. (2001) Over-expression of HSP-70 inhibits bacterial lipopolysaccharide-induced production of cytokines in human monocyte-derived macrophages. Cytokine, 16: 210–219.

Di Santo, E., Sironi, M., Mennini, T., Zinetti, M., Savoldi, G., Di Lorenzo, D. and Ghezzi, P. (1996) A glucocorticoid receptor-independent mechanism for neurosteroid inhibition of tumor necrosis factor production. Eur. J. Pharmacol., 299: 179–186.

Dorozynski, A. (2003) Chirac announces investigation into heat wave's death toll. Br. Med. J., 327: 465.

Drexler, A.M. (1995) Tumor necrosis factor: its role in HIV/AIDS. STEP Perspect., 7: 13–15.

Drobatz, K.J. and Macintire, D.K. (1996) Heat-induced illness in dogs: 42 cases (1976–1993). J. Am. Vet. Med. Assoc., 209: 1894–1899.

D'Souza, S.D., Antel, J.P. and Freedman, M.S. (1994) Cytokine induction of heat shock protein expression in human oligodendrocytes: an interleukin-1-mediated mechanism. J. Neuroimmunol., 50: 17–24.

Dubois, E.F. (1949) Why are fevers over 106°F rare? Am. J. Med. Sci., 217: 361–368.

Dubose, D.A., Basamania, K., Maglione, L. and Rowlands, J. (1983a) Role of bacterial endotoxins of intestinal origin in rat heat stress mortality. J. Appl. Physiol., 54: 31–36.

Dubose, D.A., McCreary, J., Sowders, L. and Goode, L. (1983b) Relationship between rat heat stress mortality and alterations in reticuloendothelial carbon clearance function. Aviat. Space Environ. Med., 54: 1090–1095.

Dubose, D.A., Wenger, C.B., Flinn, S.D., Judy, T.A., Dubovtsev, A.I. and Morehouse, D.H. (2003) Distribution and mitogen response of peripheral blood lymphocytes after exertional heat injury. J. Appl. Physiol., 95: 2381–2389.

Ensor, J.E., Crawford, E.K. and Hasday, J.D. (1995) Warming macrophages to febrile range destabilizes tumor necrosis factor-α mRNA without inducing heat shock. Am. J. Physiol., 269: C1140–C1146.

Ensor, J.E., Wiener, S.M., McCrea, K.A., Viscardi, R.M., Crawford, E.K. and Hasday, K.A. (1994) Differential effects of hyperthermia on macrophage interleukin-6 and tumor necrosis factor-α expression. Am. J. Physiol., 266: C967–C974.

Fairchild, K.D., Singh, I.S., Patel, S., Drysdale, B.E., Viscardi, R.M., Hester, L., Lazusky, H.M. and Hasday, J.D. (2004) Hypothermia prolongs activation of NF-kB and augments generation of inflammatory cytokines. Am. J. Physiol., 287: C422–C431.

Fargnoli, J., Kunisada, T., Fornace Jr., A.J., Schneider, E.L. and Holbrook, N.J. (1990) Decreased expression of heat shock protein 70 mRNA and protein after heat treatment in cells of aged rats. Proc. Natl. Acad. Sci. U.S.A., 87: 846–850.

Feder, M.E. and Hofmann, G.E. (1999) Heat-shock proteins, molecular chaperones, and the stress response: evolutionary and ecological physiology. Annu. Rev. Physiol., 61: 243–282.

Fink, A.L. (1999) Chaperone-mediated protein folding. Physiol. Rev., 79: 425–449.

Flanagan, S.W., Ryan, A.J., Gisolfi, C.V. and Moseley, P.L. (1995) Tissue-specific HSP70 response in animals undergoing heat stress. Am. J. Physiol., 268: R28–R32.

Follenius, M., Brandenberger, G., Oyono, S. and Candas, V. (1982) Cortisol as a sensitive index of heat intolerance. Physiol. Behav., 29: 509–513.

Fouqueray, B., Philippe, C., Amrani, A., Perez, J. and Baud, L. (1992) Heat shock prevents lipopolysaccharide-induced tumor necrosis factor-α synthesis by rat mononuclear phagocytes. Eur. J. Immunol., 22: 2983–2987.

Frisina, J.P., Gaudieri, S., Cable, T., Keast, D. and Palmer, T.N. (1994) Effects of acute exercise on lymphocyte subsets and metabolic activity. Int. J. Sports Med., 15: 36–41.

Gathiram, P., Gaffin, S.L., Brock-Utne, J.G. and Wells, M.T. (1987a) Time course of endotoxemia and cardiovascular changes in heat-stressed primates. Aviat. Space Environ. Med., 58: 1071–1074.

Gathiram, P., Wells, M.T., Brock-Utne, J.G. and Gaffin, S.L. (1987b) Antilipopolysaccharide improves survival in primates subjected to heat stroke. Circ. Shock, 23: 157–164.

Gordon, C.J. (2001) The therapeutic potential of regulated hypothermia. Emerg. Med. J., 18: 81–89.

Gotoh, K., Nonoguchi, K., Higashitsuji, H., Kaneko, Y., Sakurai, T., Sumitomo, Y., Itoh, K., Subjeck, J.R. and Fujita, J. (2004) Apg-2 has a chaperone-like activity similar to Hsp110 and is overexpressed in hepatocellular carcinomas. FEBS Lett., 560: 19–24.

Gower, D.J., Hollman, C., Lee, K.S. and Tytell, M. (1989) Spinal cord injury and the stress protein response. J. Neurosurg., 70: 605–611.

Graber, C.D., Reinhold, R.B., Breman, J.G., Harley, R.A. and Hennigar, G.R. (1971) Fatal heat stroke: circulating endotoxin and gram-negative sepsis as complications. JAMA, 216: 1195–1196.

Hadad, E., Rav-Acha, M., Heled, Y., Epstein, Y. and Moran, D.S. (2004) Heat stroke; a review of cooling methods. Sports Med., 34: 501–511.

Hales, J.R., Khogali, M., Fawcett, A.A. and Mustafa, M.K. (1987) Circulatory changes associated with heat stroke: observations in an experimental animal model. Clin. Exp. Pharmacol. Physiol., 14: 761–777.

Hall, D.M., Buettner, G.R., Matthes, R.D. and Gisolfi, C.V. (1994) Hyperthermia stimulates nitric oxide formation: electron paramagnetic resonance detection of •NO-heme in blood. J. Appl. Physiol., 77: 548–553.

Hall, D.M., Buettner, G.R., Oberley, L.W., Xu, L., Mattes, R.D. and Gisolfi, C.V. (2001) Mechanisms of circulatory and intestinal barrier dysfunction during whole body hyperthermia. Am. J. Physiol., 280: H509–H521.

Hammami, M.M., Bouchama, A., Al-Sedairy, S., Shail, E., Al Ohaly, Y. and Mohamed, G.E. (1997) Concentrations of soluble tumor necrosis factor and interleukin-6 receptors in heatstroke and heatstress. Crit. Care Med., 25: 1314–1319.

Hanagata, K., Matsukawa, T., Sessler, D.I., Miyaji, T., Funayama, T., Koshimizu, M., Kashimoto, S. and Kumazawa, T. (1995) Isoflurane and sevoflurane produce a dose-dependent reduction in shivering threshold in rabbits. Anesth. Analg., 81: 581–584.

Hart, G.R., Anderson, R.J., Crumpler, C.P., Shulkin, A., Reed, G. and Knochel, J.P. (1982) Epidemic classical heat stroke: clinical characteristics and course of 28 patients. Medicine, 61: 189–197.

Hartl, F.U. and Hayer-Hartl, M. (2002) Molecular chaperones in the cytosol: from nascent chain to folded protein. Science, 295: 1852–1858.

Hashim, I.A., Al-Zeer, A., Al-Shohaib, S., Al-Ahwal, M. and Shenkin, A. (1997) Cytokine changes in patients with heatstroke during pilgrimage to Makkah. Mediat. Inflamm., 6: 135–139.

Haveman, J., Geerdink, A.G. and Rodermond, H.M. (1996) Cytokine production after whole body and localized hyperthermia. Int. J. Hyperthermia, 12: 791–800.

Haynes, B.F. and Fauci, A. (1978) The differential effect of in vivo hydrocortisone on the kinetics of subpopulations of human peripheral blood thymus-derived lymphocytes. J. Clin. Invest., 61: 703–707.

Heidemann, S.M., Lomo, L., Ofenstein, J.P. and Samaik, A.P. (2000) The effect of heat on cytokine production in rat endotoxemia. Crit. Care Med., 28: 1465–1468.

Heinrich, P.C., Castell, J.V. and Andus, T. (1990) Interleukin-6 and the acute phase response. Biochem. J., 265: 621–636.

Hennig, J., Rzepka, U., Mai, B. and Netter, P. (1995) Suppression of HPA-axis activity by haloperidol after experimentally induced heat stress. Prog. Neuro-Psychopharmacol. Biol. Psychiat., 19: 603–614.

Herbert, J.M., Savi, P., Laplace, M.C. and Lale, A. (1992) IL-4 inhibits LPS-, IL-1β- and TNFα-induced expression of tissue factor in endothelial cells and monocytes. FEBS Lett., 310: 31–33.

Hickey, M.J., Issekutz, A.C., Reinhardt, P.H., Fedorak, R.N. and Kubes, P. (1998) Endogenous interleukin-10 regulates hemodynamic parameters, leukocyte-endothelial cell interactions, and microvascular permeability during endotoxemia. Circ. Res., 83: 1124–1131.

Hightower, L.E. (1980) Cultured animal cells exposed to amino acid analogues or puromycin rapidly synthesize several polypeptides. J. Cell Physiol., 102: 407–427.

Holbrook, N.J. and Udelsman, R. (1994) Heat shock protein gene expression in response to physiologic stress and aging. In: Morimoto R.I., Tissieres A. and Georgopoulos C. (Eds.), Heat Shock Proteins: Structure, Function and Regulation. Cold Spring Harbor Lab. Press, New York, pp. 577–593.

Hotchkiss, R., Nunnally, I., Lindquist, S., Taulien, J., Perdrizet, G. and Karl, I. (1993) Hyperthermia protects mice against the lethal effects of endotoxin. Am. J. Physiol., 265: R1447–R1457.

Hubbard, R.W., Bowers, W.D., Matthew, W.T., Curtis, F.C., Criss, R.E.L., Sheldon, G.M. and Ratteree, J.W. (1977) Rat model of acute heatstroke mortality. J. Appl. Physiol., 42: 809–816.

Hubbard, R.W., Mager, M. and Kerstein, M. (1982) Water as a tactical weapon: a doctrine for preventing heat casualties. Proceedings of the Army Science Conference; June 15–18; Deputy Chief of Staff for Research, Development and Acquisition, Department of the Army, Vol. II, pp. 125–139.

Hubbard, R.W., Matthew, W.T., Linduska, J.D., Curtis, F.C., Bowers, W.D., Leav, I. and Mager, M. (1976) The laboratory rat as a model for hyperthermic syndromes in humans. Am. J. Physiol., 231: 1119–1123.

Hutchison, K.A., Dittmar, K.D., Czar, M.J. and Pratt, W.B. (1994) Proof that Hsp70 is required for assembly of the glucocorticoid receptor into a heterocomplex with Hsp90. J. Biol. Chem., 269: 5043–5049.

Hutchison, V.H. (1961) Critical thermal maxima in salamanders. Physiol. Zool., 34: 92–125.

Hutchison, V.H. and Murphy, K. (1985) Behavioral thermoregulation in the salamander *Necturus maculosus* after heat shock. Comp. Biochem. Physiol., 82A: 391–394.

Hutter, J.J., Mestril, R., Tam, E.K.W., Sievers, R.E., Dillmann, W.H. and Wolfe, C.L. (1996) Overexpression of heat shock protein 72 in transgenic mice decreases infarct size in vivo. Circulation, 94: 1408–1411.

Hutter, M.M., Sievers, R.E., Barbosa, V. and Wolfe, C.L. (1994) Heat-shock protein induction in rat hearts: a direct correlation between the amount of heat-shock protein induced and the degree of myocardial protection. Circulation, 89: 355–360.

Hwang, T.S., Han, H.S., Choi, H.K., Lee, Y.J., Kim, Y.J., Han, M.Y. and Park, Y.M. (2003) Differential, stage-dependent expression of Hsp70, Hsp110 and Bcl-2 in colorectal cancer. J. Gastroenterol. Hepatol., 18: 690–700.

Ibuka, N. and Fukumura, K. (1997) Unpredictable deprivation of water increases the probability of torpor in the Syrian hamster. Physiol. Behav., 62: 551–556.

Inoue, Y., Shibasaki, M., Ueda, H. and Ishizashi, H. (1999) Mechanisms underlying the age-related decrement in the human sweating response. Eur. J. Appl. Physiol., 79: 121–126.

IUPS Thermal Commission. (2001) Glossary of Terms for Thermal Physiology. 3rd edn. Revised by The Commission for Thermal Physiology of the International Union of Physiological Sciences. Jap. J. Physiol., 51: 245–280.

Jaattela, M. (1999) Heat shock proteins as cellular lifeguards. Ann. Med., 31: 261–271.

Maron, M.B., Wagner, J.A. and Horvath, S.M. (1977) Thermoregulatory responses during competitive marathon running. J. Appl. Physiol., 42: 909–914.

May, L.T., Neta, R., Moldawer, L.L., Kenney, J.S., Patel, K. and Sehgal, P.B. (1993) Antibodies chaperone circulating IL-6: paradoxical effects of anti-IL-6 "neutralizing" antibodies in vivo. J. Immunol., 151: 3225–3236.

McLay, R.N., Kastin, A.J. and Zadina, J.E. (2000) Passage of interleukin-1-beta across the blood-brain barrier is reduced in aged mice: a possible mechanism for diminished fever in aging. Neuroimmunomodulation, 8: 148–153.

Miller, D.J., Yoshikawa, T.T. and Norman, D.C. (1995) Effect of age on fever response to recombinant interleukin-6 in a murine model. J. Gerontol. A Biol. Sci. Med. Sci., 50: M276–M279.

Minson, C.T., Wladkowski, S.L., Cardell, A.F., Pawelczyk, J.A. and Kenney, W.L. (1998) Age alters the cardiovascular response to direct passive heating. J. Appl. Physiol., 84: 1323–1332.

Mitchell, J.B., Dugas, J.P., McFarland, B.K. and Nelson, M.J. (2002) Effect of exercise, heat stress, and hydration on immune cell number and function. Med. Sci. Sports Exerc., 34: 1941–1950.

Mizzen, L.A. and Welch, W.J. (1988) Characterization of the thermotolerant cell. I. Effects on protein synthesis activity and the regulation of heat-shock protein 70 expression. J. Cell Biol., 106: 1105–1116.

Moldoveanu, A.I., Shephard, R.J. and Shek, P.N. (2001) The cytokine response to physical activity and training. Sports Med., 31: 115–144.

Morita, T., Saitoh, K., Takagi, T. and Maeda, Y. (2000) Involvement of the glucose-regulated protein 94 (Dd-GRP94) in starvation response of *Dictyostelium discoideum* cells. Biochem. Biophys. Res. Commun., 274: 323–331.

Morrison-Bogorad, M., Zimmerman, A.L. and Pardue, S. (1995) Heat-shock 70 messenger RNA levels in human brain: correlation with agonal fever. J. Neurochem., 64: 235–246.

Moseley, P.L., Gapen, C., Wallen, E.S., Walter, M.E. and Peterson, M.W. (1994) Thermal stress induced epithelial permeability. Am. J. Physiol., 267: C425–C434.

Naughton, M.P., Henderson, A., Mirabelli, M.C., Kaiser, R., Wilhelm, J.L., Kieszak, S.M., Rubin, C.H. and McGeehin, M.A. (2002) Heat-related mortality during a 1999 heat wave in Chicago. Am. J. Prev. Med., 22: 221–227.

Neta, R., Sayers, T.J. and Oppenheim, J.J. (1992) Relationship of TNF to interleukins. In: Aggarwal B.B. and Vilcek J. (Eds.), Tumor Necrosis Factor: Structure, Function, and Mechanism of Action. Marcel Dekker, New York, pp. 499–566.

Neumann, F.J., Ott, I., Marx, N., Luther, T., Kenngott, S., Gawaz, M., Kotzsch, M. and Schomig, A. (1997) Effect of human recombinant interleukin-6 and interleukin-8 on monocyte procoagulant activity. Arterioscler. Thromb. Vasc. Biol., 17: 3399–3405.

Neville, A.J. and Sauder, D.N. (1988) Whole body hyperthermia (41° to 42°C) induces interleukin-1 in vivo. Lymphokine Res., 7: 201–206.

Nielsen, H.B. and Pedersen, B.K. (1997) Lymphocyte proliferation in response to exercise. Eur. J. Appl. Physiol., 75: 375–379.

Nieman, D.C., Henson, D.A., Smith, L.L., Utter, A.C., Vinci, D.M., Davis, J.M., Kaminsky, D.E. and Shute, M. (2001) Cytokine changes after a marathon race. J. Appl. Physiol., 91: 109–114.

Nolan, J.P. (1981) Endotoxin, reticuloendothelial function, and liver injury. Hepatology, 1: 458–465.

Nomoto, K. and Yoshikai, Y. (1991) Heat-shock proteins and immunopathology: regulatory role of heat-shock protein-specific T cells. Springer Semin. Immunopathol., 13: 63–80.

Norman, D.C., Yamamura, R.H. and Yoshikawa, T.T. (1988) Fever response in old and young mice after injection of interleukin 1. J. Gerontol., 43: M80–M85.

Ohara, K., Furuyama, F. and Isobe, Y. (1975) Prediction of survival time of rats in severe heat. J. Appl. Physiol., 38: 724–729.

Oka, T., Oka, K. and Hori, T. (2001) Mechanisms and mediators of psychological stress-induced rise in core temperature. Psychosom. Med., 63: 476–486.

Okusawa, S., Gelfand, J.A., Ikejima, T., Connolly, R.J. and Dinarello, C.A. (1988) Interleukin 1 induces a shock-like state in rabbits: synergism with tumor necrosis factor and the effect of cyclooxygenase inhibition. J. Clin. Invest., 81: 1162–1172.

Oshima, T., Laroux, F.S., Coe, L.L., Morise, Z., Kawashi, S., Bauer, P., Grisham, M.B., Specian, R.D., Carter, P., Jennings, S., Granger, D.N., Joh, T. and Alexander, J.S. (2001) Interferon-γ and interleukin-10 reciprocally regulate endothelial junction integrity and barrier function. Microvasc. Res., 61: 130–143.

Paidas, C.N., Mooney, M.L., Theodorakis, N.G. and De Maio, A. (2002) Accelerated recovery after endotoxic challenge in heat shock-pretreated mice. Am. J. Physiol., 282: R1374–R1381.

Paleolog, E.M., Delasalle, S.A.J., Buurman, W.A. and Feldmann, M. (1994) Functional activities of receptors for tumor necrosis factor-α on human vascular endothelial cells. Blood, 84: 2578–2590.

Pals, K.L., Chang, R.T., Ryan, A.J. and Gisolfi, C.V. (1997) Effect of running intensity on intestinal permeability. J. Appl. Physiol., 82: 571–576.

Patel, R.T., Deen, K.I., Youngs, D., Warwick, J. and Keighley, M.R.B. (1994) Interleukin 6 is a prognostic indicator of outcome in severe intra-abdominal sepsis. Br. J. Surg., 81: 1306–1308.

Peloso, E.D., Florez-Duquet, M., Buchanan, J.B. and Satinoff, E. (2003) LPS fever in old rats depends on the ambient temperature. Physiol. Behav., 78: 651–654.

Peters, M., Jacobs, S., Ehlers, M., Vollmer, P., Mullberg, J., Wolf, E., Brem, G., Meyer zum Buschenfelde, K.H. and Rose-John, S. (1996) The function of the soluble interleukin 6 (IL-6) receptor in vivo: sensitization of human soluble IL-6 receptor transgenic mice towards IL-6 and prolongation of the plasma half-life of IL-6. J. Exp. Med., 183: 1399–1406.

Petersdorf, R.G. (1994) Hypothermia and hyperthermia. In: Isselbacher K.J., Braunwald E., Wilson J.D., Martin J.B., Fauci A.S. and Kasper D.L. (Eds.), Harrison's Principles of Internal Medicine. McGraw-Hill, New York, pp. 2473–2479.

Pockley, A.G., Bulmer, J., Hanks, B.M. and Wright, B.H. (1999) Identification of human heat shock protein 60 (Hsp60) and anti-Hsp60 antibodies in the peripheral circulation of normal individuals. Cell Stress Chaperones, 4: 29–35.

Pockley, A.G., Wu, R., Lemne, C., Kiessling, R., de Faire, U. and Frostegard, J. (2000) Circulating heat shock protein 60 is associated with early cardiovascular disease. Hypertension, 36: 303–307.

van der Poll, T., Buller, H.R., ten Cate, H., Wortel, C.H., Bauer, K.A., van Deventer, S.J., Hack, C.E., Sauerwein, H.P., Rosenberg, R.D. and ten Cate, J.W. (1990) Activation of coagulation after administration of tumor necrosis factor to normal subjects. N. Engl. J. Med., 322: 1622–1627.

van der Poll, T., de Jonge, E. and Levi, M. (2001) Regulatory role of cytokines in disseminated intravascular coagulation. Semin. Thromb. Hemost., 27: 639–651.

van der Poll, T., Levi, M., Hack, C.E., ten Cate, H., van Deventer, S.J., Eerenberg, A.J., de Groot, E.R., Jansen, J., Gallati, H., Buller, H.R., ten Cate, J.W. and Aarden, L.A. (1994) Elimination of interleukin 6 attenuates coagulation activation in experimental endotoxemia in chimpanzees. J. Exp. Med., 179: 1253–1259.

Pradier, O., Gerard, C., Delvaux, A., Lybin, M., Abramowicz, D., Capel, P., Velu, T. and Goldman, M. (1993) Interleukin-10 inhibits the induction of monocyte procoagulant activity by bacterial lipopolysaccharide. Eur. J. Immunol., 23: 2700–2703.

Pratt, W.B., Silverstein, A.M. and Galigniana, M.D. (1999) A model for the cytoplasmic trafficking of signaling proteins involving the Hsp90-binding immunophilins and p50cdc37. Cell Signal., 11: 839–851.

Rajdev, S., Hara, K., Kokubo, Y., Mestril, R., Dillmann, W., Weinstein, P.R. and Sharp, F.R. (2000) Mice overexpressing rat heat shock protein 70 are protected against cerebral infarction. Ann. Neurol., 47: 782–791.

Raju, S.F., Robinson, G.H. and Bower, J.O. (1973) The pathogenesis of acute renal failure in heat stroke. South Med. J., 66: 330–333.

Ramlow, J.M. and Kuller, L.H. (1990) Effect of the summer heat wave of 1988 on daily mortality in Allegheny County, PA. Public Health Rep., 105: 283–289.

Rasch, W., Samson, P., Cote, J. and Cabanac, M. (1991) Heat loss from the human head during exercise. J. Appl. Physiol., 71: 590–595.

Rea, I.M., McNerlan, S. and Pockley, A.G. (2001) Serum heat shock protein and anti-heat shock protein antibody levels in aging. Exp. Gerontol., 36: 341–352.

Redl, H., Schlag, G., Schiesser, A. and Davies, J. (1995) Thrombomodulin release in baboon sepsis: its dependence on the dose of *Escherichia coli* and the presence of tumor necrosis factor. J. Infect. Dis., 171: 1522–1527.

Riabowol, K.T., Mizzen, L.A. and Welch, W.J. (1988) Heat shock is lethal to fibroblasts microinjected with antibodies against hsp70. Science, 242: 433–436.

Ritossa, F. (1962) A new puffing pattern induced by temperature shock and DNP in *Drosophila melanogaster*: relation to chromosome puffs. Experientia, 18: 571–573.

Roberts, W.O. (2004) Death in the heat: can football heat stroke be prevented? Curr. Sports Med. Rep., 3: 1–2.

Romanovsky, A.A. and Blatteis, C.M. (1996) Heat stroke: opioid-mediated mechanisms. J. Appl. Physiol., 81: 2565–2570.

Roumen, R.M.H., Hendricks, T., Wevers, R.A. and Goris, R.J.A. (1993) Intestinal permeability after severe trauma and hemorrhagic shock is increased without relation to septic complications. Arch. Surg., 128: 453–457.

Ruszczak, Z., Detmar, M., Imcke, E. and Orfanos, C.E. (1990) Effects of rIFN alpha, beta, and gamma on the morphology, proliferation, and cell surface antigen expression of human dermal microvascular endothelial cells in vitro. J. Invest. Dermatol., 95: 693–699.

Ryan, A.J., Flanagan, S.W., Moseley, P.L. and Gisolfi, C.V. (1992) Acute heat stress protects rats against endotoxin shock. J. Appl. Physiol., 73: 1517–1522.

Ryan, A.J., Matthes, R.D., Mitros, F.A. and Gisolfi, C.V. (1994) Heat stress does not sensitize rats to the toxic effects of bacterial lipopolysaccharide. Med. Sci. Sports Exerc., 26: 687–694.

Salo, D.C., Donovan, C.M. and Davies, K.J. (1991) Hsp70 and other possible heat shock or oxidative stress proteins are induced in skeletal muscle, heart, and liver during exercise. Free Radic. Biol. Med., 11: 239–246.

Satinoff, E., Peloso, E. and Plata-Salamn, C.R. (1999) Prostaglandin E_2-induced fever in young and old Long-Evans rats. Physiol. Behav., 67: 149–152.

Satoh, J. and Kim, S.U. (1995) Cytokines and growth factors induce HSP27 phosphorylation in human astrocytes. J. Neuropathol. Exp. Neurol., 54: 504–512.

Sawdey, M.S. and Loskutoff, D.J. (1991) Regulation of murine type 1 plasminogen activator inhibitor gene expression in vivo: tissue specificity and induction by lipopolysaccharide, tumor necrosis factor-alpha, and transforming growth factor-beta. Clin. Invest., 88: 1346–1353.

Schena, M., Shalon, D., Heller, R., Chai, A., Brown, P.O. and Davis, R.W. (1996) Parallel human genome analysis: microarray-based expression monitoring of 1000 genes. Proc. Natl. Acad. Sci. U.S.A., 93: 10614–10619.

Schmid, E.F., Binder, K., Grell, M., Scheurich, P. and Pfizenmaier, K. (1995) Both tumor necrosis factor receptors, TNFR60 and TNFR80, are involved in signaling endothelial tissue factor expression by juxtracrine tumor necrosis factor alpha. Blood, 86: 1836–1841.

Schobitz, B., Pezeshki, G., Pohl, T., Hemmann, U., Heinrich, P.C., Holsboer, F. and Reul, J.M.H.M. (1995) Soluble interleukin-6 (IL-6) receptor augments central effects of IL-6 in vivo. FASEB J., 9: 659–664.

Schwager, I. and Jungi, T.W. (1994) Effect of human recombinant cytokines on the induction of macrophage procoagulant activity. Blood, 83: 152–160.

Sharif, S.F., Hariri, R.J., Chang, V.A., Barie, P.S., Wang, R.S. and Ghajar, J.B.G. (1993) Human astrocyte production of tumour necrosis factor-α, interleukin-1β and interleukin-6 following exposure to lipopolysaccharide endotoxin. Neurol. Res., 15: 109–112.

Sharma, H.S. and Dey, P.K. (1986) Probable involvement of 5-hydroxytryptamine in increased permeability of blood-brain barrier under heat stress in young rats. Neuropharmacology, 25: 161–167.

Sharma, H.S. and Dey, P.K. (1987) Influence of long-term acute heat exposure on regional blood-brain barrier permeability, cerebral blood flow and 5-HT level in conscious normotensive young rats. Brain Res., 424: 153–162.

Sharma, H.S., Westman, J., Nyberg, F., Cervos-Navarro, J. and Dey, P.K. (1994) Role of serotonin and prostaglandins in brain edema induced by heat stress: an experimental study in the young rat. Acta Neurochir. Suppl., 60: 65–70.

Shattuck, G.C. and Hilferty, M.M. (1932) Sun stroke and allied conditions in the United States. Am. J. Trop. Med., 12: 223–245.

Shattuck, G.C. and Hilferty, M.M. (1933) Cause of death from heat in Massachusetts. N. Engl. J. Med., 209: 319–329.

Shintani, F., Kanba, S., Nakaki, T., Nibuya, M., Kinoshita, N., Suzuki, E., Yagi, G., Kato, R. and Asai, M. (1993) Interleukin-1β augments release of norepinephrine, dopamine, and serotonin in the rat anterior hypothalamus. J. Neurosci., 13: 3574–3581.

Sigmund, C.D. (1993) Major approaches for generating and analyzing transgenic mice. Hypertension, 22: 599–607.

Silva, A.T., Bayston, K.F. and Cohen, J. (1990) Prophylactic and therapeutic effects of a monoclonal antibody to tumor necrosis factor-α in experimental gram-negative shock. J. Infect. Dis., 162: 421–427.

Singh, I.S., Viscardi, R.M., Kalvakolanu, I., Calderwood, S. and Hasday, J.D. (2000) Inhibition of tumor necrosis factor-α transcription in macrophages exposed to febrile range temperatures. J. Biol. Chem., 275: 9841–9848.

Skidmore, R., Gutierrez, J.A., Guirriero Jr., V. and Kregel, K.C. (1995) HSP70 induction during exercise and heat stress in rats: role of internal temperature. Am. J. Physiol., 268: R92–R97.

Smalley, B., Janke, R.M. and Cole, D. (2003) Exertional heat illness in Air Force basic military trainees. Mil. Med., 168: 298–303.

Snyder, Y.M., Guthrie, L., Evans, G.F. and Zuckerman, S.H. (1992) Transcriptional inhibition of endotoxin-induced monokine synthesis following heat shock in murine peritoneal macrophages. J. Leukoc. Biol., 51: 181–187.

Sonna, L.A., Cullivan, M.L., Sheldon, H.K., Pratt, R.E. and Lilly, C.M. (2003) Effect of hypoxia on gene expression by human hepatocytes (HepG2). Physiol. Genomics, 12: 195–207.

Sonna, L.A., Gaffin, S.L., Pratt, R.E., Cullivan, M.L., Angel, K.C. and Lilly, C.M. (2002) Effect of acute heat shock on gene expression by human peripheral blood mononuclear cells. J. Appl. Physiol., 92: 2208–2220.

Sonna, L.A., Wenger, C.B., Flinn, S., Sheldon, H.K., Sawka, M.N. and Lilly, C.M. (2004) Exertional heat injury and gene expression changes: a DNA microarray analysis study. J. Appl. Physiol., 96: 1943–1953.

Stauss, H.M., Morgan, D.A., Anderson, K.E., Massett, M.P. and Kregel, K.C. (1997) Modulation of baroreflex sensitivity and spectral power of blood pressure by heat stress and aging. Am. J. Physiol., 272: H776–H784.

Stoen, R. and Sessler, D.I. (1990) The thermoregulatory threshold is inversely proportional to isoflurane concentration. Anesthesiology, 72: 822–827.

Stojadinovic, A., Kiang, J., Smallridge, R., Galloway, R. and Shea-Donohue, T. (1995) Induction of heat-shock protein 72 protects against ischemia/reperfusion in rat small intestine. Gastroenterology, 109: 505–515.

Stott, P.A., Stone, D.A. and Allen, M.R. (2004) Human contribution to the European heat wave of 2003. Nature, 432: 610–614.

Stouthard, J.M., Levi, M., Hack, C.E., Veenhof, C.H., Romijn, H.A., Sauerwein, H.P. and van der Poll, T. (1996) Interleukin-6 stimulates coagulation, not fibrinolysis, in humans. Thromb. Haemost., 76: 738–742.

Subjeck, J.R., Sciandra, J.J. and Johnson, R.J. (1982) Heat shock proteins and thermotolerance: a comparison of induction kinetics. Br. J. Radiol., 656: 579–584.

Suzuki, K., Nakaji, S., Yamada, M., Liu, Q., Kurakake, S., Okamura, N., Kumae, T., Umeda, T. and Sugawara, K. (2003) Impact of a competitive marathon race on systemic cytokine and neutrophil response. Med. Sci. Sports Exerc., 35: 348–355.

Tatar, M., Khazaeli, A.A. and Curtsinger, J.W. (1997) Chaperoning extended life. Nature, 390: 30.

Tateda, K., Matsumoto, T., Miyazaki, S. and Yamaguchi, K. (1996) Lipopolysaccharide-induced lethality and cytokine production in aged mice. Infect. Immun., 64: 769–774.

Tissières, A., Mitchell, H.K. and Tracy, V.M. (1974) Protein synthesis in salivary glands of *Drosophila melanogaster*: relation to chromosome puffs. J. Mol. Biol., 84: 389–398.

Unno, N., Wnag, H., Menconi, M.J., Tytgat, S.H.A.J., Larkin, V., Smith, M., Morin, M.J., Chavez, A., Hodin, R.A. and Fink, M.P. (1997) Inhibition of inducible nitric oxide synthase ameliorates endotoxin-induced gut mucosal barrier dysnfunction in rats. Gastroenterology, 113: 1246–1257.

Vandentorren, S., Suzan, F., Medina, S., Pascal, M., Maulpoix, A., Cohen, J.-C. and Ledrans, M. (2004) Mortality in 13 French cities during the August 2003 heat wave. Am. J. Public Health, 94: 1518–1520.

Velasco, S., Tarlow, M., Olsen, K., Shay, J.W., McCracken Jr., G.H. and Nisen, P.D. (1991) Temperature-dependent modulation of lipopolysaccharide-induced interleukin-1 beta and tumor necrosis factor alpha expression in cultured human astroglial cells by dexamethasone and indomethacin. J. Clin. Invest., 87: 1674–1680.

Villar, J., Edelson, J.D., Post, M., Mullen, J.B.M. and Slutsky, A.S. (1993) Induction of heat stress proteins is associated with decreased mortality in an animal model of acute lung injury. Am. Rev. Respir. Dis., 147: 177–181.

Villar, J., Ribeiro, S.P., Mullen, J.B.M., Kuliszewski, M., Post, M. and Slutsky, A.S. (1994) Induction of the heat shock response reduces mortality rate and organ damage in a sepsis-induced acute lung injury model. Crit. Care Med., 22: 914–921.

Voellmy, R. (1994) Transduction of the stress signal and mechanisms of transcriptional regulation of heat shock/stress protein gene expression in higher eukaryotes. Crit. Rev. Eukaryot. Gene Expr., 4: 357–401.

Waage, A., Slupphaug, G. and Shalaby, R. (1990) Glucocorticoids inhibit the production of IL6 from monocytes, endothelial cells and fibroblasts. Eur. J. Immunol., 20: 2439–2443.

Wada, H., Ohiwa, M., Kaneko, T., Tamaki, S., Tanigawa, M., Takagi, M., Mori, Y. and Shirakawa, S. (1991a) Plasma level of tumor necrosis factor in disseminated intravascular coagulation. Am. J. Hematol., 37: 147–151.

Wada, H., Tamaki, S., Tanigawa, M., Takagi, M., Mori, Y., Deguchi, A., Katayama, N., Yamamoto, T., Deguchi, K. and Shirakawa, S. (1991b) Plasma level of IL-1β in disseminated intravascular coagulation. Thromb. Haemost., 65: 364–368.

Wada, H., Tanigawa, M., Wakita, Y., Nakase, T., Minamikawa, K., Kaneko, T., Ohiwa, M., Kageyama, S., Kobayashi, T., Noguchi, T., Deguchi, K. and Shirakawa, S. (1993) Increased plasma level of interleukin-6 in disseminated intravascular coagulation. Blood Coagul. Fibrinol., 4: 583–590.

Wakefield, E.G. and Hall, W.W. (1927) Heat injuries: a preparatory study for experimental heatstroke. JAMA, 89: 92–95.

Walters, T.J., Ryan, K.L. and Mason, P.A. (2001) Regional distribution of Hsp70 in the CNS of young and old food-restricted rats following hyperthermia. Brain Res. Bull., 55: 367–374.

Walters, T.J., Ryan, K.L., Tehrany, M.R., Jones, M.B., Paulus, L.A. and Mason, P.A. (1998) HSP70 expression in the CNS in response to exercise and heat stress in rats. J. Appl. Physiol., 84: 1269–1277.

Wang, Q., Fang, C.H. and Hasselgren, P.-O. (2001a) Intestinal permeability is reduced and IL-10 levels are increased in septic IL-6 knockout mice. Am. J. Physiol., 281: R1013–R1023.

Wang, Q. and Hasselgren, P.-O. (2002b) Heat shock response reduces intestinal permeability in septic mice: potential role of interleukin-10. Am. J. Physiol., 282: R669–R676.

Wang, X., Zou, Y., Wang, Y., Li, C. and Chang, Z. (2001b) Differential regulation of interleukin-12 and interleukin-10 by heat shock response in murine peritoneal macrophages. Biochem. Biophys. Res. Commun., 287: 1041–1044.

Watanabe, N., Tsuji, N., Akiyama, S., Sasaki, H., Okamoto, T., Kobayashi, D., Sato, T., Hagino, T., Yamauchi, N. and Niitsu, Y. (1998) Endogenous tumour necrosis factor regulates heat-inducible heat shock protein 72 synthesis. Int. J. Hyperthermia, 14: 309–317.

Watanabe, N., Tsuji, N., Akiyama, S., Sasaki, H., Okamoto, T., Kobayashi, D., Sato, T., Hagino, T., Yamauchi, N., Niitsu, Y., Nakai, A. and Nagata, K. (1997) Induction of heat shock protein 72 synthesis by endogenous tumor necrosis factor via enhancement of the heat shock element-binding activity of heat shock factor 1. Eur. J. Immunol., 27: 2830–2834.

Wexler, R.K. (2002) Evaluation and treatment of heat-related illness. Am. Fam. Physician, 65: 2307–2314.

Wieczorek, M., Swiergiel, A.H., Pournajafi-Nazarloo, H. and Dunn, A.J. (2005) Physiological and behavioral responses to interleukin-1beta and LPS in vagotomized mice. Physiol. Behav., 85: 500–511.

Wijsman, J.A. and Shivers, R.R. (1993) Heat stress affects blood-brain barrier permeability to horseradish peroxidase in mice. Acta Neuropathol., 86: 49–54.

Wilkinson, D.A., Burholt, D.R. and Shrivastava, P.N. (1988) Hypothermia following whole-body heating of mice: effect of heating time and temperature. Int. J. Hyperthermia, 4: 171–182.

Williams, R.S., Thomas, J.A., Fina, M., German, Z. and Benjamin, I.J. (1993) Human heat shock protein 70 (hsp70) protects murine cells from injury during metabolic stress. J. Clin. Invest., 92: 503–508.

Wright, G.L. (1976) Critical thermal maximum in mice. J. Appl. Physiol., 40: 683–687.

Wright, G., Knecht, E. and Wasserman, D. (1977) Colonic heating pattern and the variation of thermal resistance among rats. J. Appl. Physiol., 43: 59–64.

Wu, C. (1995) Heat shock transcription factors: structure and regulation. Annu. Rev. Cell Dev. Biol., 11: 441–469.

Xiao, X., Zuo, X., Davis, A.A., McMillan, D.R., Curry, B.B., Richardson, J.A. and Benjamin, I.J. (1999) HSF1 is required for extra-embryonic development, postnatal growth and protection during inflammatory responses in mice. EMBO J., 18: 5943–5952.

Xie, Y., Chen, C., Stevenson, M.A., Auron, P.E. and Calderwood, S.K. (2002) Heat shock factor 1 represses transcription of the IL-1β gene through physical interaction with the nuclear factor of interleukin 6. J. Biol. Chem., 277: 11802–11810.

Xu, Q. (2002) Role of heat shock proteins in atherosclerosis. Arterioscler. Thromb. Vasc. Biol., 22: 1547–1559.

Yamamoto, K. and Loskutoff, D.J. (1996) Fibrin deposition in tissues from endotoxin-treated mice correlates with decreases in the expression of urokinase-type but not tissue-type plasminogen activator. J. Clin. Invest., 97: 2440–2451.

Yamamoto, K., Shimokawa, T., Kojima, T., Loskutoff, D.J. and Saito, H. (1999) Regulation of murine protein C gene expression in vivo: effects of tumor necrosis factor-alpha, interleukin-1, and transforming growth factor-beta. Thromb. Haemost., 82: 1297–1301.

Yang, Y.-L., Lu, K.-T., Tsay, H.-J., Lin, C.-H. and Lin, M.-T. (1998) Heat shock protein expression protects against death following exposure to heatstroke in rats. Neurosci. Lett., 252: 9–12.

CHAPTER 25

Oxidative stress and ischemic injuries in heat stroke

Chen-Kuei Chang[1], Ching-Ping Chang[2,3,*], Shyun-Yeu Liu[4] and Mao-Tsun Lin[2,*]

[1]Department of Surgery, Mackay Memorial Hospital, Taipei, Taiwan; Graduate Institute of Injury Prevention and Control, Taipei Medical University and Municipal Wan-Fan Hospital, Taipei, Taiwan
[2]Department of Medical Research, Chi-Mei Medical Center, Tainan, Taiwan
[3]Department of Biotechnology, Southern Taiwan University of Technology, Tainan, Taiwan
[4]Department of Oral and Maxillofacial Surgery, National Defense Medical Center and Taipei Medical University, Taipei, Taiwan

Abstract: When rats were exposed to high environmental temperature (e.g., 42 or 43°C), hyperthermia, hypotension, and cerebral ischemia and damage occurred during heat stroke were associated with increased production of free radicals (specifically hydroxyl radicals and superoxide anions), higher lipid peroxidation, lower enzymatic antioxidant defenses, and higher enzymatic pro-oxidants in the brain of heat stroke-affected rats. Pretreatment with conventional hydroxyl radical scavengers (e.g., mannitol or α-tocopherol) prevented increased production of hydroxyl radicals, increased levels of lipid peroxidation, and ischemic neuronal damage in different brain structures attenuated with heat stroke and increased subsequent survival time. Heat shock preconditioning (a mild sublethal heat exposure for 15 min) or regular, daily exercise for at least 3 weeks, in addition to inducing overproduction of heat shock protein 72 in multiple organs including brain, significantly attenuated the heat stroke-induced hyperthermia, hypotension, cerebral ischemia and damage, and overproduction of hydroxyl radicals and lipid peroxidation. The precise function of heat shock protein 72 are unknown, but there is considerable evidence that these proteins are essential for survival at both normal and elevated temperatures. They also play a critical role in the development of thermotolerance and protection from oxidative damage associated with cerebral ischemia and energy depletion during heat stroke. In addition, Shengmai San or magnolol (Chinese herbal medicines) or hypervolemic hemodilution (produced by intravenous infusion of 10% human albumin) is effective for prevention and repair of ischemic and oxidative damage in the brain during heat stroke. Thus, it appears that heat shock protein 72 preconditioning induced by prior heat shock or regular exercise training, as well as pretreatment with Shengmai San or magnolol is able to prevent the oxidative damage during heat stroke. On the other hand, hypervolemic hemodilution, Shengmai San, or magnolol is able to treat the oxidative damage after heat stroke onset.

Keywords: heat stroke; cerebral ischemia; neuronal injury; oxidative stress; lipid peroxidation; free radical scavengers

Oxidative stress in rats with heat stroke-induced cerebral ischemia

We have performed several studies in a rat heat stroke model, showing that heat stroke is associated with cerebral ischemia, higher levels of

*Corresponding authors. Tel.: +886-6-2812811 ext. 52657; Fax: +886-6-2517850; E-mail: 891201@mail.chimei.org.tw

interleukin-1β, and increased release of dopamine and glutamate in the brain (Kao et al., 1994; Lin et al., 1995b, c; Lin, 1997, 1999). These factors are known to increase free radical production (Ikeda and Long, 1990; Halliwell et al., 1991). It is not unexpected that an excessive accumulation of cytotoxic free radicals in the brain and oxidative stress can occur during heat stroke. To test the hypothesis, experiments were performed to detect *in vivo* hydroxyl radical production, extent of lipid peroxidation, total superoxide dismutase (SOD), and catalase activity in the brain and rate of superoxide generation of submitochondrial particles in the brain of rats suffering from heat stroke compared with normothermic controls.

Adult Sprague-Dawley rats were obtained from the Animal Resource Center of the National Science Council of the Republic of China (Taipei, Taiwan). All protocols were approved by the Animal Ethics Committee of the National Yang-Ming University School of Medicine. Adequate anesthesia was maintained to abolish the corneal reflex and pain reflexes induced by tail pinching throughout all experiments (approximately 8 h) after a single intraperitoneal dose of urethane (1.4 g/kg body weight) during thermal experiments. At the end of the experiments, control rats and any rats that had survived heat stroke were killed with an overdose of urethane.

The right femoral artery and vein of rats were cannulated with polyethylene tubing (PE 50), under urethane anesthesia, for blood pressure monitoring and drug administration. The animals were positioned in a stereotoxic apparatus (Kopf model 1406; Grass Instrument Co.) to insert probes for measurement of cerebral blood flow (CBF) and concentrations of 2,3-dihydroxybenzoic acid (2,3-DHBA; Sigma Chemical Co.) or 2,5-DHBA (Sigma). Physiological monitoring included colonic temperature (T_{co}), MAP, CBF, and 2,3-DHBA, and 2,5-DHBA concentration values in the striatum. Colon temperature was monitored continuously by a thermocouple.

Animals were randomly assigned to one of the following two groups. One group of rats was exposed to an ambient temperature (T_a) of 42°C (with relative humidity of 60% in a temperature-controlled chamber). The time when both MAP and local CBF in the striatum began to decrease from their peak levels was taken as the onset of heat stroke, as shown in Fig. 1. The second group was exposed to T_a of 24°C for at least 90 min to reach thermal equilibrium and was used as control. Their T_{co} was maintained at approximately 36.5°C with the use of an electric thermal mat before the start of the heat stress experiments.

A 24-gauge stainless steel needle probe (diameter, 0.58 mm; length, 40 mm) was inserted into the right striatum with the atlas and coordinates of Paxinos and Watson (1982). At the end of the experiments, the brain was removed, fixed in 10% neutral buffered formalin, and embedded in paraffin blocks. Serial (10 μm) sections through the frontal cortex, striatum, hippocampus, or hypothalamus were stained with hematoxylin and eosin for microscopic evaluation. The extent of cerebral neuronal damage was scored on a scale of 0–3 modified from the grading system of Pulsinelli et al. (1982).

A microdialysis probe (CMA20) with a 4-mm-long dialysis membrane was vertically implanted into the right striatum with the atlas and coordinates of Paxinos and Watson (1982). The concentrations of hydroxyl radicals were measured by a modified procedure based on the hydroxylation of sodium salicylate by hydroxyl radicals, leading to the production of 2,3-DHBA and 2,5-DHBA (Lancelot et al., 1995; Obata and Yamanaka, 1995; Obata, 1997). These two compounds in dialysates were then measured by high-performance liquid chromatography with electrochemical detection.

At 20 min after the onset of heat stroke, eight rats were decapitated. The eight normothermic rats were used as a control group. The brains were removed and dissected into several parts, including the frontal cortex, striatum, hippocampus, and hypothalamus. A fluorescence assay procedure was used to measure lipid peroxidation in these areas (Kikugawa et al., 1989; Mohanakumar et al., 1994). The lipid peroxidation was determined by measuring the levels of malondialdehyde (MDA) and its dihydropyridine polymers at 356 nm excitation and 426 nm emission.

The rate of superoxide ($O_2^{\cdot-}$) generation by submitochondrial particles was measured according to

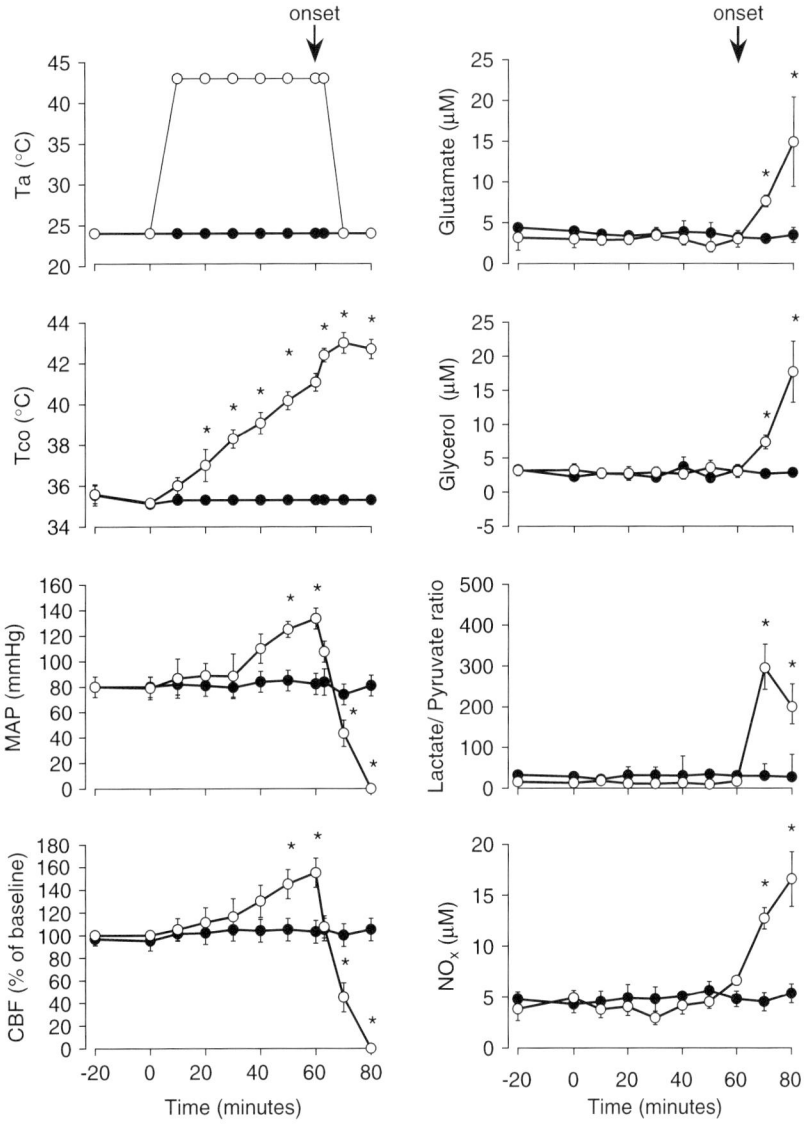

Fig. 1. Time-course changes in ambient temperature (T_a), colonic temperatures (T_{co}), mean arterial pressure (MAP), cerebral blood flow (CBF), or striatal levels of glutamate, glycerol, lactate/pyruvate ratio evaluated in normothermic control rats (○) or in rats subjected to heat stress (43°C T_a) (●). The arrow indicates time of heat stroke onset in heat stroke group. Values are means ± SEM; $n = 8$ animals per group. *$P < 0.05$ versus normothermic group.

Boveris (1984). Tissues from the brain were homogenized in 10 volumes (wt./vol.) of 0.1% Triton X. The homogenates were centrifuged at 3000 g for 5 min, and the supernatant was centrifuged at 50,000 g for 30 min. The resulting supernatant was used for measurement of the activity of SOD and catalase. The SOD activity was measured according to Misra and Fridovich (1977). Catalase activity was measured by the method of Luck (1965). Protein concentration in the aforementioned experiments was determined by the method of Lowry et al. (1951).

It was found that the values of MAP and CBF after heat stroke onset were all significantly lower than those in normothermic controls (or animals kept at room temperature) (Yang and Lin, 2002). However, the values of T_{co} and DHBA levels in the striatum, and neuronal damage score were greater. The extent of lipid peroxidation and the rate of $O_2^{\cdot -}$ generation in the brain were all greater in rats after heat stroke onset. In contrast, the values of total SOD in the brain and the catalase activity in the brain were lower (Yang and Lin, 2002).

The present results are consistent with many studies showing that reactive oxygen species (ROS) are involved in brain injuries in many neurological disorders such as cerebral ischemia and reperfusion, hypoxia, inflammation, and fever (Suzuki et al., 1985; Cao et al., 1988; Riedel and Maulik, 1999; Babior, 2000; Chan, 2001). During heat stroke, many endogenous antioxidative defenses are likely to be perturbed as a result of overproduction of oxygen radicals by cytosolic pro-oxidant enzymes and mitochondria, inactivation of detoxification systems, consumption of antioxidants, and failure to adequately replenish antioxidants in heat-injured (because of hyperthermia) and ischemic (because of ensuing hypotension) brain tissue. These ROS have been shown to be directly involved in oxidative damage with macromolecules in ischemic tissue, which leads to cell death (Chan, 2001). An increased level of dopamine, serotonin, glutamate, cytokines (Lin, 1999), or inducible nitric oxide synthase (iNOS)-dependent nitric oxide (NO) (Chang et al., 2004b) was associated with cerebral ischemia during rat heat stroke. All these transmitters may cause the increased release of ROS and result in ischemic neuronal damage in the brain (Lancelot et al., 1995; Fabian et al., 2000).

Protective effects of α-tocopherol, mannitol, and magnolol on oxidative damage and cerebral ischemia in heat stroke

As described in the aforementioned section, cerebral ischemia and damage are associated with an increased production of free radicals (specifically, hydroxyl radicals and superoxide), increased lipid peroxidation, and decreased enzymatic antioxidant defenses in the brain of heat stroke-affected rats. This raises the possibility that pretreatment with free radical scavengers will protect against heat stroke-induced cerebral ischemia and damage.

To deal with this question, the aim of the present study was to assess the effect of three free radical scavengers, namely α-tocopherol (Yamamoto et al., 1983; Suzuki et al., 1985), mannitol (Suzuki et al., 1985; Schwarz et al., 1998), and magnolol (Lo et al., 1994; Lin et al., 1995a) on heat stroke-induced cerebral ischemia and oxidative damage.

Adult Sprague-Dawley rats were obtained from the Animal Resource Center of the National Science Council of the Republic of China (Taipei, Taiwan). Adequate anesthesia was maintained to abolish the corneal reflex and pain reflexes induced by tail pinch throughout the course of all experiments (approximately 8 h) after a single intraperitoneal dose of urethane (1.4 g/kg body weight). The right femoral artery and vein of rats were cannulated with polyethylene tubing (PE 50) for the monitoring of blood pressure and drug administration, respectively. Animals were positioned in a stereotoxic apparatus to insert probes for the measurement of CBF and concentrations of 2,3-DHBA or 2,5-DHBA in the striatum. Physiological monitoring included T_{co}, MAP, and CBF.

Animals were assigned randomly to one of the following groups. One group of rats treated with normal saline (NS) (1 mL/kg body weight, intravenously, i.v.) 30 min before the start of heat stress was exposed to 42°C T_a (with 60% relative humidity in a temperature-controlled chamber). The second group of rats was treated with an i.v. dose of 10% mannitol (1 mL/kg body weight) 30 min before the start of heat stress. The third group of rats was treated with an i.p. dose of corn oil (1 mL/kg body weight) 30 min before the start of heat stress. The fourth group of rats was treated with an i.p. dose (1 mL/kg) of α-tocopherol (20 mg/mL per kg body weight) 30 min before the start of heat stress. The fifth group of rats was treated with an i.v. dose of either 20 or 40 mg/kg magnolol "0" min.

For measurement of extracellular ischemia and damage markers, as well as hydroxyl radials, in the striatum, the microdialysis probe was implanted stereotaxically into the striatum according to the atlas and coordinates of Paxinos and Watson (1982). The dialysates were sampled in microvials and were collected every 10 min in a CMA140 (Carnegie Medicine, Stockholm, Sweden) fraction collector. Aliquots of dialysates (5 μL) were injected onto a CMA600 Microdialysis analyzer for measurement of lactate, glycerol, pyruvate, and glutamate. All reagents required for analysis were obtained from CMA Microdialysis (Carnegie Medicine). The concentrations of hydroxyl radicals were measured by a modified procedure based on the hydroxylation of sodium salicylate by hydroxyl radicals, leading to production of 2,3-DHBA and 2,5-DHBA (Lancelot et al., 1995; Obata and Yamanaka, 1995; Obata, 1997).

Brains were removed and dissected into several parts, including the cortex, striatum, hippocampus, and hypothalamus. A fluorescence assay procedure was used to measure lipid peroxidation in these areas (Kikugawa et al., 1989). Lipid peroxidation was determined by measuring the levels of MDA and its dihydropyridine polymers at 356 nm excitation and 426 nm emission. The rate of superoxide ($O_2^{\cdot-}$) generation by submitochondrial particles was measured according to the method of Boveris (1984).

For measurement of intracranial pressure (ICP), a probe was inserted into the brain in a stereotaxic apparatus (Kopf model 1406; Grass Instrument, Quincy, MA, USA). The ICP was monitored with a Statham P23AC transducer (Gould Instruments, Cleveland, OH, USA) via a 20 gauge stainless-steel needle probe (diameter, 0.90 mm; length, 38 mm), which was introduced into the right lateral cerebral ventricle according to the stereotaxic coordinates of Paxinos and Watson (1982).

Pretreatment with α-tocopherol (20 mg/kg, i.v.) or mannitol (10%, i.v.) 30 min before the onset of heat exposure significantly attenuated heat stroke-induced arterial hypotension, cerebral ischemia and neuronal damage, the increased free radical formation and lipid peroxidation in the brain, and the increased plasma levels of cytokines (Niu et al., 2003). Pretreatment with α-tocopherol or mannitol resulted in a prolongation of survival time in heat stroke. In addition, as shown in both Figs. 1 and 2, after the onset of heat stroke, arterial hypotension (decreased MAP), cerebral ischemia (decreased CBF), intracranial hypertension (increased ICP), cerebral hypoperfusion (decreased cerebral perfusion pressure (CPP) = MAP–ICP), and increased levels of cellular ischemia markers (e.g., glutamate, lactate/pyruvate ratio, and NOx) and damage marker (e.g., glycerol) in striatum, and deep body temperature occurred. It was found that magnolol (20 or 40 mg/kg, i.v.) significantly attenuated the heat stroke-induced hyperthermia, arterial hypotension, intracranial hypertension, cerebral ischemia (Fig. 2) and neuronal damage, and increased free radical formation and lipid peroxidation in the brain (Chang et al., 2003). The increased levels of extracellular concentrations of ischemic markers in the striatum that occurred during heat stroke were also significantly attenuated by pretreatment with magnolol (Chang et al., 2003).

At T_a higher than those of the body, heat gain exceeds heat losses and lead to hyperthermia, cerebral vascular congestion, and edema during heat stress (Shih et al., 1984). Cerebral vascular congestion and edema result in intracranial hypertension. In contrast, after the onset of heat stroke, MAP is decreased. Both intracranial hypertension and arterial hypotension eventually lead to a reduction in CPP, as shown in the present study. A reduction of CPP to below the autoregulatory levels causes cerebral ischemia, which leads to neuronal damage in the brain. The accumulation of acidic metabolites and hypoxia resulting from cerebral ischemia initiates neuronal depolarization, which leads to enhanced release of glutamate (Yang et al., 1998a, b), dopamine (Lin et al., 1995c), serotonin (Chiu et al., 1995; Kao and Lin, 1996), and cytokines (Chiu et al., 1995; Lin et al., 1995b; Liu et al., 2000). The present results demonstrate that free radical scavengers such as magnolol, α-tocopherol, or mannitol may protect against heat stroke by maintaining appropriate levels of MAP and CPP.

In the present study, the heat stroke-induced increases in circulating cytokines (including interleukin-1β, interleukin-6, and tumor necrosis factor-alpha, TNF-α), arterial hypotension, intracranial hypertension, cerebral ischemia and damage were associated with increased hydroxyl radicals and lipid peroxidation in rat brain. Pretreatment with

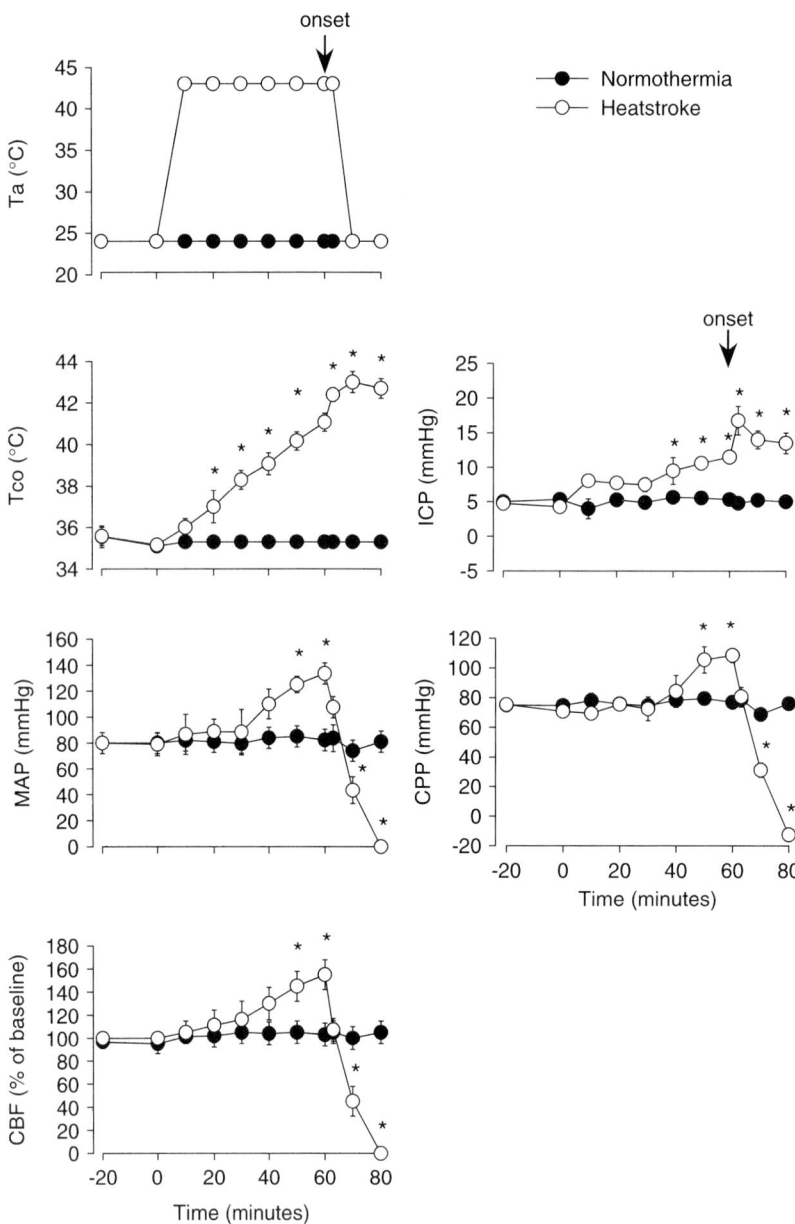

Fig. 2. Time-course change in ambient temperature (T_a), colonic temperature (T_{co}), mean arterial pressure (MAP), cerebral blood flow (CBF), intracranial pressure (ICP), and cerebral perfusion pressure (CPP) evaluated in normothermic control rats (●) or in rats subjected to heat stress (43°C T_a) (○). The arrow indicates time of heat stroke onset in heat stroke group. Values are means ± SEM; $n = 8$ animals per group. *$P < 0.05$ versus normothermic group.

mannitol, α-tocopherol, or magnolol, in addition to attenuating the elevating levels of cytokines in the serum, decreased hydroxyl radical formation, lipid peroxidation and ischemic injury in brain, and improved survival in rats with heat stroke. This demonstrates that although no one pretreatment prevents the heat stroke syndrome entirely, all three drugs can attenuate it.

Protective effects of heat shock preconditioning on oxidative stress during heat stroke

In our previous studies (Yang et al., 1998; Yang and Lin, 1999), we showed that the brain preconditioned by thermal or chemical injury increased heat shock protein (HSP) 72, and that this induction was associated with an attenuation of cerebral ischemic damage and animal death during heat stroke. Although oxidative stress may be involved in heat stroke-induced cerebral ischemia and damage, it is unknown whether HSP72 preconditioning may limit the genesis of oxidative damage during heat stroke-induced cerebral ischemia. To deal with the matter, in the present study, we compared the temporal profiles of MAP, HR, and extracellular levels of ischemia and damage markers and hydroxyl radical production in the striatum of heat stroke rats with or without heat shock pretreatment. In addition, experiments were conducted to determine the effect of heat shock pretreatment on the MDA, reduced-form glutathione (GSH), oxidized-form GSH (GSSG), total glutathione (GSH + GSSG), GSSG/GSH ratio, glutathione peroxidase (GPx) activity, and glutathione reductase (GR) activity, in the striatum of heat stroke rats with or without heat shock pretreatment.

Experiments were performed in male adult Wistar rats (weighing 250–350 g), which were obtained from the Animal Resource Center, National Science Council (Taipei, Taiwan, ROC). Four groups of animals were designated for experiments: (a) normothermic control rats (N); (b) rats without heat shock subjected to heat stroke (HS); (C) rats, 16 h after heat shock, subjected to heat stroke (HSP-16 + HS); and (d) rats, 96 h after heat shock, subjected to heat stroke (HSP-96 h + HS). In separate groups of eight animals each, animals were subjected to (a) measurement of HSP72; (b) measurement of latency; (c) measurement of survival time; (d) measurement of T_{co}, MAP, HR, brain PO_2, brain temperature (BT), and CBF; (e) measurement of T_{co}, MAP, HR, extracellular dihydroxybenzoic acid (DHBA), glutamate, and lactate/pyruvate ratio; (f) measurement of MDA; and (g) measurement of total glutathione (GSH + GSSG), and GSSG.

For heat shock treatment, rats under general anesthesia (1 h after 40 mg/kg pentobarbital sodium i.p.) were heated gently with an electric pad. The T_{co} of the heated animals was kept at 42 ± 0.5°C for 15 min. They were returned to room temperature during the recovery period. These groups of animals were subjected to heat stroke experiments at 16 or 96 h after the start of heat shock treatment.

Heat stroke was induced by exposing these groups of animals to an T_a of 43°C with a relative humidity of 60% in a temperature-controlled chamber; the moment at which MAP began to decrease from its peak level was taken as the onset of heat stroke. Immediately after the onset of heat stroke in these groups of animals, the heat exposure was switched off, and the T_a was restored to 25°C.

The animals were killed by decapitation at the end of the experiment for detection of HSP72. The striatum was quickly dissected from the brain and placed into microcentrifuge tubes and then stored at −20°C. The samples (40 mg/lane) were incubated for 5 min at 95°C in Laemmli buffer and then separated on 10% SDS-polyacrylamide discontinuous gel. After electrophoresis, the gels were processed for an immunoblotting study that was performed as described previously (Yang and Lin, 1999).

For measurements of brain PO_2, temperature, and CBF, a 100-mm-diameter thermocouple and two 230-mm fibers are attached to the oxygen probe. This combined probe measures oxygen, temperature, and microvascular blood flow. The measurement requires OxyLite and OxyFlo instruments. OxyLite 2000 (Oxford Potronix, Oxford, UK) is a two-channel device (measuring PO_2 and temperature at two sites simultaneously), and OxyFlo 2000 is a two-channel laser Doppler perfusion-monitoring instrument. The combination of these two instruments provides simultaneous tissue blood flow, oxygenation, and temperature data. After each animal, under urethane anesthesia, was placed in a stereotaxic apparatus, a combined probe was implanted into the striatum of rat brain.

For measurement of extracellular ischemia and damage markers, a microdialysis probe (4 mm in length; CMA/12, Carnegie Medicine, Stockholm, Sweden) was stereotaxically implanted into the right striatum according to the atlas and coordinates of Paxinos and Watson (1982). According to the methods described by Berger et al. (2002), an equilibrium period of 60 min without sampling was allowed after probe implantation. Aliquots of dialysates (8 μL) were injected onto a CMA600 microdialysis analyzer for measurement of glutamate, lactate, pyruvate, and glycerol.

For measurements of hydroxyl radicals, the concentrations of hydroxyl radicals were measured by a modified procedure based on the hydroxylation of sodium salicylate by hydroxyl radicals, leading to the production of 2,3-DHBA and 2,5-DHBA (Yang and Lin, 2002).

Lipid peroxidation was assessed by measuring the concentration of MDA, the end product of lipid peroxidation, with 2-thiobarlbituric acid (TBA) to form a chromophore absorbing, at 532 nm (Dahle et al., 1962). The values were expressed as nanometers of TBA-reactive substances (MDA equivalent) per milligram of protein (Ohkawa et al., 1979).

Tissues were homogenized in 5% 5-sulfosalicylic acid (1:10 W/V) at 4°C, and the supernatant were used to analysis of total and oxidized glutathione. Total GSH (GSH + GSSG) was analyzed according to the method of Tietze (1969) and oxidized GSH was determined as described by Griffith (1980).

To measure cytosolic GPx and GR activities, tissues were homogenized in 0.05 M phosphate buffer, pH 7.0. The supernatants were used for GPx and GR activity assay. The GPx and GR activities were performed with a commercial glutathione peroxidase cellular activity assay kit (Sigma, USA) and a glutathione reductase assay kit (Sigma, USA), respectively. One unit of GPx and GP activity was defined as the amount of sample required to oxidized 1 mmol of NADPH per minute based on the molecular absorbance of 6.22×10^6 NADPH.

It was found that striatal HSP72 expression was significantly increased in rats with heat stress (43°C T_a for 59 min) 16 h after heat shock preconditioning (42°C T_{co} for 15 min) (HSP-16 h + HS), but not in rats with heat stress 96 h after heat shock preconditioning (HSP-96 h + HS), in rats with heat stress but without heat shock preconditioning (HS), or in rats without heat stress and without heat shock preconditioning (N) (Wang et al., 2005a). As compared with those of the HS group, both the latency and survival time values were significantly increased in HSP-16 + HS rats but not in HSP-96 h + HS rats. Figure 3 shows the effects of heat exposure (43°C T_a for 59 min) on T_{co}, MAP, HR, and striatal levels of DHBAs, glutamate, glycerol, and lactate/pyruvate ratio in different groups of rats. As shown in this figure, 15 min after the onset of heat stroke in the HS group, all the MAP and HR values were significantly lower than those of N group ($P < 0.05$). On the other hand, the values of T_{co} and striatal levels of DHBAs, glutamate, glycerol, and lactate/pyruvate ratio were significantly higher in HS rats 15 min after the onset of heat stroke than in those of the N group ($P < 0.05$). Heat stroke-induced hyperthermia, arterial hypotension, bradycardia, and increased levels of striatal DHBAs, glycerol, glutamate, and lactate/pyruvate ratio were significantly attenuated by heat shock preconditioning 16 h before the initiation of heat exposure in the HSP-16 h + HS group. As shown in Figs. 4 and 5, 15 min after the onset of heat stroke, in the HS group, all the MDA, GSSG, and GSSG/GSH ratio values were significantly greater than those of N group ($P < 0.05$). GSG was significantly lower compared with control. Likewise, glutathione peroxidase activity and glutathione reductase activity were significantly lower in HS group 15 min after the onset of heat stroke than in those of the N group ($P < 0.05$) (Fig. 6). Heat stroke-induced increased values of MDA, GSSG, and GSSG/GSH ratio as well as the decreased values of GSH, glutathione peroxidase activity, and glutathione reductase activity were significantly attenuated by heat shock preconditioning 16 h before the initiation of heat exposure in the HSP-16 h + HS group, but not in the HSP-96 h + HS group.

Expression of HSP72 increase tolerance of cerebral neurons to ischemic damage (Kirino et al., 1991) and protects neurons against subsequent severe heat shock (47°C) (Fink et al., 1997). In our previous (Yang and Lin, 1999) and present study (Wang et al., 2005a), we induce heat shock preconditioning and showed overproduction of HSP72 in the striatum at 16 h that returned to baseline at 96 h. Prior heat shock conferred significant protection against heat stroke-induced arterial hypotension and striatal ischemia and damage at 16 h. However, 96 h after heat shock, when HSP72

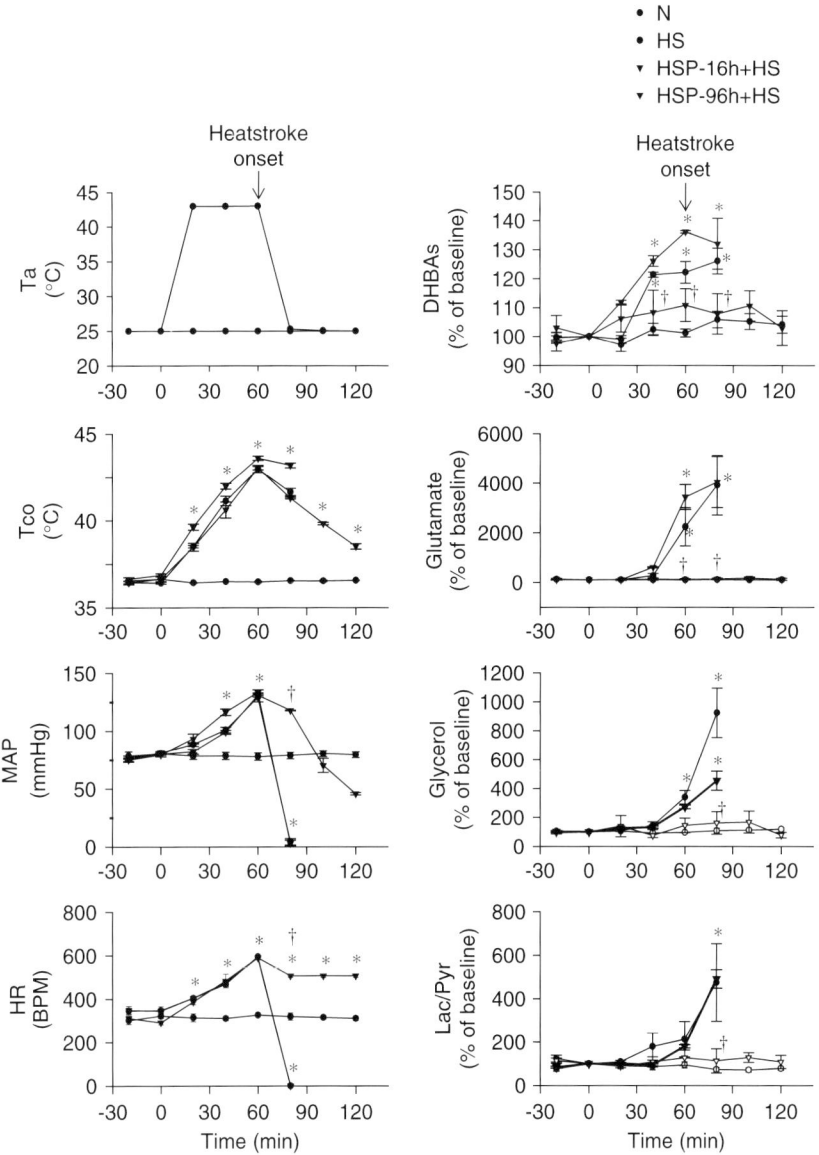

Fig. 3. Time-course changes in ambient temperature (T_a), colonic temperature (T_{co}), mean arterial pressure (MAP), HR, or striatal levels of DHBAs, glutamate, glycerol, and lactate/pyruvate, evaluated in normothermic control rats (○), in rats subjected to heat stress (43°C T_a) without heat shock preconditioning (●), in rats subjected to heat stress (43°C T_a) 16 h after heat shock preconditioning (42°C T_{co} for 15 min) (▽), or in rats subjected to heat stress (43°C T_a) 96 h after heat shock preconditioning (42°C T_{co} for 15 min) (▼). The arrow indicates time of heat stroke onset in HS group. Values are means ± SEM; $n = 8$ animals per group. *$P<0.05$ versus N group; $P<0.05$ versus HS group in Duncan multiple range test. Adapted from Wang et al. (2005a).

expression returns to basal levels, no protection is evident when compared with animal snot preconditioned with heat shock. These findings suggest that heat shock pretreatment (42°C T_{co} for 15 min) induced HSP72 overproduction in the striatum and conferred protection against circulatory shock and striatal ischemic injury during heat stroke.

The present results further demonstrated that the decreased levels of glutathione and glutathione peroxidase activity in striatum exhibited during

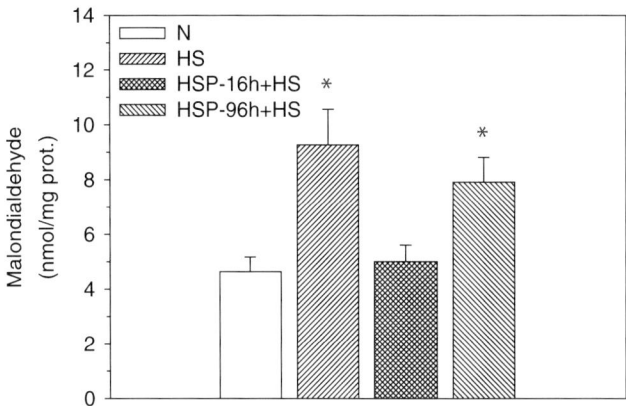

Fig. 4. The concentrations of malondialdehyde in striatal tissue homogenates in rats without heat shock pretreatment and heat stress (N), in rats subjected to heat stress (43°C T_a) without heat shock pretreatment (42°C T_{co} for 15 min) (HS), in rats subjected to heat stress (43°C T_a for 59 min) 16 h after heat shock pretreatment (42°C T_{co} for 15 min) (HSP-16 h + HS), and in rats subjected to heat stress (43°C T_a for 59 min) 96 h after heat shock pretreatment (42°C T_{co} for 15 min) (HSP-96 h + HS). Values are means ± SEM; $n = 8$ animals per group. *$P < 0.05$ versus N group in Duncan multiple-range test. Adapted from Wang et al. (2005a).

heat stroke were also ameliorated in rats 16 h but not 96 h after heat shock pretreatment. After the onset of heat stroke, ischemic injury was noted to occur in different brain structures including striatum, hypothalamus, cortex, and thalamus (Kao et al., 1994; Lin et al., 1995b; Kao and Lin, 1996). Prior heat shock induced HSP72 in different brain structures (including striatum, hypothalamus, and cortex) (Yang and Lin, 1999). In the present study (Wang et al., 2005b), the striatum was chosen as an representative region for measurement of HSP72 induction and neuronal damage during heat stroke. The precise functions of HSP72 are unknown, but there is considerable evidence that these proteins are essential for survival at both normal and elevated temperatures. They also play a critical role in the development of thermotolerance and protection from cellular damage associated with stresses such as ischemia, cytokines, and energy depletion.

Progressive exercise preconditioning protects against oxidative damage and cerebral ischemia during heat stroke

During heat stroke, rodents display hyperthermia, arterial hypotension, intracranial hypertension, and cerebral ischemia neuronal damage (Lin and Chang, 2004). However, the above-mentioned heat stroke syndromes are attenuated by induction of HSP72 in rat brain (Yang and Lin, 1999; Li et al., 2001). Other evidences have also demonstrated that HSP72 can be detected in skeletal muscle, heart, liver, and adrenal gland from rats with endurance exercise training for 7–12 weeks (wk) (Ecochard et al., 2000; Gonzalez et al., 2000; Smolka et al., 2000). This raises the possibility that pretreatment of rats with progressive exercise induces HSP72 and improves oxidative damage and cerebral ischemia and cerebral ischemia and damage during heat stroke.

To deal with the matter, we investigated the effects of heat stress on survival time, T_{co}, MAP, stroke volume (SV), total peripheral resistance (TPR), cardiac output (CO), heart rate (HR), blood gases and extent of tissue or serum TNF-α in rats with or without prior chronic exercise training. In addition, the levels of the HSP72 expression in multiple organs including hearts were determined in rats with or without prior exercise training.

Animals were randomly assigned to one of the following five groups (1) sedentary control group (SED); (2) rats 1 day after exercise training for 1 wk (1wk1D); (3) rats 1 day after exercise training for 2 wk (2wk1D); (4) rats 1 day after exercise training for 3 wk (3wk1D); and (5) rats 10 days after exercise training for 3 wk (3wk10D). Each

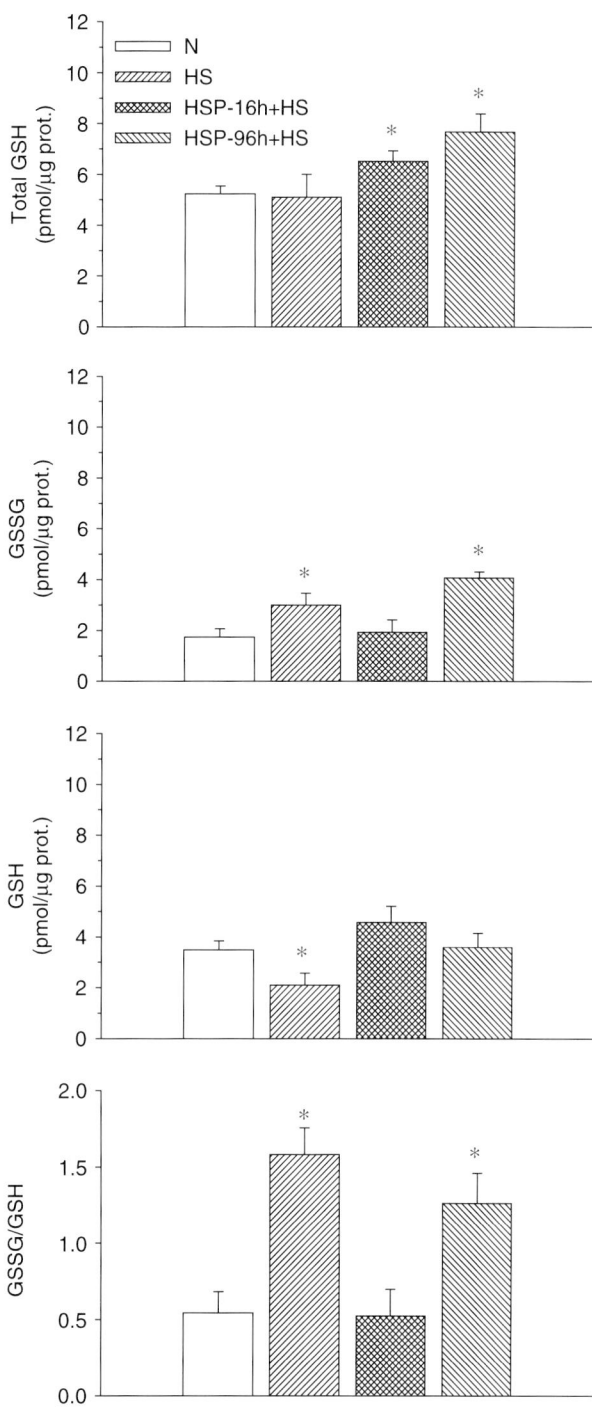

Fig. 5. The concentrations of total glutathione (GSH + GSSG), oxidation-form glutathione (GSSG), reduced-form glutathione (total GSH-GSSG), and GSSG/GSH ratio of striatal tissue homogenates in rats without heat shock pretreatment and heat stress (N), in rats subjected to heat stress (43°C T_a) without heat shock pretreatment (42°C T_{co} for 15 min) (HS), in rats subjected to heat stress (43°C T_a for 59 min) 16 h after heat shock pretreatment (42°C T_{co} for 15 min) (HSP-16 h + HS), and in rats subjected to heat stress (43°C T_a for 59 min) 96 h after heat shock pretreatment (42°C T_{co} for 15 min) (HSP-96 h + HS). Values are means ± SEM; $n = 8$ animals per group. *$P < 0.05$ versus N group in Duncan multiple-range test. Adapted from Wang et al. (2005a).

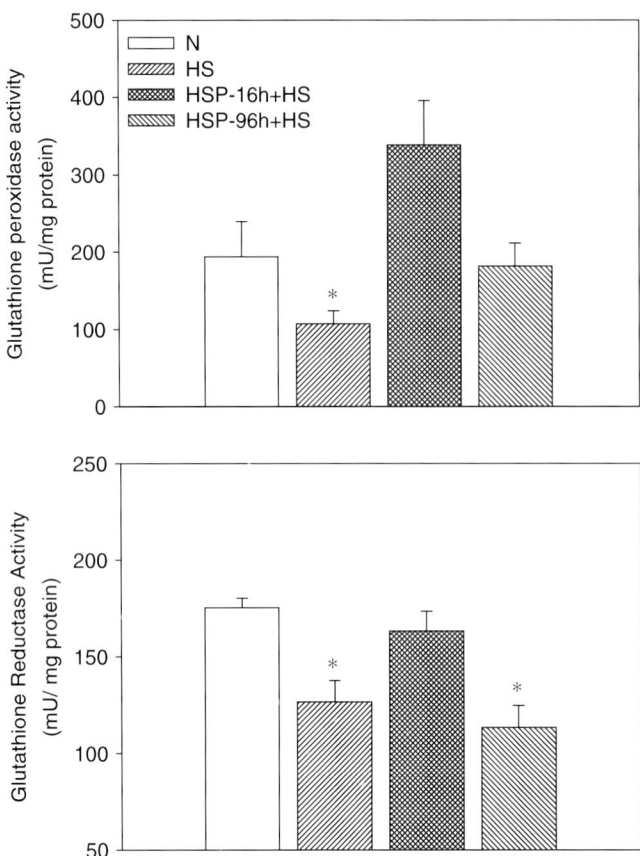

Fig. 6. The activities of glutathione peroxidase and glutathione reductase of striatal tissue homogenates in rats without heat shock pretreatment and heat stress (N), in rats subjected to heat stress (43°C T_a) without heat shock pretreatment (42°C T_{co} for 15 min) (HS), in rats subjected to heat stress (43°C T_a for 59 min) 16 h after heat shock pretreatment (42°C T_{co} for 15 min) (HSP-16 h + HS), and in rats subjected to heat stress (43°C T_a for 59 min) 96 h after heat shock pretreatment (42°C T_{co} for 15 min) (HSP-96 h + HS). Values are means ± SEM; $n = 8$ animals per group. *$P < 0.05$ versus N group in Duncan multiple-range test. Adapted from Wang et al. (2005a).

one of the above-mentioned groups were divided into two subgroups. One subgroup of rats was exposed to a T_{pad} of 43°C to obtain the latency for the onset of heat stroke. In the SED group, the latency of the onset of heat stroke was found to be 55 ± 3 min. The other subgroup of rats were exposed to a T_{pad} of 43°C for only 55 min, and then kept at 24°C until death. The survival time (interval between the onset of heat stroke and animals death) was obtained for each group.

Different group of animals was used for the different sets of experiments: (i) determination of both latency and survival time for SED, 1wk1D, 2wk1D, 3wk1D, and 3wk10D rats; (ii) determination of T_{co}, MAP, CBF, and striatal levels of glutamate, glycerol, and lactate/pyruvate ratio in SED rats kept at 24°C, SED rats kept at 43°C, or 3wk1D rats kept at 43°C; (iii) determination of MAP, ICP, and CPP in SED rats kept at 24°C, SED rats kept at 43°C, and 3wk1D rats kept at 43°C; and (iv) determination of the extent of lipid peroxidation, rats of $O_2^{\cdot -}$ generation, and ATP in tissue homogenates of striatum and the extracellular concentrations of DHBA in the striatum of SDE or 3wk1D rats. Another group of SED rats kept at 24°C was used as normothermic controls.

For determination of rat TNF-α level, blood samples and organs were taken 15 min after the onset of heat stroke. The blood samples were allowed to clot for 2 h at room temperature and then were centrifuged (2000 g, 20 min, 4°C). The supernatants were harvested. The organ samples were prepared according to previous reports (Wolf et al., 2001; Zhou et al., 2003). The organs were disintegrated in five volumes of ice-cold Ripa buffer. The homogenates were incubated on ice for 30 min and then centrifuged (15,000 g, 30 min, 4°C) twice. The supernatants were stored at −70°C until measurement. The concentrations of TNF-α in serum and organ lysates were determined using double-antibody sandwich ELISA (R & D Systems, Minneapolis, MN, USA) according to the manufacturer's instructions.

Prior exercise training for 3 wk, but not 1 or 2 wk, conferred significant protection against heat stroke-induced hyperthermia, arterial hypotension, decreased CO, decreased SV, decreased peripheral vascular resistance, and increased levels of serum or tissue TNF-α and correlated with overexpression of HSP72 in multiple organs including heart, liver, and adrenal gland. However, 10 days after 3 wk progressive exercise training, when HSP72 expression in multiple organs returned to basal values, the beneficial effects exerted by 3 wk exercise training were no more observed. Conclusion: These results strongly suggest that HSP72 preconditioning with progressive exercise training may protect against heat stroke-induced circulatory shock by reducing TNF-α overproduction (Hung et al., 2005). Furthermore, we observed the values of MAP and CBF in heat stroke SED rats which received no chronic exercise pretreatment obtained at 70 min after initiation of heat exposure were significantly lower than those of the normothermic controls (Table 1). In contrast, the values of striatal levels of glutamate, glycerol, and lactate/pyruvate ratio, and T_{co} were significantly greater at 70 min after initiation of heat exposure in SED rats. As compared with those at the same time point of heat stroke SED rats the values of MAP and CBF in 3wk1D rats were significantly greater, whereas the values of T_{co} and striatal levels of glutamate, glycerol, and lactate/pyruvate ratio were significantly lower (Table 1).

As shown in Table 2, the values of both MAP and CPP in heat stroke SED rats obtained at 70 min after initiation of heat exposure were significantly lower than those of the normothermic controls. In contrast, the values of ICP were significantly greater at 70 min after initiation of heat exposure in SED rats. When compared with those at the same time point of heat stroke SED rats, the values of both MAP and CPP in 3wk1D rats were significantly greater, whereas the values of ICP were significantly lower.

Table 1. Effects of heat exposure (43°C for 60 min) on colonic temperature (T_{co}), mean arterial pressure (MAP), and local cerebral blood flow (CBF) and extracellular concentration of glutamate, glycerol, and lactate/pyruvate ratio in the striatum of SED or 3wk1D rats

Groups/time course	HSP72 expression	T_{co} (°C)	MAP (mmHg)	CBF (% baseline)	Glutamate (μM)	Glycerol (μM)	Lactate/pyruvate
SED rats kept at 24°C	Negative						
0 min		35.6±0.2	80±4	100±5	4.5±0.6	4±2	7±2
60 min		36.1±0.3	78±4	102±6	4.8±0.7	3±1	6±3
70 min		36.0±0.2	81±3	103±4	4.2±0.4	4±1	8±3
SED rats kept at 43°C	Negative						
0 min		35.9±0.2	79±3	100±4	4.6±0.5	3±1	7±3
60 min		42.6±0.3*	130±8*	156±5	5.0±0.6	4±1	8±2
70 min		43.2±0.2*	40±5*	42±8*	7.6±0.4*	8±1*	300±43*
3wk1D rats kept at 43°C	Positive						
0 min		36.0±0.3	81±3	99±3	4.4±0.6	3±1	6±2
60 min		40.3±0.3†	133±7	158±6	4.9±0.5	4±2	8±3
70 min		41.0±0.4†	66±5†	109±5†	4.5±0.4†	3±1†	9±4†

Notes: Another group of rats kept at 24°C were used as normothermic controls. Data are means ± SEM of eight rats per group.
*$P<0.05$ compared with SED rats kept at 24°C.
†$P<0.05$ compared with SED rats kept at 43°C (ANOVA followed by Duncan's test).

Table 2. Effects of heat exposure (43°C for 60 min) on MAP, CBF in striatum, intracranial pressure (ICP), and cerebral perfusion pressure (CPP) of SED or 3wk1D rats

Groups/time course	MAP (mmHg)	ICP (mmHg)	CPP (mmHg)
SED rats kept at 24°C			
0 min	80±3	5±1	75±3
60 min	79±4	6±2	73±2
70 min	78±3	7±2	71±2
SED rats kept at 43°C			
0 min	79±2	5±2	74±2
60 min	133±7*	15±3*	118±4*
70 min	41±4*	13±2*	28±3*
3wk1D rats kept at 43°C			
0 min	82±4	4±2	78±3
60 min	136±8	5±2†	131±4†
70 min	67±6†	6±1†	61±3†

Notes: Another group of rats kept at 24°C were used as normothermic controls. Data are means ± SEM of eight rats per group.
*$P > 0.05$ compared with SED rats kept at 24°C.
†$P < 0.05$ compared with SED rat kept at 43°C (ANOVA followed by Duncan's test).

Table 3. Effects of heat exposure (43°C for 60 min) on the extent of lipid peroxidation, rates of $O_2^{\cdot -}$ generation, and ATP in tissue homogenates and the extracellular concentrations of DHBA in the striatum of SED or 3wk1D rats

Groups/time course	Lipid peroxidation (RFU)	Rate of $O_2^{\cdot -}$ generation (nmol/min per mg protein)	DHBA (% baseline)
SED rats kept at 24°C			
0 min	1.55±0.10	6.88±0.59	100±3
60 min	1.53±0.08	6.76±0.54	98±3
70 min	1.57±0.09	6.91±0.01	101±4
SED rats kept at 43°C			
0 min	1.56±0.09	6.75±0.54	100±4
60 min	2.24±0.11*	9.23±0.83*	760±25*
70 min	2.41±0.12*	10.4±0.9*	751±34*
3wk1D rats kept at 43°C			
0 min	1.58±0.11	6.73±0.53	100±3
60 min	1.59±0.09†	6.95±0.74†	156±3†
70 min	1.53±0.08†	7.03±0.86†	153±4†

Notes: Another group of SED rats kept at 24°C were used as normothermic controls. Data are means ± SEM of eight rats per group. RFC, relative fluorescence units.
*$P < 0.05$ compared with SDE rats kept at 24°C.
†$P < 0.05$ compared with SED rats kept at 43°C (ANOVA followed by Duncan's test).

It was found that the values of striatal levels of lipid peroxidation, rate of superoxide generation, and DHBA in heat stroke SED rats obtained at 70 min after initiation of heat exposure were significantly higher than those of the normothermic controls (Table 3). In contrast, the values of ATP in the striatum were significantly lower at 70 min after initiation of heat exposure is SED rats. When compared with those at the same time point of heat stroke SED rats, the values of lipid peroxidation, rate of superoxide generation, and DHBA in the striatum in 3wk1D rats were significantly lower, whereas the values of ATP in the striatum were significantly greater.

In the present series of experiments, we have first demonstrated that progressive exercise preconditioning protects oxidative damage, TNF-α overproduction, and cerebral ischemia and damage during heat stroke. We realize a positive correlation between oxidative damage or cerebral ischemia during heat stroke and HSP72 expression in rats sacrificed without heat exposure to the acute experiment. The present results demonstrated that progressive exercise for 3 wk, in addition to inducing overproduction of HSP72 in several vital organs (Hung et al., 2005), significantly attenuated the hyperthermia, arterial hypotension (due to decreased SV and peripheral vascular resistance), intracranial hypertension, TNF-α overproduction, cerebral hypoperfusion, decreased striatal levels of cellular ischemia and damage markers, and increased levels of ATP, lipid peroxidation, rate of superoxide generation and DHBA in the striatum exhibited during heat stroke. However, 10 days after termination of 3-wk exercise training, when the HSP72 levels in different vital organs return to their basal levels, the above-mentioned protections against heat exposure were no more observed. These results indicate that HSP72 preconditioning with progressive exercise training may protect against circulatory shock and cerebral ischemia by reducing oxidative stress and TNF-α overproduction during heat stroke.

Hall et al. (2001) have proposed that hyperthermia during heat stroke stimulates xanthine oxidase production of ROS and limits heat tolerance by promoting circulatory and intestinal barrier dysfunction, increased oxygen species are able to augment the tissue injury by operating with the regulation of cytokine gene expression (Cybulsky et al., 1988). Our previous results further showed that the heat stroke-induced hyperthermia, arterial hypotension, and cerebral ischemia was associated with an increased production of free radicals, higher lipid peroxidation, and lower enzymatic antioxidant defenses in multiple organs including hearts and brains (Yang and Lin, 2002; Niu et al., 2003). Pretreatment with hydroxyl radical scavengers significantly attenuated the heat stroke-induced arterial hyperthermia, hypotension, and cerebral ischemia. Regular exercise training has also been found to protect against myocardial injury by reducing lipid peroxidation (Demirel et al., 1998) and increasing cardiac antioxidant defense system (Husain and Hazelrigg, 2002). Therefore, in the present results, regular exercise training, in addition to inducing HSP overexpressing in multiple organs, may ameliorate the oxidative stress and damage and arterial hypotension during heat stroke by reducing hyperthermia.

Chinese herbal medicine, Shengmai San, is effective for improving oxidative damage and cerebral ischemia during heat stroke

Shengmai San (SMS) is routinely being used for treating coronary heart disease (Fang et al., 1987; Zhao et al., 1996). In was also found that SMS effectively suppressed the thiobarbituric acid reactive substance (TBARS) formation during reperfusion following ischemia, indicating that SMS improves the oxidative damage in the brain (Wang et al., 1999). We studied here the effect of SMS on brain oxidative and ischemic damage in heat stroke rats.

Adult Sprague-Dawley rats were assigned randomly to one of the following four major groups. One major group of rats was treated with an oral dose of distilled water (DW) (1 mL per rat) daily and consecutively for 7 days before initiation of heat stress or one single oral dose of DW (1 mL per rat) immediately after the onset of heat stroke. For heat stroke induction, animals were exposed to a T_a of 43°C (with relative humidity of 60% in a temperature-controlled chamber). At a certain point in the heat stroke group, when MAP and local CBF in the striatum of rat brains began to decrease from their peak levels, this moment was considered as the onset of heat stroke. Immediately after the onset of heat stroke, heat stress was terminated and the animals were allowed to recover at room temperature (24°C). Our results showed that the latency for the onset of heat stroke were found to be ~70 min for the vehicle-treated heat stroke group. Then, both physiological parameters and survival time were observed for up to 450 min (or at the end of experiments).

The second group of rats was treated with an oral dose of SMS (0.3–1.2 g/mL per rat, daily and consecutively for 7 days). The third group of rats was treated with one single oral dose of SMS (0.3–1.2 g/mL per rat) immediately after the start of heat stroke. Both the second and the third group of rats were exposed to heat exposure (43°C for exactly 70 min) to induce heat stroke. Again, after 70-min heat exposure, the animals were allowed to recover at room temperature (24°C). These two groups of rats were used as drug-treated heat stroke groups. The fourth major group of DW-treated rats was exposed to room temperature (24°C) and used as the normothermic controls.

Different group of animals was used for different sets of experiments: (i) measurements of MAP, ICP, CPP (CPP = MAP−ICP), and striatal CBF, PO_2, and brain temperature (T_b); (ii) determination of T_{co}, MAP, HR, and extracellular concentrations of glutamate, glycerol, lactate/pyruvate ratio in the striatum; or (iii) determination of the extent of lipid peroxidation and rate of O_2^- generation in striatal mitochondria and extracellular levels of 2,3-DHBA and 2,5-DHBA in the striatum; or (iv) determination of neuronal damage scores for different brain structures.

As mentioned in the foremore section, after the onset of heat stroke, the values of MAP, CPP, CBF, and brain partial pressure of O_2 (PO_2) were all significantly lower than those in normothermic controls. However, the values of ICP, brain and colonic temperatures, and brain levels of free radials, lipid peroxidation and rate of $O_2^{\cdot-}$ generation, and glutamate, glycerol, and lactate/pyruvate ratio were all greater in heat stroke rats pretreated

Table 4. Effects of heat exposure (43°C for 70 min) on the extent of lipid peroxidation and rate of $O_2^{\cdot-}$ generation in tissue homogenates and the extracellular concentration of DHBA in the striatum of rats treated with distilled water (DW) or Shengmai San (SMS)

Treatment	Lipid peroxidation (RFU)	Rate of $O_2^{\cdot-}$ generation (nmol/min per mg protein)	DHBA (% baseline)
1. Daily oral administration for consecutive 7 days before initiation of HE			
(1) DW (1 mL)-treated rats kept at 24°C	1.58 ± 0.11	6.91 ± 0.68	100 ± 4
(2) DW (1 mL)-treated rats kept at 43°C	2.26 ± 0.13*	9.15 ± 0.82*	726 ± 27*
(3) SMS (0.6 g in 1 mL)-treated rats kept at 43°C	1.64 ± 0.09†	7.07 ± 0.88†	150 ± 5*,†
2. Single one oral dose immediately after the onset of heat stroke			
(1) DW (1 mL)-treated rats kept at 24°C	1.54 ± 0.09	6.88 ± 0.66	100 ± 5
(2) DW (1 mL)-treated rats kept at 43°C	2.31 ± 0.14*	9.18 ± 0.72*	731 ± 29*
(3) SMS (0.6 g on 1 mL)-treated rats kept at 43°C	1.77 ± 0.10†	7.15 ± 0.56†	142 ± 4*,†

Source: From Wang et al. (2005b).
Note: Data are means ± SEM of eight rats per group. RFC, relative fluorescence units. For the determination of lipid peroxidation and rate of $O_2^{\cdot-}$ generation in the homogenates, and extracellular concentration of DHBA of striatum, samples were obtained 85 min after the initiation of heat exposure or 15 min after the onset of heat stroke or at the equivalent time after injection of DW for the rats at 24°C.
*$P<0.05$ compared with DW-treated rats at 24°C.
†$P<0.05$ compared with DW-treated rats at 43°C (ANOVA followed by Duncan's test).

or post-treated with vehicle solution compared with those of normothermic controls (Table 4). Pretreatment or past-treatment with SMS, a traditional Chinese herbal medicine, significantly reduced the arterial hypotension, intracranial hypertension, cerebral hypoperfusion and hypoxia, and increased levels of ischemia and damage markers in the brain during heat stroke (Wang et al., 2005b).

As demonstrated in the present results, the prolongation of survival in heat stroke rats with SMS treatment was found to be related to enhancement of MAP and local CBF, as well as reduction in both intracranial hypertension and cerebral neuronal damage during heat stroke. The augmentation of CBF in rats treated with SMS may be brought about by higher cerebral perfusion resulting from lower ICP (due to reduction in cerebral edema and cerebrovascular congestion) and higher MAP during heat stroke (Shih et al., 1984).

There is evidence that both circulatory shock and cerebral ischemia are associated with increased production of free radicals (specifically, hydroxyl radicals and superoxide), increased lipid peroxidation, and decreased enzymatic antioxidant defenses in the brain of heat stroke-affected rats (Yang and Lin, 2002). Pretreatment with α-tocopherol or mannitol 30 min before the onset of heat exposure significantly attenuate heat stroke-induced arterial hypotension, cerebral ischemia and neuronal damage, and the increased radical formation and lipid peroxidation in the brain (Niu et al., 2003). Results obtained here clearly stated that SMS had a strong activity to prevent the cerebral oxidative stress and neuronal damage produced by cerebral ischemia during heat stroke (Wang et al., 2005b). Pretreatment with SMS prevented both the increased production of free radicals and increased lipid peroxidation in the brain of heat stroke-affected rats. Moreover, it was found that SMS administered immediately after the onset of heat stroke was still effective both in preventing circulatory shock and free radicals accumulation and increased lipid peroxidation in the brain already damaged to considerable extent during heat stroke.

Hypervolemic hemodilution attenuates oxidative and ischemic damage during heat stroke

Evidence has accumulated to indicate that heat stroke-induced cerebral ischemia and injury is greatly attenuated by induction of hypervolemic hemodilution with intravenous administration of 10% human started either before the heat stress or right after the onset of heat stroke in rats (Chang et al., 2001). The present study attempted to ascertain whether the neuroprotective effect of hypervolemic hemodilution therapy is associated with inhibition of cerebral release of glutamate, glycerol, lactate, and free radicals after cerebral ischemia in heat stroke rats. The extracellular concentrations of glutamate, glycerol, lactate, and hydroxyl radicals in the striatum of rat brain were assessed by intracerebral microdialysis methods.

Adult Sprague-Dawley rats were assigned randomly to the following five groups: NS-pretreated rats maintained at 24°C; NS-pretreated rats maintained at 42°C; albumin-pretreated rats maintained at 42°C; NS-post-treated rats maintained at 42°C; and albumin-post-treated rats maintained at 42°C. In NS-pretreated rats at 24°C, animals were exposed to T_a of 24°C for at least 90 min to reach thermal equilibrium and injected with 0.9% NaCl solution (10 mL/kg of body weight) before the start of the experiment. In NS-pretreated rats at 42°C, animals were injected with 0.9% NaCl solution (10 mL/kg of body weight) i.v. and exposed immediately to T_a of 42°C. The moment at which MAP and local CBF began to decrease from their peak levels was taken as the onset of heat stroke. In albumin-pretreated rats at 42°C, animals received 10 mL of 10% human albumin/kg of body weight (Travenol Laboratories, Glendale, CA, USA) i.v. at the onset of heat exposure. In NS-post-treated rats at 42°C, animals received 0.9% NaCl solution (10 mL/kg of body weight) i.v. at the onset of heat stroke. In albumin-post-treated rats, at 42°C, animals received 10 mL of 10% albumin/kg of body weight i.v. at the onset of heat stroke. Immediately after the onset of heat stroke in these groups of animals, the heat exposure was switched off. In the present study, all groups of rats were exposed to 42°C until the onset of heat stroke. For determination of survival time, different physiological parameter changes or mortality rate, the heat exposure was switched off at exactly 88 min in all groups of rats. Different group of animals was used for the different sets of experiments: (i) measurement of ICP, CPP, MAP and striatal CBF, PO_2, and temperature; (ii) determination of T_{co}, MAP, HR, and extracellular concentrations of glutamate, glycerol, lactate/pyruvate ratio in the striatum; (iii) determination of latency; (iv) determination of survival time; and (v) determination of brain lipid peroxidation and rate of $O_2^{\cdot-}$ generation.

In the present study (Chang et al., 2004a), we attempted to assess the mechanisms underlying the neuroprotective effect of hypervolemic hemodilution in rat heat stroke. In anesthetized rats treated with NS immediately after the onset of heat stroke induced by a high ambient temperature (42°C), the values for MAP, ICP, CPP, CBF, brain PO_2, and striatal glutamate, glycerol, lactate/pyruvate ratio, hydroxyl radicals, and neuronal damage score were 42 ± 3 mmHg, 33 ± 3 mmHg, 9 ± 3 mmHg, 109 ± 20 blood perfusion units (BPU), 6 ± 1 mmHg, 51 ± 7 μmol/L, 24 ± 3 μmol/L, 124 ± 32, $694 \pm 22\%$ of baseline, and 2.25 ± 0.05, respectively. In animals

Table 5. Effects of heat exposure (HE; 42°C for 88 min) on the extent of lipid peroxidation and rate of $O_2^{\cdot-}$ generation in mitochondria in the tissue homogenates of corpus striatum measured at 100 min (i.e., 12 min after heat exposure was withdrawn) in rats treated with normal saline (NS) or albumin (AB) immediately before the start of HE or at the onset of heat stroke

Animal treatment	Lipid peroxidation (RFU)	Rate of $O_2^{\cdot-}$ generation (nmol/min/mg protein)
NS-pretreated rats at 24°C		
0 min after testing	1.71 ± 0.13	6.72 ± 0.73
100 min after testing	1.68 ± 0.12	6.65 ± 0.71
NS-pretreated rats at 42°C		
0 min before HE	1.66 ± 0.12	6.85 ± 0.69
100 min after HE	$2.18 \pm 0.15^*$	$9.11 \pm 0.87^*$
AB-pretreated rats at 42°C		
0 min before HE	1.73 ± 0.14	6.92 ± 0.72
100 min after HE	$1.88 \pm 0.11^{\dagger}$	$7.15 \pm 0.72^{\dagger}$
NS-post-treated rats at 42°C		
0 min before HE	1.69 ± 0.12	6.74 ± 0.69
100 min after HE	$2.21 \pm 0.16^*$	$9.23 \pm 0.85^*$
AB-post-treated rats at 42°C		
0 min before HE	1.72 ± 0.13	6.85 ± 0.69
100 min after HE	$1.79 \pm 0.10^{\dagger}$	$7.09 \pm 0.71^{\dagger}$

Source: From Chang et al. (2004a).
Note: Values are means $\pm$ SEM for eight rats per group.
*$P < 0.05$, significantly different from corresponding control values (saline-treated rats at 24°C; ANOVA followed by Duncan's test).
$^{\dagger}P < 0.05$, significantly different from corresponding control values (saline-treated rats at 42°C; ANOVA followed by Duncan's test).

treated with 10% albumin immediately after the onset of heat stroke (T_a of 42°C for 88 min), the values for MAP, ICP, CPP, CBF, brain PO_2 and striatal glutamate, glycerol, lactate/pyruvate ratio, hydroxyl radicals, and neuronal damage score were 64 ± 6 mmHg, 10 ± 2 mmHg, 54 ± 5 mmHg, 452 ± 75 BPU, 15 ± 2 mmHg, 3 ± 2 μmol/L, 4 ± 2 μmol/L, 7 ± 3, 119 ± 7% of baseline, and 0.38 ± 0.05, respectively. Apparently, the heat stroke-induced arterial hypotension, intracranial hypertension, cerebral hypoperfusion, cerebral ischemia, brain hypoxia, increased levels of striatal glutamate, glycerol, lactate/pyruvate ratio and hydroxyl radicals, and increased striatal neuronal damage scorer values were all attenuated significantly by the induction of hypervolemic hemodilution in rats immediately at the onset of heat stroke (Tables 5 and 6).

Hypervolemic hemodilution has been used to treat animals with acute focal (Cole et al., 1990) or global (Graham et al., 1978) cerebral ischemia by increasing blood flow through a reduction in the viscosity and oxygen content of the blood, caused by reductions in both the hematocrit and the fibrinogen (Grotta, 1987). Indeed, the heat stroke-induced arterial hypotension, cerebral ischemia, increased striatal dopamine overload, and increased neuronal damage score were attenuated by induction of both hypervolemic and hemodilution state with 10% human albumin either before or after the onset of heat stroke (Chang et al., 2001). Our recent results (Chang et al., 2004a) further demonstrated that the heat stroke-induced arterial hypotension, intracranial hypertension, cerebral hypoperfusion, cerebral ischemia, brain hypoxia, increased levels of striatal glutamate, glycerol, lactate/pyruvate satio and hydroxyl radicals, and increased striatal neuronal damage score values were all significantly attenuated by induction of hypervolemic hemodilution in rats immediately at the onset of heat stroke. These results show that the neuroprotective effect of hypervolemic hemodilution is related to a decrease in the elevation of glutamate, glycerol,

Table 6. Effects of heat exposure (HE; 42°C for 88 min) on mean T_{co}, MAP, and extracellular concentrations of glutamate, glycerol, lactate/pyruvate, and DHBA in the corpus striatum of rats pretreated with NS or AB immediately before the start of HE or rats post-treated with normal saline or AB 12 min after the onset of heat stroke

Animal treatment	T_{co} (°C)	MAP (mmHg)	Glutamate (μmol/L)	Glycerol (μmol/L)	Lactate/ pyruvate	DHBA (% baseline)
NS-pretreated rats at 24°C						
0 min after testing	36.2 ± 0.2	79 ± 5	0.88 ± 0.52	3 ± 1	6 ± 2	100 ± 6
100 min after testing	36.1 ± 0.3	75 ± 4	0.97 ± 0.63	4 ± 2	7 ± 3	98 ± 4
NS-pretreated rats at 42°C						
0 min before HE	36.0 ± 0.3	78 ± 4	0.95 ± 0.53	3 ± 2	6 ± 3	100 ± 5
100 min after HE	$42.6\pm0.2^*$	$2\pm2^*$	$54\pm8^*$	$22\pm4^*$	$127\pm34^*$	$705\pm26^*$
AB-pretreated rats at 42°C						
0 min before HE	36.3 ± 0.3	77 ± 3	0.93 ± 0.48	4 ± 2	7 ± 2	100 ± 6
100 min after HE	42.8 ± 0.3	$50\pm4^\dagger$	$2\pm1^\dagger$	$5\pm2^\dagger$	$8\pm4^\dagger$	$124\pm8^\dagger$
NS-post-treated rats at 42°C						
0 min before HE	36.1 ± 0.2	81 ± 4	0.85 ± 0.53	4 ± 2	7 ± 3	100 ± 5
100 min after HE	$42.7\pm0.3^*$	$4\pm3^*$	$51\pm7^*$	$24\pm3^*$	$124\pm32^*$	$694\pm22^*$
AB-post-treated rats at 42°C						
0 min before HE	36.2 ± 0.2	79 ± 4	0.91 ± 0.49	3 ± 2	6 ± 2	100 ± 6
100 min after HE	42.5 ± 0.3	$52\pm5^\dagger$	$3\pm2^\dagger$	$4\pm2^\dagger$	$7\pm3^\dagger$	$119\pm7^\dagger$

Source: From Chang et al. (2004a).
Note: Values are means $\pm$ SEM for eight rats per group.
$^*P<0.05$, significantly different from corresponding control values (saline-treated rats at 24°C; ANOVA followed by Duncan's test).
$^\dagger P<0.05$, significantly different from corresponding control values (saline-treated rats at 42°C; ANOVA followed by Duncan's test).

lactate, and free radicals in the brain exposed to heat stroke-induced cerebral ischemia/hypoxia injury.

Possible role played by oxidative damage in the pathogenesis of cerebral ischemia during heat stroke

Figure 7 depicts a scheme showing possible events between the heat exposure and animal death from heat stroke, with known or proposed interrelationship derived from the published data. After the onset of heat stroke, both arterial hypotension and intracranial hypertension induce cerebral ischemia, which leads to oxygen and nutrient deprivation and the initiation of a cascade of secondary mechanisms. This neurotoxic cascade involves overloading of glutamate, NO, oxygen free radicals, and other substances in the brain as well as derangements in normal metabolic and physiological functions. Cerebral ischemia generated ROS within the mitochondria which then signal the release of cytochrome C by mechanisms that may be related to Bcl-2 and translocation of Bax (Chan, 2001). Cytochrome C activates caspase 3 which is known to lead to nuclear DNA damage without repair, resulting in apotosis. The activation of the N-methyl-D-aspartate (NMDA) receptor and

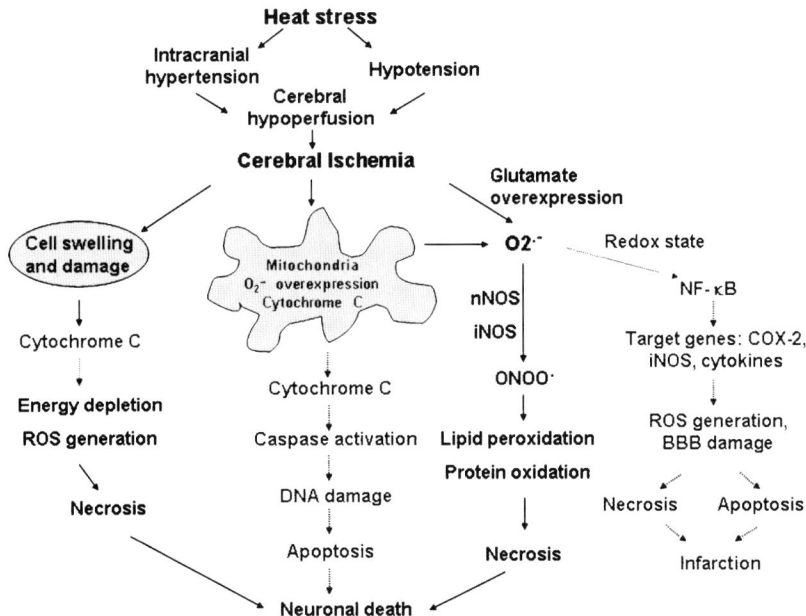

Fig. 7. A scheme depicting events occurring between the exposure of animals to a hot environment and cerebral neuronal death from heat stroke with known or proposed interrelationships derived from the published and unpublished data. Cerebral ischemia generated reactive oxygen species (ROS) within the mitochondria which then signal the release of cytochrome c by mechanisms that may be related to Bcl-2 and translocation of Bax (Chan, 2001). Cytochrome c activates caspase 3 which is known to lead to nuclear DNA damage without repair, resulting apoptosis. The activation of the N-methyl-D-aspartate receptors and formation of $O_2^{\cdot -}$ and nitric oxide (NO) may directly signal the mitochondrial release of cytochrome c or formation of peroxynitrite and subsequent hydroxyl radical formation can directly damage lipids proteins and DNA and lead to cell death, most likely necrosis. A more severe cerebral ischemia may cause mitochondria swelling and damage, which then cause the inhibition of ATP syntheses and increased ROS production, directly causing necrotic cell death. Oxygen radicals are known to alter the redox state of the cell. The major signaling pathway of oxygen radical involves the nuclear factor-kappa B (NF-κB). The activation of NF-κB causes its translocation into the nucleus and binding to the NF-κB site of many inducible genes, including but not limited to cyclooxygenase-2, inducible nitric oxide synthase, and cytokines. The expression of these genes may lead to formation of ROS and blood–brain barrier breakdown, which lead to apoptosis and/or necrosis. Adapted from Lin and Chang (2004). ICP, intracranial pressure; MAP, mean arterial pressure; CPP, cerebral perfusion pressure; CBF, cerebral blood flow.

formation of $O_2^{\cdot -}$ and NO may directly signal the mitochondrial release of cytochrome C or formation of peroxynitrite ($ONOO^-$) and subsequent hydroxyl radical formation can directly damage lipid proteins and DNA and lead to cell death most likely necrosis. A more severe cerebral ischemia may cause mitochondria swelling and damage, which then cause the inhibition of ATP synthesis and increased ROS production, directly causing necrotic cell death. Oxygen radicals are known to alter the redox state of the cell. The major signaling pathway of oxygen radicals involves the nuclear factor-kappa B (NF-κB). The activation of NF-κB causes its translocation into the nucleus and binding to the NF-κB site of many inducible genes, including but not limited to cyclooxygenase-2 (COX-2), iNOS, and cytokines. The expression of these genes may lead to formation of ROS and blood-brain barrier breakdown, which lead to apoptosis and/or necrosis. Thus, any measures which are able to attenuate oxidative damage and/or intervene the secondary neurotoxic cascades can be used to prevent and/or to treat ischemic neuronal damage in brain during heat stroke. HSP72 preconditioning induced by prior heat shock, regular exercise training, or SMS pretreatment is able to prevent the oxidative damage during heat stroke. On the other hand, hypervolemic hemodilution or SMS therapy is able to treat the oxidative damage during heat stroke.

Acknowledgments

This work was supported by grants-in-aid for special projects (NSC 91-2314-B-195-021 and NSC 92-2314-B-195-017) from the National Science Council of Republic of China (Taipei, Taiwan), and a grant from the Chi-Mei Medical Center (Tainan, Taiwan).

References

Babior, B.M. (2000) Phagocytes and oxidative stress. Am. J. Med., 109: 33–44.

Berger, C., Schabitz, W.R., Georgiadis, D., Steiner, T., Aschoff, A. and Schwab, S. (2002) Effects of hypothermia on excitatory amino acids and metabolism in stroke patients: a microdialysis study. Stroke, 33m: 519–524.

Boveris, A. (1984) Determination of the production of superoxide radicals and hydrogen peroxide in mitochondria. Methods Enzymol., 105: 429–435.

Cao, W., Carney, J.M., Duchon, A., Floyd, R.A. and Chevion, M. (1988) Oxygen free radical involvement in ischemia and reperfusion injury to brain. Neurosci. Lett., 88: 233–288.

Chan, P.H. (2001) Reactive oxygen radicals in signaling and damage in the ischemic brain. J. Cereb. Blood Flow Metab., 21: 2–14.

Chang, C.K., Chien, C.H., Chou, H.L. and Lin, M.T. (2001) The protective effect of hypervolemic hemodilution in experimental heatstroke. Shock, 16: 153–158.

Chang, C.K., Chiu, W.T., Chang, C.P. and Lin, M.T. (2004a) Effect of hypervolaemic haemodilution on cerebral glutamate, glycerol, lactate and free radicals in heatstroke rats. Clin. Sci. (Lond.), 106: 501–509.

Chang, C.P., Hsu, Y.C. and Lin, M.T. (2003) Magnolol protects against cerebral ischaemic injury of rat heatstroke. Clin. Exp. Pharmacol. Physiol., 30: 387–392.

Chang, C.P., Lee, C.C., Chen, S.H. and Lin, M.T. (2004b) Aminoguanidine protects against intracranial hypertension and cerebral ischemic injury in experimental heatstroke. J. Pharmacol. Sci., 95: 56–64.

Chiu, W.T., Kao, T.Y. and Lin, M.T. (1995) Interleukin-1 receptor antagonist increases survival in rat heatstroke by reducing hypothalamic serotonin release. Neurosci. Lett., 202: 33–36.

Cole, D.J., Drummond, J.C., Ruta, T.S. and Peckham, N.H. (1990) Hemodilution and hypertension effects on cerebral hemorrhage in cerebral ischemia in rats. Stroke, 21: 1333–1339.

Cybulsky, M.I., McComb, D.J. and Movat, H.Z. (1988) Neutrophil leukocyte emigration induced by endotoxin. Mediator roles of interleukin 1 and tumor necrosis factor alpha 1. J. Immunol., 140: 3144–3149.

Dahle, L.K., Hill, E.G. and Holman, R.T. (1962) The thiobarbituric acid reaction and the autoxidations of polyunsaturated fatty acid methyl esters. Arch. Biochem. Biophys., 98: 253–261.

Demirel, H.A., Powers, S.K., Caillaud, C., Coombes, J.S., Naito, H., Fletcher, L.A., Vrabas, I., Jessup, J.V. and Ji, L.L. (1998) Exercise training reduces myocardial lipid peroxidation following short-term ischemia-reperfusion. Med. Sci. Sports Exerc., 30: 1211–1216.

Ecochard, L., Lhenry, F., Sempore, B. and Favier, R. (2000) Skeletal muscle HSP72 level during endurance training: influence of peripheral arterial insufficiency. Pflugers Arch., 440: 918–924.

Fabian, R.H., Perez-Polo, J.R. and Kent, T.A. (2000) Electrochemical monitoring of superoxide anion production and cerebral blood flow: effect of interleukin-1 beta pretreatment in a model of focal ischemia and reperfusion. J. Neurosci. Res., 60: 795–803.

Fang, J., Jiang, J. and Luo, D.C. (1987) Effect of Shengmai San oral fluid on coronary arteriosclerotic. Chin. Med. J., 26: 403.

Fink, S.L., Chang, L.K., Ho, D.Y. and Sapolsky, R.M. (1997) Defective herpes simplex virus vectors expressing the rat brain stress-inducible heat shock protein 72 protect cultured neurons from severe heat shock. J. Neurochem., 68: 961–969.

Gonzalez, B., Hernando, R. and Manso, R. (2000) Anabolic steroid and gender-dependent modulation of cytosolic HSP70s in fast- and slow-twitch skeletal muscle. J. Steroid Biochem. Mol. Biol., 74: 63–71.

Graham, D.G., Tiffany, S.M., Bell Jr., W.R. and Gutknecht, W.F. (1978) Autoxidation versus covalent binding of quinones as the mechanism of toxicity of dopamine, 6-hydroxydopamine, and related compounds toward C1300 neuroblastoma cells in vitro. Mol. Pharmacol., 14: 644–653.

Griffith, O.W. (1980) Determination of glutathione and glutathione disulfide using glutathione reductase and 2-vinylpyridine. Anal. Biochem., 106: 207–212.

Grotta, J.C. (1987) Current status of hemodilution in acute cerebral ischemia. Stroke, 18: 689–690.

Hall, D.M., Buettner, G.R., Oberley, L.W., Xu, L., Matthes, R.D. and Gisolfi, C.V. (2001) Mechanisms of circulatory and intestinal barrier dysfunction during whole body hyperthermia. Am. J. Physiol. Heart Circ. Physiol., 280m: H509–H521.

Halliwell, B., Kaur, H. and Ingelman-Sundberg, M. (1991) Hydroxylation of salicylate as an assay for hydroxyl radicals: a cautionary note. Free Radic. Biol. Med., 10: 439–441.

Hung, C.H., Chang, N.C., Cheng, B.C. and Lin, M.T. (2005) Progressive exercise preconditioning protects against circulatory shock during experimental heat stroke. Shock, 23: 426–433.

Husain, K. and Hazelrigg, S.R. (2002) Oxidative injury due to chronic nitric oxide synthase inhibition in rat: effect of regular exercise on the heart. Biochim. Biophys. Acta, 1587: 75–82.

Ikeda, Y. and Long, D.M. (1990) The molecular basis of brain injury and brain edema: the role of oxygen free radicals. Neurosurgery, 27: 1–11.

Kao, T.Y., Chio, C.C. and Lin, M.T. (1994) Hypothalamic dopamine release and local cerebral blood flow during onset of heatstroke in rats. Stroke, 25: 2483–2487.

Kao, T.Y. and Lin, M.T. (1996) Brain serotonin depletion attenuates heatstroke-induced cerebral ischemia and cell death in rats. J. Appl. Physiol., 80: 680–684.

Kikugawa, K., Kato, T., Beppu, M. and Hayasaka, A. (1989) Fluorescent and cross-linked proteins formed by free radical and aldehyde species generated during lipid oxidation. Adv. Exp. Med. Biol., 266: 345–356.

Kirino, T., Tsujita, Y. and Tamura, A. (1991) Induced tolerance to ischemia in gerbil hippocampal neurons. J. Cereb. Blood Flow Metab., 11: 299–307.

Lancelot, E., Callebert, J., Plotkine, M. and Boulu, R.G. (1995) Striatal dopamine participates in glutamate-induced hydroxyl radical generation. Neuroreport, 6: 1033–1036.

Li, P.L., Chao, Y.M., Chan, S.H. and Chan, J.Y. (2001) Potentiation of baroreceptor reflex response by heat shock protein 70 in nucleus tractus solitarii confers cardiovascular protection during heatstroke. Circulation, 103: 2114–2119.

Lin, M.H., Chao, H.T. and Hong, C.Y. (1995a) Magnolol protects human sperm motility against lipid peroxidation: a sperm head fixation method. Arch. Androl., 34: 151–156.

Lin, M.T. (1997) Heatstroke-induced cerebral ischemia and neuronal damage. Involvement of cytokines and monoamines. Ann. N.Y. Acad. Sci., 813: 572–580.

Lin, M.T. (1999) Pathogenesis of an experimental heatstroke model. Clin. Exp. Pharmacol. Physiol., 26: 826–827.

Lin, M.T. and Chang, C.P. (2004) The neuropharmacological basis of heat intolerance and its treatment. J. Thermal. Biol., 29: 463–469.

Lin, M.T., Kao, T.Y., Chio, C.C. and Jin, Y.T. (1995b) Dopamine depletion protects striatal neurons from heatstroke-induced ischemia and cell death in rats. Am. J. Physiol., 269: H487–H490.

Lin, M.T., Kao, T.Y., Jin, Y.T. and Chen, C.F. (1995c) Interleukin-1 receptor antagonist attenuates the heat stroke-induced neuronal damage by reducing the cerebral ischemia in rats. Brain Res. Bull., 37: 595–598.

Liu, C.C., Chien, C.H. and Lin, M.T. (2000) Glucocorticoids reduce interleukin-1 concentration and result in neuroprotective effects in rat heatstroke. J. Physiol., 527: 333–343.

Lo, Y.C., Teng, C.M., Chen, C.F., Chen, C.C. and Hong, C.Y. (1994) Magnolol and honokiol isolated from *Magnolia officinalis* protect rat heart mitochondria against lipid peroxidation. Biochem. Pharmacol., 47: 549–553.

Lowry, O.H., Rosebrough, N.J., Farr, A.L. and Randall, R.J. (1951) Protein measurement with the Folin phenol reagent. J. Biol. Chem., 193: 265–275.

Luck, H. (1965) Catalase. In: Bergmeyer H. (Ed.), Methods of Enzymatic Analysis. Academic Press, New York, NY, pp. 885–894.

Misra, H.P. and Fridovich, I. (1977) Superoxide dismutase: "positive" spectrophotometric assays. Anal. Biochem., 70: 553–560.

Mohanakumar, K.P., de Bartolomeis, A., Wu, R.M., Yeh, K.J., Sternberger, L.M., Peng, S.Y., Murphy, D.L. and Chiueh, C.C. (1994) Ferrous-citrate complex and nigral degeneration: evidence for free-radical formation and lipid peroxidation. Ann. N.Y. Acad. Sci., 738: 392–399.

Niu, K.C., Lin, K.C., Yang, C.Y. and Lin, M.T. (2003) Protective effects of alpha-tocopherol and mannitol in both circulatory shock and cerebral ischaemia injury in rat heatstroke. Clin. Exp. Pharmacol. Physiol., 30: 745–751.

Obata, T. (1997) Use of microdialysis for in-vivo monitoring of hydroxyl free-radical generation in the rat. J. Pharm. Pharmacol., 49: 724–730.

Obata, T. and Yamanaka, Y. (1995) Intracranial microdialysis of salicylic acid to detect hydroxyl radical generation by monoamine oxidase inhibitor in the rat. Neurosci. Lett., 188: 13–16.

Ohkawa, H., Ohishi, N. and Yagi, K. (1979) Assay for lipid peroxides in animal tissues by thiobarbituric acid reaction. Anal. Biochem., 95: 351–358.

Paxinos, G. and Watson, C. (1982) The Rat Brain in Stereotaxic Coordinates (2nd ed.). Academic Press, New York.

Pulsinelli, W.A., Brierley, J.B. and Plum, F. (1982) Temporal profile of neuronal damage in a model of transient forebrain ischemia. Ann. Neurol., 11: 491–498.

Riedel, W. and Maulik, G. (1999) Fever: an integrated response of the central nervous system to oxidative stress. Mol. Cell. Biochem., 196: 125–133.

Schwarz, S., Schwab, S., Bertram, M., Aschoff, A. and Hacke, W. (1998) Effects of hypertonic saline hydroxyethyl starch solution and mannitol in patients with increased intracranial pressure after stroke. Stroke, 29: 1550–1555.

Shih, C.J., Lin, M.T. and Tsai, S.H. (1984) Experimental study on the pathogenesis of heat stroke. J. Neurosurg., 60: 1246–1252.

Smolka, M.B., Zoppi, C.C., Alves, A.A., Silveira, L.R., Marangoni, S., Pereira-Da-Silva, L., Novello, J.C. and Macedo, D.V. (2000) HSP72 as a complementary protection against oxidative stress induced by exercise in the soleus muscle of rats. Am. J. Physiol. Regul. Integr. Comp. Physiol., 279: R1539–R1545.

Suzuki, J., Imaizumi, S., Kayama, T. and Yoshimoto, T. (1985) Chemiluminescence in hypoxic brain-the second report: cerebral protective effect of mannitol, vitamin E and glucocorticoid. Stroke, 16: 695–700.

Tietze, F. (1969) Enzymic method for quantitative determination of nanogram amounts of total and oxidized glutathione: applications to mammalian blood and other tissues. Anal. Biochem., 27: 502–522.

Wang, J.L., Ke, D.S. and Lin, M.T. (2005a) Heat shock pretreatment may protect against heatstroke-induced circulatory shock and cerebral ischemia by reducing oxidative stress and energy depletion. Shock, 23.

Wang, N.L., Liou, Y.L., Hsu, Y.C., Lin, M.T. and Chang, C.K. (2005b) Chinese herbal medicine, Shengmai San, is effective for improving circulatory shock and oxidative damage in the brain during heatstroke. J. Pharmacol. Sci., 97: 253–265.

Wang, X., Magara, T. and Konishi, T. (1999) Prevention and repair of cerebral ischemia-reperfusion injury by Chinese herbal medicine, Shengmai San, in rats. Free Radic. Res., 31: 449–455.

Wolf, D., Schumann, J., Koerber, K., Kiemer, A.K., Vollmar, A.M., Sass, G., Papadopoulos, T., Bang, R., Klein, S.D., Brune, B. and Tiegs, G. (2001) Low-molecular-weight hyaluronic acid induces nuclear factor-kappaB-dependent resistance against tumor necrosis factor alpha-mediated liver injury in mice. Hepatology, 34: 535–547.

Yamamoto, M., Shima, T., Uozumi, T., Sogabe, T., Yamada, K. and Kawasaki, T. (1983) A possible role of lipid peroxidation in cellular damages caused by cerebral ischemia and the protective effect of alpha-tocopherol administration. Stroke, 14: 977–982.

Yang, C.Y. and Lin, M.T. (2002) Oxidative stress in rats with heatstroke-induced cerebral ischemia. Stroke, 33: 790–794.

Yang, Y.L. and Lin, M.T. (1999) Heat shock protein expression protects against cerebral ischemia and monoamine overload in rat heatstroke. Am. J. Physiol., 276: H1961–H1967.

Yang, Y.L., Lu, K.T., Tsay, H.J., Lin, C.H. and Lin, M.T. (1998a) Heat shock protein expression protects against death following exposure to heatstroke in rats. Neurosci. Lett., 252: 9–12.

Yang, Y.L., Pan, W.H., Chiu, T.H. and Lin, M.T. (1998b) Striatal glutamate release is important for development of ischemic damage to striatal neurons during rat heatstroke. Brain Res., 795: 121–127.

Zhao, M., Brong, H., Lu, B., Ihu, X., Huang, C. and Yang, J. (1996) Effect of Shengmai San on serum lipid peroxidation in acute viral myocarditis. J. Combined Tradit. Chin. West. Med., 16: 142–145.

Zhou, Z., Wang, L., Song, Z., Lambert, J.C., McClain, C.J. and Kang, Y.J. (2003) A critical involvement of oxidative stress in acute alcohol-induced hepatic TNF-alpha production. Am. J. Pathol., 163: 1137–1146.

Subject Index

AA (see arachidonic acid)
α-AR agonists 7
α4β7 integrin 144
absolute core temperatures 52–53
absolute heat production 51
accidental hypothermia 154
acclimatization to heat 113
acclimatory plasticity 373
acetaminophen 9, 17, 161, 210
　effect on T_b 18
　treatment, early 211
acetazolamide 463
acidosis 328
activation-induced cell death (AICD) 145, 146
active cooling 45
　of head 55–56
active transporters, fundamental differences in 463
active transport of Na and Cl in CSF 463
acute heat stress 374
acute inflammatory effects 247
acute intoxication 236
acute mortality 233
acute phase response 208
acute physical exercise on gene expression 328–334
　acidosis 328
　anti-inflammatory cytokine IL-6 329
　CD4 lymphocytes 329
　CD8 lymphocytes 329
　cell growth 330
　circulating IL-6 330
　differentiation 330
　"high interest" genes 330
acute stroke 204
　patients 211
acute toxicity 233
addictive drugs 220, 233
adenosine 3′, 5′–monophosphate (cyclic AMP) 466
adenosine receptor 434

adenosine triphosphate (ATP)
　decrease in intracellular stores of 326
　hydrolysis of 399
adrenaline 86
adrenergic neurons 40
adrenoceptors (AR) 7
　β2-adrenoceptor 144
　α_1 agonist 7
　α_2 agonist 7
adverse environmental conditions 219
aerobic fitness 49
afebrile humans 17
African runners 51
Ag (Silver) 245
　nanoparticles 255
age factors 118
aging 49
　and heat stroke 512–513
　impedes coping with hyperthermia 438–439
　of polytetrafluoroethylene (PTFE) fume 247
agitation 90
AH-6809 20
airflow 45, 191
air velocity 112
Al (Aluminum) 245
　nanoparticles 255
　treated nanoparticles 262
alanine aminotransferase (ALT) 492
albumin/IgG quotients 463
alcohol 483
　addicts 286
　consumption 205
　effects of 113
　heat strain and, abuse 483
　rubs 162, 210
　vasomotor reflexes depressed by 483
alcoholism 114, 482
alertness sensation 54
altered psycho-emotional states 220

Alzheimer's disease 512
amantadine 88
ambient temperature(s) 48, 64–65
   different, thermoeffector responses at 72
   drugs toxicity patterns and 68–69
   and hyperthermic response 66–68
ambient vapor pressure 112
amino acids 449
   excitatory 277
     and inhibitory 295
   neurotransmission 295
   neurotransmitters 295 (see also specific entries)
amino acid and opioid interaction in hyperthermia 311–312
   antagonized by naloxone 311
   blockade of cell death 311
   excitatory effects of opioids 311
   κ-opioid receptor 311, 312
   opioids 311
   pathophysiology of hyperthermic brain injuries 311
amino acid neurotransmitters in hyperthermia 298–301
   aspartate 298
   cell injury in hyperthermia 300
     amino acid neurotransmitter 300
     antioxidant compound H-290/51 300
     Ca^{2+} influx 300
     cell death in heat stress 300
     glutamate receptors 300
     imbalance between oxidants and antioxidants 300
     induction of oxidative stress by glutamate 300
     programmed cell death 300
   excitatory amino acids 298
     glutamate and aspartate 298–300
   GABA 298
   glutamate 298
   glycine 298
   high-performance liquid chromatography (HPLC) 298
   inhibitory amino acids 298
     GABA and glycine 300
   naloxone treatment 300–301
     cellular stress response 301
     δ-opioid receptor 301
     influence on amino acid neurotransmitters 301
     κ-opioid receptor 301
     L-type Ca^{2+} channel blocker nimodipine 301
     μ-opioid receptor 301
     opioid antagonist naloxone 301
   o-phthalaldehyde method 298
amino acid neurotransmitters in pathophysiology of hyperthermia 296–297
   axonal damage 296
   brain heating 296
   drug addiction 297
   GABA depresses axonal conduction 296
   glutamate-induced neurotoxicity 296
   glutamate release 296
   hyperthermia, intensity and severity of 296
   memory function 297
   opioid neurotransmission 297
   opioid peptides 297
   sleep 297
   stress 297
   thermoregulation 297
amino acid neurotransmitters in thermoregulation 296
   aspartate and glutamate 296
   frontal cerebral cortex 296
   γ-amino butyric acid (GABA) 296
   prostaglandin E1 (PGE1)-induced hyperthermia 296
   taurine 296
6-aminomethyl- 3-methyl-4H-1,2,4-benzothiadiazine-1,1-dioxide (TAG) 450
amitryptyline 91
amphetamine 233
   derivatives 177
amphetamine-like drugs 233
   neurotoxic effects of 236
amphetamine-like psychomotor stimulants 233
amphetamine-like psychostimulants 220
amphetamines 450
   combination in hyperthermia induced brain damage 348
   stroke and 347
amygdala stimulation 35
amygdaloidal complex, subnuclei of 427
analgesics 91
anatomical models 111
ancient Indian literature 153

anesthesia 154, 155, 158, 177, 184–185, 500
   choice of 184–185
   equithesin 185
   heat stress influences 184
   pentobarbital 177, 185
   urethane 177, 184–185
anesthetic drugs, general 238
anesthetic gases 83
anesthetics and brain temperature 177
anesthetize states 188
Ang II modulation 377
angiotensin 277, 278, 280, 282–284, 288–289, 373, 460, 463 (see also renin-angiotensin system)
   and related peptides 288–289
   system 277
   and thermoregulation 376–377
      evaporative heat loss 377
angiotensin-converting enzyme 283
angiotensin II 281
animal models 173
   of hyperthermia 178
   hypothermia in 21
animal models of heat stroke 485, 499
animal models of hyperthermia 178, 298, 467
   BBB permeability 298
   biological oxygen demand (BOD) 298
   heat-related morbidity or mortality 298
   heat stress without heat stroke 298
   hyperthermia-induced brain damage 298
   neurochemical mediators 298
   opioid peptides 298
   prostaglandins 298
   relative humidity 298
   serotonin 298
   wind velocity 298
animal models of stress 278
anoxic brain injury 164
anterior hypothalamus 3
   preoptic area of 19, 21, 39, 40
anti-apoptotic protein, bcl-2 77
anti-asialo GM1 antibody 147–148
anti-cancer agents 74
anticancer drugs 246
anticholinergics 174
antidepressants 81, 90
   tricyclic 91
antigen: antibody complexes 510
antigen presentation 140

antigen-presenting cells 140
anti-inflammatory cytokine 142
   IL-6 329
antineoplastic drugs 148
antinociception 282
antioxidants 10
anti-parkinsonian drugs 83
antipsychotic drugs 82
antipsychotics 81
antipyretic agents 17, 210
antipyretic drugs 17, 87, 161
antipyretic effects 11
antipyretic therapy 153
anti-pyrexial therapy 210
antituberculosis drug 90
anxiety 175
   in serotonin syndrome 100
apathy 153, 154
apoptosis 139, 201, 360, 435
   abnormal, due to neural tube defects 348
   caspase activation due to 350
   disruption of cytoskeleton due to 361
   drug-induced 148
   heat-induced 139, 361
   heat shock-induced 326
   immunologically induced 145–148
   JNK activation-induced 362
   lymphocytes 144, 145
   neuronal, due to 12 min exposure to 43°C 348
   non-caspase mediated apoptotic death 361
   overactivation of PARP 361
   receptor-induced 145
   stress-induced 144–145
apoptotic cascades 202
apoptotic cell death 403
arachidonic acid (AA) 4, 15, 21
   α-carbon carboxylic acid moiety of 21
   conjugates 21–22
   COX metabolism of 17
arachidonoylethanolamide (AEA) 21
2-arachidonoylglycerol (2-AG) 21
arachodonic acid 233
area postrema 188
L-arginine 10
arginine vasopressin (AVP) 463
arousal-related brain hyperthermia. 223

arousal-related temperature increase 226
Arrhenius equation 63
Arrhenius plot 74
arterial blood 238
　pressure, recovery of mean 207
　temperature 221–222, 230
arterial pressure 436
　mean arterial pressure 500
arteriosclerotic heart disease 483
arterio-venous blood temperature 35
aspartate 295, 298, 449
　glutamate and 296, 298–300
aspartate amino transferase (AST) 492
aspirin 6, 18
　non-selective COX inhibitor 6
　oral 17
astrocytes 249, 250
　derived from adult human brain 424
　　unable to induce Hsp70 424
　Hsc70 424
　Hsp70 induction in 417
　specific marker of 262
astrocytic processes 8
astrocytosis 233
asymmetrical ^{123}I-N-isopropyl-p-
　iodoamphetamine (^{123}I-IMP) 86
AT receptor blockade 377
AT1 receptors 377
AT2 receptors 377
atropine 185, 450
　in animals 90
atypical antipsychotics 89
auditory nerve-brainstem evoked responses (ABR)
　376
autoimmunity 145
autonomic nervous symptoms 86, 100
autonomic nervous system (ANS) 373
autonomic symptoms 82
autonomic thermoeffectors 74
autonomic thermoregulation 113
autoregulation and cerebral vasomotor responses
　155, 159
aversion of RF-EMFs 122–124
axillary ice packs 210
axonal damage 232
axonal translocation 247
axons, morphological changes in 154

bacterial endotoxic lipopolysaccharide 3
　(see lipopolysaccharide)
bacterial endotoxin-induced hyperthermia
　174
bacterial infection 17
bacterial invasion 220
bacterial products 17
β2-antagonists 144
barbiturates 161
baroreceptor reflex, impairment of 513
barrier permeability 461–463
　(see also blood–brain barrier)
base damage repair 139
baseline core temperature 72
basement membrane 250
Bax translocation 361
BBB (see blood–brain barrier)
BBB permeability 205
　brain edema and 191
　to Evans blue 187–188
β-blocker 95
bcl-2 77, 145
behavioral performance disruption, of RF-EMFs
　124–128
behavioral pharmacology, animal 90
behavioral responses, to heat stress 48, 53
behavioral thermoregulation 48, 113
benzodiazepines 96
　derivatives for serotonin syndrome 92
　for NMS treatment 86, 88
Bergmann glia 175, 436
17β estradiol 206
BH3 protein 145
bicuculline 453
biodistribution 247
biological oxygen demand (BOD) incubator
　179
biomarkers 10
biotinylated IL-1-receptor 4
bladder temperatures 207
bleomycin 74
blinking of eyes 185
blood 38–39, 229–231, 259, 260, 264–265,
　305–306
　circulation 223
　and hypothalamic thermoregulatory centers
　　irritation 208
　supply to brain 221–222

blood-brain barrier (BBB) 5, 159, 162, 173, 203, 210, 454
　brain fluid microenvironment 247
　breakdown of 251
　in cerebral capillaries 459
　disruption
　　exacerbation of 245
　　and microvascular function 209
　　by nanoparticles 251
　dysfunction 246
　　heat stress-induced 297
　in endothelial cells 249
　endothelium 251
　function and nanoparticles (see nanoparticles, and blood-brain barrier function)
　opening 425
　　with nanoparticles 252
　opening, reversibility of 188–190
　permeability 232
　　alterations in 349
　　5-HT in regulation of 501
　　naloxone on regional 305
　　in nanoparticles (see nanoparticles, blood-brain barrierpermeability)
　　properties of 249
　to protein tracers 209
　severely disrupted 165
　structure and function of 250
blood-brain interface 190
blood-CSF barrier (BCSFB) 188, 451
　and cell and tissue injuries in CNS 459
　in choroid plexus 459
　dysfunction of, augmented fluxes of ions, water and proteins 459
　in heat-related illnesses 467
　in hyperthermia 459
　permeability 459
　protein flux across 462
blood-CSF interfaces 461
blood-CSF vs. blood-brain barriers 461–464
blood flow, high 55
blood gases 179
blood pressure 347
　decreased 436
　maintenance of 436
　mean arterial blood pressure (MABP) 255
　in Neurolepticmalignant syndrome (NMS) 82
　recovery of mean arterial 207
　regulation 436
blood temperature 35, 486
blood volume, reduced 45
body core 35, 114, 221, 223, 224
　basal temperatures at 224
　temperature (T_c) 4, 5, 29, 31
body heat 66, 113, 114, 118, 230, 277, 449, 455, 529
　dissipation of 449
　production 234
body hyperthermia 233, 238, 395
body mass 51, 63
　index 50
body size, impact on vasomotor responses 71
body's natural defense system 185–186
body surface blood flow 436
body temperature(s) (T_b) 15, 64, 66, 233
　acetaminophen effect on 18
　after marathon run 229
　and brain temperature, homeostasis 229–231
　changes during heat exposure 180
　circadian rhythm 18
　control 64, 436
　diurnal variations of 20
　dose-dependent effects 219
　elevated 45
　increase of 450
　linear increases in 224
　lowering of basal 18
　maintenance of 17
　monitoring 233
　regulation 173
　　and MDMA 72
　　in NMS 90
　RF-EMF exposure and variation in 113
body warming, external 238
body weight loss 182
bombesin 376
bone marrow 143
brain 18
　activation during exercise 29
　activity
　　altered 47
　　and hyperthermia 47
　astrocytes derived from adult human 424
　blood supply to 221–222
　calcium metabolism 453
　cells 116–117, 220

cellular alterations in 245
circulation 223
cooling 114
critically low perfusion of 38–39
cyclooxygenase-2 (COX-2) 464
cytokines and monoamines 436
damage 174–175, 233, 277
   amphetamines combination in hyperthermia-induced 348
   hyperthermia causing 347
   hyperthermia-induced 298
   stroke and amphetamines 347
drug delivery to 251
drug release in 251
dysfunction 187, 418
   induced by nanoparticles 245
edema 153, 154, 159, 173
   BBB permeability and 191
   in heat-related illnesses 188–191
   precautions during measurement 191
endothelial cells, endocytosis within 251
energy consumption 155
fatigue 29
fluid microenvironment 247
function, influence of nanoparticles on 245
heat arrival to 221
heating 296
heat outflow from, diminished 230
heat production 234
heat removal from 230
heat treatment of tumors in 176
human 155
hyperthermia 153, 154, 165, 176–177, 190, 219, 278, 285, 286, 427, 451, 464, 467, 468, 529
   induced by meth-amphetamine (METH) and 3,4-methylenedioxymethamphetamine (MDMA) 233–236
   neurotransmission and 223
   as physiological phenomenon 220–231
   rapid brain temperature fluctuations and 224–229
hypothermia 55
injured 155
injuries 201, 207
   anoxic 164
   hyperthermic 154, 311
   hypothermic protection after 207
   hypoxic 153
   traumatic 76, 153, 161, 163
lesions 347
metabolism 230, 238
   and cerebral blood flow 223
microfluid environment 190
overheating, dangerous 219
parenchyma 251
pathology 245
   aggravation by nanoparticles 245
   in vivo rat model 246
proximal frontal cortex 252
regional 307
   influence of naloxone 307
sites 238
stem 175, 208
   reticular formation 185, 232
thermodynamic response of 36
thermorecording, development of 224
trauma 207
tumor 184
uptake of nanoparticles 248
ventral surface of 252
volume swelling of 190
water content 190–191, 210
waves 47
brain-blood temperature differentials 222
   amplified 238
brain-body temperature homeostasis 224–226
brain/hypothalamic temperature 31
brain-muscle differentials 234
brain pathology in hyperthermia 301–302
   electron microscopy 301
   endothelial cell membrane, distortion of 302
   lanthanum across the endothelial cell membrane 302
   lanthanum tracer, leakage of 302
   light microscopy 301
   loss of nucleolus 302
   naloxone 302
      neuroprotective role of 301
   nuclear membrane, irregular shape 302
   Somogyi fixative 301
   structural changes in hyperthermia 302
   synaptic damage 301
brainstem auditory evoked potential (BAEP) 397

brain temperatures 29, 34–36, 161, 204–205, 207, 211, 220, 347, 449
  affecting neuronal and neurochemical evaluations 236–238
  anesthetics and 177
  and body temperature, homeostasis of 229–231
  direct measurements of 220
  dissociation between core temperature and 206–207
  dose-dependent effects 219
  as factor affecting neural functions 223–224
  as factor for neural damage 231–233
  fluctuations, rapid 219
  fluctuations, source of 221
  increase of 418
    range of 41° to 42°C 418
  indirect measure of 230
  integrative role 223
  intraischemic 202
  local 35
  monitoring 233
  naturally occurring fluctuations of 220
  psychostimulants and 177
  rectal temperature vs. 184
  regional 176–177
  valid index of 230
branched-chain amino acid supplementation 39
β2-receptors 144
bromocriptine 83
  oral use 88
bromophenol blue 187
bronchial secretions 185
Brooks anatomical models 111
bupropion 39–40
by-product of metabolic activity 223

Ca^{2+} 76, 223, 300, 301, 305, 307, 353, 354, 356, 357, 362, 374, 378, 384, 400, 449, 450, 466
  channels 223
  imbalance 76
cabosil 267
caffeine halothane contracture tests 83
CA1 hippocampus 202
calcitonin gene-related peptide (CGRP) 277, 281, 287–288, 450
  angiotensin II 281
  cAMP response element binding protein (CREB) 281

cardiac tumor necrosis factor-alpha (TNF_α) 288
CGRP-1 281
CGRP-2 281
co-localized with SP 281
concentrations in rat brain 288
calcium release, NMS treatment 88
caloric restriction 439
calorimetry 51
camptothecin 148
cancer treatment 467
cannabinoids 177, 233
cannabinoid receptor 21 (see also CB)
  agonists 21
  signaling pathways 22
Cannabis sativa 21
capillary hydrostatic pressure 155
capsaicin receptor 21
carbon nanotubes
  administration of 247
  oxidative stress by 247
cardiac arrest 55, 164, 204
  patients, improved outcome in 211
cardiopulmonary bypass 164
cardiopulmonary resuscitation 161
cardiovascular activity 436
cardiovascular diseases 174
cardiovascular function 31
cardiovascular mechanism 46
cardiovascular strain 48
cardiovascular stress 154
cardiovascular systems 29
carotid rete 37, 114
case fatalities 153
caspases 350, 352
caspase-3 204
caspase-8 145, 148
caspase 9 148
caspase activation in heat-stressed neurons 350–353
  Ac-Asp-Glu-Val-Asp-aldehyde (DEVD-CHO) 352
  annexin V staining 350
  apoptosis 350
  caspase-3 352
    activation 352
  caspase-8 352
  caspase-9 352

caspase inhibitors Z-Val-Ala-Asp(OMe)-
fluoromethylketone(zVAD-fmk) 352
cysteine-dependent aspartate-specific proteases
350, 352
prolonged neuronal survival 352
quinoly-valyl-O-methyl-aspartyl-[2,6-difluoro
phenoxyl-methylketone (qVD-OPH) 352
signaling apoptotic pathways 352
TUNEL 350
caspase-inhibition 148
caspase-specific peptide inhibitor 147
catastrophic heat injury 31
catastrophic hyperthermia 29, 45
catechin 10
catecholamines 233
  levels 86
catechol-quinones 233
Caucasian athletes 51
Caucasian runners 51
caudate nucleus 188
cavernous sinuses 114
CB_1 antagonist, SR141716A 21
CB_2 antagonist, SR144528 21
CB_1/CB_2 agonist 21
CBF 35, 37, 38, 155, 159, 160, 165, 305, 306, 307,
312, 526, 527, 528, 529, 530, 531, 536–542
(see also cerebral blood flow)
CB_1 knockout mice 21
C57/BL6 mice 18
CB_1 receptor 21
CB_2 receptor 21
CD4 lymphocytes 329
CD8 lymphocytes 329
CD91 receptor 140
CD/1 strain study 20
cell cycle arrest 139
  cell damage 77
  cell death 311
    hyperthermic 139
    kinetics of 74
  cell growth 330
  cell injury in hyperthermia 300
    amino acid neurotransmitter 300
    antioxidant compound H-290/51 300
    Ca^{2+} influx 300
    cell death in heat stress 300
    glutamate receptors 300
    imbalance between oxidants and antioxidants
    300
    induction of oxidative stress by glutamate 300
    programmed cell death 300
cell-killing 74
  properties 139
cellular alterations in brain 245
cellular antioxidants 247
cellular breakdown 71
cellular depolarization 210
cellular fatigue 436
cellular gene expression 321
cellular homeostasis 395
cellular matrix 251
cellular mechanisms 347
  of neuronal damage from hyperthermia 350
    blebbing of processes 350
    cell swelling 350
    cellular damage 350
    cerebellar granule neurons 350
    condensed nuclei 350
    cultured neuron, killing due to long exposure
      to high temperature 350
    dorsal root ganglion neurons 350
    moderate hyperthermia 350
    producing delayed apoptotic, death in cortical
      neurons 350
    septal neurons 350
    stressing cultured striatal neurons 350
cellular membranes 139
cellular resistance, increase of 140
cellular responses 208
cellular responses to heat 325–327
  calcium concentrations, increase in 326
  cytokines 326
  cytoskeletal components disruption 326
  disruption of transcription 325
  heat shock factor 1 (HSF1) 325
  heat shock-induced apoptosis 326
  heat shock responses 325
  heat shock responses in vitro 326–327
    cell-cycle proteins p53 327
    cell-cycle proteins p21 (WAF-1) 327
    cell fate after heat shock 327
    cellular redox state 327
    cellular signaling 327
    chaperonins 327
    co-chaperons 327

Cu superoxide dismutase 327
　　degradation of proteins 327
　　degradation of proteins (ubiquitins) 327
　　disruption in RNA splicing due to
　　　temperatures 327
　　extracellular signal-regulated kinases (ERK)
　　　327
　　gene expression alteration 327
　　heme oxygenase-1 327
　　HSP 90 family 327
　　HSP 70 family lack introns 327
　　jun-N-terminal kinases (JNK) 327
　　mitogen-activated protein kinases 327
　　p38 327
　　steroid receptors 327
　　stress-activated protein kinases 327
　　Zn superoxide dismutase 327
　hydrogen ion, increase in 326
　interferons 326
　intracellular sodium, increase in 326
　intracellular stores of ATP, decrease in
　　326
　membrane permeability alteration 326
　non-lethal thermal stress 326
　prior exposure to LPS 326
　prior exposure to TNF-α 326
　pro- and anti-apoptotic pathways 326
　protein degradation, increased 326
　transcription factors, activation of 325
central blood volume, depletion of 436
central drive for exercise 45
central fatigue 29, 30–34
central nervous system 31, 33, 48, 82, 86, 92, 94,
　　95, 101, 113,139, 153, 154, 173, 175, 201,
　　202, 245, 246, 277, 295, 385, 395, 397,
　　398, 401–403, 409, 433, 434, 460, 482
　　(see also CNS)
central noradrenergic system 6
central temperature thresholds 375
central thermal afferents 45
central thermoreceptors 49
central venous pressure 436
cerebellar atrophy 175
cerebellar efferent pathways, degeneration of
　　436
cerebellar pressure cone 188
cerebellum 86, 175, 232, 403
　temperature 221

cerebral blood flow (CBF) 35, 155, 230, 233
　brain metabolism and 223
　increase in 206
　and metabolism 38–39
　regional 34
cerebral blood flow in hyperthermia, regional
　　305–307
cerebral capillaries 250
　blood-brain barrier (BBB) in 459
　endothelial cells of 249
　non-cerebral capillaries 250
cerebral circulation 176, 220
cerebral cooling 37
cerebral cortex 175
cerebral edema 163
cerebral endothelial cells 460
cerebral endothelium 249
　effect of decrease in electrical resistance 251
cerebral energy production 34–35
cerebral glucose 29
cerebral glycogen content 38
cerebral heat balance 35
cerebral heat production 35
cerebral heat release and heat production, balance
　　in 35
cerebral hypoxia 373, 436
cerebral ischemia 205, 395, 451
　gender importance 206
　hypervolemic hemodilution for 540–543
　models of 202
　oxidative damage role in pathogenesis of, during
　　heat stroke 543–544
　protective effects of free radical scavengers on,
　　in heat stroke 528–530
　Shengmai San for improvement 539–540
　transient 161
cerebral metabolic demand 155
cerebral metabolic heat production 34
cerebral metabolic rate of oxygen ($CMRO_2$) 35,
　　38, 160
cerebral metabolic uncoupling 153
cerebral metabolism 159–161, 176
cerebral microvessels 232
cerebral neuronal activity 38
cerebral neurons 417
cerebral oxygen consumption 233
cerebral oxygen uptake 38
cerebral pathophysiology 153–154

cerebral perfusion pressure 155
    reduction in 451
cerebral perturbations 30
cerebral stimulation 35
cerebral temperature 35
cerebral thermodynamic responses during exercise 34–38
cerebral vascular congestion 451
cerebral vasomotor responses and autoregulation 155, 159
cerebral vasoparalysis 155
cerebral vasoreactivity impairment 159
cerebral venous blood temperature 37
cerebral vessels, Hsp70 staining and 425–427
cerebromicrovascular endothelial cells 8
cerebrospinal fluid (see CSF)
cerebrovascular permeability 162
cerebrovascular resistance 155
c-fos 4, 6, 427
chaperoning function 139
chemical or heavy metal toxicity 397
chemical sympathectomy 6
chemotherapy 138, 154
chloral hydrate 177
chloralose 177
chlorpromazine 68, 95–96
choroidal blood vessels, lumen of 461
choroidal epithelial cells, degeneration of 459
choroidal epithelium 463
    fluid formation by 464
choroid epithelial cells 460
choroid plexus 425, 464
    blood-CSF barrier (BCSFB) in 459
    cells of 408
    hyperthermia-induced damage to 467–473
choroid plexus–CSF system 473–474
chronic destructive CNS changes 219
chronic drug use 234
chronic hepatitis C virus infection 160
chronic illness 114
chronic infectious diseases 154
chronic protein neutralization systems 502
chronic stressors 433
cingulated cortex 232
cirazoline 7
circadian rhythm 18
    COX inhibition on 18

circadian variation 113
circumventricular organs 252, 408
circumventricular organs (CVOs)
    capillaries of 460
cisplatin 148
cisterna magna 449, 463
classical heat stroke 175
climatic chamber 141
climatic warming 45
clinical brain death 164
clinical burden 174
clinical neurology 153–154
clomipramine 90–92
clonidine 7
clorgyline 90, 92
closed head injury (CHI) 385–386
clothing 112
CNS 18, 31, 34, 39–40, 86, 92, 173, 175, 202, 395
    barriers or transport interfaces 459–461
    BCSFB and cell and tissue injuries in 459
    blood-CSF barrier (BCSFB) dysfunction of, and 459
    disturbances of 114
    dysfunction, heat stress induced 295
    entry of nanoparticles, in air 249
    function at deep heating 113
    heat damage of 139
    injuries 154, 162
        hyperthermia after 201, 207, 208
    ischemia 76
    microenvironment 250
    microvessel 247
    nanoparticles in 246
    neurons 21–22
    pathology, neural damage and 220
    pattern of CNS activation 33
    serotonergic pathways in 72
    temperature increase 208
    toxicity 153, 154
cobra venom factor (CVF) 5, 6
cocaine 154, 177, 219, 233, 450
cognitive and motor function deficits in hyperthermia 308
    footprint analysis 308–310
    grid walking 308, 309
    inclined plane test 308–309

influence of naloxone on cognitive and motor
   deficits in hyperthermia 310
    blockade of opioid receptor 310
    minimum placement errors 310
   Rota-Rod performances 308, 309
cognitive deficits 245
cognitive deficits after TBI 205
cognitive dysfunction 164
cognitive effects of hyperthermia 46–47
cognitive function 46
cognitive functioning 46
cognitive impairment 47–48, 54, 245, 295
cognitive performance 48
cognitive processes 46
cognitive responses 45, 55
   heat stress to 53
cold-defense responses 20
cold environments 68
   thermoregulatory responses 22
cold exposure 20
cold-induced vasodilation (CIVD) 64
cold receptors 48
cold-sensitive (cold-activated) neurons 453
cold stress 46
cold stressing 66
cold temperature, subjective feelings of 74
collision impact 54
colloidal carrier 251
colonic temperature (T_c) 115, 126, 438
coma 114, 153, 295, 450
compensable *vs.* uncompensable heat stress
   322–323
  biophysics of heat exchange 323
  cutaneous blood flow 323
  elevation in core temperature 323
  endotoxemia 323
  heat acclimatization 323
  heat dissipation 323
  heat loads 322
  ischemia of the splanchnic circulation 323
  loss of body water 323
  magnitude of heat load 323
  outcome of heat stress 323
  reduction in circulating blood volume 323
  skin temperature 323
  translocation of bacteria 323
compensatory mechanisms, 47–48
computational dosimetry 110–111

conduction 53, 65
confusion 90
congestive heart failure 483
consciousness, altered, in NMS 82
constitutive cytosolic (c)PGES 17
constitutive expression of heat shock proteins
   399–401
constitutive microsomal PGE synthase
   (mPGES)-2 17
controlled hypothermia 207
contused brain regions 162
contusion injury, moderate 205
contusion volume
   increase 202, 205, 206
   reduction in 206
convection 53, 65, 107
convulsion 295
"cooked and coagulated" organs 492
cool environments 51
   exercise in 34
cooling blankets 210
cooling helmets 55
cooling of face 49
copulatory behavior 226–227, 228
core body temperature 224
   increase in 175
   reduction in 211
core temperature(s) 31, 35, 37, 46, 49, 52, 485
   dissociation between brain and 206–207
   heat stroke with 114
   high 51
   measurement 74
   responses 46
corpus callosum 406
cortical blood flow 207
cortical brain temperature 184
corticosterone in rodents 502
cortisol 141, 144
   in humans 502
COXs (cyclooxygenase) 6, 7, 10, 16–18, 20, 464
   fever/heat activation of 459
   inhibitors 464
COX-1 464
   knockout mice 18
   null mice 18
COX-2 464
   knockout mice 18
   null mice 18

COX enzymes
  COX-1 5, 16, 18
  COX-2 4, 5, 16–17
  COX-1 inhibitor 7
  COX-2 inhibitor 7
  COX metabolism 17
  inhibition 18
  inhibitors 17, 18
  isoforms 16
COX-independent effects, salicylates 20
COX-independent pathway 7
COX-1$^{-/-}$ mice 7
cranial compartment 190
C-reactive protein (CRP) 162
  hepatic synthesis of 498
creatine phosphokinase (CPK) 492
  serum 82
critical internal temperatures 30–34, 36, 46
critical limiting body temperature 51
critical limiting central temperature 45
critical thermal maximum 155
cross-tolerance (C-T) 373
crowded dance clubs 74
CSF 20, 83, 116, 162, 406, 449, 459
  active transport of Na and Cl in 463
  altered Na$^+$/Ca^{2+} ratio in 466
  analyses 463
  aspartate and glutamate levels in 453
  calcium concentration 453
    rise during hyperthermia 454
  cations 449
  concentration of PGE2 464
  elevation of myelin basic protein (MBP) 466
  fluid homeostasis 463
  formation 464
  GABA 449, 451
    levels 92
  heat stress stimulating Na$^+$ in 466
  homovanillic acid (HVA) 88
  hyperosmolality 472
  levels of hsp72 407
  monoamines 83
  noradrenaline (NA) level, in serotonin
      syndrome 92
  PGE levels in 466
  production rate 464
  as sensitive indicator of brain function 466
  sodium decrease in 454

substance P in 465
taurine 449
as a temperature-sensitive process 464
CSF composition
  altered 464–467
    in schizophrenia and other mental illnesses
      465–466
  in hyperthermia 466
CSF 5-HIAA 83, 88
in neuroleptic malignant syndrome (NMS) 83, 88
  in serotonin syndrome 92
Cu (Copper) 245
  nanoparticles 245, 254–255
    in air 254
    in algaecides 254
    in bactericides 254
    in fungicides 254
    human exposure 254
cutaneous circulation 113
cyclin dependent kinases (cdk) 361
  inhibitors of
    butyrolactone-1 361
    fascaplysin 361
    indirubin-30-monoxime 361
    Tat-LFG 361
cyclooxygenase enzymes (see COX enzymes)
cyclooxygenases (see COX)
cyproheptadine 96
cysteinyl LTs 20
cytochrome P-450 22
cytokines 4
  cellular responses to heat 326
  circulating 11
  and cytokine receptor knockout mice 502
  heat shock proteins and 511
  heat stroke and 481, 492–495
  IL-β 442
  inflammatory 438
    secretion, induction of 409
  injection induces heat stroke symptoms 502
  levels prior to heat exposure 497
  lymphocyte controlling 142
  modulation of DIC 490
  monoamines and 436
  neutralization 481, 502
  pro- and anti-inflammatory 494
  production 484, 512
  proinflammatory 268

pyrogenic 11, 288
  release of 208
cytokine-induced stimulation of NOS II 504
cytokine-inducing stimuli 494
cytokine-mediated heat stroke 502–503
cytokine-targeted therapy 148
cytoplasmic protein clumping 492
cytoskeletal damage to hyperthermia-induced death 361
  Bcl-2 361
  Bim 361
  c-jun 361
  cytoskeleton 142, 354, 361, 404
    activation of SAPK2/p38 mitogen-activated protein (MAP) kinase 361
    disruption of, causing apoptosis 361
    effect of hyperthermia on 361
  N-terminal kinase (JNK) 361
cytoskeletal function 210
cytoskeletal organisation 139
cytoskeletal proteolysis 165
cytosolic phospholipase A_2 (cPLA$_2$) 5
cytostatics 148
cytotoxic effects
  of hyperthermia 138–139
  of reactive oxygen species 232
cytotoxicity of lymphocytes 147
cytotoxic T-cell(s) 146
  activation 140
cytotoxic therapy for cancer treatment 467
cytotoxic T-lymphocyte 140

dantrolene 88, 96
  effects of, on MDMA-induced hyperthermia 96–97, 100
DDT 68
death 45, 69
  by hyperthermia 72
  pathways 148
  receptors 145
death-by-neglect 145
decision making 46
deep brain microthermistors 207
deep-seated tumors 154
deep venous thrombosis 205
dehydration 30–31
  and NMS 82
dehydration and hemoconcentration 489

deliberate hyperthermia 153–154, 165
delirium 114, 153, 295
demyelating peripheral neuropathy 153, 154
dentritic spines 7
depression 81, 174
depressive patients 90
desensitization of G-protein-coupled receptors 374
detection of RF-EMFs 121–122
dexamethasone 164
  replacement therapy 502
diabetes 174
diabetic neuropathy 175
diaphoresis 82, 90
diarrhea 90
diazepam 96
diclofenac 161
diffuse axonal injury 163
diffuse neuronal injury 208
dihydroxy fatty acids 15
diminished heat dissipation 154
dinitrophenol 68
discomfort 175
diseased populations 174
disseminated intravascular coagulation (DIC) 87, 490–491
  cytokine modulation of 490
disseminated intravascular coagulopathy (DIC) 397
diurnal body temperature 16–20, 22, 48
  T_b 18
  variability 48
dizziness 175
DNA damage to hyperthermia-induced death 360–361
  apoptosis
    heat-induced 361
    non-caspase mediated apoptotic death 361
    overactivation of PARP 361
  Bax translocation 361
  cyclin dependent kinases (cdk) 361
  heat-induced apoptosis 361
  inhibitors of *cdks*
    butyrolactone-1 361
    fascaplysin 361
    indirubin-30-monoxime 361
    Tat-LFG 361

PARP inhibitor DPQ (3,4-dihydro-5-[4(1-piperidinyl)butoxy]-1(2H)-isoquinolinone) 361
  p53 inhibitor pifithrin-α 361
  topoisomerase I 360
    heat-sensitive 360
  tumor suppressor p53 361
DNA internucleosomal breakages 436
dopamine (DA) 500
  agonists 88
  concentration in serotonin syndrome 94–95
  depletion 233
  precursor 88
  receptors blockade 83
  reduced striatal levels of 75
  release 526
dopamine D_2 receptors 86, 90
  blocking effect 82–83
  gene 87
  reduced affinity of 82
dopamine-hypofunction 83, 89
dopamine/noradrenalin reuptake inhibitor 39
dopaminergic D_2 activity 34
dopaminergic neurons 40
dopaminergic system 29, 39
dopamine serotonin imbalance theory 89
dorsal stratum temperature 221, 224
dorsal vagal complex 408
dorso-ventral temperature gradient 176, 177
dosimetry 70
  of RF-EMF 108
    computational 110–111
    empirical 109–110
double-strand breaks 139
doxorubicin 148
D_2 receptor antagonist haloperidol 501–502
drugs 64–66, 68–69, 74–76, 91, 148, 358
  of abuse 71, 219
  addiction 297
  alcohol addicts and 286
  anticancer 246
  development, in rodents 69
  effects of 113
  hyperthermic efficacy of 64–66
  release in brain 251
  targeting in magnetic nanoparticles 246
  thermoregulatory profile 64–66
  transport by polymer nanoparticles 251
  use 482
drug-delivery, induced 245
drug-induced alterations 219
drug-induced apoptosis 148
drug-induced heat production 234
drug-induced hyperthermia 427
drug-induced toxicity 219
drugs toxicity 64
  and hyperthermia 74–76
  magnitude *versus* duration 69–71
  patterns and ambient temperature 68–69
    homeotherms 68
    lethal endpoints 68
    non-lethal endpoints 68
    thermoregulatory set-point 68
    toxic potency 68
    Type A response 68
    Type C response 68
    xenobiotics, toxicity of 68
drug-taking behavior 219, 233
dry weight 190
dye extravasation 188
dynorphin 281, 297
dysphagia 88

ecstasy 71, 174, 233
edema 114, 395 (see also nanoparticles, brain edemaformation)
  formation 245, 246
  general sponginess and 262
  in regional brain 265
edematous swelling 232
EEG 29, 34, 47
  abnormality 88
  frequency shift 34
  stimuli induced changes in 221
EEG-RPE relation 34
effector systems 219–220
eicosanoids 15
  thermoregulatory role 15–16
electrical resistance 461
electrical responses of brain, altered 48
electrical stimulation 30
electric field vectors 108
electrochemical measurements 238
electroconvulsive therapy (ECT) 88–89
  of NMS 88–89

electroencephalogram (see EEG)
electromagnetic spectrum 108
electromyographic response 34
electron dense tracer 247
electrostatic charges 251
EL4 T-lymphoma 253
emotional factors and body temperature 113
emotional hyperthermia 223
empirical dosimetry 109–110
endocannabinoids 15, 21
endocrine disorders 277
endogenous antioxidative defenses 528
endogenous bacteria, leakage of 503
endogenous cannabinoid receptor ligands 21
endogenous cryogens 449
   down-regulation of body temperature 449
endogenous drugs 148
endogenous neuroprotectors 232
endogenous thermoregulatory mediator 16
endogenous transmitters 233
endoperoxide PGH_2 17
endoplasmatic reticulum 139
endothelial adhesion molecules 162
endothelial barrier dysfunction 503
endothelial cells 7, 17, 247
   cerebromicrovascular 8
   hyperthermia-induced destruction of 232
endothelial permeability 503
endothelium-derived relaxing factor 503
endothermy 113
endotoxemia 323, 503
   and heat stroke 506–507
endotoxemic syndrome 481
endotoxin 503
   circulating 492
   exposure 435
   levels, high 506
endovanilloids 15, 21
endovascular catheters 211
energy metabolism, impaired recovery of 165
environmental conditions 219
   brain-body temperature homeostasis under 224–226
environmental heat stress 53
environmental humidity 191
environmental overheating 219
environmental stress 46
environmental warming 229–231

enzymatic antioxidant defenses 525
enzymatic pro-oxidants 525
enzymatic protein kinases, inhibition of 210
EP 19 (see also E-prostanoid)
ependyma 460
epidemiology of heat wave mortalities 482–484
epinephrine 48, 141, 144
epithelial cells of plexus 463
EP2 knockouts 19–20
epoxide LTA_4 20
5,6-epoxyeicosatrienoic acid (5,6-EET) 22
epoxy fatty acids 15
E-prostanoid (EP) 19
   EP2 knockouts 19–20
   EP4 knockouts 19–20
   EP1 receptor antagonist 20
   EP2 receptor antagonist 20
   EP_3 receptors 5
   receptor 19
equithesin 185
errors in judgment 46
erythropoietin 382
escape behavior 114
esophageal temperature 55, 486
estradiol 425
estrogen 206
ethanol 68, 185, 434
euphoric-like state 72
Evans blue-albumin 252
   68 kDa 461
Evans blue dye (T-1824) 187
   BBB permeability to 187–188
   extravasation 165
   extravasation of 188
evaporation 53, 68
evaporative heat exchange 51
evaporative heat loss 180
   mechanisms 65, 181
evaporative water loss 64, 182
   by MDMA 72
event-related potential (ERP) 48
exacerbating inflammatory cascades 209
excessive body heating 114
excessive lymphocyte expansion 145
excitation 100
excitatory and inhibitory amino acids 295
excitatory neurotransmitter 34
excitotoxic cascades 202

excitotoxic contribution to hyperthermia-induced
    neuronal death 353–354
  activation of nitric oxide synthetase 354
  excessive glutamate release 354
  excitotoxicity 353
  febrile hyperthermia in children
    causing seizures 353–354
      producing brain damage 353–354
  impaired glutamate reuptake 354
  increasing levels of reactive oxygen species 354
  ionotropic glutamate receptor 354
    block AMPA/kainate receptors 354
    CNQX 354
    MK801 354
  overactivation of poly(ADP-ribose) polymerase
    (PARP) 354
  saxitoxin 354
  stress-induced increases in cytosolic [Ca^{2+}] 353
  tetrodotoxin 354
  voltage-dependent Na^+ channels 354
excitotoxic injury, neurons sensitivity to 210
excitotoxicity 201, 295, 353
  causing hyperthermia 353
execution 46
exercise 29, 34–38, 328–341, 534–539
  blood temperature after 230
  capacity 45
  central drive for 45
  cerebral thermodynamic responses during 34–38
  in cool environments 34
  global cerebral temperature and 35
  in heat 29, 49
  with heat stress 34
  in hot environments 36
  with hyperthermia 38
  intensity 31
  performance 40, 48
    thermal stress effect on 51
  tolerance 45
exercise-induced brain hyperthermia 177
exercise-induced hyperthermia 33
exercising muscles 34, 35
exertion, perceived 47
exertional heat illness 321–325
  adapt to heat stress 324
  cardiac output 325
  dehydration 325
  febrile illness 324

  high body fat 325
  temperate conditions 324
  viral infection 324
exertional heat illnesses 45
exertional heat injury (EHI) 323–325
exertional heat stroke (EHS) 175–176, 323–325
  patients 506
exertional hyperthermia 322
exhaustion 29, 45
exogenous hyperthermia 153
experimental hyperthermia 208
exponential cell death 140
exposure standards 54, 121, 130, 131 (see also
    safety standards, for RF-EMF)
exposure to high ambient temperature 450
external cooling processes 210
external heat stress 53
extracorporeal perfusion methods 154
extrahypothalamic sites 18
extrapyramidal symptoms 82, 91–92
extreme hyperthermia 144
extreme inflammation 145
extreme stress 145

facial cooling 55
facial fanning 37
facial skin temperature 55
facial thermoreceptor density 49
fainting 175
fanning of face 55
fans 162
Fas-associated death domain (FADD) 145
Fas death receptors to hyperthermia-induced
    damage, contribution of 360
  apoptosis 360
  caspase-3 activation 360
  death-inducing signaling complex (DISC) 360
  Fas receptor-mediated death 360
  FLIP, downregulation by heat stress 360
  heat-induced apoptotic mechanisms 360
  procaspase-8 360
Fas-receptor 145
fatigue 114
  after marathon run 229
  developed during intense exercise 230
  development of 34
  sensations 54
fats, breakdown of 492

fatty acid oxidation 15
febrigenic PGE$_2$ 17
febrigenic process 11
febrile episodes 205, 210
febrile periods 201
febrile responses 4, 6, 10, 17, 19, 162, 402
febrile state 408
febrile temperatures 17
female patients 202
fentanyl 145
fever 5, 15, 91, 201, 204, 220, 449, 450, 464, 512
    (see also high fever)
  adverse effects 208
  after stroke and neurologic mortality 208
  epidemiology of 161–162
  generation of 20
  Gram-negative bacteria 3
  induced by injection
    of E. coli 453
    of PGE$_1$ 453
    of PGE$_2$ 453
  and inflammation following CNS injury 162
  in neurologically injured patients 161
  and neurologic outcome in patients 162
    anoxic brain injury 164
    cognitive dysfunction 164
    non-traumatic subarachnoid hemorrhage 163
    stroke 162–163
    traumatic brain injury 163
  NO in 10, 11 (see also Nitric oxide)
  pathogenesis 17
  pathophysiology 17
  poor outcome in stroke patients with 208
  prevalence of 208
  producing center 3
  production 7
  prolonged high 176
  in serotonin syndrome 90
  spontaneous 153
  in subarachnoid hemorrhage 208
  as a symptom of heat stroke 488
fever-inducing substances 453
"fever-like" core temperature 489
fever-like hyperthermia 139
fever-range whole-body hyperthermia (FR-WBH) 144
fibrinolysis or anticoagulation, impairment of 490
fibroblast growth factor (FGF$_2$), basic 463

finite-difference time-domain (FDTD) method 110–111
firefighting operations 53
fires 154
flat body posture 90
flow-metabolism coupling 155
fluctuating homeostatic parameter 220
fluid percussion (F-P) brain injury 205
fluorouracil 148
fluoxetine 76, 90, 92
focal ischemia 202, 203
focal neuronal injury 208
foot-print analysis 258
  bromophenol blue 258
  stride length 258
forearm blood flow 55
forepaw treading 90
free radicals 9, 503
  formation 76, 247
  generation 201, 209
    by nanoparticles 250
  production 10, 492, 525, 526
    hyperthermia and 210
frontal cortex 47
fundic mucosa 184

GABA 86, 449
  and glycine 295
  levels 449
GABA$_A$ or GABA$_B$ receptors agonists 450
GABAergic system 86
GABA-ergic transmission, increase of 451
GABA mimetic agents 86
gamma-aminobutyric acid (see GABA)
gaseous transmitter 10
gastric lavage with cold saline 162
gastric ulceration 184
gastrointestinal (GI) barrier function
  ischemia and increased vascular permeability 503
  permeability and heat stroke 503–506
  physical and physiological disruptions of 503
γδ-T-cells 143
gender, importance 202, 206
gene expression
  by cells at barriers 463–464
  changes in PBMCs 334
  responses in EHI 334–341

genotoxicants
  bleomycin 74
  paraquat 74
Glasgow Coma Scale 163
glia 7
glial cells 7, 232
  morphological changes in 154
glial cell types 417
glial fibrillary acidic protein (GFAP) 262, 420
glial pathology 175
glial/vascular expression 417
global cerebral metabolic activity 230
global cerebral oxygen and glucose uptake, ratio between 38
global cerebral temperature 35
global dementia 347
global genomic response 373
  to heat acclimation 378–379
global ischemia 202–204
global warming 174
glucocorticoid hormones 502
glucocorticoids (GC) 148
glutamate 10, 233, 295, 449
  neurotransmission 210
  overloading of 543
  receptors 88
  release 526
glutamate-induced neurotoxicity 232
glutamatergic excitation 451
glutaminergic pathways 10
GM-CSF 409
golden hamsters, exposure to RF radiation 69
gold particles 248 (see also Au)
  silver-coated colloidal 248
  translocation of 248
G-protein-coupled receptors 19
gradual rewarming protocol 207
Gram-negative bacteria 116
granular cell layer 403
granzyme B 147
gray matter oligodendrocytes express Hsp70 420
gross motor behavior, depression of 450
guinea pigs 5
gut microbes 17

habituation, repeated 221
haloperidol 82
  in animals 90

halothane 177
handling of animals 178–179
handling stress 178
hand temperatures 47–48
headache 175
head cooling 55, 56
  active 55–56
head-injured patients 201, 211
  severe 205
head injury 162, 207
head skin temperature 37
head surface temperature 56
head temperature 54
head weaving 90
heart rate (HR) 50, 53, 113, 178, 534
  decreased 55
heat acclimation 373
  increasing HSP reserves 439
  modulates the heat shock response 439–440
heat acclimatization 323, 483
heat accumulation, progressive 229
heat balance 35, 51, 53
heat chambers 179, 180
heat collapse 482
heat conservation 8
heat cramps 277, 482
heat death 45
heat disorders 277
heat-dissipating mechanisms 175, 373
  interferring with 450
heat dissipation 49, 51, 186, 322
  diminished 154, 219
  drug-induced impairment of 234
  insufficient 450
  mechanisms 181
  potential 54
heat endurance 373
heat exchange 53
  catheters 211
  evaporative 51
  regulation 71
heat exhaustion (HE) 175, 278, 323, 482
  state 181, 182
heat exposure 512
  BOD incubator for 179
  body temperature changes during 180
  body weight loss after 182
  brain dysfunction after 187

cardiovascular response to 503
    increase in skin blood flow 503
cytokine levels prior to 497
duration of 483
guidelines 45
prolonged 489, 492
heat health warning systems 174
heat illness 435
    continuum 482
    exertional 45
    risks 50
        factors 324
    severity of 324
    syndrome 482
heat-induced apoptosis 139, 361
heat-induced brain 190, 500
    damage 191
    edema, symptoms 190
    injury 232
heat-induced cell death 139
heat-induced damage, of brain tissue 232
heat-induced hypoglycemia 489
heat-induced hypothermia 487
    paradoxical decrease of core temperature 487
heat-induced responses, age-associated alterations 512
heat-induced tissue damage 491
heating 33
heat injury, potentially 45
heat loads 54, 322
    magnitude of 323
heat loss 450
    mechanisms 18
        breakdown of 113–114
        impairment of 71
    restriction 219
heat production 20, 50, 51, 63, 64, 233, 322, 450
    body and brain 234
    cerebral heat release and, balance in 35
    from dancing in warm environment 74
    excessive 51
    internal 29
    intra-brain 219
    maximal potential 229
    MDMA effects on 72
    metabolic 115
    by organism 219
    rate of 35

heat prostration 482
heat-related deaths 153, 174–175
    cause of 176
heat-related disorders 277
heat-related illnesses 173, 174, 178
    brain edema in 188–191
heat-related morbidity 173
heat-related mortality 174
heat release 223
heat removal 34, 35
    from brain 230
    impaired 35
heat resistance 71
heat "shock" 486
heat shock elements (HSE) 402
heat shock factor 1 (HSF1) 435
heat shock factors (HSF) 139, 402
heat shock 27 kDa protein-1 (HSPB1) 333
heat shock paradigms 486
heat shock preconditioning 525
    effects of, on oxidative stress 531–534
heat shock proteins 139–140, 232, 321, 409, 417, 433–435, 507–508 (see also HSP, or Hsps)
    in cardioprotection 437
    constitutive expression of 399–401
    and cytokines 511
    cytoprotective role of 435
    expression 395, 512
        alterations in 508–510
    extracellular release of 409
    function 512
    gene expression 509
    Hsp70 417
    immunological features of 140
    inducible 72 kDa 397
    interaction with integrative physiological response 435
    72 kDa, highly inducible 399
    low molecular weight 407–408
    overexpressed 403
    protection, mechanism of: inhibition of cytokine transcription 511–512
    protective role of 434
    and protein folding 508
    secreted 407–408
    synthesis 140, 208
        inhibition of 140

and systemic integrative responses 433
and thermotolerance 439
and viral neurovirulence 408–410
heat shock response 433
heat shock response element (HSE) 435
heat shock response in animal models 328
    JNK pathways, acquisition of thermotolerance 328
    non-thermal stresses, increased tolerance to ischemia-reperfusion 328
        oxidative stress 328
heat shock responses and moderate hyperthermia 328
    exercise hyperthermia 328
    exercise-trained rats 328
    reduced mortality 328
heat shock treatment 512
heat storage 31, 46, 51
heat strain 46, 47, 53, 483
    alcohol abuse 483
    aspirin 483
    circulatory collapse and death 483
    diuretic medications 483
    high death rate 483
    hypertension 483
heat stress 17, 29, 31, 35, 45, 46, 49, 175–176, 177–178, 245, 277–278, 295, 323, 417, 449, 464, 486 (see also HS)
    acute 374
    animals handling prior to 178–179
    behavioral response to 48, 53
    classical 502
    cognitive responses to 53
    countermeasures 53–56
    effects of 45
    environment 49
    exercise with 34
    handling of animals prior to 178–179
    indices 53
    influences anesthesia 184
    mild 436
    pathophysiological states 322
    perceived 54
    perception of 48–50
    post-mortem symptoms of 184
    psychological components of 186
    responses 512
        indicators of 418
    survival in IL-6 knockout mice, decreased 503
    and thermal load 180–181
    uncompensable 51
    without heat stroke 181
heat-stress-induced hyperthermia on CSF
    calcium, sodium, potassium and magnesium 453–454
    taurine, GABA, aspartate and glutamate 451–452
heat stroke 113–115, 175–176, 324, 418, 481, 482
    aging and 512–513
    animal models of 485, 499
    bacterial translocation 484
    changes in immune status 484
    cooling method of 497
    cytokine-mediated 502–503
    cytokines in 481
    effects of heat shock preconditioning on oxidative stress 531–534
    endotoxemia and 506–507
    fever as a symptom of 488
    and found cardiovascular illness 483
    gastrointestinal (GI) permeability and 503–506
        endotoxemia with 503
    heat stress without 181
    and hyperthermia 435–437
    hypothalamic levels of IL-1β in 499
    IL-1 and 499–502
    IL-6R in 497
    leukocyte distribution 484
    military recruits with exertional 497
    morbidity, mitigation of 481
    mortalities in elderly 483
    oxidative stress and ischemic injuries in 525
    pathophysiology of 435–436
    patients, exertional 506
    permanent neurological damage in 483
    production of cytokines 484
    symptoms 502
    thermoregulatory consequences of 486–489
    victims 436
heat stroke-induced hyperthermia 525
heat stroke syndrome 435, 436
heat tolerance 504
heat transfer gradient 55
heat treatment 138
heat waves 173
    epidemiology of, mortalities 482–484

heavy metals and toxins 434
helmet use 54–55
hemispheric cerebral infarction 163
hemodynamic alterations 208
hemodynamic changes 489–490
hemorrhage 397
hemorrhagic stroke 163
hepatic synthesis of C-reactive protein (CRP) 498
Hepatitis B Virus 409
heroin 177, 219
Herpes-Simplex Virus 2 409
high blood pressure 155
high brain temperatures 220
high-dose steroids 148
high environmental temperature 525
high fever
  in NMS 82, 87
  in serotonin syndrome 92, 94
high temperature environment 90
high temperature *per se* 232
hippocampal fissure 406
hippocampal neurons 403
hippocampus (Hippo) 175, 224, 408
histaminergics 174
histopathological outcome 202, 207
homeostasis 145
  body and brain temperature 229–231
  thermal 66
homeostatic parameter 220
homovanillic acid (HVA) 83
  levels 86
horseradish peroxidase (40 kDa) 461
host-derived mediators 17
hot air 154
hot conditions 51, 229
hot-dry environment 54
hot environments 51, 154
  exercise in 36, 45
  exposure to 45
  prolonged exercise in 34
  working in 46
hot-humid environment 54
hot peppers 22
hottest summer month, July 484
  protective clothing 484
HS- and febrile-animals (see heat shock)
  calcium content increase in 449
  CSF sodium decrease in 449

Hsc70 400
  73 kDa isoform, constitutively expressed 399
HSF (see heat shock factors)
HSF1 monomer 435
HS-induced hyperthermia 449
hsp27 407 (see also HSP, or Hsps)
hsp32 407
HSP 60 139
Hsp70 139, 175, 399, 420
  accumulation 435
  in brain
    regions 441
    vessels of hyperthermic female 425
  in cerebral vascular cells 425
  in choroid plexus 425
  circulating estrogens and 425
  in CNS tissues, inducible form of 436
  cytoplasmic relocation of 420
  distribution of 417
  epitopes 140
  exogenous 406
  expression 418
    after hyperthermia 427
    arsenite-induced brain patterns 417
    in oligodendrocytes 417
  following heat stress 423
  gray matter oligodendrocytes expressed 420
  in hyperthermic brain 418
  and immune system, link between 433
  increasing plasma osmolarity and 427
  induction in astrocytes 417
  less staining in capillaries 425
  levels 433
  mRNA 420
  mRNA after heat stress 427
  neuroprotective role of 437
  nuclear presence of 423
  nuclear translocation of 420–423
  oligodendrocytes as major producers of 418–420
  protein in brain arteries 425
  SP 71 420 (see also Substance P)
  staining of endothelial nuclei 425
    indication of vascular cell stress 425
  stress-inducible form 140
hsp-70 210

Hsp72 383, 400–402, 407, 532, 538 (see also HSP, HSPs)
  accumulation in liver 439
  CSF levels of 407
  fever-induced expression of 410
  intracellular 442
  in neurons 409
  overproduction of 525
  serum levels of 407
  stress-inducible cytosolic 508
  and viral core particle 408
Hsp90 139, 401, 407
  protecting neurons from thermal stress 407
HSP100 139
Hsp27 and Hsp32 401
HSP/antigen complexes 409
hsp72-dependent, increases in viral antigenic burden 408–409
hsp72 expression 408
HSP70 family 435
  prevents cellular damage 435
  prevents protein damage 435
HSP70 gene product 440
(HSP70–HSF1) complex 435
hsp27 levels, expression after heat shock 408
hsp72-mediated enhancement of viral gene expression 409
5-HT 39, 91, 92, 94, 95, 500, 501 (see also 5-hydroxytryptamine, serotonin)
  and IL-1 in the mediation of heat injury 501
  precursors 92
  in regulation of BBB permeability 501
  re-uptake inhibitor 39
5-HT$_{1A}$ receptors 91
  agonist 92
  antagonist 96
  blocking effect 96
  hyperstimulation of 91
5-HT$_{2A}$ receptors 91
5-HT$_{2C}$ receptor 39
5-HT receptors 91
  agonists 94
  antagonists 95
human(s)
  brain, structural integrity and function of 155
  cerebral pathophysiology and clinical neurology of hyperthermiain 153–165
  exposure 46

gene expression 321
  oxygen consumption 229
  response, of MDMA 74
human heat stroke 495–499
  diabetes, cardiovascular abnormalities 496
  during the Hajj 495
  hot, arid desert environment 496
  life-threatening effects of 496
human recombinant IL-1β 450
humid conditions 229
humid environments 51, 114
HVA levels 86
hydrocephalus 207
8-hydroxy-2-(di-*n*-propylamino) tetralin (8-OH-DPAT) 92
6-hydroxydopamine (6-OHDA) 501
  increases striatal blood flow 501
5-hydroxyindoleacetic acid (5-HIAA) 83
hydroxyl radical(s) 525
  scavengers 525
5-hydroxytryptamine 39 (see also serotonin)
5-hydroxytryptamine (5-HT$_{2C}$) 39
  receptor blockade 39
hyperalgesia 282
hyperemia, decrease in 206
hyperglycemia 163
hyper- or hypoglycemia 489
hyperoxia 373
hyperpyrexia 175
hyperreflexia 90, 100
hypertension 174, 483, 507
  intracranial 451
  in serotonin syndrome 91
hyperthermia 16, 29, 31, 45, 52, 53, 113, 175, 208, 220, 321, 347, 418, 449, 450, 490, 512
  amphetamine use 347
  anti-cancer strategies increased by 348
  bacterial endotoxin-induced 174
  during Biblical times 153
  blood-cerebrospinal fluid barrier in 459
  brain
    activity and 47
    cancer treatment 348
    damage by 347
    dysfunction and 173
  breakdown of fats by 492
  cellular and molecular effects 139
  cerebrovascular effects of 155

CSF composition in 466
cytotoxic effects 138–139
developing brain and 347
developing nervous system, vulnerability to 348–349
  anencephaly 348
  anterior neural tube defects in mouse embryos 348
  cardiac development 348
  children of spina bifida, mothers of 348
  cortical gray matter, thickness reduction in 348
  cranial neural tube defects with facial cleft 348
  embryonic brain damage due to exposure to high temperatures 348
  embryonic death 348
  exencephaly 348
  growth retardation 348
  hyperthermia during pregnancy 348
  maternal hyperthermia in guinea pigs 348
  mental retardation 348
  microencephaly due to heat exposure during early neurogenesis 348
  neural tube defects in guinea pigs 348
  neural tube defects in pregnant women 348
  neuronal apoptosis due to 12 min exposure to 43°C 348
  neuronal migration 348
  neuronal progenitors 348
  teratogenic hyperthermia 348
drug-induced 427
and drug-induced toxicity 219
and drug toxicity 74–76
effect on immune system 140
effects of dantrolene on, induced by MDMA 96–97, 100
exercise with 38
extreme 144
in humans 74
hyperthermia-induced damage 347
  denaturing of nascent polypeptide chains 347
  endoplasmic reticular stress 347
  mammalian neurons 347
  mitochondrial damage 347
  in non-neuronal cells 347
  nuclear and cytoskeletal damage 347
incidence, higher 208
incidence of severe heat waves 347
incidents of, in athletes 347
increased catecholamine and cortisol release 490
increased lymphocytes 490
increases glutamate extracellular concentrations 451
induced 137
induce Hsp70 in systemic arteries 425
induction models, 450
infrared exposure to induce 186
in-vitro effects on lymphocytes 140
life-threatening 96
malignant and NMS 83, 86
management of 210–211
necrotic neuronal death 347
neuronal damage with, increased 164
neuronal injury by 347
neurotransmitter release 210
passively induced 30
postoperative 164
post-traumatic 205, 206
problems of 296
  global warming 296
  heat-related mortality 296
  heat waves, frequency and intensity of 296
prolonged therapeutic 153
provoke of 20
quantum dots in breast tissues, alteration following 246
rise in CSF calcium concentration 454
risk of neuronal injury 348
soldiers in hot environments 347
temperature and duration of exposure 347
temperature coefficients and 77
T suppressor-cytotoxic cells 490
whole-body, by RF-EMF fields 108
hyperthermia exacerbates post-ischemic neuronal death 349–350
  acetaminophen 349
  alterations in blood-brain barrier permeability 349
  anti-pyretic treatments 349
  blood platelets 349
  brain temperatures of 40°C 349
  cardiac arrhythmia 349
  cooling brain below 37°C 349
  cytoskeletal integrity 349
  elevated body temperature during stroke 349
  excitotoxic glutamate release 349

free radical production 349
hypotension 349
increased risk of pneumonia 349
ischemia-induced energy stress 349
post-ischemic hyperthermia 349
hyperthermia/heat stroke 503
hyperthermia induced brain damage 468
  combination of amphetamines 348
  fever caused by infection 348
  neuroleptic malignant syndrome
    neuroleptics producing 348
  neurological deficits 348
  pathological changes in brain 348
  physical activity 348
  potential brain injury 348
  warm environment 348
hyperthermia-induced brain pathology 295
hyperthermia-induced caspase activation 354
hyperthermia-induced damage to choroid plexus 467–473
hyperthermia-induced fatigue 29, 30
  neurohumoral responses and 39–40
hyperthermia-induced hyperventilation 38
hyperthermic brain injuries 154
  pathophysiology of 311
hyperthermic cell death 139
hyperthermic death 66, 72
hyperthermic drugs 63, 71
hyperthermic effects
  of drugs 64–66
  of MDMA 72
hyperthermic exercise 35, 38
hyperthermic-induced toxicity 63
hyperthermic induction of heat shock proteins 401–408
hyperthermic level of heat stroke development 435–436
hyperthermic mechanisms of damage 208–210
hyperthermic pre-conditioning 403, 408
hyperthermic rats 164
hyperthermic response
  after CNS injury 208
  ambient temperature and 66–68
hypervolemic hemodilution 540–543
hypocomplementemia 5
hypofibrinogenemia 397
hypofunction of dopamine 83, 89
hypomania 90, 100

hypotension 91, 349, 502, 525
hypothalami 8
hypothalamic dysfunction 163
hypothalamic hormones 29
hypothalamic levels of IL-1β in heat stroke 499
hypothalamic neurons 22
hypothalamic PGE_2 18
  concentrations 20
hypothalamic thermoregulatory centers irritation 208
hypothalamo-pituitary axis (HPA) 144
hypothalamus 17, 18, 94, 208, 232, 408
  anterior 499–500
  interleukin-1β+/ED1+ cells within 209
  thermoregulation coordination 347
hypothermia 16–18, 74
  accidental 154
  in animal 65
  bcl-2 and 77
  induced 487
  mechanisms exacerbating toxicity of drugs and other agents 77
  rapid rewarming after 207
hypothermic cardiopulmonary bypass, rewarming after 154
hypothermic effects, in mammals 21
hypothermic protection after injury 207
hypoxia, in NMS 82
hypoxia-ischemia 204
hypoxic brain injury 153

ibuprofen 18, 210
ICNIRP (see International Commission on Non-Ionizing Radiation Protection)
IEEE C95.1 (IEEE Standard for Safety Levels with Respect to Human Exposure to Radio Frequency Electromagnetic Fields, 3 kHz to 300 GHz) 107, 129–130
^{123}I-iodo-benzamide (^{123}I-IZM) 86
IL-1
  and heat stroke 499–502
  systemic injection of 502
IL-6 329
  anti-inflammatory cytokine 329
  circulating 330
  knockout mice, decreased heat stress survival in 503
  mRNA 330

IL-12 409
IL-1β 4–6
  endogenous pyrogen 449
Il-1β 409
IL-1β-induced fever 453
IL-1-induced fever
  on CSF calcium, sodium, potassium and magnesium 454–455
  on CSF taurine, GABA, aspartate and glutamate contents, effect of 453
IL-10 knockout mice 506
  treatment with an anti-TNF antibody 506
IL-6R in heat stroke 497
imipramine 91
immediate early genes, expression of specific 210
immune cells 21
immune dysfunction 490
immune regulation 144
immune system 17
  effects of hyperthermia on 137, 140–144
immunity, development of 145
immunocompromised populations 482
immunologically induced apoptosis 145–148
immunological mechanism 46
immunosuppression 144
impaired heat dissipation 220
inactivation of neuropeptides
  aminopeptidases 282
  angiotensin-converting enzyme 283
  carboxypeptidases 282
  inactivation of bradykinin 283
  metallo-endopeptidases 282
  neuropeptide degradation 283
incontinence, in NMS 82
incoordination 90
indomethacin 17, 20, 161, 211, 464
induced hyperthermia 137
  passively 30
inducible nitricoxide synthase (iNOS) 382
induction of fever and hyperthermia 450–451
induction of hsp72 408
infarct size 163
infarct volume 164, 202
infection(s)
  prevalence of 208
  from respiratory and urinary etiologies 208
infectious agents 174
infectious diseases 153

inferior vena cava 5
inflammation 201
  after CNS injury 162
  mediators of 20
inflammatory cascades 202, 210
inflammatory cell accumulation 205
inflammatory cytokines 438
  secretion, induction of 409
inflammatory mediators 162, 341–342
  inhibitor I-kappa-B 342
  NF-kappa-B 342 (see also NF-κ-B)
    dependent transcription 342
    inhibition by heat shock 342
inflammatory processes 209
inflammatory response 155
information processing 46
infrared exposure 186
infrared lamp 186
infrared radiation 154
inhalational anesthetics 83
^{123}I-N-isopropyl-p-iodoamphetamine (^{123}I-IMP) 86
injured brain 155, 161
injury cascades 208
iNOS 382 (see inducible Nitric oxide synthase)
inositol-(1,4,5)-trisphosphate 455
intact females 206
intense cycling 229
intense thirst 175
intensive care units 201
inter-endothelial routes 251
Interferon-gamma (IFN-γ) 142
interleukin-1β 526
interleukin-1β+/ED1+cells 209
interleukin(s) (IL) 162
  IL-1 142
  IL-6 142
  IL-8 142
  IL-10 142
  IL-12 142
  increased plasma levels of 490
  soluble interleukin-2 receptors (sIL-2R) 143
internal body temperatures 29, 113
internal cooling, method 211
internal temperatures 180
  high stability 219
International Commission on Non-Ionizing Radiation Protection (ICNIRP) 107, 129

inter-neuronal communication 223
intestinal barrier permeability, increased 504
intestinal blood flow 503
intestinal flora 17
intestinal oxidative and nitrosative stress 503
intestine 223
intra-brain heat accumulation 230
intra-brain heat production 219
    balance of 220
intra-brain origin 219
intra-brain thermogenesis 230
intracellular calcium 328
intracellular glutamate concentrations 210
intracellular receptors 15
intracerebral hemorrhage 153, 154, 163, 207
    fever and outcome after 208
intracerebral pressure 347
intracerebroventricularly (i.c.v.) 449
intracranial hypertension 153, 154, 451, 529
    rebound 207
intracranial pressure (ICP) 205, 529
    increased 207
intracranial volume homeostasis 205
intracytoplasmatic membranes 139
intraischemic brain temperature 202–203
intraischemic hyperthermia 209
intraportal vein injection 5
intravascular disorder, systemic 490
intravenous salicylates 17
intraventricular blood 163
intraventricular hemorrhage 208
intrinsic thermal resistance 395
in vivo microdialysis 238
iodine 259
ionic channels 223
irradiation on skin 121
irreversible neuronal injury 208
irritation 100
ischemia 397
ischemic brain injury 153
ischemic depolarizations 165
ischemic glutamate release 210
ischemic injuries, in heat stroke 525
ischemic neuronal injury 164
isokinetic contractions 29
isoleukotrienes 15
isometric contractions 29, 33
isoprostanes 15

janus kinases phosphodiesterase (JANKa) 381
jugular venous blood 35
jugular venous to arterial temperature difference 35

K^+ (Potassium) 449, 453–455
ketamine 185, 233
ketanserin (KET) 76, 96, 500
killer heat waves 173
κ-opioid receptor 311,312

labile blood pressure 100
lack of sleep 114
lanthanum 247
    chloride solutions, buffered 461
lateral cerebral ventricle 529
LDL receptor-mediated transcytosis 251
L-dopa 88
leptomeninges 188
lethal body temperature, in mammals 71
lethal catatonia 88
lethal heat stress 71
lethal hyperthermia 72
leukocyte
    adhesion, increase in 505
    E- and P-selectin reduced 506
    recruitment, inhibition of 505
leukocytosis 163
    in NMS 82
leukotrienes (LTs) 15, 20
LTB_4 20
LTC_4 20
LTD_4 20
life-threatening complications 219
life-threatening hyperthermia 96
lifetime disabilities 173
lipid
    damage 77
    mediator 4
    peroxidation 63, 76, 525
    rafts 400
lipid insoluble non-electrolytes 460
    restricted at barrier sites 460
lipid insoluble proteins 460
    restricted at barrier sites 460
lipid-soluble solutes 460

lipopolysaccharides (LPS) 142
  bacterial endotoxic 3–6
    mononuclear phagocytes 3
    pyrogenic signal 6
liposomes 245
  nanoparticles 246
lipoxins 15
lipoxygenase 21
  12-lipoxygenase enzymes 21
  15-lipoxygenase enzymes 21
  pathways 15
  products of AA 20
5-lipoxygenase, activating protein 20
5-lipoxygenase-catalyzed oxygenation of AA 20
lithium carbonate 91
lithium chloride treatment 403
liver 11, 223
  damage 174
  HSP72 accumulation in 439
  injury 492
L-NAME (N(G)-nitro-L-arginine methyl ester) treatment 504
local brain temperature 35
local hyperthermia 154
localized head cooling 211
locomotive systems 29
locoregional hyperthermia 138
long term heat acclimation (LTHA) 374
loss of sensation 450
lower critical temperature 64, 65
low molecular weight antioxidants (LMWA) 385
LPS (see lipopolysaccharide)
L-selectin 144
lumbar puncture 83
luminal and abluminal membranes 463
Luxol Fast Blue staining 262
lymphocyte apoptosis and therapeutical hyperthermia 144
  drug-induced apoptosis 148
  immunologically induced apoptosis 145–148
  stress-induced apoptosis 144–145
lymphocytes 140
  apoptosis of 145
  controlling cytokines 142
  cytotoxicity of 147
  function 140, 144–145
  in-vitro effects of hyperthermia 140
  shift 144–145
  transient impairment of, during WBH 142–143
lysergic acid diethylamide-induced hyperthermia 420
lysergic acid diethylamide (LSD) 401
lysosomes 139

macromolecular transport, vesicles for 249
magnetic field vectors 108
magnetic resonance image (MRI) 111
magnetite cationic liposome nanoparticles (MCLN) 253
magnolol 525, 528–530
major histocompatibility complex (MHC) 140
male patients 202
male rats 206
malignant catatonia 88
malignant diseases, treatment of 154
malignant hyperthermia 83, 86
malignant hyperthermic state 234
manganese nanoparticles, inhalation of 249
mannitol 461
manual dexterity, loss of 48
MAO-A inhibitor 92 (see also Monoamine oxidase)
MAO-B inhibitor 92, 94
MAP and JNK kinase pathways in hyperthermia-induced neuronal damage 361–362
  glycogen synthase kinase 3 (GSK-3) 362
  IRE1-α stress sensorprotein 362
  JNK
    activation-induced apoptosis 362
    phosphatases, inhibition by heat stress 362
  TNF receptor-associated factor 2 (TRAF2) 362
marijuana 174
maximal isometric contractions 30
maximal voluntary contraction (MVC) 31
maximum permissible exposure levels (MPEs) 108
MCP-1 (monocyte chemoattractant protein) 409
MDMA (see 3,4-methylenedioxymethamphetamine)
MDMA-induced hyperthermia 96–97, 100
mean arterial blood pressure (MABP) 255
mean arterial pressure (MAP) 500
mean body temperature 49
mechanical deformation 328
mechanical ventilation 402

medial preoptic area of hypothalamus (MPAH) 224
medial thalamic neurons 238
medulla 232
memantine 96
membrane alterations 139
membrane blebbing 139
memory loss 347
menstrual cycle 18
mental diseases 174
mental errors 54
mental function, impairments in 45
mental illness 153, 482
   cause of 176
mental performance 54
mental retardation 82
mental status change 90
meperidine 90
metabolic acidosis, in NMS 82
metabolic activation 233
metabolic alterations 208
metabolic heat dissipation 229
metabolic heat production 46, 48, 115, 322
metabolic rate 64, 74
   of MDMA 72
metabolic specialists 71
metabolic thermogenesis 66, 68
metabolism and cerebral blood flow 38–39
metastatic disease 154
METH (see meth-amphetamine)
meth-amphetamine (METH) 154, 174, 219, 233
   brain hyperthermia induced by 233–236
   intoxication 234
methamphetamines 75
3,4-methylenedioxymethamphetamine (MDMA) 63, 91, 219, 233
   autonomic thermoeffectors in 74
   body temperature regulation and 72
   brain hyperthermia induced by 233–236
   effects of dantrolene on hyperthermia induced by 96–97, 100
   effects on heat production 72
   human response 74
   hyperthermic effects of 72
   mortality and 72
   overdose exposures to 72
   rodent response 71–74
   stimulation 72
      of motor activity 72
   and therapeutic treatment of psychoses 72
   thermoeffector responses of rats with 72
   thermoregulatory mechanisms, disabling of 72
   vasoconstriction by 72
α-methyl-p-tyrosine (AMPT) 76
Mg^{2+} (Magnesium) 449, 453–455
mice, exposure to RF radiation 69
microarray studies 327–328
microdialysis, of NE 6
microglia 417
microglial marker OX-42 420
microhemorrhages 179
   in stomach 184
microperoxidase (20 kDa) 461
microsomal PGE synthase (mPGES)-1 4, 5, 17
microthermistors, deep brain 207
microtubule-associated protein-2 204
microvascular function 209
microwave field 107
middle cerebral artery mean blood velocity (MCA V_{mean}) 38
mild hyperthermia 201, 202, 205
   by RF-EMF fields 107
mild hypothermia 74, 205
military
   exercise 267
   and non-military personals 267
   operations 53
   recruits, with exertional heat stroke 497
mining operations 53
MIP-2 (macrophage-inflammatory protein-2 (MIP-2)) 409
missing antigen-receptor 145
mitochondria, irreversible damage of 236
mitochondrial damage causing hyperthermia-induced neuronal death 355–357
   ATP production 356
   Bcl-2-associated X protein (Bax) 356
   Ca^{2+-}activated protease calpain 356
   Ca^{2+} cytosolic elevation 356
      activation of Ca^{2+}-dependent proteases 357
      contributing to excitotoxic damage 357
   calcium sequestration 356
   disruption of Ca^{2+} transport 356
   increased intracellular Ca^{2+} due to heat stress 356

increased permeability of mitochondrial inner
    membrane 356
  mitochondrial depolarization 356
  mitochondrial permeability transition pore
    (MPTP) 356
  reduced mitochondrial respiration 356
mitochondrial energy metabolism 145
mitochondrial κATP dependent channels 434
mitochondrial membranes 232
mitochondrial oxygen tension 38
mitomycin 148
mitotic apparatus 138
MK-0663 7
MK 801 96
MK-886 21
moclobemide 91
modulator 11
molecular chaperone protect against hyperthermia
    359–360
  ATP for refolding proteins 359
  dorsal root ganglion neurons, protection from
    heat stress 359
  heat shock proteins (HSPs), increase in
    expression of 359
  hippocampal neurons 359
  HSP10 359
  HSP40 359
  HSP60 359
  HSP70, overexpression 359
  HSP12A 359
  HSP12B 359
  misfolding of nascent polypeptide chains 359
  neuronal populations 359
  protein stabilization within mitochondria 359
molecular chaperones 139, 417
monoamine metabolism 83
monoamine overload 347
monoamine oxidase (MAO) inhibitors 90–92
monoaminergic systems 277
monocyte stimulation by WBH 140–142
monohydroxy fatty acids 15
mononuclear phagocytes 3, 4, 251
μ-opioid peptide (MOP) receptors 281
morbidity 163
morphine 174, 177
mortality 163
  and MDMA 72
  mild hyperthermia and 205

motor activity 29
motor cortex 30
motor deficits 245
α-motor drive 29
α-motor neurons 29
motor planning 46
motor-unit recruitment 34
MPAH (see medial preoptic area of
    hypothalamus)
mucosal wall of stomach 184
multi-organ dysfunction 176, 347, 397
multiple injury mechanisms 202
multiple sclerosis (MS) 55, 462
muscle 29, 30, 31, 82, 83, 226, 278, 328–329, 441,
    534
  activity, strenuous 482
  chemoreceptors 33
  metabolism 31
  metabolite 31
  metaboreceptors 33
  relaxants 185
  rigidity 82, 91
  temperatures 46, 226, 227
  tone, decreased 450
myelin 252, 262, 263, 266, 301, 302, 471–473
  pallor 175
  pathology 175
myelin basic protein (MBP) 232, 262, 301, 303, 466
myocardial function, depressed 502
myoclonus 90, 100
myoglobinuria 82

NA (see noradrenaline)
Na$^+$ (Sodium) 449, 453–455 (see also Sodium)
Na$^+$ channels 223
naloxone 295, 297, 300, 302, 311
  neuroprotective role of 301
  treatment 300–301
    cellular stress response 301
    δ-opioid receptor 301, 311–312
    influence on amino acid neurotransmitters
      301
    κ-opioid receptor 295, 301, 311–312
    L-type Ca^{2+} channel blocker nimodipine
      301
    μ-opioid receptor 295, 301, 311–312
naltrexone 297
nanomaterials, transport of 251

nanoparticles 174, 191
  absorption 246
  administration 255
  Ag, intracerebral administration of 252
  aggravation
    of brain pathology 245
    of cell and tissue injuries 246
  aminosilane-coated iron-oxide 253
  animals treated with 245
  antioxidant defense mechanism 253
  and apolipoprotein 251
  BBB disruption by 253
  biodistribution 247
  and blood-brain barrier function 249
    disruption *in vivo* 252
    facilitation of drug transport 250–251
    and neurotoxicity 252
    surface charge of nanoparticles and BBB 251–252
  blood-brain barrier permeability 259–261, 264–265
    chronic administration of nanoparticles 259
    Evans blue 259
    intravascular tracers 259
    iodine 259
    radioiodine 259
    regional BBB permeability 259
  blood-testis barriers 252
  brain and spinal cord blood flow 259
    carbonized microspheres 259
    Cu nanoparticles-treated groups 259
    microcirculation in brain and spinal cord 259
    nanoparticles on cerebral ischemia 259
  brain edema formation 259, 262
    breakdown of BBB 259–262
    chronic administration of metal nanoparticles 262
    modification in BBB functions 262
    neurodegenerative changes 262
    protein tracers 259
    vasogenic edemaformation 262
    volume swelling 262
    water content in brainand spinal cord 259–262
  brain pathology 253, 255, 262, 267
  carbon tubes 253
  cationized 251
    brain permeability of 251
  causing brain dysfunction 254–255
    ad libitum 254
    Ag nanoparticles 255–266
    Al nanoparticles 255–266
    ambient air temperature 254
    blood gases 255
    core body temperature 254
    Cu nanoparticles 254–255
    deep visceral temperature 254
    heart rate 255
    mean arterial blood pressure (MABP) 255
    metal nanoparticles 255
    muscle temperature 254
    respiration 255
    skin temperature 254
    stress-induced fluctuations 254
    treatment with nanoparticles 255
    tympanic membrane temperature 254
    in vivo effects 254
  cell membrane damage by 253
  cellular changes in brain and spinal cord
    BBB leakage 262
    chromatolysis 262
    condensed cytoplasm 262
    degradation of myelin 262
    dense karyoplasm 262
    distortion of nerve cells 262
    distortion or loss of myelin 262
    eccentric nucleoli 262
    endothelial-glial interface 262
    general sponginess and edema 262
    glial fibrillary acidic protein (GFAP) 262
    gliosis 262
    infiltration of lanthanum 262
    Luxol Fast Blue staining 262
    myelin basic protein (MBP) 262
    nanoparticle-induced brain pathology *in vivo* 262
    non-neural cells 262
    nuclear membrane damage 262
    specific marker of astrocytes 262
  coated, transfer in brain 251
  cobalt 253
  copper (Cu) 245
  degradation 251
  effect on heat stressed animals 262, 264
    Al treated nanoparticles 262
    blood-spinal cord barrier (BSCB) 264

brain fluid microenvironment 268
cabosil 267
cardiovascular and respiratory functions in hyperthermia 262
chromatolysis 266
coagulatory disturbances 268
decrease in angle of inclined plane 264
defense planning 267
degeneration 266
diffusion of lanthanum 266
distortion of cell nucleus 266
endothelial cell membrane disruption 266
evaporative heat loss 264
exposure of airborne particles 267
functional and pathological outcome 262
heart rate 264
induction of oxidative stress 268
instillation of diesel exhaust 267
military and non-military personals 267
military exercise 267
nanoparticles within body fluid microenvironment 268
proinflammatory cytokines 268
prostration 264
reduction in body weight 264
regional brain edema formation 265
respiration 264
salivation 264
spinal cord swelling 265
stride length 264
tissue compliance or tissue pressure 265
WBH-induced gastric ulceration 264
EL4 T-lymphoma 253
enhancing tumor temperature after thermotherapy 253–254
foot-print analysis 258
bromophenol blue 258
stride length 258
grid walking 258
gross abnormal particles 255
inclined plane test 258
inducing oxidative stress and formation of free radicals 253
magnetic 253–254
in air, entry in CNS 249
antitumor immunity 253
in central nervous system (CNS) 246
dendritic cells (DC) 253

derived from metals 245
distribution 246
drug delivery to brain 251
in drug targeting 246
effect of environmental conditions on 246
engineered 247
excretion 246
free radical formation 250, 253
high tumor temperature by 253
inclusion in body fluid compartments 245
induced drug-delivery 245
inducing brain dysfunction 245
inducing hyperthermia 253
inflammatory changes caused by 249
inflammatory potential of 246
influence on brain function 245, 249
inhalation of manganese 249
inhaled from ambient air 246
intraperitoneal administration of 252
intratumoral administered 253
intravenous, intraperitoneal or intracerebral administration 252
lipid insoluble 252
magnetite cationic liposome nanoparticles (MCLN) 253
metabolism 246
of cytochrome P450s 253
microhemorrhages in stomach wall 258
gastic ulceration in stomach 258
magnitude and intensity of stressor 258
necrotic areas 253
negatively charged 251
neurodegenerative changes caused by 249
and neurotoxicity 246
opening of BBB with 252
oxidative stress induction 250
particle chemistry of 246–247
PEGylated polycyanoacrylate 251
physicochemical properties 246
polysorbate 80-coated polybutylcyanoacrylate (PBCA) 250
prostration 258
quantum dots 253
ROS production 248, 253 (see Reactive oxygen species)
Rota-Rod performance 258
salivation 258
from semiconductor materials 246

semi-solid or soft 245
size of, effect on drug delivery to brain 251
species used 247
surface charges of 252
therapeutic effects of hyperthermia 253
toxic effects of 247
toxicity 246
toxicological effects of 246
transcytosis and 247
translocation 250
   in biological system 247–248
   in neurons 248–249
transport from trachea to ganglion nodosum 249
and tumor 253–254
   temperature 253
   treatment 253
ultrafine particles 253
uptake in brain 248
vesicular transport by 250
in vitro studies 246
in vivo effects of 246
in vivo studies 246
without surfactant coating 251
nanoscale products 246
nanotechnology 250–251, 253
  recent advancement 250–251
  therapeutic stategies 253
  thermotherapy 253
nanotoxicity, acute 247
N-arachidonoyldopamine (NADA) 22
nasal regions 248
nasogastric tube 88
natural-killer lymphocyte activity (NKA) 140, 148
natural-killer lymphocyte (NK-cell) 140, 145, 147
natural killer T-cells (NKT-cells) 143, 146
natural motivated behavior 224–229
nausea 153
NE (see norepinephrine)
neck cooling 55
necrosis 138, 139
negative temperature coefficient 64
neoplastic diseases 153, 154
nerve cells, morphological changes in 154
nerve conduction velocity 113
nervous damage 449
neural activation (arousal), widespread 228

neural cell(s) 219, 232
  damage 233
neural damage 55
  associated with CNS pathology 220
  brain temperature as factor for 231–233
  non-specific marker of 232
neural dysfunction 436
neural excitability, decreased 374
neural functions and brain temperature 223–224
neural hormones 206
neural intracellular acidosis 210
neural metabolism 219
neural network 116
neural tube defects, mechanisms of formation of
  abnormal apoptosis 348
    due to fever 348
    inhibition of mitosis after hyperthermia 348
  decreased expression of Krox20 gene 349
  heat resistance by glia 349
  neuronal vulnerability, age dependence of 349
  utero hyperthermia 349
neuroactive peptides 277
neuroactive substances 238
neurochemical measurements 238
neurocognitive dysfunction, modulator of 164
neurodegeneration 417
  premature 173
neuroendocrine
  functions 297
  response to stress 441
neurogenic cells 139
neurogenic hyperthermia 211
neuroglia 418
neuroglial cells in culture 423–424
neurohumoral responses 39–40
neuroleptic malignant syndrome (NMS) 81, 82–90
  animal models of 89–90
  clinical manifestations and diagnosis 82
  differentiation of serotonin syndrome and 100–101
  dysphagia in 88
  EEG abnormality in 88
  GABA
    level 86
    mimetic agents 86
  high temperature environment 90
  lumbar puncture in 83
  pathogenesis 83

pathophysiology of 83–86
relationship between serotonin syndrome and 101–102
risk factors 82–83
   antipsychotic drugs 82
   dehydration 82
   dopamine D_2 receptor blocking effect 82–83
   haloperidol 82
   mental retardation 82
   olanzapine 82
   organic brain disorders 82
   physical exhaustion 82
   psychiatric disorder 82
   psychomotor activity 82
   quetiapine 82
   reduced D_2 receptor affinity 82
sialorrhea in 82, 88
skeletal muscle abnormality 83
treatment of 87–89
neuroleptics 81, 174
neurological deficit 295
   permanent 467
neurological intensive care unit 449
neurological symptoms 153
neurologic impairment 162
neurologic mortality 208
neuromodulation 376–377
neuromuscular mechanisms 46, 50
neurons 7, 248–249, 349–354, 417–428 (see also nerve cells)
   fluid environment of 459
   and glia 395
   sensitivity to excitotoxic injury 210
neuronal activation 328
neuronal cell bodies 7
neuronal damage 347, 451, 525
   with hyperthermia, increased 164
neuronal excitability 10
neuronal excitotoxicity 210
neuronal Hsp70 non-responsiveness 417
   to heat stress 427–428
neuronal measurements 238
neuronal pathology 175, 203
neuronal vulnerability 202
neuron specific enolase (NSE) 465
neuropathological studies 175
neuropathological vs. neurotherapeutic roles of choroid plexus–CSF system 473–474

neuropeptides 277
   biosynthesis 278–279
      in CNS 278
         neurotransmitters or neuromodulators 278
         prepropeptides 279
   in heat stress, measurement of 284
   under stressful conditions 278
neuropeptide conversion 281–282
   antinociception 282
   hyperalgesia 282
   neutral endopeptidase (NEP) 282
   nociceptin 282
   substance P endopeptidase (SPE) 282
neuropeptide receptors 278, 283–284
   adenylate cyclase/cAMP system 283
   β-funaltrexamine 284
   G-protein-coupled receptors 283
   phospholipase C/inositol phosphate system 283
   regulation of ion channels 283
neuropeptides in hyperthermia, 277–289
   angiotensin and related peptides 288–289
      fever receptor 289
      lipopolysaccharides (LPS) 288
      LPS-induced hypothermia 289
      PGE2 288
      pyrogenic cytokines 288
      stress-induced fever 288
   calcitonin gene-related peptide 287–288
      cardiac tumor necrosis factor-alpha (TNF$_\alpha$) 288
      CGRP concentrations in rat brain 288
   opioid peptides 284–286
      drug or alcohol addicts 286
      L-NAME or L-NMMA 285
      Met-enkephalin 286
      morphine-induced hyperthermia 286
      nitric oxide synthase (NOS) inhibitors 285
      opioid-related peptide nociceptin 286
   tachykinins 286–287
      heat-induced SP release 287
      nitrite/nitrate (NO$_x$) 287
      NK-1 receptor antagonist RP67580 287
      NK-2 receptor antagonist SR48968 287
neuropharmacology 76
neuroprotection 76
neuroprotective hyperthermia 203
neuroprotective temperature 202, 205
neuropsychological factors 54

neuropsychological responses 45, 46
  to hyperthermia 46–53
neuropsychological tests 48
neurorepair 250
  processes 245
neurosurgical intensive-care unit (ICU) 155
neurotoxicants 75
neurotoxic cascade 543
neurotoxic drugs 220
neurotoxicity 76, 236, 250, 251
  glutamate-induced 232
neurotransmission
  brain hyperthermia and 223
  dynamics of 238
neurotransmitter release, hyperthermia and 210
neurotransmitters 39, 76
  enhanced release of 165
  excitatory 450
  or neuromodulators 278
neurotropic viruses 395
  enhanced gene expression of 408
neutral endopeptidase (NEP) 282
neutrophils 162
NF-κ-B 140
  translocation of 409
nicotine 233
nigrostriatal dopaminergic system 501
nitric oxide (NO) 377
nitric oxide synthase (NOS) inhibitors 297
NMDA (see N-methyl-D-aspartate)
NMDA/AMPA receptors 295
N-methyl-D-aspartate (NMDA) 298
  receptor antagonists 96
  receptors 10
N-methyl-D-aspartyl (NMDA) 281
NMS (see neuroleptic malignant syndrome)
NO 10, 147, 233, 543 (see also Nitric oxide)
  cellular 504
  concentrations, increased 503
  in fever 11
  and leukocyte adhesion 505
  overproduction of 504
  production 409
  radicals 10
  release 10
  scavengers 11
  synthesis 503
NO/cGMP-signaling pathway 455

nociceptin 282
nocturnal T_b 19
non-cerebral capillaries 249
non-evaporative mechanisms 71
non-febrile T_b 17
non-febrile thermoregulation 15
non-heat related illness 497
non-selective COX inhibitor, aspirin 6
non-shivering thermogenesis 64
non-traumatic subarachnoid hemorrhage 163
noradrenaline (NA) 83, 86
norepinephrine (NE) 1, 6, 48, 141
  microinjection 6
normothermia 153
normothermic rats 164
NOS 10, 287, 300 (see also Nitric oxide synthase, enzyme)
NOS II
  cytokine-induced stimulation of 504
NO synthase, isoforms of
  endothelial, neural, and inducible NOS 10
nuclear DNA damage 139
nuclear steroid hormone receptor 145
nuclear translocation of Hsp70 420–423
nucleic acids, damage 77
nucleus accumbens (NAcc) 224
nucleus of the solitary tract (NTS) 4
NUR77 (nuclear receptor 77) 145

obesity 114
occipital cortex 86, 232, 305
occupational health 45
occupational safety 45, 46
olanzapine 82
oligodendrocyte(s) 406, 418–420
  as major producers of Hsp70 418–420
  marker lectin GS II 420
o-phthalaldehyde method 298
opiates 233
opioids 145, 148, 311
  and hyperthermia 284–286, 297–298
    activation of glutamate receptors 298
    allodynia-induced pain 298
    blockade of opioid receptors 297
    brain pathology 297
    dynorphin and cell damage 297
    dynorphin antiserum 298
    dynorphin upregulation in hyperthermia 298

glutamate release in CNS 298
heat-induced neurotoxicity 298
heat stress-induced blood-brain barrier (BBB) dysfunction 297
hyperthermia-induced brain pathology 297
interaction between opioids 298
naloxone 297
naltrexone 297
neuroendocrine functions 297
nitric oxide synthase (NOS) inhibitors 297
N-methyl-D-aspartate (NMDA) 298
temperature regulation 297
upregulation of dynorphin 297
opioid antagonist naloxone 284, 286, 295, 300–305, 307, 310–311
opioid peptides (OP) 277, 280–281
β-casein 281
β-casomorphins 281
β-endophin 280
big dynorphin 281
endomorphin-1 281
μ-opioid peptide (MOP)receptors 281
N-methyl-D-aspartyl (NMDA) 281
precursor proteins 280
nociceptin 280
orphanin 280
pro-orphanin 280
prodynorphin (ProDYN) 280
α/β-neo-endophin 280
dynorphin A 280
dynorphin B 280
proenkephalin (ProENK) 280
pro-opiomelanocortin (POMC) 280
opioid receptors 145, 295–312
multiple, blockade of 295
neuromodulatory roles 295
oral antibiotics 17
oral aspirin 17
oral temperature 54, 486
organic brain disorders 82
organism's heat production 219
organum vasculosum laminae terminalis 5
osmotic pressure gradients 190
outcome aggravation 208, 209
ovariectomized female rats 206
ovariectomized rats 206
overheating 40, 113
brain 219

oxidative currents, generation of 238
oxidative damage 528–544
hypervolemic hemodilution for 540–543
progressive exercise preconditioning protects against 534–539
protective effects of free radical scavengers on, in heat stroke 528–530
role in pathogenesis of cerebral ischemia during heat stroke 543–544
Shengmai San for improvement 539–540
oxidative injury 397
oxidative phosphorylation 63, 66
uncouplers 68
oxidative stress 10, 165, 233
heat shock preconditioning effects on 531–534
in heat stroke 525
in rats with heat stroke-induced cerebral ischemia 525–528
oxygenated fatty acids 15
oxygen consumption 35
in humans 229
oxygen delivery, inadequate 38
oxygen deprivation 381–384
oxygen free radicals 543
oxygen radicals 10
production, exaggerated 165
oxygen uptake 29

pain 115, 277
receptors 124
stimulus 185
panting 18, 65
paraffin oil 179
paraquat 74
parenteral salicylate 17
parkinsonism 83
Parkison's disease 83
paroxetine 39
PARP inhibitor DPQ (3,4-dihydro-5-[4(1-piperidinyl)butoxy]-1(2H)-isoquinolinone) 361
passive heating 186
passive hyperthermia 47
alternative method for 185–187
passively induced hyperthermia 30
pathognomonic of NMS 82
pathological brain hyperthermia 219, 232, 236
pathomechanisms 201, 208

patients with SCI 205
pediatric TBI patient, severe 208
PEGylated polycyanoacrylate nanoparticles 251
pentazocine 91
pentobarbital 177, 185
peptide biosynthesis, fundamentals of 278
peptidergic systems 277
perceived exertion 47
perceived heat stress 54
perceived temperature 48
perception 46
    of heat stress 48–50
perceptual heat strain 45
perceptual responses to thermal stimuli 48
perceptual thermal strain 47
performance/tolerance, exercise 53
perimesencephalic cisterns 163
peripheral blood mononuclear cells
    (PBMCs) 322
peripheral fatigue 31
peripheral motor neuron 30
peripheral nervous diseases 175
peripheral nervous system (PNS) 175
peripheral pyrogens, activation 10
peripheral thermal afferents 45
peripheral thermoreceptors 49
peripheral thermoregulatory effectors 375
peripheral vascular beds 45
peripheral vasoconstriction 64, 66
peripheral vasodilation 450
perivascular cells 17
periventricular inflammation 209
peroxidative products 247
peroxynitrite 233
personality changes 347
pethidine 91
PGE_2 16, 464
    antiserum 6
    CSF concentration of 464
    hypothalamic 18
PGES enzymes 18
PGE synthases (PGES) 17
pH, low 154
phagocytes, mononuclear 3
pharmacological brain hyperthermia 229, 233
pharmacology 69
pharmacotherapy 81
    of NMS 88

phasic temperature fluctuation 226
phenotypic adaptation 373
phosphoenolpyruvate carboxykinase (PEPCK)
    489
phosphoinositide (PI) 5
phospholipase A_2 511
pH-sensitive calcium channel 22
physical activity 35
    brain HSP and cytokine responses alteration
        440–442
physical exhaustion 82
physical fitness 46
physical refrigeration 162
physical stress 140–142
physical workload 53
physiological brain hyperthermia 219–232
    source of 221–223
physiological heat strain 45
physiological strain 45, 175
physiological strain index (PSI) 53–54
physiological thermoregulatory capacity 49
physiological thermoregulatory strategies 49
pia-glia 460
pilot clothing 56
p53 inhibitor pifithrin-α 361
pinocytosis 247
pipamperone 96
PI-specific phospholipase C (PI-PLC) 5
pituitary gland 34
    prolactin release from 34
pizotifen 39
plasma 5, 6, 86, 141, 142, 144, 165, 187, 239, 249,
    279, 283, 284, 356, 436, 462, 467, 469, 481,
    490, 492, 497, 499
    cytokine, elevated circulating 436
    glucose homeostasis 489
    IL-12p40 499
    membrane, extended 249, 460
    of NMS patients 86
    osmolarity, increasing 427
    proteins, slow leak of 462
    tonicity 468
    viscosity 468
plasma free fatty acids 20
plasma membranes 232
plasma PGE_2 6
plasticity of the autonomic responses 375–377
plexus-CSF-brain nexus 463

pneumonia 88
pO$_2$, low 154
POA (preoptic area) 10
polio virus, intranasal administration 248
polymer nanoparticles 250–251
　drug transport by 251
polymorphonuclear leukocytes 165, 205, 209
　increased accumulation of 209
　infiltration 165
polymorphonuclear neutrophil leukocytes 144
polysorbate 80 251
polysorbate 80-coated polybutylcyanoacrylate (PBCA) 250
　nanoparticles 250
polytetrafluoroethylene (PTFE) fume
　aging of 247
　effect of exposure 247
polyunsaturated fatty acids 15
pons 232
porcine stress syndrome (PSS) model 89
positron emission tomography (PET) 86
post-mortem symptoms of heat stress 184
postoperative hyperthermia 164
post-traumatic hyperthermia 205, 206, 209
posttraumatic stress disorder (PTSD) 465
　major depression in 465
potassium, 453–455
prazosin 7
prefrontal cortex 34
preoptic area of anterior hypothalamus 19, 21, 39, 40
preoptic temperature set point, elevation of 450
prepropeptide and precursor processing 279
　biosynthesis of neuropeptide 279
　peptidergic neuron 279
pressure autoregulation 155
pressure-flow autoregulation 153
pre-synaptic neurotransmitter release 404
presyncopal signs 38–39
prevalence of fever and infection 208
progesterone 206
programmed cell death (apoptosis) 139, 144, 148, 435
progressive hyperthermia 33–34
proinflammatory cytokines 209
pro-inflammatory cytokines 142
proinflammatory stimuli 6

prolactin 34, 465
　secretion 34
　serum concentration of 34
proliferation 330
prolonged exercise 29, 30, 34
prolonged high fever 176
prolonged therapeutic hyperthermia 153
propofol anesthesia 155
propranolol 95, 161
propylene glycol 185
prostaglandin (PG)E$_2$ 5–6
　plasma 6
　receptors 5
　thermogenic 4
prostaglandins (PGs) 15–20, 162, 298
　synthesis 459
prostanoids 17, 464
prostration 182
proteins
　damage 71, 77
　denaturation 113
　folded and misfolded 232
　general survival 140
　S100B 465
proteinase-inhibitor 9 (PI9) 147
protein kinases 210, 283, 327, 425
　enhanced inhibition of 165
　miitogen activated 148
protein misfolding, contributing to hyperthermia-induced neuronal death 358–359
　blocking protein synthesis 358
　cycloheximide 358
　denaturation of proteins, heat-induced 358
　heat-labile proteins 358
　lysosomes 358
　misfolding of proteins 358
　protein synthesis inhibitors 358
　proteosomes 358
　puromycin 358
　threshold temperature 358
proteotoxic stress by sodium arsenite 424–425
proton magnetic resonance spectroscopy 56
psychiatric diseases 277
psychiatric disorders 82, 174
psychiatric symptoms, in NMS 88
psycho-emotional stimulation 233
psychological component 45

psychological safeguards 45
psychomotor activity 82
psychomotor stimulant drugs 219
psychoses, therapeutic treatment 72
psychostimulants 174
  and brain temperature 177
psychostimulant treatment 191
psychotic symptoms, in serotonin syndrome 100–101
pulmonary ventilation 37
Purkinje cells 403
  loss of 175, 436
puromycin 358
pyloric mucosa 184
pyretic function 10
pyrexia 153, 154, 161
pyrogenic cytokines 3, 11, 450
pyrogenic doses, LPS 8
pyrogenic function 6
pyrogenic signal, of LPF 6

quantum dots 246
  in breast tissues, alteration following hyperthermia 246
quetiapine 82

rabbits, exposure to RF radiation 69
radiation 53, 65
radiation-induced cell damage 139
radiation-induced DNA fragmentation 139
radioactive leucine 420
radio frequency electromagnetic fields (RF-EMF) 107–131
  aversive properties of 122–124
  dosimetry of 108
    computational 110–111
    empirical 109–110
  effects on physiological functions and survival 114–115
  emitting devices 108
  frequency 116
    range 108
  mild hyperthermia by 107
  performance disruption of 124–128
  physiological responses of 115–117
  safety standards 107, 128–130
  thermal tissue response 115
  thermoelastic expansion 121
  thermoregulation during exposure with 112–120
  whole-body hyperthermia 108, 113–114
radioiodine 259
radiotelemetric pill 74
radiotelemetry 486
radiotherapy 138, 154
RANTES (regulated on activation, normal T expressed and secreted) 409
rapid brain hyperthermia 228–229
rapid brain temperature fluctuations 219
  brain hyperthermia and 224–229
rapid mode of cell death 138
rapid rewarming after hypothermia 207
rating of perceived exertion (RPE) 34
rats, exposure to RF radiation 69–71
reactive astrocytes 209
reactive oxygen species (ROS) 76–77, 249, 253, 528
  cytotoxic effects of 232
reactive oxygen species (ROS), contribution to hyperthermic stress 357–358
  catalase 357
  dimethylsulfoxide (DMSO) 358
  DNA damage induced by free radicals 357
  gamma-glutamate cysteine 357
  glutathione peroxidase 357
  hydrogen peroxide 357
  increased ROS production 357
    during heat stress 357
  N-aceytl-cysteine 357
  peroxynitrite scavenger MnTBAP 358
  superoxide 357
receptor-induced apoptosis 145–146
receptor-induced death 146
receptors 15, 19, 33, 86, 120, 124, 140, 283–284, 295–312, 354, 360, 450, 497
  N-methyl-D-aspartate 10
  PGE_2 5
rectal temperatures 50–53, 55, 96, 179, 184, 207, 486
  brain temperature vs. 184
  measurement 179
regional brain temperatures 176–177
regional brain water content 190–191
regional cerebral blood flow (CBF) 34, 35
regional hyperthermia 154
remifentanil 145
renal failure 87, 492
  in serotonin syndrome 95

renin-angiotensin system (RAS) 281
respiration 38, 113, 178, 255, 264, 356
respiratory center, modulation of 440
respiratory cooling, of upper airways 37
respiratory failure 87
respiratory tract infection 205
restlessness 90
restrictive gap junctions 460
reticuloendothelial (RES) system 507
rewarming
   in injury models, slow 207
     phase 207–208
     slow but not rapid 207
RF-EMF energy 114
   absorption during whole-body exposure 115
   thermoregulatory responses of humans and human volunteers to 118–120
RF-EMF exposure 108, 109, 111, 113, 114, 118–120
   behavioral effects of 120–128
RF-EMF responses, by human 118–124
   aversion 122–124
   detection of 121–122
   performance disruption 124–128
RF radiation exposure 69
   dosimetry units for 70
rhabdomyolysis 87, 492
rhesus monkeys 127
risperidone 96
ritanserin 96
RNA translation factor (eEF1A) 435
robust brain hyperthermia 230
robust hyperthermia 233
rodents 63
   drug development in 69
   extrapolation from, to human 69–71
   response of MDMA 71–74
room temperatures 238
Rota-Rod performances 258, 308, 309
running in heat 50
running speed 51
ryanodine receptor gene 87

safety standards, for RF-EMF 107, 128–130
salicylates 17
   COX-independent effects 20
saliva grooming 65

salivation 182, 258, 264
   blocking of 66
Salmonella enteritidis 8
salt balance 174
SAR 70 (see specific absorption rate)
sarcoplasmic reticulum 83
saxitoxin 354
SC-560 (Sulindac sulfide 560) 7
scalp temperature 56
schizophrenia 81
SCI 201–211 (see also spinal cord injury)
secondary hyperthermic insult 206
secondary hypoxia 207
secondary injury processes 165
selective brain cooling 35–37
selective head cooling 37–38
selective 5-HT$_2$-receptor antagonist ketanserin 500
   increase in intracerebral pressure 500
   reduction in mean arterial pressure (MAP) 500
selective serotonin reuptake inhibitors (SSRIs) 90–92
sensitive organs 232
sensory nerve endings 248
sensory stimulation 35
serotonergic activity 34, 39
serotonergic neurones 34
serotonergic pathways 72
serotonergics 174
serotonergic system 34, 39
serotonin (5-HT) 34, 39, 72, 83, 91, 298, 500
   depletion in striatal levels and MDMA 75–76
   nonspecific, antagonist 96
   receptor gene 87
serotonin behavioral syndrome 90
serotonin-fatigue hypothesis 39
serotonin syndrome 81, 90–102
   abnormal behaviors in animals with 90
   amitryptyline 91
   analgesics 91
   antidepressants 91
   benzodiazepine derivatives for 92
   causative drugs 91
   clinical symptoms and diagnosis of 90–91
   CSF NA level in 92
   differentiation of NMS and 100–101
   dopamine concentration in 94–95
   fluid replacement for 95
   imipramine 91

lithium carbonate 91
MAO inhibitors 91
3,4-methylenedioxymethamphetamine (MDMA) 91
moclobemide 91
pathogenesis of 91–92
pathophysiology 92, 94–95
pentazocine 91
pethidine 91
pharmacotherapy in animal model of 96
relationship between serotonin syndrome and 101–102
tramadol 91
treatment of 95–101
tricyclic antidepressants 91
L-tryptophan 91
serum 34, 143, 162, 180, 187, 232, 307, 397, 406–407, 470, 497, 537
  enzyme abnormalities 154
  levels of hsp72 407
  proteins 190
  zinc, elevations in 162
"set point" for body temperature 455
set point temperatures 116
severe body hyperthermia 436
severe disability 208
severe pediatric TBI patient 208
sexual behavior 226
  neuronal recordings during 228
sexually arousing stimuli, exposure to 226
Shengmai San 525, 539–540
Shigella dysenteriae 466
shivering 49, 55, 90
  thermogenesis 64
short term heat acclimation (STHA) 374
12-(S)-hydroperoxyeicosatetraenoic acids (12S-HpETE) 21
15-(S)-hydroperoxyeicosatetraenoic acids (15S-HpETE) 21
sialorrhea, in NMS 82, 88
Simian Virus 40, 409
single-photon emission computed tomography (SPECT) 86
skeletal muscle(s) 29–30
  abnormality 83
  fatigue 29

skeletal muscles
  contraction 322
  spasms of 482
skin 48–49, 66–67, 72, 109, 114, 119, 223
  basal temperatures at 224
  receptors 186
  skin temperature 37, 48–49, 55, 64–66, 74, 120, 178, 181, 323
skin blood flow 64, 66, 322, 436
  cardiovascular response to heat exposure 503
  vasoconstriction of 72
sleep 114
  EEG during 34
slow rewarming
  in injury models 207
  protocol 207
sodium arsenite 424
  causing protein denaturation 424
  causing stress response 424
  induced Hsp70 424
sodium pentobarbital 185, 238
sound stimuli 223
species' body mass 69
specific absorption rate (SAR) 109, 110
specific absorption (SA) 109
specific heat capacity of blood 35
spinal cord 175
spinal cord injury (SCI) 201, 202, 205–207
  hypothermic protection after 207
  patients with 205
splanchnic blood flow 503
  reduction in 492
splanchnic resistance 504
splanchnic vascular responses 503
splenocytes 144–145
spontaneous fever 153, 154
spontaneous intracerebral hemorrhage 208
spontaneous subarachnoid hemorrhage 449
stabile internal temperatures 219
steady-state probe topography 48
steroids 148
  high-dose 148
strenuous exercise 38, 503
stress 144
  biomarker 436
  hormones 141, 142, 144–145
  susceptibility, biomarkers of 507
  tolerance 512

stressful conditions 278, 395, 507
stress-induced apoptosis 144–145
stress-induced hyperthermia 223
stress-inducible genes 438
stressors, novel 373
stress-related transcriptional activity 402
striatal blood flow 223
striatum 86, 202
stroke 162–163, 174
    fever after, and neurologic mortality 208
    fever and outcome after 208
    patients, acute 210
    patients, poor outcome in 208
stroke volume (SV) 534
stupor 153
subarachnoid blood 163
subarachnoid hemorrhage 163, 207
    fever and outcome after 208
    fever in 208
    lead to hyperthermia 208
subdiaphragmatic vagotomy 4, 6
subdural brain temperature 37
substance P endopeptidase (SPE) 282
sub-threshold hyperthermia 397
sucrose, permeability of 252
sudomotor 45
superoxide anions 525
superoxide dismutase (SOD) 526
surface area: mass ratio 66
surface temperature 71
sustained maximal muscle contractions 29
sustained muscle contractions 30
swallowing 185
sweating 49, 65, 74
    blocking 66
    profuse, loss of electrolytes in 482
    rates 53, 55, 229
    response 49, 54
sweat secretion 113
sympathetic activity 34
sympathetic-adrenomedullary system (SAS) 144
sympathetic nervous system 92, 184–185
sympathomimetics 68
synaptic function 403
synaptic transmission 210

synergistic damaging effects of combined hyperthermia and ischemia 362–363
    brain damage increased by hyperthermia 362–363
    mitochondrial damage 362
    protein denaturation 362
        energy stress with hyperthermia 362
        synergistic damaging effects 362
        proteosome overload 362
systemic heat stress 301, 395, 397
systemic hyperprolactinemia 34
systemic hyperthermia 139, 144, 154, 159, 231–232
systemic inflammatory response 176
systemic inflammatory response syndrome (SIRS) 492
systemic norepinephrine levels 34
systemic responses to heat stroke 484–486
systemic thermoregulation 56
systemic vascular resistance 502
systolic hypotension 163

tachycardia 82, 100
tachykinins 277
tachykinins and other neuropeptides 281
    AT1 and AT2 receptors 281
    calcitonin gene-related peptide (CGRP) 281
        angiotensin II 281
        cAMP response element binding protein (CREB) 281
        CGRP-1 281
        CGRP-2 281
        co-localization with SP 281
    NK1 receptor 281
    preprotachykinins 281
    renin-angiotensin system (RAS) 281
    substance P (SP) 281–283, 286
tachypnea 82, 91
tail vasoconstriction 72
taurine 449
    antagonist 450
TBI 204–207 (see also traumatic brain injury)
temperate environment, exercise in 45
temperature see also *specific* entries
    balance 220
    and cell death 74
    elevations in, after brain injury 208
    gradient 66, 72, 221

homeostasis 220
increase 208
overshooting 207
probes 110–111
toxicity 164–165
temperature coefficients 63, 69, 375
negative 64
positive 63, 77
temperature-sensitive neurons 238
temperature thresholds (T-Tsh) 373
temporal muscles 224
testicles 232
(-)-Δ^9-tetrahydrocannabinol 21
tetrodotoxin 354
thalamus 202, 232
therapeutic hyperthermia 137
definition 176
in humans 154, 161
clinical effects 154–155
physiological effects 155, 159–160
therapeutic treatment strategies 206
thermal afferents 45, 49
input, overall 49
thermal chemosensitisation 138
thermal clamping 49
thermal comfort 49
thermal damage 115–117, 492
thresholds for 115–117
thermal discomfort 48–50
thermal effects, of AR agonists 7
thermal enhancement ratio (TER) 138
thermal environments 46
thermal gradient 55, 74, 487, 488
thermal integration sites 48
thermal isoeffect dose (TID) 138
thermalizing energy 112
thermal kinetics 63
thermal load 112, 178, 180–181, 507
heat stress and 180–181
thermal manipulation 49
thermal nociception 182–184
thermal overload 180–181
thermal perception 49–50
thermal physiology 48
thermal radiosensitisation 138
thermal receptors 124
thermal sensation 49
thermal sensitivity 45

thermal sensors 124
thermal set point 449, 450, 453
thermal stimulation 49
thermal stimulus 49
integration 48
perception of 48
perceptual responses to 48
thermal strain 48, 54
thermal stress 21, 46, 48, 49, 321, 397
effect on exercise performance 51
modulating perception of 48
sensations of 54
thermal tissue response 115
thermal tolerance 116, 155, 438
thermistor probes 179, 254
thermoafferent convergence 49
thermodynamic response of brain 36
thermoeffector(s) 64, 66
activity 64
responses of rats with MDMA 72
thermoelastic expansion 121
thermogenesis 64
metabolic 66
thermoinsensitive neurons 7–8
thermometrics 138
thermoneutral environments 18, 29, 35
exercise during 35
thermoneutral zone 20, 64
for rabbit 69
temperature regulation 64
thermoreceptors 49, 186
thermoregulation 21, 39, 45, 112–120, 220
age factors and 118
autonomic 113
behavioral 48, 113
CB_1 receptor and 21
central mechanisms of 449
endocannabinoids and 21
thermoregulatory ability 49, 63
thermoregulatory capacity 440
thermoregulatory centers, damage to 208, 486
thermoregulatory consequences of heat stroke 486–489
thermoregulatory factors 34
thermoregulatory failure 66, 69, 153–154
thermoregulatory mechanisms, disabling of 72
thermoregulatory overload 113–114
thermoregulatory profile, of drugs 64–66

thermoregulatory sensitivity 69
thermoregulatory set-point 68
thermoregulatory system 64, 436
thermosensitive cells 116
thermosensitivity 49, 55
   regional differences in 45, 49
thermosensors 115, 116
thermotolerance 373, 433
   acquired 510
   development of 321
   and HSPs 439
   and injury protection 510–511
   in spontaneously hypertensive rats 483
thermotolerance, development of 525
thermotolerant rats 438
thirst sensation 49, 54
thrombocytopenia 397
tight junctions 250, 252, 460
   choroidal 462
   permeability of 461
   structural changes in 251
   ultrastructural differences in 462
   widening of 251
tissue 115–117, 246–247, 268, 284, 323, 407, 433–434, 487–489, 491–492, 532 (see also brain tissue)
   histopathology 492
   hypoxia 328
   injury responses 491–492
   necrosis 502
   temperature 64
titanium dioxide (TiO_2) 246
T-lymphocytes (T-cells) 140, 144, 145
   activity 143
   apoptosis 145
   proliferation 143
TNF 145, 289, 362, 490–492, 496, 499, 506, 509
   α and β 496
   in human heat stroke mortality 496
   p55/p75 receptor (TNFR) knockout mice 503
TNF-α after endotoxin exposure 435
TNF-α over production 538
TNFR60 496
TNFR80 496
TNF-receptor family 146
TNKα expression 209

α-tocopherol 525
   effect on oxidative damage and cerebral ischemia 528–530
tolerance in heat 48
toll-like receptor 6
toluene 68
tonic temperature 226
topoisomerase I 360
   heat-sensitive 360
torso cooling 55, 56
total peripheral resistance (TPR) 534
toxicant exposure 74
toxicants 66, 68
toxicity 63, 64, 68–69, 74–76, 153, 154, 246, 247, 249, 252, 385
   chemical or heavy metal 397
   of drugs 64
      and ambient temperature 68–69
      of hyperthermic drugs 64
      magnitude *versus* duration 69–70
toxicology 69
toxic potency 68
tracheobronchial regions 248
tramadol 91
transcranial magnetic stimulation 30
transcytosis 247
transgenic C57BL/6 mice 409
transient cerebral ischemia 161
transient forebrain ischemia 207
transient hyperemic response 155, 159
transient receptor potential of vanilloid (TRPV)
   TRPV4 22
   TRPV1 deficient mice 22
   TRPV1 knockout mice 22
   TRPV1 null mice 22
   TRPV1 receptor 21–22
transient vasoparalysis 159
tranylcypromine 92
trauma 162, 163, 201, 205, 232, 311, 381, 467, 510
   gender importance 206
   models of 202
traumatic axonal injury (TAI) 202
traumatic brain injury (TBI) 153, 161, 163, 204–205, 373, 449
   cognitive deficits after 205
tremor 82, 90
tricyclic antidepressants 91, 148
trigeminus nerves 248

TRPV 21, 22 (see also transient receptor potential of vanilloid)
Trypan blue 187
tryptophan 34
L-tryptophan 91, 92
tubular necrosis 492
tumor 154, 176, 178, 253–254, 508
  defence 147
  immunity 140
  necrosis factor-α 249
  remission 137
  resection 207
  specific peptides 140
  suppressor p53 361
  vaccination 140
tumouricidal effects 147
tumour necrosis factor alpha (TNF-α) 4, 142, 249, 288, 360, 490, 529 (see also TNF)
tympanic membrane temperature 36, 184, 254, 255, 262, 263
tympanic temperature measurements 230
tyrosine-hydroxylase depletion 233

"ultradian" T_b oscillations 22
ultrafine particles 191
uncompensable heat stress 51
unconsciousness 450
upper critical temperature 65
urban heat islands 483
urethane (ethyl carbamate) 177, 184–185, 526
urinary tract infection 205
urination 182
urine, of NMS patients, catecholamines levels in 86

vagal afferents 4
vanilloid receptor signaling pathways 22
vanillylmandelic acid 86
vascular cascades 210
vascular cells 417
vascular endothelial growth factor (VEGF) 383
vascular endothelium 154, 175
  morphological changes in 154
vascular engorgement 155
vascular infarction 190
vasoconstriction 55
  by MDMA 72
  of skin blood flow 72

vasodilatation 18
vasogenic edema 155, 259, 262
vasomotor index 71
vasomotor reflexes, depressed by alcohol 483
vasomotor responses 55, 70–71, 155, 159, 377
  body size impact on 71
  lack of 55
vasomotor tone 48
vasoparalysis 153
  transient 159
vasospasm 164
  and poor outcome 208
vegetative survival 208
venous blood outflow 229–231
venous blood temperature 230
venous return, reduced 45
ventilation 45, 54
ventral stratum temperature 221
ventral tegmental area of midbrain, temperature of 224
ventricular space, dilated 459
ventricular walls 188
ventriculostomy catheter 207
ventromedial preoptic area (VMPO) 3, 6, 8
  PGE_2 4, 7, 9–10
vesicles for macromolecular transport 249
Vesicular Stomatitis Virus 409
vibrissa 185
viral antigenic peptides 409
viral gene expression, stress-conditioning and 408
viral infection 397
viral invasion 220
virus inoculation 174
visceral temperature 179
visual cortex 35
vital centers 190
vitamin C, transport of 463
VMPO (see ventromedial preoptic area)
voltage-gated ion channels 380
volume swelling of brain 188–190, 262
voluntary exercise tolerance 45

warm acclimation 376
warm environmental temperatures 234
warm environments 51, 66
  thermoregulatory responses 22
warm receptors 48

warm-sensitive neurons 4, 7, 238
warm-sensitive (warm-activated) neurons 453
warm stress 17, 29, 31, 35, 45, 46, 49, 175–178, 180, 245, 277–278, 295, 323, 417, 449, 464, 486
   (see also heat stress)
warm temperature, subjective feelings of 74
water content, in brain 190–191
water-heated suits 154
WAY 100635 (N-[2-[4-(2-methoxyphenyl)-1-piperazinyl]ethyl)-N-(2-pyridinyl) cyclohexanecarboxamide trihydrochloride) 96
WBH 137, 138, 140–143, 145, 146, 148, 174, 178–191, 245, 254, 264, 267–268, 467–473
   (see also whole-body hyperthermia)
weakness 175
wet bulb globe temperature (WBGT) 53
wet weight 190
whiskers, movements of 185
whole body cooling 55
whole body heating 47–48
whole-body hyperthermia (WBH) 108, 137, 154, 161, 174, 245, 408, 417, 459, 464
  affect on rats treated with nanoparticles 245
  induced stress reactions 140–142
  monocytes stimulation by 142
  in primates 499
  transient impairment of lymphocytes during 142–143
whole-body hyperthermia, models of 178, 185–186
  anesthesia, choice of 184–185
  animals handling prior to heat stress 178–179
  behavioral observations 181–182
  body temperature changes during heat exposure 45, 137, 180
  body weight loss after heat exposure 182, 310
  brain temperature *vs.* rectal temperature 184
  exposure of rats 179–180
  heat chamber 179, 180, 186
  heat exposure, observations during 180
  heat stress and thermal load 180–181
  heat stress influences anesthesia 184
  heat stress without heat stroke 181, 298
  post-mortem symptoms of heat stress 184
  rectal temperature measurement 179
  thermal nociception 182–184
whole-body oxygen consumption 233
whole-body resonance 124
Whole-body specific absorption rate (SARs) 128
widespread neural activation (arousal) 228
wild-type mice 7
WIN 55212-2[(4,5-dihydro-2-methyl-4(4-morpholinylmethyl)-1-(1-naphthalenyl-carbonyl)-6H-pyrrolo[3,2,1ij]quinolin-6-one]-induced hypothermia 21
ω-6 (*n*-6) C_{20} polyunsaturated fatty acids 15
ω-36 (*n*-3) C_{20} polyunsaturated fatty acids 15
work stoppage 121, 124, 126
worse outcomes 202, 206

xenobiotics, toxicity of 68
xylazine 185

yohimbine 7–10

zonulae occludentes 460
Z-VAD*fmk* 147